*Editors*

Duoqian Miao
Tongji University
Shanghai
China

Witold Pedrycz
University of Alberta
Edmonton Alberta
Canada

Dominik Ślęzak
University of Warsaw
Warsaw
Poland

Georg Peters
University of Applied Sciences
München
Germany

Qinghua Hu
Tianjin University
Tianjin
China

Ruizhi Wang
Tongji University
Shanghai
China

# Lecture Notes in Artificial Intelligence  8818

## Subseries of Lecture Notes in Computer Science

### LNAI Series Editors

Randy Goebel
  *University of Alberta, Edmonton, Canada*
*Yuzuru Tanaka*
  *Hokkaido University, Sapporo, Japan*
*Wolfgang Wahlster*
  *DFKI and Saarland University, Saarbrücken, Germany*

### LNAI Founding Series Editor

Joerg Siekmann
  *DFKI and Saarland University, Saarbrücken, Germany*

ISSN 0302-9743                    ISSN 1611-3349    (electronic)
ISBN 978-3-319-11739-3           ISBN 978-3-319-11740-9    (eBook)
DOI 10.1007/978-3-319-11740-9

Library of Congress Control Number: 2014950507

LNCS Sublibrary: SL 7 – Artificial Intelligence

Springer Cham Heidelberg New York Dordrecht London

Printed on acid-free paper

Springer is part of Springer Science+Business Media (www.springer.com)

More information about this series at http://www.springer.com/series/1244

Duoqian Miao · Witold Pedrycz
Dominik Ślęzak · Georg Peters
Qinghua Hu · Ruizhi Wang (Eds.)

# Rough Sets
# and Knowledge Technology

9th International Conference, RSKT 2014
Shanghai, China, October 24–26, 2014
Proceedings

 Springer

# Lecture Notes in Artificial Intelligence 8818

Subseries of Lecture Notes in Computer Science

More information about this series at http://www.springer.com/series/1244

Duoqian Miao · Witold Pedrycz
Dominik Ślęzak · Georg Peters
Qinghua Hu · Ruizhi Wang (Eds.)

# Rough Sets
# and Knowledge Technology

9th International Conference, RSKT 2014
Shanghai, China, October 24–26, 2014
Proceedings

 Springer

*Editors*

Duoqian Miao
Tongji University
Shanghai
China

Witold Pedrycz
University of Alberta
Edmonton Alberta
Canada

Dominik Ślęzak
University of Warsaw
Warsaw
Poland

Georg Peters
University of Applied Sciences
München
Germany

Qinghua Hu
Tianjin University
Tianjin
China

Ruizhi Wang
Tongji University
Shanghai
China

ISSN 0302-9743
ISBN 978-3-319-11739-3
DOI 10.1007/978-3-319-11740-9

ISSN 1611-3349 (electronic)
ISBN 978-3-319-11740-9 (eBook)

Library of Congress Control Number: 2014950507

LNCS Sublibrary: SL 7 – Artificial Intelligence

Springer Cham Heidelberg New York Dordrecht London

Printed on acid-free paper

Springer is part of Springer Science+Business Media (www.springer.com)

# Preface

This book contains the papers selected for presentation at the ninth International Conference on Rough Sets and Knowledge Technology (RSKT 2014) held at Tongji University, Shanghai, China from October 24 to 26, 2014.

The RSKT conference series was established in 2006 as a major international forum that brings researchers and industry practitioners together to discuss and deliberate on fundamental issues of knowledge processing and management, and knowledge-intensive practical solutions in the knowledge age. The past eight RSKT conferences were held in Chongqing, China (2006), Toronto, Canada (2007), Chengdu, China (2008), Gold Coast, Australia (2009), Beijing, China (2010), Banff, Canada (2011), Chengdu, China (2012), and Halifax, Canada (2013). The conferences have attracted experts from around the world to present their state-of-the-art scientific results, to nurture academic and industrial interaction, and to promote collaborative research in rough set theory and knowledge technology. The RSKT conference series so far impressively shows the immense progress that has been achieved and the wide acceptance and impact of rough sets in academia and practice. The RSKT proceedings, all published in Springer's LNCS series, have become rich reference material for researchers, students, and practitioners in rough sets and data and knowledge processing.

In total, RSKT 2014 received 162 submissions to the general conference, two special sessions, and four workshops. The special sessions were focused on

- Domain-Oriented Data-Driven Data Mining (3DM) organized by Andrzej Skowron, Guoyin Wang, and Yiyu Yao.
- Uncertainty Analysis in Granular Computing: An Information Entropy-Based Perspective organized by Duoqian Miao, Lin Shang and Hongyun Zhang.

The workshops were organized in the areas of

- Advances in Granular Computing (AGC 2014) organized by Andrzej Bargiela, Wei-Zhi Wu, William Zhu, Fan Min, Athanasios Vasilakos, and JingTao Yao.
- Big Data to Wise Decisions (First International Workshop, BD2WD 2014) organized by Andrzej Skowron, Guoyin Wang, Jiye Liang, Vijay V. Raghavan, Jie Tang, and Yiyu Yao.
- Rough Set Theory (Fifth Workshop, RST 2014) organized by Davide Ciucci, Yanyong Guan, and Marcin Wolski.
- Three-way Decisions, Uncertainty, and Granular Computing (Second International Workshop) organized by Tianrui Li, Salvatore Greco, Jerzy W. Grzymala-Busse, Ruizhi Wang, Dun Liu, and Pawan Lingras.

Following the tradition of rigorous reviewing in the RSKT conference series, each submitted paper went through a thorough review by at least two Program Committee (PC) members and domain experts; some papers even received up to five reviews. As a result, only top-quality papers were chosen for presentation at the conference including 38 regular papers (acceptance rate of 23.5%) and 40 papers for special sessions or workshops. The authors were requested to address the reviewer's comments and make revisions. Eventually, the final camera-ready submissions were carefully checked and approved by PC Chairs. Therefore, we would like to thank all the authors for submitting their best papers; without their contributions, this conference would not have been possible.

The RSKT 2014 program was further very much enriched by invited keynote speeches and plenary talks. We are most grateful to the high-profile keynote speakers, Zongben Xu, Hamido Fujita, and Thierry Denoeux, for their visionary addresses. The conference was also fortunate in bringing a group of established scholars to deliver invited talks. We thank the plenary speakers, Jerzy Grzymala-Busse, Jiming Liu, Duoqian Miao, Mikhail Moshkov, Sheila Petty, Andrzej Skowron, Shusaku Tsumoto, Guoyin Wang, Ruizhi Wang, JingTao Yao, and Yiyu Yao, for sharing their latest research insights.

RSKT 2014 was supported by many people and organizations. We thank the members of the Advisory Board and the Steering Committee for their invaluable suggestions and support throughout the organization process. We thank the Conference Chairs, Changjun Jiang, Witold Pedrycz, and Jue Wang, for the kind support, and the special session and workshop organizers for their great dedication. We express our thanks to the excellent work of Publicity Chairs, Ruizhi Wang, Lijun Sun, Hanli Wang, Pawan Lingras, and Xiaodong Yue, Local Organization Chairs, Hongyun Zhang, Cairong Zhao, Zhihua Wei, and Feifei Xu, and Webmaster, Zhifei Zhang. We thank the PC members and additional external reviewers for their thorough and constructive reviews.

We greatly appreciate the cooperation, support, and sponsorship of many institutions, companies, and organizations, including Tongji University, National Natural Science Foundation of China, International Rough Set Society, and Rough Sets and Soft Computing Society of the Chinese Association for Artificial Intelligence.

We acknowledge the use of EasyChair conference system for paper submission, review, and editing of the proceedings. We are thankful to Alfred Hofmann and the excellent LNCS team at Springer for their support and cooperation publishing the proceedings as a volume of the Lecture Notes in Computer Science.


October 2014                                          Duoqian Miao
                                                     Witold Pedrycz
                                                     Dominik Ślęzak
                                                       Georg Peters
                                                        Qinghua Hu
                                                       Ruizhi Wang

# Organization

## Organizing Committee

### Advisory Board

Bo Zhang (China)
Roman Slowinski (Poland)
Sankar K. Pal (India)
Ruqian Lu (China)
Fuji Ren (Japan)
Xindong Wu (USA)
Jian Pei (Canada)

### Steering Committee Chairs

Andrzej Skowron (Poland)
Guoyin Wang (China)
Yiyu Yao (Canada)

### Conference Chairs

Changjun Jiang (China)
Witold Pedrycz (Canada)
Jue Wang (China)

### Program Committee Chairs

Duoqian Miao (China)
Dominik Ślęzak (Poland)
Qinghua Hu (China)
Georg Peters (Germany)

### Publicity Chairs

Ruizhi Wang (China)
Lijun Sun (China)
Hanli Wang (China)
Pawan Lingras (Canada)
Xiaodong Yue (China)

**Local Organization Chairs**

Hongyun Zhang (China)
Cairong Zhao (China)
Zhihua Wei (China)
Feifei Xu (China)


# Program Committee

Aijun An (Canada)
Qiusheng An (China)
Mohua Banerjee (India)
Jan Bazan (Poland)
Theresa Beaubouef (USA)
Zbigniew Bonikowski (Poland)
Maciej Borkowski (USA)
Gianpiero Cattaneo (Italy)
Nick Cercone (Canada)
Mihir K. Chakraborty (India)
Chien-Chung Chan (USA)
Degang Chen (China)
Jiaxing Cheng (China)
Davide Ciucci (Italy)
Chris Cornelis (Belgium)
Krzysztof Cyran (Poland)
Jianhua Dai (China)
Martine De Cock (Belgium)
Dayong Deng (China)
Ivo Düntsch (Canada)
Lin Feng (China)
Yang Gao (China)
Anna Gomolinska (Poland)
Xun Gong (China)
Salvatore Greco (Italy)
Jerzy Grzymala-Busse (USA)
Shenming Gu (China)
Yanyong Guan (China)
Jianchao Han (USA)
Suqing Han (China)
Jun He (China)
Christopher Henry (Canada)
Daryl Hepting (Canada)
Joseph Herbert (Canada)
Shoji Hirano (Japan)
Qinghua Hu (China)

Xiaohua Hu (USA)
Xuegang Hu (China)
Bing Huang (China)
Masahiro Inuiguchi (Japan)
Lakhmi Jain (Australia)
Ryszard Janicki (Canada)
Jouni Jarvinen (Finland)
Richard Jensen (UK)
Xiuyi Jia (China)
Chaozhe Jiang (China)
Na Jiao (China)
Sankar K. Pal (India)
Janusz Kacprzyk (Poland)
C. Maria Keet (South Africa)
Jan Komorowski (Sweden)
Jacek Koronacki (Poland)
Abd El-monem Kozae (Egypt)
Marzena Kryszkiewicz (Poland)
Yasuo Kudo (Japan)
Aboul Ella Lassanien (Egypt)
Jingsheng Lei (China)
Henry Leung (Canada)
Daoguo Li (China)
Deyu Li (China)
Fanzhang Li (China)
Jinjin Li (China)
Longshu Li (China)
Huaxiong Li (China)
Tianrui Li (China)
Wen Li (China)
Yuefeng Li (Australia)
Jiye Liang (China)
Tsau Young Lin (USA)
Pawan Lingras (Canada)
Baoxiang Liu (China)
Caihui Liu (China)

Jian Yu (China)
Xiaodong Yue (China)
Slawomir Zadrozny (Poland)
Xianhua Zeng (China)
Bo Zhang (China)
Hongyun Zhang (China)
Ling Zhang (China)
Nan Zhang (China)
Qinghua Zhang (China)
Xianyong Zhang (China)

Xiaohong Zhang (China)
Yan-Ping Zhang (China)
Cairong Zhao (China)
Shu Zhao (China)
Caiming Zhong (China)
Ning Zhong (Japan)
Xianzhong Zhou (China)
Zhihua Zhou (China)
William Zhu (China)
Wojciech Ziarko (Canada)

## Additional Reviewers

Donald Chinn (USA)
Lynn D'Eer (Belgium)
Sofie De Clercq (Belgium)

Nele Verbiest (Belgium)
Fan Min (China)
Zehua Zhang (China)

## Sponsoring Institutions

Tongji University
National Natural Science Foundation of China
International Rough Set Society
Rough Sets and Soft Computing Society of the Chinese Association for Artificial
    Intelligence

# Highlights of Invited Presentations

## Keynote Speeches

### Research Directions in Subjective Decision Support System: Case Study on Medical Diagnosis

In decision making most approaches take into account objective criteria, however, the subjective correlation among decision makers provided as preference utility is necessary to be presented to provide confidence preference additive among decision makers reducing ambiguity and producing better utility preferences measurement for subjective criteria among decision makers. Most models in Decision support systems assume criteria as independent. Therefore, these models are ranking alternatives based on objective data analysis. Also, different types of data (time series, linguistic values, interval data, etc.) impose difficulties to perform decision making using classical multi-criteria decision making models.

Subjectiveness is related to the contextual form of criteria. Subjectiveness is contributing to provide knowledge and information on the criteria of mental existence to integrate its appropriateness relationship with the objective criteria for better understanding of incomplete decision makers' criteria setting. Uncertainty of some criteria in decision making is also considered as an other important aspect to be taken care of, in decision making and decision support system-based prediction. These drawbacks in decision making are major research challenges that are attracting wide attention, like on big data analysis for risk prediction, medical diagnosis, and other applications that are in practice more subjective to user situation and its knowledge-related context. Subjectivity is an important aspect in data analysis in considering a specific situation bounded by user preferences. Subjectiveness is a set of criteria that are filtering or reordering the objective criteria for better fitness and appropriateness in decision making based on a situation prediction.

For example, Probabilistic rough set adopts statistics to estimate the probability of a subset of objects with respect to equivalence relations. Subjective criteria are those criteria which are sensitive to the situation bounded by the mental relation to expert like sentimental relations to the expert experience or user-oriented mental interest in certain situations. In this talk I highlight subjectivity projected on Three Way decisions and explain the boundary region using reference point in Prospect Theory. This is to provide some highlight on Evidential Rough Set as an alternative perspective situated in a subjective context in real practices, like medical diagnosis. These are projected on rough set based on issues in entropy not necessarily based on Shannon's entropy theory, which in major part provides statistical-based gain (i.e., entropy) that would make quantified quality measurement not reflect on actual specific situation, as it does not reflect the communication context (subjective) and consider uncertainty as white noise. Therefore, new measuring directions are needed to qualify the gain of attribute context and its relationship.

I think quantifiable measurement of the boundary region in three-way decisions could provide a measure that can handle these drawbacks and also provide better understanding of the subjective uncertainty gain.

Subjectivity would be examined based on correlations between different contextual structures reflecting the framework of personal context, for example, in nearest neighbor-based correlation analysis fashion. Some of the attributes of incompleteness also may lead to affect the approximation accuracy. Attributes with preference-ordered domain relation properties become one aspect in ordering properties in rough approximations.

This talk provides an overview of subjectiveness in decision making, exhibited by a case study using medical diagnosis.

The Virtual Doctor System (VDS) developed by my group is a system assisting the human doctor, who is practicing medical diagnosis in the real situation and environment. The interoperability is represented by utilizing the medical diagnosis cases of medical doctors, represented in machine executable fashion based on human patient interaction with a virtual avatar resembling a real doctor. VDS is practiced as a virtual avatar interacting with the human patient based on physical views and mental view analysis. In this talk, I outline our VDS system and then discuss related issues on subjective decision making in the medical domain. Using fuzzy reasoning techniques in VDS, it has been shown is possible to provide better precision in circumstances related to partial known data and uncertainty on the acquisition of medical symptoms. We employed a combination of Fuzzy geometric aggregation for attributes for representing Physical view of VDS (Subjective attributes). We propose harmonic fuzzy reasoning in mental view in VDS projected on cognitive emotional models representation using fuzzy reasoning model. These are aligned and aggregated on the medical knowledge base using different distance functions and entropy functions. The purpose is to derive the weight of related attributes from the medical knowledge base and rank the preference order of the set of alternatives employing intuitionistic fuzzy similarity measures related to mental (Subjective) and physical (Objective) symptoms for decision making. A set of ideal solutions is provided based on simple case scenarios. The weight of mental decision making is derived based on hamming distance fuzzy operators. The alignment is to provide intelligent mapping between the mental view (Subjective) and physical view (Objective) based on fuzzy representation of each through different type of aggregation function. If the weights of some attributes are not known or partially known, then we need to predict using patient preferences by looking to the subjective risks.

These case studies are resembled through three-way decisions, by looking to an experiment related to simple case medical diagnosis. This is to classify objects into simple or not simple medical cases based on attribute reduction to search minimal attribute subset preserving medical concept of simple case, and their hierarchical structure by detecting irrelevant attributes from the knowledge bases based. Attribute reduction and classification are represented through regions, discernibility values in evidential rough set presentation. The subjective projection would provide some support means for the setting regions specifying discernible borders. These new directions are for present as well as future research.

Hamido Fujita
Iwate Prefectural University, Japan
Hfujita-799@acm.org

## Clustering in the Dempster-Shafer Framework: Comparison with Fuzzy and Rough Set Approaches

The Dempster-Shafer theory of belief is a powerful and well-founded formalism for reasoning under uncertainty. A belief function may be seen as a generalized set or as a nonadditive measure, and the theory extends both set-membership and Bayesian probability theory. In this talk, I review the basic principles of this theory and focus on its applications to clustering. There are basically two ways in which Dempster-Shafer theory can be applied to clustering. A first approach is based on the notion of credal partition, which extends the notions of hard, fuzzy, and possibilistic partitions. Several algorithms, including EVCLUS and ECM (evidential c-means), have been proposed to generate credal partitions from data by optimizing some objective criterion. The other approach is to search for space in all partition for the most plausible one, where the plausibility of a partition is computed from pairwise distances between objects. These approaches will be reviewed and contrasted with some fuzzy- and rough sets-based clustering algorithms.

Thierry Denoeux
Université de Technologie de Compiègne, France
Thierry.Denoeux@hds.utc.fr

# Plenary Talks

## Mining Incomplete Data: A Rough Set Approach

Incomplete data sets may be affected by three different kinds of missing attribute vales: lost values, attribute-concept values, and do not care conditions. For mining incomplete data sets three different generalizations of lower and upper approximations: singleton, subset, and concept are applicable. These approximations are further generalized to include a parameter interpreted as the probability. Properties of singleton, subset, and concept approximations are presented. Additionally, all these approximations are validated experimentally.

Jerzy Grzymala-Busse
University of Kansas, USA
jerzy@ku.edu

## From Big Data Analytics to Healthcare Intelligence

E-technology is among the key trends in defining and shaping the future of the healthcare sector, from personalized healthcare delivery to global health innovation. This transformation is mainly driven by the new opportunities and challenges created by *big data* analytics—giving meaning to voluminous, dynamically evolving, and often unstructured individuals health data (e.g., EHRs, genomic sequences, neuroimages, tracking device data, lifestyle choices, sentiment, and social networks), as well as those publically available open data (e.g., public health data, policies and clinical guidelines, claims, and cost data, R&D findings, demographics, and socioeconomic data). To engage in such a transformation, data analytics, data mining, and modeling technologies will play an increasingly important role in offering novel solutions to complex healthcare problems such as effective surveillance and prevention of diseases, efficient, and optimal utilization of healthcare services, to name a few.

In this talk, I discuss the key promise and challenges of big data analytics in achieving healthcare intelligence. In particular, I present some examples from our ongoing research supporting evidence-based policy-making for disease surveillance and prevention. As shown in the examples, data analytics and modeling helps characterize, in terms of tempo-spatial patterns, when, where, and how certain infectious or chronic diseases will likely develop and can be effectively prevented.

Jiming Liu
Hong Kong Baptist University, China
jiming@comp.hkbu.edu.hk

## Extensions of Dynamic Programming for Design and Analysis of Decision Trees

The aim of usual dynamic programming is to find an optimal object from a finite set of objects. We consider extensions of dynamic programming which allow us to (i) describe the set of optimal objects, (ii) count the number of these objects, (iii) make sequential optimization relative to different criteria, (iv) find the set of Pareto optimal points for two criteria, and (v) describe relationships between two criteria. The areas of application include discrete optimization, fault diagnosis, complexity of algorithms, machine learning, and knowledge representation.

In the presentation, we consider applications of this new approach to the study of decision trees as algorithms for problem solving, as a way for knowledge extraction and representation, and as predictors which, for a new object given by values of conditional attributes, define a value of the decision attribute.

The obtained results include the minimization of average depth for decision trees sorting eight elements (this question was open since 1968), improvement of upper bounds on the depth of decision trees for diagnosis of 0–1-faults in read-once combinatorial circuits, existence of totally optimal (with minimum depth and minimum number of nodes) decision trees for monotone Boolean functions with at most five variables, study of time-memory tradeoff for decision trees for corner point detection, study of relationships between number and maximum length of decision rules derived from decision trees, and study of accuracy-memory tradeoff for decision trees.

Mikhail Moshkov
King Abdullah University of Science and Technology, Saudi Arabia
`mikhail.moshkov@kaust.edu.sa`

## Remediation and Screen-Based Interfaces

This talk examines how the process of remediation works in screen-based interfaces. Remediation is a cornerstone of digital media studies and as Bolter and Grusin (1996, 1999) argue, the desire for newness or immediacy with each new visual aesthetic technology has resulted in a combinatory process that takes older visual aesthetic forms and reconfigures them with new approaches to create new forms. This is particularly true of digital media which routinely reshape or 'remediate' one another and their analog predecessors such as film, television, and photography in pursuit of new experiences of immediacy. Through a variety of digital media examples, this talk also probes issues of authenticity and interpretation in remediation of screen-based interfaces and attempts to determine if it is possible to apply the principles of granular computing to create effective knowledge processes.

Sheila Petty
University of Regina, Canada
`sheila.petty@uregina.ca`

## Foundations for Cyber-Physical Systems Based on Interactive Granular Computing

**Cyber-Physical Systems (CPS)** help us to interact with the physical world just as the Internet helps us to interact with one another. It is predicted that applications based on CPS will have enormous societal impact and bring enormous economic benefit.

We discuss a computational model for CPS based on interactive complex granules (c-granules, for short). C-granules are controlled by agents. Any c-granule of a given agent specifies a perceived structure of its local environment consisting of portions of matter (physical objects), called hunks. There are three kinds of such hunks: (i) hunks in the agent external environment creating the *hard_suit* of c-granule, (ii) internal hunks of agent creating the *soft_link* of c-granule, some of which can be represented by agent as information granules (infogranules, for short), and (iii) hunks creating the *link_suit* of c-granule and playing the role of links between hunks from the *hard_suit* and *soft_suit*. This structure is used in recording by means of infogranules the results of interactions of hunks from the local environment.

Our approach is based on the Wisdom Technology meta-equation

$$WISDOM = INTERACTIONS + ADAPTIVE\_JUDGMENT$$
$$+ KNOWLEDGE\_BASES.$$

The understanding of interactions is the critical issue of CPS treated as complex systems. Using c-granules one can model interactions of agents with the physical world and represent perception of interactions in the physical world by agents. Adaptive judgment allows agents to reason about c-granules and interactive computations performed on them. In adaptive judgment, different kinds of reasoning are involved such as deduction, induction, abduction, reasoning by analogy, or reasoning for efficiency management. In the approach, an important role is also played by knowledge bases and interactions of agents with them.

Some illustrative applications of the proposed approach related to real-life projects (e.g., respiratory failure, UAV control, algorithmic trading, sunspot classification, semantic search engine, firefighter safety) are reported. We emphasize the pivotal role of the proposed approach for efficiency management in CPS.

Andrzej Skowron
University of Warsaw, Poland
skowron@mimuw.edu.pl

## Data Mining in Hospital Information System

Twenty years have passed since clinical data were stored electronically as a hospital information system (HIS). Stored data give all the histories of clinical activities in a hospital, including accounting information, laboratory data and electronic patient records and their executed histories. Due to the traceability of all the information, a hospital cannot function without the information system, which is one of the most important infrastructures. The size of data in HIS is monotonically increasing, which can be viewed as "big data". However, reuse of the stored data has not yet been discussed in detail, except for laboratory data and accounting information to which OLAP methodologies are applied. Data mining approach just started ten years ago, but their usefulness in clinical environments has not yet been fully reported.

This talk presents the research achievements of the speaker's institution, where a scheme for innovation of hospital services based on data mining has been introduced. The goal for the hospital services is to realize the following three layers of hospital services: services for hospital management, devices for medical staff and services for patients in order to achieve the efficient hospital services. Since all the results of services were stored as data, data mining in hospital information system is very important. The first layer is called services for patients, which supports the improvement of healthcare service delivery for patients. This is a fundamental level of healthcare services in which medical staff directly gives medical services to the patients. Patient records and other results of clinical examinations support the quality of this service. The second layer is called services for medical staff, which supports decision making of medical practitioner. Patient histories and clinical data are applied to data mining techniques which give useful patterns for medical practice. Especially, detection of risk of patients, such as drug adverse effects or temporal status of chronic diseases will improve the qualities of medical services. The top layer is called services for hospital management. This level is achieved by capturing global behavior of a hospital: the bridging between microscopic behavior of medical staff and macroscopic behavior of hospital is very important to deploy medical staff in an optimal way for improving performance of the hospital. Thus, it is highly expected that data mining in data extracted from hospital information systems plays a central role in achieving the hierarchical scheme.

Research of the speaker's group applied several data mining techniques for this purpose for twenty years. The talk surveys the achievements, where the studies focus on the two aspects of hospital data, which consist of executed results and histories of services. Mainly, mining results of executed services (contents mining) is connected with medical decision making, whereas mining histories of services (history mining) is connected with service innovation. In both cases, rough set-based rule induction, decision tree induction, trajectories mining, clustering and multidimensional scaling (MDS) were adopted for analysis.

The results show that the introduced methods are not only useful for improvement of medical decision making, but also useful for capturing the characteristics of clinical activities in hospital as follows. Concerning decision making, the most important contributions are that these methods gave tools for data mining based risk management, called "risk mining". Rough set-based rule induction plays a central role in extracting rules for medical diagnosis, prevention of medical accidents and intra-hospital infection. Although extraction of medical diagnosis rule led to several important discoveries, rule induction in prevention of risky events gave more impact: interpretation of rules enabled the medial staff to reduce the number of errors in clinical environments. Furthermore, comparison of risk factors between different institutes was obtained in terms of intra-hospital infection. Trajectories mining succeeded in detecting two or three dimensional temporal evolution of diseases and classification of temporal characteristics of these diseases, which shed lights on risk factors of disease progression. Clustering and MDS successfully visualized characteristics of results obtained by rule induction and trajectories mining.

Concerning history mining, the following five contributions were focused on: first, the chronological overview of hospital activities, periodical behavior of the number of orders were visualized, which can be viewed as a "life-cycle" of a hospital. Secondly, a pattern of long-term follow up patients with respect to the number of orders by using HIS data is extracted. Thirdly, trajectories mining technique is applied to temporal analysis the number of orders. The results of clustering analysis gave two groups of clinical actions. The one was a pattern where orders are given both in wards and outpatient clinics. The other one was a pattern where orders are provided mainly in the wards. Fourth, clustering of temporal sequences of the number of orders captures the behavior of medical doctors, whose interpretation leads to improvements in service of outpatient clinic. Finally, similarity-based analysis was applied to temporal trends of the numbers of nursing orders. The results showed that nursing orders are automatically classified into two major categories, "disease-specific" and "patient-specific" ones. Furthermore, the former one was classified into three subcategories, according to the temporal characteristics. The method is now used for construction of clinical pathways, which improve the efficiency of hospital services.

Shusaku Tsumoto
Shimane University, Japan
tsumoto@med.shimane-u.ac.jp

## Multi-granularity Bidirectional Computational Cognition for Big Data Mining and Cognition

Big data processing is a challenging problem in intelligent uncertain information processing. The big data-based cognition and mining useful knowledge from big data are key issues of big data processing. Granular computing (GrC) provides some useful models and methods for dealing with this problem. Multi-granularity computational cognition might be a useful tool for intelligent cognition of big data. In this talk, multi-granularity bidirectional computational cognition between concepts' extension and intension, one of the key issues of multi-granularity computational cognition are introduced. It is also a basic problem of both cognitive computing and big data processing.

Many data mining and machine learning methods have been proposed for extracting knowledge from data in the past decades. Unfortunately, they are all unidirectional cognitive computing models for extracting knowledge from concepts' extension to intension only. In the view of granular computing, they transfer knowledge from finer granule levels to coarser granule levels. However, bidirectional computational cognition between concept's extension and intension provide bidirectional knowledge (information) transformations between intension and extension respectively. Some new research results on multi-granularity bidirectional computational cognition models based on cloud model for intelligent uncertain information processing are introduced. The relation between cloud model and gauss normal distribution model for expressing and processing uncertain concepts are analyzed.

Some interesting cognition experiments based on bidirectional computational cognition models, such as cognizing a concept over and over again, increasing cognition of a concept, multi-granularity cognition of a concept, image cognition, etc., are also introduced.

Guoyin Wang

Chongqing University of Posts and Telecommunications, China

wanggy@ieee.org

## Three-Way Decisions with Game-Theoretic Rough Sets

There are two major problems in intelligent decision making, i.e., too many options and contradictive criteria. Simplifying multi-option decision making to binary decision making, e.g., acceptance and rejection, is a commonly used approach. However, making a decision with insufficient information may have a low accuracy level and result in unexpected consequences. The recent proposed ternary or three-way decision model with a non-commitment option may shed light on such a problem. Probabilistic rough sets introduce a pair of thresholds which leads to three-way decision making. The game-theoretic rough set (GTRS) model determines a balanced, optimal threshold pair by setting up a game for trading off between different criteria. This produces to a moderate, cost-effective, or efficient level of acceptance, rejection or deferment decision. The GTRS model also provides an alternative mechanism for solving the problems of decision making with contradictive criteria.

JingTao Yao

University of Regina, Canada

jtyao@cs.uregina.ca

## Uncertainty in Three-Way Decisions and Granular Computing

Three-Way Decisions (3WD) are formulated based on the options of acceptance, rejection, and non-commitment, leading to three pair-wise disjoint classes of objects called the positive, negative, and neutral/middle/boundary regions. Three-way decisions can be motivated, interpreted, and implemented based on the notion of information granularity. When coarse-grained granules are used, it may only be possible to make a definite decision of acceptance or rejection for some objects. For other objects, it is impossible to make a definite decision due to a lack of detailed information and hence the third non-commitment option is used. Objects with a non-commitment decision may be further investigated by using fine-grained granules. In this way, multiple levels of granularity lead naturally to sequential three-way decisions. The notion of uncertainty plays a crucial role in both three-way decisions and granular computing. Compared with the commonly used two-way/binary models, the existence of uncertain information makes three-way decisions a superior model for practical decision making. Different levels of granularity are associated with different degrees of uncertainty, resulting in three-way decisions with different quality. A study of uncertainty in three-way decisions and granular computing may provide insights into the inherent relationships between the two fields.

|  |  |  |
|---|---|---|
| Duoqian Miao | Ruizhi Wang | Yiyu Yao |
| Tongji University, | Tongji University, | University of Regina, |
| China | China | Canada |
| dqmiao@tongji.edu.cn | ruizhiwang@tongji.edu.cn | yyao@cs.uregina.ca |

# Contents

## Foundations and Generalizations of Rough Sets

Generalized Dominance-Based Rough Set Model for the Dominance
Intuitionistic Fuzzy Information Systems ........................... 3
    *Xiaoxia Zhang and Degang Chen*

On Definability and Approximations in Partial Approximation
Spaces ......................................................... 15
    *Davide Ciucci, Tamás Mihálydeák, and Zoltán Ernő Csajbók*

Many-Valued Rough Sets Based on Tied Implications ................. 27
    *Moataz El-Zekey*

On the Topological Structure of Rough Soft Sets .................... 39
    *Vinay Gautam, Vijay K. Yadav, Anupam K. Singh, and S.P. Tiwari*

Knowledge Granulation in Interval-Valued Information Systems Based
on Maximal Consistent Blocks ..................................... 49
    *Nan Zhang and Xiaodong Yue*

Multi-granulation Rough Sets Based on Central Sets ................. 59
    *Caihui Liu, Meizhi Wang, Yujiang Liu, and Min Wang*

Rough Set Theory on Topological Spaces ............................ 69
    *K. Anitha*

## Attribute Reduction and Feature Selection

Co-training Based Attribute Reduction for Partially Labeled Data ..... 77
    *Wei Zhang, Duoqian Miao, Can Gao, and Xiaodong Yue*

Approximate Reduction for the Interval-Valued Decision Table ........ 89
    *Feifei Xu, Zhongqin Bi, and Jingsheng Lei*

Global Best Artificial Bee Colony for Minimal Test Cost Attribute
Reduction ....................................................... 101
    *Anjing Fan, Hong Zhao, and William Zhu*

Reductions of Intuitionistic Fuzzy Covering Systems Based on
Discernibility Matrices ........................................... 111
    *Tao Feng and Jusheng Mi*

Multi-label Feature Selection with Fuzzy Rough Sets ................. 121
    *Lingjun Zhang, Qinghua Hu, Jie Duan, and Xiaoxue Wang*

A Logarithmic Weighted Algorithm for Minimal Test Cost Attribute
Reduction.................................................... 129
    *Junxia Niu, Hong Zhao, and William Zhu*

Attribute Reduction in Object Oriented Concept Lattices Based on
Congruence Relations ......................................... 139
    *Xia Wang and Wei-Zhi Wu*

An Explicit Sparse Mapping for Nonlinear Dimensionality Reduction ... 149
    *Ying Xia, Qiang Lu, JiangFan Feng, and Hae-Young Bae*

## Applications of Rough Sets

A Web-Based Learning Support System for Rough Sets .............. 161
    *Ying Zhou and JingTao Yao*

A Parallel Matrix-Based Approach for Computing Approximations in
Dominance-Based Rough Sets Approach .......................... 173
    *Shaoyong Li and Tianrui Li*

Propositional Compilation for All Normal Parameter Reductions of a
Soft Set.................................................... 184
    *Banghe Han and Xiaonan Li*

Top-N Recommendation Based on Granular Association Rules ......... 194
    *Xu He, Fan Min, and William Zhu*

An Efficient BP-Neural Network Classification Model Based on
Attribute Reduction.......................................... 206
    *Yongsheng Wang and Xuefeng Zheng*

Hardware Accelerator Design Based on Rough Set Philosophy ......... 216
    *Kanchan S. Tiwari, Ashwin G. Kothari, and
    K.S. Sreenivasa Raghavan*

## Intelligent Systems and Applications

Properties of Central Catadioptric Circle Images and Camera
Calibration................................................. 229
    *Huixian Duan, Lin Mei, Jun Wang, Lei Song, and Na Liu*

Communication Network Anomaly Detection Based on Log File
Analysis ................................................... 240
    *Xin Cheng and Ruizhi Wang*

Using the Multi-instance Learning Method to Predict Protein-protein
Interactions with Domain Information ........................... 249
    *Yan-Ping Zhang, Yongliang Zha, Xinrui Li, Shu Zhao, and
    Xiuquan Du*

An Improved Decimation of Triangle Meshes Based on Curvature . . . . . .    260
    Wei Li, Yufei Chen, Zhicheng Wang, Weidong Zhao, and Lin Chen

A Community Detecting Algorithm Based on Granular Computing . . . . .    272
    Lu Liu, Taorong Qiu, Xiaoming Bai, and Zhongda Lin

A Robust Online Tracking-Detection Co-training Algorithm with
Applications to Vehicle Recognition . . . . . . . . . . . . . . . . . . . . . . . . . . . . .    285
    Chen Jiyuan and Wei Zhihua

An Approach for Automatically Reasoning Consistency of
Domain-Specific Modelling Language . . . . . . . . . . . . . . . . . . . . . . . . . . . .    295
    Tao Jiang, Xin Wang, and Li-Dong Huang

## Knowledge Technology

Online Object Tracking via Collaborative Model within the Cascaded
Feedback Framework . . . . . . . . . . . . . . . . . . . . . . . . . . . . . . . . . . . . . . . . . .    309
    Sheng Tian and Zhihua Wei

Optimal Subspace Learning for Sparse Representation Based Classifier
via Discriminative Principal Subspaces Alignment . . . . . . . . . . . . . . . . . . .    320
    Lai Wei

A Statistics-Based Semantic Relation Analysis Approach for Document
Clustering . . . . . . . . . . . . . . . . . . . . . . . . . . . . . . . . . . . . . . . . . . . . . . . . . . .    332
    Xin Cheng, Duoqian Miao, and Lei Wang

Multi-granulation Ensemble Classification for Incomplete Data . . . . . . . . .    343
    Yuan-Ting Yan, Yan-Ping Zhang, and Yi-Wen Zhang

Heterogeneous Co-transfer Spectral Clustering . . . . . . . . . . . . . . . . . . . . . .    352
    Liu Yang, Liping Jing, and Jian Yu

Mixed Pooling for Convolutional Neural Networks . . . . . . . . . . . . . . . . . . . .    364
    Dingjun Yu, Hanli Wang, Peiqiu Chen, and Zhihua Wei

An Approach for In-Database Scoring of R Models on DB2 for z/OS . . .    376
    Yikun Xian, Jie Huang, Yefim Shuf, Gene Fuh, and Zhen Gao

On the $FM_\alpha$-Integral of Fuzzy-Number-Valued Functions . . . . . . . . . . . . .    386
    Yabin Shao, Qiang Ma, and Xiaoxia Zhang

## Domain-Oriented Data-Driven Data Mining

Rough Classification Based on Correlation Clustering . . . . . . . . . . . . . . . . .    399
    László Aszalós and Tamás Mihálydeák

An Improved Hybrid ARIMA and Support Vector Machine Model for
Water Quality Prediction ............................................. 411
    *Yishuai Guo, Guoyin Wang, Xuerui Zhang, and Weihui Deng*

Multi-label Supervised Manifold Ranking for Multi-instance Image
Retrieval.............................................................. 423
    *Xianhua Zeng, Renjie Lv, and Hao Lian*

Nearest Neighbor Condensation Based on Fuzzy Rough Set for
Classification ........................................................ 432
    *Wei Pan, Kun She, Pengyuan Wei, and Kai Zeng*

Predicting Movies User Ratings with Imdb Attributes ................ 444
    *Ping-Yu Hsu, Yuan-Hong Shen, and Xiang-An Xie*

Feature Selection for Multi-label Learning Using Mutual Information
and GA................................................................ 454
    *Ying Yu and Yinglong Wang*

## Uncertainty in Granular Computing

Characterizing Hierarchies on Covering-Based Multigranulation
Spaces................................................................. 467
    *Jingjing Song, Xibei Yang, Yong Qi, Hualong Yu,*
    *Xiaoning Song, and Jingyu Yang*

Uncertainty Measures in Interval-Valued Information Systems ......... 479
    *Nan Zhang and Zehua Zhang*

A New Type of Covering-Based Rough Sets ......................... 489
    *Bin Yang and William Zhu*

A Feature Seletion Method Based on Variable Precision Tolerance
Rough Sets........................................................... 500
    *Na Jiao*

Incremental Approaches to Computing Approximations of Sets in
Dynamic Covering Approximation Spaces .......................... 510
    *Guangming Lang, Qingguo Li, Mingjie Cai, and Qimei Xiao*

## Advances ih Granular Computing

Knowledge Approximations in Multi-scale Ordered Information
Systems .............................................................. 525
    *Shen-Ming Gu, Yi Wu, Wei-Zhi Wu, and Tong-Jun Li*

An Addition Strategy for Reduct Construction ..................... 535
    *Cong Gao and Yiyu Yao*

Analysis of User-Weighted $\pi$ Rough k-Means ...................... 547
   *Georg Peters and Pawan Lingras*

An Automatic Virtual Calibration of RF-Based Indoor Positioning
with Granular Analysis ............................................. 557
   *Ye Yin, Zhitao Zhang, Deying Ke, and Chun Zhu*

Algebraic Structure of Fuzzy Soft Sets ............................. 569
   *Zhiyong Hong and Keyun Qin*

A Rapid Granular Method for Minimization of Boolean Functions ...... 577
   *Zehua Chen, He Ma, and Yu Zhang*

## Big Data to Wise Decisions

QoS-Aware Cloud Service Selection Based on Uncertain User
Preference ......................................................... 589
   *Bin Mu, Su Li, and Shijin Yuan*

Attribute Reduction in Decision-Theoretic Rough Set Model Using
MapReduce ......................................................... 601
   *Jin Qian, Ping Lv, Qingjun Guo, and Xiaodong Yue*

A Context-Aware Recommender System with a Cognition Inspired
Model ............................................................. 613
   *Liangliang Zhao, Jiajin Huang, and Ning Zhong*

Study on Fuzzy Comprehensive Evaluation Model of Education
E-government Performance in Colleges and Universities .............. 623
   *Fang Yu, Lijuan Ye, and Jiaming Zhong*

Parallel Attribute Reduction Based on MapReduce ................... 631
   *Dachao Xi, Guoyin Wang, Xuerui Zhang, and Fan Zhang*

Dynamic Ensemble of Rough Set Reducts for Data Classification ...... 642
   *Jun-Hai Zhai, Xi-Zhao Wang, and Hua-Chao Wang*

## Rough Set Theory

Intuitionistic Fuzzy Rough Approximation Operators Determined by
Intuitionistic Fuzzy Triangular Norms ............................. 653
   *Wei-Zhi Wu, Shen-Ming Gu, Tong-Jun Li, and You-Hong Xu*

Covering Approximations in Set-Valued Information Systems .......... 663
   *Yanqing Zhu and William Zhu*

Rough Fuzzy Set Model for Set-Valued Ordered Fuzzy Decision
System ............................................................ 673
   *Zhongkui Bao, Shanlin Yang, and Ju Zhao*

Optimal-Neighborhood Statistics Rough Set Approach with Multiple
Attributes and Criteria . . . . . . . . . . . . . . . . . . . . . . . . . . . . . . . . . . . . . . . . .   683
   *WenBin Pei, He Lin, and LingYue Li*

Thresholds Determination for Probabilistic Rough Sets with Genetic
Algorithms . . . . . . . . . . . . . . . . . . . . . . . . . . . . . . . . . . . . . . . . . . . . . . . . . . . . .   693
   *Babar Majeed, Nouman Azam, and JingTao Yao*

## Three-Way Decisions, Uncertainty, and Granular Computing

Three-Way Weighted Entropies and Three-Way Attribute Reduction . . .   707
   *Xianyong Zhang and Duoqian Miao*

Applying Three-Way Decisions to Sentiment Classification with
Sentiment Uncertainty . . . . . . . . . . . . . . . . . . . . . . . . . . . . . . . . . . . . . . . . . .   720
   *Zhifei Zhang and Ruizhi Wang*

Three-Way Formal Concept Analysis . . . . . . . . . . . . . . . . . . . . . . . . . . . . .   732
   *Jianjun Qi, Ling Wei, and Yiyu Yao*

Three-Way Decision Based on Belief Function . . . . . . . . . . . . . . . . . . . . . .   742
   *Zhan'ao Xue, Jie Liu, Tianyu Xue, Tailong Zhu, and Penghan Wang*

Semantically Enhanced Clustering in Retail Using Possibilistic
K-Modes . . . . . . . . . . . . . . . . . . . . . . . . . . . . . . . . . . . . . . . . . . . . . . . . . . . . . .   753
   *Asma Ammar, Zied Elouedi, and Pawan Lingras*

A Three-Way Decisions Clustering Algorithm for Incomplete Data . . . . .   765
   *Hong Yu, Ting Su, and Xianhua Zeng*

Sentiment Analysis with Automatically Constructed Lexicon and
Three-Way Decision . . . . . . . . . . . . . . . . . . . . . . . . . . . . . . . . . . . . . . . . . . . .   777
   *Zhe Zhou, Weibin Zhao, and Lin Shang*

A Novel Intelligent Multi-attribute Three-Way Group Sorting Method
Based on Dempster-Shafer Theory . . . . . . . . . . . . . . . . . . . . . . . . . . . . . . .   789
   *Baoli Wang and Jiye Liang*

Dynamic Maintenance of Three-Way Decision Rules . . . . . . . . . . . . . . . . .   801
   *Chuan Luo, Tianrui Li, and Hongmei Chen*

An Overview of Function Based Three-Way Decisions . . . . . . . . . . . . . . . .   812
   *Dun Liu and Decui Liang*

Multicost Decision-Theoretic Rough Sets Based on Maximal Consistent
Blocks . . . . . . . . . . . . . . . . . . . . . . . . . . . . . . . . . . . . . . . . . . . . . . . . . . . . . . . . .   824
   *Xingbin Ma, Xibei Yang, Yong Qi, Xiaoning Song, and Jingyu Yang*

A Method to Reduce Boundary Regions in Three-Way Decision
Theory .......................................................... 834
    *Ping Li, Lin Shang, and Huaxiong Li*

An Integrated Method for Micro-blog Subjective Sentence Identification
Based on Three-Way Decisions and Naive Bayes ..................... 844
    *Yanhui Zhu, Hailong Tian, Jin Ma, Jing Liu, and Tao Liang*

Probabilistic Rough Set Model Based on Dominance Relation ......... 856
    *Wentao Li and Weihua Xu*

**Author Index** ................................................ 865

# Foundations and Generalizations
# of Rough Sets

# Generalized Dominance-Based Rough Set Model for the Dominance Intuitionistic Fuzzy Information Systems

Xiaoxia Zhang* and Degang Chen**

School of Control and Computer Engineering,
North China Electric Power University, Beijing 102206, China

**Abstract.** A dominance-based rough set approach was proposed by replacing the indiscernibility relation with a dominance relation. The aim of this paper is to present a new extension of the dominance-based rough set by means of defining a new dominance relation, i.e., generalized dominance-based rough set model is proposed based on the dominance intuitionistic fuzzy information systems. To get the optimal decision rules from the existing dominance intuitionistic fuzzy information systems, a lower and upper approximation reduction and rule extraction algorithm are investigated. Furthermore, several properties of the generalized dominance-based rough set model are given, and the relationships between this model and the others dominance-based rough set models are also examined.

**Keywords:** Dominance-based rough set, generalized dominance-based rough set, attribute reduction, rule extraction.

## 1 Introduction

Rough set theory, as an extension of the set theory for the study of intelligent systems characterized by insufficient and incomplete information, was proposed by Pawlak in 1982 [1]. Rough sets have been successfully applied in pattern recognition, data mining, machine learning, and so on. A key notion in Pawlak's rough set model is equivalence relation. The equivalence classes are the building blocks for the construction of the lower and upper approximations. By replacing the equivalence relation with fuzzy relations [2], ordinary binary relations [3], tolerance relations [4], dominance relations [5], covering relations [6], and others [7], various generalized rough set models are proposed, in which the dominance-based rough set models and the fuzzy rough set theory are two types of the most important extended rough set models.

---

\* Thanks to the support by Natural Scientific Fund of Gansu Province of China (1308RJZA125).
\*\* Thanks to the support by National Natural Scientific Fundation of China (No. 71471060).

D. Miao et al. (Eds.): RSKT 2014, LNAI 8818, pp. 3–14, 2014.
DOI: 10.1007/978-3-319-11740-9_1 © Springer International Publishing Switzerland 2014

Presently, work on dominance-based rough set model progressing rapidly. To consider the ranking properties of criteria, Greco et al. [8,9] proposed a dominance-based rough set approach (DRSA) based on the substitution of the indiscernibility relation with a dominance relation. In the DRSA, condition attributes are the criteria used and classes are ranked by preferences. Therefore, the knowledge approximated is a collection of dominance classes, which are sets of objects defined by a dominance relation. In recent years, a number of studies on DRSA have been conducted [10,11,12]. By introducing the concept of DRSA into the fuzzy environment, Greco et al. also proposed the dominance-based rough fuzzy model [13]. In such model, the fuzzy target is approximated by using a dominance relation instead of an indiscernibility relation. As a further investigation, Hu et al. [14] presented an algorithm to compute the reductions of the variable precision dominance-based rough set model. Wang and Chen et al. [15] first examined the relationships between covering information systems and ordered information systems. Huang et al. [16,17] discussed dominance-based (interval-valued) intuitionistic fuzzy rough set models and their applications.

In aforementioned researches about DRSA, dominance class $[x]_{R_C^{\preceq}}$ is the set of objects dominating $x$ for each attribute of attribute set $C$. As we all known, if an object dominating $x$ for each attribute of $C$, then its comprehensive evaluation value is also better than that of $x$. Conversely, if the comprehensive evaluation value of one object is better than that of $x$, then it may not be better than $x$ for each attribute of condition attribute set $C$. However, in some real-life situations, we always consider the comprehensive evaluation value of alternatives if they have no special requirements for individual attribute value. Therefore, dominance class $[x]_{R_C^{\preceq}}$ of Huang and Wei et al. [16,17] is too restrictive to inconvenient for various practical applications. Based on this point, we introduce the notion of generalized dominance classes $[x]_C^{\preceq}$ and $[x]_C^{\succeq}$ in intuitionistic fuzzy information systems. Generalized dominance classes only consider the comprehensive evaluation value and regardless of single attribute value of alternatives, i.e, generalized dominance class $[x]_C^{\preceq}$ is the set of objects dominating $x$ with respect to comprehensive evaluation value. Thus, we propose a new dominance-based rough set model, i.e., generalized dominance-based rough set model. Meanwhile, several properties of the generalized dominance-based rough set model are given, and the relationships between this model and the others dominance-based rough set models are also examined. Furthermore, we also investigated the reductions and rule extraction of this model.

The structure of this paper is organized as follows: Section 2 briefly introduces the preliminary issues considered in the study, such as the notations of intuitionistic fuzzy set, some basic operations of intuitionistic fuzzy sets, and how to determine the dominance classes of an object with respect to its attribute values. In Section 3, we define a generalized dominance-based (intuitionistic fuzzy) rough set model and investigate the corresponding properties and the relationships between this model and the others dominance-based rough set models. Meanwhile, the attribute reduction and rule extraction of the model are also examined.

## 2   Preliminaries

Let $U$ be a nonempty and finite universe of discourse. Throughout this paper, the universe $U$ is considered to be finite. The class of all subsets of $U$ will be denoted by $P(U)$.

*Definition 2.1.* [18,19,20] Let $U$ be a nonempty and finite universe of discourse. An intuitionistic fuzzy set(IF) $A$ in $U$ is an object having the form

$$A = \{\langle x, \mu_A(x), \nu_A(x)\rangle | x \in U\},$$

where $\mu_A : U \to [0,1]$ and $\nu_A : U \to [0,1]$ satisfy $0 \le \mu_A(x) + \nu_A(x) \le 1$ for all $x \in U$, and $\mu_A(x)$ and $\nu_A(x)$ are called the degree of membership and the degree of non-membership of the element $x \in U$ to $A$, respectively. $\pi_A(x) = 1 - \mu_A(x) - \nu_A(x)$ called the degree of hesitancy of the element $x \in U$ to $A$. The complement of an IF set $A$ is defined by $\sim A = \{\langle x, \nu_A(x), \mu_A(x)\rangle | x \in U\}$.

We call $A(x) = (\mu_A(x), \nu_A(x))$ an intuitionistic fuzzy values. Especially, for any $A \in P(U)$, if $x \in A$, then $A(x) = (1, 0)$; if $x \notin A$, then $A(x) = (0, 1)$.

*Definition 2.2.* [20,21,22,23] Let $A_i(x) = (\mu_{A_i}(x), \nu_{A_i}(x))$, $i = 1, 2, \cdots, m$. Then $\oplus_{i=1}^n A_i$ and $\otimes_{i=1}^n A_i$ are defined as follows:

$$\oplus_{i=1}^n A_i = \left\{ \langle x, 1 - \prod_{i=1}^n (1 - \mu_{A_i}(x)), \prod_{i=1}^n \nu_{A_i}(x)\rangle | x \in U \right\},$$

$$\otimes_{i=1}^n A_i = \left\{ \langle x, \prod_{i=1}^n \mu_{A_i}(x), 1 - \prod_{i=1}^n (1 - \nu_{A_i}(x))\rangle | x \in U \right\}.$$

*Definition 2.3* [24] Let $A(x) = (\mu_A(x), \nu_A(x))$ and $B(x) = (\mu_B(x), \nu_B(x))$ be two intuitionistic sets, $s(A(x)) = \mu_A(x) - \nu_A(x)$ and $s(B(x)) = \mu_B(x) - \nu_B(x)$ be the scores of $A(x)$ and $B(x)$, respectively; and let $h(A(x)) = \mu_A(x) + \nu_A(x)$ and $h(B(x)) = \mu_B(x) + \nu_B(x)$ be precisions of $A(x)$ and $B(x)$, respectively, then
  (1) If $s(A(x)) < s(B(x))$, then $A(x) \prec B(x)$;
  (2) If $s(A(x)) = s(B(x))$ and $h(A(x)) = h(B(x))$, then $A(x) = B(x)$;
  (3) If $s(A(x)) = s(B(x))$ and $h(A(x)) < h(B(x))$, then $A(x) \prec B(x)$;
  (4) If $s(A(x)) = s(B(x))$ and $h(A(x)) > h(B(x))$, then $A(x) \succ B(x)$.

*Definition 2.4.* An intuitionistic fuzzy information system (IFIS) is a quadruple $S = (U, AT = C \cup D, V, f)$, where $U$ is a non-empty and finite set of objects called the universe, $C$ is a non-empty and finite set of conditional attributes, $D = \{d\}$ is a singleton of decision attribute $d$, and $C \cap D = \emptyset$. $V = V_C \cup V_D$, where $V_C$ and $V_D$ are domains of condition and decision attributes, respectively. The information function $f$ is a map from $U \times (C \cup D)$ onto $V$, such that $f(x, c) \in V_C$ for all $c \in C$ and $f(x, d) \in V_D$ for $D = \{d\}$, where $f(x, c)$ and $f(x, d)$ are intuitionistic fuzzy values, denoted by $f(x, c) = c(x) = (\mu_c(x), \nu_c(x))$, $f(x, d) = d(x) = (\mu_d(x), \nu_d(x))$.

Let $S = (U, AT = C \cup D, V, f)$ be an IFIS. For any $a \in AT$ and $x, y \in U$, denoted by

$$f(x, a) \preceq f(y, a) \Leftrightarrow (\forall a \in AT)[f(x, a) \prec f(y, a) \vee f(x, a) = f(y, a)],$$

$$f(x, a) \succeq f(y, a) \Leftrightarrow (\forall a \in AT)[f(x, a) \succ f(y, a) \vee f(x, a) = f(y, a)].$$

By using $\preceq$ and $\succeq$, we can obtain an increasing preference and a decreasing preference. If the domain of an attribute is ordered according to a decreasing or increasing preference, then the attribute is a criterion.

*Definition 2.5.* An IFIS $S = (U, AT = C \cup D, V, f)$ is called a dominance intuitionistic fuzzy information system (DIFIS) if all condition attributes are criterions, denoted by $\widetilde{S}$.

*Definition 2.6.* Let $\widetilde{S} = (U, AT = C \cup D, V, f)$ be a dominance intuitionistic fuzzy information system. Dominance relations $R_B^{\preceq}$ and $R_B^{\succeq}$ in DIFIS $\widetilde{S}$ are defined as follows:

$$R_B^{\preceq} = \{(x, y) \in U \times U | f(x, a) \preceq f(y, a), \forall\, a \in B \subseteq AT\},$$

$$R_B^{\succeq} = \{(x, y) \in U \times U | f(x, a) \succeq f(y, a), \forall\, a \in B \subseteq AT\}.$$

The dominance class $[x]_{R_B^{\preceq}}$ induced by $R_B^{\preceq}$ is the set of objects dominating $x$, i.e. $[x]_{R_B^{\preceq}} = \{y \in U | f(x, a) \preceq f(y, a), \forall\, a \in B \subseteq AT\}$, where $[x]_{R_B^{\preceq}}$ describes the set of objects that may dominate $x$, and $[x]_{R_B^{\preceq}}$ is called the dominating class with respect to $x \in U$. Meanwhile, the dominated class with respect to $x \in U$ can be defined as $[x]_{R_B^{\succeq}} = \{y \in U | f(x, a) \succeq f(y, a), \forall\, a \in B \subseteq AT\}$.

Equivalence relation $R$ in $\widetilde{S}$ is defined as follows:

$$R = \{(x, y) \in U \times U | f(x, a) = f(y, a), \forall\, a \in AT\}.$$

With respect to $R$, we can define an equivalence class of $x \in U$ with respect to $B$ as follows:

$$[x]_{R_B} = \{y \in U | f(x, a) = f(y, a), \forall\, a \in B \subseteq AT\}.$$

*Definition 2.7.* Let $\widetilde{S} = (U, AT = C \cup D, V, f)$ be a dominance intuitionistic fuzzy information system, $R_B^{\preceq}$ be dominance relation, $B \in AT$. For any $X \subseteq P(U)$, the lower and upper approximations of $X$ with respect to $R_B^{\preceq}$ are defined as follows:

$$\underline{R_B^{\preceq}}(X) = \{x \in U | [x]_{R_B^{\preceq}} \subseteq X\}, \overline{R_B^{\preceq}}(X) = \{x \in U | [x]_{R_B^{\preceq}} \cap X \neq \emptyset\}.$$

According to Definition 2.7, we obtain that the lower approximation $\underline{R_B^{\preceq}}(X)$ is the greatest definable set contained in $X$, and the upper approximation $\overline{R_B^{\preceq}}(X)$

is the least definable set containing $X$. The pair $(\underline{R_B^{\preceq}}(X), \overline{R_B^{\preceq}}(X))$ is referred to as a dominance-based rough set of $X$.

Similarly, the lower and upper approximations of $X$ with respect to $R_B^{\succeq}$ are defined as follows:

$$\underline{R_B^{\succeq}}(X) = \{x \in U | [x]_{R_B^{\succeq}} \subseteq X\}, \overline{R_B^{\succeq}}(X) = \{x \in U | [x]_{R_B^{\succeq}} \cap X \neq \emptyset\}.$$

## 3   Generalized Dominance-Based Rough Set Model

For dominance-based intuitionistic fuzzy rough set, according to the definition of dominance class

$$[x]_{R_B^{\preceq}} = \{y \in U | f(x, a) \preceq f(y, a), \forall\, a \in B \subseteq AT\},$$

we see that for any $y \in [x]_{R_B^{\preceq}}$, they require $f(x, a) \preceq f(y, a)$ for each $a \in B \subseteq AT$. In fact, this condition is too strict to use conveniently in practical applications, and sometimes it would lead to the loss of useful information in some decision making problems. From the perspective of information granularity, dominance class $[x]_{R_B^{\preceq}}$ produces a finer granulation, which will increasing the computation in large universes. This problem will be illustrated in detail in Example 3.1.

**Table 1.** A dominance intuitionistic fuzzy information system

| $x$ | $c_1$ | $c_2$ | $c_3$ | $c_4$ | $c_5$ | $d$ |
|---|---|---|---|---|---|---|
| $x_1$ | (0.4, 0.5) | (0.3, 0.5) | (0.8, 0.2) | (0.4, 0.5) | (0.7, 0.1) | (0.3, 0.6) |
| $x_2$ | (0.3, 0.5) | (0.4, 0.5) | (0.6, 0.1) | (0.4, 0.5) | (0.7, 0.3) | (0.4, 0.6) |
| $x_3$ | (0.3, 0.5) | (0.1, 0.8) | (0.8, 0.1) | (0.4, 0.5) | (0.7, 0.3) | (0.2, 0.7) |
| $x_4$ | (0.1, 0.8) | (0.1, 0.8) | (0.4, 0.5) | (0.1, 0.8) | (0.8, 0.2) | (0.2, 0.8) |
| $x_5$ | (0.7, 0.3) | (0.4, 0.5) | (0.9, 0.1) | (0.4, 0.6) | (0.8, 0.1) | (0.4, 0.6) |
| $x_6$ | (0.3, 0.6) | (0.4, 0.6) | (0.7, 0.2) | (0.5, 0.5) | (0.8, 0.2) | (0.4, 0.5) |
| $x_7$ | (0.4, 0.5) | (0.4, 0.5) | (0.8, 0.2) | (0.4, 0.5) | (0.8, 0.2) | (0.6, 0.4) |
| $x_8$ | (0.4, 0.6) | (0.4, 0.5) | (0.9, 0.1) | (0.7, 0.3) | (0.8, 0.2) | (0.6, 0.4) |
| $x_9$ | (0.4, 0.6) | (0.7, 0.3) | (0.9, 0.1) | (0.4, 0.5) | (0.9, 0.0) | (0.8, 0.2) |
| $x_{10}$ | (0.7, 0.3) | (0.7, 0.3) | (0.8, 0.2) | (0.9, 0.0) | (0.4, 0.5) | (0.8, 0.2) |

*Example 3.1* Table 1 is a computer audit risk assessment decision table with intuitionistic fuzzy attributes. In Table 1, object set $U = \{x_1, x_2, \cdots, x_{10}\}$ concludes 10 audited objects, and condition attribute set is $C = \{c_1, c_2, \cdots, c_5\}$, where $c_1$ =Better Systems Circumstance, $c_2$ =Better Systems Control, $c_3$ =Safer Finance Data, $c_4$ =Credible Auditing Software, $c_5$ =Operation Standardization; decision attribute set is $D = \{d\}$, where $d$ =Acceptable Ultimate Computer Auditing Risk.

**Table 2.** Dominating classes $[x]_{R_C^{\preceq}}$ and dominated classes $[x]_{R_C^{\succeq}}$

| $x$ | $[x]_{R_C^{\preceq}}$ | $[x]_{R_C^{\succeq}}$ | $x$ | $[x]_{R_C^{\preceq}}$ | $[x]_{R_C^{\succeq}}$ |
|---|---|---|---|---|---|
| $x_1$ | $\{x_1, x_7\}$ | $\{x_1\}$ | $x_6$ | $\{x_6, x_8\}$ | $\{x_6\}$ |
| $x_2$ | $\{x_2, x_9, x_8, x_7\}$ | $\{x_2\}$ | $x_7$ | $\{x_7\}$ | $\{x_7, x_1, x_2, x_4\}$ |
| $x_3$ | $\{x_3, x_9, x_8\}$ | $\{x_3\}$ | $x_8$ | $\{x_8\}$ | $\{x_8, x_6, x_3, x_2, x_4\}$ |
| $x_4$ | $\{x_4, x_9, x_5, x_8, x_7\}$ | $\{x_4\}$ | $x_9$ | $\{x_9\}$ | $\{x_9, x_3, x_2, x_4\}$ |
| $x_5$ | $\{x_5\}$ | $\{x_5, x_4\}$ | $x_{10}$ | $\{x_{10}\}$ | $\{x_{10}\}$ |

According to Definition 2.6, we get all the dominating classes $[x]_{R_C^{\preceq}}(x \in U)$ and dominated classes $[x]_{R_C^{\succeq}}(x \in U)$ with respect to condition attribute set $C$ as Table 2.

By Definition 2.2, for any $x_i \in U(i = 1, 2, \cdots, 10)$, $C(x_i) = \oplus_{k=1}^5 f(x_i, c_k)$ denotes the comprehensive evaluation value of $x_i$ with respect to all condition attributes $c_k, k = 1, 2, 3, 4, 5$, where $f(x_i, c_k) = c_k(x_i) = (\mu_{c_k}(x_i), \nu_{c_k}(x_i))$. According to Table 2, $C(x_i) = \oplus_{k=1}^5 f(x_i, c_k)$ are given as follows:

$$C(x_1) = \oplus_{k=1}^5 f(x_1, c_k) = \left(1 - \prod_{k=1}^5 (1 - \mu_{c_k}(x_1)), \prod_{k=1}^5 \nu_{c_k}(x_1)\right) = (0.98488, 0.0025);$$

$C(x_2) = (0.96976, 0.00375); C(x_3) = (0.97732, 0.006); C(x_4) = (0.91252, 0.0512);$
$C(x_5) = (0.99784, 0.0009); C(x_6) = (0.9874, 0.0072); C(x_7) = (0.99136, 0.005);$
$C(x_8) = (0.99784, 0.0018); C(x_9) = (0.99892, 0.000); C(x_{10}) = (0.99892, 0.000).$

By Definition 2.3, we obtain the scores $s(C(x_i))$ of the alternatives $x_i$:

$$s(C(x_1)) = 0.98238; s(C(x_2)) = 0.96601, s(C(x_3)) = 0.97132,$$

$$s(C(x_4)) = 0.86132, s(C(x_5)) = 0.99694, s(C(x_6)) = 0.98020,$$

$$s(C(x_7)) = 0.98636, s(C(x_8)) = 0.99604, s(C(x_9)) = 0.99892, s(C(x_{10})) = 0.99892.$$

Then, we rank all the objects $x_i$ by using $s(C(x_i))(i = 1, 2, \cdots, 10)$:

$$x_9 = x_{10} \succ x_5 \succ x_8 \succ x_7 \succ x_1 \succ x_6 \succ x_3 \succ x_2 \succ x_4. \tag{1}$$

Since $h(C(x_9)) = h(C(x_{10})) = 0.99892$, then (1) is the final results.

For the following discussions, let $[x]_C^{\preceq}$ be the set of objects dominating $x$ with respect to the comprehensive evaluation value of condition attribute set $C$, and $[x]_C^{\succeq}$ be the set of objects dominated $x$ with respect to the comprehensive evaluation value of condition attribute set $C$.

From Table 2, we get $[x_1]_{R_C^{\preceq}} = \{x_1, x_7\}, [x_1]_{R_C^{\succeq}} = \{x_1\}$. However, equation (1) shows $[x_1]_C^{\preceq} = \{x_1, x_9, x_{10}, x_5, x_8, x_7\}, [x_1]_C^{\succeq} = \{x_1, x_2, x_3, x_4, x_6\}$. For $x_9$, only $c_1(x_1) \succ c_1(x_9)$, except attribute $c_1$, $c_k(x_1) \prec c_k(x_9)$ hold for each $c_k \in \{c_2, c_3, c_4, c_5\}$. It shows even only one attribute value dissatisfy the dominance relation will results in rejection of that classes $[x]_{R_C^{\preceq}}$ and $[x]_{R_C^{\succeq}}$. These properties can be important especially in case of large universes, e.g. generated from dynamic processes.

To address this deficiency, we try to propose a new way to produce the dominance classes, then give a comparison between the presented dominance classes with Definition 2.6.

**Definition 3.1.** Let $\widetilde{S} = (U, AT = C \cup D, V, f)$ be a dominance intuitionistic fuzzy information system, $U = \{x_1, x_2, \cdots, x_n\}, C = \{c_1, c_2, \cdots, c_m\}, D = \{d\}$. Generalized dominace relations $\preceq_C$ and $\succeq_C$ in DIFIS $\widetilde{S}$ are defined as follows:

$$\preceq_C = \{(x_i, x_j) \in U \times U \mid \oplus_{k=1}^m f(x_i, c_k) \preceq \oplus_{k=1}^m f(x_j, c_k), \forall\, c_k \in C\},$$

$$\succeq_C = \{(x_i, x_j) \in U \times U \mid \oplus_{k=1}^m f(x_i, c_k) \succeq \oplus_{k=1}^m f(x_j, c_k), \forall\, c_k \in C\}.$$

The generalized dominance class $[x_i]_{\overline{C}}^{\preceq}$ induced by $\preceq_C$ is the set of objects dominating $x_i$, i.e. $[x_i]_{\overline{C}}^{\preceq} = \{x_j \in U \mid \oplus_{k=1}^m f(x_i, c_k) \preceq \oplus_{k=1}^m f(x_j, c_k), \forall\, c_k \in C\}$, where $[x_i]_{\overline{C}}^{\preceq}$ is called the generalized dominating class with respect to $x_i \in U$. Meanwhile, the generalized dominated class with respect to $x_i \in U$ can be defined as $[x_i]_{\overline{C}}^{\succeq} = \{x_j \in U \mid \oplus_{k=1}^m f(x_i, c_k) \succeq \oplus_{k=1}^m f(x_j, c_k), \forall\, c_k \in C\}$.

In $[x_i]_{\overline{d}}^{\preceq} = \{x_j \in U \mid f(x_i, d) \preceq f(x_j, d)\}$, $x_i$ is the minimal element that dominating itself.

All the generalized dominance classes form a covering of $U$, i.e. $U = \bigcup\limits_{i=1}^{n} [x_i]_{\overline{C}}^{\preceq}$.

**Example 3.2.** (Following Example 3.1) By Definition 3.1, the dominating classes $[x]_{R_d^{\preceq}}$ and dominated classes $[x]_{R_d^{\succeq}}$ with respect to decision attribute $d$ are given in Table 3.

**Table 3.** Dominating $[x]_{R_d^{\preceq}}$ and dominated classes $[x]_{R_d^{\succeq}}$

| $x$ | $[x]_{\overline{d}}^{\preceq}$ | $[x]_{\overline{d}}^{\succeq}$ |
|---|---|---|
| $x_1$ | $\{x_1, x_9, x_{10}, x_5, x_6, x_7, x_8, x_2\}$ | $\{x_1, x_3, x_4\}$ |
| $x_2$ | $\{x_2, x_9, x_{10}, x_5, x_6, x_7, x_8\}$ | $\{x_2, x_5, x_4, x_1, x_3\}$ |
| $x_3$ | $\{x_3, x_1, x_9, x_{10}, x_5, x_6, x_7, x_8, x_2\}$ | $\{x_3, x_4\}$ |
| $x_4$ | $\{x_4, x_9, x_{10}, x_5, x_8, x_7, x_1, x_6, x_3, x_2\}$ | $\{x_4\}$ |
| $x_5$ | $\{x_2, x_9, x_{10}, x_5, x_6, x_7, x_8\}$ | $\{x_2, x_5, x_4, x_1, x_3\}$ |
| $x_6$ | $\{x_6, x_9, x_{10}, x_8, x_7\}$ | $\{x_6, x_3, x_2, x_4, x_1, x_5\}$ |
| $x_7$ | $\{x_7, x_9, x_{10}, x_8\}$ | $\{x_7, x_1, x_6, x_3, x_2, x_4, x_5\}$ |
| $x_8$ | $\{x_8, x_9, x_{10}, x_7\}$ | $\{x_8, x_7, x_1, x_6, x_3, x_2, x_4, x_5\}$ |
| $x_9$ | $\{x_9, x_{10}\}$ | $\{x_9, x_{10}, x_5, x_8, x_7, x_1, x_6, x_3, x_2, x_4\}$ |
| $x_{10}$ | $\{x_{10}, x_9\}$ | $\{x_{10}, x_9, x_5, x_8, x_7, x_1, x_6, x_3, x_2, x_4\}$ |

By Definition 2.3, we obtain the scores $s(d(x_i))$ of the alternatives $x_i$:

$$s(d(x_1)) = -0.3, s(d(x_2)) = -0.2, s(d(x_3)) = -0.5, s(d(x_4)) = -0.6, s(d(x_5)) = -0.2,$$

$s(d(x_6)) = -0.1, s(d(x_7)) = 0.2, s(d(x_8)) = 0.2, s(d(x_9)) = 0.6, s(d(x_{10})) = 0.6.$

Then, we rank all the objects $x_i$ by using $s(d(x_i))(i = 1, 2, \cdots, 10)$ as

$$x_9 = x_{10} \succ x_7 = x_8 \succ x_6 \succ x_2 = x_5 \succ x_3 \succ x_5 \succ x_6.$$

Furthermore, by Definition 2.3, for those objects which according to the score function can not be divided, we obtain the precisions $h(d(x_i))$ of $x_i$ $(i = 2, 5, 7, 8, 9, 10)$ as follows:

$$h(d(x_2)) = 0.9, h(d(x_5)) = h(d(x_7)) = h(d(x_8)) = h(d(x_9)) = h(d(x_{10})) = 1.0.$$

Then, we rank all the objects $x_i$ by using $h(d(x_i))(i = 1, 2, \cdots, 10)$ as

$$x_9 = x_{10} \succ x_8 = x_7 \succ x_6 \succ x_5 \succ x_2 \succ x_1 \succ x_3 \succ x_4. \tag{2}$$

According to equation (2), by employing Definition 3.1, the generalized dominating classes $[x]_{\bar{d}}^{\preceq}$ and generalized dominated classes $[x]_{\bar{d}}^{\succeq}$ with respect to decision attribute $d$ are obtained in Table 4'.

**Table 4.** The generalized dominating classes $[x]_{\bar{d}}^{\preceq}$ and dominated classes $[x]_{\bar{d}}^{\succeq}$

| $x$ | $[x]_{\bar{d}}^{\preceq}$ | $[x]_{\bar{d}}^{\succeq}$ |
|---|---|---|
| $x_1$ | $\{x_1, x_9, x_{10}, x_5, x_6, x_7, x_8, x_2\}$ | $\{x_1, x_3, x_4\}$ |
| $x_2$ | $\{x_2, x_9, x_{10}, x_5, x_6, x_7, x_8\}$ | $\{x_2, x_5, x_4, x_1, x_3\}$ |
| $x_3$ | $\{x_3, x_1, x_9, x_{10}, x_5, x_6, x_7, x_8, x_2\}$ | $\{x_3, x_4\}$ |
| $x_4$ | $\{x_4, x_9, x_{10}, x_5, x_8, x_7, x_1, x_6, x_3, x_2\}$ | $\{x_4\}$ |
| $x_5$ | $\{x_2, x_9, x_{10}, x_5, x_6, x_7, x_8\}$ | $\{x_2, x_5, x_4, x_1, x_3\}$ |
| $x_6$ | $\{x_6, x_9, x_{10}, x_8, x_7\}$ | $\{x_6, x_3, x_2, x_4, x_1, x_5\}$ |
| $x_7$ | $\{x_7, x_9, x_{10}, x_8\}$ | $\{x_7, x_1, x_6, x_3, x_2, x_4, x_5\}$ |
| $x_8$ | $\{x_8, x_9, x_{10}, x_7\}$ | $\{x_8, x_7, x_1, x_6, x_3, x_2, x_4, x_5\}$ |
| $x_9$ | $\{x_9, x_{10}\}$ | $\{x_9, x_{10}, x_5, x_8, x_7, x_1, x_6, x_3, x_2, x_4\}$ |
| $x_{10}$ | $\{x_{10}, x_9\}$ | $\{x_{10}, x_9, x_5, x_8, x_7, x_1, x_6, x_3, x_2, x_4\}$ |

For dominance intuitionistic fuzzy information system (Table 1), from the equation (1) of Example 3.1, the generalized dominating classes $[x]_{\bar{C}}^{\preceq}(x \in U)$ and generalized dominated classes $[x]_{\bar{C}}^{\succeq}(x \in U)$ with respect to condition attribute set $C$ are obtained in Table 5.

According to Table 5, it is easy to get $[x_j]_{\bar{C}}^{\preceq} \subseteq [x_i]_{\bar{C}}^{\preceq}$, if $x_i \preceq x_j$.

Consequently, by using the comprehensive evaluation value of each object with respect to all attributes to produce generalized dominating and dominated classes would avoid discarding many excellent objects. However, the dominance relations of Definition 2.6 are so strict that many objects are discarded, and thus lead to the dominance classes are more finer so that it would largely increasing the computation.

**Table 5.** The generalized dominating classes $[x]_{\overline{C}}^{\preceq}$ and dominated classes $[x]_{\overline{C}}^{\succeq}$

| $x$ | $[x]_{\overline{C}}^{\preceq}$ | $[x]_{\overline{C}}^{\succeq}$ |
|---|---|---|
| $x_1$ | $\{x_1, x_9, x_{10}, x_5, x_8, x_7\}$ | $\{x_1, x_6, x_3, x_2, x_4\}$ |
| $x_2$ | $\{x_2, x_9, x_{10}, x_5, x_8, x_7, x_1, x_6, x_3\}$ | $\{x_2, x_4\}$ |
| $x_3$ | $\{x_3, x_9, x_{10}, x_5, x_8, x_7, x_1, x_6\}$ | $\{x_3, x_2, x_4\}$ |
| $x_4$ | $\{x_4, x_9, x_{10}, x_5, x_8, x_7, x_1, x_6, x_3, x_2\}$ | $\{x_4\}$ |
| $x_5$ | $\{x_5, x_9, x_{10}\}$ | $\{x_5, x_8, x_7, x_1, x_6, x_3, x_2, x_4\}$ |
| $x_6$ | $\{x_6, x_9, x_{10}, x_5, x_8, x_7, x_1\}$ | $\{x_6, x_3, x_2, x_4\}$ |
| $x_7$ | $\{x_7, x_9, x_{10}, x_5, x_8\}$ | $\{x_7, x_1, x_6, x_3, x_2, x_4\}$ |
| $x_8$ | $\{x_8, x_9, x_{10}, x_5\}$ | $\{x_8, x_7, x_1, x_6, x_3, x_2, x_4\}$ |
| $x_9$ | $\{x_9, x_{10}\}$ | $\{x_9, x_{10}, x_5, x_8, x_7, x_1, x_6, x_3, x_2, x_4\}$ |
| $x_{10}$ | $\{x_{10}, x_9\}$ | $\{x_{10}, x_9, x_5, x_8, x_7, x_1, x_6, x_3, x_2, x_4\}$ |

From Table 2, we get $[x_1]_{R_{\overline{C}}^{\preceq}} = \{x_1, x_7\}$, but according to Table 5, $[x_1]_{\overline{C}}^{\preceq} = \{x_1, x_9, x_{10}, x_5, x_8, x_7\}$. Therefore, according to the comprehensive evaluation values, $x_9, x_{10}, x_5, x_8, x_7$ are all better than $x_1$, only the individual attributes value of $x_9, x_{10}, x_5, x_8$ do not satisfy $f(x, a) \preceq f(y, a), \forall\ a \in C$. For example, for object $x_1$ and $x_5$, only $f(x_1, c_4) \succ f(x_5, c_4)$, except $c_4 =$ Credible Auditing Software, $f(x_1, c_i) \preceq f(x_5, c_i)$ hold for any $c_i \in C(i \neq 4)$. In a word, for any $x_i, x_j \in U$, $[x_i]_{R_{\overline{C}}^{\preceq}}$ contains only those objects that fully dominating $x_i$ with respect to each attribute of $C$, it is so inconvenient for decision making problems. Hence, in many practical applications, if there is no particular requirements for each attribute value of objects, we would only consider the comprehensive evaluate value of each object.

*Remark 3.1.* If $C$ is a singleton of attribute set, i.e., $|C| = 1$, then $[x_i]_{\overline{C}}^{\preceq} = [x]_{R_{\overline{C}}^{\preceq}}$.

*Theorem 3.1.* Let $\widetilde{S} = (U, AT = C \cup D, V, f)$ be a DIFIS, $U = \{x_1, x_2, \cdots, x_n\}$, $C = \{c_1, c_2, \cdots, c_m\}, D = \{d\}$. Then

(1) $\preceq_C$ and $\succeq_C$ are reflexive and transitive;

(2) $x_j \in [x_i]_{\overline{C}}^{\preceq} \Leftrightarrow [x_j]_{\overline{C}}^{\preceq} \subseteq [x_i]_{\overline{C}}^{\preceq}$; $x_j \in [x_i]_{\overline{C}}^{\succeq} \Leftrightarrow [x_j]_{\overline{C}}^{\succeq} \subseteq [x_i]_{\overline{C}}^{\succeq}$;

(3) $[x_i]_{\overline{C}}^{\preceq} = \cup\{[x_j]_{\overline{C}}^{\preceq} | x_j \in [x_i]_{\overline{C}}^{\preceq}\}$;

(4) $[x_i]_{R_{\overline{C}}^{\preceq}} \subseteq [x_i]_{\overline{C}}^{\preceq}, [x_i]_{R_{\overline{C}}^{\succeq}} \subseteq [x_i]_{\overline{C}}^{\succeq}$;

(5) $[x_i]_{\overline{C}}^{\succeq} \cap [x_i]_{\overline{C}}^{\preceq} = \{x_j \in U | \oplus_{k=1}^m f(x_i, c_k) = \oplus_{k=1}^m f(x_j, c_k), \forall\ c_k \in C\}$;

(6) $U = \bigcup_{i=1}^n [x_i]_{\overline{C}}^{\preceq} = \bigcup_{i=1}^n [x_i]_{\overline{C}}^{\succeq}$;

(7) $[x_i]_{\overline{C}}^{\preceq} = [x_j]_{\overline{C}}^{\preceq}$ or $[x_i]_{\overline{C}}^{\succeq} = [x_j]_{\overline{C}}^{\succeq} \Leftrightarrow \oplus_{k=1}^m f(x_i, c_k) = \oplus_{k=1}^m f(x_j, c_k), c_k \in C$.

*Proof.* We only prove (2) and (4), the others can be obtained easily according to Definition 3.1.

(2) "$\Rightarrow$" According to Definition 3.1, if $x_j \in [x_i]_{\overline{C}}^{\preceq}$, then $\oplus_{k=1}^m f(x_i, c_k) \preceq \oplus_{k=1}^m f(x_j, c_k)$ for any $c_k \in C$. Similarly, for any $x_l \in [x_j]_{\overline{C}}^{\preceq}$, $\oplus_{k=1}^m f(x_j, c_k) \preceq \oplus_{k=1}^m f(x_l, c_k)$ holds. Thus, $\oplus_{k=1}^m f(x_i, c_k) \preceq \oplus_{k=1}^m f(x_l, c_k), \forall\ c_k \in C$. Hence, $x_l \in [x_i]_{\overline{C}}^{\preceq}$, then $[x_j]_{\overline{C}}^{\preceq} \subseteq [x_i]_{\overline{C}}^{\preceq}$ holds.

"⇐"It is easy to get from Definition 3.1.

(4) For any $c_k \in C$, if $x_j \in [x_i]_{\overline{C}}^{\preceq}$, then $f(x_i, c_k) \preceq f(x_j, c_k)$, thus $\oplus_{k=1}^m f(x_i, c_k) \preceq \oplus_{k=1}^m f(x_j, c_k)$. Hence, $x_j \in [x_i]_{\overline{C}}^{\preceq}$, i.e. $[x_i]_{R_{\overline{C}}^{\preceq}} \subseteq [x_i]_{\overline{C}}^{\preceq}$.

Now, we introduce the notion of generalized dominance-based rough set model which is composed of the generalized dominance classes.

*Definition 3.2.* Let $\widetilde{S} = (U, AT = C \cup D, V, f)$ be a dominance intuitionistic fuzzy information system, $U = \{x_1, x_2, \cdots, x_n\}, C = \{c_1, c_2, \cdots, c_m\}, D = \{d\}$. $[x_i]_{\overline{C}}^{\preceq}$ is the generalized dominating class induced by $\preceq_C$. For any $X \in P(U)$, the lower and upper approximations of $X$ are defined as follows:

$$\underline{X_{\overline{C}}^{\preceq}} = \{x_i \in U | [x_i]_{\overline{C}}^{\preceq} \subseteq X\}, \overline{X_{\overline{C}}^{\preceq}} = \{x_i \in U | [x_i]_{\overline{C}}^{\preceq} \cap X \neq \emptyset\}.$$

$\underline{X_{\overline{C}}^{\preceq}}$ and $\overline{X_{\overline{C}}^{\preceq}}$ are called the generalized dominating lower approximation and generalized dominating upper approximation of $X$ with respect to $\preceq_C$. $(\underline{X_{\overline{C}}^{\preceq}}, \overline{X_{\overline{C}}^{\preceq}})$ is referred to as a generalized dominating-based rough set. Elements in $\underline{X_{\overline{C}}^{\preceq}}$ can be classified as members of $X$ with complete certainty using attribute set $C$, whereas elements in $\overline{X_{\overline{C}}^{\preceq}}$ can be classified as members of $X$ with only partial certainty using attribute set $C$. The class $\overline{X_{\overline{C}}^{\preceq}} - \underline{X_{\overline{C}}^{\preceq}}$ is referred to as boundary of $X$ with respect to $C$ and denoted by $BN_{\overline{C}}^{\preceq}(X)$.

Similarly, $(\underline{X_{\overline{C}}^{\succeq}}, \overline{X_{\overline{C}}^{\succeq}})$ is referred to as a generalized dominated-based rough set. The generalized dominating-based rough set and generalized dominated-based rough set are both referred to as generalized dominance-based rough set.

If $X \in IF(U)$, then

$$\underline{X_{\overline{C}}^{\preceq}}(x_i) = \min_{x_j \in [x_i]_{\overline{C}}^{\preceq}} X(x_j), \quad \overline{X_{\overline{C}}^{\preceq}}(x_i) = \max_{x_j \in [x_i]_{\overline{C}}^{\preceq}} X(x_j).$$

The operators $\underline{X_{\overline{C}}^{\preceq}}$ and $\overline{X_{\overline{C}}^{\preceq}}$ are, respectively, referred to as lower and upper generalized dominating-based intuitionistic fuzzy rough approximation operators of $x_i$. The pair $(\underline{X_{\overline{C}}^{\preceq}}, \overline{X_{\overline{C}}^{\preceq}})$ is referred to as a generalized dominating-based intuitionistic fuzzy rough set. $\underline{X_{\overline{C}}^{\preceq}}(x_i)$ is just the degree to which $x$ certainly belongs to $X$, $\overline{X_{\overline{C}}^{\preceq}}(x_i)$ is the degree to which $x$ possibly belongs to $X$.

Let $D_j = [x_j]_{\overline{d}}^{\preceq}(j = 1, 2, \cdots, m)$ be the generalized dominating classes of $x_j$ with respect to decision attribute $d$, and $(\underline{D_{1\overline{C}}^{\preceq}}, \underline{D_{2\overline{C}}^{\preceq}}, \cdots, \underline{D_{m\overline{C}}^{\preceq}})$ be the low generalized dominating rough approximations of $D_j$ with respect to condition attribute set $C$.

*Example 3.3.* By using Definition 3.1 to compute the $\underline{D_{j\overline{C}}^{\preceq}}(j = 1, 2, \cdots, 10)$ of Example 3.1, we obtain

$$\underline{D_{1\overline{C}}^{\preceq}} = \{x_1, x_5, x_6, x_7, x_8, x_9, x_{10}\}; \underline{D_{2\overline{C}}^{\preceq}} = \{x_5, x_7, x_8, x_9, x_{10}\}; \underline{D_{3\overline{C}}^{\preceq}} = U, x_2\} = U;$$

$$\underline{D_{4\overline{C}}^{\preceq}} = U, \underline{D_{5\overline{C}}^{\preceq}} = \{x_5, x_7, x_8, x_9, x_{10}\}; \underline{D_{6\overline{C}}^{\preceq}} = \underline{D_{7\overline{C}}^{\preceq}} = \underline{D_{8\overline{C}}^{\preceq}} = \underline{D_{9\overline{C}}^{\preceq}} = \underline{D_{10\overline{C}}^{\preceq}} = \{x_9, x_{10}\}.$$

*Theorem 3.2.* Let $\widetilde{S} = (U, AT = C \cup D, V, f)$ be a dominance intuitionistic fuzzy information system, $U = \{x_1, x_2, \cdots, x_n\}, C = \{c_1, c_2, \cdots, c_m\}, D = \{d\}. [x_i]_C^{\preceq}$ is the generalized dominating class induced by $\preceq_C$. For any $X, Y \in P(U)$, then

(1) $\underline{X_C^{\preceq}} \subseteq X, X \subseteq \overline{X_C^{\preceq}}$;

(2) $\sim \underline{X_C^{\preceq}} = \overline{\sim X_C^{\preceq}}, \sim \overline{X_C^{\preceq}} = \underline{\sim X_C^{\preceq}}$;

(3) $\underline{\emptyset_C^{\preceq}} = \emptyset, \overline{\emptyset_C^{\preceq}} = \emptyset; \underline{U_C^{\preceq}} = U, \overline{U_C^{\preceq}} = U$;

(4) $\underline{X \cap Y_C^{\preceq}} = \underline{X_C^{\preceq}} \cap \underline{Y_C^{\preceq}}, \overline{X \cup Y_C^{\preceq}} = \overline{X_C^{\preceq}} \cup \overline{Y_C^{\preceq}}$;

(5) $\underline{X \cup Y_C^{\preceq}} \supseteq \underline{X_C^{\preceq}} \cup \underline{Y_C^{\preceq}}, \overline{X \cap Y_C^{\preceq}} \subseteq \overline{X_C^{\preceq}} \cap \overline{Y_C^{\preceq}}$;

(6) $X \subseteq Y \Rightarrow \underline{X_C^{\preceq}} \subseteq \underline{Y_C^{\preceq}}, \overline{X_C^{\preceq}} \subseteq \overline{Y_C^{\preceq}}$.

*Proof.* The proofs of (1)-(6) can be obtained directly from Definition 3.2.

*Definition 3.3.* Let $\widetilde{S} = (U, AT = C \cup D, V, f)$ be a DIFIS and $B \subseteq C$. Then $B$ is referred to as a consistent set of $\widetilde{S}$ if $[x]_B^{\preceq} = [x]_C^{\preceq}$ for all $x \in U$. If $B$ is a consistent set and no proper subset of $B$ is a consistent set of $\widetilde{S}$, then $B$ is referred to as a reduct of $\widetilde{S}$.

For any $x_i \in U$, let $C(x_i) = \oplus_{c_k \in C} f(x_i, c_k)$ represents the comprehensive evaluation value of $x_i$ with respect to condition attribute set $C$. For any $x_i, x_j \in U$, if there exist a subset $B \subseteq C$, such that $C(x_i) \preceq C(x_j) \Rightarrow B(x_i) \preceq B(x_j)$, then $B$ is referred to as a consistent set of $\widetilde{S}$. In other words, a consistent set $B$ is a subset of $C$ where the rank results of $C(x_i)$ and $B(x_i)$ remain unchanged.

By Definition 3.3, we see that a reduction of $\widetilde{S}$ is a minimal set of condition attributes preserving the same lower approximations and upper approximations.

The knowledge hidden in a DIFIS $\widetilde{S} = (U, AT = C \cup \{d\}, V, f)$ may be discovered and expressed in the form of decision rules: $t \to s$, where $t = \wedge c(x_i), c \in C$ and $s = d(x_i), t$ and $s$ are, respectively, called the condition and decision parts of the rule.

Let $\widetilde{S} = (U, AT = C \cup \{d\}, V, f)$ be a DIFIS, $B \subseteq C$. For any generalized condition attribute dominance classes $[x_i]_C^{\preceq}$ and generalized decision attribute dominance classes $D_j = [x_j]_d^{\preceq}$, if $x_k \in \underline{D_{jB}^{\preceq}} = \{x_i \in U | [x_i]_B^{\preceq} \subseteq D_j\}$, then $\bigwedge_{c \in B} c(x_k) \Rightarrow d(x_j) \preceq d(x_k)$.

Definition 3.2 extends the definition of dominance rough approximations of $X \subseteq P(U)$ with respect to DIFIS by using generalized dominance classes.

However, once there are special requirements for the individual attribute in decision making problems, Definition 3.2 is no longer applicable. Hence, it is necessary to construct a new dominance-based rough set model to solve these problems both considering the comprehensive evaluation value and individual attribute value of each attribute, which will be investigated in our next work.

# References

1. Pawlak, Z.: Rough Sets. International Journal of Computer and Information Sciences 11, 341–356 (1982)
2. Dubois, D., Prade, H.: Rough Fuzzy Sets and Fuzzy Rough Sets. International Journal of General System 17, 191–209 (1990)

3. Yao, Y.Y.: Relational Interpretations of Neighborhood Operators and Rough Set Approximation Operators. Information Sciences 111(1), 239–259 (1998)
4. Skowron, A., Stepaniuk, J.: Tolerance Approximation Spaces. Fundamenta Informaticae 27(2), 245–253 (1996)
5. Greco, S., Matarazzo, B., Slowinski, R.: Rough Approximation of A Preference Relation by Dominance Relation. European Journal of Operational Research 117(1), 63–83 (1999)
6. William, Z., Wang, F.Y.: Reduction and Axiomization of Covering Generalized Rough Sets. Information Sciences 152, 217–230 (2003)
7. Mi, J.S., Leung, Y., Zhao, H.Y., Feng, T.: Generalized Fuzzy Rough Sets Determined by A Triangular Norm. Information Sciences 178, 3203–3213 (2008)
8. Greco, S., Matarazzo, B., Słowiński, R.: A New Rough Set Approach to Multicriteria and Multiattribute Classification. In: Polkowski, L., Skowron, A. (eds.) RSCTC 1998. LNCS (LNAI), vol. 1424, pp. 60–67. Springer, Heidelberg (1998)
9. Greco, S., Matarazzo, B., Slowinski, R.: Rough Sets Theory for Multicriteria Decision Analysis. European Journal of Operational Research 129, 1–47 (2001)
10. Kotlowski, W., Dembczynski, K., Greco, S., et al.: Stochastic Dominance-based Rough Set Model for Ordinal Classification. Information Sciences 178, 4019–4037 (2008)
11. Hu, Q.H., Yu, D., Guo, M.Z.: Fuzzy Preference based Rough Sets. Information Sciences 180, 2003–2022 (2010)
12. Liou, J.J.H., Tzeng, G.H.: A Dominance-based Rough Set Approach to Customer Behavior in The Airline Market. Information Sciences 180, 2230–2238 (2010)
13. Greco, S., Inuiguchi, M., Slowinski, R.: Fuzzy Rough Sets and Multiple-Premise Gradual Decision Rules. International Journal of Approximate Reasoning 41, 179–211 (2006)
14. Hu, Q.H., Yu, D.R.: Variable Precision Dominance based Rough Set Model and Reduction Algorithm for Preference-Ordered Data. In: Proceedings of 2004 International Conference on Machine Learning and Cybernetics, vol. 4, pp. 2279–2284. IEEE (2004)
15. Wang, C.Z., Chen, D.G., He, Q., et al.: A Comparative Study of Ordered and Covering Information Systems. Fundamenta Informaticae 122, 1–13 (2012)
16. Huang, B., Li, H.X., Wei, D.K.: Dominance-based Rough Set Model in Intuitionistic Fuzzy Information Systems. Knowledge-Based Systems 28, 115–123 (2012)
17. Huang, B., Wei, D.K., Li, H.X., et al.: Using a Rough Set Model to Extract Rules in Dominance-based Interval-Valued Intuitionistic Fuzzy Information Systems. Information Sciences 221, 215–229 (2013)
18. Mieszkowicz-Rolka, A., Rolka, L.: Fuzzy implication operators in variable precision fuzzy rough sets model. In: Rutkowski, L., Siekmann, J.H., Tadeusiewicz, R., Zadeh, L.A. (eds.) ICAISC 2004. LNCS (LNAI), vol. 3070, pp. 498–503. Springer, Heidelberg (2004)
19. Mieszkowicz-Rolka, A., Rolka, L.: Fuzzy Rough Approximations of Process Data. International Journal of Approximate Reasoning 49(2), 301–315 (2008)
20. Atanassov, K.: Intuitionistic Fuzzy Sets. Fuzzy sets and Systems 20, 87–96 (1986)
21. Atanassov, K.: More on Intuitionistic Fuzzy Set. Fuzzy sets Syst. 33, 37–45 (1989)
22. Atanassov, K.: Intuitionistic Fuzzy Sets. Fuzzy Sets and Systems 31, 343–349 (1986)
23. Atanassov, K.: Intuitionistic Fuzzy Sets: Theory and Applicatons. Physica, Heidelberg (1999)
24. Xu, Z.S.: Intuitionistic Preference Relations and Their Application in Group Decision Making. Information Sciences 177, 2363–2379 (2007)

# On Definability and Approximations
# in Partial Approximation Spaces

Davide Ciucci[1], Tamás Mihálydeák[2], and Zoltán Ernő Csajbók[3]

[1] Dipartimento di Informatica, Sistemistica e Comunicazione
Università di Milano – Bicocca
Viale Sarca 336 – U14, I–20126 Milano, Italia
ciucci@disco.unimib.it

[2] Department of Computer Science, Faculty of Informatics,
University of Debrecen
Kassai út 26, H-4028 Debrecen, Hungary
mihalydeak.tamas@inf.unideb.hu

[3] Department of Health Informatics, Faculty of Health,
University of Debrecen,
Sóstói út 2-4, H-4400 Nyíregyháza, Hungary
csajbok.zoltan@foh.unideb.hu

**Abstract.** In this paper, we discuss the relationship occurring among the basic blocks of rough set theory: approximations, definable sets and exact sets. This is done in a very general framework, named Basic Approximation Space that generalizes and encompasses previous known definitions of Approximation Spaces. In this framework, the lower and upper approximation as well as the boundary and exterior region are independent from each other. Further, definable sets do not coincide with exact sets, the former being defined "a priori" and the latter only "a posteriori" on the basis of the approximations. The consequences of this approach in the particular case of partial partitions are developed and a discussion is started in the case of partial coverings.

## 1 Introduction

Since the beginning of rough set theory (RST) and with an increasing interest, a great attention has been paid to approximations and to generalized approaches to rough sets. This lead to the definition of several new models such as dominance-based rough sets [10], covering rough sets (where nowadays more than 20 pairs of approximations are known [18,20,17]), probabilistic rough sets [19], etc... On the other hand, very few attention has been paid to some other basic notions such as definability and roughness. In particular, the existing discussions on definability regard mainly relation-based rough sets, and usually standard rough sets based on a partition of the universe.

In [21], Yao presents a discussion on definability as primitive notion and approximations as derived ones, which are needed to describe undefinable objects. His study is based on a logical approach where a set is definable if "its extension can be precisely defined by a logic formula". A topological approach is given in

D. Miao et al. (Eds.): RSKT 2014, LNAI 8818, pp. 15–26, 2014.
DOI: 10.1007/978-3-319-11740-9_2 © Springer International Publishing Switzerland 2014

[13]: in the case that approximations are based on a tolerance or a quasi ordered relation, definable sets are defined in order to form an Alexadroff topology. A discussion on vagueness and language in Pawlakian rough sets is given in [2], where different definitions of rough sets are given and different forms of vagueness outlined. A study on rough sets based on generalized binary relations is conduced in [11], where a set is considered *definable* if it is the union of the elementary granules. We will discuss more about this choice in Section 4.

To the best of our knowledge, there has never been a discussion on definability and roughness in a generic context. Indeed, some considerations can be found in [3], where some problems in the definition of exact sets as primitive notion ("a priori" attitude) are put forward and more recently in [8], where the authors deal with the problem of defining different vagueness categories in partial approximation spaces.

With this work we would like to start a discussion on definability and approximations. We will develop our considerations in a basic environment, making few assumptions on the properties that the definable sets and the approximations have to satisfy.

In Section 2, the relationship among the basic elements of rough sets is debated and the definition of Basic Approximation Spaces provided. This discussion is then developed in the case of partial partitions in Section 3, where different categories of exactness and vagueness are also introduced. Finally, a similar approach is started in the more complex case of partial coverings.

## 2    Basic Approximation Spaces

In generalized approaches to rough sets we have some ingredients that mutually interact: exact sets, rough sets, lower ($l$) and upper ($u$) approximations and the exterior or negative ($n$) and boundary ($b$) region. Usually, these four regions $l, u, b, n$ are not independent, that is $u(S) = l(S) \cup b(S)$, $n(S) = u(S)^c$ and two of them are sufficient to define the others. In other words, we have a tri-partition of the universe in lower-boundary-exterior (or also an orthopair [4]).

However, in some situations, it makes sense that this dependence is relaxed. In standard (i.e., equivalence based) RST the boundary can be defined in three equivalent ways: if $l(S) = \cup \mathfrak{B}_1$, $u(S) = \cup \mathfrak{B}_2$ (where $\mathfrak{B}_1$ and $\mathfrak{B}_2$ are given families of base sets), then

1. $b(S) = u(S) \setminus l(S)$;
2. $b(S) = \cup(\mathfrak{B}_2 \setminus \mathfrak{B}_1)$;
3. $b(S)$ is the union of those base sets, which are not subsets of $S$ and their intersections with $S$ are not empty.

In generalized theories of rough sets these definitions give different notions of boundaries, and so one has to make clear the informal notion of the boundary in order to decide the applicable version.

In the first two cases boundaries can be defined by lower and upper approximations, and so they are not independent of approximation functions. On the other hand, the third version is the closest form of the informal notion of bound-aries. Moreover, in general cases, it is independent of (cannot be defined by) lower and upper approximations. For example in membrane computing appli-cations the third version has to be used in order to represent the vicinity of a membrane (see partial approximation of multisets and its application in mem-brane computing [15,16]). Another situation where the dependence is relaxed is in partial settings. In this case, we can have that, for a set $S$, $u(S)^c \neq n(S)$, since some members in $u(S)^c$ can be outside the domain of definition.

According to the former facts, it is worth to consider all the four regions/operators $l, u, b, n$ as primitive ones and then study how they can interact. In this way we give the same importance to all of them. We do not decide, at this stage, which of them is primitive and which is derived. In particular, the boundary is often neglected, but, as also pointed out in [5], thinking in terms of lower-boundary instead of lower-upper or lower-negative (as usually done) can give a new perspective and for instance, develop new operations on rough sets.

Similarly, in standard RST, it is clear what the *exact* sets are (the sets which coincide with their approximations, that is, equivalence classes and their unions) and what the *rough* sets are (all the other sets). In general situations this can become more tricky, and we can make a further distinction between base sets, which are the building blocks (equivalence classes in the standard case), definable objects, that is the sets we can obtain by operations on base sets (union of equivalence classes) and exact sets, that are defined using the approximations.

Given all these considerations, we define a basic structure for approximations, that takes into account all these elements and some properties they should satisfy.

**Definition 2.1.** *A* Basic Approximation Space *over a set $U$ is defined as the structure $(U, \mathfrak{B}, \mathfrak{D}_{\mathfrak{B}}, \mathfrak{E}_{\mathfrak{B}}, l, b, u, n)$ such that*

1. *$\emptyset \neq \mathfrak{B} \subseteq \mathcal{P}(U)$ if $B \in \mathfrak{B}$ then $B \neq \emptyset$ (the base system)*
   *We will denote the union of all the members of the base system as $\mathbb{B} = \cup_{B \in \mathfrak{B}} B$.*
2. *$\mathfrak{B} \subseteq \mathfrak{D}_{\mathfrak{B}} \subseteq \mathcal{P}(U)$ and $\emptyset \in \mathfrak{D}_{\mathfrak{B}}$ (the definable sets)*
   *We will denote the union of all the definable sets as $\mathbb{D} = \cup_{D \in \mathfrak{D}_{\mathfrak{B}}} D$.*
3. *$\mathfrak{E}_{\mathfrak{B}} \subseteq \mathcal{P}(U)$ and $\emptyset \in \mathfrak{E}_{\mathfrak{B}}$ (the exact sets)*
4. *$l, b, u, n : \mathcal{P}(U) \mapsto \mathfrak{D}_{\mathfrak{B}}$ are, respectively, the lower, boundary, upper and negative mappings. They satisfy the following properties:*
   - *$u(\emptyset) = \emptyset$, $b(\emptyset) = \emptyset$, $n(\emptyset) = \mathbb{B}$; (normality conditions)*
   - *If $S_1 \subseteq S_2$ then $l(S_1) \subseteq l(S_2)$, $u(S_1) \subseteq u(S_2)$ and $n(S_2) \subseteq n(S_1)$ (mono-tonicity)*
   - *$l(S) \subseteq u(S)$, $l(S) \cap n(S) = \emptyset$ (weak approximation properties)*

*Remark 2.1.* The substructure $(U, \mathfrak{B}, \mathfrak{D}_{\mathfrak{B}}, l, u)$ with corresponding properties is a Generalized Approximation Space as defined in [8]. The substructure $(\mathcal{P}(U), l, u)$ is a Boolean Approximation Algebra as defined in [3].

As a trivial consequence of the above definition, we have that $l(\emptyset) = \emptyset$.

Let us make some remarks on this definition.

– The set $\mathbb{B}$ does not necessary coincide with the universe $U$, that is, we only have a partial covering. This does not mean that the objects in $U \setminus \mathbb{B}$ are *unknown*, but rather that they are *undefined*, in a similar way as Kleene used these two terms [14]. Consequently, if a name refers to an object which does not belong to $\mathbb{B}$, all propositions on this object are undefined. For instance, in Fig. 2(a) since $P_5 \notin \mathbb{B}$, for any proposition $Pr$, $Pr(P_5)$ is undefined.

- According to the actual definition of exactness, we may have that exact sets are not necessarily chosen among the definable ones (i.e., it is not required that $\mathfrak{E}_{\mathfrak{B}} \subseteq \mathfrak{D}_{\mathfrak{B}}$). If exact elements are defined according to a request based on approximation operators (such as, $l(S) = u(S)$), in the general case we cannot say neither that base sets are exact nor that exact sets are definable (or vice versa). We will see some examples in the following sections.
- The lower and upper approximations as well as the boundary and negative mappings are definable. This is due to the interpretation of *definable elements* as the *all and only* elements that given by our knowledge we can speak about.

## 3  Partial Partitions

Let us consider the simplest case, that is the sets of the base system are mutually disjoint (but do not necessarily cover the universe). We can call this situation *partial partition* or *one-layered* approximation space [9]. This case is of course a generalization of the standard setting where the partition is total.

*Remark 3.1.* It is an important case also taking into account that any partial covering can be reduced to a partial partition by considering the collection of disjoint granules:

$$gr(x) := \cap\{B \mid x \in B\} \setminus \cup\{B \mid x \notin B\}, \tag{1}$$

A similar formula was suggested in [1] to obtain a (total) partition from a (total) covering. Of course, by this operation we lose part of the semantic of the covering approach but we gain in simplicity. That is, the definable sets will change and also the approximations, that typically will become finer (that is closer to the set under approximation).

So, let us suppose to have a base system which constitutes a partial partition, we now define all the other elements of the approximation space according to Definition 2.1.

- The *definable elements* are the base sets and their unions: $\emptyset \in \mathfrak{D}_{\mathfrak{B}}$, $\mathfrak{B} \subseteq \mathfrak{D}_{\mathfrak{B}}$, if $D_1, D_2 \in \mathfrak{D}_{\mathfrak{B}}$ then $D_1 \cup D_2 \in \mathfrak{D}_{\mathfrak{B}}$. Clearly, we have that $\mathbb{B} = \mathbb{D}$.
- The *lower* and *upper approximations* are defined as usual as

$$l(S) := \cup\{B \in \mathfrak{B} \mid B \subseteq S\} \tag{2}$$
$$u(S) := \cup\{B \in \mathfrak{B} \mid B \cap S \neq \emptyset\} \tag{3}$$

- The *boundary* is $u(S) \setminus l(S)$ and the *exterior* is $n(S) = \mathbb{D} \setminus u(S)$.
- The exact sets are those we can define without ambiguity using the approximations. However, by means of the upper and lower approximations

we can define at least two different notions of exactness. The first relies entirely on the approximations: $\mathfrak{E}_\mathfrak{B} := \{S : l(S) = u(S)\} \supseteq \mathfrak{D}_\mathfrak{B}$. But the sets in $\mathfrak{E}_\mathfrak{B}$ do not generally coincide with the sets under approximation: $l(E) = E = u(E)$, indeed $E$ can contain some undefined objects. For instance, in Fig. 2(b), $l(S) = u(S) = B_4 \neq S = B_4 \cup \{P_5\}$. Thus, we can also consider a further class of exact sets, that we can name *absolutely exact*: $\mathfrak{A}_\mathfrak{B} = \{S : l(S) = S = u(S)\} \subseteq \mathfrak{E}_\mathfrak{B}$. The difference is that exact sets are defined relatively to our knowledge that can be only expressed through the approximations whereas absolutely exact sets are defined with respect to the "real" world (that we are not always allowed to know). For a discussion on the differences between these two levels (sets and approximations) and the problems arising in putting them at the same level see also [2,6].

We notice that since we are in a partial setting, besides the exterior, we also have the *undefinable exterior* $U \setminus (u(S) \cup n(S))$ that characterizes the undefinable objects (which are different from the unknown ones belonging to the boundary).

Further, we used only the union operator to build the definable sets from the base sets, but in this case we have that $B_i \cap B_j = \emptyset$ for any choice of $B_i \neq B_j$ and that $\mathbb{D} \setminus B$ (i.e., the negation of $B$ wrt definable sets) is also a definable set. As we will see, this does not generalize to the covering situation.

Finally, in this simplified situation, we have that the condition $l(S) = u(S)$ can be equivalently expressed as $(S \setminus l(S)) \cap \mathbb{D} = \emptyset$.

*Example 3.1.* Let us suppose to study the diagnosis of thyroid dysfunctions via clinical symptoms. It is an important but inexact classification problem. We deal with only hypothyroidism thyroid disorder which occurs when the thyroid is "underactive", i.e., it does not produce enough thyroid hormones. For more details about thyroid dysfunctions and their informatics considerations see [7] and the references therein. Here, the study is considerably simplified for illustrative purposes.

At the beginning, let us suppose to have at our disposal an information table containing clinical symptoms weight change, edema, bradycardia, affection which together or separately may indicate hypothyroidism (see Fig. 1). Hypothyroidism can be accurately diagnosed with laboratory tests, last column of the table is based on these results.

$U$ is the set of patients: $\{P_1, P_2, P_3, P_4, P_5, P_6, P_7, P_8, P_9, P_{10}, P_{11}, P_{12}, P_{13}\}$.

First, let us form the base system directly according to the clinical symptoms which separately may indicate hypothyroidism:

- weight change = gain: $\{P_2, P_6, P_{10}, P_{11}\}$;
- edema = yes: $\{P_6, P_7, P_8, P_9\}$;
- bradycardia = yes: $\{P_2, P_3, P_4, P_6, P_{12}, P_{13}\}$;
- affection = depression: $\{P_2, P_{12}, P_{13}\}$.

These base sets form a partial covering (Fig. 2(a)), because their union does not add up the whole set $U$, namely, patients No. 1 and No. 5 do not possess any clinical symptom which may indicate hypothyroidism. In Fig. 2, ovals contain the patients who suffer from hypothyroidism according to the laboratory tests.

| No. | Weight change | Edema | Brady-cardia | Affection | Hypothy-roidism |
|-----|---------------|-------|--------------|-----------|-----------------|
| $P_1$ | not change | no | no | normal | no |
| $P_2$ | **gain** | **no** | **yes** | **depression** | yes |
| $P_3$ | **not change** | **no** | **yes** | **normal** | yes |
| $P_4$ | loss | no | yes | normal | yes |
| $P_5$ | loss | no | no | nervousness | no |
| $P_6$ | **gain** | **yes** | **yes** | **normal** | yes |
| $P_7$ | loss | yes | no | normal | yes |
| $P_8$ | not change | yes | no | normal | no |
| $P_9$ | loss | yes | no | nervousness | no |
| $P_{10}$ | gain | no | no | normal | no |
| $P_{11}$ | gain | no | no | nervousness | no |
| $P_{12}$ | **loss** | **no** | **yes** | **depression** | yes |
| $P_{13}$ | not change | no | yes | depression | no |

Clinical symptoms which
may indicate hypothyroidism:
- Weight change = gain
- Edema = yes
- Bradycardia = yes
- Affection = depression

$U$ is the set of all patients:
$U = \{P_1, P_2, P_3, P_4, P_5, P_6, P_7,$
$\quad P_8, P_9, P_{10}, P_{11}, P_{12}, P_{13}\}$

**Fig. 1.** Information table with clinical symptoms which may indicate hypothyroidism

In Fig. 2(a), the base sets are not pairwise disjoint. Let us reduce this partial covering to a partial partition according to the formula (1) (see Fig. 2(b)):

- weight change = gain AND bradycardia = yes AND affection = depression: $B_1 = \{P_2\}$;
- bradycardia = yes OR (bradycardia = yes AND affection = depression): $B_2 = \{P_3, P_4, P_{12}, P_{13}\}$;
- weight change = gain AND edema = yes AND bradycardia = yes: $B_3 = \{P_6\}$;
- edema = yes: $B_4 = \{P_7, P_8, P_9\}$;
- weight change = gain: $B_5 = \{P_{10}, P_{11}\}$.

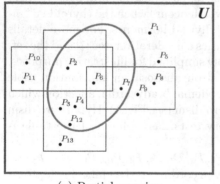

(a) Partial covering

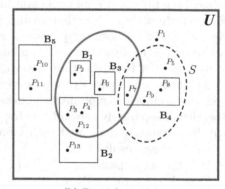

(b) Partial partition

**Fig. 2.** Partial covering and its reduction to partial partition

## 3.1 Categories of Vagueness

Up to now we have defined four different categories of elements with respect to the possibility to describe them using the available knowledge. They are:

- The base sets $\mathfrak{B}$, our bricks to build the available knowledge;
- The definable sets $\mathfrak{D} \supseteq \mathfrak{B}$, whatever we can define using the basic bricks;

- The exact sets $\mathfrak{E}$, which are given according to the lower and upper approximations, without any reference to real sets. We have that $\emptyset, \mathbb{D} \in \mathfrak{E}$;
- The absolutely exact sets $\mathfrak{A} \subseteq \mathfrak{E}$, which are given according to the lower and upper approximations *and* making reference also to real sets.

We remark that exact sets are not necessarily definable, since an exact set $S \in \mathfrak{E}$ can contain some undefined object. On the other hand, absolutely exact sets coincide with definable sets: $\mathfrak{A} = \mathfrak{D}$. In order to understand what the available knowledge (base and definable sets) and the approximations enable us to describe, it is also interesting to differentiate the situations where the lower and upper approximations are not trivial, that is the lower is not empty and the upper is different from the universe (in this case the union of definable elements $\mathbb{D}$). Of course we obtain the standard four notions of definability/roughness plus one, only definable in partial settings (see also [8]).

- The set $S$ is *roughly definable*: there exist objects that surely belong to $S$ and objects that surely do not belong to $S$. That is, $l(S) \neq \emptyset$ (therefore, $u(S) \neq \emptyset$), $u(S) \neq \mathbb{D}$ (equiv., $n(S) \neq \emptyset$).
- *Internally definable* or *externally undefinable*: there do exist objects surely belonging to $S$ and there are no objects that certainly do not belong to $S$. That is, $l(S) \neq \emptyset$, $n(S) = \emptyset$ (equiv., $u(S) = \mathbb{D}$) and $b(S) \neq \emptyset$. As a trivial consequence we have that, $u(S) \neq \emptyset$.
- *Externally definable* or *internally undefinable*: it is the opposite of the previous case, that is, there do exist objects surely not belonging to $S$ but not surely belonging to $S$. In other words, $l(S) = \emptyset$, $n(S) \neq \emptyset$ and $n(S) \neq \mathbb{D}$. Consequently, we have $u(S) = b(S)$ and they are both not empty and different from $\mathbb{D}$.
- *Totally undefinable*: no objects are known with certainty, that is $l(S) = n(S) = \emptyset$, $u(S) = \mathbb{D} = b(S)$.
- *Negatively definable*: all the definable objects do not belong to $S$, that is $l(S) = u(S) = b(S) = \emptyset$ and $n(S) = \mathbb{D}$. This can happen only in partial cases, where $S$ is made only of objects outside the definition domain $\mathbb{B}$.

## 4   Partial Covering: The "Relative" Boolean Case

It is well known that moving from partitions to covering, the possibilities to define the approximations are much more wider. See for instance [18,20,17], where almost 30 different pairs of approximations are classified and their properties studied. If to a covering we also add the possibility to be partial, these possibilities can only become wider. This holds also for the other elements, besides lower and upper approximations, of a basic approximation algebra.

### 4.1   Definable Sets

Let us address as the first important problem the definition of *definable* sets. As previously discussed, *definable* is what we are allowed to talk about, given some basic blocks, that is to say, how can we combine the elements of $\mathfrak{B}$. In case of partitions we considered the union of elements of $\mathfrak{B}$ as definable elements and this does not present particular problems since intersection of elements of $\mathfrak{B}$ are empty and the complement (with respect to $\mathbb{B}$) of definable elements are definable.

In case of covering this is no more true, and it makes a difference whether to consider only the union or also other operations. We can distinguish at least two interesting cases.

1. We consider only the union. In this case, the intersection of two base sets is not generally definable, as well as the complement. An immediate consequence is that the usual definition of boundary and negative mappings does not give us a definable set in this setting. For instance, if $b(S) = u(S) \setminus l(S) = u(S) \cap l(S)^c$, then it is not assured that $l(S)^c$ is definable, nor its intersection with the upper approximation. The justification to consider this case, a part from a purely theoretical one, is that it makes sense in some situations such as the rough set approach to membrane computing [16,15]. A study about definability and approximations in this basic situation can be found in [8].

2. On the other hand we have the most complete situation where all the set operations are allowed: union, intersection and negation (with respect to $\mathbb{B}$). That is, definable sets are a complete field of sets [12]. From a point of view of logic, this is the most natural approach. We have some facts (the base sets) which constitute our knowledge and we combine them using all the instruments available using a (partial) Boolean logic. For instance, if we have a set $B_1$ representing "the students that submitted a program written in Java", a set $B_2$ representing "the students that submitted a program written in group", then it sounds natural to ask which are "the students that submitted a paper written in Java and in group", as well as "the student that did not write a program in Java". Even if we do not have this basic information, we can build it up.

From now on, let us consider this second scenario that we can call *Relative Boolean System*. Definable elements are thus formally defined as:

- $\emptyset \in \mathfrak{D}_\mathfrak{B}$, $\mathfrak{B} \subseteq \mathfrak{D}_\mathfrak{B}$
- if $D_1, D_2 \in \mathfrak{D}$ then $D_1 \cup D_2 \in \mathfrak{D}_\mathfrak{B}$, $D_1 \cap D_2 \in \mathfrak{D}_\mathfrak{B}$, $\mathbb{B} \setminus D_1 \in \mathfrak{D}_\mathfrak{B}$.

Also in this case, we have that $\mathbb{B} = \mathbb{D}$.

## 4.2   Exact Sets

In order to define the exact sets, we have to choose a lower and an upper approximation definition among the several ones known in literature (or also define some new one). A complete analysis of this multitude of approximations is out of scope of the present work, here we just limit the discussion to the natural extension of the standard $l, u$ definitions in equations (2). So, we are in a special case with respect to [8], since we also admit intersection and negation to define the definable elements. The boundary can now be defined in (at least) two ways:

$$b(S) = u(S) \setminus l(S) \tag{4}$$
$$b_p(S) = \cup\{B \in \mathfrak{B} | B \cap S \neq \emptyset, B \nsubseteq S\} \tag{5}$$

*Example 4.1.* Let us take again the information table from Fig. 1 and the partial covering of $U$ defined in Fig.3: $\mathfrak{B} = \{B_1, B_2, B_3\}$, where $\cup \mathfrak{B} \subsetneq U$.

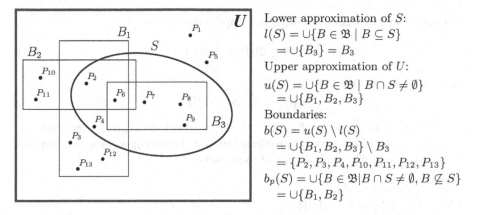

Lower approximation of $S$:
$$l(S) = \cup\{B \in \mathfrak{B} \mid B \subseteq S\}$$
$$= \cup\{B_3\} = B_3$$
Upper approximation of $U$:
$$u(S) = \cup\{B \in \mathfrak{B} \mid B \cap S \neq \emptyset\}$$
$$= \cup\{B_1, B_2, B_3\}$$
Boundaries:
$$b(S) = u(S) \setminus l(S)$$
$$= \cup\{B_1, B_2, B_3\} \setminus B_3$$
$$= \{P_2, P_3, P_4, P_{10}, P_{11}, P_{12}, P_{13}\}$$
$$b_p(S) = \cup\{B \in \mathfrak{B} | B \cap S \neq \emptyset, B \nsubseteq S\}$$
$$= \cup\{B_1, B_2\}$$

**Fig. 3.** Boundaries, lower and upper approximations in partial covering

According to Fig. 3, the following formulas can easily be checked:

$$b(S) \subsetneq b_p(S) \subsetneq u(S) \tag{6}$$
$$b_p(S) \cap l(S) = \{P_6\} \neq \emptyset \tag{7}$$
$$b_p(S) \cup l(S) = u(S) \tag{8}$$

As it can be seen in Example 4.1, in general the two definitions are not equal and we have $b(S) \subseteq b_p(S)$, see formula (6). Moreover, we notice that $b_p$ is defined independently from the lower and upper approximations. So, we no more have that $b_p(S) \cap l(S) = \emptyset$, see formula (7). On the other hand, we easily get that $b_p(S) \subseteq u(S)$ and $b_p(S) \cup l(S) = u(S)$, see formula (8). Both boundaries (as well as the lower and upper approximations) are definable since they can be obtained by standard set operations starting from elements in the base system.

Also in the case of the negative region we can have more than one definition: two based on the approximations and one independent from approximations.

$$n_l(S) = l(\mathbb{B} \setminus u(S)) \tag{9}$$
$$n(S) = \mathbb{B} \setminus u(S) \tag{10}$$
$$n_p(S) = \cup\{B \in \mathfrak{B} | B \cap S = \emptyset\} \tag{11}$$

All are definable sets and they are not equal but we only have that $n_l(S) \subseteq n(S) \subseteq n_p(S)$. Moreover, while $n_l(S) \cap u(S) = n(S) \cap u(S) = \emptyset$ we can have a non-empy intersection between $n_p(S)$ and $u(S)$. In any case, the intersection between the lower approximation and all the three negative regions is empty. So, considering also that they are monotonic, we can take any of them as defining a Basic Approximation Space, according to definition 2.1.

*Example 4.2.* Let us consider again the example 4.1 and suppose to have a further base set $B_4 = \{P_{13}, P_{14}, P_{15}\}$. So, we have that $n_l(S) = \emptyset$, $n(S) = \{P_{14}, P_{15}\}$ and $n_p(S) = B_4$.

Moreover, we can define a new upper approximation as

$$u_n(S) = \cup\{B \in \mathfrak{B} | B \cap S \neq \emptyset\} \setminus n_p(S) \tag{12}$$

and a new boundary:

$$b_n(S) = u_n(S) \setminus l(S) \tag{13}$$

and we have $u_n(S) \subseteq u(S)$ and $b_n(S) \subseteq b(S)$ and both $u_n, b_n$ are monotonic.

Now, by means of the lower approximation and the two upper approximations we can define four different notions of exact sets:

- $E_1 = \{S : l(S) = u(S)\}$
- $E_2 = \{S : l(S) = S = u(S)\}$
- $E_3 = \{S : l(S) = u_n(S)\}$
- $E_4 = \{S : l(S) = S = u_n(S)\}$

We have that both $E_1$ and $E_2$ conditions imply that $b_n(S) = b(S) = b_p(S) = \emptyset$. On the other hand, $E_3$ and $E_4$ imply that $b_n(S) = \emptyset$ but not that $b_p(S)$ nor $b(S)$ are empty. Finally, we have that $E_1 \subseteq E_3$ and $E_2 \subseteq E_4$.

## 4.3   Categories of Vagueness

The final step is to define the different categories of vagueness. Clearly, the situation is more complex with respect to the partition case, since for any category we have more than one plausible definition. Here, it follows a list of categories based on the approximations defined above.

- The *roughly definable sets*
  - $V_1$: $l(S) \neq \emptyset$, $u(S) \neq \mathbb{D}$ and $l(S) \neq u(S)$ (in order to differentiate it from $E_1$);
  - $V_2$: $l(S) \neq \emptyset$, $u_n(S) \neq \mathbb{D}$ and $l(S) \neq u_n(S)$ (in order to differentiate it from $E_3$);
- The *internally definable sets*
  - $V_3$: $l(S) \neq \emptyset$, $u(S) = \mathbb{D}$ and $l(S) \neq u(S)$;
  - $V_4$: $l(S) \neq \emptyset$, $u_n(S) = \mathbb{D}$ and $l(S) \neq u_n(S)$, and we have $V_4 \subseteq V_3$;

- The *externally definable sets*
  - $V_5$: $l(S) = \emptyset$, $u(S) \neq \mathbb{D}$ and $u(S) \neq \emptyset$;
  - $V_6$: $l(S) = \emptyset$, $u_n(S) \neq \mathbb{D}$ and $u_n(S) \neq \emptyset$;
- The *totally undefinable sets*
  - $V_7$: $l(S) = \emptyset$, $u(S) = \mathbb{D}$;
  - $V_8$: $l(S) = \emptyset$, $u_n(S) = \mathbb{D}$, and we have $V_8 \subseteq V_7$;
  - $V_9$: $l(S) = \emptyset$, $n_l(S) = \emptyset$;
- The *negatively definable sets*, $V_{10}$: $l(S) = u(S) = b(S) = \emptyset$ and $n_l(S) = \mathbb{D}$, that is, $S$ is made only of elements outside $\mathbb{D}$.

Clearly, this four categories as well as the exact sets outlined at the end of the previous section are mutually disjoint.

# 5  Conclusion and Perspectives

In this paper, lower, boundary, upper and negative regions/operators have been considered as primitive ones. It is assumed that they are independent of each other and we have studied how they can interact both in the case of a partial partition and in the case of the covering. In this last generalization we only considered one pair of lower and upper approximation. Nevertheless, starting from just a lower and an upper approximation, we arrived at defining two upper approximations $u_n(S) \subseteq u(S)$, three boundaries $b_n(S) \subseteq b(S) \subseteq b_p(S)$ and three exterior regions $n_l(S) \subseteq n(S) \subseteq n_p(S)$ and all of them can be combined to obtain a Basic Approximation Space.

This approach can give a new perspective on the generalization of rough set theory. As a further development, other approximations can be considered in the covering case and their interactions studied. Moreover, a logical approach (in the style of [21]) in the definition of *definable* and *rough* sets is worth considering.

**Acknowledgements.** The publication was supported by the TÁMOP–4.2.2.C–11/1/KONV–2012–0001 project. The project has been supported by the European Union, co-financed by the European Social Fund.

# References

1. Bianucci, D., Cattaneo, G.: Information entropy and granulation co-entropy of partitions and coverings: A summary. Transactions on Rough Sets 10, 15–66 (2009)
2. Chakraborty, M.K., Banerjee, M.: Rough sets: Some foundational issues. Fundamenta Informaticae 127(1-4), 1–15 (2013)
3. Ciucci, D.: Approximation algebra and framework. Fundamenta Informaticae 94(2), 147–161 (2009)
4. Ciucci, D.: Orthopairs: A Simple and Widely Used Way to Model Uncertainty. Fundamenta Informaticae 108(3-4), 287–304 (2011)
5. Ciucci, D.: Orthopairs in the 1960s: Historical remarks and new ideas. In: Cornelis, C., Kryszkiewicz, M., Ślęzak, D., Ruiz, E.M., Bello, R., Shang, L. (eds.) RSCTC 2014. LNCS, vol. 8536, pp. 1–12. Springer, Heidelberg (2014)
6. Ciucci, D., Dubois, D.: Three-valued logics, uncertainty management and rough sets. In: Peters, J.F., Skowron, A. (eds.) Transactions on Rough Sets XVII. LNCS, vol. 8375, pp. 1–32. Springer, Heidelberg (2014)

7. Csajbók, Z.E., Mihálydeák, T., Ködmön, J.: An adequate representation of medical data based on partial set approximation. In: Saeed, K., Chaki, R., Cortesi, A., Wierzchoń, S. (eds.) CISIM 2013. LNCS, vol. 8104, pp. 120–128. Springer, Heidelberg (2013)
8. Csajbók, Z.E., Mihálydeák, T.: From vagueness to rough sets in partial approximation spaces. In: Kryszkiewicz, M., Cornelis, C., Ciucci, D., Medina-Moreno, J., Motoda, H., Raś, Z.W. (eds.) RSEISP 2014. LNCS, vol. 8537, pp. 42–52. Springer, Heidelberg (2014)
9. Csajbók, Z.E.: Approximation of sets based on partial covering. In: Peters, J.F., Skowron, A., Ramanna, S., Suraj, Z., Wang, X. (eds.) Transactions on Rough Sets XVI. LNCS, vol. 7736, pp. 144–220. Springer, Heidelberg (2013)
10. Greco, S., Matarazzo, B., Slowinski, R.: Granular computing and data mining for ordered data: The dominance-based rough set approach. In: Meyers, R.A. (ed.) Encyclopedia of Complexity and Systems Science, pp. 4283–4305. Springer (2009)
11. Grzymala-Busse, J., Rzasa, W.: Definability and other properties of approximations for generalized indiscernibility relations. In: Peters, J.F., Skowron, A. (eds.) Transactions on Rough Sets XI. LNCS, vol. 5946, pp. 14–39. Springer, Heidelberg (2010)
12. Järvinen, J.: Approximations and rough sets based on tolerances. In: Ziarko, W.P., Yao, Y. (eds.) RSCTC 2000. LNCS (LNAI), vol. 2005, pp. 182–189. Springer, Heidelberg (2001)
13. Järvinen, J., Kortelainen, J.: A note on definability in rough set theory. In: De Baets, B., De Caluwe, R., De Tré, G., Kacprzyk, J., Zadrozny, S. (eds.) Current Issues in Data and Knowledge Engineering, Problemy Współczesnej Nauki, Teoria i Zastosowania, Informatyka, pp. 272–277. EXIT (2004)
14. Kleene, S.C.: Introduction to metamathematics. North–Holland Pub. Co., Amsterdam (1952)
15. Mihálydeák, T., Csajbók, Z.E.: Partial approximation of multisets and its applications in membrane computing. In: Lingras, P., Wolski, M., Cornelis, C., Mitra, S., Wasilewski, P. (eds.) RSKT 2013. LNCS, vol. 8171, pp. 99–108. Springer, Heidelberg (2013)
16. Mihálydeák, T., Csajbók, Z.E., Takács, P.: Communication rules controlled by generated membrane boundaries. In: Alhazov, A., Cojocaru, S., Gheorghe, M., Rogozhin, Y., Rozenberg, G., Salomaa, A. (eds.) CMC 2013. LNCS, vol. 8340, pp. 265–279. Springer, Heidelberg (2014)
17. Restrepo, M., Cornelis, C., Gomez, J.: Duality, conjugacy and adjointness of approximation operators in covering-based rough sets. International Journal of Approximate Reasoning 55, 469–485 (2014)
18. Samanta, P., Chakraborty, M.K.: Generalized rough sets and implication lattices. In: Peters, J.F., Skowron, A., Sakai, H., Chakraborty, M.K., Slezak, D., Hassanien, A.E., Zhu, W. (eds.) Transactions on Rough Sets XIV. LNCS, vol. 6600, pp. 183–201. Springer, Heidelberg (2011)
19. Yao, J., Yao, Y., Ziarko, W.: Probabilistic rough sets: Approximations, decision-makings, and applications. International Journal of Approximate Reasoning 49(2), 253–254 (2008)
20. Yao, Y., Yao, B.: Covering based rough set approximations. Information Sciences 200, 91–107 (2012)
21. Yao, Y.: A note on definability and approximations. In: Peters, J.F., Skowron, A., Marek, V.W., Orłowska, E., Słowiński, R., Ziarko, W.P. (eds.) Transactions on Rough Sets VII. LNCS, vol. 4400, pp. 274–282. Springer, Heidelberg (2007)

# Many-Valued Rough Sets Based on Tied Implications

Moataz El-Zekey

Institute for Research and Applications of Fuzzy Modeling, CE IT4Innovations
University of Ostrava, 30. dubna 22, 701 03 Ostrava 1, Czech Republic
moataz.elzekey@bhit.bu.edu.eg

**Abstract.** We investigate a general many-valued rough set theory, based on tied adjointness algebras, from both constructive and axiomatic approaches. The class of tied adjointness algebras constitutes a particularly rich generalization of residuated algebras and deals with implications (on two independently chosen posets $(L, \leq_L)$ and $(P, \leq_P)$, interpreting two, possibly different, types of uncertainty) tied by an integral commutative ordered monoid operation on $P$. We show that this model introduces a flexible extension of rough set theory and covers many fuzzy rough sets models studied in literature. We expound motivations behind the use of two lattices $L$ and $P$ in the definition of the approximation space, as a generalization of the usual one-lattice approach. This new setting increase the number of applications in which rough set theory can be applied.

**Keywords:** Many-valued rough sets, Approximation operators, Calculi of approximation Spaces, Tied implications, Tied adjointness algebras.

## 1 Introduction

The theory of rough sets [22] generalizes traditional set theory by allowing a concept (represented by a subset of a (finite) universe $V$ of interest) to be described approximately by a lower and upper bound based on the information (knowledge) on hand. This knowledge is represented by a binary relation $R$ on $V$. Usefulness of rough sets theory has been fully demonstrated by its applications (see, e.g. [26]).

When we deal with real-life problems, fuzzy structures often provide much more adequate models of information than classical structures. This, in turn, leads to fuzzy rough sets which encapsulate the related, but distinct, concepts of fuzziness and indiscernibility. These occur as a result of uncertainty in knowledge or data. Dubois and Prade [5, 6] were among the first who investigated the problem of a fuzzyfication of a rough set, and since then many papers [4, 7, 10, 15, 16, 19, 24, 23, 28–31] have focused on the refinement of this model using both constructive approaches, which propose new definitions of approximation operators (using fuzzy logical extensions of the Boolean implication and conjunction), and axiomatic approaches, which set forth a set of axioms or desirable properties, and characterize the operators that satisfy them.

In Pawlak's rough set model [22], an equivalence relation is a key and primitive notion. This equivalence relation, however, seems to be a very stringent condition that may limit the application domain of the rough set model. So various types of binary

D. Miao et al. (Eds.): RSKT 2014, LNAI 8818, pp. 27–38, 2014.
DOI: 10.1007/978-3-319-11740-9_3 © Springer International Publishing Switzerland 2014

relations have been considered to replace the indiscernibility equivalence relation. Like classical rough set, several other authors considered fuzzy rough set models based on general fuzzy relations (see, e.g. [15, 16, 23, 29–31]).

The fuzzy rough set theory, proposed by the others authors, has been made up the deficiencies of the traditional rough set theory in several aspects. However many real world systems and applications require information management components that provide support for managing imprecise and uncertain data. A single model cannot process all type of uncertainties. There are various approaches for representing and processing uncertainty in the context of different domains of applicability in the literature. For example, classical set or crisp set, Lattice-valued fuzzy set (e.g. interval-valued fuzzy set, Vague set) and rough sets model different type of uncertainty. Since these types are relevant for many applications (e.g. database applications), combining them together is of both theoretical and practical importance. The applications of this idea are manifold.

There are several trials in the literature for integrating the classical rough sets theory with lattice-valued fuzzy sets. In [10], the authors combines the interval-valued fuzzy sets and the rough sets, and studies the basic theory of the interval-valued rough fuzzy sets. They were motivated by the idea mentioned in [25] where there are both of the symbolic values, real values and possibly lattice values (e.g. interval values) of the attributes in the real life database. Paper [27] concerns the processing of imprecision and indiscernibility in relational databases using vague rough technique leading to vague rough relational database model. They utilized the notion of indiscernibility and possibility from rough set theory coupled with the idea of membership and non-membership values from vague set theory to represent uncertain information in a manner that maintains the degree of uncertainty of information for each tuple of the original database and also those resulting from queries. Dubois and Prade [5], by using an equivalence relation, were among the first who introduce lower and upper approximations in fuzzy set theory to obtain an extended notion called rough fuzzy set [5, 6]. Moreover, Dubois and Prade also pointed out that the rough fuzzy set is a special case of the fuzzy rough set in the universe in their literatures.

One important point in all of these approaches from the above paragraph is that the semantic of the information relation (which is crisp) is different from the semantic of the approximated sets (which are interval-valued fuzzy sets, vague sets and Zadah fuzzy sets, respectively, or generally lattice-valued sets). This gives rise to conclude that, from a practical point of view, using one semantic for both of information relation and approximated sets is not always convenient. Hence, more general frameworks can be obtained by involving the approximations of $L$-valued fuzzy sets based on $P$-valued fuzzy relations, where $L$ and $P$ denote two independently chosen posets $(L, \leq)$ and $(P, \leq)$, interpreting two, possibly different, types of uncertainty. Therefore, the traditional fuzzy rough set theory could not deal with such kinds of situations effectively. It is then necessary to extend the traditional fuzzy rough set theory in a general sense.

In this paper, we propose a general many-valued rough set theory, based on a new algebraic model for non-classical logics, i.e. the class of tied adjointness algebras [20, 21]. It deals with implications $\Rightarrow: P \times L \to L$ tied by an integral commutative ordered monoid operation on $P$, in the sense given in [1], where $L$ and $P$ denote two independently chosen posets, interpreting two, possibly different, types of uncertainty, and generalizes the usual one-lattice approach. The property of being tied extends to multiple-valued logic the equivalence in classical logic known as the *law of importation*. It holds for several types of implications used in fuzzy logic, among which the *residuated implications* and *S-implications* are two types. There have been many papers, both theoretical and showing usefulness of tied implications in approximate reasoning in the recent past, e.g. [1–3, 11–14, 18, 20, 21]. Tied adjointness algebras constitute a particularly rich generalization of residuated algebras and couple a strong set of theorems with a wide variety of logical connectives frequently met in the literature.

The present paper investigates a general many-valued rough sets theory, called $(L, P)$-valued rough sets, in which both the constructive and axiomatic approaches are used. In the constructive approach, based on an arbitrary $P$-valued fuzzy relation and tied adjointness algebras on $(L, P)$, a coupled pair of generalized many-valued rough approximation operators is defined. The proposed many-valued rough set theory combines the $L$-valued fuzzy set theory with the traditional fuzzy rough set theory based on $P$-valued fuzzy relations. Therefore, the traditional fuzzy rough set theory is extended and its weaknesses are overcome. It extends to $P$-valued fuzzy relations and $L$-valued fuzzy sets most of the basic notions of the rough sets of Pawlak. The connections between $P$-valued fuzzy binary relations and many-valued approximation operators are examined. The resulting $(L, P)$-valued rough sets are proper generalizations of rough fuzzy sets [5], fuzzy rough set [4–7, 10, 15, 16, 19, 23, 24, 28–31], the interval-valued rough fuzzy set model [10] and Vague Rough Sets [27]. Hence, this model introduces a flexible extension of rough set theory and covers many lattice-valued fuzzy rough sets models studied in literature. This new setting increase the number of applications in which rough set theory can be applied. In the axiomatic approach, various classes of fuzzy rough sets are characterized by different sets of axioms, axioms of fuzzy approximation operators guarantee the existence of certain types of fuzzy relations producing the same operators.

## 2 Preliminaries

### 2.1 Implications and Their Adjoints

**Definition 1.** *Let $(P, \leq_P)$ and $(L, \leq_L)$ be posets and $\Rightarrow: P \times L \to L$, $\&: P \times L \to L$, $\supset: L \times L \to P$ are binary operations on $(L, P)$. An adjoint triple $(\Rightarrow, \&, \supset)$ on $(L, P)$ is an ordered triple in which the three operations $\Rightarrow$, $\&$ and $\supset$ are mutually related by the following* adjointness *condition, $\forall a \in P$, $\forall y, z \in L$:*

$$Adjointness: \quad y \leq_L a \Rightarrow z \quad iff \quad a \& y \leq_L z \quad iff \quad a \leq_P y \supset z. \tag{1}$$

**Proposition 1.** *Let $(\Rightarrow, \&, \supset)$ be an adjoint triple on $(L, P)$. Then $\forall a \in P$, $\forall y, z \in L$:*

$$a \leq_P (a \Rightarrow z) \supset z, \quad a \& (a \Rightarrow z) \leq_L z, \tag{2}$$

$$y \leq_L a \Rightarrow (a \& y), \quad (y \supset z) \& y \leq_L z, \tag{3}$$

$$a \leq_P y \supset (a \& y), \quad y \leq_L (y \supset z) \Rightarrow z. \tag{4}$$

**Proposition 2.** *Let $(\Rightarrow, \&, \supset)$ be an adjoint triple on $(L, P)$. Then $\forall a \in P$, $\forall y, z \in L$:*

(i) *The operation* & *is monotone in each argument, and the operations* $\Rightarrow, \supset$ *are antitone in the left argument and monotone in the right argument.*

(ii) *For all indexed families* $\{a_j\}$ *in* $P$ *and* $\{z_m\}, \{y_m\}$ *in* $L$, *such that the suprema and infima in the left-hand sides exist:*

$$\sup_j a_j \Rightarrow \inf_m z_m = \inf_{j,m} (a_j \Rightarrow z_m), \tag{5}$$

$$\sup_j a_j \& \sup_m y_m = \sup_{j,m} (a_j \& y_m), \tag{6}$$

$$\sup_j y_j \supset \inf_m z_m = \inf_{j,m} (y_j \supset z_m). \tag{7}$$

(iii) *In the case when* $L = P$, $\Rightarrow = \supset$ *if and only if* & *is commutative.*

**Proposition 3.** *[18] Let* $(\Rightarrow, \&, \supset)$ *be an adjoint triple on* $(L, P)$ *with a top element* $\mathbf{1}$ *for* $(P, \leq_P)$. *Then,* $\mathbf{1}$ *is a left identity element for* $\Rightarrow$ *(i.e.* $\forall z \in L: \mathbf{1} \Rightarrow z = z$) *iff* $\mathbf{1}$ *is a left identity element for* & *iff* $\supset$ *satisfies the following comparator axiom:*
   *Comparator axiom:* $\forall (y, z) \in L^2: \quad y \supset z = \mathbf{1} \quad iff \quad y \leq_L z.$

**Definition 2.** *(cf. [17, 18]) An* adjointness algebra *is an 8-tuple* $(L, \leq_L, P, \leq_P, \mathbf{1}, \Rightarrow, \&, \supset)$, *in which* $(L, \leq_L), (P, \leq_P)$ *are two posets with a top element* $\mathbf{1}$ *for* $(P, \leq_P)$, $\mathbf{1}$ *is a left identity element for* $\Rightarrow$ *and* $(\Rightarrow, \&, \supset)$ *is an adjoint triple on* $(L, P)$.
A complete adjointness lattice *is an adjointness algebra whose two underlying posets are complete lattices.*

In adjointness algebras the operations $\Rightarrow, \&, \supset$ are called *implication, conjunction* and *comparator* ($\supset$ is called a *forcing-implication* in [1, 17]), respectively. The ordered triple $(\Rightarrow, \&, \supset)$ is called an *implication triple* on $(L, P)$ (cf. [1] and [18]).

*Remark 1.* The definition of adjointness algebras in [1] and [18] includes other axioms, namely the mixed monotonicity properties of the operations $\Rightarrow, \&$ and $\supset$. In fact, we do not need these axioms, since they follows form adjointness (see Proposition 2).

We can speak of right identity, associativity or commutativity of a conjunction & only when $(P, \leq_P)$ equals $(L, \leq_L)$, and hence, we say that $\Rightarrow, \supset, \&$ are connectives *on* $P$. If, in this case, & is a *monoid* operation with unit element $\mathbf{1}$, it is called an *object-conjunction*, see [20, Section 4] (it is also called a *tying-conjunction* in [21]). We usually denote an object-conjunction by $\otimes$, and its adjoint $\Rightarrow$ and $\supset$ by $\rightarrow$ and $\rightsquigarrow$, respectively.

We can rephrase the well-known definition (cf. [8]) of (possibly *noncommutative*) *partially ordered residuated integral monoid*, or a *porim* for short, to read

**Definition 3.** *A* partially ordered residuated integral monoid *is a special case of adjointness algebras over one posets, in which the conjunction is a* monoid *operation with unit element* $\mathbf{1}$.

Therefore, a partially ordered residuated integral monoid always takes the form $(P, \leq_P, \mathbf{1}, \rightarrow, \otimes, \rightsquigarrow)$. Hence, in a porim, both $\rightarrow$ and $\rightsquigarrow$ become simultaneously an implication and a comparator. If the poset $(P, \leq_P)$ is a lattice with the associated meet and join operations $\wedge$ and $\vee$, respectively, then the algebra $(P, \leq_P, \wedge, \vee, \mathbf{1}, \rightarrow, \otimes, \rightsquigarrow)$ is called an *integral residuated lattice*. When $\otimes$ is also commutative, $\rightarrow$ has to coincide with $\rightsquigarrow$, and this adjointness algebra becomes $(P, \leq_P, \wedge, \vee, \mathbf{1}, \otimes, \rightarrow)$. These adjointness algebras are called *commutative integral residuated lattices.*

## 2.2 Tied Adjointness Algebras

Let $\circledast : P \times L \to L$ be a binary operation on $(L, P)$. A binary operation $\star$ on $P$ is said to *tie* $\circledast$ (or $\circledast$ is *tied* by $\star$) if the following identity holds (see [1]) $\forall a, b \in P, \forall z \in L$ : $(a \star b) \circledast z = a \circledast (b \circledast z)$. The property of *being tied* can be seen as a weakened form of associativity, particularly when the operation $\circledast$ is a conjunction. $\circledast$ is said to be *faithful*[1] if for each distinct pair $a$ and $b$ in $P$ there is $z \in L$ such that $a \circledast z \neq b \circledast z$.

**Proposition 4.** *(cf. [1]) In a complete adjointness lattice* $(L, \leq_L, P, \leq_P, 1, \Rightarrow, \&, \supset)$, *the following four conditions are equivalent:*

(i) *The implication* $\Rightarrow$ *is faithful.*      (ii) *The conjunction* $\&$ *is faithful.*

(iii) *For all* $a \in P$, $\inf_{z \in L} ((a \Rightarrow z) \supset z) = a$.    (iv) *For all* $a \in P$, $\inf_{y \in L} (y \supset (a \& y)) = a$.

**Definition 4.** *(cf. [20, 21]) A* tied adjointness algebra *is an algebra*

$$\Lambda = (P, \leq_P, 1, L, \leq_L, \Rightarrow, \&, \supset, \otimes, \to)$$

*in which,* $(L, \leq_L, P, \leq_P, 1, \Rightarrow, \&, \supset)$ *is an adjointness algebra,* $(P, \leq_P, \otimes, \to, 1)$ *is a commutative* porim, *and* $\otimes$ *ties* $\Rightarrow$.

The class of all tied adjointness algebras is denoted by ADJT[2].

**Theorem 1.** *[20] Let* $\Lambda = (P, \leq_P, 1, L, \leq_L, \Rightarrow, \&, \supset, \otimes, \to)$ *be a tied adjointness algebra. Then the following properties hold in* $\Lambda$, $\forall a, b, c \in P$, $\forall x, y, z, w \in L$:

$$\otimes \text{ ties } \Rightarrow: \quad ((a \otimes b) \Rightarrow z) = (a \Rightarrow (b \Rightarrow z)) \tag{8}$$

$$\otimes \text{ ties } \&: \quad ((a \otimes b) \& z) = (a \& (b \& z)) \tag{9}$$

$$\text{Strong adjointness}: \quad (y \supset (a \Rightarrow z)) = (a \to (y \supset z)) = (a \& y \supset z) \tag{10}$$

$$\text{Exchange axiom for } \Rightarrow: \quad (a \Rightarrow (b \Rightarrow z)) = (b \Rightarrow (a \Rightarrow z)) \tag{11}$$

$$\text{Exchange axiom for } \&: \quad (a \& (b \& z)) = (b \& (a \& z)) \tag{12}$$

$$\supset \text{ is } \otimes -\text{transitive}: \quad ((x \supset y) \otimes (y \supset w)) \leq_P (x \supset w) \tag{13}$$

$$\text{Prefixing with } \Rightarrow: \quad y \supset z \leq_P ((a \Rightarrow y) \supset (a \Rightarrow z)) \tag{14}$$

$$\text{Prefixing with } \&: \quad y \supset z \leq_P ((a \& y) \supset (a \& z)) \tag{15}$$

$$\text{Prefixing with } \supset: \quad y \supset z \leq_P ((x \supset y) \to (x \supset z)) \tag{16}$$

$$\text{Suffixing with } \Rightarrow: \quad a \to b \leq_P ((b \Rightarrow z) \supset (a \Rightarrow z)) \tag{17}$$

$$\text{Suffixing with } \&: \quad a \to b \leq_P ((a \& y) \supset (b \& y)) \tag{18}$$

$$\text{Suffixing with } \supset: \quad y \supset z \leq_P ((z \supset w) \to (y \supset w)) \tag{19}$$

As we have mentioned above, we are entitled to consider a commutative residuated algebra $(P, \leq_P, \otimes, \to)$ as a particular case of tied adjointness algebra; that is, by setting

---

[1] It is said to *distinguish left arguments* in [1, 20, 18].

[2] Due to lack of space, we cannot include examples of tied adjointness algebras. Instead, the reader is referred to [1, 20, 21] for examples and for more information.

$\Rightarrow = \supset = \rightarrow$ and $\& = \otimes$. As such, the above properties of tied adjointness algebras become algebraic properties of residuated algebras.

Let $\Lambda = (P, \leq_P, 1, L, \leq_L, \Rightarrow, \&, \supset, \otimes, \rightarrow)$ be a tied adjointness algebra, and let $L^{op}$, the opposite poset of $L$, be defined by reversing the order on $L$. Then a tied adjointness algebra $\Lambda^{dual} = (P, \leq_P, 1, L^{op}, \leq_L, \Rightarrow^d, \&^d, \supset^d, \otimes, \rightarrow)$, said to be the *dual* of $\Lambda$, is obtained by taking $\Rightarrow^d = \&$, $\&^d = \Rightarrow$, $\supset^d$ is the symmetric of $\supset$ ($x \supset^d y = y \supset x$), and by keeping $P$, $\otimes$ and $\rightarrow$ unchanged. This bijection : $\Lambda \longmapsto \Lambda^{dual}$ is self-inverse. The dual of any universally valid inequality in ADJT is universally valid. This way, duality works to establish some new inferences from their duals, without new proofs. This duality principle has been made precise in [21] (see also [20]). We apply a duality principle, in this paper; through which we manage to cut down the number of proofs.

### 2.3 Lattice-Valued Fuzzy Sets and Relations

Lattice-valued fuzzy sets were proposed by Goguen [9] as a generalization of the notion of Zadehs fuzzy sets. Assume that $V$ is a nonempty universe and $L$ is a lattice, then a mapping $A : V \mapsto L$ is called an $L$-valued fuzzy set on $V$. The set of all $L$-valued fuzzy sets on $V$ is denoted by $L^V$.

Given $z \in L$, $v \in V$ and $\emptyset \neq U \subseteq V$, $L$-valued fuzzy sets $z_U$ and $z^v$ are defined by

$$z_U(u) = \begin{cases} z, & u \in U, \\ \bot, & u \notin U. \end{cases} \qquad z^v(u) = \begin{cases} z, & u = v, \\ \top, & u \neq v. \end{cases} \tag{20}$$

Where $\top$ and $\bot$ denote the top and the bottom elements of $L$, respectively. If $U = \{u\}$, then $z_U$ is denoted by $z_u$.

A $P$-valued fuzzy set $R$ on $V^2$ is called a $P$-valued fuzzy relation on $V$. $R$ is said to be *serial* if for each $u \in V$, $\sup_{v \in V} R(u, v) = 1$, *reflexive* if $R(v, v) = 1$ for all $v \in V$, *symmetric* if $R(u, v) = R(v, u)$ for all $u, v \in V$ and $\otimes$-*transitive* if $R(u, v) \otimes R(v, w) \leq_P R(u, w)$ for all $u, v, w \in V$. $R$ is called an $P$-valued fuzzy preorder if it is reflexive and transitive; $R$ is called an $P$-*valued fuzzy* $\otimes$-*similarity relation* if it is reflexive, symmetric and $\otimes$-transitive.

For all $B \in L^V$ and $a \in P$, $L$-valued fuzzy subsets $a \Rightarrow B$ and $a\&B$ of $V$ are defined as $(a \Rightarrow B)(v) = a \Rightarrow B(v)$ and $(a\&B)(v) = a\&B(v)$, respectively. For all $A, B \in L^V$, $A \subseteq B$ denotes $A(v) \leq_L B(v)$ for all $v \in V$. For any $\{A_j\}_{j \in J} \subseteq L^V$, we write $\bigcup_{j \in J} A_j$ and $\bigcap_{j \in J} A_j$ to denote the $L$-valued fuzzy sets on $V$ given by $(\bigcup_{j \in J} A_j)(u) = \sup_{j \in J} A_j(u)$ and $(\bigcap_{j \in J} A_j)(u) = \inf_{j \in J} A_j(u)$, respectively.

## 3 Generalized $(L, P)$-Valued Rough Sets

In this section, we propose generalized many-valued rough sets and investigate their properties. Assume that $V$ is a nonempty universe and $R$ is an arbitrary $P$-fuzzy relation on $V$, then the pair $(V, R)$ is called a $P$-valued fuzzy approximation space. In the sequel, we always take a complete tied adjointness algebra $\Lambda = (P, \leq_P, 1, L, \leq_L, \Rightarrow, \&, \supset, \otimes, \rightarrow)$ in which $\Rightarrow$ (equivalently, $\&$) is faithful as a basic structure, and denote the top and bottom elements of $L$ by $\top$ and $\bot$, respectively.

**Definition 5.** *Let* $(V, R)$ *be a P-valued fuzzy approximation space. The upper rough approximation operators* $\overline{R}$ *and the lower rough approximation operators* $\underline{R}$, *induced by* $R$, *are the operators on* $L^V$ *defined by: for all L-valued fuzzy set A and* $u \in V$,

$$\overline{R}(A)(u) = \sup_{v \in V}(R(u, v)\&A(v)), \tag{21}$$

$$\underline{R}(A)(u) = \inf_{v \in V}(R(u, v) \Rightarrow A(v)). \tag{22}$$

*The pair* $(\underline{R}(A), \overline{R}(A))$ *of L-valued fuzzy sets on V is called a generalized* $(L, P)$-*valued rough set of A with respect to P-valued fuzzy approximation space* $(V, R)$.

These two operators generalize many of the corresponding ones studied in the literature (see, e.g. [4–7, 10, 15, 16, 19, 23, 24, 28–31] and the references therein), whereby usually residuated lattices (either commutative or non-commutative) are employed, and only one semantic for both of the information relation and the approximated sets is assumed. Note the conjunction & in (21) needs not to be commutative nor associative (but it is tied) and that (21) and (22) are dual to each other in the sense given before.

The following dual pairs of useful properties in Lemma 1 below can be easily derived from Definition 5 and from the properties of complete tied adjointness algebras.

**Lemma 1.** *Let* $(V, R)$ *be a P-valued fuzzy approximation space. Then for all* $a \in P$, $z \in L$, $u, v \in V$, $A, B \in L^V$ *and* $\{A_j\}_{j \in J}$:

1. $\overline{R}(\perp_V) = \perp_V$ *and* $\underline{R}(\top_V) = \top_V$,
2. $A \subseteq B$ *implies* $\overline{R}(A) \subseteq \overline{R}(B)$ *and* $\underline{R}(A) \subseteq \underline{R}(B)$,
3. $\overline{R}(\bigcup_{j \in J} A_j) = \bigcup_{j \in J} \overline{R}(A_j)$ *and* $\underline{R}(\bigcap_{j \in J} A_j) = \bigcap_{j \in J} \underline{R}(A_j)$,
4. $\overline{R}(\bigcap_{j \in J} A_j) \subseteq \bigcap_{j \in J} \overline{R}(A_j)$ *and* $\bigcup_{j \in J} \underline{R}(A_j) \subseteq \underline{R}(\bigcup_{j \in J} A_j)$,
5. $\overline{R}(a\&A) = a\&\overline{R}(A)$ *and* $\underline{R}(a \Rightarrow A) = a \Rightarrow \underline{R}(A)$,
6. $z_V \subseteq \underline{R}(z_V)$ *and* $\overline{R}(z_V) \subseteq z_V$,
7. $\overline{R}(z_v)(u) = R(u, v)\&z$ *and* $\underline{R}(z^v)(u) = R(u, v) \Rightarrow z$,
8. $\overline{R}(\top_v)(u) = R(u, v)\&\top$ *and* $\underline{R}(\top_{V-\{v\}})(u) = R(u, v) \Rightarrow \perp$,
9. *If R is symmetric, then* $\overline{R}(A) \subseteq B$ *if, and only if,* $A \subseteq \underline{R}(B)$.

**Proposition 5.** *Let* $(V, R)$ *be a P-valued fuzzy approximation space. Then, R is serial iff* $\overline{R}(z_V) = z_V$ *for all* $z \in L$ *iff* $\underline{R}(z_V) = z_V$ *for all* $z \in L$.

*Proof.* It follows immediately from (5), (6) and Proposition 4.

**Proposition 6.** *Let* $(V, R)$ *be a P-valued fuzzy approximation space. Then (i) R is reflexive, iff (ii)* $\underline{R}(A) \subseteq A$ *for all* $A \in L^V$, *iff (iii)* $A \subseteq \overline{R}(A)$ *for all* $A \in L^V$.

*Proof.* (i) implies (ii) and (iii) are Obvious. We only prove (ii) implies (i). Similarly we can prove that (iii) implies (i). Assume (ii), by Lemma 1(7) and Proposition 4, we have $\forall u \in V, R(u, u) = \inf_{z \in L}((R(u, u) \Rightarrow z) \supset z) = \inf_{z \in L}(\underline{R}(z^u)(u)) \supset z) = 1$.

**Proposition 7.** *Let* $(V, R)$ *be a P-valued fuzzy approximation space. Then the following statements are equivalent:*

**(i)** *R is symmetric.*

**(ii)** $A \subseteq \underline{R}(\overline{R}(A))$ *for all* $A \in L^V$.
**(iii)** $\overline{R}(z_v)(u) = \overline{R}(z_u)(v)$ *for all* $z \in L$ *and* $u, v \in V$.
**(v)** $\overline{R}(\underline{R}(A)) \subseteq A$ *for all* $A \in L^V$.
**(vi)** $\underline{R}(z^v)(u) = \underline{R}(z^u)(v)$ *for all* $z \in L$ *and* $u, v \in V$.

*Proof.* The equivalences between items (i), (iii) and (vi) are direct consequences from Lemma 1 (7) by faithfulness of both $\Rightarrow$ and $\&$.
(i) implies (ii): By (3), for all $A \in L^V$ and $u \in V$,

$$A(u) \leq_L \inf_{v \in V} (R(u, v) \Rightarrow \sup_{w \in V} (R(v, w) \& A(w)))$$

$$= \inf_{v \in V} (R(u, v) \Rightarrow \overline{R}(A)(v)) = \underline{R}(\overline{R}(A))(u).$$

(ii) implies (i): Assume that $R$ is not symmetric, then there exist $u, v \in V$ such that $R(u, v) \neq R(v, u)$. Since $\Rightarrow$ is faithful, then there exists $z \in L$ such that $R(u, v) \Rightarrow z \neq R(v, u) \Rightarrow z$. Consider the following case: Assume $R(v, u) \Rightarrow z < R(u, v) \Rightarrow z$. It follows from Comparator axiom that $(R(u, v) \Rightarrow z) \supset (R(v, u) \Rightarrow z) < 1$. Let $A(w) = R(u, w) \Rightarrow z$, for $z \in L$. Then, by (2), the following hold

$$A(v) \supset \underline{R}(\overline{R}(A))(v) = A(v) \supset \inf_{w \in V} (R(v, w) \Rightarrow \overline{R}(A)(w))$$

$$\leq_L (R(u, v) \Rightarrow z) \supset (R(v, u) \Rightarrow \sup_{w' \in V} (R(u, w') \& (R(u, w') \Rightarrow z)))$$

$$\leq_L (R(u, v) \Rightarrow z) \supset (R(v, u) \Rightarrow z) < 1.$$

Hence, $A(v) \not\leq_L \underline{R}(\overline{R}(A))(v)$, which implies a contradiction.
The others two cases, i.e. $R(u, v) \Rightarrow z < R(v, u) \Rightarrow z$ and, $R(v, u) \Rightarrow z$ and $R(u, v) \Rightarrow z$ are incomparable can be proved in a similar way as the above case.
The equivalence between items (i) and (v) now follows by Duality.

**Proposition 8.** *Let* $(V, R)$ *be a P-valued fuzzy approximation space. Then the following statements are equivalent:*

**(i)** $R$ *is* $\otimes$-*transitive.*
**(ii)** $\overline{R}(\overline{R}(A)) \subseteq \overline{R}(A)$ *for all* $A \in L^V$.
**(iii)** $\underline{R}(A) \subseteq \underline{R}(\underline{R}(A))$ *for all* $A \in L^V$.

*Proof.* (i) implies (ii): Let $R$ be transitive. Then, by (6) and (9), we get

$$\overline{R}(\overline{R}(A))(u) = \sup_{v \in V} \sup_{w \in V} ((R(u, v) \otimes R(v, w)) \& A(w))$$

$$\leq_L \sup_{w \in V} (R(u, w) \& A(w)) = \overline{R}(A)(u).$$

(ii) implies (i): Assume (ii), then $\sup_{v \in V} (R(u, v) \& \overline{R}(A)(v)) \leq_L \overline{R}(A)(u)$. Let $A = z_w$ for $z \in L$ and $w \in V$. Then $\sup_{v \in V} ((R(u, v) \otimes R(v, w)) \& z) \leq_L R(u, w) \& z$ (by Lemma 1(7) and then (9)). From this (by adjointness) we get $R(u, v) \otimes R(v, w) \leq_L \inf_{z \in L} (z \supset (R(u, w) \& z)) = R(u, w)$ (by Proposition 4).
The equivalence between items (i) and (iii) now follows by Duality.

**Proposition 9.** *Let $(V, R)$ be a P-valued fuzzy approximation space and R be reflexive. Then, for all $A \in L^V$, $\overline{R}(\overline{R}(A)) = \overline{R}(A)$ iff $\underline{R}(\underline{R}(A)) = \underline{R}(A)$ iff R is fuzzy preorder.*

*Proof.* It follows immediately from Propositions 6 and 8.

The following useful properties follow from the results in this section and the properties of tied adjointness algebras. Their proofs trace much the same lines of the corresponding proofs given in [19] in the more special setting of commutative residuated lattices on $[0, 1]$.

**Lemma 2.** *Let $(V, R)$ be a P-valued fuzzy approximation space where R is the P-valued fuzzy $\otimes$-similarity relation. Then, for all $a \in P$, $z \in L$ and $A, B \in L^V$,*

1. *$\underline{R}(z_V) = \overline{R}(z_V) = z_V$, $A \subseteq \overline{R}(A)$ and $\underline{R}(A) \subseteq A$,*
2. *$\overline{R}(\underline{R}(A)) = \underline{R}(A)$, $\underline{R}(\overline{R}(A)) = \overline{R}(A)$, $\overline{R}(\overline{R}(A)) = \overline{R}(A)$ and $\underline{R}(\underline{R}(A)) = \underline{R}(A)$,*
3. *$A = \overline{R}(A)$ if and only if $A = \underline{R}(A)$,*
4. *$\overline{R}(a \& \underline{R}(A)) = a \& \underline{R}(A)$ and $\underline{R}(a \Rightarrow \overline{R}(A)) = a \Rightarrow \overline{R}(A)$.*

Two coarse classifications of the $L$-valued fuzzy subsets of $V$ are induced from the $P$-valued fuzzy $\otimes$-similarity relation $R$ when we replace each $L$-valued fuzzy set $A \in L^V$ by either its upper rough approximation $\overline{R}(A) \in L^V$, or its lower rough approximation $\underline{R}(A) \in L^V$. These classifications reduce $L^V$ to the smaller collection $\{A \in L^V \mid A = \overline{R}(A)\} = \{A \in L^V \mid A = \underline{R}(A)\}$.

# 4   Axiomatic Approach

In this section, we study the axiomatic characterizations of generalized rough approximation operators. We work with unary operators on $L^V$ and some axioms which guarantees the existence of certain types of $P$-valued fuzzy binary relations producing the same generalized rough approximation operators. Such an approach is useful to get insight in the logical structure of generalized $(L, P)$-valued rough sets.

Given $A, B \in L^V$, we denote by $|A \supset B|$ the $P$-valued fuzzy subset of $V$ given by, for all $u \in V$, $|A \supset B|(u) = A(u) \supset B(u)$. Given $z \in L$, we denote by $|A \supset B| \Rightarrow z_V$, the $L$-valued fuzzy subset of $V$ given by, for all $u \in V$, $(|A \supset B| \Rightarrow z_V)(u) = (A(u) \supset B(u)) \Rightarrow z$. Similarly for & in place of $\Rightarrow$. We write $\bigcap_{z \in L} |\Psi(z^v) \supset z_V|$ and $\bigcap_{z \in L} |z_V \supset \Phi(z_u)|$ to denote the $P$-valued fuzzy sets on $V$ given by $(\bigcap_{z \in L} |\Psi(z^v) \supset z_V|)(u) = \inf_{z \in L}(\Psi(z^v)(u) \supset z)$ and $(\bigcap_{z \in L} |z_V \supset \Phi(z_v)|)(u) = \inf_{z \in L}(z \supset \Phi(z_v)(u))$, respectively.

**Definition 6.** *Let $\Psi, \Phi : L^V \to L^V$ be two mappings. For all $u \in V$, $z \in L$ and $\{A_j\}_{j \in J} \subseteq L^V$, $\Phi$ is called upper rough approximation operator if it satisfies*

($\Phi 1$)  $\Phi(z_u) = (\bigcap_{x \in L} |x_V \supset \Phi(x_u)|) \& z_V$
($\Phi 2$)  $\Phi(\bigcup_{j \in J} A_j) = \bigcup_{j \in J} \Phi(A_j)$

*$\Psi$ is called lower rough approximation operator if it satisfies*

($\Psi 1$)  $\Psi(z^u) = (\bigcap_{x \in L} |\Psi(x^u) \supset x_V|) \Rightarrow z_V$

($\Psi2$) $\Psi(\bigcap_{j\in J} A_j) = \bigcap_{j\in J} \Psi(A_j)$

**Proposition 10.** *A mapping* $\Phi : L^V \to L^V$ *is an upper rough approximation operator if and only if there exists a unique P-valued fuzzy relation R on V such that* $\Phi = \overline{R}$.

*Proof.* One direction follows easily by Lemma 5 (3), (7) and Proposition 4. For the proof of the other direction, assume $\Phi$ is an upper rough approximation operator. Then we define a $P$-valued fuzzy relation $R$ as follows, for all $u, v \in V$:

$$R(u, v) = \inf_{z\in L}(z \supset \Phi(z_v)(u)). \tag{23}$$

It is easy to proof that $A = \bigcup_{v\in V}(A(v))_v$ holds for all $A \in L^V$. By Definition 6, the following hold for all $A \in L^V$ and $u \in V$,

$$\overline{R}(A)(u) = \sup_{v\in V}(R(u, v)\&A(v)) = \sup_{v\in V}(\inf_{z\in L}(z \supset \Phi(z_v)(u))\&A(v))$$

$$= \sup_{v\in V}(\Phi((A(v))_v)(u)) = \Phi(\bigcup_{v\in V}(A(v))_v)(u) = \Phi(A)(u).$$

Hence, $\overline{R}(A) = \Phi(A)$. By Proposition 4, it is obvious that $R$ is unique.

**Proposition 11.** *A mapping* $\Psi : L^V \to L^V$ *is an upper rough approximation operator if and only if there exists a unique P-valued fuzzy relation R on V such that* $\Psi = \underline{R}$.

*Proof.* Just note that it is the dual of Proposition 10, so it follows by Duality principle and by taking the dual of (23) to define a $P$-valued fuzzy relation $R$ as follows, for all $u, v \in V$:

$$R(u, v) = \inf_{z\in L}(\Psi(z^v)(u) \supset z), \tag{24}$$

and noting that, for all $A \in L^V$, $A = \bigcap_{v\in V}(A(v))^v$.

Adding more axioms to Definition 6, by Propositions 10, 11 and characterization of several classes of generalized $(L, P)$-valued rough sets in Section 3, it is easy to obtain the axiomatic characterizations of them. Here we do not list them.

The above propositions characterize lower and upper rough approximations separately. To link them together, below we introduce and characterize the notion of *coupled pair of rough approximations* (the analogous notion of the duality in classical rough set).

**Definition 7.** *Let* $\Psi, \Phi : L^V \to L^V$ *be two mappings. We call* $(\Psi, \Phi)$ *a coupled pair of lower and upper rough approximation operators if the following conditions hold, for all* $u \in V, x, z \in L$ *and* $\{A_j\}_{j\in J} \subseteq L^V$:

**(C1)** $\Phi$ *is an upper rough approximation operator*
**(C2)** $\Psi(\bigcap_{j\in J} A_j) = \bigcap_{j\in J} \Psi(A_j)$
**(C3)** $|\Phi(x_u) \supset z_V| = |x_V \supset \Psi(z^u)|$

**Proposition 12.** *Let* $\Psi, \Phi : L^V \to L^V$ *be two mappings. The pair* $(\Psi, \Phi)$ *is a coupled pair of lower and upper rough approximation operators if and only if there exists a unique P-valued fuzzy relation R on V such that* $\Psi = \underline{R}$ *and* $\Phi = \overline{R}$.

*Proof.* Assume $(\Psi, \Phi)$ is a coupled pair of lower and upper rough approximation operators. By (C1), $\Phi$ is an upper rough approximation operator, so by Proposition 10, there exists a unique $R$ such that $\Phi = \overline{R}$, where $R(u, v) = \inf_{z \in L}(z \supset \Phi(z_v)(u))$, for $u, v \in V$. By (C3), Lemma 1 (7) and strong adjointness (10), we get, for all $y \in L$,

$$y \supset \Psi(x^v)(u) = \Phi(y_v)(u) \supset x = \inf_{z \in L}(z \supset \Phi(z_v)(u)) \& y \supset x$$
$$= y \supset (\inf_{z \in L}(z \supset \Phi(z_v)(u)) \Rightarrow x)$$

Hence, for all $x \in L$ and $u, v \in V$, we have

$$\Psi(x^v)(u) = \inf_{z \in L}(z \supset \Phi(z_v)(u)) \Rightarrow x. \tag{25}$$

On the other hand, for any $A \in L^V$, it can be verified that $A = \bigcap_{v \in V}(A(v))^v$, so by (C2) and (25), we have, for all $u \in V$

$$\Psi(A)(u) = \Psi(\bigcap_{v \in V}(A(v))^v)(u) = \inf_{v \in V} \Psi((A(v))^v)(u)$$
$$= \inf_{v \in V}((\inf_{z \in L}(z \supset \Phi(z_v)(u)) \Rightarrow A(v)) = \underline{R}(A)(u).$$

Conversely, it is clear that $\overline{R}$ and $\underline{R}$ are an upper and lower rough approximation satisfying (C1) and (C2), respectively. Also, by Lemma 1 (7) and strong adjointness (10), it is easy to see that (C3) holds.

**Acknowledgments.** The author was on leave from Benha faculty of engineering, Benha university (www.bu.edu.eg), Egypt, when writing and investigating this paper. This work was supported by the European Regional Development Fund in the IT4Innovations Centre of Excellence project (CZ.1.05/1.1.00/02.0070).

# References

1. Abdel-Hamid, A., Morsi, N.: Associatively Tied Implications. Fuzzy Sets and Systems 136, 291–311 (2003)
2. Baczyński, M., Jayaram, B.: (S,N)- and R-implications: a state-of-the-art survey. Fuzzy Sets and Systems 159, 1836–1859 (2008)
3. Baczyński, M., Jayaram, B., Mesiar, R.: R-implications and the exchange principle: The case of border continuous t-norms. Fuzzy Sets and Systems 224, 93–105 (2013)
4. Boixander, D., Jacas, J., Recasens, J.: Upper and Lower Approximations of Fuzzy Sets. International Journal of General Systems 29(4), 555–568 (2000)
5. Dubois, D., Prade, H.: Rough Fuzzy Sets and Fuzzy Rough Sets. International Journal of General Systems 17, 191–209 (1990)
6. Dubois, D., Prade, H.: Putting Fuzzy Sets and Rough Sets Together. In: Słowiński, R. (ed.) Intelligent Decision Support - Handbook of Applications and Advances of the Rough Sets Theory, pp. 203–232. Kluwer Academic Publishers (1992)
7. Hu, Q., Zhang, L., Chen, D., Pedrycz, W., Yu, D.: Gaussian kernel based fuzzy rough sets: model, uncertainty measures and applications. International Journal of Approximate Reasoning 51, 453–471 (2010)

8. Galatos, N., Jipsen, P., Kowalski, T., Ono, H.: Residuated Lattices: an algebraic glimpse at substructural logics. Elsevier (2007)
9. Goguen, J.A.: L-fuzzy sets. Journal of Mathematical Analysis and Applications 18, 145–174 (1967)
10. Gong, Z., Sun, B., Chen, D.: Rough set theory for the interval-valued fuzzy information systems. Information Sciences 178, 1968–1985 (2008)
11. Jayaram, B.: On the law of Importation $(x \wedge y) \to z \equiv x \to (y \to z)$ in Fuzzy Logic. IEEE Transaction on Fuzzy Systems 16, 130–144 (2008)
12. Mas, M., Monserrat, M., Torrens, J.: The law of importation for discrete implications. Information Sciences 179(24), 4208–4218 (2009)
13. Mas, M., Monserrat, M., Torrens, J.: A characterization of (U, N), RU,QL and D-implications derived from uninorms satisfying the law of importation. Fuzzy Sets and Systems 161(10), 1369–1387 (2010)
14. Massanet, S., Torrens, J.: The law of importation versus the exchange principle on fuzzy implication. Fuzzy Sets and Systems 168, 47–69 (2011)
15. Mi, J., Zhang, W.: An axiomatic characterization of a fuzzy generalization of rough sets. Information Sciences 160, 235–249 (2004)
16. Mi, J., Leung, Y., Zhao, H., Feng, T.: Generalized fuzzy rough sets determined by a triangular norm. Information Sciences 178, 3203–3213 (2008)
17. Morsi, N.: Propositional calculus under adjointness. Fuzzy Sets and Systems 132, 91–106 (2002)
18. Morsi, N., Roshdy, E.: Issues on adjointness in multiple-valued logics. Information Sciences 176(19), 2886–2909 (2006)
19. Morsi, N., Yakout, M.: Axiomatics for Fuzzy Rough Sets. Fuzzy Sets and Systems 100, 327–342 (1998)
20. Morsi, N., Lotfallah, W., El-Zekey, M.: The Logic of Tied Implications, Part 1: Properties, Applications and Representation. Fuzzy Sets and Systems 157(5), 647–669 (2006)
21. Morsi, N., Lotfallah, W., El-Zekey, M.: The Logic of Tied Implications, Part 2: Syntax. Fuzzy Sets and Systems 157(15), 2030–2057 (2006)
22. Pawlak, Z.: Rough Sets. International Journal of Computer and Information Sciences 11(5), 341–356 (1982)
23. Pei, D.: A generalized model of fuzzy rough sets. International Journal of General Systems 34(5), 603–613 (2005)
24. Radzikowska, A.M., Kerre, E.E.: A Comparative Study of Fuzzy Rough Sets. Fuzzy Sets and Systems 126, 137–155 (2002)
25. Richard, J., Shen, Q.: Fuzzy-rough sets for descriptive dimensionality reduction. In: Proceeding of the IEEE International Conference on Fuzzy Systems, FUZZ-IEEE 2002, pp. 29–34 (2002)
26. Skowron, A., Polkowski, L.: Rough Set in Knowledge Discovery. Springer, Berlin (1998)
27. Singh, K., Thakur, S., Lal, M.: Vague Rough Set Techniques for Uncertainty Processing in Relational Database Model. Informatica 19(1), 113–134 (2008)
28. Wang, C.Y., Hu, B.Q.: Fuzzy rough sets based on generalized residuated lattices. Information Sciences 248, 31–49 (2013)
29. Wu, W., Mi, J., Zhang, W.: Generalized fuzzy rough sets. Information Sciences 151, 263–282 (2003)
30. Wu, W., Zhang, W.: Constructive and axiomatic approaches of fuzzy approximation operators. Information Sciences 159, 233–254 (2004)
31. Wu, W., Leung, Y., Mi, J.: On characterizations of (I, T)-fuzzy rough approximation operators. Fuzzy Sets and Systems 154, 76–102 (2005)

# On the Topological Structure of Rough Soft Sets

Vinay Gautam[1], Vijay K. Yadav[1], Anupam K. Singh[2], and S.P. Tiwari[1]

[1] Department of Applied Mathematics, Indian School of Mines,
Dhanbad-826004, India
{gautam.ism181,vkymaths,sptiwarimaths}@gmail.com
http://www.ismdhanbad.ac.in
[2] I.T.S Engineering Collage,
46 Knowledge Park - III, Greater Noida -201308, India
anupam09.bhu@gmail.com
http://www.its.edu.in

**Abstract.** The concept of rough soft set is introduced to generalize soft sets by using rough set theory, and then the soft topologies on soft sets are introduced.

**Keywords:** Soft approximation space, Soft topology, Soft relation, Soft closure operator.

## 1 Introduction

The soft set theory introduced by Molodtsov [18], which is assumed as a mathematical tool for dealing with uncertainties, has been developed significantly with a number of applications such as it can be applied in game theory, Riemann integration, probability theory, etc. (cf. [19]). It has also been seen that the mathematical objects such as topological spaces, fuzzy sets and rough sets can be considered as a particular types of soft sets (cf., [16,18]). Recently, so many authors have tried to develop the mathematical concepts based on soft set theory, e.g., in [2,6,27,29], rough soft sets and fuzzy soft sets ; in [8], Soft rough fuzzy sets and soft fuzzy rough sets; in [10], the algebraic structure of semi-rings by applying soft set theory; in [3], fuzzy soft group; in [13], soft BCK/BCI-algebras; in [14], the applications of soft sets in ideal theory of BCK/BCI-algebras; in [5,28,1], soft set relations and functions; in [4,7,12,25], soft topology, which itself is showing the interest of researchers in this area.

Beside soft set theory, rough set theory, firstly proposed by Pawlak [20] has now been developed significantly due to its importance for the study of intelligent systems having insufficient and incomplete information. In rough set introduced by Pawlak, the key role is played by equivalence relations. In literature (cf., [15,20,21,23], several generalizations of rough set have been made by replacing the equivalence relation by an arbitrary relation. Simultaneously, the relation of rough set with topology is also studied (cf., [15,24]).

D. Miao et al. (Eds.): RSKT 2014, LNAI 8818, pp. 39–48, 2014.
DOI: 10.1007/978-3-319-11740-9_4 © Springer International Publishing Switzerland 2014

As both the theories approaches to vagueness, it will be interesting to see the connection between both the theories. In this direction, an initiation has already been made (cf., [11,26]), in which, soft set theory is utilized to generalize the rough set model introduced by Pawlak (cf., [20]). Also, the resultant hybrid model has been applied to multicriteria group decision making (cf., [9]). It is the natural question that what will happen if rough set theory is used to generalize soft sets. This paper is toward this study. Specifically, we try to introduce the concept of rough soft set, and as topology is closely related to rough sets, we try to introduce soft topologies on soft sets with the help of rough soft sets.

## 2   Preliminaries

In this section, we collect some concepts associated with soft sets, which we will use in the next section. Throughout, $U$ denotes an universal set and $E$, the set of all possible parameters with respect to $U$. The family of all subsets of $U$ is denoted by $P(U)$.

**Definition 1.** [18] *A pair $F_A = (F, A)$ is called a **soft set** over $U$, where $A \subseteq E$ and $F : A \to P(U)$ is a map.*

In other words, a soft set $F_A$ over $U$ is a parameterized family $\{F(a) : a \in A\}$ of subsets of the universe $U$. For $\epsilon \in A$, $F(\epsilon)$ may be considered as the set of $\epsilon$-appximate elements of the soft set $F_A$.

For the universe $U$, $S(U)$ will denote the class of all soft sets over $U$.

**Definition 2.** [22] *Let $A, B \subseteq E$ and $F_A, G_B \in S(U)$. Then $F_A$ is **soft subset** of $G_B$, denoted by $F_A \subseteq G_B$, if*

**(i)** $A \subseteq B$, *and*
**(ii)** $\forall a \in A$, $F(a) \subseteq G(a)$.

**Definition 3.** [16] *$F_A$ and $G_B$ are said to be **soft equal** if $F_A \subseteq G_B$ and $G_B \subseteq F_A$. For a soft set $F_A \in S(U)$, $\widetilde{P}(F_A)$ denotes the set of all soft subsets of $F_A$.*

**Definition 4.** [17] *Let $A \subseteq E$ and $F_A \in S(U)$. Then $F_A$ is called **soft empty**, denoted by $F_\phi$, if $F(a) = \phi$, $\forall a \in A$.*

$F(a) = \phi$, $\forall a \in A$ means that there is no element in $U$ related to the parameter $a \in A$. Therefore, there is no need to display such elements in the soft sets, as it is meaningless to consider such parameters.

**Definition 5.** [17] *Let $A, B \subseteq E$ and $F_A, G_B \in S(U)$. Then the **soft union** of $F_A$ and $G_B$ is a soft set $H_C = (H, C)$, where $C = A \cup B$ and $H : C \to P(U)$ such that $\forall a \in C$,*

$$H(a) = \begin{cases} F(a) & \text{if } a \in A - B \\ G(a) & \text{if } a \in B - A \\ F(a) \cup G(a) & \text{if } a \in A \cap B \end{cases}$$

**Definition 6.** [17] *Let* $A, B \subseteq E$ *and* $F_A, G_B \in S(U)$. *Then the* **soft intersection** *of* $F_A$ *and* $G_B$ *is a soft set* $H_C = (H, C)$, *where* $C = A \cap B$ *and* $H : C \to P(U)$ *such that* $H(a) = F(a) \cap G(a)$, $\forall a \in C$.

**Definition 7.** [17] *Let* $E = \{e_1, e_2, e_3, ...e_n\}$ *be a set of parameters. Then the* **NOT** *set of* $E$ *is* $\rceil E$ *is defined by* $\rceil E = \{\rceil e_1, \rceil e_2, \rceil e_3, ...\rceil e_n\}$, *where* $\rceil e = not\ e_i$, $\forall i = 1, 2, ..., n$.

**Definition 8.** [1] *Let* $A \subseteq E$ *and* $F_A \in S(U)$. *Then the* **soft complement** *of* $F_A$ *is* $(F_A)^c$ *and defined by* $(F_A)^c = F_A^c$, *where* $F^c : A \to P(U)$ *is a map such that* $F^c(a) = U - F(a)$, $\forall a \in A$.

We call $F^c$, the soft complement function of $F$. It is easy to see that $(F^c)^c = F$ and $(F_A^c)^c = F_A$. Also, $F_\phi^c = F_E$ and $F_E^c = F_\phi$.

**Proposition 1.** [17] *Let* $F_A \in S(U)$. *Then*

**(i)** $F_A \cup F_A = F_A$, $F_A \cap F_A = F_A$
**(ii)** $F_A \cup F_\phi = F_A$, $F_A \cap F_\phi = F_\phi$
**(iii)** $F_A \cup F_E = F_E$, $F_A \cap F_E = F_A$
**(iv)** $F_A \cup F_A^c = F_E$, $F_A \cap F_A^c = F_\phi$.

**Definition 9.** [5] *Let* $A, B \subseteq E$ *and* $F_A, G_B \in S(U)$. *Then the* **cartesian product** *of* $F_A$ *and* $G_B$ *is the soft set* $H_{A \times B} = (H, A \times B)$, *where* $H_{A \times B} = F_A \times G_B$ *and* $H : A \times B \to P(U \times U)$ *such that* $H(a, b) = F(a) \times G(b)$, $\forall (a, b) \in A \times B$, *i.e.,* $H(a, b) = \{(h_i, h_j) : h_i \in F(a)\ and\ h_j \in G(b)\}$.

**Definition 10.** [5] *Let* $A, B \subseteq E$ *and* $F_A, G_B \in S(U)$. *Then a* **soft relation** *from* $F_A$ *to* $G_B$ *is a soft subset of* $F_A \times G_B$.

In other words, a soft relation from $F_A$ to $G_B$ is of the form $H_C'$, where $C \subseteq A \times B$ and $H'(a, b) = H(a, b)$, $\forall (a, b) \in C$, and $H_{A \times B} = F_A \times G_B$ as defined in Definition 9. Any subset of $F_A \times F_A$ is called a **soft relation** on $F_A$.

In an equivalent way, the soft relation $R$ on the soft set $F_A$ in the parameterized form is as follows:

If $F_A = \{F(a_1), F(a_2), ...\}$, $a_1, a_2, ... \in A$, then $_{F(a_i)}R_{F(a_j)} \Leftrightarrow F(a_i) \times F(a_j) \in R$.

**Definition 11.** [5] *A soft relation* $R$ *on a soft set* $F_A \in S(U)$ *is called*

**(i)** **soft reflexive** *if* $H'(a, a) \in R$, $\forall a \in A$,
**(ii)** **soft symmetric** *if* $H'(a, b) \in R \Rightarrow H'(b, a) \in R$, $\forall (a, b) \in A \times A$, *and*
**(iii)** **soft transitive** *if* $H'(a, b) \in R$, $H'(b, c) \in R \Rightarrow H'(a, c) \in R$, $\forall a, b, c \in A$.

Above definition can be restated as follows:

**Definition 12.** [28] *A soft relation* $R$ *on a soft set* $F_A \in S(U)$ *is called*

**(i)** **soft reflexive** *if* $F(a) \times F(a) \in R$, $\forall a \in A$,

(ii) **soft symmetric** if $F(a) \times F(b) \in R \Rightarrow F(b) \times F(a) \in R, \forall(a,b) \in A \times A$, and

(iii) **soft transitive** if $F(a) \times F(b) \in R, F(b) \times F(c) \in R \Rightarrow F(a) \times F(c) \in R$, $\forall a, b, c \in A$.

**Definition 13.** [5] Let $A \subseteq E$ and $F_A \in S(U)$. Then $[F(a)] = \{F(a') : F(a) \times F(a') \in R, \forall a, a' \in A\}$.

*Remark 1.* For $A \subseteq E$ and $F_A \in S(U)$, it can be seen that $[F(a)] = (F, A_a), a \in A$ is a soft subset of $F_A$, where $A_a = \{a' \in A : F(a) \times F(a') \in R\}$.

**Definition 14.** [7] Let $F_A \in S(U)$ and $\tau \subseteq \widetilde{P}(F_A)$. Then $\tau$ is called a **soft topology** on $F_A$ if

(i) $F_\phi, F_A \in \tau$,
(ii) for $F_{A_i} \in \widetilde{P}(F_A), i \in I$, if $F_{A_i} \in \tau$, then $\cup_{i \in I} F_{A_i} \in \tau$, and
(iii) for $F_{A_1}, F_{A_2} \in \widetilde{P}(F_A)$, if $F_{A_1}, F_{A_2} \in \tau$, then $F_{A_1} \cap F_{A_2} \in \tau$.

The pair $(F_A, \tau)$ is called **soft topological space** and soft subsets of $F_A$ in $\tau$ are called **soft open** set. The compliment of a soft open set is called a **soft closed** set.

## 3   Rough Soft Set and Soft Topology

In this section, we introduce the concept of rough soft set and introduce soft topologies on soft sets. Throughout this section, $F_A$ is a soft set over $U$.

**Definition 15.** A pair $(F_A, R)$ is called a **soft approximation space**, where $F_A \in S(U)$ and $R$ is a soft relation on $F_A$.

**Definition 16.** Let $(F_A, R)$ be a soft approximation space. Then **soft lower approximation** and **soft upper approximation** of $G_B \subseteq F_A$, are respectively, defined as:

$$\underline{apr}(G_B) = \cup_{a \in A}\{F(a) \in F_A : [F(a)] \subseteq G_B\}, \text{ and}$$

$$\overline{apr}(G_B) = \cup_{a \in A}\{F(a) \in F_A : [F(a)] \cap G_B \neq F_\phi\}.$$

The pair $(\underline{apr}(G_B), \overline{apr}(G_B))$ is called a **rough soft set**.

*Remark 2.* From above definition, it is clear that $\underline{apr}(G_B)$ and $\overline{apr}(G_B)$ are soft subsets of $F_A$.

*Example 1.* Let $U = \{u_1, u_2, u_3\}, E = \{x_1, x_2, x_3\}, A = \{x_1, x_2\}$, $F_A = \{(x_1, \{u_1, u_2\}), (x_2, \{u_2, u_3\})\}$ and $G_B \subseteq F_A$, where $G_B = \{(x_2, \{u_2, u_3\})\}$. Also, consider a soft relation $R = \{F(x_1) \times F(x_1), F(x_1) \times F(x_2), F(x_2) \times F(x_2)\}$. Then $[F(x_1)] = \{F(x_1), F(x_2)\}$ and $[F(x_2)] = \{F(x_2)\}$. It can be easily seen that $\underline{apr}(G_B) = G_B$ and $\overline{apr}(G_B) = F_A$.

**Proposition 2.** For a soft approximation space $(F_A, R)$ and $\forall G_B, H_C \subseteq F_A$,

**(i)** $\underline{apr}(F_\phi) = F_\phi = \overline{apr}(F_\phi)$;

**(ii)** $\underline{apr}(F_A) = F_A = \overline{apr}(F_A)$;

**(iii)** If $G_B \subseteq H_C$, then $\underline{apr}(G_B) \subseteq \underline{apr}(H_C)$ and $\overline{apr}(G_B) \subseteq \overline{apr}(H_C)$;

**(iv)** $\underline{apr}(G_B) = (\overline{apr}(G_B^c))^c$;

**(v)** $\overline{apr}(G_B) = (\underline{apr}(G_B^c))^c$;

**(vi)** $\underline{apr}(G_B \cap H_C) = \underline{apr}(G_B) \cap \underline{apr}(H_C)$;

**(vii)** $\underline{apr}(G_B) \cup \underline{apr}(H_C) \subseteq \underline{apr}(G_B \cup H_C)$;

**(viii)** $\overline{apr}(G_B \cup H_C) = \overline{apr}(G_B) \cup \overline{apr}(H_C)$;

**(ix)** $\overline{apr}(G_B \cap H_C) \subseteq \overline{apr}(G_B) \cap \overline{apr}(H_C)$

**Proof** $(i)$ and $(ii)$ are obvious.

$(iii)$ Let $G_B \subseteq H_C$ and $F(a) \in \underline{apr}(G_B)$, $a \in A$. Then $[F(a)] \subseteq G_B$, and so $[F(a)] \subseteq H_C$. Thus $F(a) \in \underline{apr}(H_C)$, whereby $\underline{apr}(G_B) \subseteq \underline{apr}(H_C)$. Similarly, we can show that $\overline{apr}(G_B) \subseteq \overline{apr}(H_C)$.

$(iv)$ $F(a) \in (\overline{apr}(G_B^c))^c \Leftrightarrow F(a) \notin (\overline{apr}(G_B^c)) \Leftrightarrow [F(a)] \cap G_B^c = F_\phi \Leftrightarrow [F(a)] \subseteq G(B) \Leftrightarrow F(a) \in G(B)$. Thus $\underline{apr}(G_B) = (\overline{apr}(G_B^c))^c$.

$(v)$ Similar to that of (iv).

$(vi)$ $F(a) \in \underline{apr}(G_B \cap H_C) \Leftrightarrow [F(a)] \subseteq G_B \cap H_C \Leftrightarrow [F(a)] \subseteq G_B$ and $[F(a)] \subseteq H_C \Leftrightarrow F(a) \in \underline{apr}(G_B)$ and $F(a) \in \underline{apr}(H_C) \Leftrightarrow F(a) \in \underline{apr}(G_B) \cap \underline{apr}(H_C)$. Thus $\underline{apr}(G_B \cap H_C) = \underline{apr}(G_B) \cap \underline{apr}(H_C)$.

$(vii)$ Follows as above.

$(viii)$ $F(a) \in \overline{apr}(G_B \cup H_C) \Leftrightarrow [F(a)] \cap (G_B \cup H_C) \neq F_\phi \Leftrightarrow [F(a)] \cap G_B \neq F_\phi$ or $[F(a)] \cap H_C \neq F_\phi \Leftrightarrow F(a) \in \overline{apr}(G_B)$ or $F(a) \in \overline{apr}(H_C) \Leftrightarrow F(a) \in \overline{apr}(G_B) \cup \overline{apr}(H_C)$. Thus $\overline{apr}(G_B \cup H_C) = \overline{apr}(G_B) \cup \overline{apr}(H_C)$.

$(ix)$ Follows as above.

Following example support each of proposition (i) to (ix).

*Example 2.* Let $U = \{u_1, u_2\}, E = \{x_1, x_2, x_3\}, A = \{x_1, x_2\}$. Also, let $F_A = \{(x_1, \{u_1, u_2\}), (x_2, \{u_1, u_2\})\}$ and $F_A^i, i \in I$ denotes soft subsets of $F_A$. Then all soft subsets of $F_A$ are

$F_A^1 = \{(x_1, \{u_1, u_2\}), (x_2, \{u_1\})\}$,

$F_A^2 = \{(x_1, \{u_1, u_2\}), (x_2, \{u_2\})\}$,

$F_A^3 = \{(x_1, \{u_1\}), (x_2, \{u_1, u_2\})\}$,

$F_A^4 = \{(x_1, \{u_2\}), (x_2, \{u_1, u_2\})\}$,

$F_A^5 = \{(x_1, \{u_1\}), (x_2, \{u_1\})\}$,

$F_A^6 = \{(x_1, \{u_2\}), (x_2, \{u_2\})\}$,

$F_A^7 = \{(x_1, \{u_1\}), (x_2, \{u_2\})\}$,

$F_A^8 = \{(x_1, \{u_2\}), (x_2, \{u_1\})\}$,

$F_A^9 = \{(x_1, \{u_1, u_2\})\},$
$F_A^{10} = \{(x_2, \{u_1, u_2\})\},$
$F_A^{11} = \{(x_1, \{u_1\})\},$
$F_A^{12} = \{(x_1, \{u_2\})\},$
$F_A^{13} = \{(x_2, \{u_1\})\},$
$F_A^{14} = \{(x_2, \{u_2\})\},$
$F_A^{15} = F_\emptyset,$
$F_A^{16} = F_A.$

Let $R = \{F(x_1) \times F(x_1), F(x_2) \times F(x_2), F(x_1) \times F(x_2)\}.$
By definition 13 and 16 it follows that
$[F(x_1)] = \{F(x_1), F(x_2)\}, [F(x_2)] = \{F(x_2)\}$ and
$\underline{apr}(F_A^i : i = 3, 4, 10) = \{F(x_2)\}, \underline{apr}(F_A) = F_A,$
$\underline{apr}(F_A^i : i = 1, 2, 5, 6, 7, 8, 9, 11, 12, 13, 14, 15) = F_\emptyset,$ also
$\overline{apr}(F_A^i : i = 1, 2, 3, 4, 5, 6, 7, 8, 10, 13, 14, 16) = F_A,$
$\overline{apr}(F_A^i : i = 9, 11, 12) = \{F(x_1)\}$ and $\overline{apr}(F_A^{15}) = F_\emptyset.$

**Proposition 3.** *Let $(F_A, R)$ be a soft approximation space and $R$ be soft reflexive. Then $\forall G_B \subseteq F_A,$*

**(i)** $G_B \subseteq \overline{apr}(G_B),$ *and*
**(ii)** $\underline{apr}(G_B) \subseteq G_B.$

**Proof** Follows easily from the fact that $R$ is reflexive.

*Example 3.* In Example 2, let $R = \{F(x_1) \times F(x_1), F(x_2) \times F(x_2)\}.$ Then $[F(x_1)] = \{F(x_1)\}, [F(x_2)] = \{F(x_2)\}.$ Thus $\underline{apr}(F_A^i : i = 1, 2, 9) = \{F(x_1)\}, \underline{apr}(F_A^i : i = 3, 4, 10) = \{F(x_2)\}, \overline{apr} = F_A, \underline{apr}(F_A^i : i = 5, 6, 7, 8, 11, 12, 13, 14, 15) = F_\emptyset, \overline{apr}(F_A^i : i = 1, 2, 3, 4, 5, 6, 7, 8, 16) = F_A, \overline{apr}(F_A^i : i = 9, 11, 12) = \{F(x_1)\}, \overline{apr}(F_A^i : i = 10, 13, 14) = \{F(x_2)\}, \overline{apr}(F_A^{15}) = F_\emptyset.$ Clearly, $F_A^i \subseteq \overline{apr}(F_A^i) \forall i = 1, ...16,$ and $\underline{apr}(F_A^i) \subseteq F_A^i, \forall i = 1, ...16.$

**Proposition 4.** *Let $(F_A, R)$ be a soft approximation space and $R$ be soft symmetric. Then $\forall G_B, H_C \subseteq F_A,$*

**(i)** $\overline{apr}(\underline{apr}(G_B)) \subseteq G_B,$ *and*
**(ii)** $G_B \subseteq \underline{apr}(\overline{apr}(G_B)).$

**Proof** (i) Let $F(a) \in \overline{apr}(\underline{apr}(G_B)), a \in A.$ Then $[F(a)] \cap \underline{apr}(G_B) \neq F_\phi,$ or that, there exists $F(a') \in [F(a)], a' \in A$ such that $F(a') \in \underline{apr}(G_B).$ $F(a') \in \underline{apr}(G_B),$ implying that $[F(a')] \subseteq G_B.$ Since $R$ is symmetric and $F(a) \times F(a') \in R,$ so $F(a') \times F(a) \in R.$ Thus $F(a) \in [F(a')],$ and so $F(a) \in G_B.$ Hence $\overline{apr}(\underline{apr}(G_B)) \subseteq G_B.$

(ii) Follows as above.

**Proposition 5.** *Let $(F_A, R)$ be a soft approximation space and $R$ be soft transitive. Then $\forall G_B \subseteq F_A,$*

(i) $\overline{apr}(\overline{apr}(G_B)) \subseteq \overline{apr}(G_B)$, and

(ii) $\underline{apr}(G_B) \subseteq \underline{apr}(\underline{apr}(G_B))$.

**Proof** (i) Let $F(a) \in \overline{apr}(\overline{apr}(G_B))$, $a \in A$. Then $[F(a)] \cap \overline{apr}(G_B) \neq F_\phi$, i.e., there exists $F(a') \in [F(a)]$, $a' \in A$ such that $F(a') \in \overline{apr}(G_B)$. Now, $F(a') \in \overline{apr}(G_B) \Rightarrow [F(a')] \cap G_B \neq F_\phi$, i.e., there exists $F(a'') \in [F(a')]$, $a'' \in A$ such that $F(a'') \in G_B$. But $R$ being soft transitive, $F(a') \in [F(a)]$ and $F(a'') \in [F(a')]$ implying that $F(a'') \in [F(a)]$. Thus $[F(a)] \cap G_B \neq F_\phi$, whereby $F(a) \in \overline{apr}(G_B)$. Hence $\overline{apr}(\overline{apr}(G_B)) \subseteq \overline{apr}(G_B)$.

(ii) Follows as above.

**Proposition 6.** *If a soft relation $R$ on $F_A$ is soft reflexive. Then $\tau = \{G_B \subseteq F_A : \underline{apr}(G_B) = G_B\}$ is a soft topology on $F_A$.*

**Proof** In view of Proposition 2, we only need to show that if $G_{B_i} \in \tau$, then $\cup_{i \in I} G_{B_i} \in \tau$, where $G_{B_i} \in \widetilde{P}(F_A)$, $i \in I$. For which, it is sufficient to show that $\cup_{i \in I} G_{B_i} \subseteq \underline{apr}(\cup_{i \in I} G_{B_i})$. Let $F(a) \in \cup_{i \in I} G_{B_i}$, $a \in A$. Then their exists some $j \in J$ such that $F(a) \in G_{B_j} = \underline{apr}(G_{B_j})$, i.e., $[F(a)] \subseteq G_{B_j} \subseteq \cup_{i \in I} G_{B_i}$, or that $F(a) \in \underline{apr}(\cup_{i \in I} G_{B_i})$. Thus $\cup_{i \in I} G_{B_i} \subseteq \underline{apr}(\cup_{i \in I} G_{B_i})$, whereby $\cup_{i \in I} G_{B_i} \subseteq \underline{apr}(\cup_{i \in I} G_{B_i})$. Hence $\tau$ is a soft topology on $F_A$.

**Proposition 7.** *Let $R$ be soft reflexive and soft symmetric. Then $\underline{apr}(G_B) = G_B$ if and only if $G_B^c = \overline{apr}(G_B^c)$.*

**Proof** Let $\underline{apr}(G_B) = G_B$. As, $\overline{apr}(G_B^c) \subseteq G_B^c$, we only need to show that $G_B^c \subseteq \overline{apr}(G_B^c)$. For this, let $F(a) \notin \overline{apr}(G_B^c)$, $a \in A$. Then $\exists F(b) \in F_A$ such that $F(b) \in [F(a)]$ and $F(b) \notin ((G_B)^c)$, or that, $F(b) \in G_B = \underline{apr}(G_B)$ and $F(b) \in [F(a)]$. Now, $R$ being soft symmetric and $F(b) \in [F(a)]$ so $F(a) \in [F(b)]$. Also, $F(b) \in \underline{apr}(G_B) \Rightarrow [F(b)] \subseteq G_B$. Thus $F(a) \in G_B$, or that $F(a) \notin ((G_B)^c)$, whereby $G_B^c = \overline{apr}(G_B^c)$. The converse part can be proved similarly.

Following is an easy consequence of the above proposition.

**Proposition 8.** *Let $R$ be soft reflexive and soft symmetric relation on $F_A$. Then $(F_A, \tau)$ is the soft topological space having the property that $G_B$ is soft open if and only if $G_B$ is soft closed.*

**Proof** As $R$ is soft reflexive, from Proposition 6, $\tau$ is a topology on a $F_A$. Also, $G_B$ is soft open if and only if $G_B \in \tau$ if and only if $\underline{apr}(G_B) = G_B$ if and only if $\underline{apr}(G_B))^c = (G_B)^c$ if and only if $(G_B)^c \in \tau$ if and only if $(G_B)^c$ is open if and only if $G_B$ is soft closed.

Now, we introduce the following concept of soft closure and soft interior operator on a soft set.

**Definition 17.** *A mapping $\widetilde{c} : \widetilde{P}(F_A) \to \widetilde{P}(F_A)$ is called a **soft closure operator** if $\forall G_B, G_{B_1}, G_{B_2} \in \widetilde{P}(F_A)$,*

(i) $\widetilde{c}(F_\phi) = F_\phi$,

**(ii)** $G_B \subseteq \widetilde{c}(G_B)$,
**(iii)** $\widetilde{c}(G_{B_1} \cup G_{B_2}) = \widetilde{c}(G_{B_1}) \cup \widetilde{c}(G_{B_2})$,
**(iv)** $\widetilde{c}(\widetilde{c}(G_B)) = \widetilde{c}(G_B)$.

*Remark 3.* Let $\tau = \{G_B \subseteq F_A : \widetilde{c}(G_B^c) = G_B^c\}$. Then it can be seen that $\tau$ is a soft topology on $F_A$.

**Definition 18.** *A mapping* $\widetilde{i} : \widetilde{P}(F_A) \to \widetilde{P}(F_A)$ *is called a* **soft interior operator** *if,* $\forall G_B, G_{B_1}, G_{B_2} \in \widetilde{P}(F_A)$ ,

**(i)** $\widetilde{i}(F_A) = F_A$,
**(ii)** $\widetilde{i}(G_B) \subseteq G_B$,
**(iii)** $\widetilde{i}(G_{B_1} \cap G_{B_2}) = \widetilde{i}(G_{B_1}) \cap \widetilde{i}(G_{B_2})$,
**(iv)** $\widetilde{i}(\widetilde{i}(G_B)) = \widetilde{i}(G_B)$.

*Remark 4.* Let $\tau = \{G_B \subseteq F_A : \widetilde{i}(G_B) = G_B\}$. Then it can be seen that $\tau$ is a soft topology on $F_A$.

**Proposition 9.** *If a soft relation $R$ on $F_A$ is soft reflexive and soft transitive, then $\underline{apr}$ and $\overline{apr}$ are saturated[1] soft interior and saturated soft closure operators respectively.*

**Proof** Follows from Propositions 2, 3 and 5.

Finally, we show that each saturated soft closure operator on a soft set also induces a soft reflexive and soft transitive relation as:

**Proposition 10.** *Let $\widetilde{c}$ be a saturated soft closure operator on $F_A$. Then there exists an unique soft reflexive and soft transitive relation $R$ on $F_A$ such that $\widetilde{c}(G_B) = \overline{apr}(G_B), \forall G_B \subseteq F_A$.*

**Proof** Let $\widetilde{c}$ be a saturated soft closure operator and $R$ be a soft relation on $F_A$ given by $F(a) \times F(a') \in R \Leftrightarrow F(a) \in \widetilde{c}(\{F(a')\}), a, a' \in A$. As, $\{F(a)\} \subseteq \widetilde{c}(\{F(a)\})$, $F(a) \in \widetilde{c}(\{F(a)\})$, or that, $F(a) \times F(a) \in R$. Thus $R$ is a soft reflexive relation on $F_A$. Also, let $F(a) \times F(a') \in R$ and $F(a') \times F(a'') \in R; a, a', a'' \in A$. Then $F(a) \in \widetilde{c}(\{F(a')\})$ and $F(a') \in \widetilde{c}(\{F(a'')\})$. Thus $F(a) \in \widetilde{c}(\{F(a')\})$ and $\widetilde{c}(\{F(a')\}) \subseteq \widetilde{c}(\widetilde{c}(\{F(a'')\})) = \widetilde{c}(\{F(a'')\})$, or that, $F(a) \in \widetilde{c}(\{F(a'')\})$, i.e., $F(a) \times F(a'') \in R$. Therefore $R$ is a soft transitive relation on $F_A$. Now, let $G_B \subseteq F_A$ and $F(a) \in \overline{apr}(G_B), a \in A$. Then $[F(a)] \cap G_B \neq F_\phi$, or that, $\exists F(a') \in F_A$ such that $F(a') \in [F(a)] \cap G_B$, showing that $F(a) \in \widetilde{c}(\{F(a')\})$ and $F(a') \in G_B$. Thus $F(a) \in \widetilde{c}(G_B)$, whereby $\overline{apr}(G_B) \subseteq \widetilde{c}(G_B)$. Conversely, let $F(a) \in \widetilde{c}(G_B)$. Then $F(a) \in \widetilde{c}(\cup\{F(a') : F(a') \in G_B\}) = \cup\{\widetilde{c}(\{F(a')\}) : F(a') \in G_B\}$ (as $\widetilde{c}$ is a saturated closure operator). Now, $F(a) \in \cup\{c(\{F(a')\}) : F(a') \in G_B\} \Rightarrow F(a) \in c(\{F(a')\})$, for some $F(a') \in G_B$, or that, $F(a') \in [F(a)]$, for some $F(a') \in G_B$, i.e., $[F(a)] \cap G_B \neq F_\phi$, showing that $F(a) \in \overline{apr}(G_B)$. Thus $\widetilde{c}(G_B) \subseteq \overline{apr}(G_B)$. Therefore $\widetilde{c}(G_B) = \overline{apr}(G_B)$. The uniqueness of soft relation $R$ can be seen easily.

---

[1] A soft closure operator $\widetilde{c} : \widetilde{P}(F_A) \to \widetilde{P}(F_A)$ on $F_A$ is being called here saturated if the (usual) requirement $\widetilde{c}(G_{B_1} \cup G_{B_2}) = \widetilde{c}(G_{B_1}) \cup \widetilde{c}(G_{B_2})$ is replaced by $\widetilde{c}(\cup_{i \in I} G_{B_i}) = \cup_{i \in I} \widetilde{c}(G_{B_i})$, where $G_{B_i} \in \widetilde{P}(F_A), i \in I$.

# 4 Conclusion

In this paper, we tried to introduce the concept of rough soft sets by combining the theory of rough sets and that of soft sets, as well as introduce soft topologies on a soft set induced by soft lower approximation operator. As rough soft sets are generalization of soft sets with the help of rough set theory and the rough set theory has already been established much more; so this paper opens some new directions.

# References

1. Ali, M.I., Feng, F., Xiaoyan, L., Min Won, K., Sabir, M.: On some new operations in soft set theory. Computers and Mathematics with Applications 57, 1547–1553 (2009)
2. Ali, M.I.: A note on soft sets, rough soft sets and fuzzy soft sets. Applied Soft Computing 11, 3329–3332 (2011)
3. Aygunoglu, A., Aygun, H.: Introduction to fuzzy soft groups. Computers and Mathematics with Applications 58, 1279–1286 (2009)
4. Aygunoglu, A., Aygun, H.: Some notes on soft topological spaces. Neural computing and Applications 21, 113–119 (2012)
5. Babitha, K.V., Sunil, J.J.: Soft set relations and functions. Computers and Mathematics with Applications 60, 1840–1849 (2010)
6. Broumi, S., Majumdar, P., Smarandache, F.: New operations on intuitionistic fuzzy soft sets based on second Zadeh's logical operators. International Journal of Information Engineering and Electronic business 1, 25–31 (2014)
7. Cagman, N., Karatas, S., Enginoglu, S.: Soft topology. Computers and Mathematics with Applications 62, 351–358 (2011)
8. Dan, M., Zhang, X., Qin, K.: Soft rough fuzy sets and soft fuzy rough sets. Computers and Mathematics with Applications 62, 4635–4645 (2011)
9. Feng, F.: Soft rough sets applied to multicriteria group decision making. Annals of Fuzzy Mathematics and Informatics 2, 69–80 (2011)
10. Feng, F., Jun, Y.B., Zhao, X.Z.: Soft semirings. Computers and Mathematics with Applications 56, 2621–2628 (2008)
11. Feng, F., Liu, X., Fotea, V.L., Jun, Y.B.: Soft sets and soft rough sets. Information Sciences 181, 1125–1137 (2011)
12. Georgiou, D.N., Megaritis, A.C.: Soft set theory and topology. Applied General Topology 15, 93–109 (2014)
13. Jun, Y.B.: Soft BCK/BCI-algebras. Computers and Mathematics with Applications 56, 1408–1413 (2008)
14. Jun, Y.B., Park, C.H.: Applications of soft sets in ideal theory of BCK/BCI-algebras. Information Sciences 178, 2466–2475 (2008)
15. Kondo, M.: On the structure of generalized rough sets. Information Sciences 176, 586–600 (2006)
16. Maji, P.K., Roy, A.R.: An application of soft sets in decision making problem. Computers and Mathematics with Applications 44, 1077–1083 (2002)
17. Maji, P.K., Biswas, R., Roy, A.R.: Soft set theory. Computers and Mathematics with Applications 45, 552–562 (2003)
18. Molodtsov, D.A.: Soft set theory-first results. Computers and Mathematics with Applications 37, 19–31 (1999)

19. Molodtsov, D.A., Leonov, V.Y., Kovkov, D.V.: Soft sets technique and its application. Nechetkie Sistemy i Myagkie Vychisleniya 1, 8–39 (2006)
20. Pawlak, Z.: Rough sets. International Journal of Computer and Information Sciences 11, 341–356 (1982)
21. Pawlak, Z., Skowron, A.: Rudiments of rough sets. Information Sciences 177, 7–27 (2007)
22. Pei, D., Miao, D.: From soft set to information systems. In: Proceedings of the IEEE International Conference on Granular Computing, vol. 2, pp. 617–621 (2005)
23. Polkowaski, L.: Rough sets. Mathematical Foundations. Springer, Berlin (2002)
24. Qin, K., Yang, J., Pei, Z.: Generalized rough sets based on reflexive and transitive relations. Information Sciences 178, 4138–4141 (2008)
25. Shabir, M., Naz, M.: On soft topological spaces. Computers and Mathematics with Applications 61, 1786–1799 (2011)
26. Shabir, M., Ali, M.I., Shaheen, T.: Another approach to soft rough sets. Knowledge-Based Systems 40, 72–80 (2013)
27. Sun, B., Ma, W.: Soft fuzzy rough sets and its application in decision making. Artificial Intelligence Review 41, 67–80 (2014)
28. Yang, H.L., Guo, Z.L.: Kernels and Closures of soft set relations and soft set relation mappings. Computers and Mathematics with Applications 61, 651–662 (2011)
29. Zhang, Z., Wang, C., Tian, D., Li, K.: A novel approach to interval-valued intuitionistic fuzzy soft set based decision making. Applied Mathematical Modeling 38, 1255–1270 (2014)

# Knowledge Granulation in Interval-Valued Information Systems Based on Maximal Consistent Blocks

Nan Zhang[1] and Xiaodong Yue[2]

[1]School of Computer and Control Engineering, Yantai University,
Yantai, Shandong, 264005, China
zhangnan0851@163.com
[2]School of Computer Engineering and Science, Shanghai University,
Shanghai, 200444, China
yswantfly@gmail.com

**Abstract.** Rough set theory, proposed by Pawlak in the early 1980s, is an extension of the classical set theory for modeling uncertainty or imprecision information. In this paper, we investigate partial relations and propose the concept of knowledge granulation based on the maximal consistent block in interval-valued information systems. The knowledge granulation can provide important approaches to measuring the discernibility of different knowledge in interval-valued information systems. These results in this paper may be helpful for understanding the essence of rough approximation and attribute reduction in interval-valued information systems.

**Keywords:** rough set theory, knowledge granulation, uncertainty measure.

## 1 Introduction

Rough set theory, proposed by Pawlak [1], has become a well-established mechanism for uncertainty management and reasoning [2,3,4,5,6,7,8,9]. It has a wide variety of applications in pattern recognition and artificial intelligence. As one of the most important issues in rough set theory, the knowledge measure [7] has been widely investigated.

To evaluate knowledge uncertainty, the concept of entropy was proposed by Shannon [10] in 1948. It is a very powerful mechanism for characterizing information contents in various modes and has been applied in many fields. Information measures of uncertainty of rough sets and rough relation databases were investigated by Beaubouef [11]. Miao et al. [12] proposed knowledge granularity and discernibility based on the equivalence relation in complete information systems. Liang et al. [13,14] investigated information granulation in incomplete information systems, which have been effectively applied in measuring for attribute significance, feature selection, decision-rule extracting, etc. Qian and Liang [15] proposed combination granulation with intuitionistic knowledge content nature to measure the size of information granulation in information systems. Xu et al. [16] provide the knowledge granulation to measure the discernibility of different knowledge in ordered information systems.

D. Miao et al. (Eds.): RSKT 2014, LNAI 8818, pp. 49–58, 2014.
DOI: 10.1007/978-3-319-11740-9_5 © Springer International Publishing Switzerland 2014

From a measurement-theoretic perspective, Yao et al. [17] investigated a class of measures of granularity of partitions and also introduced new measures of granularity of partitions in the paper.

The rest of this paper is organized as follows. Some preliminary concepts such as interval-valued information systems, the similarity coefficient and $\alpha$-tolerance relation are briefly recalled in Section 2. Sections 3 introduce the partial relations in an interval-valued information system. In Section 4, the knowledge granulation in an interval-valued information system based on maximal consistent block is proposed. The paper is summarized in Section 5.

## 2    Preliminaries

In this section, we will review some basic concepts related to interval-valued information system. Detailed descriptions about interval-valued information systems (IvIS) can be found in the paper [18,19].

An *Interval-valued Information System* is defined by $\zeta = (U, AT, V, f)$, where

- $U = \{u_1, u_2, ..., u_n\}$ is a non-empty finite set called the universe of discourse;

- $AT = \{a_1, a_2, ..., a_m\}$ is a non-empty finite set of $m$ attributes, such that:

  $a_k(u_i) = [l_i^k, u_i^k]$, $l_i^k \leq u_i^k$, for all $i = 1, 2, ..., n$ and $k = 1, 2, ..., m$;

- $V = \bigcup_{a_k \in AT} V_{a_k}$, $V_{a_k}$ is a domain of attribute $a_k$;

- $f : U \times AT \rightarrow V$ is called the information function such that $f(u_i, a_k) \in V_{a_k}$.

To avoid producing the large number of equivalence classes in Pawlak information systems, we give the similarity coefficient to measure the closeness degree of different interval numbers under the same attribute, as follows:

**Definition 1.** *Let* $\zeta = (U, AT, V, f)$ *be an interval-valued information system. For* $\forall u_i$, $u_j \in U, a_k \in A, A \subseteq AT$, *the similarity coefficient between* $a_k(u_i)$ *and* $a_k(u_j)$ *is defined as:*

$$\alpha_{ij}^k = \begin{cases} 0 & \left[l_i^k, u_i^k\right] \cap \left[l_j^k, u_j^k\right] = \varnothing \\ \dfrac{\min\left\{\left|u_i^k - l_j^k\right|, \left|u_j^k - l_i^k\right|, \left|u_i^k - l_i^k\right|, \left|u_j^k - l_j^k\right|\right\}}{\left|\max\left\{u_i^k, u_j^k\right\} - \min\left\{l_i^k, l_j^k\right\}\right|} & \text{otherwise} \end{cases}$$

Based on the similarity coefficient, we can get the corresponding tolerance relation as:

**Definition 2.** *Let* $\zeta = (U, AT, V, f)$ *be an interval-valued information system. For a given similarity rate* $\alpha \in [0.1]$, *and* $A \subseteq AT$. *The* $\alpha$-*tolerance relation* $T_A^\alpha$ *is expressed as follow:*

$$T_A^\alpha = \{(u_i, u_j) \in U \times U : \alpha_{ij}^k \geq \alpha, \forall a_k \in A\},$$

*where $\alpha_{ij}^k$ is similarity coefficient.*

It is clear that the relation $T_A^\alpha$ is reflexive and symmetric, but not transitive. For a $u_i \in U$, the $\alpha$-tolerance class corresponding to $T_A^\alpha$ is defined as:

**Definition 3.** *Let $\zeta = (U, AT, V, f)$ be an IvIS. $S_A^\alpha(u_i)$, called $\alpha$-tolerance class for $u_i$ with respect to $T_A^\alpha$, is given as:*

$$S_A^\alpha(u_i) = \{u_j \in U : (u_i, u_j) \in T_A^\alpha\}.$$

For $S_A^\alpha(u_i)$, $u_i \in U$, we can get the set in which any object satisfies the $\alpha$-tolerance relation $T_A^\alpha$ with the object $u_i$. To obtain a maximal set in which the objects are tolerant with each other, we introduce the $\alpha$-*Maximal Consistent Block* ($\alpha$-MCB) in interval-valued information systems as following:

In $\zeta = (U, AT, V, f)$, for $\forall u_i, u_j \in M$, $M \subseteq U$ satisfying $(u_i, u_j) \in T_A^\alpha$, then $M$ is the $\alpha$-tolerance class in an interval-valued information system. Further, if $\forall u_m \in U - M$, there exists $u_i \in M$ satisfying $(u_i, u_m) \notin T_A^\alpha$. Here, $M$ is called the $\alpha$-Maximal Consistent Block ($\alpha$-MCB). For a subset $A \subseteq AT$, $M_A^\alpha(u_i)$ is the $\alpha$-MCB with respect to object $u_i \in U$.

## 3    Partial Relations in Interval-Valued Information Systems

Let $\zeta = (U, AT, V, f)$ be an interval-valued information system, for any $A, B \subseteq AT$, then

$$S^\alpha(A) = (S_A^\alpha(u_1), S_A^\alpha(u_2), \cdots, S_A^\alpha(u_{|U|})),$$
$$S^\alpha(B) = (S_B^\alpha(u_1), S_B^\alpha(u_2), \cdots, S_B^\alpha(u_{|U|})).$$

Binary relations "$\preceq$", "$\approx$" and "$\prec$" can be defined as following:

· $S^\alpha(A) \preceq S^\alpha(B) \Leftrightarrow$ for any $i \in \{1, 2, \cdots, |U|\}$, we have $S_A^\alpha(u_i) \subseteq S_B^\alpha(u_i)$, $S_A^\alpha(u_i) \in S^\alpha(A)$ and $S_B^\alpha(u_i) \in S^\alpha(B)$, denoted by $A \preceq B$;

· $S^\alpha(A) \approx S^\alpha(B) \Leftrightarrow$ for any $i \in \{1, 2, \cdots, |U|\}$, we have $S_A^\alpha(u_i) = S_B^\alpha(u_i)$, $S_A^\alpha(u_i) \in S^\alpha(A)$ and $S_B^\alpha(u_i) \in S^\alpha(B)$, denoted by $A \approx B$;

· $S^\alpha(A) \prec S^\alpha(B) \Leftrightarrow S^\alpha(A) \preceq S^\alpha(B)$, $S^\alpha(A) \neq S^\alpha(B)$, denoted by $A \prec B$.

**Theorem 1.** *Let* $\zeta = (U, AT, V, f)$ *be an interval-valued information, if* $\mathbf{S} = \{S^{\alpha}(A) \mid \forall A \subseteq AT\}$ *and* $S^{\alpha}(A) = \{S^{\alpha}_A(u_1), S^{\alpha}_A(u_2), \cdots, S^{\alpha}_A(u_n)\}$, *then* $(\mathbf{S}, \preceq)$ *is a partial set.*

*Proof.*

For any $A, B, C \subseteq AT$,

$$S^{\alpha}(A) = (S^{\alpha}_A(u_1), S^{\alpha}_A(u_2), \cdots, S^{\alpha}_A(u_{|U|})),$$

$$S^{\alpha}(B) = (S^{\alpha}_B(u_1), S^{\alpha}_B(u_2), \cdots, S^{\alpha}_B(u_{|U|})),$$

$$S^{\alpha}(C) = (S^{\alpha}_C(u_1), S^{\alpha}_C(u_2), \cdots, S^{\alpha}_C(u_{|U|})).$$

(1)    For any $u_i \in U$, we can get $S_A(u_i) = S_A(u_i)$, thus $A \preceq A$.

(2)    Suppose $A \preceq B$ and $B \preceq A$, from the definition of $A \preceq B$, we have

$A \preceq B \Leftrightarrow$ for any $i \in \{1, 2, \cdots, |U|\}$, $S^{\alpha}_A(u_i) \subseteq S^{\alpha}_B(u_i)$, where $S^{\alpha}_A(u_i) \in S^{\alpha}(A)$ and $S^{\alpha}_B(u_i) \in S^{\alpha}(B)$;

$B \preceq A \Leftrightarrow$ for any $i \in \{1, 2, \cdots, |U|\}$, $S^{\alpha}_B(u_i) \subseteq S^{\alpha}_A(u_i)$, where $S^{\alpha}_A(u_i) \in S^{\alpha}(A)$ and $S^{\alpha}_B(u_i) \in S^{\alpha}(B)$.

Therefore, $S^{\alpha}_A(u_i) \subseteq S^{\alpha}_B(u_i) \subseteq S^{\alpha}_A(u_i)$, i.e., $S^{\alpha}_A(u_i) = S^{\alpha}_B(u_i)$. For any $\mu_i \in U$, we can have $S^{\alpha}_A(u_i) = S^{\alpha}_B(u_i)$, i.e., $A \approx B$.

(3)    Suppose $A \preceq B$ and $B \preceq C$, from the definition of $A \preceq B$, we get

$A \preceq B \Leftrightarrow$ for any $i \in \{1, 2, \cdots, |U|\}$, $S^{\alpha}_A(u_i) \subseteq S^{\alpha}_B(u_i)$, where $S^{\alpha}_A(u_i) \in S^{\alpha}(A)$ and $S^{\alpha}_B(u_i) \in S^{\alpha}(B)$;

$B \preceq C \Leftrightarrow$ for any $i \in \{1, 2, \cdots, |U|\}$, $S^{\alpha}_B(u_i) \subseteq S^{\alpha}_C(u_i)$, where $S^{\alpha}_B(u_i) \in S^{\alpha}(B)$ and $S^{\alpha}_C(u_i) \in S^{\alpha}(C)$;

Therefore, for any $i \in \{1, 2, \cdots, |U|\}$, $S^{\alpha}_A(u_i) \subseteq S^{\alpha}_B(u_i) \subseteq S^{\alpha}_C(u_i)$, namely, $S^{\alpha}_A(u_i) \subseteq S^{\alpha}_C(u_i)$. For any $\mu_i \in U$, we have $S^{\alpha}_A(u_i) \subseteq S^{\alpha}_C(u_i)$, i.e., $A \preceq C$.

Thus, $(\mathbf{S}, \preceq)$ is a partial set.

# 4    Knowledge Granulation in Interval-Valued Information Systems

In paper [17], Yao et al. proposed the generalized concept of knowledge granulation which should satisfy three conditions.

**Definition 4.** *Let* $IS = (U, AT, V, f)$ *be an information system,* $\forall A \subseteq AT$ *and* $GD$ *be a mapping from the power set of* $A$ *to the set of real numbers. We say that* $GD(A)$

*is the knowledge granulation in an information system* $IS = (U, AT, V, f)$, *if* $GD(A)$ *satisfies the following conditions:*

- **Non-negativity**: *for* $\forall A \subseteq AT$, $GD(A) \geq 0$;
- **Invariability**: *for* $\forall A, B \subseteq AT$ *and* $A \approx B$, *then* $GD(A) = GD(B)$;
- **Monotonicity**: *for* $\forall A, B \subseteq AT$ *and* $A \prec B$, *then* $GD(A) < GD(B)$.

In 2002, Miao *et al.* [12] proposed the knowledge granularity to measure the uncertainty in complete information systems as follows:

**Definition 5.** *Let* $IS = (U, AT, V, f)$ *be a complete information system,* $U / IND(A) = \{X_1, X_2, \cdots, X_m\}$, *knowledge granularity related to attribute sets* $A$ *in* $IS$ *is defined by:*

$$GD(A) = \frac{|R|}{|U|^2}$$

$$= \frac{1}{|U|^2} \sum_{i=1}^{m} |X_i|^2,$$

*where,* $\sum_{i=1}^{m} |X_i|^2$ *is the number of elements in the equivalence relation induced by*

$$\bigcup_{i=1}^{m} (X_i \times X_i).$$

**Theorem 2.** $GD(A)$ *is the knowledge granulation in* $IS = (U, AT, V, f)$ *under the Definition 4*

*Proof.*

(1)  Obviously, $GD(A) \geq 0$。

(2)  For $A, B \subseteq AT$,

$$U / IND(A) = (X_1, X_2, \cdots, X_m),$$

$$U / IND(B) = (X_1', X_2', \cdots, X_n').$$

If $A \approx B$, then $m = n$, and $X_i = X_i' \ (1 \leq i \leq n)$. Therefore,

$$GD(A) = \frac{1}{|U|^2} \sum_{i=1}^{m} |X_i|^2$$

$$= \frac{1}{|U|^2} \sum_{i=1}^{n} |X_i'|^2$$

$$= GD(B).$$

(3)  For $A, B \subseteq AT$, $A \prec B$ and $m > n$. Therefore,

$$GD(B) = \frac{1}{|U|^2} \sum_{i=1}^{n} |X_i'|^2$$

$$= \frac{1}{|U|^2} \sum_{i=1}^{n} (|X'_{i1}| + |X'_{i2}| +,...,+ |X'_{it}|)^2$$

$$> \frac{1}{|U|^2} \sum_{i=1}^{n} (|X'_{i1}|^2 + |X'_{i2}|^2 +,...,+ |X'_{it}|^2)$$

$$= \frac{1}{|U|^2} \sum_{i=1}^{m} |X_i|^2$$

$$= GD(A),$$

where, $1 \le t \le |X'_i|$. Thus, $GD(A)$ is the knowledge granulation under Definition 4.

To obtain the knowledge granulation based on the tolerance relation in interval-valued information systems, we can give another definition of knowledge granularity in complete information systems as:

$$GD(A) = \frac{1}{|U|^2} \sum_{i=1}^{m} |X_i|^2$$

$$= \frac{1}{|U|^2} \sum_{i=1}^{|U|} |X(u_i)|$$

$$= \sum_{i=1}^{|U|} \frac{|X(u_i)|}{|U| \times |U|},$$

where, $|X(u_i)|$ is the number of elements which satisfy the equivalence relation with the object $u_i$ in $IS$, and $\sum_{i=1}^{|U|} |X(u_i)| = \sum_{i=1}^{m} |X_i|^2$ is the number of elements in the equivalence relation induced by $\bigcup_{i=1}^{m} (X_i \times X_i)$. Thus, we can introduce the definition of knowledge granulation in interval-valued information systems as following:

**Definition 6.** *Let* $\varsigma = (U, AT, V, f)$ *be an interval-valued information system, for any* $A \subseteq AT$, $\alpha \in [0.1]$, *knowledge granulation related to* $A$ *in* $\varsigma$ *is given as following:*

$$GDI^{\alpha}(A) = \frac{1}{|U|^2} \sum_{i=1}^{|U|} |UM_A^{\alpha}(u_i)|$$

$$= \sum_{i=1}^{|U|} \frac{|UM_A^{\alpha}(u_i)|}{|U| \times |U|}$$

$$= \sum_{i=1}^{|U|} \frac{|S_A^{\alpha}(u_i)|}{|U| \times |U|},$$

*where* $M_A^{\alpha}(u_i) \in \xi_A^{\alpha}(u_i)$, $\xi^{\alpha}(A) = \{M_A^{\alpha}(u_1), M_A^{\alpha}(u_2),..., M_A^{\alpha}(u_n)\}$.

Based on the definition of knowledge granulation, the smaller $GDI^{\alpha}(A)$ is, the stronger knowledge discernibility is.

**Theorem 3.** $GDI^{\alpha}(A)$ *is the knowledge granulation in* $\varsigma = (U, AT, V, f)$ *under the Definition 4.*

*Proof:*

(1)  Obviously,  $GDI^\alpha(A) \geq 0$ .

(2)  For  $\forall A, B \subseteq AT$ , we have

$$S^\alpha(A) = (S_A^\alpha(u_1), S_A^\alpha(u_2), \cdots, S_A^\alpha(u_{|U|})) ,$$

$$S^\alpha(B) = (S_B^\alpha(u_1), S_B^\alpha(u_2), \cdots, S_B^\alpha(u_{|U|})) .$$

If  $A \approx B$ ,  $i \in \{1, 2, \cdots, |U|\}$ ,  then  $S_A^\alpha(u_i) = S_B^\alpha(u_i)$ ,  i.e.,  $|S_A^\alpha(u_i)| = |S_B^\alpha(u_i)|$ ,

therefore,

$$GDI^\alpha(A) = \frac{1}{|U|^2} \sum_{i=1}^{|U|} |S_A^\alpha(u_i)|$$

$$= \frac{1}{|U|^2} \sum_{i=1}^{|U|} |S_B^\alpha(u_i)|$$

$$= GDI^\alpha(B).$$

(3)  If $\forall A, B \subseteq AT$  and  $A \prec B$ , for any  $u_i \in U$ ,  then  $S_A^\alpha(u_i) \subseteq S_B^\alpha(u_i)$ , thus,

$$GDI^\alpha(A) = \frac{1}{|U|^2} \sum_{i=1}^{|U|} |S_A^\alpha(u_i)|$$

$$< \frac{1}{|U|^2} \sum_{i=1}^{|U|} |S_B^\alpha(u_i)|$$

$$= GDI^\alpha(B) \circ$$

Therefore,  $GDI^\alpha(A)$  is a knowledge granulation in interval-valued information systems under the Definition 4.

**Theorem 4 (Minimum).** *Let*  $\zeta = (U, AT, V, f)$  *be an interval-valued information system, and*  $T_A^\alpha$  *be an*  $\alpha$ *-tolerance relation. The minimum of knowledge granulation in an interval-valued information system*  $\zeta$  *is*  $1/|U|$ . *This value is achieved if and only if*  $T_A^\alpha = \check{T}_A^\alpha$ , *where*  $\check{T}_A^\alpha$  *is an unit tolerance relation, i.e.,*

$$U / \check{T}_A^\alpha = \{M_A^\alpha(u_i) = \{u_i\} : u_i \in U\}$$

$$= \{\{u_1\}, \{u_2\}, \dots, \{u_n\}\} .$$

*Proof.*

If  $T_A^\alpha = \check{T}_A^\alpha$ , then

$$GDI^\alpha(A) = \frac{1}{|U|^2} \sum_{i=1}^{|U|} |\cup M_A^\alpha(u_i)|$$

$$= \sum_{i=1}^{|U|} \frac{|S_A^\alpha(u_i)|}{|U| \times |U|}$$

$$= \sum_{i=1}^{|U|} \frac{1}{|U| \times |U|}$$

$$= \frac{1}{|U|}.$$

**Theorem 5 (Maximum).** *Let* $\zeta = (U, AT, V, f)$ *be an interval-valued information system, and* $T_A^\alpha$ *be an* $\alpha$*-tolerance relation. The maximum of knowledge granulation in an interval-valued information system* $\zeta$ *is 1. This value can be obtained if and only if* $T_A^\alpha = \hat{T}_A^\alpha$, *where* $\hat{T}_A^\alpha$ *is an universe tolerance relation, i.e.,*

$$U / \hat{T}_A^\alpha = \{M_A^\alpha(u_i) = U : u_i \in U\}$$
$$= \{U, U, ..., U\}.$$

*Proof.*

If $T_A^\alpha = \hat{T}_A^\alpha$, then

$$GDI^\alpha(A) = \frac{1}{|U|^2} \sum_{i=1}^{|U|} |\cup M|$$

$$= \sum_{i=1}^{|U|} \frac{|S_A^\alpha(u_i)|}{|U| \times |U|}$$

$$= \sum_{i=1}^{|U|} \frac{|U|}{|U| \times |U|}$$

$$= 1.$$

**Theorem 6 (Boundedness).** *Let* $\zeta = (U, AT, V, f)$ *be an interval-valued information system, and* $T_A^\alpha$ *be an* $\alpha$*-tolerance relation. The knowledge granulation in an interval- valued information system* $\zeta$ *exists the boundedness, namely,*

$$\frac{1}{|U|} \le GDI^\alpha(A) \le 1,$$

*where* $GDI^\alpha(A) = 1/|U|$ *if and only if* $T_A^\alpha = \check{T}_A^\alpha$, *and* $GDI^\alpha(A) = 1$ *if and only if* $T_A^\alpha = \hat{T}_A^\alpha$.

**Proposition 1.** *Let* $\zeta = (U, AT, V, f)$ *be an interval-valued information system, for* $0 \le \alpha < \beta \le 1$, $A \subset B \subset AT$, *we have*

(1)   $GDI^\alpha(A) > GDI^\beta(A)$;

(2)   $GDI^\alpha(A) > GDI^\alpha(B)$;

(3)   $GDI^\alpha(A) > GDI^\beta(B)$.

*Proof:*

(1)  In a given $\zeta = (U, AT, V, f)$, if $0 \leq \alpha < \beta \leq 1$, for any $A \subset AT$, then we can get

$| S_A^{\alpha}(u_i) | > | S_A^{\beta}(u_i) |$, thus, $\dfrac{1}{|U|^2} \sum\limits_{i=1}^{|U|} | S_A^{\alpha}(u_i) | > \dfrac{1}{|U|^2} \sum\limits_{i=1}^{|U|} | S_A^{\beta}(u_i) |$, $GDI^{\alpha}(A) > GDI^{\beta}(A)$;

(2)  In a given $\zeta = (U, AT, V, f)$, if $A \subset B \subset AT$, for any $0 \leq \alpha \leq 1$, then we can get

$| S_A^{\alpha}(u_i) | > | S_B^{\alpha}(u_i) |$, thus, $\dfrac{1}{|U|^2} \sum\limits_{i=1}^{|U|} | S_A^{\alpha}(u_i) | > \dfrac{1}{|U|^2} \sum\limits_{i=1}^{|U|} | S_B^{\alpha}(u_i) |$, $GDI^{\alpha}(A) > GDI^{\alpha}(B)$;

(3)  From (1) and (2), we can get $GDI^{\alpha}(A) > GDI^{\beta}(A)$ and $GDI^{\beta}(A) > GDI^{\beta}(B)$ respectively. Therefore, $GDI^{\alpha}(A) > GDI^{\beta}(B)$.

## 5   Conclusions

In the paper, partial relations and knowledge granulation based on the maximal consistent block are investigated in interval-valued information systems. The knowledge granulation can provide important approaches to measuring the discernibility of different knowledge in interval-valued information systems. These results in this paper may be helpful for understanding the essence of rough approximations and attribute reduction in interval-valued information systems.

**Acknowledgements.** This work was partially supported by the National Natural Science Foundation of China (No. 61170224, 61305052), the Natural Science Foundation of Shandong Province (No. ZR2013FQ020), the Science and Technology Development Plan of Shandong Province (No. 2012GGB01017), the Doctor Research Foundation of Yantai University (No. JS12B28).

## References

1. Pawlak, Z.: Rough sets. International Journal of Computer and Information Sciences 11, 341–356 (1982)
2. Pawlak, Z.: Rough Sets: Theoretical Aspects of Reasoning About Data. Kluwer Academic Publishers, Boston (1991)
3. Yao, Y.Y.: On Generalizing Pawlak Approximation Operators. In: Polkowski, L., Skowron, A. (eds.) RSCTC 1998. LNCS (LNAI), vol. 1424, pp. 298–307. Springer, Heidelberg (1998)
4. Yao, Y.Y., Miao, D.Q., Xu, F.F.: Granular Structures and Approximations in Rough Sets and Knowledge Spaces. In: Ajith, A., Rafael, F., Rafael, B. (eds.) Rough Set Theory: A True Landmark in Data Analysis. Springer, Berlin (2009)
5. Yao, Y.Y.: Probabilistic approaches to rough sets. Expert Systems 20, 287–297 (2003)
6. Qian, Y.H., Liang, J.Y., Yao, Y.Y., et al.: MGRS: A multi-granulation rough set. Information Sciences 6, 949–970 (2010)
7. Wang, G.Y., Zhang, Q.H., Ma, X.A., Yang, Q.S.: Granular computing models for knowledge uncertainty. Chinese Journal of Software 4, 676–694 (2011)
8. Qian, Y.H., Liang, J.Y., Dang, C.Y.: Incomplete multi-granulation rough set. IEEE Transactions on Systems, Man, and Cybernetics: Part A 2, 420–431 (2012)

9. Miao, D.Q., Zhao, Y., Yao, Y.Y., et al.: Relative reducts in consistent and inconsistent decision tables of the Pawlak rough set model. Information Sciences 24, 4140–4150 (2009)
10. Shannon, C.E.: The mathematical theory of communication. The Bell System Technical Journal 3-4, 373–423 (1948)
11. Beaubouef, T., Petry, F.E., Arora, G.: Information-theoretic measures of uncertainty for rough sets and rough relational databases. Information Sciences 109, 185–195 (1998)
12. Miao, D.Q., Fan, S.D.: The calculation of knowledge granulation and its application. Journal of Systems Engineering: Theory and Practice 24, 93–96 (2002)
13. Liang, J.Y., Shi, Z.Z.: The information entropy, rough entropy and knowledge granularity in rough set theory. International Journal of Uncertainty, Fuzziness and Knowledge-Based Systems 1, 37–46 (2004)
14. Liang, J.Y., Shi, Z.Z., Li, D., Wierman, M.: Information entropy, rough entropy and knowledge granularity in incomplete information systems. International Journal of General Systems 35, 641–654 (2006)
15. Qian, Y., Liang, J.: Combination entropy and combination granulation in incomplete information system. In: Wang, G.-Y., Peters, J.F., Skowron, A., Yao, Y. (eds.) RSKT 2006. LNCS (LNAI), vol. 4062, pp. 184–190. Springer, Heidelberg (2006)
16. Xu, W.H., Zhang, X.Y., Zhang, W.X.: Knowledge granulation, knowledge entropy and knowledge uncertainty measure in ordered information systems. Applied Soft Computing 9, 1244–1251 (2009)
17. Yao, Y.Y., Zhao, L.Q.: A measurement theory view on the granularity of partitions. Information Sciences 213, 1–13 (2012)
18. Zhang, N., Miao, D.Q., Yue, X.D.: Knowledge reduction in interval-valued information systems. Journal of Computer Research and Development 47, 1362–1371 (2010)
19. Zhang, N.: Research on Interval-valued Information Systems and Knowledge Spaces: A Granular Approach. PhD Thesis, Tongji University, China (2012)

# Multi-granulation Rough Sets
# Based on Central Sets

Caihui Liu[1], Meizhi Wang[2], Yujiang Liu[1], and Min Wang[1]

[1] Department of Mathematics and Computer Science, Gannan Normal University
Ganzhou, Jiangxi Province, P.R. China, 341000
[2] Department of Physical Education, Gannan Normal University, Ganzhou, China,
341000
liu_caihui@163.com

**Abstract.** Exploring rough sets from the viewpoint of multi-granulation
has become one of the promising topics in rough set theory, in which
lower or upper approximations are approximated by multiple binary relations. The purpose of this paper is to develop two new kinds of multi-
granulation rough set models by using concept of central sets in a given
approximation space. Firstly, the concepts of the two new models are
proposed. Then some important properties and the relationship of the
models are disclosed. Finally, several uncertainty measures of the models
are also proposed. These results will enrich the theory and application
of multi-granulation rough sets.

**Keywords:** Central sets, Measures, Multi-granulation, Rough sets.

## 1 Introduction

The concept of multi-granulation rough set was first introduced by Qian et
al. [1] in 2006. Unlike single-granulation rough sets, the approximations of a
target concept in multi-granulation rough sets are constructed by using multi-
distinct sets of information granules. Since then, a lot of researchers dedicated
to the development of it. Qian et al. [2] presented a multi-granulation rough
set model based on multiple tolerance relations in incomplete information systems. Yang et al. [3] constructed multi- granulation rough sets based on the
fuzzy approximation space. Yang et al. [4] discussed the hierarchical structures
of multi-granulation rough sets and She et al. [5] investigated the topological
and lattice-theoretic properties of multi- granulation rough sets. Liu et al. [6,7,8]
introduced multi-granulation covering rough sets and multi-granulation covering
fuzzy rough sets in a covering approximation space and some lattice-based properties are disclosed. Lin et al. [9] introduced two kinds of neighborhood-based
multi-granulation rough sets and discussed the covering reducts of the models.
Xu et al. [10,11] constructed two types of multi-granulation rough sets based
on the tolerance and ordered relations respectively. And Xu et al. [12] proposed
multi- granulation fuzzy rough set model and some measures in the model are
constructed. Taking the test cost into consideration, Yang et al. [13] proposed a

D. Miao et al. (Eds.): RSKT 2014, LNAI 8818, pp. 59–68, 2014.
DOI: 10.1007/978-3-319-11740-9_6 © Springer International Publishing Switzerland 2014

test cost sensitive multigraulation rough set model. A pessimistic multigranulation rough set model is developed by Qian et al. [14] based on "Seeking common ground while eliminating differences" strategy. Qian et al. [15] developed the multigranulation decision- theoretic rough set and proved that many existing multigranulation rough set models can be derived from the multigranulation decision-theoretic rough set framework. This paper discusses two new types of multi-granulation rough set models by employing the concept of central sets.

## 2   Preliminaries

This section reviews some basic notions such as binary relation, binary neighborhood, central sets and multi-granulation rough sets.

**Definition 1 [16].** Let $U = \{1, 2, \ldots, n\}$ be a universe of discourse and $R$ a binary relation on $U$. For any $x \in U$, the sets $N_R(x) \subseteq U$ is called the right binary neighborhood of $x$, where $N_R(x) = \{y \in U \mid y \in U \wedge (x, y) \in R\}$.

**Definition 2 [17].** Let $U$ be a universe of discourse and $R$ a binary relation on $U$. The pair $< U, C >$ is called a covering approximation space. For any $x, y \in U$, an equivalence relation $E_R$ is defined by $(x, y) \in E_R \Leftrightarrow N_R(x) = N_R(y)$. A central set of $x$ is defined as $C_R(x) = \{y \in U \mid N_R(x) = N_R(y), x, y \in U\}$.

It is easy to prove that $\pi_{E_R} = \{C_R(x) \mid x \in U\}$ is a partition of $U$.

Qian et al. first proposed the rough set model based on multi-granulation called MGRS in the reference [1].

**Definition 3 [1].** Let $K = (U, \mathbf{R})$ is a knowledge base, $\mathbf{R}$ a family of equivalence relations on $U$, for any $X \subseteq U$, $P, Q \in \mathbf{R}$, the optimistic lower and upper approximations of $X$ can be defined by the following.

$\underline{O_{P+Q}}(X) = \{x \in U \mid [x]_P \subseteq X \text{or} [x]_Q \subseteq X\}$

$\overline{O_{P+Q}}(X) = \sim \underline{P + Q}(\sim X)$

Where $\sim X$ is the complement of $X$ in $U$.

**Definition 4 [1].** Let $K = (U, \mathbf{R})$ is a knowledge base, $\mathbf{R}$ a family of equivalence relations on $U$, for any $X \subseteq U$, $P, Q \in \mathbf{R}$, the pessimistic lower and upper approximations of $X$ can be defined by the following.

$\underline{P_{P+Q}}(X) = \{x \in U \mid [x]_P \subseteq X \text{and} [x]_Q \subseteq X\}$

$\overline{P_{P+Q}}(X) = \sim \underline{P + Q}(\sim X)$

Where $\sim X$ is the complement of $X$ in $U$.

## 3   The Concepts and Properties of Two Types of Multi-granulation Rough Sets Based on Central Sets

First, this section proposes the concepts of multi-granulation rough sets on the basis of central sets. Then some important properties of the models are disclosed.

Some illustrative examples are employed to show the mechanisms of the models. For the purpose of simplicity, throughout the paper, only the case of two relations is considered in the process of defining a multi-granulation rough set model.

**Definition 5.** Let $< U, \mathbf{R} >$ be an approximation space, where $U$ is a universe of discourse and $\mathbf{R}$ a family of binary relations based on $U$. For any $X \subseteq U$ and $R_1, R_2 \in \mathbf{R}$, the optimistic lower and upper approximations of $X$ based on central sets with respect to $R_1, R_2$ are defined as follows.

$$\underline{O_{R_1+R_2}}(X) = \cup\{C_{R_1}(x) \cap C_{R_2}(x) | \, x \in U((N_{R_1}(x) \neq \emptyset, N_{R_1}(x) \subseteq X) \text{or}(N_{R_2}(x)$$
$$\neq \emptyset, N_{R_2}(x) \subseteq X))\}$$
$$\overline{O_{R_1+R_2}}(X) = \cup\{C_{R_1}(x) \cap C_{R_2}(x) | \, x \in U(N_{R_1}(x) \cap X \neq \emptyset \text{and} N_{R_2}(x) \cap X \neq \emptyset)\}$$

If $\underline{O_{R_1+R_2}}(X) = \overline{O_{R_1+R_2}}(X)$, then $X$ is said to be definable. Otherwise is called optimistic multi-granulation rough sets based on central sets with respect to $R_1, R_2$. The pair $(\underline{O_{R_1+R_2}}(X), \overline{O_{R_1+R_2}}(X))$ is called a multi-granulation rough set of $X$.

**Definition 6.** Let $< U, \mathbf{R} >$ be an approximation space, where $U$ is a universe of discourse and $\mathbf{R}$ a family of binary relations based on $U$. For any $X \subseteq U$ and $R_1, R_2 \in \mathbf{R}$, the pessimistic lower and upper approximations of $X$ based on central sets with respect to $R_1, R_2$ are defined as follows.

$$\underline{P_{R_1+R_2}}(X) = \cup\{C_{R_1}(x) \cap C_{R_2}(x) | \, x \in U((N_{R_1}(x) \neq \emptyset, N_{R_1}(x) \subseteq X) \text{and}(N_{R_2}(x)$$
$$\neq \emptyset, N_{R_2}(x) \subseteq X))\}$$
$$\overline{P_{R_1+R_2}}(X) = \cup\{C_{R_1}(x) \cap C_{R_2}(x) | \, x \in U(N_{R_1}(x) \cap X \neq \emptyset \text{or} N_{R_2}(x) \cap X \neq \emptyset)\}$$

If $\underline{P_{R_1+R_2}}(X) = \overline{P_{R_1+R_2}}(X)$, then $X$ is said to be definable. Otherwise is called pessimistic multi-granulation rough sets based on central sets with respect to $R_1, R_2$. The pair $(\underline{P_{R_1+R_2}}(X), \overline{P_{R_1+R_2}}(X))$ is called a multi-granulation rough set of $X$.

**Example 1.** Given an approximation space $< U, \mathbf{R} >$, where $U = \{1, 2, 3, 4, 5\}$, $R_1, R_2 \in \mathbf{R}$ and $R_1 = \{(1, 3), (1, 5), (2, 3), (2, 5), (3, 1), (3, 2), (4, 1), (4, 2), (5, 3), (5, 5)\}$, $R_2 = \{(1, 1), (1, 2), (1, 3), (2, 1), (2, 2), (2, 3), (3, 3), (3, 4), (3, 5), (4, 3), (4, 4), (4, 5), (5, 3), (5, 4), (5, 5)\}$. Given a subset $X = \{2, 3\}$ of $U$, according to Definition 5 and 6, the following results can be obtained.
For $R_1$:

$$N_{R_1}(1) = N_{R_1}(2) = N_{R_1}(5) = \{3, 5\},$$
$$N_{R_1}(3) = N_{R_1}(4) = \{1, 2\},$$
$$\pi_{E_{R_1}} = \{\{1, 2, 5\}, \{3, 4\}\}.$$

For $R_2$:

$$N_{R_2}(1) = N_{R_2}(2) = \{1, 2, 3\},$$
$$N_{R_2}(3) = N_{R_2}(4) = N_{R_2}(5) = \{3, 4, 5\},$$
$$\pi_{E_{R_2}} = \{\{1, 2\}, \{3, 4, 5\}\}.$$

Then

$$\underline{O_{R_1+R_2}}(X) = \emptyset,$$
$$\overline{O_{R_1+R_2}}(X) = (\{1, 2, 5\} \cap \{1, 2\}) \cup (\{1, 2, 5\} \cap \{1, 2\}) \cup (\{3, 4\} \cap \{3, 4, 5\}) \cup$$
$$(\{3, 4\} \cap \{3, 4, 5\}) \cup (\{1, 2, 5\} \cap \{3, 4, 5\})$$

$$= \{1, 2, 3, 4, 5\};$$
$$\underline{P_{R_1+R_2}}(X) = \emptyset,$$
$$\overline{P_{R_1+R_2}}(X) = (\{1,2,5\} \cap \{1,2\}) \cup (\{1,2,5\} \cap \{1,2\}) \cup (\{3,4\} \cap \{3,4,5\}) \cup$$
$$(\{3,4\} \cap \{3,4,5\}) \cup (\{1,2,5\} \cap \{3,4,5\})$$
$$= \{1,2,3,4,5\}.$$

Example 1 shows how to calculate the approximations for a given concept in the given approximation space. Although the example shows that the outputs of optimistic and pessimistic lower and upper approximations are the same, they are different to each other. We will show this later in the paper.

**Proposition 1.** Suppose that $< U, \mathbf{R} >$ is an approximation space, $R_1, R_2 \in \mathbf{R}$. For any $X \subseteq U$, the optimistic multi-granulation rough approximations may not satisfy the following two properties.
(1) $\underline{O_{R_1+R_2}}(X) \subseteq X$
(2) $X \subseteq \overline{O_{R_1+R_2}}(X)$

But, here we must make it clear that $\underline{O_{R_1+R_2}}(X) \subseteq \overline{O_{R_1+R_2}}(X)$.

**Example 2.** (Continued from Example 1) Let $X = \{3, 5\}$, according to the definition, we have that
$$\underline{O_{R_1+R_2}}(X) = (\{1,2,5\} \cap \{1,2\}) \cup (\{1,2,5\} \cap \{1,2\}) \cup (\{1,2,5\} \cap \{3,4,5\})$$
$$= \{1,2,5\}$$
$$\overline{O_{R_1+R_2}}(X) = (\{1,2,5\} \cap \{1,2\}) \cup (\{1,2,5\} \cap \{1,2\}) \cup (\{1,2,5\} \cap \{3,4,5\})$$
$$= \{1,2,5\}$$
Therefore,
$$\underline{O_{R_1+R_2}}(X) = \{1,2,5\} \not\subseteq X = \{3,5\},$$
$$X = \{3,5\} \not\subseteq \overline{O_{R_1+R_2}}(X) = \{1,2,5\},$$
But we have that
$$\underline{O_{R_1+R_2}}(X) = \{1,2,5\} \subseteq \overline{O_{R_1+R_2}}(X) = \{1,2,5\}.$$

Example 2 verifies the results in the Proposition 1.

**Proposition 2.** Suppose that $< U, \mathbf{R} >$ is an approximation space, $R_1, R_2 \in \mathbf{R}$. For any $X \subseteq U$, the pessimistic multi-granulation rough approximations satisfy $\underline{P_{R_1+R_2}}(X) \subseteq X \subseteq \overline{P_{R_1+R_2}}(X)$.
**Proof.** It can be easily proved according to Definition 6.

**Proposition 3.** Suppose that $< U, \mathbf{R} >$ is an approximation space, $R_1, R_2 \in \mathbf{R}$. For any $X \subseteq U$, its lower and upper approximations based on central sets with respect to $R_1, R_2$ satisfy the following properties.
(1) $\underline{O_{R_1+R_2}}(U) = U, \overline{O_{R_1+R_2}}(U) = U, \underline{P_{R_1+R_2}}(U) = U, \overline{P_{R_1+R_2}}(U) = U$
(2) $\underline{O_{R_1+R_2}}(\emptyset) = \emptyset, \overline{O_{R_1+R_2}}(\emptyset) = \emptyset, \underline{P_{R_1+R_2}}(\emptyset) = \emptyset, \overline{P_{R_1+R_2}}(\emptyset) = \emptyset$
(3) $\overline{P_{R_1+R_2}}(\underline{P_{R_1+R_2}}(X)) = \underline{P_{R_1+R_2}}(\overline{X})$
(4) $\overline{P_{R_1+R_2}}(X) = \overline{P_{R_1+R_2}}(\overline{P_{R_1+R_2}}(X))$
**Proof.** Properties(1) and (2) is straightforward according Definition 5 and 6, here we only give the proofs of (3) and (4).

(3): On the one hand, according to Proposition 2, we have that $\underline{P_{R_1+R_2}}(X) \subseteq X$, therefore, $\underline{P_{R_1+R_2}}(\underline{P_{R_1+R_2}}(X)) \subseteq \underline{P_{R_1+R_2}}(X)$;

On the other hand, by the Proposition 1 in [1], we have that

$$\underline{P_{R_1+R_2}}(\underline{P_{R_1+R_2}}(X)) = \underline{P_{R_1}}(\underline{P_{R_1+R_2}}(X)) \cup \underline{P_{R_2}}(\underline{P_{R_1+R_2}}X)$$
$$= \underline{P_{R_1}}(\underline{P_{R_1}}(X) \cup \underline{P_{R_2}}(X)) \cup \underline{P_{R_2}}(\underline{P_{R_1}}(X) \cup \underline{P_{R_1}}(X))$$
$$\supseteq \underline{P_{R_1}}\underline{P_{R_1}}(X) \cup \underline{P_{R_2}}\underline{P_{R_2}}(X)$$
$$= \underline{P_{R_1}}(X) \cup \underline{P_{R_2}}(X)$$
$$= \underline{P_{R_1+R_2}}(X)$$

Thus, $\underline{P_{R_1+R_2}}(\underline{P_{R_1+R_2}}(X)) = \underline{P_{R_1+R_2}}(X)$ holds.

(4): It can be proved similarly as (3).

**Proposition 4.** Suppose that $< U, \mathbf{R} >$ is an approximation space, $R_1, R_2 \in \mathbf{R}$. For any $X \subseteq U$, the optimistic multi-granulation rough approximations may not satisfy the following two properties.

(1) $\underline{O_{R_1+R_2}}(\underline{O_{R_1+R_2}}(X)) = \underline{O_{R_1+R_2}}(X)$

(2) $\overline{O_{R_1+R_2}}(X) = \overline{O_{R_1+R_2}}(\overline{O_{R_1+R_2}}(X))$

**Example 3.** (Continued from Example 2) From Example 2, we know that $\underline{O_{R_1+R_2}}(X) = \overline{O_{R_1+R_2}}(X) = \{1,2,5\}$, but according to Definition 5, we can calculate the following results.

$\underline{O_{R_1+R_2}}(\underline{O_{R_1+R_2}}(X)) = \emptyset$,

$\overline{O_{R_1+R_2}}(\overline{O_{R_1+R_2}}(X)) = \{1,2,3,4,5\}$

Therefore,

$\underline{O_{R_1+R_2}}(\underline{O_{R_1+R_2}}(X)) = \emptyset \neq \underline{O_{R_1+R_2}}(X) = \{1,2,5\}$,

$\overline{O_{R_1+R_2}}(\overline{O_{R_1+R_2}}(X)) = \{1,2,3,4,5\} \neq \overline{O_{R_1+R_2}}(X) = \{1,2,5\}$.

**Proposition 5.** Suppose that $< U, \mathbf{R} >$ is an approximation space, $R_1, R_2 \in \mathbf{R}$. For any $X \subseteq U$, we have that $\overline{O_{R_1+R_2}}(\sim X) =\sim \underline{O_{R_1+R_2}}(X)$, but $\underline{O_{R_1+R_2}}(\sim X) =\sim \overline{O_{R_1+R_2}}(X)$ may not be satisfied.

**Example 4.** (Continued from Example 2) One can calculate $\sim X = \{1,2,4\}$, then the following outputs could be obtained.

$\underline{O_{R_1+R_2}}(\sim X) = \emptyset$,

$\overline{O_{R_1+R_2}}(\sim X) = (\{3,4\} \cap \{3,4,5\}) \cup (\{3,4\} \cap \{3,4,5\})$
$= \{3,4\}$.

Therefore,

$\underline{O_{R_1+R_2}}(\sim X) = \emptyset \neq \{3,4\} =\sim \overline{O_{R_1+R_2}}(X)$,

$\overline{O_{R_1+R_2}}(\sim X) = \{3,4\} =\sim \underline{O_{R_1+R_2}}(X)$.

Example 4 is a proof example for Proposition 5.

**Proposition 6.** Suppose that $< U, \mathbf{R} >$ is an approximation space, $R_1, R_2 \in \mathbf{R}$. For any $X \subseteq U$, we have $\underline{P_{R_1+R_2}}(\sim X) =\sim \overline{P_{R_1+R_2}}(X)$ and $\overline{P_{R_1+R_2}}(\sim X) =\sim \underline{P_{R_1+R_2}}(X)$.

**Proof.** According to Definition 6, we have that

$\underline{P_{R_1+R_2}}(\sim X) = \cup\{ C_{R_1}(x) \cap C_{R_2}(x)|\, x \in U((N_{R_1}(x) \neq \emptyset, N_{R_1}(x) \subseteq (\sim X))$

$and(N_{R_2}(x) \neq \emptyset, N_{R_2}(x) \subseteq (\sim X)))\}$

$=\sim \cup\{C_{R_1}(x) \cap C_{R_2}(x)|\, x \in U(N_{R_1}(x) \cap X = \emptyset \text{or} N_{R_2}(x) \cap X = \emptyset)\}$

$=\sim \overline{P_{R_1+R_2}}(X)$

$\overline{P_{R_1+R_2}}(\sim X) = \cup\{C_{R_1}(x) \cap C_{R_2}(x)|\, x \in U(N_{R_1}(x) \cap (\sim X) \neq \emptyset \text{or} N_{R_2}(x) \cap (\sim X) \neq \emptyset)\}$

$=\sim \cup\{C_{R_1}(x) \cap C_{R_2}(x)|\, x \in U((N_{R_1}(x) \neq \emptyset, N_{R_1}(x) \subseteq X \text{and}(N_{R_2}(x) \neq \emptyset, N_{R_2}(x) \subseteq X))\}$

$=\sim \underline{P_{R_1+R_2}}(X)$

**Example 5.** (Continued from Example 2) We have $\sim X = \{1, 2, 4\}$, then we can get the following results.

$\underline{P_{R_1+R_2}}(\sim X) = \emptyset,$

$\overline{P_{R_1+R_2}}(\sim X) = (\{1, 2, 5\} \cap \{1, 2\}) \cup (\{1, 2, 5\} \cap \{1, 2\}) \cup (\{3, 4\} \cap \{3, 4, 5\}) \cup$
$(\{3, 4\} \cap \{3, 4, 5\}) \cup (\{1, 2, 5\} \cap \{3, 4, 5\})$

$= \{1, 2, 3, 4, 5\}$

Therefore,

$\underline{P_{R_1+R_2}}(\sim X) = \emptyset =\sim \overline{P_{R_1+R_2}}(X),$

$\overline{P_{R_1+R_2}}(\sim X) = \{1, 2, 3, 4, 5\} =\sim \underline{P_{R_1+R_2}}(X).$

**Proposition 7.** Suppose that $< U, \mathbf{R} >$ is an approximation space, $R_1, R_2 \in \mathbf{R}$. Then, for arbitrary $X \subseteq U$, the following properties hold.

(1) If $X \subseteq Y$, then $O_{R_1+R_2}(X) \subseteq O_{R_1+R_2}(Y)$

   If $X \subseteq Y$, then $\underline{P_{R_1+R_2}}(X) \subseteq \underline{P_{R_1+R_2}}(Y)$

(2) If $X \subseteq Y$, then $\overline{O_{R_1+R_2}}(X) \subseteq \overline{O_{R_1+R_2}}(Y)$

   If $X \subseteq Y$, then $\overline{P_{R_1+R_2}}(X) \subseteq \overline{P_{R_1+R_2}}(Y)$

(3) $O_{R_1+R_2}(X \cap Y) \subseteq O_{R_1+R_2}(X) \cap O_{R_1+R_2}(Y)$

   $\underline{P_{R_1+R_2}}(X \cap Y) \subseteq \underline{P_{R_1+R_2}}(X) \cap \underline{P_{R_1+R_2}}(Y)$

(4) $\overline{O_{R_1+R_2}}(X \cup Y) \supseteq \overline{O_{R_1+R_2}}(X) \cup \overline{O_{R_1+R_2}}(Y)$

   $\overline{P_{R_1+R_2}}(X \cup Y) \supseteq \overline{P_{R_1+R_2}}(X) \cup \overline{P_{R_1+R_2}}(Y)$

(5) $O_{R_1+R_2}(X \cup Y) \supseteq O_{R_1+R_2}(X) \cup O_{R_1+R_2}(Y)$

   $\underline{P_{R_1+R_2}}(X \cup Y) \supseteq \underline{P_{R_1+R_2}}(X) \cup \underline{P_{R_1+R_2}}(Y)$

(6) $\overline{O_{R_1+R_2}}(X \cap Y) \subseteq \overline{O_{R_1+R_2}}(X) \cap \overline{O_{R_1+R_2}}(Y)$

   $\overline{P_{R_1+R_2}}(X \cap Y) \subseteq \overline{P_{R_1+R_2}}(X) \cap \overline{P_{R_1+R_2}}(Y)$

**Proof.** First, we give the proofs of (1) and (2). According to Definition 5 and 6, if $X \subseteq Y$, then

$O_{R_1+R_2}(X) = \cup\{C_{R_1}(x) \cap C_{R_2}(x)|\, x \in U((N_{R_1}(x) \neq \emptyset, N_{R_1}(x) \subseteq X \text{or}(N_{R_2}(x) \neq \emptyset, N_{R_2}(x) \subseteq X))\}$

$\subseteq \cup\{C_{R_1}(x) \cap C_{R_2}(x)|\, x \in U((N_{R_1}(x) \neq \emptyset, N_{R_1}(x) \subseteq Y \text{or}(N_{R_2}(x) \neq \emptyset, N_{R_2}(x) \subseteq Y))\}$

$= O_{R_1+R_2}(Y)$

$\underline{P_{R_1+R_2}}(X) = \underline{\cup\{C_{R_1}(x) \cap C_{R_2}(x)|\, x \in U((N_{R_1}(x) \neq \emptyset, N_{R_1}(x) \subseteq X \text{and}(N_{R_2}(x)}$
$\neq \emptyset, N_{R_2}(x) \subseteq X))\}$

$\subseteq \cup\{C_{R_1}(x) \cap C_{R_2}(x)|\, x \in U((N_{R_1}(x) \neq \emptyset, N_{R_1}(x) \subseteq Y \text{and}(N_{R_2}(x)$
$\neq \emptyset, N_{R_2}(x) \subseteq Y))\}$

$= \underline{P_{R_1+R_2}}(Y)$

$$\overline{O_{R_1+R_2}}(X) = \cup\{C_{R_1}(x) \cap C_{R_2}(x)|\, x \in U(N_{R_1}(x) \cap X \neq \emptyset \text{and} N_{R_2}(x) \cap X \neq \emptyset)\}$$
$$\subseteq \cup\{C_{R_1}(x) \cap C_{R_2}(x)|\, x \in U(N_{R_1}(x) \cap Y \neq \emptyset \text{and} N_{R_2}(x) \cap Y \neq \emptyset)\}$$
$$= \overline{O_{R_1+R_2}}(Y)$$
$$\overline{P_{R_1+R_2}}(X) = \cup\{C_{R_1}(x) \cap C_{R_2}(x)|\, x \in U(N_{R_1}(x) \cap X \neq \emptyset \text{or} N_{R_2}(x) \cap X \neq \emptyset)\}$$
$$\subseteq \cup\{C_{R_1}(x) \cap C_{R_2}(x)|\, x \in U(N_{R_1}(x) \cap Y \neq \emptyset \text{or} N_{R_2}(x) \cap Y \neq \emptyset)\}$$
$$= \overline{P_{R_1+R_2}}(Y)$$

(3) As we know, $X \cap Y \subseteq X$ and $X \cap Y \subseteq Y$. Then, according to (1), we have
$$\underline{O_{R_1+R_2}}(X \cap Y) \subseteq \underline{O_{R_1+R_2}}(X) \text{ and } \underline{O_{R_1+R_2}}(X \cap Y) \subseteq \underline{O_{R_1+R_2}}(Y)$$
Therefore,
$$\underline{O_{R_1+R_2}}(X \cap Y) \cap \underline{O_{R_1+R_2}}(X \cap Y) \subseteq \underline{O_{R_1+R_2}}(X) \cap \underline{O_{R_1+R_2}}(Y)$$
That is, $\underline{O_{R_1+R_2}}(X \cap Y) \subseteq \underline{O_{R_1+R_2}}(X) \cap \underline{O_{R_1+R_2}}(Y)$.

$\underline{P_{R_1+R_2}}(X \cap Y) \subseteq \underline{P_{R_1+R_2}}(X) \cap \underline{P_{R_1+R_2}}(Y)$ can be proved similarly.

(4) As we know, $X \subseteq X \cup Y$ and $Y \subseteq X \cup Y$. Then, according to (2), we have
$$\overline{O_{R_1+R_2}}(X \cup Y) \supseteq \overline{O_{R_1+R_2}}(X) \text{ and } \overline{O_{R_1+R_2}}(X \cap Y) \supseteq \overline{O_{R_1+R_2}}(Y)$$
Therefore,
$$\overline{O_{R_1+R_2}}(X \cup Y) \cup \overline{O_{R_1+R_2}}(X \cup Y) \supseteq \overline{O_{R_1+R_2}}(X) \cup \overline{O_{R_1+R_2}}(Y)$$
That is, $\overline{O_{R_1+R_2}}(X \cup Y) \supseteq \overline{O_{R_1+R_2}}(X) \cup \overline{O_{R_1+R_2}}(Y)$.

One can prove $\overline{P_{R_1+R_2}}(X \cup Y) \supseteq \overline{P_{R_1+R_2}}(X) \cup \overline{P_{R_1+R_2}}(Y)$ in a similar way.

(5) Obviously, we have $X \subseteq X \cup Y$ and $Y \subseteq X \cup Y$. Then, according to (1), the following are satisfied.
$$\underline{O_{R_1+R_2}}(X \cup Y) \supseteq \underline{O_{R_1+R_2}}(X) \text{ and } \underline{O_{R_1+R_2}}(X \cap Y) \supseteq \underline{O_{R_1+R_2}}(Y)$$
Therefore,
$$\underline{O_{R_1+R_2}}(X \cup Y) \cup \underline{O_{R_1+R_2}}(X \cup Y) \supseteq \underline{O_{R_1+R_2}}(X) \cup \underline{O_{R_1+R_2}}(Y)$$
That is, $\underline{O_{R_1+R_2}}(X \cup Y) \supseteq \underline{O_{R_1+R_2}}(X) \cup \underline{O_{R_1+R_2}}(Y)$.

$\underline{P_{R_1+R_2}}(X \cup Y) \supseteq \underline{P_{R_1+R_2}}(X) \cup \underline{P_{R_1+R_2}}(Y)$ can be proved in a similar way.

(6) Noting that $X \cap Y \subseteq X$ and $X \cap Y \subseteq Y$, by employing (2), we have
$$\overline{O_{R_1+R_2}}(X \cap Y) \subseteq \overline{O_{R_1+R_2}}(X) \text{ and } \overline{O_{R_1+R_2}}(X \cap Y) \subseteq \overline{O_{R_1+R_2}}(Y)$$
Therefore,
$$\overline{O_{R_1+R_2}}(X \cap Y) \cap \overline{O_{R_1+R_2}}(X \cap Y) \subseteq \overline{O_{R_1+R_2}}(X) \cap \overline{O_{R_1+R_2}}(Y)$$
That is, $\overline{O_{R_1+R_2}}(X \cap Y) \subseteq \overline{O_{R_1+R_2}}(X) \cap \overline{O_{R_1+R_2}}(Y)$.
We can prove $\overline{P_{R_1+R_2}}(X \cap Y) \subseteq \overline{P_{R_1+R_2}}(X) \cap \overline{P_{R_1+R_2}}(Y)$ by using the similar method.

**Proposition 8.** Suppose that $< U, \mathbf{R} >$ is an approximation space, $R_1, R_2 \in \mathbf{R}$. Then, for arbitrary $X \subseteq U$, the following property holds.
$$\underline{P_{R_1+R_2}}(X) \subseteq \underline{O_{R_1+R_2}}(X) \subseteq \overline{O_{R_1+R_2}}(Y) \subseteq \overline{P_{R_1+R_2}}(Y)$$
**Proof.** It is straightforward according to Definition 5 and 6.

## 4  Uncertainty Measures

Rough set theory is an efficient tool for handling vagueness and uncertainty. In this section, some measures to characterize the vagueness and uncertainty of the new model are introduced.

**Definition 7[18].** Let $S = (U, AT)$ be an information system. For $A \subseteq AT$, $X \subseteq U$, the rough membership of $x$ in $X$ is defined as $u_X^A(x) = \frac{|[x]_A \cap X|}{|[x]_A|}$.

The rough membership defined in Definition 7 can only evaluate the uncertainty in classical rough set model. Next, a new rough membership is proposed, which can be used to handle the uncertainty of neighborhood rough sets.

**Definition 8.** Suppose that $< U, \mathbf{R} >$ is an approximation space, $R \in \mathbf{R}$ and $X \subseteq U$. The neighborhood rough membership of $x$ is defined as $u_X^R(x) = \frac{|N_R(x) \cap X|}{|N_R(x)|}$.

Based on Definition 8, a rough membership for multi-granulation rough sets based on central sets is defined as follows.

**Definition 9.** Suppose that $< U, \mathbf{R} >$ is an approximation space, $R_1, R_2 \in \mathbf{R}$. Then for any $X \subseteq U$, the rough membership of $x$ can be defined as $u_X^{\mathbf{R}}(x) = \frac{u_X^{R_1}(x) + u_X^{R_2}(x)}{2}$

**Example 6.**(Continued from Example 1) According to Definition 9, the following results can be obtained.

$$u_X^{R_1}(1) = u_X^{R_1}(2) = u_X^{R_1}(5) = \frac{|\{3,5\} \cap \{2,3\}|}{|\{3,5\}|} = \frac{1}{2}$$

$$u_X^{R_1}(3) = u_X^{R_1}(4) = \frac{|\{1,2\} \cap \{2,3\}|}{|\{1,2\}|} = \frac{1}{2}$$

$$u_X^{R_2}(1) = u_X^{R_2}(2) = \frac{|\{1,2,3\} \cap \{2,3\}|}{|\{1,2,3\}|} = \frac{2}{3}$$

$$u_X^{R_2}(3) = u_X^{R_2}(4) = u_X^{R_2}(5) = \frac{|\{3,4,5\} \cap \{2,3\}|}{|\{3,4,5\}|} = \frac{1}{3}$$

Then

$$u_X^{\mathbf{R}}(1) = \frac{u_X^{R_1}(1) + u_X^{R_2}(1)}{2} = \frac{\frac{1}{2} + \frac{2}{3}}{2} = \frac{7}{12}$$

$$u_X^{\mathbf{R}}(2) = \frac{u_X^{R_1}(2) + u_X^{R_2}(2)}{2} = \frac{\frac{1}{2} + \frac{2}{3}}{2} = \frac{7}{12}$$

$$u_X^{\mathbf{R}}(3) = \frac{u_X^{R_1}(3) + u_X^{R_2}(3)}{2} = \frac{\frac{1}{2} + \frac{1}{3}}{2} = \frac{5}{12}$$

$$u_X^{\mathbf{R}}(4) = \frac{u_X^{R_1}(4) + u_X^{R_2}(4)}{2} = \frac{\frac{1}{2} + \frac{1}{3}}{2} = \frac{5}{12}$$

$$u_X^{\mathbf{R}}(5) = \frac{u_X^{R_1}(5) + u_X^{R_2}(5)}{2} = \frac{\frac{1}{2} + \frac{1}{3}}{2} = \frac{5}{12}$$

**Proposition 9.** Suppose that $< U, \mathbf{R} >$ is an approximation space, $R_1, R_2 \in \mathbf{R}$. Then, for arbitrary $X \subseteq U$, $0 < u_X^{\mathbf{R}}(x) \leq 1$ holds.

**Proof.** It can be easily proved according to definitions 8 and 9.

# 5    Conclusion

In this paper, two new kinds of multi-granulation rough set models have been proposed by using the central set of elements. Some meaningful properties of the models have been discussed in details and several examples were given to explain the concepts and properties in the paper. Finally, some uncertainty measures are discussed. The further research may include how to establish decision method based on the theory of multi-granulation rough set and how to employ the proposed models for decision-making in the context of uncertainty.

**Acknowledgements.** This work was supported by the China National Natural Science Foundation of Youth Science Foundation under Grant No.: 61305052.

# References

1. Qian, Y.H., Liang, J.Y.: Rough set method based on multi-granulations. In: Proceedings of 5th IEEE International Conference on Cognitive Informatics, Beijing, China, July 17-19, vol. 1, pp. 297–304 (2006)
2. Qian, Y.H., Liang, J.Y., Dang, C.Y.: Incomplete mutigranulation rough set. IEEE Trans. Syst. Man Cybern. A 40(2), 420–431 (2010)
3. Yang, X.B., Song, X.N., Dou, H.L., Yang, J.Y.: Multi-granulation rough set: from crisp to fuzzy case. Ann. Fuzzy Mathe. Inf. 1, 55–70 (2011)
4. Yang, X.B., Qian, Y.H., Yang, J.Y.: Hierarchical structures on multigranulation spaces. J. Comput. Sci. Technol. 27, 1169–1183 (2012)
5. She, Y.H., He, X.L.: On the structure of the multigranulation rough set model. Knowl.-Based Syst. 36, 81–92 (2012)
6. Liu, C.H., Miao, D.Q.: Covering rough set model based on multi-granulations. In: Proceedings of the 13th International Conference on Rough Sets, Fuzzy Sets, Data Mining and Granular Computing, Moscow, Russia, June 25-27, pp. 87–90 (2011)
7. Liu, C.H., Wang, M.Z.: Covering fuzzy rough set based on multi-granulations. In: Proceedings of the International Conference on Uncertainty Reasoning and Knowledge Engineering, Bali, Indonesia, August 4-7, vol. 2, pp. 146–149 (2011)
8. Liu, C.H., Miao, D.Q., Qian, J.: On multi-granulation covering rough sets. Int. J. Approx. Reason. 55, 1404–1418 (2014)
9. Lin, G.P., Qian, Y.H., Li, J.J.: NMGRS: Neighborhood-based multi granulation rough sets. Int. J. Approx. Reason. 53, 1080–1093 (2012)
10. Xu, W.H., Sun, W., Zhang, X., Zhang, W.X.: Multiple granulation rough set approach to ordered information systems. Int. J. Gen. Syst. 41, 475–501 (2012)
11. Xu, W.H., Wang, Q.R., Zhang, X.T.: Multi-granulation rough sets based on tolerance relations. Soft Comput. 17(7), 1241–1252 (2013)
12. Xu, W.H., Wang, Q.R.: Multi-granulation fuzzy rough sets. J. Intell. Fuzzy Syst., doi:10.3233 /IFS-1 30818 (2013)
13. Yang, X.B., Qi, Y.S., Song, X.N., Yang, J.Y.: Test cost sensitive multigranulation rough set: Model and minimal cost selection. Inf. Sci. 250, 184–199 (2013)
14. Qian, Y.H., Li, S.Y., Liang, J.Y., Shi, Z.Z., Wang, F.: Pessimistic rough set based decisions: A multigranulation fusion strategy 264, 196–210 (2014)
15. Qian, Y.H., Zhang, H., Sang, Y.L., Liang, J.L.: Multigranulation decision- theoretic rough sets, Int. J. Approx. Reason. 55(1), 225–237 (2014)

16. Lin, T.Y.: Neighborhood systems and relational database. Abstract. In: Proc CSC 1988, p. 725 (Februray 1988)
17. Chen, Z.H., Lin, T.Y., Xie, G.: Knowledge Approximations in Binary Relation: Granular Computing Approach. Int. J. Intell. Syst. 28(9), 843–864 (2013)
18. Pawlak, Z.: Rough sets. Int. J. Comp. Inf. Sci. 11, 341–365 (1982)

# Rough Set Theory on Topological Spaces

K. Anitha

Department of Mathematics, S.A.Engineering College, Chennai-600 077,
Tamil Nadu, India
subramanianitha@yahoo.com

**Abstract.** We consider that Rough Sets that arise in an Information System from the point of view of Topology. The main purpose of this paper is to show how well known topological concepts are closely related to Rough Sets and generalize the Rough sets in the frame work of Topological Spaces. We presented the properties of Quasi-Discrete topology and $\Pi_0$-Roughsets.

**Keywords:** Upper and Lower Approximations, Reducts, Discernibility Matrix, Rough Topology, $\Pi_0$-Roughsets.

## 1 Introduction

A Rough Set first described by Polish Computer Scientist Zdzislaw Pawlak is a formal approximation of Crisp Set in terms of a pair of a sets which give the Lower and Upper approximation of the original set.(en.wikipedia.org/wiki/Rough-Set). Rough set Theory based on the assumption that every object in the universe we associate some information. The basics of Roughset theory is based on similarity relation in which objects are having same information.

Rough Set Theory is the extension of conventional set theory that supports approximations is decision making. A Rough Set itself is a approximation of vague concepts by a pair of Crisp Sets called Lower and Upper approximation. The Lower approximation is the set of objects with certainty for belong to the subset of interest where as upper approximation is the set of all objects that are possibly belongs to the subset.

## 2 Terminologies of Rough Set Theory

An Information System can be represented as table of data, consisting of Objects in Rows and Attributes in Columns. For example in a Medical Data Set Patients might be represented as Objects (Rows), their Medical History such as Blood Pressure, Height, Weight etc. as Attributes (Columns). Attributes are also called as Variables or Features.

An Information System is a pair $IS = (U, A)$ where $U$ is a non-empty finite set of objects called Universe and $A$ is a non-empty finite set of attributes such that a: $U \longrightarrow V_a$ for every $a \in A$. The set $V_a$ is called the value set of 'a'.

D. Miao et al. (Eds.): RSKT 2014, LNAI 8818, pp. 69–74, 2014.
DOI: 10.1007/978-3-319-11740-9_7 © Springer International Publishing Switzerland 2014

## 3   Decision System

From the above Information System we can give the classification of Patients for finding the Patient is ill or healthy. Such Classification is known as Decision System and the corresponding attribute is called Decision Attribute.

## 4   Indiscernibility

Rough Set is a pair $(U, R)$ Where $U$ is Universe of Objects and is an Equivalence Relation on $R$. For each object $x \in U$ we can construct the set $[x]_R = \{y \in U : (x, y) \in R\}$ and they are called Equivalence Class. The Set of all Equivalence Classes is called Quotient set and which is defined by $U/R = \{[x]_R : x \in U\}$. For each $x \in U$ we can find a unique value $a_R(x) = [x]_R$ and construct a function $a_R : U \to U/R$ and this function $a_R$ is called Feature or Variable or Attribute of objects in the universe $U$ with respect to $R$. Two objects $x, y \in U$ are said to be Indiscernible if and only if $a_R(x) = a_R(y)$, and the relation is called Indiscernibility relation denoted by $IND_a$.

For any $P \subseteq A$ there is an equivalence relation $IND(P)$ which is defined as

$$IND(P) = \{(x, y) \in U^2 / \forall a \in P, a(x) = a(y)\} \tag{1}$$

which represents two objects are equivalent if and only if their attribute values are same in $P$.

## 5   Upper Approximation and Lower Approximation

Let $Y$ be a Rough Set and $Y \subset U$ can be approximated using only the information contained within $R$ by constructing the Lower and Upper approximation of $R$. Equivalence classes contained within $Y$ belong to the Lower Approximation $(\underline{R}Y)$. Equivalence classes within $X$ and along its border form the Upper Approximation $(\overline{R}Y)$. They are expressed as

$$\underline{R}Y = \{x \in U : [x]_R \subseteq Y\} \to \text{Lower Approximation.}$$
$$\overline{R}Y = \{x : [x]_R \cap Y \neq \emptyset\} \to \text{Upper Approximation}$$

*Remark 1.*

     *i.* $\underline{R}Y \subseteq Y \subseteq \bar{R}Y$

     *ii.* If $\underline{R}Y = \overline{R}Y$ then $Y$ is Exact

     *iii.* $\underline{R}(X \cap Y) = \underline{R}X \cap \underline{R}Y$

     *iv.* $\underline{R}X \subseteq X$

     *v.* $\underline{R}(\underline{R}X) = \underline{R}X$

     *vi.* $\underline{R}(\emptyset) = \emptyset$

$$vii.\ \overline{R}\,(X \cup Y) = \overline{R}X \ \cup \ \overline{R}Y$$
$$viii.\ X \subseteq \overline{R}X$$
$$ix.\ \overline{R}\,(\overline{R}X) = \overline{R}X$$
$$x.\ \overline{R}\,(U) = U.$$

## 6   Decision System

A Decision system is a triple $A_d = (U, A, d)$ *where* $(U, A)$ is an Information System, $d$ is a Distinguished attribute called the *Decision Attribute* which is defined by the function $d : U \to V_d$ on the universe $U$ into the value set $V_d$ and the $d$ induces the partition of the universe $U$ into equivalence classes of the $IND_d$. From this Decision System $A_d$, and a set $B \subseteq A$ of Attributes we can define the Positive Region as [7]

$$POS_B\,(d) = \{\ x \ \in U\ :\ \exists i \ \in \{1, 2, \ldots k\,(d)\}.\ [x]_B \subseteq\ X_i \ \} \tag{2}$$

where $\{1, 2, \ldots k\,(d)\}.[x]_B \subseteq X_i$ are Decision Classes.

## 7   Reduct

Reduct is the process of finding the Minimal Subset for the original Data Set without information loss. Hence Reduct is a subset which is defined as a minimal subset R of the initial attribute set $c$ such that for a given set of attributes D,

$$\gamma_R(D) = \gamma_c(D)$$

That is $R$ is a minimal subset if

$$\gamma_{R-\{a\}}(D) \neq \ \gamma_c(D), \forall a \in R$$

which means that no attributes can be removed from the subset without affecting dependency degree. A given data set may have many reduct sets, and the collection of all reducts is denoted by

$$R_{ALL} = \left\{X/X \subseteq C, \gamma_R\,(D) = \gamma_c\,(D), \gamma_{R-\{a\}}\,(D) \neq \gamma_c\,(D), \forall a \in X\right\}.$$

The intersection of all the sets in $R_{ALL}$ is called "CORE". The condition for locate a single element of the reduct set is $R_{MIN} \subseteq R_{ALL}$.

## 8   Finding Reduct Through Discernibility Matrix

Let $A$ be an information system with $n$ objects and $k$ attributes. The Discernibility Matrix of $A$ is a Symmetric Square matrix $M_A$ of order $n$ which is defined by [2]

$$C_{ij} = \{a \ \in A \ : a\,(x_i) \neq\ a\,(x_j)\} \ \text{where}\ i, j = 1, 2, 3, \ldots n \tag{3}$$

and $f_A\,(a_1^*, a_2^*, \ldots a_k^*) \ : \ \wedge\{\wedge\ C_{ij}^* \ : 1 \leq j,\ C_{ij} \neq \emptyset\}$ Where $f_A$ is the discernibility function.

**Theorem 1.** *A Conjunction $\bigwedge_{j=1}^{m} a_{i_j}^{*}$ is a Prime Implicants of the Discernibility function $f_A$ if and only if the set $\{ a_{i_1}, a_{i_2}, \ldots\ldots a_{i_m} \}$ is a $RED(A)$ where $RED(A)$ is Reducts of the Information System $A$.*

*Proof.* Let us consider a set $B \subseteq A$ of attributes and define the value $v_B$ as

$$v_B = \begin{cases} v_B\,(a^{*}) = 1 \; if \; a \in B \\ v_B\,(a^{*}) = 1 \; if \; a \notin B \end{cases}$$

Let us assume that $f_A$ takes the value 1 under $v_B$
$\implies$ Normal Disjunctive form and Dual of $f_A$ takes the value 1.
$\implies \{ a_{i_1}, a_{i_2}, \ldots\ldots a_{i_k} \} \subseteq B$
$\implies$ Minimal Set $B$ with the property $IND_A = IND_B$ is of the form
$\{a_{i_1}, a_{i_2}, \ldots\ldots a_{i_k}\}$ for some $k$, which is $RED\,(A)$.

**Proposition 1.** *The Set $IND(P) = \{(x,y) \in U^2 // \forall a \in P, a(x) = a(y)\}$ forms Partition of $U$.*

The Partition of $U$, determined by $IND(P)$ is denoted by $U/IND(P)$, which is the set of equivalence classes generated by $IND(P)$. The Partition characterizes a topological space which is called Approximation Space $AS = (U, \mathcal{P})$ where $U$ is the Universal Set and $\mathcal{P}$ is an Equivalence Relation and Equivalence classes of $\mathcal{P}$ are called Granules, Blocks or Elementary Sets. $\mathcal{P}_x \subseteq U$ denote equivalence class containing $x \in U$.

Using the above notations we can express Rough Membership Function which is defined as

$$\eta_X^{\mathcal{P}}(x) = \frac{|\mathcal{P}_X \bigcap \mathrm{X}|}{|\mathcal{P}_X|}, x \in U.$$

# 9    Toplogical Properties of Rough Set Theory

The Rough Set Topology is generated by its equivalence classes in which every Open set is Closed and is known as Clopen Topology or Quasi-Discrete Topology. [5]

**Definition 1.** *A Topology for $U$ is a collection $\tau$ of subsets of $U$ satisfying the following conditions:*
*i. $\emptyset \in \tau$ and $U \in \tau$.*
*ii. The arbitrary union of members of each sub-collection of $\tau$ is a member of $\tau$.*
*iii. The finite intersection of members of $\tau$ is a member of $\tau$.*

# 10    Rough Topology

**Definition 2.** *Let $(U, \tau)$ be a Topological Space and and let $R(\tau)$ be the equivalance relation on $P(U)$ defined as $(A, B) \in R\,(\tau) \Leftrightarrow \{A^I = B^I \text{ and } A^C = B^C\}$. A Topology $\tau$ is Quasi-Discrete if the sets in $\tau$ are Clopen.*

*Remark 2.* i. The members of $\tau$ are called Open Sets.

ii. The complement of *Open Sets* are Called *Closed Sets*.

iii. The Interior of $X \in U$ is the largest Open Subset of $X$ and it is denoted as $X^I$

iv. The Closure of $X \in U$ is the smallest closed subset that includes $X$ and it is denoted as $X^C$

v. A Topological Rough Set in $(U, \tau)$ is the Element of the Quotient set $P(U)/R(\tau)$.

*Remark 3.* Let $X^C$, $X^I$ and $X^B$ be Closure, Interior and Boundary Points respectively. Then the following axioms are hold

i. If $X^B = \emptyset$, then $X$ is Exact otherwise it is said to be Rough.

ii. If $X^C = X = X^I$, then $X$ is Totally Definable.

iii. If $X^C \neq X, X^I \neq X$ then $X$ Undefinable.

iv. If $X^B \neq X, X^C = X$ then $X$ is Externally Definable.

v. If $X^B = X^C \neq X$ then $X$ is Internally Definable.

**Proposition 2.** *If A is Exact in $(U, \tau)$ and $\tau \subset \tau'$ then A is Exact with respect to $\tau'$*

*Proof.* If A is Exact means then $X^B = \emptyset$ in $(U, \tau)$. Since $\tau \subset \tau', X^B = \emptyset$ in $(U, \tau')$ and hence $A$ is Exact with respect to $\tau'$

**Proposition 3.** *If $(U, \tau)$ is a Quasi-Discrete Topology and $\tau \subset \tau'$ then each Exact Set in $\tau$ is Exact in $\tau'$ iff $X^C$ in $(U, \tau) = X^C$ in $(U, \tau'), \forall X \in \tau'$*

## 11   $\Pi_0$ - Rough Sets

Let $U$ be the Universe and $P = \{[x]_{R_n} : x \in X, \ n = 1, 2, 3, \ldots\}$ be the family of all equivalence classes of all relations $R_n$ as open base from $U$. Given $R_n$ we can construct $\Pi_n$ the topology obtained by taking the family $P_n = \{[x]_{R_n} : x \in X, \ n = 1, 2, 3, \ldots\}$ as a open base, where $P \subseteq P_n$, from this concept we construct $T = Cl(\Pi_0) X$, $T = U/(Int\Pi_0)X$

**Definition 3.** *$\Pi_0$ – Rough Sets are defined as a pair $(Q, T)$,*
*where $Q = Cl(\ \Pi_0\ ) X, \ T = U/(Int\ \Pi_0\ )X$.*

*Remark 4.* A pair $(Q, T)$ of $\Pi_0$ Closed Sub Sets in $U$ satisfies the conditions $Q = Cl(\ \Pi_0\ ) X, T = U/(Int\ \Pi_0\ )X$ with a rough subset $X \subseteq U$ if and only if $Q, T$ satisfy the following conditions

i. $Q \bigcup T = U$

ii. $Q \bigcap T \neq \emptyset$

iii. $Q \bigcap T$ does not contain any point $x$ such that the singleton $\{x\}$ is $\Pi_0$-Open.

## 12   **Rough Set Metric**

The Metric Topology on $R_{\Pi_0}$ is defined by

$$d_n(x, y) = \begin{cases} 1 \ when \ \ [x]_n \neq [y]_n \\ 0 \qquad\qquad otherwise \end{cases}$$

where $d_n : UXU \to R^+$ and $R^+$ is the set of Non-Negative Real Numbers.

**Proposition 4.** *The Metric Topology Induced by d coincides with the topology $\Pi_0$*

*Proof.* Let $B = x_n^t$ for $y \in B$ and $r < 10^{-n}$, the Open Ball $B(y, r)$. From the definition of metric we have for any $z \in B(y, r)$, $[z]_n = [y]_n$ otherwise $(y, z) \geq \frac{1}{9} .10^{-n+1}$ hence $[z]_n = x_n^t \Rightarrow B(y, r) \subseteq x_n^t \Rightarrow x_n^t$ is open. Conversly, for an open ball $B(x, s)$, let $r = s - d(x, y)$ and select $n$, such that $10^{-n} < r$. For $z \in [y]_n$ we have $(y, z) \leq \frac{1}{9} .10^{-n+1}$. Then By Triangle Inequality $d(z, x) < s \Rightarrow [y]_n \subseteq B(x, s)$. Hence $B(x, s)$ is open in $\Pi_0$.

# 13  Conclusion

In this paper we have discussed Quasi-Discrete topology, Metric Topology, $\Pi_0$. – Rough Sets with their significant properties and relation between Metric Topology on $\Pi_0$. – Rough Sets. Finally we conclude such significant Topological properties of Rough Set Theory will be useful in Datamining, Information Analysis and Knowledge Processing.

# References

1. Abraham, A., Falćon, R., Bello, R. (eds.): Rough Set Theory. A True Landmark in Data Analysis. Springer
2. Kozae, A.M., Abo Khadra, A.A., Medhat, T.: Rough Set Theory for Topological Spaces. International Journal of Approximate Reasoning 40, 35–43 (2005)
3. Stadler, B.M.R., Stadler, P.F.: Basic Properties of Closure Spaces 42, 577–585
4. Vlach, M.: Algebraic and Topological Aspects of Rough Set Theory. In: Fourth International Workshop on Computational Intelligence and Applications. IEEE
5. Thuan, N.D.: Covering Rough Sets From a Topological Point of View. IACSIT International Journal of Computer Theory and Engineering 1, 1793–8201 (2009)
6. Raghavan, R., Tripathy, B.K.: On Some Topological Properties of Multigranular Rough Sets 2, 536–543 (2011)
7. Gnilka, S.: On Extended Topologies I: Closure Operators. Ann. Soc. Math. Pol. Math. 34, 81–94 (1994)
8. Iwinski, T.: Algebraic Approach to Rough Sets. Bull, Polish Academy of Science and Math. 35, 673–683 (1987)
9. Zhu, W.: Relationship between generalized Rough Sets Based on Binary Relation and Covering. Information Sciences 179, 210–225 (2009)
10. Zhu, W.: Generalized Rough Sets Based on Relations. Information Sciences 177, 4997–5011 (2007)
11. Pawlak, Z.: Some Issues on Rough Sets. In: Peters, J.F., Skowron, A., Grzymała-Busse, J.W., Kostek, B.z., Swiniarski, R.W., Szczuka, M.S. (eds.) Transactions on Rough Sets I. LNCS, vol. 3100, pp. 1–58. Springer, Heidelberg (2004)

# Attribute Reduction
and Feature Selection

# Co-training Based Attribute Reduction
# for Partially Labeled Data

Wei Zhang[1,2,3], Duoqian Miao[1,3], Can Gao[4], and Xiaodong Yue[5]

[1] School of Electronics and Information Engineering, Tongji University, Shanghai 201804
[2] School of Computer Science and Technology,
Shanghai University of Electric Power, Shanghai 200090
[3] The Key Laboratory of Embedded System and Service Computing,
Ministry of Education, Tongji University, Shanghai 201804
[4] Zoomlion Heavy Industry Science And Technology Development Co.,Ltd.,
Changsha 410013
[5] School of Computer Engineering and Science, Shanghai University, Shanghai 200444

**Abstract.** Rough set theory is an effective supervised learning model for labeled data. However, it is often the case that practical problems involve both labeled and unlabeled data. In this paper, the problem of attribute reduction for partially labeled data is studied. A novel semi-supervised attribute reduction algorithm is proposed, based on co-training which capitalizes on the unlabeled data to improve the quality of attribute reducts from few labeled data. It gets two diverse reducts of the labeled data, employs them to train its base classifiers, then co-trains the two base classifiers iteratively. In every round, the base classifiers learn from each other on the unlabeled data and enlarge the labeled data, so better quality reducts could be computed from the enlarged labeled data and employed to construct base classifiers of higher performance. The experimental results with UCI data sets show that the proposed algorithm can improves the quality of reduct.

**Keywords:** Rough Sets, Co-training, Incremental Attribute Reduction, Partially Labeled Data, Semi-supervised learning.

## 1 Introduction

Since the initial work of Pawlak [1,2], rough set theory, as an effective approach to dealing with imprecise, uncertain and incomplete information, has been applied to many domains, such as pattern recognition, artificial intelligence, machine learning, economic forecast, knowledge acquisition and data mining [3-7].

Attribute reduction is one of the most fundamental and important notions in rough set theory, which provides a minimum subset of attributes that preserves the same descriptive ability as the entire set of attributes [1].So there is a variety of research and application on the issue [8,9]. These researches mainly concentrate on the attribute reduction of labeled data.

While in many practical applications (e.g. web-page classification, speech recognition, natural language parsing, spam filtering, video surveillance, protein 3D structure

D. Miao et al. (Eds.): RSKT 2014, LNAI 8818, pp. 77–88, 2014.
DOI: 10.1007/978-3-319-11740-9_8 © Springer International Publishing Switzerland 2014

prediction, etc.), labeled data may be difficult to obtain because they require human annotators, special devices, or expensive and slow experiments. Moreover, the unlabeled data are available in large quantity and easy to collect [10]. A partially labeled data consists of a few labeled data and large numbers of unlabeled data. So the problem arises of how to get the attribute reduct of partially labeled data. It is outside the realm of traditional rough set theory. If only labeled data are considered, it is difficult to select high quality reduct due to the scarcity of the labeled data. Or the partially labeled data are transformed into labeled data, such as reference [11]. It attaches a special pseudo-class symbol which differs from that of all labeled data to every unlabeled data and unlabeled data will have a class symbol. This method exploits the unlabeled data but would introduce some irrelevant attributes.

The aim of the paper is to capitalize on the unlabeled data to improve the quality of attribute reducts from few labeled data, which could make rough sets more practicably.

The main idea is to get higher quality attribute reducts during the process of two classifiers co-training iteratively. So firstly section 2 presents the preliminary knowledge on co-training and incremental updating algorithm for attribute reduction based on discernibility matrix. Section 3 proposes the attribute reduction algorithm for partially labeled data. The theoretical analysis and the experimental results are given in section 4 and section 5, respectively. Finally, Section 6 concludes the paper and indicates the future work.

## 2     Preliminaries

In rough set theory, an information system( $IS$ ) is denoted $IS = (U, A, V, f)$ ,where $U$ is a finite nonempty set of objects $\{x_1, x_2, ..., x_n\}$, A is a finite nonempty set of attributes to characterize the objects $\{a_1, a_2, ..., a_m\}$ ,To be more specific, $IS$ is also called a decision table if $A = C \cup D$ , where $C$ is the set of condition attributes and $D$ is one of class attributes, the V is the union of attribute domains such that $V = \bigcup_{a \in A} V_a$ for $V_a$ denoting the value domain of attribute a, and $f : U \times A \to V$ is an information function which associates a unique value of each attribute with every object belonging to U, such that for any $a \in A$ and $u \in U, f(u, a) \in V_a$ .Also, $(U, A, V, f)$ can be written more simply as $(U, A)$ .

### 2.1     Incremental Updating Algorithm for Attribute Reduction

**Definition 1.** For any $x_i, x_j \in U(x_i \neq x_j)$, $x_i$ and $x_j$ are consistent if $C_i = C_j$ implies $D_i = D_j$ ;otherwise, $x_i$ and $x_j$ are inconsistent. Condition attributes $R \subseteq C$ , $x_i$ and $x_j$ are R-consistent if $R_i = R_j$ implies $D_i = D_j$ ;otherwise, $x_i$ and $x_j$ are R-inconsistent.

**Definition 2.** If $F = \{X_1, X_2, ..., X_n\}$ is a family of non-empty sets(classification), then $\underline{R}F = \{\underline{R}X_1, \underline{R}X_2, ..., \underline{R}X_n\}$. The quality of approximation of $F$ by $R$ is defined as $\gamma_R(F) = (\sum card \underline{R}X_i) / card U$.

**Definition 3.** The indiscernibility relation $IND(R), R \subseteq C$, is defined as $IND(R) = \{x_i, x_j \in U^2 : \forall a \in C, a(x_i) = a(x_j)\}$.

**Definition 4.** The set of $x_i (x_i \in U)$ which is equivalent with $x(x \in U)$ employing $IND(R)$ is denoted by $[x(U)]_R$.

Many existing algorithms mainly aim at the case of stationary information system or decision table, very little work has been done in updating of an attribute reduction. Reference [12] introduces an incremental updating algorithm(named algorithm-IUAARI )for attribute reduction based on discernibility matrix in the case of updating, which only inserts a new row and column, or deletes one row and updates corresponding column when updating the discernibility matrix. Attribute reduction can be effectively updated by utilizing the old attribute reduction. Main idea is stated as follows. Details are referenced to reference [12].

Two objects are discernible if their values are different in at least one attribute. Skowron and Rauszer suggested a matrix representation for storing the sets of attributes that discern pairs of objects, called a discernibility matrix [13]. For decision table, reference[12] considers consistent and inconsistent cases, redefines the matrix $M$ as follows:

$$
m_{ij} = \left\{
\begin{array}{l}
\{a \in C : f(x_i, a) \neq f(x_j, a)\}, f(x_i, D) \neq f(x_j, D), x_i, x_j \in U_1 \\
\{a \in C : f(x_i, a) \neq f(x_j, a)\}, x_i \in U_1, x_j \in U_2{}' \\
/
\end{array}
\right.
$$

$$
U_1 = \bigcup_{i=1}^{k} \underline{C}X_i(U), U_2 = U - U_1, U_2{}' = delrep(U_2)
$$

function $delrep(U_2)$ is defined as follows:

begin
  $U_2{}' = \varnothing$;
  for $\forall x \in U_2$ do
if not $\exists y \in U_2{}'$ makes $\forall a \in C, f(x, a) = f(y, a)$ and $f(x, D) \neq f(y, D)$ then
  $U_2{}' = U_2{}' \cup \{x\}$;
return $U_2{}'$
end

According to analysis of matrix updating on the new added data, algorithm IUAARI considers three cases to get new reduct efficiently, described as follows:

---

**Algorithm IUAARI** (Incremental Updating Algorithm of Attribute Reduction for Inserting)

**Input:** (1) $U_1 = \bigcup\limits_{i=1}^{k} \underline{C} X_i(U), U_2 = U - U_1, U_2' = delrep(U_2)$, discernibility matrix $M_1$

(2) $red \in R(U)$  //  $R(U)$ is a set of all reducts for $U$

(3) An added data $x$

**Output:** $redUpdated$  $(redUpdated \in R(U \cup \{x\}))$

(1)   Compute updated $M_1(x)$ and $core(U \cup \{x\})$

(2)   $redUpdated = \emptyset$;

    ① if $x$ is not consistent with $y(y \in U_2')$ then $redUpdated = red$, go to (3);

    ② if $x$ is not consistent with $y(y \in U_1)$ then

        $redUpdated = redUpdated \cup core$;

            if $|[y(U)]|_{red} == 1$ then

                select attributes in $(red - core)$ to construct $redUpdated$ for $U \cup \{x\}$;

            else

                select attributes in $(C - core)$ to construct $redUpdated$ for $U \cup \{x\}$;

            go to (3);

    ③ if $x$ is $red - $consistent with $y(y \in U_1)$ then $redUpdated = red$, go to (3);

        else{

            $redUpdated = redUpdated \cup core$;

            select attributes in $(C - core)$ to construct $redUpdated$ for $U \cup \{x\}$;

            go to (3);

        }

(3)   **return** $redUpdated$

---

## 2.2     Co-training

Co-training [14] proposed by Blum and Mitchell is an important semi-supervised learning model. It is a classical semi-supervised style and has been used in many applications successfully, such as web-page categorization, image retrieval and intrusion detection [15]. The standard co-training assumes that there exist two sufficient and redundant sets of attributes or views that describe the data. Two base classifiers are first trained on the initial labeled data using two attribute sets respectively. Then, alternately, one classifier labels some confident unlabeled data and adds those data with predicted labels to the training set of the other one. The classifiers are iteratively retrained until the predefined stopping criterion is met. But in many practical applications, there is only a single natural set of attributes, and the assumption for two sufficient and redundant attribute sets is difficultly satisfied.

Then two kinds of ideas are proposed as follows. One is to meet the conditions as far as possible, such as two view based on random splitting, mutual independence measures and genetic algorithm optimization, etc. [16-21]. In this way, it might be not stable of view splitting or not sufficient. The other is to relax the conditions, such as different classifiers based on resampling of a single attribute set [22-25]. In fact, it only employs a single view.

Usually, there are a number of reducts for a given partially labeled data. Each reduct preserves the discriminating power of the original data. Therefore, it is sufficient to train a good classifier. Moreover, different reducts describe the data in different views. So after transformation of partially labeled data to decision table by attaching a special pseudo-class symbol which differs from that of all other data to every unlabeled data, reference [11,26] employ two diverse attribute reducts on the transformed decision table as two views to train the base classifiers of co-training. In this way, however, the reducts might include irrelevant attributes as mentioned above.

## 3    Co-training Based Attribute Reduction Algorithm

Formally, a partially labeled data is denoted as $PS = (U = L \cup N, A = C \cup D, V, f)$ ,where $L$ —is a nonempty and finite set of labeled data. $N$ —is a nonempty and finite set of unlabeled data.

The number of initial labeled data $L$ is very limited. It is difficult to get high quality reduct from it. But lots of unlabeled data $N$ could be considered to capitalize on to improve the quality of attribute reducts. According to the condition of co-training, the original attributes are split into two sufficient and diverse attribute sets as two views. And two base classifiers are trained on them. The two base classifiers co-train iteratively. In every round, the base classifiers learn from each other on the unlabeled data and enlarge the labeled data, so better quality reducts could be computed on the enlarged labeled data and employed to construct base classifiers of higher performance. It goes by iteration in this way until the classifiers don't change any more. The structure of co-training based attribute reduction for partially labeled data is shown in Fig. 1.

So firstly, how to split the original attributes into the two sufficient and diverse views? Usually, there are a number of reducts for a given labeled data. Each reduct preserves the discriminating power of the original data. Therefore, it is sufficient to train a good classifier. Moreover, different reducts describe the data in different views, which means the classifying rules induced by the different reducts would be different, so the classifiers are different. Therefore, we firstly compute two diverse reduct to train the base classifiers of co-training. The detailed procedure is shown in Algorithm TDRCA.

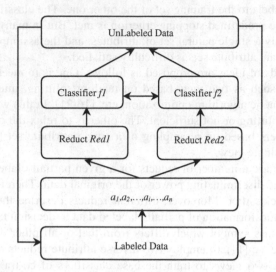

**Fig. 1.** Framework of the incremental attribute reduction for partially labeled data

---

**Algorithm TDRCA** (Two Diverse Reducts Construction Algorithm)

**Input:**    $PS = (U = L \cup N, A = C \cup D, V, f)$.

**Output:** Two diverse reducts: $red1, red2$.

(1) $LS = (U = L, A = C \cup D, V, f)$, Compute the discernibility matrix $M$ of $LS$;

(2) perform matrix absorption on $M$;

(3) Add the attributes in the singleton set of $M$ to $Core$ and remove the attributes that do not appear in $M$ from $C$;

(4) $red1 = core$, $C_1 = C - core$; sort the attributes in $C_1$ by frequency ascending-ly and add them to Clist;

   **while** $M$ contains non-singleton subsets **do**

     Select the left-most frequent attribute $a$ in $C_1$,

     $red1 = red1 \cup \{a\}$,

     Perform matrix deletion on $M$ to remove matrix elements which includes $a$;

   **end while**

(5) $red2 = core$, $C_2 = C - core - red1$;

   **while** $M$ contains non-singleton subsets **do**

     Select the left-most frequent attribute $a$ in $C_2$,

     $red2 = red2 \cup \{a\}$,

     Perform matrix deletion on $M$ to remove matrix elements which includes $a$;

     if( $C_2 = \varnothing$ ) put the attributes of $red1 - core$ in $C_2$;

   **end while**

(6) **return** $red1, red2$

Algorithm TDRCA firstly gets an optimal reduct $red1$ based on discernibility matrix. And to get the second reduct $red2$, the following strategy is employed. The core attributes are added into $red2$. In order to get as different as possible reduct with the $red1$, we prioritize the attributes that don't be included in $red1$. If these attributes are not enough to construct a reduct, the attriubutes in $red1-core$ are be taken into account. In this way, we could get two as different reducts as possible.

We treat the two reducts as the views for co-training to get base classifiers. As analyzed above, the two reducts not only keep the discriminating power of the original data but also describe the initial labeled data from different points of view. Then during the process of co-training, unlabeled data are exploited to enlarge labeled data set, two classifiers are retrained and better reducts are gotten iteratively as depicted in Fig1.It can be formulated by Algorithm CTARAPLD.

---

**Algorithm CTARAPLD** ( Co-training Based Attribute Reduction Algorithm for Partially Labeled Data)

**Input:**  $PS = (U = L \cup N, A = C \cup D, V, f)$.

**Output:**  Reduct $red1, red2$ .

(1) On $L$ ,decompose condition attribute set $C$ into two diverse reducts $red1$ and
  $red2$ by Algorithm TDRCA;

(2)  $L_1 = L_2 = L, N_1 = N_2 = N$ ,Train two base classifiers f1 and f2 on $L_1, L_2$ in
  reducts $red1$ and $red2$ respectively;

(3) **while**( $N_1 \neq \varnothing$ or $N_2 \neq \varnothing$ ) **do**

  ① Add $Samples1$ ( $Samples1 \in N_1$ ) with the class symbols predicted by $f_2$
  to the training set $L_1$ and $N_1 = N_1 - Samples1$ . Compute $red1$ on updated
  $L_1$ by Algorithm IUAARI.

  ② Add $Samples2$ ( $Samples2 \in N_2$ ) with the class symbols predicted by $f_1$
  to the training set $L_2$ and $N_2 = N_2 - Samples2$ . Compute $red2$ on updated
  $L_2$ by Algorithm IUAARI .

  ③ Check the updating of reducts on enlarged $L_1, L_2$ .
  If (no updating in reducts)
    Retrain classifiers f1 and f2 on updated $L_1, L_2$ ;
  else
    Retrain classifiers f1 and f2 on updated $L_1, L_2$ in updated $red1$ and $red2$
  respectively;
  **end while**

(4) **return** $red1, red2$ .

---

By algorithm TDRCA, two diverse reducts are gotten as views, in which two base classifiers are trained on the initial labeled data respectively. For any unlabeled data, two classifiers predict it and result will be one of the following cases—only one classifier is predictable, both of classifiers are predictable or unpredictable. The classifiers could learn from each other in the first case. Unlabeled data with predicted class symbol are exploited to enlarge labeled data. In this way, reduct could be updated to more close with true reduct of all training data labeled. After the classifiers are retrained on the enlarged labeled data, their classification performance are boosted and unpredictable data before would be predictable. The classifiers will learn from each other once again. In an ideal situation, every unlabeled data would be labeled a class symbol and the reduct may be very close to the true one of all training data labeled.

# 4    Empirical Analysis

## 4.1    Benchmark Data Sets and Experiment Parameters

Five UCI data sets [27] are used in the experiments. The detailed information of these data sets is shown in Table 1. Data set "Iono", "Wine" and "WPBC" are described by continuous attributes. We use the principle of equal frequency [28] to discretize continuous data. For each data set, 10-fold cross validation is employed for evaluation. In each fold, the training set is randomly partitioned into labeled set $L$ and unlabeled set $N$ under a given label rate ( $\alpha$ ), which can be computed by the size of $L$ over the size of $L \cup N$.

**Table 1.** UCI data sets

| Data Set | #Attributes | #Instances | #Classes |
|---|---|---|---|
| Ionosphere (Iono) | 34 | 351 | 2 |
| Wine | 13 | 178 | 3 |
| Wisconsin Diagnostic Breast Cancer (WDBC) | 30 | 569 | 2 |
| Wisconsin Prognostic Breast Cancer (WPBC) | 32 | 198 | 2 |
| tic-tac-toe(TTT) | 9 | 958 | 2 |

In the every round of iteration , class symbols ratio of selected unlabeled data will match the one of data in the underlying data distribution of initial labeled data.

## 4.2    Validation of the Classifiers' Generalization in Reducts

For the evaluation of reduct quality, we collect the cardinality of redcuts and accuracy of classifiers trained in reducts by the following ways:

(a) Compute the reduct on the labeled data under the labeled rate $\alpha$ =100%, which is showed in table 2.
(b) Without exploiting the unlabeled data, compute the reduct only on the labeled data under the labeled rate $\alpha$ =10%, which is showed in table 3.

(c) Transform the partially labeled data into labeled data in the way attaching a pseudo-class symbol which is discernible from all other data. Compute the reduct on the transformed data, which is showed in table 4.

(d) By the algorithm CTARAPLD proposed in the paper, compute the reduct on the partially labeled data under the labeled rate $\alpha$=10%, which is showed in table 5. The cardinality of reduct with best classification accuracy is recorded in table 6.

We apply CART and Bayes , two popular classification learning algorithm to evaluate the quality of the reducts. For each data set under a specific label rate, 10-fold cross validation is applied, and the results are averaged. Table 2-5 show the average cardinality of reducts and the average accuracy of the learned classifiers in the different cases as described above. In table 2-5, N1, N2 denote the average cardinality of original attribute set and computed reduct, respectively. Accuracy1 and Accuracy2 means the average accuracy of classifiers in original attribute set and computed reduct, respectively.

**Table 2.** Cardinality of reducts and accuracy of classifiers on data under $\alpha$=100%

| Data Set | #Attributes | | CART | | Bayes | |
|---|---|---|---|---|---|---|
| | N1 | N2 | Accuracy1 | Accuracy2 | Accuracy1 | Accuracy2 |
| Wine | 13 | 5 | 0.8786 | 0.9092 | 0.9424 | 0.9315 |
| Iono | 34 | 8 | 0.8221 | 0.8386 | 0.8097 | 0.6983 |
| WDBC | 30 | 8 | 0.9342 | 0.9350 | 0.9250 | 0.9388 |
| WPBC | 32 | 7 | 0.6667 | 0.6869 | 0.5814 | 0.7169 |
| TTT | 9 | 8 | 0.8213 | 0.7612 | 0.6258 | 0.6521 |
| Avg. | 23.6 | 7.2 | 0.8246 | 0.8262 | 0.7768 | 0.7875 |

**Table 3.** Cardinality of reducts and accuracy of classifiers on labeled data under $\alpha$=10%

| Data Set | #Attributes | | CART | | Bayes | |
|---|---|---|---|---|---|---|
| | N1 | N2 | Accuracy1 | Accuracy2 | Accuracy1 | Accuracy2 |
| Wine | 13 | 3 | 0.8786 | 0.6980 | 0.9424 | 0.7237 |
| Iono | 34 | 6 | 0.8221 | 0.7281 | 0.8097 | 0.7215 |
| WDBC | 30 | 3 | 0.9342 | 0.8757 | 0.9250 | 0.8753 |
| WPBC | 32 | 5 | 0.6667 | 0.7113 | 0.5814 | 0.6907 |
| TTT | 9 | 6 | 0.8213 | 0.7269 | 0.6258 | 0.6472 |
| Avg. | 23.6 | 4.6 | 0.8246 | 0.7480 | 0.7768 | 0.7317 |

**Table 4.** Cardinality of reducts and accuracy of classifiers on transformed labeled data under $\alpha$=10%.

| Data Set | #Attributes | | CART | | Bayes | |
|---|---|---|---|---|---|---|
| | N1 | N2 | Accuracy1 | Accuracy2 | Accuracy1 | Accuracy2 |
| Wine | 13 | 9 | 0.8786 | 0.8047 | 0.9424 | 0.8832 |
| Iono | 34 | 13 | 0.8221 | 0.7978 | 0.8097 | 0.8117 |
| WDBC | 30 | 11 | 0.9342 | 0.9063 | 0.9250 | 0.8918 |
| WPBC | 32 | 7 | 0.6667 | 0.6874 | 0.5814 | 0.7291 |
| TTT | 9 | 8 | 0.8213 | 0.7612 | 0.6258 | 0.6521 |
| Avg. | 23.6 | 9.6 | 0.8246 | 0.7915 | 0.7768 | 0.7936 |

**Table 5.** Cardinality of reducts and accuracy of classifiers on partially labeled data under $\alpha$=10%.

| Data Set | #Attributes | | CART | | Bayes | |
|---|---|---|---|---|---|---|
| | N1 | N2 | Accuracy1 | Accuracy2 | Accuracy1 | Accuracy2 |
| Wine | 13 | 5 | 0.8786 | 0.8376 | 0.9424 | 0.8605 |
| Iono | 34 | 7 | 0.8221 | 0.7954 | 0.8097 | 0.7303 |
| WDBC | 30 | 6 | 0.9342 | 0.9253 | 0.9250 | 0.9282 |
| WPBC | 32 | 6 | 0.6667 | 0.6975 | 0.5814 | 0.7204 |
| TTT | 9 | 8 | 0.8213 | 0.7761 | 0.6258 | 0.6502 |
| Avg. | 23.6 | 6.4 | 0.8246 | 0.8064 | 0.7768 | 0.7779 |

**Table 6.** Cardinality of reduct with the best classification accuracy in Table 5

| Data Set | #Attributes | CART | | Bayes | |
|---|---|---|---|---|---|
| | N1 | N2 | Best Accuracy2 | N2 | Best Accuracy2 |
| Wine | 13 | 4 | 0.9000 | 5 | 0.8977 |
| Iono | 34 | 7 | 0.8332 | 7 | 0.7827 |
| WDBC | 30 | 5 | 0.9360 | 5 | 0.9419 |
| WPBC | 32 | 6 | 0.7056 | 6 | 0.7381 |
| TTT | 9 | 8 | 0.7950 | 8 | 0.6553 |
| Avg. | 24 | 6.0 | 0.8340 | 6.2 | 0.8031 |

It is clear from these data in table 2 that reduction could reduce the cardinality of attribute set and retain or improve the classification accuracy if we have lots of labeled data. However, if number of labeled data is very limited, such as labeled rate 10% in table 3, accuracy of classifiers trained in the reduct from the initial labeled data would decrease much.

Compare table 3 to table 4 and table 5, we can see the performance of each algorithm is boosted with exploiting unlabeled data. In table 3, the classifier is trained only on the initial labeled data. The cardinality of reduct is fewer and the classification accuracy on almost all data is lower than that in table 4 and table 5 which capitalize on unlabeled data to enlarge labeled data.

Classification accuracy of algorithm in table 5 ( proposed algorithm in this paper) is comparable to that in table 4. Moreover, the former retain fewer attributes than later. As analyzed above, every unlabeled data is labeled pseudo-class symbol discernible from all other data, which may bring irrelevant attributes and decrease the classification accuracy. Like data "WDBC" in table 5, the classification accuracy 92.53, 92.82 by CART, Bayes respectively are very close to that in original attribute set and have 6 attributes, while in table 4, the classification accuracy is lower and have 11 attributes. It shows proposed algorithm improves the quality of reduct.

# 5    Conclusions

Traditional rough set theory is not applicable for partially labeled data. Based on co-training for capitalizing on unlabeled data to enlarge labeled data, this paper proposes a co-training based attribute reduction algorithm to deal with partially labeled data. The experiments on UCI data sets prove the effectiveness. Accuracy of classifiers trained in the reduct computed on the partially labeled data is much higher than that on the initial labeled data. The quality of reduct is much improved.

In order to further investigate the effectiveness of proposed algorithm, we will perform the experiments under other different label rates. In addition, unlabeled data misclassified by the classifiers will decrease the quality of the reduct, how to deal with it is also our work in the future.

**Acknowledgements.** The work is supported by the National Natural Science Foundation of China (Serial No. 61075056, 61273304, 61202170, 61103067) and the Specialized Research Fund for the Doctoral Program of Higher Education of China  (Serial No. 20130072130004).

# Reference

1. Pawlak, Z.: Rough sets. International Journal of Computer & Information Sciences 11(5), 341–356 (1982)
2. Pawlak, Z.: Rough sets: Theoretical aspects of reasoning about data. Kluwer Academic Publishers. Dordrecht & Boston (1991)
3. Liu, Q.: Rough sets and rough reasoning. Academic Pub., Beijing (2001)
4. Wang, G.Y.: Rough set theory and knowledge acquisition. Xi'an Jiaotong University Press, Xi'an (2001)
5. Zhang, W.X., Wu, W.Z., Liang, J.Y., et al.: Rough set theory and methods. Science and Technology Press, Beijing (2001)
6. Polkowski, L.: Rough sets: Mathematical foundations. Springer Science & Business (2013)
7. Miao, D.Q., Li, D.G.: Rough Set Theory, Algorithms and Applications. Tsinghua University Press, Beijing (2008)

8. Miao, D.Q., Zhao, Y., Yao, Y.Y., et al.: Relative reducts in consistent and inconsistent deci-sion tables of the Pawlak rough set model. Information Sciences 179(24), 4140–4150 (2009)
9. Thangavel, K., Pethalakshmi, A.: Dimensionality reduction based on rough set theory: A review. Applied Soft Computing 9(1), 1–12 (2009)
10. Xiaojin, Z.: Semi-supervised learning literature survey. Computer Sciences TR 1530. De-partment of Computer Sciences, University of Wisconsin (2008)
11. Miao, D.Q., Gao, C., Zhang, N., et al.: Diverse reduct subspaces based co-training for partial-ly labeled data. International Journal of Approximate Reasoning 52(8), 1103–1117 (2011)
12. Yang, M.: An incremental updating algorithm for attribute reduction based on improved dis-cernibility matrix. Chinese Journal of Computers 30(5), 815–822 (2007)
13. Skowron, A., Rauszer, C.: The discernibility matrices and functions in information sys-tems. In: Intelligent Decision Support. Theory and Decision Library, vol. 11, pp. 331–362. Springer, Netherlands (1992)
14. Blum, A., Mitchell, T.: Combining labeled and unlabeled data with co-training. In: Pro-ceedings of the 11th Annual Conference on Computational Learning Theory, pp. 92–100. ACM, New York (1998)
15. Zhu, X.J., Goldberg, A.B.: Introduction to semi-supervised learning. Synthesis Lectures on Artificial Intelligence and Machine Learning 3(1), 1–130 (2009)
16. Nigam, K., Ghani, R.: Analyzing the effectiveness and applicability of co-training. In: Proceedings of the Ninth International Conference on Information and Knowledge Man-agement, pp. 86–93. ACM, New York (2000)
17. Feger, F., Koprinska, I.: Co-Training using RBF nets and different feature splits. In: Inter-national Joint Conference on Neural Networks, pp. 1878–1885. IEEE, Piscataway (2006)
18. Wang, J., Luo, S.W., Zeng, X.H.: A random subspace method for co training. Acta Elec-tronica Sinica 36(12A), 60–65 (2008)
19. Tang, H.L., Lin, Z.K., Lu, M.Y., et al.: An advanced co-training algorithm based on mu-tual independence and diversity measures. Journal of Computer Research and Develop-ment. 45 (11),1874-1881 (2008)
20. Salaheldin, A., El Gayar, N.: New feature splitting criteria for co-training using genetic al-gorithm optimization. In: El Gayar, N., Kittler, J., Roli, F. (eds.) MCS 2010. LNCS, vol. 5997, pp. 22–32. Springer, Heidelberg (2010)
21. Yaslan, Y., Cataltepe, Z.: Co-training with relevant random subspaces. Neurocomput-ing 73(10), 1652–1661 (2010)
22. Goldman, S.A., Zhou, Y.: Enhancing Supervised Learning with Unlabeled Data. In: Pro-ceedings of the 17th International Conference on Machine Learning, pp. 327–334. Morgan Kaufmann, San Francisco (2000)
23. Zhou, Y., Goldman, S.: Democratic co-learning. In: The 16th IEEE International Confe-rence on Tools with Artificial Intelligence, pp. 594–602. IEEE, Piscataway (2004)
24. Zhou, Z.H., Li, M.: Tri-training: Exploiting unlabeled data using three classifiers. IEEE Transactions on Knowledge and Data Engineering 17(11), 1529–1541 (2005)
25. Li, M., Zhou, Z.H.: Improve computer-aided diagnosis with machine learning techniques using undiagnosed samples. IEEE Transactions on Systems, Man, and Cybernetics: Sys-tems 37(6), 1088–1098 (2007)
26. Gao, C., Miao, D.Q., Zhang, Z.F., et al.: A Semi-Supervised rough set model for classifi-cation based on active learning and co-training. Pattern Recognition and Artificial Intelli-gence 25(5), 745–754 (2012)
27. Blake, C., Merz, C.J.: UCI Repository of machine learning databases, http://archive.ics.uci.edu/ml/datasets.html
28. Øhrn, A., Komorowski, J.: ROSETTA–A Rough Set Toolkit for Analysis of Data. In: 5th International Workshop on Rough Sets and Soft Computing, pp. 403–407 (2007)

# Approximate Reduction for the Interval-Valued Decision Table

Feifei Xu*, Zhongqin Bi, and Jingsheng Lei

College of Computer Science and Technology
University of Shanghai Electric Power
2103 Pingliang Road, Shanghai, China
xufeifei1983@hotmail.com

**Abstract.** Many specific applications for electric power data, such as load forecasting and fault diagnosis, need to consider data changes during a period of time, rather than one record, to determine their decision classes, because the class label of only one record is meaningless. Based on the above discussion, interval-valued rough set is introduced. From the algebra view, we define the related concepts and prove the properties for the interval-valued reduction based on dependency, and present the corresponding heuristic reduction algorithm. In order to make the algorithm to achieve better results in practical applications, approximate reduction is introduced. To evaluate the proposed algorithm, we experiment on six months' operating data of one 600MW unit in some power plant. Experimental results show that the algorithm proposed in this article can maintain a high classification accuracy with the proper parameters, and the numbers of objects and attributes can both be greatly reduced.

**Keywords:** Interval-value; approximate reduction; dependency; the decision table.

## 1 Introduction

Rough set theory[1] is a powerful mathematical tool to deal with the uncertainty, as well as probability theory, fuzzy sets and evidence theory. It has been successfully applied in many fields of science and engineering. As a relatively new soft computing method, rough set theory has already attracted more and more attention in recent years. Its effectiveness has been demonstrated in various application areas, and it has become one of the hot areas in current artificial intelligence theories and application researches[2–16]. There are many similar characteristics among rough sets, probability theory, fuzzy sets and evidence theory. Compared

* Project supported by Natural Science Foundation of China (No.61272437, No.61305094) and the Research Award Fund for Outstanding Young Teachers in Higher Education Institutions, Shanghai (No.sdl11003) and the Innovation Program of Shanghai Municipal Education Commission (No.12YZ140,No.14YZ131) and the Natural Science Foundation of Shanghai (No.13ZR1417500).

D. Miao et al. (Eds.): RSKT 2014, LNAI 8818, pp. 89–100, 2014.
DOI: 10.1007/978-3-319-11740-9_9 © Springer International Publishing Switzerland 2014

to the latter three, rough sets do not need any priori knowledge of the data. The knowledge can be obtained by the theory itself, while probability theory, fuzzy sets and evidence theory need the information of probability, membership and assignment of probability respectively.

Most of the data on electric power are composed of continuous-valued attributes. Different from the traditional classification methods, the classification problems in many applications in the data on electric power are no longer considering a single record of data, but using the form of data blocks as research objects. This is because it is meaningless to determine the class label by relying solely on one record of data. We should consider the characteristics of the data for a period of time in order to determine which category of the data segment belongs to. Take load forecasting for example, we can not say that the load is low or high for only one record of data. We need to find the similar data segment with the data segment to be predicted, and determine its load value. Therefore, the classification analysis should be based on the blocks of data. In order to build the classification model for the data on electric power efficiently, the data blocks are approximately represented by interval values, namely, the data block can be approximately described by using its maximum and minimum value(non-numerical condition attributes can convert to numeric attributes). Then attribute reduction strategy of the interval-valued decision table can be designed, and the classification model can be established. There has been some research on interval-valued attribute reduction[17–20], however, the computing complexity is too high.

For the data on electric power are mainly continuous, and many classification problems for the power data should consider the data blocks as object units, the data block is described as the interval-valued form, on which the heuristic reduction is discussed. The related concepts and properties proof of dependency based interval-valued attribute reduction are given, and the corresponding algorithm is presented. To enhance the practicability of the proposed algorithm, approximate reduction is introduced. We use the real power data to test the proposed algorithm. The results are analyzed and discussed. Experimental results show that the proposed algorithm can maintain a high classification accuracy when choosing proper interval length.

The rest of the paper is organized as follows: Section 2 describes the basic concepts in interval-valued information systems. In section 3, the related concepts of interval-valued reduction based on positive region are given, and some of the properties are proved. Then, a heuristic algorithm of the interval-valued decision table based on dependency is proposed. The experiments are conducted in section 4. Conclusion and future work are raised in section 5.

## 2   Preliminaries

This section describes the related concepts and properties in interval-valued information systems. Current research on interval-valued information systems are mostly based on the information systems without classification labels[18–20]. In this paper, considering the characteristics of the power data, we discuss the situation of conditional attributes with interval values and the decision attribute as the class labels.

**Definition 1.** *Suppose an interval-valued decision table $DT =< U, C \cup D, V, f >$. The attribute set $C \cup D$ is nonempty and finite, including conditional attributes $C = \{a_1, a_2, ..., a_h\}$ and decision attribute $D = \{d\}$. $V = V_C \cup V_D$, where $V_C$ is the set of all values of conditional attributes and $V_D$ represents the set of values of the decision attribute. $f : U \times C \to V_C$ is a mapping of interval values. $f : U \times D \to V_D$ is a mapping of single value.*

Table 1 is an interval-valued decision table[17], where the universe is $U = \{u_1, u_2, ..., u_{10}\}$, the conditional attributes $C = \{a_1, a_2, a_3, a_4, a_5\}$, and the decision attribute $D = \{d\}$. The conditional attribute value $f(a_k, u_i) = [l_i^k, u_i^k]$ is an interval value, eg. $f(a_2, u_3) = [7.03, 8.94]$. The decision attribute value $d(u_i)$ is a single value, eg. $d(u_3 = 2)$.

**Table 1.** An Interval-valued Decision Table

| $U$ | $a_1$ | $a_2$ | $a_3$ | $a_4$ | $a_5$ | $d$ |
|---|---|---|---|---|---|---|
| $u_1$ | [2.17,2.96] | [5.32,7.23] | [3.35,5.59] | [3.21,4.37] | [2.46, 3.59] | 1 |
| $u_2$ | [3.38,4.50] | [3.38,5.29] | [1.48,3.58] | [2.36,3.52] | [1.29,2.42] | 2 |
| $u_3$ | [2.09,2.89] | [7.03,8.94] | [3.47,5.69] | [3.31,4.46] | [3.48,4.61] | 2 |
| $u_4$ | [3.39,4.51] | [3.21,5.12] | [0.68,1.77] | [1.10,2.26] | [0.51,1.67] | 3 |
| $u_5$ | [3.70,4.82] | [2.98,4.89] | [1.12,3.21] | [2.07,3.23] | [0.97,2.10] | 2 |
| $u_6$ | [4.53,5.63] | [5.51,7.42] | [3.50,5.47] | [3.27,4.43] | [2.49,3.62] | 2 |
| $u_7$ | [2.03,2.84] | [5.72,7.65] | [3.68,5.91] | [3.47,4.61] | [2.53,3.71] | 1 |
| $u_8$ | [3.06,4.18] | [3.11,5.02] | [1.26,3.36] | [2.25,3.41] | [1.13,2.25] | 3 |
| $u_9$ | [3.38,4.50] | [3.27,5.18] | [1.30,3.40] | [4.21,5.36] | [1.11,2.23] | 1 |
| $u_{10}$ | [1.11,2.26] | [2.51,3.61] | [0.76,1.85] | [1.30,2.46] | [0.42,1.57] | 4 |

In crisp rough sets, equivalence relation is used to partition the universe. However, in interval-valued decision tables, it is hard to partition the universe reasonably by using the equivalence classes formed by the objects with same interval values.

**Definition 2.** *Suppose an interval-valued decision table $DT =< U, C \cup D, V, f >$. $a_k \in C$. $f(a_k, u_i) = [l_i^k, u_i^k]$, where $l_i^k \le u_i^k$. $l_i^k = u_i^k$ means object $u_i$ is a constant in attribute $a_i$. For any $u_i$ and any conditional attribute $a_k$, $l_i^k = u_i^k$, then the decision table is a traditional decision table. The similarity between $u_i$ and $u_j$ on attribute $a_k$ is defined:*

$$r_{ij}^k = \begin{cases} 0 & [l_i^k, u_i^k] \cap [l_j^k, u_j^k] = \not\subset \\ \frac{card([l_i^k, u_i^k] \cap [l_j^k, u_j^k])}{card(max\{u_i^k, u_j^k\} - min\{l_i^k, l_j^k\})} & [l_i^k, u_i^k] \cap [l_j^k, u_j^k] \ne \not\subset \end{cases}$$

$card()$ means the length of the interval. Obviously, $0 \le r_{ij}^k \le 1$. If $r_{ij}^k = 0$, the conditional attribute values of $f(a_k, u_i)$ and $f(a_k, u_j)$ are separate. If $0 < r_{ij}^k < 1$, the conditional attribute values of $f(a_k, u_i)$ and $f(a_k, u_j)$ are properly including. If $r_{ij}^k = 1$, it is indisernable between the conditional attribute value $f(a_k, u_i)$ and $f(a_k, u_j)$.

The similarities of conditional attributes describe the degrees of equivalence between the objects in interval-valued decision tables.

**Definition 3.** *Suppose an interval-valued decision table $DT =< U, C \cup D, V, f >$. Given a threshold $\lambda \in [0, 1]$ and any attribute subset $A \subseteq C$. We can define a binary relation $R_A^\lambda$ on $U$: $R_A^\lambda = \{(x_i, x_j) \in U \times U : r_{ij}^k > \lambda, \forall a_k \in A\}$ as the $\lambda$-tolerance relation on $A$.*

**Property 1.** Suppose an interval-valued decision table $DT =< U, C \cup D, V, f >$. Given a threshold $\lambda \in [0, 1]$ and any attribute subset $A \subseteq C$. $R_A^\lambda$ is reflexive and symmetric, but not transitive.

**Property 2.** Suppose an interval-valued decision table $DT =< U, C \cup D, V, f >$. Given a threshold $\lambda \in [0, 1]$ and any attribute subset $A \subseteq C$. $R_A^\lambda = \bigcap\limits_{a_k \in A} R_{a_k}^\lambda$ .

Denote $R_A^\lambda(u_i)$ as the $\lambda$-tolerance class of interval object $u_i$ on attribute set $A$. Take table 1 for an example, when $\lambda = 0.7$ and $A = a_1$, according to the definitions 2 and 3, $R_{\{a_1\}}^{0.7}(u_1) = \{u_1, u_3, u_7\}, R_{\{a_1\}}^{0.7}(u_2) = \{u_2, u_4, u_9\}, R_{\{a_1\}}^{0.7}(u_3) = \{u_1, u_3, u_7\}, R_{\{a_1\}}^{0.7}(u_4) = \{u_2, u_4, u_9\}, R_{\{a_1\}}^{0.7}(u_5) = \{u_5\}, R_{\{a_1\}}^{0.7}(u_6) = \{u_6\}, R_{\{a_1\}}^{0.7}(u_7) = \{u_1, u_3, u_7\}, R_{\{a_1\}}^{0.7}(u_8) = \{u_8\}, R_{\{a_1\}}^{0.7}(u_9) = \{u_2, u_4, u_9\}, R_{\{a_1\}}^{0.7}(u_{10}) = \{u_{10}\}$. On account of property 2, if $A$ is composed of multiple attributes, we can firstly calculate the $\lambda$-tolerance classes of interval-valued objects on each attribute, then compute the $\lambda$-tolerance classes of multiple attributes by intersection operation.

**Definition 4.** *Suppose an interval-valued decision table $DT =< U, C \cup D, V, f >$. Given a threshold $\lambda \in [0, 1]$ and any attribute subset $A \subseteq C$, $X \subseteq U$. The upper and lower approximate operators of $X$ on $A$ can be defined as:*

$$\overline{R}_A^\lambda(X) = \{u_i \in U, R_A^\lambda(u_i) \cap X \neq \not\subset\}, \underline{R}_A^\lambda(X) = \{u_i \in U, R_A^\lambda(u_i) \subseteq X\}$$

The above definitions and properties are not referred to the decision attribute.

## 3  Interval-Valued Heuristic Reduction Method Based on Dependency

[17] presented a reduction algorithm in interval-valued decision table based on discernibility function. However, the computational complexity of the algorithm is too high. In this section, from the algebra view, the concepts and properties of the heuristic reduction are given, and the corresponding algorithm is put forward.

### 3.1  Concepts and Properties

In the light of definition 4, we can define the lower and upper approximate operators of decision attribute on interval-valued conditional attributes as follows.

**Definition 5.** *Suppose an interval-valued decision table* $DT =< U, C \cup D, V, f >$. *Given a threshold* $\lambda \in [0,1]$. *A partition* $\{\psi_1, \psi_2, ..., \psi_l\}$ *of* $U$ *is derived by the decision classes of* $D$. *For any conditional attribute subset* $A \subseteq C$, *the upper and lower approximate operators of* $D$ *on* $A$ *can be defined as:*

$$\overline{R}_A^\lambda(D) = \bigcup_{i=1}^l \overline{R}_A^\lambda(\psi_i), \underline{R}_A^\lambda(D) = \bigcup_{i=1}^l \underline{R}_A^\lambda(\psi_i),$$

where $\overline{R}_A^\lambda(X) = \{u_i \in U, R_A^\lambda(u_i) \cap X \neq \varnothing\}, \underline{R}_A^\lambda(X) = \{u_i \in U, R_A^\lambda(u_i) \subseteq X\}$. $R_A^\lambda(u_i)$ represents the $\lambda$-tolerance class of interval object $u_i$ on attribute set $A$.

The lower approximation of decision attribute $D$, named positive region, can be denoted as $POS_A^\lambda(D)$. The size of positive region reflects the degree of separation of classification problem in a given attribute space. The bigger the positive region is, the less overlapping region of tolerance class becomes. To measure the significance of attributes, we define $\lambda$-dependency of decision attribute $D$ on interval-valued conditional attributes $A$ as

$$\gamma_A^\lambda(D) = \frac{|\underline{R}_A^\lambda(D)|}{|U|},$$

where $|\bullet|$ means the set base. $0 \leq \gamma_A^\lambda(D) \leq 1$ represents the ratio of the number of the objects whose tolerance classes completely contained in the decision classes to the number of universe, in accordance with $A$. Obviously, the bigger the positive region is, the stronger the dependency of decision attribute $D$ to conditional attributes $A$ will be.

**Property 3.** Given an interval-valued decision table $DT =< U, C \cup D, V, f >$ and a threshold $\lambda$, if $B \subseteq A \subseteq C$ and $u_i \in POS_B^\lambda(D)$, $u_i \in POS_A^\lambda(D)$.

**Proof.** Suppose $u_i \in \underline{R}_B^\lambda(D_j)$, where $D_j$ represents the objects with decision attribute value equivalent to $j$, namely, $R_B^\lambda(u_i) \subseteq D_j$. Since $B \subseteq A \subseteq C, R_A^\lambda(u_i) \subseteq R_B^\lambda(u_i)$. Therefore, $R_A^\lambda(u_i) \subseteq R_B^\lambda(u_i) \subseteq D_j$. Then $u_i \in POS_A^\lambda(D)$.

**Property 4.** $\gamma_A^\lambda(D)$ is monotonous. If $A_1 \subseteq A_2 \subseteq ... \subseteq C$, $\gamma_{A_1}^\lambda(D) \leq \gamma_{A_2}^\lambda(D) \leq ... \leq \gamma_C^\lambda(D)$.

**Proof.** From property 3, we know $\forall u_i \in POS_{A_1}^\lambda(D)$. Then we have $u_i \in POS_{A_2}^\lambda(D), ..., u_i \in POS_C^\lambda(D)$. There may exist $u_j \notin POS_{A_1}^\lambda(D)$, but $u_j \in POS_{A_2}^\lambda(D), ..., u_j \in POS_C^\lambda(D)$. Then we have $|POS_{A_1}^\lambda(D)| \leq |POS_{A_2}^\lambda(D)| \leq ... \leq |POS_C^\lambda(D)|$. According to the definition $\gamma_A^\lambda(D) = \frac{|POS_A^\lambda(D)|}{|U|}$, then we have $\gamma_{A_1}^\lambda(D) \leq \gamma_{A_2}^\lambda(D) \leq ... \leq \gamma_C^\lambda(D)$.

**Definition 6.** *Suppose an interval-valued decision table* $DT =< U, C \cup D, V, f >$, $\lambda \in [0,1]$, $A \subseteq C$. *For* $\forall a_k \in A$, *if* $\gamma_{A-\{a_k\}}^\lambda(D) < \gamma_A^\lambda(D)$, *we say attribute* $a_k$ *is necessary to* $A$. *Otherwise, if* $\gamma_{A-\{a_k\}}^\lambda(D) = \gamma_A^\lambda(D)$, *we say attribute* $a_k$ *is redundant to* $A$. *If* $\forall a_k \in A$ *is necessary, the attribute set is independent.*

If $\gamma_{A-\{a_k\}}^\lambda(D) = \gamma_A^\lambda(D)$, it indicates that the positive regions of decision tables will not be changed when $a_k$ is reduced. That is the discernabilities of each classes keeping unchanged, namely, $a_k$ does not bring any contribution to classification. So $a_k$ is redundant. Conversely, if $a_k$ is reduced, the positive region gets smaller. It shows the discernabilities of each classes worse. At this point, $a_k$ can not be reduced.

**Definition 7.** *Suppose an interval-valued decision table* $DT =< U, C \cup D, V, f >$, $\lambda \in [0,1]$, $A \subseteq C$. *We say* $A$ *is a* $\lambda$-*reduct of* $C$, *if* $A$ *satisfies:*
$(1) \gamma_A^\lambda(D) = \gamma_C^\lambda(D)$;
$(2) \forall a_k \in A, \gamma_{A-\{a_k\}}^\lambda(D) < \gamma_A^\lambda(D)$.

In this definition, condition (1) states that $\lambda$-reduct can not lower the discernabilities of decision tables. That is $\lambda$-reduct has the same discernability with all conditional attributes. Condition (2) accounts for no redundant attribute existed in the reduct. Each attribute should be necessary. The definition is consistent with crisp rough sets in formal. However, the model defines $\lambda$-reduct in the interval-valued space, while crisp rough set theory is defined in the discrete space.

**Definition 8.** *Suppose an interval-valued decision table* $DT =< U, C \cup D, V, f >$, $\lambda \in [0,1]$. $A_1, A_2, ..., A_s$ *denote all the* $\lambda$-*reducts in the decision table. Define* $Core = \bigcap_{i=1}^{s} A_i$ *as the core of the decision table.*

**Definition 9.** *Given an interval-valued decision table* $DT =< U, C \cup D, V, f >$ *and* $\epsilon(\epsilon \geq 0)$. *If* $|\gamma_C^\lambda(D) - \gamma_A^\lambda(D)| \leq \epsilon(A \subseteq C)$, *and* $|\gamma_C^\lambda(D) - \gamma_B^\lambda(D)| > \epsilon(\forall B \subset A)$, *we call* $A$ *an* $\epsilon$-*approximate reduction of* $DT$.

### 3.2 $\lambda$-reduction in the Interval-Valued Decision Table Based on Dependency

If we want to find all the $\lambda$-reducts of an interval-valued decision table, we need to compute $2^h - 1$ attribute subsets and determine whether they meet the conditions of $\lambda$-reduct, where $h$ is the number of conditional attributes. The calculation is not tolerable. In this paper, based on the notion of dependency, a heuristic reduction algorithm is proposed. The algorithm complexity is greatly reduced. Dependency describes the contribution to classification of conditional attributes, therefore, it can be used as the evaluation of attribute importance.

**Definition 10.** *Suppose an interval-valued decision table* $DT =< U, C \cup D, V, f >$, $\lambda \in [0,1]$, $A \subseteq C$. $\forall a_k \in C - A$, *the significance of* $a_k$ *to* $C$ *is defined:*

$$SIG(a_k, A, D) = \gamma_{A \cup \{a_k\}}^\lambda(D) - \gamma_A^\lambda(D)$$

With the definition of significance, we can construct the greedy algorithm for interval-valued reduction. The algorithm sets the empty set as the starting point. Then calculate the attribute importance of all the remaining attributes and select the attribute with highest importance, and add to the set of $\lambda$-reduct, until all the remaining attribute importance equals to 0, namely, adding any new attribute, the dependence degree is no longer changing. Forward search algorithm can guarantee the important attribute added to the $\lambda$-reduct, so there is no loss of important features. Backward search algorithm is difficult to guarantee the result, because for a large number of redundant attributes in the interval-valued decision table, even if the important attribute is removed, it will not lower the the ability to discernablility of the decision table. Therefore, it may ultimately retain a large number of attributes with weak distinguished ability, but as a whole is still able to maintain the ability to discernablility of the original data, rather than a few features with strong distinguishing ability. The reduction algorithm for an interval-valued decision table based on the dependency is described in algorithm 1.

**Algorithm 1.** $\lambda$-Reduction for the Interval-valued Decision Table Based on Dependency (RIvD)

   **Input:** $DT = <U, C \cup D, V, f>, \lambda$
   **Output:** $\lambda$-reduct $red$
   **Step1.** Let $red = \emptyset$;
   **Step2.** For all $a \in C$, compute the $\lambda$-tolerance classes $R_{\{a\}}^{\lambda}$ on attribute $a$;
   **Step3.** For any $a_k \in C - red$, compute $SIG(a_k, red, D) = \gamma_{red \cup \{a_k\}}^{\lambda}(D) - \gamma_{red}^{\lambda}(D)$;//Define $\gamma_{\emptyset}^{\lambda}(D) = 0$
   **Step4.** Select $a_i$, which satisfies $SIG(a_i, red, D) = \max_k(SIG(a_k, red, D))$;
   **Step5.** If $SIG(a_i, red, D) > 0, red = red \cup \{a_i\}$, goto Step3; Otherwise, return $red$, end.

Suppose the number of conditional attributes $C$ is $h$, and the number of interval objects is $n$, the time complexity of the algorithm is $O(n^2 + hn)$.

In order to solve the problems in real life, we cannot use too strict reduction conditions, so the reduction condition $SIG(a_i, red, D) > 0$ can be improved by $0 < SIG(a_i, red, D) < \varepsilon$. $\epsilon$ needs to be set in advance on the basis of the specific data. The improvement will be more close to reality and more practical to a certain extent. The value of $\epsilon$ will directly affect the classification results, as well as the application of the algorithm. If the value of $\epsilon$ is too small, the conditional attributes will be selected too many, so as to influence the practicability of the algorithm; If the value of $\epsilon$ is too large, it will lead to the selected conditional attributes too few and influence the algorithm accuracy.

## 4   Experiments

In power plants, production data have strong regularity by time. Decision making can be provided for the operations, maintenances and accident treatments of power plants by analyzing historical data through power stations, and finding the operation rules for power stations. Existing data mining techniques made lots of attempts in power production data, and obtained some results. However, the existing data mining methods did not pay much attention to the characteristics of operating data. Without a judgement of a steady state, the data was directly mined. Due to lack of the clear working condition, the result of data mining is not so good as the actual operation data. In this section, according to the production characteristics of power plants, we classify the production data to steady state or unsteady state and build the classification model. The accuracies of the classification results and running time for building a classification model are used to evaluate the effectiveness of the proposed algorithm.

## 4.1   Experimental Data

In this section, a 600MW unit in some power plant is used for the experiments. The data acquisition frequency of the plant is each 1 minute, namely, each minute produces one record of data. The data of first half 2012 year are used as the experimental object. Removing the maintenance downtime of the unit, a total of 107184 records are produced. The data have 427 attributes. Removing the keyword ID and data retention time by the system automatically generating, a total of 425 attributes (all numeric) are produced. We use the formula of steady state judgement to classify the operating data, forming a decision attribute. Then we get a large decision table.

To evaluate the performance of the algorithm, we design a variety of data interval division, eg. each 10 minutes, 20 minutes,..., 90 minutes for an interval. In the process of interval division, if an interval corresponds to different decision classes, the data with the same decision class constitute to a small interval, and the next interval starts from another different decision class.

## 4.2   Experimental Environment

All experiments are run on the workstation of Intel Xeon(R) Processor (Four Cores, 2.5GHz, 16GRAM), JAVA programming. In order to guarantee the fairness of the experimental comparison, we use ten fold cross validation to estimate the accuracy of classification.

## 4.3   Evaluation Criteria

Since the value of $\epsilon$ affects only the length of the selected subset of attributes (i.e. the number of the selected attributes), it does not affect the sequence of selecting the attribute according to the importance, so this paper does not discuss the value of $\epsilon$. To construct the classification models on power data, in addition to considering the running time of the algorithm, we also should consider the average classification accuracy. This paper mainly studies the interval-valued heuristic reduction, when evaluating the accuracy of the algorithm, the test data and training data are divided into blocks in accordance with the same time. Then for each attribute, record the maximum and minimum values, and get the data blocks by using interval values. Selecting the data block with highest similarity in training data, let it be the decision class label of the test data block. Compared with the real decision class of the test data, calculate the percentage of correctly classifying. The whole process is still using ten fold cross validation to calculate the average accuracy rate of classification.

## 4.4    The Selection and Set of Parameters

The algorithm 1 is performed on the data set. Record the running time of different interval length and different number of selected attributes. Figure 1 indicates the relation between the running time and the number of selected attributes when $\lambda = 0.7$ and the interval length by 10 minutes, 20 minutes,...,90 minutes. Figure 2(a) shows the running time when $\lambda = 0.5, 0.7, 0.9$ and the number of selected attributes equals to 3. Figure 2(b) expresses the running time changing along with the interval length when assigning different $\lambda$ and the number of selected attributes=6.

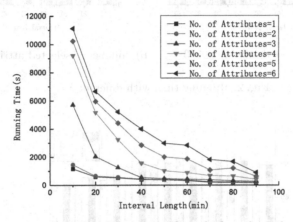

**Fig. 1.** The running time with different number of selected attributes when $\lambda = 0.7$

From figure 1, with the increase of the interval length, data objects reduce greatly, so is the running time. But with the increase of the interval length, the coincidence degree of intervals is also increasing. It's easy to result the number of attributes in $\lambda$-tolerance classes increasing. Adding an attribute, the intersection operation for tolerance classes is increased, so the running time of the algorithm is not linearly changing. At some point, especially when the interval length is longer, with the increase of interval length, the running time does not decrease as the objects decrease. This may be because although the interval length increases, the number of attributes in $\lambda$-tolerance classes is increasing, the running time is then increased.

Figure 3 reveals the average classification accuracy of different attribute subsets selected when $\epsilon = 0.01$.

From figure 3, When the interval length is more than 1 hours, while the running time is reduced, the average classification accuracies drop significantly. This is mainly because if the data interval length is too large, the data interval values cannot express the characteristics of data blocks; Similarly, if the data interval length is too small, not only makes the running time longer, but also the information of data blocks is insufficient. It can also lead to low classification

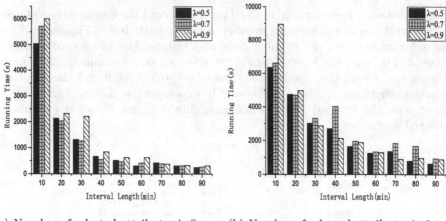

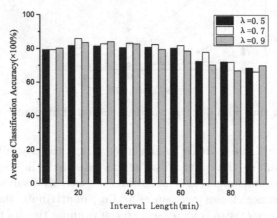

(a) Number of selected attributes is 3    (b) Number of selected attributes is 6

**Fig. 2.** Running time with different $\lambda s$

**Fig. 3.** The average classfication accuracy with different $\lambda s$ when $\epsilon = 0.01$

accuracy. Therefore, for the interval-valued reduction algorithm, selection of the interval length has a great effect on the algorithm results. We should set the length of intervals based on different applications.

Therefore, the proposed interval-valued reduction algorithm is suitable to deal with continuous distribution data, and cannot handle greatly changed data type.

## 5    Conclusions and Future work

Based on the characteristics of classification problems in power data, an interval-valued reduction algorithm was proposed by using dependency. We experimented it on the power data with the application in the judging stable state, and achieved good results. Since the applications of power data mostly need to consider a data segment changing rather than a record of data, by dividing the data set into intervals can not only significantly reduce the data objects, reduce the difficulty of data analysis, but also meet the specific application of power data. Meanwhile, the attribute reduction reduces the dimension of the data set, but does not affect the classification ability of the entire data set. The data can be greatly reduced, thereby, the difficulty in data analysis is reduced. From the experimental results, the proposed algorithm is effective, and provides a new way to interval-valued reduction method.

In future, we will focus on the following several aspects: the parameters will be discussed in detail. We will give more reasonable and effective selection method through more experiments; We will further apply the method into load prediction and fault diagnosis in the power data, and validate the effectiveness of the algorithm in the round.

# References

1. Pawlak, Z.: Rough sets. International Journal of Compute and Information Science 11(4), 341–356 (1982)
2. Wang, G.Y., Yao, Y.Y., Yu, H.: A survey on rough set theory and applications. Chinese Journal of Computers 32(7), 1229–1246 (2009) (in Chinese with English abstract)
3. Mac Parthalain, N., Jensen, R., Shen, Q.: Rough and fuzzy-rough methods for mammographic data analysis. Intelligent Data Analysis-An International Journal 14(2), 225–244 (2010)
4. Zhu, W.: Generalized rough sets based on relations. Information Sciences 177(22), 4997–5011 (2007)
5. Zhang, W.X., Wu, W.Z., Liang, J.Y., et al.: Rough set theory and method. Science Press, Beijing (2001)
6. Mi, J.S., Wu, W.Z., Zhang, W.X.: Constructive and Axiomatic Approaches of Theory of Rough Sets. PRAI 15(3), 280–284 (2002) (in Chinese with English abstract)
7. Zhu, W.: Topological approaches to covering rough sets. Information Sciences 177(6), 1499–1508 (2007)
8. Zhang, W.X., Yao, Y.Y., Liang, Y.: Rough Set and Concept Lattice. Xi'an Jiaotong University Press, Xi'an (2006)
9. Qian, Y., Liang, J., Yao, Y., et al.: MGRS: A multi-granulation rough set. Information Sciences 180(6), 949–970 (2010)
10. Suyun, Z., Tsang, E., Degang, C.: The Model of Fuzzy Variable Precision Rough Sets. IEEE Transactions on Fuzzy Systems 17(2), 451–467 (2009)
11. Huang, B., et al.: Dominance relation-based fuzzy-rough model and its application to audit risk evaluation. Control and Decision 24(6), 899–902 (2009) (in Chinese with English abstract)
12. Zhang, D.B., Wang, Y.N., Huang, H.X.: Rough neural network modeling based on fuzzy rough model and its application to texture classification. Neurocomputing 72(10-12), 2433–2443 (2009)
13. Xu, F.F., Miao, D.Q., Wei, L.: Fuzzy-rough attribute reduction via mutual information with an application to cancer classification. Computers Mathematics with Applications 57(6), 1010–1017 (2009)

14. Hu, Q.H., Zhao, H., Yu, D.R.: Efficient Symbolic and Numerical Attribute Reduction with Rough Sets. PRAI 6, 732–738 (2008) (in Chinese with English abstract)
15. Liang, J.Y., Qian, Y.H., Pedrycz, W., et al.: An efficient accelerator for attribute reduction from incomplete data in rough set framework. Pattern Recognition 44(8), 1658–1670 (2011)
16. Wang, W.H., Zhou, D.H.: An Algorithm for Knowledge Reduction in Rough Sets Based on Genetic Algorithm. Journal of System Simulation 13, 91–94 (2001) (in Chinese with English abstract)
17. Zhang, N., Miao, D.Q., Yue, X.D.: Approaches to Knowledge Reduction in Interval-Valued Information Systems. Journal of Computer Research and Development 47(8), 1362–1371 (2010) (in Chinese with English abstract)
18. Chen, Z.C., Qin, K.Y.: Attribute Reduction of Interval-valued Information System Based on the Maximal Tolerance Class. Fuzzy Systems and Mathematics 23(6), 126–132 (2009) (in Chinese with English abstract)
19. Guo, Q., Liu, W.J., Jiao, X.F., Wu, L.: A Novel Interval-valued Attribution Reduction Algorithm Based on Fuzzy Cluster. Fuzzy Systems and Mathmatics 27(1), 149–153 (2013) (in Chinese with English abstract)
20. Gong, W.L., Li, D.Y., Wang, S.G., Cheng, L.T.: Attribute Reduction of Interval-Valued Information System Based on Fuzzy Discernibility Matrix. Journal of Shanxi University (Natural Science Edition) 34(3), 381–387 (2011) (in Chinese with English abstract)

# Global Best Artificial Bee Colony for Minimal Test Cost Attribute Reduction

Anjing Fan, Hong Zhao*, and William Zhu

Lab of Granular Computing,
Minnan Normal University, Zhangzhou 363000, China
hongzhaocn@163.com

**Abstract.** The minimal test cost attribute reduction is an important component in data mining applications, and plays a key role in cost-sensitive learning. Recently, several algorithms are proposed to address this problem, and can get acceptable results in most cases. However, the effectiveness of the algorithms for large datasets are often unacceptable. In this paper, we propose a global best artificial bee colony algorithm with an improved solution search equation for minimizing the test cost of attribute reduction. The solution search equation introduces a parameter associated with the current global optimal solution to enhance the local search ability. We apply our algorithm to four UCI datasets. The result reveals that the improvement of our algorithm tends to be obvious on most datasets tested. Specifically, the algorithm is effective on large dataset Mushroom. In addition, compared to the information gain-based reduction algorithm and the ant colony optimization algorithm, the results demonstrate that our algorithm has more effectiveness, and is thus more practical.

**Keywords:** Cost-sensitive learning, Minimal test cost, Attribute reduction, Granular computing, Biologically-inspired algorithm.

## 1 Introduction

Cost-sensitive learning is one of the most active and important research areas in machine learning and data mining. In conventional data mining, attribute reduct tries to maximize the accuracy or minimize the error rate in general. In real-world applications, one should pay cost for obtaining a data item of an attribute. It is important to take the test cost account into attribute reduct [1,2]. The minimal test cost attribute reduction [9] is an important problem in cost-sensitive learning. This problem is not a simple extension of existing attribute reduction problems, it is a mandatory stage in dealing with the test cost issue. The problem is a task to select an attribute subset with minimal test cost. The performance of the minimal test cost attribute reduction is the test cost, which is independent of the performance of attribute reduction.

In the recent years, some algorithms are proposed to deal with the minimal test cost attribute reduction problem, such as ant colony optimization algorithm (ACO) [3] and information gain-based λ-weighted reduction (λ-weighted) algorithm [4]. However, the

---

* Corresponding author.

D. Miao et al. (Eds.): RSKT 2014, LNAI 8818, pp. 101–110, 2014.
DOI: 10.1007/978-3-319-11740-9_10 © Springer International Publishing Switzerland 2014

effectiveness of these algorithms for large datasets is often needed to improve. To deal with this problem, artificial bee colony (ABC) algorithm be considered. The ABC algorithm is a biologically-inspired optimization algorithm, it is able to produce high quality solutions with fast convergence. Due to its simplicity and easy implementation, the ABC algorithm has captured much attention and has been applied to solve many practical optimization problems [14].

In this paper, we propose a global best artificial bee colony (GABC) algorithm for the minimal test cost attribute reduction problem. The GABC algorithm is inspired by the ABC algorithm. The ABC algorithm [5] is proposed to optimize continuous functions. Although it has fewer control parameters, it shows competitive performance compared with other population-based algorithms. However the algorithm cannot be effective using the individual information to optimize search method, so the traditional artificial bee colony algorithm is good at exploration but poor at exploitation. As we know, the exploitation is determined by the solution search equation. In this paper, the GABC algorithm improves the solution to balance the exploration and exploitation ability of the ABC algorithm. The GABC algorithm induces a parameter Lb into the improved solution search equation. The parameter Lb value is mainly composed of the fitness of the global optimal solution.

We evaluate the performance of our algorithm on four UCI (University of California Irvine) datasets [6,7], which serve the machine learning community. Since there is no cost settings for attribute on the four datasets, we use Normal distribution to generate test cost for datasets. The viability and effectiveness of the GABC algorithm are tested on four datasets. The results demonstrate the good performance of the GABC algorithm in solving the minimal test cost attribute reduction problem when compared with the $\lambda$-weighted algorithm, ACO algorithm and ABC algorithms. Experiments are undertaken by an open source software called Coser (cost-sensitive rough sets) [8].

The rest of the paper is organized as follows. Section 2 presents attribute reduction in cost-sensitive learning and discusses the problem of the minimal test cost attribute reduction. Section 3 analyzes the parameters of the GABC algorithm for getting optimal. Section 4 presents the experimental results and the comparison results. Finally, conclusions and recommendations for future studies are drawn in Section 5.

## 2   Preliminaries

In this section, we present some basic notions for the minimal test cost attribute reduction problem. The one conveyed is the test cost independent decision system, the other we proposed is the minimal test cost attribute reduction problem.

### 2.1   Test Cost Independent Decision System

In this paper, datasets are fundamental for the minimal test cost attribute reduction problem. We consider the datasets with a test cost independent decision system [4]. A test-cost-sensitive decision system is defined as follows.

**Definition 1.** *[4] A test cost independent decision system (TCI-DS) $S$ is the 6-tuple:*

$$S = (U, C, d, \{V_a | a \in C \cup \{d\}\}, \{I_a | a \in C \cup \{d\}\}, tc), \tag{1}$$

where $U$ is a finite set of objects called the universe, $C$ is the set of attributes, $d$ is the decision class, $V_a$ is the set of values for each $a \in C \cup \{d\}$, $I_a : U \rightarrow V_a$ is an information function for each $a \in C \cup \{d\}$, and $tc : C \rightarrow R^+ \cup \{0\}$ is the test cost function for each $a \in C$.

Here test costs are independent of one another. A test cost function can be represented by a vector $tc = [tc(a_1), tc(a_2), ..., tc(a_{|C|})]$. It is easy to calculate the test cost for an attribute subset $B$ (any $B \subseteq C$), which is counted as follows: $tc(B) = \sum_{a \in B} tc(a)$.

**Table 1.** A clinical decision system

| Patient | Headache | Temperature | Lymphocyte | Leukocyte | Eosinophil | Heartbeat | Flu |
|---------|----------|-------------|------------|-----------|------------|-----------|-----|
| $x_1$ | yes | high | high | high | high | normal | yes |
| $x_2$ | yes | high | normal | high | high | abnormal | yes |
| $x_3$ | yes | high | high | high | normal | abnormal | no |
| $x_4$ | no | high | normal | normal | high | normal | no |

An exemplary decision system is given by Table 1. The attributes of this decision system are symbolic. Here $C = \{$Headache, Temperature, Lymphocyte, Leukocyte, Eosinophil, Heartbeat $\}$, $\{d\} = \{$Flu$\}$, $U = \{x_1, x_2, x_3, x_4\}$, and the corresponding test cost of attributes is represented by a vector $tc = [12, 5, 15, 20, 15, 10]$.

## 2.2 The Minimal Test Cost Attribute Reduction Problem

Attribute reduction plays an important role in rough sets [11]. We review the reduction based on positive region [12].

**Definition 2.** *[13] Any $B \subseteq C$ is called a decision relative reduction (or a reduction for brevity) of S if and only if:*
1.$POS_B(\{d\}) = POS_C(\{d\})$;
2.$\forall a \in B, POS_{B-\{a\}}(\{d\}) \neq POS_C(\{d\})$.

In applications, a number of reductions sometimes are needed. However, in most applications, only one reduction is needed. Since there may exist many reductions, an optimization metric is needed. In this paper, the test cost is taken into account in attribute reduction problem. Naturally, the test cost of the attribute reduction is employed as a metric in our work. In other words, we are interested in the attribute reduction with minimal test cost. We define reductions of this type as follows.

**Definition 3.** *[13] Let $S$ be a TCI-DS and Red $(S)$ be the set of all reductions of S. Any $R \in Red(S)$ where $tc(R) = min\{tc(R^{'}) \mid R \in Red(S)\}$ is called a minimal test cost attribute reduction.*

As indicated in Definition 3, the set of all minimal test cost attribute reductions is denoted by $MTR(S)$. The optimal objective of our paper is $MTR$ problem.

# 3   Algorithm

This section introduces the global best artificial bee colony (GABC) algorithm in detail. Similar to the artificial bee colony (ABC) [14] algorithm, our algorithm consists food sources and three groups of bees: employed bees, onlookers and scouts. The ABC algorithm [15] is composed of two main steps: recruit an optimal good source and abandon a bad source. The process of artificial bees seeking good food sources equal the process of finding the minimal test cost attribute reduction.

In GABC algorithm, let one employed bee is on one food source and the number of employed bees or onlookers equal the number of food sources. The position of a food source represents an attribute subset and it is exploited by one employed bee or one onlookers. The number of food sources is set to 1.5 times number of attributes. Employed bees search new foods and remember the food source in their memory, and then pass the food information to onlookers. The onlookers tend to select good food sources from those foods founded by the employed bees, then further search the foods around the selected food source. The scouts are translated from a few employed bees, which abandon their food sources and search new ones.

As well known that both exploration and exploitation are necessary for the ABC algorithm. In the algorithm, the exploration refers to the ability to investigate the various unknown regions in the solution space to discover the global optimum. While the exploitation refers to the ability to apply the knowledge of the previous good solutions to find better solutions. In practice, to achieve good optimization performance, the two abilities should be well balanced. As we know, a new candidate solution is given by the following solution search equation in the artificial bee colony algorithm:

$$v_{ij} = x_{ij} + \phi_{ij}(x_{ij} - x_{kj}). \tag{2}$$

In Equation (2), we can know that the coefficient $\phi_{ij}$ is an uniform random number in [0, 1] and $x_{kj}$ is a random individual in the population, therefore, the solution search dominated by Equation (2) is random enough for exploration. However, alternatively the new candidate solution is generated by moving the old solution towards another solution selected randomly from the population. That is to say, the probability that the randomly selected solution is a good solution is the same as that the randomly selected solution is a bad one, so the new candidate solution is not promising to be a solution better than the previous one. To sum up, the solution search equation described by Equation (2) is good at exploration but poor at exploitation.

By taking advantage of the information of the global best solution to guide the search of candidate solutions, we rebuild the ABC algorithm to improve the exploitation. The GABC algorithm as follows.

**Step 1.** Create an initial food source position, and calculate the fitness value of the food source.

Food sources initialization is a crucial task in the ABC algorithm because it can affect the convergence speed and the quality of finding optimal solution. We replace the random select attribute subset with an attribute subset containing core attribute and satisfying the position region constraint [16].

The fitness value of the food source is defined as the reciprocal of the corresponding test cost. The fitness equation as follows:

$$fitness = \frac{1}{1 + tc},\tag{3}$$

where $tc$ is the test cost of an attribute subset selected.

After initialization, the GABC algorithm enters a loop of operations: updating feasible solutions by employed bees, selecting feasible solutions by onlooker bees, and avoiding suboptimal solutions by scout bees.

**Step 2.** Produce new solution $v_{ij}$ for the employed bees by Equation (5) and evaluate it by Equation (3).

The best solution in the current population is a very useful source which can be used to improve the convergence speed. We introduce a parameter Lb that associates with the current global optimal solution. The equation of Lb is conveyed as follows:

$$Lb = fitness_i / global fitness,\tag{4}$$

where fitness $_i$ is the $i$-th iteration fitness of food source, and global fitness stand for the fitness of current global optimal food source. As can be seen from Equation (4), Lb is a positive real number, typically less than 1.0.

Through the analysis, we propose a new solution search equation as follows:

$$v_{ij} = x_{ij} + \phi_{ij}(x_{ij} - x_{kj}) + Lb(g_i - x_{ij}),\tag{5}$$

where $k \in \{1, 2, 3, ..., SN\}$ and $j \in \{1, 2, 3, ..., D\}$ are randomly chosen indexes. $k$ is different from $i$. $SN$ is the number of the attribute, $D$ is the number of the food source. $\phi_{ij}$ is a random value in $[0, 1]$. $v_{ij}$ and $x_{kj}$ is a new feasible solution that is modified from its previous solution $x_{ij}$, $g_i$ is the best solution that explored in the history used to direct the movement of the current population.

When Lb takes 0, Equation (5) is identical to Equation (2). We can get a new solution better than the old one, then turn the new solution to be an old one in the next iteration. Apply the greedy selection process for the employed bees.

**Step 3.** Calculate the probability values $P_i$ for the solution $v_{ij}$ by Equation (6).

Produce the new solution $u_{ij}$ for onlooker bee by Equation (5), and evaluate it by Equation (3). Where $u_{ij}$ is produced from the solutions $v_{ij}$ depending on $P_i$. An onlooker bee chooses a food source depending on the probability values $P_i$ associated with that food source.

The equation of calculating the probability values $P_i$ is shown as follows:

$$P_i = \frac{fitness_i}{\sum\limits_{n=1}^{SN} fitness_n},\tag{6}$$

where $fitness_i$ is the fitness value of the $i$-th solution, $SN$ is the food number.

Apply the greedy selection process for the onlookers.

**Step 4.** When a food source can not improve further through limit cycles, the food source is abandoned for a scout bee. The food source is replaced with a new randomly solution produced by Equation (5).

The limit is an important control parameter of the GABC algorithm for abandonment. This step avoid the algorithm falling into suboptimal solutions. The Steps 2, 3 and 4 are repeated until the running generation reaches the maximal number of iteration.

The GABC algorithm deletes redundant attributes of each food source in inverted order with the positive region constraint. Through the above steps, an attribute reduction with minimal test cost has been produced, it is the final solution.

## 4  Experiments

To test the performance of the GABC algorithm, an extensive experimental evaluation and comparison with the ABC, the $\lambda$-weighted [4] and the ACO [3] algorithms are provided based on four datasets as follows. The four datasets are shown in Table 2. The finding optimal factor (FOF) [4] is used as comparison criteria in this paper.

### 4.1  Data Settings

In our experiments, there are four UCI datasets used to test. These are Zoo, Voting, Tic-tac-toe and Mushroom. The information of the four datasets is summarized in Table 2. On the four datasets, attributes are no test cost settings, so we apply Normal distribution to generate random test cost in [1, 10].

**Table 2.** Database information

| Name | Domain | $|U|$ | $|C|$ | $D = \{d\}$ |
|---|---|---|---|---|
| Zoo | Zoology | 101 | 16 | Type |
| Voting | Society | 435 | 16 | Vote |
| Tic-tac-toe | Game | 958 | 9 | Class |
| Mushroom | Botany | 8124 | 22 | Class |

### 4.2  Experiment Results

In experiment, each algorithm is undertaken with 100 different test cost settings on four datasets. The experiments reveal the performance of the GABC algorithm through analyzing parameters: limit, iteration and Lb. Nextly, we investigate the impact of the three parameters on the GABC algorithm.

Figure 1 presents solutions along iterations and limits for the four datasets. It can be observed that the evolution curves of the GABC algorithm reach higher FOF much faster. Thus, it can be concluded that overall the GABC algorithm outperforms well. It can be found from Figure 1, the FOF of a large limit(i.e., 60 or 80) is superior to the FOF of a small limit (i.e., 20 or 30) on most datasets, this rule also applies to the iteration.

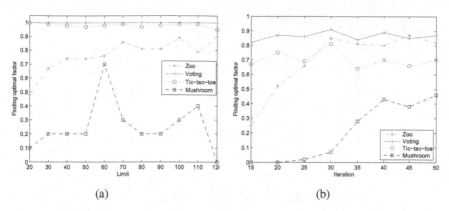

**Fig. 1.** Finding optimal factor on four datasets: (a) Limit, (b) Iteration

In Figure 1, we investigate the impact of parameter of limit and iterations on the GABC algorithm.

1) When the maximal iteration is set to 300, we let the parameter Lb be 0.8. As can be seen, we obtain better value of limit on the Mushroom dataset when limit is 60. For the other three test datasets, better results are obtained when limit is 110.

2) When limit is set to 110 and the parameter Lb is kept in 0.8. We can obtain that better value of iteration on the Mushroom dataset is 40. For the other three test datasets, better results are obtained when iteration is 30. The performance on parameter iteration is likely sensitive to the number of attributes.

3) As can be seen, when the value of limit is increased, the FOF is also improved. This trend also applies to the parameter of iteration. Figure 1(b) shows the parameter of iteration is needed to converge towards the optimal solution for the GABC algorithm, which same to the parameter of limit. We observe that the values of limit and iteration can greatly influence the experimental results.

In order to reveal the impact of control parameter Lb, we conduct experiments for our algorithm, where Lb in [0, 1] with 0.2 stepsize and use the competition approach [4] to improve the results. In Figure 2, when the value of Lb is set to 1, we can obtain a good result on Mushroom dataset. For the other three test datasets, better results are obtained when Lb is around 0.8. As can be seen, when the values of Lb are increased, the values of FOF are also improved. Therefore, the selective Lb is set at 0.8 for all the datasets tested. We can observe that the values of Lb also have effect on the results.

This can be explained by the basic principle of the ABC algorithm. The parameter Lb in Equation (5) plays an important role in balancing the exploration and exploitation of the candidate solution search. When Lb increases from zero to a certain value, the exploitation of Equation (5) will also increase correspondingly.

### 4.3   Comparison Results

In the following, we illustrate the advantage of the GABC algorithm compared with the $\lambda$-weighted algorithm, the ACO algorithm and the ABC algorithm. The limit of the GABC and ABC algorithms is set to 100, and the iteration is 40.

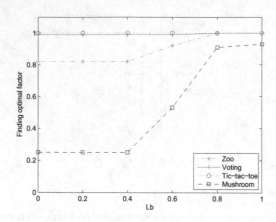

**Fig. 2.** Finding optimal factor for Lb value on four datasets

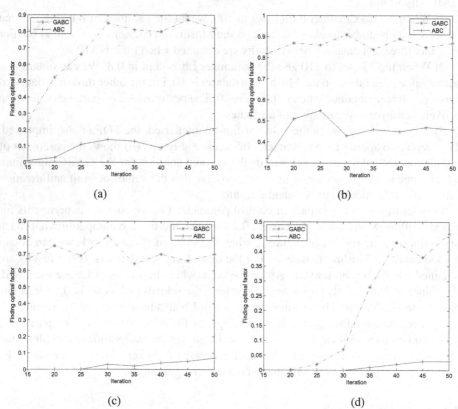

**Fig. 3.** Finding optimal factor for iteration on four datasets: (a) Zoo, (b) Voting, (c) Tic-tac-toe, (d) Mushroom

**Table 3.** Finding optimal factor of three algorithms with the competition approach

| Datasets | $\lambda$-weighted | ACO | GABC |
|----------|--------------------|-----|------|
| Zoo | 0.833 | 0.987 | **1.000** |
| Voting | **1.000** | **1.000** | **1.000** |
| Tic-tac-toe | 0.408 | **1.000** | **1.000** |
| Mushroom | 0.176 | 0.958 | **0.970** |

Figure 3 presents the FOF for iteration on the four different datasets. Table 3 draws the FOF for three algorithms on the four different datasets by competition approach. The best results are marked in bold in table. The results in Table 3 and Figure 3 further demonstrate that the GABC algorithm is a great algorithm since it generates significantly better results than the $\lambda$-weighted algorithm, ACO algorithm and ABC algorithm for datasets.

The results show that the FOF of the other two algorithms produced are acceptable results on most datasets. However, the performances of the two algorithms are shortly on Mushroom dataset. The results of the $\lambda$-weighted algorithm are especially obvious. For example, the FOF is only 17.6% of the $\lambda$-weighted algorithm on the Mushroom dataset. However, it is 97% of our algorithm on the Mushroom dataset.

In summary, the GABC algorithm can produce an optimal reduction in general. The algorithm has the highest performance among the three algorithms for all four datasets.

## 5  Conclusions

In this paper, we have developed the global based artificial bee colony algorithm to cope with the minimal test cost attribute reduction problem. The algorithm has been improved by introducing the parameter Lb based on global optimal. We have demonstrated the effectiveness of the GABC algorithm and provided comparisons with two other algorithms. The results have shown that the GABC algorithm possesses superior performance in finding optimal solution as compared to the other algorithms. In the future, we will improve the stability and efficiency of the global based artificial bee colony algorithm.

**Acknowledgments.** This work is in part supported by the Zhangzhou Municipal Natural Science Foundation under Grant No. ZZ2013J03, the Key Project of Education Department of Fujian Province under Grant No. JA13192, the National Science Foundation of China under Grant Nos. 61379049 and 61379089, and the Postgraduate Research Innovation Project of Minnan Normal University under Grant No. YJS201438.

## References

1. Fumera, G., Roli, F.: Cost-sensitive learning in support vector machines. In: Proceedings of VIII Convegno Associazione Italiana per L' Intelligenza Artificiale (2002)
2. Ling, C.X., Yang, Q., Wang, J.N., Zhang, S.C.: Decision trees with minimal costs. In: Proceedings of the 21st International Conference on Machine Learning, p. 69 (2004)

3. Xu, Z., Min, F., Liu, J., Zhu, W.: Ant colony optimization to minimal test cost reduction. In: 2012 IEEE International Conference on Granular Computing (GrC), pp. 585–590. IEEE (2012)
4. Min, F., He, H.P., Qian, Y.H., Zhu, W.: Test-cost-sensitive attribute reduction. Information Sciences 181, 4928–4942 (2011)
5. Karaboga, D., Akay, B.: A comparative study of artificial bee colony algorithm. Applied Mathematics and Computation 214(1), 108–132 (2009)
6. Johnson, N., Kotz, S.: Continuous distributions. J. Wiley, New York (1970) ISBN: 0-471-44626-2
7. Johnson, R., Wichern, D.: Applied multivariate statistical analysis, vol. 4. Prentice Hall, Englewood Cliffs (1992)
8. Min, F., Zhu, W., Zhao, H., Xu, Z.L.: Coser: Cost-senstive rough sets (2012), http://grc.fjzs.edu.cn/~{}fmin/coser/
9. Min, F., Zhu, W.: Minimal cost attribute reduction through backtracking. In: Kim, T.-H., et al. (eds.) DTA/BSBT 2011. CCIS, vol. 258, pp. 100–107. Springer, Heidelberg (2011)
10. Susmaga, R.: Computation of minimal cost reducts. In: Raś, Z.W., Skowron, A. (eds.) ISMIS 1999. LNCS, vol. 1609, pp. 448–456. Springer, Heidelberg (1999)
11. Zhu, W.: A class of fuzzy rough sets based on coverings. In: Proceedings of Fuzzy Systems and Knowledge Discovery, vol. 5, pp. 7–11 (2007)
12. Pawlak, Z.: Rough sets and intelligent data analysis. Information Sciences 147(12), 1–12 (2002)
13. Min, F., Zhu, W.: Attribute reduction of data with error ranges and test costs. Information Sciences 211, 48–67 (2012)
14. Gao, W., Liu, S.: Improved artificial bee colony algorithm for global optimization. Information Processing Letters 111(17), 871–882 (2011)
15. Banharnsakun, A., Achalakul, T., Sirinaovakul, B.: The best-so-far selection in artificial bee colony algorithm. Applied Soft Computing 11(2), 2888–2901 (2011)
16. Cai, J., Ding, H., Zhu, W., Zhu, X.: Artificial bee colony algorithm to minimal time cost reduction. J. Comput. Inf. Systems 9(21), 8725–8734 (2013)

# Reductions of Intuitionistic Fuzzy Covering Systems Based on Discernibility Matrices

Tao Feng[1],[*] and Jusheng Mi[2]

[1] School of Sciences, Hebei University of Science and Technology, Shijiazhuang, China
`fengtao_new@163.com`
[2] College of Mathematics and Information Science, Hebei Normal University,
Shijiazhuang, China

**Abstract.** In stead of intuitionistic fuzzy covering, a new intuitionistic fuzzy binary relation is proposed using a set in an intuitionistic fuzzy covering, and correspondingly the intuitionistic fuzzy approximation space is obtained. Then a novel discernibility matrix is defined which is based on the intuitionistic fuzzy binary relation we defined. And then reductions of intuitionistic fuzzy covering information systems are studied by remaining the intuitionistic fuzzy binary relation unchanged.

**Keywords:** Intuitionistic fuzzy covering, reduction, discernibility matrix, similarity degree.

## 1 Introduction

The theory of rough sets [21] is an extension of set theory for the study of intelligent systems characterized by insufficient and incomplete information. Now, rough set theory has generated classification results and decision rules from datasets [12,17,22,31]. A key definition in Pawlak's rough set model is the indiscernibility relation, called equivalence relation. By the indiscernibility relation, the universe of discourse can be divided into many indiscernibility classes. And indiscernibility classes are basic information granules for constructions of lower and upper approximations[36,37]. The attribute reduction plays an important role in pattern recognition and machine learning. And there are several methods of reductions based on different springboards[30], among which the discernibility matrix is a very useful and feasible method. For example, in order to find three types of reducts (the assignment reduct, the distribution reduct and the maximum distribution reduct), Zhang et al. utilized discernibility matrices with respect to those reducts and obtained the corresponding Boolean functions, called discernibility functions [23,30]. Meanwhile, many scholars discussed reductions of various information systems[16]. And some of them research the attribute reductions using covering rough sets, for example, Zhu proposed the concept of reductions of coverings, and studied the axiomatic characterization of the lower approximation operator in a covering approximation space [40,41]. Since fuzzy

---

[*] Corresponding author.

D. Miao et al. (Eds.): RSKT 2014, LNAI 8818, pp. 111–120, 2014.
DOI: 10.1007/978-3-319-11740-9_11 © Springer International Publishing Switzerland 2014

sets can also be approximated by coverings [7,40,41], Wang etc studied a novel method of attribute reductions of covering decision systems[28]. Fuzzy covering approximations were also studied in [6].

On the other hand, as a more general case of fuzzy sets [39], the concept of intuitionistic fuzzy (IF, for short) sets [1], originally proposed by Atanassov, is an important tool dealing with imperfect and imprecise information. An intuitionistic fuzzy set (IFS, for short) gives membership degree and non-membership degree to describe the level of an element belonging to a set. Hence, handling imperfect and imprecise information is more flexible and effective by intuitionistic fuzzy sets. In IF decision systems, operations of intuitionistic fuzzy set's membership functions and non-membership functions usually use dual triangle norms to fit different situations. More recently, rough set approximations were introduced into IF situations by many authors. They respectively proposed the concept of IF rough sets in which the lower and upper approximations are both IF sets [19,24,25,29]. Then definitions and properties of approximations of IF sets were investigated with respect to an arbitrary IF binary relation in which the universe of discourse may be infinite [29]. Recently, due to its greater flexibility in handling vagueness or uncertainty, IF set theory had been widely used in many areas, such as medical diagnosis [5], pattern recognition[15], clustering analysis[33], and decision making[13,34,38]. As far as reliability field is concerned, a lot of work had been done by many researchers to develop and enrich IF set theory [4,11,10,14,18]. And notice that an IF covering approximation space, which enlarges application fields of the intuitionistic fuzzy rough set theory, is also a generalization of IF approximation spaces[8]. Because the structure of IF covering approximation spaces is relatively complex, we pay more attention to reductions of IF covering (decision) approximation spaces using discernibility matrices.

Along this line of research, we study the reductions of IF covering information systems by discernibility matrices. In Section 2, we introduce definitions of induced IF covering lower and upper approximation operators. Then in section 3 an IF binary relation is constructed based on an IF covering information system. And the reductions are studied by using discernibility matrices remaining the IF binary relation unchanged. In section 4, we define similarity degree between IF covering systems and IF decision attributes, which is used to construct a discernibility matrix for reducing IF covering decision systems. We then conclude the paper with a summary and an outlook for further researches in Section 5.

## 2    Basic Definitions and Properties of IF Covering Approximation Operators

**Definition 1.** *[7] Let $U$ be a nonempty fixed set. An intuitionistic fuzzy (IF, for short) set $A$ is an object having the form*

$$A = \{\langle x, \mu_A(x), \gamma_A(x) \rangle : x \in U\},$$

where $\mu_A : U \to [0,1]$ and $\gamma_A : U \to [0,1]$ denote the degree of membership (namely $\mu_A(x)$) and the degree of nonmembership (namely $\gamma_A(x)$) of each element $x \in U$ belonging to set $A$, respectively, and $0 \le \mu_A(x) + \gamma_A(x) \le 1$ for each $x \in U$. The family of all IF subsets of $U$ is denoted by $IF(U)$.

**Definition 2.** *[7] Let $U$ be a nonempty set, and IF sets $A$ and $B$ be in the forms $A = \{\langle x, \mu_A(x), \gamma_A(x)\rangle : x \in U\}$, $B = \{\langle x, \mu_B(x), \gamma_B(x)\rangle : x \in U\}$. Then*

1. $A \subseteq B$ iff $\mu_A(x) \le \mu_B(x)$ and $\gamma_A(x) \ge \gamma_B(x)$, $\forall x \in U$;
2. $A = B$ iff $A \subseteq B$ and $B \subseteq A$;
3. $\sim A = \{\langle x, \gamma_A(x), \mu_A(x)\rangle : x \in U\}$;
4. $A \cap B = \{\langle x, \mu_A(x) \wedge \mu_B(x), \gamma_A(x) \vee \gamma_B(x)\rangle : x \in U\}$;
5. $A \cup B = \{\langle x, \mu_A(x) \vee \mu_B(x), \gamma_A(x) \wedge \gamma_B(x)\rangle : x \in U\}$.

*Where $\sim A$ is the complement IF set of $A$. $\widehat{\langle \alpha, \beta\rangle} = \{\langle x, \alpha, \beta : x \in U\rangle\}$ is a constant IF set.*

Let $U$ be a non-empty and finite set, called the universe of discourse. In the following, we review the properties of IF approximation spaces.

An IF binary relation $R$(IFR, for short) on $U$ is an IF subset of $U \times U$ [27], namely, $R$ is given by

$$R = \{\langle (x,y), \mu_R(x,y), \gamma_R(x,y)\rangle : (x,y) \in U \times U\},$$

where $\mu_R : U \times U \to [0,1]$ and $\gamma_R : U \times U \to [0,1]$ satisfy the condition $0 \le \mu_R(x,y) + \gamma_R(x,y) \le 1$ for all $(x,y) \in U \times U$. We denote the set of all IF binary relations on $U$ by $IFR(U \times U)$.

Meanwhile, we introduce a special IF set (IF singleton set) $1_{\{y\}} = \{\langle x, \mu_{1_{\{y\}}}(x), \gamma_{1_{\{y\}}}(x)\rangle : x \in U\}$ for $y \in U$ as follows

$$\mu_{1_{\{y\}}}(x) = \begin{cases} 1, \text{ if } x = y; \\ 0, \text{ if } x \neq y. \end{cases} \quad \gamma_{1_{\{y\}}}(x) = \begin{cases} 0, \text{ if } x = y; \\ 1, \text{ if } x \neq y. \end{cases}$$

If a family of IFSs $\mathcal{C} = \{\{\langle x, \mu_{A_i}(x), \gamma_{A_i}(x)\rangle : x \in U\} : i \in J\}$ of $U$ satisfies the conditions $\bigcup\{\{\langle x, \mu_{A_i}(x), \gamma_{A_i}(x)\rangle : x \in U\} : i \in J\} = \widehat{\langle 1,0\rangle}$, and $\forall i \in J$, $\{\langle x, \mu_{A_i}(x), \gamma_{A_i}(x)\rangle\} \neq \widehat{\langle 0,1\rangle}$, then $\mathcal{C}$ is called an IF covering of $U$. A finite subfamily of an IF covering $\mathcal{C}$ of $U$ is called a finite subcovering of $\mathcal{C}$, if it is also an IF covering of $U$. If $x \in U$ and $C$ is an IFS, $\mu_C(x) = 1$ and $\gamma_C(x) = 0$, then we say that $C$ covers $x$ [8].

**Definition 3.** *[8] Let $\mathcal{B} = \{B_i : i = 1, \ldots, m\}$ and $\mathcal{C} = \{C_j : j = 1, \ldots, n\}$ be two IF coverings of $U$. Then $\mathcal{B}$ is called finer than $\mathcal{C}$ (or $\mathcal{C}$ is coarser than $\mathcal{B}$), denoted by $\mathcal{B} \prec \mathcal{C}$, if the following conditions hold:*
*(1) $m \ge n$*
*(2) $\forall B_i \in \mathcal{B}$, there exists an IF set $C_i \in \mathcal{C}$ such that $B_i \subseteq C_j$.*

Let $U$ be a non-empty and finite set, $\mathcal{C} = \{A_i : i \in J\}$ an IF covering of $U$. In the following, we study the IF covering approximation space. In Section 3, we mainly discuss the uncertainty degree of an IF set in a new IF covering approximation space. In Section 4, we study the reductions of an IF covering.

**Definition 4.** *[8] Suppose $U$ is a finite and nonempty universe of discourse, and $\mathcal{C} = \{C_1, \ldots, C_n\}$ is an IF covering of $U$. For every $x \in U$, $C_x = \cap\{C_i : C_i \in \mathcal{C}, \mu_{C_i}(x) = 1, \gamma_{C_i}(x) = 0\}$ , then $Cov(\mathcal{C}) = \{C_x : x \in U\}$ is also an IF covering of $U$, which is called the induced IF covering of $U$ based on $\mathcal{C}$.*

Obviously, every element in $Cov(\mathcal{C})$ cannot be written by the union of other elements in $Cov(\mathcal{C})$.

**Definition 5.** *[8] Let $U$ be a finite and nonempty universe of discourse, $\mathcal{C}$ an IF covering of $U$. For any $A \in IF(U)$, the induced IF covering upper and lower approximations of $A$ w.r.t. $Cov(\mathcal{C})$, denoted by $\overline{\mathcal{C}}(A)$ and $\underline{\mathcal{C}}(A)$, are two IF sets and are defined, respectively, as follows:*

$$\overline{\mathcal{C}}(A) = \{\langle x, \mu_{\overline{\mathcal{C}}(A)}(x), \gamma_{\overline{\mathcal{C}}(A)}(x)\rangle : x \in U\}; \tag{1}$$

$$\underline{\mathcal{C}}(A) = \{\langle x, \mu_{\underline{\mathcal{C}}(A)}(x), \gamma_{\underline{\mathcal{C}}(A)}(x)\rangle : x \in U\}. \tag{2}$$

*Where*
$\mu_{\overline{\mathcal{C}}(A)}(x) = \bigvee_{y \in U}[\mu_{C_x}(y) \wedge \mu_A(y)], \gamma_{\overline{\mathcal{C}}(A)}(x) = \bigwedge_{y \in U}[\gamma_{C_x}(y) \vee \gamma_A(y)];$
$\mu_{\underline{\mathcal{C}}(A)}(x) = \bigwedge_{y \in U}[\gamma_{C_x}(y) \vee \mu_A(y)], \gamma_{\underline{\mathcal{C}}(A)}(x) = \bigvee_{y \in U}[\mu_{C_x}(y) \wedge \gamma_A(y)].$
*$\overline{\mathcal{C}}, \underline{\mathcal{C}} : IF(U) \to IF(U)$ are referred to as induced IF covering upper and lower rough approximation operators w.r.t. $Cov(\mathcal{C})$, respectively. The pair $(U, Cov(\mathcal{C}))$ is called an induced IF covering approximation space. $\forall A \in IF(U)$, if $A = \underline{\mathcal{C}}(A)$, then $A$ is inner definable; if $A = \overline{\mathcal{C}}(A)$, then $A$ is outer definable. $A$ is definable iff $\underline{\mathcal{C}}(A) = \overline{\mathcal{C}}(A)$.*

## 3    Reductions of IF Covering Information Systems

In an IF covering system, first we try to construct an IF approximation space. If $U$ is a nonempty and finite universe of discourse, $\mathcal{C}$ is an IF covering of $U$, $(U, \mathcal{C})$ is an IF covering information system. $\forall C \in \mathcal{C}$, we define an IFR $\forall x, y \in U$, $R_C(x, y) = (\mu_{R_C}(x, y), \gamma_{R_C}(x, y))$ as

$$\mu_{R_C}(x, y) = \begin{cases} 1 - 2K(x, y), & K(x, y) \leq \frac{1}{2}; \\ 0, & \text{otherwise}; \end{cases}$$

$$\gamma_{R_C}(x, y) = \begin{cases} 2((\mu_C(x) - \mu_C(y))^2 + (\gamma_C(x) - \gamma_C(y))^2), & K(x, y) \leq \frac{1}{2}; \\ 1, & \text{otherwise}. \end{cases}$$

Where $K(x, y) = (|\mu_C(x) - \mu_C(y)| + |\gamma_C(x) - \gamma_C(y)|)^2, \forall x, y \in U$.

**Proposition 1.** *Let $U$ be a nonempty and finite universe of discourse, $(U, \mathcal{C})$ an IF covering information system. $C \in \mathcal{C}$, $x, y \in U$, then $R_C$ is an IFR which has the following properties:*

1. $R_C(x, x) = (1, 0)$.
2. $R_C(x, y) = R_C(y, x)$.

3. When $K(x,y) \leq \frac{1}{2}$, if $\mu_C(x) = \mu_C(y)$ or $\gamma_C(x) = \gamma_C(y)$, then $\gamma_{R_C}(x,y) = 1 - \mu_{R_C}(x,y)$.
4. When $K(x,y) > \frac{1}{2}$, $R_C(x,y) = (0,1)$.

*Proof.* Obvious.

By Proposition 1, $R_C$ is a reflexive and symmetric IFR. If the difference between $x$ and $y$ with respect to $C$ is big, then $R_C(x,y) = (0,1)$.

Thus, IF binary relation $R_C$ generated by IF covering information system $(U, \mathcal{C})$ is defined by $R_{\mathcal{C}} = \bigcap_{C \in \mathcal{C}} R_C$. And $R_{\mathcal{C}}$ is also a reflexive and symmetric IFR.

Then $(U, R_{\mathcal{C}})$ is an IF approximation space. And if $\mathcal{B} \subseteq \mathcal{C}$, then $R_{\mathcal{B}} = \bigcap_{C \in \mathcal{B}} R_C$ satisfying:

1. $R_{\mathcal{C}} \subseteq R_{\mathcal{B}} \Leftrightarrow R_{\mathcal{C}}(x,y) \leq R_{\mathcal{B}}(x,y), \forall x, y \in U$
2. $R_{\mathcal{C}} = R_{\mathcal{B}} \Leftrightarrow R_{\mathcal{C}}(x,y) = R_{\mathcal{B}}(x,y), \forall x, y \in U$

**Example** $U = \{x_1, x_2, x_3, \ldots, x_6\}$ is the set of 6 houses for sale. And there are many customers to evaluate the prices of the 6 houses from 4 characteristics, where $C_1$ is the quality of habitation, $C_2$ is the price comparing favorably with those of others, $C_3$ is the exterior looking, and $C_4$ is the intimate structure. And for every characteristic, customers give different evaluations, thus we integrate all evaluations giving the membership degree of every house and nonmembership degree of every house. Let $\mathcal{C} = \{C_1, C_2, C_3, C_4\}$, then $(U, \mathcal{C})$ is an IF covering information system, where

| $U$ | $x_1$ | $x_2$ | $x_3$ | $x_4$ | $x_5$ | $x_6$ |
|---|---|---|---|---|---|---|
| $C_1(x_i)$ | $(1,0)$ | $(0.7, 0.1)$ | $(0.9, 0)$ | $(1,0)$ | $(0.8, 0.1)$ | $(0.3, 0.6)$ |
| $C_2(x_i)$ | $(0,1)$ | $(1,0)$ | $(0.6, 0.4)$ | $(0, 0.6)$ | $(1,0)$ | $(1,0)$ |
| $C_3(x_i)$ | $(0, 0.5)$ | $(1,0)$ | $(1,0)$ | $(0.1, 0.6)$ | $(1,0)$ | $(0.8, 0.1)$ |
| $C_4(x_i)$ | $(0.1, 0.6)$ | $(1,0)$ | $(0.8, 0)$ | $(0.5, 0.3)$ | $(1,0)$ | $(1,0)$ |

Then

$$R_{C_1} = \begin{pmatrix} (1,0) & (0.68, 0.2) & (0.98, 0.02) & (1,0) & (0.82, 0.1) & (0,1) \\ (0.68, 0.2) & (1,0) & (0.82, 0.1) & (0.68, 0.2) & (0.98, 0.02) & (0,1) \\ (0.98, 0.02) & (0.82, 0.1) & (1,0) & (0.98, 0.02) & (0.92, 0.04) & (0,1) \\ (1,0) & (0.68, 0.2) & (0.98, 0.02) & (1,0) & (0.82, 0.1) & (0,1) \\ (0.82, 0.1) & (0.98, 0.02) & (0.92, 0.04) & (0.82, 0.1) & (1,0) & (0,1) \\ (0,1) & (0,1) & (0,1) & (0,1) & (0,1) & (1,0) \end{pmatrix}$$

$$R_{C_2} = \begin{pmatrix} (1,0) & (0,1) & (0,1) & (0.68, 0.32) & (0,1) & (0,1) \\ (0,1) & (1,0) & (0,1) & (0,1) & (1,0) & (1,0) \\ (0,1) & (0,1) & (1,0) & (0,1) & (0,1) & (0,1) \\ (0.68, 0.32) & (0,1) & (0,1) & (1,0) & (0,1) & (0,1) \\ (0,1) & (1,0) & (0,1) & (0,1) & (1,0) & (1,0) \\ (0,1) & (1,0) & (0,1) & (0,1) & (1,0) & (1,0) \end{pmatrix}$$

$$
R_{C_3} = \begin{pmatrix}
(1,0) & (0,1) & (0,1) & (0.92,0.04) & (0,1) & (0,1) \\
(0,1) & (1,0) & (1,0) & (0,1) & (1,0) & (0.82,0.1) \\
(0,1) & (1,0) & (1,0) & (0,1) & (1,0) & (0.82,0.1) \\
(0.92,0.04) & (0,1) & (0,1) & (1,0) & (0,1) & (0,1) \\
(0,1) & (1,0) & (1,0) & (0,1) & (1,0) & (0.82,0.1) \\
(0,1) & (0.82,0.1) & (0.82,0.1) & (0,1) & (0.82,0.1) & (1,0)
\end{pmatrix}
$$

$$
R_{C_4} = \begin{pmatrix}
(1,0) & (0,1) & (0,1) & (0.02,0.5) & (0,1) & (0,1) \\
(0,1) & (1,0) & (0.92,0.08) & (0.28,0.5) & (1,0) & (1,0) \\
(0,1) & (0.92,0.08) & (1,0) & (0.68,0.2) & (0.92,0.08) & (0.92,0.08) \\
(0.02,0.5) & (0.28,0.5) & (0.68,0.2) & (1,0) & (0.28,0.5) & (0.28,0.5) \\
(0,1) & (1,0) & (0.92,0.08) & (0.28,0.5) & (1,0) & (1,0) \\
(0,1) & (1,0) & (0.92,0.08) & (0.28,0.5) & (1,0) & (1,0)
\end{pmatrix}
$$

Thus

$$
R_C = \begin{pmatrix}
(1,0) & (0,1) & (0,1) & (0.02,0.5) & (0,1) & (0,1) \\
(0,1) & (1,0) & (0,1) & (0,1) & (0.98,0.02) & (0,1) \\
(0,1) & (0,1) & (1,0) & (0,1) & (0,1) & (0,1) \\
(0.02,0.5) & (0,1) & (0,1) & (1,0) & (0,1) & (0,1) \\
(0,1) & (0.98,0.02) & (0,1) & (0,1) & (1,0) & (0,1) \\
(0,1) & (0,1) & (0,1) & (0,1) & (0,1) & (1,0)
\end{pmatrix}
$$

Obviously, $R_C(x_1, x_3) \neq C_{x_1}(x_3)$ and $R_C(x_1, x_3) \neq C_{x_3}(x_1)$. Thus, the new IF binary relation is different to the set induced by the IF covering defined in Definition 4. Using the IF binary relation $R_C$, we can construct IF approximation operators as follows: $A \in IF(U)$,

$$
\overline{R_C}(A) = \{\langle x, \mu_{\overline{R_C}(A)}(x), \gamma_{\overline{R_C}(A)}(x)\rangle : x \in U\};
$$

$$
\underline{R_C}(A) = \{\langle x, \mu_{\underline{R_C}(A)}(x), \gamma_{\underline{R_C}(A)}(x)\rangle : x \in U\}.
$$

Where $\mu_{\overline{R_C}(A)}(x)$, $\gamma_{\overline{R_C}(A)}(x)$, $\mu_{\underline{R_C}(A)}(x)$, $\gamma_{\underline{R_C}(A)}(x)$ are similar to the Definition in [27].

**Proposition 2.** *Let $U$ be a nonempty and finite universe of discourse. $C$ is an IF covering of $U$. $R_C$ is the IF binary relation on $U \times U$, then*

1. $\overline{R_C}(1_{\{x\}})(y) = \overline{R_C}(1_{\{y\}})(x)$
2. $\underline{R_C}(1_{U-\{x\}})(y) = \underline{R_C}(1_{U-\{y\}})(x)$

*Proof.* (1) By the definition of $\overline{R_C}$, we have

$$
\begin{aligned}
\mu_{\overline{R_C}(1_{\{x\}})}(y) &= \bigvee\nolimits_{z \in U}[\mu_{R_C}(y,z) \wedge \mu_{1_{\{x\}}}(z)] \\
&= \bigvee\nolimits_{z \neq x}[\mu_{R_C}(y,z) \wedge \mu_{1_{\{x\}}}(z)] \vee [\mu_{R_C}(y,x) \wedge \mu_{1_{\{x\}}}(x)] \\
&= \bigvee\nolimits_{z \neq x}[\mu_{R_C}(y,z) \wedge (0,1)] \vee [\mu_{R_C}(y,x) \wedge \mu_{1_{\{x\}}}(x)] \\
&= \mu_{R_C}(y,x).
\end{aligned}
$$

$$\gamma_{\overline{R_C}(1_{\{x\}})}(y) = \bigwedge_{z\in U}[\gamma_R(y,z) \vee \gamma_{1_{\{x\}}}(z)]$$
$$= \bigwedge_{z\neq x}[\gamma_{R_C}(y,z) \vee \gamma_{1_{\{x\}}}(z)] \wedge [\gamma_{R_C}(y,x) \vee \gamma_{1_{\{x\}}}(x)]$$
$$= \bigwedge_{z\neq x}[\gamma_{R_C}(y,z) \vee (1,0)] \wedge [\gamma_{R_C}(y,x) \vee \gamma_{1_{\{x\}}}(x)]$$
$$= \gamma_{R_C}(y,x).$$

Similarly, we have $\overline{R_C}(1_{\{y\}})(x) = R_C(x,y)$. Thus, $\overline{R_C}(1_{\{x\}})(y) = \overline{R_C}(1_{\{y\}})(x)$.

(2) By the definition of $\underline{R_C}$, we have

$$\mu_{\underline{R_C}(1_{\{x\}})}(y) = \bigwedge_{z\in U}[\gamma_{R_C}(y,z) \vee \mu_{1_{U-\{x\}}}(z)]$$
$$= \bigwedge_{z\neq x}[\gamma_{R_C}(y,z) \vee \mu_{1_{U-\{x\}}}(z)] \wedge [\gamma_{R_C}(y,x) \vee \mu_{1_{U-\{x\}}}(x)]$$
$$= \bigwedge_{z\neq x U}[\gamma_{R_C}(y,z) \vee (1,0)] \wedge [\gamma_{R_C}(y,x) \wedge \mu_{1_{U-\{x\}}}(x)]$$
$$= \gamma_{R_C}(y,x).$$

$$\gamma_{\underline{R_C}(1_{U-\{x\}})}(y) = \bigvee_{z\in U}[\mu_{R_C}(y,z) \wedge \gamma_{1_{U-\{x\}}}(z)]$$
$$= \bigvee_{z\neq x U}[\mu_{R_C}(y,z) \wedge \gamma_{1_{U-\{x\}}}(z)] \vee [\mu_{R_C}(y,x) \wedge \gamma_{1_{U-\{x\}}}(x)]$$
$$= \bigwedge_{z\neq x U}[\mu_{R_C}(y,z) \vee (0,1)] \wedge [\mu_{R_C}(y,x) \vee \gamma_{1_{U-\{x\}}}(x)]$$
$$= \mu_{R_C}(y,x).$$

Similarly, we have $\underline{R_C}(1_{U-\{y\}})(x) = R_C(x,y)$. Thus, $\underline{R_C}(1_{U-\{x\}})(y) = \underline{R_C}(1_{U-\{y\}})(x)$.

In the following, we discuss reductions of IF covering information systems remaining the IF binary relation unchanged.

**Definition 6.** *If $(U,C)$ is an IF covering information system, $R_C$ is an IF binary relation generated by $(U,C)$. $B \subseteq C$, if $R_B = R_C$, then $B$ is called a consistent set of $C$ in $(U,R_C)$. Moreover, if $\forall B' \subset B$, $R_{B'} \neq R_C$, then $B$ is a reduct of $C$.*

Let $Red(C) = \{B, B$ is a reduct of $(U,C)\}$. The intersection of all the reducts is called the core set of $Cov(U,C)$. It should be pointed out that the core set can be an empty set in some cases.

Let $U$ be a nonempty and finite universe of discourse, $C$ an IF covering of $U$. $\forall x_i, x_j \in U$, we define

$$\sigma_{ij} = \{C_\alpha^\beta : C \in C, \alpha = \mu_{R_C}(x_i, x_j), \beta = \gamma_{R_C}(x_i, x_j)\}.$$

Note that $\sigma_{ij} = \sigma_{ji}$ and $\sigma_{ii} = \{C_\alpha^\beta : C \in C, \alpha = 1, \beta = 0\}$.

**Definition 7.** *Given an IF covering information system $(U,C)$, where $U$ is a nonempty and finite set, $C$ is a finite IF covering of $U$. The discernibility matrix is defined as follows:*

$$M_{n\times n} = (M_{ij})_{n\times n} = \begin{pmatrix} M_{11} & M_{12} & \dots & M_{1n} \\ M_{21} & M_{22} & \dots & M_{2n} \\ \vdots & \vdots & \ddots & \vdots \\ M_{n1} & M_{n2} & \dots & M_{nn} \end{pmatrix}$$

*where $M_{ij}$ is defined as:*

1. *If* $\exists C \in \mathcal{C}$ *such that* $R_C(x_i, x_j) = \bigcap_{C \in \mathcal{C}} \{R_C(x_i, x_j)\}$, *then*

$$M_{ij} = \{C \in \mathcal{C} : (R_C(x_i, x_j) = \bigcap_{C \in \mathcal{C}} \{R_C(x_i, x_j)\}) \wedge (R_C(x_i, x_j) \neq (0, 1))\}.$$

2. *Otherwise,*

$$M_{ij} = \{C_i \wedge C_j : (\mu_{R_{C_i}}(x_i, x_j) = \bigwedge_{C \in \mathcal{C}} \mu_{R_C}(x_i, x_j)) \wedge (\gamma_{R_{C_j}}(x_i, x_j) = \bigvee_{C \in \mathcal{C}} \gamma_{R_C}(x_i, x_j))\}.$$

Note that if $\forall x_i, x_j \in U$, $\sigma_{ij} = \{C_1^0 : \forall C \in \mathcal{C}\}$, then $M_{ij} = \emptyset$.

It is easy to find that $M_{ij}$ is the set of subsets of an IF covering which has the maximum IF discernibility of $x_i$ and $x_j$ and satisfies:

1. $M_{ij} = M_{ji}$,
2. $M_{ii} = \emptyset$.

**Proposition 3.** *Let* $(U, \mathcal{C})$ *be a finite IF covering information system,* $\mathcal{B} \subseteq \mathcal{C}$, *then* $R_{\mathcal{B}} = R_{\mathcal{C}}$ *if and only if* $\mathcal{B} \cap M_{ij} \neq \emptyset$, $\forall M_{ij} \neq \emptyset$.

**Theorem 1.** *Given a finite IF covering information system* $(U, \mathcal{C})$, *if* $C \in \mathcal{C}$ *is an element in the core of* $\mathcal{C}$, *then there exists a* $M_{ij}$ *such that* $M_{ij} = \{C\}$ *or* $M_{ij} = \{C \wedge C_i : C_i \in \mathcal{C}\}$.

*Proof.* It is obvious.

**Definition 8.** *Let* $(U, \mathcal{C})$ *be a finite IF covering information system. A discernibility function* $f$ *of* $(U, \mathcal{C})$ *is a Boolean function of* $m$ *Boolean variables* $C_1^*, C_2^*, \ldots, C_m^*$ *corresponding to the elements of the IF covering* $\{C_1, C_2, \ldots, C_m\}$ *respectively, and defined as:*

$$f(C_1^*, C_2^*, \ldots, C_m^*) = \bigwedge \{\bigvee M_{ij} : M_{ij} \in M_{n \times n}\}$$

*where* $\bigvee M_{ij}$ *is the disjunction of the element* $a \in M_{ij}$, *and* $\bigwedge$ *denotes conjunction.*

## 4    Concluding Remarks

In this paper, we defined a novel pair of IF upper and lower approximations based on an IF covering and discussed their properties and the similarity degree between an intuitionistic fuzzy covering system and an intuitionistic fuzzy decision covering system. Then, we defined the reductions of an IF covering approximation space. Finally we discussed the reductions of IF covering systems using discernibility matrices. In the future, we will pay more attention to the study of uncertainty and reductions of IF decision systems.

**Acknowledgements.** This paper is supported by the National Natural Science Foundation of China (No. 61300121, 61170107), by the Natural Science Foundation of Hebei Province (No. A2013208175) and by the Doctoral Starting up Foundation of Hebei University of Science and Technology (QD201228).

# References

1. Atanassov, K.: Intuitionistic fuzzy sets. Fuzzy Sets Systems 20, 87–96 (1986)
2. Atanassov, K.: Review and new results on intuitionistic fuzzy sets. IMMFAIS 1 (1998)
3. Burillo, P., Bustince, H.: Entropy on intuitionistic fuzzy sets and on interval-valued fuzzy sets. Fuzzy Sets and Systems 78, 305–316 (1996)
4. Chakrabortty, S., Pal, M., Nayak, P.K.: Intuitionistic fuzzy optimization technique for Pareto optimal solution of manufacturing inventory models with shortages. European Journal of Operational Research 228, 381–387 (2013)
5. De, S.K., Biswas, R., Roy, A.: An application of intuitionistic fuzzy sets in medical diagnosis. Fuzzy Sets and Systems 117, 209–213 (2001)
6. Deng, T.Q., Chen, Y.M., Xu, W.L., Dai, Q.H.: A novel approach to fuzzy rough sets based on a fuzzy covering. Information Sciences 177, 2308–2326 (2007)
7. Feng, T., Mi, J.S., Wu, W.Z.: Covering-based generalized rough fuzzy sets. In: Wang, G.-Y., Peters, J.F., Skowron, A., Yao, Y. (eds.) RSKT 2006. LNCS (LNAI), vol. 4062, pp. 208–215. Springer, Heidelberg (2006)
8. Feng, T., Mi, J.S.: Uncertainty and reduction in intuitionistic fuzzy covering based rough sets. Journal of Zhejiang Ocean University 5, 481–489 (2010)
9. Feng, T., Zhang, S.-P., Mi, J.-S., Li, Y.: Intuitionistic fuzzy topology space based on fuzzy rough sets. In: Proceedings of the 2009 International Conference on Machine Learning and Cybernetics, vol. 2, pp. 706–711 (2009)
10. Garg, H.: An approach for analyzing fuzzy system reliability using particle swarm optimization and intuitionistic fuzzy set theory. Journal of Multiple- Valued Logic and Soft Computing 21, 335–354 (2013)
11. Garg, H., Rani, M., Sharma, S.P., Vishwakarma, Y.: Intuitionistic fuzzy optimization technique for solving multi-objective reliability optimization problems in interval environment. Expert Systems with Applications 41, 3157–3167 (2014)
12. Hamiderza, R.K., Karim, F.: An improved feature selection method based on ant colony optimization (ACO) evaluated on face recognition system. Applied Mathematics and Computation 205(2), 716–725 (2008)
13. He, Y.D., Chen, H.Y., Zhou, L.G., Han, B., Zhao, Q.Y., Liu, J.P.: Generalized intuitionistic fuzzy geometric interaction operators and their application to decision making. Expert Systems with Applications 41(5), 2484–2495 (2014)
14. Huang, B., Wei, D.K., Li, H.X., Zhuang, Y.L.: Using a rough set model to extract rules in dominance-based interval-valued intuitionistic fuzzy information systems. Information Sciences 221, 215–229 (2013)
15. Li, D.F., Cheng, C.: New similarity measures of intuitionistic fuzzy sets and application to pattern recognitions. Pattern Recognition Letters 23, 221–225 (2002)
16. Li, M., Shang, C.X., Feng, S.Z., Fan, J.P.: Quick attribute reduction in inconsistent decision tables. Information Sciences 254, 155–180 (2014)
17. Liu, G.L., Zhu, K.: The relationship among three types of rough approximation pairs. Knowledge-Based Systems 60, 28–34 (2014)
18. Mahapatra, G.S., Mitra, M., Roy, T.K.: Intuitionistic fuzzy multi-objective mathematical programming on reliability optimization model. International Journal of Fuzzy Systems 12(3), 259–266 (2010)
19. Nanda, S., Majumda, S.: Fuzzy rough sets. Fuzzy Sets and Systems 45, 157–160 (1992)
20. Niu, C.Y., Yang, Y., Jin, L.: New entropy for intuitionistic fuzzy sets. Computer Engineering and Applications 45(34), 32–34 (2009)

21. Pawlak, Z.: Rough sets. International Journal of Computer and Information Sciences 11, 341–356 (1982)
22. Pawlak, Z., Skowron, A.: Rudiments of rough sets. Information Sciences 177, 3–27 (2007)
23. Qian, Y.H., Liang, J., Pedrycz, W., Dang, C.: Positive approximation: an accelerator for attribute reduction in rough set theory. Artificial Intelligence 174, 597–618 (2010)
24. Rizvi, S., Naqvi, H.J., Nadeem, D.: Rough intuitionistic fuzzy set. In: Proceedings of the 6th Joint Conference on Information Sciences (JCIS), Durham, NC, pp. 101–104 (2002)
25. Samanta, S.K., Mondal, T.K.: Intuitionistic fuzzy rough sets and rough intuitionistic fuzzy sets. Journal of Fuzzy Mathematics 9, 561–582 (2001)
26. Sujit, K.D., Adrijit, G., Shib, S.S.: An interpolating by pass to Pareto optimality in intuitionistic fuzzy technique for a EOQ model with time sensitive backlogging. Applied Mathematics and Computation 230(1), 664–674 (2014)
27. Supriya, K.D., Ranjit, B., Akhil, R.R.: Some operations on intuitionistic fuzzy sets. Fuzzy Sets and Systems 114, 477–484 (2000)
28. Wang, C.Z., He, Q., Chen, D.G., Hu, Q.H.: A novel method for attribute reduction of covering decision systems. Information Sciences 254, 181–196 (2014)
29. Wu, W.Z., Zhou, L.: Topological structures of intuitionistic fuzzy rough sets. In: Proceedings of the Seventh International Conference on Machine Learning and Cybernetics, Kunming, pp. 618–623 (2008)
30. Wu, W.Z., Zhang, W.X., Li, H.Z.: Knowledge acquisition in incomplete fuzzy information systems via rough set approach. Expert Systems 20, 280–286 (2003)
31. Xie, G., Zhang, J.L., Lai, K.K., Yu, L.: Variable precision rough set for group decision-making: an application. Approximation Reasoning 49, 331–343 (2008)
32. Xu, W.H., Li, Y., Liao, X.W.: Approaches to attribute reductions based on rough set and matrix computation in inconsistent ordered information systems. Knowledge-Based Systems 27, 78–91 (2012)
33. Xu, Z.S., Chen, J., Wu, J.J.: Clucluster algorithm for intuitionistic fuzzy sets. Information Sciences 178, 3775–3790 (2008)
34. Xu, Z.S.: A method based on distance measure for interval-valued intuitionistic fuzzy grioup decision making. Information Sciences 180, 181–191 (2010)
35. Xu, Z.S., Yager, R.R.: Some geometric aggregation operators based on intuitionistic fuzzy sets. International Journal of General Systems 35, 417–433 (2006)
36. Yao, Y.Y.: Information granulation and rough set approximation. International Journal of Intelligent Systems 16, 87–104 (2001)
37. Yao, Y.Y.: On generalizing rough set theory, Rough Sets. In: Wang, G., Liu, Q., Yao, Y., Skowron, A. (eds.) RSFDGrC 2003. LNCS (LNAI), vol. 2639, pp. 44–51. Springer, Heidelberg (2003)
38. Ye, J.: Fuzzy cross entropy of interval-valued intuitionistic fuzzy sets and its optimal decision-making method based on the weights of alternatives. Expert Systems with Applications 38(5), 6179–6183 (2011)
39. Zadeh, L.A.: Fuzzy sets. Information and Control 8, 338–353 (1965)
40. Zhu, W., Wang, F.Y.: Reduction and axiomization of covering generalized rough sets. Information Sciences 152, 217–230 (2003)
41. Zhu, W.: Relationship between generalized rough sets based on binary relation and covering. Information Sciences 179(15), 210–225 (2009)

# Multi-label Feature Selection with Fuzzy Rough Sets

Lingjun Zhang, Qinghua Hu, Jie Duan, and Xiaoxue Wang

School of Computer Science and Technology, Tianjin University, Tianjin, 300072, China
huqinghua@tju.edu.cn

**Abstract.** Feature selection for multi-label classification tasks has attracted attention from the machine learning domain. The current algorithms transform a multi-label learning task to several binary single-label tasks, and then compute the average score of the features across all single-label tasks. Few research discusses the effect in averaging the scores. To this end, we discuss multi-label feature selection in the framework of fuzzy rough sets. We define a novel dependency functions with three fusion methods if the fuzzy lower approximation of each label has been calculated. A forward greedy algorithm is constructed to reduce the redundancy of the selected features. Numerical experiments validate the performance of the proposed method.

## 1 Introduction

Multi-label learning is a common issue in many practical applications, where a sample is associated with more than one class label simultaneously. For example, a patient may be diagnosed with many several kinds of diseases; there are multiple objects in an images, and a gene is related with various functions.

To deal with multi-label learning, two kinds of strategies, called problem transformation and algorithm adaption, are developed. Binary relevance (BP), label power set (LP) and pruned problem transformation (PPT) are well-known problem transformation methods. BP [1] transforms the multi-label learning problem into several binary single-label tasks, so this method cut up the relationship between labels and easily generate unbalanced data. LP [2] considers the correlations by taking each distinct label set as a new class. However, this method easily lead to lower prediction rate and computation cost since the number of new classes is increased exponentially with the increase of label. Furthermore, many new labels may be associated with a few samples which lead to class unbalance problem. PPT [3] abandons new class with too small number of samples or assign these samples with new labels, while this irreversible process could result in class information loss. Algorithm adaption methods [4] solve the label overlapping and improve prediction results by adapting or extending existing single-label algorithms, rather than transforming the data. In this paper, we focus on the problem transformation method.

Just as the curse of dimensionality occurs in single-label classification problems, there are also redundant or irrelevant attributes in multi-label tasks [5-7]. Feature selection, by removing the superfluous features, can improve prediction performance. Zhang and Zhou [8] proposed a dimensionality reduction method, multi-label Dimensionality Reduction via Dependence Maxiization (MDDM), with two kinds of projection

D. Miao et al. (Eds.): RSKT 2014, LNAI 8818, pp. 121–128, 2014.
DOI: 10.1007/978-3-319-11740-9_12 © Springer International Publishing Switzerland 2014

strategies. Zhang and Jose [9] proposed feature selection mechanisms, which are incorporated into the multi-label naive Bayes classifiers (MLNB) to improve the classification performances. In addition, there are many feature evaluation methods based on the strategies of data transformation in recent years. Newton Spolaor and Everton Alvares Cherman [10] proposed four multi-label feature selection methods by using ReliefF (RF) and Information Gain (IG) as feature evaluation measures for each label and combining with approaches BR and LP. Gauthier Doquir and Michel Verleysen [11] proposed the multi-label feature selection approach by using the multivariate mutual information criterion and combining with PPT.

Fuzzy rough sets, which are mathematical tool for describing incomplete and fuzzy data, have been widely discussed in attribute reduction and feature selection [12-14], and good performance was reported in literature [15-17]. However, there is not report on fuzzy rough sets based multi-label feature selection. In this paper, we define novel dependency functions with three different fusion methods if the fuzzy lower approximation of each label has been calculated for multi-label classification. Some extended properties are discussed. And then we developed a feature selection algorithm, which uses the fuzzy dependency to evaluate the quality of features, and takes the forward greedy search strategy. We analyze the semantics of different fusion methods and show some experiments to test the proposed algorithm.

## 2   Related Work

In this section, we introduce some basic concepts and notations about binary relevance and fuzzy rough sets.

Binary relevance decomposes a multi-label problem into $L$ independent binary classification problems. A sample is labeled as positive when it is associated with this label; otherwise, it is negative.

**Table 1.** Multi-label data set

| $U$ | $c_1$ | $c_2$ | $\cdots$ | $c_M$ | $d_1$ | $d_2$ | $\cdots$ | $d_L$ |
|---|---|---|---|---|---|---|---|---|
| $x_1$ | $x_1^{c_1}$ | $x_1^{c_2}$ | $\cdots$ | $x_1^{c_M}$ | $x_1^{d_1}$ | $x_1^{d_2}$ | $\cdots$ | $x_1^{d_L}$ |
| $x_2$ | $x_2^{c_1}$ | $x_2^{c_2}$ | $\cdots$ | $x_2^{c_M}$ | $x_1^{d_1}$ | $x_1^{d_2}$ | $\cdots$ | $x_2^{d_L}$ |
| $\vdots$ | $\vdots$ | $\vdots$ | $\ddots$ | $\vdots$ | $\vdots$ | $\vdots$ | $\cdots$ | $\vdots$ |
| $x_N$ | $x_N^{c_1}$ | $x_N^{c_2}$ | $\cdots$ | $x_N^{c_M}$ | $x_N^{d_1}$ | $x_N^{d_2}$ | $\cdots$ | $x_N^{d_L}$ |

Let $\langle U, C, D \rangle$ be a multi-label classification problem, where $U = \{x_1, x_2, \ldots, x_N\}$ is a nonempty set composed of $N$ samples, $C = \{c_1, c_2, \ldots, c_M\}$ is a feature set with $M$ features, and $D = \{d_1, d_2, \ldots, d_L\}$ is a label set with $L$ labels. Table 1 give a multi-label data set. Furthermore, $\langle U, C, D \rangle$ degenerate to single-label classification if $L = 1$. Each sample $x$ can be described by a $M$-dimensional vector, and $x^c$ is the feature value of $x$ on $c$. The label set of each sample $x$ can be represented as a $L$-dimensional vector, and $x^d$ is the label value of $x$ on $d$. where $x^d \in \{1, -1\}$. If sample $x$ is associated with

**Table 2.** $L$ Binary single-label data sets

| $U$ | $C$ | $d_1$ | $U$ | $C$ | $d_2$ | $U$ | $C$ | $d_L$ |
|-----|-----|-------|-----|-----|-------|-----|-----|-------|
| $x_1$ | $\cdots$ | $x_1^{d_1}$ | $x_1$ | $\cdots$ | $x_1^{d_2}$ | $x_1$ | $\cdots$ | $x_1^{d_L}$ |
| $x_2$ | $\cdots$ | $x_1^{d_1}$ | $x_2$ | $\cdots$ | $x_1^{d_2}$ | $x_2$ | $\cdots$ | $x_1^{d_L}$ |
| $\vdots$ | $\cdots$ | $\vdots$ | $\vdots$ | $\cdots$ | $\vdots$ | $\vdots$ | $\cdots$ | $\vdots$ |
| $x_N$ | $\cdots$ | $x_N^{d_1}$ | $x_N$ | $\cdots$ | $x_N^{d_2}$ | $x_N$ | $\cdots$ | $x_N^{d_L}$ |

label $d$, then $x^d = 1$; otherwise, $x^d = -1$. The $L$ single-label binary data sets after BR transformation are given in Table 2.

Given a single-label binary classification problem $\langle U, C, d \rangle$, and $R_B$ is a fuzzy equivalence relation on $U$ induced by feature subset $B$. For $\forall x \in U$, we associate a fuzzy equivalence class $[x]_{R_B}$ with $x$ in terms of feature subset $B$. The membership function of $\forall y \in U(y \neq x)$ to $[x]_{R_B}$ is defined as $[x]_{R_B}(y) = R_B(x, y)$. $U$ is divided into two subset $A_+, A_-$ by label $d$. If $x^d = 1$, then $A_+(x) = 1$; If $x^d = -1$, $A_+(x) = 0$. The membership of sample $x$ to the lower approximation of $d$ in term of $B$ is defined as follows [13]:

$$\underline{R_B}d(x) = \underline{R_B}A_+(x) \cup \underline{R_B}A_-(x) = \inf_{x^d=1, y^d=-1}(1 - R(x, y)), \qquad (1)$$

It is easy to get that the membership of $x$ to the lower approximation of $d$ is determined by the closest sample out of $d$ if $x$ belongs to $d$, while the membership of $x$ to the lower approximation of $d$ is zero if $x$ is not associated with $d$.

## 3  Multi-label Feature Evaluation

Now we give new definitions of feature dependency for multi-label feature selection.

### 3.1  Multi-label Feature Evaluation

**Definition 1.** Given a multi-label problem $\langle U, C, D \rangle$, $R_B$ is a fuzzy equivalence relation on $U$ induced by feature subset $B \subseteq C$. The membership of sample $x$ to the lower approximation of $D$ in term of $B$ is defined as three forms:

$$\underline{R_B}D(x)_1 = max_{\{d|x^d=1\}}(\underline{R_B}d(x)), \qquad (2)$$

$$\underline{R_B}D(x)_2 = min_{\{d|x^d=1\}}(\underline{R_B}d(x)), \qquad (3)$$

$$\underline{R_B}D(x)_3 = mean_{\{d|x^d=1\}}(\underline{R_B}d(x)), \qquad (4)$$

where $\underline{R_B}d(x)$ is the membership of sample $x$ to the lower approximation of all label $d$ associated with $x$ in term of $B$.

**Definition 2.** Let $F$ be a fuzzy subset, the cardinality of $F$ is defined as $|F| = \sum_{x \in U} F(x)$, where $F(x)$ is the membership of $x$ to $F$.

**Definition 3.** Given a multi-label problem $\langle U, C, D \rangle$, $R_B$ is a fuzzy equivalence relation on $U$ induced by feature subset $B \subseteq C$. The dependency function of $D$ in term of $B$ is defined as:

$$\gamma_B(D) = \frac{|R_B D|}{|U|} = \frac{\sum_{x \in U} R_B D(x)}{|U|}. \tag{5}$$

The dependency between the label and feature reflect the approximation ability of fuzzy equivalence class induced by feature to the label. In multi-label classification problem, the approximation ability of feature is related to multiple label. The result in each label should be combined to obtain the final result. The first fusion method is the loosest requirement, which selects the best feature from the best situation by $max$ function. The second fusion method is the most strict requirement, which selects the best feature from the worst situation by $min$ function. The third fusion method selects the best feature from the average situation by $mean$ function.

## 3.2    Multi-label Feature Selection

Feature evaluation and search strategy are two keys in constructing an algorithm for feature selection. We take the forward greedy algorithm as the search strategy and use feature dependency as heuristic knowledge.

*Multi-label Feature Selection Algorithm Based on Fuzzy Rough set*

```
Input: Multi-label classification table <U,C,D>,
       where |U|=N, |C|=M, |D|=L
Output: Feature_select
Forward greedy search strategy based on feature dependency
   Dependency function:
      for feature c=1:C
         RD=zeros(N,L)
         for label d=1:L
            X+=find(U(:,d)==1)
            X-=find(U(:,d)~=1)
            Generate fuzzy relation matrix Rd(X-,X+)
            RD(X+,d)=min(1-Rd)
         end
       switch
        case 'min'
         dependency(c)=sum(min(RD'))
        case 'max'
         dependency(c)=sum(max(RD'))
         case 'mean'
         dependency(c)=sum(mean(RD'))
       end
     end
   end
 Feature_select
end
```

We only consider the similarity between samples from different class on each label. The time complexity in evaluating $M$ features is $O(LNlogNM)$ if there are $N$ samples and $L$ labels. Therefore, the complexity of selecting features with a forward greedy search strategy is $O(LM^2NlogN)$.

## 4    Experimental Study

In order to test the effectiveness of our methods, we design a set of comparative experiments, and the effects of the methods on determination are analyzed and evaluated in this section.

The details of two multi-label data sets used in the experiments are given in Table 3, where $N$, $M$, $L$ stand for the numbers of samples, features, and labels, respectively. We download the data sets from the Mulan's repository (*http://mulan.sourceforge.net*).

**Table 3.** Details of two multi-label data sets

| name | sample | feature | label | cardinality | density | distinct |
|------|--------|---------|-------|-------------|---------|----------|
| birds | 645 | 260 | 19 | 1.014 | 0.053 | 133 |
| emotion | 593 | 72 | 6 | 1.869 | 0.311 | 27 |

For each dataset, we compare our proposed method, i.e.,FRS1, FRS2 and FRS3 with existing method MLNB. To validate the effectiveness of our approah, we calculate the following evaluation metrics: Average Precision (AP), Coverage (CV), Hamming Loss (HL), and One Error (OE) [4]. We utilize ML-kNN as the classifier where parameter $k$ and $\mu$ of ML-kNN is set to be 10 and 1, respectively. The results of original feature space are taken as an evaluation benchmark.

Note that, MLNB is the wrapper approach which is directly addressed in multi-label data set. Furthermore,for each dataset, we compare the second approach (FRS2) our proposed with ReliefF-BR and MDDM, which are a classical filter feature selection and dimension reduction respectively. The ReliefF-BR approach combined ReliefF and BR method. The former is implemented by Weka, and the latter is the one available in Mulan. The threshold value for ReliefF is set to be 0.01. The kernel for MDDMproj and MDDMspc is linear kernel. For comparative reasons, the number of feature subset of these methods is set to be equal to the number of feature subset after the FR2 method. Because of the limited passage, only the result of AP is presented. The result of AP in original feature space is used as a baseline to evaluate the feature selection methods.

Table 4 compares that the classification results on birds dataset in different feature spaces. We observe that FRS2 are the best than all other methods in AP and better than most methods in other performance criteria. FRS3 has the fewest feature than other methods, while its performances is slightly worse than other methods. It is different from our imagine that FRS1 has more feature than FRS3 and the performances is better. The reason for FRS1's slow convergence speed is that FRS1 selects a inadequate feature for the first iteration, which can not be discard. Similar results are obtained on emotion dataset as shown in Table 5, in Which our methods are learly better than benchmark except FRS3. We can see that MLNB has the fewest feature with the lowest performance.

**Table 4.** Four performance criteria on *birds* dataset when ML-kNN is used

| methods | feature | AP (↓) | CV (↑) | HL (↑) | OE (↑) |
|---------|---------|--------|--------|--------|--------|
| Original | 260 | 0.6949 | 3.3994 | 0.0536 | 0.3901 |
| MLNB | 131 | 0.6969 | 3.4583 | 0.0520 | 0.3653 |
| FRS3 | 35 | 0.6894 | 3.5294 | 0.0523 | 0.3808 |
| FRS1 | 103 | 0.6965 | 3.5666 | 0.0525 | 0.3684 |
| FRS2 | 107 | 0.7136 | 3.4025 | 0.0554 | 0.3529 |

**Table 5.** Four performance criteria on *emotion* dataset, where ML-kNN is used

| methods | feature | AP (↓) | CV (↑) | HL (↑) | OE (↑) |
|---------|---------|--------|--------|--------|--------|
| Original | 72 | 0.7808 | 1.9158 | 0.2137 | 0.3317 |
| MLNB | 25 | 0.7529 | 2.0743 | 0.2450 | 0.3762 |
| FRS3 | 35 | 0.7803 | 1.9802 | 0.2120 | 0.3119 |
| FRS1 | 46 | 0.7881 | 1.9257 | 0.2269 | 0.3119 |
| FRS2 | 50 | 0.7932 | 1.9208 | 0.2186 | 0.2921 |

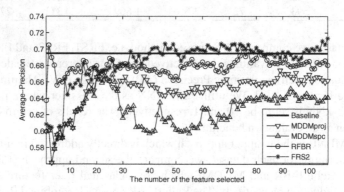

**Fig. 1.** AP on *birds* dataset if different numbers of features are selected(ML-kNN)

Fig. 1 shows AP on birds dataset when ML-kNN is used as a classifier. It can be seen that the performance of FR2 is better than MLNB and original feature space by getting the highest result with fewest feature. $MDDM_{proj}$ and $MDDM_{sps}$ seem not to be suitable for the birds dataset. RfBR gets the highest precision in the first two feature, then relatively bad results are obtained until the feature number increase to 76. In contrast, the result of FRS2 at the first is lower than RFBR. However, the highest result of FRS2 is better than that of RFBR.

In Fig. 2, we can found that the performance of FR2 is worse than other methods at the first, while the highest result is larger than two MDDM methods and original feature space. RF-BR is best in emotion dataset.

The experimental results demonstrate that our methods have their comparative advantages in some aspects. Comparing with other approaches, the major contributions of this research lies in its suggestions for different method to combine single-label

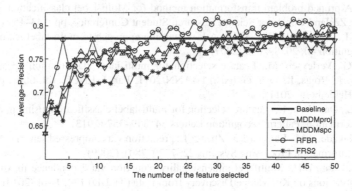

**Fig. 2.** AP on *emotion* dataset if different numbers of features are selected(ML-kNN)

problems to multi-label problem for feature selection. Moreover, our methods produce higher precision performance with fewer features than other methods in birds dataset. Therefore, this indicates that the dependency function is able to find the useful subsets of features for multi-label classification.

## 5 Conclusion

In this work, we proposed the dependency function combined with BR problem transformation methods as feature evaluation based on fuzzy rough sets used for multi-label feature selection. we got some important conclusions through discussing the different meaning of three fusion methods, which are used for collecting feature score on each label, and the results of experiment illustrated the effectiveness of the method. However, the exploration of the article is just a preliminary tries to solve multi-label problem by using fuzzy rough sets, further research is needed in the future on following aspects: Firstly, BR method do not consider the correlations of labels, how to take the correlations into account for feature selection. Secondly, how to solve the label overlapping directly and improve prediction results by adapting or extending our algorithms. Finally, how to combine multi-label problem with other problems exist in real applications, for example, multi-modal problem.

**Acknowledgments.** This work is supported by National Natural Science Foundation of China under Grants 61222210.

## References

1. Boutell, M.R., Luo, J., Shen, X., Brown, C.M.: Learning multi-label scene classification. Pattern Recognition 37, 1757–1771 (2004)
2. Tsoumakas, G., Katakis, I., Vlahavas, I.: Mining multi-label data. J.: Data mining and knowledge discovery handbook, pp. 667–685. Springer, US (2010)

3. Read, J.: A pruned problem transformation method for Multi-label classification. In: Proc. 2008 New Zealand Computer Science Research Student Conference, pp. 143–150 (2008)
4. Zhang, M.L., Zhou, Z.H.: ML-KNN: A lazy learning approach to multi-label learning. Pattern Recognition 40, 2038–2048 (2007)
5. Doquire, G., Verleysen, M.: Feature selection for multi-label classification problems. In: Cabestany, J., Rojas, I., Joya, G. (eds.) IWANN 2011, Part I. LNCS, vol. 6691, pp. 9–16. Springer, Heidelberg (2011)
6. Jaesung, L., Dae-Won, K.: Feature selection for multi-label classification using multivariate mutual information. Pattern Recognition Letters 34, 349–357 (2013)
7. Hsu, D., Kakade, S.M., Langford, J., Zhang, T.: prediction via compressed sensing. Advances in Neural Information Processing Systems 22, 7720–7801 (2009)
8. Zhang, Y., Zhou, Z.H.: Multilabel dimensionality reduction via dependence maximization. ACM Transactions on Knowledge Discovery from Data (TKDD) 122, 1–14 (2010)
9. Zhang, M.L., Jose, M.P.: Feature selection for multi-label naive Bayes classification. Information Sciences 179, 3218–3229 (2009)
10. Newton, S., Everton, A.C.: A Comparison of Multi-label Feature Selection Methods using the Problem Transformation Approach. Electronic Notes in Theoretical Computer Science 292, 135–151 (2013)
11. Gauthier, D., Michel, V.: Mutual information-based feature selection for multilabel classification. Neurocomputing 122, 148–155 (2013)
12. Hu, Q.H., Zhang, L.: Gaussian Kernel Based Fuzzy Rough Sets: Model, Uncertainty Measures and Applications. J. Approximate Reasoning 51, 453–471 (2010)
13. Hu, H., Yu, D.R., Pedrycz, W., Chen, D.G.: Kernelized fuzzy rough sets and their applications. IEEE Trans. on TKDE 23, 1649–1667 (2011)
14. Yeung, D.S., Chen, D.G., Tsang, E.C.: On the generalization of fuzzy rough sets. IEEE Trans. on Fuzzy systems 13, 343–361 (2005)
15. Qian, Y., Wang, Q., Cheng, H.: Fuzzy-rough feature selection accelerator. Fuzzy Sets and Systems (2014)
16. Jensen, R., Tuson, A., Shen, Q.: Finding rough and fuzzy-rough set reducts with SAT. Information Sciences 255, 100–120 (2014)
17. Maji, P., Garai, P.C.: Simultaneous Feature Selection and Extraction Using Fuzzy Rough Sets. In: Proceedings of the Second International Conference on Soft Computing for Problem Solving, pp. 115–123. Springer, India (2014)

# A Logarithmic Weighted Algorithm for Minimal Test Cost Attribute Reduction

Junxia Niu, Hong Zhao*, and William Zhu

Lab of Granular Computing,
Minnan Normal University, Zhangzhou 363000, China
hongzhaocn@163.com

**Abstract.** Minimal test cost attribute reduction is an important problem in cost-sensitive learning since it reduces the dimensionality of the attributes space. To address this issue, many heuristic algorithms have been used by researchers, however, the effectiveness of these algorithms are often unsatisfactory on large-scale datasets. In this paper, we develop a logarithmic weighted algorithm to tackle the minimal test cost attribute reduction problem. More specifically, two major issues are addressed with regard to the logarithmic weighted algorithm. One relates to a logarithmic strategy that can suggest a way of obtaining the attribute reduction to achieve the best results at the lowest cost. The other relates to the test costs which are normalized to speed up the convergence of the algorithm. Experimental results show that our algorithm attains better cost-minimization performance than the existing a weighted information gain algorithm. Moreover, when the test cost distribution is Normal, the effectiveness of the proposed algorithm is more effective for dealing with relatively medium-sized datasets and large-scale datasets.

**Keywords:** Cost-sensitive learning, Granular computing, Attribute reduction, Test cost, Logarithmic weighted algorithm.

## 1 Introduction

Cost-sensitive learning is one of the most challenging problems in both data mining and machine learning. The research work on cost-sensitive learning has attracted considerable attention in different data mining domains, such as rough sets [2,3,4], decision trees [5], artificial neural networks [6] and bayes networks [7]. Cost-sensitive learning deals with the problem of learning from decision systems relative to a variety of costs. Test costs is one of the most important types of cost in cost-sensitive learning. It is the measurement cost of determining the value of an attribute exhibited by an object. The test costs can be money, time, or other types of resources.

Test-cost-sensitive attribute reduction is one of the most fundamental problems in cost-sensitive learning. It aims at finding an attribute reduction with the lowest test cost, meanwhile maintaining enough information of original decision system. This issue is called the minimal test cost attribute reduction (MTR) [9] problem. In many applications, such as the medical examination, where patients need to do a lot of tests.

---

* Corresponding author.

D. Miao et al. (Eds.): RSKT 2014, LNAI 8818, pp. 129–138, 2014.
DOI: 10.1007/978-3-319-11740-9_13 © Springer International Publishing Switzerland 2014

In general, we tend to undertake only a part of tests to save resources. The MTR method provides a good solution in the case. Specifically, a $\lambda$-weighted information gain (WIG) algorithm [2] has been designed to deal with MTR problem. This approach produces good results in small-sized datasets. Unfortunately, it often does not find the optimal solution on large-scale datasets. For example, on Mushroom dataset, when the test cost distribution is Normal, the finding optimal factor (FOF) [2] is only 18%.

In this paper, we propose a logarithmic weighted (LW) algorithm for the MTR problem. The major contributions of proposed algorithm are fourfold. First, a heuristic function is constructed by a logarithmic strategy, which is based on information gain of attributes, test costs and a user-specified logarithmic $\delta$. Particularly, $\delta$ is used to increase the influence of the test costs. The larger the test cost is, the smaller the heuristic function is. In this case, the smaller test cost take more advantage in the choice. Second, from the heuristic function and different $\delta$ settings, a number of candidate reductions with the maximal fitness level in each $\delta$ setting environment are produced. Third, a competition method is used to select a best reduction among these candidate reductions. Fourth, test costs are normalized to speed up the convergence of the algorithm. Experimental results show that our algorithm can generate a minimal test cost attribute reduction in most cases.

Experiments are undertaken with open source software cost-sensitive rough sets (COSER) [21]. Four open datasets from the University of California-Irvine (UCI) library are employed to study the effectiveness of our algorithm. Two representative distributions, namely Uniform and Normal, are employed to generate test costs. We adopt three measures to evaluate the performance of the reduction algorithms from a statistical viewpoint. Experimental results indicate that the LW algorithm outperforms the WIG algorithm in terms of finding optimal factor (FOF), maximal exceeding factor (MEF) and average exceeding factor (AEF), proposed in [2]. For example, on large-scale dataset such as Mushroom, when the test cost distribution is Normal, the FOF of the LW algorithm is 42%, compared with only 18% of the WIG algorithm. With the same distribution, on medium-sized dataset such as Tic-tac-toe dataset, the FOF of the LW algorithm is 95%, compared with only 40% of the WIG algorithm.

The rest of the paper is organized as follows. Section 2 introduces the basic knowledge of test-cost-independent decision systems and the MTR problem. Section 3 presents a logarithmic weighted algorithm and the competition approach. Section 4 states the experimental process and shows the comparison results. Finally, Section 5 concludes in this paper and outlines further research.

## 2   Preliminaries

This section reviews basic knowledge involving test-cost-independent decision systems and the minimal test cost attribute reduction.

### 2.1   Test-cost-independent Decision Systems

Most supervised learning approaches are based on decision systems (DS). A DS is often represented by a decision table. A decision table is often denoted as $S = (U, At = C \cup$

$D, \{V_a|a \in At\}, \{I_a|a \in At\})$, where $U$ is a finite nonempty set of objects called the universe, $At$ is a finite nonempty set of attributes, $C$ is the set of conditional attributes describing the objects, and $D$ is the set of decision attributes that indicates the classes of objects, $V_a$ is the set of values for each $a \in At$, and $I_a: U \rightarrow V_a$ is an information function for each $a \in At$. We often denote $\{V_a|a \in At\}$ and $\{I_a|a \in At\}$ by $V$ and $I$, respectively. Cost-sensitive decision systems are more general than DS. We consider the simplest through most widely used model [4] as follows.

**Definition 1.** *[4] A test-cost-independent decision system (TCI-DS) is the 5-tuple:*

$$S = (U, At, V, I, c), \tag{1}$$

*where $U, At, V$, and $I$ have the same meanings as in a DS, and c: $C \rightarrow R^+ \cup \{0\}$ is the test cost function. Here the attribute test cost function can be represented by a vector $c = [c(a_1), c(a_2), \dots, c(a_{|C|})]$.*

For example, a TCI-DS is represented by Table 1 and a test cost vector, where $U = \{x_1, x_2, x_3, x_4, x_5, x_6\}$, $C=\{$ Muscle pain, Temperature, Snivel, Heartbeat$\}$, $D=\{$ Flu$\}$, and let the cost vector be c=[25, 40, 20, 50]. In other words, if the doctor select test Temperature and Heartbeat, each with cost is $40 and $50, the total test cost would be $40 + $50 = $90, which indicates that test costs are independent of one another. If any element in c is 0, a TCI-DS coincides with a DS. Therefore, this type of decision system is called a test-cost-independent decision system.

**Table 1.** A decision table for medical treatment

| Patient | Muscle pain | Temperature | Snivel | Heartbeat | Flu |
|---------|-------------|-------------|--------|-----------|-----|
| $x_1$ | Yes | Normal | No | Abnormal | No |
| $x_2$ | No | High | Yes | Normal | Yes |
| $x_3$ | No | High | Yes | Abnormal | Yes |
| $x_4$ | Yes | Normal | Yes | Normal | Yes |
| $x_5$ | Yes | High | No | Abnormal | No |
| $x_6$ | Yes | High | Yes | Normal | Yes |

## 2.2 The Minimal Test Cost Attribute Reduction

Attribute reduction is an important problem of rough set theory. There are many rough set models to address the attribute reduction problem from different perspectives, such as covering-based [10,11], decision-theoretical [3], variable-precision [12], dominance-based [13], and neighborhood [14,15]. Many definitions of relative reductions exist [14,16] for rough set models. This paper employs the definition based on the positive region.

**Definition 2.** *An attribute set $R \subseteq C$ is a region reduction of $C$ with respect to $D$ if it satisfies the following two conditions:*
*1. $POS_R(D) = POS_C(D)$, and*
*2. $\forall a \in R, POS_{R-\{a\}}(D) \subset POS_C(D)$.*

Most existing reduction problems aim at finding the minimal reduction. For example, let the set of all relative reductions of $S$ be $Red(S)$, and let $B \in Red(S)$ be a minimal reduction if and only if $|B|$ is minimal. Because the test cost is the key point of this article, we are interested in reductions with the minimal test cost. For simplicity, the minimal test cost attribute reduction (MTR) can be expressed as $c(B) = min\{c(B')|B' \in Red(S)\}$. In fact, the MTR problem is more general than the minimal reduction problem. When all tests have the same cost, the MTR problem coincides with the minimal reduction problem.

## 3   The Logarithmic Weighted Algorithm

Attribute reduction is an important concept in rough set theory. However, for large dataset, finding a minimal test cost attribute reduction (MTR) is NP-hard. Consequently, we design a logarithmic weighted (LW) heuristic algorithm to tackle the MTR problem.

The relative concepts of the information gain [20] and the information entropy are important feasible heuristic information for our algorithm, we first introduce the following concepts. The attribute set $P$ is an equivalent relation of domain $U$ on the cluster, which is composed of $k$ class objects, namely $\{X_1, X_2, \ldots, X_k\}$, $X_i \in U/IND(P)$. Corresponding to the probability are $p(X_1), p(X_2), \ldots, p(X_k)$, then information entropy of $P$ is often denoted as $H(P) = -\sum_{i=1}^{k} p(X_i) log p(X_i)$. For simplicity, the conditional entropy of the attribute set $Q$ respect to attribute set $P$ is often represented $H(Q|P) = -\sum_{i=1}^{n} p(X_i) \sum_{j=1}^{m} p(Y_j|X_i) log p(Y_j|X_i)$, where $U/IND(P) = \{X_1, X_2, \ldots, X_n\}$, $U/IND(Q) = \{Y_1, Y_2, \ldots, Y_m\}$, $p(Y_j|X_i) = \frac{X_i \cap Y_j}{|X_i|}$.

**Definition 3.** [2] Let $R \subseteq C$, $a_i \in C - R$. The information gain of $a_i$ respect to $R$ is

$$f_e(R, a_i) = H(D|R) - H(D|R \cup \{a_i\}). \tag{2}$$

To speed up the convergence of algorithm, we employ the logarithmic normalization method. In fact, there are a number of normalization approaches. For simplicity, we use the method which is based on the 10 logs base normalization method. In our algorithm, since datasets do not provide the test cost, we apply Uniform and Normal distributions to generate random test costs in [1, 100]. Since $lg^1 = 0$, $lg^{100} = 2$, therefore, the value of test cost are normalized from their value into a range from 0 to 2. Let $R \subseteq C$, $a_i \in R$, $c^*(a_i) = lg^{c(a_i)}$, where $c^*(a_i)$ is the normalized cost of attribute $a_i$, $c(a_i)$ is the test cost of attribute $a_i$. Finally, we propose a LW heuristic information function:

$$f(R, a_i, c^*(a_i)) = f_e(R, a_i)(1 + c^*(a_i) \times lg^\delta), \tag{3}$$

where $c^*(a_i)$ is the normalized cost of attribute $a_i$, and $0 < \delta \leq 1$ is a user-specified parameter. If $\delta = 1$, the test costs are not considered in the evaluation. If $\delta > 0$, tests with lower cost have bigger significance. Different $\delta$ settings can adjust the significance of test cost.

The algorithm framework is listed in Algorithm 1 containing three main steps. First, lines 1 and 2 initialize the basic variables saving the minimal cost reduction. The key

---

**Algorithm 1.** An addition-deletion logarithmic weighted algorithm

---

**Input**: $S = (U, At = C \cup D, V, I, c)$
**Output**: A reduction with minimal test cost
**Method**: lw-reduction

1. $R = \emptyset$;
2. $MC = +\infty$;
   //Addition
3. $CA = C$;
4. **while** $(POS_R(D) \neq POS_C(D))$ **do**
5.    **for** (each $a \in CA$) **do**
6.       Compute $f(R, a, c^*(a))$;
7.    **end for**
8.    Select $a'$ with the maximal $f(R, a', c^*(a'))$;
9.    $R = R \cup \{a'\}$; $CA = CA - \{a'\}$;
10. **end while**
    //Deletion
11. $CD = R$; //sort attributes in $CD$ from respective test cost in a descending order;
12. **while** $(CD \neq \emptyset)$ **do**
13.    $CD = CD - \{a'\}$, where $a'$ is the first element of $CD$;
14.    **if** $(POS_{R-\{a'\}}(D) = POS_R(D))$ **then**
15.       $R = R - \{a'\}$;
16.    **end if**
17. **end while**
    //Select the reduction with minimal test cost
18. **for** ($k$=0; $k < |R|$; $k$++) **do**
19.    **if** $(c(R_k) < MC)$ **then**
20.       $MC = c(R_k)$
21.    **end if**
22. **end for**
23. $Return(R_k, MC)$

---

code of this framework is listed in lines 3 and 10, and the attribute significance function is defined to select best attributes. In this process, the algorithm adds the current best attribute $a$ to $R$ from the heuristic function $f(R, a_i, c^*(a_i))$ until $R$ becomes a super reduction. Then, Lines 11 and 18 are the process of deleting attribute, where the algorithm deletes the attribute $a$ from $R$ guaranteeing $R$ with the current minimal total cost. Finally, lines 18 and 23 indicate that the attribute subset with minimal test cost is selected as the result.

To obtain better results, the competition approach has been discussed in [2]. In the new applications, it is still valid because there is no universally optimal $\delta$. In the approach, reductions compete against each other with only one winner with different $\delta$ values. For simplicity, a reduction with minimal test cost, which can be obtained by $\delta \in \Lambda$. Let $R_\delta$ be the reduction by Algorithm 1 using the heuristic information, and let $|\Lambda|$ be the set of user-specified logarithm $\delta$ values. The competition approach is represented as $c_\Lambda = \min_{\delta \in \Lambda} c(R_\delta)$.

# 4    Experiments

In this section, we try to answer the following questions by experimentation.
1. Is the LW algorithm efficient?
2. Is there an optimal setting of $\delta$ for any dataset?
3. Does the LW algorithm outperform the previous approach?

## 4.1    Data Generation

Experiments are carried out on four standard datasets obtained from the UCI [22] repository: Mushroom, Tic-tac-toe, Voting and Zoo. Each dataset should contain exactly one decision attribute and have no missing value. On the four datasets, attributes are no test cost settings. For statistical purposes, different test cost distributions correspond to different applications. The Uniform distribution and Normal distribution are common and meet the most of the cases, hence this paper use these two kinds of distribution to generate random test cost in [1, 100].

## 4.2    Experiment Settings

To test the quality of the proposed algorithm, we set $\delta = 0.03, 0.15, \ldots, 0.99$. The algorithm runs 100 times with different test cost distributions and different $\delta$ settings on four datasets. From different test cost distributions, the performance of the algorithm is different. The finding optimal factor (FOF), maximal exceeding factor (MEF) and average exceeding factor (AEF) [2] are used as comparison criteria. The detail of definition can be seen in [2]. Fig. 1 shows FOF of the LW algorithm on different datasets with different test cost distributions.

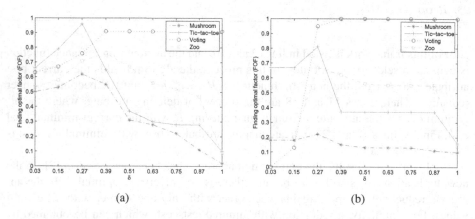

(a)                                                    (b)

**Fig. 1.** The FOF of the LW algorithm on different datasets with different test cost distributions: (a) Uniform distribution; (b) Normal distribution

From the results we observe the following.

1) When $\delta = 1$, it is observed that there is a good performance only on the Voting dataset. However, the performance of the algorithm is very poor based on other three datasets. This is because when $\delta = 1$, from Equation (3), we have $f(R, a_i, c^*(a_i)) = f_e(R, a_i)(1 + c^*(a_i) \times lg^\delta) = f_e(R, a_i)$. In other words, the algorithm degrades to an entropy-based reduction algorithm without taking into account test costs, and the result obtained is a minimal reduction rather than a minimal test cost reduction.

2) Obviously there is a tendency that with increasing $\delta$, the performance of the LW algorithm increases until it reaches a maximum, after which it decreases slowly.

3) It can be seen that there is no universally optimal parameter setting of $\delta$ that is valid for any dataset, however, there is a tradeoff for $\delta = \delta^*$=0.27 might be a rational setting if no further information is available for all datasets we tested.

## 4.3 Comparison of Three Approaches

For the LW algorithm, $\delta$ is a key parameter. Three approaches can be obtained with different setting of $\delta$. We compare the performance for the three methods based on the LW algorithm. The first approach, which is implemented by setting $\delta = 1$, called the non-weighted approach. The second approach, called the best $\delta$ approach by considering $\delta = \delta^*$, which chooses the best parameter $\delta$ value as depicted in Fig. 1. The third approach is the competition approach based $\Lambda$ as discussed in Section 3. Three approaches are based on the same datasets.

**Table 2.** Results for $\delta = 1$, $\delta = \delta^*$ with the optimal setting, and $\delta = \Lambda$ with a number of choices

| Dataset | Distribution | FOF | | | MEF | | | AEF | | |
|---------|--------------|-----|-----|-----|-----|-----|-----|-----|-----|-----|
| | | $\delta = 1$ | $\delta = \delta^*$ | $\delta \in \Lambda$ | $\delta = 1$ | $\delta = \delta^*$ | $\delta \in \Lambda$ | $\delta = 1$ | $\delta = \delta^*$ | $\delta \in \Lambda$ |
| Mushroom | Uniform | 0.020 | 0.620 | 0.810 | 5.171 | 0.641 | 0.265 | 0.796 | 0.0780 | 0.0173 |
| | Normal | 0.110 | 0.220 | 0.420 | 0.062 | 0.515 | 0.056 | 0.021 | 0.1860 | 0.0121 |
| Tic-tac-toe | Uniform | 0.100 | 0.960 | 1.000 | 0.444 | 0.026 | 0.000 | 0.121 | 0.0004 | 0.0000 |
| | Normal | 0.160 | 0.810 | 0.950 | 0.025 | 0.007 | 0.007 | 0.007 | 0.0005 | 0.0001 |
| Voting | Uniform | 0.910 | 0.910 | 1.000 | 0.088 | 0.088 | 0.000 | 0.003 | 0.0032 | 0.0000 |
| | Normal | 1.000 | 1.000 | 1.000 | 0.000 | 0.000 | 0.000 | 0.000 | 0.0000 | 0.0000 |
| Zoo | Uniform | 0.090 | 0.710 | 0.930 | 1.202 | 0.534 | 0.155 | 0.260 | 0.0354 | 0.0036 |
| | Normal | 0.140 | 0.670 | 0.850 | 0.044 | 0.202 | 0.016 | 0.013 | 0.0489 | 0.0011 |

General three approaches results are depicted in Table 2, from which we observe the following.

1) When $\delta = 1$, the non-weighted approach only performs well on the Voting dataset. However, the results are very poor on the other three datasets. In other words, the non-weighted approach is not suitable for the minimal test cost attribute reduction problem.

2) By considering $\delta = \delta^*$, the best $\delta$ approach obtains optimal results in most cases. However, from the real point of view, it is different to guess the best value of $\delta$.

3) In general, the competition approach is a simple and effective method to improve the performance of the algorithm, specifically on datasets where the optimal reduction is hard to find. For example, on Mushroom dataset, the FOF for the competition approach is 0.81, but the FOF for the best $\delta$ is only 0.62.

## 4.4    Comparison with Previous Heuristic Algorithm

The LW heuristic algorithm is developed to deal with the minimal test cost reduction problem. We illustrate the advantage of the LW algorithm compared with the $\lambda$-weighted information gain (WIG) algorithm from effectiveness. Since two different algorithms have different parameters, we compare the results of the competition approach on four datasets with different test costs distributions. Fig. 2(a) shows competition approach results of two algorithms with Normal distribution. Fig. 2(b) indicates the improving efficiency ratio of two algorithms with different distributions. To show the performance of the new approach more intuitive, we adopt Table 3 to express it. From the results we observe the following.

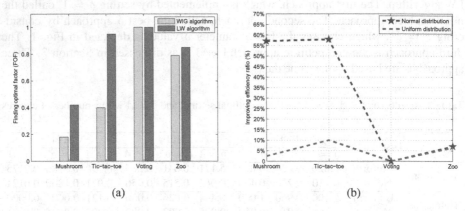

(a)                                    (b)

**Fig. 2.** Competition approach results: (a) Finding optimal factor; (b) Improving efficiency ratios

1) From Fig. 2(a) we can see that on the Voting dataset, two algorithms have the same performance. However, our algorithm has the highest performance on the other three datasets, meanwhile it can produce better results. For example, on Tic-tac-toe dataset, the FOF is only 40% for the WIG algorithm. In correspondence, it is 95% for the LW algorithm. In addition, on the large datasets, the improvement tends to be more obvious. Such as on Mushroom dataset, the FOF for our algorithm is 42%. However, it is only 18% for the WIG algorithm.

2) The improving efficiency ratios of two algorithms on four datasets as show in Fig. 2(b). In larger datasets, the improvement of LW algorithm tends to be quite significant. For example, on Mushroom dataset, when the test cost distribution are Normal and Uniform, the improving efficiency ratio reaches 57% and 2%, respectively. On Tic-tac-toe dataset, it reaches 58% and 10%, respectively.

3) The MEF and AEF are showed in Table 3. It is clear that the LW algorithm significantly outperforms the WIG algorithm. For example, on Mushroom dataset, the MEF on the Normal distribution is 2.7% for the WIG algorithm. But it is only 1.1% for our algorithm. In short, the performance of our algorithm is better than the WIG algorithm in finding success rate, the average performance and the worst performance.

**Table 3.** Results of two algorithms with the competition approach

| Dataset | Distribution | WIG algorithm | | | LW algorithm | | |
|---|---|---|---|---|---|---|---|
| | | FOF | MEF | AEF | FOF | MEF | AEF |
| Mushroom | Uniform | 0.790 | 0.3354 | 0.0228 | 0.810 | 0.2659 | **0.0173** |
| | Normal | 0.180 | 0.2777 | 0.1109 | **0.420** | 0.1109 | **0.0121** |
| Tic-tac-toe | Uniform | 0.900 | 0.2412 | 0.0010 | **1.000** | **0.0000** | **0.0000** |
| | Normal | 0.400 | 0.0127 | 0.0034 | **0.950** | **0.0057** | **0.0001** |
| Voting | Uniform | **1.000** | **0.0000** | **0.0000** | **1.000** | **0.0000** | **0.0000** |
| | Normal | **1.000** | **0.0000** | **0.0000** | **1.000** | **0.0000** | **0.0000** |
| Zoo | Uniform | 0.870 | 0.1556 | 0.0046 | 0.930 | 0.1556 | **0.0036** |
| | Normal | 0.790 | 0.0120 | 0.0012 | 0.850 | 0.0120 | 0.0011 |

## 5    Conclusions

In this paper, a logarithmic weighted algorithm for minimal test cost attribute reduction has been proposed. It adopts a logarithmic strategy in the addition and deletion stages of algorithm. Experimental results show that our algorithm can produce better results than the $\lambda$-weighted information gain algorithm, and our algorithm is also more obvious for dealing with relatively medium-sized datasets and large-scale datasets. With regard to future research, on the one hand, we will design more sophisticated approaches to produce better results. On the other hand, we will also apply it to similar problems such as an error range based attribute reduction or other models such as an interval-valued decision system.

**Acknowledgments.** This work is in part supported by the Zhangzhou Municipal Natural Science Foundation under Grant No. ZZ2013J03, the Key Project of Education Department of Fujian Province under Grant No. JA13192, and the National Science Foundation of China under Grant Nos. 61379049 and 61379089.

## References

1. Yang, Q., Wu, X.: 10 challenging problems in data mining research. International Journal of Information Technology 5(04), 597–604 (2006)
2. Min, F., He, H., Qian, Y., Zhu, W.: Test-cost-sensitive attribute reduction. Information Sciences 181(22), 4928–4942 (2011)

3. Yao, Y.Y., Zhao, Y.: Attribute reduction in decision-theoretic rough set models. Information Sciences 178(17), 3356–3373 (2008)
4. Min, F., Liu, Q.: A hierarchical model for test-cost-sensitive decision systems. Information Sciences 179, 2442–2452 (2009)
5. Turney, P.: Cost-sensitive classification: Empirical evaluation of a hybrid genetic decision tree induction algorithm. Journal of Artificial Intelligence Research (JAIR) 2 (1995)
6. Zhou, Z., Liu, X.: Training cost-sensitive neural networks with methods addressing the class imbalance problem. IEEE Transactions on Knowledge and Data Engineering 18(1), 63–77 (2006)
7. Chai, X., Deng, L., Yang, Q., Ling, C.: Test-cost sensitive naive bayes classification. In: Fourth IEEE International Conference on Data Mining, ICDM 2004, pp. 51–58. IEEE (2004)
8. Pawlak, Z.: Rough sets. International Journal of Computer and Information Sciences 11, 341–356 (1982)
9. He, H., Min, F.: Accumulated cost based test-cost-sensitive attribute reduction. In: Kuznetsov, S.O., Ślęzak, D., Hepting, D.H., Mirkin, B.G. (eds.) RSFDGrC 2011. LNCS (LNAI), vol. 6743, pp. 244–247. Springer, Heidelberg (2011)
10. Zhu, W., Wang, F.: Reduction and axiomization of covering generalized rough sets. Information Sciences 152, 217–230 (2003)
11. Zhu, W.: Topological approaches to covering rough sets. Information Sciences 177(6), 1499–1508 (2007)
12. Ziarko, W.: Variable precision rough set model. Journal of Computer and System Sciences 46(1), 39–59 (1993)
13. Greco, S., Matarazzo, B., Słowiński, R., Stefanowski, J.: Variable consistency model of dominance-based rough sets approach. In: Ziarko, W.P., Yao, Y. (eds.) RSCTC 2000. LNCS (LNAI), vol. 2005, pp. 170–181. Springer, Heidelberg (2001)
14. Hu, Q., Yu, D., Liu, J., Wu, C.: Neighborhood rough set based heterogeneous feature subset selection. Information Sciences 178(18), 3577–3594 (2008)
15. Zhao, H., Min, F., Zhu, W.: Cost-sensitive feature selection of numeric data with measurement errors. Journal of Applied Mathematics 2013, 1–13 (2013)
16. Qian, Y., Liang, J., Pedrycz, W., Dang, C.: Positive approximation: An accelerator for attribute reduction in rough set theory. Artificial Intelligence 174(9), 597–618 (2010)
17. Min, F., Du, X., Qiu, H., Liu, Q.-H.: Minimal attribute space bias for attribute reduction. In: Yao, J., Lingras, P., Wu, W.-Z., Szczuka, M.S., Cercone, N.J., Ślęzak, D. (eds.) RSKT 2007. LNCS (LNAI), vol. 4481, pp. 379–386. Springer, Heidelberg (2007)
18. Min, F., Zhu, W.: Attribute reduction with test cost constraint. Journal of Electronic Science and Technology of China 9(2), 97–102 (2011)
19. Yao, Y., Zhao, Y., Wang, J.: On reduct construction algorithms. In: Wang, G.-Y., Peters, J.F., Skowron, A., Yao, Y. (eds.) RSKT 2006. LNCS (LNAI), vol. 4062, pp. 297–304. Springer, Heidelberg (2006)
20. Wang, G., Yu, H., Yang, D.: Decision table reduction based on conditional information entropy. Chinese Journal of Computers 2(7), 759–766 (2002)
21. Min, F., Zhu, W., Zhao, H., Pan, G.: Coser: Cost-sensitive rough sets (2011), http://grc.fjzs.edu.cn/~{}fmin/coser/
22. Blake, C.L., Merz, C.J.: UCI repository of machine learning databases (1988), http://www.ics.uci.edu/~{}mlearn/mlrepository/

# Attribute Reduction in Object Oriented Concept Lattices Based on Congruence Relations

Xia Wang and Wei-Zhi Wu

School of Mathematics, Physics and Information Science,
Zhejiang Ocean University, Zhoushan, Zhejiang, 316022, P.R. China
bblylm@126.com

**Abstract.** This paper studies a new definition and an approach to attribute reduction in an object oriented concept lattice based on congruence relations. Firstly, dependence space based on the object oriented concept lattice is researched to obtain the relationship among object oriented concept lattices and the corresponding congruence relations. Then the notion of attribute reduct in this paper, resembling that in rough set theory, is defined to find minimal attribute subsets which can preserve all congruence classes determined by the attribute set. Finally, an approach of discernibility matrix is presented to calculate all attribute reducts. It is shown that attribute reducts can also keep all object oriented extents and their original hierarchy in the object oriented concept lattice.

**Keywords:** Object oriented concept lattice, rough set, attribute reduction, congruence relation, formal context.

## 1 Introduction

Formal concept analysis [1, 26] and rough set theory [13, 14] are two effective and complementary mathematical tools to analyze data. The relationship and combination of formal concept analysis and rough set theory are studies to provide new approaches to data analysis. In [3, 15, 16, 28] rough set approximation operators are introduced into formal concept analysis by considering different types of definability. On the other hand, the notion of formal concept and formal concept lattice can also be introduced into rough set, such as the object and property oriented formal concept lattices introduced by Yao [29, 30] respectively based on approximation operators.

One of the key problems of formal concept analysis and rough set theory is knowledge reduction which can make knowledge representation of database more succinct, knowledge hiding in database clearer, and adaptability of rule sets for decision tables better. Many types of approaches to knowledge reduction in rough set theory[4, 9, 10, 14, 17–20, 31, 32] have been proposed, as well as in the area of formal concept analysis [1, 5–8, 11, 21–25, 27, 33, 34]. Moreover, almost all of the approaches to knowledge reduction in rough set theory are based on binary relations such as equivalence relations, partial ordering relations and so on which

D. Miao et al. (Eds.): RSKT 2014, LNAI 8818, pp. 139–148, 2014.
DOI: 10.1007/978-3-319-11740-9_14 © Springer International Publishing Switzerland 2014

define on the object (attribute) sets or the object (attribute) power sets. For formal concept analysis, most of the approaches consider knowledge reduction from the viewpoint of the extents of formal concepts and their hierarchy. In [1], the reducible attribute and reducible object were proposed from the viewpoint of shortening lines or rows. In [33, 34], an attribute reduction approach was presented to find minimal attribute sets which can determine all extents and their original hierarchy in the concept lattice. And attribute reduction in a consistent formal decision context was also investigated in [25]. Then, in [8] the approach was generalized to attribute reduction in the attribute oriented concept lattices and the object oriented concept lattices. Wang et al. [21, 22] provided another approach to attribute reduction, which only required to preserve all extents of ∧−irreducible elements. Subsequently the approach was extended to attribute reduction in object oriented concept lattices and property oriented concept lattices. Wu et al. [27] studied attribute reduction in formal contexts from the viewpoint of keeping granular structure of concept lattices. In [7] an efficient post-processing method was shown to prune redundant rules by virtue of the property of Galois connection, which inherently constrains rules with respect to objects. In [11] a Boolean approach was formulated to calculate all reducts of a formal context via the use of discernibility function. Wang et al. [23] developed an approach to attribute reduction in a formal context and a consistent formal decision context based on congruence relations. The approach in a formal context is to find minimal attribute sets which can preserve all original congruence classes. Four types of approaches to attribute reduction in inconsistent formal decision contexts were defined in [24]. In [5], methods for attribute reduction were studied by an order-preserving mapping between the set of all the extents of the condition concept lattice and that of the decision concept lattice. In [6], methods of approximate concept construction were presented for an incomplete formal context.

The purpose of this paper is mainly to study the notion and approaches to attribute reduction in object oriented concept lattices. Basic definitions and properties of formal concept analysis are recalled in Section 2. In Section 3, dependence space is introduced into the object oriented concept lattice, and relationships among object oriented concept lattices and the corresponding congruence relations are also discussed. In Section 4, an approach to attribute reduction in object oriented concept lattices are proposed using discernibility matrices to obtain all attribute reducts in Section 4. Finally, we conclude the paper in Section 5.

## 2   Preliminaries

In this section, we recall some basic notions and properties about formal concept analysis which will be used in this paper.

**Definition 1.** [1] A formal context $(U, A, I)$ consists of object set $U$ and attribute set $A$, and a relation $I \subseteq U \times A$. The elements of $U$ are called objects and the elements of $A$ are called attributes of the formal context.

For any $X \subseteq U$ and $B \subseteq A$, Y.Y. Yao defined two pairs of dual operators:

$$X^{\square} = \{a \in A \mid \forall\, x \in U,\ xIa \Rightarrow x \in X\}, \tag{1}$$
$$X^{\diamond} = \{a \in A \mid \exists\, x \in U,\ xIa \wedge x \in X\}, \tag{2}$$
$$B^{\square} = \{x \in U \mid \forall\, a \in A,\ xIa \Rightarrow a \in B\}, \tag{3}$$
$$B^{\diamond} = \{x \in U \mid \exists\, a \in A,\ xIa \wedge a \in B\}. \tag{4}$$

**Definition 2.** [30] Let $(U, A, I)$ be a formal context and $B \subseteq A$. The formal context $(U, B, I_B)$ is called a subcontext of $(U, A, I)$, where $I_B = I \cap (U \times B)$.

For any $B \subseteq A$, let $^{\square B}, ^{\diamond B}$ stand for the operator in the subcontext $(U, B, I_B)$. Clearly, for any $X \subseteq U, X^{\square A} = X^{\square}$, $X^{\square B} = X^{\square A} \cap B$, $X^{\diamond A} = X^{\diamond}$ and $X^{\diamond B} = X^{\diamond A} \cap B$.

**Definition 3.** [30] An object oriented concept of a formal context $(U, A, I)$ is a pair $(X, B)$ with $X \subseteq U, B \subseteq A, X^{\square} = B$ and $B^{\diamond} = X$. We call $X$ the extent and $B$ the intent of the object oriented concept $(X, B)$.

For any two object oriented concepts $(X_1, B_1)$ and $(X_2, B_2)$, Y.Y. Yao defined two operators meet and join as follows:

$$(X_1, B_1) \wedge (X_2, B_2) = ((B_1 \cap B_2)^{\diamond}, (B_1 \cap B_2)), \tag{5}$$

$$(X_1, B_1) \vee (X_2, B_2) = ((X_1 \cup X_2), (X_1 \cup X_2)^{\square}). \tag{6}$$

The set of all object oriented concepts of $(U, A, I)$ is denoted by $L_O(U, A, I)$ and is called the object oriented concept lattice of the formal context $(U, A, I)$.

**Property 1.** [30] Let $(U, A, I)$ be a formal context, $X, X_1, X_2$ be object sets, and $B, B_1, B_2$ be attribute sets, then

(1) If $X_1 \subseteq X_2$, then $X_1^{\square} \subseteq X_2^{\square}$, and $X_1^{\diamond} \subseteq X_2^{\diamond}$.
   If $B_1 \subseteq B_2$, then $B_1^{\square} \subseteq B_2^{\square}$, and $B_1^{\diamond} \subseteq B_2^{\diamond}$;
(2) $X^{\square \diamond} \subseteq X \subseteq X^{\diamond \square}$, and $B^{\square \diamond} \subseteq B \subseteq B^{\diamond \square}$;
(3) $X^{\square} = X^{\square \diamond \square}$, and $B^{\square} = B^{\square \diamond \square}$;
   $X^{\diamond} = X^{\diamond \square \diamond}$, and $B^{\diamond} = B^{\diamond \square \diamond}$;
(4) $(X_1 \cap X_2)^{\square} = X_1^{\square} \cap X_2^{\square}$, and $(B_1 \cap B_2)^{\square} = B_1^{\square} \cup B_2^{\square}$;
(5) $(X_1 \cup X_2)^{\diamond} = X_1^{\diamond} \cup X_2^{\diamond}$, and $(B_1 \cup B_2)^{\diamond} = B_1^{\diamond} \cup B_2^{\diamond}$.

# 3   Dependence Space Based on an Object Oriented Concept Lattice

An information system is a triple $(U, A, F)$, where $U$ is the finite set of objects and $A$ is the finite set of attributes, $F$ is a set of functions between U and A.

In [12], Novotný defined a congruence relation on the attribute power set $\mathcal{P}(A)$ and a dependence space in information systems.

**Definition 4.** [12] Let $(U, A, F)$ be an information system. $\mathcal{K}$ is an equivalence relation on $\mathcal{P}(A)$. Then, $\mathcal{K}$ is called a congruence relation on $(\mathcal{P}(A), \bigcup)$, whenever it satisfies the following condition: if $(B_1, C_1) \in \mathcal{K}, (B_2, C_2) \in \mathcal{K}$, then $(B_1 \cup B_2, C_1 \cup C_2) \in \mathcal{K}$.

**Definition 5.** [12] Let $A$ be a finite nonempty set, $\mathcal{K}$ a congruence relation on $(\mathcal{P}(A), \bigcup)$. Then the ordered pair $(A, \mathcal{K})$ is said to be a dependence space.

**Definition 6.** [2] An interior operator is a mapping $int : 2^U \to 2^U$ such that for all $X, Y \subseteq U$, $X \subseteq Y \subseteq U \Rightarrow int(X) \subseteq int(Y)$; $int(X) \subseteq X$ and $int(X) = int(int(X))$.

Let $(U, A, I)$ be a formal context and $B \subseteq A$. Developed by Novotný's idea, we define a binary relation on the object power set $\mathcal{P}(U)$ as follows:

$$R^{\Box B} = \{(X, Y) \in \mathcal{P}(U) \times \mathcal{P}(U) | X^{\Box B} = Y^{\Box B}\}. \tag{7}$$

It is easy to prove that $R^{\Box B}$ is a congruence relation on $(\mathcal{P}(U), \bigcup)$. That is, $(U, R^{\Box B})$ is a dependence space. We define $[X]_{R^{\Box B}} = \{Y \in \mathcal{P}(U) | (X, Y) \in R^{\Box B}\}$, the congruence class determined by $X$ with respect to the congruence relation $R^{\Box B}$. Then we define $int_{R^{\Box B}}(X) = \bigcap\{Y | Y \in [X]_{R^{\Box B}}\}$.

By Property 1 and the definition of $int_{R^{\Box B}}(X)$, we have the following Lemma 1 immediately.

**Lemma 1.** Let $(U, A, I)$ be a formal context. For any $X, Y, Z \in \mathcal{P}(U)$ and $B \subseteq A$, the following statements hold: (1) $(int_{R^{\Box B}}(X), X) \in R^{\Box B}$; (2) $int_{R^{\Box B}}$ is an interior operator; (3) $int_{R^{\Box B}}(X) = X^{\Box B \Diamond B}$.

By Lemma 1 (4) and Property 1 (3), $(int_{R^{\Box B}}(X), X^{\Box}) \in L_O(U, B, I_B)$ for any $X \subseteq U$ and $B \subseteq A$. On the other hand, if $X \in L_{OU}(U, B, I_B)$, then $X = int_{R^{\Box B}}(X)$, where $L_{OU}(U, A, I) = \{X | (X, B) \in L_O(U, A, I)\}$. That is $L_{OU}(U, B, I_B) = \{int_{R^{\Box B}}(X) | X \subseteq U\}$.

**Lemma 2.** Let $(U, A, I)$ be a formal context. For any $X, Y, Z \in \mathcal{P}(U)$ and $B \subseteq A$, if $X \subseteq Y \subseteq Z$ and $(X, Z) \in R^{\Box B}$, then $(X, Y) \in R^{\Box B}$ and $(Y, Z) \in R^{\Box B}$.

Since $R^{\Box B}$ is a congruence relation, Lemma 2 can be easily proved by the definition of congruence relation.

**Lemma 3.** Let $(U, A_1, I_1)$ and $(U, A_2, I_2)$ be two formal contexts with the same object set. For any $X \subseteq U$, if $L_{OU}(U, A_2, I_2) \subseteq L_{OU}(U, A_1, I_1)$, then we have (1) $int_{R^{\Box A_1}}(int_{R^{\Box A_2}}(X)) = int_{R^{\Box A_2}}(X)$, and (2) $int_{R^{\Box A_2}}(X) \subseteq int_{R^{\Box A_1}}(X)$.

*Proof.* (1) Since $int_{R^{\Box A_2}}(X) \in L_{OU}(U, A_2, I_2)$ for any $X \subseteq U$, if $L_{OU}(U, A_2, I_2) \subseteq L_{OU}(U, A_1, I_1)$, then we obtain that $int_{R^{\Box A_2}}(X) \in L_{OU}(U, A_1, I_1)$ which implies that $(int_{R^{\Box A_2}}(X))^{\Box A_1 \Diamond A_1} = int_{R^{\Box A_2}}(X)$. Therefore, (i) is concluded considering that $int_{R^{\Box A_1}}(int_{R^{\Box A_2}}(X)) = (int_{R^{\Box A_2}}(X))^{\Box A_1 \Diamond A_1}$.

(2) Since $int_{R^{\Box A_1}}$ is an interior operator, we have $int_{R^{\Box A_1}}(int_{R^{\Box A_2}}(X)) \subseteq int_{R^{\Box A_1}}(X)$. Thus, $int_{R^{\Box A_2}}(X) \subseteq int_{R^{\Box A_1}}(X)$ follows directly from (i).           †

The following Theorem 1 shows us the relationships among object oriented concept lattices and the corresponding congruence relations.

**Theorem 1.** Let $(U, A_1, I_1)$ and $(U, A_2, I_2)$ be two formal contexts with the same object set. Then we have $L_{OU}(U, A_2, I_2) \subseteq L_{OU}(U, A_1, I_1)$ if and only if $R^{\Box A_1} \subseteq R^{\Box A_2}$.

*Proof.* Sufficiency. Assume $L_{OU}(U, A_2, I_2) \not\subseteq L_{OU}(U, A_1, I_1)$, then there exists $X \in L_{OU}(U, A_2, I_2)$ such that $X \notin L_{OU}(U, A_1, I_1)$. Thus, $int_{R^{\Box A_1}}(X) \subsetneq X = int_{R^{\Box A_2}}(X)$ is concluded. Since $R^{\Box A_1} \subseteq R^{\Box A_2}$ if and only if $[X]_{R^{\Box A_1}} \subseteq [X]_{R^{\Box A_2}}$ for any $X \subseteq U$, we have $int_{R^{\Box A_2}}(X) \subseteq int_{R^{\Box A_1}}(X)$, which is a contradiction to $int_{R^{\Box A_1}}(X) \subsetneq int_{R^{\Box A_2}}(X)$. Consequently, $L_{OU}(U, A_2, I_2) \subseteq L_{OU}(U, A_1, I_1)$.

Necessity. Assume $R^{\Box A_1} \not\subseteq R^{\Box A_2}$, then there exits $X \subseteq U$ such that $[X]_{R^{\Box A_1}} \not\subseteq [X]_{R^{\Box A_2}}$. Thus, there exists $Y \in [X]_{R^{\Box A_1}}$ such that $Y \notin [X]_{R^{\Box A_2}}$. We prove it from two cases: $X \in L_{OU}(U, A_1, I_1)$ and $X \notin L_{OU}(U, A_1, I_1)$.

Firstly, we suppose $X \in L_{OU}(U, A_1, I_1)$. Since $Y \in [X]_{R^{\Box A_1}}$ and $Y \notin [X]_{R^{\Box A_2}}$, we obtain $X = int_{R^{\Box A_1}}(Y) \subsetneq Y$. Combining with $int_{R^{\Box A_2}}(Y) \subseteq int_{R^{\Box A_1}}(Y)$ by Lemma 3 (2), we have $int_{R^{\Box A_2}}(Y) \subseteq X \subsetneq Y$. Due to Lemma 2, $(Y, X) \in R^{\Box A_2}$, which is a contradiction to $Y \notin [X]_{R^{\Box A_2}}$. Therefore, $[X]_{R^{\Box A_1}} \subseteq [X]_{R^{\Box A_2}}$ holds. That is $R^{\Box A_1} \subseteq R^{\Box A_2}$.

Secondly, we suppose $X \notin L_{OU}(U, A_1, I_1)$. According to the above discussions, we have $[int_{R^{\Box A_1}}(X)]_{R^{\Box A_1}} \subseteq [int_{R^{\Box A_1}}(X)]_{R^{\Box A_2}}$ due to $int_{R^{\Box A_1}}(X) \in L_{OU}(U, A_1, I_1)$. Since $Y \in [X]_{R^{\Box A_1}}$, it is evident that $int_{R^{\Box A_1}}(X) \subseteq Y$ and $Y \in [int_{R^{\Box A_1}}(X)]_{R^{\Box A_2}}$. Combining with $int_{R^{\Box A_2}}(X) \subseteq int_{R^{\Box A_1}}(X)$, we have $int_{R^{A_2}}(X) \subseteq Y$. Since $int_{R^{\Box A_2}}$ is an interior operator and $Y \in [int_{R^{\Box A_1}}(X)]_{R^{\Box A_2}}$, we obtain $int_{R^{\Box A_2}}(Y) = int_{R^{\Box A_2}}(int_{R^{\Box A_1}}(X)) \subseteq int_{R^{\Box A_2}}(X)$. Thus, $int_{R^{\Box A_2}}(Y) \subseteq int_{R^{\Box A_2}}(X) \subseteq Y$. By Lemma 2, $(Y, int_{R^{\Box A_2}}(X)) \in R^{\Box A_2}$ holds. That is, $(Y, X) \in R^{\Box A_2}$, which is a contradiction to $Y \notin [X]_{R^{\Box A_2}}$. Therefore, $[X]_{R^{\Box A_1}} \subseteq [X]_{R^{\Box A_2}}$ is concluded. That is $R^{\Box A_1} \subseteq R^{\Box A_2}$.           †

## 4   The Notion and Approaches to Attribute Reduction in Object Oriented Concept Lattices

In this section, we develop the notion of attribute reduction in an object oriented concept lattice based on the congruence relations and then define an approach to attribute reduction.

**Definition 7.** Let $(U, A, I)$ be a formal context. For any $B \subseteq A$, if $R^{\Box A} = R^{\Box B}$, then $B$ is called a consistent set of the object oriented concept lattice $L_O(U, A, I)$. Further, for any $b \in B$ if $R^{\Box A} \neq R^{\Box B-\{b\}}$, then $B$ is called an attribute reduct of $L_O(U, A, I)$.

Definition 6 shows that consistent sets preserve all original congruence classes determined by the attribute set. By Theorem 1 and Definition 6, we can obtain the following result directly.

**Theorem 2.** Let $(U, A, I)$ be a formal context. For any $B \subseteq A$ and $b \in B$, we have the following two statements:

(1) $B$ is a consistent set if and only if $L_{OU}(U, A, I) = L_{OU}(U, B, I_B)$.

(2) $B$ is an attribute reduct if and only if $L_{OU}(U, A, I) = L_{OU}(U, B, I_B)$, $L_{OU}(U, A, I) \neq L_{OU}(U, B - \{b\}, I_{B-\{b\}})$.

For convenience, we use $R^{\Box a}$ instead of $R^{\Box\{a\}}(a \in A)$.

**Definition 8.** Let $(U, A, I)$ be a formal context. For any $X_i, X_j \subseteq U$, we define

$$D([X_i]_{R^{\Box A}}, [X_j]_{R^A}) = \{a \in A | (X_i, X_j) \notin R^{\Box a}\}. \tag{8}$$

Then $D([X_i]_{R^{\Box A}}, [X_j]_{R^{\Box A}})$ is called the discernibility attribute set between $[X_i]_{R^{\Box A}}$ and $[X_j]_{R^{\Box A}}$, and $\mathcal{D} = (D([X_i]_{R^{\Box A}}, [X_j]_{R^{\Box A}}) | X_i, X_j \in \mathcal{P}(U))$ is called the discernibility matrix.

**Theorem 3.** Let $(U, A, I)$ be a formal context. For any $X_i, X_j \subseteq U$, we have $D([X_i]_{R^{\Box A}}, [X_j]_{R^{\Box A}}) = B_i \cup B_j - B_i \cap B_j$, where $(int_{R^A}(X_i), B_i) \in L_O(U, A, I)$ and $(int_{R^{\Box A}}(X_j), B_j) \in L_O(U, A, I)$.

*Proof.* Since for any $X_i, X_j \subseteq U$, $(int_{R^{\Box A}}(X_i), B_i) \in L_O(U, A, I)$ and $(int_{R^{\Box A}}(X_j), B_j) \in L_O(U, A, I)$, we obtain

$$a \in D([X_i]_{R^{\Box A}}, [X_j]_{R^{\Box A}}) \Leftrightarrow (X_i, X_j) \notin R^{\Box a}$$

$$\Leftrightarrow X_i^{\Box a} \neq X_j^{\Box a}$$

$$\Leftrightarrow (int_{R^{\Box A}}(X_i))^{\Box a} \neq (int_{R^{\Box A}}(X_j))^{\Box a}$$

$$\Leftrightarrow a \in B_i \cup B_j - B_i \cap B_j,$$

The proof is completed.                                                      †

**Property 2.** Let $(U, A, I)$ be a formal context. For any $X_i, X_j, X_k \subseteq U$ and $B \subseteq A$, the following properties hold:

(1) $D([X_i]_{R^{\Box A}}, [X_i]_{R^{\Box A}}) = \emptyset$.

(2) $D([X_i]_{R^{\Box A}}, [X_j]_{R^{\Box A}}) = D([X_j]_{R^{\Box A}}, [X_i]_{R^{\Box A}})$.

(3) $D([X_i]_{R^{\Box A}}, [X_j]_{R^{\Box A}}) \subseteq D([X_i]_{R^{\Box A}}, [X_k]_{R^{\Box A}}) \cup D([X_k]_{R^{\Box A}}, [X_j]_{R^{\Box A}})$.

(4) $D([X_i]_{R^{\Box B}}, [X_j]_{R^{\Box B}}) = D([X_i]_{R^{\Box A}}, [X_j]_{R^{\Box A}}) \cap B$.

**Theorem 4.** Let $(U, A, I)$ be a formal context. For any nonempty attribute set $B$, we have that $B$ is a consistent set if and only if for any $D([X_i]_{R^{\Box A}}, [X_j]_{R^{\Box A}}) \neq \emptyset$, $B \cap D([X_i]_{R^{\Box A}}, [X_j]_{R^{\Box A}}) \neq \emptyset$.

*Proof.* Sufficiency. Since for any $D([X_i]_{R^{\square A}}, [X_j]_{R^{\square A}}) \neq \emptyset$, $B \cap D([X_i]_{R^{\square A}}, [X_j]_{R^{\square A}}) \neq \emptyset$, we have if $D([X_i]_{R^{\square A}}, [X_j]_{R^{\square A}}) \neq \emptyset$, then $D([X_i]_{R^{\square B}}, [X_j]_{R^{\square B}}) \neq \emptyset$. On the other hand, we have that if $D([X_i]_{R^{\square B}}, [X_j]_{R^{\square B}}) \neq \emptyset$, then $D([X_i]_{R^{\square A}}, [X_j]_{R^{\square A}}) \neq \emptyset$ by Property 2 (4).

Necessity. If $B$ is a consistent set, then $R^{\square A} = R^{\square B}$ which implies $[X_i]_{R^{\square A}} = [X_j]_{R^{\square B}}$ for any $X_i, X_j \subseteq U$. Thus we have if $[X_i]_{R^{\square A}} \cap [X_j]_{R^{\square A}} = \emptyset$, then $[X_i]_{R^{\square B}} \cap [X_j]_{R^{\square B}} = \emptyset$. Hence if $D([X_i]_{R^{\square A}}, [X_j]_{R^{\square A}}) \neq \emptyset$, then $D([X_i]_{R^{\square B}}, [X_j]_{R^{\square B}}) \neq \emptyset$. Since $D([X_i]_{R^{\square B}}, [X_j]_{R^{\square B}}) = D([X_i]_{R^{\square A}}, [X_j]_{R^{\square A}}) \cap B$, we have $B \cap D([X_i]_{R^{\square A}}, [X_j]_{R^{\square A}}) \neq \emptyset$.    †

**Example 1.** Table 1 gives a formal context $(U, A, I)$ with $U = \{1, 2, 3, 4, 5\}$ and $A = \{a, b, c, d, e\}$. Table 2 shows the corresponding discernibility matrix $\mathcal{D}$ of $(U, A, I)$ and Fig. 1 gives the object oriented concept lattice $L(U, A, I)$.

**Table 1.** A formal context $(U, A, I)$

| $U$ | $a$ | $b$ | $c$ | $d$ | $e$ |
|---|---|---|---|---|---|
| 1 | 0 | 1 | 0 | 1 | 0 |
| 2 | 1 | 0 | 1 | 0 | 1 |
| 3 | 1 | 1 | 0 | 0 | 1 |
| 4 | 0 | 1 | 1 | 1 | 0 |
| 5 | 1 | 0 | 0 | 0 | 1 |

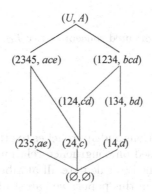

**Fig. 1.** The object oriented concept lattices $L_O(U, A, I)$

According to Theorem 4 and Table 2, $B_1 = \{a, b, c, d\}$ and $B_2 = \{b, c, d, e\}$ are the attribute reducts of $(U, A, I)$. Fig. 2 shows the object oriented concept lattices $L_O(U, B_1, I_{B_1})$ and $L_O(U, B_2, I_{B_2})$ respectively. It is easy to see that the three object oriented concept lattices $L_O(U, A, I)$, $L_O(U, B_1, I_{B_1})$ and $L_O(U, B_2, I_{B_2})$ are isomorphic to each other.

**Table 2.** The discernibility matrix $\mathcal{D}$ of $(U, A, I)$

| | $U$ | $\{2345\}$ | $\{1234\}$ | $\{235\}$ | $\{124\}$ | $\{134\}$ | $\{24\}$ | $\{14\}$ | $\emptyset$ |
|---|---|---|---|---|---|---|---|---|---|
| $U$ | $\emptyset$ | $\{b,d\}$ | $\{a,e\}$ | $\{b,c,d\}$ | $\{a,b,e\}$ | $\{a,c,e\}$ | $\{a,b,d,e\}$ | $\{a,b,c,e\}$ | $A$ |
| $\{2345\}$ | | $\emptyset$ | $\{a,b,d,e\}$ | $\{c\}$ | $\{a,d,e\}$ | $A$ | $\{a,e\}$ | $\{a,c,d,e\}$ | $\{a,c,e\}$ |
| $\{1234\}$ | | | $\emptyset$ | $A$ | $\{b\}$ | $\{c\}$ | $\{b,d\}$ | $\{c,d\}$ | $\{b,c,d\}$ |
| $\{235\}$ | | | | $\emptyset$ | $\{a,c,d,e\}$ | $\{a,b,d,e\}$ | $\{a,c,e\}$ | $\{a,d,e\}$ | $\{a,e\}$ |
| $\{124\}$ | | | | | $\emptyset$ | $\{b,c\}$ | $\{d\}$ | $\{c\}$ | $\{c,d\}$ |
| $\{134\}$ | | | | | | $\emptyset$ | $\{b,c,d\}$ | $\{b\}$ | $\{b,d\}$ |
| $\{24\}$ | | | | | | | $\emptyset$ | $\{c,d\}$ | $\{c\}$ |
| $\{14\}$ | | | | | | | | $\emptyset$ | $\{d\}$ |
| $\emptyset$ | | | | | | | | | $\emptyset$ |

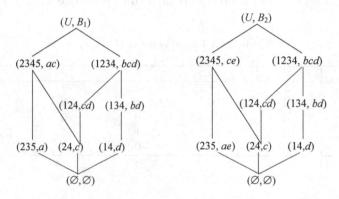

**Fig. 2.** The left is the object oriented concept lattice $L_O(U, B_1, I_{B_1})$, and the right is $L_O(U, B_2, I_{B_2})$.

## 5   Conclusion

This paper has developed notions and approaches to attribute reduction in object oriented concept lattices based on congruence relations. Discernibility matrices have been subsequently defined to calculate all attribute reducts. Basing on the reduction method proposed in this paper, we can study knowledge reduction in inconsistent formal decision contexts based on object oriented concept lattices in further research.

**Acknowledgments.** This work was supported by the Natural Science Foundation of China (Nos. 61202206, 61272021, 61075120, 61173181, 11071284, 60673096 and 11071281) and The Natural Science Foundation of Zhejiang Province (Nos. LZ12F03002, LY14F030001).

# References

1. Ganter, B., Wille, R.: Formal Concept Analysis: Mathematical Foundations. Springer, New York (1999)
2. Gediga, G., Düntsch, I.: Modal-style operators in qualitative data analysis. In: Proceedings of the 2002 IEEE International Conference on Data Mining, pp. 155–162 (2002)
3. Hu, K., Sui, Y., Lu, Y.-C., Wang, J., Shi, C.-Y.: Concept approximation in concept lattice. In: Cheung, D., Williams, G.J., Li, Q. (eds.) PAKDD 2001. LNCS (LNAI), vol. 2035, pp. 167–173. Springer, Heidelberg (2001)
4. Kryszkiewicz, M.: Comparative study of alternative type of knowledge reduction in insistent systems. International Journal of Intelligent Systems 16, 105–120 (2001)
5. Li, J.H., Mei, C.L., Lv, Y.J.: Knowledge reduction in formal decision contexts based on an order-preserving mapping. International Journal of General Systems 41(2), 143–161 (2012)
6. Li, J.H., Mei, C.L., Lv, Y.J.: Incomplete decision contexts: approximate concept construction, rule acquisition and knowledge reduction. International Journal of Approximation Reasoning 54(1), 149–165 (2013)
7. Liu, H.W., Liu, L., Zhang, H.J.: A fast pruning redundant rule method using Galois connection. Applied Soft Computing 11, 130–137 (2011)
8. Liu, M., Shao, M.W., Zhang, W.X., Wu, C.: Reduction method for concept lattices based on rough set theory and its application. Computers and Mathematics with Applications 53, 1390–1410 (2007)
9. Mi, J.S., Wu, W.Z., Zhang, W.X.: Approaches to knowledge reduction based on variable precision rough set model. Information Sciences 159, 255–272 (2004)
10. Mi, J.-S., Wu, W.-Z., Zhang, W.-X.: Approaches to approximation reducts in inconsistent decision tables. In: Wang, G., Liu, Q., Yao, Y., Skowron, A. (eds.) RSFDGrC 2003. LNCS (LNAI), vol. 2639, pp. 283–286. Springer, Heidelberg (2003)
11. Mi, J.S., Leung, Y., Wu, W.Z.: Approaches to attribute reduction in concept lattices induced by axialities. Knowledge-Based Systems 23, 504–511 (2010)
12. Novotný, M.: Dependence spaces of information system. In: Orlowska, E. (ed.) Incomplete Information: Rough Set Analysis, pp. 193–246. Physica-Verlag, Heidelberg (1998)
13. Pawlak, Z.: Rough sets. International Journal of Computer and Information Sciences 11, 341–356 (1982)
14. Pawlak, Z.: Rough Sets: Theoretical Aspects of Reasoning About Data. Kluwer Academic Publishers, Boston (1991)
15. Saquer, J., Deogun, J.S.: Formal Rough Concept Analysis. In: Zhong, N., Skowron, A., Ohsuga, S. (eds.) RSFDGrC 1999. LNCS (LNAI), vol. 1711, pp. 91–99. Springer, Heidelberg (1999)
16. Saquer, J., Deogun, J.: Concept approximations based on rough sets and similarity measures. Int. J. Appl. Math. Comput. Sci. 11, 655–674 (2001)
17. Skowron, A., Rauszer, C.: The discernibility matrices and functions in information systems. In: Slowinski, R. (ed.) Intelligent Decision Support, Handbook of Applications and Advances of the Rough Set Theory, pp. 331–362. Kluwer Academic Publisher, Dordrecth (1991)
18. Slezak, D.: Approximate reducts in decision tables. In: Proc. 6th Conf. Int. on Information Processing and Management of Uncertainty in Knowledge-Based Systems, IPMU 1996, Granada, Spain, pp. 1159–1164 (1996)

19. Slowinski, R., Stefanowski, J., Greco, S., Matarazzo, B.: Rough set based on processing of inconsistent information in decision analysis. Control Cybernet. 29(1), 379–404 (2000)
20. Stefanowski, J.: On rough set based approaches to induction of decision rules. In: Polkowski, L., Skowron, A. (eds.) Rough Sets in Knowledge Discovery, pp. 500–529. Physica, Heidelberg (1998)
21. Wang, X., Ma, J.-M.: A novel approach to attribute reduction in concept lattices. In: Wang, G.-Y., Peters, J.F., Skowron, A., Yao, Y. (eds.) RSKT 2006. LNCS (LNAI), vol. 4062, pp. 522–529. Springer, Heidelberg (2006)
22. Wang, X., Zhang, W.X.: Relations of attribute reduction between object and property oriented concept lattices. Knowledge-Based Systems 21(5), 398–403 (2008)
23. Wang, X.: Approaches to attribute reduction in concept lattices based on rough set theory. International Journal of Hybrid Information Technology 5(2), 67–79 (2012)
24. Wang, X., Wu, W.Z.: Approximate reduction in inconsistent formal decision contexts. In: Lin, T.Y., Hu, X.H., Wu, Z.H. (eds.) Proceedings of 2012 IEEE International Conference on Granular Computing, Grc 2012, Hangzhou, China, pp. 616–621 (2012)
25. Wei, L., Qi, J.J., Zhang, W.X.: Attribute reduction theory of concept lattice based on decision formal contexts. Science in China Series F-Information Science 51(7), 910–923 (2008)
26. Wille, R.: Restructuring lattice theory: An approach based on hierarchies of concepts. In: Rival, I. (ed.) Ordered Sets, pp. 445–470. Reidel, Dordrecht (1982)
27. Wu, W.Z., Leung, Y., Mi, J.S.: Granular computing and knowledge reduction in formal contexts. IEEE Transactions Knowledge and Data Engineering 21(10), 1461–1474 (2009)
28. Yao, Y.Y.: Rough Set Approximations in Formal Concept Analysis. In: Dick, S., Kurgan, L., Pedrycz, W., Reformat, M. (eds.) Proceedings of 2004 Annual Meeting of the North American Fuzzy Information Processing Society, pp. 73–78. IEEE (2004)
29. Yao, Y.Y.: Concept lattices in rough set theory. In: Proceedings of 2004 Annual Meeting of the North American Fuzzy Information Processing Society, pp. 796–801. IEEE Computer Society, Washington, D.C. (2004)
30. Yao, Y.: A comparative study of formal concept analysis and rough set theory in data analysis. In: Tsumoto, S., Słowiński, R., Komorowski, J., Grzymała-Busse, J.W. (eds.) RSCTC 2004. LNCS (LNAI), vol. 3066, pp. 59–68. Springer, Heidelberg (2004)
31. Zhang, W.X., Mi, J.S., Wu, W.Z.: Approaches to knowledge reductions in inconsistent systems. International Journal of Intelligent Systems 21(9), 989–1000 (2003)
32. Zhang, W.X., Leung, Y., Wu, W.Z.: Information Systems and Knowledge Discovery. Science Press, Beijing (2003)
33. Zhang, W.X., Wei, L., Qi, J.J.: Attribute reduction theory and approach to concept lattice. Science in China Series F-Information Science 48(6), 713–726 (2005)
34. Zhang, W.X., Qiu, G.F.: Uncertain Decision Making Based on Rough Sets. Tsinghua University Publishing House (2005)

# An Explicit Sparse Mapping
# for Nonlinear Dimensionality Reduction

Ying Xia, Qiang Lu, JiangFan Feng, and Hae-Young Bae

Research Center of Spatial Information System, Chongqing
University of Posts and Telecommunications, Chongqing 400065, China
{xiaying,fengjf}@cqupt.edu.cn, csluqiang@gmail.com,
hybae@inha.ac.kr

**Abstract.** A disadvantage of most nonlinear dimensionality reduction methods is that there are no explicit mappings to project high-dimensional features into low-dimensional representation space. Previously, some methods have been proposed to provide explicit mappings for nonlinear dimensionality reduction methods. Nevertheless, a disadvantage of these methods is that the learned mapping functions are combinations of all the original features, thus it is often difficult to interpret the results. In addition, the dense projection matrices of these approaches will cause a high cost of storage and computation. In this paper, a framework based on L1-norm regularization is presented to learn explicit sparse polynomial mappings for nonlinear dimensionality reduction. By using this framework and the method of locally linear embedding, we derive an explicit sparse nonlinear dimensionality reduction algorithm, which is named sparse neighborhood preserving polynomial embedding. Experimental results on real world classification and clustering problems demonstrate the effectiveness of our approach.

**Keywords:** Nonlinear dimensionality reduction, sparse representation.

## 1 Introduction

Great quantities of high-dimensional data are confronted in data processing. The high dimensionality of data will lead to the curse of dimensionality, thus effective dimensionality reduction techniques are demanded. Nonlinear dimensionality reduction has become a kind of important dimensionality reduction technique due to its capacity of obtaining a low intrinsic dimensionality of high-dimensional data [1, 2]. In recent years, nonlinear dimensionality reduction methods, such as locally linear embedding [1], isometric mapping [2], Laplacian eigenmap [3], maximum variance unfolding [4] achieved effective performance in comprehensive experiments. However, these methods have a disadvantage that there are no explicit mappings from the high-dimensional feature space to the low-dimensional representation space. Thus the dimensionality of new-come high-dimensional data could not be reduced quickly. This restricts the application of these nonlinear dimensionality reduction methods in many practical tasks such as pattern recognition and classification.

D. Miao et al. (Eds.): RSKT 2014, LNAI 8818, pp. 149–157, 2014.
DOI: 10.1007/978-3-319-11740-9_15 © Springer International Publishing Switzerland 2014

In recent years, some linear or nonlinear approaches such as locality preserving projections [5] and neighborhood preserving polynomial embedding [6], have been proposed to get approximate explicit mappings for nonlinear dimensionality reduction methods. These approaches retain some intrinsic structures of high-dimensional data set in the dimensionality reduction and achieve effective performance in experiments. A disadvantage of these approaches is, however, that the learned projective functions are combinations of all the original features, thus it is often difficult to interpret the results. Besides, these methods cannot get sparse translation matrices for explicit mappings, thus this will affect the performance of computation and storage. For instance, if we use a linear mapping method like locality preserving projections, it will need 100 million expensive floating-point multiplications to project a feature from 100 thousand dimensionalities to 1000 dimensionalities. Moreover, storage of the projection matrix in floating-point format is 400 million. The high cost is unaffordable in many real scenarios such as mobile applications or on embedded devices.

Several methods have been introduced to learn sparse mappings. The sparse principal component analysis method is proposed to produce modified principal components with sparse loadings in [7]. A spectral regression method is proposed for sparse subspace learning in [8]. This method casts the problem of learning the projective functions into a linear regression framework. It can obtain sparse mappings for subspace learning by using a $l1$-based linear regression. These methods get sparse mappings with the assumption that there exists a linear mapping between the high-dimensional data and their low-dimensional representations. Nevertheless, this linearity assumption may be too restrictive for nonlinear dimensionality reduction.

In order to solve the problem above, a two-step framework to obtain sparse polynomial mappings for nonlinear dimensionality reduction is proposed in this paper. Firstly, a nonlinear dimensionality reduction method could be applied to get low-dimensional representations of high-dimensional input data. Secondly, polynomial regression based on L1-norm regularization is applied to get a sparse polynomial mapping from high-dimensional input data to their low-dimensional representations. We can acquire sparse polynomial mappings for nonlinear dimensionality reduction by using this approach. In this paper, we use the method of locally linear embedding at the first step of the framework to derive a sparse nonlinear dimensionality reduction algorithm, which is named sparse neighborhood preserving polynomial embedding. Experiments have been conducted to demonstrate the effectiveness of the proposed approach.

## 2     Sparse Polynomial Mapping Framework

In this section, we present a framework to learn a sparse polynomial mapping from high-dimensional data samples to their low-dimensional representations.

This framework can be divided into two steps. In the first step, a nonlinear dimensionality method is used to get the low-dimensional representations of high-dimensional input data. In the second step, a sparse polynomial mapping which directly maps high-dimensional data to their low-dimensional representations is learned by using the $l1$-based simplified polynomial regression presented below. In the training phase, a sparse polynomial mapping is learned by this two-step framework. In the testing phase, we get the low-dimensional representations by directly projecting high-dimensional data using the sparse polynomial mapping which is learned in the training phase.

High-dimensional input data are denoted by $X = [x_1, x_2, ..., x_n], x_i \in \mathbb{R}^m$ where n is the number of the input data. Their low-dimensional representations are denoted by a matrix $Y = [y_1, y_2, ..., y_d]$ where the low-dimensional representation $y^{(i)} \in \mathbb{R}^d (d \ll m)$ is the transpose of the $i$th row of $Y$.

It has been presented in [9] that many nonlinear dimensionality reduction methods, including locally linear embedding [1], isometric mapping [2] and Laplacian eigenmap [3], can be cast into the framework of graph embedding. By this framework, learning the low-dimensional representations of the high-dimensional data is reduced to solving the following optimization problem:

$$\min_{y^{(i)}} \frac{1}{2} \sum_{i,j=1}^{n} W_{ij} \| y^{(i)} - y^{(j)} \|^2$$

$$\text{s.t. } \sum_{i=1}^{N} D_i y^{(i)} y^{(i)^T} = I \tag{1}$$

where $W_{ij} (i, j = 1, 2, ..., n)$ are weights which are defined by the input data samples, and $D_i = \sum_{j=1}^{n} W_{ij}$, and $I$ is an identity matrix. With some simple algebraic calculation, (1) is equivalent to

$$\min_{Y} \text{ tr}(Y^T (D - W)Y)$$

$$\text{s.t. } Y^T DY = I \tag{2}$$

where $D$ is a diagonal matrix whose diagonal entity is $D_i$, and $W$ is a symmetrical matrix whose entity is $W_{ij}$. The optimal solutions $y_k \in \mathbb{R}^n (k = 1, 2, ..., d)$ are the eigenvectors of the following generalized eigenvalue problem corresponding to the $d$ smallest eigenvalues:

$$(D - W)y = \lambda Dy \tag{3}$$

Once $y_k (k = 1, 2, ..., d)$ are calculated, the low-dimensional representation $y^{(i)} = (y_1(i), ..., y_d(i))^T$.

It is worth noting, however, that the computational complexity exponentially increases as the polynomial degree increases in the computing stage. Thus we use a simplified polynomial in the following by removing the crosswise items of the polynomial. We assume that the $k$th component $y_i^k$ of $y^{(i)}$ is a polynomial of degree $p$ in $x_i$ in the following manner:

$$y_i^k = v_k^T X_p^{(i)}$$

where $v_k$ is the vector of polynomial coefficients, and $X_p^{(i)}$ is defined by

$$X_p^{(i)} = \begin{pmatrix} \overbrace{x_i \otimes x_i \otimes \cdots \otimes x_i}^{p} \\ \vdots \\ x_i \otimes x_i \\ x_i \end{pmatrix}$$

where $\otimes$ stands for the Kronecker product defined on matrices. For two matrices $A = (a_{ij})$ and B, $A \otimes B$ is a block matrix whose $(i,j)$th block is $a_{ij}B$ .

An explicit sparse polynomial mapping can be learned by the sparse polynomial mapping framework. This framework uses the following simplified polynomial regression with the $L_1$ penalty. Due to the nature of the $L_1$ penalty, some coefficients will be shrunk to exact zero if the penalty parameter is large enough [10]. Thus it produces a sparse model, which is exactly what we want.

Low-dimensional representations $y^{(i)} (i = 1, 2, ..., n)$ of the high-dimensional data samples $x_i (i = 1, 2, ..., n)$ are computed by solving the generalized eigenvalue problem (3). With a $L_1$ penalty on $v_k$ which is the vector of polynomial coefficients, we have

$$\min_{v_k} \sum_{k=1}^{d} \sum_{i=1}^{n} ((y_i^k - v_k^T X_p^{(i)})^2 + \lambda \| v_k \|_1) \qquad (4)$$

which is named the $l_1$-based simplified polynomial regression, where $\lambda$ is a penalty parameter, and $\| \cdot \|_1$ is the $L_1$ norm.

The optimization problem (4) can be solved by an efficient coordinate descent algorithm [11] initialized by the value obtained in a previous iteration. Since the $L_1$ penalty is applied, the sparse coefficient vector of the polynomial mapping can be acquired. The sparsity of the vector of polynomial coefficients can be also controlled by tuning the parameter $\lambda$ . The larger the value of parameter $\lambda$ is, the higher the sparsity of the coefficients. The sparse coefficients shall bring great convenience to subsequent computation and storage.

# 3    Sparse Neighborhood Preserving Polynomial Embedding

It has been presented in [9] that most nonlinear dimensionality reduction methods, including locally linear embedding [1], isometric mapping [2] and Laplacian eigenmap [3], can be cast into the framework of graph embedding with different weights. Thus different sparse polynomial mapping algorithms could be derived by the framework of sparse polynomial mapping proposed in Section 2 with different weights. In this section, we derive an explicit sparse nonlinear dimensionality reduction algorithm, named sparse neighborhood preserving polynomial embedding (SNPPE). It is obtained by the above framework with weights defined in a way same to the locally linear embedding method [1].

Given $n$ input data points $x_1,...,x_n$ in $\mathbb{R}^m$, we construct a weighted graph with $n$ nodes, one for each sample, and a set of edges connecting neighboring samples. The sparse polynomial mapping is derived by solving the generalized eigenvalue problem (3) and the optimization problem (4). The algorithmic procedure of SNPPE is formally stated below.

Step 1. **Constructing the adjacency graph**: Put an edge between data points $x_i$ and $x_j$ if $x_i$ among $k$ nearest neighbors of $x_j$ or $x_j$ is among $k$ nearest neighbors of $x_i$.

Step 2. **Choosing the weights**: Compute the weights $W_{ij}$ that best reconstruct each data point $x_i$ from its neighbors, minimizing the cost in equation $E(W) = \sum_i \| x_i - \sum_j W_{ij} x_j \|^2$ by constrained linear fits.

Step 3. **Computing the low-dimensional representations**: The low-dimensional representations $y^{(i)} (i = 1, 2,..., n)$ of the high-dimensional data points can be computed by solving the generalized eigenvalue problem (3).

Step 4. **Solving the optimization problem**: The optimal solutions $v_k (k = 1, 2,..., d)$ of the optimization problem (4) can be computed by the pathwise coordinate descent algorithm in [11].

In the step 2, the weight matrix $W = (W_{ij})$ has a solution given by $R_i = M^{-1}I / (I^T M^{-1} I)$ where $R_i$ is a column vector formed by the $k$ nonzero entries in the $i$th row of $W$, and $I$ is a column vector of all ones. The $(j,l)$th entry of the $k \times k$ matrix $M$ is $(x_j - x_i)^T (x_l - x_i)$, where $x_j$ and $x_l$ are among the $k$ nearest neighbors of $x_i$.

## 4    Experiments

In this section, we describe an experimental evaluation of the proposed sparse neighborhood preserving polynomial embedding algorithm for nonlinear dimensionality reduction. The face recognition and face clustering are carried out using the proposed algorithm and compared algorithms in the experiments.

### 4.1    Datasets

Two face databases were used in this experiment. They are the AR database [12] and the PIE database [13], respectively. The AR database contains over 4000 color images corresponding to 126 people's faces. These images include front view of faces with different expressions, illumination conditions, and occlusions (sun glasses and scarf). In the implementation, we use a subset of the AR database, which contains 1400 face images corresponding to 100 people, where each individual has 14 different images. The PIE database contains more than 40,000 facial images of 68 individuals. These images were obtained across different poses, with different expressions, and under variable illumination conditions. In this experiment, we select face images of near frontal poses and use all the images under different illumination conditions and facial expressions.   A subset including 2040 images with 30 images per individual is acquired finally. All the face images in this experiment are manually aligned and cropped. These face images are resized to a size of $32 \times 32$ pixels, and the gray level values are rescaled to the range from 0 to 1.

## 4.2     Experimental Settings

The polynomial degree of the proposed algorithm is set as 2 in the experiments. For these two databases, we randomly choose half of the images per class for training, and the remaining for test. The training images are used to learn explicit mappings. The testing images are mapped into lower dimensional subspace by these explicit mappings. For simplicity, recognitions are implemented by using nearest neighbor classifier in the reduced feature space. 5-fold cross validation has been used in the experiment to choose the best penalty parameter. As a baseline, we also give the recognition rates of the classifier using the raw data without dimensionality reduction. In practice, 10 training/test splits are randomly chosen. The average recognition accuracies over these splits are illustrated below.

## 4.3     Recognition Result

The proposed sparse neighborhood preserving polynomial embedding algorithm (SNPPE) is an unsupervised dimensionality reduction method. We compare it with principal component analysis (PCA) [14], Sparse PCA [7], and neighborhood preserving polynomial embedding (NPPE) [6]. The recognition rates on AR and PIE are illustrated in Fig. 1(a) and Fig. 1(b), respectively. As can be seen, the performance of the PCA, Sparse PCA, NPPE and SNPPE algorithms varies with the number of dimensionalities. We demonstrate the best results with the standard deviations obtained by these algorithms in Table 1. The sparsity of the projection matrix of each mapping is also shown, and it is calculated as the ratio of the number of zero entries and the total number of entries. As can be seen, the sparsity for PCA and NPPE are both zero, while the sparsity for Sparse PCA and the proposed SNPPE are very high. Comparing with second best method, SNPPE achieves 7.6% and 13.4% relative improvements on AR and PIE, respectively. The performances of the proposed algorithm overtake the compared algorithms.

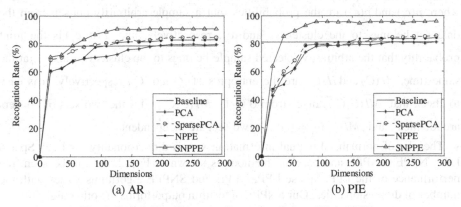

**Fig. 1.** Recognition Rates vs. dimensionality on AR and PIE

**Table 1.** Recognition Results on AR and PIE

| Method | AR | | | PIE | | |
|---|---|---|---|---|---|---|
| | accuracy (%) | dim | sparsity (%) | accuracy (%) | dim | sparsity (%) |
| Baseline | $78.32 \pm 0.3$ | 1024 | -- | $80.98 \pm 0.7$ | 1024 | -- |
| PCA | $79.36 \pm 0.4$ | 246 | 0 | $80.88 \pm 0.7$ | 230 | 0 |
| Sparse PCA | $84.36 \pm 0.5$ | 215 | $88.65 \pm 0.4$ | $85.00 \pm 0.6$ | 208 | $92.16 \pm 0.5$ |
| NPPE | $82.62 \pm 0.6$ | 185 | 0 | $83.53 \pm 0.6$ | 115 | 0 |
| SNPPE | $90.80 \pm 0.2$ | 135 | $92.68 \pm 0.3$ | $96.36 \pm 0.7$ | 93 | $94.21 \pm 0.6$ |

## 4.4    Clustering Result

The use of our proposed algorithm for face clustering is investigated on PIE database. Face clustering is an unsupervised task, and we compare our algorithm SNPPE with locality preserving projections (LPP) [5], Sparse LPP [8], and neighborhood preserving polynomial embedding (NPPE) [6]. K-means is chosen as our clustering algorithm. It is performed in the reduced spaces which are reduced by using the above algorithms. For the baseline method, K-means is performed in the original feature space. The clustering result is evaluated by comparing the obtained label of each face image with that provided by the ground truth. The normalized mutual information ($\overline{MI}$) is used to measure the clustering performance [8]. Let $C$ denote the set of clusters acquired from the ground truth, and $C'$ obtained from an algorithm. Their normalized mutual information metric $MI(C, C')$ is defined as follows:

$$\overline{MI(C, C')} = \frac{\sum_{c_i \in C, c'_j \in C'} p(c_i, c'_j) \cdot \log_2 \frac{p(c_i, c'_j)}{p(c_i) \cdot p(c'_j)}}{\max(H(C), H(C'))}$$

where $p(c_i)$ and $p(c_j')$ are the probabilities that a sample arbitrarily chosen from the data set belongs to the clusters $c_i$ and $c_j'$, respectively, and $p(c_i, c_j')$ is the joint probability that the arbitrarily selected sample belongs to the clusters $c_i$ and $c_j'$ at the same time. $H(C)$ and $H(C')$ are the entropies of $C$ and $C'$, respectively. It is easy to check that $\overline{MI}(C, C')$ ranges from 0 to 1. $\overline{MI}(C, C') = 1$ if the two sets of clusters are identical, and $\overline{MI}(C, C') = 0$ if the two sets are independent.

The plot of normalized mutual information versus dimensionality for LPP, Sparse LPP, NPPE, SNPPE and baseline methods is shown in Fig. 2. As can be seen, the performance of the LPP, Sparse LPP, NPPE and SNPPE algorithms varies with the number of dimensionalities. Our SNPPE algorithm outperforms the other algorithms.

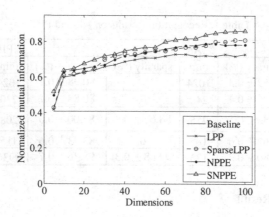

**Fig. 2.** Normalized mutual information vs. dimensionality on PIE dataset

# 5   Conclusion

In this paper, an explicit sparse polynomial mapping framework for nonlinear dimensionality reduction is proposed. It is developed from the graph embedding and the simplified polynomial regression with L1-norm regularization. An explicit sparse polynomial mapping from high-dimensional input data to their low-dimensional representations can be obtained by this framework. Thus the dimension of a new-come data sample can be quickly reduced by the learned explicit mapping, and the cost of computation and storage can be decreased considerably.

Through the proposed framework, we can derive new sparse nonlinear dimensionality reduction algorithms. In practice, we develop a new algorithm named sparse neighborhood preserving polynomial embedding using this framework. Experimental results on face recognition and face clustering show effectiveness of the proposed approach.

**Acknowledgments.** We would like to thank to the reviewers for their helpful comments. This work was financially supported by the Natural Science Foundation of China (41201378), Natural Science Foundation Project of Chongqing CSTC (cstc2012jjA40014), and Doctoral Startup Foundation of Chongqing University of Posts and Telecommunications (A2012-34).

# References

1. Roweis, S.T., Saul, L.K.: Nonlinear dimensionality reduction by locally linear embedding. Science 290(5500), 2323–2326 (2000)
2. Tenenbaum, J.B., de Silva, V., Langford, J.C.: A global geometric framework for nonlinear dimensionality reduction. Science 290(5500), 2319–2323 (2000)
3. Belkin, M., Niyogi, P.: Laplacian eigenmaps for dimensionality reduction and data representation. Neural Computation 15(6), 1373–1396 (2003)
4. Weinberger, K.Q., Saul, L.K.: Unsupervised learning of image manifolds by semidefinite programming. International Journal of Computer Vision 70(1), 77–90 (2006)
5. He, X., Niyogi, P.: Locality preserving projections. In: Advances in Neural Information Processing Systems, vol. 16, pp. 37–45. The MIT Press, Cambridge (2004)
6. Qiao, H., Zhang, P., Wang, D., Zhang, B.: An Explicit Nonlinear Mapping for Manifold Learning. IEEE Transactions on Cybernetics 43(1), 51–63 (2013)
7. Zhou, H., Hastie, T., Tibshirani, R.: Sparse principle component analysis. Journal of Computational and Graphical Statistics 15(2), 265–286 (2006)
8. Cai, D., He, X., Han, J.: Spectral regression: A unified approach for sparse subspace learning. In: Proceedings of the 7th IEEE International Conference on Data Mining, pp. 73–82 (2007)
9. Yan, S., Xu, D., Zhang, B., Zhang, H., Yang, Q., Lin, S.: Graph embedding and extensions: A general framework for dimensionality reduction. IEEE Transactions on Pattern Analysis and Machine Intelligence 24(1), 40–51 (2007)
10. Tibshirani, R.: Regression shrinkage and selection via the lasso. Journal of the Royal Statistical Society, Series B (Methodological), 267–288 (1996)
11. Friedman, J., Hastie, T., Tibshirani, R.: Regularization paths for generalized linear models via coordinate descent. Journal of Statistical Software 33(1), 1–22 (2010)
12. Martinez, A., Benavente, R.: The AR face database. CVC Tech. Report #24 (1998)
13. Sim, T., Baker, S., Bsat, M.: The CMU Pose, Illumination, and Expression Database. IEEE Transactions Pattern Analysis and Machine Intelligence 25(12), 1615–1618 (2003)
14. Jolliffe, I.: Principal component analysis. John Wiley & Sons, Ltd. (2005)

# Applications of Rough Sets

# A Web-Based Learning Support System for Rough Sets

Ying Zhou and Jing Tao Yao

Department of Computer Science, University of Regina,
Regina, Saskatchewan, Canada S4S 0A2
{zhou227y,jtyao}@cs.uregina.ca

**Abstract.** Web-based learning is gaining popularity due to its convenience, ubiquity, personalization, and adaptation features compared with
traditional learning environments. The learning subjects of Web-based
learning systems are mostly for popular sciences. Little attention has
been paid for learning cutting edge subjects and no such systems have
been developed for rough sets. This paper presents the design principle, system architectures, and prototype implementation of a Web-based
learning support system named *Online Rough Sets* (ORS). The system is
specifically designed for learning rough sets in a student-centered learning environment. Some special features, such as adaptation, are emphasized in the system. The ORS has the ability of adaptation to student
preference and performance by modifying the size and order of learning
materials delivered to each individual. Additionally, it predicts estimated
learning time of each topic, which is helpful for students to schedule their
learning paces. A demonstrative example shows ORS can support students to learn rough sets rationally and efficiently.

## 1 Introduction

Web-based learning is obtaining popularity and provides many benefits compared with traditional learning environments [6]. It is learning via electronic
media on the Web [2]. Some Web-based learning systems have been built to
facilitate teaching and learning of various subjects.

In post-secondary education, there are some Web-based learning systems for
computer science subjects, such as, artificial intelligence, network security, and
information retrieval [3,7,8,10,12,13,15]. Rough set theory has gained significant
attention as an intelligent tool dealing with uncertainty [18]. However, there are
no Web-based learning systems for rough sets in the literature. The purpose of
this paper is to design and prototype an effective Web-based learning system for
rough sets.

Web-based learning support systems (WLSS) provide a reasonable framework
to realize Web-based learning. The WLSS sustain teaching as well as learning
by using the Web as an interface, repository and resource center [17]. They offer
learning environments which can be accessed by learners regardless of their time
and places [17]. Well-designed WLSS can supply student-centered education [11].

D. Miao et al. (Eds.): RSKT 2014, LNAI 8818, pp. 161–172, 2014.
DOI: 10.1007/978-3-319-11740-9_16 © Springer International Publishing Switzerland 2014

A student-centered learning approach for adults, andragogy, may be suitable for rough sets learners, as most of them are adults. Andragogy aims to help adult learners according to nature and characteristics of adults [14]. This approach is driven more by needs, preference, and performance of students rather than those of teachers.

In this paper, we present a Web-based learning support system named Online Rough Sets (ORS). We combine WLSS with andragogy to support education, identify a set of rough sets concepts including their dependency information as structured learning materials, and collect a set of concept-specific quizzes. The design of ORS mainly depends on a learning model which is the realization of the learning process of andragogy. The learning model has the mechanism to generate adapted learning materials for each student. With this model, learning materials are tailored and sorted according to each student's preference. Learning materials delivered to a student will be automatically adjusted for review purpose when the student achieves poor performance on any quiz towards a learning material. Moreover, the ORS helps students schedule learning paces and improve learning efficiency by providing estimated learning time for each concept.

## 2    Web-Based Learning Support Systems and Rough Sets

This section provides some background and related works on which this paper is based.

### 2.1    Web-Based Learning Systems for Computer Science Subjects

E-learning refers to the use of electronic media with information and communication technology in education [16]. It is defined as learning through electronic means. Web-based learning is an alternative name of e-learning when the delivery method (i.e., the Web) is emphasized [16].

There are some Web-based learning systems for computer science subjects reported in the literature. For theory of computation, a weak ontology is used to organize learning materials [13], while a robot is utilized to simulate details of finite state automata [7]. For artificial intelligence, an interface is developed to manipulate input and output of prolog-programmed scripts through the Web [12]. For robotic, a simulator connecting with a real robot is built to perform experiments [10]. For cryptography in network security, the system generates adapted learning materials after a student querying on a particular concept [15]. For system analysis, a 3D role-play game supports students to learn the waterfall development model by themselves [3]. For fuzzy information retrieval system, a simulator is developed to visualize steps of weighted queries [8].

Although these learning systems have various advantages, there are still some limitations. We list a few common ones in the following: (1) no adaptation to performance is made; (2) no systems offer queries on multiple concepts; (3) learning materials of systems are organized in a fixed manner except the system for cryptography in network security; and (4) some systems are only simulators or interfaces for the purpose of displaying steps clearly.

## 2.2 Web-Based Learning Support Systems

The WLSS are computerized systems that support teaching and learning by making use of the Web [11]. Fan suggests WLSS should consider four major functions, namely, the complexity of learning support, the adaptability support, the interaction support, and the assessment support [5]. Well-designed WLSS are able to offer education with a student-centered learning style [11]. In this type of education, students choose and study learning materials based on their background knowledge, while instructors act as supporting roles [6].

Andragogy aims to help adults learn better. Its learning process involves eight steps: (1) preparing learners; (2) establishing a climate; (3) involving learners in mutual planning; (4) diagnosing learning needs; (5) forming learning objectives; (6) designing plans with learners; (7) carrying out plans; and (8) evaluating outcomes [9]. The andragogy approach can be integrated in WLSS which have the flexibility to provide a student-centered type of education.

## 2.3 Rough Sets

Rough set theory was proposed by Pawlak in 1980s [18]. The fundamental idea of rough sets is the lower and upper approximations of a set. Recent development of rough sets includes some models based on the probabilistic rough sets [19], such as the decision-theoretic rough set model and the game-theoretic rough set model.

Complex knowledge can be separated into a network of prerequisites [4]. Learners should master relevant prerequisites before moving to another higher level knowledge [4]. As the knowledge of rough sets can be represented by a series of formulas, it is possible to decompose them into a set of concepts with dependency. With notes, assignments, and exams of the "Rough Sets & Applications" course offered in the University of Regina, it is possible to identify the knowledge as a set of concepts, dependency, and concept-specific quizzes.

Table 1 shows the 68 concepts that we have identified in five groups. Dependency exists among related concepts, and Fig. 5 (a) partly shows some dependent relationships. Every concept accompanies its own quizzes. For example, the question for *Granule-based rough sets* may be "Please choose the right formula which represents the granule-based rough sets, given $E$ is an equivalence relation on $U$ and $C \subseteq U$." The candidate answers for student to select are Equation (1) and Equation (2).

$$\underline{apr}(C) = \{x | x \in U, \forall_{y \in U}[xEy \Rightarrow y \in C]\},$$
$$\overline{apr}(C) = \{x | x \in U, \exists_{y \in U}[xEy \wedge y \in C]\}. \tag{1}$$

$$\underline{apr}(C) = \bigcup\{[x] | [x] \in U/E, [x] \subseteq C\},$$
$$\overline{apr}(C) = \bigcup\{[x] | [x] \in U/E, [x] \cap C \neq \emptyset\}. \tag{2}$$

**Table 1.** Concepts of rough sets

| Group | Concept | Group | Concept |
|---|---|---|---|
| Basic | Information table | General | Deletion-based reduct |
| | Equivalence relation (ER) | | Advantage of relative reduct |
| | Concept in an information table | | Matrix absorption |
| | Definable set | | Addition-deletion-based reduct |
| | Family of definable set | | Addition-based reduct |
| | Partition induced by ER | | Powerset-based reduct |
| | Family of undefinable set | Advanced | PRS |
| | Property of definable set | | DTRS |
| | Family of concept | | GTRS |
| | Partition of universe | | Development of PRS |
| | Property of ER | | Equivalence relation rule |
| | Approximation on definable set | | Element-based rough sets |
| | Property of definable set 2 | | Granule-based rough sets |
| | Decision table | | Subsystem-based rough sets |
| | Discernibility matrix (DM) | | Relative attribute reduct |
| | Non-equivalence relation | | Reduct |
| | Semantic Meaning of ER | | Duality |
| | Equivalence class (EC) | | Reduct using DM |
| | Description of EC | Application | Bayesian theorem on PRS |
| | Element of DM | | Bayesian decision procedure on PRS |
| | Property of approximation | Others | Concept |
| | Three regions of rough sets | | Logic connective |
| General | Simplification | | Means of formula |
| | Minimum DM | | Logic language |
| | Boolean algebra | | Power set |
| | Degree of overlap | | Set operator |
| | ER based on BR | | De Morgan's laws |
| | Description 2 of EC | | Lattice |
| | Interpretation of $(\alpha, \beta)$ | | Binary relation (BR) |
| | Attribute reduct | | One-to-one correspondence |
| | Non-reduct-attribute | | Bayesian theorem |
| | Core-attribute | | Probabilistic independence assumption |
| | Reduct-attribute | | Bayesian decision procedure |
| | Monotonicity property | | Nash equilibrium |

# 3    Online Rough Sets Development

This section details the development processes of ORS including requirement analysis, system design, and implementation.

## 3.1    Requirement Analysis

The aim of ORS is to support students to learn rough sets through the Web in a student-centered learning environment. The learning materials of ORS are a set of concepts with dependency as presented in Section 2.3, and each concept is accompanied with a set of concept-specific quizzes.

The learning model of ORS is the realization of the eight-step learning process as described in Section 2.2. The learning model supports learners to study those dependent concepts from bottom to top with some personalization. It also ensures that prerequisites of a concept must be studied again when learners fail to pass the concept's quizzes. This is done by adding a loop between *Forming learning objectives* and *Evaluating outcomes*. Fig. 1 shows the learning model and its emphasis (i.e., preference, performance, and estimated time).

There are three categories of users, namely, expert, tutor, and student. Experts are responsible for defining learning materials, dependency, and their characters. Tutors help students customize these learning materials. Students define their preference and study the personalized learning materials.

A *concept* is a unit of learning materials defined by experts in ORS. A *range* defined by tutors points to several concepts, and is used to help students choose their learning objectives. A *goal* indicates several concepts selected by a student from either concepts or ranges. An item of *experience* is a group of concepts which appear in a goal and have been mastered by the student. An item of *preference* contains prioritized constraints based on concept characters (e.g., the difficulty, the media type), and is used to customize learning sequences for everyone.

Learning materials delivered to each individual are organized in a *study plan*. The *study plan* contains a learning sequence and estimated time for learning each concept. The learning sequence guides a student to learn concepts one by one sequentially. The estimated time helps a student allocate proper time to learn a concept, and thus helps the student reduce interruption caused by time shortage in learning this concept. The estimated time can improve learning efficiency, since learners require more time to complete a task if it has been interrupted [1].

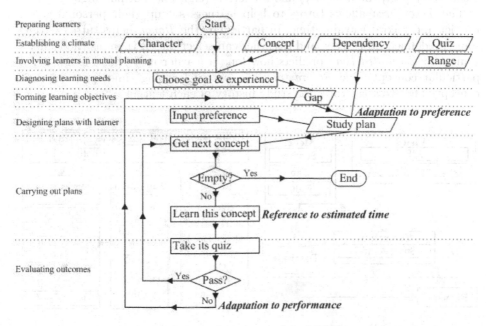

**Fig. 1.** The learning model of ORS

Adaptation to preference means that ORS forms a learning sequence based on dependency and student preference. Adaptation to performance means that ORS modifies a learning sequence by adding prerequisites of a concept into this sequence when the student achieves poor performance on this concept.

## 3.2   System Design

Model-view-controller has been widely used in designing web applications and conforms to the multi-layer architecture of WLSS. The technical architecture of ORS is extended from model-view-controller as shown in Fig. 2. The *Model* component is divided into three layers, namely, the *Service*, the *Dao* and the *POJO*. Each class of the *POJO*, or Plain Old Java Object, represents a table in the *Database*. The *Dao*, or Data Access Object, contains classes which operate data on a single class of the *POJO*. The *Service* encapsulates business logic among multiple classes of the *Dao* and serves the *Controller*.

Fig. 3 shows the functional architecture of ORS, which includes four main functional components, namely, the *Expert area*, the *Tutor area*, the *Student area*, and the *Public management area*.

The *Expert area* is designed to input concepts, dependency, and quizzes by experts. In particular, the *Concept* is used to create, read, update, and delete (CRUD) a concept, which contains the name, description, links of research papers, quizzes, and values on characters (e.g., the low on the difficulty character). The *Dependency* CRUD dependent relationships among concepts. The *Quiz* CRUD a question and candidate answers for each concept. The *Character* CRUD a character (e.g., the difficulty) and its details (e.g., low, normal, high).

The *Tutor area* allows tutors to help learners set up their personal goals, predict and record learning time. In particular, the *Range* helps tutors CRUD some predefined objectives, and a learner can select one of them as the leaner's goal. The *Estimated time* predicts how long a learner may spend to master a particular concept. The *Record time* logs actual learning time spent on each concept. The *Archive* stores the study history of each learner.

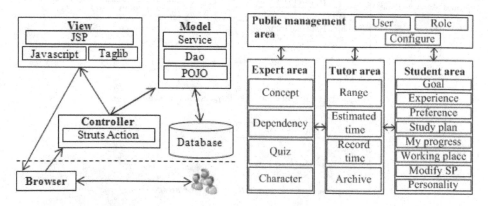

**Fig. 2.** Technical architecture          **Fig. 3.** Functional architecture

The *Student area* provides learners the interaction interfaces with ORS. In particular, the *Personality* is used to maintain student personal information (e.g., age, education level). The *Goal* CRUD a goal from ranges or concepts. The *Experience* allows a student to mark those concepts which he has mastered in his goal initially. The *Preference* allows students to have some preference (e.g., from low to high on the difficulty character) on the learning sequence without breaking dependency among concepts. The *Study plan* generates a linear study plan based on a goal, experience, and preference. The *My progress* displays how many concepts students have learned and how many remaining concepts. The *Working place* displays content of a concept and quizzes associated with it.

## 3.3 Implementation

One major function of ORS is the *Study plan*, which is the core of the adaptation. The *Study plan* uses Algorithm 1 to find a learning sequence for a particular leaner initially. A *Gap* contains concepts which are in a learner's goal and not in the experience. The algorithm assigns all elements of a *Gap* to a learning sequence $\{x_1, x_2, ...x_n\}$ regarding dependency and preference. Initially, the algorithm finds a concept with the highest priority among those concepts which belong to the *Gap* and whose prerequisites do not belong to the *Gap*. Priorities of concepts are determined by the function $h(c)$ based on preference. Subsequently, that concept is assigned to the sequence and is removed from the *Gap*. The dependent relationships which treat that concept as a prerequisite are also removed from the dependency.

---

**Algorithm 1.** A best-first search for finding a learning sequence

---

**Data**: NAC is a priority queue with h(c) as the comparator ;
       LOR is useful dependency among concepts ;
**Result**: a learning sequence
$PO = \{x_1, x_2, ...x_n\}$;                              // initialization
$i \leftarrow 0$;
$NAC.add(\{c3|c3 \in Gap \;\wedge\; \neg\exists_{(c1,c2)\in LOR}(c3 = c2)\})$ ;
**while** $NAC \neq Nil$ **do**
    $i \leftarrow i + 1$;
    $c1 \leftarrow NAC.popup()$;
    $x_i \leftarrow c1$;
    **while** $existAsParent(LOR, c1)$ **do**         // when c1 is a parent
        $c2 \leftarrow findChildByParent(LOR, c1)$; // return first child
        LOR.remove((c1,c2));
        **if** $!existAsChild(LOR, c2)$ **then**        // if c2 has no parent
           | NAC.popin(c2);                      // add into NAC
        **end**
    **end**
**end**
**return** $PO$

---

Moreover, the algorithm repeats the finding, assigning, and removing steps based on the updated *Gap* and dependency until the *Gap* is empty. Finally, each element of the learning sequence has been assigned a unique concept.

If a student achieves poor performance in learning a concept by taking its quizzes, for example, the *Gap* will be updated to prerequisites of this concept. A new learning sequence will be generated by invoking the algorithm again. The study plan will be updated based on the new learning sequence.

Another major function of ORS is the *Estimated time*. It estimates the learning time for each student to improve learning efficiency. The estimated time of a study plan is calculated on concept by concept basis as shown in Fig. 4.

For the *previous learners* who have finished learning tasks, their time spent on each concept is recorded as well as their personality (e.g., age, education level, gender). For the *learner A* who starts learning, the *Cluster analysis* identifies the actual learning time of those previous learners having similar personality with this learner. In the actual time, each concept has a series of data distributed on different dates. For each concept, the *Moving average* calculates the average time from the top N most recent pieces of the concept's actual time (the N is configurable). Finally, the estimated time of the *learner A* is the result of matching between the learning sequence and those averages.

Java and Javascript are the two programming languages implementing ORS. In particular, two Java-based open source systems, appfuse and weka, are used. Appfuse offers the project skeleton and weka provides the APIs of its clustering algorithms to support the *Cluster analysis*. Other related technologies include Struts, Hibernate, DWR, and Dojo. The system is deployed in Tomcat and the database used is MySQL.

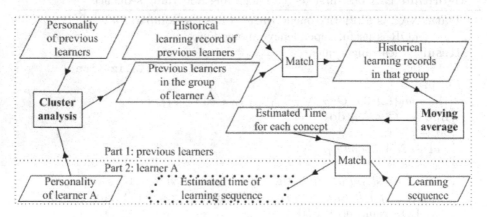

**Fig. 4.** Predict the estimated time for a new learner A

## 4    A Demonstrative Example

This section demonstrates how a student gets an adapted study plan based on the student's preference and performance.

## 4.1 Planning with Preference and Estimated Time

Suppose a student wants to learn a specific concept, called *Concept in an information table*. When the student chooses *Concept in an information table* in ORS, the related 7 concepts are automatically selected based on the dependency. Those concepts, the dependency, and their difficulty levels on the difficulty character are shown in Fig. 5 (a). We assume that the student has known *Logic connective* and marks it as the experience. What the student needs to learn are the 6 concepts presented in Fig. 5 (b).

Adaptation to preference is highlighted in Fig. 5 (c). Before applying the preference, there are multiple available learning sequences. After the student selecting and configuring the character of difficulty as preference (e.g., from low to high), there is only one learning sequence left.

A study plan is generated by ORS for the student as shown in Table 2. The learning sequence and the estimated time are presented in different columns. The learning sequence helps the student learn rationally. The estimated time is helpful to schedule the student's learning paces and improve learning efficiency, since it can help the student decrease interruption caused by time shortage in learning a concept.

## 4.2 Learning Sequences Adapted to Performance

Based on the above study plan, we assume that the student has learned the concepts of *Set operator* and *Information table* successfully. The student progress is shown in Fig. 6. Subsequently, he starts to learn *Logic language* but fails to pass its quizzes.

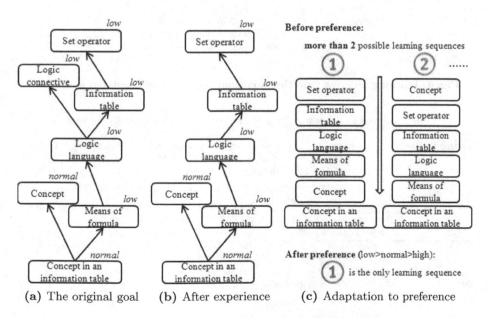

(a) The original goal    (b) After experience    (c) Adaptation to preference

**Fig. 5.** Effect of experience and preference on a goal

**Table 2.** Study plan including a learning sequence and estimated time

| Learning Sequence | Concept | Estimated Time (minutes) |
|---|---|---|
| 001 | Set operator | 12 |
| 002 | Information table | 18 |
| 003 | Logic language | 27 |
| 004 | Means of formula | 12 |
| 005 | Concept | 16 |
| 006 | Concept in an information table | 9 |
| Total: | | 94 |

In this case, adaptation to performance is made due to the failure. The ORS will notify and direct the student to learn the prerequisite concepts of *Logic language* again, namely, *Information table* and *Logic connective*. These two concepts are added into the study plan as shown in Fig. 7.

This is your current progress. Please continue to finish your job. Or, you could delete it.

Have finished:

| Learning Sequence | Concept | Created Date | Updated Date | Time Spent | Status |
|---|---|---|---|---|---|
| 001 | Set operator | 19-02-2014 | 19-02-2014 | 12 | 2 |
| 002 | Information table | 19-02-2014 | 19-02-2014 | 18 | 2 |

Need to learn:

| Learning Sequence | Concept | Created Date | Updated Date | Estimated Time | Status |
|---|---|---|---|---|---|
| 003 | Logic language | 19-02-2014 | 19-02-2014 | 27 | 1 |
| 004 | Means of formula | 19-02-2014 | 19-02-2014 | 12 | 0 |
| 005 | Concept | 19-02-2014 | 19-02-2014 | 16 | 0 |
| 006 | Concept in an information table | 19-02-2014 | 19-02-2014 | 9 | 0 |
| Total Estimated Time: | | | | 64.0 | |

**Fig. 6.** Learning progress reflecting how many concepts are mastered and left

This is your current progress. Please continue to finish your job. Or, you could delete it.

Have finished:

| Learning Sequence | Concept | Created Date | Updated Date | Time Spent | Status |
|---|---|---|---|---|---|
| 001 | Set operator | 19-02-2014 | 19-02-2014 | 12 | 2 |
| 002 | Information table | 19-02-2014 | 19-02-2014 | 18 | 2 |

Need to learn:                    Add dependencies of "Logic language into study plan

| Learning Sequence | Concept | Created Date | Updated Date | Estimated Time | Status |
|---|---|---|---|---|---|
| 003.001 | Information table | 19-02-2014 | 19-02-2014 | 14 | 0 |
| 003.002 | Logic connective | 19-02-2014 | 19-02-2014 | 4 | 0 |
| 003.999 | Logic language | 19-02-2014 | 19-02-2014 | 32 | 1 |
| 004 | Means of formula | 19-02-2014 | 19-02-2014 | 12 | 0 |
| 005 | Concept | 19-02-2014 | 19-02-2014 | 16 | 0 |
| 006 | Concept in an information table | 19-02-2014 | 19-02-2014 | 9 | 0 |
| Total Estimated Time: | | | | 87.0 | |

**Fig. 7.** Adaptation to performance

From this example, the student can learn rough sets rationally and efficiently with the adapted study plan, which is updated according to the preference and performance in the whole learning process.

## 5   Conclusion

Web-based learning support systems offer Web-based learning environments for teachers and students in education. A student-centered learning style is suitable to learners of rough sets in terms of their age. *Online Rough Sets* is the WLSS for rough sets with the student-centered learning approach.

We identified a set of rough sets concepts with dependency and collected a set of concept-specific quizzes. The concepts can be used as structured learning materials for the subject, while a quiz is used to assess whether or not its related concept has been mastered by students.

We developed a learning model from andragogy to make adaptation to each student, and integrated it into the WLSS. The system generates an adapted study plan for each student consisting of a learning sequence and estimated time. The learning sequence can guide students learn concepts from bottom to top according to dependency, preference, and performance. The estimated time helps students schedule learning paces and improve learning efficiency.

*Online Rough Sets* supports students to learn the subject in a student-centered learning environment through the Web anytime and anywhere. In addition, it addresses some limitations found in existing systems to provide personalized study plans to help students learn rationally and efficiently.

## References

1. Bailey, B.P., Konstan, J.A.: On the need for attention-aware systems: Measuring effects of interruption on task performance, error rate, and affective state. Computers in Human Behavior 22(4), 685–708 (2006)
2. Chen, C.M., Lee, H.M., Chen, Y.H.: Personalized e-learning system using item response theory. Computers & Education 44(3), 237–255 (2005)
3. Cheng, C.H., Su, C.H.: A game-based learning system for improving student's learning effectiveness in system analysis course. Procedia - Social and Behavioral Sciences 31, 669–675 (2012)
4. Corbett, A.T., Anderson, J.R.: Knowledge decomposition and subgoal reification in the act programming tutor. In: Proceedings of Artificial Intelligence and Education, vol. 95, Charlottesville (1995)
5. Fan, L.: Web-based learning support system. In: Yao, J.T. (ed.) Web-based Support Systems. AIKP, vol. XXII, pp. 81–95. Springer, London (2010)
6. Fan, L., Yao, Y.Y.: Web-based learning support systems. In: Proceedings of the WI/IAT 2003 Workshop on Applications, Products and Services of Web-Based Support Systems, pp. 43–48. Saint Mary's University, Halifax (2003)
7. Hamada, M., Sato, S.: A game-based learning system for theory of computation using lego nxt robot. Procedia Computer Science 4, 1944–1952 (2011)

8. Herrera-Viedma, E., López-Herrera, A., Alonso, S., Moreno, J., Cabrerizo, F., Porcel, C.: A computer-supported learning system to help teachers to teach fuzzy information retrieval systems. Information Retrieval 12(2), 179–200 (2009)

9. Holton, E.F., Swanson, R.A., Naquin, S.S.: Andragogy in practice: Clarifying the andragogical model of adult learning. Performance Improvement Quarterly 14(1), 118–143 (2001)

10. Hong, S.-H., Park, J.-H., Kwon, K.H., Jeon, J.W.: A distance learning system for robotics. In: Shi, Y., van Albada, G.D., Dongarra, J., Sloot, P.M.A. (eds.) ICCS 2007, Part III. LNCS, vol. 4489, pp. 523–530. Springer, Heidelberg (2007)

11. Kim, D.W., Yao, J.T.: A web-based learning support system for inquiry-based learning. In: Yao, J.T. (ed.) Web-based Support Systems. AIKP, vol. XXII, pp. 125–143. Springer, London (2010)

12. Kozak, M.M.: Teaching artificial intelligence using web-based applications. J. Comput. Sci. Coll. 22(1), 46–53 (2006)

13. Kumar, S., Haider, M.T.U.: Designing ontology-based academic course for theory of computation to facilitate personalized e-learning system. International Journal of Application or Innovation in Engineering and Management (IJAIEM) 2(3), 374–378 (2013)

14. Philip, O.O.: First, there was pedagogy and then came andragogy. Einstein J. Biol. Med. 21(2), 83–87 (2005)

15. Takahashi, Y., Abiko, T., Negishi, E., Itabashi, G., Kato, Y., Takahashi, K., Shiratori, N.: An ontology-based e-learning system for network security. In: 19th International Conference on Advanced Information Networking and Applications, vol. 1, pp. 197–202 (2005)

16. Vogel, J.J., Vogel, D.S., Cannon-Bowers, J., Bowers, C.A., Muse, K., Wright, M.: Computer gaming and interactive simulations for learning: A meta-analysis. Journal of Educational Computing Research 34(3), 229–243 (2006)

17. Yao, J.T., Kim, D.W., Herbert, J.P.: Supporting online learning with games. In: Proc. SPIE 6570, Data Mining, Intrusion Detection, Information Assurance, and Data Networks Security, Orlando, vol. 6570, pp. 65700G-1–65700G-11 (2007)

18. Yao, J.T, Zhang, Y.: A scientometrics study of rough sets in three decades. In: Lingras, P., Wolski, M., Cornelis, C., Mitra, S., Wasilewski, P. (eds.) RSKT 2013. LNCS, vol. 8171, pp. 28–40. Springer, Heidelberg (2013)

19. Yao, Y.Y.: Probabilistic rough set approximations. International Journal of Approximate Reasoning 49(2), 255–271 (2008)

# A Parallel Matrix-Based Approach for Computing Approximations in Dominance-Based Rough Sets Approach

Shaoyong Li and Tianrui Li*

School of Information Science and Technology, Southwest Jiaotong University,
Chengdu, 610031, China
meterer@163.com, trli@swjtu.edu.cn

**Abstract.** Dominance-based Rough Sets Approach (DRSA) is a useful tool for multi-criteria classification problems solving. Parallel computing is an efficient way to accelerate problems solving. Computation of approximations is a vital step to find the solutions with rough sets methodologies. In this paper, we propose a matrix-based approach for computing approximations in DRSA and design the corresponding parallel algorithms on Graphics Processing Unit (GPU). A numerical example is employed to illustrate the feasibility of the matrix-based approach. Experimental evaluations show the performance of the parallel algorithm.

**Keywords:** Rough sets, Dominance relation, Approximations, Parallel computing, GPU.

## 1 Introduction

Dominance-based Rough Set Approach (DRSA) [6] proposed by Greco et al. is an excellent mathematic tool to aid multi-criteria classification problems solving, which has been successfully applied in many fields [13].

Since computation of approximations is a vital step to problems solving with rough sets methodologies, accelerating this computation has attracted many scholars in rough sets society recently. There are many significant incremental approaches for updating approximations in rough sets methodologies under dynamic data environment, which can reduce the computational time spent on updating approximations by trying to avoid the unnecessary computation with the previous results [2–5, 7–11, 14, 15, 17, 19]. Parallelization of algorithms is a popular technique to speed up the computational process. To accelerate computation of approximations, some studies on parallel algorithms for computing rough sets' approximations have done, e.g., Zhang et al. proposed a parallel algorithm for computing approximations of rough sets under the indiscernible relation [16]. Followed that, they compared the parallel algorithms of computing approximations in rough sets on different MapReduce runtime systems [18].

---

* Corresponding author.

D. Miao et al. (Eds.): RSKT 2014, LNAI 8818, pp. 173–183, 2014.
DOI: 10.1007/978-3-319-11740-9_17 © Springer International Publishing Switzerland 2014

The Graphics Processing Unit (GPU) has been an important platform for science computing because of its low cost and massive parallel processing power. GPU computing is able to achieve the highest performance for data-parallel problems. At present, parallel computing with GPU has been applied widely in many fields [12]. Parallelization of the algorithm for computing approximations in DRSA may be helpful to reduce the computational time. This paper aims to investigate a matrix-based approach for computing approximations in DRSA in order to design the corresponding parallel algorithm on GPU.

The remainder of this paper is organized as follows. We present some basic notions of DRSA and modify the definition of $P$-generalized decision in Section 2. We introduce a matrix-based approach for computing approximations in DRSA and give a numerical example to illustrate this method in Section 3. In Section 4, the CPU-based algorithm and the GPU-based parallel algorithm are designed and experimental evaluations show GPU-based parallel algorithm dominates the CPU-based algorithm. This paper ends with conclusions and further research topics in Section 5.

## 2   Preliminaries

At first, we briefly review some basic notions of DRSA [6].

A decision system is denoted by $S = (U, A, V, f)$, where $U$ is a finite set of objects, called the universe; $A = C \cup \{d\}$, $C$ is a set of condition attributes and $d$ is a decision attribute. $V$ is regarded as the domain of all attributes. $f : U \times A \to V$ is an information function such that $f(x, a) \in V_a$, $\forall a \in A$ and $x \in U$, where $V_a$ is the domain of attribute $a$.

In DRSA, the attributes with preference-ordered domains in the decision systems are called as the criteria. There is a preference relation on the universe with respect to each of criteria, e.g., $\forall x, y \in U$, if $f(x, a) \geq f(y, a)$, then $x \succeq_a y$ means " $x$ is at least as good as $y$ with respect to the criterion $a$ ". $\succeq_a = \{(x, y) \in U \times U \mid f(x, a) \geq f(y, a)\}$ is the preference relation with respect to $a$.

Given $P \neq \emptyset$ and $P \subseteq C$, if $x \succeq_a y$ for all $a \in P$, then $x$ dominates $y$ with respect to $P$, which is denoted by $xD_Py$.

$$D_P = \{(x, y) \in U \times U \mid f(x, a) \geq f(y, a), \forall a \in P\}$$

is the dominance relation with respect to $P$.

The basic knowledge granules in DRSA are two types of sets as follows:

- A set of objects dominating $x$, called $P$-dominating set of the object $x$, $D_P^+(x) = \{y \in U \mid yD_Px\}$;
- A set of objects dominated by $x$, called $P$-dominated set of the object $x$, $D_P^-(x) = \{y \in U \mid xD_Py\}$.

$U$ is divided by $d$ into a family of equivalence classes with preference-ordered, called decision classes. Let $\boldsymbol{Cl} = \{Cl_n, n \in T\}$ be a set of decision classes,

$T = \{1, \cdots, t\}$. $\forall r, s \in T$ such that $r > s$, the objects from $Cl_r$ are preferred to the objects from $Cl_s$. In DRSA, the concepts to be approximated are an upward union and a downward union of classes such that

$$Cl_n^{\geq} = \bigcup_{n' \geq n} Cl_{n'}, \quad Cl_n^{\leq} = \bigcup_{n' \leq n} Cl_{n'}, \quad \forall n, n' \in T.$$

$x \in Cl_n^{\geq}$ means "$x$ belongs to at least class $Cl_n$", and $x \in Cl_n^{\leq}$ means "$x$ belongs to at most class $Cl_n$".

The lower and upper approximations of $Cl_n^{\geq}$ and $Cl_n^{\leq}$ are defined respectively as follows:

$$\underline{P}(Cl_n^{\geq}) = \{x \in U \mid D_P^+(x) \subseteq Cl_n^{\geq}\} \tag{1}$$

$$\overline{P}(Cl_n^{\geq}) = \{x \in U \mid D_P^-(x) \cap Cl_n^{\geq} \neq \emptyset\} \tag{2}$$

$$\underline{P}(Cl_n^{\leq}) = \{x \in U \mid D_P^-(x) \subseteq Cl_n^{\leq}\} \tag{3}$$

$$\overline{P}(Cl_n^{\leq}) = \{x \in U \mid D_P^+(x) \cap Cl_n^{\leq} \neq \emptyset\} \tag{4}$$

The lower and upper approximations partition the universe into three regions: positive region, negative region and boundary region as follows:

$$\begin{cases} POS_P(Cl_n^{\geq}) = \underline{P}(Cl_n^{\geq}) \\ NEG_P(Cl_n^{\geq}) = U - \overline{P}(Cl_n^{\geq}) \\ BN_P(Cl_n^{\geq}) = \overline{P}(Cl_n^{\geq}) - \underline{P}(Cl_n^{\geq}) \end{cases}, \quad \begin{cases} POS_P(Cl_n^{\leq}) = \underline{P}(Cl_n^{\leq}) \\ NEG_P(Cl_n^{\leq}) = U - \overline{P}(Cl_n^{\leq}) \\ BN_P(Cl_n^{\leq}) = \overline{P}(Cl_n^{\leq}) - \underline{P}(Cl_n^{\leq}) \end{cases}$$

Next, we modify the $P$-generalized decision in our previous work [8] for this study as follows:

$$\delta_P(x_i) = \langle l_i, u_i \rangle$$

where $l_i = n$ if $d_n = min\{f(y, d) \in V_d \mid y \in D_P^+(x_i)\}$ and $u_i = n$ if $d_n = max\{f(y, d) \in V_d \mid y \in D_P^-(x_i)\}$. $min(\bullet)$ and $max(\bullet)$ are the minimum and maximum of a set, respectively.

With the modified $P$-generalized decision, we can rewrite the definition of approximations in DRSA as follows:

$$\underline{P}(Cl_n^{\geq}) = \{x_i \in U \mid l_i \geq n\} \tag{5}$$

$$\overline{P}(Cl_n^{\geq}) = \{x_i \in U \mid u_i \geq n\} \tag{6}$$

$$\underline{P}(Cl_n^{\leq}) = \{x_i \in U \mid u_i \leq n\} \tag{7}$$

$$\overline{P}(Cl_n^{\leq}) = \{x_i \in U \mid l_i \leq n\} \tag{8}$$

## 3 A Matrix-Based Approach for Computing Approximations in DRSA

As we known, there is a preference relation $\succeq_a$ for each criterion $a \in P$ in DRSA. In our previous work [8], we defined a matrix

$$R^a = \begin{pmatrix} r_{1,1}^a & \cdots & r_{1,|U|}^a \\ \vdots & \ddots & \vdots \\ r_{|U|,1}^a & \cdots & r_{|U|,|U|}^a \end{pmatrix}$$

to present $\succeq_a$, where

$$r_{i,j}^a = \begin{cases} 1, & f(x_i, a) \geq f(x_j, a) \\ 0, & f(x_i, a) < f(x_j, a) \end{cases}, \quad 1 \leq i, j \leq |U|.$$

$|\bullet|$ is the cardinality of a set. To present the dominance relation with respect to the set of criteria $P$, we defined a dominance matrix

$$R^P = \begin{pmatrix} \phi_{1,1}^P & \cdots & \phi_{1,|U|}^P \\ \vdots & \ddots & \vdots \\ \phi_{|U|,1}^P & \cdots & \phi_{|U|,|U|}^P \end{pmatrix}$$

where $\phi_{i,j}^P = \sum\limits_{a \in P} r_{i,j}^a$. If $\phi_{i,j}^P = 0$, then $x_i$ is dominated by $x_j$ strictly with respect to $P$. If $0 < \phi_{i,j}^P < |P|$, then there is not a dominance relation between $x_i$ and $x_j$ with respect to $P$. If $\phi_{i,j}^P = |P|$, then $x_i$ dominates $x_j$ with respect to $P$. Hence we redefined the definitions of the $P$-dominating and $P$-dominated sets of $x_i$ as follows:

$$D_P^+(x_i) = \{x_j \in U \mid \phi_{j,i}^P = |P|\} \tag{9}$$

$$D_P^-(x_i) = \{x_j \in U \mid \phi_{i,j}^P = |P|\} \tag{10}$$

Here we do not care the cases about $0 \leq \phi_{i,j}^P < |P|$ and transform the matrix $R^P$ into a boolean matrix

$$B = \begin{pmatrix} b_{1,1} & \cdots & b_{1,|U|} \\ \vdots & \ddots & \vdots \\ b_{|U|,1} & \cdots & b_{|U|,|U|} \end{pmatrix}$$

where

$$b_{i,j} = \begin{cases} 1, & \phi_{i,j}^P = |P|; \\ 0, & \text{Otherwise.} \end{cases}$$

Let $M_d = (c_1, c_2, \cdots, c_{|U|})$ be a vector to store the values of objects corresponding to the decision attribute $d$, where $c_i = f(x_i, d)$, $1 \leq i \leq |U|$. Next, we define two sparse matrices to prepare for computing the $P$-generalized decisions of all objects on the universe as follows:

$$M_u = \begin{pmatrix} m_{1,1}^u & \cdots & m_{1,|U|}^u \\ \vdots & \ddots & \vdots \\ m_{|U|,1}^u & \cdots & m_{|U|,|U|}^u \end{pmatrix}, \quad M_l = \begin{pmatrix} m_{1,1}^l & \cdots & m_{1,|U|}^l \\ \vdots & \ddots & \vdots \\ m_{|U|,1}^l & \cdots & m_{|U|,|U|}^l \end{pmatrix}$$

where

$$m_{i,j}^u = \begin{cases} b_{i,j} \times c_j, & b_{i,j} = 1 \\ *, & b_{i,j} = 0 \end{cases}, \quad m_{i,j}^l = \begin{cases} b_{j,i} \times c_i, & b_{j,i} = 1 \\ *, & b_{j,i} = 0 \end{cases}$$

$*$ indicates an empty value in the sparse matrices. In fact, $b_{i,j} = 1$ reflects that $x_i$ dominates $x_j$, then the $i$-th row in the matrix $M_u$ gives the decision attribute values of objects in $P$-dominated set of $x_i$. Analogously, the $i$-th row in the matrix $M_l$ gives the decision attribute values of objects in $P$-dominating set of $x_i$.

Now we define two $1 \times |U|$ matrices, namely, $M_{max} = \{u_1, u_2, \cdots, u_{|U|}\}$ and $M_{min} = \{l_1, l_2, \cdots, l_{|U|}\}$, to present the $P$-generalized decisions, where $u_i = n$ if $d_n = max(M_u(i,:))$ and $l_i = n$ if $d_n = min(M_l(i,:))$. $M_u(i,:)$ and $M_l(i,:)$ are the $i$-th row in the matrices $M_u$ and $M_l$, respectively.

With $M_{max}$ and $M_{min}$, we can obtain approximations in DRSA by equations (5-8). Following is an example to validate the usefulness of the matrix-based approach discussed above.

*Example 1.* Given a decision table $S = (U, C \cup d, V, f)$ as shown in Table 1, where $U = \{x_1, x_2, \cdots, x_8\}$, $C = \{a_1, a_2\}$, $V_{a_1} = \{50, 65, 70, 80, 90\}$, $V_{a_2} = \{50, 60, 75, 80, 90\}$, $V_d = \{d_1, d_2, d_3\} = \{1, 2, 3\}$.

**Table 1.** A decision table

| $U$ | $a_1$ | $a_2$ | $d$ ‖ | $U$ | $a_1$ | $a_2$ | $d$ |
|-----|-------|-------|-------|-----|-------|-------|-----|
| $x_1$ | 50 | 75 | 2 | $x_5$ | 80 | 90 | 2 |
| $x_2$ | 65 | 50 | 1 | $x_6$ | 90 | 80 | 3 |
| $x_3$ | 70 | 75 | 1 | $x_7$ | 80 | 80 | 3 |
| $x_4$ | 50 | 60 | 1 | $x_8$ | 90 | 90 | 3 |

At first, we compute the preference matrices of Table 1 as follows:

$$
R^{a_1} = \begin{pmatrix} 1 0 0 1 0 0 0 0 \\ 1 1 0 1 0 0 0 0 \\ 1 1 1 1 0 0 0 0 \\ 1 0 0 1 0 0 0 0 \\ 1 1 1 1 1 0 1 0 \\ 1 1 1 1 1 1 1 1 \\ 1 1 1 1 1 0 1 0 \\ 1 1 1 1 1 1 1 1 \end{pmatrix}, \quad
R^{a_2} = \begin{pmatrix} 1 1 1 1 0 0 0 0 \\ 0 1 0 0 0 0 0 0 \\ 1 1 1 1 0 0 0 0 \\ 1 1 0 1 0 0 0 0 \\ 1 1 1 1 1 1 1 1 \\ 1 1 1 1 0 1 1 0 \\ 1 1 1 1 0 1 1 0 \\ 1 1 1 1 1 1 1 1 \end{pmatrix}
$$

Let $P = \{a_1, a_2\}$, then we have a dominance matrix

$$
R^P = \begin{pmatrix} 2 1 1 2 0 0 0 0 \\ 1 2 0 1 0 0 0 0 \\ 2 2 2 2 0 0 0 0 \\ 1 1 0 2 0 0 0 0 \\ 2 2 2 2 2 1 2 1 \\ 2 2 2 2 1 2 2 1 \\ 2 2 2 2 1 1 2 0 \\ 2 2 2 2 2 2 2 2 \end{pmatrix}
$$

Next, we transform $R^P$ into a boolean matrix

$$B = \begin{pmatrix} 1 & 0 & 0 & 1 & 0 & 0 & 0 & 0 \\ 0 & 1 & 0 & 0 & 0 & 0 & 0 & 0 \\ 1 & 1 & 1 & 1 & 0 & 0 & 0 & 0 \\ 0 & 0 & 0 & 1 & 0 & 0 & 0 & 0 \\ 1 & 1 & 1 & 1 & 1 & 0 & 1 & 0 \\ 1 & 1 & 1 & 1 & 0 & 1 & 1 & 0 \\ 1 & 1 & 1 & 1 & 0 & 0 & 1 & 0 \\ 1 & 1 & 1 & 1 & 1 & 1 & 1 & 1 \end{pmatrix}$$

Since $M_d = \{2, 1, 1, 1, 2, 3, 3, 3\}$, we compute two sparse matrices $M_u$ and $M_l$ as follows:

$$M_u = \begin{pmatrix} 2 & * & * & 1 & * & * & * & * \\ * & 1 & * & * & * & * & * & * \\ 2 & 1 & 1 & 1 & * & * & * & * \\ * & * & * & 1 & * & * & * & * \\ 2 & 1 & 1 & 1 & 2 & * & 3 & * \\ 2 & 1 & 1 & 1 & * & 3 & 3 & * \\ 2 & 1 & 1 & 1 & * & * & 3 & * \\ 2 & 1 & 1 & 1 & 2 & 3 & 3 & 3 \end{pmatrix} , \quad M_l = \begin{pmatrix} 2 & * & 1 & * & 2 & 3 & 3 & 3 \\ * & 1 & 1 & * & 2 & 3 & 3 & 3 \\ * & * & 1 & * & 2 & 3 & 3 & 3 \\ 2 & * & 1 & 1 & 2 & 3 & 3 & 3 \\ * & * & * & * & 2 & * & * & 3 \\ * & * & * & * & * & 3 & * & 3 \\ * & * & * & * & 2 & 3 & 3 & 3 \\ * & * & * & * & * & * & * & 3 \end{pmatrix}$$

Then we have $M_{max} = \{2, 1, 2, 1, 3, 3, 3, 3\}$ and $M_{min} = \{1, 1, 1, 1, 2, 3, 2, 3\}$. By equations (5-8), we can obtain approximations in DRSA of Table 1 as follows:

$$\underline{P}(Cl_2^{\geq}) = \{x_5, x_6, x_7, x_8\}, \overline{P}(Cl_2^{\geq}) = \{x_1, x_3, x_5, x_6, x_7, x_8\};$$
$$\underline{P}(Cl_3^{\geq}) = \{x_6, x_8\}, \qquad \overline{P}(Cl_3^{\geq}) = \{x_5, x_6, x_7, x_8\};$$
$$\underline{P}(Cl_1^{\leq}) = \{x_2, x_4\}, \qquad \overline{P}(Cl_1^{\leq}) = \{x_1, x_2, x_3, x_4\};$$
$$\underline{P}(Cl_2^{\leq}) = \{x_1, x_2, x_3, x_4\}, \overline{P}(Cl_2^{\leq}) = \{x_1, x_2, x_3, x_4, x_5, x_7\};$$

Obviously, $\underline{P}(Cl_1^{\geq}) = \overline{P}(Cl_1^{\geq}) = \underline{P}(Cl_3^{\leq}) = \overline{P}(Cl_3^{\leq}) = U$.

## 4   Algorithms and Experimental Evaluations

In this section, we design two algorithms based on the matrix-based approach discussed in the previous section, CPU-based algorithm and GPU-based parallel algorithm. Then we compare the computational time taken by each of two algorithms and analyze the performance of GPU-based parallel algorithm.

### 4.1   CPU-Based Algorithm and GPU-Based Parallel Algorithm of Computing Approximations in DRSA

We design a CPU-based algorithm for computing approximations in DRSA with reference to the proposed matrix-based approach at first.

Then, we design a GPU-based parallel algorithm for computing approximations in DRSA according to the proposed matrix-based approach.

---

**Algorithm 1.** A CPU-based algorithm for computing approximations in DRSA

---

**Input:**
The information table $S = (U, A, V, f)$ **Output:**
Approximations in DRSA.

```
1  begin
2  |    table ← S;                                    // Evaluate table with S.
3  |    M_c ← table.condition_attri_value;
4  |    M_c' ← transpose(M_c);           // Evaluate M_c' with the transposition of M_c.
5  |    M_d ← table.decision_attri_value;
6  |    for n = 1 → m do                            // Compute all preference matrices.
7  |    |    R^{a_n} ← ge(M_c(n, :), M_c'(:, n));
8  |    end
9  |    R^P ← sum(R^{a_n}, 1, m);                // Compute the dominance matrix R^P.
10 |    B ← eq(|P|, R^P);                    // Transform R^P into a boolean matrix B.
11 |    M_u ← B · M_d;            // Multiplies arrays B and M_d element by element and
                                   returns the result in M_u.
12 |    M_l ← transpose(B) · M_d;
13 |    M_l ← nonzero(M_l);                      // Replace all "0" in M_l with "NaN".
14 |    M_{max} ← max(M_u);
15 |    M_{min} ← min(M_l);
16 |    for n = 2 → m do ;                        // Compute approximations.
17 |
18 |    |    \underline{P}(Cl_n^≥) ← {x_i | i ∈ ge(n, M_{min})};
19 |    |    \overline{P}(Cl_n^≥) ← {x_i | i ∈ ge(n, M_{max})};
20 |    |    \underline{P}(Cl_{n-1}^≤) ← {x_i | i ∈ le(n, M_{max})};
21 |    |    \overline{P}(Cl_{n-1}^≤) ← {x_i | i ∈ le(n, M_{min})};
22 |    end
23 |    Output the results.
24 end
```

---

## 4.2 Analysis of the Performance of the Two Algorithms

In this subsection, we introduce some experiments to test the performance of the proposed two algorithms. In our experiments, the host machine is Intel(R)Xeon(R)CPU E5620 2.40GHz; the GPU device is Tesla C2050. The programs is coded in Matlab (R2013b) and a data set, EEG Eye State, is downloaded from UCI [1].

The aim of our experiments is to investigate that the GPU-based parallel algorithm (Algorithm 2) dominates the CPU-based one (Algorithm 1), and study whether two factors we care about, the number of objects and the number of criteria, can influence the performance of GPU-based parallel algorithm. Hence, we make the following two experiments.

In the first experiment, we select 14 condition criteria and 1000, 2000, $\cdots$, 6000 objects randomly to be experimental data sets, respectively. On these data sets, we execute Algorithms 1 and 2, respectively. The time taken by each of two algorithms is listed in Table 2. With the increasing number of objects, the trends of the computational time taken by each of two algorithms and the speedup between Algorithms 1 and 2 are shown in Fig. 1(a) and Fig. 1(b), respectively.

---

**Algorithm 2.** A GPU-based parallel algorithm for computing approximations in DRSA

---

**Input:**
The information table $S = (U, A, V, f)$ **Output:**
Approximations in DRSA.

1 **begin**
2     $table \leftarrow S$;
3     $M_c \leftarrow gpuArray(table.condition\_attri\_value)$;          // Convert an array to a
      gpuArray with data stored on the GPU device.
4     $M_c' \leftarrow gpuArray(transpose(M_c))$;
5     $M_d \leftarrow gpuArray(table.decision\_attri\_value)$;
6     $R^P \leftarrow parfun(@sum, parfun(@ge, M_c, M_c'), 3)$;          // parfun is the function
      executed on GPU in parallel and @sum is the handle of the function
      sum.
7     $B \leftarrow parfun(@eq, |P|, R^P)$;
8     $M_u \leftarrow B \cdot M_d$;
9     $M_l \leftarrow transpose(B) \cdot M_d$;
10    $M_l \leftarrow parfun(@nonzero, M_l)$;
11    $M_{max} \leftarrow max(M_u)$;
12    $M_{min} \leftarrow min(M_l)$;
13    **for** $n = 2 \rightarrow m$ **do** ;          // gather can transfer gpuArray to local workspace.
14
15        $\underline{P}(Cl_n^{\geq}) \leftarrow \{x_i \mid i \in gather(ge(n, M_{min}))\}$;
16        $\overline{P}(Cl_n^{\geq}) \leftarrow \{x_i \mid i \in gather(ge(n, M_{max}))\}$;
17        $\underline{P}(Cl_{n-1}^{\leq}) \leftarrow \{x_i \mid i \in gather(le(n, M_{max}))\}$;
18        $\overline{P}(Cl_{n-1}^{\leq}) \leftarrow \{x_i \mid i \in gather(le(n, M_{min}))\}$;
19    **end**
20    Output the results.
21 **end**

---

**Table 2.** A comparison of the time taken by Algorithms 1 and 2 vs. the number of objects

| Objects | Criteria | $t_1$ | $t_2$ | $S_p$ |
|---------|----------|-------|-------|-------|
| 1000 | 14 | 0.0549 | 0.0159 | 3.5273 |
| 2000 | 14 | 0.2452 | 0.0425 | 5.7694 |
| 3000 | 14 | 0.5298 | 0.0872 | 6.0757 |
| 4000 | 14 | 0.9846 | 0.1451 | 6.7857 |
| 5000 | 14 | 1.4527 | 0.2295 | 6.3298 |
| 6000 | 14 | 2.0871 | 0.3284 | 6.3553 |

In Table 2, $t_1$ and $t_2$ present the computational time taken by Algorithms 1 and 2, respectively. $S_p = t_1/t_2$ presents the speedup. The x-coordinates in Fig. 1(a) and (b) pertain to the size of the object set. The y-coordinates in Fig. 1(a) and (b) concern the computational time and the speedup, respectively. From Table 2 and Fig. 1(a), we can see that the computational time taken by Algorithms 1 and 2 raises with the increasing size of the object set. The computational time taken by Algorithm 2 is always less than that of its counterpart. From Fig. 1(b) and the fifth column in Table 2, we can see that the trend of the speedup grows

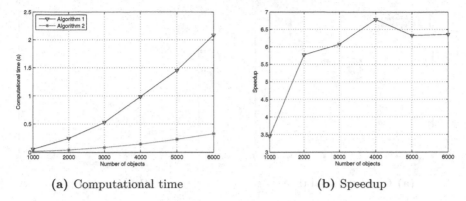

(a) Computational time                    (b) Speedup

**Fig. 1.** The trends of the computational time taken by each of two algorithms and the speedup vs. the number of objects

with the enlarging size of the object set. Hence, the results of the first experiment prove that Algorithm 2 dominates Algorithm 1 and the performance of Algorithm 2 is positively related to the size of the object set.

In the second experiment, we select 7000 objects and 2, 4, $\cdots$, 12 condition criteria randomly to be experimental data sets, respectively. We also execute Algorithms 1 and 2 on these data sets respectively. The time taken by each of two algorithms is shown in Table 3. With the increasing number of criteria, Fig. 2(a) and Fig. 2(b) illustrate the trends of the computational time taken by each of two algorithms and the speedup between Algorithms 1 and 2, respectively.

**Table 3.** A comparison of the time taken by Algorithms 1 and 2 vs. the number of criteria

| Objects | Criteria | $t_1$ | $t_2$ | $S_p$ |
|---------|----------|-------|-------|-------|
| 7000 | 2 | 1.9079 | 0.1351 | 14.1221 |
| 7000 | 4 | 2.0791 | 0.1864 | 11.1540 |
| 7000 | 6 | 2.2901 | 0.2379 | 9.6263 |
| 7000 | 8 | 2.4246 | 0.2902 | 8.3550 |
| 7000 | 10 | 2.9063 | 0.3412 | 8.5179 |
| 7000 | 12 | 3.4698 | 0.3932 | 8.8245 |

From Table 3 and Fig. 2(a), we can see that the computational time taken by Algorithms 1 and 2 raises with the increasing size of the criterion set. The computational time taken by Algorithm 1 is always more than that of Algorithm 2. From Fig. 2(b) and the fifth column in Table 3, we can see that the trend of the speedup decreases with the enlarging size of the attribute set. Hence, the results of the first experiment testify that Algorithm 2 dominates Algorithm 1 and the performance of Algorithm 2 is negatively related to the size of the criterion set.

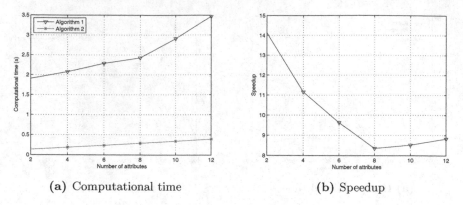

(a) Computational time    (b) Speedup

**Fig. 2.** The trends of the computational time taken by each of two algorithms and the speedup vs. the number of criteria

## 5    Conclusions and Future Work

In this paper, we proposed a matrix-based approach for computing approximations in DRSA and designed two corresponding algorithms: CPU-based algorithm and GPU-based parallel algorithm. This work is helpful to accelerate approximations computation in DRSA so that it improves the efficiency of applying DRSA to solve multi-criteria classification problems. By the numerical example and experimental evaluations, we can obtain the following conclusions: (1) The proposed matrix-based approach can be used in computing approximations in DRSA. (2) The GPU-based parallel algorithm dominates the CPU-based algorithm. (3) The performance of the GPU-based parallel algorithm is positively related to the size of the object set while negatively related to the number of criteria. In the future, we will improve the proposed GPU-based algorithm to weaken the influence from the size of the criterion set.

**Acknowledgements.** This work is supported by the National Science Foundation of China (Nos. 61175047, 61100117, 71201133) and NSAF (No. U1230117), Youth Social Science Foundation of the Chinese Education Commission (10YJCZH117, 11YJC630127) and the Fundamental Research Funds for the Central Universities (SWJTU11ZT08).

## References

1. Bache, K., Lichman, M.: UCI machine learning repository (2013), http://archive.ics.uci.edu/ml
2. Chan, C.: A rough set approach to attribute generalization in data mining. Information Sciences 107, 177–194 (1998)
3. Chen, H., Li, T., Qiao, S., Ruan, D.: A rough set based dynamic maintenance approach for approximations in coarsening and refining attribute values. International Journal of Intelligent Systems 25, 1005–1026 (2010)

4. Chen, H., Li, T., Ruan, D., Lin, J., Hu, C.: A rough-set-based incremental approach for updating approximations under dynamic maintenance environments. IEEE Transactions on Knowledge and Data Engineering 25(2), 274–284 (2013)
5. Cheng, Y.: The incremental method for fast computing the rough fuzzy approximations. Data & Knowledge Engineering 70, 84–100 (2011)
6. Greco, S., Matarazzo, B., Slowinski, R.: Rough sets theory for multicriteria decision analysis. European Journal of Operational Research 129, 1–47 (2001)
7. Li, S., Li, T., Liu, D.: Dynamic maintenance of approximations in dominance-based rough set approach under the variation of the object set. International Journal of Intelligent Systems 28(8), 729–751 (2013)
8. Li, S., Li, T., Liu, D.: Incremental updating approximations in dominance-based rough sets approach under the variation of the attribute set. Knowledge-Based Systems 40, 17–26 (2013)
9. Li, T., Ruan, D., Geert, W., Song, J., Xu, Y.: A rough sets based characteristic relation approach for dynamic attribute generalization in data mining. Knowledge-Based Systems 20, 485–494 (2007)
10. Luo, C., Li, T., Chen, H.: Dynamic maintenance of approximations in set-valued ordered decision systems under the attribute generalization. Information Sciences 257, 210–228 (2014)
11. Luo, C., Li, T., Chen, H., Liu, D.: Incremental approaches for updating approximations in set-valued ordered information systems. Knowledge-Based Systems 50, 218–233 (2013)
12. Navarro, C.A., Hitschfeld-Kahler, N., Mateu, L.: A survey on parallel computing and its applications in data-parallel problems using GPU architectures. Communications in Computational Physics 15(2), 285–329 (2014)
13. Pawlak, Z., Skowron, A.: Rough sets: Some extensions. Information Sciences 177, 28–40 (2007)
14. Qian, Y., Liang, J., Pedrycz, W., Dang, C.: Positive approximation: An accelerator for attribute reduction in rough set theory. Artificial Intelligence 174, 597–618 (2010)
15. Zhang, J., Li, T., Ruan, D.: Neighborhood rough sets for dynamic data mining. International Journal of Intelligent Systems 27, 317–342 (2012)
16. Zhang, J., Li, T., Ruan, D.: A parallel method for computing rough set approximations. Information Sciences 194, 209–223 (2012)
17. Zhang, J., Li, T., Ruan, D.: Rough sets based matrix approaches with dynamic attribute variation in set-valued information systems. International Journal of Approximate Reasoning 53, 620–635 (2012)
18. Zhang, J., Wong, J., Li, T., Pan, Y.: A comparison of parallel large-scale knowledge acquisition using rough set theory on different MapReduce runtime systems. International Journal of Approximate Reasoning 55, 896–907 (2014)
19. Zhang, J., Li, T., Chen, H.: Composite rough sets for dynamic data mining. Information Sciences 257, 81–100 (2014)

# Propositional Compilation for All Normal Parameter Reductions of a Soft Set

Banghe Han and Xiaonan Li

School of Mathematics and Statistics,
Xidian University,
Xi'an, 710071, China

**Abstract.** This paper proposes a method for compiling all the normal parameter reductions of a soft set into a conjunction of disjunctive normal form, which is generated by parameter boolean atomic formulas. A subset of parameter set is a normal parameter reduction if and only if the characteristic function of its complementary set is a model of this proposition. Three rules for simplifying this job are developed and combined.

**Keywords:** Soft set, normal parameter reduction, disjunctive normal form.

## 1 Introduction

In this paper we deal with the normal parameter reduction (a kind of feature selection) in soft set which is a special 0-1 valued information system. Soft set was initiated as a new mathematical tool for dealing with uncertainties and vagueness by Molodtsov [1] in 1999. The theory of soft sets has potential applications in various fields like game theory, operations research, decision making and so on. Although soft set can be represented as an information system and parameters behave as primitive attributes having values 0 or 1 [2], the parameter reduction in soft sets and attribute reduction in traditional information systems are not the same [3]. In soft set the classification scheme is based on additive model. Two objects are in the same class if and only if they have the same sum of parameter values. For example, the parameters in soft set may represent different voters. One of two candidates holds an advantage if and only if he has bigger number of affirmative votes. While in information systems, two objects are different if and only if they have at least one different attribute value. This means that objects can be distinguished by a single attribute. For example two clothes are different if they have different colors.

Many researchers have made contributions to parameter reduction theory of soft sets. [4] proposed the concept of reduct-soft-set. Actually the definition of the reduct there is the same with traditional information systems or rough sets. Chen et al. [3] pointed out that problems tackled by attributes reduction in rough set theory and parameters reduction in soft set theory are different. Then [3] presented a new notion of parameterization reduction in soft sets. This notion was compared with the concepts of attribute reduction in rough set theory. After

D. Miao et al. (Eds.): RSKT 2014, LNAI 8818, pp. 184–193, 2014.
DOI: 10.1007/978-3-319-11740-9_18 © Springer International Publishing Switzerland 2014

this kind of parameter reduction, only the set of optimal choices remains the same. So this concept has the problem of suboptimal choice. In order to overcome this problem the concept of normal parameter reduction was introduced in [5]. An algorithm for normal parameter reduction was also developed in [5]. But the algorithm involves a great amount of computation. Ma et al. [6] pointed out an important property of normal parameter reduction of soft sets. Then this property was used for reducing the workload for finding candidate parameter sets. Another method of reduction of parameters was proposed by Ali [7]. This method is very much similar to reduction of attributes in case of rough sets. Instead Ali proposed to delete parameters only one at each time in order to avoid the heavy searching work. Gong et al [8] developed parameters reduction concepts in bijective soft set decision system under fuzzy environments.

[9] proposed some useful simplification methods which transforms the discernibility matrix into a simper form. Although the parameter reduction of soft sets is different from the attribute reduction of information tables, we can use these good ideas in the discernibility function theory of rough sets or information systems. In this paper we want to compile all the normal parameter reductions of soft set into a proposition of parameter boolean variables. As far as the authors know, propositional representation work of normal parameter reductions of soft sets has not been well studied yet. This proposition is expected to provide an implicit representation of normal parameter reductions. Once we need one or some or all of them we can use pruning techniques to get them.

## 2    Preliminaries

In this paper, suppose $U = \{u_1, u_2, \cdots, u_n\}$ is a finite set of objects, $E$ is a set of parameters. $\wp(U)$ means the powerset of $U$, $|A|$ means the cardinality of set $A$.

**Definition 1 (Soft set).** A soft set over $U$ is a pair $S = (F, A)$, where
   (i) $A$ is a subset of $E$;
   (ii) $F : A \to \wp(U)$, $\forall e \in A$, $F(e)$ means the subset of $U$ corresponding with parameter $e$. We also use $F(e)(u) = 1$ ($F(e)(u) = 0$) to mean than $u$ is (not) an element of $F(e)$.

**Definition 2 (Choice value function).** Let $S = (F, A)$ be a soft set over $U$. The function $\sigma_S : U \to \mathbb{N}$ defined by $\sigma_S(u) = |\{e \in A | u \in F(e)\}|$ is called the choice value function of $S$.

   When the underlying soft set $S = (F, A)$ is explicit, $A_1 \subseteq A$, we also write $\sigma_{A_1}(u)$ to mean $|\{e \in A_1 | u \in F(e)\}|$.

**Example 1.** Table 1 represents a soft set $S = (F, A)$ over $U = \{u_1, u_2, \cdots, u_6\}$, where $A = \{e_1, e_2, \cdots, e_7\}$. The function $F$ is decided by the columns indexed by $e_j$, $j = 1, 2, \cdots, 7$, $F(e_j)(u_i) = 1$ if and only if the value in the $u_i$ row and $e_j$ column is equal to 1.

**Definition 3 (Normal parameter reduction).** For soft set $S = (F, A)$ over $U$, $B \subseteq A$, $B \neq \emptyset$, if the constraint $\sum_{e \in A-B} F(e)(u_1) = \cdots = \sum_{e \in A-B} F(e)(u_n)$ is satisfied, then $B$ is called a normal parameter reduction of $S$.

**Table 1.** Tabular representation for a soft set $S = (F, A)$ and $\sigma_S$

|       | $e_1$ | $e_2$ | $e_3$ | $e_4$ | $e_5$ | $e_6$ | $e_7$ | $\sigma_S$ |
|-------|-------|-------|-------|-------|-------|-------|-------|------------|
| $u_1$ | 0     | 0     | 1     | 1     | 1     | 1     | 1     | 5          |
| $u_2$ | 1     | 0     | 1     | 1     | 1     | 0     | 0     | 4          |
| $u_3$ | 0     | 1     | 1     | 1     | 0     | 1     | 0     | 4          |
| $u_4$ | 0     | 0     | 0     | 0     | 0     | 1     | 1     | 2          |
| $u_5$ | 1     | 0     | 1     | 0     | 0     | 0     | 0     | 2          |
| $u_6$ | 1     | 0     | 1     | 0     | 0     | 0     | 0     | 2          |

Our definition of normal parameter reduction here is different from that of [5], we do not require minimal property.

For soft set $S = (F, A)$ over $U$, denote the set of all normal parameter reductions of $S$ by $NPR(S)$.

**Example 2.** Consider the soft set $S = (F, A)$ in Table 1. Let $B_1 = A - \{e_4\}$, then for soft set $(F, B_1)$, we have $\sum_{e \in A - B_1} F(e)(u_2) = 1, \sum_{e \in A - B_1} F(e)(u_4) = 0$. Thus $B_1 \notin NPR(S)$. Let $B_2 = A - \{e_1, e_2, e_7\}$, then for soft set $(F, B_2)$, we have $\sum_{e \in A - B_2} F(e)(u_i) = 1, i = 1, 2, \cdots, |U|$. Thus $B_2 \in NPR(S)$.

In this paper $E$ is also considered as boolean atomic propositions when we are concerned with parameter reductions of soft set $S = (F, A)$. $F(E)$ means the set of propositions generated by $E$ with logical connectives $\neg$ (negation), $\vee$ (disjunction), $\wedge$ (conjunction). It is assumed that readers are familiar with basic concepts in propositional logic such as literal, model and tautology. Suppose $Q_1 \subseteq F(E), Q_2 \subseteq F(E)$, denote $\bigwedge Q_1 = \wedge_{q_1 \in Q_1} q_1, \bigvee Q_1 = \vee_{q_1 \in Q_1} q_1, Q_1 \bigwedge Q_2 = \{q_1 \wedge q_2 | q_1 \in Q_1, q_2 \in Q_2\}, Q_1 \bigvee Q_2 = \{q_1 \vee q_2 | q_1 \in Q_1, q_2 \in Q_2\}$.

## 3    Compilation of All Normal Parameter Reductions of Soft Set into a Proposition of Parameter Boolean Variables

In this section we will show how to compile all the normal parameter reductions of a soft set into a proposition of parameter boolean variables.

Suppose $S = (F, A)$ is a soft set over $U$. In this paper without specific explanation we always assume $\sigma(u_n) = \min\{\sigma(u_i) | i = 1, 2, \cdots, n\}$. Define a $2 \times (n-1)$ matrix $[M_{i,j}]_S$ as follows: $\forall j = 1, 2, ..., n-1$,

$$M_{i,j} = \begin{cases} D_{j \leftarrow n}, i = 1; \\ D_{n \leftarrow j}, i = 2, \end{cases} \tag{1}$$

where $D_{j \leftarrow n} = \{e \in A | F(e)(u_j) = 1, F(e)(u_n) = 0\}, D_{n \leftarrow j} = \{e \in A | F(e)(u_n) = 1, F(e)(u_j) = 0\}$.

**Example 3.** Consider the soft set $S$ given in Table 1. $|U| = 6$. By computing $D_{j \leftarrow 6}, D_{6 \leftarrow j}, j = 1, 2, \cdots, 5$, we have $[M_{i,j}]_S$ as follows:

$$[M_{i,j}]s = \begin{pmatrix} \{e_4,e_5,e_6,e_7\} & \{e_4,e_5\} & \{e_2,e_4,e_6\} & \{e_6,e_7\} & \emptyset \\ \{e_1\} & \emptyset & \{e_1\} & \{e_1,e_3\} & \emptyset \end{pmatrix}. \tag{2}$$

First of all, assume $|U| = 2$. $D_{1\leftarrow2} = \{e \in A | F(e)(u_1) = 1, F(e)(u_2) = 0\}$, $D_{2\leftarrow1} = \{e \in A | F(e)(u_2) = 1, F(e)(u_1) = 0\}$. Then we have $M_{1,1} = D_{1\leftarrow2}$, $M_{2,1} = D_{2\leftarrow1}$ and

$$[M_{i,j}]s = \begin{pmatrix} D_{1\leftarrow2} \\ D_{2\leftarrow1} \end{pmatrix}. \tag{3}$$

**Theorem 1.** Suppose $S = (F, A)$ is a soft set over $U$. $U = |2|$. $B \subseteq A$, then $B$ is a normal parameter reduction of $S$ if and only if $|(A - B) \cap D_{1\leftarrow2}| = |(A - B) \cap D_{2\leftarrow1}|$.

**Definition 4.** Suppose $S = (F, A)$ is a soft set over $U$. $U = |2|$. By our assumption $|D_{1\leftarrow2}| \geq |D_{2\leftarrow1}|$. Define a set of propositions $Q_{(1,2)}$ and a proposition $P_{(1,2)}$ as follows:

• If $|D_{2\leftarrow1}| = 0$, then define $Q_{(1,2)} = \{\bigwedge_{e \in D_{1\leftarrow2}} \neg e\}$, proposition $P_{(1,2)} = \bigwedge_{e \in D_{1\leftarrow2}} \neg e$. Particularly, when $|D_{1\leftarrow2}|$ is also equal to 0, $Q_{(1,2)} = \{\top\}$, $P_{(1,2)} = \top$.

• If $|D_{2\leftarrow1}| > 0$, $\forall j \in \{0, 1, \cdots, |D_{2\leftarrow1}|\}$, There are $C^j_{|D_{1\leftarrow2}|}$ $(C^j_{|D_{2\leftarrow1}|})$ methods for choosing an arbitrary subset of $D_{1\leftarrow2}$ $(D_{2\leftarrow1})$ with $j$ elements. Each method corresponds with a subset of $D_{1\leftarrow2}$ $(D_{2\leftarrow1})$. Denote the set of all these corresponding subsets of $D_{1\leftarrow2}$ $(D_{2\leftarrow1})$ by $D^{(j)}_{1\leftarrow2}$ $(D^{(j)}_{2\leftarrow1})$. $\forall H_1 \in D^{(j)}_{1\leftarrow2}$, $\forall H_2 \in D^{(j)}_{2\leftarrow1}$, $|H_1| = |H_2|$, define proposition

$$\bigwedge_{e \in D_{1\leftarrow2} \cup D_{2\leftarrow1}} e^{\lambda(e)}, \tag{4}$$

where $\lambda(e) \in \{0, 1\}$, $\lambda(e) = 1$ if and only if $e \in H_1 \cup H_2$, if $\lambda(e) = 1$, $e^{\lambda(e)} = e$; otherwise $e^{\lambda(e)} = \neg e$. Denote the set of all propositions generated by this way as $Q^{(j)}_{(1,2)}$. Denote $Q_{(1,2)} = \bigcup_{j=0,1,\cdots,|D_{2\leftarrow1}|} Q^{(j)}_{(1,2)}$. Define proposition $P_{(1,2)} = \bigvee Q_{(1,2)}$.

**Lemma 1.** Suppose $S = (F, A)$ is a soft set over $U$. $|U| = 2$. $\pi$ is a valuation from $D_{1\leftarrow2} \cup D_{2\leftarrow1}$ to $\{0, 1\}$, $\forall H_1 \in D^{(j)}_{1\leftarrow2}$, $\forall H_2 \in D^{(j)}_{2\leftarrow1}$, $\pi$ is model of the proposition $\bigwedge_{e \in D_{1\leftarrow2} \cup D_{2\leftarrow1}} e^{\lambda(e)}$ defined in expression (4) if and only if $\pi$ is equal to $\lambda$.

**Definition 5.** Given $B \subseteq A$, define a variable assignment $\pi_B : A \to \{0, 1\}$ by $\pi_B(e) = 1$ if and only if $e \in A - B$.

**Lemma 2.** Suppose $S = (F, A)$ is a soft set over $U$. $|U| = 2$. $B \subseteq A$, then the number of parameter variables in $D_{1\leftarrow2}$ $(D_{2\leftarrow1})$ taking value 1 with respect to the valuation $\pi_B$ is equal to $|(A - B) \cap D_{1\leftarrow2}|$ $(|(A - B) \cap D_{2\leftarrow1}|)$.

**Lemma 3.** Suppose $S = (F, A)$ is a soft set over $U$. $|U| = 2$. $B \subseteq A$, then

$$D_{1\leftarrow2} \cup D_{2\leftarrow1} = ((A - B) \cap (D_{1\leftarrow2})) \cup ((A - B) \cap (D_{2\leftarrow1})) \cup (B \cap (D_{1\leftarrow2} \cup D_{2\leftarrow1}))$$

**Theorem 2.** Suppose $S = (F, A)$ is a soft set over $U$. $|U| = 2$. $B \subseteq A$, then $B$ is a normal parameter reduction of $S$ if and only if $\pi_B$ is a model of $P_{(1,2)}$.

**Proof. (Sufficiency)** If $\pi_B$ is a model of $P_{(1,2)}$. Since $P_{(1,2)} = \bigvee Q_{(1,2)}$, $Q_{(1,2)} = \bigcup_{j=0,1,\cdots,|D_{1\leftarrow2}|} Q_{(1,2)}^{(j)}$, then $\pi_B$ is a model of one conjunction of literals $L \in Q_{(1,2)}^{(j)}$, $j \in \{0, 1, \cdots, |D_{1\leftarrow2}|\}$. It is easy to verify that the number of parameter variables in $D_{1\leftarrow2}$ taking value 1 (i.e., $|(A-B) \cap D_{1\leftarrow2}|$) is equal to that of $D_{2\leftarrow1}$ (i.e., $|(A-B) \cap D_{2\leftarrow1}|$). By Theorem 1, $B$ is a normal parameter reduction.

(**Necessity**) If $B$ is a normal parameter reduction, by Theorem 1 we have $|(A-B) \cap D_{1\leftarrow2}| = |(A-B) \cap D_{2\leftarrow1}|$. Let $j = |(A-B) \cap D_{1\leftarrow2}|$, $H_1 = (A-B) \cap D_{1\leftarrow2}$, $H_2 = (A-B) \cap D_{2\leftarrow1}$. Construct one conjunction of literals $L \in Q_{(1,2)}^{(j)}$ in the way like expression (4). Since $(H_1 \cup H_2) \subseteq A - B$, by Definition 5 we have $\forall e \in H_1 \cup H_2$, $\pi_B(e) = 1$. By Lemma 3, we have $(D_{1\leftarrow2} \cup D_{2\leftarrow1}) - (H_1 \cup H_2) = (B \cap (D_{1\leftarrow2} \cup D_{2\leftarrow1}))$, therefore $\forall e \in (D_{1\leftarrow2} \cup D_{2\leftarrow1}) - (H_1 \cup H_2)$, by Definition 5 $\pi_B(e) = 0$. Thus we have $\pi_B = \lambda$. By Lemma 1 $\pi_B$ is a model of $L$. Since $P_{(1,2)} = \bigvee Q_{(1,2)}$, $Q_{(1,2)} = \bigcup_{j=0,1,\cdots,|D_{1\leftarrow2}|} Q_{(1,2)}^{(j)}$, $\pi_B$ is a model of $P_{(1,2)}$.

Now we come to the situation when $|U| > 2$. For each pair $i \in \{1, 2, \cdots, |U| - 1\}$, we can construct proposition $P_{(i,|U|)}$. According to the transitivity property of the relation "$=$", it is easy to verify the following theorem.

**Theorem 3.** Suppose $S = (F, A)$ is a soft set over $U$. $|U| > 2$. $B \subseteq A$, then $B$ is a normal parameter reduction if and only if $\pi_B$ is a model of $P = \bigwedge_{i<|U|} P_{(i,|U|)}$.

**Example 4.** Consider the soft set $S$ given in Table 1. $|U| = 6$.

(i) $D_{6\leftarrow1} = \{e_1\}$, $D_{1\leftarrow6} = \{e_4, e_5, e_6, e_7\}$. $|D_{1\leftarrow6}| = 4$, $|D_{6\leftarrow1}| = 1$, thus $Q_{(1,6)}^{(0)} = \{\neg e_4 \wedge \neg e_5 \wedge \neg e_6 \wedge \neg e_7 \wedge \neg e_1\}$, $Q_{(1,6)}^{(1)} = \{e_4 \wedge \neg e_5 \wedge \neg e_6 \wedge \neg e_7 \wedge e_1, \neg e_4 \wedge e_5 \wedge \neg e_6 \wedge \neg e_7 \wedge e_1, \neg e_4 \wedge \neg e_5 \wedge e_6 \vee \neg e_7 \wedge e_1, \neg e_4 \wedge \neg e_5 \wedge \neg e_6 \wedge e_7 \wedge e_1\}$, $Q_{(1,6)} = \bigcup_{j=0,1} Q_{(1,6)}^{(j)} = \{\neg e_4 \wedge \neg e_5 \wedge \neg e_6 \wedge \neg e_7 \wedge \neg e_1, e_4 \wedge \neg e_5 \wedge \neg e_6 \wedge \neg e_7 \wedge e_1, \neg e_4 \wedge e_5 \wedge \neg e_6 \wedge \neg e_7 \wedge e_1, \neg e_4 \wedge \neg e_5 \wedge e_6 \wedge \neg e_7 \wedge e_1, \neg e_4 \wedge \neg e_5 \wedge \neg e_6 \wedge e_7 \wedge e_1\}$. So $P_{(1,6)} = (\neg e_4 \wedge \neg e_5 \wedge \neg e_6 \wedge \neg e_7 \wedge \neg e_1) \vee (e_4 \wedge \neg e_5 \wedge \neg e_6 \wedge \neg e_7 \wedge e_1) \vee (\neg e_4 \wedge e_5 \wedge \neg e_6 \wedge \neg e_7 \wedge e_1) \vee (\neg e_4 \wedge \neg e_5 \wedge e_6 \wedge \neg e_7 \wedge e_1) \vee (\neg e_4 \wedge \neg e_5 \wedge \neg e_6 \wedge e_7 \wedge e_1)$.

(ii) $D_{6\leftarrow2} = \emptyset$, $D_{2\leftarrow6} = \{e_4, e_5\}$, $|D_{6\leftarrow2}| = 0$, $Q_{(2,6)} = \{\neg e_4 \wedge \neg e_5\}$. So $P_{(2,6)} = \neg e_4 \wedge \neg e_5$.

(iii) $D_{3\leftarrow6} = \{e_2, e_4, e_6\}$, $D_{6\leftarrow3} = \{e_1\}$. $|D_{6\leftarrow3}| = 1, |D_{3\leftarrow6}| = 3$, thus $Q_{(3,6)}^{(0)} = \{\neg e_2 \wedge \neg e_4 \wedge \neg e_6 \wedge \neg e_1\}$, $Q_{(3,6)}^{(1)} = \{e_2 \wedge \neg e_4 \wedge \neg e_6 \wedge e_1, \neg e_2 \wedge e_4 \wedge \neg e_6 \wedge e_1, \neg e_2 \wedge \neg e_4 \wedge e_6 \wedge e_1\}$, $Q_{(3,6)} = \bigcup_{j=0,1} Q_{(3,6)}^{(j)} = \{\neg e_2 \wedge \neg e_4 \wedge \neg e_6 \wedge \neg e_1, e_2 \wedge \neg e_4 \wedge \neg e_6 \wedge e_1, \neg e_2 \wedge e_4 \wedge \neg e_6 \wedge e_1, \neg e_2 \wedge \neg e_4 \wedge e_6 \wedge e_1\}$. So $P_{(3,6)} = (\neg e_2 \wedge \neg e_4 \wedge \neg e_6 \wedge \neg e_1) \vee (e_2 \wedge \neg e_4 \wedge \neg e_6 \wedge e_1) \vee (\neg e_2 \wedge e_4 \wedge \neg e_6 \wedge e_1) \vee (\neg e_2 \wedge \neg e_4 \wedge e_6 \wedge e_1)$.

(iv) $D_{4\leftarrow6} = \{e_6, e_7\}$, $D_{6\leftarrow4} = \{e_1, e_3\}$. $|D_{6\leftarrow4}| = 2, |D_{4\leftarrow6}| = 2$, thus $Q_{(4,6)}^{(0)} = \{\neg e_6 \wedge \neg e_7 \wedge \neg e_1 \wedge \neg e_3\}$, $Q_{(4,6)}^{(1)} = \{e_6 \wedge \neg e_7 \wedge e_1 \wedge \neg e_3, e_6 \wedge \neg e_7 \wedge \neg e_1 \wedge e_3, \neg e_6 \wedge e_7 \wedge e_1 \wedge \neg e_3, \neg e_6 \wedge e_7 \wedge \neg e_1 \wedge e_3\}$, $Q_{(4,6)}^{(2)} = \{e_6 \wedge e_7 \wedge e_1 \wedge e_3\}$, $Q_{(4,6)} = \bigcup_{j=0,1,2} Q_{(4,6)}^{(j)} = \{\neg e_6 \wedge \neg e_7 \wedge \neg e_1 \wedge \neg e_3, e_6 \wedge \neg e_7 \wedge e_1 \wedge \neg e_3, e_6 \wedge \neg e_7 \wedge \neg e_1 \wedge e_3, \neg e_6 \wedge e_7 \wedge e_1 \wedge \neg e_3, \neg e_6 \wedge$

$e_7 \wedge \neg e_1 \wedge e_3, e_6 \wedge e_7 \wedge e_1 \wedge e_3\}$. So $P_{(4,6)} = (\neg e_6 \wedge \neg e_7 \wedge \neg e_1 \wedge \neg e_3) \vee (e_6 \wedge \neg e_7 \wedge e_1 \wedge \neg e_3) \vee (e_6 \wedge \neg e_7 \wedge \neg e_1 \wedge e_3) \vee (\neg e_6 \wedge e_7 \wedge e_1 \wedge \neg e_3) \vee (\neg e_6 \wedge e_7 \wedge \neg e_1 \wedge e_3) \vee (e_6 \wedge e_7 \wedge e_1 \wedge e_3)$.

(v) $D_{5 \leftarrow 6} = \emptyset$, $D_{6 \leftarrow 5} = \emptyset$, so $P_{(5,6)} = \top$.

(vi) Finally we have $P = P_{(1,6)} \wedge P_{(2,6)} \wedge P_{(3,6)} \wedge P_{(4,6)} \wedge P_{(5,6)}$.

We see that elements in $Q_{(i,|U|)}$ are conjunctions of literals of parameter boolean variables. $\forall 1 \leq i < |U|$, take one conjunction of literals $L_i$ from $Q_{(i,|U|)}$, do conjunction operation and then get a new conjunction of literals $L_1 \wedge L_2 \wedge \cdots \wedge L_{|U|-1}$. $B$ is a normal parameter reduction if and only if $\pi(B)$ is a model of one of these new conjunction of literals. The total number of these new conjunction of literals is equal to $\prod_{i<|U|} |Q_{(i,|U|)}|$. We write $\prod_{i<|U|} |Q_{(i,|U|)}|$ as $NCL$(number of conjunction of literals). Then it is easy to verify that the number of parameter reductions of soft set $S$ is no bigger than $NCL$. Note that the conjunction of literals $L_1 \wedge L_2 \wedge \cdots \wedge L_{|U|-1}$ corresponds with a normal parameter reduction if and only if $L_1 \wedge L_2 \wedge \cdots \wedge L_{|U|-1}$ contains no complementary literals.

**Example 5.** Continue with Example 4 and consider the soft set $S$ given in Table 1, where $|U| = 6$.

(i) $NCL = 5 \times 1 \times 4 \times 6 \times 1 = 120$.

(ii) Let $L_1 = \neg e_4 \wedge \neg e_5 \wedge \neg e_6 \wedge \neg e_7 \wedge \neg e_1 \in Q_{(1,6)}$, $L_2 = \neg e_4 \wedge \neg e_5 \in Q_{(2,6)}$, $L_3 = \neg e_2 \wedge \neg e_4 \wedge \neg e_6 \wedge \neg e_1 \in Q_{(3,6)}$, $L_4 = \neg e_6 \wedge \neg e_7 \wedge \neg e_1 \wedge \neg e_3 \in Q_{(4,6)}$, $L_5 = \top \in Q_{(5,6)}$, then we get a trivial normal parameter reduction $B = A$ of $S$.

(iii) Let $L_1 = \neg e_4 \wedge \neg e_5 \wedge e_6 \wedge \neg e_7 \wedge e_1 \in Q_{(1,6)}$, $L_2 = \neg e_4 \wedge \neg e_5 \in Q_{(2,6)}$, $L_3 = \neg e_2 \wedge \neg e_4 \wedge e_6 \wedge e_1 \in Q_{(3,6)}$, $L_4 = e_6 \wedge \neg e_7 \wedge e_1 \wedge \neg e_3 \in Q_{(4,6)}$, $L_5 = \top \in Q_{(5,6)}$, then we get a normal parameter reduction $B = \{e_1, e_6\}$ of $S$.

(iv) Let $L_1 = \neg e_4 \wedge \neg e_5 \wedge \neg e_6 \wedge e_7 \wedge e_1 \in Q_{(1,6)}$, $L_2 = \neg e_4 \wedge \neg e_5 \in Q_{(2,6)}$, $L_3 = e_2 \wedge \neg e_4 \wedge \neg e_6 \wedge e_1 \in Q_{(3,6)}$, $L_4 = \neg e_6 \wedge e_7 \wedge e_1 \wedge \neg e_3 \in Q_{(4,6)}$, $L_5 = \top \in Q_{(5,6)}$, then we get a normal parameter reduction $B = \{e_1, e_2, e_7\}$ of $S$.

(v) Let $L_1 = \neg e_4 \wedge \neg e_5 \wedge \neg e_6 \wedge e_7 \wedge e_1 \in Q_{(1,6)}$, $L_2 = \neg e_4 \wedge \neg e_5 \in Q_{(2,6)}$, $L_3 = e_2 \wedge \neg e_4 \wedge \neg e_6 \wedge e_1 \in Q_{(3,6)}$, $L_4 = e_6 \wedge \neg e_7 \wedge e_1 \wedge \neg e_3 \in Q_{(4,6)}$, $L_5 = \top \in Q_{(5,6)}$, then we generate no normal parameter reduction of $S$ since there exist complementary literals $\neg e_7$ and $e_7$.

If we want to output all the normal parameter reductions of $S$, we can use pruning techniques. Note that there may exist complementary literals in $L_1 \wedge L_2 \wedge \cdots \wedge L_i$ after we choose $L_1, L_2, \cdots, L_i, i < |U| - 1$. Then it is unnecessary to choose $L_{i+1}, \cdots, L_{|U|-1}$.

## 4    Simplification Rules for Compilation of Normal Parameter Reductions of Soft Sets

In this section we will introduce some simplification rules for compiling normal parameter reductions of soft sets into proposition of parameter boolean variables. This is very important because after simplification we can solve our compilation work or output all the normal parameter reductions more efficiently.

Suppose $S = (F, A)$ is a soft set over $U = \{u_1, u_2, \cdots, u_n\}$. $[M_{i,j}]_S$ is the matrix defined by expression (1). Since the work in Section 3 is actually based on $[M_{i,j}]_S$, it suffices to give simplification rules for $[M_{i,j}]_S$.

## 4.1 Core-Based Rule

**Definition 6 (Core of normal parameter reduction of soft sets).** For soft set $S = (F, A)$ over $U$, we call $\bigcap_{B \in NPR(S)} B$ the core of normal parameter reductions of $S$. Denote $Core = \bigcap_{B \in NPR(S)} B$.

**Lemma 4.** Suppose $S = (F, A)$ is a soft set over $U$. $\forall i \neq n$, if $M_{2,i} = \emptyset$ and $M_{1,i} \neq \emptyset$, then

$$M_{1,i} \subseteq Core. \tag{5}$$

**Definition 7 (Simplification Algorithm 1 by Core-based Rule).** Suppose $S = (F, A)$ is a soft set over $U$. $Core^* = \emptyset$. $\forall i \neq n$, if $M_{2,i} = \emptyset$ and $M_{1,i} \neq \emptyset$, then refresh $Core^* = Core^* \cup M_{1,i}$, and simplify $[M_{i,j}]_S$ by eliminating all the elements of $M_{1,i}$ from $[M_{i,j}]_S$.

**Example 6.** Consider the $[M_{i,j}]_S$ in Example 3, note that $M_{2,2} = \emptyset$, $M_{1,2} \neq \emptyset$. By applying the *Algorithm 1*, $Core^*$ is refreshed as $\{e_4, e_5\}$, and $[M_{i,j}]_S$ becomes

$$[M_{i,j}]_S = \begin{pmatrix} \{e_6, e_7\} & \emptyset & \{e_2, e_6\} & \{e_6, e_7\} & \emptyset \\ \{e_1\} & \emptyset & \{e_1\} & \{e_1, e_3\} & \emptyset \end{pmatrix}. \tag{6}$$

**Theorem 4.** Suppose $S = (F, A)$ is a soft set over $U$. Suppose proposition $P_1$ and $P_2$ are the compilation results with respect to $[M_{i,j}]_S$ before and after using *Algorithm 1*, $P_1$ is logically equivalent to $(\bigwedge_{e \in Core^*} \neg e) \wedge P_2$.

**Proof.** Assume $M_{1,1} \neq \emptyset$, $M_{2,1} = \emptyset$. $\forall \pi : A \to \{0, 1\}$:

(1) If $\pi(P_1) = 1$, then there exist $L_i \in Q_{(i,|U|)}$ $(i = 1, 2, \cdots, |U| - 1)$ such that $\pi$ is a model of $L_1 \wedge L_2 \wedge \cdots \wedge L_{|U|-1}$. By Section 3, we know that $L_1$ must be $\bigwedge_{e \in M_{2,1}} \neg e$. Thus $\forall e \in W_{1,1}$, $\pi(e) = 0$. $\forall j \in \{2, 3, \cdots, |U| - 1\}$, $L_j$ can't have literals $e$ $(e \in M_{1,2})$ appearing in them. Construct $L_i^* \in Q_{(i,n)}^*$ (here, we use $Q_{(i,n)}^*$ to mean the set of propositions corresponding with $Q_{(i,n)}$ after using Algorithm 1) as follows: $L_1^* = \top$. When $i \geq 2$, we get $L_i^*$ by removing all literals $\neg e$ $(e \in M_{1,1})$ appearing in $L_i$. Since $\pi$ is a model of $L_1 \wedge L_2 \wedge \cdots \wedge L_{|U|-1}$, $\pi$ is a model of $(\bigwedge_{e \in M_{1,1}} \neg e) \wedge L_1^* \wedge L_2^* \wedge \cdots \wedge L_{|U|-1}^*$. Hence $\pi$ is a model of $(\bigwedge_{e \in Core^*} \neg e) \wedge P_2$.

(2) If $\pi((\bigwedge_{e \in Core^*} \neg e) \wedge P_2) = 1$, then there exist $L_i^* \in Q_{(i,|U|)}^*$ $(i = 1, 2, \cdots, |U| - 1)$ such that $\pi$ is a model of both $L_1^* \wedge L_2^* \wedge \cdots \wedge L_{|U|-1}^*$ and $\bigwedge_{e \in M_{1,1}} \neg e$. Construct $L_i \in Q_{(i,n)}$ as follows: $L_1 = \bigwedge_{e \in M_{1,1}} \neg e$. When $i \geq 2$, we get $L_i$ by adding all literals $\neg e$ $(e \in M_{1,1} \cap (M_{1,i} \cup M_{2,i}))$ to $L_i$ in a conjunctive way. Since $\pi$ is a model of $L_1^* \wedge L_2^* \wedge \cdots \wedge L_{|U|-1}^*$ and $\bigwedge_{e \in M_{1,1}} \neg e$, $\pi$ is a model of $L_1 \wedge L_2 \wedge \cdots \wedge L_{|U|-1}$. Hence $\pi$ is a model of $P_1$.

**Corollary 1.** Suppose $S = (F, A)$ is a soft set over $U = \{u_1, u_2, \cdots, u_n\}$. After running Algorithm 1, we have $Core^* \subseteq Core$.

## 4.2    Inclusion-Based Rule

**Definition 8 (Simplification Algorithm 2 by Inclusion-based Rule).**
Given $[M_{i,j}]_S$, $\forall i \neq j$, $i,j \in \{1,2,\cdots,|U|-1\}$
(1) if $M_{1,i} \cup M_{2,i} \neq \emptyset$, $M_{1,i} \subseteq M_{1,j}$ and $M_{2,i} \subseteq M_{2,j}$, then
   • when $|M_{1,j} - M_{1,i}| \geq |M_{2,j} - M_{2,i}|$, refresh $M_{1,j}$ as $M_{1,j} - M_{1,i}$ and $M_{2,j}$
as $M_{2,j} - M_{2,i}$;
   • when $|M_{1,j} - M_{1,i}| < |M_{2,j} - M_{2,i}|$, refresh $M_{2,j}$ as $M_{1,j} - M_{1,i}$ and $M_{1,j}$
as $M_{2,j} - M_{2,i}$;
(2) if $M_{1,i} \cup M_{2,i} \neq \emptyset$, $M_{2,i} \subseteq M_{1,j}$ and $M_{1,i} \subseteq M_{2,j}$, then
   • when $|M_{2,j} - M_{1,i}| \geq |M_{1,j} - M_{2,i}|$, refresh $M_{1,j}$ as $M_{2,j} - M_{1,i}$ and $M_{2,j}$
as $M_{1,j} - M_{2,i}$.
   • when $|M_{2,j} - M_{1,i}| < |M_{1,j} - M_{2,i}|$, refresh $M_{2,j}$ as $M_{2,j} - M_{1,i}$ and $M_{1,j}$
as $M_{1,j} - M_{2,i}$.

**Example 7.** Consider the $[M_{i,j}]_S$ in expression (6), note that $M_{1,1} \subseteq M_{1,4}$,
$M_{2,1} \subseteq M_{2,4}$. By applying the inclusion-based rule, $[M_{i,j}]_S$ is refreshed as:

$$[M_{i,j}]_S = \begin{pmatrix} \{e_6, e_7\} & \emptyset & \{e_2, e_6\} & \{e_3\} & \emptyset \\ \{e_1\} & \emptyset & \{e_1\} & \emptyset & \emptyset \end{pmatrix}. \tag{7}$$

**Theorem 5.** Suppose $S = (F, A)$ is a soft set over $U$. Suppose proposition $P_1$
and $P_2$ are the compilation results with respect to $[M_{i,j}]_S$ before and after using
*Algorithm 2*, $P_1$ is logically equivalent to $P_2$.

## 4.3    Diagonal-Based Rule

**Definition 9 (Simplification Algorithm 3 by Diagonal-based Rule).**
Given $[M_{i,j}]_S$, $\forall i \neq j$, $i,j \in \{1,2,\cdots,|U|-1\}$. When $M_{(1,i)} \cap M_{2,j} = \emptyset$,
$M_{2,j} \neq \emptyset$, $M_{2,i} \cap M_{1,j} = \emptyset$, $M_{2,i} \neq \emptyset$, compute $a = \sum_{k=0}^{|M_{(,i}|}(C_{|M_{1,i}|}^k \times C_{|M_{2,i}|}^k)$,
$b = \sum_{k=0}^{|M_{2,j}|}(C_{|M_{1,j}|}^k \times C_{|M_{2,j}|}^k)$, $d_1 = |(M_{1,i} \cup M_{2,j}) - (M_{1,i} \cup M_{2,j}) \cap (M_{1,j} \cup M_{2,j})|$,
$d_2 = |(M_{1,j} \cup M_{2,i}) - (M_{1,i} \cup M_{2,j}) \cap (M_{1,j} \cup M_{2,i})|$. $d = \min(d_1, d_2)$, $c = \sum_{k=0}^{d} C_{d_1}^k \times C_{d_2}^k$. If $b - c \geq a - c$, $b - c > 0$
   • when $d_1 \geq d_2$, refresh $M_{1,j}$ as $(M_{1,i} \cup M_{2,j}) - (M_{1,i} \cup M_{2,j}) \cap (M_{1,j} \cup M_{2,i})$
and $M_{2,j}$ as $(M_{1,j} \cup M_{2,i}) - (M_{1,i} \cup M_{2,j}) \cap (M_{1,j} \cup M_{2,i})$.
   • when $d_1 < d_2$, refresh $M_{2,j}$ as $(M_{1,i} \cup M_{2,j}) - (M_{1,i} \cup M_{2,j}) \cap (M_{1,j} \cup M_{2,i})$
and $M_{1,j}$ as $(M_{1,j} \cup M_{2,i}) - (M_{1,i} \cup M_{2,j}) \cap (M_{1,j} \cup M_{2,i})$.

**Example 8.** Consider the $[M_{i,j}]_S$ in expression (7), by using Algorithm 1, we
get $Core^* = \{e_3, e_4, e_5\}$ and

$$[M_{i,j}]_S = \begin{pmatrix} \{e_6, e_7\} & \emptyset & \{e_2, e_6\} & \emptyset & \emptyset \\ \{e_1\} & \emptyset & \{e_1\} & \emptyset & \emptyset \end{pmatrix}. \tag{8}$$

Note that in expression (8), $M_{1,1} \cap M_{2,3} = \emptyset$, $M_{2,3} \neq \emptyset$, $M_{2,1} \cap M_{1,3} = \emptyset$,
$M_{2,1} \neq \emptyset$, compute $a = \sum_{k=0}^{|M_{2,1}|} C_{|M_{(1,1)}|}^k \times C_{|M_{2,1}|}^k = 3$, $b = \sum_{k=0}^{|M_{2,3}|} C_{|M_{1,3}|}^k \times$

$C^k_{|M_{2,3}|} = 3$, $d_1 = |(M_{1,1} \cup M_{2,3}) - (M_{1,1} \cup M_{2,3}) \cap (M_{1,3} \cup M_{2,1})| = 1$, $d_2 = |(M_{1,3} \cup M_{2,1}) - (M_{1,1} \cup M_{2,3}) \cap (M_{1,3} \cup M_{2,1})| = 1$. $d = \min(d_1, d_2) = 1$, $c = \sum_{k=0}^{d} C^k_{d_1} \times C^k_{d_2} = 2$. Thus $b - c \geq a - c$, $b - c = 1 > 0$. $d_1 \geq d_2$. Therefore

$$M_{1,3} := (M_{1,1} \cup M_{2,3}) - (M_{1,1} \cup M_{2,3}) \cap (M_{1,3} \cup M_{2,1}) = \{e_7\},$$

$$M_{2,3} := (M_{1,3} \cup M_{2,1}) - (M_{1,1} \cup M_{2,3}) \cap (M_{1,3} \cup M_{2,1}) = \{e_2\}.$$

We have the refreshed $[M_{i,j}]_S$ as follows:

$$[M_{i,j}]_S = \begin{pmatrix} \{e_6, e_7\} & \emptyset & \{e_7\} & \emptyset & \emptyset \\ \{e_1\} & \emptyset & \{e_2\} & \emptyset & \emptyset \end{pmatrix}. \tag{9}$$

With expression (9), we can construct propositions in the way introduced in Section 3: $Q_{(1,6)} = \{\neg e_6 \wedge \neg e_7 \wedge \neg e_1, e_6 \wedge \neg e_7 \wedge e_1, \neg e_6 \wedge e_7 \wedge e_1\}$, $Q_{(2,6)} = \top$, $Q_{(4,6)} = \top$, $Q_{(5,6)} = \top$, $Q_{(3,6)} = \{\neg e_7 \wedge \neg e_2, e_7 \wedge e_2\}$. Note that we have $\{e_3, e_4, e_5\} \subseteq Core$, thus the normal parameter reductions of the soft set $S$ in Table 1 can be compiled into proposition $P = \neg e_3 \wedge \neg e_4 \wedge \neg e_5 \wedge (\bigvee Q_{(1,6)}) \wedge (\bigvee Q_{(3,6)})$.

**Theorem 6.** Suppose $S = (F, A)$ is a soft set over $U = \{u_1, u_2, \cdots, u_n\}$. Assume $\sigma(u_n) = \min\{\sigma(u_i) | i = 1, 2, \cdots, n\}$. Suppose proposition $P_1$ and $P_2$ are the compilation results with respect to $[M_{i,j}]_S$ before and after using *Algorithm 3*, $P_1$ is logically equivalent to $P_2$.

## 4.4  Simplification Algorithm 4 of $[M_{i,j}]_S$ by Combining the Three Rules

Now we give our Algorithm 4 by combining the three rules in Table 2. We announce that we set *core\** empty initially only at the first time we use Simplification Algorithm 1. Since $[M_{i,j}]_S$ has finite elements, **Algorithm 4** can be terminated in finite steps.

**Table 2.** Simplification Algorithm 4 of $[M_{i,j}]_S$ by combining the three rules

| |
|---|
| **Step 1**: input $[M_{i,j}]_S$ |
| **Step 2**: use Algorithm 1, refresh $[M_{i,j}]_S$ |
| **Step 3**: use Algorithm 2, if $[M_{i,j}]_S$ can be changed, turn to Step 2; otherwise turn to Step 4. |
| **Step 4**: use Algorithm 3, if $[M_{i,j}]_S$ can be changed, turn to Step 2; otherwise turn to Step 5. |
| **Step 5**: output $[M_{i,j}]_S$ |

By Theorems 4-6 we have the following corollary.

**Corollary 2.** Suppose $S = (F, A)$ is a soft set over $U$. Compute the $[M_{i,j}]_S$ defined by expression (1). Suppose propositions $P_1$ and $P_2$ are the compilation results with respect to $[M_{i,j}]_S$ before and after using Algorithm 4, then $P_1$ is logically equivalent to $(\bigwedge_{e \in Core^*} \neg e) \wedge P_2$.

# 5    Conclusion and Future Work

In this paper we have succeeded in compiling all the normal parameter reductions of soft set into a proposition of parameter boolean variables. This proposition actually provides an implicit representation of normal parameter reductions, once we need one or some or all of them we can use pruning techniques to get them. This propositional structure of parameter reductions allows us to make reasoning about the soft set very conveniently. There exists some potential work for the near future. How can we output one or some of the normal parameter reductions of soft set by $P$ as quickly as possible? For instance, we may be interested in the minimal parameter reductions of soft sets [5].

**Acknowledgements.** This work is supported by the Fundamental Research Funds for the Central Universities (Grant No. K5051370012), the National Natural Science Foundation of China (Grant No.s 61261047 and 61202178), and the horizontal subject (Grant No. HX01120716157) of Xidian University.

# References

1. Molodtsov, D.: Soft set theory– First results. Computers and Mathematics with Applications 37, 19–31 (1999)
2. Pei, D., Miao, D.: From soft sets to information systems. In: The Proceeding of 2005 IEEE International Conference on Granular Computing, IEEE GrC 2005, pp. 617–621. IEEE Press (2005)
3. Chen, D., Tsang, E.C.C., Yeung, D.S., Wang, X.: The parameterization reduction of soft sets and its applications. Computers and Mathematics with Applications 49 (5-6), 757-763 (2005)
4. Maji, P.K., Roy, A.R., Biswas, R.: An application of soft sets in a decision making problem. Computers and Mathematics with Applications 44, 1077–1083 (2002)
5. Kong, Z., Gao, L., Wang, L., Li, S.: The normal parameter reduction of soft sets and its algorithm. Computers and Mathematics with Applications 56(12), 3029–3037 (2008)
6. Ma, X., Sulaiman, N., Qin, H., Herawana, T., Zain, J.M.: A new efficient normal parameter reduction algorithm of soft sets. Computers and Mathematics with Applications 62(2), 588–598 (2011)
7. Ali, M.I.: Another view on reduction of parameters in soft sets. Applied Soft Computing 12, 1814–1821 (2012)
8. Gong, K., Wang, P., Xiao, Z.: Bijective soft set decision system based parameters reduction under fuzzy environments. Applied Mathematical Modelling 37(6), 4474–4485 (2013)
9. Yao, Y.Y., Zhao, Y.: Discernibility matrix simplification for constructing attribute reducts. Information Sciences 179, 867–882 (2009)

# Top-N Recommendation
# Based on Granular Association Rules

Xu He[1], Fan Min[2,1], and William Zhu[1]

[1] Lab of Granular Computing, Minnan Normal University, Zhangzhou 363000, China
[2] School of Computer Science, Southwest Petroleum University, Chengdu 610500, China
minfanphd@163.com

**Abstract.** Recommender systems are popular in e-commerce as they provide users with items of interest. Existing top-$K$ approaches mine the $K$ strongest granular association rules for each user, and then recommend respective $K$ types of items to her. Unfortunately, in practice, many users need only a list of $N$ items that they would like. In this paper, we propose confidence-based and significance-based approaches exploiting granular association rules to improve the quality of top-$N$ recommendation, especially for new users on new items. We employ the confidence measure and the significance measure respectively to select strong rules. The first approach tends to recommend popular items, while the second tends to recommend special ones to different users. We also consider granule selection, which is a core issue in granular computing. Experimental results on the well-known MovieLens dataset show that: 1) the confidence-based approach is more accurate to recommend items than the significance-based one; 2) the significance-based approach is more special to recommend items than the confidence-based one; 3) the appropriate setting of granules can help obtaining high recommending accuracy and significance.

**Keywords:** Granular computing, recommender system, granule association rules, confidence, significance.

## 1 Introduction

As the rapidly growing amount of information available on the Internet and the improving of e-commerce, recommender systems have become more and more popular [1]. Recommender systems help relieve users of massive information overload, and provide users the items in which they are interested. To date, many methods, which are popular in KDD area, have been proposed for recommender systems, such as content-based filtering method [2], collaborative filtering method [3] and model-based algorithm [4]. Recently, researchers have studied the cold-start recommendations [5], including new user, new item as well as new user and new item [6].

Granular association rules mining [7,8] is a new approach to deal with the cold-start problems for recommender systems [5,9]. This approach generates rules with three measures to reveal connections between granules in two universes. A complete example of granular association rules might be "male students rate action movies released in 2013 with a probability of 25%; 50% users are male students and 40% movies are

D. Miao et al. (Eds.): RSKT 2014, LNAI 8818, pp. 194–205, 2014.
DOI: 10.1007/978-3-319-11740-9_19 © Springer International Publishing Switzerland 2014

action ones released in 2013." Here 50%, 40%, and 25% are the *source coverage*, the *target coverage* and the *confidence*, respectively.

Granule selection is a core issue in granular computing [10,11,12,13,14,15]. A granule is also a concept. For example, "action movies released in 2013" is a granule, "movies released in 2013" is a coarser granule, and "comedy action movies released in 2013" is a finer granule. Existing top-$K$ approaches [16] mine the $K$ strongest granular association rules for each user, and then recommend respective $K$ types of items to her. However, fine recommended granules or coarse recommended granules can generate different number of recommended items. Unfortunately, in a real world, many users need only a list of $N$ items that they would like. Therefore, it is necessary to study how to form the $N$ items for the users with the appropriate setting of granules.

In this paper, we propose confidence-based and significance-based approaches exploiting granular association rules to improve the quality of top-$N$ recommendation, especially for new users on new items. First, the new users are matched by granular association rules, which satisfy source coverage, target coverage and confidence of rule thresholds. Second, we sort the rules based on the confidence measure and the significance measure. The confidence-based approach ranks the rules according to the confidence measure. The significance-based approach ranks them according to the significance measure. The confidence-based approach tends to recommend popular items, while the significance-based approach tends to recommend special ones to different users. So items matched by rules that have a higher confidence (significance) are ranked first. Third, granule selection is considered in the item recommending process. Recommend items are generated for users using the appropriate recommended granules. Fourth, we choose the first $N$ highest ranked items as the recommendation list.

Experiments are undertaken on the MovieLens dataset [17] using our open source software Grale [18]. First, with some appropriate settings, the top-$N$ cold-start recommendation made by the confidence-based approach on average is more accurate than those made by the significance-based one. Second, the loss in significance is more pronounced for the confidence-based approach than for the significance-based one. Third, there is a tradeoff between the granule number and the accuracy, and the appropriate setting of granules can help obtaining high recommending accuracy and significance.

## 2 Preliminaries

In this section, we review some preliminary definitions such as information granule, many-to-many entity-relationship system (MMER) [7,8] and granular association rules with three measures [16].

### 2.1 The Data Model

Now we first revisit the definitions of information systems. [8].

**Definition 1.** $S = (U, A)$ *is an information system, where* $U = \{x_1, x_2, \ldots, x_n\}$ *is the set of all objects,* $A = \{a_1, a_2, \ldots, a_m\}$ *is the set of all attributes, and* $a_j(x_i)$ *is the value of* $x_i$ *on attribute* $a_j$ *for* $i \in [1..n]$ *and* $j \in [1..m]$.

Two information systems are listed in Tables 1(a) and 1(b), respectively. In Table 1(b), 1 indicates *true*, and 0 indicates *false*.

Information granule is defined by Yao and Deng [19] as follows:

**Definition 2.** *A granule is a triple*

$$G = (g, i(g), e(g)), \tag{1}$$

*where $g$ is the name assigned to the granule, $i(g)$ is a representation of the granule, and $e(g)$ is a set of objects that are instances of the granule.*

*Example 1.* In Tables 1(a) and 1(b), users and movies can be described by information granules. "male student" and "action movie" is two granules $g_1$ and $g_2$. $i(g_1) = \{Gender, Occupation\}$, and $e(g_1) = \{u_1, u_4\}$. $i(g_2) = \{Action\}$, and $e(g_2) = \{m_1, m_2, m_5\}$.

In an information system, any $A' \subseteq A$ induces an equivalence relation [20,21,22,23]

$$E_{A'} = \{(x, y) \in U \times U | \forall a \in A', a(x) = a(y)\}, \tag{2}$$

and partitions $U$ into a number of disjoint subsets called *blocks* or *granules*. The granule containing $x \in U$ is

$$E_{A'}(x) = \{y \in U | \forall a \in A', a(y) = a(x)\}. \tag{3}$$

According to Equation (3), $(A', x)$ determines a granule in an information system. Hence $g = g(A', x)$ is a natural name to the granule. $i(g)$ can be formalized as the conjunction of respective attribute-value pairs, i.e.,

$$i(g(A', x)) = \bigwedge_{a \in A'} \langle a : a(x) \rangle. \tag{4}$$

$e(g)$ is given by

$$e(g(A', x)) = E_{A'}(x). \tag{5}$$

Let $x \in U$ and $A'' \subset A' \subseteq A$, we have

$$e(g(A', x)) \subseteq e(g(A'', x)). \tag{6}$$

Consequently, we say that $g(A', x)$ is *finer* than $g(A'', x)$, and $g(A'', x)$ is *coarser* than $g(A', x)$.

*Example 2.* "male student" is a granule, "student" is a coarser granule, and "Chinese male student" is a finer granule.

Now we review the definitions of binary relations and many-to-many entity relationship systems [8].

**Definition 3.** *Let $U = \{x_1, x_2, \ldots, x_n\}$ and $V = \{y_1, y_2, \ldots, y_k\}$ be two sets of objects. Any $R \subseteq U \times V$ is a binary relation from $U$ to $V$.*

An example of binary relation is given by Table 1(c), where $U$ is the set of users as indicated by Table 1(a), and $V$ is the set of movies as indicated by Table 1(b). A binary relation can be viewed as an information system. However, in order to save space, it is more often stored in the database as a table with two foreign keys.

**Definition 4.** *A many-to-many entity-relationship system (MMER) is a 5-tuple $ES = (U, A, V, B, R)$, where $(U, A)$ and $(V, B)$ are two information systems, and $R \subseteq U \times V$ is a binary relation from $U$ to $V$.*

An example of MMER is given by Table 1.

**Table 1.** A many-to-many entity-relationship system

(a) User

| User-id | Age | Gender | Occupation | Country |
|---------|-----|--------|------------|---------|
| $u_1$ | 21 | M | student | China |
| $u_2$ | 43 | F | writer | Canada |
| $u_3$ | 23 | M | writer | Canada |
| $u_4$ | 22 | M | student | China |

(b) Movie

| Movie-id | Release-decade | Action | Adventure | Animation | Western |
|----------|----------------|--------|-----------|-----------|---------|
| $m_1$ | 2013 | 1 | 0 | 0 | 1 |
| $m_2$ | 2013 | 1 | 1 | 0 | 0 |
| $m_3$ | 2014 | 0 | 1 | 1 | 1 |
| $m_4$ | 2014 | 0 | 1 | 0 | 1 |
| $m_5$ | 2014 | 1 | 0 | 1 | 1 |

(c) Rates

| User-id \ Movie-id | $m_1$ | $m_2$ | $m_3$ | $m_4$ | $m_5$ |
|--------------------|-------|-------|-------|-------|-------|
| $u_1$ | 0 | 1 | 0 | 1 | 0 |
| $u_2$ | 1 | 0 | 0 | 1 | 1 |
| $u_3$ | 0 | 1 | 0 | 0 | 1 |
| $u_4$ | 0 | 0 | 1 | 1 | 1 |

## 2.2 Granular Association Rules with Three Measures

Now we discuss the means for connecting users and items. A *granular association rule* [8] is an implication of the form

$$(GR): \bigwedge_{a \in A'} \langle a : a(x) \rangle \Rightarrow \bigwedge_{b \in B'} \langle b : b(y) \rangle, \tag{7}$$

where $A' \subseteq A$ and $B' \subseteq B$.

According to Equations (4) and (5), the set of objects meeting the left-hand side of the granular association rule is

$$LH(GR) = E_{A'}(x); \tag{8}$$

while the set of objects meeting the right-hand side of the granular association rule is

$$RH(GR) = E_{B'}(y). \tag{9}$$

Let us look at an example granular association rule "male students rate action movies released in 2013 with a probability of 25%; 50% users are male students and 40% movies are action ones released in 2013." Here 50%, 40%, and 25% are the source coverage, the target coverage, and the confidence, respectively. Formally, the *source coverage* of $GR$ is

$$scov(GR) = |LH(GR)|/|U|; \tag{10}$$

while the *target coverage* of $GR$ is

$$tcov(GR) = |RH(GR)|/|V|. \tag{11}$$

The *confidence* of $GR$ is the probability that a user chooses an item, namely

$$conf(GR) = \frac{|(LH(GR) \times RH(GR)) \cap R|}{|LH(GR)| \times |RH(GR)|}. \tag{12}$$

# 3 Top-N Recommendation Based on Granular Association Rules

Personalized recommendation and diverse recommendation are two key issues of recommendation systems. In the real world, top-$N$ recommendation more in line with the actual needs. At the same time, we also consider the cold-start problem [5,9] in recommender systems. In this section, we discuss the evaluation metrics and two different approaches based on granular association rules to improve the quality of top-$N$ recommendation, especially both of new user and item.

## 3.1 Evaluation Metrics

First, we learn the two evaluation metrics, namely accuracy and significance.

**Accuracy Metric.** The performance of the recommender is evaluated mainly by the recommendation accuracy. We first look the following definition.

**Definition 5.** *Let* $ES = (U, A, V, B, R)$ *be an MMER,* $\emptyset \subset X \subseteq U$ *and* $\emptyset \subset Y \subseteq V$. *The accuracy of recommending* $Y$ *to* $X$ *is*

$$acc(X, Y) = \begin{cases} 0 & X = \emptyset \text{ or } Y = \emptyset; \\ \frac{|(X \times Y) \cap R|}{|X| \times |Y|} & \text{otherwise.} \end{cases} \tag{13}$$

The consideration of empty sets is for the situation where some users receive no recommendation. Such situation is common for new user and/or new item cases.

A recommender can be viewed a function $RC : U \to 2^V$. $RC(x)$ is the set of items recommended to user $x \in U$. According to Definition 5, the accuracy of $RC$ on $x$ is

$$acc(x, RC) = acc(\{x\}, RC(x)) = \frac{|(\{x\} \times RC(x)) \cap R|}{|RC(x)|}. \tag{14}$$

The recommender may recommend different number of items to different user, we would like to compute the number of total recommendations divided by the number of successful ones. Therefore we propose the following definition.

**Definition 6.** *The accuracy of RC on $X \subseteq U$ is*

$$acc(X, RC) = \frac{\sum_{x \in X} |(\{x\} \times RC(x)) \cap R|}{\sum_{x \in X} |RC(x)|}. \tag{15}$$

**Significance Metric.** In many applications, many popular items with highly accurate results are recommended. For example, suggesting movies that won an Oscar award to people. Though being highly accurate, it will not astonish people due to common sense. For this reason, we would like to recommend special items to people. We propose the following definitions.

**Definition 7.** *Let $ES = (U, A, V, B, R)$ be an MMER, $\emptyset \subset X \subseteq U$ and $\emptyset \subset Y \subseteq V$. The significance of recommending $Y$ to $X$ is*

$$sig(X, Y) = acc(X, Y)/acc(U, Y). \tag{16}$$

Here $acc(U, Y)$ indicate the popularity of item set $Y$. In other words, there is a penalty on popular items such that they are not recommended to everyone.

The same term can be employed to evaluate the quality of a rule.

**Definition 8.** *The significance of a granular association rule $GR$ is*

$$sig(GR) = sig(LH(GR), RH(GR)). \tag{17}$$

The quality of the recommender $RC$ can be evaluated as follows.

**Definition 9.** *The significance of recommender $RC$ on $x \in U$ is*

$$sig(x, RC) = sig(\{x\}, RC(x)) = acc(\{x\}, RC(x))/acc(U, RC(x)); \tag{18}$$

*while the significance of $RC$ over $X \subseteq U$ is*

$$sig(X, RC) = \sum_{x \in X} sig(x, RC)/|X|. \tag{19}$$

We will employ $sig(GR)$ to build the rule set, and $sig(X, RC)$ to evaluate the performance of the recommender. For example, men like to see the adventure film with a probability of 30%, Chinese men like to see the adventure film with a probability of 60%. So that the significance of recommending the adventure film for Chinese men over the set of men can be expressed as $sig(X, RC) = 0.6/0.3 = 2.0$.

## 3.2 Mining Top-N Recommendation Items

We propose confidence-based and significance-based approaches exploiting granular association rules to improve the quality of top-$N$ recommendation, especially for new

---

**Algorithm 1.** Rule set construction

---

**Input**: The training set $ES = (U, A, V, B, R)$, a minimal source coverage threshold $ms$, a minimal target coverage threshold $mt$.

**Output**: Source granules, target granules, and the rule set, all stored in the memory.

**Method**: training

1. $SG(ms) = \{(A', x) \in 2^A \times U | \frac{|E_{A'}(x)|}{|U|} \geq ms\};$
2. $TG(mt) = \{(B', y) \in 2^B \times V | \frac{|E_{B'}(y)|}{|V|} \geq mt\};$
3. **for** each $g \in SG(ms)$ **do**
4.    **for** each $g' \in TG(mt)$ **do**
5.       $GR = (i(g) \Rightarrow i(g'));$
6.       compute $conf(GR);$
7.    **end for**
8. **end for**

---

users on new items. The whole process is divided into two aspects: the rule set constructing stage and the recommending stage.

In the first stage, a rule set is constructed as shown by Algorithm 1. To store the rule set, we need an array to store $SG(ms)$, an array to store $TG(mt)$, and a matrix with size $|SG(ms)| \times |TG(mt)|$ to store confidence of all rules. Here $SG(ms)$ ($GT(ms)$) is the set of all source (target) granules satisfying the source (target) coverage threshold.

In the second stage, top-$N$ items are respectively recommended new item for new user. The process is divided into four steps as follows.

**Step 1:** A new user is matched by granular association rules, which satisfy source coverage, target coverage and confidence of rule thresholds.

**Step 2:** Comparing different rules matching the user, top-$K$ recommended granules are obtained according to two different approaches. We sort the rules based on the confidence measure and the significance measure. The confidence-based approach ranks the rules according to the confidence measure. The significance approach ranks them according to the significance measure. The confidence-based approach tends to recommend popular items, while the significance-based approach tends to recommend special ones to different users. So the new items matched by rules that have a higher confidence (significance) are ranked first.

**Step 3:** Granule selection is considered in the item recommending process. Different granules have different coverage of items, which in turn have different number of recommended items. The finer granule we choose, the less recommended items we obtain. On the contrary, the coarser granule we choose, the more recommended items we obtain. Note that if a new user is matched by multiple source granules, we select the rule that has the highest confidence (significance).

**Step 4:** Recommend items are generated for each user using the appropriate recommended granules, and then we choose the first $N$ highest ranked items from as the recommendation list.

# 4    Experiments

In this section, we try to answer the following questions through experimentation.

1. How does the performance change for different threshold settings?
2. How does the performance change for the different number of recommendations?
3. How does the performance change for the different recommended granules values?
4. Does the significance-based approach outperform the confidence-based one in terms of accuracy and significance?

## 4.1    Dataset

The MovieLens dataset [17] is widely used in recommender systems (see, e.g., [5,6]). It is assembled by the GroupLens project (http://www.grouplens.org). We downloaded the data set from the Internet Movie Database [17]. The database schema is listed as follows.

- User (<u>userID</u>, age, gender, occupation)
- Movie (<u>movieID</u>, release-year, genre)
- Rates (<u>userID, movieID</u>)

We use the version with 943 users and 1,682 movies. The data are preprocessed to cope with Definition 4 as follows. The original Rate relation contains the rating of movies with 5 scales, while we only consider whether or not a user has rated a movie. The user age is discretized to 9 intervals as indicated by the data set [24]. Since there are few movies before 1970s and too many movies after 1990, the release year is discretized to 3 intervals: before 1970s, 1970s-1980s, and 1990s. The genre is a multi-valued attribute. Therefore we scale it to 18 boolean attributes and deal with it using the scaling-based approach proposed in [25].

## 4.2    Results

The following experiments are undertaken to answer the questions raised at the beginning of the section one by one. Here we randomly select 60% data (565 users and 1009 movies) as the training set, and the remaining data (378 users and 673 movies) as the testing set, where the users and movies are all regard as cold start ones. We deal with the cold-start recommendations through granular association rule mining. Each experiment is repeated 100 times with different sampling, and the average value is computed. We introduce two parameters $N$ and $K$, where $N$ is the number of recommended items, and $K$ is the number of recommended granules.

**Performance of top-N Recommendation for Thresholds.** Here we set $N = 10$ and $K = 15$. Fig. 1(a) compares the accuracy of the confidence-based and significance-based approaches as $ms(mt)$ increases. The chart reveals that coverage has detrimental effects on confidence-based approach. However, the accuracy of recommendation keeps stable through significance-based approach. The confidence-based approach is

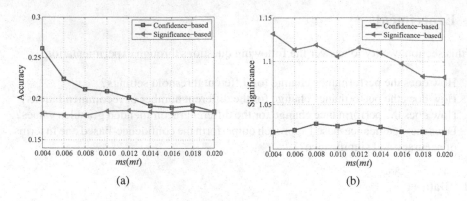

Fig. 1. Accuracy and significance comparison: (a) accuracy; (b) significance

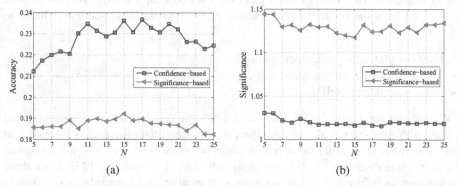

Fig. 2. $K = 10$: Performance of the recommender for different $N$ values

better than the significance-based one at first. At last, we observe that they have almost identical shape.

Fig. 1(b) compares the significance of the confidence-based and significance-based approaches as $ms(mt)$ increases. We observe that the significance decreases in general through the significance-based approach, while keeps stable through the confidence-based approach. However, the significance-based approach is always better than the confidence-based one. Moreover, the significance of the confidence-based approach keep stable and is always close to 1. From this viewpoint, high accuracy is mainly due to the fact that popular items are often recommended.

**Performance of Recommendation for Different $N$ Values.** Set $ms = mt = 0.01$ and $K = 10$. Let $N$ increases from 5 to 25, and the accuracy and significance of these two approaches are compared, as shown in Figs. 2(a) and 2(b).

We can observe that the top-$N$ cold-start recommendation made by the confidence-based approach on average is more accurate than those made by the significance-based approach, as depicted in Fig. 2(a). However, the loss in significance is more pronounced for the confidence-based approach than for the significance-based one, as depicted in Fig. 2(b).

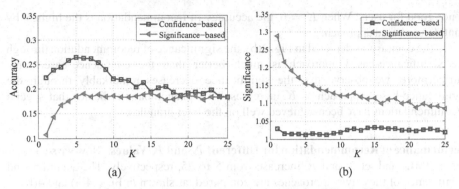

**Fig. 3.** $N = 10$: Performance of the recommender for different $K$ values

**Performance of Recommendation for Different $K$ Values.** Set $ms = mt = 0.01$ and $N = 10$. Let $K$ increases from 1 to 25, and the accuracy and significance of these two approaches are compared, as shown in Figs. 3(a) and 3(b).

Fig. 3(a) shows the accuracy of recommendation through confidence-based approach outperforms than significance-based one. We also observe that there is a tradeoff between the granule number and the accuracy. When the recommended granule is too finer, we only obtain less recommended items, which is not enough for users. When the recommended granule is too coarser, we obtain too much recommended items, which

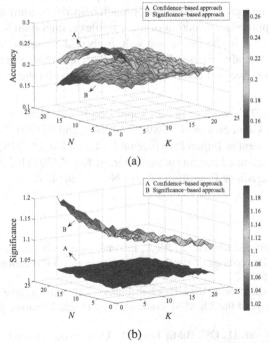

**Fig. 4.** Performance of the recommender for different $N$ and $K$ values

is not good for users. When $K = 6$, the accuracy of recommendation is the highest by confidence-based approach.

On the contrary, as depicted in Fig. 3(b), the significance of recommendation through the significance-based approach is much better than the confidence-based one. Furthermore, we observe that the significance decreases remarkably through the significance-based approach as $K$ increases. This phenomenon indicates that special recommendations have been achieved well on the finer granules.

**Performance of Recommendation for Different $N$ and $K$ Values.** Now we set $ms = mt = 0.01$, and let $N$ and $K$ increase from 5 to 25, respectively. The accuracy and significance of these two approaches are compared, as shown in Figs. 4(a) and 4(b).

Fig. 4(a) shows that confidence-based approach outperforms the significance-based one at the beginning. In the end, we observe that they have almost similar shape. When $N$ changes from 9 to 14 and $K$ changes from 5 to 8, we can obtain higher accuracy through the confidence-based approach. Fig. 4(b) shows that the significance-based approach always outperforms confidence-based one. When $N$ changes from 15 to 25 and $K$ changes from 5 to 8, we can obtain higher significance through the significance-based approach.

# 5 Conclusions

In this paper, we proposed confidence-based and significance-based approaches exploiting granular association rules to improve the quality of top-$N$ recommendation. Experimental results indicate that the confidence-based approach helped to recommend higher accuracy items and the significance-based approach helped to recommend higher significance items. The appropriate selection of the granules is important to the performance of the recommendation. In the future, we will analyze the computational complexity of our recommendation algorithm, and design a more effective algorithm for granule selection. Furthermore, we will compare our approaches with other related ones.

**Acknowledgments.** This work is in part supported by the National Science Foundation of China under Grant Nos. 61379089, 61379049, and 61170128, the Key Project of Education Department of Fujian Province under Grant No. JA13192, the Zhangzhou Municipal Natural Science Foundation under Grant No. ZZ2013J03, the Minnan Normal University Postgraduate Education Project No.YJS201437.

# References

1. Sarwar, B., Karypis, G., Konstan, J., Riedl, J.: Analysis of recommendation algorithms for e-commerce. In: Proceedings of the 2nd ACM Conference on Electronic Commerce, pp. 158–167. ACM (2000)
2. Mooney, R.J., Roy, L.: Content-based book recommending using learning for text categorization. In: Proceedings of the 5th ACM Conference on Digital Libraries, pp. 195–204. ACM (2000)
3. Goldberg, D., Nichols, D., Oki, B.M., Terry, D.: Using collaborative filtering to weave an information tapestry. Communications of the ACM 35, 61–70 (1992)

4. Bobadilla, J., Ortega, F., Hernando, A., Gutiérrez, A.: Recommender systems survey. Knowledge-Based Systems 46, 109–132 (2013)
5. Schein, A.I., Popescul, A., Ungar, L.H., Pennock, D.M.: Methods and metrics for cold-start recommendations. In: SIGIR 2002, pp. 253–260 (2002)
6. He, X., Min, F., Zhu, W.: Parametric rough sets with application to granular association rule mining. Mathematical Problems in Engineering 2013, 1–13 (2013)
7. Min, F., Hu, Q.H., Zhu, W.: Granular association rules on two universes with four measures (2013), http://arxiv.org/abs/1210.5598
8. Min, F., Hu, Q.H., Zhu, W.: Granular association rules with four subtypes. In: Proceedings of the 2011 IEEE International Conference on Granular Computing, pp. 432–437 (2012)
9. Guo, H.: Soap: Live recommendations through social agents. In: 5th DELOS Workshop on Filtering and Collaborative Filtering, pp. 1–17 (1997)
10. Bargiela, A., Pedrycz, W.: Granular computing: An introduction. Kluwer Academic Publishers, Boston (2002)
11. Yao, J.T., Yao, Y.Y.: Information granulation for web based information retrieval support systems. In: Proceedings of SPIE, vol. 5098, pp. 138–146 (2003)
12. Yao, Y.Y.: Granular computing: basic issues and possible solutions. In: Proceedings of the 5th Joint Conference on Information Sciences, vol. 1, pp. 186–189 (2000)
13. Yao, J.T.: Recent developments in granular computing: A bibliometrics study. In: IEEE International Conference on Granular Computing, pp. 74–79 (2008)
14. Zhu, W., Wang, F.: Reduction and axiomization of covering generalized rough sets. Information Sciences 152(1), 217–230 (2003)
15. Wu, W.Z., Leung, Y., Mi, J.S.: Granular computing and knowledge reduction in formal contexts. IEEE Transactions on Knowledge and Data Engineering 21(10), 1461–1474 (2009)
16. Min, F., Zhu, W.: Mining top-k granular association rules for recommendation (2013), http://arxiv.org/abs/1305.4801
17. Internet movie database, http://movielens.umn.edu
18. Min, F., Zhu, W., He, X.: Grale: Granular association rules (2013), http://grc.fjzs.edu.cn/~fmin/grale/
19. Yao, Y., Deng, X.: A granular computing paradigm for concept learning. In: Ramanna, S., Jain, L.C., Howlett, R.J. (eds.) Emerging Paradigms in ML and Applications. SIST, vol. 13, pp. 307–326. Springer, Heidelberg (2013)
20. Pawlak, Z.: Rough sets. International Journal of Computer and Information Sciences 11, 341–356 (1982)
21. Skowron, A., Stepaniuk, J.: Approximation of relations. In: Ziarko, W. (ed.) Proceedings of Rough Sets, Fuzzy Sets and Knowledge Discovery, pp. 161–166 (1994)
22. Zadeh, L.: Towards a theory of fuzzy information granulation and its centrality in human reasoning and fuzzy logic. Fuzzy Sets and Systems 19, 111–127 (1997)
23. Zhu, W.: Generalized rough sets based on relations. Information Sciences 177(22), 4997–5011 (2007)
24. He, X., Min, F., Zhu, W.: A comparative study of discretization approaches for granular association rule mining. In: Proceedings of the 2013 Canadian Conference on Electrical and Computer Engineering, pp. 725–729 (2013)
25. Min, F., Zhu, W.: Granular association rules for multi-valued data. In: Proceedings of the 2013 Canadian Conference on Electrical and Computer Engineering, pp. 799–803 (2013)

# An Efficient BP-Neural Network Classification Model Based on Attribute Reduction

Yongsheng Wang[1,2] and Xuefeng Zheng[1,2]

[1] School of Computer and Communication Engineering, University of Science and
Technology Beijing, Beijing, China
[2] Beijing Key Laboratory of Knowledge Engineering for Materials Science,
Beijing, China
yongsh_wang@126.com

**Abstract.** Classification is an important issue in data mining and
knowledge discovery, and the attribute reduction has been proven to be
effective in improving the classification accuracy in many applications. In
this paper, we first apply rough set theory to reduce irrelative attribute
and retain the important attributes, and the input neuron based on the
important attributes can simplify the structure of BP-neuron network
and improve classification accuracy. Then an efficient BP-neural network
classification model based on attribute reduction is developed for high-
dimensional data analysis. Finally, the experimental results demonstrate
the efficiency and effectiveness of the proposed model.

**Keywords:** Classification model, Attribute reduction, Neural Network,
Rough sets.

## 1 Introduction

At present, there are a lot of knowledge discovery tool to use in classification
applications. Knowledge discovery is not a single technique, some commonly
used techniques are: Case-Based Reasoning, Neural Networks, Decision Trees,
Genetic Algorithms, Fuzzy Sets and Rough Sets [1-4]. Data mining relates to
other areas, including machine learning, cluster analysis, regression analysis,
and neural networks. This model normally uses a predetermined set of features.
A machine learning algorithm of data mining generates a number of models
capturing relationships between the input features and the decisions produced by
cluster analysis. Neural networks are generally preferred for their generalization
ability, and the neural network training time is to long, it denotes the drastic
increase in computational complexity and classification error with data having a
great number of dimensions. Decision trees obviously outperform neural networks
in terms of interpretability.

The so-called curse of attributes pertinent to many knowledge discovery and
data mining algorithms, denotes the drastic increase in computational timing and
classification error with data having a large of attributes in massive data. Beside
this motivation, the concept of a rough set, introduced in [5,6], proved to be
efficient mathematical tool that can be used for both attribute reduction. It helps

D. Miao et al. (Eds.): RSKT 2014, LNAI 8818, pp. 206–215, 2014.
DOI: 10.1007/978-3-319-11740-9_20 © Springer International Publishing Switzerland 2014

us to find out the relevant optimal attribute subset to classify objects without reducing the classification performance. Rough set theory applies the unclear relation and data pattern comparison based on the notion of an information system, where the data is uncertain, fuzzy, incomplete or inconsistent. Rough set-based attribute reduction is crucial in data pre-processing, many redundant or irrelevant attribute may lead to poor performance of the learning algorithms. Selecting relevant attributes are essential to improve the classification accuracy. In past decade years, many researchers proposed a lot of attribute reduction approach based on the different model in varies application. The key object of rough set extract relative attributes from original attributes. Each attribute is independent of the other attributes. We can extract necessary attributes from the original attribute set used in the real-world information system.

Neural Networks is the most widely used model among the artificial intelligence techniques, the mechanism of neural network is come to human brain action when a pattern that had no output neuron associated with it was given as an input neuron. The efficient neural network framework is design by Section [7], it is applied to many manipulate real world classification problem. On the real world classification problems, neural network had been used widely in many kinds of application fields. For example, in the commerce field, the neural network is applied in the prediction of Stock market [8]; the neural network also is applied to discover the damage kind and location in the cable-stayed Bridge from the massive noisy data. And in e-commerce fields, the neural network is used to estimate the population of specific e-commerce [9]. The neural network is a classic type of neural network framework using a kernel-based approximation to form an estimate of the probability density function of categories in a large classification data.

In this paper, we will present an efficient BP-neural network classification model based on attribute reduction for high-dimensional data analysis. Rough set theory provides an efficient mathematical tool that can be used for both attribute reduction. It helps us to find out the relevant optimal attribute sets to classify objects without reducing the classification performance. Neural networks is competitive technique which are considered to be the most efficient tools in many pattern classification applications, so we first apply rough set theory to reduce irrelative attribute and retain the important attributes from data, and the input neuron based on the important attributes simplify the structure of neuron networks and improve classification accuracy in classification problem. Finally, results of experimental analysis are included to quantify the efficiency and effectiveness of the proposed model.

The paper is organized as follows. In Section 2, some preliminary knowledge about rough set theory and neural network are briefly reviewed. In Section 3, an efficient BP-neural network classification model based on attribute reduction is proposed. Section 4 shows experimental results are included to quantify the efficiency and effectiveness of the proposed model. We then conclude the paper with Section 5.

## 2  Preliminaries

### 2.1  Rough Sets

Rough set theory provides an efficient mathematical tool that can be used for both attribute reduction and rule extraction. It helps us to find out the relative optimal attribute sets without deterioration of classification accuracy [10,11]. The idea of attribute reduction has encouraged some researchers in improving the computation effectiveness of rough set theory in massive real world dataset, including medicine, control analysis, fault-diagnosis, web categorization, and economic prediction, and so on. Rough set theory has been used successfully in a number of applications such as data mining and knowledge discovery.

In rough sets theory, an information system is defined as a family of sets $S = (U, A)$, where $U$ is a non-empty, finite set of instances and is also called the universe, and $A$ is a non-empty, finite set of attributes. For each attribute $a \in A$, a mapping $a : U \to V_a$ is determined by a given decision table, where $V_a$ is the value domain of attribute $a$. If $A = C \cup D$ and $C \cap D == \emptyset$, such that $C$ is non-empty, finite condition attributes and $D$ is referred to as decision attributes, then it be a complete decision system.

**Definition 1.** *(Indiscernibility relation) Let $S = (U, C \cup D)$ be a complete decision system. Given a non-empty subset $B \subseteq A$ determines an indiscernibility relation: $R_B = \{(x, y) \in U \times U | a(x) = a(y), \forall a \in B\}$.*

The indiscernibility relation $R_B$ partitions $U$ into some equivalence classes given by $U/R_B = \{[x]_B | x \in U\}$, where $[x]_B$ denotes the equivalence class determined by $x$ with respect to $B$.

**Definition 2.** *(Lower and upper approximation) Let $S = (U, C \cup D)$ be a complete decision system. Given an equivalence relation $R$ on the universe $U$ and a subset $X \subseteq U$, one can define a B- lower approximation of $X$ is defined as $\underline{R}(X) = \{x \in U | [x]_R \subseteq X\}$. And an B-upper approximation of $X$ is defined as $\overline{R}(X) = \{x \in U | [x]_R \cap X \neq \}$.*

**Definition 3.** *(Positive region) Let $S = (U, C \cup D)$ be a complete decision system, $\forall P \subseteq C$, if a partition of the universe $U$ with respect to the decision attribute $D$ is $U/IND(D)=\{D_1, D_2, \cdots, D_m\}$ and another partition $U/IND(P)= \{P_1, P_2, \cdots, P_t\}$ defined by condition attribute $P$, then a positive region is defined as $POS_P(D) = \bigcup_{D_j \in U/IND(D)} \underline{P}(D_j)$.*

Reduct is the minimal subset of attributes that enables the same classification of elements of the universe as the whole set of attributes.

**Definition 4.** *(Attribute reduction) Let $S = (U, C \cup D)$ be a complete decision system, for a non-empty subset $P \subseteq C$, an attribute reduct $P$ based on positive region of $C$ with respect to $D$ is defined as*

1) $POS_P(D) = POS_C(D)$, 2) for $\forall a \in P$, there is $POS_{P-\{a\}}(D) \neq POS_C(D)$.

**Definition 5.** *(Core attributes) Let $S = (U, C \cup D)$ be a complete decision system, if Red are called an all attribute reduction set based on positive region of $C$ with respect to $D$, then core based on positive region is defined as $Core = \cap Red$.*

the core is the most important subset of attributes, for none of its elements can be removed without affecting the classification power of attributes.

**Definition 6.** *(Binary discernibility matrix) Let $S = (U, C \cup D)$ be a complete and consistent decision system. Let $M$ be the binary discernibility matrix of complete decision system, each element $M((i,j), k)$ indicates the discernibility between two objects $u_i$ and $u_j$ with different decision class by a single condition attribute $c_k$, which is defined as:*

$$M((i,j),k) = \begin{cases} 1 & f(x_i, c_k) \neq f(x_i, c_k) \, and \, f(x_i, D) \neq f(x_i, D) \\ 0 & else. \end{cases}$$

## 2.2   Neural Network

Neural network is available to build model a nonlinear mapping between the matrix $R = (r_{ij})_{m \times n}$ and the vector $C = (c_1, c_2, \cdots, c_m)$. Adjusting a set parameter of the model modifies this mapping. In order to construct the neural network process model. In this paper, here the neural network model consists of three layers; three layers include input layer, hidden layer and output layer. The number of nodes in the input and output layers can be created at $m$ and one in the neural network respectively. we assume the number of nodes in hidden layer is $l$. The connecting strength between the nodes in the former layer and the later layer is represented by an adjustable weight $w$.

The training processing procedure is applied to a feed forward network with a single hidden layer. We assume that there are $n$ decision variables in the optimization-simulation problem and that the neural has $m$ hidden neurons with a bias term in each hidden neuron and an output neuron. Note that the weights in the neural network are numbered sequentially starting with the first input to the first hidden neuron. Therefore, the weights for all the inputs to the first hidden neuron are $w_{i1}$ to $w_{in}$. The bias term for hidden neuron is $w_{i(n+1)}$. We test the activation functions for hidden neuron:

$$a_i = tanh((2(w_{i(n+1)} + \sum_{j=1}^{n} w_{(i-1)(n+1)+j} c_j)).$$

The activation function for the output layer is defined as follows:

$$u_{pj} = w_{m(n+2)+1} + \sum_{i=1}^{n} w_{m(n+1)+i} a_i.$$

Given the above network architecture and activation functions, we also test two schemes for optimizing $w$. The first scheme is the application of the training procedure. The second scheme consists of applying the training method to the set of weights associated with hidden nodes and then using linear regression to find the weight associated with the output nodes. In other words, using the training method to find the best set of values for $w_{m1}$ to $w_{m(n+1)}$, and then apply linear regression to minimize the sum of squares associated with $u_{pj}$. The advantage of the second scheme is that the number of weights that the training procedure needs to adjust is reduced by $m + 1$. On the other hand, the disadvantage is that the regression model needs to be solved each time any of the first $m(n+1)$ weights is change in order to calculate the mean squared error.

In the neural network learning scheme, the calculated output nodes in the output layer,$u_{ij}$ are compared with the expected output nodes $a_j$ to find the error, before the error signals are propagated backward through the network. The error function $E$ is defined as:

$$E = \frac{1}{2n}\sum_{i=1}^{m}\sum_{j=1}^{n}(u_{ij} - a_j)^2.$$

## 3    The Classification Model

The main advantage of rough sets is that it is good at attribute reduction and the strategy dealing of compressed data, and the problem is it is sensitive to noise data. Neural networks is the most popular neural networks, whose main advantage is high accuracy and non-sensitive with noise, but its disadvantage is that the redundant and irrelevant data can easily cause over-training of neural network, besides, networks scale and the amount of training samples influence on the speed of network training and training time are main problem. As to the advantages and the disadvantages of rough set theory and neural networks. The BP-neural network classification model based on attribute reduction (called AR-BPNN) can be showed in Figure 1.

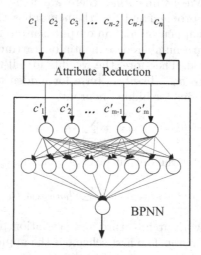

**Fig. 1.** The BP-Neural Network Classification Model Based on Attribute Reduction

In this section, an efficient classification model is proposed based the neural network and attribute reduction, The proposed model overcame rough sets sensitive to noise data; on the other hand, it reduced the training computation time of neural network, and improved efficiency much of computation.

### 3.1   A Quick Rough Set-Based Attribute Reduction Algorithm

Given an attribute set size n in decision system, the task of attribute reduction (AR) can be seen as a search for an optimal attribute subset through the competing $2^n$ candidate subsets. The definition of what an optimal subset is may vary depending on the problem to be solved. Although an exhaustive method may be used for this purpose, this is quite impractical for most large data sets. Usually attribute reduction algorithms involve heuristic or random search strategies in an attempt to avoid this prohibitive complexity. In this section, based on binary discernibility matrix, a quick rough set-based attribute reduction algorithm is proposed to solve the attribute reduction problem.

---

**Algorithm 1.** A Quick Attribute Reduction Algorithm

---

**Input:** A complete decision system $S = (U, C \cup D)$;
**Output:** An attribute reduction $Red$.
**Begin**

1. Initialize $Core \leftarrow \emptyset$; $Red \leftarrow \emptyset$;
2. Compute the discernibility matrix $M$ of the decision system $S = (U, C \cup D)$;
3.    if $c_k \in M((i,j),k)$ and $|M((i,j),k)| = 1$, then $Core = Core \cup \{c_k\}$;
4.    delete the matrix element including core attribute $Core$ in $M$;
5. Let $Red \leftarrow Core$;
6. Compute the frequency of each attribute in $M$;
7. If the frequency of attribute $a$ is the most in $M$, then let $Red = Red \cup \{a\}$; and delete the matrix element including $a$ ;
8. If the $M$ is not empty, then turn to step 6; else turn to Step 9;
9. //For some attributes are redundant attributes in $Red$,
10.    if $POS_{Red-\{b\}} = POS_{Red}(D)$, then delete the attribute $b$ from the $Red$; else, reserve the attribute $b$;
11. Output the attribute reduct $Red$.

**End**

---

**Time Complexity Analysis.** Steps 2-4 are to compute core attributes in complete decision system, the time complexity of Steps 2-4 is $O(|U|^2|C|)$. Steps 6-8 are to add some improtant conditional attributes to reduct, the time complexity of Steps 6-8 is $O(U|^2|C - Core|^2)$. Step 10 are to delete some redundant features in reduct $Red$, its time complexity is $O(|U|^2|Red|)$. Therefore, the time complexity of Algorithm 1 is $max\{O(U|^2|C - Core|^2), O(|U|^2|Red|)\}$.

## 3.2   Classification Algorithm Based on AR-BPNN

The most important advantage of neural network is that the models make no assumptions about the properties of the data.In this section, based on the above attribute reduction algorithm, we employ a three layers network to construct the AR-BPNN classification model.

---

**Algorithm 2.** Classification Algorithm Based on AR-BPNN

---

**Input:** A complete decision system $S = (U, C \cup D)$;
**Output:** Classification model;
**Begin**

1. According to the Algorithm 1, Compute the attribute reduct $Red$
2. If $|Red|=L$, let the number neurons of neural network in input layer is $L$
3. According to the type of prediction, determine the number of output neurons $V$;
4. Determine the number of hidden layer neurons, $N = \sqrt{L+V} + z$, where $z$ is a random number between 1 to 5;
5. To train the reduced sample data obtained by Step 1 embedded to neural network model shown in Fig 1, and compute and output the error $E$;
6. Adjust the weight and threshold of each neuron in accordance with the error reverse propagation algorithm according to the error of the output layer;
7. If errors meet the target value,terminate the training, else turn to Step6;

**End**

---

## 4   Experimental Results

In this section, we will illustrate the efficiency and effectiveness of the proposed model. We have downloaded six real data sets from UCI Repository of Machine Learning databases[12], which are described in Table 1. All the experiments are carried out on a PC with Windows XP, Core2 CPU2.93 GHz and memory 2GB. Algorithms are coded in C++ and the software being used is Microsoft C++ 6.0.

There are two main objectives to carry out the experiments. One is to validate the feasibility of rough set-based attribute reduction, and the other is to validate the efficiency of the proposed classification algorithm. The proposed algorithm is validated from two aspects: feature subset size, and classification accuracy. With six UCI data sets shown in Table 1, a series of experiments are conducted for evaluating the proposed classification algorithm. For the data sets with numerical attributes, we use the data tool Rosetta to discrete them.

**Table 1.** A description of six data sets

| ID | Data sets | Objects | Attributes | Classes |
|----|-----------|---------|------------|---------|
| 1 | Dermatology | 366 | 34 | 6 |
| 2 | Breast Cancer | 699 | 10 | 2 |
| 3 | Vehicle | 846 | 18 | 4 |
| 4 | Mfeat-factors | 2000 | 216 | 10 |
| 5 | Satimage | 6435 | 36 | 6 |
| 6 | Mushroom | 8124 | 22 | 2 |

The experimental results in terms of subset size of attribute reduction by the proposed attribute reduction algorithm are shown in Table 2.

**Table 2.** Subset Sizes of Attribute Reduction

| ID | Data sets | Attributes | Reduct |
|----|-----------|------------|--------|
| 1 | Dermatology | 34 | 11 |
| 2 | Breast Cancer | 10 | 6 |
| 3 | Vehicle | 18 | 10 |
| 4 | Mfeat-factors | 216 | 21 |
| 5 | Satimage | 36 | 19 |
| 6 | Mushroom | 22 | 5 |

It is easy to see that from Table 2 that the size of attribute subsets reduced by the proposed attribute reduction algorithm is much small than original attributes in most data sets, The experimental results demonstrate that the validity of the proposed attribute reduction Algorithm, which can reduce data dimensions effectively. In what follows, the classification performance of the proposed classification model (called AR-BPNN), along with a comparison with classification based on original attributes (called BPNN) [7,9], is demonstrated on the six UCI data sets using the classification accuracy. In the experiments, for the purpose of illustrating the quality of subsets discovered, we follow a 10-fold cross-validation strategy to evaluate the algorithms. Each data set is divided into two disjoint parts by random sampling: one for training and the other for test. In each fold, for the proposed classification model, we use the attribute reduction algorithms to reduce the training set. In this experiment, let the learning rate $\eta = 0.15$, inertial correction coefficient $\alpha = 0.85$, error function $E \leq 0.05$, the maximum number of training is 400 times. We train the training samples, when the network training is completed, we test the corresponding new datasets as test samples, the comparison of the average classification accuracy of the two classifiers is shown in Tables 3.

**Table** 3. Comparison of Classification Accuracy of BPNN and AR-BPNN

| ID | Data sets | BPNN | AR-BPNN |
|----|-----------|------|---------|
| 1 | Dermatology | 92.25% | 90.89% |
| 2 | Breast Cancer | 83.01% | 85.72% |
| 3 | Vehicle | 80.58% | 81.94% |
| 4 | Mfeat-factors | 77.42% | 83.65% |
| 5 | Satimage | 87.11% | 89.40% |
| 6 | Mushroom | 96.39% | 100.0% |

From the results reported in Tables 3, it can be seen that the rough set-based attribute reduction techniques are shown to improve the classification accuracy when compared to the results from unreduced data set in most cases. This indicates that classification accuracy may be increased as a result of feature selection through the removal of irrelevant and redundant attributes. In additional, in the experimental process of training and prediction, the computational time of BPNN classifier based on original attribute is longer than that of AR-BPNN classifier. The following conclusion can be drawn from the experimental results that it is feasible and effective to classification by using AR-BPNN classifier with a reasonably compact attribute subset, which is appropriate for high-dimensional data.

## 5    Conclusions

In this paper, we present the efficient classification model based on neural network pre-processing with rough sets. At first, based on binary discernibility matrix, a quick rough set-based attribute reduction algorithm is proposed to obtain the attribute reduct in high-dimensional data. Then reduced data can reduce the number of the input neuron effectively and simplify the structure of neuron networks. The experimental results shows that, compared with the general neural network model, the classification accuracy of the proposed hybrid model is efficient and feasible.

**Acknowledgments.** This work was supported in part by the Natural Science Foundation of China (No.61163025), and the 2012 Ladder Plan Project of Beijing Key Laboratory of Knowledge Engineering for Materials Science (No. Z121101002812005).

## References

1. Jagielska, I., Matthews, C., Whitfort, T.: An Investigation into the Application of Neural Network, Fuzzy Logic, Genetic Algorithms, and Rough Set to Automated Knowledge Acquisition for Classification Problems. Neurocomputing 24, 37–54 (1999)

2. Zhou, Z.H., Wu, J., Tang, W.: Ensembling neural networks: Many could be better than all. Artificial Intelligence 137, 239–263 (2002)
3. Liang, J.Y., Wang, F., Dang, C.Y., Qian, Y.H.: A group incremental approach to feature selection applying rough set technique. IEEE Transactions on Knowledge and Data Engineering 26, 294–308 (2014)
4. Pawlak, Z., Skowron, A.: Rudiments of rough sets. Information Sciences 177, 3–27 (2007)
5. Wang, F., Liang, J.Y., Dang, C.Y.: Attribute reduction for dynamic data sets. Applied Soft Computing 13, 676–689 (2013)
6. Zhang, J.B., Li, T.R., Ruan, D.: Rough sets based matrix approaches with dynamic attribute variation in set-valued information systems. International Journal of Approximate Reasoning 53, 620–635 (2012)
7. Setiono, R., Liu, H.: Neural-Network Feature Selector. IEEE Transactions on Neural Networks 8, 554–662 (1997)
8. Zhang, Y.D., Lenan, W.: Stock market prediction of S&P 500 via combination of improved BCO approach and BP neural network. Expert Systems with Applications 36, 8849–8854 (2012)
9. Khashei, M., Bijari, M.: An artificial neural network model for timeseries forecasting. Expert Systems with Applications 37, 479–489 (2012)
10. Chen, D.G., Zhao, S.Y., Zhang, L.: Sample pair selection for attribute reduction with rough set. IEEE Transactions on Knowledge and Data Engineering 24, 2080–2093 (2012)
11. Own, H.S., Abraham, A.: A new weighted rough set framework based classification for Egyptian NeoNatal Jaundice. Applied Soft Computing 12, 999–1005 (2012)
12. UCI Machine Learning Repository, http://archive.ics.uci.edu/ml/datasets

# Hardware Accelerator Design
# Based on Rough Set Philosophy

K.S. Tiwari[1], A.G. Kothari[2], and K.S. Sreenivasa Raghavan[3]

[1] MESCOE, Pune-01, India
kanchan.s.tiwari@gmail.com
[2] VNIT, Nagpur, India
agkothari72@rediffmail.com
[3] VNIT, Nagpur, India
sreenivas.raghavan7@gmail.com

**Abstract.** This paper presents a design of hardware accelerator for algorithms of rough set theory. A hardware implementation of incremental reduct generation and rule induction is proposed in this paper. Incremental reduct generation algorithm is based on simplified discernibility matrix. The design has been simulated and implemented with Xilinx Artix 7 Field Programmable Gate Array (FPGA) and verified using post synthesis simulation in Xilinx .The hardware accelerator designed is generic and easily reconfigurable due to use of FPGA.The maximum design frequency achieved is 152 MHz. The proposed hardware accelerator is used for the smart grid application. The hardware accelerator extracts important features from the database of the smart grid and generates rules using them. It automates the systems, making it more reliable and less prone to human decision making. It is worth noting that the performance of the hardware accelerator becomes more visible when dealing with larger data sets.

**Keywords:** Hardware accelerator, Discernibility matrix, FPGA, Reduct, Rough set theory, Xilinx, Smart grid.

# 1    Introduction

Rough set theory (RST), developed by Z.Pawlak, is a powerful mathematical tool for dealing with vagueness and uncertainty emphasized in decision-making. Data mining is a discipline that has an important contribution to data analysis, discovery of new meaningful knowledge, and autonomous decision-making. The rough set theory offers a viable approach for decisions and rule extraction from data, which traditional optimization methods and machine learning find difficult. Since more than quarter century, researchers are working on finding efficient reduct, rule generation algorithms, and have developed standard software for RST such as RSES, ROSE, ROSETTA, etc. for benchmarking their results. Interested readers can refer [1] for a detailed review. Recently, there is also focus on hardware mapping of RST algorithms on FPGA [2-6]. The main advantage of mapping a rough set algorithm on FPGA is

D. Miao et al. (Eds.): RSKT 2014, LNAI 8818, pp. 216–225, 2014.
DOI: 10.1007/978-3-319-11740-9_21 © Springer International Publishing Switzerland 2014

inherent speed advantage of FPGA over software-based methods. Handling ever-increasing amount of data generates increasing demand for computing power in terms of both speed and memory. General Purpose Processors do not give optimum performance for all types of applications, even if multicore technology is used. Dedicated hardware, accelerators, is specialized in computing one type of processing task and they complement conventional architectures. By using traditional microprocessors, it is also very difficult to exploit the inherent parallelism in rough set algorithms fully. Currently the complexity of programmable hardware has been evolving to the phase where large high-speed digital systems can be implemented on a single programmable logic chip.

The aim of proposed research work is to use FPGA for accelerating computationally intensive applications. In current scenario, the proposed work is more suitable for various real time applications. Literature survey shows that there is a huge gap between hardware and software implementation of rough set algorithms, as the theory itself is relatively young with respect to its counterpart. The motivation for implementing rough set algorithms in hardware stems from the fact that they are very CPU intensive while they are also intrinsically parallel algorithms and the basic operations of rough set algorithms can execute in a pipelining fashion. Our architecture incorporates built-in hardcore resources like dual port BRAMs, Multipliers etc. in order to create an efficient hardware-based rough set algorithm.

Rough set algorithm processes data by removing redundant attributes, and extracting meaningful rules from it. These rules are used for classification or decision making purpose. Currently, there is tremendous research in the data analytics particularly in the area of dynamic knowledge i.e. where the knowledge or data (either attributes or objects) is added to database during run time (using software). It provides the most valuable data (usually about past events) for good decision-making, but there is lack of hardware acceleration. This paper presents design of generic hardware accelerator, which effectively handles larger databases. It selects important features using incremental algorithm and generates rules using them in if-then form. Its use is in decision-making support system. The incremental algorithm is an important technique for added-in data without re-computing the original information in the dynamic decision table. Literature survey on hardware design of RST algorithms shows instances of handling of static databases; however, no related work for dynamic database is found.

This work bridges the gap between software and hardware implementation of RST. RST is a powerful soft computing theory, and can be very well mapped on FPGA. All time critical tasks can be effectively managed by such dedicated hardware in less amount of time.

This paper is divided into following section: in section 2, blocks of hardware accelerator are discussed. Section 3 elaborates the application of proposed hardware accelerator for smart grids. Section 4 presents simulation results and synthesis report of the design while in section 5 research work is concluded. For rough set preliminaries, readers can refer [7-8].

## 2 Proposed VLSI Implementation

In this section, we describe the VLSI implementation of hardware accelerator for dynamic database. The details of algorithms used can be found in [9]. The block diagram for computing reduct for dynamic system is shown in fig.1

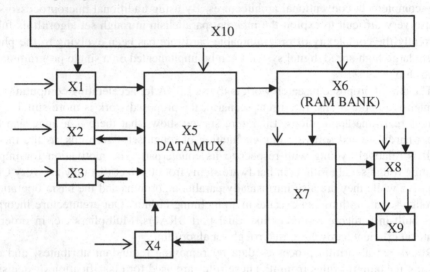

**Fig. 1.** Block diagram of proposed system

| Lines | Description |
|-------|-------------|
| ⟶ | These lines indicate "Control Signals" to each block |
| ⟶ | These lines indicate "Data" from RAM Block (Component X6) |
| ⟶ | These lines are divided into bidirectional & unidirectional data bus |

**Component X1:** This block loads the data into RAM (RAM1, RAM2).RAM1 and RAM2 present positive and negative region elements respectively. In the event of either adding new data to RAM, appropriate control signals along with address of the RAM location are given. The control signals are load, load4, load3- which when asserted adds the data in the appropriate RAM i.e. either positive and negative region. The data is transferred on rising edge of clock. Two assert signals are used for informing processor that the operation is complete (Adding of data to RAM).

**Component X2:** This block takes input data from RAM, computes the discernibility matrix elements (mij), and puts the calculated values ($m_{ij}$) on the address bus CLISTRAM1, CLISTRAM2. The control signal is start1. The data is read from RAM on rising edge and the computed discernibility matrix elements ($m_{ij}$) put on the address bus at the falling edge of the clock. An auto increment mode is utilized in the block which updates the next address to be read during falling edge of clock.

**Component X3:**  This block performs the reduct update, when the new object has same attributes as positive region elements but different decision levels. It computes discernible matrix element $m_{ij}$ and performs the require action of incrementing or decrementing count by providing appropriate $m_{ij}$. The control signals are p3enble,clck etc.

**Component X4:** This block takes in the new object, which is neither positive nor negative region element from RAM1. It forms the discernible matrix element using the definition of $m_{ij}$, which is given in [9]. The computed $m_{ij}$ are used as addresses for storing new count value i.e. incrementing. The control signals are p4enble, clck etc.

**Component X5:** This component routes the data from components X1, X2, X3, X4 to RAM for reading from RAM as well as writing the data to RAM. If we try to access the RAM for data without this datamux (X5), then it leads to multiple drivers' issue. Control Signals used are start1, start3, start4, load3, load4, load, add3, delete3, add4.

**Component X6:** This component is RAM BANK. In this block, three RAM's blocks are used, two of which are DUAL PORT RAM (RAM1 for positive region, CLISTRAM for recording the frequency of each discernible matrix element $m_{ij}$) and other one is SINGLE PORT RAM (RAM2 for negative region elements). Control signals used are readenable, writeenable. The width of data bus is RAM1, RAM2 are 19 bits wide while CLISTRAM has width of 16 bits. RAMs are implemented using IP Cores for low power and efficient use of FPGA resources. Timing diagram of dual port RAM is shown in Fig. 2.

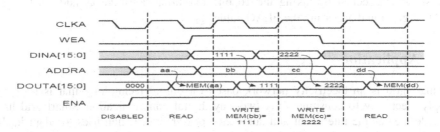

**Fig. 2.** Timing diagram of Dual port RAM (Taken from [10])

**Component X7:** This block reads the value of count obtained from CLISTRAM. The count values are examined for following. When the process is creation of discernibility matrix elements for static database, then count is incremented by one. If the process is updating block, corresponding control signals are enabled and the value of count is either incremented (if new $m_{ij}$ is being added to discernibility matrix) or decremented ($m_{ij}$ is removed from discernibility matrix). For instance, if the new element being added is neither positive region nor negative element, then it implies that, it is new element and it might lead to new discernibility matrix element. The control signals used are start1, add3, add4, delete3.

**Component X8:** This block forms the basis of reduct update algorithm when new discernibility elements are added. Control signals are add3, add4. When add3 or add4 is asserted, discernibility matrix element that has count value as 0 are ANDed with reduct, if the resulting value is null or ∅, then reduct should be updated else reduct remains same. Xilinx primitives are used to implement this block.

**Component X9:** This block is part of reduct update algorithm when discernibility elements are deleted. Control signals used are delete3. When del3 is asserted, new discernibility elements are checked with reduct, to check for superfluous elements. If any superfluous attribute is found, it is removed from the previous reduct and updated. Xilinx primitives are used to implement this block.

**Component X10:** This component is responsible for separating the incoming data stream of 19 bits wide into positive and negative elements and feeding it to the system. In this block, we denote the status of an element in a flag register. There are two flag registers, POSFLAG, NEGFLAG (table 1) which denotes whether the element is positive or negative or neither of them. Since the system has 16 bit conditional attributes and 3 bit decision attributes, therefore number of possible 16 bit conditional attribute pattern can be $65536(2^{16})$.

**Table 1.** Flag registers

| POSFLAG | NEGFLAG | STATUS |
|---------|---------|--------|
| 1 | 1 | Error. |
| 1 | 0 | Positive element |
| 0 | 1 | Negative element |
| 0 | 0 | Neither positive nor negative |

The status is checked by taking the 16 bit conditional attribute as address to this FLAG register and accessing the FLAG value.

# 3    Application

A smart grid is an electricity network based on digital technology that is used to supply electricity to consumers via two-way digital communication. Smart grid has been defined as the intelligent and automated power grid, which tries to adapt itself from time to time, and in different situations. It monitors and controls power flow between power plant and consumers and all the points in between. Smart grids allows us to integrate all sources of energy on a common platform i.e. integration of natural resources based power supplies like wind, geo-thermal, ocean currents with thermal energy based plants and nuclear based energy source. It can be used for routing power e.g. if consumer is not using the power supply supplied by plant, then consumer can acts as power plant and route the supply to power deficient areas thus equalizing the loads. Many government institutions around the world have been encouraging the use of smart grids for their potential to control and deal with global warming, emergency resilience and energy independence scenarios.

## 3.1    Rough Set Theory in Smart Grids

This basic motivation behind using hardware accelerator can be attributed to the case study of demand forecasting of power grid as mentioned in [11]. In this case, the results generated from software simulation were infeasible as it was found to have taken lots of time for processing the data (32 days). In our case study, few real time parameters are considered for decision-making process. A bottom to top approach-hierarchy is used in managing these controllers. An incremental approach for find important attributes is more effective in smart grid as database needs to be updated at regular intervals. Decision attributes will give decisions in the form of priorities like shifting of feeders, reversal of role (i.e. consumer becomes producer and route power to other part where there is scarcity of power) etc.

Attributes considered are [12 -14]:

1. Temperature of surrounding.
2. Season. (Since during summer the need for power is more).
3. Feeder failure rate.
4. Past history of replacement, repairs, test and power quality factor events.
5. Past history of failures.
6. Heat generated at feeders.
7. Power consumption.

Power companies are beginning to switch from reactive maintenance plans (fix when something goes wrong) to proactive maintenance plans (fix potential problems before they happen). To accommodate this condition, attributes pertaining to feeder failure rates and temperature should be taken into account. Even it might happen that new replaced components might be replaced at a faster rate than old components, so as to take this situation into account, attribute relating to past replacement of part, repairs must be added. The decision attributes are again split in two parts:

- Global attributes. (Whose values are constantly monitored and action is taken if any discrepancy is found)
- Local attributes. (Required at local level so that they can be taken care lower level of hierarchy).Local attributes consist of variables like repairs, replacements. If these are verified in specific amount of time, then the decision pertaining to global attributes are set, which shows a high level of problem.

Let the region (City or part of the state) be divided in J regions where J is an integer. Each of the J regions is divided in K sub regions. Each part has RST controller installed in its region. The RST controller samples data at regular instants to reflect or update any important changes and give decision in the form of priorities. The RST controller in each part has the attributes as defined above. In case of any feeder breakdown, the priority of RS T controller changes and decision is given such that the power, the failed feeder was carrying is now transferred routed to other feeder using decision bit "shifting of feeders". If the amount of power of any sub regions is less than threshold or more than threshold, a consumer from same sub regions or different region will act as producer and routes the power to the sub regions which is in need. Fault in wire generally carries more current, thus generating more heat at feeder; it leads to failure of that feeder. To take pre-emptive action, the RST controller informs the global controller about the difference in feeder temperature by sending its decision bit as '1'. This decision bit is now conditional attribute for RST controller at global level.

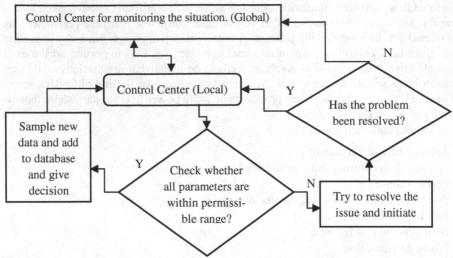

**Fig. 3.** Hierarchy of RST controllers using bottom to top approach

Control center are set of resources, which monitor the status of a state or city or region or part. The hierarchy shown in fig.4 shows the division of work among the control center. The aggregate of data from sub regions are collected at region level data from region are collected at city level and so on. If we take instance of sub region level of a region, then global variables of sub region level are local variables at regional level. In this way, efficiency of the power distribution as well as power transmission and maintenance increases.

## 4    Results

In table 2, device utilization summary used by hardware accelerator is mentioned. It is evident from these tables that the device resources are not fully utilized so more logic can be accommodated.

**Table 2.** Device Utilization summary

| Number of BUFFERS used | 5 out of 32    - 15% |
|---|---|
| Number of External IOBs | 146 out of 172 -    84% |
| Number of LOCed IOBs | 0 out of 146    -  0% |
| Number of OLOGICs | 17 out of 180   -    9% |
| Number of RAMB18 | 1 out of 26    -  3% |
| Number of RAMB36 | 17 out of 26    -  65% |
| Number of Slices | 375 out of 3120  - 12% |
| Number of Slice Registers | 492 out of 12480  - 3% |
| Number used as Flip Flops | 398 |
| Number used as Latches | 94 |

## TIMING INFORMATION OF OVERALL SYSTEM (Operating Frequency):

Minimum period: 6.575ns (Maximum Frequency: 152.080MHz)
Minimum input arrival time before clock: 15.279ns
Maximum output required time after clock: 3.040ns.

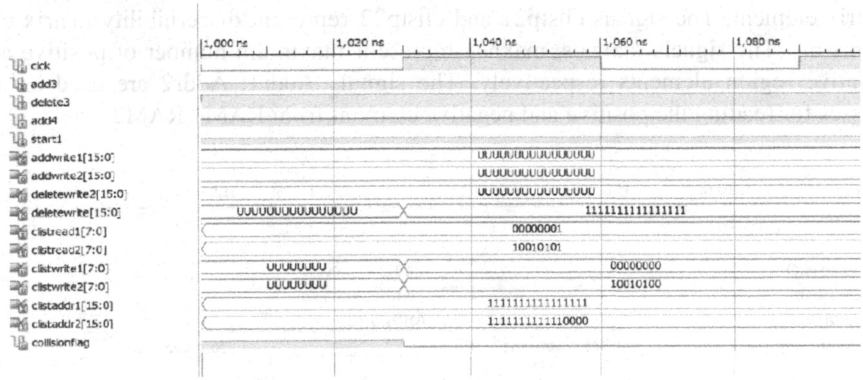

**Fig. 4.** Addcontroller unit waveform

Add controller unit (fig. 4) is responsible for incrementing the count of each discernibility matrix element. When either add3 or add4 is asserted, it passes that discernibility matrix element whose count is zero to reduct update unit. Similarly, when delete3 is asserted, it passes that discernibility matrix element whose count is zero after decrementing by one, to reduct update unit. In the fig.4, we observe two data buses- clistread1 and clistread2, which read the count value. The other two data buses- clistwrite1 and clistwrite2 are used for writing the count value in CLISTRAM, whenever new discernibility matrix element $m_{ij}$ is being added.

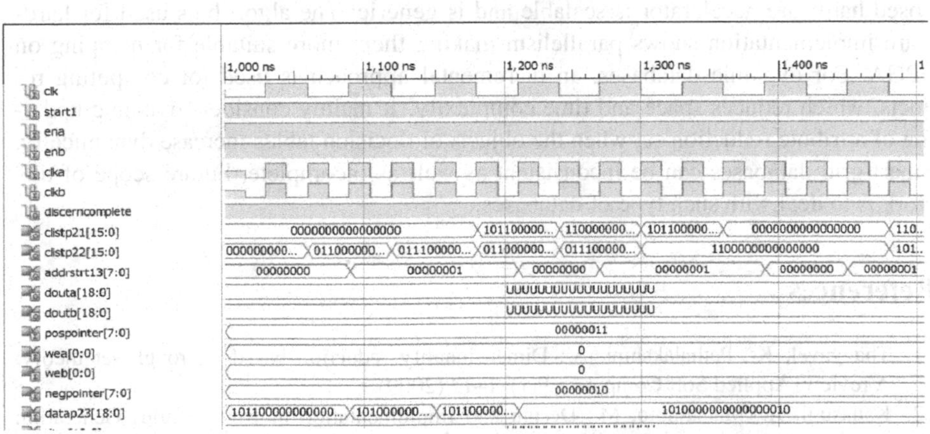

**Fig. 5.** Discernibility matrix unit waveform

Discernibility matrix unit (fig.5) computes the discernibility matrix elements $m_{ij}$. This unit reads the data from RAM BANK (on the data bus namely-datap21, datap23, datap22- positive and negative region elements respectively) during rising edge of clock and during falling edge of clock, it computes the discernibility matrix elements $m_{ij}$. These $m_{ij}$ are used as addresses, which hold the frequency of each discernibility matrix element. The signals clistp21 and clistp22 represent discernibility matrix elements $m_{ij}$. The signals maxpos, maxneg represent maximum number of positive and negative region elements respectively. The signals Addr1, Addr2 are used as addresses for reading the positive and negative element from RAM1,RAM2.

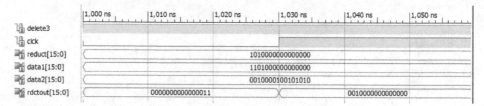

**Fig. 6.** Reduct update block waveform

Reduct update block updates reduct based on the algorithm given in [9]. In fig.6, the reduct is being observed for the case, when the discernibility matrix element $m_{ij}$ is being removed from discernibility matrix. The rdctout represents the final updated reduct.

## 5    Conclusion

The hardware implementation of the rough set algorithms is in development phase and its usage in real life will gain more importance in due course of time. Such hardware can be effectively used for any application involving decision making. The proposed hardware accelerator is scalable and is generic. The algorithms used for hardware implementation shows parallelism making them more suitable for mapping on FPGA. For dynamic databases, an incremental approach is used for computing reducts, which reduces space and time complexity. It mainly considers dynamic updating of attribute reduction set when the objects of decision tables increase dynamically. Smart grid databases can be inconsistent as well as incomplete. Future scope of this work is to deal with such type of databases.

## References

1. Thangavel, K., Pethalakshmi, A.: Dimensionality reduction based on rough set theory: A review. Applied Soft Computing 9(1), 1–12 (2009)
2. Kanasugi, A., Matsumoto, M.: Design and Implementation of Rough Rules Generation from Logical Rules on FPGA Board. In: Kryszkiewicz, M., Peters, J.F., Rybiński, H., Skowron, A. (eds.) RSEISP 2007. LNCS (LNAI), vol. 4585, pp. 594–602. Springer, Heidelberg (2007)

3. Pawlak, Z.: Elementary Rough Set Granules: Toward a Rough Set Processor. In: Pal, S.K., et al. (eds.) Rough-Neural Computing Cognitive Technologies, pp. 5–13. Springer (2004)
4. Kopczynski, M., Stepaniuk, J.: Rough Set Methods and Hardware Implementations. Zeszyty Naukowe Politechniki Białostockiej. Informatyka 8, 5–17 (2011)
5. Lewis, T., Perkowski, M., Jozwiak, L.: Learning in hardware: Architecture and implementation of FPGA based rough set machine
6. Tiwari, K.S., Kothari, A.G., Keskar, A.G.: Reduct Generation from Binary Discernibility Matrix: An Hardware Approach. International Journal of Future Computer and Communication 1(3), 270–272 (2012)
7. Zadeh, L.A.: Foreword. Applied Soft Computing 1(1), 1–2 (2001)
8. Pawlak, Z.: Rough sets,Theoretical Aspects of Reasoning about Data. Kluwer Academic Publishers, Dordrecht (1991)
9. Guan, L.: An Incremental Updating Algorithm of Attribute Reduction Set in Decision Tables. In: 2009 Sixth International Conference on Fuzzy Systems and Knowledge Discovery (2009)
10. Xilinx Block Memory Generator v7.3 Product Guide
11. Aman, S., Yin, W., Simmhan, Y., Prasanna, V.: Machine Learning for Demand Forecasting in Smart Grid
12. Bryant, R.E., Hensel, C., Katz, R.H., Gianchandani, E.P.: From Data to Knowledge to Action: Enabling the Smart Grid. In: Computing Community Consortium
13. Wu, L., Kaiser, G., Rudin, C., Waltz, D., Anderson, R., Boulanger, A., Salleb-Aouissi, A., Dutta, H., Pooleery, M.: Evaluating Machine Learning for Improving Power Grid Reliability. In: ICML 2011 Workshop on Machine Learning for Global Challenge (2011)
14. Rudin, C., et al.: Machine Learning for the New York City Power Grid. IEEE Transactions on Pattern Analysis and Machine Intelligence
15. Aman, S., Yin, W., Simmhan, Y., Prasanna, V.: Machine Learning for Demand Forecasting in Smart Grid

# Intelligent Systems and Applications

# Properties of Central Catadioptric Circle Images and Camera Calibration

Huixian Duan\*, Lin Mei, Jun Wang, Lei Song, and Na Liu

R & D Center of Cyber-Physical Systems, The Third Research Institute of Ministry
of Public Security, Bisheng Road, Shanghai, China
hxduan005@163.com

**Abstract.** Camera calibration based on circle images has unparalleled advantages in many fields. However, due to the large distortion, catadioptric camera calibration from circles remains a challenging and open problem. Under central catadioptric camera, circles in a scene are projected to quartic curves on the image plane. Except for the sufficient and necessary conditions that must be satisfied by paracatadioptric circle image, the properties of the antipodal image points and the absolute conic are both very important for catadioptric camera calibration. In this paper, we study the properties of the antipodal image points on paracatadioptric circle image, that is the criterion conditions for the antipodal image points on circle image. What's more, we show the image of the absolute conic, and derive the constraint equations about the intrinsic parameters of central catadioptric camera. Finally, we discuss on the central catadioptric camera calibration using circle images.

**Keywords:** Central catadioptric camera, image processing, the absolute conic, antipodal images, machine learning.

## 1 Introduction

Many applications in computer vision, such as robot navigation and virtual reality, a camera with a quite large field of view is required. A conventional camera has a very limited field of view. One effective way to enhance the field of view of a camera is to combine the camera with mirrors, which is referred to as catadioptric image formation. Catadioptric system can be classified into two groups, central and noncentral, based on the uniqueness of an effective viewpoint [1], [2]. Uniqueness of an effective viewpoint is desirable because it allows the mapping of any part of the scene to a perspective plane without parallax. as if it was taken with a perspective camera whose focus is the effective viewpoint. Baker and Nayar [1] introduced that a central catadioptric system can be built by setting a parabolic mirror in front of an orthographic camera, or a hyperbolic, elliptical, planar mirror in front of a perspective camera, where the single viewpoint constraint can be fulfilled via a careful alignment of the mirror and the camera.

---

\* Project supported by National Science and Technology Support Projects of China (No.2012BAH07B01).

D. Miao et al. (Eds.): RSKT 2014, LNAI 8818, pp. 229–239, 2014.
DOI: 10.1007/978-3-319-11740-9_22 © Springer International Publishing Switzerland 2014

In central catadioptric systems, camera calibration is also an important task in order to extract metric information from 2D images. In the literature, the calibration methods can be classified into the following five categories. The first category [3,4,5] requires a 3D/2D calibration pattern with control points. The second category [2,6,7,8,9] only makes use of the properties of line images. The third category [10,11] is based on the sufficient and necessary conditions of para-catadioptric circle images. The fourth category [12,13,14] is based on the properties of sphere images. The fifth category [15] only uses point correspondence in multiple views, without needing to know either the 3D location of space points or camera locations.

Camera calibration from circles has great advantages. However, due to large distortion, catadioptric camera calibration from circle images has many difficulties and lacks of studies. Based on the projection of a line complex, Sturm and Barreto[16] proved that the central catadioptric projection of a quadric is a quartic curve. According to the imaging process under central catadioptric model, Duan and Wu [17] derived the algebraic expression of a circle image and provided a unified imaging theory of different geometric elements. What's more, Duan et.al [11] gave the conditions that must be satisfied by paracatadioptric circle image, and proposed camera calibration method based on one circle image. In this paper, we study the properties of the antipodal image points on paracatadioptric circle image. And we show the projection of the absolute conic, and derive the constraint equations about the intrinsic parameters of central catadioptric camera. Finally, we discuss on the central catadioptric camera calibration using circle images.

This paper is organized as follows: Section 2 is some preliminaries. Section 3 study properties of the antipodal image points on paracatadioptric circle image and the projection of the absolute conic under central catadioptric camera. Section 4 discusses catadioptric camera calibration based on circle images. Finally, Section 5 presents some concluding remarks.

## 2    Preliminaries

A bold letter denotes a vector or a matrix. Without special explanation, a vector is homogenous coordinates. In the following, we briefly review the image formation for central catadioptric camera introduced in [18], the antipodal image points and their properties proposed in [8] and the absolute conic introduced in [19].

### 2.1    Paracatadioptric Circle Image

Let the intrinsic parameter matrix of the pinhole camera be

$$\mathbf{K}_c = \begin{pmatrix} r_c f_c & s & u_0 \\ 0 & f_c & v_0 \\ 0 & 0 & 1 \end{pmatrix}$$

where $r_c$ is the aspect ratio, $f_c$ is the focal length, $(u_0, v_0, 1)^T$ denoted as $\mathbf{p}$ is the principal point, and $s$ is the skew factor.

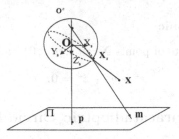

**Fig. 1.** The image formation of a point under central catadioptric camera

As shown in Fig.1, under the central catadioptric camera model, a space point
point **X** is projected to its catadioptric image by

$$\mathbf{m} = \lambda \mathbf{K}\left(\frac{\mathbf{RM} + \mathbf{t}}{\|\mathbf{RM} + \mathbf{t}\|} + \xi \mathbf{e}\right). \tag{1}$$

Where $\lambda$ is a scalar, $(\mathbf{R}, \mathbf{t})$ are a $3 \times 3$ rotation matrix and a 3-vector of
translation, $\mathbf{K}$ is the intrinsic matrix, $\|\ \|$ denotes the norm of vector in it, $\mathbf{e} =
(0, 0, 1)^T$, and $\xi$ is the mirror parameter that is the distance from $\mathbf{O}$ to $\mathbf{O}^c$. The
mirror is a paraboloid if $\xi = 1$, an ellipsoid or hyperboloid if $0 < \xi < 1$, and a
plane if $\xi = 0$.

## 2.2   The Antipodal Image Points

Under paracatadioptric camera, Wu et al. [8] gave the definition of antipodal
image points and studied their properties as follows:

**Definition 1.** $\{\mathbf{m}, \mathbf{m}'\}$ is called a pair of antipodal image points if they could
be images of two end points of a diameter of the viewing sphere(See Fig.2).

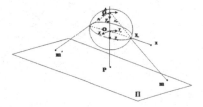

**Fig. 2.** $\{\mathbf{m}, \mathbf{m}'\}$ is a pair of antipodal image points.

**Proposition 3.** If $\{\mathbf{m}, \mathbf{m}'\}$ is a pair of antipodal image points under paracatadioptric camera, we have:

$$\frac{1}{\mathbf{m}^T \varpi \mathbf{m}} \mathbf{m} + \frac{1}{\mathbf{m}'^T \varpi \mathbf{m}'} \mathbf{m}' = \mathbf{p}. \tag{2}$$

where $\varpi = \mathbf{K}_c^{-T} \mathbf{K}_c^{-1}$, and $\mathbf{p}$ is the principal point.

## 2.3   The Absolute Conic

The absolute conic consists of points $\mathbf{X} = (x, y, z, 0)^T$ at infinity such that:

$$x^2 + y^2 + z^2 = 0.$$

# 3   Properties of Paracatadioptric Circle Images

In this section, we study the properties of antipodal image points on paracatadioptric circle image, that is the criterion conditions for the antipodal image points on circle images. What's more, we derive the algebraic equations of the absolute conic under central catadioptric camera and obtain the constraint equations for intrinsic parameters.

## 3.1   Properties of Antipodal Image Points on Paracatadioptric Circle Image

Generally, under paracatadioptric camera, the projection of a circle is a quartic curve (See Fig.3). Duan and Wu[17] derived the algebraic expression of circle image and pointed out that the image of a circle under paracatadioptric camera is a quartic curve consisting of two closed curves, one of which is visible in the image plane and the other is invisible. Now, we shall show the properties of the antipodal image points on paracatadioptric circle image. It follows from Section 2.2 that, the image point, antipodal image point and the principal point $\mathbf{p}$ are collinear. Thus, we can calculate the antipodal image point.

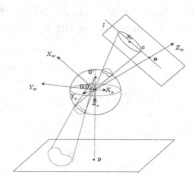

**Fig. 3.** The image formation of a circle

Suppose that $\mathbf{m}_1$ is extracted from the visible part of the circle image. Then $\mathbf{m}_1$ and the principal point $\mathbf{p}$ determinate a line $L$. If $L$ intersects the circle at two points, we know that they are a pair of antipodal points. In general case, $L$ intersects the visible part of the circle image at $\mathbf{m}_1, \mathbf{m}_2$ and the invisible part of the circle image at $\mathbf{m}_3, \mathbf{m}_4$.

It follows from the definition of the antipodal image point that if $m_2$ and $m_1$ lie on the same side of the principal point $p$, then $m_3, m_4$ lie on the other side of $p$. Moreover, the two points $m_3, m_4$ are the antipodal points of $m_1$ and $m_2$ respectively. How can we determinate which of $m_3, m_4$ is the antipodal point of $m_1$ and which is that of $m_2$?

By the image formation of paracatadioptric camera (see Fig.1), we know that, all the space point lie on the front of the virtual optical center $O^c$, that is, the formation of paracatadioptric camera has quasi-affine invariance [20], and hence the lines from $O^c$ have invariant adjacent relations. Based on this property, Theorem1 gives the determination conditions for the antipodal image points on circle images.

**Theorem 1.** *Let $m_1$ be the image point extracted from the circle image. Suppose that the line $L$ determined by $m_1$ and the principal point $p$ intersects circle image (a quartic curve) at four different points $m_1 m_2 m_3 m_4$. Suppose $m_2$ and $m_1$ lie on the same side of $p$, and the point $m_2$ is adjacent to the principal point $p$ and the point $m_1$ is nonadjacent to $p$. Then $\{m_1, m_3\}\{m_2, m_4\}$ are two pairs of antipodal image points if and only if, for each pair of $\{m_1, m_3\}$ and $\{m_2, m_4\}$, one point of the pair is adjacent to the principal point $p$ and the other is not. For instance, if $m_1(m_2)$ is adjacent to $p$, then $m_3(m_4)$ is nonadjacent to $p$.*

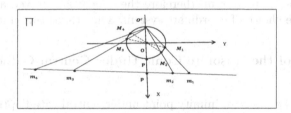

**Fig. 4.** The antipodal image points on paracatadioptric camera image

*Proof.* Virtual optical center $O^c$ and the line $L$ determine a plane $\Pi$ which intersects the viewsphere at a unit circle $c$. By the image formation of paracatadioptric camera based on circle, we can connect the center of the viewsphere and each point on the circle in the space to obtain a oblique cone, which intersects the plane $\Pi$ at two lines. The lines intersect the unit circle $c$ at four points denoted by $M_i, i = 1, 2, 3, 4$. Then the ray passing through $O^c$ and $m_i$ intersects the unit circle $c$ at $M_i = (X_{si}, Y_{si})^T, i = 1, 2, 3, 4$.

Suppose the ray passing through $O^c$ and the principal point $p$ of the camera intersects the unit circle $c$ at $P$.

We set the Euclidean coordinate as Fig.4.: the center of the viewsphere $O$ is taken as the origin, the line passing through the points $O$ and $p$ is the $X$ axis and the line perpendicular to $X$ axis is the $Y$ axis. Here we assume the circle passes through the visible part of projection on circle image of the virtual optical center $O^c$ if $Y$ coordinate of the projective point on viewsphere of the circle.

Since $\mathbf{m}_2$ and $\mathbf{m}_1$ lie on the same side of $\mathbf{p}$, that is, the points $\mathbf{m}_1, \mathbf{m}_2$ lie on the visible part of the circle image, then we have $Y_{s1} > 0, Y_{s2} > 0$.

$\Rightarrow$   Since the point $\mathbf{m}_2$ and $\mathbf{p}$ are adjacent and $\mathbf{m}_1$ and $\mathbf{p}$ are nonadjacent, we have $\mathbf{m}_2$ lies between $\mathbf{m}_1$ and $\mathbf{p}$, and therefore, the ray $\mathbf{O}^c\mathbf{m}_2$ lies between the ray $\mathbf{O}^c\mathbf{m}_1$ and $\mathbf{O}^c\mathbf{P}$. Thus we have $\angle\mathbf{m}_2\mathbf{O}^c\mathbf{p} < \angle\mathbf{m}_1\mathbf{O}^c\mathbf{p}$. Since $\{\mathbf{m}_1, \mathbf{m}_3\}$ and $\{\mathbf{m}_2, \mathbf{m}_4\}$ are two pairs of antipodal image points, it follows from the definition of antipodal image point that the line segment connecting $\mathbf{M}_1$ and $\mathbf{M}_3$ is a diameter of $\mathbf{c}$, and the line segment connecting $\mathbf{M}_2$ and $\mathbf{M}_4$ is a diameter of $\mathbf{c}$. Then we have $\angle\mathbf{m}_1\mathbf{O}^c\mathbf{m}_3 = \angle\mathbf{m}_1\mathbf{O}^c\mathbf{p} + \angle\mathbf{p}\mathbf{O}^c\mathbf{m}_3 = 90^o$, $\angle\mathbf{m}_2\mathbf{O}^c\mathbf{m}_4 = \angle\mathbf{m}_2\mathbf{O}^c\mathbf{p} + \angle\mathbf{p}\mathbf{O}^c\mathbf{m}_4 = 90^o$. Since $\angle\mathbf{m}_2\mathbf{O}^c\mathbf{p} < \angle\mathbf{m}_1\mathbf{O}^c\mathbf{p}$, we have $\angle\mathbf{m}_3\mathbf{O}^c\mathbf{p} < \angle\mathbf{m}_4\mathbf{O}^c\mathbf{p}$. Therefore the ray $\mathbf{O}^c\mathbf{m}_3$ lie between $\mathbf{O}^c\mathbf{m}_4$ and $\mathbf{O}^c\mathbf{p}$, that is, the point $\mathbf{m}_3$ is adjacent to the principal point $\mathbf{p}$, and the point $\mathbf{m}_4$ is nonadjacent to $\mathbf{p}$.

$\Leftarrow$   Assume the antipodal image points of $\mathbf{m}_1$ and $\mathbf{m}_2$ are $\mathbf{m}_1^{'}$ and $\mathbf{m}_2^{'}$, respectively. Since the point $\mathbf{m}_1$ is nonadjacent to the principal point $\mathbf{p}$ and $\mathbf{m}_2$ is adjacent to $\mathbf{p}$, then the proof above implies that $\mathbf{m}_1^{'}$ is adjacent to $\mathbf{p}$ and $\mathbf{m}_2^{'}$ is nonadjacent to $\mathbf{p}$. Since for each pair of $\{\mathbf{m}_1, \mathbf{m}_3\}$ and $\{\mathbf{m}_2, \mathbf{m}_4\}$, one point of the pair is adjacent to the principal point $\mathbf{p}$ and the other is not, that $\mathbf{m}_1$ is adjacent to $\mathbf{p}$ and $\mathbf{m}_2$ is nonadjacent to $\mathbf{p}$ implies that $\mathbf{m}_3$ is adjacent to $\mathbf{p}$ and $\mathbf{m}_4$ is nonadjacent to $\mathbf{p}$. Thus we know $\mathbf{m}_3$ ($\mathbf{m}_4$) the antipodal image point of $\mathbf{m}_1$ ($\mathbf{m}_2$, respectively).

Furthermore, the adjacent relation of the points $\mathbf{m}_i, i = 1, 2, 3, 4$ doesn't depend on the coordinate system, therefore the conclusion proved above is independent with the choice of coordinate system. Thus the theorem follows.

## 3.2   Imaging of the Absolute Conic Under Central Catadioptric Camera

First, we study the image of infinity point under central catadioptric camera. A line intersects the infinite plane at a infinity point. Since parallel lines intersect the infinite plane at the same infinity point, we can calculate the image of infinity point by the intersections of the image of the parallel lines under central catadioptric camera.

The intrinsic parameter matrix is

$$\mathbf{K}_c = \begin{pmatrix} r_c f_c & s & u_0 \\ 0 & f_c & v_0 \\ 0 & 0 & 1 \end{pmatrix}.$$

As in Fig.5., let $L$ be a space line. The lines $L_1$ and $L$ are parallel. Suppose the direction of $L$ and $L_1$ is $\mathbf{l} = (l_1, l_2, l_3)^T$. Then they intersect the infinite plane at the same infinity point $\mathbf{D} = (l_1, l_2, l_3, 0)^T$. Thus we can determine the image of the infinity point through the image of $L$ and $L_1$ under central catadioptric camera.

By the image formation of lines under central catadioptric camera, the center $\mathbf{O}$ of the viewsphere together with $L$ and $L_1$ determine two planes $\pi$ and $\pi_1$ respectively. We denote the normal vectors by $\mathbf{n}\mathbf{n}_1$, respectively. Then the two

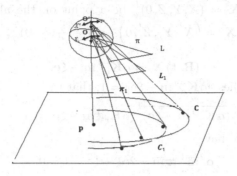

**Fig. 5.** The image formation of parallel lines under central catadioptric camera

planes $\pi$ and $\pi_1$ intersect at a line with direction $\mathbf{n} \times \mathbf{n}_1$. Since $\mathbf{n} \perp \ln_1 \perp \mathbf{l}$, we have $\mathbf{n} \times \mathbf{n}_1$ is parallel to $\mathbf{l}$. Thus the parametric equation of the intersection line is

$$(x, y, z) = \lambda \mathbf{l}. \tag{3}$$

It follows from the image formation formula Eq(1) of point under central catadioptric camera that, the images two lines $L$ and $L_1$ under central catadioptric camera have two intersections:

$$\mathbf{P}_\pm = (\frac{\lambda_\pm (r_c f_c l_1 + s l_2)}{\lambda_\pm l_3 + \xi}, \frac{\lambda_\pm (f_c l_2)}{\lambda_\pm l_3 + \xi})^T. \tag{4}$$

where $\lambda_\pm = \pm \frac{1}{\sqrt{l_1^2 + l_2^2 + l_3^2}}$. Then $\mathbf{p}_\pm$ is the image of infinity point $\mathbf{D}$ under central catadioptric camera. Thus, the image of infinity point $\mathbf{D}$ can be contained in the image formation of point under central catadioptric camera. Following from this, We study the image of the absolute conic under central catadioptric camera.

The absolute conic $\mathbf{\Omega}_\infty$ is a curve on the infinite plane in the projective space, and its image under perspective camera doesn't vary as the location position varies. Therefore, the conic has an important role in camera calibration. Now, we shall show the image of absolute conic under central catadioptric camera.

**Theorem 2.** *Set* $\mathbf{K}_c^{-T}\mathbf{K}_c^{-1} = \varpi$. *Let* $\mathbf{m}$ *be a point(complex point) on the image of the absolute conic. Denote by* $\overline{\mathbf{m}}$ *the conjugate complex point of* $\mathbf{m}$. *Then*
  *(i)   if* $0 < \xi < 1$, *the image of the absolute conic under catadioptric camera is:*

$$\xi^4(\overline{\mathbf{m}}^T \varpi \mathbf{m} - \mathbf{m}^T \varpi \mathbf{m})^2 + 2\xi^2(\overline{\mathbf{m}}^T \varpi \mathbf{m} - \mathbf{m}^T \varpi \mathbf{m})(\mathbf{m}^T \varpi \mathbf{m} - 2) + (\mathbf{m}^T \varpi \mathbf{m})^2 = 0. \tag{5}$$

  *(ii)   if* $\xi = 1$, *the image of the absolute conic under catadioptric camera is :*

$$4(\mathbf{m} - \overline{\mathbf{m}})^T \varpi \mathbf{m} + (\overline{\mathbf{m}}^T \varpi \mathbf{m})^2 = 0. \tag{6}$$

  *(iii)   if* $\xi = 0$, *the image of the absolute conic under catadioptric camera is:*

$$\mathbf{m}^T \varpi \mathbf{m} = 0. \tag{7}$$

*Proof.*    (*i*)    Let $\mathbf{X} = (X, Y, Z, 0)^T$ be a point on the absolute conic with $\| (\mathbf{R}, \mathbf{t}) \mathbf{X} \| = d$. Let $\mathbf{X}' = \left( X', Y', Z', 0 \right)^T = \left( \frac{X}{d}, \frac{Y}{d}, \frac{Z}{d}, 0 \right)^T$. Then by Eq(1), we have

$$(\mathbf{R}, \mathbf{t}) \mathbf{X}' = \alpha \mathbf{K}_c^{-1} \mathbf{m} - \xi \mathbf{e}.$$

$\| (\mathbf{R}, \mathbf{t}) \mathbf{X}' \| = 1$ implies $\| \alpha \mathbf{K}_c^{-1} \mathbf{m} - \xi \mathbf{e} \| = 1$, that is,

$$\overline{(\alpha \mathbf{K}_c^{-1} \mathbf{m} - \xi \mathbf{e})}^T (\alpha \mathbf{K}_c^{-1} \mathbf{m} - \xi \mathbf{e}) = 1.$$

By $\overline{\mathbf{m}}^T \mathbf{K}_c^{-T} \mathbf{e} = 1$, we have

$$\alpha^2 \overline{\mathbf{m}}^T \boldsymbol{\varpi} \overline{\mathbf{m}} - 2\alpha \xi + \xi^2 - 1 = 0. \tag{8}$$

According to the definition of absolute conic, we know $X^2 + Y^2 + Z^2 = 0 \Leftrightarrow X'^2 + Y'^2 + Z'^2 = 0$. Then

$$(\alpha \mathbf{K}^{-1} \mathbf{m} - \xi \mathbf{e})^T (\alpha \mathbf{K}^{-1} \mathbf{m} - \xi \mathbf{e}) = 0 \Rightarrow \alpha^2 \mathbf{m}^T \boldsymbol{\varpi} \mathbf{m} - 2\alpha \xi + \xi^2 = 0. \tag{9}$$

Subtract Eq(8) from Eq(9) and obtain

$$\alpha^2 = \frac{1}{\overline{\mathbf{m}}^T \omega \overline{\mathbf{m}} - \mathbf{m}^T \boldsymbol{\varpi} \mathbf{m}}$$

Substitute $\alpha^2$ in Eq(9) or Eq(8), we obtain the image of the absolute conic under central catadioptric camera.

(*ii*)    Substitute $\xi = 1$ in Eq(5), we have

$$(\overline{\mathbf{m}}^T \boldsymbol{\varpi} \mathbf{m} - \mathbf{m}^T \boldsymbol{\varpi} \mathbf{m})^2 + 2(\overline{\mathbf{m}}^T \boldsymbol{\varpi} \mathbf{m} - \mathbf{m}^T \boldsymbol{\varpi} \mathbf{m})(\mathbf{m}^T \boldsymbol{\varpi} \mathbf{m} - 2) + (\mathbf{m}^T \boldsymbol{\varpi} \mathbf{m})^2 = 0.$$

Expand the above formula, and then we obtain the image of the absolute conic under catadioptric camera.

(*iii*)    Substitute $\xi = 0$ in Eq(5), then we have $(\mathbf{m}^T \boldsymbol{\varpi} \mathbf{m})^2 = 0$. Thus we obtain the image of the absolute conic under catadioptric camera, which is equivalent to our conclusions.

The points on the absolute conic are complex, and then the points on its image are also complex. Let $\mathbf{m} = (a + bi, c + di, 1)^T = \mathbf{r} + i\mathbf{I}$, where $\mathbf{r} = (a, c, 1)^T$ is the real part of $\mathbf{m}$, $\mathbf{I} = (b, d, 0)^T$ is the imaginary part of $\mathbf{m}$ and $i = \sqrt{-1}$. Then we have the following corollary.

**Corollary 1.** (*i*)    *If $0 < \xi < 1$, the formula Eq(5) holds if and only if its real and imaginary parts are 0. Then*

$$\xi^2 ((\mathbf{r}^T \boldsymbol{\varpi} \mathbf{r} + \mathbf{I}^T \boldsymbol{\varpi} \mathbf{I}) + (\xi^2 - 1)(\mathbf{r}^T \boldsymbol{\varpi} \mathbf{r} - \mathbf{I}^T \boldsymbol{\varpi} \mathbf{I}))^2 - 8\xi^2 \mathbf{I}^T \boldsymbol{\varpi} \mathbf{I} = 0.$$

$$\mathbf{r}^T \boldsymbol{\varpi} \mathbf{I}((\xi^2 - 1)(1 - 2\xi^2) \mathbf{I}^T \boldsymbol{\varpi} \mathbf{I} + (1 - \xi^2) \mathbf{r}^T \boldsymbol{\varpi} \mathbf{r} + 2\xi^2) = 0. \tag{10}$$

(*ii*)    *If $\xi = 1$, the formula Eq(6) holds if and only if its real and imaginary parts are 0. Then*

$$(\mathbf{r}^T \boldsymbol{\varpi} \mathbf{r} + \mathbf{I}^T \boldsymbol{\varpi} \mathbf{I})^2 - 8\mathbf{I}^T \boldsymbol{\varpi} \mathbf{I} = 0.$$

$$\mathbf{r}^T \boldsymbol{\varpi} \mathbf{I} = 0. \tag{11}$$

This corollary gives the constraint equations of the intrinsic parameters of camera and we can use them to determine the intrinsic parameters of the catadioptric camera.

# 4   Discussion on Catadioptric Camera Calibration from Circle Images

In this section, we mainly discuss the catadioptric camera calibration using circle images. The intrinsic parameters of the catadioptric camera contains the mirror parameter $\xi$, the aspect ratio $r_c$ the focal length $f_c$ the skew factor $s$ and the principal point $\mathbf{p}$.

In the previous sections, we study the properties of paracatadioptric circle image and the image of the absolute conic under catadioptric camera. In the following, we discuss the catadioptric camera calibration from circle images in theory:

(1) Theorem1 gives the criterion conditions of the antipodal image points on paracatadioptric circle image. Firstly, we estimate the paracatadioptric circle image using the method proposed in [11]. Then, the antipodal image points on the circle image can be calculated by Theorem1. Finally, we can obtain the intrinsic parameters of paracatadioptric camera by the calibration method based on antipodal image points in [21].

(2) From Theorem2 and Corollary1, we derive the image of the absolute conic under catadioptric camera and the constraint equations of the intrinsic parameters of camera. Thus, if we know the image of the absolute conic under catadioptric camera, $\varpi$ can be obtained by Corollary1, and the intrinsic parameters of camera can be estimated by the Cholesky decomposition.

(3) Since circular points lie on the viewsphere, the image $\mathbf{m}$ of one circular point under catadioptric camera is equivalent with the image under virtual perspective camera. Then, $\mathbf{m}$ still satisfies:

$$\mathbf{m}^T \varpi \mathbf{m} = 0. \tag{12}$$

That is, if we have the images of circular points under catadioptric camera, we can straightly estimate $\varpi$ by Eq(12), and then obtain intrinsic parameters of camera by Cholesky decomposition.

It follows from the above conclusions that the circle images can be used to theoretically estimate the intrinsic parameters of catadioptric camera. However, in practice, the catadioptric camera calibration based on circles has the following problems:

(1) The projection of a circle is a quartic curve under catadioptric camera, and is very sensitive to noise. That is, the estimation accuracy of catadioptric circle image is easy to be influenced by noise.

(2) Under catadioptric camera, two circle images have 16 intersections. It is still an open problem how to find out the image of circular points among these intersections.

## 5   Conclusion

In this paper, we study properties of the antipodal images on paracatadioptric circle image, that is the criterion conditions of the antipodal image points on circle image. In addition, we derive the image of the absolute conic under the central catadioptric camera, and obtain the constraint equations about the intrinsic parameters. Finally, we analyse catadioptric camera calibration based on circle images and the problems existed in the practice. In the future work, we will try to solve the problems existed in the practice, and present the calibration method of catadioptric camera based on circle images.

# References

1. Baker, S., Nayer, S.: A Theory of Single-viewpoint Catadioptric Image Formation. International Journal Computer Vision 35(2), 175–196 (1999)
2. Geyer, C., Daniilidis, K.: Catadioptric Camera Calibration. In: International Conference on Computer Vision, vol. 1, pp. 398–404. IEEE Press, Corfu (1999)
3. Scaramuzza, D., Martinelli, A., Siegwart, R.: A flexible technique for accurate omnidirectional camera calibration and structure from motion. In: International Conference on Computer Vision Systems, pp. 45–52. IEEE Press, New York (2006)
4. Deng, X., Wu, F., Wu, Y.: An easy calibration method for central catadioptric cameras. Acta Automatica Sinica 33, 801–808 (2007)
5. Bastanlar, Y., Puig, L., Sturm, P., Barreto, J.: Dlt-like calibration of central catadioptric cameras. In: Workshop on Omnidirectional Vision, Camera Networks and Non-Classical Cameras. Marseille (2008)
6. Geyer, C., Daniilidis, K.: Paracatadioptric camera calibration. Transactions on Pattern Analysis and Machine Intelligence 24, 687–695 (2002)
7. Barreto, J., Araujo, H.: Geometry properties of central catadioptric line images and application in calibration. Transactions on Pattern Analysis and Machine Intelligence 27, 1327–1333 (2005)
8. Wu, F., Duan, F., Hu, Z., Wu, Y.: A new linear algorithm for calibrating central catadioptric cameras. Pattern Recognition 41, 3166–3172 (2008)
9. Duan, F., Wu, F., Zhou, M., Deng, X., Tian, Y.: Calibrating effective focal length for central catadioptric cameras using one space line. Pattern Recognition Letters 33, 646–653 (2012)
10. Duan, H., Li, G., Li, C., Tan, Y.: A fitting method of paracatadioptric circle image. Chinese Journal of Computers 35, 2063–2071 (2012)
11. Duan, H., Mei, L., Shang, Y., Hu, C.: Calibrating focal length for paracatadioptric camera from one circle iamge. In: International Conference on Computer Vision Theory and Application, pp. 56–63. INSTICC Press, Lisbon (2014)
12. Ying, X., Hu, Z.: Catadioptric camera calibration using geometric invariants. Transactions on Pattern Analysis and Machine Intelligence 26, 1260–1271 (2004)
13. Duan, H., Wu, Y.: Paracatadioptric camera calibration using sphere images. In: International Conference on Image Processing, pp. 649–652. IEEE Press, Brussels (2011)
14. Duan, H., Wu, Y.: A calibration method for paracatadioptric camera from sphere images. Pattern Recognition Letters 33, 677–684 (2012)
15. Kang, S.: Catadioptric self-calibration. In: Conference on Computer Vision and Pattern Recognition, vol. 1, pp. 201–207 (2000)
16. Sturm, P., Barreto, J.P.: General imaging geometry for central catadioptric cameras. In: Forsyth, D., Torr, P., Zisserman, A. (eds.) ECCV 2008, Part IV. LNCS, vol. 5305, pp. 609–622. Springer, Heidelberg (2008)

17. Duan, H., Wu, Y.: Unified imaging of geometric entities under catadioptric camera and camera calibration. Journal of Computer-Aided Design and Computer Graphics 23, 891–898 (2011)
18. Geyer, C., Daniilidis, K.: Catadioptric projective geometry. International Journal Computer Vision 45, 223–243 (2001)
19. Hartley, R., Zisserman, A.: Multiple View Geometry in Computer Vision. Cambridge University Press, Cambridge (2000)
20. Semple, J.G., Kneebone, G.T.: Algebraic Projective Geometry. Claredon Press, Oxford (1998)
21. Wu, F., Duan, F., Hu, Z., Wu, Y.: A new linear algorithm for calibration central catadioptric cameras. Pattern Recognition 41(10), 3166–3172 (2008)

# Communication Network Anomaly Detection
# Based on Log File Analysis

Xin Cheng and Ruizhi Wang

Department of Computer Science and Technology, Tongji University, Shanghai, China
cx1227@gmail.com, ruizhiwang@tongji.edu.cn

**Abstract.** Communication network today are becoming larger and increasingly complex. Failure in communication systems will cause loss of critical data and even economic losses. Therefore, detecting failures and diagnosing their root-cause in a timely manner is essential. Fast and accurate detection of these failures can accelerate problem determination, and thereby improve system reliability. Today log files have been paid attention on system and network failure detection, but it is still a challenging task to build an efficient model to detect anomaly from log files. To this effect, we propose a novel approach, which aims to detect frequent patterns from log files to build the normal profile, and then to identify the anomalous behaviour in log files. The experimental results demonstrate that our approach is an efficient way for anomaly detection with high accuracy and few false positives.

## 1 Introduction

Communication network today plays an important role in daily life, and it consists of thousand of network elements. At these scales, communication network failures are common and may even indicate serious impending failures. When a communication network has some failures, the operators would like to solve the problem quickly, but actually that is not simple, because it needs all kinds of tools to troubleshoot and diagnose problems. However, a most useful tool building in almost every software has been ignored, the log file, which can used to record the behavior of system, including normal behavior and abnormal behavior.

Log files now are playing more and more important role in System and network management [6][12]. Because log files reflect the developers original ideas about what events are valuable to report, including errors, execution tracing, or statistics about the programs internal state. Hence the analysis of log files will be an efficient way for anomaly detection.

Log file analysis techniques can be categorized into fault detection and anomaly detection [7][9]. In the case of fault detection, the domain expert creates a database of fault message patterns in which a human expert enumerates a set of rules, consisting of regular expressions and responses to take when matching messages are encountered. The difficulty of writing and maintaining regular expressions for monitoring is proportional to the number of types of messages present, and the rate at which this set of message types change (for example, additions of new devices or software, or changes in user behavior). In the case of anomaly detection, a system profile is created which reflects

D. Miao et al. (Eds.): RSKT 2014, LNAI 8818, pp. 240–248, 2014.
DOI: 10.1007/978-3-319-11740-9_23 © Springer International Publishing Switzerland 2014

normal system activity. If messages are logged that do not fit the profile, an alarm is raised. With this approach, previously unknown fault conditions are detected, but on the other hand, creating the system profile by hand is time-consuming and error-prone.

As logs are too large to examine manually, various methods have been employed. The most popular way is mining frequent pattern from log files [8][11]. But mining patterns from raw logs is often difficult, that is because there is no event type in log file lines, fortunately, it is possible to drive event types from log file lines, since very often the events of the same type correspond to a certain line pattern. For example,

*[.AddPermArp.] add arp entry ip: 80.168.5.48, mac:0:e0:fc:fc:5:30*
*[.AddPermArp.] add arp entry ip: 80.168.5.49, mac: 0:e0:fc: fc:5:31*
*[.AddPermArp.] add arp entry ip: 80.168.5.50,mac:0:e0:fc:fc:5:32*
*[.AddPermArp.] add arp entry ip: 80.168.5.51,mac:0:e0:fc:fc:5:33*

We can get the event type "*[.AddPermArp.] add arp entry ip: \*, mac \**" from these lines. Thus, we can apply frequent pattern mining on building event log models.

The rest of the paper proceeds as follows. In Section 2. we review the background and related work on frequent pattern mining, and we will describe our approach in Section 3. The experiment result will be presented in Section 4, then concluding in Section 5.

## 2    Frequent Pattern Mining

Frequent patterns are itemsets, subsequences, or substructures that appear in a data set with frequency no less than a user-specified threshold [3]. For example, a set of items, such as milk and bread that appear frequently together in a transaction data set is a frequent itemset. A subsequence, such as buying first a PC, then a digital camera, and then a memory card, if it occurs frequently in a shopping history database, is a (frequent) sequential pattern. A substructure can refer to different structural forms, such as subgraphs, subtrees, or sublattices, which may be combined with itemsets or subsequences. If a substructure occurs frequently in a graph database, it is called a (frequent) structural pattern.

Finding frequent patterns plays an essential role in mining associations, correlations, and many other interesting relationships among data. Moreover, it helps in data indexing, classification, clustering, and other data mining tasks as well. Thus, frequent pattern mining has become an important data mining task and a focused theme in data mining research. Frequent pattern mining was first proposed by Agrawal [1] for market basket analysis in the form of association rule mining. It analyses customer buying habits by finding associations between the different items that customers place in their shopping baskets. For instance, if customers are buying milk, how likely are they going to also buy cereal (and what kind of cereal) on the same trip to the supermarket. Such information can lead to increased sales by helping retailers do selective marketing and arrange their shelf space.

The first algorithm for patterns mining was proposed by Agrawal and Srikant, called Apriori [1]. The essence of the Apriori algorithm is among frequent k itemsets: A k-itemset is frequent only if all of its sub-itemsets are frequent. This implies that frequent itemsets can be mined by first scanning the database to find the frequent 1-itemsets, then using the frequent 1-itemsets to generate candidate frequent 2-itemsets, and check against the database to obtain the frequent 2-itemsets. This process iterates until no more frequent k-itemsets can be generated for some k. In many cases, the Apriori algorithm significantly reduces the size of candidate sets using the Apriori principle. However, it can suffer from two-nontrivial costs: generating a huge number of candidate sets; and repeatedly scanning the database and checking the candidates by pattern matching.

Han et al. [4] devised an FP-growth method that mines the complete set of frequent itemsets without candidate generation. FP-growth works in a divide-and-conquer way. The first scan of the database derives a list of frequent items in which items are ordered by frequency descending order. According to the frequency-descending list, the database is compressed into a frequent-pattern tree, or FP-tree, which retains the itemset association information. The FP-tree is mined by starting from each frequent length-1 pattern (as an initial suffix pattern), constructing its conditional pattern base (a sub database, which consists of the set of prefix paths in the FP-tree co-occurring with the suffix pattern), then constructing its conditional FP-tree, and performing mining recursively on such a tree. The pattern growth is achieved by the concatenation of the suffix pattern with the frequent patterns generated from a conditional FP-tree. The FP-growth algorithm transforms the problem of finding long frequent patterns to searching for shorter ones recursively and then concatenating the suffix. It uses the least frequent items as a suffix, offering good selectivity. Performance studies demonstrate that the method substantially reduces search time.

Since the first proposal of this new data mining task and its associated efficient mining algorithms, there have been hundreds of follow-up research publications, on various kinds of extensions and applications, ranging from scalable data mining methodologies, to handling a wide diversity of data types, various extended mining tasks, and a variety of new applications.

# 3 Methodology

In this work, we aim to enable the automation of operators' task of analyzing the log file to detect and present the network anomaly. In fact, important information is buried in the millions of lines of free-text logs, which can be used to automatically detect network problems. To analyze logs automatically, we need to create high quality features which can be better understandable by a machine learning algorithm, and it depends on the nature of log files.

## 3.1 The Nature of Log File

The nature of the log file data plays an important role when designing an efficient mining algorithm. By inspecting some log files from communication network, some properties are discovered from raw logs.

First, we discovered that there are strong correlations between words occurred frequently. Thats because the log file is generated from a standard format, e.g.,

*printf ([.AddPermArp.] add arp entry ip:, ipaddress,mac:,macaddress);*

When this message was logged many times, the constant strings become frequent words which occur frequently together.

Secondly, the frequent items of log files also have strong correlation. If items are event types, strong correlation between event types often exist, e.g, when a node $A$ would like to setup connection with a node $B$, $A$ need send a message to check whether $B$ is ready now, and $B$ should reply a message to $A$. In log files, it displays as follows:

*Node \*: connection to Node \* is prepared*
*Node \*: Node \* is ready*

When a standard process is logged, the events of the process should record together and frequently.

## 3.2  Line Pattern Detecting

We would like to design a fast and efficient algorithm to detect line patterns from raw log files. The algorithm relied on the nature of log files.

Our choice is the employment of data clustering algorithm. Clustering algorithms aim at dividing the set of objects into groups (clusters), where objects in each cluster are similar to each other (and as dissimilar as possible to objects from other clusters). Objects that do not fit well to any of the clusters detected by the algorithm are considered to form a special cluster of outliers. When log file lines are viewed as objects, clustering algorithms are a natural choice, because line patterns form natural clusters lines that match a certain pattern are all similar to each other, and generally dissimilar to lines that match other patterns. After the clusters (event types) have been identified, association rule algorithms can be applied for detecting temporal associations between event types. However, note that log file data clustering is not merely a preprocessing step. A clustering algorithm could identify many line patterns that reflect normal system activity and that can be immediately included in the system profile, which can be further used to analyze by using the association rule algorithms.

We take the whole log file as data space, and the data in the data space is each line of log files. The properties of data are each word from the relative line. Our algorithm consists of three steps. The first step is mining frequent words, then we will build cluster candidate by the frequent words collected at the first step. Finally, clusters are selected from the candidates.

The first step likes the word count; we take a pass over the whole log, and record every word, position and its occurrence times. If the occurrence of a word is more than a specified threshold defined by user, well take the word as frequent. This step is very close to Apriori, since frequent words can be viewed as frequent 1-itemsets.

Secondly, we need build the cluster candidates table based on the frequent words that we get at the first step. When a line is found to have more than one frequent word, its a cluster candidate. If this line does not existed in candidate table, it will be inserted with the count value as 1. If this line has existed in candidate table, just its count value will be incremented. The cluster candidate will be inserted as a region with the set of fixed attributes $(i_1, v_1), (i_2, v_2), ..., (i_m, v_m)$, $i_1, ..., i_m$ is the word position and $v_1, ..., v_m$ is the word. For example, if a line in log is *"Node 6 is prepared"*, and there exist the attribute (1, "Node"), (3,"is") and (4,"prepared"), then the three attributes will be a cluster candidate.

The last step of the algorithm is generate clusters from candidate table, all candidates with count value are greater than the threshold value are taken as the cluster. Actually, each cluster corresponds to a certain line pattern,e.g., the cluster with attributes ((1, "Node"), (3,"is") , (4,"prepared")) correspond to the line pattern *"Node * is prepared"*. Thus, we have got all the line patterns through cluster algorithm.

### 3.3   Frequent Pattern Mining

At Section 3.2, we have mined the line patterns by using clustering algorithm. Each pattern represents an event in log files, and strong correlation exists among events. In this session, we will discuss how to discover the correlation by using pattern mining.

We call some events as normal behavior, that need guarantees (1) these events should cover common cases and (2) these events should occur in a short time. So we define a frequent pattern is the events always occur together in a certain time frequently. In our approach, the time, segment flag and event sequence information are combined together to capture the normal behavior from log files. Moreover,

Firstly, if the event has time-stamp, we should consider the duration is less than $T_{max}$, which is defined by users. However, in some cases, log files may not contain time-stamps, but developers may record some segment flags, for example, *"Process * is starting"* as signal of process start, *"Process * is finished"* as signal of process end, so our first step is observing these flags and get the relevant events, and represent the relevant events by the line patterns.

Secondly, we scan through each event until we find an event followed by a time gap more than 10 times the duration since the start of the sequence. Also, represent the relevant events by the line patterns.

Finally, we prefer a pattern that can represent all events for a standard process. We called the pattern as domain pattern. We use two criteria to select the dominant pattern. (1) we start with the medoid of all sessions considered. By definition, the medoid has the minimal aggregated distance from all other data points, which indicates that it is a good representation of all data points. Intuitively, a medoid is similar to the centroid (or mean) in the space, except that the medoid must be an actual data point. Criterion 1 guarantees that the selected dominant session is a good representation of the sequences examined. (2) we require the sequence to have a minimal support of $0.2M$ from all $M$ event traces. If the medoid does not meet this minimal support, we choose the next closest session (data point) that does. Criterion 2 guarantees that the selected session is dominant and representative. The selection criteria are robust over a wide range of minimal support values because the normal traces are indeed in the majority in the log files.

### 3.4   Anomaly Detection

We use anomaly detection methods to find unusual patterns in logs. In this way, we can automatically find log segments that are most likely to indicate problems. We have investigated a variety of such methods and have found that Principal Component Analysis (PCA) [2][5] combined with term-weighting techniques from information retrieval yields excellent anomaly detection results on both feature matrices, while requiring little parameter tuning.

PCA is a statistical method that captures patterns in high-dimensional data by automatically choosing a set of coordinatesʇthe principal componentsʇthat reflect covariation among the original coordinates. We use PCA to separate out repeating patterns in feature vectors, thereby making abnormal message patterns easier to detect. PCA has runtime linear in the number of feature vectors, so the anomaly detection can scale to large log files.

As with frequent pattern mining, the goal of PCA is to discover the statistically dominant patterns and thereby identify anomalies inside data. PCA can capture patterns in high-dimensional data by automatically choosing a (small) set of coordinates (the principal components) that reflect covariation among the original coordinates. Once we estimate these patterns from the archived and periodically updated data, we use them to transform the incoming data to make abnormal patterns easier to detect. PCA detection has a model estimation phase. In the modeling phase, PCA captures the dominant pattern in a transformation matrix $PP^T$, where $P$ is formed by the top principal components chosen by PCA algorithm. Then the abnormal component of each message count vector $y$ is computed as:

$$y_a = (I - PP^T)y, \tag{1}$$

Here, $y_a$ is the projection of y onto the abnormal subspace. The squared prediction error $SPE = ||y_a||^2$ (squared length of vector $y_a$) is used for detecting abnormal events: we mark vector as abnormal if

$$SPE = ||y_a||^2 > Q_\alpha \tag{2}$$

where $Q$ denotes the threshold statistic for the SPE residual function at the $(1 - \alpha)$ confidence level [6]. Due to limitations of space, we refer readers unfamiliar with these techniques to the work [10] for details. The choice of the confidence parameter for anomaly detection has been studied in previous work, and we follow standard recommendations in choosing $= 0.001$ in our experiments.

## 4   Experiment and Evaluation

To validate the performance trends observed through experimentation, we proceed with experimentation over the network server logs from the real world. The system experts recommended using log files belonging to four specific periods, during which anomalous operations took place within the system. A minimum efficacy of 65% and a 1% maximum for false positives are established after consultation with the aforementioned experts, two system administrators who work on the network system. Our experimental data set is the network server log as depicted in Table 1.

### 4.1   Experimental Results

Table 2 shows the results of applying our process compared against those from the manual analysis performed by the system administrators. The accuracy of our approach

**Table 1.** Characteristics of Data Sets

| Data sets | Size | Lines | Total number of different words |
|---|---|---|---|
| Network server log 1 | 325.3 MB | 1,657,148 lines | 401,843 |
| Network server log 2 | 278.9 MB | 1,179,027 lines | 392,217 |
| Network server log 3 | 362.7 MB | 1,985,361 lines | 439,328 |
| Network server log 4 | 413.5 MB | 2,375,359 lines | 481,329 |

**Table 2.** Results for the analysis of the network system log files

| Data sets | Total Anomaly | Detected Anomaly | Efficiency | False Positive |
|---|---|---|---|---|
| Network server log 1 | 94 | 77 | 81.9% | 0.13% |
| Network server log 2 | 65 | 43 | 66.2% | 0.32% |
| Network server log 3 | 78 | 72 | 92.3% | 0.25% |
| Network server log 4 | 81 | 65 | 80.2% | 0.49% |

**Table 3.** Classification of the Detected Anomaly

| Type of Anomaly Detection | Percentage |
|---|---|
| Single board loading failed | 22.81% |
| Single process call failed | 30.65% |
| Unspecified protocol error | 38.17% |
| PDTCH synchronization failed | 8.37% |

for anomaly detection is always over 66%, with a mean value of 80.2%. The number of false positives in all the cases is smaller than 0.49%, with a minimum value of 0.13%. Table 3 shows the classification of the types of anomaly detected by using our procedure applied to the academic management database. The false positive cases are not included in this table. The most common abnormalities are the following: single board loading failed (about 23%), single process call failed when making a phone call (about 31%), unspecified protocol error, which is used to report a protocol error event only when no other cause in the protocol error class applies (about 38%), and PDTCH synchronization failed (about 8%).

To better understand the experimental result, the sample frequent patterns that have been discovered with our approach from the original log files are illustrated in Fig. 1, and Fig. 2 depicts sample anomalous log file lines that we discovered when the anomaly detection approach was applied to one of our experimental datasets (the network server log file 4 from Table 1).

## 4.2   Discussion

From the experiment results on the four databases of real systems, we can observe that our approach achieves higher values than the minimum values suggested by the experts in all cases. The average of accuracy in the detection of anomaly is near 80%, with a minimum of 66% and a maximum of 92.3%. In term of the percentage of false positives,

*Dec 18 \* sshd[\*]: connect from \**

*Dec 18 \* sshd[\*]: log: Connection from \* port \**

*Dec 18 \* \* log: \* \* \* \**

*Dec 18 \* sshd[\*]: connect from 1\**

*Dec 18 \* sshd[\*]: log: Connection from 1\* port \**

*Dec 18 \* sshd[\*]: log: \* authentication for \*accepted.*

*Dec 18 \* sshd[\*]: log: Closing connection to 1\**

**Fig. 1.** Sample frequent patterns

*Task \* deal \* failed! Because \*.*

*FILE: \* , LINE: \* , NUM= \* , ucTrafficType is out-of-bounds, ucTrafficType = \**

*FILE: \* , LINE: \* , NUM= \*,DCH AM UL GetRlcPduSize fail, ulL2CfgCcbIndex=\**

*syslog-ng[\*\*\*]: Error accepting AF_UNIX connection, opened connections: 300, max: 300*

*[syslog.notice] [\*\*\*]: Error connecting to remote host (\*\*\*), reattempting in 60 seconds*

**Fig. 2.** Sample anomalous log file lines

there is no cases whose value is greater than 1%, which illustrates that our approach can also meet this requirement.

In summary, our approach can help operators notice the abnormal behaviors in log files, which can greatly improve the efficiency of finding the root cause of network system anomaly.

## 5    Conclusion and Future work

Log files contain a lot of information and it is often necessary to use an automated analysis technique to mine this information. But the log files have an inherent variability due to the entangling of constant message types and variable parameter types. In this paper, we propose a approach to anomaly detection by the analysis of the log files. We get the normal patterns from log files and then perform PCA-based anomaly detection. Based on these experimental results, we can conclude that our proposed approach can be considered to be a significant success, allowing us to develop a process and to apply it on the real log file analysis and thus to facilitate the system auditor's job.

As future work, we plan to do some research on the mining of rare patterns from log files, since this might be more efficient on revealing anomalous events that represent unexpected behavior.

**Acknowledgement.** This work was supported by the National Natural Science Foundation of China (No. 61075056, 61273304), the Specialized Research Fund for the Doctoral Program of Higher Education of China (No. 20130072130004) and the Fundamental Research Funds for the Central Universities. This work was also partially supported by the National Natural Science Fund of China (No. 61203247) and the Fundamental Research Funds for the Central Universities (No. 2013KJ010).

# References

1. Agrawal, R., Srikant, R., et al.: Fast algorithms for mining association rules. In: Proc. 20th Int. Con. Very Large Data Bases, VLDB, vol. 1215, pp. 487–499 (1994)
2. Dunia, R., Qin, S.J.: Multi-dimensional fault diagnosis using a subspace approach. In: American Control Conference. Citeseer (1997)
3. Han, J., Cheng, H., Xin, D., Yan, X.: Frequent pattern mining: current status and future directions. Data Mining and Knowledge Discovery 15(1), 55–86 (2007)
4. Han, J., Pei, J., Yin, Y.: Mining frequent patterns without candidate generation. In: ACM SIGMOD Record, vol. 29, pp. 1–12. ACM (2000)
5. Jackson, J.E., Mudholkar, G.S.: Control procedures for residuals associated with principal component analysis. Technometrics 21(3), 341–349 (1979)
6. Lakhina, A., Crovella, M., Diot, C.: Diagnosing network-wide traffic anomalies. In: ACM SIGCOMM Computer Communication Review, vol. 34, pp. 219–230. ACM (2004)
7. Makanju, A.A., Zincir-Heywood, A.N., Milios, E.E.: Clustering event logs using iterative partitioning. In: Proceedings of the 15th ACM SIGKDD International Conference on Knowledge Discovery and Data Mining, pp. 1255–1264. ACM (2009)
8. Vaarandi, R.: A breadth-first algorithm for mining frequent patterns from event logs. In: Aagesen, F.A., Anutariya, C., Wuwongse, V. (eds.) INTELLCOMM 2004. LNCS, vol. 3283, pp. 293–308. Springer, Heidelberg (2004)
9. Vaarandi, R., et al.: A data clustering algorithm for mining patterns from event logs. In: Proceedings of the 2003 IEEE Workshop on IP Operations and Management (IPOM), pp. 119–126 (2003)
10. Xu, W., Huang, L., Fox, A., Patterson, D., Jordan, M.I.: Detecting large-scale system problems by mining console logs. In: Proceedings of the ACM SIGOPS 22nd Symposium on Operating Systems Principles, pp. 117–132. ACM (2009)
11. Xu, W., Huang, L., Fox, A., Patterson, D., Jordan, M.I.: Online system problem detection by mining patterns of console logs. In: Ninth IEEE International Conference on Data Mining, ICDM 2009, pp. 588–597. IEEE (2009)
12. Yamanishi, K., Maruyama, Y.: Dynamic syslog mining for network failure monitoring. In: Proceedings of the Eleventh ACM SIGKDD International Conference on Knowledge Discovery in Data Mining, pp. 499–508. ACM (2005)

# Using the Multi-instance Learning Method to Predict Protein-Protein Interactions with Domain Information

Yan-Ping Zhang[1,2], Yongliang Zha[1,2,], Xinrui Li[1,2], Shu Zhao[1,2], and Xiuquan Du[1,2]

1. Key Laboratory of Intelligent Computing and Signal Processing of Ministry of Education, Anhui University, Hefei 230601, Anhui Province, P.R. China
2. School of Computer Science and Technology, Anhui University, Hefei 230601, P.R. China
dxqllp@163.com

**Abstract.** Identifying protein-protein interactions (PPIs) can help us to know the protein function and is critical for understanding the mechanisms of proteome. Recently, lots of computational methods such as the domain-based approach have been developed for predicting the protein-protein interactions. The conventional domain-based methods usually need to infer the interacting domain pairs from already known interacting sets of proteins, and then to predict the PPIs. However, it is difficult to provide the detailed information that which of the domain pairs will actually interact for the PPIs prediction. Therefore, it is of great importance to develop a new computational model which can ignore the information whether a domain pair is interacting or not. In this paper, we propose a novel method using multi-instance learning (MIL) for predicting protein-protein interactions based on the domain information. Firstly, the domain pairs of two proteins were composed. Then, we use the amino acid composition feature encoding method to encode the domain pairs. Finally, two multi-instance learning methods were used for training the data. The experiment results demonstrate that the proposed method is effective.

**Keywords:** Protein-protein interactions (PPIs); domain; multi-instance learning (MIL).

## 1    Introduction

Proteins perform a vast array of functions within living organisms, including catalyzing metabolic reactions, replicating DNA, responding to stimuli and transporting molecules from one location to another. According to research, proteins usually cooperate with other proteins to achieve a particular function and associate to form stable protein complexes. Protein-protein interactions (PPIs) are vital to many biochemical processes and play a major role in cellular events [1-2]. So, identifying protein-protein interactions (PPIs) helps us to know the protein function and is critical for understanding the mechanisms of proteome. The prediction of protein-protein interactions has become one of the important topics in the field of molecular biology and bioinformatics.

The most reliable methods for studying protein-protein interactions are experimental methods. Previously the detection of protein-protein interactions was limited to labor-intensive experimental techniques such as co-immunoprecipitation or affinity

D. Miao et al. (Eds.): RSKT 2014, LNAI 8818, pp. 249–259, 2014.
DOI: 10.1007/978-3-319-11740-9_24 © Springer International Publishing Switzerland 2014

chromatography. Then the high-throughput experimental methods such as the yeast two-hybrid and methods based on mass spectrometry are also available. However, these methods are not often applicable because of time-consuming and needing lots of money which make it hard to investigate all the possible interactions [3-4]. For this reason, it is of great practical significance to develop the reliable computational methods for detecting the protein-protein interactions.

In the past decade, a wide range of computational methods had been developed. All these methods can be usually divided into four parts: structure-based approach, sequence-based method, annotation-based and domain-based method. Correspondingly, Zhang et al. used a structure-based approach to infer protein-protein interactions [5].You et al. presented a novel hierarchical PCA-EELM model to predict protein-protein interactions [6]. Zahiri et al. introduced a novel evolutionary based on feature extraction algorithm for PPIs prediction [7]. The domain-based approach is one of the effective methods for prediction of protein-protein interactions. Multiple studies have shown that domain-domain interactions from different experiments are more consistent than their corresponding protein-protein interactions [8]. So, it is quite reliable to use the domains and their interactions for prediction of the protein-protein interactions [9].In the past few years, lots of researchers focused on the protein-protein interactions prediction by using the domain information. Roslan et al. utilized shared interacting domain patterns and Gene Ontology information [10], Binny et al. used the domain–domain associations and Jang et al. predicted protein-protein interactions based on the multi-domain collaboration [11-12].

Despite the recent advances, the existing domain-based methods usually need to infer the interacting domain pairs from already known interacting sets of proteins, and then to predict the PPIs. However, it is difficult to provide the detailed information that which of domain pairs will really interact for the PPIs prediction. Therefore, it is of great importance to develop a new computational model which can ignore the information whether a domain pair is interacting or not.

Multi-instance learning (MIL) method [13-15] is an extension of the standard supervised learning setting. Since multi-instance problems extensively exist but are unique to those addressed by previous learning frameworks, multi-instance learning was regarded as a new learning framework [14]. Mei et al. had introduced the multi-instance learning method to predict the protein sub-cellular localization [16].The researchers treat the protein sequence as several structural domains, and then use the MIL method to capture protein sequence local information and structural domain boundary partition information. Motivated by this, we propose a novel method using multi-instance learning (MIL) for predicting protein-protein interactions based on the domain information. The results of the experiment demonstrate that the proposed method is effective.

# 2    Multi-instance Learning

## 2.1    Multiple-Instance Learning Scheme

Multiple-instance learning (MIL) is a scheme of semi-supervised learning for problems with incomplete knowledge concerning the labels of the training data. Multi-instance

Learning studies the ambiguity in input space or instance space, where an object has many alternative input descriptions, i.e. instances. The term multi-instance learning was coined by Dietterich et al. when they were investigating the problem of drug activity prediction [13-15].

In multi-instance learning, the training set is composed of many bags each of which contains many instances. A bag is positively labeled if it contains at least one positive instance; otherwise it is labeled as a negative bag. The task is to learn some concept from the training set for correctly labeling unseen bags [14].

Let $X$ be the input space and $Y = \{+1, -1\}$ be the label space. The MIL task can be seen as a function: $f_{MIL} : 2^X \rightarrow Y$. The dataset $X_i = \{X_{i1}, X_{i2}, ..., X_{in_i}\}$ is consisting of a set of bags and its labels, each $X_i \subseteq X$ is a set of $n_i$ instances $\{(X_1, y_1), (X_2, y_2), ..., (X_n, y_n)\}$. For every $X_i$, if $y_i = 1$, it has $c(X_{ij}) = 1$ for atleast one $j$ and if $y_i = 0$, it has $c(X_{ij}) = 0$ for all $j$, where $c(X_{ij})$ is the label of the instance. The framework of MIL is given in Fig. 1.

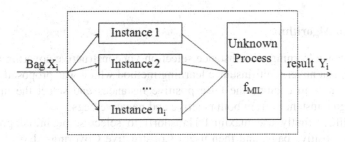

**Fig. 1.** The framework of Multi-Instance learning (MIL)

Since its introduction, a wide variety of new algorithms have been developed and well-known supervised learning algorithms extended to learn MI concepts, such as the Diverse Density, Citation-kNN and Bayesian-kNN, Relic, EM-DD, BP-MIP, MI SVMs and multi-instance ensembles [14]. In this paper, two different MIL algorithms were used to test our method.

## 2.2    Bagging_C_kNN Algorithm

In this research, we solve the problem using a modified version of the Citation-kNN[17] algorithm, which is one of the well-known MIL solutions. The Citation-kNN algorithm is a nearest neighbor style algorithm, which borrows the notion of citation of scientific references in the way that a bag is labeled through analyzing not only its neighboring bags but also the bags that regard the concerned bag as a neighbor [18].

In standard k-nearest neighbor algorithm, each object, or instance, is regarded as a feature vector in the feature space. For two different feature vectors, i.e. a and b, the distance between them can be written as:

$$Dist(a,b) = \| a - b \| \qquad (1)$$

But if the goal is to discriminate the bags, the formula must be extended. Suppose two different bags, i.e. $A=\{a_1, a_2, ..., a_m\}$ and $B=\{b_1, b_2, ..., b_n\}$. The problem of measuring the distance between different bags is in fact the problem of measuring the distance between different feature vector sets. Formally, such a distance metric can be written as:

$$Dist(A, B) = \underset{\substack{1 \leq i \leq m \\ 1 \leq j \leq n}}{MIN}(Dist(a_i, b_j)) = \underset{a \in A}{MIN} \underset{b \in B}{MIN} \parallel a - b \parallel \tag{2}$$

Bagging employs bootstrap sampling to generate several training sets from the original training set and then trains component learners, i.e. multiple versions of the base learner, from each generated training set. The predictions of the component learners are combined via majority voting. A method called Bagging_C_kNN [18] which build multi-instance ensembles for Citation-kNN learner to solve multi-instance problems.

## 2.3    MilCa Algorithm

Multiple-instance learning with instance selection via constructive covering algorithm (MilCa) [19] is a novel multi-instance learning method which was proposed by Zhang et al. MilCa aims to exclude the false positive instances and select the high representative degree instances from both positive and negative bags.

In the MilCa, firstly use maximal Hausdorff to select some initial positive instances from positive bags, and then use a Constructive Covering Algorithm (CCA) [20] to restructure the structure of the original instances of negative bags. Then an inverse testing process is employed to exclude the false positive instances from positive bags and to select the high representative degree instances ordered by the number of covered instances from training bags. The outline of these steps can be found as follows:

Input: Training bags $X = \{X_1^+, ..., X_{m^+}^+, X_1^-, ..., X_{m^-}^-\}$
Output: the set of high degree representative instances
RSI
Begin
Step 1: Label all instances of positive(negative) training bags with +1(-1)
Step 2: Select $m^+$ initial instances (RSI+) from positive bags via Eq.(3)

$$x_j^+ = \arg d(X_j^+, X_{ins}^-) \tag{3}$$

Step 3: Obtain a negative cover set $C^-$ via CCA, the instances in $RSI^+$ and $X_{ins}^-$.
Step 4: Use $C^-$ to do an inverse testing process and select the instances from positive bags that are not covered by $C^-$ via Eq. (4).

$$RSI^+ = \{x_i^+ \mid x_i^+ \in x_{ins}^+, x_i^+ \notin x_{ins}^-\} \tag{4}$$

Step 5: Construct a cover set via CCA, the instances in $RSI^+$ and $m^+$. Then obtain the new $RSI^+$ with $m^+$ instances and $RSI^-$ with $m^-$ instances according to Eq. (4) and Eq. (5).

$$RSI^- = \{x_i^- \mid x_i^- \in C_{ins}^-\} \tag{5}$$

Step 6: Form the high degree representative set of instance: $RSI = RSI^+ \cup RSI^-$

End.

---

Finally, a similarity measure function is used to convert the training bag into a single sample. For $RSI = \{x_1^+, x_2^+, ..., x_{m^+}^+\} \cup \{x_1^-, x_2^-, ..., x_{m^-}^-\}$ and the numbers of instances in the corresponding covers are $no_i^+$ and $no_j^-$. The similarity between $x_k (x_k \in RSI)$ and the $i$-th bag $X_i$ is:

$$s(x_k, X_i) = \begin{cases} \dfrac{no_k^+}{\sum\limits_{p=1}^{m^+} no_p^+} \cdot \min\limits_{x_k \in RSI^+} \exp(\dfrac{d(x_k^+, x_{ij})^2}{2\sigma^2}) \\[4mm] \dfrac{no_k^-}{\sum\limits_{p=1}^{m^-} no_p^-} \cdot \min\limits_{x_k \in RSI^-} \exp(\dfrac{d(x_k^-, x_{ij})^2}{2\sigma^2}) \end{cases} \tag{6}$$

Then, employ the exponential function and the parameter of $\sigma$ to adjust the similarity measure function. If $x_k \in RSI^+$, then a positive bag should be similar to the instance $x_k$ highly. Otherwise the positive bag has a low similarity with the instance. An embedding function $\Phi$ is defined, which converts a bag $X_i$ to a $(m^+ + m^-)$ dimensional sample:

$$\Phi(X_i) = [s(x_1^+, X_i), ..., s(x_{m^+}^+, X_i), s(x_1^-, X_i), ..., s(x_{m^-}^-, X_i)]^T \tag{7}$$

The converted single sample is also labeled as the label of $X_i$. And CCA is again used to classification for the converted samples.

# 3    Method

Multi-instance learning method had been widely used for scene or image classification, and the research had shown that multi-instance learning specializes in dealing with the problem which has natural structural partition. In the paper [16], this method also had been used to predict the protein subcellular localization. Motived by this, we introduced the multi-instance learning method into our research on PPIs. As for a protein, it is formed by domains. So, we can use the theory of multi-instance learning method based on the domain pair information to predict the PPIs.

## 3.1   Instance Extraction

Protein sequence is a long string consisting of 20 amino acids. A protein can range from ten to thousands of amino acids in length. With the evolution of the protein, the conservative of amino acid residues are highly different. Usually, only a few conserved areas of protein sequence determine the protein structure and function. These conserved sequence regions are called domains. As shown in Fig. 2, the protein P39722 has four domains.

**Fig. 2.** The **domains of protein P39722** (Miro domain [7,119], EF_assoc_2 domain [235,332], EF_assoc_1 domain [368,438], Miro domain [450,563])

The researchers who focus on the domain-based prediction of protein-protein interactions have delineated that a domain is a fundamental unit of biological functions, and the domain-domain interaction is indispensable for the PPIs [21]. It has proved that the real interaction between the proteins is the interaction between the domain pairs. When a domain on one protein and a domain on the other, we can think that a domain pair (DP) is made up of these two domains. It is difficult for previous domain based method to know which domain pairs in two proteins induce the PPIs. Here we utilize the theory of multi-instance learning which is especially useful to solve this problem.

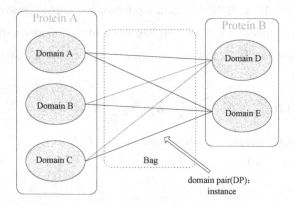

**Fig. 3.** A **MIL bag model** (the domain pair between two proteins was treated as a instance, and all the domain pairs were composed as a bag. If the bag was labeled as interaction, there were at least one interact DPs. Otherwise the bag was labeled as no interaction.)

Firstly, we should compose the distances of the datasets using the domain pair information. Suppose that there are two proteins $\{P,Q\}$, the domains of protein $P$ are $P=\{d_p^1, d_p^2, ..., d_p^m\}$ , and the domains of protein $Q$ are $Q=\{d_q^1, d_q^2, ..., d_q^n\}$. The number of domains is $m$ and $n$. A domain pair is composed by a domain randomly selected from $P$ and a domain selected from $Q$. Therefore, $m \times n$ domain pairs are composed. For example, there are $\{a,b\} \in P$ and $\{c,d,e\} \in Q$ .In this case, there are $2 \times 3 = 6$ domain pairs, they are $\{ac, ad, ae, bc, bd, be\}$. As shown in Fig. 3.In the method, we regard the domain pair (DP) between two proteins as an instance, and treat the interaction between these two proteins as a bag.

## 3.2    Feature Encoding

Amino Acid Composition (AAC) is one of the most widely-used feature encoding methods. Protein sequence is composed by 20 amino acids and the length of the sequence is noted as L. Then the number of each amino acid in the sequence which note as $A=\{A_1, A_2, ..., A_{20}\}$ is calculated. The composition of amino acid is $X=\{X_1, X_2, ..., X_{20}\}=A/L=\{A_1/L, A_2/L, ..., A_{20}/L\}$.

Here we also use the Amino Acid Composition method to encode the sequence of the domain pair. A domain pair has two domains, so, an instance has 40 features. And the feature of the bag is made up of the features of the instances, has 40×m×n dimension.

## 3.3    MIL model for PPIs

When two proteins interact, there is at least one domain pair which is real interacting. When there is no protein-protein interaction, we can know that all the domain pairs do not interact. Although we do not know which DPs induce the PPIs, multi-instance learning is especially useful to deal with this problem. Through lots of trainings, commonness (interactive domain pairs) was found. When a new bag (test interaction data) was input, the instances were scanned. If at least one instance was in the commonness, the method labeled the bag interaction, otherwise it do not interact.

The steps of the proposed method can be summarized as follow:

Step1: For every protein pair in the datasets, compose the DPs as instances.
Step2: Use the AAC to encode the instances and label the protein pair as a bag.
Step3: Apply a multi-instance learning algorithm to training the data.

# 4    Experiments

We evaluated our method on the datasets which we composed. Two MIL algorithms are applied for training. Here several criteria are used to evaluate the performance. We compare our results with some baselines to show how much improvement we could achieve.

## 4.1   Data Collection

For training, we got the protein-protein interaction pairs from one of the most popular databases DIP (http://dip.doe-mbi.ucla.edu/dip/Main.cgi). In our experiment, we chose the S.cerevisiae protein pairs that are derived from the DIP core dataset as positive examples. The reliability of this core subset has been tested by two methods, expression profile reliability and paralogous verification method. And the domain information for the proteins was extracted from Pfam (http://pfam.sanger.ac.uk/ ).

In our experiment, the interacting protein pairs were got from the dataset Scere20131031, and the pfam information was extracted from the latest dataset Pfam27.0.Firstly, the protein IDs were obtained to got the sequence from Swiss-prot. Then the proteins whose length of sequence is less than 50 were removed. The remaining proteins were scanned by Pfam27.0 and the domain information was noted for them. The proteins that have no pfam-A domains were manually removed.

The negative datasets play an important role in the model. Since the non-interacting pairs were not readily available, a strategy for constructing the negative dataset called Prcp has been described by Shen and colleagues [22] in detail. Protein pairs were generated by randomly pair of the remained proteins as above. If a protein pair appeared in the Scere20131031, it is a positive bag. The remaining interacting protein pairs were non-interacting pairs. All the positive bags in the protein pairs form the positive dataset. And five negative datasets were generated by randomly selecting equal number of positive dataset from the non-interacting pairs. Thus, five subsets were prepared.

## 4.2   Evaluation of Performance

To measure the performance of the proposed method, several evaluation criteria which are described below are used. Firstly, we should count the number of true positives (TP), false positives (FP), true negatives (TN) and false negatives (FN).TP is the number of true PPIs that are predicted correctly, FP is the number of true non-interacting pairs that are predicted to be PPIs, TN is the number of true non-interacting pairs that are predicted correctly and FN is the number of true PPIs that are predicted to be non-interacting pairs. Then we use these to compute the criteria:

$$Accuracy = \frac{TP + TN}{TP + TN + FP + FN}$$

$$Sensitivity = \frac{TP}{TP + FN}$$

$$Precision = \frac{TP}{TP + FP}$$

$$F - score = \frac{2 * sensitive * precision}{sensitive + precision}$$

$$cc = \frac{TP * TN + FN * FP}{\sqrt{(TP + FN)(TN + FP)(TP + FP)(TN + FN)}}$$

Sensitivity measures how accurately the classifier can classify out of all positive data. Precision measures how precise the data classified as positive by this classifier. Accuracy of a classifier measures the estimated probability of correct predictions. F-score takes into account both sensitivity and precision are a useful measure of overall performance. The result of cc is a number between -1 and +1 showing complete agreement, -1 complete disagreement and 0 showing that the prediction was uncorrelated with the results.

## 4.3     Experimental Results and Discussion

We evaluated the performance of the proposed approach using two MIL algorithms on some criteria which are described above. Bagging_C_kNN and MilCa respectively correspond to the traditional famous and current novel algorithm. So, they were chosen to train the dates. Five datasets were composed by using five different negative subsets. Then five results were generated from the five sets of data, and the performance of the model was evaluated by the average value of these results. In order to compare, we also implemented several experiments which used some other machine learning methods including NaiveBayes, AdaBoostM1, RandomTree, and SVM. To reduce the bias of training and testing data, in the experiment, a 10-fold cross validation technique is adopted. Table 1 gives the average prediction results achieved by all these comparative methods and the proposed approach.

It can be observed from Table 1 that the proposed approach shows a good prediction accuracy of 67% and 66% which is better than all other comparative methods. To better investigate the prediction ability of our method, we also calculated the values of Sensitive, Precision, F-Score and cc. All these criteria are widely used to evaluate the performance in protein-protein interactions prediction. From Table 1, we can see that our model gives the best prediction performance with an average precision value of 68%, F-score value of 66% and 65%. Further, it also can be seen in Table 1 that the value of sensitive by MilCa is worse than NaiveBayes and SVM method, and the value of sensitive by Bagging_C_kNN is better than other methods except Naive-Bayes. Specially, the cc value of 34% and 32% is much better than other methods. From the results, we can conclude that our method is an effective method.

**Table 1.** The prediction result generated by several methods

| Methods | Sensitive | Precision | F-score | Accuracy | cc |
|---------|-----------|-----------|---------|----------|-----|
| NaiveBayes | **0.67** | 0.57 | 0.62 | 0.59 | 0.18 |
| AdaBoostM1 | 0.63 | 0.57 | 0.60 | 0.58 | 0.16 |
| RandomTree | 0.61 | 0.58 | 0.60 | 0.59 | 0.18 |
| SVM | 0.65 | 0.65 | 0.65 | 0.64 | 0.29 |
| Bagging_C_kNN | 0.65 | **0.68** | **0.66** | **0.67** | **0.34** |
| MilCa | 0.63 | **0.68** | **0.65** | **0.66** | **0.32** |

In addition, as we know, sensitivity measures the fraction of interactions that are identified, and the precision measures the fraction of the predicted interactions that are actually interactions. So the worse value of sensitive perhaps due to the negative datasets. Because our negative dataset were composed by randomly pair. Correlation coefficient measures that how well the predicted class labels correlate with the actual class labels and it is good for measuring the performance. In the table, we can find that our value is much better than comparative methods. All the analysis shows that the proposed method outperforms other comparative methods for predicting protein-protein interactions.

## 5     Conclusion

In this paper, we developed a new method to predict the protein-protein interactions. The domain pairs of two proteins composed for feature encoding which were regarded as instances, and the interaction between these two proteins is treated as a bag. Finally, two multi-instance learning algorithms were used for training. Unlike conventional domain-based methods, the new model can ignore the information whether a domain pair is interacting or not. In our research, we just use the most basic method to compose the domain pairs, and the domain pairs are endowed equal status. More MIL algorithms could be used to train the dates. Further studies are required to address these issues.

**Acknowledgements.** This Research is partially supported by the grants of the National Natural Science Foundation of China, No. 61175046 and 61203290, supported by the Doctoral Start-up Funds of Anhui University under Grant No.33190078, supported by the outstanding young backbone teachers training under Grant NO. 02303301 and supported by the Graduate's Academic Innovation Research Project of Anhui University.

## References

1. Shi, M.G., et al.: Predicting protein–protein interactions from sequence using correlation coefficient and high-quality interaction dataset. Amino Acids 38(3), 891–899 (2010)
2. Guo, Y., Yu, L., Wen, Z., et al.: Using support vector machine combined with auto covariance to predict protein–protein interactions from protein sequences. Nucleic Acids Research 36(9), 3025–3030 (2008)
3. Skrabanek, L., Saini, H.K., Bader, G.D., et al.: Computational prediction of protein–protein Interactions. Molecular Biotechnology 38(1), 1–17 (2008)
4. Yu, J., Fotouhi, F.: Computational approaches for predicting protein–protein interactions: A survey. Journal of Medical Systems 30(1), 39–44 (2006)
5. Zhang, Q.C., Petrey, D., Deng, L., et al.: Structure-based prediction of protein-protein interactions on a genome-wide scale. Nature 490(7421), 556–560 (2012)
6. You, Z.H., Lei, Y.K., Zhu, L., et al.: Prediction of protein-protein interactions from amino acid sequences with ensemble extreme learning machines and principal component analysis. BMC Bioinformatics 14(suppl. 8), S10 (2013)

7. Zahiri, J., Yaghoubi, O., Mohammad-Noori, M., et al.: PPIevo: Protein–protein interaction prediction from PSSM based evolutionary information. Genomics 102(4), 237–242 (2013)
8. Memi, V., Wallqvist, A., Reifman, J.: Reconstituting protein interaction networks using parameter-dependent domain-domain interactions. BMC Bioinformatics 14(1), 154 (2013)
9. Wojcik, J., Schächter, V.: Protein-protein interaction map inference using interacting domain profile pairs. Bioinformatics 17(suppl. 1), S296–S305 (2001)
10. Roslan, R., Othman, R.M., Shah, Z.A., et al.: Utilizing shared interacting domain patterns and Gene Ontology information to improve protein–protein interaction prediction. Computers in Biology and Medicine 40(6), 555–564 (2010)
11. Binny, P.S., Saha, S., Anishetty, R., et al.: A matrix based algorithm for protein–protein interaction prediction using domain–domain associations. Journal of Theoretical Biology 326, 36–42 (2013)
12. Jang, W.H., Jung, S.H., Han, D.S.: A computational model for predicting protein interactions based on multidomain collaboration. IEEE/ACM Transactions on Computational Biology and Bioinformatics (TCBB) 9(4), 1081–1090 (2012)
13. Ray, S., Scott, S., Blockeel, H.: Multi-instance learning. In: Encyclopedia of Machine Learning, pp. 701–710 (2010)
14. Zhou, Z.H.: Multi-instance learning: A survey. Department of Computer Science and Technology. Nanjing University (2004)
15. Gärtner, T., Flach, P.A., et al.: Multi-Instance Kernels. In: Proceedings of the 19th International Conference on Machine Learning, Sydney, Australia, pp. 179–186 (2002)
16. Mei, S.Y., Fei, W.: Structural Domain Based Multiple Instance Learning for Predicting Gram-Positive Bacterial Protein Subcellular Localization. In: International Joint Conference, pp. 195–200. IEEE (2009)
17. Wang, J., Zucker, J.D.: Solving multiple-instance problem: A lazy learning approach. In: Proceedings of the 17th International Conference on Machine Learning, San Francisco, pp. 1119–1125 (2000)
18. Zhou, Z.-H., Zhang, M.-L.: Ensembles of multi-instance learners. In: Lavrač, N., Gamberger, D., Todorovski, L., Blockeel, H. (eds.) ECML 2003. LNCS (LNAI), vol. 2837, pp. 492–502. Springer, Heidelberg (2003)
19. Zhang, Y.P., Zhang, H., et al.: Multiple-Instance Learning with Instance Selection via Constructive Covering Algorithm. Tsinghua Science and Technology 19 (2014)
20. Zhang, L., Zhang, B.: A geometrical-representationMcCulloch-Neural model and its application. IEEETransactions on Neural Networks 10, 925–929 (1999)
21. Jang, W.H., Jung, S.H., Han, D.S.: A computational model for predicting protein interactions based on multidomain collaboration. IEEE/ACM Transactions on Computational Biology and Bioinformatics (TCBB) 9(4), 1081–1090 (2012)
22. Shen, J., Zhang, J., et al.: Predicting protein–protein interactions based only on sequences information. Proceedings of the National Academy of Sciences 104(11), 4337–4341 (2007)

# An Improved Decimation of Triangle Meshes Based on Curvature

Wei Li[1,2], Yufei Chen[1,2,*], Zhicheng Wang[1,2], Weidong Zhao[1,2], and Lin Chen[1,2]

[1]Research Center of CAD, Tongji University, Shanghai, China
[2]The Engineering Research Center for Enterprise Digital Technology,
Ministry of Education, Tongji University, Shanghai, China
`april337@163.com`

**Abstract.** This paper proposes an improved decimation of triangle meshes based on curvature. Mesh simplification based on vertex decimation is simple and easy for implementation. But in previous mesh simplification researches based on vertex decimation, algorithms generally focused on the distance error between the simplified mesh and the original mesh. However, a high quality simplified mesh must have low approximation error and preserve geometric features of the original model. According to this consideration, the proposed algorithm improves classical vertex decimation by calculating the mean curvature of each vertex and considering the change of curvature in local ring. Meanwhile, this algorithm wraps the local triangulation by a global triangulation. Experimental results demonstrate that our approach can preserve the major topology characteristics and geometric features of the initial models after simplifying most vertices, without complicated calculation. It also can reduce the influence from noises and staircase effects in the process of reconstruction, and result in a smooth surface.

**Keywords:** Mesh simplification; vertex decimation; geometric feature; curvature.

## 1 Introduction

Nowadays, the polygonal meshes have become a popular graphics primitive for computer graphics application because of their mathematical simplicity. Any three points in a three dimensional space can determine a plane, and triangles have the advantages of flexibility and efficiency. Therefore, the triangle has become a basic element of mesh models in many applications. With the development of science and technology, the data set used for multiple application areas, such as medical imaging, virtual reality, Computer Graphics, visualization, is becoming more and more complex. A mesh could have millions of elements when we deal with these data sets. More complex polygonal models obtained from 3D acquisition techniques such as laser scanning could be reproduced more accurately, but the drawbacks of the complexity are too long time on reconstruction and redundant data for subsequent processing. Especially in applications of medical imaging, the time means the patient's life. For these rea-

D. Miao et al. (Eds.): RSKT 2014, LNAI 8818, pp. 260–271, 2014.
DOI: 10.1007/978-3-319-11740-9_25 © Springer International Publishing Switzerland 2014

sons, mesh simplification has become an extremely important topic on computer graphics research in the last years.

The goal of mesh simplification is to reduce the complexity but keeping as possible as high fidelity of the original model [1]. Having this goal, a multitude of mesh simplification algorithms were developed during the time. For multiple classification methods of those algorithms, we can refer the reader to [2]. According to the way of structuring the reduced model, simplification algorithms can be categorized into two classes: refinement and decimation [3]. The former begin with a simple initial model, and then gradually add details until it reached the requirement of the approximation error. Because of the difficulty of structuring the initial approximation meshes for complex 3D mesh models, this method is infrequent in the mesh simplification. The method based on decimation begins with the original model, and then gradually delete some geometrical elements. These methods simplify the original model by removing some geometric elements such vertices, edges, triangles. Accordingly, the decimation method can be classified into three different approaches according to the difference of the selected objects [4]: removal of vertex [5], removal of edge [6] and removal of triangle [7].

In 1992, Schroeder [5] described an algorithm based on vertex decimation. His method iteratively selects a vertex removal, removes all adjacent faces, and retriangulates the resulting hole [8]. It measures the distance from the vertex to average plane by its adjacent triangles, then uses the distance to decide the order in which vertices are removed [9]. This method has a simple implementation, saves calculation time, takes up less memory and preserves the topology of the original mesh. However, the local error metric will produce accumulated error after much iteration. So it generates low-quality approximated models [2], [4], [9]. Rossignac and Borrel [10] proposed a vertex clustering method to remove vertices in meshes. This method assigns a weight to each vertex on the input mesh by its perceptual importance, then subdivides the mesh into a three-dimensional grid, and finally, all the vertices in a given grid cell are clustered to the position of the vertex with maximum weight. This method tends to be very fast but the visual appearance of the final mesh is relatively inaccurate to define [9].

Edge collapse is based on the iterative contraction of vertex pairs. The fundamental operation is to iteratively merge two neighboring vertices to the same position [11]. Progress mesh (PM) [6] representation is a kind of iterative edge collapse based mesh simplification method developed by Hoppe in 1996, in which an energy function to describe the complexity and fidelity of mesh is used to track simplification quality. PM use the edge collapse operator to construct a progressive mesh, and measures the distance from the proposed new triangles to a set of sample points from the original mesh to decide which edge to collapse. This method obtains high quality results. However, it is not easy to implement and use because of the complexity of calculating energy function [4], [12]. Garland and Heckbert [13] proposed a Quadric Error Metric (QEM) based the vertex pair-collapse operator, which can be considered the topology modifying variant of the edge-collapse operator. The key of QEM algorithm is to find an error metric as the cost function at each vertex, and the vertex pair which has the minimum cost is contracted at each iteration step. However, because this error measurement is solely based on the Euclidian distance between geometric positions, simplification may preserve geometric features only to a certain extent [14].

Triangle collapse simplifies meshes by removing selected triangles, which is a continuation of edge collapse. Hamann [7] weighted triangles by the product of equiangularity and curvature, then sorted triangles by their weights and collapsed them successively. Tran S.Gieng et al. [15] proposed a similar simplification algorithm, which weighted triangles by the product of area, curvature and et al. These algorithms preserve shape better, but have a higher complexity.

Considering the calculation time and complexity, obviously, the algorithm based on vertex decimation, which is simple and easy for implementation, is more suitable for applications of medical imaging. However, the traditional vertex decimation selects a candidate for removal based on the Euclidian distance between local geometric positions. This will lead to two drawbacks: one is accumulated error, the other one is the difficulty of preserving geometric features. In this paper, we present a modified mesh simplification based on vertex decimation, for efficiently simplifying triangle meshes with a threshold defined of curvature. This method improves classical vertex decimation by calculating the mean curvature of each vertex and considering the change of curvature in local ring. Different from the local triangulation in traditional algorithms, our method suggests a global triangulation. Experimental results demonstrated that our approach can preserve the major topology characteristics and geometric features of the initial models after simplifying most vertices, without complicated calculation. It also can reduce the influence from noises and staircase effects in the process of reconstruction, and result in a smooth surface. With the proper preferences, this algorithm will strongly improve quality of the approximated model.

The structure of the paper is organized as follows: Section 2 reviews the related work on mesh generation and vertex decimation. After that, Section 3 presents our algorithm followed by the experimental results in Section 4. Finally, Section 5 draws the conclusion and discusses the future work.

## 2    Related Work

### 2.1    Mesh Generation

The triangular mesh is one of the popular representations for free-form surfaces. Compared with splines, triangular meshes are convenient for visualization and description of complicated shapes of arbitrary topology [16]. With the development of laser scanners or other 3D data acquisition equipments, triangular mesh models are easily obtained. 3D data models are directly obtained by laser scanners, computer vision systems or medical imaging devices to model visually actual objects. Then we can obtain triangular meshes by an algorithm that extracts isosurfaces from 3D volume data. One of these methods is Marching Cubes [17], which is an effective surface construction algorithm that generates many triangles. Marching Cubes provide a method that defines voxel and generating isosurfaces accurately. The voxel is defined as a data unit that consists of eight vertices between adjacent layers. The algorithm determines a threshold of the surface, and then calculates gradients of each vertex. Cubes which contain the surface will be found out by comparing the gradient with the threshold. Finally, these surfaces are obtained through the interpolation. Now,

Marching Cubes, which matured gradually from medical applications, apply to multiple areas.

A typical medical computed tomography (CT) or magnetic resonance (MR) scanner produces over 100 slices at a resolution of 512 by 512 pixels each [18,19]. For these sampled data sets, isosurface extraction using marching cubes can produce hundreds of thousands of triangles or more. Obviously, we have trouble storing and rendering models of this size. This is the direct reason for mesh simplification.

## 2.2    Vertex Decimation

Vertex Decimation was first proposed by Schroeder [5] in 1992. Although it has some improvement in recent years, the main idea of this algorithm is unchanged. The classical vertex decimation is briefly reviewed here.

This decimation algorithm checks all vertices in the mesh model over and over again. During a pass, each vertex which conform to the appointed decimation criteria and all triangles that use the vertex are deleted. But removing a vertex result a hole in the mesh. So a local triangulation for the resulting hole is necessary. The vertex removal process repeats until some termination condition is met. The termination condition is usually defined as a percent reduction of the initial mesh, or a maximum error. This algorithm mainly consists of three steps. The first step is describing the local vertex geometry and topology. The second is evaluating the decimation criteria, and the last is triangulating the resulting hole. Fig. 1 shows a simple explanation of vertex decimation.

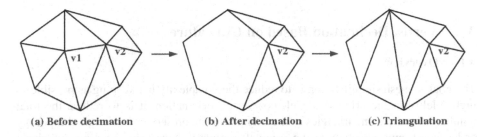

    (a) Before decimation             (b) After decimation            (c) Triangulation

**Fig. 1.** Vertex decimation

The outcome of first step determines whether the vertex is a potential candidate for removal, and if it is, which decimation criteria to use. The classical vertex decimation divides vertices in the mesh into five classifications: simple, complex, boundary, interior edge, or corner vertex. Different decimation criteria apply to different classifications of vertices. The fundamental decimation criterion is based on vertex distance to plane or vertex distance to edge. A simple vertex, which is surrounded by a whole cycle of triangles, use the distance to plane as a criterion. Simple vertices that are within the specified distance to the average plane will be removed (see Fig. 2(a)). If a simple vertex is used by a triangle not in the complete cycle of triangles, or if some edge is not used by two triangles, then this vertex is complex. Complex vertices are not removed from the mesh. Boundary vertices that are on the boundary of triangle meshes use the distance to edge as a criterion (see Fig. 2(b)). If the distance to edge is less than the specified value, this vertex can be removed. Moreover, a simple vertex can be further classified as an interior edge or corner vertex. For a detailed explanation, we can refer the reader to [5]. Removing a vertex and its adjacent triangles results one hole (two holes for interior edge vertex). These holes must be triangulated. But because the hole is generally non-planar, the two-dimensional algorithm to construct the triangulation is not suitable. Every hole is star-shaped, so the algorithm chooses a triangulation scheme based on recursive loop splitting. Each loop to be triangulated is divided into two halves along a split line defined from two non-adjacent vertices. The loop splitting processes again, until only three vertices remain in each loop. A loop of three vertices forms a triangle, which may be added to the mesh, and end the recursion process.

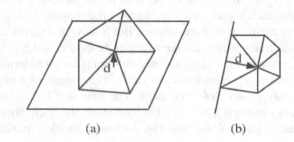

(a)                                    (b)

**Fig. 2.** Vertex distance to plane or to edge

# 3    Vertex Decimation Based on Curvature

## 3.1    Overview

The goal of mesh simplification is to reduce the complexity but keeping as possible as high fidelity of the original model. For vertex decimation, it is to reduce the total number of vertices and triangles in the triangle mesh, preserving the original topology and a good approximation to the original geometry. Moreover, keeping the visual feature characteristic is also very important for the mesh simplification.

Having this goal, we present a modified mesh simplification based on vertex decimation. We use curvature as the metric to instead of distance. This will preserve geometric features of the original model better without complicated calculation. Meanwhile, we also propose to consider the change of curvature in local area, and wrap the local triangulation by a global triangulation. By such means, geometric features of the model will receive adequate attention, and the influence from noises and staircase effects in reconstruction will be reduced.

The flowchart of our algorithm is given in Fig. 3. For an input of triangular mesh M with a vertex set $D = \{d_1, d_2, ..., d_m\}$ and a triangle set $T = \{ t_1, t_2, ..., t_n \}$, firstly, we calculate the mean curvature $h$ of each vertex in the mesh. If $h_i$ is less than a specified value of low curvature $c_l$, this vertex will be considered as a candidate for removal.

If $h_i$ is greater than a specified value of high curvature $c_h$, we calculate the average of mean curvatures (denoted by $c_i$) of neighbor vertices. If $c_i$ is high or low sufficiently, this vertex will be considered as a candidate for removal too. Afterwards, we obtain a simplified vertex set D' by removing candidates from D. Finally, triangulating D' result in the output. The process can be expressed as Table 1.

**Table 1.** Procedure of our algorithm

| Vertex Decimation Based on Curvature |
| --- |
| Input: triangular mesh M |
| FOR $d_i$ in D |
|     Calculating $h_i$ |
|     IF $h_i > c_l$ |
|         IF $h_i < c_h$ |
|             Add $d_i$ into D' |
|         ELSE |
|             Calculating $c_i$ |
|             IF $c_i$ is not high or low suffi- ciently |
|                 Add $d_i$ into D' |
|             END |
|         END |
|     END |
| END |
| Triangulating D' |
| Output: simplified model M' |

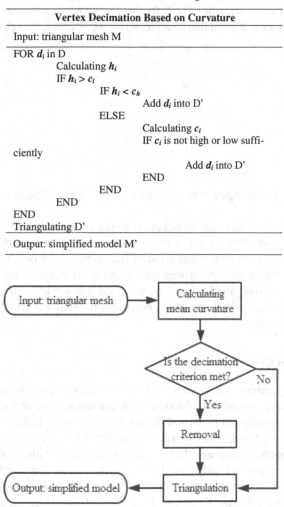

**Fig. 3.** The flowchart of our algorithm

## 3.2 Describing Meshes

We can only directly obtain coordinates of all vertices in a 3D data model. The triangle mesh is obtained by an algorithm that extracts isosurfaces from 3D volume data. An isosurface consist of information of vertices and triangles. Geometrically, triangle mesh consists of a set of vertices. A set of triangles can be used to describe a piecewise linear surface by connecting subset of the vertices together in an order [12]. Generally, each triangle is defined by three numbers of vertices which form this triangle. In other words, a triangle mesh is indicated as a set of vertices and a set of triangles, and the set of triangles is shown by vertices. Fig. 4 shows an example. In Fig. 4, we presume that numbers of vertices which form this triangle $t_1$ successively are 1, 2 and 3. The triangle $t_1$ can be defined by the sequence (1, 2, 3). In this way, we can conveniently obtain coordinates of vertices that form this triangle, when we access a triangle.

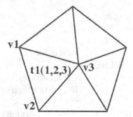

**Fig. 4.** A triangle can be indicated as three numbers of vertices.

Moreover, neighbor vertices and adjacent triangles of each vertex in the mesh are continually used in our algorithm. For this reason, it is desirable to maintain a list of the triangles that use each vertex and a list of the vertices that neighbor each vertex. We can work out the neighbor vertices and adjacent triangles of each vertex expediently by the set of triangles. In this way, the local geometric relationships between vertices are described clearly. In addition, edges are implicitly defined as vertex pairs in the triangle definition.

### 3.3    Decimation Criteria

Many classical algorithms use some fundamental parameters as the decimation criteria, like distance, area, volume etc. Methods that use distance error metric can control the error between the original mesh and the resulting, but it is difficult to preserve the major characteristics. Some related experiments [4], [12], [20] show that many 3D models contain geometric features that reflect geometrical details on the models, and if the simplification method only use distance error metric (or similar metric), the features are missing or not well respected, which implies that triangles may not be dense enough for the features. To ensure the quality of simplified models, it is necessary to develop traditional decimation criteria. Recently, shape error metric has received increasing attention to preserve geometric features in high-performance mesh simplification [21]. Shape error metric can improve the precision of the object features efficiently, but it is difficult to find a good shape description about the object features [8]. In our method, we adopt decimation criteria based on curvature.

Curvatures describe the amount by which the shape deviates from a plane [16]. A high curvature value means a great tortuosity. Generally, geometric features exist in the area that with a great tortuosity. In other words, a surface region that contains geometric features would have higher curvature values while the non-feature regions would have lower curvature values. For a vertex on a triangle mesh, there are countless curvatures that defined in various directions. Taking all possible directions then the maximum and minimum values of these curvatures at a vertex are called the principal curvatures, $k_1$ and $k_2$. The mean curvature $h$ is the average of these two principal curvatures at the vertex, and thus serves as a good metric for geometric features of the 3D model. For a triangle mesh, the mean curvature at a vertex $d_i$ can be calculated by using the information of its 1-ring neighborhood [16], [22], [23]:

$$h = \left\| \frac{1}{4A} \sum_{j=1}^{val_i} (\cot \alpha_j + \cot \beta_j)(d_j - d_i) \right\| \tag{1}$$

where A is the sum of the area of the triangle faces adjacent to $d_i$, $\alpha_j$, $\beta_j$ are the two angles opposite to the edge connecting $d_i$ and $d_j$ as shown in Fig. 5, and $val_i$ is the valence of $d_i$ (the number of adjacent triangles).

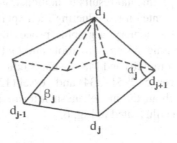

**Fig. 5.** The 1-ring neighborhood of vertex $d_i$

We let the mean curvatures calculated at each vertex serve as a measure for determining whether this vertex would be removed. If the mean curvature of a vertex is less than the specified value $c_l$, the vertex can be removed.

Nevertheless, solely using mean curvatures is not always desirable. For example, meshes may contain some noisy points with high mean curvatures. But obviously, these noisy points should be removed too. There is an important characteristic of the noisy point that can be used to advantage. That is high curvatures of its neighbor vertices with high mean curvature of itself. In addition, the vertex on the edge of staircase effects is also a kind of points with high mean curvature which should be removed. For vertices with higher mean curvatures, we can calculate the average $c_i$ of the mean curvatures of their neighbor vertices by using (2).

$$c_i = \frac{1}{n_i} \sum_{j=1}^{n_i} h_j \tag{2}$$

Here $h_j$ is the mean curvature of each neighbor vertex of $d_i$, and $n_i$ is the number of neighbor vertices. A vertex $d_i$ with a higher $h_i$ and a higher $c_i$ could be considered as a noisy point, and a vertex with higher $h_i$ and a lower $c_i$ could be considered as a staircase point. The noisy points and staircase points would be removed in our algorithm. In this way, we can obtain smoother surface.

### 3.4 Triangulation

Removing a vertex from the triangular mesh generate a hole (see Fig. 1(b)). In traditional algorithms, the hole will be filled up by a local triangulation. However, the local triangulation may produce long and narrow triangles, which are bad for the quality of simplified model, and cause an incoordination between new triangles and initial triangles. Furthermore, the closure problem needs further discussion [5], [24]. Our method suggests a global triangulation based on Delaunay Triangulation [11], [25] to instead of the local triangulation. In this way, we can obtain coordinating triangle meshes. Meanwhile, the closure of surface could be ensured.

## 4    Results and Discussions

This section presents our experimental results to demonstrate the proposed algorithm. The computer system implemented for conducting the experiment has a PC with Intel Core i5-2450M, 2.5 GHz, 4G memory, a graphic processing unit NVIDIA GeForce 610M, and Windows 7 operating system. The tools and platforms used in this project are Matlab, Vtk(visualization toolkit) and MS Visual Studio. The experimental graphic models in this work are liver (512×512×34) and vessel (125×68×322). After isosurface extraction, the details about the model mesh information are listed in Table 2. The original graphic model is illustrated in Fig. 6.

Table 2. Data information of the original models

| Model | Vertices | Triangles |
|---|---|---|
| liver | 103476 | 206948 |
| vessel | 87903 | 174352 |

Fig. 7 shows the simplification results on the models using our algorithm with different decimation ratio. The details about the corresponding reduced mesh information are listed in Table 3. The data size of models can be reduced by our algorithm, without complicated calculation. Even after simplifying most vertices, the major topology characteristics and geometric features of the initial models are still preserved. Moreover, noises and staircase effects in the process of reconstruction are avoided to a large extent.

Mesh simplification can save much time on model reconstruction. In our experiments, reconstructing the original mesh of liver model spend 5.71 seconds. After decimating 96% of the vertices, reconstructing reduced mesh model only spend 0.34 seconds, which achieves real-time demand.

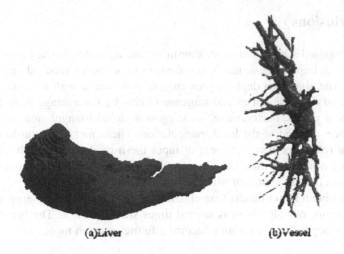

(a)Liver                                    (b)Vessel

**Fig. 6.** Original graphic model

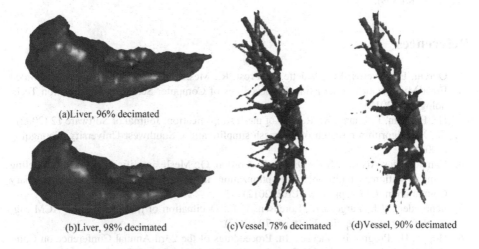

(a)Liver, 96% decimated

(b)Liver, 98% decimated          (c)Vessel, 78% decimated     (d)Vessel, 90% decimated

**Fig. 7.** Simplification results

**Table 3.** Data information of the simplification results

| Model | | Vertices | Triangles |
|---|---|---|---|
| liver | 96% decimated | 4005 | 8404 |
| | 98% decimated | 2246 | 4894 |
| vessel | 78% decimated | 19795 | 38864 |
| | 90% decimated | 9642 | 19150 |

## 5    Conclusions

This paper proposed a modified mesh simplification algorithm based on vertex deci-mation aiming at improving surface simplification for complex medical graphic mod-el rendering. The algorithm deploys geometric feature metric with mean curvature of each vertex, and decrease noises and staircase effects by the average of mean curva-ture of neighbor points. Furthermore, we suggest a global triangulation of the whole reserved vertices instead of the local triangulation. These measures employed in our algorithm can retain geometric features of input mesh models and result a relatively smooth surface after simplification. The associated experiment demonstrates that our algorithm is an efficient approach with high performance.

Even if we obtain better results than classical vertex decimation in terms of quality of approximations, our algorithm is several times slower than it. The better tradeoff between efficiency and accuracy may become a future research topic.

**Acknowledgements.** This work was supported by the Natural Science Foundation of China (No. 61103070) and Program for Young Excellent Talents in Tongji University (No. 2013KJ008).

## References

1. Ovreiu, E., Riveros, J.G., Valette, S., Prost, R.: Mesh Simplification Using a Two-Sided Error Minimization. International Proceedings of Computer Science & Information Tech-nology 50 (2012)
2. He, H., Tian, J., Zhang, X.: Review of mesh simplification. Journal of Software 12 (2002)
3. Fu, X.: Algorithm research of 3D mesh simplification. Southwest University, Chongqing (2008)
4. Campomanes-Alvarez, B.R., Damas, S., Cordón, O.: Mesh simplification for 3D modeling using evolutionary multi-objective optimization. In: 2012 IEEE Congress on Evolutionary Computation (CEC), pp. 1–8. IEEE (2012)
5. Schroeder, W.J., Zarge, J.A., Lorensen, W.E.: Decimation of triangle meshes. ACM Sig-graph Computer Graphics 26(2), 65–70 (1992)
6. Hoppe, H.: Progressive meshes. In: Proceedings of the 23rd Annual Conference on Com-puter Graphics and Interactive Techniques, pp. 99–108. ACM (1996)
7. Hamann, B.: A data reduction scheme for triangulated surfaces. Computer Aided Geome-tric Design 11(2), 197–214 (1994)
8. Jun, L., Shi, J.: A Mesh Simplification Method Based on Shape Feature. In: 2006 8th In-ternational Conference on Signal Processing, vol. 2 (2006)
9. Zhao, Y., Liu, Y., Song, R., Zhang, M.: A Retinex theory based points sampling method for mesh simplification. In: 2011 7th International Symposium on Image and Signal Processing and Analysis (ISPA), pp. 230–235. IEEE (2011)
10. Rossignac, J., Borrel, P.: Multi-resolution 3D approximations for rendering complex scenes. Springer, Heidelberg (1993)
11. Xin, S.Q., Chen, S.M., He, Y., et al.: Isotropic Mesh Simplification by Evolving the Geodesic Delaunay Triangulation. In: 2011 Eighth International Symposium on Voronoi Diagrams in Science and Engineering (ISVD), pp. 39–47. IEEE (2011)

12. Wang, J., Wang, L.R., Li, J.Z., Hagiwara, I.: A feature preserved mesh simplification algorithm. Journal of Engineering and Computer Innovations 6, 98–105 (2011)
13. Garland, M., Heckbert, P.S.: Surface simplification using quadric error metrics. In: Proceedings of the 24th Annual Conference on Computer Graphics and Interactive Techniques, pp. 209–216. ACM Press/Addison-Wesley Publishing Co. (1997)
14. Wei, J., Lou, Y.: Feature preserving mesh simplification using feature sensitive metric. Journal of Computer Science and Technology 25(3), 595–605 (2010)
15. Gieng, T.S., Hamann, B., Joy, K.I., Schussman, G.L., Trotts, I.J.: Smooth hierarchical surface triangulations. In: Proceedings of the 8th Conference on Visualization 1997, pp. 379–386. IEEE Computer Society Press (1997)
16. Wang, Y., Zheng, J.: Curvature-guided adaptive T-spline surface fitting. Computer-Aided Design 45(8), 1095–1107 (2013)
17. Lorensen, W.E., Cline, H.E.: Marching cubes: A high resolution 3D surface construction algorithm. ACM Siggraph Computer Graphics 21(4), 163–169 (1987)
18. Chen, Y., Wang, Z., Hu, J., Zhao, W., Wu, Q.: The domain knowledge based graph-cut model for liver ct segmentation. Biomedical Signal Processing and Control 7(6), 591–598 (2012)
19. Chen, Y., Zhao, W., Wu, Q., Wang, Z., Hu, J.: Liver segmentation in CT images for intervention using a graph-cut based model. In: Yoshida, H., Sakas, G., Linguraru, M.G. (eds.) Abdominal Imaging 2011. LNCS, vol. 7029, pp. 157–164. Springer, Heidelberg (2012)
20. Qing, D., Chen, J., Yu, H., Wang, Z.: Mesh simplification method based on vision feature. In: IET International Communication Conference on Wireless Mobile and Computing (CCWMC 2011), pp. 398–402. IET (2011)
21. Jian, W., Hai-Ling, W., Bo, Z., Ni, J.: An Efficient Mesh Simplification Method in 3D Graphic Model Rendering. In: 2013 Seventh International Conference on Internet Computing for Engineering and Science (ICICSE), pp. 55–59. IEEE (2013)
22. Desbrun, M., Meyer, M., Schröder, P., Barr, A.H.: Implicit fairing of irregular meshes using diffusion and curvature flow. In: Proceedings of the 26th Annual Conference on Computer Graphics and Interactive Techniques, pp. 317–324. ACM Press/Addison-Wesley Publishing Co. (1999)
23. Sullivan, J.M., Schröder, P.: Discrete differential geometry. Birkhäuser, Basel (2008)
24. Pan, Z., Zhou, K., Shi, J.: A new mesh simplification algorithm based on triangle collapses. Journal of Computer Science and Technology 16(1), 57–63 (2001)
25. Thomas, D.M., Yalavarthy, P.K., Karkala, D., Natarajan, V.: Mesh simplification based on edge collapsing could improve computational efficiency in near infrared optical tomographic imaging. IEEE Journal of Selected Topics in Quantum Electronics 18(4), 1493–1501 (2012)

# A Community Detecting Algorithm
# Based on Granular Computing

Lu Liu, Taorong Qiu, Xiaoming Bai, and Zhongda Lin

Department of Computer; Nanchang University, Nanchang, 330031, China

**Abstract.** Detecting the community structure of social network is really a very challenging and promising research in the world today.Granular Computing ,which can simplify the solution of problem by generating granules and implementation in different granularity spaces, is a kind of intelligent information processing model to simulate the human thinking. In this paper, a model of mining community structure based on granular computing is proposed through improving the similarity between nodes, that is, to design a corresponding mining algorithm by decomposing the problem in different granularity spaces so as to realize the structure detecting. The experimental results on three classic data sets show that the mining algorithm presented in this paper is reasonable.

**Keywords:** Social network, Community Detecting, Similarity, Granular Computing, Rough Set.

## 1 Introduction

Tribalization in the internet is the main trend in the future [1]. The sociality reflected in social networks is a community structure, in which nodes are joined together in tightly-knit groups between which there are only looser connections. It is a long history to research the community structure. A more useful approach taken by social network analysis with the set of techniques is known as hierarchical clustering, which can fall into two broad classes: Divisive method and Agglomerative method. The process of it can be represented using a tree of the type called "dendrograms". Agglomerative algorithm,in essence, is to find the most similar nodes in the network every time and add an edge between the two nodes. It is also known as the bordered method. While in contrast, divisive method is trying to find the least similar nodes and remove the edge between them. GN Algorithm is the most representative one of divisive methods [2]. It is a kind of mining method based on edge splitting through a significant concept called Edge Betweenness. In 2004, Clauset, Newman and Moore et al. proposed a new greedy algorithm, CNM Algorithm [3]. CNM Algorithm is considered as a kind of improved Newman algorithm. By calculating the max information gains, the groups are separated from one another and so reveal the underlying community structures. Soon afterwards, Newman studied many efficient approaches to finding community structure [4-7]. There are still some optimization algorithms, such as EO Algorithm [8] and other traditional spatial clustering methods, such as K-Means used for community

D. Miao et al. (Eds.): RSKT 2014, LNAI 8818, pp. 272–284, 2014.
DOI: 10.1007/978-3-319-11740-9_26 © Springer International Publishing Switzerland 2014

detecting [9]. In addition,considering the overlapping of community structures, Palla et al. introduced the concept of faction and presented a clique percolation algorithm in the literature[10].

Granular computing(GrC) [11-13], the basic idea of which is to reduce the problem complexity, is a tool to simulate human thinking problems in the field of intelligent information processing. For the feature of a community that nodes close connected can be divided into a same community, it is of great use to combine GrC with community detecting.In recent years, combinations with granularity and clustering gain wide attentions of universal scholars. However, it is still in the beginning stage to exploit GrC in social network. Ronald R.Yager introduced the granular computing theory into the social network analysis and provided a soft definition of community, in which, semantic concept is represented by fuzzy set theory and the relationship among individuals in social network is described by the set relation so that problems in social network can be solved by granularity in the way of human cognition [12-13]. However, the question is that it is just a simple paradigm to analysis social network with GrC. There is still lack of application in the real data sets. On the basic of Yager's research, a formal description of community is given in literature [14].

In this paper we present a class of new algorithms for network clustering,i.e.,the discovery of community structure in networks,taking advantage of the granular computing based on rough set model. Our discussion focuses primarily on how to define the information granule and how to generate the granular space in the network, both of which are basic problems in granular computing. By this mean, the community mining algorithm is transformed into problem solving under different granular spaces.

## 2     Closeness Between Nodes and Its Improvement

Studies of community detecting are based on the assumption that the individuals in the same community are joined together in tightly-knit groups between which there are only looser connections. Therefore, how to measure of the similarity between nodes became one of the main factors that decided to corporate results. It is inclined for sociologist to define the similarity between nodes based on the structural equivalence [15]. Nodes which have exactly the same neighbor are considered to be a structural equivalence. However, equivalent structure is rare in the real social network for distance between two nodes in complex network is not the space length, but the number of the edge in shortest path. Literature [14] puts forward a close degree formula based on the "small world "features, namely "six degrees of separation":

$$Close(v_i, v_j) = \frac{pathLen(v_i, v_j)}{2 \times 10^{pathLen(v_i,v_j)-2}}$$   (2.1)

In the above equation, $pathLen(v_i, v_j)$ is the shortest path length between $v_i$ and $v_j$.

But, formula(2.1) performs poor in the network with small average shortest path. As shown in Figure 1, there are two distinct community structures which are respectively formed by nodes 1,2,3,4 and 5,6,7,8 and its average shortest path(shortest path between two nodes divided by the sum of all possible number of edges in the network) is 1.8. When the formula(2.1) is used to calculate the closeness, node 5 is erroneously divided into the right association for the closeness between them is 1.

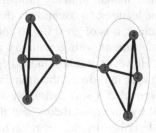

**Fig. 1.** A Simple network with two community structures

t is for this reason that the following modifications are given on the basis of formula(2.1)

***Definition 2-1 Close between nodes.*** The close between nodes can be defined as follows:

$$Close(v_i, v_j) = \begin{cases} 1 & pathLen(v_i, v_j) < 2 \\ \dfrac{4 \times pathLen(v_i, v_j)}{10^{pathLen(v_i, v_j)-1}} & otherwise \end{cases} \quad (2.2)$$

In which, $pathLen(v_i, v_j)$ denotes the shortest path between node $v_i$ and node $v_j$, $Close(v_i, v_j) \in [0\,1]$. $Close(v_i, v_j)$ is proportional to the closeness. The greater the value of $Close(v_i, v_j)$ is, the closer the relationship is. For a given $G(V, E)$, if $\exists e_{ij} \in E$ ,then $Close(v_i, v_j) = 1$. $Close(v_i, v_j)$ has reflexivity and symmetry in the unweighted and undirected graph.

The improved $Close(v_i, v_j)$ can describe the structure better. As in the previous example, node 5 is closer to nodes 6,7,8 ( $Close(5,6) = Close(5,7) = Close(5,8) = 1$ ) than to nodes 1,2,3 ( $Close(5,1) = Close(5,2) = Close(5,3) = 0.8$ ). After such improvement, node 5 can be separated easily from the community as shown in the right part of the network in Figure 1 containing node 1,2,3,4. Further more, the community structure generated with it satisfies the condition of the internal nodes are more compact than that of intercommunal.

# 3    Model of Community Detecting Based on Granular Computing

## 3.1    Model Design of Mining Community Based on Granular Computing

***Definition 3-1 Indistinguishability between nodes*** For a given Graph $G(V, E)$ , the indistinguishability $R \subset V \times V$ on $V$ can be defined as: for any $v_i$ and $v_j$ ,

$v_i \in V, v_j \in V, i \neq j$ , if $Close(v_i, v_j) = 1$ exists, then there is $v_i R v_j$ , or $(v_i, v_j) \in R$ .

Due to the specification that $Close(v_i, v_j)$ meets reflexivity and symmetry, so does the indistinguishable relationship $R$ .

**Definition 3-2 Basic Structure Granule and Basic Cover Granule of Graph.** For a given Graph $G(V, E)$, suppose $SubG(V_{SubG}, E_{SubG})$ be a subgraph of $G$ , if the two conditions are satisfied:

① nodes in set $V_{SubG}$ are indistinguishable;

② there is no other subgraph $SG(V_{SG}, E_{SG})$ in $G$ which meets the conditions that for $\forall v \in V_{SubG}$ , $\forall e \in E_{SubG}, v \in V_{SG}$ , $e \in E_{SG}$ exists;

Then $SubG(V_{SubG}, E_{SubG})$ is called as a basic structure granular of $G$ , $SG$ for short. $V_{SubG}$ is denoted as a basic cover granule.

**Definition 3-3 Basic Granular Space of Graph.** For a given graph $G(V, E)$, a basic granular space is defined as the set of all basic structure granules of $G$ , denoted as $BGS = (SG_1, SG_2, ..., SG_k)$ , which meets that for any $SG_i = (V_{SG_i}, E_{SG_i})$ , $SG_j = (V_{SG_j}, E_{SG_j})$ , $i \neq j$ there is $V_{SG_i} \not\subset V_{SG_j} \wedge V_{SG_j} \not\subset V_{SG_i}$ and $\bigcup_{i=1}^{k} V_{SG_i} = V$ .

**Definition 3-4 Basic Cover of Graph.** For a given Graph $G(V, E)$ , $BGS = \{SG_1, SG_2, ..., SG_k\}$ is the basic granular space , among which, $SG_i = (BG_i, E_{SG_i})$, $i = 1, ... k$ , then $BC = \{BG_1, BG_2, ..., BG_k\}$ is the basic cover of $G$ .

**Definition 3-5 Cover of Graph.** For a given Graph $G(V, E)$ , if the set $Cover = \{C_1, C_2, ..., C_m\}$ meets $C_i \subseteq V, i = 1, 2, ..., m$ and $\bigcup_{i=1}^{m} C_i = V$ $\wedge \exists C_j, C_h, C_j \cap C_h \neq \phi$ , then $Cover$ can be referred as a cover of $G$ , among which, $C_i, i = 1, 2, ..., m$ is called a cover granule of $G$ .

Obviously, the basic cover granule $BG_i$ surely is the cover granule of $G$ , but not vice versa.

**Definition 3-6 $\beta$_Similar Granular Set of Cover Granule** For a given Graph $G(V, E)$ , $Cover = \{C_1, C_2, ..., C_m\}$ is supposed to be the cover of $G$ and $0 \leq \beta < 0.5$, then for any cover granule $C_i \in C$ , the $\beta$_similar granular set of the cover granule $C_i$ can be defined as :

$$G_\beta(C_i) = \left\{ C_j \in C \mid C_i \overset{\beta}{\supseteq} C_j, \ i, j \in \{1..m\} \right\} \qquad (3.1)$$

or,

$$G_\beta(C_i) = \{C_j \in C \mid c(C_i, C_j) \leq \beta\} \tag{3.2}$$

It can be seen from the above definition that the cover granule is coarsen by getting the $\beta$_similar granular set and therefore transformations under different granularities are achieved.

**Definition 3-7 Minimal Cover of Graph.** For a given Graph $G(V, E)$, suppose $0 \leq \beta < 0.5$ and $Cover = \{C_1, C_2, ..., C_m\}$ be a cover of $G$, for $\forall C_i \in C$, if $G_\beta(C_i) = \{C_i\}$, then $Cover = \{C_1, C_2, ..., C_m\}$ is called a minimal cover of $G(V, E)$.

**Definition 3-8 Graph Granular Space.** A graph granular space can be defined as: $GS = \{G(V, E), MC\}$, among which, $G(V, E)$ is the original structure and $MC$ is the minimal cover of $G$.

**Definition 3-9 Node Granular Belongingnes.s** For a given Graph $G(V, E)$ and a cover of $G$, $Cover = \{C_1, C_2, ..., C_m\}$, the belongingness of node $u$ ($u \in V$) with a cover granule $C$ ($C \in Cover$) can be defined as following:

$$G\_Belongingness(u, C) = \frac{\sum_{i}^{n_C} Close(u, v_i)}{n_C} \tag{3.3}$$

There is also $G\_Belongingness(u, C) \in [0\ 1]$.

Since a cover granule can be seen as the community structure under a certain level, so the belongingness between nodes and cover granules reflects the closeness between them. It is also a criterion to estimate whether a node can be divided into a certain community. The greater $G\_Belongingness(u, C)$ is, the more possible $u$ is belong to $C$.

**Definition 3-10 Granular Compactness.** For a given Graph $G(V, E)$ and a cover $Cover = \{C_1, C_2, ..., C_m\}$ on $G$, the closeness between any two cover granules $C_i, C_j \in Cover$ can be expressed as:

$$GC(C_i, C_j) = \frac{\sum_{k=1}^{n_{C_i}} G\_Belongingness(u_k, C_j)}{n_{C_i}} \tag{3.4}$$

In (3.5), $u_k \in C_i$ and $n_{C_i}$ is the number of $C_i$. The properties of $GC(C_i, C_j)$ are as following:

① Symmetry : $GC(C_i, C_j) = GC(C_j, C_i)$ ;

② Commutativity : if $GC(C_i, C_k) \geq \beta$ ,then $GC(C_i \cup C_j, C_k) \geq \beta$ , $GC(C_i \cup C_k, C_j) \geq \beta$ and $GC(C_k \cup C_j, C_i) \geq \beta$ exists.

The proof process in detail is omitted in order to save space.

On the grounds of the above introduction,the corresponding algorithm is proposed in the following.

## 3.2   Community Mining Algorithm Based on Granular Computing(CGCC Algorithm)

A appropriate initial clustering granularity not only can reduce the computational complexity but also can improve the accuracy of clustering. In our community mining algorithm, the basic coverage of the graph is chosen as the initial community structure. Algorithm 1 introduced in the following realizes the granulation for the first time, that is to construct the basic cover of network.

**Algorithm 1.** Graph granulating --Generating a basic cover of Graph

---

**Input:** Graph $G(V,E)$
**Output:** A basic cover of $G$   $BC = \{BG_1, BG_2,..., BG_k\}$

---

**Step1:**Initialization $BC = \varnothing$ ;
**Step2:**Compute $C_c(v_i)$ of nodes in $G$, and put them in order from large to small;
   **Step3:**For a node $v_i$ ( $v_i \in V$ ),if $v_i \notin \cup BC_j$ $(j = 1,2,...k)$, get all the basic cover granule         $BG_i$ which contains $v_i$ according to the definition of Indiscernibility,go to Step4;else go
      to Step 6;
**Step4:**If $|BG_i| > 3$ ,then go to Step5;else, go to Step6;
**Step5:**Put $BG_i$ into $BC$. if there are many granules which satisfies the condition, then put all
      of them into $BC$ ;
**Step6:**Let $BG_i = \{v_i\}$, and put $BG_i$ into $BC$ ;
**Step7:**Choose the next node $v_{i+1}$ ,repeat Step3-Step6 until all nodes have been visited;
**Step8:**The end.

---

After algorithm 1, a basic cover of  graph $G$  is obtained, namely getting a basic structure of granular space. Each basic cover granule can be regarded as the kernel of initial community. In the second step, the closeness centrality is used to measure the centralization which can reflect the global topological structure.

There are, however, also many high overlapping basic cover granules, namely one node may belong to different basic cover granules at the same time. In order to solve this problem, algorithm 2 will obtain a minimal coverage of the graph by getting the $\beta$_similar granule set.

**Algorithm 2.** Granular coarsening--Obtaining the minimal coverage of graph

---

**Input:** A basic coverage of $G(V, E)$, $BC = \{BG_1, BG_2, ..., BG_k\}$ and a given
    Threshold $\beta$
**Output:** A minimal Coverage of $G$, $MC = \{MC_1, MC_2, ..., MC_n\}$

---

**Step1:** Initialization $MC = \varnothing$, $OC = BC$; put $C_i$ in order with the value $|BG_i|$
    from small to large, denoted as $OC = (C_1, C_2, ..., C_k)$;
**Step2:** If $OC \neq \phi$, go to Step3, else go to Step8;
**Step3:** $i = 1$;
**Step4:** Get the $i^{th}$ cover granule $C_i \in OC$, calculate $G_\beta(C_i)$;
**Step5:** If $G_\beta(C_i) = \{C_i\}$, put $C_i$ into $MC$ and $OC = OC - \{C_i\}$, then go
to Step 7; else

go to Step6;
**Step6:** $C_i = \bigcup E_i$, $OC = OC - \{E_i\}$, among which $E_i \in G_\beta(C_i)$;
**Step7:** If there are still cover granules that are not traversed, then process $i = i + 1$
and go to

Step4; else put $OC$ in order according to the value of $|C_i|$ from
small to large and

go to Step2;
**Step8:** The end.

---

Thus, the initialization phrase is over. In the next stage, closeness is calculated between cover granules and integration will happen between cover granules with highest similarity, if the closeness is over the threshold given.

It follows that the community mining algorithm based on granular computing proposed can be divided into two parts: the first one is the original granulation which can get a basic structure granular space by generating the basic cover granule of a graph and granule coarsening according to the β_similar granule; the second part is granular integration, in which, cover granules with certain similarities will be integrated.

The algorithm proposed for identifying communities is stated in detail as follows:

**Algorithm 3.** CGCC-Algorithm

---

**Input:** The original Data Set
**Output:** A set of communities $Clique = \{C_1, C_2, ..., C_k\}$

---

**Step1:**Read the original data set and Get the graph structure of network

**Step2:**Gain a basic cover of original data set $BC = \{BG_1, BG_2, ..., BG_k\}$ according to

Algorithm 1;

**Step3:**Obtain a minimal coverage $MC = \{MC_1, MC_2, ..., MC_n\}$ according to algorithm 2;

**Step4:**Choose an appropriate threshold $\alpha$ by experiment;

**Step5:**Put $MC_i$ in order according to the value of $|MC_i|$ from small to large, denoted as

$$MC = \{MC_1, MC_2, ..., MC_n\};$$

**Step6:**Calculate the closeness between cover granules according to formula(4.5), if

$$Max(GC(MC_i, MC_j)) \geq \alpha, MC_i = MC_i \cup MC_j \text{ exists;}$$

**Step7:** $MC = MC/\{MC_j\}$;

**Step8:**Repeat Step6-Step7 until there is no qualified communities to be associated with;

**Step9:**The end .

Apparently, communities under different granules can be got when $\alpha$ takes various values.

# 4    Test and Comparison

## 4.1    Experiment Environment

Experiments are performed on a loaded Intel(R)Core(TM)i5-3210MCPU T5250@2.50 GHz,4G of RAM,Windows 8(64 bit System) and implemented by Python2.7.

### 4.1.1    NetworkX Introduction

NetworkX [17] is a Python language software package for the creation, manipulation, and study of the structure, dynamics, and functions of complex networks. It is easy to load and store networks in standard and nonstandard data formats, generate many types of random and classic networks, analyze network structure, build network models, design new network algorithms, draw networks and much more. NetworkX has various features,for instance,it contains some date structures for graphs as well as standard graph algorithms.Also it is open source and well tested .

### 4.1.2    Graph Model Language(GML)

GML(Graph Model Language) [18], is portable file format for graphs based on ASCII representation, also as Graph Meta Language. It is a universal file format to represent graphs. Data sets used in this paper are all GML files.

#### 4.1.3    Data Sets

Three different sizes of real-world data sets for which the community structures are already known, including Zachary Karate Club [19], Dolphin Social Network [20] and College Football Network [21], are used to carried out simulation experiments. The number of nodes accurate classified is taken to evaluate the test result. Features of three data sets are shown as follows.

**Table 1.** Features of Data Set

| Data Set | Number of nodes | Number of edges | Number of real communitie |
|---|---|---|---|
| Zachary Karate Club | 34 | 78 | 2 |
| Dolphin Social Network | 62 | 159 | 2 |
| College Football Network | 115 | 613 | 12 |

### 4.2    Evaluation Index

Modularity and Normalized Mutual Information(NMI) are usually used to evaluate the community divided results. Module, used to analyze the community of the unknown network data, shows the differences between the edge inside and expectations. However, in many cases, community gained when the modularity takes the maximum value is often inconsistent with the reality, which is in fact the combination of several real communities [22]. NMI is applied in information theory. It is to measure the algorithm through entropy. In this paper, the number of nodes correctly classified is adopted as the evaluation index because the real community structures of data set employed in our experiments are already known. We make some analysis on the three data sets between CGCC and GN,CNM.

### 4.3    Analysis on The Three Data Sets With CGCC

We tested CGCC algorithm on the three classical data sets whose community structure were known beforehand. Figure 2-Figure4 show the optimal community structures obtained by CGCC.

Figure 2 illustrates the application results of our algorithm to the Zachary Karate Club. Vertices are drawn as different color according to the primary division detected. We find that the CGCC algorithm splits the network into two strong communities, which consists of 16 nodes and 18 nodes respectively.

As it is shown in Figure 3, Dolphin network is divided into two communities. The blue nodes (a total of 42 nodes ) make up the first community and the other community is formed by the remaining nodes colored in green. It turns out that the results are in excellent agreement with expectations.

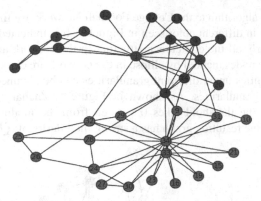

**Fig. 2.** Community Structure on Zachary Karate Club with CGCC

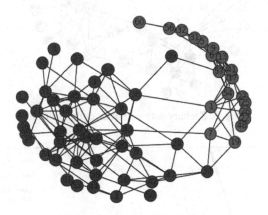

**Fig. 3.** Community Structure on Dolphin Social Network with   CGCC

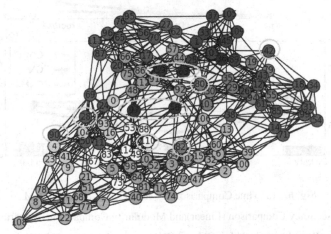

**Fig. 4.** Community Structure on College Football Network by CGCC

(10 of 115 nodes are divided incorrectly marked by ellipse )

Applying CGCC algorithm to the College Football Network, we find 12 well-defined communities drawn in different color,seen in Figure 4. It is indicated that CGCC algorithm identifies nearly all the conference structure in the network and the accuracy is about 91%. Only 10 nodes marked by ellipse in the network are classified incorrectly.

Various communities under different granularities can be obtained by adjusting the threshold between granularities. As shown in Figure 5, Zachary is divided into 4 communities when the threshold takes 0.8. Seen from the modularity, the optimal community split of the resulting graph has a strong modularity of $Q = 0.419$.

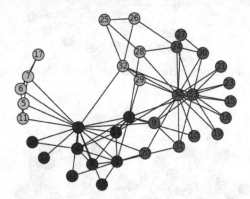

**Fig. 5.** Zachary data with four communities

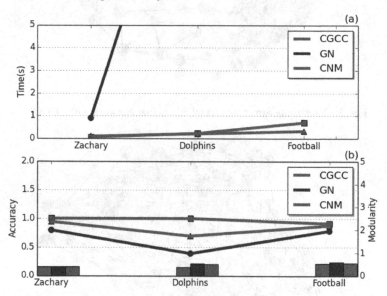

**Fig. 6.**   (a) Time Comparison between CGCC and GN,CNM

(b)Accuracy Comparison (Linear)and Modularity Comparison(Histogram)

(Results of algorithm GN and CNM referred to paper [22-25])

## 4.4      Analysis on The Three Sets Between CGCC and GN,CNM

Figure 6(a) shows the run time of CGCC, GN and CNM when the thresholds take the best value respectively,which are selected after repeated experiments. In spite of this the accuracy and modularity are also depicted in Figure 6(b).

It is indicated in Figure 6 that CGCC preforms more perfectly than GN and CNM on the three sets, and at the same time, CGCC has higher time efficiency compared with GN. However, though it is not good enough to feed College Football Network into CGCC than CNM, both algorithms give reasonably high time performance on the other two data sets.

# 5      Conclusion

In this paper, a model of community mining and corresponding algorithm are designed based on granular computing by improving the similarity between nodes, which is extended to be similarity between nodes and cover granules, also between cover granules. We have tested our method on three real-world networks with well-documented structured and find the results to be in excellent agreement with expectations. A compare analysis with GN and CNM is also given. Test results indicated that CGCC algorithm is effective and feasible. It is not only able to detect the known community structure with a high degree of success but also can obtain higher time efficiency.

A number of extensions or improvements of our method may be possible. The first one is to analysis the relationship between the average depth and the threshold value in order to choose the optimal or quasi optimal value; Secondly, it is hopeful to apply this algorithm in dynamic network. At present, it is still impractical for huge graphs, as a result, it is necessary to improve the speed of the algorithm.

**Acknowledgements.**   This work was supported by National Natural Science Fund Project (No. 61070139), the Science and technology Project of Jiangxi Provincial Department of Education (No. GJJ14134, GJJ14143) .

# Reference

1. Tan, X.T.: Tribalization: modes of life and relation to their environment in the internet[OL] (June 10, 2013), http://www.huxiu.com/article/21125/1.html
2. Newman, M.E.J., Girvan, M.: Finding and evaluating community structure in networks. Physical Review E 699(22): 026113(1-15) (2004)
3. Clauset, A., Newman, M.E.J., Moore, C.: Finding community structure in very large networks. Physical Review E 70(6): 066111(1-6) (2004)
4. Karrer, B., Levina, E., Newman, M.E.J.: Robustness of community structure in networks. Physical Review E 77(4): 046119(1-9) (2008)
5. Ball, B., Karrer, B., Newman, M.E.J.: Efficient and principled method for detecting communities in networks. Physical Review E 84(3): 036103 (1-3) (2011)
6. Karrer, B., Newman, M.E.J.: Stochastic blockmodels and community structure in networks. Physical Review E 83(1): 016107 (1-10) (2011)

7. Nadakuditi, R.R., Newman, M.E.J.: Graph Spectra and the Detectability of Community Structure in Networks. Physical Review Letters 108(18): 188701(5) (2012)

8. Duch, J., Arenas, A.: Community detection in complex networks using extreme optimization. Physical Review E 72(2): 027104(1-4) (2005)

9. Zhao, F.X., Xie, F.D.: Detecting community in complex networks using K-means cluster algorithm. Application Research of Computers 26(6), 2041–2043 (2009)

10. Palla, G., Derényi, I., Farkas, I., Vicsek, T.: Uncovering the overlapping community structure of complex networks in nature and society. Nature 435(7043), 814–818 (2005)

11. Wang, G.Y., Miao, D.Q., Yao, Y.Y., et al.: Cloud Model Granular Computing. Science Press, Beijing (2012)

12. Yager, R.R.: Intelligent social network analysis using granular computing. International Journal of Intelligent System 23, 1197–1219 (2008)

13. Yager, R.R.: Granular Computing for Intelligent Social Network Modeling and Cooperative Decisions. In: Proc. 4th IEEE International Conference Intelligent Systems 2008(1): 1_3-1_7

14. Bastani, S., Jafarabad, A.K., Zarandi, M.H.F.: Fuzzy Models for Link Prediction in Social Networks. International Journal of Intelligent System 28, 768–786 (2013)

15. Wang, L., Dai, G.Z.: Scale-free Properties, phenomena and Its Control in Complex Network. Science Press, Beijing (2009)

16. Huang, B.: Rough sets-based theory & application for knowledge acquisition in incomplete information system. Nanjing University of Science & Technology, Nanjing (2004)

17. High-productivity software for complex networkx[OL], http://networkx.github.io/

18. Himsolt, M.: GML: A portable Graph File Format[OL]. at http://www.uni-passau.de/Graphlet/GML

19. Zachary, W.W.: An information flow model for conflict and fission in small groups. Journal of Anthropological Research 33(4), 452–473 (1977)

20. Lusseau, D.: The emergent properties of a dolphin social network. Proceedings of the Royal Society of London. Series B: Biological Sciences 270, S186–S188

21. Girvan, M., Newman, M.E.J.: Community structure in social and biological networks. Proceedings of the National Academy of Sciences of the United States of America 99(12), 7821–7826 (2002)

22. Chen, Y.X.: Complex network community detection based on multi-objective generic algorithm. Lanzhou University, Lanzhou (2012)

23. Luo, L., Zhang, S.W., Chen, T.: Detecting the community by integrating the dense sub graph and edge clustering coefficient. Electronic Design Engineering 21(18), 36–40 (2013)

24. Xing, H.W.: Complex structure detection algorithm research in complex networks. Hunan University, Changsha (2011)

25. Liu, Y.: A study of feature analysis and community mining in social networks. University of Electronic Science and Technology of China, Chengdu (2013)

# A Robust Online Tracking-Detection Co-training Algorithm with Applications to Vehicle Recognition

Chen Jiyuan[1,2] and Wei Zhihua[1,2]

[1] Department of Computer Science and Technology,Tongji University
[2] Key Laboratory of Embedded System and Service Computing,
Ministry of Education
Cao'an Road, 4800,Shanghai, China, 201804

**Abstract.** Focusing on the vehicle tracking task in a video, we propose an Online Tracking-Detection Co-Training schema that integrates detecting and tracking results in a co-training style. The tracker follows the object from frame to frame and its trajectory is used in one of feature view in co-training process. The detector recognizes the patches including given object in current frame and corrects the tracker in broken frame. Our proposed model is verified through experiments on reality videos including some challenging situations.

**Keywords:** Co-training, detection, tracking, on-line, vehicle recognition.

## 1 Introduction

Object detection and tracking attract many researchers as its widely applications. However, its performance is affected by many factors such as light, weather, occlusion and so on. These limit application of the state-of-the-art methods in industry. Therefore, object detecting and tracking in reality conditions is still a challenging task. Recently, detection based on semi-supervised learning attracts much attention which learns from both labeled and unlabeled objects.The unlabeled data can improve the performance of classifier[1]. The vehicle which locates in consecutive frames defines as a trajectory which represents relationship between patches in adjacent frames. The information extracted from tracking trajectory is useful for the detection task.

The paper proposes a novel co-training detection-tracking scheme learning from both labeled and unlabeled objects. Relationship between objects is used in two feature views of classification algorithm to mutually rectify errors by constraint. Online learning is integrated to improve the global performance. Experiments on real traffic videos are performed to testify the robustness of the algorithm.

The rest of the paper is organized as follows. In section 2, basic problem is defined and various semi-supervised detection algorithms are investigated. In section 3, online tracking-detection co-training scheme is proposed.Experiments on real world traffic videos affecting by various external factors are conducted to analyzes the effectiveness of proposed algorithm.

D. Miao et al. (Eds.): RSKT 2014, LNAI 8818, pp. 285–294, 2014.
DOI: 10.1007/978-3-319-11740-9_27 © Springer International Publishing Switzerland 2014

## 2    Related Works

Many tracking algorithms have been proposed based on motion analysis. For example, CAMSHIFT is widely used for this task by conducting continuous Mean Shift[11] in videos. Henriques[15] takes sub-windows induces circulant-structure and use the Fast Fourier Transform to incorporate information from all sub-windows without iterating over them.A common technique is blurring the image for smoothing the objective function. The disadvantage of blurring is losing the original information from the image. L. Sevilla-Lara[14] proposes a new tracking method which is build an image descriptor using distribution fields, which is a representation that allows smoothing the objective function without destroying information about pixel values.

Recently, detection formulated as a binary classification problem has received a lot of concerns due to its promising results. Semi-supervised learning is popular for this task because it include unlabeled data into training set to enhance the performance. Frequently used semi-supervised learning methods include Self-Learning[2] and Co-training[3][4] etc. Self-training and co-training adapt to the independent assumption of different feature views[1]. For the problem of small labeled set, online learning is often integrated to co-training process which online updates the model initially trained from original trained set [3][4][5].

In semi-supervised learning, more feasible strategy for labeling unlabeled examples is guided by some prior knowledge or supervision information[1]. The basic idea is combining detection and tracking[6] where the detector serves as initial model for semi-supervised learning.Tracking could learn some information underlying examples, i.e. estimating object location in frame-by-frame fashion. The target can be viewed as a single labeled instance and the video as unlabeled datasets. Some authors perform self-training and co-training for tracking object[7][8][9]. The approach predicts the position of the objects with a tracker and updates the model with positive examples that are close to the selected patch and negative examples that are far from it. The strategy is able to adapt the tracker to new appearances and background, but breaks down when the tracker makes a mistake. In order to avoiding above problem, Kalal proposed Tracking-Learning-Detection by using P-N learning[10]. This approach integrates tracker and detection and made them correcting the cost function.

## 3    Online Tracking-Detection Co-training Scheme

The goal of the Online Tracking-Detection Co-Training scheme is improving the original tracking by using co-training. During the transmission from monitor to the server, some bits may lost and it will lead to some frame broken. Because broken frames cannot be tracked by the tracker, a heuristic search algorithm is designed to solve this problem. In this section, the framework of our model is introduced first. The detection with online co-training method is then described. Finally, the scheme to identify and correct the error is designed for improving the algorithm reliability.

## 3.1   Framework

The Online Tracking-Detection Co-Training framework is shown in Fig.3.1. The process should mark the target in the first frame since each frame which would be processed by our model requires the given initial target object $B_I$, which is an input rectangle and given by human. Then, the model would work in two ways simultaneously: the tracking component and the detection component.

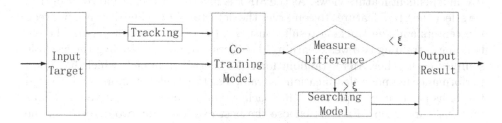

**Fig. 1.** Online Tracking-Detection Co-Training scheme

The task of the basic tracking component is obtaining the trajectory of target object $B_T$ in the consecutive frames. CAMSHIFT[11] is chosen in our framework due to its high efficiency. The detection is identifying whether the images in test set is similar to ones in training set.Some convincing examples learned by co-training process will be added into training set. In co-training, one feature view adopts the SIFT[12] and the other view considers tracking trajectory.We measure their difference by a predefined threshold. If the result is less than given threshold , we will consider these models offer the fit results. Otherwise, the algorithm will be re-initialized by applying heuristic search to relabel the target. The detector can measure which sub-rectangle is fittest and take it as final output $B_O$.

## 3.2   Online Tracking-Detection Co-training

This section investigates how the tracking and object detection components of the Online Tracking-Detection Co-Training algorithm work.

**Tracking**
The task of the tracking component determines the centroid of the target object in the present frame. CAMSHIFT is a fast tracking algorithm and we decide to use it in tracking component. The process of CAMSHIFT is chosen to initialize the target from the last frame, transforms the frame from the RGB space to the HSV space, and calculates the colorful histogram to select the slide window $X$. Then it calculates the centroid $(x_c, y_c)$ by the formulas (1) and (2):

$$x_c = \frac{\Sigma_x \Sigma_y x I(x,y)}{\Sigma_x \Sigma_y I(x,y)} \tag{1}$$

$$y_c = \frac{\Sigma_x \Sigma_y y I(x,y)}{\Sigma_x \Sigma_y I(x,y)} \tag{2}$$

$(x_c, y_c)$ is the coordinate of the search slide window $X$, and $I(x, y)$ is the pixel of the window $X$. The CAMSHIFT algorithm will adjust the size of the slide window and move the center of the window to the centroid $(x_c, y_c)$.

## Detection
Co-training is designed for detection which check each patches in frames and determine which one is matched. The co-training process trains two classifiers in two independent feature views. As the SIFT is invariant to scale and rotation[12], we select the SIFT features to construct the detector $f_1(x)$. The distance between the centers and the tracking result construct the other detector $f_2(x)$. The co-training is training two weak learning classifiers stronger by labeling the credible samples each other. Here, random forest[13] classifier model, which has good performance in image classification, is trained as the weak detectors by the initial target as positive example and some background patches as negative ones. This part has three steps. First we choose the best examples for two detector in the $K + 1$th turn. This process is shown in formula (3) and (4).

$$Y_P^+ = \{x_{s_i} \| x_{s_i} \epsilon X_{k+1}, 1 \leq i \leq m^P, max_{args_i} \Pi_{i=1}^{(m^P)} max\left(f_1^k(x_{s_i}), f_2^k(x_{s_i})\right)\} (3)$$

$$Y_N^+ = \{x_{s_i} \| x_{s_i} \epsilon X_{k+1}, 1 \leq i \leq m^N, max_{args_i} \Pi_{(i=1)}^{(m^N)} max\left(f_1^k(x_{s_i}), f_2^k(x_{s_i})\right)\} (4)$$

The $X_k$ means the unlabeled set in the $k$th turn, $x_i$ is one sample in the dataset, $s_i$ and $t_i$ is the sample index, $m^P$ and $m^N$ are the new samples adding to the labeled sample set, $max_{arg(s_i)}$ means choosing the best value for the target function to get the maximum, $Y_P^+$ and $Y_N^+$ means the positive and negative examples in the $k + 1$th turns which will be added into the labeled dataset. In order to keep the dataset size for making the co-training model speed up and reduce the effects on bad cases, we choose some examples to remove from the model. This process is shown in formula (5) and (6).

$$Y_P^- = \{x_{t_i} \| x_{t_i} \epsilon X_k^P, 1 \leq i \leq n^P, min_{argt_i} \prod_{i=1}^{n^P} max\left(f_1^k(x_{t_i}), f_2^k(x_{t_i})\right)\} \quad (5)$$

$$Y_N^- = \{x_{t_i} \| x_{t_i} \epsilon X_k^N, 1 \leq i \leq n^N, min_{argt_i} \prod_{i=1}^{n^N} max\left(f_1^k(x_{t_i}), f_2^k(x_{t_i})\right)\} \quad (6)$$

Where, $n^P$ and $n^N$ are the samples removing from the labeled sample.$min_{argt_i}$ means choosing the best value for the target function to get the minimum. $Y_P^-$ and $Y_N^-$ means the positive and negative examples in the $k + 1$th turns which will be removed from the labeled dataset. Finally we update the labeled dateset as the formula (7) and (8) shown.

$$X_{k+1}^P = X_k^P + Y_P^+ - Y_P^- \tag{7}$$
$$X_{k+1}^N = X_k^N + Y_N^+ - Y_N^- \tag{8}$$

The $X_k$ means the unlabeled set in the $k$th turn. $Y_P^+$ and $Y_N^+$ means the positive and negative examples in the $k + 1$th turns which will be added into the labeled dataset. $Y_P^-$ and $Y_N^-$ means the positive and negative examples in the $k + 1$th turns which will be removed from the labeled dataset. The pseudo code of the Tracking-Detection Co-Training model is described in Fig.3.2:

```
//The tracking and object detection code
Input : the input rebound box;
Output : the predicted object;
Procedure :
1.  Init_Camshift();
2.  Init_Detection();
3.  For(i = 1; i < frame_num; i + +)
4.     Result = Camshift(Input);
5.     X = RandomGetPat(Result, m, n)
6.     L1 = Classifier f₁(X);
7.     L2 = Classifier f₂(X);
8.     PushMostSamples(L1, X, P, N, g₁(x));
9.     PushMostSamples(L2, X, P, N, g₂(x));
10.    Update(f₁(x), L1)
11.    Update(f₂(x), L2);
12. EndFor
```

**Fig. 2.** The pseudo code of the Tracking-Detection Co-Training model

## 3.3 Correction Scheme

After tracking and detection, we get the target $B_T$ and $B_O$ by two models. In normal conditions, the tracking and detecting have the good performance and their results will be similar. It is assumed that the distance between the input and the model results satisfying the formula (9).

$$\|B_T - B_O\| < \delta \ and \ \|B_I - B_O\| < \delta \ and \ \|B_T - B_I\| < \delta \tag{9}$$

The predicted object close to the input, we consider the constant $\delta$ as the bound among the input $B_I$, the tracking result $B_T$ and the detected object $B_O$. When the formula (9) is satisfied, we consider the tracking result as the final result. However, if some frames are broken, the formula (9) will not be satisfied. The heuristic search is searching each nodes, calculating the possibility for image patches by the co-training model. In our model, we determine the Breadth-First search as the searching strategy because it can expand the neighbors. The Gaussian Mixture Model is chosen to generate the initial nodes. For each loop, a node from the candidate list is expanded by scanning the image patches which centroid is near this one. We retain the $K$ most likely nodes which is calculating the possibility by the co-training model to the candidate list. After running $M$ loops, the searching process is finished and the corrected object is obtained.

# 4  Experiments and Discussion

## 4.1  Experiment Setting

Our algorithm is applied to track vehicles in videos captured by monitor and the camera in-car. The information is listed in Table 1. As assuming the result should be sent back to the monitor, the videos are transformed into 2 frames per second. Our model is compared with three algorithms: CAMSHIFT[11],TLD[10] and the Circulant Structure model[15].Performance is evaluated by three metrics: average precision, average recall and runtime. Average Precision is the percent of the bounding box overlapped the ground truth one larger than 50%. Average Recall is the percent of the ground truth bounding box that the area overlapped is larger than 50%. Runtime is interval from the first frame to the last frame.

**Table 1.** Experiment videos information

| Test Video | Frame number | Character | Capture method |
|---|---|---|---|
| Monitor videos | 780 | Complicated background | station monitor |
| Clips1 | 158 | Object size change | Moving camera in-car |
| Clips2 | 288 | Vibration and rotation | Moving camera in-car |

## 4.2  Experiment Results on Monitor-Captured Videos

The monitor-captured video is recorded by the monitor. Some packages lose and some frames are broken due to various reason. The test video has 780 frames, and the frames between 79th and 94th frame are broken. The results by CAMSHIFT, TLD and the Circulant Structure model are presented in the Fig.4.1(a), Fig.4.1(b), Fig.4.1(c).Our model result as shown in Fig.4.1(d).

As the Fig.4.1 shown, CAMSHIFT cannot work on the video which has some broken frames.We could find that the TLD result is much larger than the target. Fig 4.1(c) shows that the algorithm cannot update the classifier perfectly. In our scheme, final recognition results are focusing on object with more suitable bounding even if some broken frames existing as shown.

## 4.3  Performance Testing with External Influence

The videos captured by the camera in moving cars are tested to observe the algorithm performance in different interference factors.

**Performance Test in Size Changing Condition.** For Clips1, there are 158 frames.The result of CAMSHIFT, TLD, the Circulant Structure and our model are represented in Fig. 4.2(a), Fig. 4.2(b), Fig. 4.2(c) and Fig. 4.2(d).

The labeled targets in CAMSHIFT are larger than reality. TLD detects targets a little bigger bounding box. Circulant Structure algorithm tracks one part patches of the target. Our model could improve the result by co-training scheme.

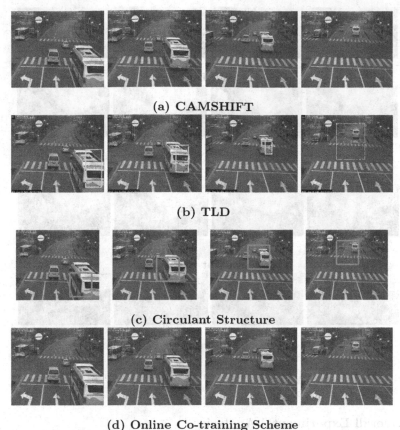

(a) CAMSHIFT

(b) TLD

(c) Circulant Structure

(d) Online Co-training Scheme

**Fig 4.1.** Experiment on the dataset of Monitor-Captured videos

**Performance Test in Rotation Condition.** In Clips2, there are 288 frames. The result of CAMSHIFT, TLD, the Circulant Structure and the Online Co-Training Scheme are represented in Fig.4.3(a), Fig.4.3(b), Fig.4.3(c) and Fig.4.3(d).

As the Fig.4.3 show, the CAMSHIFT cannot track the target when the camera was rotated. TLD and our model can mark the object correctly. However, the object detected by TLD is wider than the target. Our algorithm could successfully detect the rotated objects and its size is similar to original one. Since the size of target keeps as the first frame, the Circulant Structure algorithm can detect the target correctly and track it during the frame sequence.

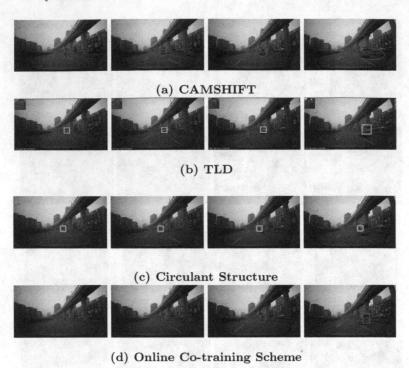

(a) CAMSHIFT

(b) TLD

(c) Circulant Structure

(d) Online Co-training Scheme

**Fig 4.2.** Experiment in the dataset of Clips2

## 4.4    Overall Experimental Analysis

The overall experiment results are shown in Table.2. For our model, the average precision or the average recall shows the best performance and the runtime is shown as the following table. For external effect brought from size change (clips1) or rotation (clips2), TLD also shows good performance in precision and recall, but it will take more time when it runs. Our model is obviously better than TLD when some frames are lost or broken, but it need more time for running program as the table shown. Additionaly, the result on clips2 indicates that the proposed algorithm is not affected by rotated videos. TLD, on the contrary, is slightly below its average level. The performance of the Circulant Structure algorithm are better than CAMSHIFT, but not good as our algorithm in each dataset because they cannot change the rebound box size when object is changed.

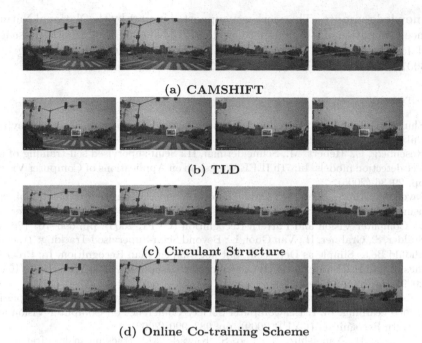

(a) CAMSHIFT

(b) TLD

(c) Circulant Structure

(d) Online Co-training Scheme

**Fig 4.3.** Experiment in the dataset of Clips3

**Table 2.** Overall Experiment results

| Test Video | Precision/Recall/Runtime | | | |
| | CAMSHIFT | TLD | Circulant Structure | Online TD Co-Training |
|---|---|---|---|---|
| Monitor-Captured videos | 0.54/0.47/147.3s | 0.62/0.93/319.0s | 0.63/0.78/457.8s | 0.91/0.93/699.3s |
| Clips1 | 0.83/0.79/39.3s | 0.93/0.88/70.1s | 0.84/0.77/112.5s | 0.95/0.98/127.3s |
| Clips2 | 0.74/0.58/51.2s | 0.86/0.82/101.3s | 0.94/0.94/179.6s | 1.00/0.97/213.4s |

## 5    Conclusion

In this paper, we proposed co-training scheme to integrate the traditional tracking algorithm with the detection to improve the object recognition performance. Correcting algorithm for broken frames is also designed to advance the reliability of proposed algorithm. However, the runtime of our model can be slower than other algorithm in complex environment. Comparing with the other tracking algorithms, our model has better performance.Future work mainly focus on optimizing the algorithm efficiency and choosing better model to make the algorithm fast.

**Acknowledgements.** This work was supported in part by the National Natural Science Foundation of China under Grants 61273304 and 61202170, the Research Fund for the Doctoral Program of Higher Education of China under grants 20130072130004.

# References

1. Zhu, X., Goldberg, A.: Introduction to semi-supervised learning. Morgan Claypool Publishers, USA (2009)
2. Rosenberg, C., Hebert, M., Schneiderman, H.: Semi-supervised self-training of object detection models. In: 7th IEEE Workshop on Applications of Computer Vision, pp. 29–36 (2005)
3. Javed, O., Ali, S., Shah, M.: Online detection and classification of moving objects using progressively improving detectors. In: IEEE Computer Society Conference on Computer Vision and Pattern Recognition (CVPR 2005), pp. 696–701 (2005)
4. Stalder, S., Grabner, H., Van Gool, L.: Beyond Semi-Supervised Tracking: Tracking Should Be as Simple as Detection, but not Simpler than Recognition. In: Proceedings of 12th IEEE International Conference on Computer Vision Workshops (ICCV 2009), pp. 1409–1416 (2009)
5. Babenko, B., Yang, M.H., Belongie, S.: Visual Tracking with Online Multiple Instance Learning. In: IEEE Computer Society Conference on Computer Vision and Pattern Recognition (CVPR 2009), pp. 983–990 (2009)
6. Li, Y., Ai, H., Yamashita, T., Lao, S., Kawade, M.: Tracking in low frame rate video: A cascade particle filter with discriminative observers of different life spans. In: 2007 IEEE Computer Society Conference on Computer Vision and Pattern Recognition (CVPR 2007), pp. 1–8 (2007)
7. Grabner, H., Bischof, H.: On-line boosting and vision. In: 2006 IEEE Computer Society Conference on Computer Vision and Pattern Recognition (CVPR 2006), pp. 260–267 (2006)
8. Avidan, S.: Ensemble tracking. IEEE Trans. Pattern Anal. Mach. Intell. 29(2), 261–271 (2007)
9. Yu, Q., Dinh, T.B., Medioni, G.: Online tracking and reacquisition using co-trained generative and discriminative trackers. In: Forsyth, D., Torr, P., Zisserman, A. (eds.) ECCV 2008, Part II. LNCS, vol. 5303, pp. 678–691. Springer, Heidelberg (2008)
10. Kalal, Z., Mikolajczyk, K., Matas, J.: Tracking-Learning-Detection. IEEE Trans. Pattern Anal. Mach. Intell. 34(7), 1409–1422 (2012)
11. Comaniciu, D., Ramesh, V., Meer, P.: Real-Time Tracking of Non-Rigid Objects using Mean Shift. In: Proceeding of IEEE Conference on Computer Vision and Pattern Recognition (CVPR 2000), pp. 142–149 (2000)
12. Lowe, D.G.: Distinctive Image Features from Scale-Invariant Key points. Int. J. Comput. Vision 60(2), 91–110 (2004)
13. Breiman, L.: Random Forests. J. Mach. Learn. 45(1), 5–32 (2001)
14. Sevilla-Lara, L., Learned-Miller, E.: Distribution fields for tracking. In: 2012 IEEE Conference on Computer Vision and Pattern Recognition (CVPR), pp. 1910–1917. IEEE (2012)
15. Henriques, J.F., Caseiro, R., Martins, P., Batista, J.: Exploiting the circulant structure of tracking-by-detection with kernels. In: Fitzgibbon, A., Lazebnik, S., Perona, P., Sato, Y., Schmid, C. (eds.) ECCV 2012, Part IV. LNCS, vol. 7575, pp. 702–715. Springer, Heidelberg (2012)

# An Approach for Automatically Reasoning Consistency of Domain-Specific Modelling Language

Tao Jiang, Xin Wang, and Li-Dong Huang

Yunnan Minzu University, 134 Yi Er Yi Avenue, Kunming, P.R. China
{jtzwyinternet73,wxkmyn}@163.com, lidonghuang@gmail.com

**Abstract.** Domain-Specific Modeling Language (DSML) defined by informal way cannot precisely represent its structural semantics, so properties of models such as consistency cannot be systematically analyzed and verified. In response, the paper proposes an approach for automatically reasoning consistency of DSML. Firstly, we establish a formal framework for DSML based on first-order logic; and then, an automatic mapping mechanism for formalizing DSML is defined; based on this, we present our method for verifying consistency of DSML and its models based on first-order logical inference; finally, the automatic mapping engine for formalizing DSML and its models is designed to show the feasibility of our formal method.

**Keywords:** Domain-Specific Modeling Language (DSML), structural semantics, consistency verification, automatic mapping.

## 1 Introduction

As a Model-Driven Development methodology for the specific domain, DSM [1] focuses on simplicity, practicability and flexibility. As a modeling language for DSM, DSMLs play an important role in system design and modeling of specific areas. Meta-language for building DSMLs is called Domain-specific Metamodeling Language (DSMML).

Semantics of DSML can be divided into structural semantics [2] and behavioral semantics. The former describes static semantic constraints between modeling elements, focusing on the static structural properties; the latter concerns analysis of execution semantics of domain models, focusing on the dynamic behavior. Although structural semantics is very important, research in structural semantics is not as extensive and deep as behavioral semantics'.

There are several problems that have not been solved well for DSML, which include mathematical description for metamodeling and model transformations, tool-independence formal specification, analysis techniques of properties of models based on formalization and automatic mapping mechanism and so on. Here are some examples that illustrate this. The precise structural semantics of the mature metaprogrammable Generic Modeling Environment (GME) [3] depend on the implementation of complex tools. Standards such as the UML superstructure [4] and Meta-Object

D. Miao et al. (Eds.): RSKT 2014, LNAI 8818, pp. 295–306, 2014.
DOI: 10.1007/978-3-319-11740-9_28 © Springer International Publishing Switzerland 2014

Facility (MOF) [5] do not provide sufficiently precise formal definitions of the DSML process.

The paper proposes an approach for automatically reasoning consistency of DSML based on first-order logical inference, based on this, design of corresponding formalization automatic mapping engine for DSML are introduced to show the application of formalization.

## 2    Related Works

Within the domain-specific language community, graph-theoretic formalisms have received the most research attention. The majority of work focuses on model transformations based on graph, but analysis and verification of properties of models has not received the same attention. In DSM domain, most of the DSMLs are defined and verified using informal way, for example, GME developed by Vanderbilt [3] university uses expanded OCL to check model, and MetaEdit+ of MetaCase Company [6] uses fixed declarative rules to verify model properties.

There are much typical work on formalization of modeling language, such as Andre's formalization and verification of UML class diagram based on ADT [7], Malcolm Shroff's formalization and verification of UML class diagram based on Z [8], and Jackson.E.K's formalization of DSML based on Horn logic [9] and so on. In these appoaches, formalization of metamodeling language has not been considered, and automatic translation mechanism from metamodels to the corresponding formal semantic domain has not been established too, so they have lower level of automated analysis. But using our method, we can implement automatically reasoning on consistency of DSML and its models by automatically mapping XML format DSML and its models to the corresponding first-order logic system based on our automatic mapping mechanism for formalizing DSML. This is very important for formalizing DSML and verifying models.

## 3    Formalization of XMML

We design a DSMML named extensible markup language based meta-modeling language (XMML) [10]. XMML is divided into four layers: metamodeling language layer used to define different DSMLs where XMML is located, DSML layer used to build concrete domain application models, domain application model layer used to make corresponding source codes of target system by code generator, and target application system layer [10].

We require that element of XMML is called metamodeling element and element of DSML built based on XMML is called domain modeling element and domain object built based on DSML is called domain model element. Among them, metamodeling element is also called metatype and type of model element is modeling element and type of modeling element is metatype [10].

Metamodeling elements of XMML are divided into two types: entity type and association type, the former is used to describe modeling entities in metamodel and the latter concerns relationships between modeling entities. Metamodeling element of entity type consists of four types such as model type, entity type, reference entity type and relationship type. Metamodeling element of association type includes the following five types: role assignment association, model containment relationship, attachment relationship, reference relationship, and refinement relationship. We finish formalization of structural semantics of XMML based on the above nine metatypes. Details of definition and formalization can be seen in [11].

Once set of predicate symbols denoting metamodeling elements and constraint axioms set denoting constraints over all metamodels are derived, we finish formalization XMML based on first-order logic. Based on this, we can create formalized system of XMML called $T_X$. Because we can prove logical consistency of $T_X$, XMML must have metamodels that can be satisfied, thus it is meaningful to discuss properties of metamodels built based on XMML.

# 4    A Formal Framework for Automatically Reasoning Consistency of DSML

## 4.1    A Framework for Formalizing DSML Based on First-order Logic

In our framework, structural semantics of DSML are characterized by a metamodel built based on XMML, so once a metamodel is formalized, we finish formalization of DSML.

A metamodel $M$ can be regarded as composition of the following five parts: a set $S$ of predicate symbols denoting modeling elements, an extended set $S_C$ of predicate symbols used to derive properties, constraint formulas set $F$ denoting constraints over all models, a set $O$ of constants denoting public properties, a set $\Omega$ of terms symbols denoting model elements used to build model. Among them, $S_C$ and $O$ may be empty, union of $S$, $S_C$ and $O$ form symbols set, and $F$ is defined by first-order predicate formulas based on symbols set. So a metamodel $M$ is defined as follows.

**Definition 1.** ($M$) a metamodel $M$ is a 5-tuple of the form $< S, S_C, \Omega, O, F >$ consisting of $S$, $S_C$, $\Omega$, $O$ and $F$.

Formalization of XMML differs from formalization of a metamodel. To the former, due to uniqueness of XMML, we can formalize XMML by artificial derivation and proving; to the latter, since both metamodels built based XMML and models built based on DSML are many and varied, an automatic mapping mechanism for formalizing any metamodel and its models has to be created. Reference to institution method for specification language in the literature [12], and combined with first-order logic theory [13], we establish a description framework for formalizing any metamodel and its models.

To build a metamodel formalized system based on first-order logic, the key is to establish a first-order language symbol set and a group of constraint axioms based on symbol set. We can derive a first-order language symbol set $\sum$ from abstract syntax of a metamodel, which consists of a set of constants and a set of predicate symbols, so the mapping from metamodels to $\sum$ called signature mapping $L$ has to be firstly established. Set of constraint axioms can be derived from structural semantics of a metamodel, which consists of type constraints of model elements and relationship between them and so on, thus we then establish the axiom mapping $A_\Sigma$ based on $\Sigma$. Upon completion of formalizing any metamodel, we can analyze logical consistency of itself. In addition, to verify consistency of models built based on any metamodel, the mapping from a model to a group of first-order logic statements based on $\Sigma$ called model mapping $S_\Sigma$ has to be established to determine whether a model as an interpretation of a metamodel formalized system satisfies a metamodel. So a framework for formalizing any metamodel consists of signature mapping $L$, axiom mapping $A_\Sigma$, model mapping $S_\Sigma$, metamodel consistency verification and determination of satisfaction of a model to a metamodel and so on. The architecture of the framework is shown in Fig 1.

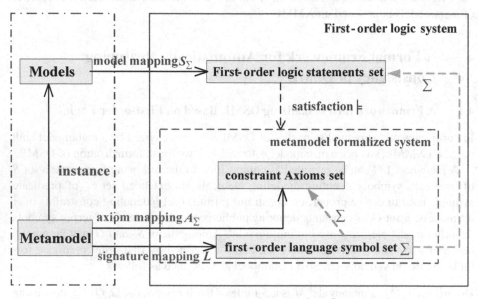

**Fig. 1.** A formal framework for metamodel based on first-order logic

Various semantic constraint relationships between modeling elements of entity type have to be analyzed to regard them as the basis for establishing mapping rules for formalizing any metamodel.

## 4.2    Automatic Mapping Mechanism for Formalizing Metamodels

We illustrate our formal mapping mechanism of metamodels based on first-order logic by software architecture metamodel $W_{SA}$ shown in Fig 2 as an example.

The metamodel $W_{SA}$ consists of modeling elements of entity type such as *SoftwareArchitecture*, *Component*, *Connection*, *Interface* and modeling elements of association type such as *AttchInfToCom* and *AttchInfToCon* denoting attachment relationship, *ComRefSA* denoting refinement relationship and *InfAssociation* denoting association relationship, denoting that software architecture consists of component and connection, and it also builds constraints over domain models that interfaces have to be attached to a component or a connection and components or connections can only be interconnected through interfaces and any component can be refined into a new software architecture model.

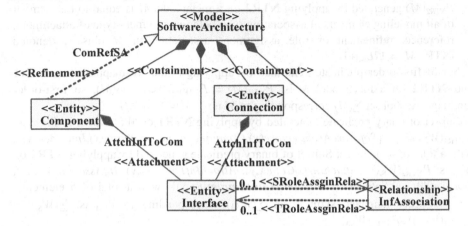

**Fig. 2.** Software architecture metamodel $W_{SA}$

Two modeling elements of attachment type and one modeling element of refinement type will be mapped directly by their name, so they are marked metamodel instance types called *AttchInfToCom*, *AttchInfToCon* and *ComRefSA* separately. But names of metamodel instance types of role assignment association and model containment relationship have not been marked because they need to be dynamically generated when they are mapped.

### 4.3    Signature Mapping

Via signature mapping, a group of predicates denoting element type and relationship between elements in the model can be derived as the basis of other mappings.

1.  Signature mapping rule **NTR1**(mapping of entity type): for every modeling element of entity type named *ME* and belonging to meta-type of model, entity, or reference entity in the metamodel, the corresponding unary predicate *ME(x)* is defined to represent that type of model element *x* is *ME*; the number of elements in unary predicate set $P_{NTR1}(M)$ generated by applying **NTR1** on a metamodel *M* is the number of all modeling elements of entity type belonging to meta-type of model, entity, or reference entity in *M*, denoted $|NTR_1(M)| = |EE_M|$;

2.  Signature mapping rule **NTR2**(mapping of association type): for each association of attachment, reference or refinement from entity type $ME_x$ to entity type $ME_y$ with *ML* as the association edge name,   a binary predicate *ML(x, y)* is defined to represent the relation between instance *x* of $ME_x$ and instance *y* of $ME_y$ is *ML*; for each role assignment association built by modeling element *ML* of relationship type, a binary predicate *ML(x, y)* is defined to represent that there exist explicit as-

sociation between entity type element $x$ at source side and entity type element $y$ at target side; for each modeling element $MC$ of model type, a binary predicate $MCContainment$ $(x, y)$ is defined to represent that entity type element $x$ is included in model type element $y$; the number of elements in binary predicate set $P_{NTR2}(M)$ generated by applying **NTR2** on a metamodel $M$ is equal to the number of all modeling elements of association type belonging to meta-type of attachment, reference, refinement or role assignment association in $M$ plus 1, denoted $|NTR_3(M)| = |RE_M|+1$.

Set of first-order predicate generated by applying Signature mapping rule **NTR1** and **NTR2** on a metamodel $M$ is: $P_{NTR}(M) = P_{NTR1}(M) \cup P_{NTR2}(M)$, so first-order language symbol set $\Sigma(M)$ corresponding to $M$ is: $\Sigma(M)= P_{NTR}(M)$.

Subset of unary predicate generated by applying **NTR1** on $W_{SA}$ shown in Fig 3 is: $P_{NTR1}(W_{SA})=$ {$SoftwareArchitecture(x),Component(x),Connection(x),Interface(x)$}, with a total of 4 elements; Subset of binary predicate generated by applying **NTR2** on $W_{SA}$ is: $P_{NTR2}(W_{SA})=\{AttachInfToCom(x,y), AttachInfToCon(x,y), InfAssociation(x,y),$ $ComRefSA(x,y),SoftwareArchitecture\textbf{Containment}(x,y)\}$, with a total of 5 elements. So first-order language symbol set $\Sigma(W_{SA})$ corresponding to $W_{SA}$ is: $\Sigma(W_{SA}) = P_{NTR1}(W_{SA}) \cup P_{NTR2}(W_{SA})$.

## 4.4    Axiom Mapping

Due to uniqueness of XMML, first-order language symbol set generated via $T_X$ is fixed, thus the corresponding constraint axioms set is unique too. In contrast, metamodels built based on XMML have many and they vary, and different metamodels contain different instance types and their structures differ from each other, thus their first-order language symbol sets $\Sigma(M)$ and constraint axioms generated via mapping vary, so constraint formulas in the axiom mapping rules is actually a group of axiom schema generated by extracting common constraint relationships to represent their structure forms, and constraint formulas corresponding to a metamodel have to be generated by replacing related symbols of axiom schema with its own and evolving related axiom schema according to structural differences.

Here we build axiom mapping rules from classification of entity type elements, constraint relation of association type elements and multiple constraint of association relation based on text concrete syntax of XMML [10].

1.  classification of entity type elements

XMML is one of typed metamodeling languages, and the metamodels built based XMML are well-typed. Correspondingly, DSML based XMML on is also one of typed modeling languages, so the models built based DSML are well-typed too. On the basis of the relevant literatures [12], we characterize typed constraints of DSML from completeness and uniqueness of classification of entity type elements.

(1)    Axiom mapping rule **ATR$_1$**(Completeness of Classification): Assume that $ME_1$, $ME_2$, ..., $ME_n$ are all all modeling elements of entity type belonging to metatype of model, entity or reference entity defined in one metamodel, a formula of classification completeness in the form of $\forall x.ME_1(x) \vee ME_2(x) \vee \cdots \vee ME_n(x)$ is defined; the number of elements in constraint axiom set $A_{ATR1}(M)$ generated by applying

**ATR₁** on a metamodel $M$ is 1, denoted $|ATR_1(M)| = 1$.

Completeness constraint axiom set containing only one element generated by applying **ATR₁** on $W_{SA}$ is: $A_{ATR1}(W_{SA})=\{$

$\forall x.$ *SoftwareArchitecture(x)* $\lor$ *Component(x)* $\lor$ *Connection* (x) $\lor$ *Interface* (x)$\}$

(2) Axiom mapping rule **ATR₂**(Uniqueness of Classification): Assume that $ME_1$, $ME_2$, ..., $ME_n$ are all all modeling elements of entity type belonging to metatype of model, entity or reference entity defined in one metamodel, a formula of classification uniqueness in the form of $\forall x. ME_i \rightarrow \neg ME_j$ is defined on any two different elements $ME_i$ and $ME_j$; the number of elements in constraint axiom set $A_{ATR2}(M)$ generated by applying **ATR₂** on a metamodel $M$ is equal to combination number produced by taking any two elements from the $EE_M$, denoted $|ATR_2(M)| = \dfrac{|EE_M| \times (|EE_M|-1)}{2}$, among them, $EE_M$ denotes a set constituted by all all modeling elements of entity type belonging to metatype of model, entity or reference entity defined in $M$.

Uniqueness constraint axiom set containing six elements generated by applying **ATR₂** on $W_{SA}$ is: $A_{ATR2}(W_{SA})= \{$

$\forall x.SoftwareArchitecture(x) \rightarrow \neg Component(x),$

$\forall x.SoftwareArchitecture(x) \rightarrow \neg Connection(x), \cdots$

$\forall x. Connection(x) \rightarrow \neg Interface(x)\}$, with a total of 6 elements.

**ATR₁** and **ATR₂** make it explicit that a model as an instance of any metamodel must have its all model elements of entity type completely and uniquely classified by the metamodel instance type.

2. constraint relation of association type elements

(1) Axiom mapping rule **ATR₃**(constraint of implicit association type): for every modeling element named $MC$ and belonging to meta-type of attachment relationship in the metamodel, the corresponding binary predicate $MC(x, y)$ is derived, among them, $x$ represents an instance of included modeling element and $y$ or $z$ represents an instance of host modeling element; if modeling elements of entity type that are included in modeling element $ME$ by attachment relationship $MC$ are $NE_1$, $NE_2$, ..., $NE_n$, a formula of type constraint in the form of $\forall x,y.MC(x,y) \rightarrow (NE_1(x) \lor NE_2(x) \lor \cdots \lor NE_n(x)) \land ME(y)$ and another formula of semantic constraint in the form of $\forall x,y,z.MC(x,y) \land MC(x,z) \rightarrow (y=z)$ are defined. For every modeling element named $MR$ and belonging to meta-type of reference or refinement relationship from modeling element of entity $ME_1$ to $ME_2$ in the metamodel, the corresponding binary predicate $MR(x, y)$ is derived, among them, $x$ represents an instance of $ME_1$ and $y$ represents an instance of $ME_2$, a formula of type constraint in the form of $\forall x,y.MR(x,y) \rightarrow ME_1(x) \land ME_2(y)$ and another formula of semantic constraint in the form of $\forall x,y,z.MR(x,y) \land MR(x,z) \rightarrow (y=z)$ are defined. For all modeling element belonging to meta-type of model containment relationship, constraint formulas generated is similar to the fomer; the number of elements in constraint axiom set of implicit association $A_{ATR3}(M)$ generated by applying **ATR₃** on a metamodel $M$ is equal to the number $CE_M$ of all modeling elements belonging to meta-type of implicit association

except for model containment in $M$ multiplied by 2 plus 2, denoted $|ATR_5(M)| = |CE_M| \times 2 + 2$.

Implicit association constraint axiom set generated by applying $\mathbf{ATR_3}$ on $W_{SA}$ is: $A_{ATR3}(W_{SA}) = \{ \ \forall x,y.\ AttachInfToCom(x,y) \rightarrow Interface(x) \land Component(y),..., \ \forall x,y,z.\ SoftwareArchitectureContain\text{-}ment(x,y) \land SoftwareArchitectureContainment(x,z) \rightarrow (y=z) \}$, with a total of 8 elements.

(2) Axiom mapping rule $\mathbf{ATR_4}$(constraint of explicit association type): For every role assignment association between modeling elements of entity type established by modeling element of relationship type $ML$ in the metamodel, the corresponding binary predicate $ML(x, y)$ is derived to represent an association from entity type element $x$ at source side to entity type element $y$ at target side; if modeling elements of entity type at source side that are connected to $ML$ are $SE_1$, $SE_2$, ..., $SE_n$ and modeling elements of entity type at target side that are connected to $ML$ are $TE_1$, $TE_2$, ..., $TE_m$, a formula of type constraint of role assignment association in the form of $\forall x,y.\ ML(x,y) \rightarrow (SE_1(x) \lor SE_2(x) \lor \cdots \lor SE_n(x)) \land (TE_1(y) \lor TE_2(y) \lor \cdots \lor TE_n(y))$ is defined; the number of elements in constraint axiom set of role assignment association $A_{ATR4}(M)$ generated by applying $\mathbf{ATR_4}$ on a metamodel $M$ is equal to the number $LE_M$ of all modeling elements belonging to meta-type of relationship type in $M$ used to establish explicit association, denoted $|ATR_4(M)| = |LE_M|$.

Role assignment association constraint axiom set generated by applying $\mathbf{ATR_4}$ on $W_{SA}$ is: $A_{ATR4}(W_{SA}) = \{ \ \forall x,y. InfAssociation(x,y) \rightarrow Interface(x) \land Interface(y) \}$

3. multiple constraint of association relation

It is pointed out in UML documents that multiplicity at the endpoints of association constrains the number of instances at the other end. Multiple concept of association relation in XMML is similar to UML's.

(1) Axiom mapping rule $\mathbf{ATR_5}$(multiple constraint of instances at the source end): For every role assignment association from entity type element at source side to entity type element at target side established by modeling element of relationship type $ML$ in the metamodel, the corresponding binary predicate $ML(x, y)$ is derived to represent an association from entity type element $x$ at source side to entity type element $y$ at target side; assume that $Smul$ is multiplicity value at the source end set in $ML$, different forms of axiom set is derived according to various value.

(a) If $SMul = 0..1$: $\forall x,y,z.ML(y,x) \land ML(z,x) \rightarrow (y=z)$

(b) If $SMul = 1$: $\exists x,y.\ ML(x,y)$ , $\forall x,y,z.ML(y,x) \land ML(z,x) \rightarrow (y=z)$

(c) If $SMul = 1..*$: $\exists x,y.\ ML(x,y)$

(d) If $SMul = 2..*$: $\exists\, x,y,z.\ ML(y,x) \land ML(z,x) \land (y \neq z)$

(e) If $SMul = 0..2$: $\forall x,y,z,u.ML(y,x) \land ML(z,x) \land ML(u,x) \rightarrow (y=z) \lor (y=u) \lor (u=z)$

(f) If $SMul = 2$: $\exists\, x,y,z.\ ML(y,x) \land ML(z,x) \land (y \neq z)$
$\forall x,y,z,u.ML(y,x) \land ML(z,x) \land ML(u,x) \rightarrow (y=z) \lor (y=u) \lor (u=z)$

Axiom set of multiple constraint of instances at the source end generated by applying $\mathbf{ATR_5}$ on $W_{SA}$ is: $A_{ATR5}(W_{SA}) = \{$
$\forall x,y,z.\ InfAssociation(y,x) \land InfAssociation(z,x) \rightarrow (y=z) \}$

(2) Axiom mapping rule $\mathbf{ATR_6}$(multiple constraint of instances at the target end):

Similar to the former rule **ATR$_5$**, it is omitted here.

The number of elements in axiom set of multiple constraint of association relation $A_{ATR5-6}(M)$ generated by applying **ATR$_5$** and **ATR$_6$** on a metamodel $M$ is: $|ATR_{5-6}(M)|$ = $|MV_M|+|SingleV_M|\times2$, among them, $MV_M$ denotes a set constituted by all multiplicity at source end and multiplicity at target end which are not single values and are established by modeling element of relationship type in $M$, $SingleV_M$ denotes a set constituted by all multiplicity at source end and multiplicity at target end which are single values in $M$.

Constraint axioms set $A_\Sigma(M)$ based on first-order language symbol set $\sum(M)$ corresponding to $M$ is union of the above six constraint axioms subsets, denoted $A_\Sigma(M)$

$$=\bigcup_{i=1}^{6} A_{ATR_i}(M).$$

## 4.5    Consistency Reasoning of Metamodel and Its Models

First-order predicate set $P_{NTR}(M)$ generated by applying signature mapping rule **NTR1** and **NTR2** on a metamodel $M$ as a group of predicate symbols and constraint axioms set $A_\Sigma(M)$ based on $P_{NTR}(M)$ generated by applying axiom mapping rule **ATR$_1$-ATR$_6$** on a metamodel $M$ as a group of constraint axioms are all added to first-order logic formalized system called $Q$ predicate calculus [13] to form formalized system of metamodel $M$ based on $Q$ – metamodel formalized system called $T_Q(M)$. Reference to the definition of $T_X$ in the literature [14], $T_Q(M)$ is defined as follows.

**Definition 2.** $(T_Q(M))$ $T_Q(M)$ consists of formal language $L(M)$ and deduction structure defined based on $L(M)$. Except some modification of symbols set $S_Q(M)$ and adding of a set of constraints  axioms, $T_Q(M)$ is entirely based on predicate calculus Q. Its symbols set $S_Q(M)$ is union of logical link symbol set $L_Q(M)$ and collection of individual variables $V_Q(M)$ and Individual constant symbol set $C_Q(M)$ and predicate symbols set $P_Q(M)$, i.e. $S_Q(M)= L_Q(M) \cup V_Q(M) \cup C_Q(M) \cup P_Q(M)$, among them, $L_Q(M)$ and $V_Q(M)$ are same as Q's, $C_Q(M)=\Phi, P_Q(M) = P_{NTR}(M)$. Union of $C_Q(M)$ and $P_Q(M)$ is called set of non-logical symbols $N_L(M)$, i.e. $N_L(M)= C_Q(M) \cup P_Q(M)$. Axioms set $A(M)$ of $T_Q(M)$ is composed of logical axioms set $A_L(M)$ of Q and constraint axioms set $A_\Sigma(M)$, i.e. $A(M)= A_L(M) \cup A_\Sigma(M)$.

The semantic interpretation of $T_Q(M)$ is a model built based on metamodel $M$, universe of discourse of interpretation is the set of all entity model elements and constants contained in the model.

Once metamodel and model are formalized based on first-order logic, we can imple-ment analysis and verification of properties such as logical consistency of meta-model and model based on first-order logical inference.

It is impossible to find a true interpretation for constraint axiom set $A_\Sigma(M)$ of every $T_Q(M)$ in order to prove semantic consistency of $T_Q(M)$, so we can only prove logical consistency of different constraint axiom $A_\Sigma(M)$ based on automatic theorem prover. Reference to the literature [13], the following definition is available.

**Definition 3.** (logical consistency of metamodel). A metamodel is logically consistent in the domain iff the constraint axiom set $A_\Sigma(M)$ of metamodel formalized system called $T_Q(M)$ is proved to be logically consistent in the automatic theorem prover; a metamodel is logically inconsistent in the domain iff the constraint axiom set $A_\Sigma(M)$ of metamodel formalized system called $T_Q(M)$ is proved to be contradictory in the automatic theorem prover, denoted $A_\Sigma(M) \vdash False$.

Taking predicate symbol set $P_{NTR}(W_{SA})$ generated via signature mapping on $W_{SA}$ as predicate definition part of automatic theorem prover SPASS [23], and taking constraint axiom set $A_\Sigma(W_{SA})$ generated via axiom mapping on $W_{SA}$ as axiom part of SPASS, we find that there is no contradiction in $A_\Sigma(W_{SA})$ by proving based on SPASS, so $W_{SA}$ is logically consistent.

If metamodel formalized system called $T_Q(M)$ is proved to logically consistent, then metamodel $M$ must have a interpretation that can be satisfied, so the domain based on metamodel $M$ is not empty, thus it is meaningful to discuss other properties based on metamodel $M$ in the domain. From the point of view of formalization, a legal model is an interpretation that satisfies all constraint formulas of $A\Sigma(M)$, so the relationship that model satisfies metamodel is equivalent to the relationship that the interpretation of $T_Q(M)$ satisfies $T_Q(M)$. By equivalence of satisfaction relationship and logical consistency, reference to the literature [12], we can get determination method of model consistency.

**Inference 1 (Logical consistency of model).** If union of constraint axiom set $A_\Sigma(M)$ of $T_Q(M)$ and set of first-order predicate statements $T_L(S)$ generated via model $S$ is logically consistent, then the model $S$ is consistent; instead, if union of constraint axiom set $A_\Sigma(M)$ of $T_Q(M)$ and set of first-order predicate statements $T_L(S)$ generated via model $S$ is logically inconsistent, denoted $A_\Sigma(M) \cup T_L(S) \vdash False$, then the model $S$ is inconsistent.

# 5     Design of *LMapMSS*

Based on signature mapping, axiom mapping and model mapping, reference to the literature [12], formalization automatic mapping engine for model and metamodel called *LMapMSS* (*Logic Mapping of Metamodels and models Based on structural semantics*) is designed and implemented to finish automatic translation from XML format model and metamodel built based on XMML concrete syntax scheme to the corresponding first-order logic system in SPASS [15] format, thus we can realize automatic process of verification of consistency of metamodel and its model.

*LMapMSS* consists of formalization automatic mapping module for metamodel called *MapMBD* (*Mapping of Metamodel Based on Domain*) and formalization automatic mapping module for model called *MapSBD* (*Mapping of inStances Based on Domain*). *MapMBD* implements formalization of any metamodel via signature mapping and axiom mapping to finish verification of logical consistency of $T_Q(M)$. *MapSBD* implements formalization of any model via model mapping based on formalization of metamodel to finish verification of logical consistency of model.

Based on .net 2.0 platform, by using C#.net as development language, we implement the corresponding prototype system for *MapMBD* and *MapSBD* and integrate them in the modeling environment of Archware, thus it becomes possible for Archware to validate metamodel and model built based on it. Running interface of *MapMBD* is shown in Fig 3, its left window shows XML format document of metamodel produced by *Archware* and the corresponding first-order logic system in SPASS format generated by translation of *MapMBD* is showed in right window.

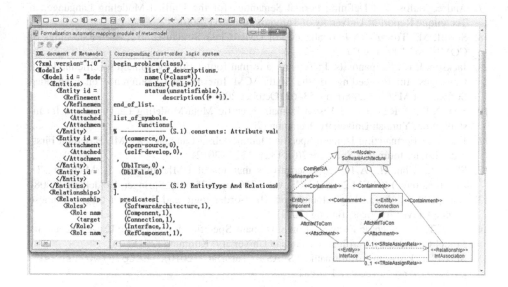

**Fig. 3.** Running interface of *LMapMSS*

## 6    Conclusions

DSML defined in the informal way cannot precisely describe its structural semantics, which makes it difficult to systematically reason on properties of its models such as consistency. In response, the paper proposes an approach for automatically reasoning consistency of DSML, and then we illustrate our approach by a classic case, based on this, design and implementation of corresponding automatic mapping engine for formalizing DSML and its models are introduced to show the feasibility of our method.

**Acknowledgements.** This work was supported by the National Natural Science Foundation of China under Grant No. 61363022.

## References

1. Steven, K., Juha-Pekka, T.: Domain-specific modeling: Enabling full code genera-tion. John Wiley & Sons, New Jersey (2008)
2. Jackson, E.K., Sztipanovits, J.: Formalizing the Structural Semantics of Domain-Specific Modeling Languages. Journal of Software and Systems Modeling 8(4), 451–478 (2008)

3. ISIS Vanderbilt University, GME Users Manual (2013),
   http://www.isis.vanderbilt.edu/Projects/gme/
4. Object Management Group, Unified Modeling Language: Superstructure version 2.4.1
   (2011), http://www.omg.org/spec/UML/2.4.1/Superstructure/PDF
5. Object Management Group, Meta Object Facility Specification version 2.4.1 (2011),
   http://www.omg.org/spec/MOF/2.4.1
6. MetaCase, MetaEdit+ Version 5.0 User's Guide (2014),
   http://www.metacase.com/support/50/manuals/meplus/Mp.html
7. Andreopoulos, W.: Defining Formal Semantics for the Unified Modeling Language, in
   Technique Report of University of Toronto, Toronto (2000)
8. Shroff, M.: Towards A Formalization of UML Class Structures in Z. In: Proceedings of
   COMPSAC 1997 (1997)
9. Jackson, E.K., Sztipanovits, J.: Towards a formal foundation for domain specific modeling
   languages. In: Proceedings of the Sixth ACM International Conference on Embedded
   Software (EMSOFT 2006), pp. 53–62 (October 2006)
10. Sun, X.P.: A Research of Visual Domain-Specific Meta-Modeling Language and Its In-
    stantiation, Yunnan University, Kunming (2010)
11. Jiang, T.: Formalizing Domain-Specific Metamodeling Language XMML Based on First-
    order Logic. Journal of Software 7(6), 1321–1328 (2012)
12. Shan, L., Zhu, H.: A formal descriptive semantics of UML. In: Liu, S., Maibaum, T.,
    Araki, K. (eds.) ICFEM 2008. LNCS, vol. 5256, pp. 375–396. Springer, Heidelberg (2008)
13. Cheng, M.Z., Yu, J.: Logic foundation—first-order logic and first-order theory. Chinese
    People University Press, Beijing (2003)
14. Jiang, T.: Research on Formalization of Domain-Specific Metamodeling Lan-guage and Its
    Model Consistency Verification, Yunnan University, Kunming (2011)
15. Max-Planck-Institut Informatik, SPASS Tutorial (2010), http://www.spass-
    prover.org/tutorial.html

# Knowledge Technology

# Online Object Tracking via Collaborative Model within the Cascaded Feedback Framework

Sheng Tian and Zhihua Wei

College of Electronic and Information Engineering, Tongji University

**Abstract.** Generative and discriminative models are commonly used in object tracking algorithms. However, the limitation of using these models lies in the fact proved by a large number of experiments that a single model is easily influenced by external factors, such as occlusion and illumination variation. To address this issue, in this paper based on a collaborative model within the cascaded feedback framework, we propose an online object tracking algorithm where an adaptive generative model has been developed which can adapt to the dynamic background. Experimentally, we show that our algorithm is able to outperform the state-of-the-art trackers on the various benchmark videos.

**Keywords:** Collaborative model, tracking, cascaded feedback.

## 1 Introduction

Object tracking plays a critical role in many vision applications such as video surveillance, human-machine interaction and robot perception [1]. In recent years much progress has been made and some assumptions about the target (e.g., restricted fast movement between two frames) can be allowed, however, tracking arbitrary object in a dynamic background is still a challenging problem. It is hard to develop a robust and efficient algorithm to deal with all factors, such as large appearance change caused by illumination variation, occlusion, perspective transformation and fast movement.

Given the initial state of a target object in a frame, the tracker can derive the parameters of the model via the prior knowledge about the target. In the next frame, generative and discriminative models can be used to predict the position of the target object. To be more specific, generative models can be used in appearance modeling for the object and make predictions using reconstruction error. An object can be represented by various features such as Haar-like features [2,9,10], holistic or local histograms [6,12], intensity [5], texture [17] or optical flow [18]. Some methods (e.g., PCA [5], integration or selection [2,4]) can be applied to improve the robustness against occlusion or illumination variation. In this paper, considering simplicity and efficiency, we adopt intensity values for representation. Experiments show that the simple feature can also have a good effect.

Compared with appearance models which can be used to find the most similar region of the target object, discriminative methods treat tracking as binary classification problem which aims at distinguishing the target object from the

D. Miao et al. (Eds.): RSKT 2014, LNAI 8818, pp. 309–319, 2014.
DOI: 10.1007/978-3-319-11740-9_29 © Springer International Publishing Switzerland 2014

background [2,8,9,10,11,17]. Furthermore, generative and discriminative models have been integrated in several frameworks (e.g., co-training [16], cascade [14], TLD [21]).

In this paper, we propose a simple but effective framework in which we can firstly exploit a discriminative model to detect the most possible region, then use a generative model to make local correction. The generative model built in this framework can adaptively modify its appearance weight to adapt to the surrounding background using the confidence value generated by the discriminative model. For the discriminative model established in the framework, we can train it online to adapt to appearance variations of the target object and the background. Experimental studies on various challenging videos show that the proposed algorithm can outperform the state-of-the-art trackers.

The rest of the paper is organized as follows. In Section 2, we review the relevant models and explain the origin of our idea; in Section 3, we introduce our tracking framework and two models we used; in Section 4, we present qualitative and quantitative results on a number of challenging videos. We conclude in Section 5.

## 2    Related Work

Sparse representation has aroused widespread concern in the field of vision problems [3,13,15,19]. Despite this success, there is a major limitation of real-time tracking - high computational cost. The appearance models based on sparse representation usually have to generate a lot of sparse vectors from trivial patches in target object region, which limits the search scope. In [19], particle filter is adopted to search potential position. Unfortunately, it is hard to balance the number of the particles, which means redundant particles increase computational time, but on the other hand, small particles cause drift. In [3], the target object is located by mean-shift. However, one potential disadvantage of this approach is the phenomenon that object location is easy to drift. In addition, occlusion is also one of the most challenging problems. Commonly, the tracker can update its model in a small rate to deal with occlusion. Zhong et al. [19] propose a weighted histogram which ignores the elements with large reconstruction error. In this way the approach may ignore some important elements, which can lead to a poor and unstable appearance representation. In light of the above analysis, we propose an effective method with the ability to dynamically update appearance structure.

Although discriminative methods can distinguish the target from the background, the positive samples still have a certain ambiguity which can result in imprecise location. Babenko et al. [2] try to treat the positive samples as a bag. However, the proposed algorithm weakens positive samples, so that similar positive samples cannot be properly classified. Therefore, an intuitive idea occurs to combine a discriminative model with an appearance model. Recently, Wu et al. [20] evaluate some established online tracking algorithms. They show in this evaluation that, a discriminative model, named CSK [8], is proved to have a very high speed of detecting and a continuous output structure. Due to these factors, we try to integrate it with an appearance model. Moreover, several experiments are designed to show that the integration can improve tracking accuracy.

# 3    Proposed Algorithm

In this section, we present the new algorithm in details. First, our framework will be introduced so that under this new framework the generative model and discriminative model can influence each other to generate a better output. Next, two models adopted in the subsequent articles will be described.

## 3.1    Framework

Using CSK and Kernel Sparsity-based Appearance Model(KSAM, which will be established in the third part of this section) we propose a collaborative model within the cascaded feedback framework (See Fig. 1). The process should start with marking the target in the first frame. To each new frame, the CSK tracker firstly generates a hot region, the region with high probability of the target object, by cropping out an image patch centered in the position which has been estimated in the previous frame. Given the feedback from CSK, KSAM will be activated to find the most similar candidate within the hot region. If the maximum likelihood generated by KSAM is less than the threshold , estimated position is the same one derived from CSK. At last, CSK and KSAM are updated in this new estimated position.

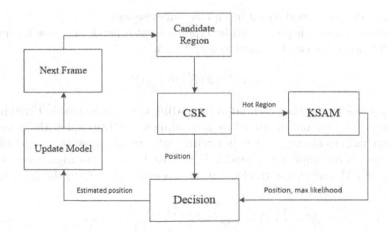

**Fig. 1.** The collaborative model within the cascaded feedback framework

## 3.2  Discriminative Model (CSK)

The CSK tracker exploits circulant matrix to train the classifier with all samples collected around the estimated position of the target. The key for its outstanding speed is that it uses Fast Fourier Transform (FFT) which converts convolutional operations to element-wise operations, and it can compute all the responses simultaneously to generate a confidence map. Here we provide a brief overview of this approach [8].

A classifier can be trained by estimating the parameters that minimize the cost function. CSK makes use of Regularized Least Squares (RLS) in Eq. 1

$$\min_{w} \sum_{i=1}^{m} \|y - <\phi(x_i), w>\|^2 + \lambda <w, w> \tag{1}$$

which has a simple closed-form solution in Eq. 2,

$$\alpha = (K + \lambda I)^{-1} y \tag{2}$$

where $\phi$ is the mapping to the Hilbert space induced by the kernel $K$ in Eq.2, $\lambda$ controls the amount of regularization, $K$ is the kernel matrix, and the solution $w$ is implicitly represented by the vector $\alpha$.

In consideration of all cyclic shifts in a gray-scale image patch centered on the target, the kernel matrix can be represented as a circulant structure if it is a unitarily invariant kernel. Then we can obtain a new solution by using Fourier Transform:

$$\alpha = F^{-1}\left(\frac{F(y)}{F(K) + \lambda}\right) \tag{3}$$

where the division is performed in an element-wise way.

The detection step is performed by cropping out a patch in a new frame, and the confidence value can be calculated as Eq. 4.

$$y = F^{-1}\left(F(K) \odot F(\alpha)\right) \tag{4}$$

The target position is then estimated by finding the translation that maximized the score $y$. Since the output structure is continuous, we can exploit the generated confidence map to obtain a hot region with high probability of the target object, as the input of our appearance model. For more details of the algorithm, see the reference [8]. Moreover, the tracker can be trained online using the function (5) in [7]

$$A^p = F(\alpha) = \frac{\sum_{i=1}^{p} \beta_i F(K^i) F(y^i)}{\sum_{i=1}^{p} \beta_i F(K^i)\left(F(K^i) + \lambda\right)} \tag{5}$$

where all extracted appearances of the target from the first frame till the current frame p are taken into consideration, the weights $\beta_i$ are set using a learning rate parameter $\gamma$. The total model and the object appearance are updated using (6)

$$A^p = A^p_N / A^p_D$$

$$A^p_N = (1 - \gamma)\, A^{p-1}_N + \gamma F(y^p) F(K^p_N) \tag{6a}$$

$$A^p_D = (1 - \gamma)\, A^{p-1}_D + \gamma F(K^p)\, (F(K^p) + \lambda) \tag{6b}$$

$$x^p = (1 - \gamma)\, x^{p-1} + \gamma x^p. \tag{6c}$$

Here the numerator $A^p_N$ and denominator $A^p_D$ are updated separately.

## 3.3   Generative Model (KSAM)

Motivated by the success of sparse coding for image classification [13,15] and object tracking [3,19], we propose a generative model based on sparse coding and exploit a dynamic Gaussian kernel to modify the weight of coefficient vectors for the adaptation to the surrounding background. We call it Kernel Sparsity-based Appearance Model (KSAM).

One image patch can be split into several trivial patches by sliding windows in the normalized image. By Eq. 7, we can get a set of sparse coefficient vector $\alpha$

$$\min_{a_i} \|x_i - D\alpha_i\|_2^2 + \lambda \|\alpha_i\|_1 \tag{7}$$

where the dictionary $D$ is generated from $k$-means cluster centers which consist of the most representative pattern of the target object [19]. Considering the spatial information of the image patch, we can write the sparse coefficient vector in matrix form:

$$A = \begin{pmatrix} \alpha_{11} & \cdots & \alpha_{1n} \\ \vdots & \ddots & \vdots \\ \alpha_{m1} & \cdots & \alpha_{mn} \end{pmatrix} \tag{8}$$

where $m$ and $n$ are determined by the size of sliding window and image patch. Matrix $A$ can be changed into its linear form by a learning rate $\lambda$:

$$A^p = (1 - \lambda)\, A^{p-1} + \lambda A^p \tag{9}$$

The reconstruction error can be calculated in Eq. 10.

$$rerr^p = \sum_{r,c} \left\| x^p_{r,c} - D\alpha^{p-1}_{r,c} \right\|^2 \tag{10}$$

were $(r, c) \in \{1...m\} \times \{1...n\}$.

Typically, the rectangle representation of the target object can lead to the ambiguity between the object and the background (See Fig. 2). Fortunately, it is not a big question to discriminative models because they can distinguish the object and the background by positive and negative samples. However, the ambiguity has a negative effect on appearance model, especially when occlusion occurs.

In order to solve the problem of the ambiguity, we try to apply a Gaussian kernel. Further, to avoid losing the information of the initial object, we extend (10) by a similar function (See Eq. 11)

**Fig. 2.** The ambiguity between the object and the background

$$rerr^p_{new} = \sum_{r,c} K(r,c) \left\| x^p_{r,c} - D\alpha^{p-1}_{r,c} \right\|^2 \left(1 - sim\left(\alpha^p_{r,c}, \alpha^1_{r,c}\right)\right). \qquad (11)$$

$$K(r,c) = exp\left(-\frac{1}{2h^2}\left((r-cen)^2 + (c-cen)^2\right)\right) \qquad (12)$$

$$sim(\alpha,\beta) = \frac{\sum_i min(a_i, b_i)}{\sum_i b_i} \qquad (13)$$

Here $cen$ in (12) is the center of the image patch, $h$ is the bandwidth of the Gaussian kernel. $a_i$ and $b_i$ in (13) are the coefficient elements of the vector $\alpha$ and $\beta$.

Moreover, according to the max confidence value generated by CSK, the information whether the target object can be easy to be distinguished with the background will be obtained. For doing this, we modify the bandwidth $h$ in Eq. 12 to change the weight of the object appearance (See Eq.14)

$$h = \sigma \times max\_response + base. \qquad (14)$$

Here $base$ is the benchmark value determined by the initial state of the target object, and $\sigma$ is the weighing coefficient which controls the change rate of $h$. Max response is derived by CSK.

Given the reconstruction error of the candidate, the likelihood can be calculated by:

$$likelihood_{candidate} = exp\left(-\frac{rerr^2_{candidate}}{2\sigma^2}\right) \qquad (15)$$

## 4   Experiments and Discussion

### 4.1   Experiment Setting

In order to evaluate the performance of our algorithm, we make experiments on some challenging image sequences listed in Table 1. These sequences cover most challenging problems in object tracking: abrupt motion, deformation, illumination variation,background cluster etc. For comparison, we select seven

**Table 1.** List of the test sequences (IV: Illumination Variation, SV: Scale Variation, OCC: Occlusion, DEF: Deformation, MB: Motion Blur, FM: Fast Motion, IPR: In-Plane Rotation, OPR: Out-of-Plane Rotation, BC: Background Clutters)

| Sequence | Frame Number | Attribute |
|----------|--------------|-----------|
| bolt | 350 | OCC, DEF, IPR, OPR |
| deer | 71 | MB, FM, IPR, BC |
| david | 350 | OCC, DEF,OPR,BC |
| faceocc | 892 | OCC |
| tiger | 354 | IV, OCC, DEF, MB, FM, IPR, OPR |
| soccer | 392 | IV, SV, OCC, MB, FM, IPR, OPR, BC |
| skating | 400 | IV, SV, OCC, DEF, OPR, BC |
| coke | 291 | IV, OCC, FM, IPR, OPR, BC |

**Table 2.** Precision plot based on [20] (threshold = 20 pixels)

| Sequence | MIL | SCM | CSK | IVT | LSK | Struck | TLD | Our |
|----------|-----|-----|-----|-----|-----|--------|-----|-----|
| bolt | 0.014 | 0.031 | 0.034 | 0.014 | 0.977 | 0.02 | 0.306 | 0.597 |
| deer | 0.127 | 0.028 | 1 | 0.028 | 0.338 | 1 | 0.732 | 1 |
| david | 0.738 | 0.496 | 0.659 | 0.754 | 0.476 | 0.337 | 0.111 | 1 |
| faceocc | 0.221 | 0.933 | 0.947 | 0.645 | 0.122 | 0.575 | 0.203 | 0.872 |
| tiger | 0.095 | 0.126 | 0.255 | 0.080 | 0.418 | 0.175 | 0.446 | 0.338 |
| soccer | 0.191 | 0.268 | 0.135 | 0.174 | 0.120 | 0.253 | 0.115 | 0.370 |
| skating | 0.130 | 0.768 | 0.988 | 0.108 | 0.698 | 0.465 | 0.318 | 0.873 |
| coke | 0.151 | 0.430 | 0.873 | 0.131 | 0.258 | 0.949 | 0.684 | 0.787 |

state-of-the-art algorithms with the same initial position of the target. These algorithms include the MIL [2], SCM [19], CSK [8], IVT [5], LSK [3], Struck [11] and TLD [21]. We present some representative results in the next part of this section, together with the analysis of the experimental results respectively from the perspectives of qualitative and quantitative.

Values of the parameters involved are presented as follows. The variable $\lambda$ in Eq. 5 is fixed to be 0.01 and the learning rate $\gamma$ in Eq. 6 is fixed to be 0.075. The size of sliding window for KSAM is $6 \times 6$, and the target and candidate object are resized and normalized to $32 \times 32$. The row number and column number of dictionary D in Eq. 11 are 36 and 50. The variable $\lambda$ in Eq. 9 and $\sigma$ in Eq. 11 are fixed to be 0.025 and 0.8. The variable $\sigma$ in Eq. 14 is fixed to be 2, but the variable *base* in Eq. 13 ranges from 2 to 4.

## 4.2 Experiment Results

We evaluate the above-mentioned algorithms using the precision plot as well as the success plot described in [20]. The results are shown in Table 2 (using the score for the threshold = 20 pixels) and Table 3 (using the success rate value for the threshold = 0.2). Figure 3 shows the precision plot in various situations.

**Table 3.** Precision plot based on [20] (threshold = 20 pixels)

| Sequence | MIL | SCM | CSK | IVT | LSK | Struck | TLD | Our |
|---|---|---|---|---|---|---|---|---|
| bolt | 0.014 | 0.027 | 0.034 | 0.014 | 1 | 0.020 | 0.306 | 0.603 |
| deer | 0.225 | 0.155 | 1 | 0.042 | 0.352 | 1 | 0.761 | 1 |
| david | 0.738 | 0.500 | 0.675 | 0.746 | 0.528 | 0.353 | 0.115 | 1 |
| faceocc | 1 | 1 | 1 | 1 | 1 | 1 | 0.952 | 1 |
| tiger | 0.232 | 0.261 | 0.456 | 0.132 | 0.948 | 0.244 | 0.593 | 0.607 |
| soccer | 0.250 | 0.398 | 0.214 | 0.237 | 0.122 | 0.281 | 0.176 | 0.582 |
| skating | 0.238 | 1 | 0.888 | 0.113 | 0.710 | 0.550 | 0.248 | 0.873 |
| coke | 0.385 | 0.574 | 0.969 | 0.155 | 0.371 | 0.959 | 0.804 | 0.873 |

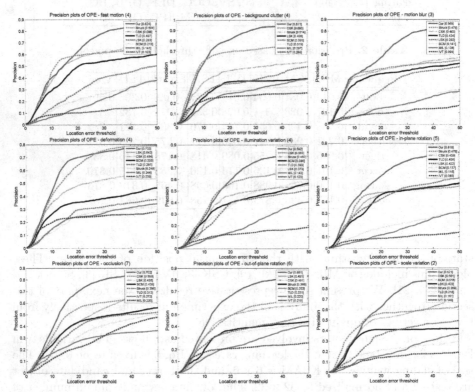

**Fig. 3.** The precision plot in various situations. A comparative study of the proposed algorithm with other seven state-of-the-art methods on challenging sequences.

## 4.3 Experiment Analysis

In terms of the experimental results, our algorithm exhibits an excellent performance. For our tracker based on CSK, the results of our algorithm for bolt, david, tiger and soccer are better than the result of CSK. It is noted, however, that for coke, skat-ing and faceocc, our tracker is not as good as CSK because the appearance model fails to update the appearance correctly for big occlusion. Figure 3 shows some tracking results on sequences.

(a) Tracking results on sequence bolt.

(b) Tracking results on sequence coke.

(c) Tracking results on sequence david.

(d) Tracking results on sequence deer.

(e) Tracking results on sequence faceocc.

(f) Tracking results on sequence skating.

(g) Tracking results on sequence soccer.

(h) Tracking results on sequence tiger.

TLD —— IVT —— CSK —— SCM —— Our —— Struck —— LSK —— MIL

**Fig. 4.** Sample tracking results of evaluated algorithms

For the sequence coke, it is obvious that our tracker lost the target object from frame 260. Owing to complete occlusion for a period of time, our tracker learned the wrong appearance leading to wrong estimate. In the sequence david and deer, our tracker achieved perfect performance which can track the object completely. In the sequence bolt, most of evaluated trackers lost the target with similar color of the object and the background between frame 10 and frame 20. For appearance model, our tracker succeeded in tracking the target object, but lost target from frame 200 due to the sprinting of the object in front of the finish line which leaded to fast motion and large scale change in the appearance. In the frame 350 for sequence skating, there was illumination variation and a low contrast between the foreground and the background . Our tracker failed at that frame but CSK and SCM can track the object to the end. For the sequence soccer, in spite of partial occlusion and low contrast between the foreground and the background, our tracker performed better than other trackers. For the sequence tiger, when the target was covered by the leaf in the frame 230, all the trackers failed. In average, Figure 3 shows that our algorithm has a better performance than other algorithms in various situations.

## 5  Conclusion

In this paper, we have presented a cascaded feedback scheme for the combination of the discriminative model and generative model so as to improve the object tracking performance. Through the experiments, our algorithm has a good performance, but it cannot fully adapted to the types of problems. In particular, if an object is completely occluded for a long time or experiences with a large appearance change in a short time, our tracker will lose the target object. Furthermore, the variable base in Eq. 14 can be adjusted according to the initial state of the object and can be sensitive to the scale variation. Further work will be done to deal with these problems to get a more robust tracking algorithm.

**Acknowledgments.** Our work was supported in part by the National Natural Science Foundation of China #61273304 and #61202170, the Research Fund for the Doctoral Program of Higher Education Grant #20130072130004.

## References

1. Yilmaz, A., Javed, O., Shah, M.: Object Tracking: A Survey. ACM Computing Surveys 38(4), 1–45 (2006)
2. Babenko, B., Yang, M.-H., Belongie, S.: Visual tracking with online multiple instance learning. In: IEEE Conference on Computer Vision and Pattern Recognition, pp. 983–990. IEEE (2009)
3. Liu, B., Huang, J., Yang, L., Kulikowsk, C.: Robust tracking using local sparse appearance model and k-selection. In: IEEE Conference on Computer Vision and Pattern Recognition, pp. 1313–1320. IEEE (2011)
4. Collins, R.T., Liu, Y., Leordeanu, M.: Online selection of discriminative tracking features. IEEE Transactions on Pattern Analysis and Machine Intelligence 27(10), 1631–1643 (2005)

5. Ross, D., Lim, J., Lin, R.-S., Yang, M.-H.: Incremental learning for robust visual tracking. Computer Vision–IJCV 77(1-3), 125–141 (2008)

6. Comaniciu, D., Member, V.R., Meer, P.: Kernel-based object tracking. IEEE Transactions on Pattern Analysis and Machine Intelligence 25(5), 564–575 (2003)

7. Danelljan, M., Shahbaz Khan, F., Felsberg, M., et al.: Adaptive Color Attributes for Real-Time Visual Tracking. Computer Vision and Pattern Recognition (2014)

8. Henriques, J.F., Caseiro, R., Martins, P., Batista, J.: Exploiting the circulant structure of tracking-by-detection with kernels. In: Fitzgibbon, A., Lazebnik, S., Perona, P., Sato, Y., Schmid, C. (eds.) ECCV 2012, Part IV. LNCS, vol. 7575, pp. 702–715. Springer, Heidelberg (2012)

9. Grabner, H., Bischof, H.: Online boosting and vision. In: IEEE Computer Society Conference on Computer Vision and Pattern Recognition, vol. 1, pp. 260–267. IEEE (2006)

10. Grabner, H., Leistner, C., Bischof, H.: Semi-supervised on-line boosting for robust tracking. In: Forsyth, D., Torr, P., Zisserman, A. (eds.) ECCV 2008, Part I. LNCS, vol. 5302, pp. 234–247. Springer, Heidelberg (2008)

11. Hare, S., Saffari, A., Torr, P.H.S.: Struck: Structured output tracking with kernels. In: IEEE International Conference on Computer Vision–ICCV, pp. 263–270. IEEE (2011)

12. He, S., Yang, Q., Lau, R.W.H., et al.: Visual tracking via locality sensitive histograms. In: IEEE Conference on Computer Vision and Pattern Recognition, pp. 2427–2434. IEEE (2013)

13. Wright, J., Ma, Y., Maral, J., Sapiro, G., Huang, T., Yan, S.: Sparse representation for computer vision and pattern recognition. Proceedings of the IEEE 98(6), 1031–1044 (2010)

14. Santner, J., Leistner, C., Saffari, A., Pock, T., Bischof, H.: PROST:Parallel robust online simple tracking. In: IEEE Conference on Computer Vision and Pattern Recognition, pp. 723–730. IEEE (2010)

15. Wright, J., Yang, A.Y., Ganesh, A., Sastry, S.S., Ma, Y.: Robust face recognition via sparse representation. IEEE Transactions on Pattern Analysis and Machine Intelligence 31(2), 210–227 (2009)

16. Yu, Q., Dinh, T.B., Medioni, G.G.: Online tracking and reacquisition using co-trained generative and discriminative trackers. In: Forsyth, D., Torr, P., Zisserman, A. (eds.) ECCV 2008, Part II. LNCS, vol. 5303, pp. 678–691. Springer, Heidelberg (2008)

17. Avidan, S.: Ensemble tracking. IEEE Transactions on Pattern Analysis and Machine Intelligence 29(2), 261–271 (2007)

18. Brox, T., Bruhn, A., Papenberg, N., Weickert, J.: High AccuracyOptical Flow Estimation Based on a Theory for Warping. In: Proc. European Conf. Computer Vision, pp. 25–36 (2004)

19. Zhong, W., Lu, H., Yang, M.-H.: Robust Object Tracking via Sparsity-based Collaborative Model. In: IEEE Conference on Computer Vision and Pattern Recognition, pp. 1838–1845. IEEE (2012)

20. Wu, Y., Lim, J., Yang, M.-H.: Online object tracking: A benchmark. In: IEEE Conference on Computer Vision and Pattern Recognition, pp. 2411–2418. IEEE (2013)

21. Kalal, Z., Matas, J., Mikolajczyk, K.: P-N Learning: Bootstrapping Binary Classifiers by Structural Constraints. In: IEEE Conference on Computer Vision and Pattern Recognition, pp. 49–56. IEEE (2010)

# Optimal Subspace Learning for Sparse Representation Based Classifier via Discriminative Principal Subspaces Alignment

Lai Wei

Department of Computer Science, Shanghai Maritime University, Haigang Avenue 1550, Shanghai, P.R. China
weilai@shmtu.edu.cn

**Abstract.** Sparse representation based classifier (SRC) has been successfully applied in different pattern recognition tasks. Based on the analyses on SRC, we find that SRC is a kind of nearest subspace classifier. In this paper, a new feature extraction algorithm called discriminative principal subspaces alignment (DPSA) is developed according to the geometrical interpretations of SRC. Namely, DPSA aims to find a subspace wherein samples lie close to the hyperplanes spanned by the their homogenous samples and appear far away to the hyperplanes spanned by the their heterogenous samples. Different from the existing SRC-based feature algorithms, DPSA does not need the reconstruction coefficient vectors computed by SRC. Hence, DPSA is much more efficient than the SRC-based feature extraction algorithms. The face recognition experiments conducted on three benchmark face images databases (AR database, the extended Yale B database and CMU PIE) demonstrate the superiority of our DPSA algorithm.

**Keywords:** Feature extraction, nearest subspace classifier, sparse representation based classifier, linear regression based classification.

## 1 Introduction

Feature extraction is often required as a preliminary stage in many data processing applications. In the past few decades, many feature extraction methods have been proposed[1,2,3]. Particularly, the subspace learning[2,3] based feature extraction methods have become some of the most popular ones.

Two most well-known subspace learning methods are principal component analysis (PCA)[4] and linear discriminant analysis (LDA)[5]. Because of their high efficiencies, PCA and LDA have been widely applied for different practical feature extraction applications. Moreover, a family of manifold learning-related methods have aroused a lot of interest. For example, He et al. proposed two well-known algorithms, locality preserving projections (LPP)[6] and neighborhood preserving projection (NPE)[7]. Yan et al. presented a marginal Fisher analysis (MFA)[8] algorithm and formulated many existing subspace learning algorithms into a unified graph-embedding framework.

D. Miao et al. (Eds.): RSKT 2014, LNAI 8818, pp. 320–331, 2014.
DOI: 10.1007/978-3-319-11740-9_30 © Springer International Publishing Switzerland 2014

The goals of these mentioned methods are to find subspaces wherein specific statistical properties of data sets can be well preserved. However, in pattern recognition fields, after the extracted features are obtained, a proper classifier needs to be chosen. If the chosen classifier is appropriate for the extracted features, good recognition performances can be achieved. Hence, feature extraction algorithms designed according to the given classifiers are required. Yang et al. proposed a local mean based nearest neighbor discriminant analysis (LMNN-DA)[9] on the basis of the local mean based nearest neighbor (LM-NN) classifier[10]. Based on the recently proposed linear regression-based classification method (LRC)[11], Chen et al proposed a reconstructive discriminant analysis (RDA) method[12]. As reported in [9,12], LMNN-DA, HOLDA and RDA have achieved promising results in biometrics recognition and handwriting digits recognition.

Recently, sparse representation based classifier (SRC)[13] and its variation (collaborative representation based classifier (CRC)[14]) have achieved excellent results in face recognition. Hence, SRC-related feature extraction algorithms have attracted many attentions. Qiao et al. constructed a kind of $l_1$-norm affinity graph based on SRC and presented a sparsity preserving projections (SPP) method[15]. However, SPP is an unsupervised algorithm and it aims to find a subspace in which the sparse reconstructive relationship of original data points will be preserved by the low dimensional embeddings. Based on the classification rules of SRC, Cui et al. introduced the class information of data sets into SPP and devised a sparse maximum margin discriminant analysis (SMMDA)[17]. Lu et al. presented a discriminant sparsity neighborhood preserving embedding (DSNPE)[16] whose objective function is much similar to RDA. We can find that there exist two drawbacks in the two algorithms: firstly, they use the reconstruction coefficient vectors computed in the original space. Because the reconstruction relationship of the samples in the original space may not be in accord with that in the obtained low-dimensional subspace, SRC can not be guaranteed to achieve better results in the projected subspaces. Secondly, because SRC is used to compute the reconstruction coefficient vectors, the computation cost of the two algorithms is very expensive.

In this paper, based on the analysis on SRC and LRC, we find that SRC can be regarded as an extension of nearest subspace classifier. Hence, we propose a new supervised feature extraction algorithm, termed discriminative principal subspaces alignment (DPSA), to find a subspace wherein samples lie close to the hyperplanes spanned by the their homogenous samples and appear far away to the hyperplanes spanned by the their heterogenous samples.

## 2    Sparse Representation Based Classifier (SRC) and Linear Regression Classification (LRC)

Although SRC and LRC have different reconstruction strategies, they share the similar classification rules: a query image is identified to the class with the minimum reconstruction error. Let us introduce SRC firstly.

## 2.1  Sparse Representation Based Classifier (SRC)

SRC has a compact mathematical expression. Suppose we have $C$ classes of subjects. Denote $\mathbf{X}^i \in R^{D \times n_i}$ as the dataset of the $i$th class, and each column of $\mathbf{X}^i$ is a sample of class $i$. The entire training set is defined as $\mathbf{X} = [\mathbf{X}^1, \mathbf{X}^2, \cdots, \mathbf{X}^C] \in R^{D \times n}$, where $n = \sum_i n_i$. Once a query image $\mathbf{y}$ comes, we represent $\mathbf{y}$ by using the training samples, namely $\mathbf{y} = \mathbf{X}\mathbf{a}$, where $\mathbf{a} = [\mathbf{a}^1, \mathbf{a}^2, \cdots, \mathbf{a}^C]$ and $\mathbf{a}^i$ is the reconstruction representation vector associated with class $i$. If $\mathbf{y}$ is from the $i$th class, usually $\mathbf{y} = \mathbf{X}^i \mathbf{a}^i$ holds well. This implies that most coefficients in $\mathbf{a}^j (j \neq i)$ are nearly zeros and only $\mathbf{a}^i$ has significant nonzero entries. Then the sparest solution $\mathbf{a}^*$ can be sought by solving the following optimization problem:

$$\mathbf{a}^* = \arg \min \|\mathbf{a}\|_0 \ s.t. \ \mathbf{y} = \mathbf{X}\mathbf{a} \tag{1}$$

where $\| \cdot \|_o$ is the $l_0$-norm, which counts the number of nonzero entries in a vector.

Solving $l_0$ optimization problem (1), however, is NP hard. Fortunately, recent research efforts reveal that for certain dictionaries, if the solution $\mathbf{a}$ is sparse enough, finding the solution of the $l_0$ optimization problem is equivalent to a $l_1$ optimization problem[18]:

$$\mathbf{a}^* = \arg \min \|\mathbf{a}\|_1 \ s.t. \ \mathbf{y} = \mathbf{X}\mathbf{a} \tag{2}$$

In practice, Eq. (2) is usually transferred to a regularization problem, which is also called Lasso[19]:

$$\min_{\mathbf{a}} \|\mathbf{y} - \mathbf{X}\mathbf{a}\|_2^2 + \lambda \|\mathbf{a}\|_1 \tag{3}$$

Once $\mathbf{a}^*$ is obtained, SRC computes the residuals $r^i(\mathbf{y}) = \|\mathbf{y} - \mathbf{X}^i (\mathbf{a}^*)^i\|_2$ for each class $i$. If $r^l = \min\{r^i(\mathbf{y}) | i = 1, 2, \cdots, C\}$, $\mathbf{y}$ is assigned to class $l$.

Recently, Zhang et al. pointed out that the use of collaborative representation was more crucial than the $l_1$-sparsity of $\mathbf{a}$[13]. And the $l_2$-norm regularization on $\mathbf{a}$ can do a similar job to $l_1$-norm regularization but with much less computational cost. So they proposed the collaborative representation based classifier(CRC) method which tried to solve the following problem:

$$\min_{\mathbf{a}} \|\mathbf{y} - \mathbf{X}\mathbf{a}\|_2^2 + \lambda \|\mathbf{a}\|_2^2 \tag{4}$$

The optimal solution of Eq. (4) can be computed efficiently, namely $\mathbf{a}^* = (\mathbf{X}^T\mathbf{X} + \lambda\mathbf{I})^{-1}\mathbf{X}^T\mathbf{y}$. The classification rules of CRC is much similar to that of SRC, but the residuals in CRC is defined as $r^i(\mathbf{y}) = \|\mathbf{y} - \mathbf{X}^i\mathbf{a}^i\|_2 / \|\mathbf{a}^i\|_2$.

## 2.2  Linear Regression Classification (LRC)

LRC is devised based on the assumption that samples from a specific class lie in a linear subspace. Then for a query sample, LRC uses training samples from different classes to linearly represent it, namely $\mathbf{y} = \mathbf{X}^i\mathbf{a}^i$, where $\mathbf{a}^i$ is the reconstruction coefficient vector obtained by using the training samples belonging to class $i$. Then by using Least-squares estimation, we can obtain $\mathbf{a}^i = ((\mathbf{X}^i)^T\mathbf{X}^i)^{-1}(\mathbf{X}^i)^T\mathbf{y}$. Finally, LRC computes the residuals $r^i(\mathbf{y}) = \|\mathbf{y} - \mathbf{X}^i\mathbf{a}^i\|_2$ for each class $i$ and assigns $\mathbf{y}$ to the class with corresponding minimal residual.

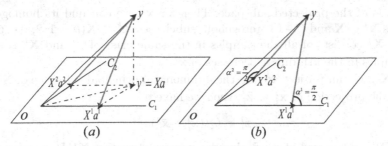

**Fig. 1.** The geometric interpretations of SRC, CRC and LRC. (a) SRC (CRC); (b) LRC

## 2.3    Analysis on SRC and LRC

We use Fig. 1 to illustrate the geometric interpretations of SRC and LRC. Fig.1 (a) is used to explain the classification rules of SRC and Fig. 1(b) shows the classification method of LRC. Here we suppose the training set $\mathbf{X} = [\mathbf{X}^1, \mathbf{X}^2]$ and $\mathbf{y}$ is a query sample. $C^1$ and $C^2$ are the subspaces spanned by the samples belonging to $\mathbf{X}^1$ and $\mathbf{X}^2$ respectively. In Fig. 1(a), $\mathbf{a}$ is computed by SRC and $\mathbf{a}^i = \delta_i(\mathbf{a})(i = 1, 2)^1$. The residuals $r^1_{SRC}(\mathbf{y})$ and $r^2_{SRC}(\mathbf{y})$ obtained by SRC are defined as the magnitude of the cyan line and blue line respectively.

In LRC, $r^i_{LRC}$ is defined as the distance between $\mathbf{y}$ and the subspace spanned by $\mathbf{X}^i$. Hence, in Fig. 1(b), the angles $\alpha^1$ and $\alpha^2$ are shown to be $\pi/2$. Thus LRC is a kind of nearest subspace classification method[11]. By comparing Fig. 1(a) and Fig. 1(b), we can find that SRC can be regarded as an extension of the nearest subspace classifier.

## 3    Discriminative Principal Subspaces Alignment (DPSA)

In this section, we propose a new supervised feature extraction algorithm termed discriminative principal subspaces alignment (DPSA)which can fit SRC and LRC well.

### 3.1    Motivation

From the descriptions in Section 2.3, we know that SRC is an extension of nearest subspace classifier. Hence, if SRC can achieve better recognition accuracies in a projected space, query samples should lie close to the subspaces spanned by the their homogenous samples and appear far away to the subspaces spanned by the their heterogenous samples. Suppose a training set $\mathbf{X}$, a training sample $\mathbf{x}^j_i \in \mathbf{X}(i = 1, 2, \cdots, n, j = 1, 2, \cdots, C)$ and its low dimensional projection $\mathbf{z}^j_i = \mathbf{P}^T \mathbf{x}^j_i$, where $\mathbf{P} \in R^{D \times d}$ is the required transformation matrix. $d$ is the

---

[1] $\delta_i$ is the characteristic function that selects the coefficients associated with the $i$th class.

dimension of the projected subspace. Then for $\mathbf{x}_i^j$, we can find its homogenous subclass $\mathbf{X}^j \subset \mathbf{X}$ and $C-1$ heterogenous subclasses $\mathbf{X}^k \subset \mathbf{X}(k = 1, 2, \cdots, C$ and $k \neq j)$. $\mathbf{X}^j$ consists of all the samples in the same class of $\mathbf{x}_i^j$ and $\mathbf{X}^k$ contains the samples in the $k$th different classes of $\mathbf{x}_i^j$.

For $\mathbf{X}^j$, we can project $\mathbf{x}_i^j$ into the $d$-dimension principal subspace of $\mathbf{X}^j$. We denote the projection of $\mathbf{x}_i^j$ as $\tilde{\mathbf{z}}_i^j$, then we have

$$\tilde{\mathbf{z}}_i^j = \mathbf{Q}^j (\mathbf{x}_i^j - \bar{\mathbf{x}}^j) \tag{5}$$

where $\mathbf{Q}^j$ is composed of $d$ left singular vectors of matrix $\mathbf{X}^j \mathbf{H}^j$ corresponding to its $d$ largest singular values. $\bar{\mathbf{x}}^j$ is the mean of $\mathbf{X}^j$ and $\mathbf{H}^j = \mathbf{I}^j - \mathbf{e}^j (\mathbf{e}^j)^T / n_j$, $\mathbf{I}^j \in R^{n_j \times n_j}$ is an identity matrix and $\mathbf{e}^j$ is a $n_j$-dimension column vector with all entities equal to 1. Define $\tilde{\mathbf{Z}}^j = [\tilde{\mathbf{z}}_1^j, \tilde{\mathbf{z}}_2^j, \cdots, \tilde{\mathbf{z}}_{n_i}^j]$, then $\tilde{\mathbf{Z}}^j = \mathbf{Q}^j \mathbf{X}^j \mathbf{H}^j$.

Meanwhile, we can obtain another $d$-dimension embedding of $\mathbf{x}_i^j$ (denoted as $\mathbf{z}_i^j$) by using DPSA, namely $\mathbf{z}_i^j = \mathbf{P}^T \mathbf{x}_i^j$. Then the following equation holds

$$\mathbf{z}_i^j - \bar{\mathbf{z}}^j = f(\tilde{\mathbf{z}}_i^j) + \varepsilon_{wi}^j \tag{6}$$

where $\varepsilon_{wi}^j = \varepsilon_{wi}^{jt} + \varepsilon_{wi}^{jr}$ is the within-class residual error. $\varepsilon_{wi}^{jt}$ represents the within-class translation error and $\varepsilon_{wi}^{jr}$ indicates the within-class reconstruction error. $\bar{\mathbf{z}}^j$ is the mean of $\mathbf{Z}^j = [\mathbf{z}_1^j, \mathbf{z}_2^j, \cdots, \mathbf{z}_{n_j}^j]$. Suppose $\mathbf{E}_w^j = [\varepsilon_{w1}^j, \varepsilon_{w2}^j, \cdots, \varepsilon_{wn_j}^j] = \mathbf{E}_w^{jt} + \mathbf{E}_w^{jr}$, then

$$\mathbf{Z}^j \mathbf{H}^j = f(\tilde{\mathbf{Z}}^j) + \mathbf{E}_w^j \tag{7}$$

Because $\mathbf{z}_i^j$ and $\tilde{\mathbf{z}}_i^j$ are all the d-dimensional embeddings of $\mathbf{x}_i^j$, they can be regarded as lying in a same $d$-dimension space. Hence, we have $f(\tilde{\mathbf{Z}}_i^j) = \mathbf{L}^j \tilde{\mathbf{Z}}_i^j$. $\mathbf{L}^j$ is an affine matrix which is used to translate $\tilde{\mathbf{z}}_i^j$ in the $d$-dimension space. In order to minimize the translation error vector $\mathbf{E}_w^{jt}$, the optimal $\mathbf{L}^j$ can be obtained by solving the following problem

$$\min_{\mathbf{L}^j} \| \mathbf{Z}^j \mathbf{H}^j - \mathbf{L}^j \tilde{\mathbf{Z}}^j \|_2^2 \tag{8}$$

then $\mathbf{L}^j = \mathbf{Z}^j \mathbf{H}^j (\tilde{\mathbf{Z}}^j)^+$. $(\tilde{\mathbf{Z}}^j)^+$ is the Moore-Penrose generalized inverse of $\tilde{\mathbf{Z}}^j$. Substitute $\mathbf{L}^j$ into Eq. (8), then the within-class translation error $\mathbf{E}_w^{tj}$ can be minimized. Hence $\mathbf{E}_w^j \approx \mathbf{E}_w^{rj}$. Then we have

$$\mathbf{E}_w^{rj} = \mathbf{Z}^j \mathbf{H}^j - \mathbf{Z}^j \mathbf{H}^j (\tilde{\mathbf{Z}}^j)^+ \tilde{\mathbf{Z}}^j \tag{9}$$

Actually, $\| \mathbf{E}_w^{rj} \|^2$ can be regarded as the sum of all the distances from $\mathbf{z}^j (j = 1, 2, \cdots, n_j)$ to the $d$-dimension principal subspace of $\mathbf{X}^j$. DPSA aims to minimize $\mathbf{E}_w^{rj}$ for arbitrary $j$, this goal can be achieved by the following optimization problem:

$$\min \| \mathbf{E}_w^r \|_2^2 = \min \sum_{j=1}^{C} \| \mathbf{E}_w^{rj} \|_2^2$$

$$= \min \sum_{j=1}^{C} \| \mathbf{Z}^j \mathbf{H}^j (\mathbf{I} - (\tilde{\mathbf{Z}}^j)^+ \tilde{\mathbf{Z}}^j) \|_2^2 \tag{10}$$

where $\mathbf{E}_w^r$ is the total within-class reconstruction error. let $\mathbf{W}_w^j = \mathbf{H}^j(\mathbf{I} - (\tilde{\mathbf{Z}}^j)^+\tilde{\mathbf{Z}}^j)$, $\mathbf{Z} = [\mathbf{Z}^1, \mathbf{Z}^2, \cdots, \mathbf{Z}^C]$ and $\mathbf{S} = [\mathbf{S}^1, \mathbf{S}^2, \cdots, \mathbf{S}^C]$. $\mathbf{S}^j$ is the $0 - 1$ selection matrix such that $\mathbf{Z}\mathbf{S}^j = \mathbf{Z}^j$. Because $\mathbf{Z} = \mathbf{P}^T\mathbf{X}$, Eq. (10) can be expressed as follows:

$$\min \|\mathbf{E}_w^r\|_2^2 = \min \sum_{j=1}^C Tr\big(\mathbf{Z}\mathbf{S}^j\mathbf{W}_w^j(\mathbf{W}_w^j)^T(\mathbf{S}^j)^T(\mathbf{Z}^j)^T\big)$$

$$= \min Tr\Big(\mathbf{P}^T\mathbf{X}\big(\sum_{j=1}^C \mathbf{S}^j\mathbf{W}_w^j(\mathbf{W}_w^j)^T(\mathbf{S}^j)^T\big)\mathbf{X}^T\mathbf{P}\Big)$$

$$= \min Tr\big(\mathbf{P}^T\mathbf{X}\mathbf{S}_w\mathbf{X}^T\mathbf{P}\big) \tag{11}$$

where $\mathbf{S}_w = \sum_{j=1}^C \mathbf{S}^j\mathbf{W}_w^j(\mathbf{W}_w^j)^T(\mathbf{S}^j)^T$ and $Tr(\cdot)$ is the trace of a matrix.

On the other hand, for one of the heterogenous subclasses $\mathbf{X}^k$ of $\mathbf{x}_i^j$, we also find the $d$-dimension projection of $\mathbf{x}_i^j$ (denoted as $\tilde{\mathbf{z}}_i^{jk}$) in the principal subspace of $\mathbf{X}^k$ by using $\mathbf{Q}^k$, where $\mathbf{Q}^k$ is composed of $d$ left singular vectors of matrix $\mathbf{X}^k\mathbf{H}^k$ corresponding to its $d$ largest singular values. Then by following the methodology described above, we can obtain an affine matrix $\mathbf{L}^{jk} = \mathbf{Z}^j\mathbf{H}^j(\tilde{\mathbf{Z}}_i^{jk})^+$ and the $k$th between-class reconstruction error

$$\|\mathbf{E}_b^{rjk}\|_2^2 = \|\mathbf{Z}^j\mathbf{H}^j(\mathbf{I} - (\tilde{\mathbf{Z}}^{jk})^+\tilde{\mathbf{Z}}^{jk})\|_2^2 \tag{12}$$

$\|\mathbf{E}_b^{rjk}\|^2$ is the sum of all the distances from $\mathbf{z}^j(j = 1, 2, \cdots, n_j)$ to the $d$-dimension principal subspace of $\mathbf{X}^k$. We hope that $\mathbf{E}_b^{rjk}$ can be maximized in the projected space for every pairs of $j$ and $k$, hence

$$\max \|\mathbf{E}_b^r\|_2^2 = \max \sum_{j=1}^C \sum_{k=1, k\neq j}^C \|\mathbf{E}^{rjk}\|_2^2$$

$$= \max Tr\Big(\mathbf{P}^T\mathbf{X}\big(\sum_{j=1}^C \mathbf{S}^j\big(\sum_{k=1, k\neq j}^C \mathbf{W}_b^{jk}(\mathbf{W}_b^{jk})^T\big)(\mathbf{S}^j)^T\big)\mathbf{X}^T\mathbf{P}\Big)$$

$$= \max Tr\big(\mathbf{P}^T\mathbf{X}\mathbf{S}_b\mathbf{X}^T\mathbf{P}\big) \tag{13}$$

where $\mathbf{S}_b = \sum_{j=1}^C \mathbf{S}^j\big(\sum_{k=1, k\neq j}^C \mathbf{W}_b^{jk}(\mathbf{W}_b^{jk})^T\big)(\mathbf{S}^j)^T$, $\mathbf{W}_b^{jk} = \mathbf{H}^j(\mathbf{I} - (\tilde{\mathbf{Z}}^{jk})^+\tilde{\mathbf{Z}}^{jk})$. Finally, the objective function of DPSA is defined as follows:

$$J_{DPSA} = \max \frac{\|\mathbf{E}_b^r\|_2^2}{\|\mathbf{E}_w^r\|_2^2} = \max \frac{Tr(\mathbf{P}^T\mathbf{X}\mathbf{S}_b\mathbf{X}^T\mathbf{P})}{Tr(\mathbf{P}^T\mathbf{X}\mathbf{S}_w\mathbf{X}^T\mathbf{P})} \tag{14}$$

Generally, we add the constraint $\mathbf{P}^T\mathbf{X}\mathbf{S}_w\mathbf{X}^T\mathbf{P} = \mathbf{I}$, such that the extracted features are uncorrelated. Then the column vectors of the optimal projection matrix $\mathbf{P}$ can be chosen as the $d$ largest generalized eigenvectors of $\mathbf{X}\mathbf{S}_b\mathbf{X}^T\mathbf{p} = \lambda\mathbf{X}\mathbf{S}_w\mathbf{X}^T\mathbf{p}$.

## 4    Experiments

In this section, we will use three face image databases (AR database[20], the extend Yale B face database[21], CMU PIE face database[22]) to verify the effectiveness of our proposed DPSA. We choose five representative algorithms including PCA, LDA, MFA (manifold learning-related algorithm), SMMDA(SRC-based algorithm), RDA (LRC-based algorithm) for comparisons. We use these feature extraction algorithms to find the low-dimensional subspaces, then SRC and LRC are used for classification in the projected spaces. The classification accuracy is used as the criterion to compare the performances of corresponding feature extraction algorithms.

Moreover, several problems need to be pointed out. Firstly, we use Eq. (3) to solve $l_1$-minimization problems in SRC. The regularization parameter $\lambda$ is set to 0.001. And MATLAB function: LeastR in the SLEP package[23] to solve the $l_1$-minimization problems. Secondly, in MFA, the two neighborhood parameters are both set to 6. Thirdly, in RDA, the number of between-class nearest subspaces $K$ should be chosen. In our experiments, we set $K = C - 1$. Fourthly, all the experiments are carried out using MATLAB on a 1.5 GHz machine with 4GB RAM.

### 4.1    Information of Three Face Databases

We briefly introduce the three face image databases as follows.

AR database consists of over 4000 face images of 126 individuals. For each individual, 26 pictures were taken in two sessions (separated by two weeks) and each section contains 13 images. We use a non-occluded subset (14 images per subject) of AR, which consists of 50 male and 50 female subjects. Some sample images of one person are shown in Fig. 2(a).

The extended Yale B face database contains 38 human subjects and around 64 near frontal images under different illuminations per individual. In our experiments, each image is resized to $32 \times 32$ pixels. Some sample images of one person are shown in Fig. 2(b).

The CMU PIE face database contains 68 subjects with over 40000 faces. Here we use a subset containing images of pose C27 (a frontal pose) of 68 persons, each with 49 images. All images are resized to be $64 \times 64$ pixels. Some sample images of one person are shown in Fig. 2(c).

To avoid overfitting, we first perform PCA and reduce the dimension of images data to be 200 before implementing the evaluating algorithms.

### 4.2    Experiments on AR Database

For AR face database, the recognition rate curves of each algorithm versus the variation of dimensions are shown in Fig. 3. From Fig. 3, we firstly can find that SRC outperforms LRC on the corresponding subspaces obtained by the feature extraction algorithms. DPSA is superior to PCA, LDA, MFA and RDA. We illustrate the average training time of each algorithm in Table 1. It can be

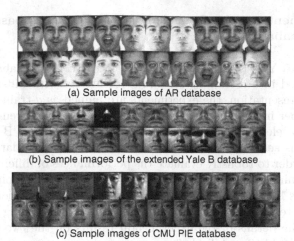

(a) Sample images of AR database

(b) Sample images of the extended Yale B database

(c) Sample images of CMU PIE database

**Fig. 2.** Sample images of three face databases (a) AR database; (b) Yale B database; (c) CMU PIE database

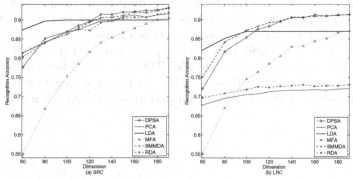

**Fig. 3.** Recognition accuracy curves of each method versus variation of dimensions on AR database. (a) The results obtained by SRC; (b) The results obtained by LRC.

found that the training time of SMMDA is about ten times more than that of DPSA. The maximal recognition rates obtained by each algorithm and their corresponding dimensions are also reported in Table 1.

**Table 1.** Maximal recognition rates (%), corresponding dimensions and CPU time for training on AR face database

| Classifier | Algorithms | | | | | |
|---|---|---|---|---|---|---|
| | DPSA | PCA | LDA | MFA | SMMDA | RDA |
| SRC | **93.13(190)** | 91.56(190) | 89.99(99) | 90.98(186) | 92.99(180) | 91.85(190) |
| LRC | **92.99**(160) | 72.10(190) | 86.98(99) | 87.43(176) | 92.13(180) | 73.24(105) |
| Time(s) | 46.27 | 0.03 | 0.18 | 0.21 | 463.18 | 14.42 |

### 4.3    Experiments on the Extended Yale B Face Database and CMU PIE Database

For the extended Yale B face database and CMU PIE face database, the image set is partitioned into a training set and a testing set with different numbers. For ease of representation, the experiments are named as $q$-train, which means that $q$ images per individual are selected for training and the remaining images for testing. $q$ is selected as 10 and 20 when the extended Yale B face database is used, and $q$ is selected as 10 and 15 when CMU PIE face database is used. Moreover, in order to robustly evaluate the performances of different algorithms in different training and testing conditions, we select images randomly and repeat the experiments 10 times in each condition. The average recognition rates of each method across ten tests on the two databases are shown in Fig. 4 and Fig. 5 respectively. The details of the experimental results on the two databases are summarized in Table 2 and Table 3.

From Fig. 4 and Fig. 5, we can get several conclusions:

1) DPSA outperforms PCA, LDA, MFA and RDA. Namely, SRC and LRC can achieve better performances in the subspaces obtained by DPSA.

2) Compared to DPSA, SMMDA achieves competitive results. In the experiments conducted on the extended Yale B database, we can see that DPSA outperforms SMMDA in relative low-dimensional subspaces (e.g. $d \leq 90$) when SRC is used. This may imply that the subspaces obtained by DPSA are more suitable for SRC than that obtained by SMMDA. However, when LRC is applied, DPSA is only slightly better than SMMDA. In the experiments conducted on CMU PIE database, DPSA constantly outperforms SMMDA when SRC is used.

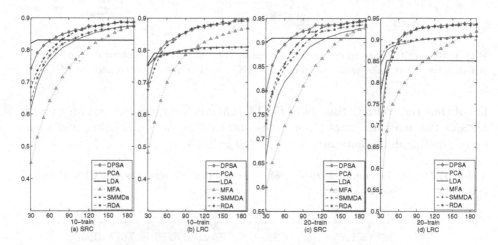

**Fig. 4.** Recognition accuracy curves of each method versus variation of dimensions on the extended Yale B face database.(a) and (b) show the results obtained by SRC and LRC on 10-train respectively, (c) and (d) show the results obtained by SRC and LRC on 20-train respectively.

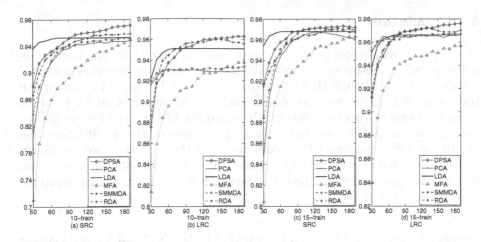

**Fig. 5.** Recognition accuracy curves of each method versus variation of dimensions on CMU PIE face database. (a) and (b) show the results obtained by SRC and LRC on 10-train respectively, (c) and (d) show the results obtained by SRC and LRC on 15-train respectively.

**Table 2.** Maximal recognition rates (%), corresponding dimensions and CPU time for training on the extended Yale B face database

|  | Classifier | Algorithms | | | | | |
|---|---|---|---|---|---|---|---|
|  |  | DPSA | PCA | LDA | MFA | SMMDA | RDA |
| 10-train | SRC | **88.78(170)** | 87.57(190) | 83.06(37) | 87.62(190) | 88.61(170) | 87.34(180) |
|  | LRC | **89.75(190)** | 81.05(180) | 79.08(37) | 86.85(190) | 89.18(190) | 81.18(190) |
|  | Time(s) | 2.68 | 0.08 | 0.09 | 0.17 | 167.45 | 14.42 |
| 20-train | SRC | **94.95(190)** | 92.90(180) | 90.87(37) | 93.32(190) | 94.50(190) | 93.71(190) |
|  | LRC | **93.89(170)** | 90.84(180) | 85.13(37) | 91.99(190) | 93.56(170) | 90.84(190) |
|  | Time(s) | 13.19 | 0.12 | 0.60 | 0.93 | 383.24 | 27.184 |

**Table 3.** Maximal recognition rates (%), corresponding dimensions and CPU time for training on CMU PIE face database

|  | Classifier | Algorithms | | | | | |
|---|---|---|---|---|---|---|---|
|  |  | DPSA | PCA | LDA | MFA | SMMDA | RDA |
| 10-train | SRC | **97.13(190)** | 94.79(190) | 95.28(67) | 94.94(190) | 95.85(190) | 95.21(180) |
|  | LRC | **96.30(190)** | 92.94(190) | 95.09(67) | 93.77(190) | 95.96(150) | 93.36(190) |
|  | Time(s) | 19.54 | 0.02 | 0.14 | 0.17 | 334.63 | 18.43 |
| 20-train | SRC | **97.33(170)** | 96.75(110) | 96.80(67) | 96.28(190) | 96.97(140) | 97.23(130) |
|  | LRC | **97.62(190)** | 96.62(190) | 96.62(67) | 95.67(190) | 97.18(130) | 96.71(190) |
|  | Time(s) | 44.11 | 0.02 | 0.18 | 0.26 | 492.71 | 65.79 |

And When LRC is used, DPSA is obviously better than SMMDA in relatively high-dimensional subspaces (e.g. $d > 120$).

In addition, from Table 2 and Table 3, we can find that the CPU time of SMMDA for training on the two databases are much larger than that of DPSA. Therefore, based on all the experiments, we can conclude that DPSA is an efficient feature extraction algorithm for SRC and LRC.

# 5  Conclusion

In this paper, we propose a novel feature extraction algorithm, termed discriminative principal subspaces alignment (DPSA), based on the geometrical view of SRC and LRC. Although DPSA and the existing SRC-based (LRC-based) feature extraction algorithms have similar goals, the structure of DPSA is totally different. Compared to the existing SRC-based (or LRC-based) feature extraction algorithms, DPSA is not faithfully designed based on the classification rules of SRC or LRC. Namely, DPSA does not need the reconstruction coefficient vectors computed by SRC or LRC. Hence, DPSA is much efficient and robust. Finally, the experiments on three benchmark face images databases prove the effectiveness of DPSA.

**Acknowledgments.** This work is supported by the National Science Foundation of China (No. 61203240) and the Innovation Program of Shanghai Municipal Education Commission (14YZ102).

# References

1. Fukunaga, K.: Introduction to Statistical Pattern Recognition, 2nd ed., Academic Press (1990)
2. Cheng, H., Vu, K., Hua, K.A.: SubSpace projection: a unified framework for a class of partition-based dimension reduction techniques. Information Sciences 179, 1234–1248 (2009)
3. Li, X., Tao, D.: Subspace learning. Neurocomputing 73, 10539–10540 (2010)
4. Turk, M.A., Pentland, A.P.: Face recognition using eigenfaces. In: IEEE Computer Society Conference on Computer Vision and, Pattern Recognition, pp. 586–591 (1991)
5. Belhumeur, P.N., Hespanha, J.P., Kriegman, D.J.: Eigenfaces vs. fisherfaces: recognition using class specific linear projection. IEEE Transactions on Pattern Analysis and Machine Intelligence 19, 711–720 (1997)
6. He, X., Yan, S., Hu, Y., Niyogi, P., Zhang, H.J.: Face recognition using laplacianfaces. IEEE Transaction on Pattern Analysis and Machine Intelligence 27(3), 328–340 (2005)
7. He, X., Cai, D., Yan, S., Zhang, H.: Neighborhood Preserving Embedding. In: ICCV 2005, pp. 1208–1213 (2005)
8. Yan, S., Xu, D., Zhang, B., Zhang, H., Yang, Q., Lin, S.: Graph embedding and extension: A general framework for dimensionality reduction. IEEE Trans. Pattern Anal. Mach. Intell. 29(1), 40–51 (2007)
9. Yang, J., Zhang, L., Yang, J.Y., Zhang, D.: From classifiers to discriminators: A nearest neighbor rule induced discriminant analysis. Pattern Recognition 44(7), 1387–1402 (2011)
10. Mitani, Y., Hamamoto, Y.: A local mean-based nonparametric classifier. Pattern Recognition Letters 27(10), 1151–1159 (2006)
11. Naseem, I., Togneri, R., Bennamoun, M.: Linear Regression for Face Recognition. IEEE Transaction on Pattern Analysis and Machine Intelligence 32(11), 2106–2112 (2010)

12. Chen, Y., Jin, Z.: Reconstructive discriminant analysis: A feature extraction method induced from linear regression classification. Neurocomputing 87, 41–50 (2012)
13. Wright, J., Yang, A., Ganesh, A., Sastry, S., Ma, Y.: Robust face recognition via sparse representation. IEEE Transaction on Pattern Analysis and Machine Intelligence 31(2), 210–227 (2009)
14. Zhang, L., Yang, M., Feng, X.: Sparse Representation or Collaborative Representation Which Helps Face Recognition. In: ICCV 2011 (2011)
15. Qiao, L.S., Chen, S.C., Tan, X.Y.: Sparsity preserving projections with applications to face recognition. Pattern Recognition 43(1), 331–341 (2010)
16. Lu, G.F., Jin, Z., Zou, J.: Face recognition using discriminant sparsity neighborhood preserving embedding. Knowledge-Based Systems 31, 119–127 (2012)
17. Cui, Y., Zheng, C.H., Yang, J., Sha, W.: Sparse maximum margin discriminant analysis for feature extraction and gene selection on gene expression data. Comput. Biol. Med. 43(7), 933–941 (2013)
18. Donoho, D.: For most large underdetermined systems of linear equations the minimal $l_1$-norm solution is also the sparsest solution, Commun. Pure Appl. Math. 59(6), 797–829 (2006)
19. Tibshirani, R.: Regression shrinkage and selection via the lasso, Journal of the Royal Statistical Society. Series B (Methodological) 58(1), 267–288 (1996)
20. Martinez, A.M., Benavente, R.: The AR face database, CVC, Univ. Autonoma Barcelona, Barcelona, Spain, Technical Report (June 24, 1998)
21. Lee, K.C., Ho, J., Driegman, D.: Acquiring linear subspaces for face recognition under variable lighting, IEEE Trans. Pattern Anal. Mach. Intell. 27(5), 684–698 (2005)
22. Sim, T., Baker, S., Bsat, M.: The CMU pose, illumination, and expression database, IEEE Trans. Pattern Anal. Mach. Intell. 25(12), 1615–1618 (2003)
23. Liu, J., Ji, S., Ye, J.: SLEP: A Sparse Learning Package. Available (2008), http://www.public.asu.edu/jye02/Software/SLEP/

# A Statistics-Based Semantic Relation Analysis Approach for Document Clustering

Xin Cheng, Duoqian Miao, and Lei Wang

Department of Computer Science and Technology, Tongji University, Shanghai, China
cx1227@gmail.com, maioduoqian@163.com

**Abstract.** Document clustering is a widely research topic in the area of machine learning. A number of approaches have been proposed to represent and cluster documents. One of the recent trends in document clustering research is to incorporate the semantic information into document representation. In this paper, we introduce a novel technique for capturing the robust and reliable semantic information from term-term co-occurrence statistics. Firstly, we propose a novel method to evaluate the explicit semantic relation between terms from their co-occurrence information. Then the underlying semantic relation between terms is also captured by their interaction with other terms. Lastly, these two complementary semantic relations are integrated together to capture the complete semantic information from the original documents. Experimental results show that clustering performance improves significantly by enriching document representation with the semantic information.

## 1   Introduction

Document clustering aims to organize the documents into groups according to their similarity. The traditional approaches are mostly based on Bags of words (BOW) model, which represents the documents with the terms and their frequency in the document. However, this model has the limitation that it assumes the terms in the document are independent thus regardless of the semantic relationship between them. It considers the documents are dissimilar if no overlapped terms exist, even though they describe the same topic.

To overcome the disadvantage of BOW model, a lot of approaches have been proposed to capture the semantic relation between terms to enhance document clustering. Generally, there are two directions to explore the semantic relation between terms: knowledge-based approach and statistics-based approach [3][6][7][13]. The knowledge-based approach measures the semantic relation between terms using the background knowledge which is constructed from ontology, such as WordNet [12] and Wikipedia [6]. Although the incorporation of the background information into BOW model has shown an improvement in document clustering, this approach has the limitation that the coverage of the ontology is limited, even for WordNet or Wikipedia. Besides, the context information has been overlooked to compute the semantic relation between terms. The statistics-based approach captures the semantic relation between terms based on term co-occurrence information, which evaluates the semantic relation between terms from the significance of their co-occurrence pattern. The most previous statistics-based

D. Miao et al. (Eds.): RSKT 2014, LNAI 8818, pp. 332–342, 2014.
DOI: 10.1007/978-3-319-11740-9_31 © Springer International Publishing Switzerland 2014

approaches only capture the explicit semantic relation between terms from their co-occurrence information, but the underlying relation has been overlooked, which is also essential for capturing the complete semantic relation between terms. Besides, the synonymous and ambiguous terms could not be accurately handled in the previous approaches, and that would affect the accuracy of semantic relation evaluation in a certain degree.

In this paper, we propose a novel approach to capture the semantic relation between terms based on both the explicit and implicit relations between terms. It firstly captures the explicit relation between terms from their co-occurrence information, and then the implicit semantic relation is revealed by their interaction with other terms. Meanwhile, Wikipedia is exploited to handle the synonymous and ambiguous terms. Lastly, the explicit and implicit semantic relations are integrated to capture the complete semantic information from the original documents, and then we extends the original BOW model with the semantic information for document clustering.

The rest of the paper is organized as follows. Section 2 presents the background of document clustering problem and reviews some related work. Section 3 proposes a novel approach for mining the semantic relation between terms and analyzing the semantic information of the original documents. The experimental results are discussed in Section 4, and the conclusion and future work will be describe in Section 5.

## 2    Related Work

Document clustering is an unsupervised approach to group the similar document together, and most document clustering approaches are based on the BOW (Bag of Words) model, which assumes that the terms in the document are independent. However, the terms are always related to each other, and the related information between them could be hierarchical relationship, compound word relation and synonym relation etc.

The semantic relation between terms was first introduced by Wong for document representation [14], and then many approaches are proposed to measure the relation between terms. Some approaches have been proposed to explore the semantic relation between terms with background knowledge, like WordNet and Wikipedia. In [2], they proposed to measure the relatedness between terms not by the exact term matching, but by their semantic relation, which is measured based on the semantic information in WordNet. However, WordNet has the limited coverage because it is manually built. In [7], Wikipedia, the largest electronic encyclopaedia, was exploited for document clustering. They construct a proper semantic matrix based on the semantic relation between terms from the underlying structural information in Wikipedia, and then they incorporated the semantic matrix into traditional document similarity measure.

Another direction of term relation measure is based on the statistical information. Examples of such work like the generalized vector space model (GVSM), which was proposed by Wong et al. [14], captures the semantic relation between terms in an explicit way by using their co-occurrence information. It simply utilizes the document-term matrix $W^T$ as the semantic matrix S, and then each document vector is projected as $d' = d * W^T$. The corresponding kernel between two document vectors is expressed as $k'(d_i, d_j) = d_i W^T W d_j$. The entry in matrix $W^T W$ reflects the similarity between terms which is measured by their frequency of co-occurrence across the document collection, which means two terms are similar if they frequently co-occur in the same document. Holger et al. [1] uses term co-occurrence patterns to estimate term dependency. It integrates the semantic information into document representation for calculation of the document similarity. The empirical results confirm that it improves the performance of document retrieval for particular document collections. Argyris et al. [8] take the local distance of the co-occurrence terms into consideration while computing the relation between terms. They exploit the relation between terms in the local context, and then combined all the local relation together to constitute the global relation matrix.

# 3   Methodology

The BOW model exploits each term in document as document features, so it cannot model efficiently the rich semantic information of documents. To capture the accurate similarity between documents, its essential to build a high quality document representation which could reserve the semantic information from the original documents. A lot of work have proposed that if two terms co-occur in the same document, they are relational in a certain degree [5][8][11]. However, they just consider the explicit relation of terms in the same document, but the underlying relation between them has been overlooked, which is also essential to capture the robust and reliable relation between terms. In our approach, a novel approach is proposed to capture the relation between terms, which identifies the relation between terms by not only themselves, but also their interaction with other terms.

In our work, we propose a novel semantic analysis model. This model capitalizes on both the explicit relation and implicit relation to compute the semantic relation between terms. The key points of the proposed model are: (a) it computes the semantic relation between each pair of terms using their co-occurrence information as the explicit relation; (b) it further constructs semantic links between terms by considering their interaction with other terms as the implicit relation; and (c) it combines the explicit and implicit relations together to compute the semantic relation for each pair of terms. Using this model, the semantic relation between terms can be captured more precisely, which can be integrated into document representation to enhance the quality of document representation.

## 3.1   The Semantic Relation Analysis between Terms

The first step of our approach for measuring the semantic relation between terms is to explore the explicit semantic relation. In most of the previous approaches, the relation between terms is simply estimated by considering the co-occurrence frequency but overlooking the discriminative power of terms, which will lead to the incorrect estimation of the relation between terms. In this work, the $tfidf$ scheme is used to measure the relation between terms which is based not only on the frequency of terms but also on their discriminative ability. Firstly, we introduce the definition of the explicit relation between terms:

**Definition 1.** *Let D be a document collection, two terms $t_i$, $t_j$ are considered to be explicitly related only if they co-occur in the same document. To evaluate the explicit relation between two terms, we propose an efficient measure which is defined as:*

$$Relation_{exp}(t_i, t_j) = \frac{1}{|H|} \sum_{d_x \in H} w_{xi} w_{xj} / (w_{xi} + w_{xj} - w_{xi} w_{xj}) \quad (1)$$

Where $w_{xi}$ and $w_{xj}$ are the *tfidf* values of term $t_i$, $t_j$ in the document $d_x$, and $H$ denotes the documents where $t_i$ and $t_j$ co-occur.

With the explicit relation between terms, the quality of document representation can be enhanced by integrating the explicit relation into document representation. However, the underlying relation between terms cannot be discovered from term co-occurrence information. In the following, we will introduce a novel approach to capture the implicit relation between terms:

**Definition 2.** *Let D be a document collection, two terms $t_i, t_j$ are from different documents $(t_i \in d_m, t_j \in d_n)$, if there is a term $t_s$ co-occur with them in the respective documents, they are considered as being linked by term $t_s$, and they are implicitly related.*

Fig. 1 shows an example of term implicit relation analysis, two terms $t_i$ and $t_j$ are from different document, and $t_{s1}, t_{s2}$ are the co-occurrence terms with them in the respective documents. Terms $t_i$ and $t_j$ are not related based on the explicit relation analysis, but they are considered to be relational using the implicit relation analysis because they co-occur with the same terms in the respective documents. Therefore, we define the calculation of the implicit relation between terms as follows:

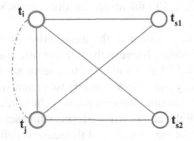

**Fig. 1.** An example of the implicit relation analysis

**Definition 3.** *Let D be a document set, a pair of terms $(t_i, t_j)$ are from different documents, the relation between $t_i$ and $t_j$ can be linked by $t_s$ which is the same co-occurrence terms with $t_i$ and $t_j$ in the respective documents. The implicit relation between $t_i$ and $t_j$, by their interaction with their co-occurrence term $t_s \in S$, is defined as:*

$$Relation_{imp}(t_i, t_j) = \frac{1}{|S|} \sum_{t_s \in S} \frac{\min((Relation_{exp}(t_i, t_s), Relation_{exp}(t_j, t_s))}{\sum_{t_x \in T}(Relation_{exp}(t_x, t_s)}, \quad (2)$$

Where $Relation_{exp}(t_i, t_s)$, $Relation_{exp}(t_j, t_s)$ represent the explicit relation of the term $t_i$ and $t_j$ with term $t_s$ in the respective documents, and $S$ is the term collection which $t_i$ and $t_j$ co-occur with, $T$ is the term collection of this corpus.

**Term Sense Disambiguation.** It is essential to measure whether an ambiguous term takes the same sense in different documents. That is because if two terms co-occur with an ambiguous term and it takes different sense in each document, then they could not be considered that they co-occur with the same term, which means the co-occurrence term could not be taken as the link term.

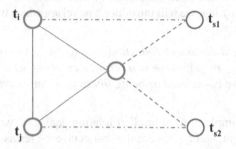

**Fig. 2.** An example of the relation with equivalent terms

As Fig. 2 demonstrates, term $t_i$ and $t_j$ co-occurrence with the same term $t_s$, but $t_s$ takes different sense $t_{s1}$ and $t_{s2}$ in the respective documents, so $t_i$ and $t_j$ could not be linked by the term $t_s$.

To alleviate this problem, we explore the intersection of their surrounding text to disambiguate the sense of terms, because the context information is an indication of the sense of each term, and the terms with the same sense should appear in the similar contexts. The sense similarity can be evaluated by two main steps: context information extraction and similarity evaluation. We first identify the context information from the co-occurrence matrix, as all the co-occurrence terms with each term is considered to be the context information. Then the similarity of the sense is defined as:

$$sim(s_1, s_2) = (|N(s_1) \cap N(s_2))|)/(|N(s_1)| + |N(s_2)|) \tag{3}$$

Where $N(s_i)$ represents all the co-occurrence terms with term $s_i$, and $N(s_1) \cap N(s_2)$ is the common co-occurrence terms between $s_1$ and $s_2$. In our approach, if $sim(s_1, s_2) < 0.5$, term $t_s$ is considered as an ambiguous term, which means terms $t_i$ and $t_j$ can not be linked by $t_s$.

**Mapping of Equivalent Terms.** In some cases, two terms are similar even same in sense but differs in spelling. For example, "disk" and "disc", "motor" and "engine", "BBC" and "British Broadcasting Corporation", and they should be taken as the same term because they are just the alternative names, alternative spellings or abbreviations of the same thing.

Like in Fig. 3, $t_i$ co-occurs with $t_{s1}$ while $t_j$ co-occurs with the term $t_{s2}$, $t_{s1}$ and $t_{s2}$ are not same in appearance, like "Car" and "Automobile", but they have the same meaning of term $t_s$, then it is intuitive that $t_i$ and $t_j$ should be considered as being related as they co-occur with the same term $t_s$.

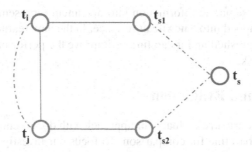

**Fig. 3.** An example of the relation with polysemous words

To solve this problem, its essential to map the equivalent terms to the identical expression. In our paper, we take Wikipedia, which has been proved to be an efficient thesaurus, as background knowledge to solve this problem. In Wikipedia, the redirect hyperlinks group the terms that have the same sense together and link to the identical concept, and they are very useful as an additional source of synonyms. Hence, if two terms link to the indexical concept, they are considered as being the link term between $t_i$ and $t_j$.

The explicit relation discovers the relation between terms by using their co-occurrence statistics and the implicit relation discovers the relation between terms by using their interaction with other terms. To capture the complete semantic relation between terms, we integrate the explicit and implicit relations together to measure the semantic relation between terms in this section.

**Definition 4.** *Let D be a document collection, terms $t_i$ and $t_j$ appear in this document collection, then the semantic relation between $t_i$ and $t_j$ is defined as:*

$$Relation(t_i, t_j) = Relation_{exp}(t_i, t_j) \cdot Relation_{imp}(t_i, t_j)), \qquad (4)$$

where $Relation_{exp}(t_i, t_j)$ is explicit relation between $t_i$ and $t_j$, and $Relation_{imp}(t_i, t_j)$ is the implicit relation between $t_i$ and $t_j$.

In our approach, the co-occurrence statistics are modeled with the integration of explicit and implicit relations. In this sense, our approach has the advantage of capturing the complete semantic relation between terms from term co-occurrence statistics. Furthermore, the semantic relation matrix can be constructed which reflects the semantic relation between each pair of terms, and then it can be used to project the original document representation into a new feature space with better discriminative ability.

## 3.2  The Document Semantic Analysis

Based on the proposed semantic relation analysis, the semantic matrix $S$ can be further constructed whose elements reflect the semantic relation between each pair of terms.

With the semantic matrix $S$, the original documents can be mapped into a new feature space, which reserves the semantic information from the original documents.

$$d : d \mapsto d' = d * S, \tag{5}$$

By integrating the semantic information into document representation, the original documents can be mapped into a new feature space. In the new feature space, the documents are well distinguished and it can further improve the performance of the related document analysis task.

# 4    Experiment and Evaluation

In this section, we empirically evaluate our approach with document clustering, and the BOW is used as the baseline for comparison. To focus our investigation on the representation rather than the clustering method, we used the standard k-means algorithm in the experiments.

## 4.1  Data Sets

To validate our strategy, we conduct experiments on four document collections. D1 is the subset of 20 Newsgroups while D2 is the mini-newsgroup version, D3 is the subsets of Reuters 21578, and D4 is the WebKB document collection. The detailed information of these document collections is described as follows:

**Table 1.** Characteristics of Data Sets

| Data sets | Name | Classes | $m$ | $n$ | $n_{avg}$ |
|-----------|------|---------|-----|-----|-----------|
| D1 | 20 newsgroup | 5 | 1864 | 16516 | 76 |
| D2 | 20 newsgroup | 20 | 1989 | 24809 | 55 |
| D3 | Reuters21578 | 8 | 2091 | 8674 | 33 |
| D4 | WebKB | 4 | 4087 | 7769 | 32 |

1. The first data set (D1) is a subset of 20 Newsgroups(20NG), which is a widely used data set for document clustering [9]. It consists 1864 newsgroup documents across 5 classes.
2. The second data set (D2) is the mini-newsgroups version, which has $1,989$ documents across all 20 classes in 20-newsgroups.
3. The third data set (D3) is a subset derived from the popular Reuters-21578 document collection [10] which has $2,091$ documents belonging to 8 classes (acq, crude, earn, grain, interest, money-fx, ship, trade).
4. The last data set (D4) is WebKB [4]. It consists of 4087 web pages and manually classified into 4 categories.

## 4.2  Evaluation Criteria

Cluster quality is evaluated by four criterions: purity, rand index, F1-measure and normalized mutual information.

Purity is a simple and transparent way to measure the quality of clustering. The purity of a cluster is computed by the ratio between the size of the dominant class in the cluster and the size of cluster. $purity(c_i) = \frac{1}{|c_i|} \max_j |c_j|$. Then the overall purity can be expressed as the weighted sum of all individual cluster purity:

$$purity = \frac{|c_i|}{N} \sum_{i=1}^{n} purity(c_i), \qquad (6)$$

Rand Index (RI) measures the clustering quality by the percentage of the true positive and true negative decisions in all decisions during clustering:

$$RI = ((TP + TR))/((TP + TR + FP + FR)) \qquad (7)$$

where TP (true positive) denotes that two similar documents are assigned to the same cluster; TN (true negative) denotes that two dissimilar documents are assigned to different clusters; FP (false positive) denotes that two dissimilar documents are assigned to the same cluster, and FN (false negative) denotes that two similar documents are assigned to different clusters.

F1-measure considers both the precision and recall for clustering evaluation:

$$F1 = ((precision * recall))/((precision + recall)) \qquad (8)$$

where $precision = TP/(TP + FP), recall = TP/(TP + FN)$.

Normalized mutual information (NMI) is a popular information theoretic criterion for evaluating clustering quality. It is computed by dividing the Mutual Information between the entropy of the clusters and the label of dataset:

$$NMI(C, L) = (I(C; L))/((H(C) + H(L))/2) \qquad (9)$$

where $C$ is a random variable for cluster assignments, $L$ is a random variable for the pre-existing classes on the same data. $I(C; L)$ is the mutual information between the clusters and the label of the dataset, and $H(C)$ and $H(L)$ is the entropy of $C$ and $L$.

## 4.3  Performance Evaluation

Table 2 shows the performance of our proposed approach on each dataset compared with two other approaches: the classic BOW model and the GVSM model, and the classic BOW model is taken as the baseline for comparison. For these quality measures, higher value in [0, 1] indicates better clustering solution. We can observe that our approach achieves significant improvement in all quality measures. Compared with the base line, our proposed approach has achieved 10.4%, 22.7%, 11.1% and 19.4% average improvement. Compared to GVSM model, our approach also achieves 7.4%, 16.9%, 8.8% and 15.5% average improvement. The experimental results demonstrate the benefit of integrating both the explicit and implicit probabilistic relation between

**Table 2.** Document Clustering Results by Using K-means

| Data | Purity | | | RI | | | F1-measure | | | NMI | | |
|------|------|------|------|------|------|------|------|------|------|------|------|------|
| Sets | *BOW* | *GVSM* | *CRM* | *BOW* | *GVSM* | *CRM* | *BOW* | *GVSM* | *CRM* | *BOW* | *GVSM* | *CRM* |
| D1 | 0.293 | 0.325 | **0.413** | 0.340 | 0.461 | **0.541** | 0.356 | 0.351 | **0.417** | 0.139 | 0.158 | **0.403** |
| D2 | 0.125 | 0.114 | **0.189** | 0.447 | 0.447 | **0.760** | 0.123 | 0.122 | **0.197** | 0.207 | 0.198 | **0.325** |
| D3 | 0.740 | 0.775 | **0.821** | 0.669 | 0.691 | **0.817** | 0.594 | 0.567 | **0.749** | 0.421 | 0.447 | **0.597** |
| D4 | 0.431 | 0.495 | **0.581** | 0.357 | 0.448 | **0.604** | 0.455 | 0.478 | **0.505** | 0.094 | 0.216 | **0.312** |

terms into document representation. Although the GVSM model is assisted by the proposed semantic smoothing, which takes into account the local contextual information associated with term occurrence, it overlooks the underlying semantic relation between terms. Compared to the GVSM model, our proposed approach considers both the explicit and implicit relations between terms, which can capture more reliable semantic relation between terms.

An interesting point to stress according to Table 2 is that larger gains are obtained in the document collections which are harder to classify, where the baseline does not perform well. For example, for the D1 and D2 collections, which are more difficult to obtain good clustering results using only bag-of-words representation. By integrating the semantic information captured with our approach into document representation, the clustering results have been significantly improved.

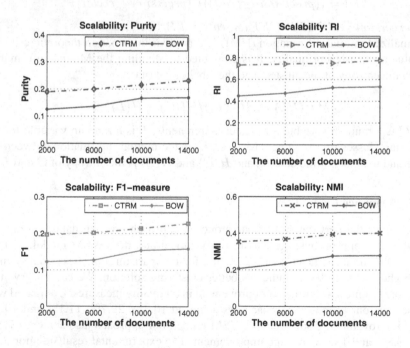

**Fig. 4.** The impact of corpus size

Besides, even in the cases where the performance of baseline is good and improvements consequently tend to be more limited, we also achieve statistically significant gains. Likewise, for D3, we still achieves 8.1%, 14.8%, 15.5% and 17.6% gains.

### 4.4   The Impact of Corpus Size

In this subsection, we analyze the effect of corpus size on the semantic relation analysis of our approach. To show the effect of corpus size, we conduct a set of experiments on the document collection 20-newsgroups by increasing the number of documents from 2, 000 to 14, 000 at increments of 4, 000.

The experimental results are shown in Fig. 4. It is interesting to note that our approach achieves significant gains compared to the baseline on the small collection with 2, 000 documents. Meanwhile, with the increase in the document collection size, the performance of our approach shows a slightly higher improvement over the baseline. In summary, the experimental results show that our strategy augments performance on different sizes of document collection, even on the small document collection, and the improved performance is stable with the increasing size of document collection.

## 5   Conclusion and Future Work

This paper presents a novel approach for the semantic relation analysis. In this approach, the semantic relation between terms is measure based on both the explicit and implicit relations. The experiment results indicate that our approach can significantly improve the performance of document clustering.

In the future, we will work on three aspects to improve our approach: (1) the independence test is essential to determine whether two terms co-occur together more often than by chance; (2) the optimal integration of the explicit and implicit relations can be further improved; (3) the reduction of time complexity is worthy further analysis.

**Acknowledgement.**   This work was supported by the National Natural Science Foundation of China (No. 61075056, 61273304), the Specialized Research Fund for the Doctoral Program of Higher Education of China (No. 20130072130004) and the Fundamental Research Funds for the Central Universities.

## References

1. Billhardt, H., Borrajo, D., Maojo, V.: A context vector model for information retrieval. Journal of the American Society for Information Science and Technology 53(3), 236–249 (2002)
2. Budanitsky, A., Hirst, G.: Evaluating wordnet-based measures of lexical semantic relatedness. Computational Linguistics 32(1), 13–47 (2006)
3. Bullinaria, J.A., Levy, J.P.: Extracting semantic representations from word co-occurrence statistics: A computational study. Behavior Research Methods 39(3), 510–526 (2007)
4. Craven, M., DiPasquo, D., Freitag, D., McCallum, A., Mitchell, T., Nigam, K., Slattery, S.: Learning to extract symbolic knowledge from the world wide web. In: Proceedings of the 15th National Conference on Artificial Intelligence (1998)

5. Figueiredo, F., Rocha, L., Couto, T., Salles, T., Gonçalves, M.A., Meira Jr, W.: Word co-occurrence features for text classification. Information Systems 36(5), 843–858 (2011)
6. Gabrilovich, E., Markovitch, S.: Computing semantic relatedness using wikipedia-based explicit semantic analysis. In: IJCAI, vol. 7, pp. 1606–1611 (2007)
7. Hu, X., Zhang, X., Lu, C., Park, E.K., Zhou, X.: Exploiting wikipedia as external knowledge for document clustering. In: Proceedings of the 15th ACM SIGKDD International Conference on Knowledge Discovery and Data Mining, pp. 389–396 (2009)
8. Kalogeratos, A., Likas, A.: Text document clustering using global term context vectors. Knowledge and Information Systems 31(3), 455–474 (2012)
9. Lang, K.: Newsweeder: Learning to filter netnews. In: Proceedings of the Twelfth International Conference on Machine Learning, pp. 170–178 (1995)
10. Lewis, D.D.: Reuters-21578 text categorization test collection, distribution 1.0 (1997), http://www.research.att.com/~lewis/reuters21578.html
11. Burgess, C., Lund, K.: Modelling parsing constraints with high-dimensional context space. Language and cognitive processes 12(2-3), 177–210 (1997)
12. Miller, G.A.: Wordnet: a lexical database for english. Communications of the ACM 38(11), 39–41 (1995)
13. Wang, P., Hu, J., Zeng, H.J., Chen, Z.: Using wikipedia knowledge to improve text classification. Knowledge and Information Systems 19(3), 265–281 (2009)
14. Wong, S.K.M., Ziarko, W., Wong, P.: Generalized vector spaces model in information retrieval. In: SIGIR 1985. pp. 18–25. ACM (1985)

# Multi-granulation Ensemble Classification
# for Incomplete Data

Yuan-Ting Yan, Yan-Ping Zhang*, and Yi-Wen Zhang

School of Computer Science and Technology, Anhui University,
Key Laboratory of Intelligent Computing and
Signal Processing of Ministry of Education,
Hefei, Anhui Province, 230601, China
zhangyp2@gmail.com

**Abstract.** A new learning algorithm is introduced that can deal with
incomplete data. The algorithm uses a multi-granulation ensemble of
classifiers approach. Firstly, the missing attributes tree (MAT) was constructed according to the missing values of samples. Secondly, the incomplete dataset was projected into a group of data subsets based on
MAT, those data subsets were used as the training sets for the neural
network. Based on bagging algorithm, each data subset was used to generate a group of classifiers and then using classifier ensemble to get the
final prediction on each data subset. Finally, we adopt the conditional
entropy as the weighting parameter to overcome the precision insuffi-
ciency of dimension based algorithm. Numerical experiments show that
our learning algorithm can reduce the influence of missing attributes
for classification results, and it is superior in performance to algorithm
compared.

**Keywords:** Incomplete data, multi-granulation, missing attribute tree,
neural network, classifier ensemble.

## 1 Introduction

Common classification algorithm requires that the number of attributes of testing samples is equal to the number of attributes of samples during the learning
process. In many real word applications, missing values are often inevitable.
There are various reasons for their existence, such as equipment errors, data loss
and manual data input, etc. Therefore, classification should be capable to deal
with incomplete data. The missing value problem has long been recognized as an
important practical issue and has been intensively investigated [1-3]. The simplest way of dealing with incomplete data is to ignore the samples with missing
values [4]. However, this method will not be able to use all of the information,
and it could also discard some important information. It is not practical when
the dataset contains a relatively large number of samples with missing values.

---

* Corresponding author.

D. Miao et al. (Eds.): RSKT 2014, LNAI 8818, pp. 343–351, 2014.
DOI: 10.1007/978-3-319-11740-9_32 © Springer International Publishing Switzerland 2014

Another common method is the imputation method[5-8]. This method often cannot satisfy the Missing at Random assumption and lead to data bias [9].

Ensemble method was applied to deal with incomplete data by Krause and Polikar [10], sub classifier was trained on random feature subset and then the ensemble learning can be conducted. However, this method can not ensure that all samples can be trained because of the random select with feature subset. Chen and Jiang [11-12] propose a noninvasive procedure ensemble learning method. This method uses neural network as the base classifier, and does not need any assumption about distribution or other strong assumptions. But this method believes that the generalization ability of sub classifier is related to the number of missing attributes. It does not consider the existence of redundant attributes and the difference influence degree of attributes.

This paper provides a classification algorithm of incomplete dataset by considering the influence degree of attributes. The missing attributes tree (MAT) is constructed according to the missing values of samples. Then the incomplete dataset is divided into a group of data subsets based on MAT. After that, the neural network is trained on these data subsets based on bagging algorithm [13], and ensemble learning can be implemented by using conditional entropy as the weighting parameter. With the point view of granular computing[14-18], the algorithm in this paper is called multi-granulation ensemble classification algorithm for incomplete data (for short, MGEI).

The study is organized as follows. Some basic concepts of information entropy are briefly reviewed in Section 2. In section 3, we establish the missing attributes tree (MAT) of incomplete dataset and then project the incomplete dataset into a group of data subsets based on MAT. After that, we propose the multi-granulation ensemble classification algorithm for incomplete data (MGEI algorithm). Experiments on datasets from UCI Machine Learning Repository in section 4 show that MGEI outperform the dimension based algorithm in terms of prediction accuracy. Finally, Section 5 concludes this paper by bring some remarks and discussion.

## 2  Information Entropy

Information entropy was first considered by C.E Shannon in the field of data communication [19]. It was used to evaluate the degree of information from a given random event. We introduce the concept about information entropy for discrete variables as follows.

**Definition 2.1.** Let $X$ be a random variables, $p(x)$ represents the probability of variable $X$ values $x$. The entropy of $X$ is defined as

$$H(X) = - \sum_{x \in X} p(x) \log_b P(x) \tag{1}$$

Where $b$ is the base number, different value of $b$ represent different unit. Entropy is typically measured in bits ($b = 2$) or nats ($b = e$). Entropy is related to the event's probability directly, uniform probability distribution leads to greater entropy. Suppose $X$ have $n$ different values, $H(X) = \log n$ if and only if the probability of each value are equal.

**Definition 2.2.** Given two random variables $X$ and $Y$, $(X, Y)$ is the joint random variable and $p(x, y)$ is the corresponding probability distribution. The joint entropy of $X$ and $Y$ defined as follows

$$H(X, Y) = -\sum_{x \in X} \sum_{y \in Y} p(x, y) \log p(x, y) \tag{2}$$

Joint entropy is a measure of the uncertainty associated with a set of variables.

**Definition 2.3.** For two variables $X$ and $Y$, suppose $Y$ is known, the conditional entropy $X$ to $Y$ is defined as

$$H(X|Y) = -\sum_{x \in X} \sum_{y \in Y} p(x, y) \log p(x|y) \tag{3}$$

$H(X|Y)$ describes the degree of correlation between $X$ and $Y$. The greater the correlation degree between $X$ and $Y$, the smaller the conditional entropy is. On the contrary, the smaller the correlation degree, the greater the conditional entropy is.

## 3 New Algorithm for Incomplete Dataset (MGEI)

Ensemble learning algorithm is generally getting a group of sample sets through sampling. Then these sample sets are used to training the classifiers. However, because of the existence of the missing values, this kind of method for incomplete datasets cannot be applicable. In this section, we will propose a multi-granulation ensemble algorithm to deal with incomplete datasets based on information entropy.

### 3.1 Construction of MAT

Missing values are various in dataset. We first construct the MAT according to the missing values as follows:

First, traversing the dataset, and dividing the samples with the same missing attributes into the same sample set. Then, according to the number of missing attributes, the sample sets were sorted in ascending order. Hence, we got some sample set $X^{mset}$, where $mset$ denote the missing attributes set. Finally, establishing MAT (illustrated in Fig.1) based on the inclusion relationship of missing attributes set.

For any node in MAT, the missing attributes set corresponding to the node is included in its child node's missing attributes set. Note that in order to avoid a loop here, we sorted the node in ascending order from left to right. For example, attributes set (1), (2) and (1, 2) satisfies $(1) \subset (1, 2) \wedge (2) \subset (1, 2)$. We make the following rules: Let $X^{(1,2)}$ be the child node of $X^{(1)}$, the child node of $X^{(1)}$ satisfy $X^{(1,i)}, i > 1$.Similarly, let $X^{(2,j)}, j > 2$ be the child nodes of $X^{(2)}$ and $X^{(1,...,k_m,l)}, l > k_m + 1$ be the child nodes of $X^{(1,...,k_m)}$.

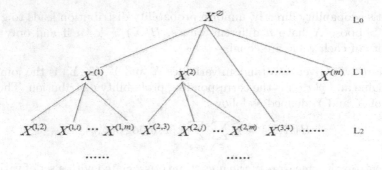

**Fig. 1.** Diagram of MAT. $L_0$ denotes the level corresponding to the sample with no missing attributes. Similarly, $L_i$ represents the level corresponding to the sample with $i$ missing attributes.

## 3.2   Maximizing Data Subset

In order to make full use of the original dataset information and to improve the classification performance, the size of data subset is maximized as follows:

For arbitrary sets $X^{mset_1} \subset X^{mset}$ and $X^{mset_2} \subset X^{mset}$, if $mset_1 \subset mset_2$, then we update $X^{mset_2}(X^{mset_2} = X^{mset_2} \bigcup X^{mset_1})$. Join the samples in $X^{mset_1}$ into $X^{mset_2}$, and set these samples' missing attributes set as $mset_2$. A new MAT is obtained after the maximize operation from top to bottom in the MAT. Note that, for any node of MAT, the set inclusion relationship is only judged between the node of the upper level and the current node. Fig.2 gives the maximize operation of MAT from section 3.1.

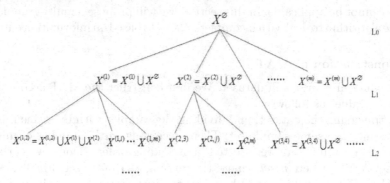

**Fig. 2.** Illustrate of maximization operation

$X^{\varnothing}$ is the root of MAT, the maximization operation is initialized from the second level ($L_2$). Because $\varnothing$ is the subset of any nonempty set, for each $X^{(i)}$ in $L_2$, update $X^{(i)}$ ($X^{(i)} = X^{(i)} \bigcup X^{\varnothing}$). Similarly, the above operation is repeated for each node in level $L_j$. e.g., $(1) \subset (1,2) \wedge (2) \subset (1,2)$, so update $X^{(1,2)}$, $X^{(1,2)} = X^{(1,2)} \cup X^{(1)} \cup X^{(2)}$. Similarly, a new MAT is obtained by updating all the nodes in MAT.

### 3.3  Constructing Sub Classifier

We use the maximized data subset as the training set for the neural network. In order to improve the prediction performance, each data subset is used to train several sub classifiers based on bagging algorithm. Then we adopt classifier ensemble to improve the prediction accuracy, this process is called sub-ensemble in this study. Formula (4) gives the majority voting method of each sub-ensemble that the weight of each sub classifier is equal.

$$y_{jm} = \arg\max_{C_k} \sum_{i=1}^{N} \sum_{tc_i=c_k} \omega^k \tag{4}$$

Where $y_{jm}$ denotes the prediction category of sample $j$ by using classifiers ensemble on the classifiers that trained on data subsets $m$, $C_k$ represents the category value, $N$ is the number of bootstrap replicates of data subsets, $tc_i$ denotes the prediction result on sub classifier $i$, $\omega^k$ represents the total weight of the prediction results $C_k$. Here the weight of each sub classifier is equal to $1/N$.

After sub-ensemble process, we got a classifier set and the size of this classifier set is equal to the number of data subsets from section 3.2. These classifiers are used to predict the testing sample at different granularity according to its missing attributes set. In order to obtain the final prediction of test sample, we need to determine the final category from the results of the sub-ensemble process. The conditional entropy of each missing attributes set is calculated on data subset $X^{\varnothing}$. Smaller conditional entropy indicates that the influence of the missing attributes set is greater, so the generalization ability of the corresponding classifier is more likely to be poor. From this point of view, we can directly use the conditional entropy as the weighting parameter to the corresponding classifier.

Note that the number of samples is different in different data subsets, and generally, more learning sample could lead to better performance of the sub classifier. In addition, the accuracy of sub classifier could also be different, higher prediction accuracy means better performance. Moreover, the influence degree of different missing attributes is also different. Considering the various factors, we use formula (5) to calculate the weight of the corresponding classifiers.

$$\omega_i = \frac{Acc_i |X^{mset_i}| H_i}{\sum Acci |X^{mset_i}| H_i} \tag{5}$$

Where $Acc_i$ denotes the prediction accuracy of classifier $h_i$, $|X^{mset_i}|$ denotes the size of $X^{mset_i}$, $H_i$ denotes the conditional entropy category attribute set to $mset_i$.

### 3.4  Multi-granulation Prediction

The maximization process in section 3.2 also shows the prediction process. Suppose a testing sample $x$ satisfies that the missing value of $x(x^{mvset})$ is included in $X^{(1)}$. And $x$ should be tested on the sub classifier corresponds to $X^{(1)}$.

Then $x$ should be tested on the sub classifier corresponds to $X^{(1,2)}$, because of the set inclusion relationship $x^{mvset} \subset (1,2)$.

For a testing sample, if the sample's missing attributes set contained in the missing attributes set of some data subsets, then, choose the corresponding classifier as the testing classifier of the sample. There may have several sub classifiers satisfy the above set inclusion condition. It could lead to some different predict results of one sample. For this case, the final prediction result $y_j$ of sample $j$ is determined by the weighted voting as (6)

$$y_j = \arg \max_{C_k} \sum_{tc_i^j = C_k} \omega^j \tag{6}$$

## 4   Experiments

We use a three-layered BP neural network as the base classifier. The number of input nodes ($id$) of the neural network is determined by the number of missing attributes, and the number of output nodes ($od$) is determined by the number of category in the dataset. According to the geometric pyramid rule, the number of hidden nodes is $\sqrt{id * od}$. We evaluate the accuracy using ten-fold cross validation approach where a given dataset is randomly partitioned into ten folds of equal size. The algorithm was tested on a variety of datasets from UCI machine learning repository [20]. In order to reduce variation of the estimate, 20 experiments for each algorithm were performed. Table 1 gives the details about the eight testing datasets. Note that CVR is the abbreviations of Congressional Voting Records. And $|subsets|$ denotes the number of data subsets of each dataset, $|X^{\varnothing}|$ denotes the size of $X^{\varnothing}$.

**Table 1.** Summary of Datasets

| Dataset | Size | dimension | $|classes|$ | $|subsets|$ | $|X^{\varnothing}|$ |
|---|---|---|---|---|---|
| adult | 32561 | 14 | 2 | 5 | 27175 |
| arrhythmia | 452 | 279 | 13 | 7 | 65 |
| bands | 540 | 39 | 2 | 64 | 240 |
| bridges | 108 | 13 | 8 | 14 | 63 |
| credit approval | 690 | 15 | 2 | 8 | 588 |
| dermatology | 366 | 34 | 6 | 2 | 325 |
| CVR | 435 | 16 | 2 | 73 | 207 |
| lung cancer | 32 | 56 | 2 | 3 | 25 |

In order to determine the number of bootstrap replicates in the sub-ensemble stage, we ran four datasets using 1, 5, 10, 15, 20 and 30 replicates to study the relationship between bootstrap replicates and classification accuracy. The results appear in Table 2. The experiments show that 10 bootstrap replicates are reasonable, the improvement of classification accuracy on all four datasets are nearly unchanged.

**Table 2.** Bagged Classification Accuracy(%)

| No.Replicates | 1 | 5 | 10 | 20 | 30 |
|---|---|---|---|---|---|
| arrhythmia | 0.596061 | 0.65197 | 0.687424 | 0.687556 | 0.687584 |
| bridges | 0.563651 | 0.576984 | 0.597751 | 0.597824 | 0.597846 |
| credit approval | 0.833333 | 0.83913 | 0.84058 | 0.84058 | 0.84058 |
| dermatology | 0.938889 | 0.966667 | 0.970244 | 0.970273 | 0.970276 |

The proposed algorithm (MGEI) is compared with the algorithm (based on the dimension) in [11]. The classification performance of these two algorithms is compared in table 5. The results can be divided into two parts

**Table 3.** Test Performance Comparison (%)

| Dataset Name | MGEI | dimension |
|---|---|---|
| adult | 0.8469745 | 0.8469009 |
| arrhythmia | 0.6874242 | 0.5984648 |
| bands | 0.5726813 | 0.5767564 |
| bridges | 0.5977248 | 0.5652856 |
| credit approval | 0.8398553 | 0.8368116 |
| dermatology | 0.97037 | 0.9294444 |
| CVR | 0.9547621 | 0.9415227 |
| lung cancer | 0.8166668 | 0.6700002 |

(a) The performance is slightly worse on dataset 'bands'. The result could be caused by the following reasons: First, the size of data subset $X^{\varnothing}$ is small or the number of classes in $X^{\varnothing}$ is smaller than the real number of classes in 'adult'. Then the conditional entropy cannot reflect the real influence of the missing attributes very well. Second, using the number of available attributes as the weighting parameter may be just closer to the real influence of the missing attributes than the conditional entropy. The performance of these two methods is nearly equal on two datasets (adult and credit approval). The reason could be that the prediction results of the sub classifiers are stable and the accuracy is higher in the method with no sub-ensemble. This lead to the final prediction accuracy of the two methods is similar.

(b) Five datasets (arrhythmia, bridges, dermatology, CVR and lung cancer) have better performance than the comparison algorithm. Two reasons could lead to the superior of performance on prediction accuracy. First, the sub-ensemble process improves the prediction accuracy on each granulation, and it also makes the prediction results more stable on each granulation. Second, the conditional entropy on $X^\varnothing$ are more effective than the dimension of the available attributes to measure the influence of missing attributes set. Generally speaking, our method MGEI is superior to the method in [11].

## 5    Conclusion and Discussion

We presented the MGEI algorithm, employing the ensemble of classifiers to realize the classification of incomplete dataset. The algorithm first construct the MAT of the incomplete dataset, then dividing the dataset into a group of data subsets based on MAT. After that, the algorithm generates multiple classifiers , each trained on a data subset based on bagging algorithm. When a sample of unknown label that has missing values is presented to the algorithm, the algorithm picks all the classifiers that did not use the missing values in its training. Then use the conditional entropy as the weighting parameter to determine the importance of each classifier. After that, these classifiers are combined through weighted majority voting.

Some datasets drawn from UCI machine learning repository were used to evaluate the proposed algorithm. Comparative experiments show that the performance of the proposed algorithm has better performance than the algorithm based on dimension. Moreover, for some datasets, the improvement on prediction accuracy is surprising (arrhythmia, dermatology and lung-cancer).

Because the conditional entropy was computed on the data subset $X^\varnothing$, so the performance of MGEI is closely related to the size of sub dataset $X^\varnothing$. More complete samples will make the conditional entropy more close to the real impact of missing attributes, and it is more likely to lead to better performance. When the dataset is lack of complete samples, the performance of the algorithm remains need to be improved. In addition, this paper focus on neural network ensemble, considering other classification methods to deal with incomplete dataset is also our future work.

**Acknowledgments.** This work was supported by National Natural Science Foundation of China (Nos.61175046 and 61073117), Natural Science Foundation of Anhui Province (No.1408085MF132).

# References

1. Wiberg, T.: Computation of principal components when data are missing. In: Second Symp. Computational Statistics, pp. 229–236 (1976)
2. Tipping, M.E., Bishop, C.M.: Probabilistic principal component analysis. Journal of the Royal Statistical Society: Series B (Statistical Methodology) 61(3), 611–622 (1999)
3. Geng, X., Smith-Miles, K., Zhou, Z.H., Wang, L.: Face image modeling by multi-linear subspace analysis with missing values. IEEE Transactions on Systems, Man and Cybernetics, Part B: Cybernetics 41(3), 881–892 (2011)
4. Allison, P.D.: Missing Data. Sage Publications, Thousand Oaks (2001)
5. Feng, H.H., Chen, G.S., Yin, C., Yang, B.R., Chen, Y.M.: A SVM regression based approach to filling in missing values. In: Knowledge-Based Intelligent Information and Engineering Systems, pp. 581–587 (2005)
6. Farhangfar, A., Kurgan, L.A., Pedrycz, W.A.: A novel framework for imputation of missing values in databases. IEEE Transaction on Systems, Man and Cybernetics 37(5), 692–709 (2007)
7. Farhangfar, A., Kurgan, L.A., Jennifer, D.: Impact of imputation of missing values on classification error for discrete data. Pattern Recognition 41(12), 3692–3705 (2008)
8. Luengo, J., Sáez, J.A., Herrera, F.: Missing data imputation for fuzzy rule-based classification systems. Soft Computing 16(5), 863–881 (2012)
9. Little, R.J., Rubin, D.B.: Statistical Analysis with Missing Data. Wiley, New York (2002)
10. Krause, S., Polikar, R.: An ensemble of classifiers approach for the missing feature problem. In: Proceedings of the International Joint Conference on Neural Networks, pp. 553–558. IEEE Press (2003)
11. Jiang, K., Chen, H.X., Yuan, S.M.: Classification for incomplete data using classifier ensembles. In: International Conference on Neural Networks and Brain, vol. 1, pp. 559–563. IEEE Press (2005)
12. Chen, H.X., Du, Y.P., Jiang, K.: Classification of incomplete data using classifier ensembles. In: International Conference on Systems and Informatics, pp. 2229–2232. IEEE Press (2012)
13. Breiman, L.: Bagging predictors. Machine learning 24(2), 123–140 (1996)
14. Yao, Y.Y., Zhang, N., Miao, D.Q.: Set-theoretic Approaches to Granular Computing. Fundamenta Informaticae 115(2), 247–264 (2012)
15. Yao, Y.Y., Yao, J.T.: Granular computing as a basis for consistent classification problems. In: Proceedings of PAKDD, vol. 2, pp. 101–106 (2002)
16. Qian, Y.H., Liang, J.Y., Dang, C.Y.: Incomplete multigranulation rough set. IEEE Transactions on Systems, Man and Cybernetics, Part A: Systems and Humans 40(2), 420–431 (2010)
17. Zhu, W.: Generalized rough sets based on relations. Information Sciences 177(22), 4997–5011 (2007)
18. Xu, J., Shen, J., Wang, G.: Rough set theory analysis on decision subdivision. In: Tsumoto, S., Słowiński, R., Komorowski, J., Grzymała-Busse, J.W. (eds.) RSCTC 2004. LNCS (LNAI), vol. 3066, pp. 340–345. Springer, Heidelberg (2004)
19. Shannon, C.E.: A mathematical theory of communication. Bell System Technical Journal 27, 379–423, 623–656 (1948)
20. UCI Repository of machine learning databases for classification, http://archive.ics.uci.edu/ml/datasets.html

# Heterogeneous Co-transfer Spectral Clustering

Liu Yang[1,2], Liping Jing[1,*], and Jian Yu[1]

[1] Beijing Key Lab of Traffic Data Analysis and Mining,
Beijing Jiaotong University, Beijing, China
[2] College of Mathematics and Computer Science,
Hebei University, Baoding, Heibei, China
{11112091,lpjing,jianyu}@bjtu.edu.cn

**Abstract.** With the rapid growth of data collection techniques, it is very common that instances in different domains/views share the same set of categories, or one instance is represented in different domains which is called co-occurrence data. For example, the multilingual learning scenario contains documents in different languages, the images in the social media website simultaneously have text descriptions, and etc. In this paper, we address the problem of automatically clustering the instances by making use of the multi-domain information. Especially, the information comes from heterogeneous domains, i.e., the feature spaces in different domains are different. A heterogeneous co-transfer spectral clustering framework is proposed with three main steps. One is to build the relationships across different domains with the aid of co-occurrence data. The next is to construct a joint graph which contains the inter-relationship across different domains and intra-relationship within each domain. The last is to simultaneously group the instances in all domains by applying spectral clustering on the joint graph. A series of experiments on real-world data sets have shown the good performance of the proposed method by comparing with the state-of-the-art methods.

**Keywords:** Heterogeneous feature spaces, co-transfer learning, spectral clustering, canonical correlation analysis.

## 1 Introduction

Clustering algorithms are generally used in an unsupervised fashion. They are presented with a set of instances that can be grouped according to some similarity or distance strategies, and perform well when they are provided with a large amount of data. However, in practice, it is hard or expensive to collect ample data in some views, for example, the annotation information of the social media. In this case, most existing methods will lead to poor performance. In order to solve this problem, various learning strategies have been proposed such as semi-supervised learning [22], transfer learning [17,12,20], and etc, to make use of the information from different domains. These methods can be roughly divided into

---

* Corresponding author.

D. Miao et al. (Eds.): RSKT 2014, LNAI 8818, pp. 352–363, 2014.
DOI: 10.1007/978-3-319-11740-9_33 © Springer International Publishing Switzerland 2014

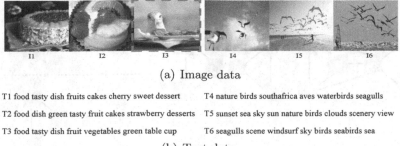

(a) Image data

T1 food tasty dish fruits cakes cherry sweet dessert      T4 nature birds southafrica aves waterbirds seagulls

T2 food dish green tasty fruit cakes strawberry desserts  T5 sunset sea sky sun nature birds clouds scenery view

T3 food tasty dish fruit vegetables green table cup       T6 seagulls scene windsurf sky birds seabirds sea

(b) Text data

**Fig. 1.** Instance examples in heterogenous domains (a) image domain and (b) text domain from the NUS-WIDE dataset. The first three images and documents belong to "food", and the last three images and documents belong to "bird".

two categories. One deals with the multi-domain data within same feature space, such as semi-supervised learning. The other concerns multi-domain data across heterogenous feature spaces as shown in Fig.1. In this paper, we focus on the analysis with heterogenous data.

As demonstrated in Fig.1, six images in Fig.1(a) and six documents in Fig.1(b) are all related with two categories "birds" and "food". When clustering the images in Fig.1(a), it is a little difficult to decide their correct categories. Because the essential structure between images is not very clear as shown in Fig.2(a) (each line indicates that two nodes has relatively high similarity). Especially, the image $I_3$ (marked by the dotted line) may be misclassified into the other category. On the other hand, the document collection has a clear structure as shown in Fig.2(b). Based on this observation, some researchers tried to combine the image processing and text mining tasks to enhance performance of the singe task, and proposed multi-view learning [2,11,3], heterogenous transfer learning [21,23] and co-transfer learning [15,16]. Among them, multi-view learning prefers to the situation where each instance appears in all views, i.e., each instance is represented in all feature spaces. Heterogenous transfer learning aims to improve the performance of the learning task in one target domain. Co-transfer learning framework makes use of the co-transferring knowledge across different domains to simultaneously handle multi-learning tasks, however, the existing methods are only designed for supervised learning rather than unsupervised learning.

In real applications, the following situations are ubiquitous: a) the instances in different domains are different but they share the same set of categories, b) there is no labeled data in each domain, c) the data from different domains are expected to be simultaneously handled. In this case, multi-view learning cannot work for a), co-transfer learning cannot be applied in b) and heterogenous transfer learning cannot perform on c). Thus, in this paper, we propose a Heterogeneous Co-Transfer Spectral Clustering framework (HCTSC) to simultaneously handle these situations with the aid of co-occurrence data. Like heterogenous transfer learning [21] and co-transfer learning [15], we assume that it is possible

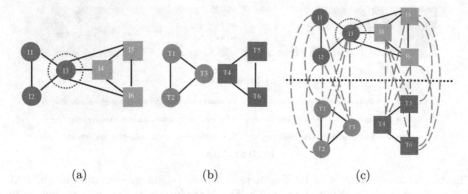

**Fig. 2.** Demonstration of graph constructed on different data, a) on image data, b) on text data, and c) on both image and text data via the proposed method HCTSC. The instances having same category label are marked with same shape (circle for "food" and square for "birds"). The solid line indicates linkage based on the single domain data, the dotted line is the new linage after merging two tasks via HCTSC.

to obtain co-occurrence information between different domains so that we can build a bridge to transfer knowledge across domains. For instance, the images co-occur with texts on the same web page [6], and the linkage between texts and images in social networks [18,23], can be used for co-occurrence information. HCTSC makes use of the co-occurrence data to form a bridge to connect two separate clustering tasks and improve the final performance of all tasks.

HCTSC framework contains three main steps: the first step is to build the relationship across heterogenous domains. We utilize canonical correlation analysis technique [10] to obtain two mapping functions from two feature spaces to one common space, then, re-represent the two-domain data via the mapping functions. The second step is to construct a joint graph which contains the inter-relationship across heterogenous domains and intra-relationship in each domain (as shown in Fig.2(c)). Note that we ignore the edges indicating weak relationships between instances in this figure. Finally, the clustering results of all instances in different domains can be simultaneously obtained by applying the spectral clustering method [13] on the joint graph. The joint graph (such as Fig.2(c)) contains more information than the two separate graphs (Fig.2(a) and Fig.2(b)). For example, image $I_3$ links more edges to the circle part in Fig.2(c), then it leads to a higher probability to correctly cluster $I_3$ into the circle part.

The main contribution of this paper is to propose a Heterogeneous Co-Transfer Spectral Clustering (HCTSC) framework. It has the following merits: (i) HCTSC has ability to co-transfer knowledge across different heterogeneous domains. (ii) It can simultaneously perform more than one unsupervised learning (clustering) tasks. (iii) It outperforms the state-of-the-art algorithms in the real-world benchmark data sets. The rest of this paper is organized as follows. In Section 2, the related work is described. In Section 3, a new heterogeneous co-transfer spectral clustering framework is presented. In Section 4, a series of experimental results are listed to illustrate the performance of HCTSC. Finally, some concluding remarks are given in Section 5.

## 2   Related Work

Traditional clustering algorithms assume that there are a large amount of instances for better performance. However, it is hard or expensive to collect enough instances in some real applications. In this case, traditional clustering algorithms will lead to poor performance as shown in Fig.2(a). Many transfer learning strategies have been proposed to solve this problem by using the information from other heterogenous domains/views, such as multi-view learning [2,11,3], heterogenous transfer learning [21,23] and co-transfer learning [15,16].

Multi-view clustering methods [2,11,3] have been proposed to improve the performance over traditional single-view clustering by utilizing more information from other views. The multi-view clustering requires that each instance should be represented in all views, but in our clustering tasks, there is only one view in one clustering task.

Various heterogenous transfer learning strategies [21,23,1,19] have been proposed and developed to perform learning tasks across different domains. In [21], Yang et. al proposed to combine the source domain and the target domain in one annotation-based Probabilistic Latent Semantic Analysis (aPLSA) model for image clustering. In [23], Zhu et al. considered the Heterogenous Transfer Learning problem by using unlabeled text data for helping Image Classification via the HTLIC model. Correlational Spectral Clustering (CSC) [1] made use of the kernel canonical correlation analysis based on the co-occurrence data to project the original data in the kernel space. Singh et al. [19] identified the latent bases with the co-occurrence data by the Collective Non-negative Matrix Factorization (CNMF) model, and mapped all instances to the latent bases space. These heterogenous transfer learning methods only consider to migrate the knowledge from the source domains to one target domain. While our heterogenous co-transfer spectral clustering framework can transfer knowledge across different domains simultaneously.

Co-transfer learning [15,16] makes use of transferring knowledge cross different domains and performs supervised learning simultaneously via the knowledge learned from co-occurrence data. It focuses on how to use labeled data in multiple different domains to enhance the performances of all tasks simultaneously. The difference of the co-transfer learning and our model is that it bases on the labeled data for classification while our model is unsupervised for clustering tasks.

## 3   Heterogenous Co-transfer Spectral Clustering

The Heterogenous Co-Transfer Spectral Clustering (HCTSC) algorithm is a technique which combines all instances in heterogenous domains to one joint graph. In this section, we show how to learn the inter-relationships of instances from different domains, and construct the joint graph with all instances. Spectral clustering is used on the joint graph to obtain the clustering results. The last part is the extended HCTSC model for more than two domains.

### 3.1   Data Representation

Given $n_1$ and $n_2$ instances $\{\mathbf{x}_i\}_{i=1}^{n_1}$ and $\{\mathbf{y}_j\}_{j=1}^{n_2}$ in two heterogenous domains respectively (where $\mathbf{x}_i \in \mathbb{R}^{m_1}$ and $\mathbf{y}_j \in \mathbb{R}^{m_2}$), we have two data matrices $X = [\mathbf{x}_1, \cdots, \mathbf{x}_{n_1}] \in \mathbb{R}^{m_1 \times n_1}$ and $Y = [\mathbf{y}_1, \cdots, \mathbf{y}_{n_2}] \in \mathbb{R}^{m_2 \times n_2}$. The goal is to partition $\{\mathbf{x}_i\}_{i=1}^{n_1}$ and $\{\mathbf{y}_j\}_{j=1}^{n_2}$ into $c$ clusters respectively. Traditional clustering methods, such as k-means and spectral clustering, handle these two tasks separately. If there are not enough data in each clustering task, the performance may be not satisfied. For example, if we use the spectral clustering on the graphs (Fig.2(a) and 2(b)) respectively, the image $I_3$ in Fig.2(a) is misclassified.

We assume that although $\mathbf{x}_i$ and $\mathbf{y}_j$ are in different domains, they share the same categories. Under this situation, the two separate tasks can be combined and accomplished to enhance performance of each single task. Because $\mathbf{x}_i$ and $\mathbf{y}_j$ come from heterogenous domains, there should be some information to link the instances in all domains. The co-occurrence data (such as the images co-occur with texts on the same web page, and the linkage between texts and images) can be easily obtained in real application [6,21,23], they can be used to be a bridge to link two separated tasks. We assume that there is a co-occurrence data set $O = \{A, B\}$ which contains $n_o$ instances described in two domains ($A \in \mathbb{R}^{m_1 \times n_o}$ and $B \in \mathbb{R}^{m_2 \times n_o}$), where $A$ and $X$, $B$ and $Y$ are in the same domain respectively. We expect to combine two clustering tasks together by using the co-occurrence data to improve the single clustering performance.

### 3.2   Inter-Relationships across Heterogenous Domains

The instance interactions from heterogenous domains are studied in this subsection. The instances $\mathbf{x}_i$ and $\mathbf{y}_j$ are in different feature spaces with different dimensional feature vectors, the relationship cannot be computed directly by using traditional similarity or distance methods. $\mathbf{x}_i$ and $\mathbf{y}_j$ should be firstly mapped to the common space, and then the similarity matrix can be computed in the new space. Canonical correlation analysis (CCA) is general technique which can learn the projections by utilizing the co-occurrence data to simultaneously map the data from all feature spaces to one common space [10]. So we use CCA to learn the mapping matrices $W_a$ and $W_b$ based on the set $O = \{A, B\}$.

$$\max_{W_a, W_b} \ W_a^T A B^T W_b$$
$$\text{subject to} \quad W_a^T A A^T W_a = I, W_b^T B B^T W_b = I \tag{1}$$

where $W_a \in \mathbb{R}^{m_1 \times r}$, $W_b \in \mathbb{R}^{m_2 \times r}$, and $r$ is the dimension of the common space. Then $\mathbf{x}_i$ and $\mathbf{y}_j$ can be mapped to the new $r$-dimensional space by $\mathbf{p}_i = \mathbf{x}_i \times W_a$ and $\mathbf{q}_j = \mathbf{y}_j \times W_b$ respectively.

## 3.3   Joint Graph Construction

Given $\{\mathbf{x}_i\}_{i=1}^{n_1}$ in the $m_1$-dimensional feature space, the similarity of $\mathbf{x}_i$ and $\mathbf{x}_k$ can be calculated. In this paper, we use Gaussian kernel function [7] to yield the non-linear version of similarity in the intrinsic manifold structure of data.

Then, we construct the $n_1$-by-$n_1$ nearest neighbor graph $G^{(1,1)}$ whose points indicate all instances and edges denote the similarities with the neighbor instances following [9]. The $(i, k)$-th entry of $G_{i,k}^{(1,1)}$ is given as follows,

$$G_{i,k}^{(1,1)} = \begin{cases} \exp(-||\mathbf{x}_i - \mathbf{x}_k||^2/\sigma^2), & k \in \mathcal{N}_i \text{ or } i \in \mathcal{N}_k \\ 0, & \text{otherwise} \end{cases} \tag{2}$$

where $\mathcal{N}_i$ and $\mathcal{N}_k$ are the index sets of $\alpha$ nearest-neighbors of point $\mathbf{x}_i$ and $\mathbf{x}_k$ respectively. The neighborhood size ($\alpha$) for the graph construction is self-tuned as in [4]. $\sigma$ is a positive parameter to control the linkage in the manifold, and like [4], the average of $\alpha$ distance values is used for the parameter $\sigma$.

Similarly, given $\{\mathbf{y}_j\}_{j=1}^{n_2}$ in the $m_2$-dimensional feature space, we can also compute the nearest neighbor graph $G^{(2,2)} \in \mathbb{R}^{n_2 \times n_2}$ via (2) by replacing $\mathbf{x}$ with $\mathbf{y}$. The graphs $G^{(1,1)}$ and $G^{(2,2)}$ represent the intra-relationships within the same space.

Meanwhile, we can construct an $n_1$-by-$n_2$ matrix $G^{(1,2)}$ with its $(i, j)$-th entry given by the similarity of $\mathbf{x}_i$ and $\mathbf{y}_j$ by replacing $\mathbf{x}_i$ with $\mathbf{p}_i$ and $\mathbf{x}_k$ with $\mathbf{q}_j$ in (2). The graph $G^{(1,2)}$ describes the inter-relationship across different spaces. $G^{(1,2)}$ is not necessarily to be symmetric, but it has $[G^{(1,2)}]_{i,j} = [G^{(2,1)}]_{j,i}$, i.e., $G^{(1,2)}$ is the transpose of $G^{(2,1)}$.

$$G = \begin{bmatrix} G^{(1,1)} & G^{(1,2)} \\ (G^{(1,2)})^T & G^{(2,2)} \end{bmatrix}. \tag{3}$$

Finally, by using the graphs $G^{(1,1)}$, $G^{(2,2)}$ and $G^{(1,2)}$, we construct the joint nearest neighbor graph $G$ via (3). The joint graph $G$ contains the inter-relationship across different domains and the intra-relationship within the same domain.

## 3.4   Spectral Clustering on the Joint Graph

After constructing the joint graph $G$, we can exploit the properties of the Laplacian of the graph by $L = D - G$, where $D$ is a diagonal matrix with $D_{i,i} = \sum_j G_{i,j}$. The first smallest $c$ eigenvectors of the graph Laplacian can be obtained, and then k-means method can be used on them to assign each instance in the graph to one of the clusters [14].

Here we briefly outline the Heterogenous Co-Transfer Spectral Clustering framework in Algorithm 1. Step 1 is to learn the inter-relationship across heterogenous domains, Steps 2-3 are to construct the joint graph and Steps 4-7 are the procedures of spectral clustering on the joint graph. More details of them are in Section 3.2, 3.3 and 3.4 respectively.

---

**Algorithm 1.** The Heterogenous Co-Transfer Spectral Clustering algorithm

---

**Input:**
   Two data matrices $X$ and $Y$, co-occurrence data set $O = \{A, B\}$.

**Output:**
   Clustering results of $X$ and $Y$.

1. Learn the inter-relationships across heterogenous domains by using CCA via (1) on $O = \{A, B\}$.
2. Construct the nearest neighbor graphs $G^{(1,1)}, G^{(2,2)}, G^{(1,2)}$ via (2).
3. Combine $G^{(1,1)}, G^{(2,2)}, G^{(1,2)}$ to form the $(n_1 + n_2) * (n_1 + n_2)$ joint graph via (3).
4. Compute the Laplacian matrix by $L = D - G$, where $D_{i,i} = \sum_j G_{i,j}$.
5. Decompose $L$ to obtain the eigenvector matrix denoted by $U$.
6. Select the first $c$ columns of $U$, and then divide them into the first $n_1$ and the other $n_2$ rows to form an $n_1 \times c$ matrix $U^{(1)}$ and an $n_2 \times c$ matrix $U^{(2)}$ respectively.
7. Run the k-means algorithm on $U^{(1)}$ and $U^{(2)}$ respectively and assign each instance to the corresponding cluster.

---

### 3.5   Generation with Multiple Clustering Tasks

We remark that the above HCTSC model can be generalized for the co-transfer spectral clustering with data in $K$ domains. The extended HCTSC model can simultaneously implement $K$ clustering tasks, one task for one domain. The joint graph is constructed as shown in (4).

$$
G = \begin{bmatrix} G^{(1,1)} & G^{(1,2)} & \cdots & G^{(1,K)} \\ G^{(2,1)} & G^{(2,2)} & \cdots & G^{(2,K)} \\ \cdots & \cdots & \cdots & \cdots \\ G^{(K,1)} & G^{(K,2)} & \cdots & G^{(K,K)} \end{bmatrix}, \tag{4}
$$

where $\{\{G^{(i,j)}\}_{i=1}^K\}_{j=1}^K$ is the nearest neighbor graph set based on the $i$-th and $j$-th domains and the similarities with the neighbor instances in the graph can be computed via (2).

The clustering on joint graph $G$ can be implemented with Steps 4-7 in Algorithm 1. Note that in Step 6, the first $c$ columns of $U$ should be divided into $K$ parts $\{U^{(i)}\}_{i=1}^K$, where each part $U^{(i)}$ is an $n_i \times c$ matrix. The final clustering results can be obtained by applying k-means on each matrix $U^{(i)}$.

## 4   Experimental Results

In this section, a series of experiments are conducted to demonstrate the effectiveness of the proposed HCTSC algorithm.

### 4.1   Methodology

The proposed HCTSC algorithm is tested on two real-world data sets. The first data set is NUS-WIDE [5] which comes from Flickr and contains 81 categories.

We follow [15] to construct 45 binary tasks by selecting 10 categories (bird, boat, flower, food, rock, sun, tower, toy, tree, car). For each task, 600 images, 1,200 texts and 1,600 co-occurred image and text pairs are sampled. 500 visual words and 1,000 text terms are extracted to build the image and document features. The second dataset is Cross-language Data which is crawled from Google and Wikipedia [16], it contains two categories of text documents about birds and animals. Among them, there are 3,415 English documents in the first domain, 2,511 French documents in the second domain, 3,113 Spanish documents in the third domain, and 2,000 co-occurred English-French-Spanish documents. 5,000 important terms extracted from each domain are used to form a feature vector.

The proposed HCTSC is compared with six methods. Among them, k-means [8] and spectral clustering (SC) only use the instances in single domain, the other four methods (aPLSA [21], CNMF [19], HTLIC [23] and CSC [1]) are based on transfer learning by making use of the co-occurrence data. In aPLSA, CNMF and HTLIC, there is a parameter which controls the transferred weight from source domain to target domain within the range [0-1]. To obtain the best performance for them, different values from 0 to 1 with the step 0.1 have been tried, and the best results are recorded in the following experiments.

The clustering result is evaluated by comparing the obtained label of each instance with the label provided by the data set. The performance is measured by the accuracy with averaging 10 trials. The larger the accuracy, the better the clustering performance.

## 4.2   Experiment 1: NUS-WIDE Data

Since aPLSA, CNMF, HTLIC and CSC only consider to migrate the knowledge from one source domain to one target domain, the first experiment records the results of all methods on one clustering task.

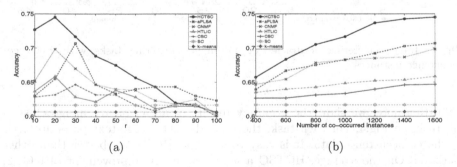

(a)                                    (b)

**Fig. 3.** Effect of (a) number of canonical variables or bases $r$ and (b) co-occurred data size on image clustering task

In order to demonstrate how to select proper value of $r$, which is the number of canonical variables in HCTSC and CSC or the number of bases in aPLSA,

CNMF and HTLIC, we randomly select one binary image clustering task (with 600 images, 300 for birds and 300 for sun, and 1,600 image-text pairs) as an example. Fig.3(a) gives the image clustering performance in terms of varying the number of $r$. We can see that HCTSC can get the best performance with 20 bases which is much smaller than the number of original image features (500). Then, the computational complexity of clustering can be efficiently reduced with the new image representation. In the experiments, we set $r = 20$.

Usually, the performance is affected by the co-occurrence data size (i.e.,$n_o$) on transfer learning. We run one image clustering task (birds vs sun) with different sizes (400-1,600) as an example, and the accuracies are recorded in Fig. 3(b). As we can see, when the number of co-occurrence instances increases, the accuracies of HCTSC, aPLSA, CNMF, HTLIC and CSC increase as well. It indicates that more co-occurrence instances make the representation more precise and helpful for the clustering data in transfer learning. The performances of transfer learning methods are better than k-means and spectral clustering which do not transfer knowledge from other domains. As expected, the proposed HCTSC outperforms other transfer learning methods ( aPLSA, CNMF, HTLIC and CSC). The reason is that in our model the text information is used in both the image representing and clustering processes, while other methods only use it for the image re-representation.

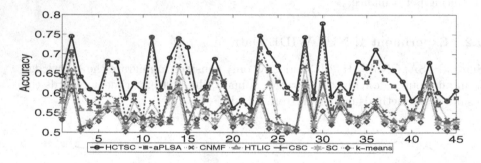

**Fig. 4.** Image clustering accuracies of 45 image clustering tasks with 1,600 co-occurrence instances

Furthermore, we show the accuracy results of all 45 binary image clustering tasks in Fig.4. For each task, the related 1,600 image-text pairs are used as the co-occurrence data. It is easy to see that HCTSC is better than other methods. On the average, HCTSC achieves accuracy improvement of 0.0362, 0.0747, 0.0952, 0.0987, 0.1083 and 0.1188 (0.6502 versus 0.6200, 0.5815, 0.5610, 0.5575, 0.5479 and 0.5374) against aPLSA, CNMF, HTLIC, CSC, SC and k-means on all image clustering tasks respectively. The corresponding win-tie-loss values with pairwise t-tests at 0.03 significance level for HCTSC against aPLSA, CNMF, HTLIC, CSC, SC and k-means are 33-9-3, 39-5-1, 43-2-0, 42-3-0, 45-0-0 and 45-0-0 respectively. This indicates that the image clustering benefits from the knowledge transferred from text domain. Because HCTSC can utilize the text information more effective and reasonable, it gains better performance than other methods.

## 4.3   Experiment 2: Cross-Language Data

In this experiment, we evaluate the proposed HCTSC for cross-language clustering. Because aPLSA, CNMF, HTLIC and CSC only consider one source domain and one target domain, we show the results of clustering English documents as an example. We cluster all 3,415 English documents with the aid of 2,511 French and 2,000 co-occurred English-French documents. The clustering results of all methods with $r$=30 are shown in Fig.5. It is clear that HCTSC gains the best performance by using the joint graph which combines the information of two domains. The knowledge from French domain is applied for the re-representation of documents and the clustering procedure. HCTSC benefits from taking full advantage of the French information.

As described in Section 3.5, HCTSC can simultaneously handel multiple clustering tasks, we conduct the experiments on three languages data and the clustering results are shown in Table 1. Because other transfer learning methods cannot deal with all three clustering tasks, we do not give their results in this table. We list the results of k-means in the last row of Table 1 for comparison. The first row is the results of instances in single domain (i.e. spectral clustering), the second and third rows are the results in two domains, and the fourth row is the results in all three domains. Obviously, the best performance happens when using all instances in three domains, and the performance is the worst with

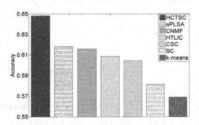

**Fig. 5.** Clustering results of English documents by using French knowledge

**Table 1.** Three cross language clustering results

| English | | French | | Spanish | |
|---|---|---|---|---|---|
| English only | 0.5816 | French only | 0.5881 | Spanish only | 0.5789 |
| English + French | 0.6481 | French + English | 0.6457 | Spanish + English | 0.6432 |
| English + Spanish | 0.6458 | French + Spanish | 0.6505 | Spanish + French | 0.6492 |
| All three | 0.6711 | All three | 0.6725 | All three | 0.6708 |
| k-means | 0.5690 | k-means | 0.5798 | k-means | 0.5654 |

the data only in one domain. The results indicate that documents in different domains help each other for single language clustering task.

## 5    Conclusions

In this paper, a heterogeneous co-transfer spectral clustering framework has been proposed to simultaneously complete more than one clustering task in different domains with the aid of co-occurrence data. The inter-relationship across different domains is learned by canonical correlation analysis via the co-occurrence data, the next is to construct a joint graph which contains the inter-relationship across different domains and intra-relationship within the same domain, and at last, the spectral clustering has been used to simultaneously group all instances in the joint graph. The proposed algorithm can be used to perform effectively in heterogenous co-transfer learning across different feature spaces like image-text and cross-language clustering problems. As for the future work, we consider how to automatically identify the transferred weights across different domains.

**Acknowledgement.** This work was supported in part by the National Natural Science Foundation of China under Grant 61375062, Grant 61370129, and Grant 61033013, the Ph.D Programs Foundation of Ministry of Education of China under Grant 20120009110006, the Opening Project of State Key Laboratory of Digital Publishing Technology, the Fundamental Research Funds for the Central Universities under Grant 2014JBM029 and Grant 2014JBZ005, the Program for Changjiang Scholars and Innovative Research Team (IRT 201206), the Beijing Committee of Science and Technology of China under Grant Z131110002813118, the Planning Project of Science and Technology Department of Hebei Province under Grant 13210347, and the Project of Education Department of Hebei Province under Grant QN20131006.

## References

1. Blaschko, M., Lampert, C.: Correlational spectral clustering. In: Proceedings of the Computer Vision and Pattern Recognition (2008)
2. Blum, A., Mitchell, T.: Combining labeled and unlabeled data with co-training. In: Proceedings of the Annual Conference on Computational Learning Theory, pp. 92–100 (1998)
3. Chaudhuri, K., Kakade, S., Livescu, K., Sridharan, K.: Multi-view clustering via canonical correlation analysis. In: Proceedings of the International Conference on Machine Learning, pp. 129–136 (2009)
4. Chen, W., Song, Y., Bai, H., Lin, C., Chang, E.: Parallel spectral clustering in distributed systems. IEEE Transactions on Pattern Analysis and Machine Intelligence 33(3), 568–586 (2011)
5. Chua, T., Tang, J., Hong, R., Li, H., Luo, Z., Zheng, Y.: NUS-WIDE: a real-world web image database from National University of Singapore. In: Proceedings of the ACM International Conference on Image and Video Retrieval (2009)

6. Dai, W., Chen, Y., Xue, G., Yang, Q., Yu, Y.: Translated learning: transfer learning across different feature spaces. In: Proceedings of the Advances in Neural Information Processing System, pp. 299–306 (2008)
7. Gartner, T.: A survey of kernels for structured data. ACM SIGKDD Explorations Newsletter 5(1), 49–58 (2003)
8. Hartigan, J., Wong, M.: A k-Means clustering algorithm. Applied Statistics 28(1), 100–108 (1979)
9. He, X., Yan, S., Hu, Y., Niyogi, P., Zhang, H.: Face recognition using laplacianfaces. IEEE Transactions on Pattern Analysis and Machine Intelligence 27(3), 328–340 (2005)
10. Hotelling, H.: Relations between two sets of variates. Biometrika 28, 321–377 (1936)
11. Kakade, S., Foster, D.: Multi-view regression via canonical correlation analysis. In: Proceedings of the Annual Conference on Computational Learning Theory, pp. 82–96 (2007)
12. Lu, Z., Zhu, Y., Pan, S., Xiang, E., Wang, Y., Yang, Q.: Source free transfer learning for text classification. In: Proceedings of the Association for the Advancement of Artificial Intelligence (2014)
13. Luxburg, U.: A tutorial on spectral clustering. Statistics and Computing 14(4), 395–416 (2007)
14. Ng, A., Jordan, M., Weiss, Y.: On spectral clustering: analysis and an algorithm. In: Advances in Neural Information Processing Systems (2002)
15. Ng, M., Wu, Q., Ye, Y.: Co-transfer learning via joint transition probability graph based method. In: Proceedings of the International Workshop on Cross Domain Knowledge Discovery in Web and Social Network Mining, pp. 1–9 (2012)
16. Ng, M., Wu, Q., Ye, Y.: Co-transfer learning using coupled markov chains with restart. IEEE Intelligent Systems (2013)
17. Pan, S., Yang, Q.: A survey on transfer learning. IEEE Transactions on Knowledge and Data Engineering 22, 1345–1359 (2010)
18. Qi, G., Aggarwal, C., Huang, T.: Towards semantic knowledge propagation from text corpus to web images. In: Proceedings of the International Conference on World Wide Web, pp. 297–306 (2011)
19. Singh, A., Gordon, G.: Relational learning via collective matrix factorization. In: Proceedings of the ACM SIGKDD International Conference on Knowledge Discovery and Data Mining, pp. 650–658 (2008)
20. Tan, B., Zhong, E., Ng, M., Yang, Q.: Mixed-transfer: transfer learning over mixed graphs. In: Proceedings of SIAM International Conference on Data Mining (2014)
21. Yang, Q., Chen, Y., Xue, G., Dai, W., Yu, Y.: Heterogeneous transfer learning for image clustering via the social Web. In: Proceedings of the Joint Conference of the Annual Meeting of the ACL and the International Joint Conference on Natural Language Processing of the AFNLP, pp. 1–9 (2009)
22. Zhu, X.: Semi-supervised learning literature survey. Technical Report 1530, Computer Science, University of Wisconsin-Madison (last modified on July 19, 2008)
23. Zhu, Y., Chen, Y., Lu, Z., Pan, S., Xue, G., Yu, Y., Yang, Q.: Heterogeneous transfer learning for image classification. In: Proceedings of the Association for the Advancement of Artificial Intelligence (2011)

# Mixed Pooling for Convolutional Neural Networks

Dingjun Yu, Hanli Wang*, Peiqiu Chen, and Zhihua Wei

Key Laboratory of Embedded System and Service Computing, Ministry of Education,
Tongji University, Shanghai, P.R. China
hanliwang@tongji.edu.cn

**Abstract.** Convolutional Neural Network (CNN) is a biologically inspired trainable architecture that can learn invariant features for a number of applications. In general, CNNs consist of alternating convolutional layers, non-linearity layers and feature pooling layers. In this work, a novel feature pooling method, named as mixed pooling, is proposed to regularize CNNs, which replaces the deterministic pooling operations with a stochastic procedure by randomly using the conventional max pooling and average pooling methods. The advantage of the proposed mixed pooling method lies in its wonderful ability to address the over-fitting problem encountered by CNN generation. Experimental results on three benchmark image classification datasets demonstrate that the proposed mixed pooling method is superior to max pooling, average pooling and some other state-of-the-art works known in the literature.

**Keywords:** Convolutional neural network, pooling, regularization, model average, over-fitting.

## 1 Introduction

Since its first introduction in the early 1980's [1], the Convolutional Neural Network (CNN) has demonstrated excellent performances for a number of applications such as hand-written digit recognition [2], face recognition [3], etc. With the advances of artificial intelligence, recent years have witnessed the growing popularity of deep learning with CNNs on more complicated visual perception tasks.

In [4], Fan *et al.* treat human tracking as a learning problem of estimating the location and the scale of objects and employ CNNs to reach this learning purpose. Cireşan *et al.* [5] propose an architecture of multi-column CNNs which

---

* Corresponding author (H. Wang, E-mail: hanliwang@tongji.edu.cn). This work was supported in part by the National Natural Science Foundation of China under Grants 61102059 and 61472281, the "Shu Guang" project of Shanghai Municipal Education Commission and Shanghai Education Development Foundation under Grant 12SG23, the Program for Professor of Special Appointment (Eastern Scholar) at Shanghai Institutions of Higher Learning, the Fundamental Research Funds for the Central Universities under Grants 0800219158, 0800219270 and 1700219104, and the National Basic Research Program (973 Program) of China under Grant 2010CB328101.

D. Miao et al. (Eds.): RSKT 2014, LNAI 8818, pp. 364–375, 2014.
DOI: 10.1007/978-3-319-11740-9_34 © Springer International Publishing Switzerland 2014

can be accelerated by Graphics Processing Unit (GPU) for image classification and amazing performances are achieved on a number of benchmark datasets. In [6], a 3D CNN model is designed for human action recognition, in which both the spatial and temporal features are mined by performing 3D convolutions. krizhevsky *et al.* [7] train a very large CNN for the ImageNet visual recognition challenge [8] and achieve an astonishing record-breaking performance in 2012.

Despite the aforementioned encouraging progresses, there are still several problems encountered by CNNs such as the over-fitting problem due to the high capacity of CNNs. In order to address this issue, several regularization techniques have been proposed, such as weight decay, weight tying and augmentation of training sets [9]. These regularization methods allow the training of larger capacity models than would otherwise be possible, which are able to achieve superior test performances as compared with smaller un-regularized models [10].

Another promising regularization approach is Dropout which is proposed by Hinton *et al.* [11]. The idea of Dropout is to stochastically set half the activations in a hidden layer to zeros for each training sample. By doing this, the hidden units can not co-adapt to each other, and they must learn a better representation for the input in order to generalize well. Dropout acts like a form of model averaging over all possible instantiations of the model prototype, and it is shown to deliver significant gains in performance in a number of applications.

However, the shortcoming of Dropout is that it can not be generally employed for several kinds of CNN layers, such as the convolutional layer, non-linearity layer and feature pooling layer. To overcome this defect, a generalization of Dropout, called DropConnect, is proposed in [12]. Instead of randomly selecting activations within the network, DropConnect sets a randomly selected subset of weights to zeros. As compared to Dropout, better performances have been achieved by DropConnect in certain cases. In [10], another type of regularization for convolutional layers, named stochastic pooling, is proposed to enable the training of larger models for weakening over-fitting. The key idea of stochastic pooling is to make the pooling process in each convolutional layer a stochastic process based on multinomial distribution.

In this work, similar to [10], a novel type of pooling method, termed as mixed pooling, is proposed in order to boost the regularization performance for training larger CNN models. Inspired by Dropout (that randomly sets half the activations to zeros), the proposed mixed pooling method replaces the conventional deterministic pooling operations with a stochastic procedure, randomly employing the max pooling and average pooling methods during the training of CNNs. Such a stochastic nature of the proposed mixed pooling method helps prevent over-fitting to some extent. Experiments are performed to verify the superiority of the proposed mixed pooling method over the traditional max pooling and average pooling methods.

The rest of this paper is organized as follows. Section 2 provides a background review of CNNs. The proposed mixed pooling method is introduced in Section 3. In Section 4, the comparative experimental results are presented. Finally, Section 5 concludes this paper.

## 2    Review of Convolutional Neural Networks

A brief review of CNNs is presented herein which is useful to elicit the proposed mixed pooling method. In general, CNNs are representatives of the multi-stage Hubel-Wiesel architecture [13], which extract local features at a high resolution and successively combine these into more complex features at lower resolutions. The loss of spatial information is compensated by an increasing number of feature maps in higher layers.

A powerful CNN is composed of several feature extraction stages, and each stage consists of a convolutional layer, a non-linear transformation layer and a feature pooling layer. The convolutional layer takes inner product of the linear filter and the underlying receptive field followed by a nonlinear activation function at every local portion of the input. Then, the non-linear transformation layer performs normalization among nearby feature maps. Finally, the feature pooling layer combines local neighborhoods using an average or maximum operation, aiming to achieve invariance to small distortions. An example of a two-stage CNN with the aforementioned three layers is shown in Fig. 1 for illustration.

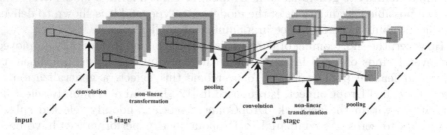

**Fig. 1.** An example of a two-stage CNN. An input image is passed through a convolutional layer, followed by non-linear transformation layer and pooling layer to extract low-level features in the first stage. Then, these three layers are applied again in the second stage to extract high-level features.

### 2.1    Convolutional Layer

The aim of the convolutional layer is to extract patterns found within local regions of the input images that are quite common in natural images [10]. Generally speaking, the convolutional layer generates feature maps by linear convolutional filters followed by nonlinear activation functions, such as ReLU [14], sigmoid, tanh, etc. In this layer, the $k$th output feature map $y_k$ can be calculated as follows:

$$y_k = f(w_k * x), \tag{1}$$

where $x$ denotes the input image, $w_k$ stands for the convolutional filter associated with the $k$th feature map, $*$ indicates the 2D convolution operator which is used to calculate the inner product of the filter template at every location in the input image, and $f(\cdot)$ is the nonlinear activation function.

## 2.2    Non-linear Transformation Layer

It has been shown in [15] that using a rectifying non-linear transformation layer is an effective way to further improve the CNN performance for visual recognition tasks. This layer usually performs local subtractive or divisive operations for normalization, enforcing a kind of local competition between features at the same spatial location in different feature maps. There are usually two kinds of non-linear transformations. One is the local response normalization [11], which yields the normalized output $y_{kij}$ at the position $(i, j)$ in feature map $k$ as

$$y_{kij} = \frac{x_{kij}}{\left(1 + \frac{\alpha}{N} \cdot \sum_{l=k-\frac{N}{2}}^{k+\frac{N}{2}} (x_{lij})^2\right)^{\beta}}, \tag{2}$$

where the sum runs over $N$ adjacent feature maps at the same spatial location, and the parameters of $\alpha$ and $\beta$ can be determined using a validation set.

Another is the local contrast normalization [15] with the normalized output $y_{kij}$ produced with the following formula.

$$y_{kij} = \frac{x_{kij}}{\left(1 + \frac{\alpha}{M_1 \cdot M_2} \sum_{p=i-\frac{M_1}{2}}^{i+\frac{M_1}{2}} \sum_{q=j-\frac{M_2}{2}}^{j+\frac{M_2}{2}} (x_{kpq} - m_{kij})^2\right)^{\beta}}, \tag{3}$$

where the local contrast is computed within a local $M_1 \times M_2$ region with the center at $(i, j)$, and $m_{kij}$ is the mean of all $x$ values within the above $M_1 \times M_2$ region in the $k$th feature map as computed as

$$m_{kij} = \frac{1}{M_1 \cdot M_2} \cdot \sum_{p=i-\frac{M_1}{2}}^{i+\frac{M_1}{2}} \sum_{q=j-\frac{M_2}{2}}^{j+\frac{M_2}{2}} x_{kpq}. \tag{4}$$

## 2.3    Feature Pooling Layer

The purpose of pooling is to transform the joint feature representation into a more usable one that preserves important information while discarding irrelevant details. The employment of pooling layer in CNNs aims to achieve invariance to changes in position or lighting conditions, robustness to clutter, and compactness of representation. In general, the pooling layer summarizes the outputs of neighboring groups of neurons in the same kernel map [7]. In the pooling layer, the resolution of the feature maps is reduced by pooling over local neighborhood on the feature maps of the previous layer, thereby enhancing the invariance to distortions on the inputs.

In CNNs, there are two conventional pooling methods, including max pooling and average pooling. The max pooling method selects the largest element in each pooling region as

$$y_{kij} = \max_{(p,q) \in \mathcal{R}_{ij}} x_{kpq}, \tag{5}$$

where $y_{kij}$ is the output of the pooling operator related to the $k$th feature map, $x_{kpq}$ is the element at $(p, q)$ within the pooling region $\mathcal{R}_{ij}$ which represents a local neighborhood around the position $(i, j)$. Regarding the average pooling method, it takes the arithmetic mean of the elements in each pooling region as

$$y_{kij} = \frac{1}{|\mathcal{R}_{ij}|} \sum_{(p,q) \in \mathcal{R}_{ij}} x_{kpq}, \tag{6}$$

where $|\mathcal{R}_{ij}|$ stands for the size of the pooling region $\mathcal{R}_{ij}$.

## 3   Proposed Mixed Pooling

### 3.1   Motivation

As mentioned before, the max pooling and average pooling methods are two popular choices employed by CNNs due to their computational efficiency. For instance, the average pooling method is used in [15] which obtains an excellent image classification accuracy on the Caltech101 dataset. In [7], the max pooling method is successfully applied to train a deep 'convnet' for the ImageNet competition. Although these two kinds of pooling operators can work very well on some datasets, it is still unknown which will work better for addressing a new problem. In another word, it is a kind of empiricism to choose the pooling operator.

On the other hand, both the max pooling and average pooling operators have their own drawbacks. About max pooling, it only considers the maximum element

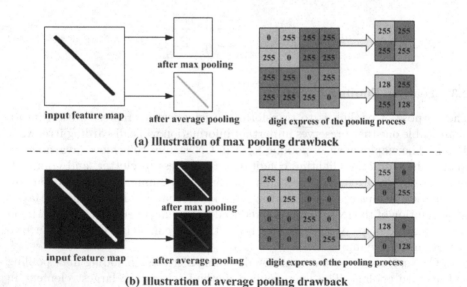

(a) Illustration of max pooling drawback

(b) Illustration of average pooling drawback

**Fig. 2.** Toy example illustrating the drawbacks of max pooling and average pooling.

and ignores the others in the pooling region. Sometimes, this will lead to an unacceptable result. For example, if most of the elements in the pooling region are of high magnitudes, the distinguishing feature vanishes after max pooling as shown in Fig. 2(a). Regarding average pooling, it calculates the mean of all the elements within the pooling region. This operator will take all the low magnitudes into consideration and the contrast of the new feature map after pooling will be reduced. Even worse, if there are many zero elements, the characteristic of the feature map will be reduced largely, as illustrated in Fig. 2(b).

It is well known that images in the nature world are ever-changing, and it is of high possibility that the defective aspects of max pooling and average pooling (as shown in Fig. 2) will have negative effects in applying pooling layers to CNNs. Therefore, as a solution, we consider to replace the deterministic pooling operation with a stochastic procedure, which randomly employs the local max pooling and average pooling methods when training CNNs. This is the proposed mixed pooling method to be introduced next.

### 3.2    Pooling Scheme

The proposed mixed pooling is inspired by the random Dropout [11] and DropConnect [12] methods. As mentioned before, when training with Dropout, a randomly selected subset of activations are set to zeros within each layer. While for DropConnect, it instead sets a randomly selected subset of weights within the network to zeros. Both of these two techniques have been proved to be powerful for regularizing neural networks.

In this work, the proposed mixed pooling method generates the pooled output with the following formula.

$$y_{kij} = \lambda \cdot \max_{(p,q) \in \mathcal{R}_{ij}} x_{kpq} + (1 - \lambda) \cdot \frac{1}{|\mathcal{R}_{ij}|} \sum_{(p,q) \in \mathcal{R}_{ij}} x_{kpq}, \tag{7}$$

where $\lambda$ is a random value being either 0 or 1, indicating the choice of using the max pooling or average pooling. In another word, the proposed method changes the pooling regulation scheme in a stochastic manner which will address the problems encountered by max pooling and average pooling to some extent.

### 3.3    Back Propagation

As usual, CNN layers are trained using the back propagation algorithm. For error propagation and weight adaptation in fully connected layers and convolutional layers, the standard back propagation procedure is employed. For the pooling layer, the procedure is a little bit different. As noted in [2], the pooling layers do not actually do any learning themselves. Instead, they just reduce the dimension of the networks. During forward propagation, an $N \times N$ pooling block is reduced to a single value. Then, this single value acquires an error computed from back propagation. For max pooling, this error is just forwarded to where it comes from because other units in the previous layer's pooling blocks do not contribute to it. For average pooling, this error is forwarded to the whole pooling block by dividing $N \times N$ as all units in the block affect its value.

In mixed pooling, it is also needed to locate where the error comes from so that it can modify the weights correctly. The proposed mixed pooling randomly apply the max pooling and average pooling during forward propagation. For this reason, the pooling history about the random value $\lambda$ in Eq. (7) must be recorded during forward propagation. Then, for back propagation, the operation is performed depending on the records. Specifically, if $\lambda = 1$, then the error signals are only propagated to the position of the maximum element in the previous layer; otherwise, the error signals will be equally divided and propagated to the whole pooling region in the previous layer.

### 3.4   Pooling at Test Time

When the proposed mixed pooling is applied for test, some noises will be introduced into CNNs' predictions, which is also found in [10]. In order to reduce this kind of noise, a statistical pooling method is used. During the training of CNNs, the frequencies of using the max pooling and average pooling methods related to the $k$th feature map are counted as $F_{max}^k$ and $F_{avg}^k$. If $F_{max}^k \geq F_{avg}^k$, then the max pooling method is applied in the $k$th feature map; otherwise, the average pooling method is used. In this sense, the proposed statistical pooling at the test time can be viewed as a form of model averaging.

## 4   Experimental Results

### 4.1   Overview

The proposed mixed pooling method is evaluated on three benchmark image classification datasets, including CIFAR-10 [16], CIFAR-100 [16] and the Street View House Number (SVHN) dataset [17], with a selection of images from CIFAR-10 and SVHN as shown in Fig. 3. The proposed method is compared with the max pooling and average pooling methods for demonstrating the performance improvement. In the experiments, the CNNs are generated from the raw RGB values of the image pixels. As a regularizer, the data augmentation technique [18] is applied for CNN training, which is performed by extracting 24×24 sized images as well as their horizontal reflections from the original 32×32 image and then training CNNs on these extracted images. Another regularizer applied in this work is the weight decay technique as used in [7].

In this work, the publicly available cuda-convnet [19] package is used to perform experiments with a single NVIDIA GTX 560TI GPU. Currently, the CNNs are trained using stochastic gradient descent approach with a batch size of 128 images and momentum of 0.9. Therefore, the update rule for weight $w$ is

$$v_{i+1} = 0.9v_i + \epsilon \left\langle \frac{\partial L}{\partial w} |_{w_i} \right\rangle_i, \tag{8}$$

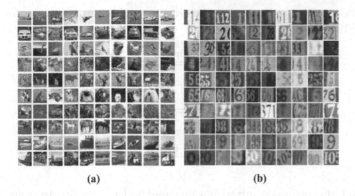

(a)                                          (b)

Fig. 3. A selection of images we evaluated. (a) CIFAR-10 [16]. (b) SVHN [17].

$$w_{i+1} = w_i + v_{i+1}, \tag{9}$$

where $i$ is the iteration index, $v$ is the momentum variable, $\epsilon$ is the learning rate, and $\left\langle \frac{\partial L}{\partial w}|_{w_i} \right\rangle_i$ is the average over the $i$th batch of the derivative of the objective with respect to $w_i$.

## 4.2   CIFAR-10

CIFAR-10 [16] is a collection of natural color images of 32×32 pixels. It contains 10 classes, each of them having 5,000 samples for training and 1,000 for testing. The CIFAR-10 images are highly varied, and there is no standard viewpoint or scale at which the objects appear. Except for subtracting the mean activity of the training set, the CIFAR-10 images are not preprocessed.

A two-stage CNN model is trained in this work, with each stage consisting of a convolutional layer, a local response normalization layer and a pooling layer. All the convolutional layers have 64 filter banks and use a filter size of 5×5. Local response normalization layers follow the convolutional layers, with $N = 9$, $\alpha = 0.001$ and $\beta = 0.75$ (as used in Eq.(2)), which normalize the output at each location over a subset of neighboring feature maps. This typically helps training by suppressing extremely large outputs allowed by the rectified linear units and helps neighboring features communicate with each other. Additionally, all of the pooling layers that follow local response normalization layers summarize a 3×3 neighborhood and use a stride of 2. Finally, two locally connected layers and a softmax layer are used as classifier at the end of the entire network.

We follow the common experimental protocol for CIFAR-10, which is to choose 50,000 images for training and 10,000 images for testing. The network parameters are selected by minimizing the error on a validation set consisting of the last 10,000 training examples.

The comparative results are shown in Table 1, where the test accuracy results of several state-of-the-art approaches are cited for illustration besides the max pooling, average pooling and mixed pooling methods. From the results, it can

**Table 1.** Comparative classification performances with various pooling methods on the CIFAR-10 dataset

| Method | Training error (%) | Accuracy (%) |
|---|---|---|
| 3-layer Convnet [11] | - | 83.4% |
| 10-layer DNN [5] | - | 88.79% |
| Stochastic pooling [20] | - | 84.87% |
| Max pooling | 3.01% | 88.64% |
| Average pooling | 4.52% | 86.25% |
| Mixed pooling | 6.25% | **89.20%** |

be seen that the proposed mixed pooling method is superior to other methods in terms of the test accuracy although it produces larger training errors than that of max pooling and average pooling. This indicates that the proposed mixed pooling outperforms max pooling and average pooling to address the over-fitting problem. As observed from the results, a test accuracy of 89.20% is achieved by the proposed mixed pooling method which is the best result which we are aware of without using Dropout. In addition, the features which are learnt in the first convolutional layer by using different pooling methods are shown in Fig. 4, where it can observed that the features learnt with the proposed mixed pooling method contains more information than that of max pooling and average pooling.

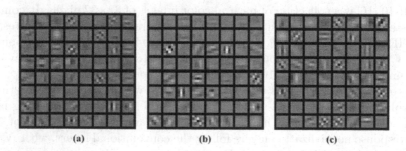

**Fig. 4.** Visualization of 64 features learnt in the first convolutional layer on the CIFAR-10 dataset. The size of each feature is 5×5×3. (a) Features learnt with max pooling. (b) Features learnt with average pooling. (c) Features learnt with mixed pooling.

### 4.3    CIFAR-100

The CIFAR-100 dataset [16] is the same in size and format as the CIFAR-10 dataset, but it contains 100 classes. That is to say, each class in CIFAR-100 has 500 images to train and 100 images to test. We preprocess the data just like the way we have done for the CIFAR-10 dataset, and the same CNN structure as used for CIFAR-10 is applied to CIFAR-100. The only difference is that the last softmax layer outputs 100 feature maps. The comparative results are shown

**Table 2.** Comparative classification performances with various pooling methods on the CIFAR-100 dataset

| Method | Training error (%) | Accuracy (%) |
|---|---|---|
| Learnable pooling regions [21] | - | 56.29% |
| Stochastic pooling [20] | - | 57.49% |
| Max pooling | 5.42% | 59.91% |
| Average pooling | 14.61% | 55.99% |
| Mixed pooling | 25.71% | **61.93%** |

in Table 2, where it can be observed that the proposed mixed pooling method outperforms the other methods in terms of test accuracy.

## 4.4 SVHN

Finally, we also perform experiments on the SVHN dataset [17]. SVHN consists of images of house numbers collected by Google Street View. There are 73,257 digits in the training set, 26,032 digits in the test set and 531,131 additional examples as an extra training set. We follow [22] to build a validation set which contains 400 samples per class from the training set and 200 samples per class from the extra set. The remaining digits of the training and extra sets are used for training. The local contrast normalization operator is applied in the same way as used in [20]. The comparative results are presented in Table 3, which demonstrate the superiority of the proposed mixed pooling method over the others.

**Table 3.** Comparative classification performances with various pooling methods on the SVHN dataset

| Method | Training error (%) | Accuracy (%) |
|---|---|---|
| Lp-pooling Convnet [22] | - | 95.10% |
| 64-64-64 Stochastic pooling [20] | - | 96.87% |
| 64-64-64 Max pooling | 2.03% | 96.61% |
| 64-64-64 Average pooling | 2.41% | 96.14% |
| 64-64-64 Mixed pooling | 3.54% | **96.90%** |

## 4.5 Time Performance

To further illustrate the advantage of the proposed mixed pooling method, the time consumption performances are illustrated in Fig. 5 with two testing scenarios evaluated for max pooling, average pooling, stochastic pooling [10] and mixed pooling, where nine epoches are tested. From Fig. 5, it can be seen that the computational complexity of mixed pooling is almost the same as that of average pooling and max pooling, and far lower than that of stochastic pooling.

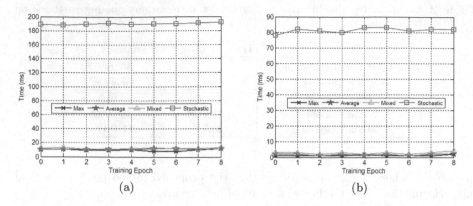

(a)                                    (b)

**Fig. 5.** Time performance comparison among max, average, stochastic and mixed pooling. (a) Time consumption when feature map size is $28 \times 28$ and pooling size is $2 \times 2$. (b) Time consumption when feature map size is $14 \times 14$ and pooling size is $2 \times 2$.

## 5  Conclusion

In this paper, a novel pooling method called mixed pooling is proposed, which can be combined with any other forms of regularization such as weight decay, Dropout, data augmentation, and so on. Comparative experimental results demonstrate that the proposed mixed pooling method is superior to the traditional max pooling and average pooling methods to address the over-fitting problem and improve the classification accuracy. With the proposed method, we achieve the start-of-the-art performances on the CIFAR-10, CIFAR-100 and SVHN datasets as compared with other approaches that do not employ Dropout. Furthermore, the proposed method requires negligible computational overheads and no hyper-parameters to tune, thus can be widely applied to CNNs.

## References

1. Fukushima, K.: Neocognitron: A self-organizing neural network model for a mechanism of pattern recognition unaffected by shift in position. Biological Cybernetics 36(4), 193–202 (1980)
2. LeCun, Y., Boser, B., Denker, J.S., Henderson, D., Howard, R.E., Hubbard, W., Jackel, L.D.: Backpropagation applied to handwritten zip code recognition. Neural computation 1(4), 541–551 (1989)
3. Lawrence, S., Giles, C.L., Tsoi, A.C., Back, A.D.: Face recognition: A convolutional neural-network approach. IEEE Transactions on Neural Networks 8(1), 98–113 (1997)
4. Fan, J., Xu, W., Wu, Y., Gong, Y.: Human tracking using convolutional neural networks. IEEE Transactions on Neural Networks 21(10), 1610–1623 (2010)
5. Cireşan, D., Meier, U., Schmidhuber, J.: Multi-column deep neural networks for image classification. In: CVPR, pp. 3642–3649 (2012)

6. Ji, S., Xu, W., Yang, M., Yu, K.: 3D convolutional neural networks for human action recognition. IEEE Transactions on Pattern Analysis and Machine Intelligence 35(1), 221–231 (2013)
7. Krizhevsky, A., Sutskever, I., Hinton, G.E.: ImageNet classification with deep convolutional neural networks. In: NIPS, vol. 1 (2012)
8. Deng, J., Dong, W., Socher, R., Li, L.-J., Li, K., Fei-Fei, L.: ImageNet: A large-scale hierarchical image database. In: CVPR, pp. 248–255 (2012)
9. Montavon, G., Orr, G.B., Müller, K.R. (eds.): Neural networks: tricks of the trade, 2nd edn. Spinger, San Francisco (2012)
10. Zeiler, M.D.: Hierarchical convolutional deep learning in computer vision. PhD thesis, ch. 6, New York University (2014)
11. Hinton, G.E., Srivastava, N., Krizhevsky, A., Sutskever, I., Salakhutdinov, R.R.: Improving neural networks by preventing co-adaptation of feature detectors. arXiv preprint arXiv:1207.0580 (2012)
12. Wan, L., Zeiler, M.D., Zhang, S., LeCun, Y., Fergus, R.: Regularization of neural networks using DropConnect. In: ICML, pp. 1058–1066 (2013)
13. LeCun, Y., Kavukcuoglu, K., Farabet, C.: Convolutional networks and applications in vision. In: ISCAS, pp. 253–256 (2010)
14. Nair, V., Hinton, G.E.: Rectified linear units improve restricted boltzmann machines. In: ICML, pp. 807–814 (2010)
15. Jarrett, K., Kavukcuoglu, K., Ranzato, M., LeCun, Y.: What is the best multi-stage architecture for object recognition? In: ICCV, pp. 2146–2153 (2009)
16. Krizhevsky, A.: Learning multiple layers of features from tiny images. Technical Report TR-2009, University of Toronto (2009)
17. Netzer, Y., Wang, T., Coates, A., Bissacco, A., Wu, B., Ng, A.Y.: Reading digits in natural images with unsupervised feature learning. In: NIPS Workshop on Deep Learning and Unsupervised Feature Learning, vol. 2011 (2011)
18. Simard, P.Y., Steinkraus, D., Platt, J.C.: Best practices for convolutional neural networks applied to visual document analysis. In: ICDAR, pp. 958–962 (2003)
19. Krizhevsky, A.: cuda-convnet., http://code.google.com/p/cuda-convnet/
20. Zeiler, M.D., Fergus, R.: Stochastic pooling for regularization of deep convolutional neural networks. arXiv preprint arXiv:1301.3557 (2013)
21. Malinowski, M., Fritz, M.: Learnable pooling regions for image classification. arXiv preprint arXiv:1301.3516 (2013)
22. Sermanet, P., Chintala, S., LeCun, Y.: Convolutional neural networks applied to house numbers digit classification. In: ICPR, pp. 3288–3291 (2012)

# An Approach for In-Database Scoring of R Models on DB2 for z/OS

Yikun Xian[1], Jie Huang[1,*], Yefim Shuf[2], Gene Fuh[3], and Zhen Gao[1]

[1] School of Software Engineering, Tongji University, Shanghai, China
[2] IBM T. J. Watson Research Center, NY, U.S.
[3] IBM GCG STG Laboratory, Beijing, U.S.
{082878xyk,huangjie,gaozhen}@tongji.edu.cn
{yefim,fuh}@us.ibm.com

**Abstract.** Business Analytics is comprehensively used in many enterprises with large scale of data from databases and analytics tools like R. However, isolation between database and data analysis tool increases the complexity of business analytics, for it will cause redundant steps such as data migration and engender latent security problem. In this paper, we propose an in-database scoring mechanism, enabling application developers to consume business analytics technology. We also validate the feasibility of the mechanism using R engine and IBM DB2 for z/OS. The result evinces that in-database scoring technique can be applicable to relational databases, largely simplify the process of business analytics, and more importantly, keep data governance privacy, performance and ownership.

**Keywords:** Business Analytics, In-Database Scoring, DB2 for z/OS, R.

## 1   Introduction

Business Analytics technology (BA) prevails in various industries for developing new insights and planning business performance through historical data, statistical models or quantitative methods [5]. Data manipulation and modeling management are two major sections during the whole process of BA. However, in traditional BA procedure, the limitation is evitable that analytics models are usually created in one environment while transactional data is stored in other different places. This may engender latent problems comprising heavy dependency of manual work, inconsistency of data migration, complexity of model scoring process and security problem as well.

One alternate solution is to integrate data and model into single environment. This implies two possible methods: 1) make database accessible to modeling tool; 2) enable database to score models. The first method is already implemented by Oracle R Enterprise that makes it much easier for data scientists to train and score models on R, the data analysis engine, which provides an interface to manipulate data in remote Oracle databases. However, although it does simplify procedures for data scientists,

---

* Corresponding author.

D. Miao et al. (Eds.): RSKT 2014, LNAI 8818, pp. 376–385, 2014.
DOI: 10.1007/978-3-319-11740-9_35 © Springer International Publishing Switzerland 2014

the problem remains unsolved that it is costly to leverage scoring results and integrate trained models into the existing system. Thus, the second method will be better because it can not only provide convenience for developers to consume models, but also largely avoid a series of problems brought by data migration.

In this paper, we propose a general in-database scoring mechanism that aims 1) to make BA technology 'consumable' by simplifying the process of BA and 2) to control the security risk by minimizing data migration. To validate feasibility of the mechanism, we choose R and IBM DB2 for z/OS as our experimental platform. The reason is R has been the most popular data mining tool since 2010 [6], and IBM mainframe has been widely used in many banks or large enterprises for a long period. In the experiment, we implement an interface for in-database scoring on DB2 with the help of some mainframe features. The interface is exposed as a DB2 stored procedure, which allows users to do both model scoring and data manipulation directly through SQL statements. Architecture of the whole process, mechanism of R model storage, and design of communication between R and stored procedure are explicitly described later in this paper.

## 2     Related Work

Architecture and implementation of in-database scoring technology is associated with database features and analytics tools. To some extent, several databases have already been enhanced with analytics function, but focus on different points.

Oracle R Enterprise (ORE) allow R users to do model scoring and operate Oracle data in R [8, 9, and 10]. It extends the database with the R library of statistical functionality and pushes down computations to the database. It provides a database-centric environment for end-to-end analytical processes in R. The public interface of ORE is the R APIs through which R users work solely within R for data analysis and trigger in-database scoring in Oracle database. However, as mentioned before, application developers who manipulate databases are still isolated from business analytics or it still costs too much for them to consume models.

Sybase IQ is a column-based relational database system used for business analytics and data warehouse [12]. It mainly focuses on analyzing large amount of data in a low-cost and high-availability environment. The in-database scoring here happens in two kinds of ways: in-process and out-process. In-process means all shared libraries and user's functions are loaded into Sybase IQ process in the memory and execute altogether. This can definitely keep high performance, but will also incur security risks and robustness risks. On the opposite, out-process causes lower security and robustness risks, but lead to lower performance.

SAS in-database processing integrates SAS solutions, SAS analytic processes, and third-party database management systems [11]. With SAS in-database processing, users can run scoring models, SAS procedures and formatted SQL queries inside the database. However, the solution mainly aims at offering an integrated analytics environment and is quite different from what we propose to solve.

Other work includes using PMML as the bridge between the data science lab and the warehouse [2]. Predictive models can be embedded into the database and automatically be turned into SQL functions. However, models supported by PMML are not rich enough, unlike original data analysis engine such as R.

# 3     In-Database Scoring Mechanism

## 3.1     General Model Definition

Let's denote the general form of model $m$ as a four-element tuple:

$$M = (X, Y, f_M, \Theta) \tag{1}$$

where $X = \{X_1, X_2, \ldots, X_m\}, m \in \mathbb{N}^+$, denoted as the set of model input variables, $Y = \{Y_1, Y_2, \ldots, Y_n\}, n \in \mathbb{N}^+$, denoted as the set of model output variables, $f$ is the formula or algorithm of the model and $\Theta = \{\theta_1, \theta_2, \ldots, \theta_k\}, k \in \mathbb{N}^+$, denoted as the set of model parameters. The model can also be written as function style:

$$Y = f_M(X, \Theta) \tag{2}$$

With observation data of input variables $x$ and observation data of output variables $y$, parameters $\Theta$ of model $m$ can be trained into estimated parameters $\widehat{\Theta}$:

$$\widehat{\Theta} = G_M(f_M | x, y) \tag{3}$$

where $G_M$ is the function or algorithm for training model $M$. Therefore, a trained model $\widehat{M}$ can denoted as:

$$\widehat{M} = (X, Y, f_M | \widehat{\Theta}) \tag{4}$$

It is obvious that $f_M$ along with estimated parameters $\widehat{\Theta}$ here is the *scoring function*. Let's define it as $f_s$ such that:

$$Y = f_M(X | \widehat{\Theta}) = f_s(X) \tag{5}$$

So given new input observation data $x'$, predicted values of output variables equals to:

$$\hat{y} = f_s(x') \tag{6}$$

## 3.2     Analysis of Scoring Function

For the training process in equation (3), it is usually done beforehand by data scientists with their own data analysis engines. Thus application developers who can now easily consume these trained model only care about three things according to equation (5), namely input variables $X$, output variables $Y$, and scoring function $f_s$.

In most cases, the output $\hat{y}$ represents values of prediction or labels of classification, which is determined by scoring function $f_s$ and new data $x'$. From equation (6) we know that as long as we save information of scoring function $f_s$, information of every input variable $X_1 \ldots X_m$ and every output variable $Y_1 \ldots Y_n$, and new input observation data $x'$, the scoring job can be done easily.

## 3.3     Process of In-Database Scoring

According to features of common relational database along with general models defined in section 3.1, we first define some useful data structures as following:

**Definition 1** (Data Analysis Engine): Data Analysis Engine (DAE) is a real software or tool for model training and model scoring. Let's define DAE $e_i$ as:

$$e_i = (\{(key, value)\}), e_i \in E \tag{7}$$

where each pair of $(key, value)$ represents attributes of DAE in real scenario, such as 'location', 'access_token', etc., and $E$ is the universal set of all DAE.

**Definition 2** (Model Structure): Model Structure (MS) is the general description framework of a model. Let's define MS $m_i$ as:

$$m_i = (e_i, mro(m_i), msf(m_i), \{(key, value)\}), m_i \in M \tag{8}$$

where $d_i$ is the DAE that matches features of the model (support for scoring at least), $mro(m_i)$, Model Result Object of Model i, represents the result of trained model ($f_M$ and $\hat{\Theta}$), $msf(m_i)$, Model Scoring Function of Model i, represents the program to run scoring function ($f_s$), and $M$ is the universal set of all models.

**Definition 3** (Model Parameter Structure): Model Parameter Structure (MPS) is the general description framework of a model variable. Let's define MPS $p_i$ as:

$$p_i = (m_i, io, order, \{(key, value)\}), order \in \mathbb{N}^+, p_i \in P \tag{9}$$

where $m_i$ is the MS that this variable belongs to, $io$ is the identifier that flags whether the variable is input or output, $order$ is the order of variable in the connected model, and $P$ is the universal set of all model variables.

**Definition 4** (Model Data Structure): Model Data Structure (MDS) is the general description framework of observation data of a variable. Let's define MDS $d_i$ as:

$$d_i = (x_1, \dots, x_m, y_1, \dots, y_n) \tag{10}$$

Notice that $x_1, \dots, x_m$ and $y_1, \dots, y_n$ are observation values of model variables strictly in the same order as those defined in $P$.

**Algorithm 1** (Process of Model Scoring): According to equation (6), input of modeling scoring is a set of model $m_i$ and model data $d_i$ where $x_1, \dots, x_m$ are real observation data while $y_1, \dots, y_n$ are empty because those are the result this scoring. Pseudo code is shown as following:

```
INITIALIZE database
INPUT <m_i, d_i> ARRAY
FOREACH <m_i, d_i> IN ARRAY
    VALIDATE m_i ∈ M
    Find DAE e_i SUCH THAT e_i ∈ E and e_i IN m_i
    Find all p_1, ... p_k SUCH THAT p_1, ... p_k ∈ P and p_1, ... p_k IN m_i
    VALIDATE k = m + n  FOR d_i = (x_1, ..., x_m, y_1, ..., y_n)
    SCORE < m_i, e_i, p_1, ... p_k, x_1, ..., x_m >
    CALCUALTE AND SAVE < y_1, ..., y_n >
END FOREACH
OUTPUT < y_1, ..., y_n > ARRAY
```

# 4    In-Database Scoring with DB2/z and R

## 4.1    Architecture

The architecture of in-database scoring module for DB2 for z/OS is based on IBM mainframe, which is divided into several logical partitions (LPARs), each supporting different operating systems like z/OS and z/Linux. In most cases, DB2 is mostly deployed on z/OS by enterprise customers due to its high availability, scalability and adaptability while z/Linux is a good choice to run analytics engine like R. Along with factors and features, as Figure 1 shown, the whole module mainly consists of three parts: the DB2 stored procedure as public interface for database user, the internal model-related tables on DB2 for z/OS, R server and model files on z/Linux.

On z/OS side, DB2 stored procedure, as the only public interface for in-database scoring, is responsible for execution of whole scoring process, including interaction with DB2 user, reading and updating table data and communication with R server. So with this interface, DB2 users can easily access to the R engine to score models and need not care about how the procedure works. Detailed implementation of stored procedure will be discussed later.

Corresponding to data structures defined in equation 7, 8, 9, and 10, four tables are required in this architecture, MODEL table for R model related information, MODELPRMT table for R model input and output parameters information and SERVER table for R server information. These three tables should be setup as DB2 system tables, and the only permission for regular users is reading to prevent arbitrary modification. SCORE DATA table is a temporary table for saving input and output data and is created by DB2 regular users who can import scoring input data from other tables and get scoring result accordingly.

On z/Linux side, R server acts as analytics engine and listens to a particular port, waiting for model scoring requests. There can be several z/Linux systems, each with one R server, so multiple scoring process may happen at the same time, but executed by different R servers for workload balancing. For each individual scoring, R server should load two files from local file system, Model Result Object (MRO) and Model Scoring Function (MSF). The previous one is a serialized R object file, which comes from a trained R model of built-in model classes or lists. The second one is R script that consists of a function to call MRO, score test data and return result string. The two files should be provided by R users by publishing R models into database.

## 4.2    Workflow

The first step is to deploy R models into mainframe by R users through an R interface. As Figure 2 shows, it will complete three things: upload MRO and MSF files and save in R server, fill in model related information in three system tables, and create a temporary data table structure that can be duplicated by DB2 users to store scoring data. Notice that MRO and MSF files are prepared by R users beforehand, and related model information should also be necessarily provided.

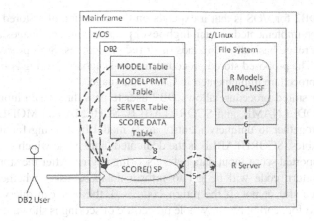

**Fig. 1.** Architecture of In-Database Scoring

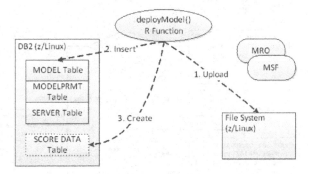

**Fig. 2.** Workflow of Deploy Model

When the stored procedure is called by the DB2 user (as Figure 1 shows), R model and server related information will be looked for in three system tables. Meanwhile, scoring input data will be read from SCORE DATA table and prepared in a particular format. Depending on these clues, the program can then locate R server and model files, passing the data to R server for the following in-database scoring. The analytics process is executed in R just like normal circumstances, but the result will be passed to the stored procedure internal program and saved to SCORE DATA table. So this data table plays a role of medium that temporarily loads scoring input data, and allows database users to access prediction result after the stored procedure execution succeeds.

## 4.3    Stored Procedure Interface and Implementation

Different from other regular stored procedure, the scoring stored procedure here should complete not only table data manipulation but also network communication with R server. CRUD (Create/Read/Update/Delete) for database table can be performed in SQL statements, but for network communication or even complex text process, these stored procedures can be hardly capable (Oracle PL/SQL is one choice). One significant advantage of DB2 for z/OS is that it expands on the definition of a stored procedure by allowing it to be implemented in such high-level programming languages as C/C++ and Java etc. So there are basically two kinds of stored procedures: SQL procedure or external procedure. The proposed stored procedure for In-Database Scoring is finally adopted as an external procedure implemented in C/C++ language.

The scoring stored procedure allows DB2 user to pass three main input parameters: SCHEMA, MODELNAME, and SCORETABLE. SCHEMA and MODELNAME can be combined together to uniquely identify an R model by looking for information in three system tables. SCORETABLE is the dedicated table name which should be firstly created and imported score input data before scoring begins. After the stored procedure completes, a return code will be get to inform database users whether the scoring process is successful or not. If the return code is 0, the user can query the specified SCORETABLE for scoring result. Whole procedure of scoring is shown in Figure 3.

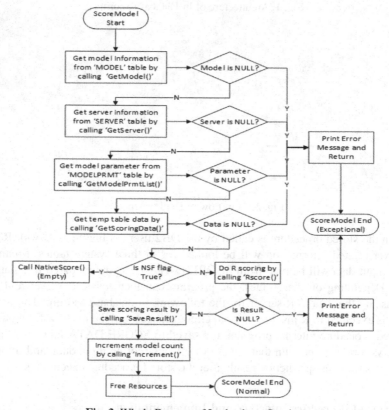

**Fig. 3.** Whole Process of In-database Scoring

## 4.4    Query Optimization

To improve SQL query performance, the embedded SQL instead of ODBC is used to fetch data from table, For tables like MODEL, SERVER and MODELPRMT, the primary key  can uniquely identify a single record, and table attributes are always fixed, so static embedded SQL that begins with 'EXEC SQL' and followed by a query statement is adopted. However, for SCORETABLE, the number of attributes is always changeable, as there can be different input variables and output variables according to different models. This causes the 'SELECT' statement cannot be fixed in source C/C++ code. A feasible solution is to adopt dynamic embedded SQL assisted by SQLCA (SQL Communication Area) and SQLDA (SQL Descriptor Area). The structure of SQLDA consists of 16 bytes SQLDA header and several 44 bytes SQLVARs, which represents the same number of attributes to be fetched from table. Table data is stored as void* array in the member 'sqldata' of SQLVAR._

## 4.5    Communication between Database and R

Since R is an analytics engine for statistical computing and graphics, an extra module is required to make R as a server that can listen to request for scoring. 'Rserve' is such an enhanced R package. It is a TCP/IP server integrated with R which allows other programs to use facilities of R. Thus, the main task turns to using C socket API to communicate with 'Rserve'.Three things will be executed during scoring process: pass scoring data to R, load MRO and MSF file into R engine and run scoring function coded in MSF. The scoring data will be constructed as an R command string which defines an array of numbers or strings.

## 4.6    Scenario and Experimental Result

We have tested three basic models in the experiment: linear regression for prediction, k-means for clustering and neural network for classification. The experimental platform is on DB2 for z/OS v9.

Before using R scoring function in DB2 for z/OS, following steps are needed to be done beforehand.

1. MRO (model result object) and MSF (model scoring function) are needed to be uploaded to z/Linux server.
2. The R model information has to be inserted into relative DB2z tables.
3. Model input and output temporary table has to be created.

Take a real scenario of k-means as example. Suppose an insurance company wants to cluster customers into several groups with three factors: age, income, historical insurance. We need to prepare two MRO and MSF files. Suppose data analyst in the company has trained a K-Means Clustering model with R using function 'kmeans()', which accepts three Xs (age, income, historic insurance) as independent variables and returns one Y as dependent variable. The returned object of function 'kmeans()' should be assigned to a variable that will then be saved to a serialized MRO file with function 'save()'. For MSF script, the first argument of function should be MRO and rests are all input variables (X). The order of these input variables should be carefully ordered. All these Xs can each be a single value or an array (vector) of values, but each X should be in the same size. The return value of function is simply a make-up string. The string is composed of all output data (Y) calculated from trained models and separated by some delimiters. Each set (row) of Ys are separated by semicolons while each cell (column) of Ys are separated by commas. Furthermore, related information of the model should be inserted into the three system tables: MODEL, MODELPRMT and SERVER. Temporary table is supposed to be created with accordance to Xs and Y and filled with several records of test data Xs, but Y left empty.

For three models in the experiment, detailed parameters are shown in table 1.

**Table 1.** Parameter of Tested Models

| Model Name | Linear Regression | K-Means | Nerual Network |
|---|---|---|---|
| Number of X | 4 | 3 | 2 |
| Data Type of X | Float | String | Float |
| Number of Y | 1 | 1 | 1 |
| Data Type of Y | Float | String | Float |
| Row of Data | 1000 | 1000 | 1000 |
| Data Table | lrdata | kmdata | nndata |

Then, we can call the stored procedure named 'R.SCORE()' to do scoring. For each model, the stored procedure program will sequentially read model information from MODEL table, MODELPRMT table, SERVER table and input data of X from TEMP DATA tables (here are 'lrdata', 'kmdata' and 'nndata'). 1000 row of input data will be formatted as a whole string and passed to R server after the z/Linux is positioned by respective model information. Once the scoring (prediction, clustering and classification) is executed by R, stored procedure program will fetch output data of Y from R server and insert into TEMP DATA tables.

Finally, we simply use SELECT SQL statement to read scored data from three tables ('lrdata', 'kmdata' and 'nndata') successfully. We can find that all manual operations are completed with SQL statements inside DB2 without help of any other tools.

## 5    Conclusion and Future Work

We have proposed a feasible in-database scoring mechanism validated by an experiment that it's successfully implemented on DB2 for z/OS. With this technology, users like application developers can directly consume business analytics like prediction and classification with SQL statements inside database where transactional data is stored. The architecture and approach is quite general and can also be applied to other databases. Meanwhile, to improve performance of fetching data from table in C program, we leverage static embedded SQL together with dynamic embedded SQL assisted with SQLCA, SQLDA and some other DB2 features rather than traditional ODBC.

The future work includes the grand vision of evolving this work eventually into a full-equipped cloud infrastructure where BA service (hardware, middleware, software, etc.) will be offered as a service to accelerate the adoption of the technology by unlimited population of application developers and end users.

# References

1. Conn, S.S.: OLTP and OLAP data integration: a review of feasible implementation methods and architectures for real time data analysis. In: Proceedings of the SoutheastCon, pp. 515–520. IEEE (2005)
2. Das, K.K., Fratkin, E., Gorajek, A.: Massively Parallel In-Database Predictions using PMML. In: PMML 2011 (2011)
3. ALzain, M.A., Pardede, E.: Using Multi Shares for Ensuring Privacy in Database-as-a-Service. In: System Sciences (HICSS), pp. 1–9 (2011)
4. Davidson, G.S., Boyack, K.W., Zacharski, R.A., Helmreich, S.C., Cowie, J.R.: Data-Centric Computing with the Netezza Architecture. SANDIA REPORT, SAND2006-1853, Unlimited Release, Printed (April 2006)
5. Chen, H., Chiang, R.H.L., Storey, V.C.: Business intelligence and analytics: From big data to big impact. MIS Quarterly (11), 1–24 (2012)
6. Revolutions, R users: Be counted in Rexer's, Data Miner Survey (January 30, 2013)
7. Hornick, M.: Quick! Swap those models – I've got a better one (August 12, 2013)
8. Oracle White Paper, Big Data Analytics - Advanced Analytics in Oracle Database (March 2013)
9. Oracle White Paper, Bringing R to the Enterprise - A Familiar R Environment with Enterprise-Caliber Performance, Scalability, and Security (May 2013)
10. Hornick, M.: Senior Manager, Development, Session 2: Oracle R Enterprise 1.3 Transparency Layer (2012)
11. SAS Documentation, SAS® 9.4 In-Database Products User's Guide Second Edition SAS (2013)
12. Neugebauer, A.: SYBASE IQ 15 In-Database Analytics Option
13. Urbanek, S.: Rserve - A Fast Way to Provide R Functionality to Applications, Proceedings of the 3rd International Workshop on Distributed Statistical Computing (DSC 2003), March 20-22 (2003)

# On the $FM_\alpha$-Integral of Fuzzy-Number-Valued Functions

Yabin Shao[1], Qiang Ma[2], and Xiaoxia Zhang[3]

[1] College of Science,
Chongqing University of Posts and Telecommunications, Chongqing, 400065, China
[2] Information Management Center,
Northwest University for Nationalities Network, Lanzhou, 730030 China
[3] Mathematics and Information Science,
Northwest Normal University, Lanzhou, 730070, China

**Abstract.** In this paper, we define the $FM_\alpha$ integrals of fuzzy-number-valued functions and discuss its properties. Especially, we give two examples which show that the $FM_\alpha$ integrable function is not fuzzy McShane integrable, and the fuzzy Henstock integrable function is not $FM_\alpha$ integrable function. As the main outcomes, we prove that a fuzzy-number-valued function $f : I_0 \to E^n$ is $FM_\alpha$ integrable on $I_0$ if and only if there exists an $ACG_\alpha$ function $F$ such that $F' = f$ almost everywhere on $I_0$.

**Keywords:** Fuzzy number, fuzzy-number-valued functions, $FM_\alpha$ integrals.

## 1 Introduction

In real analysis, the Henstock integral is designed to integrate highly oscillatory functions which the Lebesgue integral fails to do. It is well-known that the Henstock integral includes the Riemann, improper Riemann, Lebesgue and Newton integrals [1,2]. Though such an integral was defined by Denjoy in 1912 and also by Perron in 1914, it was difficult to handle using their definitions. But with the Riemann-type definition introduced more recently by Henstock[1] in 1963 and also independently by Kurzweil[3], the definition is now simple and furthermore the proof involving the integral also turns out to be easy. For more detailed results about the Henstock integral, we refer to [2]. Wu and Gong [4,5] have combined the above theories and fuzzy sets theory [6] and discussed the fuzzy Henstock integrals of fuzzy-number-valued functions which extended Kaleva integration [7]. In order to complete the theory of fuzzy calculus and to meet the solving need of transferring a fuzzy differential equation into a fuzzy integral equation, we [8] have defined the strong fuzzy Henstock integrals and discussed some of their properties and the controlled convergence theorem.

It is well-known [9] that a real function $f : [a, b] \to R$ is $C-$integrable on $[a, b]$ if and only if there exists an $ACG_c$ function F such that $F' = f$ almost everywhere on $[a, b]$. In this paper, we shall continuously extend fuzzy integration theory. Using the definition of $M_\alpha$ partition [10], we will define the fuzzy $M_\alpha$ integrals of fuzzy-number-valued functions and discuss the properties of the

D. Miao et al. (Eds.): RSKT 2014, LNAI 8818, pp. 386–396, 2014.
DOI: 10.1007/978-3-319-11740-9_36 © Springer International Publishing Switzerland 2014

fuzzy $M_\alpha-$ integral. Especially, we give two examples which show that the $FM_\alpha$ integrable function is not fuzzy McShane integrable, and the fuzzy Henstock integrable function is not $FM_\alpha$ integrable function. As the main outcomes, we prove that a fuzzy-number-valued function $f : [a, b] \to E^n$ is $FM_\alpha$ integrable on $[a, b]$ if and only if there exists an $ACG_\alpha$ function $F$ such that $F' = f$ almost everywhere on $[a, b]$.

The rest of the paper is organized as follows: in Section 2, we will recall some basic results of fuzzy numbers space. In section 3, we discuss the properties of the fuzzy $M_\alpha$-integral of fuzzy-number-valued functions. And in Section 4, we present some concluding remarks and future research.

## 2   Preliminaries

Let us denote by $E^n$ the class of fuzzy subsets of the real axis $u : \mathbb{R} \to [0, 1]$, satisfying the following properties:

(1) $u$ is normal, i.e. there exists $x_0 \in \mathbb{R}$ with $u(x_0) = 1$;

(2) $u$ is a convex fuzzy set (i.e $u(tx + (1 - t)y) \geq \min\{u(x), u(y)\}, \forall t \in [0, 1], x, y \in \mathbb{R}$);

(3) $u$ is semicontinuous on $\mathbb{R}$;

(4) $\overline{\{x \in \mathbb{R} : u(x) > 0\}}$ is compact, where $\overline{A}$ denotes the closure of $A$.

Then $E^n$ is called the space of fuzzy numbers. For $0 < \lambda \leq 1$, denote $[\tilde{u}]_\lambda = \{x \in \mathbb{R} : u(x) \geq \lambda\}$ and $[u]_0 = \{x \in \mathbb{R} : u(x) > 0\}$. Then it is well-known that for any $\lambda \in [0, 1], [\tilde{u}]_\lambda$ is a bounded closed interval. For $\tilde{u}, \tilde{v} \in \mathbb{R}_\mathcal{F}$ and $k \in \mathbb{R}$, the sum $\tilde{u} + \tilde{v}$ and the product $k \cdot \tilde{u}$ are defined by $[\tilde{u} + \tilde{v}]_\lambda = [\tilde{u}]_\lambda + [\tilde{v}]_\lambda$, $[k \cdot \tilde{u}]_\lambda = k \cdot [\tilde{u}]_\lambda, \forall \lambda \in [0, 1]$, where $[\tilde{u}]_\lambda + [\tilde{v}]_\lambda = \{x + y : x \in [\tilde{u}]_\lambda, y \in [\tilde{u}]_\lambda\}$ means the usual addition of two intervals and $k \cdot [\tilde{u}]_\lambda = \{\lambda x, x \in [\tilde{u}]_\lambda\}$ means the usual product between a scalar and a subset of $\mathbb{R}$.

**Lemma 1 ([11]).** *Let $\tilde{u} \in \mathbb{R}_\mathcal{F}$ and $[\tilde{u}]_\lambda = [u_\lambda^-, u_\lambda^+]$. Then the following conditions are satisfied:*

*(1) $u_\lambda^-$ is a bounded left continuous non-decreasing function on $(0, 1]$;*

*(2) $u_\lambda^+$ is a bounded left continuous non-increasing function on $(0, 1]$;*

*(3)$u_\lambda^-$ and $u_\lambda^+$ are right continuous at $\lambda = 0$;*

*(4)$u_1^- \leq u_1^+$.*

*Conversely, if a pair of function $a(\lambda)$ and $b(\lambda)$ satisfy condition $(1) - (4)$, then there exists a unique $\tilde{u} \in \mathbb{R}_\mathcal{F}$ such $[\tilde{u}]_\lambda = [a(\lambda), b(\lambda)]$ for each $\lambda \in [0, 1]$.*

Let $A$ and $B$ be two nonempty bounded subset of $R^n$. The distance between $A$ and $B$ is defined by the Hausdorff metric[12]:

$$d_H(A, B) = \max\{\sup_{a \in A} \inf_{b \in B} \parallel a - b \parallel, \sup_{b \in B} \inf_{a \in A} \parallel b - a \parallel\}.$$

Define $D : E^n \times E^n \to [0, \infty)$

$$D(u, v) = \sup\{d_H([u]^\alpha, [v]^\alpha) : \alpha \in [0, 1]\},$$

where $d$ is the Hausdorff metric defined in $P_k(R^n)$. Then it is easy see that $D$ is a metric in $E^n$. Using the results [13], we know that

(1) $(E^n, D)$ is a complete metric space,

(2) $D(u + w, v + w) = D(u, v)$ for all $u, v, w \in E^n$,

(3) $D(\lambda u, \lambda v) = |\lambda| D(u, v)$ for all $u, v, w \in E^n$ and $\lambda \in R$.

## 3   Properties of the Fuzzy $M_\alpha$-integral

Throughout this paper, $I_0 = [a, b]$ is a compact interval in R. Let $P$ be a finite collection of interval-point pairs $\{(I_i, \xi_i)\}_{i=1}^n$, where $\{I_i\}_{i=1}^n$ are non-overlapping subintervals of $I_0$ and let $\delta$ be a positive function on $I_0$, i.e. $\delta : I_0 \to R^+$. We say that $P = \{(I_i, \xi_i)\}_{i=1}^n$ is

(1) a partial tagged partition of $I_0$ if $\cup_{i=1}^n \subset I_0$);

(2) a tagged partition of $I_0$ if $\cup_{i=1}^n = I_0$.

**Definition 1 ([1,2]).** *Let $\delta(x)$ be a positive function defined on the interval $[a, b]$. A MacShane division $P = \{[x_{i-1}, x_i], \xi_i\}$ is said to be $\delta$−fine if the following conditions are satisfied:*

*(1) $a − x_0 < x_1 < \cdots < x_n = b$;*

*(2) $[x_{i-1}, x_i] \subset (\xi_i − \delta(\xi_i), \xi_i + \delta(\xi_i))$.*

**Definition 2 ([1,2]).** *Let $\delta(x)$ be a positive real function on a closed set $[a, b]$. A Henstock division $P = \{[x_{i-1}, x_i], \xi_i\}$ is said to be $\delta$-fine, if the following conditions are satisfied:*

*(1) $a = x_1 < x_2 < \cdots < x_n = b$;*

*(2) $\xi_i \in [x_{i-1}, x_i] \subset (\xi_i − \delta(\xi_i), \xi_i + \delta(\xi_i))$.*

**Definition 3 ([10]).** *Let $\delta(x)$ be a positive real function on a closed set $[a, b]$. A $M_\alpha$ division $P = \{[x_{i-1}, x_i], \xi_i\}$ is said to be $\delta$-fine, if the following conditions are satisfied:*

*(1) for a constant $\alpha > 0$ if it is a $\delta$−fine McShane partition of $I_0$;*

*(2) the $\sum_{i=1}^n dist(\xi_i, I_i) < \alpha$, where $dist(\xi_i, I_i) = \inf\{|t − \xi_i| : t \in I_i\}$.*

Given a $\delta$−fine partition $P = \{[x_{i-1}, x_i], \xi_i\}$ we write

$$S(f, P) = \sum_{i=1}^n f(\xi_i)|I_i|$$

for integral sums over $P$, whenever $f : I_0 \to E^n$.

**Definition 4.** *Let $\alpha > 0$ be a constant. A function $f : I_0 \to E^n$ is $FM_\alpha$-integrable if there exists a fuzzy number $A$ such that for each $\varepsilon > 0$ there is a positive function $\delta : I_0 \to R^+$ such that*

$$D(S(f, P), A) < \epsilon$$

*for each $\delta$-fine $M_\alpha$-partition $P = \{(I_i, \xi_i)\}_{i=1}^n$ of $I_0$. A is called the $FM_\alpha$-integral of $f$ on $I_0$, and we write $A = (FM_\alpha) \int_{I_0} f$.*

In the following, we shall give some properties of the $FM_\alpha$-integral.

**Theorem 1.** *A function $f : I_0 \to E^n$ is $FM_\alpha$-integrable if and only if for each $\varepsilon > 0$ there is a positive function $\delta : I_0 \to R^+$ such that*

$$D(S(f, P_1), S(f, P_2)) < \epsilon$$

*for any $\delta$-fine $M_\alpha$-partitions $P_1$ and $P_2$ of $I_0$.*

*Proof.* Assume that $f : I_0 \to E^n$ is $FM_\alpha$-integrable on $I_0$. For each $\varepsilon > 0$ there is a positive function $\delta : I_0 \to R^+$ such that

$$D\left(S(f, P), \int_{I_0} f\right) < \frac{\epsilon}{2}.$$

for each $\delta$-fine $M_\alpha$-partition $D$ of $I_0$. If $P_1$ and $P_2$ are $\delta$-fine $M_\alpha$-partitions, then

$$D(S(f, P_1), S(f, P_2)) \le D\left(S(f, P_1), \int_{I_0} f\right) + D\left(\int_{I_0} f, S(f, P_2)\right)$$

$$< \frac{\epsilon}{2} + \frac{\epsilon}{2} = \epsilon.$$

Conversely, assume that for each $\varepsilon > 0$, there is a positive function $\delta : I_0 \to R^+$ such that $D(S(f, P_m), S(f, P_k)) < \epsilon$ for any $\delta$-fine $M_\alpha$-partitions $P_m, P_k$ of $I_0$. For each $n \in N$, choose $\delta_n : I_0 \to R^+$ such that $D(S(f, P_1), S(f, P_2)) < \frac{1}{n}$ for any $\delta_n$-fine $M_\alpha$-partitions $P_1$ and $P_2$ of $I_0$. Assume that $\{\delta_n\}$ is decreasing. For each $n \in N$, let $P_n$ be a $\delta_n$-fine $M_\alpha$-partition of $I_0$. Then $\{S(f, P_n)\}$ is a Cauchy sequence. Let $L = \lim_{n \to \infty} S(f, P_n)$ and let $\epsilon > 0$. Choose $N$ such that $\frac{1}{N} < \frac{\epsilon}{2}$ and $D(S(f, P_n), L) < \frac{\epsilon}{2}$ for all $n \ge N$. Let D be a $\delta_N$-fine $M_\alpha$-partition of $I_0$. Then

$$D(S(f, P), L) \le D(S(f, P), S(f, P_N)) + D(S(f, P_N), L)$$

$$< \frac{1}{N} + \frac{\epsilon}{2} < \frac{\epsilon}{2} + \frac{\epsilon}{2} = \epsilon.$$

Hence $f$ is $FM_\alpha$-integrable on $I_0$, and $\int_{I_0} f = L$.

We can easily get the following theorems.

**Theorem 2.** *A fuzzy-number-valued function $f : I_0 \to E^n$. Then*
  (1) *If $f$ is $FM_\alpha$-integrable on $I_0$, then $f$ is $FM_\alpha$-integrable on every subinterval of $I_0$.*
  (2) *If $f$ is $FM_\alpha$-integrable on each of the intervals $I_1$ and $I_2$, where $I_1$ and $I_2$ are non-overlapping and $I_1 \bigcup I_2 = I_0$, then $f$ is $FM_\alpha$-integrable on $I_0$ and $\int_{I_1} f + \int_{I_2} f = \int_{I_0} f$.*

**Theorem 3.** *Let $f$ and $g$ be $FM_\alpha$-integrable functions on $I_0$. Then*
  (1) *$kf$ is $FM_\alpha$-integrable on $I_0$ and $\int_{I_0} kf = k \int_{I_0} f$ for each $k \in R$,*
  (2) *$f + g$ is $FM_\alpha$-integrable on $I_0$ and $\int_{I_0}(f + g) = \int_{I_0} f + \int_{I_0} g$.*

**Theorem 4.** *Let* $f : I_0 \to E^n$ *be* $FM_\alpha$-*integrable on* $I_0$. *Let* $\varepsilon > 0$. *Suppose that* $\delta$ *is a positive function on* $I_0$ *such that*

$$D(S(f, D), \int_{I_0} f) < \frac{\epsilon}{2}.$$

*for each* $\delta$-*fine* $M_\alpha$-*partition* $P = \{(I, \xi)\}$ *of* $I_0$. *If* $P' = \{(I_i, \xi_i)\}_{i=1}^m$ *is a* $\delta$-*fine partial* $M_\alpha$-*partition of* $I_0$, *then*

$$D(S(f, P'), \sum_{i=1}^m \int_{I_i} f(\xi_i)) < \epsilon.$$

*Proof.* Assume that $P' = \{(I_i, \xi_i)\}_{i=1}^m$ is an arbitrary $\delta$-fine partial $M_\alpha$-partition of $I_0$. Let $\overline{I_0 - \cup_{i=1}^m I_i} = \cup_{j=1}^k I_j'$.

Let $\eta > 0$. Since $f$ is $FM_\alpha$-integrable on $I_j'$, there exists a positive function $\delta_j : I_j' \to R^+$ such that

$$D(S(f, P_j), \int_{I_j'} f) < \frac{n}{k}.$$

for each $\delta_j$-fine $M_\alpha$-partition of $I_j'$.

Assume that $\delta_j(\xi) \le \delta(\xi)$ for all $\xi \in P_0$. Let $P_0 = P' + P_1 + P_2 + \cdots + P_k$. Then $P_0$ is a $\delta$-fine $M_\alpha$-partition of $I_0$ and we have

$$D(S(f, P_0), \int_{I_0} f) = D(S(f, P') + \sum_{j=1}^k S(f, P_j), \int_{I_0} f) < \epsilon.$$

Consequently, we obtain

$$D(S(f, P'), \sum_{i=1}^m \int_{I_i} f)$$

$$= D(S(f, P_0), \sum_{j=1}^k S(f, P_j)) - D(\int_{I_0} f, \sum_{j=1}^k \int_{I_j} f)$$

$$\le D(S(f, P_0), \int_{I_0} f) + \sum_{j=1}^k D(S(f, P_j), \int_{I_j} f)$$

$$< \epsilon + \frac{k\eta}{k} = \epsilon + \eta.$$

Since $\delta > 0$ was arbitrary, we have $D(S(f, P'), \sum_{i=1}^m \int_{I_i} f) \le \epsilon$.

Now we recall the definition of the derivative of a fuzzy-number-valued function.

**Definition 5 ([15]).** *Let* $f : (a, b) \to E^n$ *and* $x_0 \in (a, b)$. *We say that* $f$ *is differentiable at* $x_0$, *if there exists an element* $f'(t_0) \in E^n$, *such that*

(1) for all $h > 0$ sufficiently small, there exists $f(x_0 + h) -_H f(x_0), f(x_0) -_H f(x_0 - h)$ and the limits (in the metric $D$)

$$\lim_{h \to 0} \frac{f(x_0 + h) -_H f(x_0)}{h} = \lim_{h \to 0} \frac{f(x_0) -_H f(x_0 - h)}{h} = f'(x_0)$$

or

(2) for all $h > 0$ sufficiently small, there exists $f(x_0) -_H f(x_0 + h), f(x_0 - h) -_H f(x_0)$ and the limits

$$\lim_{h \to 0} \frac{f(x_0) -_H f(x_0 + h)}{-h} = \lim_{h \to 0} \frac{f(x_0 - h) -_H f(x_0)}{-h} = f'(x_0)$$

or

(3) for all $h > 0$ sufficiently small, there exists $f(x_0 + h) -_H f(x_0), f(x_0 - h) -_H f(x_0)$ and the limits

$$\lim_{h \to 0} \frac{f(x_0 + h) -_H f(x_0)}{h} = \lim_{h \to 0} \frac{f(x_0 - h) -_H f(x_0)}{-h} = f'(x_0)$$

or

(4) for all $h > 0$ sufficiently small, there exists $f(x_0) -_H f(x_0 + h), f(x_0) -_H f(x_0 - h)$ and the limits

$$\lim_{h \to 0} \frac{f(x_0) -_H f(x_0 + h)}{-h} = \lim_{h \to 0} \frac{f(x_0) -_H f(x_0 - h)}{h} = f'(x_0)$$

($h$ and $-h$ at denominators mean $\frac{1}{h} \cdot$ and $-\frac{1}{h} \cdot$, respectively).

**Theorem 5.** If the function $F : I_0 \to E^n$ is differentiable on $I_0$ with $f(\xi) = F'(\xi)$ for each $\xi \in I_0$, then the fuzzy-number-valued function $f : I_0 \to E^n$ is $FM_\alpha$-integrable.

*Proof.* By the definition of derivative, for each $\xi \in I_0$ there is a positive function $\delta : I_0 \to R^+$ such that

$$D(S(f, P_j), \int_{I'_j} f) < \frac{n}{k}.$$

for each $\delta_j$-fine $M_\alpha$-partition of $I'_j$.

Assume that $\delta_j(\xi) \leq \delta(\xi)$ for all $\xi \in P_0$. Let $P_0 = P' + P_1 + P_2 + \cdots + P_k$. Then $P_0$ is a $\delta$-fine $M_\alpha$-partition of $I_0$ and we have

$$D(\frac{F(\zeta) - F(\xi)}{\zeta - \xi}, f(\xi)) < \frac{\alpha}{2(\epsilon + |I_0|)}$$

for all $\zeta \in I_0$ with $|\zeta - \xi| < \delta(\xi)$. Assume that $P = \{(I_i, \xi_i)\}_{i=1}^n$ is a $\delta$-fine $M_\alpha$-partition of $I_0$. Then we have

$$D(\sum_{i=1}^n [f(\xi)|I_i| - F(I_i)], \tilde{0}) \leq \sum_{i=1}^n D(f(\xi)|I_i|, F(I_i))$$

$$< \frac{\alpha}{\epsilon + |I_0|} \sum_{i=1}^n (dist(I_i, \xi_i) + |I_i|)$$

$$< \frac{\alpha}{\epsilon + |I_0|} (\alpha + |I_0|) = \epsilon.$$

Hence $f : I_0 \to E^n$ is $FM_\alpha$-integrable on $I_0$.

**Definition 6.** *Let $\alpha > 0$ be a constant. Let $F : I_0 \to E^n$ and let $E$ be a subset of $I_0$.*

*(a) $F$ is said to be $ACG_\alpha$ on $E$ if for each $\epsilon > 0$ there is a constant $\eta > 0$ and a positive function $\delta : I_0 \to R^+$ such that $D(\sum_i F(I_i), \tilde{0}) < \epsilon$ for each $\delta$-fine partial $M_\alpha$-partition $P = \{(I_i, \xi_i)\}$ of $I_0$ satisfying $\xi_i \in E$ and $\sum_i |I_i| < \eta$.*

*(b) $F$ is said to be $ACG_\alpha$ on $E$ if $E$ can be expressed as a countable union of sets on each of which $F$ is $ACG_\alpha$.*

**Theorem 6.** *If a fuzzy-number-valued function $f : I_0 \to E^n$ is $FM_\alpha$-integrable on $I_0$ with the primitive $F$, then $F$ is $ACG_\alpha$ on $I_0$.*

*Proof.* By the definition of $FM_\alpha$-integral and the Theorem 4, for each $\epsilon > 0$ there is a positive function $\delta : I_0 \to R^+$ such that

$$D(\sum_{i=1}^n f(\xi)|I_i|, \sum_{i=1}^n F(I_i)) \le \epsilon.$$

for each $\delta$-fine partial $M_\alpha$-partition $P = \{(I_i, \xi_i)\}_{i=1}^n$ of $I_0$.

Assume that $E_n = \{\xi \in I_0 : n - 1 \le D(f(\xi), \tilde{0}) < n\}$ for each $n \in N$. Then we have $I_0 = \cup E_n$. To show that $F$ is $ACG_\alpha$ on each $E_n$, fix $n$ and take a $\delta$-fine partial $M_\alpha$-partition $P_0 = \{(I_i, \xi_i)\}$ of $I_0$ with $\xi_i \in E_n$ for all $i$. If $\sum_i |I_i| < \frac{\eta}{n}$, then

$$D(\sum_i F(I_i), \tilde{0}) \le D(\sum_i F(I_i), \sum_i F(I_i)f(\xi)) \cdot |I_i| + D(\sum_i f(\xi)|I_i|, \tilde{0})$$

$$< D(\sum_i F(I_i), \sum_i f(\xi))|I_i| + \sum_i D(f(\xi), \tilde{0}) \cdot |I_i|$$

$$< \epsilon + n \sum_i |I_i| < 2\epsilon.$$

Now we recall the definitions of the McShane and Henstock integrals of fuzzy-number-valued functions [5,14].

**Definition 7 ([5]).** *A function $f : I_0 \to E^n$ is fuzzy Henstock integrable if there exists a real number $A$ such that for each $\epsilon > 0$ there is a positive function $\delta : I_0 \to R^+$ such that*

$$D(S(f, P), A) < \epsilon$$

*for each $\delta$-fine Henstock partition $P = \{(I_i, \xi_i)\}_{i=1}^n$ of $I_0$.*

**Definition 8 ([14]).** *A function $f : I_0 \to E^n$ is fuzzy McShane integrable if there exists a real number $A$ such that for each $\epsilon > 0$ there is a positive function $\delta : I_0 \to R^+$ such that*

$$D(S(f, P), A) < \epsilon$$

*for each $\delta$-fine McShane partition $P = \{(I_i, \xi_i)\}_{i=1}^n$ of $I_0$.*

Since every Henstock partition is an $M_\alpha$-partition and every $M_\alpha$- partition is a McShane partition, we get the following theorem.

**Theorem 7.** *Let $f : I_0 \to E^n$ be a fuzzy valued function.*
*(a) If $f$ is fuzzy McShane integrable on $I_0$, then $f$ is $FM_\alpha$-integrable on $I_0$.*
*(b) If $f$ is $FM_\alpha$-integrable on $I_0$, then $f$ is fuzzy Henstock integrable on $I_0$.*

A function $f : I_0 \to E^n$ is $FM_\alpha$-integrable on $I_0$ if and only if there exists on $ACG_\alpha$ function $F$ on $I_0$ such that $F' = f$ almost everywhere on $I_0$. To prove this , we need the following two lemmas.

**Lemma 2.** *Suppose that $f : I_0 \to E^n$ and let $E \subseteq [a, b]$. If $\mu(E) = 0$, then for each $\epsilon > 0$ there exists a positive function $\delta$ on $E$ such that $S(|f|, P) < \epsilon$ for every $\delta$-fine partial $M_\alpha$-partition $P = \{(I_i, \xi_i)\}_{i=1}^n$ of $[a, b]$ with $\xi_i \in E$.*

*Proof.* For each $n$, let $E_n = \{x \in E : n - 1 \leq |f(x)| < n\}$ and let $\epsilon > 0$. Then $E = \cup E_n$. Since $\mu(E_n) = 0$ for each $n$, we can choose an open set $O_n \supseteq E_n$ with $\mu(O_n) < \frac{\epsilon}{n \cdot 2^n}$.

Define $\delta(x) = \rho(x, O_n^c)$ for $x \in E_n$. Suppose that $P$ is a $\delta$-fine partial $M_\alpha$-partition of $[a, b]$. Let $P_n$ be a subset of $P$ that has tags in $E_n$ and let $\pi = \{n \in Z^+ : P_n \neq \emptyset\}$. Then

$$S(|f|, P) = \sum_{n \in \pi} S(|f|, P_n) \leq \sum_{n \in \pi} n \cdot |I_i|$$

$$\leq n \cdot \mu(O_n) < n \cdot \frac{\epsilon}{n \cdot 2^n} = \epsilon.$$

**Lemma 3.** *Suppose that $F : I_0 \to R$ is $ACG_\alpha$ on $I_0$ and let $E \subseteq I_0$. If $\mu(E) = 0$, then for each $\epsilon > 0$ there exists a positive function $\delta$ on $E$ such that $D(\sum_{i=1}^n F(I_i), \tilde{0}) < \epsilon$ for every $\delta$-fine partial $M_\alpha$-partition $P = \{(I_i, \xi_i)\}_{i=1}^n$ of $I_0$ with $\xi_i \in E$ for all $i = 1, 2, \cdots, n$.*

*Proof.* Let $E = \cup_{n=1}^\infty E_n$ where $F$ is $ACG_\alpha$ on each $E_n$. Let $\epsilon > 0$. For each $n$, there exists a positive function $\delta_n : E_n \to R^+$ and a positive number $\eta_n > 0$ such that $|\sum_{i=1}^n |F(I_i)| < \frac{\epsilon}{2^n}$ for each $\delta_n$-fine partial $M_\alpha$-partition of $I_0$ with $\xi_n \in E_n$ and $|\sum_{i=1}^n |I_i| < \eta_n$. For each $n$, choose an open set $O_n \supseteq E_n$ and $\mu(O_n) < \eta_n$. Define $\delta(x) = \min\{\delta_n(x), \rho(x, O_n^c)\}$ for $x \in E_n$. Suppose that $P = \{(I_i, \xi_i)\}$ is a $\delta$-fine partial $M_\alpha$-partition of $I_0$ with $\xi_i \in E$. Let $P_n$ be subset of $P$ that has tags in $E_n$ and note that $(P_n) \sum_{i=1}^n |I_i| < \mu(O_n) < \eta_n$. Hence,

$$\sum_{i=1}^n D(F(I_i), \tilde{0}) \leq \sum_n (P_n) \sum_{i=1}^n D(F(I_i), \tilde{0}) < \sum_n \frac{\epsilon}{2^n}.$$

**Theorem 8.** *If a fuzzy-number-valued function $f : I_0 \to E^n$ is $FM_\alpha$-integrable on $I_0$ if and only if there is an $ACG_\alpha$ function $F$ on $I_0$ such that $F' = f$ almost everywhere on $I_0$.*

*Proof.* Suppose that $f$ is $FM_\alpha$-integrable on $I_0$ and let $F(x) = \int_a^x f$ for each $x \in I_0$. Then by Theorem 7, $F$ is $ACG_\alpha$ on $I_0$. Since $f$ is fuzzy Henstock integrable on $I_0$, $F' = f$ almost everywhere on $I_0$.

Conversely, suppose that there is an $ACG_\alpha$ function $F$ such that $F = f'$ almost everywhere on $I_0$. Let $E = \{x \in I_0 : F'(x) \neq f(x)\}$ and let $\epsilon > 0$. Then $\mu(E) = 0$. For each $x \in I_0 - E$, choose $\delta(x) > 0$ such that

$$D(F(y) - F(x), f(x)(y - x)) < \frac{\epsilon}{6(\alpha + |I_0|)} |y - x|$$

whenever $|y - x| < \delta(x)$ and $y \in I_0$ By Lemma 2 and Lemma 3, we can find $\delta(x) > 0$ on $E$ such that $D(\sum f(\xi)|I_i|, \tilde{0}) < \frac{\epsilon}{3}$ and $D(\sum F(I_i), \tilde{0}) < \frac{\epsilon}{3}$, whenever $P = \{(I_i, \xi_i)\}$ is a $\delta$-fine $M_\alpha$-partial partition of $I_0$ with $\xi_i \in E$.

Suppose that $P = \{(I_i, \xi_i)\}$ is a $\delta$-fine $M_\alpha$-partial partition of $I_0$. Let $P_1$ be the subset of $D$ that has tags in $E$ and let $P_2 = P - P_1$ then

$$\begin{aligned}
D((P) \sum f(\xi)|I_i|, (P) \sum F(I_i)) &= D((P_2) \sum f(\xi)|I_i|, (P_2) \sum F(I_i)) \\
&\quad + D((P_1) \sum f(\xi)|I_i|, \tilde{0}) \\
&\quad + D((P_1) \sum F(I_i), \tilde{0}) \\
&\leq (P_2) \sum D(f(\xi)|I_i|, F(I_i)) + \frac{\epsilon}{3} + \frac{\epsilon}{3} \\
&\leq \frac{\epsilon}{3(\alpha + |I_0|)} \sum (dist(I_i, \xi_i) + |I_i|) + \frac{2\epsilon}{3} \\
&\leq \frac{\epsilon}{3(\alpha + |I_0|)}(\alpha + |I_i|) + \frac{2\epsilon}{3} \\
&= \frac{\epsilon}{3} + \frac{2\epsilon}{3} = \epsilon.
\end{aligned}$$

Hence $f$ is $FM_\alpha$-integrable on $I_0$.

The following examples show that the converse of Theorem 7 is not true.

*Example 1.* Let $f$ be a fuzzy-number-valued function defined by

$$f(x) = \begin{cases} \tilde{A}(2x \sin \frac{1}{x^2} - \frac{2}{x} \cos \frac{1}{x^2}), & 0 < x \neq 1, \\ 0, & x = 0. \end{cases}$$

where fuzzy number $\tilde{A}$ defined as following:

$$\tilde{A}(s) = \begin{cases} s, & 0 \leq s \leq 1, \\ 2 - s, & 1 < s \leq 2, \\ 0, & others. \end{cases}$$

Then it is easy to show that the primitive of $f$ is

$$F(x) = \begin{cases} \tilde{A} \cdot x^2 \sin \frac{1}{x^2}, & 0 < x \neq 1, \\ 0, & x = 0. \end{cases}$$

Since $F(x)$ is differentiable and $F'(x) = f(x)$ everywhere on $[0,1]$, $f(x)$ is $M_\alpha$-integrable from Theorem 5. But $F(x)$ is not absolutely continuous on $[0,1]$ and therefore $f(x)$ is not fuzzy McShane integrable on $[0,1]$.

*Example 2.* The fuzzy-number-valued function $F$ defined by

$$F(x) = \begin{cases} \tilde{A} \cdot x \sin \frac{1}{x^2}, & 0 < x \neq 1, \\ 0, & x = 0. \end{cases}$$

is differentiable nearly everywhere on $[0,1]$. Obviously, the function $F'$ is fuzzy Henstock integrable on $[0,1]$. But we can show that $F$ is not $ACG_\alpha$ on $[0,1]$. To show this, suppose that $F$ is $ACG_\alpha$. Then there exists a set $E \subseteq [0,1]$ such that $0 \in E$ and $F$ is $ACG_\alpha$ on $E$.

For $\epsilon = \frac{\alpha}{2}$, there exist a positive function $\delta : [0,1] \to R^+$ and a positive number $\eta > 0$ such that $D(\sum F(I_i), \tilde{0}) < \frac{\alpha}{2}$, whenever $P = \{(I_i, x_i)\}_{i=1}^n$ is a $\delta$-fine partial $M_\alpha$-partition of $[0,1]$ with $x_i \in E$ and $\sum_{i=1}^n |I_i| < \eta$.

Let $a_n = \frac{1}{\sqrt{(2n+\frac{1}{2})\pi}}$ and $b_n = \frac{1}{\sqrt{2n\pi}}$ for each positive integer $n$. Then $a_n < b_n < 1$ and $\sum_{n=1}^\infty a_n = \infty$. Choose a $\delta$-fine partial partition $P = \{([a_i, b_i], 0) : N \leq i \leq M\}$ such that $\frac{\alpha}{2} < \sum_{i=N}^M a_i < \alpha$ and $b_N < \min\{\delta(0), \eta\}$. Then $0 \in E$, $\sum_{i=N}^M (b_i - a_i) < \eta$, and $\sum_{i=N}^M dist(0, [a_i, b_i]) = \sum_{i=N}^M a_i < \alpha$.

Hence, $P$ is a $\delta$-fine $M_\alpha$-partial partition of $[0,1]$. But we have

$$D(\sum_{i=N}^M F(b_i), \sum_{i=N}^M F(a_i)) = \sum_{i=N}^M D(F(b_i), F(a_i)) = \sum_{i=N}^M a_i > \frac{\alpha}{2}.$$

This contradiction shows that $F$ is not $ACG_\alpha$ on $[0,1]$. Hence, $F'$ is not $FM_\alpha$-integrable on $[0,1]$.

# 4 Conclusion

The aim of this paper is attempt to extend the theory of the nonabsolute fuzzy integration theory. We provide a minimal constructive integration process of Riemann type which includes the Keleva integral and also integrates the derivatives of differentiable functions follows by Theorem 8. In the future research, we shall discuss the convergence theorems of such integration.

**Acknowledgements.** Thanks to the support by National Natural Science Foundation of China (No. 11161041) and Fundamental Research Fund for the Central Universities (No. ZYJ2012004) and the Science-Technology Foundation for Middle-aged and Young Scientist of Northwest University for Nationalities (No. X2009-012).

# References

1. Henstock, R.: Theory of Integration. Butterworth, London (1963)
2. Lee, P.: Lanzhou Lectures on Henstock Integration. World Scientific, Singapore (1989)
3. Kurzweil, J.: Generalized Ordinary Differential Equations and Continuous Dependence on a Parameter. Czechoslovak Math. J. 7, 418–446 (1957)
4. Wu, C., Gong, Z.: On Henstock Intergrals of Interval-Valued and Fuzzy-Number-Valued Functions. Fuzzy Sets and Syst. 115, 377–391 (2000)
5. Wu, C., Gong, Z.: On Henstock Intergrals of Fuzzy-valued Functions (I). Fuzzy Sets and Syst. 120, 523–532 (2001)
6. Zadeh, L.A.: Fuzzy sets. Info. and Cont. 8, 338–353 (1965)
7. Kaleva, O.: Fuzzy Diffrential Equations. Fuzzy Sets and Syst. 24, 301–319 (1987)
8. Gong, Z., Shao, Y.: The Controlled Convergence Theorems for the Strong Henstock Integrals of Fuzzy-Number-Valued Functions. Fuzzy Sets and Syst. 160, 1528–1546 (2009)
9. Bongiorno, B., Di Piazza, L., Preiss, D.: A constructive minimal integral which includes Lebesque integrable functions and derivatives. J. London Math. Soc. 62, 117–126 (2000)
10. Park, J.M., Ryu, H.W., Lee, H.K.: The $M_\alpha-$ integral. Chungcheong J. Math. 23, 99–108 (2010)
11. Goetschel, R., Voxman, W.: Elementary of fuzzy caculus. Fuzzy Sets and Syst. 18, 31–43 (1986)
12. Dubois, D., Prade, H.: Towards Fuzzy Differential Calculus, Part 1. Integration of fuzzy mappings. Fuzzy Sets and Syst. 8, 1–17 (1982)
13. Diamond, P., Kloeden, P.: Metric Space of Fuzzy Fets: Theory and Applications. World Scientific, Singapore (1994)
14. Gong, Z., Wu, C.: The McShane integral of fuzzy-valued functions. Southeast Asian Bull. of Math. 24, 365–373 (2000)
15. Bede, B., Gal, S.: Generalizations of the Differentiability of Fuzzy-Number-Valued Functions with Applications to Fuzzy Differential Equation. Fuzzy Sets and Syst. 151, 581–599 (2005)

# Domain-Oriented Data-Driven Data Mining

# Rough Classification Based on Correlation Clustering

László Aszalós and Tamás Mihálydeák

University of Debrecen
laszalos@unideb.hu,
mihalydeak.tamas@inf.unideb.hu

**Abstract.** In this article we propose a two-step classification method. At the first step it constructs a tolerance relation from the data, and at second step it uses correlation clustering to construct the base sets, which are used at the classification of the objects. Besides the exposition of the theoretical background we also show this method in action: we present the details of the classification of the well-known iris data set. Moreover we frame some open question due this kind of classification.

## 1 Introduction

As more and more parts of our lives are recorded and documented by computers, it became possible to generate easily a numerous different kinds of statistics. Some of these statistics was possible to be done before the computer era with huge human resources. Using computers made these calculations cheap and further enabled us to make even more complicated analyses. The process of analysing data from different perspectives and summarising it into useful information is called *data mining*. This term is new, but the methods of analyses are usually not as we have known and applied them for decades. The most important is to find the correlations and patterns in the possibly huge data.

The term data mining involves six type of task, but in this article we show only two of them:

- **Clustering**: discovering groups and structures in the data that are *similar* in some sense, without using any special structure of data.
- **Classification**: generalizing the structure to the new data.

The clustering and classification originates from the first half of the twentieth century [5], so we cannot wonder why so many different approaches exists, and why different methods developed according to the different demands. We cite here only some of them related to the rough set theory: [6,7,9,11,14,17]. In this article we will show a clustering and classification which based on rough sets and correlation clustering. We remark that although the aim is similar, but the approach differs from [12]. There Pawlak uses categorical-value attributes, while in our example (the iris data set) the attributes are real-valued. Later Pawlak and Skowron used discretization of real valued attributes to get an information

D. Miao et al. (Eds.): RSKT 2014, LNAI 8818, pp. 399–410, 2014.
DOI: 10.1007/978-3-319-11740-9_37 © Springer International Publishing Switzerland 2014

table with categorical data [13]. They constructed a set of basic cuts on each attribute. In this paper we construct a partial similarity relation on real valued attributes, and use all of them together, hence we have no cuts. Moreover we do not generate rules for classification as [15,16] proposed.

Correlation clustering was invented in 2004 [2], and diverged from the traditional clustering methods: it uses a tolerance relation and not the concept of distance. Moreover this clustering has a cost function which assigns an integer to each partition. This number marks the disagreements between the original relation and its partition. The solution of the correlation clustering is a partition with minimal disagreements. This kind of clustering uses only the relation as its input, thus it is not necessary to give the number of clusters, which is common at other methods. Correlation clustering have many social, economic, physical, biological and computer science applications [3,10,18]. Though, this method also has a drawback: it is NP-complete, so it is very hard to find the partition with minimal disagreements, however its approximations usually give acceptable results in practice.

In typical cases we have no tolerance relation describing the similarity of objects, that is another drawback of our approach. Section 3 suggest a solution to this problem. Before that — in Section 2 — we show how a classification can be constructed from a tolerance relation. In Section 4 we shall discuss our approach and compare it with other well-known methods. Next, we demonstrate this method step-by-step, using the iris data set, which is a standard data at classification. Finally we conclude our work and suggest further research areas.

## 2   Second Step: Classification Based on a Relation

Bansal at al. defined correlation clustering for complete signed graphs [2]. Here $G = (V, E)$ is a graph and function $s : E \to \{+, -\}$ is the sign of edges. Sign + and $-$ denotes the similarity/dissimilarity of the nodes of the edges. We always treat a node similar to itself. This signed graph defines a relation: $uRv$ iff $s((u, v)) = +$ or $u = v$. It is obvious, that this relation $R$ is tolerance relation: it is reflexive and symmetric.

In real life we cannot decide in all cases, whether the similarity of two objects holds or not. In the next section, we will go into details concerning this problem. Here, we simply use a partial tolerance relation. The definition above remains the same, but the graph is not necessarily complete.

As mentioned before, correlation clustering assigns a number to each partition, denoting the disagreement of the partition and the relation. This counts the number of cases when two (by the relation) similar objects are in different clusters (of the partition) and when two dissimilar objects are in the same cluster. The correlation clustering selects a partition, where this disagreement is minimal.

It is possible to check that in many cases there does not exist a unique solution for correlation clustering. There are more partition with the same minimal disagreement. We have two options: to choose one of them or to combine them. We have used the latter option: we have taken the intersections of the clusters of the optimal (minimal) partitions. (This so called *rough clustering* is presented in detail in [1].) These intersections are the system of **base sets** $\mathcal{B}$ of the rough set theory. Using this base sets we can build the lower and upper approximations of objects: the lower approximation of an object is the intersection of clusters containing it. Similarly the upper approximation of an object is the union of the clusters containing it. It is easy to check that the lower and upper approximations of an object are the unions of the base sets, and the upper approximation of an object is the upper approximation of all the objects that are in its lower approximation.

This base set generated from the tolerance relation is the key for classification. The tolerance relation was given on the training set. This means that the objects in this set are identified. Hence we can identify the base sets by its dominant (most frequent) types. In some seldom cases there is no dominant type in some set: there are same number from two or more kind of objects. Then we can select from the following possibilities:

- not to classify objects that will belong to this set,
- to choose randomly from the most frequent types,
- to use multi-type — describing the most frequent types — as the result.

There is one doubt: how can we classify the new objects? One assumption is that the tolerance relation can be extended to these new objects automatically. In this case we can calculate the level of attraction-repulsion of the new object $x$ and the base set $b$:

$$A(x, b) = \#\{y \mid y \in b, \ s((x, y)) = +\} - \#\{y \mid y \in b, \ s((x, y)) = -\}$$

Next we need to search the base set $b'$ for which $A(x, b') = \max_{b \in \mathcal{B}} A(x, b)$, and if $A(x, b') > 0$, then type of $x$ will be the type of $b'$. Without formulae: the new element belongs to that base set, where its elements are mostly similar to it, and the type of the new object is the type of this set.

Of course if we use the last option and the resulting type is a multi-type, or for the new object $x$ there are exists base sets $b'$ and $b''$, that $0 < A(x, b') = A(x, b'') = \max_{b \in \mathcal{B}} A(x, b)$ then the new object can multi-type, and hence not just the classifying process is rough, but its result, too. We think that this rough outcome would be profitable at medical applications, but in this article we use only the crisp outcome.

## 3    First Step: Construct a Near Optimal Relation

From the previous section we know how we can classify objects if we have a tolerance relation. Unfortunately, we usually have no such kind of relation, so we need to construct it somehow. The data describing the objects can be different: they could be continuous (heigh, weight), discrete (number of children, phonenumbers) and categorical (gender, marital status, occupation). In some cases the continuous data are discretized, and the categorial data are numerized (encoded).

These transformations are common approaches, but in these cases we cannot treat these codes as numbers, namely we cannot take into account the difference in the codes of divorced, widowed and single status.

In our database we have four continuous data to describe the different kinds of sizes of flowers, and a categorical to give their types. At classifying a new object we can only use its continuous data. Hence to express the similarity, we need an acceptance and a rejection region. If we have no other region, then this becomes unusable in the practice. For example, based on difference on number of hairbreadth we cannot decide the baldness of a man. We cannot say for any $n$, that if the difference is $n$, they are similar, and if the difference is $n + 1$, then they are dissimilar.

We need a border region, which softens the similarity measure. This can be done with two parameters, as follows. If the difference of numbers of hairbreadth is hundred, we can say that the two men are similar, according to baldness. Meanwhile if this difference is more than 1.5 million, we treat them dissimilar. Of course these parameters (100 and 1,500,000) can vary. What else we need is to near these parameters to each other to fit our idea about baldness. Unfortunately we cannot agree about the exact values of this parameters, because a teenager girl and elderly man have different ideas about baldness.

If we have a learning/training set, we are in a better position. Namely each object has a type, and these types can be used to rate the parameters. The process is the following: for a pair of parameters — $d_1$ and $d_2$, where $d_1 < d_2$ — we generate a tolerance relation. If the distance of the two objects is smaller than $d_1$, then the relation holds between the two objects. If the distance of the two objects is larger than $d_2$, then the relation does not hold between the two objects. Otherwise (the distance is between $d_1$ and $d_2$) the relation is not defined for these objects. Applying the correlation clustering with this tolerance relation for the learning set gives a partition.

Our task is to rate this partition (and its parameters). Our aim is for the clusters of the partition that they are homogeneous, and non of them is a singleton. Singletons are rejected as the objects of the learning sets are not outliers. In an extreme case — when both parameters are very small — each cluster of the partition is a singleton, each object is similar to only itself. We wish to exclude this case.

To calculate the dissimilarity of a cluster we need to count the objects in minority. For example if a cluster contains 5 objects of type A, 9 objects of type B and 3 objects of type C, then type B objects are in majority, objects of type A and C are in minority, so the dissimilarity of this cluster is $5 + 3 = 8$. To get the dissimilarity of the partition we need to sum of dissimilarity of its clusters. So we are able to compare two pair of parameters by dissimilarity of partitions they generate.

Remember, that our aim is to find the optimal values of the parameters. This is nothing else that a global optimization problem. It is very close to the non-linear programming (NLP) problems. The main difference is that the cost function in this case is $f : \mathbb{R}^2 \to \mathbb{N}$ and not $f : \mathbb{R}^n \to \mathbb{R}$. Of course the standard

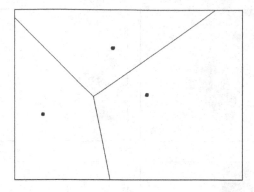

**Fig. 1.** Voronoi diagram of three object

mathematical methods used to solve NLP problems are useless here. Hence, we cannot solve such problems, only get a good approximation of the optimal solution. In such case we can use (meta)heuristic and stochastic methods. We tested the particle swarm optimization and the simulated annealing. They gave similar results, but the latter was a bit faster, so we used it for the iris data set.

The optimization produces a pair of parameters, which are near optimal, and based on them it is possible to generate a tolerance relation. This relation produces a system of base sets, and based on similarity of the new objects and the base sets the new object can be classified, as we have seen in the previous section.

## 4    Discussion

The classification on similarity to base sets could remind the reader to the nearest neighbour classification (1-NN), or to its graphical representation: to the Voronoi diagram (Fig 1). This method is the special case of the $k$-nearest neighbour classification ($k$-NN), which is a well-known and frequently used classification method.

There are similarities, e.g. both methods assign the new object to the *nearest* object/base set, but the distance used to determine the nearest object is completely different. Moreover at the $k$-NN method the outliers are classified (to the closest class), but in our case they are not. Each base set has several levels of attraction denoted with different shade of the grey at the first three pictures on Fig. 2. Here the dark grey parts denote the regions of attraction, while the light gray and white denote repulsion. At these pictures we applied the traditional euclidean distance for $d_1$ and $d_2$ to simplify the pictures, but they still remained complicated. The same picture with our distances is untraceable.

The fourth picture shows the border of attraction. The most important here is that these regions are finite, hence the outliers will not be classified. Fig. 1 shows

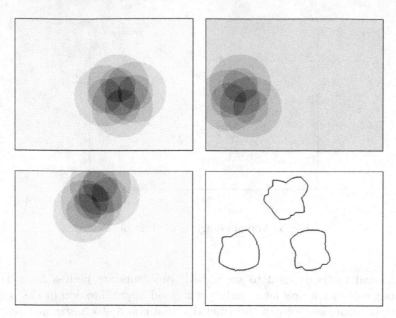

**Fig. 2.** Stucture of the attraction-repulsion effect of base sets and the attractive regions

that some regions of the Voronoi diagram are infinite, and this is the reason for classifying outliers.

# 5   Classification of the Iris Data Set

## 5.1   Flowers and Their Similarity

For testing our ideas we used the famous database [4] from 1936 (see Table 1). This data set contains 3 classes, 50 instances each, where each class refers to a type of iris plant. The data of the flowers are the length and width of sepal and petal, plus the class/type of it. As we have real data from the nature, we can assume that they follow normal distribution. As we work with distances, it is reasonable to normalize these data to get standard normal distribution in each dimension.

We have many options for distance of two *normalized* flowers. For similarity we have chosen the Chebyshev-distance: $d(x, y) = \max_i(|x_i - y_i|)$. If $d(x, y) < d_1$, then $x$ and $y$ are similar, where $x = (x_1, x_2, x_3, x_4)$ and $y = (y_1, y_2, y_3, y_4)$.

For dissimilarity we used an other *metric*: $d'(x, y) = \min_i(|x_i - y_i|)$. If $d'(x, y) > d_2$, then $x$ and $y$ are dissimilar. The minimum function warrants that in this case the points are really far away from each other. Unfortunately $d'$ is not a real metric, it does not satisfy the triangle-inequality, not even the identity of indiscernibles.

**Table 1.** Portion of the iris data set

| No. | Sepal | | Petal | | Type |
| | length | width | length | width | |
| --- | --- | --- | --- | --- | --- |
| #1 | 5.1 | 3.5 | 1.4 | 0.2 | I. setosa |
| #2 | 4.9 | 3.0 | 1.4 | 0.2 | I. setosa |
| #3 | 4.7 | 3.2 | 1.3 | 0.2 | I. setosa |
| ⋮ | ⋮ | ⋮ | ⋮ | ⋮ | ⋮ |
| #51 | 7.0 | 3.2 | 4.7 | 1.4 | I. versicolor |
| #52 | 6.4 | 3.2 | 4.5 | 1.5 | I. versicolor |
| #53 | 6.9 | 3.1 | 4.9 | 1.5 | I. versicolor |
| ⋮ | ⋮ | ⋮ | ⋮ | ⋮ | ⋮ |
| #101 | 6.3 | 3.3 | 6.0 | 2.5 | I. virginica |
| #102 | 5.8 | 2.7 | 5.1 | 1.9 | I. virginica |
| #103 | 7.1 | 3.0 | 5.9 | 2.1 | I. virginica |
| ⋮ | ⋮ | ⋮ | ⋮ | ⋮ | ⋮ |

## 5.2 Upper Level Optimization: Simulated Annealing

Fig 3. shows the process of rough classification. At the prototype software in the first step the upper level approximation had done with simulated annealing. Here the acceptance probability function was the standard: the cooling rate $\alpha = 0.95$ and the initial temperature $T = 2.0$. The number of trials at the initial temperature was 20, and each cooling increased by 5 until it reached the 200. At picking the neighbours we used random variables with standard normal distribution, and the values were divided with the actual number of trials, so we narrowed the neighbourhoods. This approach is also not common. As we have an integer-value step function on a continuous two-dimensional domain, at the beginning we need big steps (big neighbourhood), to discover the search place, and later we need little steps (small neighbourhood) to prevent it from becoming a maverick. As the process frequently ends at non-optimal regions, we recorded the best parameters during the search.

## 5.3 Lower Level Optimization: Greedy and Max-min Conflicts Methods

At clustering, our method starts with a greedy algorithm: it puts the next object to the most attractive set/cluster. It is a linear complexity method, but does not give an optimal partition. Hence the outcome of this greedy algorithm was treated as input of the max-min conflict method. At the min-conflict method [8] the randomly chosen object gets the state where it generates the minimal number of conflicts. This is a very effective algorithm in artificial intelligence. We use its variant. Here the object to be put somewhere else is not selected

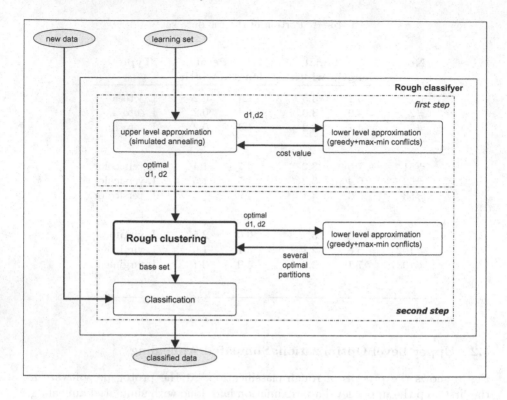

**Fig. 3.** The scheme of the main process

randomly, but by its conflicts: the most conflicting object is selected. Next, it is checked whether it can be put some state where it generates less, more precisely, minimal number of conflicts. If yes, then the object is moved, and the process starts from the beginning. Otherwise we forget this object, and the next most conflicting object is selected. By memorizing the number of conflicts we can reduce the calculations, and could get a fast, and a good optimization method. This method is parameter-free.

## 5.4 Base Sets and Results

The ordering of the objects at greedy method is not negligible, different ordering could produce different initial partitions. We utilize this property at searching the optimal partitions. We have shuffled the learning set 500 times, and applied the greedy–max-min conflicts algorithm for it, and filtered out the best partitions. The shuffling and the greedy algorithm together gives different starting points for the max-min conflicts algorithm, so it is possible to hope of getting near optimal partitions. The construction of the base set and the classification is parameterless, so based on the best partitions we unambiguously get the base set.

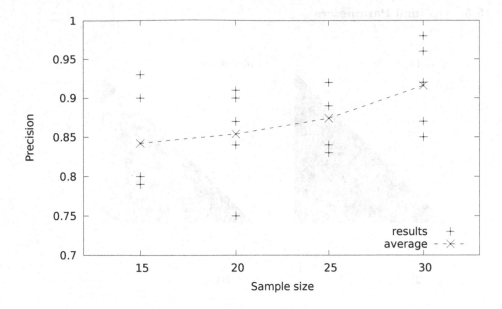

**Fig. 4.** Results of the rough classification

The attraction is a discrete function. It can occur that two (or more) nearby base sets attract with the same *force* a new object. Our software uses a conventional maximum function: it classifies to the fists base set, although it can be assigned it to any of them.

Fig. 4 shows the result of several tests. We have selected $n$ objects (denoted on $x$-axes) randomly from each class. After the construction of the system of the base sets, we have checked all the objects from the data set which do not belong to the learning set. The $y$-axes denotes the rate of correct classification. There is a big variance between the results, because an unsuccessful optimization of the parameters gives unfavourable classification result. Hence we draw on the figure the average, too. This shows a solid improvement in the results, but there are better classification methods; almost all published methods are above 95 percent. Fig. 4 shows that this kind of perfection is reachable for our method, too. For this, we need more attention on optimization of the parameters. By us, this is the key of the success.

One class of the iris data set is linearly separable, and the other two just touch each other. The latter could be a problem for our method, because the nearby objects may be added to a cluster if it is in the learning set, or it will classify them to the foreign base sets.

## 5.5  Optimal Parameters

$n = 15$                                   $n = 20$

$n = 25$                                   $n = 30$

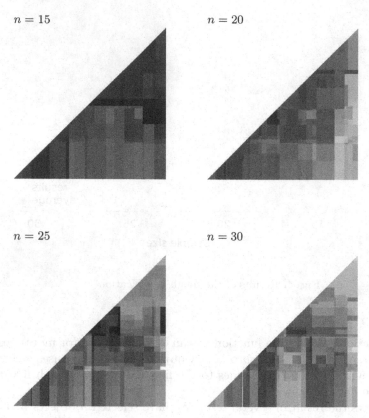

**Fig. 5.** Contour maps of the cost function used at optimization of parameters

Different learning sets mean different optimal parameters. Based on 15, 20, 25 and 30 instances from each class we generated the pictures on Fig. 5, respectively. According the value of $f(d1, d2)$ the pixel at $(d2, d1)$ was coloured. Here the different shades of grey mean different values. From left to right the pictures become brighter, it means if we have more objects in the learning set, the dissimilarity of the best partition becomes higher. We cannot prove, however we assume, that the regions $R_i = \{(x, y) \mid f(x, y) = i\}$ are (topologically) open set, i.e. there exist an $\varepsilon > 0$, such that if $d((x, y), (u, v)) < \varepsilon$ and $(x, y) \in R_i$, then $(u, v) \in R_i$, too.

In the picture, neighbouring regions are where the difference between values is bigger than 1. It is not clear whether this is the result of the poor resolution, and taking a fine enough grid would mean no such jumps in the function values; or the clustering gave not optimal values (the clustering is NP-complete task, so we have only approximations of the optimum); or there really exists such a jump in the function value.

# 6   Conclusion and Further Work

In this article we have presented a two-step classification method. At first step it constructs a near optimal tolerance relation by applying a global optimization method, where the cost function is calculated with the help of correlation clustering. At second step, by using the optimal tolerance relation it constructs a system of base sets by intersecting the optimal clusters. Finally the classification of new objects is obtained by the *nearest* base sets.

During this process there are many options. We choose the most obvious, natural ones. It would be worth to analyse the others, too. For example, the classification is crisp in the sense that it assigns only a type to an object. In our case, if in the *nearest* base set there is no majority type, we reject the classification of the object. It could define a soft classification, by allowing several types in such cases.

We have chosen fast clustering and optimizing methods. The learning set was small, so the complexity of this methods was not so interesting. We would like to test our approach on a big data sets, too. A detailed comparison of the presented clustering method with other clustering-based classifications can give such results which should make the suggested algorithm stronger.

The distances based on maximum and minimum function enable us to classify objects with missing values. This can be the base of a different research direction.

**Acknowledgments.** The publication was supported by the TÁMOP-4.2.2.C-11/1/KONV-2012-0001 project. The project has been supported by the European Union, co-financed by the European Social Fund.

**Source Code.** The formal algorithm can be found in the prototype, written in Python 3 which contains around 600 LoC and is under continuous development. It is available from the authors. The software of rough clustering — we used at classification — can be found at http://morse.inf.unideb.hu/~aszalos/roughclusters/

# References

1. Aszalós, L., Mihálydeák, T.: Rough clustering generated by correlation clustering. In: Ciucci, D., Inuiguchi, M., Yao, Y., Ślęzak, D., Wang, G. (eds.) RSFDGrC 2013. LNCS, vol. 8170, pp. 315–324. Springer, Heidelberg (2013)
2. Bansal, N., Blum, A., Chawla, S.: Correlation clustering. Machine Learning 56(1-3), 89–113 (2004)
3. Becker, H.: A survey of correlation clustering. Advanced Topics in Computational Learning Theory, pp. 1–10 (2005)
4. Fisher, R.A.: The use of multiple measurements in taxonomic problems. Annals of Eugenics 7(7), 179–188 (1936)
5. Jain, A.K.: Data clustering: 50 years beyond k-means. Pattern Recognition Letters 31(8), 651–666 (2010), http://linkinghub.elsevier.com/retrieve/pii/S0167865509002323

6. Kumar, P., Krishna, P.R., Bapi, R.S., De, S.K.: Rough clustering of sequential data. Data & Knowledge Engineering 63(2), 183–199 (2007), http://linkinghub. elsevier.com/retrieve/pii/S0169023X07000055

7. Lingras, P., Peters, G.: Applying rough set concepts to clustering. In: Peters, G., Lingras, P., Ślęzak, D., Yao, Y. (eds.) Rough Sets: Selected Methods and Applications in Management and Engineering, pp. 23–37. Springer, London (2012), http://www.springerlink.com/index/10.1007/978-1-4471-2760-4$_$2

8. Minton, S., Johnston, M.D., Philips, A.B., Laird, P.: Minimizing conflicts: a heuristic repair method for constraint satisfaction and scheduling problems. Artificial Intelligence 58(1), 161–205 (1992)

9. Mitra, S., Pedrycz, W., Barman, B.: Shadowed c-means: Integrating fuzzy and rough clustering. Pattern Recognition 43(4), 1282–1291 (2010), http:// linkinghub.elsevier.com/retrieve/pii/S0031320309003732

10. Néda, Z., Sumi, R., Ercsey-Ravasz, M., Varga, M., Molnár, B., Cseh, G.: Correlation clustering on networks. Journal of Physics A: Mathematical and Theoretical 42(34), 345003 (2009), http://www.journalogy.net/Publication/18892707/correlation-clustering-on-networks

11. Parmar, D., Wu, T., Blackhurst, J.: MMR: an algorithm for clustering categorical data using rough set theory. Data & Knowledge Engineering 63(3), 879–893 (2007), http://linkinghub.elsevier.com/retrieve/pii/S0169023X07001012

12. Pawlak, Z.: Rough classification. International Journal of Man-Machine Studies 20(5), 469–483 (1984)

13. Pawlak, Z., Skowron, A.: Rough sets and boolean reasoning. Information Sciences 177(1), 41–73 (2007)

14. Peters, G., Weber, R., Nowatzke, R.: Dynamic rough clustering and its applications. Applied Soft Computing 12(10), 3193–3207 (2012), http://linkinghub.elsevier.com/retrieve/pii/S1568494612002517

15. Shen, Q., Chouchoulas, A.: A rough-fuzzy approach for generating classification rules. Pattern Recognition 35(11), 2425–2438 (2002)

16. Tsai, Y.C., Cheng, C.H., Chang, J.R.: Entropy-based fuzzy rough classification approach for extracting classification rules. Expert Systems with Applications 31(2), 436–443 (2006)

17. Yang, L., Yang, L.: Study of a cluster algorithm based on rough sets theory. In: Proceedings of the Sixth International Conference on Intelligent Systems Design and Applications, ISDA 2006, vol. 1, pp. 492–496. IEEE Computer Society, Washington, DC (2006), http://dx.doi.org/10.1109/ISDA.2006.253

18. Zimek, A.: Correlation clustering. ACM SIGKDD Explorations Newsletter 11(1), 53–54 (2009)

# An Improved Hybrid ARIMA and Support Vector Machine Model for Water Quality Prediction

Yishuai Guo[1,2], Guoyin Wang [2,*], Xuerui Zhang [2], and Weihui Deng [2]

[1] Chongqing Key Laboratory of Computational Intelligence, Chongqing University of Posts and Telecommunications, Chongqing 400065, China
guoyishuai@cigit.ac.cn
[2] Institute of Electronic Information & Technology, Chongqing Institute of Green and Intelligent Technology, Chinese Academy of Sciences, Chongqing 401122, China
Wangguoyin@cigit.ac.cn

**Abstract.** Traditionally, the hybrid ARIMA and support vector machine model has been often used in time series forecasting. Due to the unique variability of water quality monitoring data, the hybrid model cannot easily give perfect forecasting. Therefore, this paper proposed an improved hybrid methodology that exploits the unique strength in predicting water quality time series problems. Real data sets of water quality provided by the Ministry of Environmental Protection of People's Republic of China during 2008-2014 were used to examine the forecasting accuracy of proposed model. The results of computational tests are very promising.

**Keywords:** ARIMA, Support vector machine, Time series forecasting, Water quality prediction.

## 1    Introduction

The water quality problem is a subject ongoing concern. Deterioration of water quality has initiated serious management efforts in many countries [1]. Most acceptable ecological and water related decisions are difficult to make without careful modeling, prediction and analysis of river water quality for typical development scenarios [2]. Accurate predictions of future phenomena are the lifeblood of optimal water resources management in a watershed. So far, two kind of approach have been proposed for water quality prediction [3]. One kind is the based on the mechanism of movement, physical, chemical and other factors in the water and has been widely employed in different basins [4]. But the mechanistic models usually need complete observed data and mechanism knowledge, of which are difficult to get [5]. Another kind is the models based on statistics and artificial intelligence. The rapid development of artificial intelligence provides us with more approaches for regression and better accuracy under varies situations [6-7].  For example, the support vector machine (SVM)

---

* Corresponding author.

D. Miao et al. (Eds.): RSKT 2014, LNAI 8818, pp. 411–422, 2014.
DOI: 10.1007/978-3-319-11740-9_38 © Springer International Publishing Switzerland 2014

[8-10] has been widely used for prediction and forecasting in water resources and environmental engineer.

Computer science and statistics have improved modeling approaches for discovering patterns found in water resources time series data [1]. Much effort has been devoted over the past several decades to the development and improvement of time series prediction models. One of the most important and widely used time series model is the autoregressive integrated moving average (ARIMA) model [11].

Before 2005, most of the studies reported above were simple applications of using traditional time series approaches and support vector machine [12-13]. Recently, there have been several studies suggesting hybrid models, combining the ARIMA model and support vector machine [14-17]. However, many of the real-life time series are extremely complex to be modeled using simple approaches especially when high accuracy is required. This study presents an improved hybrid model of ARIMA and SVMs to solve the water quality prediction problem.

## 2    Hybrid Model in Forecasting

### 2.1    ARIMA Model

In an autoregressive integrated moving average model (ARIMA), the future value of a variable is assumed to be a linear function of several past observations and random errors [18]. In an ARIMA model, the future value of a variable is supposed to be a linear combination of past values and past errors, expressed as follows

$$y_t = \theta_0 + \phi_1 y_{t-1} + \ldots + \phi_p y_{t-p} + \varepsilon_t - \theta_1\varepsilon_{t-1} - \theta_2\varepsilon_{t-2} - \ldots - \theta_q\varepsilon_{t-q}, \tag{1}$$

Where $y_t$ is the value of observations and $\varepsilon_t$ is the random error at time $t$, $\phi_i$ and $\theta_j$ are the coefficients, $p$ and $q$ are integers that are often referred to as autoregressive and moving average polynomials, respectively. Basically, this method has three phases: model identification, parameter estimation and diagnostic checking.

For example, the ARIMA (1, 0, 1) can be represented as follows

$$y_t = \theta_0 + \phi_1 y_{t-1} + \varepsilon_t - \theta_1\varepsilon_{t-1}. \tag{2}$$

The residuals are modeled by the ARIMA can be represented as follows

$$\varepsilon_{1t} = y_t - y_{1t}. \tag{3}$$

Where $\varepsilon_{1t}$ is the error of ARIMA model at time $t$; $y_t$ is the value of observations at time $t$; $y_{1t}$ is the value of prediction of ARIMA at time $t$.

The ARIMA model is basically a data-oriented approach that is adapted from the structure of the data themselves.

## 2.2    Support Vector Machine

The support vector machine (SVM) was proposed by Vapnik [19]. Based on the structured risk minimization (SRM) principle, SVM seeks to minimize an upper bound of the generalization error instead of the empirical error as in other neural networks. Additionally, the SVMs models generate the regress function by applying a set of high dimensional linear functions. The SVM regression function is formulated as follows

$$y = w\varphi(x) + b,$$    (4)

Where $\varphi(x)$ is called the feature, which is nonlinear mapped from the input space $x$. The coefficients $w$    and $b$    are estimated by minimizing

$$R(C) = C\frac{1}{N}\sum_{i=1}^{N}L_\varepsilon(d_i, y_i) + \frac{1}{2}\|w\|^2$$    (5)

$$L\varepsilon(d, y) = \begin{cases} |d - y| - \varepsilon & |d - y| \ge \varepsilon \\ 0 & others \end{cases},$$    (6)

Where both $C$ and $\varepsilon$ are prescribed parameters. The first term $L\varepsilon(d, y)$    is called the $\varepsilon$ -intensive loss function. The $d_i$ is the actual water quality data in the $i$th period. This function indicates that errors below $\varepsilon$ are not penalized.

The term $C(1/N)\sum_{i=1}^{N}L\varepsilon(d_i, y_i)$ is the empirical error. The second term, $\frac{1}{2}\|w\|^2$ measures the flatness of the function. $C$ evaluates the trade-off between the empirical risk and the flatness of the model. Introducing the positive slack variables $\zeta$ and $\zeta *$, which represent the distance from the actual values to the corresponding boundary values of ε-tube. Equation 5 is transformed to the following constrained formation: Minimize:

$$R(w, \zeta, \zeta*) = \frac{1}{2}ww^T + C*(\sum_{i=1}^{N}(\zeta_i + \zeta*_i))$$    (7)

Subjected to:

$$w\varphi(x_i) + b_i - d_i \le \varepsilon + \zeta^*_i$$    (8)

$$d_i - w\varphi(x_i) - b_i \le \varepsilon + \zeta_i$$    (9)

$$\zeta_i, \zeta^*_i \ge 0,$$    (10)

$$i = 1, 2\dots N.$$

Finally, introducing Lagrangian multipliers and maximizing the dual function of Equation 7 changes Equation 7 to the following form:

$$R(\alpha_i - \alpha^*_i) = \sum_{i=1}^{N}d_i(\alpha_i - \alpha^*_i) - \varepsilon\sum_{i=1}^{N}(\alpha_i - \alpha^*_i) - \frac{1}{2}\sum_{i=1}^{N}\sum_{j=1}^{N}(\alpha_i - \alpha^*_i)$$

$$\times(\alpha_j - \alpha h_j)K(x_i, x_j)$$    (11)

With the constraints

$$\sum_{i=1}^{N}(\alpha_i - \alpha^*_i) = 0, \tag{12}$$

$$0 \le \alpha_i \le C, \tag{13}$$

$$0 \le \alpha^*_i \le C, \tag{14}$$

$$i=1, 2 \ldots N.$$

In Equation 11, $\alpha_i$ and $\alpha^*_i$ are called Lagrangian multipliers. They satisfy the equalities,

$$\alpha_i * \alpha^*_i = 0,$$

$$f(x, \alpha, \alpha^*) = \sum_{i=1}^{l}(\alpha_i - \alpha^*_i)K(x, x_i) + b \tag{15}$$

Here, $K(x_i, x_j)$ is called the kernel function. The value of the kernel is equal to the inner product of two vectors $x_i$ and $x_j$ in the feature space $\varphi(x_i)$ and $\varphi(x_j)$, such that $K(x_i, x_j) = \varphi(x_i) * \varphi(x_j)$. Any function that satisfying Mercer's condition can be used as the Kernel function.

The Gaussian kernel function

$$K(x_i, x_j) = \exp(-|x_i - x_j|^2 / (2\sigma^2))$$

is specified in this study. The SVM was employed to estimate the nonlinear behavior of the forecasting data set because Gaussian kernels tend to give good performance under general smoothness assumptions.

## PSO for Parameter Optimization

In this paper, we use the PSO (Particle Swarm Optimization) algorithm to optimize the parameters of the SVM. The necessary optimization parameters of SVM are best-$g$, best-$mse$ and best-$c$.

PSO is a population-based search method that exploits the concept of social sharing of information. In the following, we will describe briefly the main concepts of the basic PSO algorithm.

Each particle in the solution space to two points close to at the same time, the first point is in the whole particle swarm all particles in the search process to achieve the optimal solution, is called a global optimal solution g-best. Another point is each particle in the process of all previous dynasties search own to achieve the optimal solution; this solution is called an individual optimal solution p-best.

Each particle represents a point in n-dimensional space, using a particle $i$ represents individual optimal ($i$-th particle fitness value corresponding to the minimum solution) the $i$-th particle is represented as $x_i = [x_{i1}, x_{i2}, ..., x_{in}]$; globally optimal solution (the swarm in the history of the search process to adapt the minimum value corresponding solution) is expressed as $pbest_i = [p_{i1}, p_{i2}, ..., p_{in}]$, and the $k$-th iteration $x_i$ correction amount (velocity of particles moving particles) is expressed as: $v_i^k = [v_{i1}^k, v_{i2}^k, ..., v_{im}^k]$.

Each particle are updated according to their speed and position:

$$v_{id}^k = w_i v_{id}^{k-1} + c_1 rand_1(p_{id}^{k-1} - x_{id}^{k-1}) + c_2 rand_2(g_d^{k-1} - x_{id}^{k-1}) \tag{16}$$

$$x_{id}^k = x_{id}^{k-1} + v_{id}^k \qquad (17)$$

Where: $k$ denotes the $k$-th iteration; $i = 1, 2, ..., m$; $d = 1, 2, ..., n$; $m$ is the number of particles in particle swarm; $n$ is the dimension of the solution vector; $n$ for the dimensions of the solution vector; $c1$ and $c2$ are acceleration factor, respectively, are two normal number; $rand1$ and $rand2$ are two independent random number between [0, 1]; $w_i$ as the momentum coefficient, adjust its size can change the search ability strong and the weak.

## 2.3    The Hybrid Methodology

Both the ARIMA and the SVM models have different capabilities to capture data characteristics in linear or nonlinear domains, so the hybrid model proposed in this study is composed of the ARIMA component and the SVMs component [1]. Thus, the hybrid model can model linear and nonlinear patterns with improved overall forecasting performance. The hybrid model $Z_t$ can then be represented as follows

$$Z_t = Y_t + N_t, \qquad (18)$$

Where $Y_t$ is the linear part and $N_t$ is the nonlinear part of the hybrid model. Both $Y_t$ and $N_t$ are estimated from the data set. $\tilde{Y}_t$ is the forecast value of the ARIMA model at time $t$. Let $\varepsilon_t$ represent the residual at time $t$ as obtained from the ARIMA model; then

$$\varepsilon_t = Z_t - \tilde{Y}_t \qquad (19)$$

The residuals are modeled by the SVM can be represented as follows

$$\varepsilon_t = f(\varepsilon_{t-1}, \varepsilon_{t-2}, ..., \varepsilon_{t-n}) + \Delta t \qquad (20)$$

Where $f$ is a nonlinear function modeled by the SVM; $\varepsilon_t$ obtained by Equation 3; $\Delta t$ is the random error. Therefore, the combined forecast is

$$\tilde{Z}_t = \tilde{Y}_t + \tilde{N}_t \qquad (21)$$

Notably, $\tilde{N}_t$ is the forecast value of Equation 17.

## 2.4    The Improved Hybrid Methodology

The behavior of water quality can't easily be captured; even the hybrid methodology cannot give perfect perfection to every river. In our experiment, there is a strong correlation between $\varepsilon_t$ sequence and $Y_t$ sequence. Realization of an experiment, we fed row set and error set simultaneously input support vector machine, try to get a more accurate prediction results and the result is what we want. Therefore, we present an improve hybrid strategy which has a greater capacity in the use of non-linear portion of the information data for forecasting water quality.

The $f_1$ and $f_2$ are modeled by the SVM can be represented as follows:

$$Z_t = f_1(y_{t-1}, y_{t-2}, ..., y_{t-n}) + f_2(\varepsilon_{t-1}, \varepsilon_{t-2}, ..., \varepsilon_{t-n}) + \Delta \varepsilon * \qquad (22)$$

Where $f_1$ and $f_2$ are nonlinear function modeled by the SVM; $\varepsilon_t$ obtained by Equation 3; $y_t$ is the value of observations at time t; $\Delta \varepsilon *$ is the random error. In the formula $f_1$, enter the sequence of previous observations and using PSO-SVM to get the next point in time predictions; in the formula $f_2$, arima error sequence to enter the formula 3, and the use of PSO-SVM to get arima prediction error next time algorithms to predict. The parameter optimization of $f_1$ and the parameter optimization of $f_2$ do not interfere with each other.

Therefore, the combined forecast can be list as:

$$\widetilde{Z}_t = \widetilde{F}_1 + \widetilde{F}_2 \tag{23}$$

Where $\widetilde{F}_1$ is the forecast value of prediction of observation and $\widetilde{F}_2$ is the forecast value of prediction of error.

# 3    Forecasting of Water Quality

Six sets of water quality monitoring data in this study to examine the performance of the proposed model. The monthly data of the water quality were provided by Ministry of Environmental Protection of People's Republic of China during 2008-2014. Monthly data of water quality (from Oct. 21, 2011 to Jun. 2014) were used as a testing data set. The remaining data of water quality in the same set were used as a testing data set. In this study, only one-step-ahead forecasting is considered. One-step-ahead forecasting can prevent problems associated with cumulative errors from the previous period for out-of-estimation sample forecasting [20-21].

Table 1 lists the corresponding periods.

**Table 1.** Data sets of water quality monitoring data

| River section | Parameter | Training data set | Testing data set |
|---|---|---|---|
| Zhutuo | PH | 46 | 20 |
| Jinzi | PH | 43 | 20 |
| Yuxi | PH | 45 | 20 |
| Zhutuo | DO | 46 | 20 |
| Fengshouba | DO | 47 | 20 |
| Jinzi | DO | 43 | 20 |

Four indices, MAE (mean absolute error), MSE (mean square error), MAPE (mean absolute percent error), and RMSE (root mean square error), were used as measures of forecasting accuracy. The indices are shown as follows

$$MAE = \frac{1}{N}\sum_{t=1}^{N} |di - zt| \tag{24}$$

$$MAPE = \frac{100}{N}\sum_{t=1}^{N} |\frac{d_i - z_t}{d_t}| \tag{25}$$

$$MSE = \frac{1}{N}\sum_{t=1}^{N} (d_t - z_t)^2 \tag{26}$$

$$RMSE = \left\{ \frac{1}{N} \sum_{t=1}^{N} (d_t - z_t)^2 \right\}^{0.5}$$    (27)

Where $N$ is the number of forecasting periods, $d_i$ is the actual water quality monitoring data at period $t$, and $z_t$ is the forecasting water quality data at period $t$.

In this study, the ARIMA model has three phases: model identification, parameter estimation, and diagnostic checking. Table 2 shows the most appropriate model for water quality prediction of different river section. For the ARIMA models, three parameters: $p$, $d$ and $q$, were adjusted based on the test sets. For the SVM models, three parameters: $g$, $mse$ and $c$, were adjusted based on the test sets. The parameter sets with the lowest values of MSE were selected for use in the best fitted model. For the hybrid models, ARIMA served as a preprocessor to filter the linear pattern of data sets. Then, the error terms from ARIMA were fed into the SVM in the hybrid models. The SVM was conducted to reduce the error function from the ARIMA. For the improved hybrid models, the raw data of water quality and error term entry support vector machine to obtain more accurate predictions.

The three parameters ($g$, $mse$ and $c$) of SVMs were adjusted. Table 2 lists suitable parameters for different models.

**Table 2.** Parameter of different models

| River section / Models | Parameter | ARIMA Models | SVM Models | Hybrid Models | Improved hybrid Models |
|---|---|---|---|---|---|
| Zhutuo | PH | (3,0,12) | c=2.8 g=0 mse=13.1 | c=0.5 g=2 mse=15.8 | c=8 g=0 mse=12.9 |
| Jinzi | PH | (12,1,10) | c=11.3 g=0 mse=8.5 | c=181 g=0 mse=13.3 | c=16 g=0 mse=8 |
| Yuxi | PH | (10,0,12) | c=14.8 g=0.1 mse=14.8 | c=1.1 g=0 mse=14.2 | c=2 g=0 mse=15 |
| Zhutuo | DO | (12,0,10) | c=11.3 g=0 mse=13.1 | c=5.6 g=0 mse=11.5 | c=32 g=0 mse=6.9 |
| Fengshouba | DO | (8,0,11) | c=181 g=0 mse=9.1 | c=2.8 g=0 mse=10.5 | c=724 g=0 mse=6.5 |
| Jinzi | DO | (3,0,9) | c=8 g=0 mse=9.8 | c=8 g=0 mse=14.2 | c=16 g=0 mse=7 |

Table 3 compares the forecasting results of different models.

Our experimental results as follows. The predictions from improved hybrid model were compared with those obtained from the ARIMA, SVM and hybrid models traditional time series approaches. The ARIMA, SVM and hybrid models produced acceptable results for hydrogen ion concentration, but the models poorly represented the pattern of water quality data for dissolved oxygen. The results from the hybrid model indicated that the modeling approach gave more reliable predictions of water hydrogen ion concentration and dissolved oxygen time series data. The accuracy measures MAE, MAPE, MSE and RMSE demonstrated that the improved hybrid model provided much better accuracy over the ARIMA, SVM and hybrid models for water quality predictions.

**Table 3.** comparison of forecasting indices

|  | Parameter | MAE | MAPE | MSE | RMSE |
|---|---|---|---|---|---|
| ARIMA model |  |  |  |  |  |
| (1)Zhutuo | PH | 0.2443 | 3.2292 | 0.1126 | 0.3356 |
| (2)Jinzi | PH | 0.1395 | 1.7771 | 0.0368 | 0.1918 |
| (3)Yuxi | PH | 0.1567 | 1.9956 | 0.0425 | 0.2062 |
| (4)Zhutuo | DO | 0.6483 | 7.6599 | 0.7036 | 0.8388 |
| (5)Fengshouba | DO | 0.4935 | 5.6264 | 0.4547 | 0.6743 |
| (6)Jinzi | DO | 1.2344 | 14.103 | 2.0909 | 1.4459 |
| SVM model |  |  |  |  |  |
| (1)Zhutuo | PH | 0.2341 | 3.0909 | 0.0925 | 0.3041 |
| (2)Jinzi | PH | 0.0973 | 1.2403 | 0.0222 | 0.1491 |
| (3)Yuxi | PH | 0.1399 | 1.7763 | 0.329 | 0.1816 |
| (4)Zhutuo | DO | 0.4667 | 5.5781 | 0.4434 | 0.6659 |
| (5)Fengshouba | DO | 0.55 | 6.3018 | 0.4463 | 0.668 |
| (6)Jinzi | DO | 0.847 | 9.8101 | 1.0583 | 1.0287 |
| Hybrid    model |  |  |  |  |  |
| (1)Zhutuo | PH | 0.222 | 2.9511 | 0.1037 | 0.3221 |
| (2)Jinzi | PH | 0.1288 | 1.6373 | 0.0302 | 0.1738 |
| (3)Yuxi | PH | 0.1478 | 1.8927 | 0.0433 | 0.2081 |
| (4)Zhutuo | DO | 0.556 | 6.4149 | 0.5847 | 0.7646 |
| (5)Fengshouba | DO | 0.4467 | 5.0885 | 0.3189 | 0.5647 |
| (6)Jinzi | DO | 1.0132 | 11.863 | 1.7036 | 1.3052 |
| Improved hybrid    model |  |  |  |  |  |
| (1)Zhutuo | PH | **0.2075** | **2.7483** | **0.797** | **0.2824** |
| (2)Jinzi | PH | **0.0942** | **1.1939** | **0.016** | **0.1266** |
| (3)Yuxi | PH | **0.1364** | **1.7329** | **0.0286** | **0.1691** |
| (4)Zhutuo | DO | **0.4077** | **4.8291** | **0.09675** | **0.598** |
| (5)Fengshouba | DO | **0.3321** | **3.9279** | **0.2649** | **0.5147** |
| (6)Jinzi | DO | **0.6716** | **7.8146** | **0.8075** | **0.8986** |

Fig.1, Fig.2, Fig.3, Fig.4, Fig.5 and Fig.6 make point-to-point comparisons of observation values and predicted values. As the result show, the prediction value of improve hybrid method is more closed to the observation value than other methods.

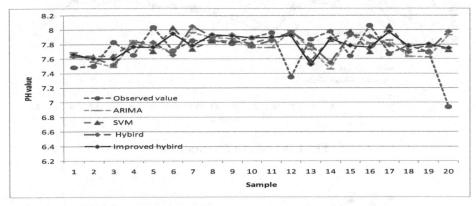

**Fig. 1.** PH value of Zhutuo section

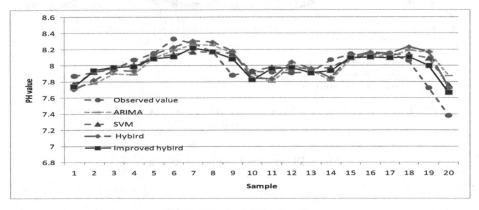

**Fig. 2.** PH value of Jinzi section

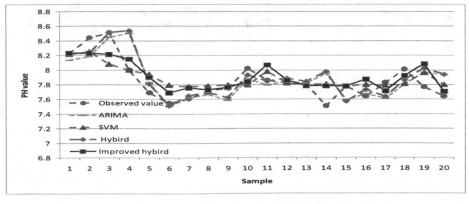

**Fig. 3.** PH value of Yuxi section

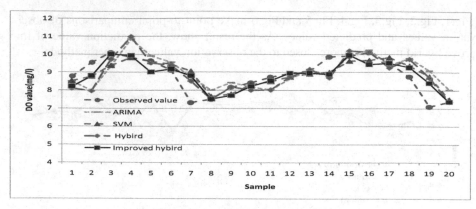

**Fig. 4.** DO value of Zhutuo section

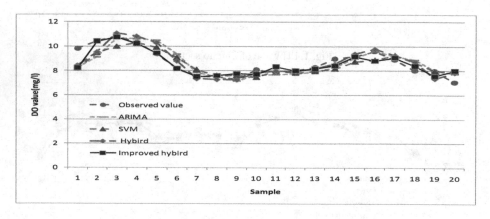

**Fig. 5.** DO value of Fengshouba section

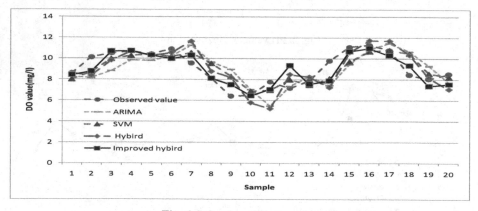

**Fig. 6.** DO value of Jinzi section

# 4 Conclusion

The proposed improved hybrid algorithm can be used for the Yangtze, Jialing and Fujiang and other similar rivers for predicting water quality data of monthly time step to detect water quality severity with respect to hydrogen ion concentration and dissolved oxygen time series data in future. The improved hybrid model developed for the Yangtze, Jialing and Fujiang can be employed for the development of a water quality emergency management plan so as to ensure sustainable water resources management in the basin.

But, we found that the improved hybrid model is not good at predicting the trip point. So if we want to have a more accurate prediction of water quality, we can think about how to get the more accurate prediction of trip point.

**Acknowledgements.** This work is supported by the National Science and Technology Major Project (NO.2014ZX07104-006), the National Natural Science Foundation of China under Grant (No.61272060), the Hundred Talents Program of CAS (NO.Y21Z110A10).

# References

1. Ömer Faruk, D.: A hybrid neural network and ARIMA model for water quality time series prediction. J. Engineering Applications of Artificial Intelligence 23(4), 586–594 (2010)
2. Najah, A., Elshafie, A., Karim, O.A., et al.: Prediction of Johor River water quality parameters using artificial neural networks. European Journal of Scientific Research 28(3), 422–435 (2009)
3. Krenkel, P.: Water quality management. Elsevier (2012)
4. Hawkins, C.P., Olson, J.R., Hill, R.A.: The reference condition: predicting benchmarks for ecological and water-quality assessments. Journal of the North American Benthological Society 29(1), 312–343 (2010)
5. Zou, X., Wang, G., Gou, G., Li, H.: A Divide-and-Conquer Method Based Ensemble Regression Model for Water Quality Prediction. In: Lingras, P., Wolski, M., Cornelis, C., Mitra, S., Wasilewski, P. (eds.) RSKT 2013. LNCS, vol. 8171, pp. 397–404. Springer, Heidelberg (2013)
6. Han, H.G., Chen, Q., Qiao, J.F.: An efficient self-organizing RBF neural network for water quality prediction. J. Neural Networks 24(7), 717–725 (2011)
7. Mahapatra, S.S., Nanda, S.K., Panigrahy, B.K.: A Cascaded Fuzzy Inference System for Indian river water quality prediction. J. Advances in Engineering Software 42(10), 787–796 (2011)
8. Yunrong, X., Liangzhong, J.: Water quality prediction using LS-SVM and particle swarm optimization. In: IEEE Second International Workshop on Knowledge Discovery and Data Mining, WKDD 2009, pp. 900–904 (2009)
9. Huang, W., Huang, F., Song, J.: Water Quality Retrieval and Performance Analysis Using Landsat Thermatic Mapper Imagery Based on LS-SVM. Journal of Software 6(8), 1619–1627 (2011)
10. Singh, K.P., Basant, N., Gupta, S.: Support vector machines in water quality management. J. Analytica Chimica Acta 703(2), 152–162 (2011)

11. Zhang, G.P.: Time series forecasting using a hybrid ARIMA and neural network model. J. Neuro Computing 50, 159–175 (2003)
12. Yunrong, X., Liangzhong, J.: Water quality prediction using LS-SVM and particle swarm optimization. In: IEEE Second International Workshop on Knowledge Discovery and Data Mining, WKDD 2009, pp. 900–904 (2009)
13. Singh, K.P., Basant, N., Gupta, S.: Support vector machines in water quality management. J. Analytica Chimica Acta 703(2), 152–162 (2011)
14. He, Y., Zhu, Y., Duan, D.: Research on hybrid ARIMA and support vector machine model in short term load forecasting. In: IEEE Sixth International Conference on Intelligent Systems Design and Applications, ISDA 2006, vol. 1, pp. 804–809 (2006)
15. Nie, H., Liu, G., Liu, X., et al.: Hybrid of ARIMA and SVMs for short-term load forecasting. J. Energy Procedia 16, 1455–1460 (2012)
16. Zhu, B., Wei, Y.: Carbon price forecasting with a novel hybrid ARIMA and least squares support vector machines methodology. J. Omega 41(3), 517–524 (2013)
17. Ming, W., Bao, Y., Hu, Z., et al.: Multistep-Ahead Air Passengers Traffic Prediction with Hybrid ARIMA-SVMs Models. The Scientific World Journal 2014 (2014)
18. Box, G.E.P., Jenkins, G.M., Reinsel, G.C.: Time series analysis: forecasting and control. John Wiley & Sons (2013)
19. Vapnik, V.: The nature of statistical learning theory. Springer (2000)
20. Reisen, V.A., Lopes, S.: Some simulations and applications of forecasting long-memory time-series models. Journal of Statistical Planning and Inference 80(1), 269–287 (1999)
21. Sivakumar, B., Jayawardena, A.W., Fernando, T.: River flow forecasting: use of phase-space reconstruction and artificial neural networks approaches. Journal of Hydrology 265(1), 225–245 (2002)

# Multi-label Supervised Manifold Ranking
## for Multi-instance Image Retrieval

Xianhua Zeng, Renjie Lv, and Hao Lian

Chongqing Key Lab of Computational Intelligence,Chongqing University of Posts and
Telecommunications, Chongqing 400065, China
College of Computer Science and Technology, Chongqing University of Posts and
Telecommunications, Chongqing 450401899, China
xianhuazeng@gmail.com

**Abstract.** Current manifold ranking is mainly used in single-instance image
retrieval without considering the prevailing semantic ambiguity problem. This
paper introduces multi-instance technique and supervised information to image
retrieval based on manifold ranking, and proposes a Multi-label Supervised
Manifold Ranking algorithm (MSMR) for multi-instance image retrieval. The
divergence between images is modified by using the multi-label information of
training samples. Our method can solve partly the 'input ambiguity problem' in
the feature extraction stage and the 'output ambiguity problem' in the output
stage. Compared with the traditional Expectation Maximization Diverse Density
(EMDD) and Citation-kNN algorithm on Corel Image Set, the multi-instance
image retrieval experimental results show that the average precision rate of our
algorithm has be enhanced .

**Keywords:** Manifold ranking, Multi-label learning, Multi-instance learning,
Image retrieval.

## 1    Introduction

Manifold Ranking (*MR*) is the semi-supervised learning method which is used to find
the sorted structure of the data by using manifold learning. This is an image retrieval
ranking method based on topology sorting with the feedback of users [1],[2]. Currently,
the manifold ranking algorithms for image retrieval are mainly applied to processing
the single-instance problem.   However, the image retrieval exits the "*input ambiguity
problem*" in the feature extraction stage and the "*output ambiguity problem*" in the
output stage of image retrieval [3].

In order to solve the semantic ambiguity problem of images, Zhou et al. proposed the
multi-label and multi-instance learning framework on the basis of multi-sample
learning, and investigated the ambiguity of the retrieval objects from the input space
and output space [3],[4].   Manifold ranking technique has extensively used in image
retrieval applications, and can used naturally to solve those multi-instance image
retrieval problems. So we consider multi-label information into defining the supervised
and weighted similarity between images, and propose a multi-label supervised
manifold ranking (MSMR) algorithm for multi-instance image retrieval. Expectation

D. Miao et al. (Eds.): RSKT 2014, LNAI 8818, pp. 423–431, 2014.
DOI: 10.1007/978-3-319-11740-9_39 © Springer International Publishing Switzerland 2014

Maximization Diverse Density (*EMDD*) [5] is a classical multi-instance learning method. In addition, *Citation-kNN*[6], [7]can be used as the multi-label learning algorithm. Some comparison experiments on Corel Image Database show the performance of our algorithm.

## 2   Manifold Ranking-Based Image Retrieval

The main idea of Manifold Ranking-Based Image Retrieval (MRBIR) [8] is as follows: First of all, computing a neighborhood matrix in the Euclidean space. Then the sorting values of queried samples can be passed to other nodes of the neighborhood graph combined with users' feedback, and takes the new sorting values as relevance ranking scores of the searched images. Finally, these retrieved images can be sorted based on the relevance ranking values of them.

   *MRBIR* is the earliest manifold ranking algorithm for image retrieval. However, *MRBIR* is only effective to the training set. For the queried images which outside of the database, *MRBIR* needs to reconstruct the neighbor matrix, which needs to take a lot of computing time. At the same time, each query of online retrieval iterate more times. So Generalized Manifold Ranking-Based Image Retrieval (*G-MRBIR*) [9] is also proposed.

   But, these manifold ranking algorithms are still mainly used in single-instance image retrieval without considering the prevailing semantic ambiguity problem. So, in next section, we will introduce the multi-label supervised information under multi-instance learning framework into manifold ranking.

## 3   Multi-label Supervised Manifold Ranking for Multi-instance Image Retrieval

### 3.1   Construction of Multi-instance Image Package and Similarity Measure

For multi-instance image retrieval, we firstly need to take an image to describe into multi-instance image package, before image retrieved in. In this paper, we use the same method as literature [10]. A given image is divided into sub-blocks, and compute statistical features of each block. Then, taking these features as the vectors for representing the given image. In following experiments, each image is described by 9 attribute vectors with 15 dimensions,   including 3 dimensions from (R, G, B) values of the mean of the central region block, 12 dimensions from (R,G,B) values of difference between central region block and four adjacent blocks.

   In traditional manifold ranking-based image retrieval, an image is thought as a single instance and the Euclidean distance is used as computing the similarity between images. Under the multi-instance learning framework, image features are no longer a single combination feature set. However, the features of each image   consist of multiple attribute vectors and each vector is an instance of multi-instance image packages. So, we need a different method to describe the similarity between multi-instance image packages. Hausdorff distance metric is a suitable way to describe the distance between packages in multi-instance learning [11]. Hausdorff distance is also known as the maximum-minimum distance, which can represent the distance between the two subsets, that is,

**Define** 1. The **k-th** Hausdorff distance between the two subsets $A = \{a_1, a_2, \cdots, a_m\}$ and $B = \{b_1, b_2, \cdots, b_n\}$ can be defined as:

$$h_k(A, B) = \max_{a_k \in A} \; \min_{b_k \in B} \| a_k - b_k \| \tag{1}$$

Obviously, the first Hausdorff distance is the maximum distance from element s of sub-set $A$ to subset $B$. The rest of Hausdorff distances can be gained by other elements in each subset.

### 3.2 Multi-label Supervised Manifold Ranking for Multi-instance Image Retrieval

In actual image retrieval, an image may have multiple labels. For example, an image may be related with 'mountain' and 'oceans', with corresponding to two labels.

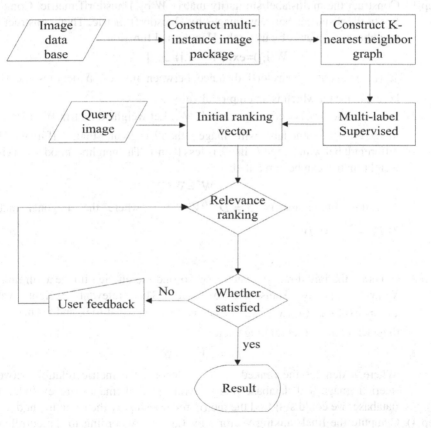

**Fig. 1.** The flow chart of MSMR in multi-instance image retrieval

The problem of multiple attribute characteristics of data is called as multi-label problem. During image retrieval, images' multi-label information can be used as supervision information to optimize the distances between images with multi-label features. In particular, for the images which are labeled by the same classes, we could add the supervision information into constructing nearest neighbor graph, that is, multiplying the distances between images by a value which is less than 1 to make distances smaller. On the other hand, for images from different classes, we multiply the distances between images by a value which is larger than 1 to increase the distances. In all, these operations to make the distances between related images are decreased, while the distances between irrelevant images are increased. This trick is more suitable with users' expectations. So we propose Multi-label Supervised Manifold Ranking algorithm (MSMR) for multi-instance Image Retrieval, as shown the basic processing in Fig.1.

Let $X=\{x_1, x_2,...,x_q, x_{q+1},...,x_n\}$ denote the image database, $x_1,...,x_q$ denote $q$ sample images, and $x_{q+1},...,x_n$ denote $n$-$q$ images to be searched. The detail steps of MSMR algorithm is described as follows:

Step 1) Construct the multi-label similarity matrix $W$ by Hausdorff metric. Compute firstly the distances between images by Hausdorff metric. Then construct the neighborhood matrix by using the Heat kernel function:

$$W(i,j)=\exp[-h_k^2(i,j)/2\delta^2] \qquad (2)$$

$h_k(i,j)$ denotes Hausdorff distance between images, $\delta$ denotes the Heat kernel constant which is an empirical value .

Step 2) Compute the supervised weight matrix $W^*$. Let weighted matrix $W_\lambda(i,j)=\lambda$. If image $i$ has the same label with image $j$, then $\lambda$ is greater than 1. If image $i$ has different label with image $j$, then $\lambda$ is less than 1. The neighborhood supervised weight matrix can be computed:

$$W^*=W*W_\lambda \qquad (3)$$

And the similar matrix $S = D^{-1/2}WD^{-1/2}$, where the diagonal matrix $D(i,i) = \sum_j W(i, j)$.

Step 3) Generate the initial ranking vector $f_0$. To find the queried image $x_0$ in images $X=\{x_1,...,x_n\}$ by Hausdorff distance. The relevant weight value $e(i) = \exp[-d^2(x_0, x_1)/2\delta^2]$. The rest $e(i)$ is equal to 0, and normalizes $e$ to generate the initial ranking vector.

$$f_0 = a(I - aS)^{-1}e \qquad (4)$$

Where $a$ denotes the ranked score, $S$ denotes the metric relation between queried image and database image. Even queried images are excluded the database, we could still find the metric relationship by the same method.

Step 4) Compute the final ranking vector $f$ by Eq. (9). According to the correlation feedback $y$ from users, updates the ranking vector $f$. If related with queried image, then $y_i=1$, else $y_i = 0$. The update of the ranking vector $f$ is

$$f(t+1)=aSf(t)+(1-a)y \qquad (5)$$

Where $aSf(t)$ denotes the ranked values which are from near points. And $(1-a)y$ denotes the ranked values which are from the feedback labels of users. The generation formula of $f(t)$

$$f(t) = (aS)^t y + (1-a)\sum_{i=0}^{t-1}(aS)^t y \qquad (6)$$

Where $S = D^{-1/2}SD^{-1/2}$, and diagonal matrix $D_{ii} = \sum_{j=1}^{n}W_{ij}$,

$P = D^{-1}W = D^{-1/2}SD^{-1/2}$. Obviously, $P \in [-1,1]$ and $S \in [-1,1]$, and $\lim_{t\to\infty}(aS)^t = 0$, so we get the following conclusion:

$$f = \lim_{t\to\infty} f(t) = (1-a)(I - aS)^{-1}y \qquad (7)$$

And because $(1-a)$ does not affect the convergence and sorting relation of the result, so

$$f = (I - aS)^{-1}y \qquad (8)$$

If the feedback image from users is relevant with queried image, then the $i^{th}$ feedback $y_i^* = 1$. If the feedback is not relevant with queried image, then $y_i^* = -1$. And the other un-label image can be set to $y_i^* = 0$. So, the positive correlation vector $y^+$ and the negative correlation vector $y^-$ are defined respectively as follows:

$$y_i^+ = \begin{cases} 1 & ,y_i^*=1 \\ 0 & ,y_i^*=0,-1 \end{cases} \qquad y_i^- = \begin{cases} -1 & , y_i^*=-1 \\ 0 & , y_i^*=0,1 \end{cases}$$

According to Eq.(8), the positive ranking vector $f^+ = (I - aS)^{-1}y^+$ and negative ranking vector $f^- = (I - aS)^{-1}y^-$. So the final manifold ranking vector $f$ is computed as follows:

$$f = \eta f_0 + (1-\eta)(I - aS)^{-1}y^+ + (I - aS)^{-1}y^- \quad (0 < \eta < 1) \qquad (9)$$

Where $\eta = \exp(-\eta_y)$, $\eta_y$ is the number of 1 in $y+$, and $f_0 = a(I - aS)^{-1}y^-$ is the initial manifold ranking vector.

Step 5) Output the corresponding images with high ranking values in vector $f$. According to the final manifold ranking vector $f$ and the images with corresponding to higher values are fed to users.

Step 6)    If users do not satisfy with the result, then return to Step 3. Else stop.

# 4    Experiments

In the following experiments, we use 2000 ambiguous natural scene images [12]. These images are divided into 5 categories, which are labeled as 'desert', 'mountain', 'sea', 'sunset', 'forest', respectively. 1000 images of them come from Corel image data set and the rest from Internet. Meanwhile, their multi-label information is shown in Table 1:

**Table 1.** Multi-labels information of 2000 images

| Label | Images | Label | Images | Label | Images | Label | Images |
|-------|--------|-------|--------|-------|--------|-------|--------|
| d | 340 | d +m | 19 | m+su | 19 | d+m+su | 1 |
| m | 268 | d+s | 5 | m+t | 108 | d+su+t | 3 |
| s | 341 | d+su | 21 | s+su | 172 | m+s+t | 6 |
| su | 216 | d+t | 20 | s+t | 14 | m+su+t | 1 |
| t | 378 | m+s | 38 | su+t | 28 | s+su+t | 4 |

In table 1, *d* denotes (desert), *m* denotes (mountains), *s* denotes (sea), *su* denotes (sunset), *t* denotes (trees), and 22% of these images have multi-labels.

We use 10 fold cross-validations to test 2000 sample images set, and select 40 images of each class every time . The other 1800 images are used as a training set. In this experiment some important parameters are set as follows: Heat-kernel Constant $\sigma=100$, $k=20$ , the equilibrium constant $a=0.8$, supervision weights $\lambda=2$, cited several $r$ =4. We label the first 20 images of each feedback, and calculate the average precision of 2000 images by iterative methods.

Without multi-label supervision information, we compared proposed method with the classical multi-instance learning method of Citation-kNN.

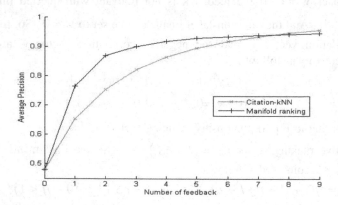

**Fig. 2.** the average precision plot of Citation-kNN and Manifold ranking for multi-instance image retrieval

In Fig.2, the average precision based on manifold ranking and multi-instance learning for image retrieval are higher than Citation-kNN when the number of feedback is smaller than 7.  But the average precision decreases lightly when the number of feedback is larger than 7. As previously mentioned, we are more concerned about the situation of the first several feedbacks in the process of the actual retrieval, especially in the first three feedbacks. In Fig.2, the average precision is enhanced about 10% by using manifold ranking in the first three feedbacks.

We also conduct experiments to compare the average retrieval rate after the second feedback with different number of feedback labels (10, 20, 30), as shown in Table 2:

**Table 2.** The average precision after different number of feedback labels

| Different number of feedback labels | Citation-kNN | Manifold Ranking |
|---|---|---|
| 10 | 0.83225 | 0.89585 |
| 20 | 0.821 | 0.8988 |
| 30 | 0.809167 | 0.901117 |

From Table 2, in the case of different number of feedback labels, the average rate which based on Manifold Ranking for multi-instance image retrieval is higher than *Citation-kNN*. Meanwhile, comparing with *EMDD* multi-instance image retrieval, the results have demonstrated once again the performance of manifold ranking under the multi-instance framework for image retrieval in Table 3.

**Table 3.** The average precision of three algorithms in the first several feedbacks

| Different Algorithm | EMDD | Citation-kNN | Manifold Ranking |
|---|---|---|---|
| Without feedback | 0.47925 | 0.47925 | 0.47925 |
| The first feedback | 0.52125 | 0.653825 | 0.763975 |
| The second feedback | 0.56925 | 0.753825 | 0.86845 |

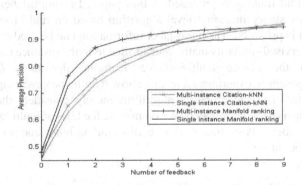

**Fig. 3.** The average precision plot of manifold ranking

The following comparative experiments are completed under the situation of combining with multi-labels as supervision information after adding supervision information, as experimental results shown in Fig.3, the average precision of our method is better than Citation-kNN in the first several feedbacks. However, it is worth noting that Citation-kNN raise less than 1%, while the proposed   MSMR algorithm for multi-instance image retrieval increase about 10%. This reason is to that manifold ranking consider global manifold structure, while the referenced set of Citation-kNN always considered local. So, global supervision information is more suitable for manifold ranking.

Meanwhile, the comparison experiment results based on manifold ranking and multi-instance image retrieval in case of presence or absence of supervision information are shown in Table 4:

**Table 4.** The average precision of manifold ranking after using supervision information

| Feedback numbers | Unsupervised | Semi-supervised | Supervised |
|---|---|---|---|
| Without feedback | 0.47925 | 0.47925 | 0.47925 |
| First feedback | 0.763975 | 0.7922 | 0.8852 |
| Second feedback | 0.86845 | 0.894775 | 0.959875 |
| Third feedback | 0.8988 | 0.9246 | 0.97565 |

As can be shown in Table 4, in the case of semi-supervised or supervised condition, the average precision is all enhanced. It demonstrates that the combination of multi-labels to supervised manifold ranking can perform well in multi-instance image retrieval once more.

## 5    Conclusion

Under the multi-instance learning framework, a novel image retrieval algorithm combining manifold ranking is proposed in this paper. For multi-label problems in reality, the multi-instance image retrieval algorithm based on multi-label supervised manifold ranking is proposed. The multi-label information can be used as constructing the weighted supervised similarity matrix and applied to multi-instance image retrieval. Compared with the classic multi-instance learning algorithms (EMDD and Citation-kNN algorithm), experimental results show that the proposed algorithm has a better performance. But the proposed algorithm merely considers the multi-label information as supervision information. We do not use the kind of multi-labels learning to predict multi-label sets of image. It is a difficult problem and we will further investigate it in the future.

**Acknowledgments.** The authors would like to thank the anonymous reviewers for their help. This work was supported by the National Natural Science Foundation of China (No. 61379114,61203308) and the Chongqing Natural Science Foundation (No. CSTC-2010BB2406).

## References

1. Liu, L., Wei, J., Ma, Q.L.: State-of-the-Art on Image Retrieval Based on Manifold Learning. Journal of Beijing Jiaotong University 34(5), 164–171 (2010)
2. Zhou, D.Y., Ohvier, B., Thomas, N.L., et al.: Ranking on Data Manifolds. Advances in Neural Information Processing Systems (2003)
3. Zhou, Z.H., Zhang, M.L., Huang, S.J., et al.: Multi-instance multi-label learning. Artificial Intelligence 176(1), 2291–2320 (2012)

4. Dai, H.B., Zhang, M.L., Zhou, Z.H.: A Multi-Instance Learning Based Approach to Image Retrieval. PR & AI 19(2), 179–185 (2006)
5. Wang, J., Zucker, J.D.: Solving the multiple-instance problem: A lazy learning approach. In: Proceedings of the 17th International Conference on Machine Learning, San Francisco, CA, pp. 1119–1125 (2000)
6. Zhou, Z.H., Zhang, M.L.: Neural networks for multiple-instance learning. Technical Report, AI Lab, Computer Science & Technology Department, Nanjing University, Nanjing, China (August 2002)
7. Jiang, L., Cai, Z., Wang, D., et al.: Bayesian citation-KNN with distance weighting. International Journal of Machine Learning and Cybernetics 5(2), 193–199 (2014)
8. He, J.R., Li, M.J., Zhang, H.J., et al.: Manifold-Ranking Based Image Retrieval. In: Proceedings of the 12th Annual ACM International Conference on Multimedia, pp. 9–16 (2004)
9. He, J.R., Li, M.J., Zhang, H.J., et al.: Generalized Manifold Ranking Based Image Retrieval. IEEE Trans. Image Process 15(10), 3170–3177 (2006)
10. Maron, O., Lozano, P.T.: A framework for multiple-instance learning. Advances in Neural Information Processing Systems, 570-576 (1998)
11. Jesorsky, O., Kirchberg, K.J., Frischholz, R.W.: Robust face detection using the hausdorff distance. In: Bigun, J., Smeraldi, F. (eds.) AVBPA 2001. LNCS, vol. 2091, pp. 90–95. Springer, Heidelberg (2001)
12. Zhang, M.L., Zhou, Z.H.: ML-KNN: A lazy learning approach to multi-label learning. Pattern Recognition 40(7), 2038–2048 (2007)

# Nearest Neighbor Condensation Based on Fuzzy Rough Set for Classification

Wei Pan[1,2], Kun She[1], Pengyuan Wei[1], and Kai Zeng[1]

[1] School of Computer Scicence & Engineering, University of Electronic Science
and Technology of China, Chengdu, Sichuan, 611731, P.R. China
[2] School of Computer Scicence & Engineering, China West Normal University,
Nanchong, Sichuan, 637009, P.R. China

**Abstract.** This work introduces a novel algorithm, called Condensation rule based on Fuzzy Rough Sets (FRSC), based on the FCNN rule together with fuzzy rough sets theory, to compute training-set-consistent subset for the nearest neighbor decision rule. In combination with fuzzy rough set theory, the FRSC rule improves the performance of FCNN rule. Two variants, named as FRSC1 and FRSC2, of the FRSC rule, are presented. The FRSC1 rule is suitable for small data set and the FRSC2 adapts to larger data sets. Compared with the FCNN rule, the FRSC1 rule requires much less time cost and gets smaller subset for small data set. For medium-size data set, less than 5000 samples, the FRSC2 rule has better time performance than FCNN rule.

**Keywords:** nearest neighbor rule, training-set-consistent subset, fuzzy rough set, lower approximation, classification.

## 1  Introduction

The nearest neighbor (NN) decision rule assigns to an unclassified sample point the classification of the nearest of a set of previously classified points [2]. The k-NN decision rule selects k nearest samples and   assigns to a sample the class label with the maximum number $k_i$ of samples, where sum($k_i$) is equal to $k$, $i=1,2,3,...l$, $l$ is the number of classes. When $k$ is equal to 1, the $k$-NN rule becomes the NN rule。 In this work, we denote the NN rule and $k$-NN rule with NN rule.

The naïve implementation of the NN rule needs to store all the previously classified data and compare each sample point to be classified with each stored point. A serious drawback, for this decision rule, is the complexity in search of the nearest neighbors among the N available training samples. Brute-force searching amounts to operations proportional to $kN(O(kN))^2$. The problem becomes particularly severe in high-dimensional feature spaces and large data set. A training-set-consistent subset may retain better classification accuracy than the entire training set. Unfortunately, computing a minimum-cardinality training-set-consistent subset for the NN rule turns out to be intractable [23]. To reduce the computational burden, a number of efficient searching schemes have been suggested, aiming to select the training-set-consistent

D. Miao et al. (Eds.): RSKT 2014, LNAI 8818, pp. 432–443, 2014.
DOI: 10.1007/978-3-319-11740-9_40 © Springer International Publishing Switzerland 2014

subset, such as condensed nearest neighbor(CNN), modified condensed nearest neighbor (MCNN), structural risk minimization using the NN rule (NNSRM), reduced nearest neighbor rule (RNN), fast condensed nearest neighbor rule (FCNN), and so on. Among these techniques, the FCNN rule is a novel order-independent algorithm aiming to select a training-set-consistent subset, which classifies the remaining data correctly through the NN rule. In particular, the FCNN rule is scalable on large multidimensional training sets.

The concept of rough sets was originally proposed by Pawlak [1] as a mathematical approach to handle imprecision, vagueness, and uncertainty in data analysis [4]. The traditional rough sets (TRS) theory will have difficulty in handling real-valued attributes. One way to solve this problem is the use of fuzzy rough sets. In fuzzy rough sets a fuzzy similarity relation is employed to characterize the degree of similarity between two objects instead of the equivalence relation used in the crisp rough sets. The degree of similarity for two objects takes values on the unit interval. If the degree of similarity is 1, then they are indiscernible. They are discernible if the degree of similarity degree is 0. If the degree of similarity takes a value between 0 and 1, then these two objects are similar to a certain degree.

The nearest neighbor condensation aims to select samples close to the decision boundary. In rough set theory, a sample with less lower approximation is more close to the decision boundary. Combination these two theories is an effective way for nearest neighbor condensation. In this work, we purpose a method based on the FCNN rule together with fuzzy rough sets theory, to reduce samples number.

The rest of this paper is organized as follows. In Section 2, some theoretical background, including fast condensed nearest neighbor (FCNN) rule and fuzzy rough set theory is described. In Section 3, the nearest neighbor condensation rule based on fuzzy rough set (FRSC) is illustrated, and two variants and its main properties are stated. And the Section 4 analyses the experimental results and compares the FRSC rule with the FCNN rule. Finally, in Section 5, the advantage sand disadvantages are drawn in the conclusion of this work.

## 2    Theoretical Background

### 2.1    Nearest Neighbor Rule

The nearest neighbor decision rule assigns to an unclassified sample point the classification of the nearest of a set of previously classified points. The algorithm for the nearest neighbor rule is summarized as follows. Given an unknown sample $x$ and a distance measure, then:

First, out of the N training samples, identify the $k$ nearest neighbors, regardless of the class label, where $k$ is chosen to be odd for a two class problem, and in general not to be a multiple of the number of classes M.

Second, out of these $k$ samples, identify the number of samples, $k_i$, that belong to class $w_i$, $i = 1,2,3,...,$M. Obviously, sum($k_i$) = $k$.

Finally, assign sample $x$ to the class $w_i$ with the maximum number $k_i$ of samples.

The simplest version of the algorithm is for $k = 1$, known as the nearest neighbor(NN) rule. In other words, an unclassified sample is assigned to the class of the nearest neighbor. For this decision rule, no explicit knowledge of the underlying distributions the data is needed.

## 2.2    Fast Condensed Nearest Neighbor Rule

The fast condensed nearest neighbor rule(FCNN) is introduced by Fabrizio Angiulli [2]. Fabrizio Angiulli showed that FCNN is a novel algorithm for the computation of a training-set-consistent subset for the NN rule. We denote by $T$ a labeled training set from a metric space, and denote by $S$ the training-set-consistent subset of $T$. The algorithm works in an incremental manner, it works as follows.

First, the consistent subset $S$ is initialed to some seed elements from each class label of the training set $T$. In general, the seeds employed are the centroids of the classes in training set $T$, denoted as $Centroids(T)$.

During each iteration, for each point $p$ in $S$, a representative element $q$ in $T$ the nearest neighbor of which is $p$ in S but having a different class label is added to $S$.

If during an iteration, no new element can be added to $S$, that is, when $T$ is correctly classified using $S$, then $S$ is a training-set-consistent subset of $T$, and the algorithm terminates, returning the set $S$.

The representative element $q$, in the algorithm, is denoted as $rep(p, Voren(p,S,T))$, where $Voren(p,S,T)$ is the set of voronoi enemies of $p$ in $T$ with respect to $S$.

According to different selection of representative element, FCNN rule has 4 variants, called as FCNN1, FCNN2, FCNN3, and FCNN4, respectively. In particular, the FCNN1 and FCNN2 rules augment the subset with all such representatives, whereas the FCNN3 and FCNN4 rules select only a representative per iteration. The FCNN1 (respectively, FCNN2) and the FCNN3 (respectively, FCNN4) rules are based on the same definition of a representative.

```
Algorithm FCNN rule
Input: A training set T
Output: A training set consistent subset S of T
Method:
        S=Φ;
        ΔS=Centroids(T);
        While (ΔS≠Φ){
        S=S∪ΔS;
            ΔS=Φ;
        for each (p∈S)
            ΔS=ΔS∪{rep(p,Voren(p,S,T))};
        }
    return(S);
```

a.    FCNN rule

```
Algorithm: FRSC
Input: Training set T;
Output: Training set consistent subset S of T
Method:
        [l,c] = size(T);
        LA = lower_app(T);
        T = sortbyLA(T, LA);
        S = [];
        m = rand * c;
        dS = getSamples(T, m);
        while ~isempty(dS)    {
            S = S U dS;
            dS = [];
            dS = getvoren(S, T);
            T = T – dS;   }
    return(S);
```

b.    FRSC rule

Fig. 1. The FCNN rule and   FRSC rule

FCNN3 and FCNN4 are more careful in selecting the points for insertion and, hence, FCNN3 (respectively, FCNN4) returns a subset smaller than FCNN1 (respectively, FCNN2). On the contrary, FCNN1 and FCNN2 execute few iterations and are noticeably faster, with FCNN2 being the fastest. FCNN1 is slightly slower than FCNN2, and it requires more iterations. However, together with the FCNN3, it is likely to select points very close to the decision boundary and hence may return a subset smaller than that of FCNN2.

## 2.3  Rough Sets and Fuzzy Rough Sets

Let $U$ denote a finite and nonempty set called the universe. Suppose $R \in U \times U$ is an equivalence relation on $U$, i.e., $R$ is reflexive, symmetric, and transitive. Elements in the same equivalence class is said to be indistinguishable. $(U, R)$ is called an approximation space. Given an arbitrary set $X \subseteq U$, the lower approximation is the greatest definable set contained in $X$, and the upper approximation is the least definable set containing $X$. They can be computed by the following equivalent formulas.

$$\underline{arp}_R X = \{x : [x]_R \subseteq X\}$$

$$\overline{apr}_R X = \{x : [x]_R \subseteq X\}$$

In the same way that crisp equivalence relation is central to rough set, fuzzy equivalence relation is central to fuzzy rough sets. The concept of crisp equivalence classes can be extended by the inclusion of a fuzzy similarity relation $R$ on the universe. Let $U$ be a nonempty universe. A fuzzy binary relation $R$ on $U$ is called a fuzzy similarity relation if $R$ is reflexivity($R(x,x)$=1), symmetric ($R(x,y)=\mu R(y,x)$), and transitivity($R(x,z) \geq (R(x,y) \wedge \mu R(y,z))$).

$$\underline{R}(X)(x) = \inf_{y \in U} \max\{1 - R(x, y), X(y)\}$$

$$\overline{R}(X)(x) = \sup_{y \in U} \min\{R(x, y), X(y)\}$$

The smaller the lower approximation is, the lower the certainty of the sample belonging to the class is. In rough set theory, a sample with less lower approximation is more close to the decision boundary.

# 3    Nearest Neighbor Condensation Based on Fuzzy Rough Sets

We start by giving some preliminary definitions.

**Definition 1:** Given a labeled training set $T$ from a metric space, Let $S$ be a subset of $T$, we denote by $l(p)$ the label associated with $p$, the rule $NN(p,S)$ assigns the sample $p$ the label of its nearest neighbor in $S$.   We say $S$ is training-set-consistent subset of $T$ if for each $p$ in $T$-$S$, $l(p) = NN(p,S)$.

**Definition 2[2]:** Let $S$ be a subset of $T$ and let $p$ be an element of $S$. We denote by $Vor(p,S,T)$ the set$\{q \in T | p=NN(q,S)\}$. At the same time, we denote by $Voren(p,S,T)$ the set of *Voronoi* enemies of $p$ in $T$ with respect to $S$, defined as $\{q \in Vor(p,S,T) | l(q) \neq l(p)\}$.

**Theorem 1[2]:** $S$ is a training-set-consistent subset of $T$ for the $NN$ rule if and only if for each element $p$ of $S$, $Voren(p,S,T)$ is empty.

## 3.1    FRSC Algorithm

Different from the FCNN rule, FRSC rule initializes the consistent subset $S$ with $m$ seed elements from $T$ with minimal lower approximation, where $m$ is a random integer, larger a little than the class number c of $T$.

At first, the algorithm computes the lower approximation of every training sample in $T$ to the class which the sample itself is included in, and sorts them by lower approximation ascending. And then, the algorithm works in an incremental manner. It initializes the consistent subset $S$ with the $m$ samples with the minimal lower approximate value. During iterations, the set $S$ is augmented until the stop condition, that $T$ is empty, is reached. During every iteration, a subset $dS$ of $T$, is selected according to parameter $m$.

```
Algorithm: FRSC2
Input: Training set T
Output: Training set consistent subset S of T
Method:
    [l,c] = size(T);
    T1 = T;
    T =compact(T);
    Tag = 1;
    S = [];
    while (tag>0){
        D = getclasses(T);
        T = lower_app(T, D);
        T = sortbyLA(T);
        S = [];
        m = rand * c;
        dS = getSamples(T, m);
        while ~isempty(dS)
        {
            S = S U dS;
            dS = getvoren(S, T);
            T = T – dS;
        }
        z = k_nn_classifier(S,T1);
        tag = length(find(T1(:,end)~=z'));
        T = T1(find(T1(:,end)~=z'));
    }
    return(S);
```

a.    The frsc2 rule

```
Algorithm: COMPACT
Input: Training set T
Output: Compacted samples set Y
Method:
    D = Classes(T);.
    [r,c]= size(T);
    for each d in D
        Td = samples(T,d);
        C(d) = centroids(Td);
    end
    Cd = zeros(length(D),length(D));
    for m = 1 : D-1
        for n = 2 : D
            if m<n
                Cd(m,n)= distance(C(m,:), C(n,:));
                Cd(n,m) = Cd(m,n);
            end
        end
    end
    Y = [];
    for each d in D
        Td = T(find(T(:,end)==d),1:end);
        maxd = max(Cd(d,:));
        Ci = find(Cd(d,:)==max(Cd(d,:)));
        dist = distance(Td(:,1:end-1), C(Ci,:));
        Y = [Y; Td((dist<maxd), :)];
    end
```

b.    The compact algorithm

**Fig. 2.** The frsc2 rule and the compact algorithm

In the algorithm, we compute the lower approximation of every samples in $T$ using the function $lower\_app(T)$ , sort those samples in $T$ with function $sortbyLA(T, LA)$, and get the set of *Voronoi* enemies in $T$ with respect to $S$ with the function $getvoren(S, T)$.    If during an iteration, no new elements can be added to $S$, that is,the function $getvoren(S, T)$ return null, the algorithm terminates, and return the set $S$.

The FRSC rule has the following properties. The FRSC rule terminates in a finite times, computes a training-set-consistent subset, and is order independent..

**Theorem 2:** The FRSC rule requires at most $|T|^2+|T|.|S|$    distance computations using $O(|T|)$ space.

***Proof.*** The algorithm of FRSC rule employs two arrays $T$ and $S$, having size $|T|$ and $|S|$, with $|S| \leq |T|$, so it has space complexity $O(|T|)$.

The time complexity of the algorithm is composed of two parts, computing the lower approximation and computing a training-set-consistent subset. It takes less than $|T|^2$ distance computations to compute the lower approximations of $T$. While using the FCNN1 rule, computing the training-set-consistent subset requires at most $|T|.|S|$ distance computations [2]. Hence, $|T|^2 + |T|.|S|$ is an upper bound to the number of distance computations required of the FRSC rule.

During computing the lower approximations, the distances between each pair of samples are calculated. For small data sets, we may store the distances. Hence, we do not need compute the distances between samples in $S$ and $T$-$S$ and can use them directly. In this case, the distance computation becomes $|T|^2$. We name this variant of FRSC as FRSC1.

The FRSC1 rule requires $|T|^2$ space storing the distances. For larger data sets, it is unable to store all the distances of every pair of samples. In most cases, the size of $S$ is far less than that of $T$, so the time complexity of FRSC rule is mainly from computing the lower approximations. Specially, with the rising $|T|$, the time cost of FRSC increases exponentially.

### 3.2 Extension to Larger Data

As discussed above, the time complexity of the lower approximation increases exponentially. As far as a big data set is concerned, it will take a very long time to compute the lower approximations, usually insufferable. Just as the time complexity of the FRSC rule is mainly from computing the lower approximation, decreasing the size of samples to compute lower approximations is an effective approach to decrease the time complexity. Here, we put forward another variant of FRSC, named as FRSC2, which can decrease the number of distance computations. Instead of computing the lower approximations of all samples in $T$, the FRSC2 rule filters those samples far from the bounder with the algorithm *compact*.

The *compact* algorithm starts by getting those classifications and the centroids of each class in training set $T$. We denote by *classes*($T$) the classifications in training set $T$, and denote by *centroids*($Td$) the centroids of samples set $Td$. For the sake of simplicity, usually, we use the average of all samples belonging to the same class as the centroids. For every class, the distances from its centroid to other class centroids are computed, and the maximum distance *maxd* and the corresponding classification $C_i$ are kept. Only those samples are remained whose distance to the centroid of $C_i$ are smaller than *maxd* and the other samples are filtered. If the samples take on normal distribution, about two third of all samples are filtered. Hence, the work computing the lower approximations decreases to only one ninth.

### 3.3 Extension to *k*-NN

The nearest neighbor decision rule assigns to an unclassified sample point the classification of the nearest of a set of previously classified points. It is the simplest version of the $k$-NN rule for $k = 1$. In $k$-NN case, a new object is assigned the class $wi$ with the maximum number $k_i$ of samples. More importantly, the $k$-NN rule provides a good estimate of the Bayes error and its error probability asymptotically approaches the Bayes error. Just as the FCNN, the FRSC rule is very easy to extend to $k$-NN situation by using the following notion of consistency: A subset $S$ of $T$ is said to be a $k$-training-set-consistent subset of $T$ if for each $p \in (T-S)$, $l(p)=NNk(p,S)$ [2], where $l(p)$ is the class label of sample $p$, $NNk(p,S)$ is the class label of $k$-NN of $p$ in $S$.

# 4    Experimentation Results

In this section, we present experimental results using four data sets, three library data sets and one real-life data set, the Forest Cover Type. The Forest Cover Type data set, that contains forest cover type data from US Forest Service (USFS) Region 2 Resource Information System (RIS) data, is composed of 581,012 tuples associated with 30×30 meter cells. The data is partitioned into 7 classes. The data contains binary (0 or 1) columns of data for qualitative independent variables, which make the computation of distance and lower approximation complex. For the sake of simplicity, we filter those binary columns and remain only the first 10 conditional attributes and the decision attribute, the last column. In different experiments, we select some parts of the data set according to the purpose of experiment.

**Table 1.** Data sets used for experiments

| Data set | Number of samples | Number of attributes | Number of classifications |
|---|---|---|---|
| Data set 1 | 100 | 2 | 2 |
| Data set 2 | 500 | 3 | 3 |
| Data set 3 | 1000 | 2 | 2 |
| Forest Cover | 581012 | 10 | 7 |

**Table 2.** Experiments for different $k$

| $k$ | Data set | Data size | Subset size | Accuracy |
|---|---|---|---|---|
| 1 | Data set 1 | 100 | 6 | 95.1% |
| | Data set 2 | 500 | 10 | 99.6% |
| | Data set 3 | 1000 | 11 | 99.9% |
| | | 12240 | 22 | 100% |
| | Forest Cover | 4080 | 23 | 99.9% |
| | | 1020 | 20 | 99.9% |
| 5 | Data set 1 | 100 | 9 | 94.9% |
| | Data set 2 | 500 | 9 | 99.8% |
| | Data set 3 | 1000 | 16 | 99.8% |
| | | 12240 | 71 | 100% |
| | Forest Cover | 4080 | 61 | 99.9% |
| | | 1020 | 57 | 99.9% |
| 9 | Data set 1 | 100 | 14 | 98.0% |
| | Data set 2 | 500 | 11 | 96.5% |
| | Data set 3 | 1000 | 20 | 99.7% |
| | | 12240 | 121 | 100% |
| | Forest Cover | 4080 | 122 | 99.9% |
| | | 1020 | 86 | 99.8% |

## 4.1    Experimental Results for k-NN

Table 2 summaries the experiments results with k=1, 5 and 9, respectively. We perform experiments on data set 1, data set 2, data set 3 and Forest Cover Type data set. For the sake of simplicity, we extract 12240 records, with class label 4 or 5, from the real-life data set, Forest Cover Type data set. And then, we pick up 6120 records and 3060 records randomly from the subset. For the three library data sets, we execute the FRSC1 rule, and we perform the FRSC2 rule on Forest Cover Type data set. In order to get the accuracy of classification, we classify those samples in the total data set using the k-NN classifier.

It is shown in table 2 that the subset size increases with k increasing. For data set 1, the subset size increase from 6 to 14, while k increases from 1 to 9. For the other data set, we can obtain similar conclusion from table 2.

We can see from experiments results that the more records in the training set, the greater the subset. When the data size is 100, the subset size is 6. When the number of data is 500 and 1000, the subset size is 10 and 11, respectively, for k=1. For k=5 and 9, there exist similar results.

At the same time, however, we must notice that the size of subset does not change monotonously with increasing k. While the size of training set becomes larger, the record number in the subset increases all the time. The subset size is 10 while k=1 for data set 2, but the subset size is 9 while k=5. The subset size of data set 1, with 100 records, is 14, but the subset size of data 2, with 500 records, is 11, for k=9.

We can find that the accuracy is very high. The accuracy fluctuates slightly with the data size and the value of k. The accuracy is higher a little while k becomes larger or there are more samples in the training set. When we extract 12240 records from data set Forest Cover Type, the accuracy is 100%, but when the number of extracted data decreases to 4080 and 1020, the accuracy is high up to 99.9%. Even the lowest accuracy is 95.1% when we perform an experiment on data set 1 with k=1.

## 4.2    Comparison with FCNN

As stated in above section, among FCNN,   CNN, MCNN, NNSRM, RNN and other methods computing the training set consistent subset, the FCNN rule is better. Hence, we only compare the FRSC rule with FCNN rule.

Table 3 summarizes FRSC1 rule with FCNN rule in data set 1, data set 2 and data set3. Fig 3 shows the distribution of data set 1. From Table 3, we can see that FRSC1 rule has better performance than FCNN rule. The subset size by FCRS1 rule is a bit smaller than that by the FCNN rule. It takes less time to perform the FRSC1 algorithm than to perform the FCNN1 rule. For data set 1 with 100 samples, the execution time by FRSC1 is 0.0159 seconds, while the FCNN1 costs 0.3038 seconds. For data set 2 with 500 samples, the execution time by FRSC1 is 0.1740 seconds, while the FCNN1 costs 1.6199 seconds. For data set 3 with 1000 samples, the execution time by FRSC1 is 0.6811 seconds, while the FCNN1 costs 1.6032 seconds. Meanwhile, we find also that the difference of execution time becomes smaller and smaller with the ascending data size. The execution time of FRSC1 rule becomes about 2.5 times of FCNN for data set 3 with 1000 samples, while the difference is about 20 times for data set 1 with 100 samples. Fig 3 shows the distribution of samples in subset. We can find that those samples in the subset by FRSC1 are more close to the boundary than FCNN. It results in smaller subset size.

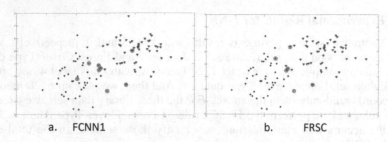

a.    FCNN1                                    b.        FRSC

**Fig. 3.** Example of training set consistent subsets computed by the FCNN1 and FRSC

**Table 3.** Small data set:Compare FRSC with FCNN1

| Data set | Data size | Method | Subset size | Time |
|----------|-----------|--------|-------------|------|
| data set 1 | 100 | FCNN | 8 | 0.3038 |
|          |     | FRSC1 | 6 | 0.0159 |
| data set 2 | 500 | FCNN | 13 | 1.6199 |
|          |     | FRSC1 | 10 | 0.1740 |
| data set 3 | 1000 | FCNN | 13 | 1.6032 |
|          |      | FRSC1 | 11 | 0.6811 |

**Table 4.** Small data set:Compare FRSC1 with FCNN1

| Data set | Data size | Method | Subset size | Time |
|----------|-----------|--------|-------------|------|
| data set 1 | 100 | FCNN1 | 9 | 0.0551 |
|          |     | FRSC1 | 6 | 0.0152 |
| data set 2 | 500 | FCNN1 | 15 | 0.4319 |
|          |     | FRSC1 | 9 | 0.1740 |
| data set 3 | 1000 | FCNN1 | 16 | 0.9286 |
|          |      | FRSC1 | 11 | 0.6561 |

**Table 5.** Large data set:Compare FRSC1 with FRSC2

| Data set | Data set size | Subset size | Time |
|----------|---------------|-------------|------|
| FRSC 1 | 1000 | 11 | 3.8930 |
|        | 3060 | 26 | 51.00 |
|        | 6120 | 27 | 314.8 |
|        | 12240 | 22 | 3924 |
| FRSC 2 | 1000 | 11 | 0.6887 |
|        | 3060 | 34 | 5.15 |
|        | 6120 | 35 | 23.03 |
|        | 12240 | 26 | 105 |

**Table 6.** Large data set:Compare FRSC2 with FCNN1

| Data set | Data set size | Subset size | Time |
|----------|---------------|-------------|------|
| FRSC 2 | 1000 | 12 | 0.6887 |
|        | 3060 | 34 | 5.15 |
|        | 6120 | 35 | 23.03 |
|        | 12240 | 22 | 105 |
| FCNN 1 | 1000 | 11 | 0.8435 |
|        | 3060 | 34 | 6.09 |
|        | 6120 | 41 | 14.7 |
|        | 12240 | 44 | 32 |

For FCNN1 rule, the difference from FRSC1 is much smaller from Table IV. For data set 1, the execution time of FCNN1 is only about 3 times of FRSC1, much less than 20 times of FCNN. While the data size becomes 1000 for data set 3, the execution time of FCNN1 is only about 1.5 times of that of FRSC1. Furthermore, with the increasing data size, the execution time of FRSC1 will be more than that of FCNN1 inevitably.

The FRSC2 rule needs less execution time than FRSC1 rule, which is shown in Table 5. The disparity increases quickly with the ascending data size. But the subset size of FRSC2 is larger a bit than FRSC1. For large data set, we use FRSC2 rule instead of FRSC1 rule.

Table 6 represents the comparison between FRSC2 and FCNN1 on Forest Cover Type data set. While the data set size is less than 3000, the FRSC2 rule needs less execution time. While there are more than 5000 samples in the data set, the execution time of FRSC2 becomes more than FCNN1, which is the same to FCNN1. We can find out the reason from the FCNN rule. The time complexity of the FCNN rule is composed of two parts, computing the lower approximation and computing a training-set-consistent subset. It takes less than $|T|2$ distance computations to compute the lower approximations of T. The time of computing the other part is at most $|T|.|S|$, much less than the first part. For large data set, the execution time is mainly caused by computing the lower approximation. So, reducing the complexity of lower approximation becomes the key factor, which is our next work also.

# 5    Conclusions

This work introduces a novel algorithm, called the FRSC rule, for computing a training-set-consistent subset for the NN rule.

The algorithm starts by computing the lower approximation of every sample in training set T, and sorts them by lower approximation ascending. And then, the algorithm works in an incremental manner. It initializes the consistent subset S with the m samples with the minimal lower approximate value. During each iteration, the set S is augmented until the stop condition, that T is empty, is reached. During each iteration, a subset $dS$ of $T$ is selected and added to set $S$ until the consistence is achieved. In most cases, the size of $S$ is far less than that of $T$, the time complexity of FRSC rule is mainly from computing the lower approximations.

Two variants of the basic method are presented, called FRSC1 and FRSC2. The FRSC1 rule requires less distance computations and the FRSC2 can deal with larger data set. Instead of computing the lower approximations of all samples in T, the FRSC2 rule compacts the data set by filtering those samples far from the boundary.

About the FRSC rule, some strengths and weaknesses can be summarized as follows:

The FRSC rule has better performance than FCNN rule on small data set. It requires less execution time and returns a smaller subset. Ordered by the lower approximation, those samples close to the boundary are selected into subset $S$, that can achieve the consistence more quickly.

For larger data set, the FRSC2 rule does better than FRSC1. The FRSC2 rule needs less execution time.

In dealing with big data set, more than 5000 samples, the FRSC rule has no superiority to FCNN rule, which is presented by the comparison of FCNN and FRSC. The reason is that the time complexity of the lower approximation increases exponentially. Reducing the complexity of lower approximation becomes the key factor, which is our next work also.

To conclude, this work presents a novel condensation algorithm for the *NN* rule, which has better performance than FCNN rule on small and medium-sized data sets.

**Acknowledgement.** The reported research was supported in part by an operating grant from natural science foundation of education department of Sichuan province. And the author would like to thank the anonymous reviewers for their useful comments.

# References

1. Pawlak, Z.: Rough sets. Int. J. Comput. Inf. Sci. 11(5), 341–356 (1982)
2. Angiulli, F.: Fast nearest neighbor condensation for large data sets classification. IEEE Trans. Knowledge and Data Engineering 19(11), 1450–1464 (2007)
3. Kryszkiewicz, M.: Rough set approach to incomplete information systems. Inf. Sci. 112, 39–49 (1998)
4. Pawlak, Z.: Rough Sets: Theoretical Aspects of Reasoning About Data. Kluwer, Norwell (1991)
5. Tsumoto, S.: Automated extraction of medical expert system rules from clinical databases based on rough set theory. Inf. Sci. 112, 67–84 (1998)
6. Skowron, A., Polkowski, L.: Rough Sets in Knowledge Discovery, vol. 1, 2. Springer, Berlin (1998)
7. Jensen, R., Shen, Q.: Fuzzy-rough attribute reduction with application to web categorization. Fuzzy Sets and Systems 141, 469–485 (2004)
8. Yeung, D.S., Chen, D., Tsang, C.C., Lee, J.W.T., Xizhao, W.: On the generalization of fuzzy rough sets. IEEE Trans. Fuzzy Systems 13(3) (July 2005)
9. Aha, D.W.: "Editorial," Artificial Intelligence Rev. Special Issue on Lazy Learning 11(5-5), 7–10 (1997)
10. Aha, D.W., Kibler, D., Albert, M.K.: Instance-Based Learning Algorithms. Machine Learning 6, 37–66 (1991)
11. Alpaydin, E.: Voting over Multiple Condensed Nearest Neighbors. Artificial Intelligence Rev. 11, 115–132 (1997)
12. Angiulli, F.: Fast Condensed Nearest Neighbor Rule. In: Proc. 22nd Int'l Conf. Machine Learning (ICML 2005), pp. 25–32 (2005)
13. Bay, S.: Combining Nearest Neighbor Classifiers through Multiple Feature Subsets. In: Proc. 15th Int'l Conf. Machine Learning, ICML 1998 (1998)
14. Bay, S.: Nearest Neighbor Classification from Multiple Feature Sets. Intelligent Data Analysis 3, 191–209 (1999)
15. Bhattacharya, B., Kaller, D.: Reference Set Thinning for the k-Nearest Neighbor Decision Rule. In: Proc. 14th Int'l Conf. Pattern Recognition, ICPR 1998 (1998)

16. Brighton, H., Mellish, C.: Advances in Instance Selection for Instance-Based Learning Algorithms. Data Mining and Knowledge Discovery 6(2), 153–172 (2002)
17. Cover, T.M., Hart, P.E.: Nearest Neighbor Pattern Classification. IEEE Trans. Information Theory 13(1), 21–27 (1967)
18. Dasarathy, B.: Minimal Consistent Subset (MCS) Identification for Optimal Nearest Neighbor Decision Systems Design. IEEE Trans. Systems, Man, and Cybernetics 24(3), 511–517 (1994)
19. Dasarathy, B.: Nearest Unlike Neighbor (NUN): An Aid to Decision Confidence Estimation. Optical Eng. 34, 2785–2792 (1995)
20. Devijver, P., Kittler, J.: On the Edited Nearest Neighbor Rule. In: Proc. Fifth Int'l Conf. Pattern Recognition (ICPR 1980), pp. 72–80 (1980)
21. Devroye, L.: On the Inequality of Cover and Hart in Nearest Neighbor Discrimination. IEEE Trans. Pattern Analysis and Machine Intelligence 3, 75–78 (1981)
22. Fukunaga, K., Hostetler, L.D.: k-Nearest-Neighbor Bayes-Risk Estimation. IEEE Trans. Information Theory 21, 285–293 (1975)
23. Gates, W.: The Reduced Nearest Neighbor Rule. IEEE Trans. Information Theory 18(3), 431–433 (1972)
24. Hart, P.E.: The Condensed Nearest Neighbor Rule. IEEE Trans. Information Theory 14(3), 515–516 (1968)
25. Karaçali, B., Krim, H.: Fast Minimization of Structural Risk by Nearest Neighbor Rule. IEEE Trans. Neural Networks 14(1), 127–134 (2003)
26. Ritter, G.L., Woodruff, H.B., Lowry, S.R., Isenhour, T.L.: An Algorithm for a Selective Nearest Neighbor Decision Rule. IEEE Trans. Information Theory 21, 665–669 (1975)
27. Wilfong, G.: Nearest Neighbor Problems. Int'l J. Computational Geometry & Applications 2(4), 383–416 (1992)
28. Wilson, D.L.: Asymptotic Properties of Nearest Neighbor Rules Using Edited Data. IEEE Trans. Systems, Man, and Cybernetics 2, 408–420 (1972)
29. Wilson, D.R., Martinez, T.R.: Reduction Techniques for Instance-Based Learning Algorithms. Machine Learning 38(3), 257–286 (2000)
30. Ziarko, W.P.: Rough sets, fuzzy sets and knowledge discovery. In: Workshop in Computing, London, U.K. (1994)
31. Chanas, S., Kuchta, D.: Further remarks on the relation between rough and fuzzy sets. Fuzzy Sets Syst. 47, 391–394 (1992)
32. Dubois, D., Prade, H.: Rough fuzzy sets and fuzzy rough sets. Int. J. Gen. Syst. 17(2-3), 191–209 (1990)
33. Yao, Y.Y.: A comparative study of fuzzy sets and rough sets. J. Inf. Sci. 109, 227–242 (1998)
34. Radzikowska, A.M., Kerre, E.E.: A comparative study of fuzzy rough sets. Fuzzy Sets Syst. 126, 137–155 (2002)
35. Yao, Y.Y.: Constructive and Algebraic methods of the theory of rough sets. Inf. Sci. 109, 21–47 (1998)

# Predicting Movies User Ratings with Imdb Attributes

Ping-Yu Hsu, Yuan-Hong Shen, and Xiang-An Xie

Department of Business Administration, National Central University, Jhongli City,
Taoyuan County, Taiwan (R.O.C.)
pyhsu@mgt.ncu.edu.tw
{edwardshen1976,ho2009}@gmail.com

**Abstract.** In the era of Web 2.0, consumers share their ratings or comments easily with other people after watching a movie. User rating simplified the procedure which consumers express their opinions about a product, and is a great indicator to predict the box office [1-4]. This study develops user rating prediction models which used classification technique (linear combination, multiple linear regression, neural networks) to develop. Total research dataset included 32968 movies, 31506 movies were training data, and others were testing data. Three of research findings are worth summarizing: first, the prediction absolute error of three models is below 0.82, it represents the user ratings are well-predicted by the models; second, the forecast of neural networks prediction model is more accurate than others; third, some predictors profoundly affect user rating, such as writers, actors and directors. Therefore, investors and movie production companies could invest an optimal portfolio to increase ROI.

**Keywords:** User rating, prediction model, classification, linear combination, convex combination, neural networks, multiple linear regression, stepwise regression, IMDb.

## 1 Introduction

Since the 20th century, movies have been an essential and important recreation to human beings. According to a survey by Motion Picture Association of America (MPAA), global box office for all films released in each country around the world reached $35.9 billion in 2013, up 4% over 2012's total [5]. More than 4000 movies were produced within one year in the whole world, and only the top 5 movies box office exceeded US$100 million and these movies gained 14% of the gross box office[6]. It is a winner-take-most industry. In early 21st century, the production cost of one movie already reaches US$65 million while the advertisement and marketing budget also reaches US$35 million [6]. An investigation with 281 movies produced in the period from 2001 to 2004 pointed out, the return on investment (ROI) of movies ranges from −96.7% up to over 677%, an average ROI at −27.2% [7]. The rigorous circumstances exposed movie investors and production companies to higher financial risks. However, to the industry practitioners, forecasting the box office of a specific movie is a difficult mission because of some uncertain characteristics. Therefore, the industry practitioners relied heavily on traditional wisdom and simple empirical rules

D. Miao et al. (Eds.): RSKT 2014, LNAI 8818, pp. 444–453, 2014.
DOI: 10.1007/978-3-319-11740-9_41 © Springer International Publishing Switzerland 2014

to make their decision in the past [8]. Most movie investments seem like a gambling. Therefore, the engagement of a forecast in box office is an imperative and challenging study issue to scholars and the industry practitioners.

User rating is a kind of Word of Mouth (WOM), it simplified the procedure which consumers express their opinions about a product. User rating is highly important to a certain product or service, because it reflects the wisdom of crowds. Undoubtedly user rating is a great indicator to predict the future sales performance of a product. Movie industry specialists agree that it is a key success factor of movie and help movie production company and investor gain a financial success [1-4].

The purpose of this study was to develop an accurate user rating prediction models which based on the early information. We used classification technique of data mining to develop prediction model. First, developed one learning algorithm to identify a linear combination (convex combination) model that best fits the relationship between the attribute set and class label of the input data. Second, testing data was used to estimate the accuracy of the model. In the meantime, we employed other techniques (multiple linear regression and neural networks) to develop comparison model. The results of comparison model were used to illustrate how effective these attributes are.

## 2    Related Works

### 2.1    Internet Movie Database (IMDb) Voting (User Rating)

IMDb (www.imdb.com) is the largest movie database in the world. The service was launched in 1990. The website had 2.8 million titles (includes episodes) and 5.9 million personalities in its database on May 2014.

IMDb registered users can rate every movie in the website (rating scale from 1 to 10). User can rate one movie as many times as they want but each rating overwrite the previous rating for the same movie. The rating shown in IMDb is not an average rating of the original data by every voting user but a kind of weighted average of an undisclosed calculation method. IMDb applies various filters to screen the original data, the objective is to present a more representative rating which is immune from abuse by subsets of individuals who have combined together with the aim of influencing (either up or down) the ratings of specific movies; IMDb keeps the mystery of rating calculation method, without disclosing whether/when/how to perform a weight for certain ratings, to provide a more objective rating.

### 2.2    Movie Box Office and User Rating Prediction

During the past 20 years, marketing scholars have developed some prediction models and decision support tools to increase the accuracy of forecast. One mainstream in which is to use multiple linear regression, by making the box office of movie as the dependent variable while the independent variable as the predictors with an impact on box office forecast, to establish a forecast model [1, 9-15]. [16] points out some production and marketing characteristic factors influence the financial performance of a movie. [17] used neural networks in predicting the financial performance of a movie. They compared their prediction model with models that used other statistical techniques; it is found the model built by neural networks do a better job of predicting box office.

### 2.3    Linear Combination (Convex Combination) and Prediction

Linear combination model is a decision rule for deriving a linear combination that predicts some criterion of interest. This method is intuitive and easy to understand to decision makers[18]. A linear combination is constructed from a set of terms by multiplying each term by a constant and adding the results. The constants were considered as weights when the linear combination model was used for decision-making or predictive purposes. Given a finite number of predictor variables $x_1, x_2, ..., x_n$, a linear combination of these predictor variables (independent variables) is a criterion variables (dependent variables) of the form.

$$w_1 x_1 + w_2 x_2 + \cdots + w_n; \text{ where the constant } w_i \geq 0 \text{ and } \sum w_i = 1; \ i = 1, 2, ..., n$$

A proper linear combination model is a linear equation which predictor variables are given optimal weights to optimize the relationship between the prediction and the criterion [19]. However, some authors pointed that it is a misunderstanding to interpret the weights as measures of the importance [18, 20]. The value of weight is dependent on the range of predictor variable values; in other words, a weight of a predictor variable can be different by increasing or decreasing the range of observed value of predictor variable. In this study, all the observed values (score) of predictor variable were average user rating which comes from IMDb user voting. Furthermore, the average user rating is interval scale and ranges from 1 to 10.

## 3    Data Preprocessing

### 3.1    Data Collection

Data for this study were collected from IMDb. We collected the user rating and attributes of all movies released from 2002 to 2012. We obtained a data set of 32968 movies. The data set consists of attributes: actors, as known, country, directors, episodes, film locations, genres, IMDb id, IMDb URL, language, plot, plot simple, poster, rated, rating, rating count, release date, runtime, title, type, writers, year, opening weekend, gross, filming dates, budget, weekend gross, copyright holder. In this study, the structure of data set is listed below. Attributes are factors that related to movies (e.g., user rating, genre, actor). Element is a subgroup of attribute (e.g., action is one kind of genre). In other words, at the high level are the attributes which can be defined in terms of more elements.

## 3.2    Data Cleaning, Transformation and Reduction

For the purpose of our analysis, we need to remove or reduce the noise and missing values from test data. This step reduce confusion to derive more useful classification rules[21]. We remove the irrelevant, weakly relevant or redundant attributes according to previous research conclusion and IT scholars' opinions. Besides, runtime is conti-nuous type data. For research purpose, we convert runtime to categorical nominal type data and divide runtime data into four groups. Table 1 presents the attributes (inde-pendent variables) which we used in this study.

**Table 1.** Description of selected attributes

|   | Attributes | Attribute Types | Number of element | Literature |
|---|---|---|---|---|
| 1 | genres | Categorical nominal | 24 | [11, 12, 16, 17] |
| 2 | directors | Categorical nominal | 8,880 | [1, 11, 15] |
| 3 | actors | Categorical nominal | 98,116 | [1, 11, 13, 15-17] |
| 4 | writers | Categorical nominal | 13,447 | [7] |
| 5 | country | Categorical nominal | 117 | [22] |
| 6 | film_locations | Categorical nominal | 1,220 | - |
| 7 | runtime | Categorical nominal | 4 | [16] |

## 3.3    Calculate the Score of Elements and Attributes

In this study, element is a quantifiable indicator of the extent to user rating. We col-lected movies that related to a certain element, and then we averaged user rating of the movies. The average user rating is the score of element. For example, Ang Lee is a director of Brokeback Mountain (2005), Hulk (2003), Talking Woodstock (2009). The user ratings of these movies are 7.6, 5.7, and 6.6. The score of Ang Lee is (7.6 + 5.7 + 6.6) / 3 = 6.63.

As noted in the previous section, element is a subgroup of attribute. We calculated the score of attribute after we had calculated the element score. We averaged the score of elements that belong to a certain attribute, and then the result was the attribute score. For example, there are five elements (animation, action, adventure, family, and mys-tery) which belong to the genre of The Adventures of Tintin (2011). The elements scores are 5.89, 5.97, 5.88, 6.87, 6.29, the genre score of The Adventures of Tintin is (5.89 + 5.97 + 5.88 + 6.87 + 6.29) / 5 = 6.18. The other attributes (actors, writers, country, film locations, runtime) use the same method to calculate the score.

# 4    Develop Prediction Models

We collected the user rating and attributes of all movies released from 2002 to 2012. Total dataset is including 32968 movies, 31,506 movies were used to be training data, and others were testing data. In section 3.3, we calculated all attribute scores and element scores, and then used training data to generate prediction rules. The rules can be used to predict future data. Methods used to develop the prediction models are represented below:

## 4.1    Linear Combination (Convex Combination) Model with Enumerating Value

The predicted user rating is derived by a linear combination of the scores of the attributes. The attributes may have different weights in deriving the predicted user rating. The computation method is as follows:

Predicted user rating =
$$w_1 \times Score_{genres} + w_2 \times Score_{directors} + w_3 \times Score_{actors} + w_4 \times Score_{writers} + w_5 \times Score_{country} + w_6 \times Score_{f\_location} + w_7 \times Score_{runtime} \qquad (1)$$

Weights: $w_1, w_2, ..., w_7 \in [0, 1]$; $w_1 + w_2 + w_3 + w_4 + w_5 + w_6 + w_7 = 1$

To find the optimal line combination, we tested all combinations of $w_1, w_2, ..., w_7$ by enumerating the values systematically in increments of 0.01 range from 0 to 1. When the accumulated difference between predicted user rating and actual user rating is the smallest, it can be considered as an optimal line combination. We use algorithm 1 and algorithm 2 to find the optimal weight combination. The algorithm and prediction model is as follows:

```
ALGORITHM 1: List all linear combinations
OUTPUT:
weight
PROGRAM:
For w1 = 0 To 1 Step 0.01
For w2 = 0 To 1 Step 0.01
    For (…)
      If w1 + w2 + w3 + w4 + w5 + w6 + w7 = 1 Then
         weight(weightcount, 1) = w1
         weight(weightcount, 2) = w2
         (…)
         weightcount ++
      End If
    End For
  End For
End For
```

```
ALGORITHM 2: Calculate the forecast error to yield the optimal
linear combination
INPUT:
movie
GenresScore, DirectorsScore, ActorsScore, WritersScore,
CountryScore,
LocationsScore, RuntimeScore
Genres, Directors, Actors, Writers, Country, Locations,
Runtime
weight
OUTPUT:
BestWeight
BestError
PROGRAM:
BestError = infinite
For i = 0 To weight.count-1
 For Each m In movie
     Error = Abs(m.Score - (GenresScore*weight(i,1) + Di-
rectorsScore*weight(i,2) +
     ActorsScore*weight(i,3) + WritersScore*weight(i,4) +
CountryScore*weight(i,5) +
     LocationsScore*weight(i,6)) +
RuntimeScore*weight(i,7)))
     If  Error < BestError Then
         BestError = Error
   BestWeight = i
     End If
   End For
End For
```

$$\text{Predicted user rating} = 0.05 \times \text{Score}_{\text{genres}} + 0.05 \times \text{Score}_{\text{directs}} + 0.15 \times \text{Score}_{\text{actors}} + 0.75 \times \text{Score}_{\text{writers}}$$

## 4.2    Multiple Linear Regression Model

In this section, we use multiple linear regression analysis to yield another user rating prediction model. User rating is dependent variable and other attribute (predictor variables) are independent variables. We applied stepwise regression technique to select predictor variables. In each step, we included a significant variable (at the 5% level) that brought the highest increase in adjusted $R^2$. After each variable inclusion step, we removed any previously included variable if the variable is no longer significant (at the 10% level). We stopped adding variables when the adjusted $R^2$ did not increase when additional variables were no longer significant. In this study, all variables were included in the regression model. We listed the result of the 7th step in stepwise regression procedure in table 2.

**Table 2.** Results of stepwise regression

| Step | Coefficient | | Standard Coefficient | t-value | p-value |
|---|---|---|---|---|---|
| | Beta | Standard Error | Beta | | |
| constant | -0.632 | 0.054 | | -11.804 | .000 |
| writers | 0.409 | 0.008 | 0.384 | 49.113 | .000 |
| actors | 0.556 | 0.009 | 0.432 | 61.934 | .000 |
| directors | 0.192 | 0.008 | 0.181 | 25.060 | .000 |
| 7  runtime | 0.028 | 0.005 | 0.014 | 5.894 | .000 |
| country | -0.031 | 0.006 | -0.012 | -4.944 | .000 |
| genres | -0.028 | 0.006 | -0.012 | -4.768 | .000 |
| film_ location | -0.012 | 0.003 | -0.009 | -3.929 | .000 |

The prediction model is as follows:

$$\text{Predicted user rating} = -0.632 + 0.409 \times \text{Score}_{writers} + 0.556 \times \text{Score}_{actors} + 0.192 \times \text{Score}_{directors} + 0.028 \times \text{Score}_{runtime} - 0.031 \times \text{Score}_{country} - 0.028 \times \text{Score}_{genres} - 0.012 \times \text{Score}_{file\_location} + \varepsilon$$

### 4.3    Neural Networks Model

Neural networks is a massive parallel distributed processor made up of simple processing units[23]. Neural networks is composed of several interconnected nodes and links. It modifies its interconnection weights by apply a set of training data. The attribute scores is the input vector and the corresponding output is actual user rating. The prediction model was developed using a commercial software product called SQL Server Business Intelligence Development Studio.

It is difficult to interpret the meaning behind the interconnection weights and hidden layer in the networks [21]. Due to the poor interpretability, the result of neural networks was used to illustrate how effective these attributes are.

## 5    Conclusion and Future Works

### 5.1    Forecast Accuracy

In order to test the forecast accuracy of the prediction models, we use 1,462 movies to be testing data which we obtained from IMDb. For testing the forecast accuracy, we used the testing data to calculate the predicted user rating from three prediction models which we developed in chapter 4. Then, we calculate the difference (Prediction absolute error; PAE) between the forecast value and actual user rating. The smaller value of PAE is, the better forecast accuracy is. We calculate the percentage of the appearance frequency in the different PAE area accounting for all testing data, to be used to compare the forecast accuracy of the three kinds of method.

PAE = |Predicted user rating − Actual user rating|

$$\text{Average PAE} = \frac{\sum_1^n \text{PAE}_n}{n}$$

$$\text{Percent of PAE between a and b} = \frac{a \le \text{number of PAE} < b}{\text{total number of testing data}} \quad 0 \le a < b \le 10$$

**Table 3.** Comparison predicted absolute error between the linear combination method, multiple linear regression and neural networks prediction models

| Prediction model | Average PAE | PAE | | | |
|---|---|---|---|---|---|
| | | $0 \le \text{PAE} < 1$ | $1 \le \text{PAE} < 2$ | $2 \le \text{PAE} < 3$ | $3 \le \text{PAE} < 4$ |
| Linear combination | 0.7347 | 72.73% | 24.45% | 2.19% | 0.31% |
| Multiple linear regression | 0.8186 | 67.08% | 28.21% | 4.08% | 0.31% |
| Neural networks | 0.6973 | 76.8% | 18.5% | 4.39% | 0.31% |

As shown in Table 3, the average PAE of the linear combination is 0.7347, lower than the average PAE 0.8186 of multiple linear regression, while the average PAE of neural networks is only 0.6973, as the method with the lowest average PAE. The PAE percentage of the linear combination lower than 1 is 72.73%, higher than the 67.08% of multiple linear regression by 5.65%, while the PAE of neural networks lower than 1 is 76.8%, higher than 72.73% of the linear combination by 4.07%. The results of paired t-test were also indicated that there is a significant difference between the PAE of neural networks and the PAE of multiple linear regression. However, there is no significant difference between the PAE of neural networks and the PAE of linear combination. As mentioned above, it can be seen that the forecast performance of using neural networks prediction model to be greater than or equal to the linear combination prediction model, while the forecast performance of the neutral networks prediction model is better than that of multiple linear regression model.

## 5.2    Conclusion and Future Work

A proper weight combination forecast equation is obtained in this study to solve the linear combination (convex combination) by enumerating value systematically; meanwhile, multiple linear regression and neural networks applied to the development of forecast models. The result indicated that using neural networks is superior to the optimal weight combination provided in this study, while the forecast performance of the linear combination is better than multiple linear regression. These findings are in line with previous studies [17]. It is noteworthy that if we only focus on the PAE lower than 2, the linear combination model is the great ratio 97.18% (72.73%+ 24.45%), that is, 97.18% testing data predicted user rating error lower than 2 when we used linear combination prediction model. The PAE of neutral networks model is 95.3% in the same condition.

In Table 4, we listed the weights of linear combination and standard coefficient of multiple linear regression. Writers, actors and directors profoundly affect user rating. A writer is in charge of such core elements as the scheme, characters, scene, and structure of the whole movie; a good screen scripts can find an echo in everyone's heart. On the contrary, a poor screen script hardly gains the favor even under sufficient resources of various aspects. The next important factor is actors. The actor selection of a movie production company is extremely important. Most studies considered star as one of the covariates with box office performance. However, directors and genres also account for considerable influence on user rating. Therefore, before investors and a movie production company prepare to shoot a movie, they may well consider the favorable portfolios of the audience from such aspects of writers, actors, directors and genres to acquire a higher anticipated user rating. Once the anticipated user rating is reached, the increase of movie revenue will take place.

**Table 4.** Comparison between weights of the linear combination and standard coefficient of multiple linear regression

| | Weight/ Standard Coefficient | | | | | | |
| --- | --- | --- | --- | --- | --- | --- | --- |
| | Genres | Directors | Actors | Writers | Country | Film_locations | Runtime |
| Linear combination | 0.05 | 0.05 | 0.15 | 0.75 | 0 | 0 | 0 |
| Multiple linear regression | -0.012 | 0.181 | 0.432 | 0.384 | -0.012 | -0.09 | 0.014 |

While our results are encouraging, there are still many improvements to be made. We consider that there are many factors with impact on user rating which are not explored. Some potential endogenous relationships exist among the factors [4], it is recommended that more studies of these questions could be performed.

This study uses enumerating value systematically to find out the proper weight combination; such kind of method highly consumes time and computer resources. The time performance of linear combination method is about 450 minutes, on the contrary, multiple linear regression is about 4 seconds and neural networks is about 8 seconds. Further research might adopt the other algorithms of solving a convex combination to reduce the calculation time and resources.

# References

1. Elberse, A., Eliashberg, J.: Demand and supply dynamics for sequentially released products in International markets: The case of motion pictures. Marketing Science 22(3), 329–354 (2003)
2. Reinstein, D.A., Snyder, C.M.: The influence of expert reviews on consumer demand for experience goods: A case study ofmovie critics. The Journal of Industrial Economics 53(1), 27–51 (2005)

3. Eliashberg, J., Shugan, S.M.: Film critics: Influencers or predictors? Journal of Marketing 61(2), 68–78 (1997)
4. Basuroy, S., Chatterjee, S., Ravid, S.A.: How critical are critical reviews? The box office effects of film critics, star power, and budgets. Journal of Marketing 67(4), 103–117 (2003)
5. Motion Picture Association of America, I., Theatrical Market Statistics Report, Motion Picture Association of America, Inc. p. 31 (2013)
6. Motion Picture Association of America, I., MPAA Economic Review (2004)
7. Eliashberg, J., Hui, S.K., Zhang, Z.J.: From story line to box office: A new approach for green-lighting movie scripts. Management Science 53(6), 881–893 (2007)
8. Eliashberg, J., Elberse, A., Leenders, M.A.A.M.: The motion picture industry: Critical issues in practice, current research, and new research directions. Marketing Science 25(6), 638–661 (2006)
9. Jones, J.M., Ritz, C.J.: Incorporating distribution into new product diffusion models. International Journal of Research in Marketing 8(2), 91–112 (1991)
10. Krider, R.E., Weinberg, C.B.: Competitive dynamics and the introduction of new products: The motion picture timing game. Journal of Marketing Research 35(1), 1–15 (1998)
11. Ainslie, A., Drèze, X., Zufryden, F.: Modeling movie life cycles and market share. Marketing Science 24(3), 508–517 (2005)
12. Zufryden, F.S.: Linking advertising to box office performance of new film releases: A marketing planning model. Journal of Advertising Research 36, 29–42 (1996)
13. Ravid, S.A.: Information, blockbusters, and stars: A study of the film industry. The Journal of Business 72(4), 463–492 (1999)
14. Hennig-Thurau, T., Houston, M.B., Sridhar, S.: Can good marketing carry a bad product? Evidence from the motion picture industry. Marketing Letters 17(3), 205–219 (2006)
15. Litman, B.R., Kohl, L.S.: Predicting financial success of motion pictures: The '80s experience. Journal of Media Economics 2(2), 35–50 (1989)
16. Simonton, D.K.: Cinematic success criteria and their predictors: The art and business of the film industry. Psychology & Marketing 26(5), 400–420 (2009)
17. Sharda, R., Delen, D.: Predicting box-office success of motion pictures with neural networks. Expert Systems with Applications 30(2), 243–254 (2006)
18. Malczewski, J.: On the use of weighted linear combination method in GIS: Common and best practice approaches. Transactions in GIS 4(1), 5–22 (2000)
19. Dawes, R.M.: The robust beauty of improper linear models in decision making. American Psychologist 34(7), 571–582 (1979)
20. Johnson, J.W., LeBreton, J.M.: History and use of relative importance indices in organizational research. Organizational Research Methods 7(3), 238–257 (2004)
21. Han, J., Kamber, M.: Data Mining: Concepts and Techniques, 2nd edn. Morgan Kaufmann, San Francisco (2006)
22. Lee, F.L.F.: Cultural discount and cross-culture predictability: Examining the box office performance of American movies in Hong Kong. Journal of Media Economics 19(4), 259–278 (2006)
23. Kantardzic, M.: Data Mining: Concepts, Models, Methods, and Algorithms. IEEE Press, Piscataway (2003)

# Feature Selection for Multi-label Learning Using Mutual Information and GA

Ying Yu[1,2,*] and Yinglong Wang[2]

[1] Software School, East China Jiaotong University, Nanchang 330045, P.R. China
[2] Software School, Jiangxi Agricultural University, Nanchang 330045, P.R. China

**Abstract.** As in the traditional single-label classification, the feature selection plays an important role in the multi-label classification. This paper presents a multi-label feature selection algorithm MLFS which consists of two steps. The first step employs the mutual information to complete the local feature selection. Based on the result of local selection, GA algorithm is adopted to select the global optimal feature subset and the correlations among the labels are considered. Compared with other multi-label feature selection algorithms, MLFS exploits the label correlation to improve the performance. The experiments on two multi-label datasets demonstrate that the proposed method has been proved to be a promising multi-label feature selection method.

**Keywords:** multi-label, feature selection, GA, mutual information.

## 1 Introduction

In the traditional single-label learning, each instance is only associated with one semantic label. However, in many real-world problems, one instance usually have multiple semantic labels simultaneously. Nowadays, multi-label learning has been applied to many domains, such as the image and video annotation[1], sentiment analysis[2] and text categorization[3]. Several application domains of multi-label learning (e.g. text categorization, gene expression) involve data with large numbers of features. Similar to the single-label learning, the multi-label learning also suffers from the curse of dimensionality[4].

Contrary to the single-label classification which assumes the classes are mutually exclusive, the multi-label classification allows different classes to overlap. Therefore, a key challenge of multi-label dimensionality reduction is how to exploit the label correlations. In this paper, we propose a supervised feature selection algorithm called Multi-Label Feature Selection(MLFS) for multi-label data set based on mutual information and genetic algorithm. MLFS consists of two steps which respectively selects the features from local and global perspective in turn and exploits the label correlation to improve the performance. We use this method as a preprocessing step and achieve encouraging results on the multi-label data sets.

* This paper is supported by the 2009 Natural Science Fund of Jiangxi Agricultural University and the Jiangxi Provincial natural science fund (No.20132BAB201045).

D. Miao et al. (Eds.): RSKT 2014, LNAI 8818, pp. 454–463, 2014.
DOI: 10.1007/978-3-319-11740-9_42 © Springer International Publishing Switzerland 2014

The rest of this paper is organized as follows. Section 2 briefly reviews the related studies about multi-label dimensionality reduction. In Section 3, th feature selection algorithm MLFS is proposed, which is based on mutual information and genetic algorithm. Section 4 illustrates the effectiveness of MLFS through some experiments. Finally, Section 5 concludes the studies.

## 2    Related Works

Numerous studies have been devoted to the dimensionality reduction, most of them focus on the single-label learning. Recently, several dimensionality reduction algorithms also have been proposed for multi-label learning. These approaches could be organized into two categories:

1) feature selection

Most papers propose a previous transformation of multi-label data set to one or more single-label data sets and then use the existing feature selection methods. One of the most popular approaches, especially in text categorization, uses the BR transformation in order to evaluate the discriminative power of each feature with respect to each label. Subsequently the obtained scores are aggregated in order to obtain an overall ranking. Common aggregation strategies include taking the maximum or a weighted average of the obtained scores[5]. The LP transformation was used in reference [2], while the copy, copy-weight, select-max, select-min and ignore transformations are used in reference [6]. After problem transformation, the filter approach is usually applied to the single-label data for which many methods have been proposed. To this end, importance measures, such as Information Gain[6] and Chi-square[2], have been the most frequently used.

2) feature extraction

Feature extraction methods construct new features out of the original ones either using class information (supervised) or not (unsupervised).Unsupervised methods, such as principal component analysis and latent semantic indexing (LSI) are obviously directly applicable to multi-label data. In reference [7], the authors directly apply LSI based on singular value decomposition in order to reduce the dimensionality of the text categorization problem. In reference [8], the PCA is used to extract the feature space. Supervised feature extraction methods for single-label data, such as linear discriminant analysis (LDA), require modification prior to their application to multi-label data. In reference [9], an feature extraction approach called class balanced linear discriminant analysis is proposed. The key idea of this method is to define a within-class scatter matrix and a between-class scatter matrix for multi-label learning. An algorithm called Multi-label informed Latent Semantic Indexing (MLSI) [10] is a modified version of the LSI method, which preserves the information of data and meanwhile captures the correlations among multiple labels. In addtion, an supervised multi-label feature extraction algorithm called Multi-label Dimensionality reduction via Dependence Maximization (MDDM) was proposed in reference [11], which adopts the Hilbert-Schmidt independence criterion to measure the dependence between the features and the corresponding labels due to its simplicity and elegant theoretical peoperties.

## 3    The MLFS Method

First of all, we present the formal notation that we use throught. Let $\mathcal{X} = R^d$ denote the input space and $L = \{l_1, l_2, ..., l_m\}$ denote the finite set of possible labels. An instance is represented as a vector of features values $x = [x_1, x_2, ..., x_d](x \in \mathcal{X})$. $F$ is the set of $d$ features.

When assigning the labels to the instances, most of the existing multi-label classification algorithms make decisions for different labels not according to different feature sets but according to the same feature set. By analysising the multi-label data sets, we find that the essential features for each label, namely those features that have the greatest discriminative power for the specified label, are not necessarily identical, or even completely different. For a specified label, several features maybe relevant, while others may be irrelevant, redundant or useless. Taking the text categorization for example, in order to judge whether the text belongs to the political class, those features related with the government, the president and the war should be paid more attention to. In order to judge whether the text belongs to the entertainment class, those features which associated with the star, the film should be considered. Those irrelevant and redundant features not only increase the dimensions of the feature space but also reduce the learning efficiency. Furthermore, the irrelevant and redundant features would produce noise and interfere with learning, which would affect the construction of classification model.

The negative factors mentioned above have been confirmed by some works. As for the nearest neighbor algorithm, reference [12] indicates that the computational complexity of the algorithm and the number of the needed instances exponentially increase with the growth of the number of the irrelevant features. When the number of the irrelevant features increases, the number of the instances needed by the decision tree algorithm also shows an exponential growth in the XOR conditions. In addition, the performance of the Bayesian classifier is also affected by the redundant features. Therefore, in order to decrease the effect of the adverse factors, the dimension of feature space should be reduced. MLFS consists of two steps. The first step is to select the local optimal feature subset for each label using dependence maximization and eliminate the irrelevant and redundant features. The second step is to summarize the result of the first step and then the optimizaition method GA is used to find the global optimal feature subset which meets the criterion.

In the first step, the forward search strategy is adopted to obtain an ordered feature sequence for each label using the dependence maximization as a metric. Then the first $q$ features are selected as the optimal feature subsets $S$. Formula 1 gives the dependence maximization between feature subset $S$ and label $l$.

$$maxI(S; l), S \subseteq F \tag{1}$$

In the begining, $S$ is an empty set. Then feature is selected in turn into $S$ based on the principle of maximizing dependency . When the number of the feature in $S$ is equal to $q$, the process terminates. It can be seen from the Formula 1 that the high-dimensional joint probability density is difficult to compute when the number of the selected features is large. So we use the *Max-Relevance and Min-Redundancy* [13] to replace the *Max-Dependency* which is an equivalent form of maximization dependency.

$$max_{f_i \in F-S_{q-1}}[I(f_i; l) - \frac{1}{q-1} \sum_{f_j \in S_{q-1}} I(f_i; f_j)] \tag{2}$$

In Formula 2, $I(f_i; l)$ means max-relevance, namely the dependency between the new selected feature and the label is largest, while the second part means the min-redundancy, namely the reduncancy between the new selected feature and the features in $S$ is smallest. When selecting features for each label, the number of features should not be too much but be controlled within a reasonable range. Because in the multi-label system, there exists correaltions among labels. So an essential feature subset for a label may also has discriminating power for other labels. The feature subset obtained from the first setp is removed many redundant and irrelevant information. However, it is an local result and it still has redundant information for the whole.

Genetic Algorithm (GA)[14] is an famous evolutionary algorithm in the artificial intelligence, which employs the heuristic search strategy. GA is a feature selection method of wrapper style, embedded with the learning algorithm to evaluate the selected features. We use GA to select the globally optimal feature subset. Let the learning algorithm run on the selected subset using ten cross-validation. The subset whose evaluations(average precison) are best is regarded as the final output $\hat{S}$. The fitness function of GA is defined as Formula 3. *avgprec* is the average precision. the greater the average presion, the better the feature subset. $\hat{S}$ is a result of joint action by all labels. So the second step considers the correlation among labels. The pseudo code of MLFS is shown in Fig 1.

$$Fitness(\hat{S}) = \frac{1}{10} \sum_{i=1}^{10} avgprec \tag{3}$$

The computational complexity of the algorithm is decided by the Step 1 and Step 4. In Step 1, the complexity of calculating the dependency is $O(m \times d)$ and the time complexity of the sorting of the dependency is $O(m \times d^2)$. So the total computational complexity of Step 1 is $O(m \times d^2)$. The computational complexity of GA in Step 4 could divided into three parts: the calculation of fitness value, the crossover and the mutation. The complexity of the calculation of fitness depends on the fitness function and here it is depends on the complexity of classification algorithm. The complexity of the crossover and the mutation depends on the number of the iterations and the size of the feature subset obtained from the Step 1.

---

**Algorithm 1** Multi-label feature selection algorithm MLFS

Input: $T$ , $q$

Output: $\hat{S}$

Step 1 For each label $l_t (1 \le t \le m)$ in $L$, the feature sequence is generated according

to the dependency maximization. Then the first $q$ features are selected as the

feature subset $S_j$.

Step 2 All the subsets $S_t (1 \le t \le m)$ are merged together to generate the set $S$

$(S = \cup S_i)$ and the repeated features are removed from $S$.

Step 3 Copy from the training set $T$ to $T'$ and $T'$ only retains the features included

in the subset $S$.

Step 4 : Initialize the GA algorithm and the function defined in Formula 3 is treated as

the evaluation function.

Step 5 : GA is run on $T'$ and the feature set whose fitness value is biggest is the

optimal feature set $\hat{S}$ 。

---

**Fig. 1.** pseudo code of MLFS

# 4    Experiments

To test the effectiveness of MLFS presented in this paper, we apply it to two
multi-label datasets which come from the the open source Mulan library[15] and
Table 1 shows their associated properties.

**Table 1.** Multi-label datasets used for experiments

| name | instances | attribute | labels | cardinality | density |
|------|-----------|-----------|--------|-------------|---------|
| medical | 978 | 1449 | 45 | 1.245 | 0.028 |
| Corel5k | 5000 | 499 | 374 | 3.522 | 0.009 |

We use MLFS-1 (the first step of MLFS) and MLFS to select features and
obtain two feature subsets. Then MLkNN [16] runs on three datasets which
respectively have three different feature sets, namely the two selected feature
subsets and the original feature set. The number of the nearest neighbors is
set as 10. The experimental results of ten-fold cross-validation in terms of five
metrics [16], namely *Hamming loss, average precision, coverage, one-error* and
*ranking loss*, are shown in Fig 3 to Fig 12. Fig 2 shows the number of the features
selected by MLFS-1 and MLFS.

It can be seen from Fig 2 to Fig 12 that the performance of MLFS-1 is better
than other two algorithms on medical. when compared with MLkNN, MLFS-1
not only has higher precision but also involves fewer features and MLFS has
poorest performance on medical. Perhaps the second step removes too much
useful information. While the performances of three algorithms on Corel5k are
exactly opposite. MLFS is best among three algorithms and the number of the

selected features is small. It is obviously that the feature selection removes many irrelevant and redundant features.

## 5    Conclusion

This paper presents a multi-label feature selection algorithm which consists of two steps. The first step employs the mutual information to evaluate the improtance of the features for each labels and completes the local feature selection. Based on the result of the first step, GA algorithm is adopted to select the global optimal feature subset. Experiments on two multi-label datasets show that MLFS could remove irrelevant features. However, it is not clear whether they are also helpful for other kinds of multi-label learning methods and whether there are better choices than dependency maximization and GA for this purpose. These are interesting issues worth further investigation.

| q | Initial features | MLFS_1 | MLFS | | q | Initial features | MLFS_1 | MLFS |
|---|---|---|---|---|---|---|---|---|
| 1 | 1449 | 43 | 11 | | 1 | 499 | 241 | 72 |
| 2 | 1449 | 82 | 29 | | 2 | 499 | 364 | 186 |
| 3 | 1449 | 118 | 48 | | 3 | 499 | 428 | 211 |
| 4 | 1449 | 152 | 66 | | 4 | 499 | 462 | 263 |
| 5 | 1449 | 184 | 74 | | 5 | 499 | 479 | 218 |
| 6 | 1449 | 219 | 87 | | 6 | 499 | 489 | 279 |
| 7 | 1449 | 246 | 104 | | 7 | 499 | 495 | 249 |
| 8 | 1449 | 279 | 116 | | 8 | 499 | 497 | 289 |
| 9 | 1449 | 314 | 124 | | 9 | 499 | 499 | 281 |
| 10 | 1449 | 346 | 142 | | 10 | 499 | 499 | 283 |
| 11 | 1449 | 366 | 170 | | 11 | 499 | 499 | 266 |
| 12 | 1449 | 395 | 184 | | 12 | 499 | 499 | 277 |
| 13 | 1449 | 423 | 209 | | 13 | 499 | 499 | 240 |
| 14 | 1449 | 446 | 208 | | 14 | 499 | 499 | 240 |
| 15 | 1449 | 471 | 213 | | 15 | 499 | 499 | 294 |
| 16 | 1449 | 494 | 216 | | 16 | 499 | 499 | 240 |
| 17 | 1449 | 516 | 221 | | 17 | 499 | 499 | 240 |
| 18 | 1449 | 537 | 236 | | 18 | 499 | 499 | 240 |
| 19 | 1449 | 561 | 269 | | 19 | 499 | 499 | 240 |
| 20 | 1449 | 583 | 261 | | 20 | 499 | 499 | 287 |

(a)  The number of the initial features and the remained features on medical

(b)  The number of the initial features and the remained features on Corel5k

**Fig. 2.** the comparison of the number of the features

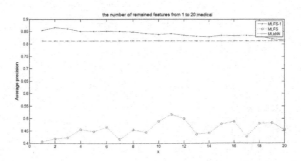

**Fig. 3.** average precision on medical

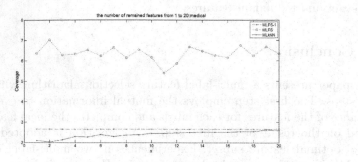

**Fig. 4.** coverage precision on medical

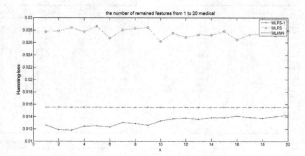

**Fig. 5.** Hamming-loss on medical

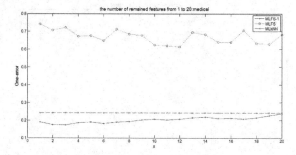

**Fig. 6.** one-error on medical

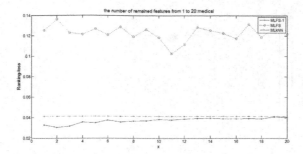

**Fig. 7.** ranking-loss on medical

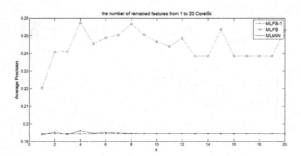

**Fig. 8.** average precision on Corel5k

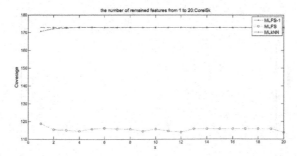

**Fig. 9.** coverage on Corel5k

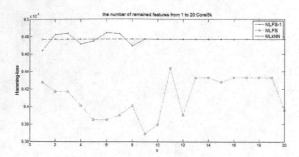

**Fig. 10.** Hamming-loss on Corel5k

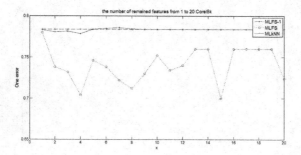

**Fig. 11.** one-error on Corel5k

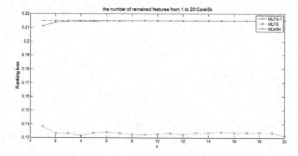

**Fig. 12.** ranking-loss on Corel5k

# References

1. Wang, J., Zhao, Y., Wu, X., et al.: A transductive multi-label learning approach for video concept detection. Pattern Recognition 44(10), 2274–2286 (2011)
2. Trohidis, K., Tsoumakas, G., Kalliris, G., et al.: Multi-Label Classification of Music into Emotions. ISMIR 8, 325–330 (2008)
3. Zhang, M.L., Zhou, Z.H.: Multilabel neural networks with applications to functional genomics and text categorization. IEEE Transactions on Knowledge and Data Engineering 18(10), 1338–1351 (2006)
4. Bellman, R.E.: Adaptive Control Process: A Guided Tour. Princeton University Press, Princeton (1961)
5. Yang, Y., Pedersen, J.O.: A comparative study on feature selection in text categorization. ICML 97, 412–420 (1997)
6. Chen, W., Yan, J., Zhang, B., et al.: Document transformation for multi-label feature selection in text categorization. In: The Seventh IEEE International Conference on Data Mining, pp. 451–456. IEEE (2007)
7. Gao, S., Wu, W., Lee, C.H., et al.: A MFoM learning approach to robust multiclass multi-label text categorization. In: The Twenty-First International Conference on Machine Learning, p. 42. ACM (2004)
8. Zhang, M.L., Pea, J.M., Robles, V.: Feature selection for multi-label naive Bayes classification. Information Sciences 179(19), 3218–3229 (2009)
9. Wang, H., Ding, C.H.Q., Huang, H.: Multi-Label Classification: Inconsistency and Class Balanced K-Nearest Neighbor. In: The Twenty-Fouth AAAI Conference on Artifical Intelligence, pp. 1264–1266 (2010)
10. Yu, K., Yu, S., Tresp, V.: Multi-label informed latent semantic indexing. In: The 28th Annual International ACM SIGIR Conference on Research and Development in Information Retrieval, pp. 258–265. ACM (2005)
11. Zhang, Y., Zhou, Z.H.: Multilabel dimensionality reduction via dependence maximization. ACM Transactions on Knowledge Discovery from Data (TKDD) 4(3), 14 (2010)
12. Langley, P., Simon, H.A.: Applications of machine learning and rule induction. Communications of the ACM 38(11), 54–64 (1995)
13. Peng, H., Long, F., Ding, C.: Feature selection based on mutual information criteria of max-dependency, max-relevance, and min-redundancy. IEEE Transactions on Pattern Analysis and Machine Intelligence 27(8), 1226–1238 (2005)
14. Goldberg, D.E.: Genetic algorithms in search, optimization, and machine learning. Addison-wesley, Reading Menlo Park (1989)
15. Tsoumakas, G., Katakis, I.: Multi-label classification: An overview. International Journal of Data Warehousing and Mining 3, 1–13 (2007)
16. Zhang, M.L., Zhou, Z.H.: ML-KNN: A lazy learning approach to multi-label learning. Pattern Recognition 40(7), 2038–2048 (2007)

# Uncertainty in Granular Computing

# Characterizing Hierarchies on Covering-Based Multigranulation Spaces

Jingjing Song[1], Xibei Yang[1,2,*], Yong Qi[3], Hualong Yu[1],
Xiaoning Song[1], and Jingyu Yang[2,4]

[1] School of Computer Science and Engineering, Jiangsu University of Science and
Technology, Zhenjiang, Jiangsu 212003, P.R. China
[2] Key Laboratory of Intelligent Perception and Systems for High-Dimensional
Information, Nanjing University of Science and Technology, Ministry of Education,
Nanjing Jiangsu 210094, P.R. China,
yangxibei@hotmail.com
[3] School of Economics and Management, Nanjing University of Science and
Technology, Nanjing, Jiangsu 210094, P.R. China
[4] School of Computer Science and Technology, Nanjing University of Science and
Technology, Nanjing, Jiangsu 210094, P.R. China

**Abstract.** Hierarchy plays a fundamental role in the development of the
Granular Computing(GrC). In many practical applications, the granules
are formed in a family of the coverings, which can construct a Covering-
based Multigranulation Space(CBMS). It should be noticed that the
hierarchies on Covering-based Multigranulation Spaces has become a
necessity. To solve such problem, the concepts of the union knowledge
distance and the intersection knowledge distance are introduced into the
CBMS, which can be used to construct the knowledge distance lattices.
According to the union knowledge distance and the intersection knowledge distance, two partial orderings can be derived, respectively. The
example shows that the derived partial orderings can compare the finer
or coarser relationships between two different Covering-based Multigranulation Spaces effectively. The theoretical results provide us a new way
to the covering based granular computing.

**Keywords:** covering-based multigranulation space, granular computing, granular structure, hierarchical structure.

## 1 Introduction

Granular computing(GrC) [22] was firstly proposed by Zadeh, which can be considered as a structured way of thinking, structured problem solving and paradigm
of information processing [20]. It can also be considered as a label of a new field
of multi-disciplinary study, dealing with theories, methodologies, techniques, and
tools that make use of granules in the process of complex problem solving [18].

---

* Corresponding author.

D. Miao et al. (Eds.): RSKT 2014, LNAI 8818, pp. 467–478, 2014.
DOI: 10.1007/978-3-319-11740-9_43 © Springer International Publishing Switzerland 2014

In recent years, many researchers paid attentions to the development of GrC. For example, Yao [20] proposed a triarchic theory of GrC, which includes multi-level, multi-view [1] and granular computing triangle; Saberi et al. [11] applied the granular computing to credit scoring modeling, which is a pioneer in examining the concept of granularity for selecting the optimum size of testing and training group in machine learning area; Yao et al. [19] reviewed foundations and schools of research and elaborated current developments in GrC research; Zhang et al. [23] discussed the double-quantitative approximation space of precision and grade and then tackled the fusion problem, which can further conduct double-quantification studies on granular computing; Hu et al. [2] proposed a novel learning approach which combines fuzzy logical designing with machine learning to construct a granular computing system. Presently, GrC provides an effective tool for problem solving in many different fields [4, 6, 15], GrC research is moving into the mainstream of computer science [19].

Information granulation will adopt various strategies according to people's requirements [10]. It is necessary to research the hierarchies on information granulations. Different levels of abstraction may represent different granulated views of our understanding of a real world problem [3], i.e., the hierarchy will help us to solve problems in different views. Generally speaking, hierarchy reflects the finer or coarser relationships among information granulations. For example, Liang et al. [5] researched the difference among the information granulations from the view of knowledge distance [9]; Qian et al. [7] characterized the hierarchies on granular structures with the cardinality when the set-inclusion does not exists among information granules; Yang et al. [17] constructed the algebraic lattices based on the set distance and knowledge distance to characterize the hierarchies on information granules.

It should be noticed that the hierarchies of granular computing mentioned are based on the equivalence relations. However, in many practical problems, the equivalence relation is hard to be acquired, it is often the covering [12–14, 24, 25] or even a family of coverings [17]. We often need to describe concurrently a target concept from some independent environments [8]. From this point of view, it is practical to research the hierarchies on a family of coverings. As a family of coverings can construct a multigranulation space, in the context of the paper, we call it Covering-based Multigranulation Space(CBMS). To further push the development of GrC, the study of hierarchies on Covering-based Multigranualtion Spaces has become a necessity. This is what will be discussed in this paper. The objective of this study is to characterize the hierarchies on Covering-based Multigranulation Spaces use a knowledge distance approach.

## 2    Preliminaries

### 2.1    Information Granulations

*Definition 1.* [21] Let $U$ be the universe of discourse, a subset $g \in 2^U$ is called a granule, where $2^U$ is the power set of $U$.

*Definition 2.* [21] Let $U$ be the universe of discourse, suppose that $G \subseteq 2^U$ is a nonempty family of subsets of $U$, the pair $(G, \subseteq)$ is called a granular structure, where $\subseteq$ is the set-inclusion relation.

*Remark 1.* Though the granular structure is defined by set-inclusion relation $\subseteq$, which is a special example of partial ordering. That is to say, a granular structure can be defined on a partial ordering.

Let $U \neq \emptyset$ be the universe, $\boldsymbol{R}$ is a family of the equivalence relations (reflexive, symmetric, transitive) on $U$, then the pair $KB = (U, \boldsymbol{R})$ is called the Pawlak knowledge base[20]. If $P \neq \emptyset$ and $P \subseteq \boldsymbol{R}$, then $IND(P) = \cap P$ (intersection of all equivalence relations in $P$) is also an equivalence relation. $U/IND(P)$ is a family of the equivalence classes, which are generated from the equivalence relation $IND(P)$. In the viewpoint of granular computing, according to Definition 1, the set of Pawlak information granules $G(P) = \{[x]_P : \forall x \in U\}$ is regarded as a Pawlak information granulation over $U$ where $[x]_P = \{y \in U : (x, y) \in IND(P)\}$. In the context of the paper, the set of all the Pawlak information granulations over $U$ is denoted by $PG(U)$.

In Pawlk's knowledge base, an equivalence relation $U/IND(P)$ can induce a partition. The definition of the partition is as Definition 3 shows.

*Definition 3.* Let $U$ be the universe of discourse, $P = \{p_1, p_2, ..., p_m\}$ is a partition on $U$, if and only if $\bigcup_{i=1}^m p_i = U$ and $\forall i \neq j, p_i \cap p_j = \emptyset$.

In the context of the paper, the pair $(U, P)$ is regarded as a Pawlak approximation space.

As we all know, in reality, the granules are formed in equivalence relation are hard to be obtained, it is often formed in covering, in which two granules may not be disjoint to each other. The covering can be defined as Definition 4 shows.

*Definition 4.* Let $U$ be the universe of discourse, $C$ is a covering on $U$, if and only if $\bigcup_{i=1}^m = U$ where $\forall c_i = C$.

By Definition 3 and 4, it is obvious that the partition is a special case of the covering. Similar to the Pawlak approximation space, the pair $(U, C)$ is regarded as a covering approximation space.

Let $U$ be the universe of discourse, $F = \{C_1, C_2, ..., C_m\}$ is a family of the coverings on $U$. Similar to the Pawlak knowledge base, the pair $KF = (U, F)$ is called the covering knowledge base. Similar to the Pawlak information granulation, let $U$ be the universe, $C$ is a covering on $U$, the set of covering information granules $G(C) = \{N(x) : \forall x \in U\}$ is regarded as a covering information granulation over $U$ where $N(x) = \cap \{c_i : c_i \in C \wedge x \in c_i\}$ is the neighborhood of $x$. The set of all the covering information granulations over $U$ is denoted by $CG(U)$.

## 2.2  Knowledge Distance on Pawlak Information Granulation

It should be noticed that in $PG(U)$, there are two special information granulations: one is the coarsest information granulation, it is denoted by $\sigma = \{U : \forall x \in U\}$. In this situation, the knowledge we obtained is the least, each object in the universe can not be distinguished; the other one is the finest information granulation, it is denoted by $\omega = \{\{x\} : \forall x \in U\}$. In this situation, the knowledge we obtained is the most, each object in the universe can be distinguished.

*Definition 5.* [9] Let $KB = (U, \boldsymbol{R})$ be the knowledge base, $\forall G(P), G(Q) \in PG(U)$, three operators can be defined as

$$G(P) \cap G(Q) = \{[x]_{P \cap Q} : \forall x \in U, [x]_{P \cap Q} = [x]_P \cap [x]_Q\}; \tag{1}$$
$$G(P) \cup G(Q) = \{[x]_{P \cup Q} : \forall x \in U, [x]_{P \cup Q} = [x]_P \cup [x]_Q\}; \tag{2}$$
$$\wr G(P) = \{\ [x]_P : \forall x \in U, \wr[x]_P = \{x\} \cup (U - [x]_P)\}. \tag{3}$$

By Definition 5, we know that the three operators can be seen as intersection operation, union operation and complement operation in-between two Pawlak information granulation $G(P)$ and $G(Q)$, respectively.

*Theorem 1.* [9] Let $KB = (U, \boldsymbol{R})$ be the knowledge base,

1. $(K(U), \cap, \cup)$ is a lattice;
2. $(K(U), \cap, \cup)$ is a distributive lattice;
3. $(K(U), \cap, \cup, \wr)$ is a complement lattice.

To characterize the relationship among information granulations, we introduce the knowledge distance for measuring the difference between two information granulations on the same knowledge base in the following[9].

*Definnition 6.* [5, 9] Let $U$ be the universe of discourse, $\forall G(P), G(Q) \in PG(U)$, the knowledge distance between Pawlak information granulations $G(P)$ and $G(Q)$ is denoted by $D(G(P), G(Q))$ such that

$$D(G(P), G(Q)) = \frac{1}{|U|} \sum_{x \in U} \frac{|[x]_P \oplus [x]_Q|}{|U|}, \tag{4}$$

where $|X|$ is the cardinal number of the set $X$, $[x]_P \oplus [x]_Q$ is the symmetric difference of $[x]_P$ and $[x]_Q$, i.e., $[x]_P \oplus [x]_Q = ([x]_P - [x]_Q) \cup ([x]_Q - [x]_P)$. Obviously, $0 \le D(G(P), G(Q)) \le 1 - \frac{1}{|U|}$ holds. $D(G(P), G(Q)) = 0$ if and only if $G(P) = G(Q)$, while $D(G(P), G(Q)) = 1 - \frac{1}{|U|}$ if and only if $G(P) = \wr G(Q)$.

*Theorem 2.* [9] Let $U$ be the universe of discourse, $\forall G(P), G(Q), G(R) \in PG(U)$, the following properties about knowledge distance hold:

1. Positive: $D(G(P), G(Q)) \ge 0$;
2. Symmetric: $D(G(P), G(Q)) = D(G(Q), G(P))$;
3. Triangle inequalities:
   (1) $D(G(P), G(Q)) + D(G(P), G(R)) \ge D(G(Q), G(R))$;
   (2) $D(G(P), G(Q)) + D(G(Q), G(R)) \ge D(G(P), G(R))$;
   (3) $D(G(P), G(R)) + D(G(Q), G(R)) \ge D(G(P), G(Q))$.

By Theorem 2, we can see that $(PG(U), D)$ is a distance space.

# 3  Hierarchies on Covering-Based Multigranulation Space

Multigranulation space is a core concept in granular computing theory. In this section, we may discuss the hierarchies on Covering-based Multigranulation Spaces. Similar to the covering information granulation, in CBMS, the multi-covering information granules of the object $x$ in the CBMS is denoted by $F(x) = \{c_i \in C : x \in c_i \land C \in F\}$. By Section 2.1, we can see that $F(x) = \{N(x) : C \in F\}$. The multi-covering information granulation over $U$ is denoted by $MS(F) = \{F(x) : \forall x \in U\}$, which constitutes a CBMS. The collection of all the Covering-based mulatigranulation spaces over $U$ is denoted by $MG(U)$.

## 3.1  Knowledge Distance on Covering-Based Multigranulation Spaces

To research the knowledge distances between two Covering-based Multigranulation Spaces, it is a necessity to discuss the knowledge distances between two Covering-based Multigranulation Spaces in terms of one certain object. The knowledge distances between two Covering-based Multigranulation Spaces in terms of the object can be defined as Definition 7 shows.

*Definnition 7.* Let $U$ be the universe of discourse, $\forall F_1, F_2 \in MG(U), \forall x \in U$, the knowledge distances between Covering-based Multigranulation Spaces $F_1$ and $F_2$ in terms of the object $x$ will be defined as follow:

$$D_x^{\cup}(F_1, F_2) = \frac{|\bigcup_{N_{1i} \in F_1(x)} \bigcup_{N_{2j} \in F_2(x)} (N_{1i} \oplus N_{2j})|}{|U|}, \tag{5}$$

$$D_x^{\cap}(F_1, F_2) = \frac{|\bigcap_{N_{1i} \in F_1(x)} \bigcap_{N_{2j} \in F_2(x)} (N_{1i} \oplus N_{2j})|}{|U|}, \tag{6}$$

where $\bigcup_{N_{1i} \in F_1(x)}$ denotes the union of the multi-covering information granules of the object $x$ in the CBMS $F_1$, $\bigcap_{N_{1i} \in F_1(x)}$ denotes the intersection of the multi-covering information granules of the object $x$ in the CBMS $F_1$. Therefore, the two formulas can be seen as two knowledge distances with different combinations of the symmetric difference in terms of Covering-based Multigranulation Spaces $F_1$ and $F_2$.

To facilitate our discussions, $\forall F_1, F_2 \in MG(U), D_x^{\cup}(F_1, F_2)$ is called the union knowledge distance in terms of the object $x$ and $D_x^{\cap}(F_1, F_2)$ is called the intersection knowledge distance in terms of the object $x$.

*Theorem 3.* Let $U$ be the universe of discourse, $\forall F_1, F_2, F_3 \in MG(U), \forall x \in U$, the following properties about knowledge distance hold:

1. Positive: $D_x^{\cup}(F_1, F_2) \geq 0, D_x^{\cup}(F_1, F_2) \geq 0$;
2. Symmetric: $D_x^{\cup}(F_1, F_2) = D_x^{\cup}(F_2, F_1), D_x^{\cap}(F_1, F_2) = D_x^{\cap}(F_2, F_1)$;

3. Triangle inequalities:
(1) $D_x^{\cup}(F_1, F_2) + D_x^{\cup}(F_1, F_3) \geq D_x^{\cup}(F_2, F_3)$,
$D_x^{\cap}(F_1, F_2) + D_x^{\cap}(F_1, F_3) \geq D_x^{\cap}(F_2, F_3)$;
(2) $D_x^{\cup}(F_1, F_3) + D_x^{\cup}(F_2, F_3) \geq D_x^{\cup}(F_1, F_2)$,
$D_x^{\cap}(F_1, F_3) + D_x^{\cap}(F_2, F_3) \geq D_x^{\cap}(F_1, F_2)$;
(3) $D_x^{\cup}(F_1, F_2) + D_x^{\cup}(F_2, F_3) \geq D_x^{\cup}(F_1, F_3)$,
$D_x^{\cap}(F_1, F_2) + D_x^{\cap}(F_2, F_3) \geq D_x^{\cap}(F_1, F_3)$.

*Proof.* We only prove the situation of union knowledge distance $D_x^{\cup}(F_1, F_2)$, the proof of intersection knowledge distance $D_x^{\cap}(F_1, F_2)$ is similar to the situation of $D_x^{\cup}(F_1, F_2)$. The properties of positive and symmetric can be obtained directly by Definition 7. In the following, we only prove the triangle inequalities.

1). If there exists equivalent Covering-based Multigranulation Spaces in $F_1, F_2$ and $F_3$. Suppose that $F_1 = F_2$, by Definition 7, it is obvious that the triangle inequalities hold.

2). If the Covering-based Multigranulation Spaces in $F_1, F_2$ and $F_3$ are different from each other. According to the basic set theory, we know that $\forall X, Y, Z$ are three finite set, $(Y \oplus Z) \subseteq (X \oplus Y) \cup (X \oplus Z)$ holds. By Definition 7, we have

$$D_x^{\cup}(F_1, F_2) + D_x^{\cup}(F_1, F_3) =$$

$$\frac{|\bigcup_{N_{1i} \in F_1(x)} \bigcup_{N_{2j} \in F_2(x)} (N_{1i} \oplus N_{2j})|}{|U|} + \frac{|\bigcup_{N_{1i} \in F_1(x)} \bigcup_{N_{3k} \in F_3(x)} (N_{1i} \oplus N_{3k})|}{|U|} \geq$$

$$\frac{|(\bigcup_{N_{1i} \in F_1(x)} \bigcup_{N_{2j} \in F_2(x)} (N_{1i} \oplus N_{2j})) \cup (\bigcup_{N_{1i} \in F_1(x)} \bigcup_{N_{3k} \in F_3(x)} (N_{1i} \oplus N_{3k}))|}{|U|} \geq$$

$$\frac{|\bigcup_{N_{1i} \in F_1(x)} \bigcup_{N_{2j} \in F_2(x)} \bigcup_{N_{3k} \in F_3(x)} ((N_{1i} \oplus N_{2j}) \cup ((N_{1i} \oplus N_{3k})))|}{|U|} \geq$$

$$\frac{|\bigcup_{N_{2j} \in F_2(x)} \bigcup_{N_{3k} \in F_3(x)} (N_{2j} \oplus N_{3k})|}{|U|} = D_x^{\cup}(F_2, F_3).$$

Similarity, it is not difficult to prove the other two triangle inequalities.

By Theorem 3, we can see that $(MG(U), D_x^{\cup})$ and $(MG(U), D_x^{\cap})$ are distance spaces. By Definition 7, the knowledge distances between two Covering-based Multigranulation Spaces can be defined as Definition 8 shows.

*Definition 8.* Let $U$ be the universe of discourse, $\forall F_1, F_2 \in MG(U), \forall x \in U$, the knowledge distances between Covering-based Multigranulation Spaces $F_1$ and $F_2$ will be defined as following:

$$D^{\cup}(F_1, F_2) = \frac{1}{|U|} \sum_{x \in U} D_x^{\cup}(F_1, F_2), \tag{7}$$

$$D^{\cap}(F_1, F_2) = \frac{1}{|U|} \sum_{x \in U} D_x^{\cap}(F_1, F_2). \tag{8}$$

It not difficult to find that $D^{\cup}(F_1, F_2)$ is the mean value of the union knowledge distances $D_x^{\cup}(F_1, F_2)$ of all the objects in the universe $U$. $D^{\cap}(F_1, F_2)$ is the mean value of the union knowledge distances $D_x^{\cap}(F_1, F_2)$ of all the objects in the universe $U$.

To facilitate our discussions, $D^{\cup}(F_1, F_2)$ denotes the union knowledge distance on Covering-based Multigranulation Spaces $F_1$ and $F_2$, $D^{\cap}(F_1, F_2)$ denotes the intersection knowledge distance on covering-based multigranulation spaces $F_1$ and $F_2$. On the basis of Theorem 3, Theorem 4 is easy to be obtained.

*Theorem 4.* Let $U$ be the universe of discourse, $\forall F_1, F_2, F_3 \in MG(U), \forall x \in U$, the following properties about knowledge distances hold:

1. Positive: $D^{\cup}(F_1, F_2) \geq 0, D^{\cup}(F_1, F_2) \geq 0$;
2. Symmetric: $D^{\cup}(F_1, F_2) = D^{\cup}(F_2, F_1), D^{\cap}(F_1, F_2) = D^{\cap}(F_2, F_1)$;
3. Triangle inequalities:
   (1) $D^{\cup}(F_1, F_2) + D^{\cup}(F_1, F_3) \geq D^{\cup}(F_2, F_3)$,
   $D^{\cap}(F_1, F_2) + D^{\cap}(F_1, F_3) \geq D^{\cap}(F_2, F_3)$;
   (2) $D^{\cup}(F_1, F_3) + D^{\cup}(F_2, F_3) \geq D^{\cup}(F_1, F_2)$,
   $D^{\cap}(F_1, F_3) + D^{\cap}(F_2, F_3) \geq D^{\cap}(F_1, F_2)$;
   (3) $D^{\cup}(F_1, F_2) + D^{\cup}(F_2, F_3) \geq D^{\cup}(F_1, F_3)$,
   $D^{\cap}(F_1, F_2) + D^{\cap}(F_2, F_3) \geq D^{\cap}(F_1, F_3)$;

*Proof.* It is easy to prove by the results on Theorem 3 and Definition 8.

## 3.2   The Algebraic Structures of Knowledge Distance on CBMS

Similar to the situation of $PG(U)$, there are also two special information granulations in $CG(U)$: one is the coarsest information granulation $\sigma = \{U : \forall x \in U\}$, the other one is the finest information granulation $\omega = \{\{x\} : \forall x \in U\}$. To discuss the algebraic structures of knowledge distances on Covering-based Multigranulation Spaces, a frame of reference should be used, in the context of the paper, the finest information granulation $\omega$ is selected.

*Definnition 9.* Let $U$ be the universe of discourse, $\forall F_1, F_2 \in MG(U), \forall x \in U$, two operators on the union knowledge distance on Covering-based Multigranulation Spaces $F_1$ and $F_2$ can be defined as following:

$$D_x^{\cup}(F_1, \omega) \wedge D_x^{\cup}(F_2, \omega) = \min\{D_x^{\cup}(F_1, \omega), D_x^{\cup}(F_2)\}; \tag{9}$$
$$D_x^{\cup}(F_1, \omega) \vee D_x^{\cup}(F_2, \omega) = \max\{D_x^{\cup}(F_1, \omega), D_x^{\cup}(F_2)\}. \tag{10}$$

*Definnition 10.* Let $U$ be the universe of discourse, $\forall F_1, F_2 \in MG(U), \forall x \in U$, two operators on the intersection knowledge distance on Covering-based Multigranulation Spaces $F_1$ and $F_2$ can be defined as following:

$$D_x^{\cap}(F_1, \omega) \wedge D_x^{\cap}(F_2, \omega) = \min\{D_x^{\cap}(F_1, \omega), D_x^{\cap}(F_2)\}; \tag{11}$$
$$D_x^{\cap}(F_1, \omega) \vee D_x^{\cap}(F_2, \omega) = \max\{D_x^{\cap}(F_1, \omega), D_x^{\cap}(F_2)\}. \tag{12}$$

In Section 2.1, we know that $MG(U)$ denotes the collection of all the Covering-based Multigranulation Spaces over $U$, $\forall x \in U, DF_\omega^\cup(x) = \{D_x^\cup(F,\omega) : \forall F \in MG(U)\}$ denotes the set of the union knowledge distances on the multi-covering information granulations and the finest information granulation in terms of the object $x$, $DF_\omega^\cup(U) = \{DF_\omega^\cup(x) : \forall x \in U\}$ denotes the set of the union knowledge distances on the multi-covering information granulations and the finest information granulation of all the objects on the universe.

It is not difficult to know that $\forall x \in U, DF_\omega^\cap(x) = \{D_x^\cap(F,\omega) : \forall F \in MG(U)\}$ denotes the set of the intersection knowledge distances on the multi-covering information granulations and the finest information granulation in terms of the object $x$, $DF_\omega^\cap(U) = \{DF_\omega^\cap(x) : \forall x \in U\}$ denotes the set of the intersection knowledge distances on the multi-covering information granulations and the finest information granulation of all the objects on the universe.

**Theorem 5.** Let $U$ be the universe of discourse, $\forall x \in U$, $(DF_\omega^\cup(x), \wedge, \vee)$ is a distributive lattice, which can be regarded as the union knowledge distance lattice.

*Proof.* By Definition 9, it is a trivial to prove this theorem.

**Theorem 6.** Let $U$ be the universe of discourse, $\forall x \in U$, $(DF_\omega^\cap(x), \wedge, \vee)$ is a distributive lattice, which can be regarded as the intersection knowledge distance lattice.

*Proof.* By Definition 10, it is a trivial to prove this theorem.

### 3.3  Hierarchies on Covering-Based Mulatigranulation Spaces

In general, a partial ordering can be induced by " $\wedge$ " and " $\vee$ " operators in a lattice, which can be used to judge the finer or coarser relationships between two Covering-based Mulatigranulation Spaces.

*Definnition 11.* Let $U$ be the universe of discourse, $\forall F_1, F_2 \in MG(U)$,
1) the partial ordering induced by the union knowledge distance lattice $(DF_\omega^\cup(x), \wedge, \vee)$ is

$$D_x^\cup(F_1,\omega) \preceq_1 D_x^\cup(F_2,\omega) \Leftrightarrow D_x^\cup(F_1,\omega) \wedge D_x^\cup(F_2,\omega) = D_x^\cup(F_1,\omega)$$
$$\text{or } D_x^\cup(F_1,\omega) \vee D_x^\cup(F_2,\omega) = D_x^\cup(F_2,\omega), \quad (13)$$

2) the partial ordering induced by the intersection knowledge distance lattice $(DF_\omega^\cap(x), \wedge, \vee)$ is

$$D_x^\cap(F_1,\omega) \preceq_1 D_x^\cap(F_2,\omega) \Leftrightarrow D_x^\cap(F_1,\omega) \wedge D_x^\cap(F_2,\omega) = D_x^\cap(F_1,\omega)$$
$$\text{or } D_x^\cap(F_1,\omega) \vee D_x^\cap(F_2,\omega) = D_x^\cap(F_2,\omega). \quad (14)$$

It should be noticed that if $D_x^\cup(F_1,\omega) \preceq_1 D_x^\cup(F_2,\omega)$ and $D_x^\cup(F_2,\omega) \preceq_1 D_x^\cup(F_1,\omega)$ hold at the same time, then $D_x^\cup(F_1,\omega) = D_x^\cup(F_2,\omega)$. However, it does not mean that $F_1$ and $F_2$ are equivalent, it means that they have the same union distance with the finest information granulation $\omega$. if $D_x^\cap(F_1, \omega) \preceq_2 D_x^\cap(F_2, \omega)$ and $D_x^\cap(F_2, \omega) \preceq_2 D_x^\cap(F_1, \omega)$ hold at the same time, then $D_x^\cap(F_1, \omega) = D_x^\cap(F_2, \omega)$. However, it does not mean that $F_1$ and $F_2$ are equivalent, it means that they have the same intersection distance with the finest information granulation $\omega$.

The partial orderings defined on the Definition 11 can be used to characterize the hierarchies on Covering-based Multigranulation Spaces, i.e., $\forall F_1, F_2 \in MG(U)$,

$$1) \quad F_1 \trianglelefteq_1 F_2 \Leftrightarrow \forall x \in U, D_x^\cup(F_1, \omega) \preceq_1 D_x^\cup(F_2, \omega); \tag{15}$$

$$2) \quad F_1 \trianglelefteq_2 F_2 \Leftrightarrow \forall x \in U, D_x^\cap(F_1, \omega) \preceq_2 D_x^\cap(F_2, \omega). \tag{16}$$

*Theorem 7.* Let $U$ be the universe of discourse, $\forall F_1, F_2 \in MG(U)$,

$$1) \quad F_1 \trianglelefteq_1 F_2 \Rightarrow \sum_{x \in U} |\cup F_1(x)| \le \sum_{x \in U} |\cup F_2(x)|; \tag{17}$$

$$2) \quad F_1 \trianglelefteq_2 F_2 \Rightarrow \sum_{x \in U} |\cap F_1(x)| \le \sum_{x \in U} |\cap F_2(x)|. \tag{18}$$

*Proof.* We only prove the Eq.(17), the proof of Eq.(18) is similar to the proof of Eq.(17).

$$F_1 \trianglelefteq_1 F_2 \Leftrightarrow$$
$$\forall x \in U, D_x^\cup(F_1, \omega) \preceq_1 D_x^\cup(F_2, \omega) \Leftrightarrow$$
$$\forall x \in U, D_x^\cup(F_1, \omega) \wedge D_x^\cup(F_2, \omega) = D_x^\cup(F_1, \omega) \text{ or}$$
$$D_x^\cup(F_1, \omega) \vee D_x^\cup(F_2, \omega) = D_x^\cup(F_2, \omega) \Leftrightarrow$$
$$\forall x \in U, D_x^\cup(F_1, \omega) \le D_x^\cup(F_2, \omega) \Leftrightarrow$$
$$\forall x \in U, \frac{|\bigcup_{N_{1i} \in F_1(x)} (N_{1i} \oplus \{x\})|}{|U|} \le \frac{|\bigcup_{N_{2j} \in F_2(x)} (N_{2j} \oplus \{x\})|}{|U|} \Leftrightarrow$$
$$\forall x \in U, |\cup F_1(x)| \le |\cup F_2(x)| \Rightarrow$$
$$\sum_{x \in U} |\cup F_1(x)| \le \sum_{x \in U} |\cup F_2(x)|.$$

*Example 1.* Suppose that $U = \{x_1, x_2, x_3, x_4, x_5, x_6, x_7, x_8\}$ is the universe, two Covering-based Multigranulation Spaces are given by $F_1 = \{C_1, C_2\}$ and $F_2 = \{C_3, C_4\}$ such that $C_1 = \{\{x_1, x_2, x_3, x_4, x_8\}, \{x_4, x_7, x_8\}, \{x_5, x_6\}\}, C_2 = \{\{x_1, x_2, x_3\}, \{x_2\}, \{x_5, x_6\}, \{x_7\}, \{x_4, x_7, x_8\}\}, C_3 = \{\{x_1, x_2, x_3, x_4, x_5, x_7, x_8\}, \{x_1, x_2, x_3, x_4, x_5, x_6, x_7, x_8\}\}, C_4 = \{\{x_1, x_2, x_3, x_4, x_6\}, \{x_4, x_6\}, \{x_7\}, \{x_4, x_5, x_6, x_8\}\}$.

$\forall x \in U$, we first compute the neighborhood of $x$.
$N_1(x_1) = \{x_1, x_2, x_3, x_4, x_8\}, N_1(x_2) = \{x_1, x_2, x_3, x_4, x_8\}$,
$N_1(x_3) = \{x_1, x_2, x_3, x_4, x_8\}, N_1(x_4) = \{x_4, x_8\}, N_1(x_5) = \{x_5, x_6\}$,
$N_1(x_6) = \{x_5, x_6\}, N_1(x_7) = \{x_4, x_7, x_8\}, N_1(x_8) = \{x_4, x_8\}$;

$N_2(x_1) = \{x_1, x_2, x_3\}, N_2(x_2) = \{x_2\}, N_2(x_3) = \{x_1, x_2, x_3\},$
$N_2(x_4) = \{x_4, x_7, x_8\}, N_2(x_5) = \{x_5, x_6\}, N_2(x_6) = \{x_5, x_6\},$
$N_2(x_7) = \{x_7\}, N_2(x_8) = \{x_4, x_7, x_8\};$
$N_3(x_1) = \{x_1, x_2, x_3, x_4, x_5, x_7, x_8\}, N_3(x_2) = \{x_1, x_2, x_3, x_4, x_5, x_7, x_8\},$
$N_3(x_3) = \{x_1, x_2, x_3, x_4, x_5, x_7, x_8\}, N_3(x_4) = \{x_1, x_2, x_3, x_4, x_5, x_6, x_7, x_8\},$
$N_3(x_5) = \{x_1, x_2, x_3, x_4, x_5, x_7, x_8\}, N_3(x_6) = \{x_1, x_2, x_3, x_4, x_5, x_6, x_7, x_8\},$
$N_3(x_7) = \{x_1, x_2, x_3, x_4, x_5, x_7, x_8\}, N_3(x_8) = \{x_1, x_2, x_3, x_4, x_5, x_7, x_8\};$
$N_4(x_1) = \{x_1, x_2, x_3, x_4, x_6\}, N_4(x_2) = \{x_1, x_2, x_3, x_4, x_6\},$
$N_4(x_3) = \{x_1, x_2, x_3, x_4, x_6\}, N_4(x_4) = \{x_4, x_6\}, N_4(x_5) = \{x_4, x_5, x_6, x_8\},$
$N_4(x_6) = \{x_4, x_6\}, N_4(x_7) = \{x_7\}, N_4(x_8) = \{x_4, x_5, x_6, x_8\}.$

By Eq.(5) in Definition 7, $\forall x \in U$,
$D_{x_1}^{\cup}(F_1, \omega) = \frac{1}{2}, D_{x_2}^{\cup}(F_1, \omega) = \frac{1}{2}, D_{x_3}^{\cup}(F_1, \omega) = \frac{1}{2}, D_{x_4}^{\cup}(F_1, \omega) = \frac{1}{4}, D_{x_5}^{\cup}(F_1, \omega) = \frac{1}{8}, D_{x_6}^{\cup}(F_1, \omega) = \frac{1}{8}, D_{x_7}^{\cup}(F_1, \omega) = \frac{1}{4}, D_{x_8}^{\cup}(F_1, \omega) = \frac{1}{4};$
$D_{x_1}^{\cup}(F_2, \omega) = \frac{7}{8}, D_{x_2}^{\cup}(F_2, \omega) = \frac{7}{8}, D_{x_3}^{\cup}(F_2, \omega) = \frac{7}{8}, D_{x_4}^{\cup}(F_2, \omega) = \frac{7}{8}, D_{x_5}^{\cup}(F_2, \omega) = \frac{7}{8}, D_{x_6}^{\cup}(F_2, \omega) = \frac{7}{8}, D_{x_7}^{\cup}(F_2, \omega) = \frac{3}{4}, D_{x_8}^{\cup}(F_2, \omega) = \frac{7}{8};$

By Eq.(7) in Definition 8, $D^{\cup}(F_2, \omega) = \frac{5}{16}, D^{\cup}(F_2, \omega) = \frac{27}{32}.$

From the computation, we can see that $\forall x \in U, D_x^{\cup}(F_1, \omega) \wedge D_x^{\cup}(F_2, \omega) = D_x^{\cup}(F_1, \omega)$, by Eq.(13) in Definition 11, $D_x^{\cup}(F_1, \omega) \preceq_1 D_x^{\cup}(F_2, \omega)$ holds obviously, and then $F_1 \unlhd_1 F_2$, we can say that CBMS $F_1$ is finer than $F_2$.

By further computation,

$$\sum_{x \in U} | \cup F_1(x)| = 28, \sum_{x \in U} | \cup F_2(x)| = 62,$$

which verifies the Eq.(17) in Theorem 7.

Similarly, by Eq.(6) in Definition 7, $\forall x \in U$,
$D_{x_1}^{\cap}(F_1, \omega) = \frac{1}{4}, D_{x_2}^{\cap}(F_1, \omega) = 0, D_{x_3}^{\cap}(F_1, \omega) = \frac{1}{4}, D_{x_4}^{\cap}(F_1, \omega) = \frac{1}{8}, D_{x_5}^{\cap}(F_1, \omega) = \frac{1}{8}, D_{x_6}^{\cap}(F_1, \omega) = \frac{1}{8}, D_{x_7}^{\cap}(F_1, \omega) = 0, D_{x_8}^{\cap}(F_1, \omega) = \frac{1}{4};$
$D_{x_1}^{\cap}(F_2, \omega) = \frac{3}{8}, D_{x_2}^{\cap}(F_2, \omega) = \frac{3}{8}, D_{x_3}^{\cap}(F_2, \omega) = \frac{3}{8}, D_{x_4}^{\cap}(F_2, \omega) = \frac{1}{8}, D_{x_5}^{\cap}(F_2, \omega) = \frac{1}{4}, D_{x_6}^{\cap}(F_2, \omega) = \frac{1}{8}, D_{x_7}^{\cap}(F_2, \omega) = 0, D_{x_8}^{\cap}(F_2, \omega) = \frac{1}{8};$

By Eq.(8) in Definition 8, $D^{\cap}(F_2, \omega) = \frac{9}{64}, D^{\cap}(F_2, \omega) = \frac{15}{64}.$

From the computation, we can see that $\forall x \in U, D_x^{\cap}(F_1, \omega) \wedge D_x^{\cap}(F_2, \omega) = D_x^{\cap}(F_1, \omega)$, by Eq.(14) in Definition 11, $D_x^{\cap}(F_1, \omega) \preceq_2 D_x^{\cap}(F_2, \omega)$ holds obviously, and then $F_1 \unlhd_2 F_2$, we can say that CBMS $F_1$ is finer than $F_2$.

By further computation,

$$\sum_{x \in U} | \cap F_1(x)| = 17, \sum_{x \in U} | \cap F_2(x)| = 23,$$

which verifies the Eq.(18) in Theorem 7.

## 4    Conclusions

Hierarchy is one of the key issues in granular computing. In many practical applications, one often needs to select the corresponding hierarchy to solve problems. In this paper, to research the hierarchies on Covering-based Multigranulation

Spaces, the concept of the union knowledge distance and the intersection knowledge distance are introduced into the Covering-based Multigranulation Spaces, which can be used to construct the knowledge distance lattices, respectively. Based on the proposed knowledge distances, two partial orderings can be derived to characterize the hierarchies on Covering-based Multigranulation Spaces. The example shows the method is feasible and effective. This method provides us a new perspective to the study of covering based granular computing.

In the following works, the reducts of CBMS based on the knowledge distance will be discussed and some more applications on knowledge distance will be explored.

**Acknowledgments.** This work is supported by the Natural Science Foundation of China (Nos. 61100116, 61272419, 61305058), Natural Science Foundation of Jiangsu Province of China (Nos. BK2011492, BK2012700, BK20130471), Qing Lan Project of Jiangsu Province of China, Postdoctoral Science Foundation of China (No. 2014M550293), Key Laboratory of Intelligent Perception and Systems for High-Dimensional Information (Nanjing University of Science and Technology), Ministry of Education (No. 30920130122005), Natural Science Foundation of Jiangsu Higher Education Institutions of China (Nos. 13KJB520003, 13KJD520008), Funding of Jiangsu Innovation Program for Graduate Education (No. SJLX_0495).

# References

1. Chen, Y.H., Yao, Y.Y.: Multiview Intelligent Data Analysis Based on Granular Computing. In: Proceedings of 2006 IEEE International Conference on Granular Computing, pp. 281–286 (2006)
2. Hu, H., Pang, L., Tian, D.P., Shi, Z.Z.: Perception Granular Computing in Visual Haze-free Task. Expert Systems with Applications 41, 2729–2741 (2014)
3. Hobbs, J.R.: Granularity. In: Proceedings of the Ninth International Joint Conference on Artificial Intelligence, pp. 432–435 (1985)
4. Li, H.Y., Yang, S., Liu, H.: Study of Qualitative Data Cluster Model Based on Granular Computing. AASRI Procedia 4, 329–333 (2013)
5. Liang, J.Y., Li, R., Qian, Y.H.: Distance: A More Comprehensible Perspective for Measures in Rough Set Theory. Knowledge-Based Systems 27, 126–136 (2012)
6. Liu, Y., Song, H.Z.: Study on Constructing Support Vector Machine with Granular Computing. Procedia Engineering 15, 3098–3102 (2011)
7. Qian, Y.H., Dang, C.Y., Liang, J.Y.: Partial Ordering of Information Granulations: A Further Investigation. Expert Systems 29, 3–24 (2012)
8. Qian, Y.H., Liang, J.Y.: Rough Set Method Based on Multigranulations. In: Proceedings of the 5th International Conference on Cognitive Informatics, pp. 297–304 (2006)
9. Qian, Y.H., Liang, J.Y., Dang, C.Y.: Knowledge Structure, Knowledge Granulation and Knowledge Distance in a Knowledge Base. International Journal of Approximate Reasoning 50, 174–188 (2009)
10. Qian, Y.H., Zhang, H., Li, F.J., Hu, Q.H., et al.: Set-based Granular Computing: A Lattice Model. International Journal of Approximation Reasoning 55, 834–852 (2014)

11. Saberi, M., Mirtalaie, M.S., Hussain, F.K., Azadeh, A., et al.: A Granular Computing-based Approach to Credit Scoring Modeling. Neurocomputing 122, 100–115 (2013)
12. Wang, L.J., Yang, X.B., Yang, J.Y., Wu, C.: Relationships among Generalized Rough Sets in Six Coverings and Pure Reflexive Neighborhood System. Information Sciences 207, 66–78 (2012)
13. Wang, S.P., Zhu, W., Zhu, Q.X., Min, F.: Four Matroidal Structures of Covering and Their Relationships with Rough Sets. International Journal of Approximate Reasoning 54, 1361–1372 (2013)
14. Wang, S.P., Zhu, W., Zhu, Q.X., Min, F.: Characteristic Matrix of Covering and its Application to Boolean Matrix Decomposition. Information Sciences 263, 186–197 (2014)
15. Wu, W.Z., Leung, Y., Mi, J.S.: Granular Computing and Knowledge Reduction in Formal Contexts. IEEE Transactions on Knowledge and Data Engineering 21, 1461–1474 (2009)
16. Yang, X.B., Qian, Y.H., Yang, J.Y.: Hierarchical Structures on Multigranulation Spaces. Journal of Computer Science and Technology 27, 1169–1183 (2012)
17. Yang, X.B., Qian, Y.H., Yang, J.Y.: On Characterizing Hierarchies of Granulation Structures via Distances. Fundamenta Informaticae 123, 365–380 (2013)
18. Yao, J.T.: A Ten-year Review of Granular Computing. In: Proceedings of 2007 IEEE International Conference on Granular Computing, pp. 734–739 (2007)
19. Yao, J.T., Vasilakos, A.V., Pedrycz, W.: Granular Computing: Perspectives and Challenges. IEEE Transactions on Cybernetics 43, 1977–1989 (2013)
20. Yao, Y.Y.: Perspectives of Granular Computing. In: Proceedings of 2005 IEEE International Conference on Granular Computing, pp. 85–90 (2005)
21. Yao, Y.Y., Zhang, N., Miao, D.Q., Xu, F.F.: Set-theoretic Approaches to Granular Computing. Fundamenta Informaticae 115, 247–264 (2012)
22. Zadeh, L.A.: Fuzzy Sets and Information Granulation. North-Holland, Amsterdam (1979)
23. Zhang, X.Y., Miao, D.Q.: Quantitative Information Architecture, Granular Computing and Rough Set Models in the Double-quantitative Approximation Space of Precision and Grade. Information Sciences 268, 147–168 (2014)
24. Zhu, W.: Topological Approaches to Covering Rough Sets. Information Sciences 177, 1499–1508 (2007)
25. Zhu, W., Wang, F.Y.: The Fourth Type of Covering-based Rough Sets. Information Sciences 201, 80–92 (2012)

# Uncertainty Measures in Interval-Valued Information Systems

Nan Zhang[1] and Zehua Zhang[2]

[1] School of Computer and Control Engineering, Yantai University,
Yantai, Shandong, 264005, China
zhangnan0851@163.com
[2] College of Computer Science and Technology, Taiyuan University of Technology,
Taiyuan, Shanxi, 030024, China
zehua.zzh@gmail.com

**Abstract.** Rough set theory is a new mathematical tool to deal with vagueness and uncertainty in artificial intelligence. Approximation accuracy, knowledge granularity and entropy theory are three main approaches to uncertainty research in classical Pawlak information system, which have been widely applied in many practical issues. Based on uncertainty measures in Pawlak information systems, we propose rough degree, knowledge discernibility and rough entropy in interval-valued information systems, and investigate some important properties of them. Finally, the relationships between knowledge granulation, knowledge discerniblity and rough degree have been also discussed.

**Keywords:** Upper and lower approximations, rough sets, uncertainty measures.

## 1 Introduction

Rough set theory, proposed by Pawlak in 1982 [1], can describe knowledge through set-theoretic approach based on equivalence relations for the universe of discourse. It provides a theoretical foundation for inference reasoning about data analysis and has extensive applications in areas of intelligent computing, pattern recognition and artificial intelligence [2-13].

In many applications, classical equivalence relation in rough set theory is restrictive for many issues. To address these problems, some extended binary relations in information systems have been researched in recent years. By taking the incomplete information systems as set-valued information systems, Kryszkiewicz [14] introduced knowledge reduction and rule acquisition in incomplete information systems. Leung et al. [15] introduced method of knowledge reduction based on the $\alpha$-tolerance relation in interval-valued information systems. Aiming at the problems in knowledge reduction in interval-valued information systems, Zhang et al. [16,17] investigated the knowledge reduction based on maximal consistent blocks in both interval-valued information and decision systems.

In the fields of uncertainty measures in generalized rough set theory, Huang et al. [12] proposed rough entropy based on the generalized binary relation. Liang et al.

D. Miao et al. (Eds.): RSKT 2014, LNAI 8818, pp. 479–488, 2014.
DOI: 10.1007/978-3-319-11740-9_44 © Springer International Publishing Switzerland 2014

[18] investigated the knowledge granulation, information entropy and rough entropy by the tolerance relation in incomplete information systems. Considering uncertainty measures in incomplete information systems, Xu et al. [19] introduced the knowledge granulation, knowledge entropy and knowledge uncertainty measure based on the dominance relation in ordered information systems. Dai et al. [20] proposed an extended conditional entropy in interval-valued decision systems. In ref. [21], $\theta$-accuracy and $\theta$-roughness are given in interval-valued information systems, which are generalizations of the concepts accuracy and roughness for the equivalence relation based rough set model. The uncertainty measure, called the $\theta$-rough degree, is also proposed in the paper.

The rest of this paper is organized as follows. Some preliminary concepts such as interval-valued information systems, rough approximations and relative properties are briefly recalled in Section 2. Sections 3 introduce rough degree in an interval-valued information system. In Section 4, the knowledge discernibility based on knowledge granulation is presented in an interval-valued information system. Knowledge rough entropy is proposed in Section 5. The relationships among proposed measures are discussed in Section 6. Finally, the paper is summarized in Section 7.

## 2　Rough Approximation in Interval-Valued Information Systems

In this section, we will briefly recall some basic conceptions [16,17] about interval-valued information systems and rough approximations.

*An interval-valued   information system* is defined by $\zeta = (U, AT, V, f)$, where

- $U = \{u_1, u_2, ..., u_n\}$ is a non-empty finite set called the universe of discourse;

- $AT = \{a_1, a_2, ..., a_m\}$ is a non-empty finite set of $m$ attributes, such that:

  $a_k(u_i) = [l_i^k, u_i^k]$, $l_i^k \leq u_i^k$, for all $i = 1, 2, ..., n$ and $k = 1, 2, ..., m$;

- $V = \bigcup_{a_k \in AT} V_{a_k}$, $V_{a_k}$ is a domain of attribute $a_k$;

- $f : U \times AT \to V$ is called the information function such that $f(u_i, a_k) \in V_{a_k}$.

Let $\zeta = (U, AT, V, f)$ be an interval-valued information system. For a given similarity rate $\alpha \in [0.1]$, and $A \subseteq AT$. The $\alpha$-*tolerance relation* $T_A^\alpha$ is expressed as

$$T_A^\alpha = \{(u_i, u_j) \in U \times U : \alpha_{ij}^k \geq \alpha, \forall a_k \in A\},$$

where, $\alpha_{ij}^k$ is the similarity degree of different interval numbers under the same attribute $a_k \in AT$.

By taking of $T_A^\alpha$, we can defined $\alpha$ - *tolerance class* as following:

$$S_A^\alpha(u_i) = \{u_j \in U : (u_i, u_j) \in T_A^\alpha\}.$$

To get a maximal set in which the objects can satisfy with each other, $\alpha$ - maximal consistent block is proposed in interval-valued information systems. By taking the $\alpha$-maximal consistent block $M_A^\alpha(u_i)$ as the basic "granule" in universe, we can obtain the finer upper and lower approximations in interval-valued information systems as following:

$$\overline{apr}_A^\alpha(X) = \{u_i \in U : M_A^\alpha(u_i) \cap X \neq \varnothing\}$$
$$= \cup\{M_A^\alpha(u_i) : M_A^\alpha(u_i) \cap X \neq \varnothing\},$$
$$\underline{apr}_A^\alpha(X) = \{u_i \in U : M_A^\alpha(u_i) \subseteq X\}$$
$$= \cup\{M_A^\alpha(u_i) : M_A^\alpha(u_i) \subseteq X\}.$$

The following properties related to rough approximations in $\zeta = (U, AT, V, f)$ are presented as:

(1)  $\underline{apr}_A^\alpha(X) \subseteq X \subseteq \overline{apr}_A^\alpha(X)$;

(2)  $\underline{apr}_A^\alpha(U) = \overline{apr}_A^\alpha(U) = U$;

(3)  $\underline{apr}_A^\alpha(\varnothing) = \overline{apr}_A^\alpha(\varnothing) = \varnothing$;

(4)  $\overline{apr}_A^\alpha(\overline{apr}_A^\alpha(X)) = \underline{apr}_A^\alpha(\overline{apr}_A^\alpha(X))$,

   $\underline{apr}_A^\alpha(\underline{apr}_A^\alpha(X)) = \overline{apr}_A^\alpha(\underline{apr}_A^\alpha(X))$;

(5)  $\overline{apr}_A^\alpha(X) = \sim \underline{apr}_A^\alpha(\sim X)$,  $\underline{apr}_A^\alpha(X) = \sim \overline{apr}_A^\alpha(\sim X)$;

(6)  $\underline{apr}_A^\alpha(X) \subseteq \underline{apr}_A^\alpha(Y)$,  $\overline{apr}_A^\alpha(X) \subseteq \overline{apr}_A^\alpha(Y)$;

(7)  $\underline{apr}_A^\alpha(X \cap Y) = \underline{apr}_A^\alpha(X) \cap \underline{apr}_A^\alpha(Y)$,

   $\underline{apr}_A^\alpha(X \cup Y) \supseteq \underline{apr}_A^\alpha(X) \cup \underline{apr}_A^\alpha(Y)$;

(8)  $\overline{apr}_A^\alpha(X \cup Y) = \overline{apr}_A^\alpha(X) \cup \overline{apr}_A^\alpha(Y)$,

   $\overline{apr}_A^\alpha(X \cap Y) \subseteq \overline{apr}_A^\alpha(X) \cap \overline{apr}_A^\alpha(Y)$.

## 3    Rough Degree in Interval-Valued Information Systems

Let $\zeta = (U, AT, V, f)$ be an interval-valued information system, for any $X \subseteq U$, $A \subseteq AT$, $\alpha \in [0,1]$, *approximation accuracy* is defined as:

$$\mu_A^\alpha(X) = \frac{|\underline{apr}_A^\alpha(X)|}{|\overline{apr}_A^\alpha(X)|}$$
$$= \frac{|\underline{apr}_A^\alpha(X)|}{|U| - |\underline{apr}_A^\alpha(\sim X)|}.$$

$\overline{apr}_A^{\,\alpha}(X) = \{u_i \in U : M_A^\alpha(u_i) \cap X \neq \varnothing\}$ is the upper approximation operator of set $X$,

$\underline{apr}_A^{\,\alpha}(X) = \{u_i \in U : M_A^\alpha(u_i) \subseteq X\}$ is the lower approximation operator of set $X$, and $|S|$ stands for the number of elements in subset $S \subseteq U$.

Based on the approximation accuracy in interval-valued information systems, we give the concept of rough degree.

*Rough degree* in interval-valued information systems is defined as following:

$$\rho_A^\alpha(X) = 1 - \mu_A^\alpha(X)$$

$$= 1 - \frac{|\underline{apr}_A^{\,\alpha}(X)|}{|\overline{apr}_A^{\,\alpha}(X)|}$$

$$= \frac{|bnr_A^\alpha(X)|}{|\overline{apr}_A^{\,\alpha}(X)|},$$

where $bnr_A^\alpha(X) = \overline{apr}_A^{\,\alpha}(X) - \underline{apr}_A^{\,\alpha}(X)$. Thus, we can get the following properties:

(1)    The bigger $\rho_A^\alpha(X)$ is, the more uncertainty is;

(2)    If $\rho_A^\alpha(X) = 0$, i.e., $\overline{apr}_A^{\,\alpha}(X) = \underline{apr}_A^{\,\alpha}(X)$, the set $X$ is an accurate set, and the rough degree is 0; if $0 < \rho_A^\alpha(X) < 1$, i.e., $\overline{apr}_A^{\,\alpha}(X) \neq \underline{apr}_A^{\,\alpha}(X)$, the rough degree $\in (0,$ $1)$; if $\rho_A^\alpha(X) = 1$, i.e., $\underline{apr}_A^{\,\alpha}(X) = 0$, the set $X$ is completely a rough set, and the rough degree is 1.

**Property 1.** *Let* $\zeta = (U, AT, V, f)$ *be an interval-valued information system, for* $0 \leq \alpha < \beta \leq 1$, $A \subset B \subseteq AT$, *we have*

(1)    $\mu_A^\alpha(X) < \mu_A^\beta(X)$, $\rho_A^\alpha(X) > \rho_A^\beta(X)$;

(2)    $\mu_A^\alpha(X) < \mu_B^\alpha(X)$, $\rho_A^\alpha(X) > \rho_B^\alpha(X)$;

(3)    $\mu_A^\alpha(X) < \mu_B^\beta(X)$, $\rho_A^\alpha(X) > \rho_B^\beta(X)$.

*Proof*:

(1)    In a given $\zeta = (U, AT, V, f)$, if $0 \leq \alpha < \beta \leq 1$, for any $A \subset AT$, then we can get the finer description for set $X \subseteq AT$, $|\underline{apr}_A^{\,\alpha}(X)| < |\underline{apr}_A^{\,\beta}(X)|$ and

$\overline{apr}_A^{\,\alpha}(X) > \overline{apr}_A^{\,\beta}(X)$, thus, $\dfrac{|\underline{apr}_A^{\,\alpha}(X)|}{|\overline{apr}_A^{\,\alpha}(X)|} < \dfrac{|\underline{apr}_A^{\,\beta}(X)|}{|\overline{apr}_A^{\,\beta}(X)|}$, namely, $\mu_A^\alpha(X) < \mu_A^\beta(X)$. For

$\rho_A^\alpha(X) = 1 - \mu_A^\alpha(X)$, we have $\rho_A^\alpha(X) > \rho_A^\beta(X)$.

(2)    In a given $\zeta = (U, AT, V, f)$, if $0 \leq \alpha \leq 1$, for any $A \subset B \subset AT$, then we can get the finer description for set $X \subseteq AT$, $|\underline{apr}_A^{\,\alpha}(X)| < |\underline{apr}_B^{\,\alpha}(X)|$ and

$\overline{apr}_A^\alpha(X) > \overline{apr}_B^\alpha(X)$ , thus, $\dfrac{|\ apr_A^\alpha(X)\ |}{|\ \overline{apr}_A^\alpha(X)\ |} < \dfrac{|\ apr_B^\alpha(X)\ |}{|\ \overline{apr}_B^\alpha(X)\ |}$ , namely, $\mu_A^\alpha(X) < \mu_B^\alpha(X)$ . For

$\rho_A^\alpha(X) = 1 - \mu_A^\alpha(X)$ , we have $\rho_A^\alpha(X) > \rho_B^\alpha(X)$ .

(3)   From (1) and (2), we can get $\mu_B^\alpha(X) < \mu_B^\beta(X)$ and $\mu_A^\alpha(X) < \mu_B^\alpha(X)$ respectively. Therefore, $\mu_A^\alpha(X) < \mu_B^\beta(X)$ . For $\rho_A^\alpha(X) = 1 - \mu_A^\alpha(X)$ , we have $\rho_A^\alpha(X) > \rho_B^\beta(X)$ .

# 4    Knowledge Discernibility in Interval-Valued Information Systems

**Definition 1.** *Let* $\zeta = (U, AT, V, f)$ *be an interval-valued information system, for any* $A \subseteq AT$ , *knowledge discernibility related to* $A$ *in* $\zeta$ *is given as following:*

$$DSI^\alpha(A) = \sum_{i=1}^{|U|} \frac{1}{|U|}(1 - \frac{|S_A(u_i)|}{|U|}),$$

*where* $|S_A^\alpha(u_i)| = |\bigcup M|$ , $M \in \xi_A^\alpha(u_i)$ *and* $\xi^\alpha(A) = \{M_A^\alpha(u_1), M_A^\alpha(u_2), ..., M_A^\alpha(u_n)\}$ .

**Theorem 1 (Minimum).** *Let* $\zeta = (U, AT, V, f)$ *be an interval-valued information system, and* $T_A^\alpha$ *be an* $\alpha$ *-tolerance relation. The minimum of knowledge discernibility in an interval-valued information system* $\zeta$ *is 0. This value is achieved if and only if* $T_A^\alpha = \widehat{T}_A^\alpha$ , *where* $\widehat{T}_A^\alpha$ *is an universe tolerance relation, i.e.,*

$$U / \widehat{T}_A^\alpha = \{M_A^\alpha(u_i) = U : u_i \in U\}$$
$$= \{U, U, ..., U\}.$$

*Proof.*

If $T_A^\alpha = \widehat{T}_A^\alpha$ , then

$$DSI^\alpha(A) = \sum_{i=1}^{|U|} \frac{1}{|U|}(1 - \frac{|S_A(u_i)|}{|U|})$$
$$= \sum_{i=1}^{|U|} \frac{1}{|U|}(1 - \frac{|U|}{|U|})$$
$$= 0.$$

**Theorem 2 (Maximum).** *Let* $\zeta = (U, AT, V, f)$ *be an interval-valued information system, and* $T_A^\alpha$ *be an* $\alpha$ *-tolerance relation. The maximum of knowledge discernibility in an interval-valued information system* $\zeta$ *is* $1 - 1/|U|$ . *This value can be obtained if and only if* $T_A^\alpha = \breve{T}_A^\alpha$ , *where* $\breve{T}_A^\alpha$ *is an unit tolerance relation, i.e.,*

$$U \,/\, \breve{T}_A^\alpha = \{M_A^\alpha(u_i) = \{u_i\} : u_i \in U\}$$
$$= \{\{u_1\}, \{u_2\}, ..., \{u_n\}\} \,.$$

*Proof.*

If $T_A^\alpha = \breve{T}_A^\alpha$, then

$$DSI^\alpha(A) = \sum_{i=1}^{|U|} \frac{1}{|U|}(1 - \frac{|S_A(u_i)|}{|U|})$$
$$= \sum_{i=1}^{|U|} \frac{1}{|U|}(1 - \frac{1}{|U|})$$
$$= 1 - \frac{1}{|U|} \,.$$

**Property 2 (Boundedness).** *Let* $\zeta = (U, AT, V, f)$ *be an interval-valued information system, and* $T_A^\alpha$ *be an* $\alpha$-*tolerance relation. The knowledge discernibility in an interval-valued information system* $\zeta$ *exists the boundedness, namely,*

$$0 \le DSI^\alpha(A) \le 1 - \frac{1}{|U|} \,,$$

*where* $DSI^\alpha(A) = 1 - 1/|U|$ *if and only if* $T_A^\alpha = \breve{T}_A^\alpha$, *and* $DSI^\alpha(A) = 0$ *if and only if* $T_A^\alpha = \hat{T}_A^\alpha$.

The concept of knowledge discernibility can describe discernible ability of knowledge in interval-valued information systems intuitively. The smaller $DSI^\alpha(A)$ is, the fewer knowledge discernibility is.

**Property 3.** *Let* $\zeta = (U, AT, V, f)$ *be an interval-valued information system, for* $0 \le \alpha < \beta \le 1$, $A \subset B \subset AT$, *we have*
(1)    $DSI^\alpha(A) < DSI^\beta(A)$;
(2)    $DSI^\alpha(A) < DSI^\alpha(B)$;
(3)    $DSI^\alpha(A) < DSI^\beta(B)$;

*Proof:*
(1)    In a given IvIS $\zeta = (U, AT, V, f)$, if $0 \le \alpha < \beta \le 1$, for any $A \subset AT$, we can get

$$|S_A^\alpha(u_i)| > |S_A^\beta(u_i)| \quad , \qquad \text{then} \qquad \sum_{i=1}^{|U|} |S_A^\alpha(u_i)| > \sum_{i=1}^{|U|} |S_A^\beta(u_i)| \quad , \qquad \text{therefore}$$

$$\frac{|U|}{|U|} - \frac{1}{|U|^2} \sum_{i=1}^{|U|} |S_A^\alpha(u_i)| < \frac{|U|}{|U|} - \frac{1}{|U|^2} \sum_{i=1}^{|U|} |S_A^\beta(u_i)|, \quad \text{namely,} \quad DSI^\alpha(A) < DSI^\beta(A);$$

(2)  In a given IvIS  $\zeta = (U, AT, V, f)$ , if  $0 \le \alpha \le 1$ , for any  $A \subset B \subset AT$ , we can get

$$| S_A^\alpha(u_i) | > | S_B^\alpha(u_i) | \quad , \quad \text{then} \quad \sum_{i=1}^{|U|} | S_A^\alpha(u_i) | > \sum_{i=1}^{|U|} | S_B^\alpha(u_i) | \quad , \quad \text{therefore}$$

$$\frac{|U|}{|U|} - \frac{1}{|U|^2} \sum_{i=1}^{|U|} | S_A^\alpha(u_i) | < \frac{|U|}{|U|} - \frac{1}{|U|^2} \sum_{i=1}^{|U|} | S_A^\beta(u_i) | , \quad \text{namely,} \quad DSI^\alpha(A) < DSI^\alpha(B);$$

(3)     From   (1)   and   (2),   we   can   get   $DSI^\alpha(B) < DSI^\beta(B)$   and
$DSI^\alpha(A) < DSI^\alpha(B)$  respectively. Therefore,  $DSI^\alpha(A) < DSI^\beta(B)$ .

# 5    Knowledge Rough Entropy in Interval-valued Information Systems

**Definition 2.** *Let*  $\zeta = (U, AT, V, f)$  *be an interval-valued information system,*  $T_A^\alpha$  *be an*  $\alpha$ *-tolerance relation,*  $U / T_A^\alpha = \{S_A^\alpha(u_i) : u_i \in U\}$ . *Knowledge rough entropy in interval-valued information systems, which is denoted by*  $E^\alpha(A)$ , *is defined by:*

$$E^\alpha(A) = \sum_{i=1}^{|U|} \frac{1}{|U|} \cdot \log_2 | S_A^\alpha(u_i) | .$$

**Theorem 3 (Minimum).** *Let*  $\zeta = (U, AT, V, f)$  *be an interval-valued information system, and*  $T_A^\alpha$  *be an*  $\alpha$ *-tolerance relation. The minimum of knowledge rough entropy in an interval-valued information system*  $\zeta$  *is 0. This value is achieved if and only if*  $T_A^\alpha = \breve{T}_A^\alpha$ , *where*  $\breve{T}_A^\alpha$  *is an unit tolerance relation i.e.,*

$$U / \breve{T}_A^\alpha = \{M_A^\alpha(u_i) = \{u_i\} : u_i \in U\}$$
$$= \{\{u_1\}, \{u_2\}, ..., \{u_n\}\} .$$

*Proof.*

If  $T_A^\alpha = \breve{T}_A^\alpha$ , then

$$E^\alpha(A) = \sum_{i=1}^{|U|} \frac{1}{|U|} \cdot \log_2 | S_A^\alpha(u_i) |$$
$$= \sum_{i=1}^{|U|} \frac{1}{|U|} \cdot \log_2 1$$
$$= 0 .$$

**Theorem 4 (Maximum).** *Let*  $\zeta = (U, AT, V, f)$  *be an interval-valued information system, and*  $T_A^\alpha$  *be an*  $\alpha$ *-tolerance relation. The maximum of knowledge rough* *entropy in an interval-valued information system* $\zeta$ *is* $\log_2 |U|$. *This value can be obtained if and only if* $T_A^\alpha = \hat{T}_A^\alpha$, *where* $\hat{T}_A^\alpha$ *is an universe tolerance relation, , i.e.,*

$$U / \hat{T}_A^\alpha = \{M_A^\alpha(u_i) = U : u_i \in U\}$$
$$= \{U, U, ..., U\}.$$

*Proof.*

If $T_A^\alpha = \hat{T}_A^\alpha$, then

$$E^\alpha(A) = \sum_{i=1}^{|U|} \frac{1}{|U|} \cdot \log_2 |S_A^\alpha(u_i)|$$

$$= \sum_{i=1}^{|U|} \frac{1}{|U|} \cdot \log_2 |U|$$

$$= \log_2 |U|.$$

**Property 4 (Boundedness).** *Let* $\zeta = (U, AT, V, f)$ *be an interval-valued information system, and* $T_A^\alpha$ *be an* $\alpha$-*tolerance relation. The knowledge rough entropy in an interval-valued information system* $\zeta$ *exists the boundedness, namely,*

$$0 \le E^\alpha(A) \le \log_2 |U|,$$

*where* $E^\alpha(A) = 0$ *if and only if* $T_A^\alpha = \check{T}_A^\alpha$, *and* $E^\alpha(A) = \log_2 |U|$ *if and only if* $T_A^\alpha = \hat{T}_A^\alpha$.

**Property 5.** *Let* $\zeta = (U, AT, V, f)$ *be an interval-valued information system, for* $0 \le \alpha < \beta \le 1$, $A \subset B \subset AT$, *we have*

(1)    $E^\alpha(A) > E^\beta(A)$;

(2)    $E^\alpha(A) > E^\alpha(B)$;

(3)    $E^\alpha(A) > E^\beta(B)$;

*Proof:*

(1)    In a given IvIS $\zeta = (U, AT, V, f)$, if $0 \le \alpha < \beta \le 1$, for any $A \subset AT$, then we can get $|S_A^\alpha(u_i)| \rhd |S_A^\beta(u_i)|$, thus, $\sum_{i=1}^{|U|} \frac{1}{|U|} \cdot \log_2 |S_A^\alpha(u_i)| \rhd \sum_{i=1}^{|U|} \frac{1}{|U|} \cdot \log_2 |S_A^\beta(u_i)|$, namely, $E^\alpha(A) > E^\beta(A)$.

(2)    In a given IvIS $\zeta = (U, AT, V, f)$, if $0 \le \alpha \le 1$, for any $A \subset B \subset AT$, then we can get $|S_A^\alpha(u_i)| \rhd |S_B^\alpha(u_i)|$, thus, $\sum_{i=1}^{|U|} \frac{1}{|U|} \cdot \log_2 |S_A^\alpha(u_i)| \rhd \sum_{i=1}^{|U|} \frac{1}{|U|} \cdot \log_2 |S_B^\alpha(u_i)|$, namely, $E^\alpha(A) > E^\alpha(B)$.

(3)   From (1) and (2), we can get $E^\alpha(A) > E^\beta(A)$ and $E^\beta(A) > E^\beta(B)$ respectively. Therefore, $E^\alpha(A) > E^\beta(B)$.

## 6    Relationships among the Three Uncertainty Measures

We have proposed three concepts (rough degree, knowledge discernibility and rough entropy) for uncertainty measures in interval-valued information systems. We give the relations among them as following:

**Theorem 5.** *Let* $\zeta = (U, AT, V, f)$ *be an interval-valued information system, and* $T_A^\alpha$ *be an* $\alpha$*-tolerance relation. Then,*

$$\rho_A^\alpha(X) + \mu_A^\alpha(X) = 1.$$

Considering knowledge granulation $GDI^\alpha(A) = \sum_{i=1}^{|U|} \dfrac{|S_A^\alpha(u_i)|}{|U| \times |U|}$ in interval-valued information systems, we can have:

**Theorem 6.** *Let* $\zeta = (U, AT, V, f)$ *be an interval-valued information system, and* $T_A^\alpha$ *be an* $\alpha$*-tolerance relation. Then,*

$$DSI^\alpha(A) + GDI^\alpha(A) = 1.$$

## 7    Conclusions

Based on uncertainty measures in Pawlak information systems, we proposed the rough degree, knowledge discernibility and rough entropy in interval-valued information systems, and also investigate some important properties of them. Finally, the relationships between knowledge granulation, knowledge discerniblity, approximation accuracy and rough degree have been also discussed in the paper.

**Acknowledgements.** This work was partially supported by the National Natural Science Foundation of China (No. 61170224, 61305052), the Natural Science Foundation of Shandong Province (No. ZR2013FQ020), the Science and Technology Development Plan of Shandong Province (No. 2012GGB01017), the Doctor Research Foundation of Yantai University (No. JS12B28).

## References

1. Pawlak, Z.: Rough sets. International Journal of Computer and Information Sciences 11, 341–356 (1982)
2. Pawlak, Z.: Rough Sets-Theoretical Aspects of Reasoning About Data. Kluwer Academic Publishers, Boston (1991)

3. Miao, D.Q., Zhao, Y., Yao, Y.Y., et al.: Relative reducts in consistent and inconsistent decision tables of the Pawlak rough set model. Information Sciences 24, 4140–4150 (2009)
4. Yao, Y.Y., Zhang, N., Miao, D.Q., Xu, F.F.: Set-theoretic approaches to granular computing. Fundamenta Informaticae 115, 247–264 (2012)
5. Liu, C.H., Miao, D.Q., Zhang, N.: Graded rough set model based on two universes and its properties. Knowledge-Based Systems 33, 65–72 (2012)
6. Qian, J., Miao, D.Q., Zhang, Z.H.: Knowledge reduction algorithms in cloud computing. Chinese Journal of Computers 12, 2332–2343 (2011)
7. Yao, Y.Y., Wong, S.K.M.: A decision theoretic framework for approximating concepts. International Journal of Man-Machine Studies 6, 793–809 (1992)
8. Yao, Y.Y., Zhao, Y.: Attribute reduction in decision-theoretic rough set models. Information Sciences 17, 3356–3373 (2008)
9. Wang, G.Y.: Rough reduction in algebra view and information view. International Journal of Intelligent Systems 6, 679–688 (2003)
10. Qian, Y.H., Liang, J.Y., Dang, C.Y.: Interval ordered information systems. Computer and Mathematics with Applications 8, 1994–2009 (2008)
11. Qian, Y.H., Liang, J.Y., Wang, F.: A new method for measuring the uncertainty in incomplete information systems. International Journal of Uncertainty, Fuzziness and Knowledge-Based Systems 6, 855–880 (2009)
12. Huang, B., Zhou, X.Z., Shi, Y.C.: Entropy of knowledge and rough set based on general binary relation. Journal of Systems Engineering: theory and Practice 24, 93–96 (2004)
13. Liang, J.Y., Shi, Z.Z.: The information entropy, rough entropy and knowledge granulation in rough set theory. International Journal of General Systems 1, 37–46 (2004)
14. Kryszkiewicz, M.: Comparative study of alternative types of knowledge reduction in inconsistent systems. International Journal of Intelligent Systems 1, 105–120 (2001)
15. Leung, Y., Fischer, M., Wu, W.Z., Mi, J.S.: A rough set approach for the discovery of classification rules in interval-valued information systems. International Journal of Approximate Reasoning 2, 233–246 (2008)
16. Zhang, N., Miao, D.Q., Yue, X.D.: Knowledge reduction in interval-valued information systems. Chinese Journal of Computer Research and Development 47, 1362–1371 (2010)
17. Zhang, N.: Research on Interval-valued Information Systems and Knowledge Spaces: A Granular Approach. PhD Thesis, Tongji University, Shanghai, China (2012)
18. Liang, J.Y., Shi, Z.Z., Li, D.Y., Wireman, M.J.: The information entropy, rough entropy and knowledge granulation in incomplete information systems. International Journal of General Systems 1, 641–654 (2006)
19. Xu, W.H., Zhang, X.Y., Zhang, W.X.: Knowledge granulation, knowledge entropy and knowledge uncertainty measure in ordered information systems. Applied Soft Computing 9, 1244–1251 (2009)
20. Dai, J.H., Wang, W.T., Xu, Q., Tian, H.W.: Uncertainty measurement for interval-valued decision systems based on extended conditional entropy. Knowledge Based Systems 27, 443–450 (2012)
21. Dai, J.H., Wang, W.T., Mi, J.S.: Uncertainty measurement for interval-valued information systems. Information Sciences 251, 63–78 (2013)

# A New Type of Covering-Based Rough Sets

Bin Yang and William Zhu

Lab of Granular Computing,
Minnan Normal University, Zhangzhou 363000, China
williamfengzhu@gmail.com

**Abstract.** As a technique for granular computing, rough sets deal with the vagueness and granularity in information systems. Covering-based rough sets are natural extensions of the classical rough sets by relaxing the partitions to coverings and have been applied for many fields. In this paper, a new type of covering-based rough sets are proposed and the properties of this new type of covering-based rough sets are studied. First, we introduce a concept of inclusion degree into covering-based rough set theory to explore some properties of the new type of covering approximation space. Second, a new type of covering-based rough sets is established based on inclusion degree. Moreover, some properties of the new type of covering-based rough sets are studied. Finally, a simple application of the new type of covering-based rough sets to network security is given.

**Keywords:** Granular computing, covering, rough set, neighborhood, inclusion degree, network security.

## 1 Introduction

Rough set theory, proposed by Pawlak [11,12] in 1982, is a useful tool to deal with the vagueness and granularity in information systems. Pawlak rough set theory is built on equivalence relations. However, an equivalence relation imposes restrictions and limitations on many applications [2,5,6,9,18,22]. Thus, one of the main directions of research in rough set theory is naturally the generalization of rough set approximations. Meanwhile, many extensions have been made by replacing equivalence relations with notions such as arbitrary binary relations [26,25], fuzzy relations [4,8,10], and coverings [27,28,29] of the universal sets. For instance, Zakowski [24] established covering-based rough set theory by exploiting coverings of universal sets. The study on covering-based rough sets is very necessary and important. Particularly, in recent years, with the fast development of science and technology, how to use the effective mathematical tools to address practical problems has become more and more essential.

In this paper, we define a new concept, namely, inclusion degree and then propose a new type of covering-based rough sets based on inclusion degree. First, the definition of inclusion degree is proposed and its some properties are explored. Moreover, through inclusion degree, some problems of covering-based rough sets can be addressed. Second, we propose a new type of covering-based rough sets, and then study some properties of this type of covering-based rough sets. Finally, the applications of this new type of covering-based rough sets are given.

D. Miao et al. (Eds.): RSKT 2014, LNAI 8818, pp. 489–499, 2014.
DOI: 10.1007/978-3-319-11740-9_45 © Springer International Publishing Switzerland 2014

The remainder of this paper is organized as follows: In Section 2, some basic concepts and properties of covering-based rough sets are introduced. In Section 3, A new type of covering-based rough sets are proposed by means of inclusion degree. Moreover, some properties of the new type of covering-based rough sets are explored. In Section 4, some applications of this new type of covering-based rough sets are given. Section 5 concludes this paper.

## 2    Basic Definitions

The concepts of partition and covering are the basis of classical rough sets and covering-based rough sets, respectively. And covering is the basis of the concept of neighborhood as well. So we introduce the two concepts at first.

**Definition 1.** *(Partition [14]) Let U be a universe of discourse and* $\mathbf{P}$ *be a family of subsets of U.* $\mathbf{P}$ *is called a partition of U if the following conditions hold: (1)* $\emptyset \notin \mathbf{P}$*; (2)* $\cup \mathbf{P} = U$*; (3) for any* $K, L \in \mathbf{P}$*,* $K \cap L = \emptyset$*. Every element of* $\mathbf{P}$ *is called a partition block.*

In the following discussion, unless stated to the contrary, the universe of discourse U is considered to be finite and nonempty.

**Definition 2.** *(Covering [14]) Let U be a universe of discourse and* $\mathbf{C}$ *be a family of subsets of U. If* $\emptyset \notin \mathbf{C}$ *and* $\cup \mathbf{C} = U$*,* $\mathbf{C}$ *is called a covering of U. Every element of* $\mathbf{C}$ *is called a covering block.*

It is clear that a partition of $U$ is certainly a covering of $U$. So the concept of covering is an extension of the concept of partition. Neighborhood [1,13,23] is a concept used widely in covering based rough sets. It is defined as follows.

**Definition 3.** *(Neighborhood [1]) Let* $\mathbf{C}$ *be a covering of U. For any* $x \in U, \cap \{K \in \mathbf{C} : x \in K\}$ *is denoted as* $N_{\mathbf{C}}(x)$ *and called the neighborhood of x.*

It is clear that $x \in N_{\mathbf{C}}(x)$. The following proposition gives an important property of neighborhoods.

**Theorem 1.** *([15]) Let* $\mathbf{C}$ *be a covering of U. For any* $x, y \in U$*, if* $y \in N_{\mathbf{C}}(x)$*, then* $N_{\mathbf{C}}(y) \subseteq N_{\mathbf{C}}(x)$*.*

If $y \in N_{\mathbf{C}}(x)$ and $x \in N_{\mathbf{C}}(y)$, by the above proposition, we have $N_{\mathbf{C}}(x) = N_{\mathbf{C}}(y)$. All the neighborhoods induced by a covering of a universe form a set family. This set family is still a covering of the universe. This type of set families have been studied by many scholars [13,15]. However, both the term and the mark of it are not identical. In this paper, we call it covering of neighborhoods and cite the mark proposed by Wang et al. [15].

After the concept of neighborhood has been given, we can introduce the concept of neighborhoods.

**Definition 4.** *(Covering of neighborhoods [15]) Let* $\mathbf{C}$ *be a covering of U.* $\{N_{\mathbf{C}}(x) : x \in U\}$ *is denoted as* $Cov(\mathbf{C})$ *and called the covering of neighborhoods induced by* $\mathbf{C}$*.*

There is an important property of neighborhoods presented by the following proposition.

**Theorem 2.** *([15]) For any $N_C(x) \in Cov(\mathbf{C})$, $N_C(x)$ is not a union of other blocks in $Cov(\mathbf{C})$.*

By the definition of $Cov(\mathbf{C})$, we see that $Cov(\mathbf{C})$ is also a covering of universe $U$. In particular, if $\mathbf{C}$ is a partition, we have that $Cov(\mathbf{C}) = \mathbf{C}$.

## 3 A New Type of Covering-Based Rough Sets

In this section, we introduce an inclusion degree into covering-based rough sets. Based on this, a new type of covering-based rough sets are proposed and the properties are studied. Firstly, we propose a new concept, namely, inclusion degree and then study its properties.

**Definition 5.** *Let $\mathbf{C}$ be a covering of $U$. For any $x \in U$ and $X \subseteq U$, we call*

$$\sigma_{\mathbf{C}}^X(x) = \frac{|X \cap N_C(x)|}{|N_C(x)|}$$

*is the inclusion degree of $x$ to $X$ with respect to $\mathbf{C}$.*

*Example 1.* Let $\mathbf{C} = \{\{a, b\}, \{a, c\}, \{b, d\}\}$ be a covering of $U = \{a, b, c, d\}$. Then $N_C(a) = \{a\}, N_C(b) = \{b\}, N_C(c) = \{a, c\}, N_C(d) = \{b, d\}$. Suppose $X = \{b, c\}$ and $Y = \{a, b, d\}$. Then $\sigma_{\mathbf{C}}^X(a) = 0, \sigma_{\mathbf{C}}^X(b) = 1, \sigma_{\mathbf{C}}^X(c) = 0.5, \sigma_{\mathbf{C}}^X(d) = 0.5; \sigma_{\mathbf{C}}^Y(a) = 1, \sigma_{\mathbf{C}}^Y(b) = 1, \sigma_{\mathbf{C}}^Y(c) = 0.5, \sigma_{\mathbf{C}}^Y(d) = 1; \sigma_{\mathbf{C}}^U(a) = \sigma_{\mathbf{C}}^U(b) = \sigma_{\mathbf{C}}^U(c) = \sigma_{\mathbf{C}}^U(d) = 1; \sigma_{\mathbf{C}}^{\emptyset}(a) = \sigma_{\mathbf{C}}^{\emptyset}(b) = \sigma_{\mathbf{C}}^{\emptyset}(c) = \sigma_{\mathbf{C}}^{\emptyset}(d) = 0.$

Based on the definition of inclusion degree, some properties of inclusion degree are proposed as follows:

**Proposition 1.** *Let $\mathbf{C}$ be a covering of $U$. Then*

*(1)* $0 \leq \sigma_{\mathbf{C}}^X(x) \leq 1$;
*(2)* *if* $X \subseteq Y$, *then* $\sigma_{\mathbf{C}}^X(x) \leq \sigma_{\mathbf{C}}^Y(x)$;
*(3)* *if* $x \in N_C(y)$ *and* $y \in N_C(x)$, *then* $\sigma_{\mathbf{C}}^X(x) = \sigma_{\mathbf{C}}^X(y)$;
*(4)* $N_C(x) \subseteq X \Leftrightarrow \sigma_{\mathbf{C}}^X(x) = 1$;
*(5)* $N_C(x) \cap X \neq \emptyset \Leftrightarrow \sigma_{\mathbf{C}}^X(x) \neq 0$;
*(6)* $\bigcup_{z \in X} N_C(z) = X \Leftrightarrow \sigma_{\mathbf{C}}^X(z) = 1$ *for all* $z \in X$;
*(7)* $\sigma_{\mathbf{C}}^X(x) + \sigma_{\mathbf{C}}^Y(x) = \sigma_{\mathbf{C}}^{(X \cup Y)}(x) + \sigma_{\mathbf{C}}^{(X \cap Y)}(x)$;
*(8)* $\sigma_{\mathbf{C}}^X(x) = 1 - \sigma_{\mathbf{C}}^{(\sim X)}(x)$;

*are hold, where* $X, Y \subseteq U$ *and* $x, y \in U$.

*Proof.* It is easy to prove this proposition by the above definition of inclusion degree.

*Example 2.* (Continued from Example 1) Suppose $X_1 = \{b\}, X_2 = \{b, c, d\}, X_3 = \{a, b, d\}$. Then $\sigma_{\mathbf{C}}^{X_1}(a) = 0, \sigma_{\mathbf{C}}^{X_1}(b) = 1, \sigma_{\mathbf{C}}^{X_1}(c) = 0, \sigma_{\mathbf{C}}^{X_1}(d) = 0.5; \sigma_{\mathbf{C}}^{X_2}(a) =$

$0, \sigma_{\mathbf{C}}^{X_2}(b) = 1, \sigma_{\mathbf{C}}^{X_2}(c) = 0.5, \sigma_{\mathbf{C}}^{X_2}(d) = 1; \sigma_{\mathbf{C}}^{X_3}(a) = 1, \sigma_{\mathbf{C}}^{X_3}(b) = 1, \sigma_{\mathbf{C}}^{X_3}(c) = 0.5, \sigma_{\mathbf{C}}^{X_3}(d) = 1.$ It is easy to find that $\sigma_{\mathbf{C}}^{X_1}(a) = \sigma_{\mathbf{C}}^{X_2}(a), \sigma_{\mathbf{C}}^{X_1}(a) < \sigma_{\mathbf{C}}^{X_3}(a); \sigma_{\mathbf{C}}^{X_1}(b) = \sigma_{\mathbf{C}}^{X_2}(b), \sigma_{\mathbf{C}}^{X_1}(b) = \sigma_{\mathbf{C}}^{X_3}(b); \sigma_{\mathbf{C}}^{X_1}(c) < \sigma_{\mathbf{C}}^{X_2}(c), \sigma_{\mathbf{C}}^{X_1}(c) < \sigma_{\mathbf{C}}^{X_3}(c); \sigma_{\mathbf{C}}^{X_1}(d) < \sigma_{\mathbf{C}}^{X_2}(d), \sigma_{\mathbf{C}}^{X_1}(d) < \sigma_{\mathbf{C}}^{X_3}(d).$ In addition, $X_1 = \bigcup_{x \in X_1} N_{\mathbf{C}}(x)$ and $X_3 = \bigcup_{x \in X_3} N_{\mathbf{C}}(x).$

According to the definition of inclusion degree, it is easy to see the inclusion degree of $x$ to $X$ with respect to $\mathbf{C}$ is determined by $\mathbf{C}$. In fact, some properties of $\mathbf{C}$ can also be explored by inclusion degree.

**Proposition 2.** *Let* $\mathbf{C} = \{K_1, K_2, \ldots, K_s\}$ *be a covering of* $U$. $\mathbf{C}$ *is a partition of* $U$ *if and only if, for any* $x \in U$, *there exists an* $1 \leq i \leq s$ *such that* $\sigma_{\mathbf{C}}^{K_i}(x) = 1$ *and* $\sigma_{\mathbf{C}}^{K_j}(x) = 0, j = 1, 2, \ldots, i-1, i+1, \ldots s.$

*Proof.* ($\Rightarrow$): If $\mathbf{C}$ is a partition of $U$, then there exists $K_i(i \in \{1, 2, \ldots, s\})$ such that $N_{\mathbf{C}}(x) = K_i$, and $N_{\mathbf{C}}(x) \bigcap K_j = \emptyset (j = 1, 2, \ldots, i-1, i+1, \ldots s).$ Therefore $\sigma_{\mathbf{C}}^{K_i}(x) = 1, \sigma_{\mathbf{C}}^{K_j}(x) = 0.$

($\Leftarrow$): If there exist $K_i, K_j \in \mathbf{C}$ such that $K_i \bigcap K_j \neq \emptyset$, then there exists $x_1 \in K_i \bigcap K_j$ such that $\sigma_{\mathbf{C}}^{K_i}(x_1) \neq 0$ and $\sigma_{\mathbf{C}}^{K_j}(x_1) \neq 0.$ $(i \neq j, i, j \in \{1, 2, \ldots, s\}).$ It is a contradiction.

The above proposition shows a necessary and sufficient condition for covering to be a partition from the viewpoint of inclusion degree.

*Example 3.* (Continued from Example 1) We have
$\sigma_{\mathbf{C}}^{\{a,b\}}(a) = 1, \sigma_{\mathbf{C}}^{\{a,b\}}(b) = 1, \sigma_{\mathbf{C}}^{\{a,b\}}(c) = \frac{1}{2}, \sigma_{\mathbf{C}}^{\{a,b\}}(d) = 0; \sigma_{\mathbf{C}}^{\{a,c\}}(a) = 1, \sigma_{\mathbf{C}}^{\{a,c\}}(b) = 0, \sigma_{\mathbf{C}}^{\{a,c\}}(c) = 1, \sigma_{\mathbf{C}}^{\{a,c\}}(d) = 0.$ Since $\sigma_{\mathbf{C}}^{\{a,b\}}(c) = \frac{1}{2}$ and $\sigma_{\mathbf{C}}^{\{a,c\}}(c) = 1$, then $\mathbf{C}$ is not a partition of $U$.

Neighborhood is an important concept in covering-based rough sets. That under what condition neighborhoods form a partition is a meaningful issue induced by this concept. Many scholars have paid attention to this issue and presented some necessary and sufficient conditions. In the following proposition, a new necessary and sufficient condition for neighborhoods to form a partition is presented from the viewpoint of inclusion degree.

**Proposition 3.** *Let* $\mathbf{C}$ *be a covering of* $U$. $Cov(\mathbf{C})$ *is a partition of* $U$ *if and only if, for any* $x, y \in U, \sigma_{\mathbf{C}}^{N_{\mathbf{C}}(x)}(y) = \sigma_{\mathbf{C}}^{N_{\mathbf{C}}(y)}(x) = 1$ *or* $\sigma_{\mathbf{C}}^{N_{\mathbf{C}}(x)}(y) = \sigma_{\mathbf{C}}^{N_{\mathbf{C}}(y)}(x) = 0.$

*Proof.* ($\Rightarrow$): For all $x, y \in U$, if $Cov(\mathbf{C})$ is a partition of $U$, then $N_{\mathbf{C}}(x) = N_{\mathbf{C}}(y)$ or $N_{\mathbf{C}}(x) \bigcap N_{\mathbf{C}}(y) = \emptyset.$ If $N_{\mathbf{C}}(x) = N_{\mathbf{C}}(y)$ for any $x, y \in U$, then $\sigma_{\mathbf{C}}^{N_{\mathbf{C}}(x)}(y) = \frac{|N_{\mathbf{C}}(x) \bigcap N_{\mathbf{C}}(y)|}{|N_{\mathbf{C}}(y)|} = \frac{|N_{\mathbf{C}}(x) \bigcap N_{\mathbf{C}}(y)|}{|N_{\mathbf{C}}(x)|} = \sigma_{\mathbf{C}}^{N_{\mathbf{C}}(y)}(x) = 1.$ If $N_{\mathbf{C}}(x) \bigcap N_{\mathbf{C}}(y) = \emptyset$, then $\sigma_{\mathbf{C}}^{N_{\mathbf{C}}(x)}(y) = \sigma_{\mathbf{C}}^{N_{\mathbf{C}}(y)}(x) = 0.$

($\Leftarrow$): For any $x, y \in U$, if $\sigma_{\mathbf{C}}^{N_{\mathbf{C}}(x)}(y) = \sigma_{\mathbf{C}}^{N_{\mathbf{C}}(y)}(x) = 1$, then $N_{\mathbf{C}}(x) = N_{\mathbf{C}}(y);$ if $\sigma_{\mathbf{C}}^{N_{\mathbf{C}}(x)}(y) = \sigma_{\mathbf{C}}^{N_{\mathbf{C}}(y)}(x) = 0$, then $N_{\mathbf{C}}(x) \bigcap N_{\mathbf{C}}(y) = \emptyset.$ Hence, $Cov(\mathbf{C})$ is a partition of $U$.

The above propositions and examples show some properties and applications of inclusion degree. In the following subsection, a new type of covering-based rough sets will be proposed from the perspective of inclusion degree.

**Definition 6.** *Let* $CAS = (U, \mathbf{C})$ *be a covering approximation space. For any* $0 \leq \beta < \alpha \leq 1, X \in \mathcal{P}(U)$. *Then the covering lower and upper approximations of* $X$ *about* $CAS$ *with parameter* $\alpha$ *and* $\beta$ *as follows, respectively.*

$$\underline{\mathbf{C}}^{\alpha}(X) = \{x \in U : \sigma_{\mathbf{C}}^{X}(x) \geq \alpha\}, \overline{\mathbf{C}}^{\beta}(X) = \{x \in U : \sigma_{\mathbf{C}}^{X}(x) > \beta\}.$$

*Example 4.* Let $\mathbf{C} = \{\{a, c\}, \{a, b, c\}, \{b, c, e\}, \{b, d, e, f\}\}$ be a covering of $U = \{a, b, c, d, e, f\}$. Then $N_{\mathbf{C}}(a) = \{a, c\}$, $N_{\mathbf{C}}(b) = \{b\}$, $N_{\mathbf{C}}(c) = \{c\}$, $N_{\mathbf{C}}(d) = \{b, d, e, f\}$, $N_{\mathbf{C}}(e) = \{b, e\}$, $N_{\mathbf{C}}(f) = \{b, d, e, f\}$. Suppose $X = \{a, b, e, f\}$ and $Y = \{b, c, d\}$. Then

$$\sigma_{\mathbf{C}}^{X}(a) = 0.5, \sigma_{\mathbf{C}}^{X}(b) = 1, \sigma_{\mathbf{C}}^{X}(c) = 0, \sigma_{\mathbf{C}}^{X}(d) = 0.75,$$
$$\sigma_{\mathbf{C}}^{X}(e) = 1, \sigma_{\mathbf{C}}^{X}(f) = 0.75; \sigma_{\mathbf{C}}^{Y}(a) = 0.5, \sigma_{\mathbf{C}}^{Y}(b) = 1,$$
$$\sigma_{\mathbf{C}}^{Y}(c) = 1, \sigma_{\mathbf{C}}^{Y}(d) = 0.5, \sigma_{\mathbf{C}}^{Y}(e) = 0.5, \sigma_{\mathbf{C}}^{Y}(f) = 0.5.$$

Suppose $\beta = 0, \alpha = 0.5$. Then

$$\underline{\mathbf{C}}^{0.5}(X) = \{a, b, d, e, f\}, \overline{\mathbf{C}}^{0}(X) = \{a, b, d, e, f\};$$
$$\underline{\mathbf{C}}^{0.5}(Y) = \{a, b, c, d, e, f\}, \overline{\mathbf{C}}^{0}(Y) = \{a, b, c, d, e, f\}.$$

Suppose $\beta = 0.4, \alpha = 0.75$. Then

$$\underline{\mathbf{C}}^{0.75}(X) = \{b, d, e, f\}, \overline{\mathbf{C}}^{0.4}(X) = \{a, b, d, e, f\};$$
$$\underline{\mathbf{C}}^{0.75}(Y) = \{b, c\}, \overline{\mathbf{C}}^{0.4}(Y) = \{a, b, c, d, e, f\}.$$

For any $0 \leq \beta < \alpha \leq 1$,

$$\underline{\mathbf{C}}^{\alpha}(\emptyset) = \emptyset, \overline{\mathbf{C}}^{\beta}(\emptyset) = \emptyset; \underline{\mathbf{C}}^{\alpha}(U) = U, \overline{\mathbf{C}}^{\beta}(U) = U.$$

The above definition shows a new type covering-based rough sets with parameter $\alpha$ and $\beta$. Then we study some properties of this covering-based rough sets. Firstly, the positive region, boundary region and negative region of $X \in \mathcal{P}(U)$ in $CAS = (U, \mathbf{C})$ with parameter $\alpha$ and $\beta$ could be given as follows, respectively.

$$pos(X, \alpha) = \{x \in U : \sigma_{\mathbf{C}}^{X}(x) \geq \alpha\}, bn(X, \alpha, \beta) = \{x \in U : \beta < \sigma_{\mathbf{C}}^{X}(x) < \alpha\},$$
$$neg(X, \beta) = \{x \in U : \sigma_{\mathbf{C}}^{X}(x) \leq \beta\} = U - \overline{\mathbf{C}}^{\beta}(X).$$

If $\underline{\mathbf{C}}^{\alpha}(X) = \overline{\mathbf{C}}^{\beta}(X)$, then $X$ is called a definable set in covering approximation space $CAS = (U, \mathbf{C})$. Otherwise, $X$ is called a covering-based rough set. Meanwhile, the relationships among covering lower and upper approximations, positive region, boundary region and negative region of $X$ in $CAS = (U, \mathbf{C})$ are obtained.

**Proposition 4.** *Let* $CAS = (U, \mathbf{C})$ *be a covering approximation space. For any* $0 \leq \beta < \alpha \leq 1, X \in \mathcal{P}(U)$, *the following relationships hold.*

$$\overline{\mathbf{C}}^{\beta}(X) = pos(X, \alpha) \bigcup bn(X, \alpha, \beta), \text{ or } bn(X, \alpha, \beta) = \overline{\mathbf{C}}^{\beta}(X) - pos(X, \alpha).$$

*Example 5.* (Continued from Example 4) We have

$$pos(X, 0.5) = \{a, b, d, e, f\}, bn(X, 0.5, 0) = \emptyset, neg(X, 0) = \{c\};$$
$$pos(Y, 0.5) = \{a, b, c, d, e, f\}, bn(Y, 0.5, 0) = \emptyset, neg(Y, 0) = \emptyset;$$
$$pos(X, 0.75) = \{b, d, e, f\}, bn(Y, 0.75, 0.4) = \{a\}, neg(Y, 0.4) = \{c\};$$
$$pos(Y, 0.75) = \{b, c\}, bn(Y, 0.75, 0.4) = \{a, d, e, f\}, neg(Y, 0.4) = \emptyset.$$

Like many other types of covering-based rough sets on $CAS = (U, \mathbf{C})$, we can also present the basic properties of covering approximation operators by the constructed method.

**Proposition 5.** *Let $CAS = (U, \mathbf{C})$ be a covering approximation space. For any $0 \le \beta < \alpha \le 1, X, Y \in \mathcal{P}(U)$, the covering lower approximation operator and upper approximation operator satisfies the following properties.*

*(1) $\underline{\mathbf{C}}^\alpha(\emptyset) = \overline{\mathbf{C}}^\beta(\emptyset) = \emptyset, \underline{\mathbf{C}}^\alpha(U) = \overline{\mathbf{C}}^\beta(U) = U,$*

*(2) $\underline{\mathbf{C}}^\alpha(X) =\sim \overline{\mathbf{C}}^{(1-\alpha)}(\sim X), \overline{\mathbf{C}}^\beta(X) =\sim \underline{\mathbf{C}}^{(1-\beta)}(\sim X),$*

*(3) $\underline{\mathbf{C}}^\alpha(X \cap Y) \subseteq \underline{\mathbf{C}}^\alpha(X) \cap \underline{\mathbf{C}}^\alpha(Y), \overline{\mathbf{C}}^\beta(X \cap Y) \subseteq \overline{\mathbf{C}}^\beta(X) \cap \overline{\mathbf{C}}^\beta(Y),$*

*(4) $\underline{\mathbf{C}}^\alpha(X \cup Y) \supseteq \underline{\mathbf{C}}^\alpha(X) \cup \underline{\mathbf{C}}^\alpha(Y), \overline{\mathbf{C}}^\beta(X \cup Y) \supseteq \overline{\mathbf{C}}^\beta(X) \cup \overline{\mathbf{C}}^\beta(Y),$*

*(5) If $X \subseteq Y$, then $\underline{\mathbf{C}}^\alpha(X) \subseteq \underline{\mathbf{C}}^\alpha(Y), \overline{\mathbf{C}}^\beta(X) \subseteq \overline{\mathbf{C}}^\beta(Y),$*

*(6) If $\alpha_1 \le \alpha_2, \beta_1 \le \beta_2$, then $\underline{\mathbf{C}}^{\alpha_2}(X) \subseteq \underline{\mathbf{C}}^{\alpha_1}(X), \overline{\mathbf{C}}^{\beta_2}(X) \subseteq \overline{\mathbf{C}}^{\beta_1}(X),$*

*(7) If $\alpha \le min\{\frac{1}{|N_{\mathbf{C}(x)}|} : x \in U\}$, then $X \subseteq \underline{\mathbf{C}}^\alpha(X), X \subseteq \overline{\mathbf{C}}^\beta(X),$*

*(8) $bn(X, \alpha, \beta) = \emptyset \Leftrightarrow \underline{\mathbf{C}}^\alpha(X) = \overline{\mathbf{C}}^\beta(X).$*

*Proof.* It is easy to prove the results by the above definition of the covering lower and upper approximation operators.

The above proposition shows some important properties of the new type of covering-based rough sets. In fact, it is clear that positive region will increase with parameter $\alpha$ decrease, negative region will increase with parameter $\alpha$ increase and boundary region will dwindle for this new type of covering-based rough sets. In the following examples we study another properties of this new type covering-based rough sets.

*Example 6.* (Continued from Example 4) We have $\underline{\mathbf{C}}^{0.5}(X \cap Y) = \underline{\mathbf{C}}^{0.5}(\{b, e\}) = \{b\} \subset \{a, b, d, e, f\} = \underline{\mathbf{C}}^{0.5}(X) \cap \underline{\mathbf{C}}^{0.5}(Y), \overline{\mathbf{C}}^0(X \cap Y) = \overline{\mathbf{C}}^0(\{b\}) = \{b, d, e, f\} \subset \{a, b, d, e, f\} = \overline{\mathbf{C}}^0(X) \cap \overline{\mathbf{C}}^0(Y)$. It is easy to see $\underline{\mathbf{C}}^\alpha(X \cap Y) = \underline{\mathbf{C}}^\alpha(X) \cap \underline{\mathbf{C}}^\alpha(Y)$ and $\overline{\mathbf{C}}^\beta(X \cap Y) = \overline{\mathbf{C}}^\beta(X) \cap \overline{\mathbf{C}}^\beta(Y)$ are not hold for any $0 \le \beta < \alpha \le 1, X, Y \in \mathcal{P}(U)$.

*Example 7.* (Continued from Example 4) Suppose $X = \{b\}$ and $Y = \{e\}$. Then $\underline{\mathbf{C}}^{0.5}(X) = \{b, e\}, \overline{\mathbf{C}}^{0.25}(X) = \{b, e\}; \underline{\mathbf{C}}^{0.5}(Y) = \{e\}, \overline{\mathbf{C}}^{0.25}(Y) = \{e\}; \underline{\mathbf{C}}^{0.5}(X \cup Y) = \{b, d, e, f\}, \overline{\mathbf{C}}^{0.25}(X \cup Y) = \{b, d, e, f\}$. It is easy to see
$$\underline{\mathbf{C}}^{0.5}(X) \cup \underline{\mathbf{C}}^{0.5}(Y) \subset \underline{\mathbf{C}}^{0.5}(X \cup Y), \overline{\mathbf{C}}^{0.25}(X) \cup \overline{\mathbf{C}}^{0.25}(Y) \subset \overline{\mathbf{C}}^{0.25}(X \cup Y).$$
Therefore $\underline{\mathbf{C}}^\alpha(X \cup Y) = \underline{\mathbf{C}}^\alpha(X) \cup \underline{\mathbf{C}}^\alpha(Y)$ and $\overline{\mathbf{C}}^\beta(X \cup Y) = \overline{\mathbf{C}}^\beta(X) \cup \overline{\mathbf{C}}^\beta(Y)$ are not hold for any $0 \le \beta < \alpha \le 1, X, Y \in \mathcal{P}(U)$.

In following, we give the roughness and precision of the new type of covering-based rough sets.

**Definition 7.** *Let* $CAS = (U, \mathbf{C})$ *be a covering approximation space. For any* $0 \le \beta < \alpha \le 1, X \in \mathcal{P}(U)$, *we define the roughness* $\rho_{\mathbf{C}}(X, \alpha, \beta)$ *and precision* $\eta_{\mathbf{C}}(X, \alpha, \beta)$ *of* $X$ *according to the parameter* $\alpha, \beta$ *about the covering approximation space* $CAS$ *as follows:*

$$\eta_{\mathbf{C}}(X, \alpha, \beta) = \frac{|\underline{\mathbf{C}}^{\alpha}(X)|}{|\overline{\mathbf{C}}^{\beta}(X)|}, \rho_{\mathbf{C}}(X, \alpha, \beta) = 1 - \frac{|\underline{\mathbf{C}}^{\alpha}(X)|}{|\overline{\mathbf{C}}^{\beta}(X)|}.$$

*Example 8.* (Continued from Example 4) We have $\eta_{\mathbf{C}}(X, 0.5, 0) = 1, \rho_{\mathbf{C}}(X, 0.5, 0) = 0$; $\eta_{\mathbf{C}}(X, 0.75, 0.4) = 0.8, \rho_{\mathbf{C}}(X, 0.75, 0.4) = 0.2$; $\eta_{\mathbf{C}}(Y, 0.5, 0) = 1, \rho_{\mathbf{C}}(X, 0.5, 0) = 0$; $\eta_{\mathbf{C}}(Y, 0.75, 0.4) = \frac{1}{3}, \rho_{\mathbf{C}}(X, 0.75, 0.4) = \frac{2}{3}$.

According to the roughness and precision of $X$, we can easy to know that $bn(X, \alpha, \beta) = \emptyset$ if and only if $\eta_{\mathbf{C}}(X, \alpha, \beta) = 1, \rho_{\mathbf{C}}(X, \alpha, \beta) = 0$. Meanwhile, the following proposition is obviously from the above definition.

**Proposition 6.** *Let* $CAS = (U, \mathbf{C})$ *be a covering approximation space. For any* $0 \le \beta < \alpha \le 1, X \in \mathcal{P}(U)$, *the roughness* $\rho_{\mathbf{C}}(X, \alpha, \beta)$ *and precision* $\eta_{\mathbf{C}}(X, \alpha, \beta)$ *are satisfied the following properties:*

*(1)* $0 \le \eta_{\mathbf{C}}(X, \alpha, \beta) \le 1, 0 \le \rho_{\mathbf{C}}(X, \alpha, \beta) \le 1$,
*(2)* $\rho_{\mathbf{C}}(X, \alpha, \beta)$ *is not decreased about* $\alpha$, *and not increased about* $\beta$,
*(3)* $\eta_{\mathbf{C}}(X, \alpha, \beta)$ *is not increased about* $\alpha$, *and not decreased about* $\beta$.

*Proof.* It is easy to prove this proposition by Definition 7.

According to the above definitions and propositions, a new type of covering-based rough sets have been established from the viewpoint of inclusion degree. Meanwhile, some properties of this covering-based rough sets are explored.

## 4    Applications of the New Type of Covering-Based Rough Sets

In [3], Ge has investigated separations in covering approximation spaces and gave some characterizations of these separations and some relations among these separations. As an application of these results, investigations on network security were converted into investigations on separations in covering approximation spaces by taking covering approximation spaces as mathematical models of networks. In this section, we introduce the application of the new type of covering-based rough sets into network security based on [3]. First, some definitions and lemmas should be introduced.

**Definition 8.** *(Network [3]) A network is a pair* $(V, \mathcal{B})$, *where* $\mathcal{B}$ *is a family of servers and* $V$ *is a set of their users, such that each server in* $\mathcal{B}$ *provides its service to some users in* $V$ *and each user in* $V$ *accepts some services from some servers in* $\mathcal{B}$.

**Definition 9.** *(S$_1$−security [3]) Let* $(V, \mathcal{B})$ *be a network.*

$S_1-$security: *For each pair of distinct users $x$ and $y$ in $V$ , there are servers $E_1$ and $E_2$ in $\mathcal{B}$ providing their services to $x$ and $y$ but not providing their services to $y$ and $x$, respectively.*

**Lemma 1.** *([3]) Let $(V, \mathcal{B})$ be a network. For each server $E$ in $\mathcal{B}$, let $V_E$ be a set of some users in $V$ such that $x$ is a user in $V_E$ if and only if $E$ provides its service to $x$. Put $U$ is the abstract set of $V$ and $K_E$ is the abstract set of $V_E$ for each $E \in \mathcal{B}$. Then $\mathbf{C} = \{K_E : E \in \mathcal{B}\}$ is a covering of $U$, and so $(U, \mathbf{C})$ is a covering approximation space.*

*Remark 1.* Let $(V, \mathcal{B})$ and $(U, \mathbf{C})$ be stated as Lemma 1. Then $(U, \mathbf{C})$ is called a covering approximation space induced by $(V, \mathcal{B})$.

**Definition 10.** *($S_1-$space [3]) Let $(U, \mathbf{C})$ be a covering approximation space. Then $(U, \mathbf{C})$ is an $S_1 -$ space if and only if $N_{\mathbf{C}}(x) = \{x\}$ for all $x \in U$.*

**Lemma 2.** *([3]) Let $(V, \mathcal{B})$ be a network and $(U, \mathbf{C})$ be a covering approximation space induced by $(V, \mathcal{B})$. Then $(V, \mathcal{B})$ has $S_1-$security if and only if $(U, \mathbf{C})$ is an $S_1-$space.*

Now, we give a simple application of the new type of covering-based rough sets into network security.

**Proposition 7.** *Let $(V, \mathcal{B})$ be a network and $(U, \mathbf{C})$ be a covering approximation space induced by $(V, \mathcal{B})$. Then the following are equivalent.*

*(1) $(V, \mathcal{B})$ has $S_1-$security.*
*(2) $\sigma_{\mathbf{C}}^{\{x\}}(x) \equiv 1$ for any $x \in U$.*
*(3) $\underline{\mathbf{C}}^I(X) = X$ for any $X \in \mathcal{P}(U)$.*

*Proof.* According to Definition 10 and Lemma 2, we only need to prove (2), (3) and

(4)  $N_{\mathbf{C}}(x) = \{x\}$ for any $x \in U$

are equivalent.

(2)$\Rightarrow$ (4): If $\sigma_{\mathbf{C}}^{\{x\}}(x) \equiv 1$ for any $x \in U$, then $N_{\mathbf{C}}(x) \subseteq \{x\}$. Hence $N_{\mathbf{C}}(x) = \{x\}$ for any $x \in U$. (4)$\Rightarrow$ (2): It is straightforward. Therefore, (2) and (4) are equivalent.

(3)$\Leftrightarrow$ (4): It is easy to prove by Definitions 5, 6 and Proposition 1.

This completes the proof of this proposition.

The above proposition shows some sufficient and necessary conditions for a network has $S_1-$security from the viewpoint of the new type of covering-based rough sets. A successfully application of this new type of covering-based rough sets into network is to a course examination network in Soochow University [3].

*Example 9.* A Course Examination Network $(V, \mathcal{B})$.

1. Conditions of Network $(V, \mathcal{B})$.
    (a) $(V, \mathcal{B})$ can be applied to a course examination for a group consisting of 30 students in Soochow University.

(b) In order to prevent cheats in a course examination, it is necessary to guarantee that every student can only acquire his/her own exam questions in the network $(V, \mathcal{B})$. Therefore, the designer of $(V, \mathcal{B})$ designs that every student $x$ must accept services from all servers in $\mathcal{B}_x$, where $\mathcal{B}_x = \{E \in \mathcal{B} : E$ provides its service to $x\}$. Obviously, it is sufficient if $(V, \mathcal{B})$ has $S_1$-security.

2. Establishment of Network $(V, \mathcal{B})$.
   (a) Put $V$ is the set of 30 users: $V = \{x_1, x_2, \ldots, x_{30}\}$.
   (b) Put $\mathcal{B}$ is the set 10 servers: $\mathcal{B} = \{E_1, E_2, \ldots, E_{10}\}$.
   (c) For each $E_k \in \mathcal{B}(k = 1, 2, \ldots, 10)$, the designer of $(V, \mathcal{B})$ designs $V_{E_k} = \{x \in V : E_k$ provides its service to $x\}$ as follows.
   $$V_{E_1} = \{x_1, x_2, x_3, x_4, x_5, x_6\}, V_{E_2} = \{x_6, x_7, x_8, x_9, x_{10}, x_{11}\},$$
   $$V_{E_3} = \{x_{11}, x_{12}, x_{13}, x_{14}, x_{15}, x_{16}\}, V_{E_4} = \{x_{16}, x_{17}, x_{18}, x_{19}, x_{20}, x_{21}\},$$
   $$V_{E_5} = \{x_{21}, x_{22}, x_{23}, x_{24}, x_{25}, x_{26}\}, V_{E_6} = \{x_{26}, x_{27}, x_{28}, x_{29}, x_{30}, x_1\},$$
   $$V_{E_7} = \{x_2, x_7, x_{12}, x_{17}, x_{22}, x_{27}\}, V_{E_8} = \{x_3, x_8, x_{13}, x_{18}, x_{23}, x_{28}\},$$
   $$V_{E_9} = \{x_4, x_9, x_{14}, x_{19}, x_{24}, x_{29}\}, V_{E_{10}} = \{x_5, x_{10}, x_{15}, x_{20}, x_{25}, x_{30}\}.$$
   Thus, a network $(V, \mathcal{B})$ is established.

3. Conversion of Network $(V, \mathcal{B})$.
   By Lemma 1, we convert the network $(V, \mathcal{B})$ to a covering approximation space $(U, \mathbf{C})$ as follows, where $(U, \mathbf{C})$ is induced by $(V, \mathcal{B})$.
   (a) $U$ is the abstract set of $V$: $U = \{x_1, x_2, \ldots, x_{30}\}$.
   (b) $\mathbf{C}$ is a covering of $U$: $\mathbf{C} = \{K_1, K_2, \ldots, K_{10}\}$.
   $$K_1 = \{x_1, x_2, x_3, x_4, x_5, x_6\}, K_2 = \{x_6, x_7, x_8, x_9, x_{10}, x_{11}\},$$
   $$K_3 = \{x_{11}, x_{12}, x_{13}, x_{14}, x_{15}, x_{16}\}, K_4 = \{x_{16}, x_{17}, x_{18}, x_{19}, x_{20}, x_{21}\},$$
   $$K_5 = \{x_{21}, x_{22}, x_{23}, x_{24}, x_{25}, x_{26}\}, K_6 = \{x_{26}, x_{27}, x_{28}, x_{29}, x_{30}, x_1\},$$
   $$K_7 = \{x_2, x_7, x_{12}, x_{17}, x_{22}, x_{27}\}, K_8 = \{x_3, x_8, x_{13}, x_{18}, x_{23}, x_{28}\},$$
   $$K_9 = \{x_4, x_9, x_{14}, x_{19}, x_{24}, x_{29}\}, K_{10} = \{x_5, x_{10}, x_{15}, x_{20}, x_{25}, x_{30}\}.$$
   Thus, a covering approximation space $(U, \mathbf{C})$ is obtained, which is induced by $(V, \mathcal{B})$.

4. Security of Network $(V, \mathcal{B})$.
   (a) It is not difficult to check that $\sigma_{\mathbf{C}}^{\{x_i\}}(x_i) = 1$ for each $i = 1, 2, \ldots, 30$. By Proposition 7, $(U, \mathbf{C})$ has $S_1$-security.
   (b) It is not difficult to compute that $\underline{\mathbf{C}}^1(X) = X$ for each $X \in \mathcal{P}(U)$. By Proposition 7, $(U, \mathbf{C})$ has $S_1$-security.

According to the analysis of the above example, it is easy to know that the conclusions are more scientific, reasonable and suitable to the reality when applying the new type of covering-based rough sets to the reality. Therefore, the decision of the reality will be more exactly in the practice.

## 5   Conclusions

In this paper, a new type of covering-based rough sets have been proposed from the viewpoint of inclusion degree. Meanwhile, some applications of this covering-based rough sets were introduced. First, a new concept, namely, inclusion degree was defined and some properties of it were studied. Moreover, some properties of the coverings were represented based on the inclusion degree. Second, based on the other types of covering-based rough sets, a new type of covering-based rough sets have been defined from the viewpoint of inclusion degree. Moreover, some properties of the new type of covering-based rough sets have been studied. Finally, a simple application of the new type of covering-based rough sets into network security was given. Though much research has been conducted in this paper, there are still many interesting issues worth studying.

1. Relationships between this new type of covering-based rough sets and other types of covering-based rough sets.
2. Dependency of lower and upper approximation operations of this new type of covering-based rough sets.
3. Axiomatic system for approximation operations of this new type of covering-based rough sets.
4. Topological and matroidal properties of this new type of covering-based rough sets.

In further work, we will conduct more specific research alone the lines of the above four issues, especially axiomatic systems and matroidal properties [7,16,17,19,20,21] for this new type of covering-based rough sets.

**Acknowledgments.** This work is in part supported by the National Natural Science Foundation of China under Grant Nos. 61170128, 61379049, and 61379089, the Fujian Province Foundation of Higher Education under Grant No. JK2012028, the Key Project of Education Department of Fujian Province under Grant No. JA13192, and the Minnan Normal University Postgraduate Education Project No.YJS201435.

# References

1. Bonikowski, Z., Bryniarski, E., Wybraniec-Skardowska, U.: Extensions and intentions in the rough set theory. Information Sciences 107, 149–167 (1998)
2. Dai, J., Xu, Q.: Approximations and uncertainty measures in incomplete information systems. Information Sciences 198, 62–80 (2012)
3. Ge, X.: An application of covering approximation spaces on network security. Computers and Mathematics with Applications 60, 1191–1199 (2010)
4. Hu, B.Q., Wong, H.: Generalized interval-valued fuzzy rough sets based on interval-valued fuzzy logical operators. International Journal of Fuzzy Systems 15, 381–391 (2013)
5. Järvinen, J.: On the structure of rough approximations. In: Alpigini, J.J., Peters, J.F., Skowron, A., Zhong, N. (eds.) RSCTC 2002. LNCS (LNAI), vol. 2475, pp. 123–130. Springer, Heidelberg (2002)
6. Kondo, M.: On the structure of generalized rough sets. Information Sciences 176, 589–600 (2006)
7. Li, X., Liu, S.: Matroidal approaches to rough set theory via closure operators. International Journal of Approximate Reasoning 53, 513–527 (2012)
8. Liu, G.: Generalized rough sets over fuzzy lattices. Information Sciences 178, 1651–1662 (2008)
9. Liu, G.: Closures and topological closures in quasi-discrete closure. Applied Mathematics Letters 23, 772–776 (2010)
10. Ma, Z., Hu, B.: Topological and lattice structures of $\mathcal{L}$-fuzzy rough sets determined by lower and upper sets. Information Sciences 218, 194–204 (2013)

11. Pawlak, Z.: Rough sets. International Journal of Computer and Information Sciences 11, 341–356 (1982)
12. Pawlak, Z.: Rough sets: theoretical aspects of reasoning about data. Kluwer Academic Publishers, Boston (1991)
13. Qin, K., Gao, Y., Pei, Z.: On covering rough sets. In: Yao, J., Lingras, P., Wu, W.-Z., Szczuka, M.S., Cercone, N.J., Ślęzak, D. (eds.) RSKT 2007. LNCS (LNAI), vol. 4481, pp. 34–41. Springer, Heidelberg (2007)
14. Rajagopal, P., Masone, J.: Discrete mathematics for computer science. Saunders College, Canada (1992)
15. Wang, C., Chen, D., Sun, B., Hu, Q.: Communication between information systems with covering based rough sets. Information Sciences 216, 17–33 (2012)
16. Wang, J., Zhu, W.: Applications of matrices to a matroidal structure of rough sets. Journal of Applied Mathematics 2013, article ID 493201, 9 pages (2013)
17. Wang, S., Zhu, Q., Zhu, W., Min, F.: Matroidal structure of rough sets and its characterization to attribute reduction. Knowledge-Based Systems 35, 155–161 (2012)
18. Wei, W., Liang, J., Qian, Y.: A comparative study of rough sets for hybrid data. Information Sciences 190, 1–16 (2012)
19. Yang, B., Zhu, W.: Matroidal structure of generalized rough sets based on symmetric and transitive relations. In: CCECE 2013, pp. 1–5 (2013)
20. Yang, B., Lin, Z., Zhu, W.: Covering-based rough sets on eulerian matroids. Journal of Applied Mathematics 2013, article ID 254797, 8 pages (2013)
21. Yao, H., Zhu, W., Wang, F.Y.: Secondary basis unique augmentation matroids and union minimal matroids. International Journal of Machine Learning and Cybernetics (2014), doi:10.1007/s13042–014–0237–1
22. Yao, Y.: Constructive and algebraic methods of theory of rough sets. Information Sciences 109, 21–47 (1998)
23. Yao, Y.: Relational interpretations of neighborhood operators and rough set approximation operators. Information Sciences 111, 239–259 (1998)
24. Zakowski, W.: Approximations in the space $(u, \pi)$. Demonstratio Mathematica 16, 761–769 (1983)
25. Zhang, J., Li, T., Ruan, D., Gao, Z., Zhao, C.: A parallel method for computing rough set approximations. Information Sciences 194, 209–223 (2012)
26. Zhang, Y., Li, J., Wu, W.: On axiomatic characterizations of three pairs of covering based approximation operators. Information Sciences 180, 274–287 (2010)
27. Zhu, W., Wang, F.: Reduction and axiomization of covering generalized rough sets. Information Sciences 152, 217–230 (2003)
28. Zhu, W., Wang, F.: On three types of covering rough sets. IEEE Transactions on Knowledge and Data Engineering 19, 1131–1144 (2007)
29. Zhu, W.: Relationship among basic concepts in covering-based rough sets. Information Sciences 179, 2478–2486 (2009)

# A Feature Seletion Method Based on Variable Precision Tolerance Rough Sets

Na Jiao

Department of Information Science and Technology, East China University of
Political Science and Law, Shanghai 201804, P.R. China
zdx.jn@163.com

**Abstract.** Feature selection is an important notions in rough sets. This
paper presents a method combining tolerance relation together with
rough sets. There is noise data in practical data sets. This paper investigates the feature selection method based on variable precision tolerance
rough sets. The parameter was discussed and the parameter interval was
described. With the change of the parameter value, the feature selection
was different. The efficiency of the proposed method can be illustrated
by an experiment with standard dataset from UCI database.

**Keywords:** Rough set theory, tolerance relation, variable precision,
parameter interval, dependency degree of feature.

## 1 Introduction

Rough sets, proposed by Pawlak [1], is a valid mathematical tool to deal with
imprecise, uncertain, and vague information. Researchers have proposed various
methods for feature selection [2-8].These methods can be generally divided into
three categories which are methods based on discernibility matrix [2], methods
based on positive region [2] and methods based on information entropy [3].

Feature selection methods based equivalence relation are restricted to the
requirement that all data must be discrete. Existing methods [4] are to discretize
the data sets and replace original data values with crisp values. Discretization
ignores their discrimination. This may cause information loss. A better choice
to solve the problem may be the use of tolerance rough set theory [5]. Tolerance
rough set theory can avoid the information loss caused by the discretization
process and maximize the ability of classification data set.

Ziarko proposed a variable precision rough set model, the processing strategy
is put forward when the error rate is lower than the threshold [6].Katzberg and
Ziarko described the asymmetric boundary variable precision rough set model,
which is more general and widens the scope of application of variable precision
rough set.

The main motivation of this study is to design a method that is from a practical point of view rather than the perspective of theory. We present the feature
selection method based on variable precision tolerance rough sets. The characteristics of parameter were analyzed. The relationship between the classification

D. Miao et al. (Eds.): RSKT 2014, LNAI 8818, pp. 500–509, 2014.
DOI: 10.1007/978-3-319-11740-9_46 © Springer International Publishing Switzerland 2014

quality and parameter interval was described, and the parameter value was extended to interval range. Experiments can be done using standard dataset from UCI database. Experimental results show that the proposed method and the related theory in this paper are effective.

The remainder of this paper is organized as follows. The next section deals with some preliminary concepts and properties regarding rough sets. The notion of variable precision tolerance rough sets is introduced in Section 3. Section 4 introduces feature selection method based on variable precision tolerance rough sets. A simple example is presented in Section 5. Section 6 presents experimental results on one benchmark data set. Finally, the work is concluded in Section 7.

## 2   Preliminaries

### 2.1   Basic Definitions

In this section, we review some basic concepts about traditional rough sets.

**Definition 1.** A decision table is defined as $DT = \langle U, C \cup D, V, f \rangle$, where $U$ is a non-empty finite set of objects; $C$ is a set of all conditional features and $D$ is a set of decision features; $V = \bigcup_{a \in C \cup D} V_a$, $V_a$ is a set of feature values of feature $a$; and $f : U \times (C \cup D) \to V$ is an information function. Let $B \subseteq C \cup D$, $B$ induces an equivalence (indiscernibility) relation on $U$ as shown:

$$IND(B) = \{(x, y) \in U \times U | \forall a \in B, f(x, a) = f(y, a)\}. \tag{1}$$

The family of all equivalence classes of $IND(B)$, i.e., the partition induced by $B$, is denoted as:

$$U/IND(B) = \{[x]_B : x \in U\}, \tag{2}$$

where $[x]_B$ is the equivalence class containing $x$. All the elements in $[x]_B$ are equivalent (indiscernible) with respect to $B$.

**Definition 2.** Let $X \subseteq U$ and $B \subseteq C$, the lower and upper approximations of $X$ with respect to $B$, denoted by $\underline{B}X$ and $\overline{B}X$, respectively, are defined as:

$$\underline{B}X = \cup \{[x]_B \in U/IND(B) | [x]_B \subseteq X\}, \tag{3}$$

$$\overline{B}X = \cup \{[x]_B \in U/IND(B) | [x]_B \cap X \neq \emptyset\}. \tag{4}$$

**Definition 3.** Let $X \subseteq U$, the positive, negative and boundary regions of $D$ with respect to $B \subseteq C$, respectively, are denoted as:

$$POS_B(D) = \cup_{X \in U/IND(D)} \underline{B}X, \tag{5}$$

$$NEG_B(D) = U - \cup_{X \in U/IND(D)} \overline{B}X, \tag{6}$$

$$BND_B(D) = \cup_{X \in U/IND(D)} \overline{B}X - \cup_{X \in U/IND(D)} \underline{B}X. \tag{7}$$

**Definition 4.** The degree of dependency of $D$ on $B \subseteq C$ can be defined as:

$$\gamma_B(D) = |POS_B(D)| / |U|. \tag{8}$$

## 3   Tolerance Rough Set Theory

Feature selection methods based equivalence relation are restricted to the requirement that all data must be discrete. Discretization ignores their discrimination. This may cause information loss. Tolerance rough set theory can avoid the information loss caused by the discretization process and maximize the ability of classification data set. In order to deal with real-valued data, we employ a similarity relation. This allows a relaxation in the way equivalence classes are considered.

### 3.1   Similarity Measures

In this approach, suitable similarity measure, given in [5], is described in Definition 5.

**Definition 5.** Given a decision table $DT = \langle U, C \cup D, V, f \rangle$, let $a \in C \cup D$ and $x, y \in U$, the similarity measure, given in [5], is defined as:

$$F_a(x, y) = \frac{|a(x) - a(y)|}{|a_{\max} - a_{\min}|}, \tag{9}$$

where $a_{\max}$ and $a_{\min}$ denote the maximum and minimum values respectively for feature $a$. The smaller the similarity measure between two objects, the greater the change that they belong to the same class.

**Definition 6.** Let $B \subseteq C \cup D$ and $\tau \in (0, 0.5]$, the overall similarity measure of $x$ and $y$ with respect to $B$ is defined as:

$$F_{B,\tau}(x, y) = \left\{ (x, y) \mid \frac{\sum\limits_{a \in B} F_a(x, y)}{|B|} \leq \tau \right\}, \tag{10}$$

where $\tau \in (0, 0.5]$ is a global similarity threshold.

**Definition 7.** Let $B \subseteq C \cup D$ and $\tau \in (0, 0.5]$, the similarity relation of $\tau$ of $U$ on $B$ can be defined as:

$$F_{B,\tau} = \{(x, y) \in U \times U | \forall a \in B, (x, y) \in F_{B,\tau}(x, y)\}. \tag{11}$$

The partition induced by $F_{B,\tau}$, is denoted as:

$$U/F_{B,\tau} = \{F_{B,\tau}(x) : x \in U\}, \tag{12}$$

where $F_{B,\tau}(x)$ denotes the similarity class containing $x$.

**Definition 8.** Let $X \subseteq U$ and $\tau \in (0, 0.5]$, the lower and upper approximations of $\tau$ of $X$ with respect to $B \subseteq C$, denoted by $\underline{B_\tau}X$ and $\overline{B_\tau}X$, respectively, are defined as:

$$\underline{B_\tau}X = \{x | F_{B,\tau}(x) \subseteq X\}, \tag{13}$$

$$\overline{B_\tau}X = \{x | F_{B,\tau}(x) \cap X \neq \emptyset\}. \tag{14}$$

The tuple $\langle \underline{B_\tau}X, \overline{B_\tau}X \rangle$ is called a tolerance-based rough set.

**Definition 9.** Let $X \subseteq U$ and $\tau \in (0, 0.5]$, the positive, negative and boundary regions of $\tau$ of $D$ with respect to $B \subseteq C$, respectively, are denoted as:

$$POS_{B,\tau}(D) = \cup_{X \in U/F_{D,\tau}} \underline{B_\tau}X, \tag{15}$$

$$NEG_{B,\tau}(D) = U - \cup_{X \in U/F_{D,\tau}} \overline{B_\tau}X, \tag{16}$$

$$BND_{B,\tau}(D) = \cup_{X \in U/F_{D,\tau}} \overline{B_\tau}X - \cup_{X \in U/F_{D,\tau}} \underline{B_\tau}X. \tag{17}$$

**Definition 10.** Let $\tau \in (0, 0.5]$, the dependency degree of $\tau$ of $D$ on $B \subseteq C$ can be defined as:

$$\gamma_{B,\tau}(D) = |POS_{B,\tau}(D)| / |U|. \tag{18}$$

The tolerance-based degree of dependency $\gamma_{B,\tau}(D)$, can be used to gauge the significance of feature subsets.

# 4 Feature Selection Method Based on Variable Precision Tolerance Rough Sets

Noise data is very difficult to avoid in many practical applications. In order to solve this contradiction, allowing the noise data or error, we introduce feature selection method based on variable precision tolerance rough sets.

## 4.1 Variable Precision Tolerance Rough Set Model

**Definition 11.** Let $X_i \in U/F_{C,\tau}$ $(i = 1, 2, \cdots |U/F_{C,\tau}|)$, $Y_j \in U/F_{D,\tau}$ $(j = 1, 2, \cdots |U/F_{D,\tau}|)$ and $\tau \in (0, 0.5]$, $\beta$ - lower and $\beta$ - upper approximations of $Y_j$ with respect to $X_i$, denoted by $\underline{X_{i,\tau}^\beta}Y_j$ and $\overline{X_{i,\tau}^\beta}Y_j$, respectively, are defined as:

$$\underline{X_{i,\tau}^\beta}Y_j = \left\{ X_i | \frac{|X_i \cap Y_j|}{|X_i|} \geq \beta \right\}, \tag{19}$$

$$\overline{X_{i,\tau}^\beta}Y_j = \left\{ X_i | \frac{|X_i \cap Y_j|}{|X_i|} > 1 - \beta \right\}. \tag{20}$$

$\beta$ - lower and $\beta$ - upper approximations of $Y_j$ with respect to $B \subseteq C$, are defined as:

$$\underline{B_\tau^\beta} Y_j = \cup \left\{ X_i | \frac{|X_i \cap Y_j|}{|X_i|} \geq \beta, X_i \in U/F_{B,\tau} \right\}, \tag{21}$$

$$\overline{B_\tau^\beta} Y_j = \cup \left\{ X_i | \frac{|X_i \cap Y_j|}{|X_i|} > 1 - \beta, X_i \in U/F_{B,\tau} \right\}, \tag{22}$$

where $0.5 < \beta \leq 1$.

**Definition 12.** Let $X_i \in U/F_{C,\tau}$ $(i = 1, 2, \cdots |U/F_{C,\tau}|)$ , $Y_j \in U/F_{D,\tau}$ $(j = 1, 2, \cdots |U/F_{D,\tau}|)$, $\tau \in (0, 0.5]$ and $\beta \in (0.5, 1]$, $\beta$ - positive, $\beta$ - negative and $\beta$ - boundary regions of $\tau$ of $D$ with respect to $B \subseteq C$, respectively, are denoted as:

$$POS_{B,\tau}^\beta (D) = \cup_{Y_j \in U/F_{D,\tau}} \underline{B_\tau^\beta} Y_j, \tag{23}$$

$$NEG_{B,\tau}^\beta (D) = U - \cup_{Y_j \in U/F_{D,\tau}} \overline{B_\tau^\beta} Y_j, \tag{24}$$

$$BN_{B,\tau}^\beta (D) = \cup_{Y_j \in U/F_{D,\tau}} \overline{B_\tau^\beta} Y_j - \cup_{Y_j \in U/F_{D,\tau}} \underline{B_\tau^\beta} Y_j. \tag{25}$$

**Definition 13.** Let $\tau \in (0, 0.5]$ and $\beta \in (0.5, 1]$, $\beta$ - dependency degree of $\tau$ of $D$ on $B \subseteq C$ is defined as:

$$\gamma_{B,\tau}^\beta (D) = \left| POS_{B,\tau}^\beta (D) \right| / |U|. \tag{26}$$

## 4.2  Feature Selection Method Based on Variable Precision Tolerance Rough Sets

The detailed procedure of feature selection method based on variable precision tolerance rough sets is described as follows.

**Algorithm 1.** Feature Selection Method based on Variable Precision Tolerance Rough Sets (DDTRS)

**Step 1.** Set $\tau$, calculate the similarity measure, the overall similarity measure and the similarity relation according to definition 5, 6, 7.

**Step 2.** Set $\beta$, calculate $\beta$ - lower , $\beta$ - upper approximations, $\beta$ -positive, $\beta$ -negative and $\beta$ -boundary regions of $\tau$ of $D$ with respect to $C$.

**Step 3.** Set $\mathrm{Red} = C$.

**Step 4.** For $a \in C$, If $\gamma_{P-\{a\},\tau}^\beta (D) == \gamma_{C,\tau}^\beta (D)$, go to step 5. Else go to step 6.

**Step 5.** $\mathrm{Red} = \mathrm{Red} - \{a\}$ , go to step 4.

**Step 6.** Return $\mathrm{Red}$.

### 4.3 Parameter Analysis of Feature Selection Method Based on Variable Precision Tolerance Rough Sets

The inclusion degree, the parameter cut-off point and the property are introduced in the subsection.

**Definition 14.** Given a decision table $DT = \langle U, C \cup D, V, f \rangle$, let $X_i \in U/F_{C,\tau}$ $(i = 1, 2, \cdots |U/F_{C,\tau}|)$, $Y_j \in U/F_{D,\tau}$ $(j = 1, 2, \cdots |U/F_{D,\tau}|)$, the inclusion degree of $X_i$ with respect to $Y_j$ is defined as:

$$ID(X_i, Y_j) = \begin{cases} \frac{|X_i \cap Y_j|}{|X_i|} & \text{if } |X_i| > 0 \\ 0 & \text{if } |X_i| = 0 \end{cases}$$

**Definition 15.** Let $X_i \in U/F_{C,\tau}$ $(i = 1, 2, \cdots |U/F_{C,\tau}|)$, $Y_j \in U/F_{D,\tau}$ $(j = 1, 2, \cdots |U/F_{D,\tau}|)$, the parameter cut-off point of $X_i$ with respect to $U/F_{D,\tau}$ is defined as:

$$\alpha_i = Max\left(ID(X_i, Y_j)\right)(j = 1, 2, \cdots |U/F_{D,\tau}|) \tag{27}$$

**Property 1.** Suppose $\alpha_i$ is the parameter cut-off point of $X_i$ with respect to $U/F_{D,\tau}$, $i = 1, 2, \cdots |U/F_{C,\tau}|$ and $\alpha_1 < \alpha_2 ... < \alpha_m ... < \alpha_{|U/F_{C,\tau}|}$. If given $\alpha_m$, then $POS^{\beta}_{B,\tau}(D)$ for every $\beta \, (0.5 < \alpha_m < \beta \leq \alpha_{m+1})$ is the same.

**Proof.** Suppose $0.5 < \alpha_m < \beta \leq \alpha_{m+1}$, with respect to $\forall \beta$ and $\forall i \leq m$, if the parameter cut-off point of $X_i$ with respect to $U/F_{D,\tau}$ is $\alpha_i < \beta$, then there is not $Y_p \in U/F_{D,\tau}$ which satisfies $\frac{|X_i \cap Y_p|}{|X_i|} \geq \beta$, therefore $\underline{B^{\beta}_{\tau}}Y_j = \emptyset$.

With respect to $\forall i \geq m+1$, if the parameter cut-off point of $X_i$ with respect to $U/F_{D,\tau}$ is $0.5 \leq \beta \leq \alpha_i$, then there is $Y_p \in U/F_{D,\tau}$ which satisfies $\frac{|X_i \cap Y_p|}{|X_i|} \geq \beta$, therefore $\underline{B^{\beta}_{\tau}}Y_j = X_i$.

## 5 A Simple Example

To illustrate the operation of feature selection method based on variable precision tolerance rough sets, it is applied to a simple example dataset in Table 1, which contains three real-valued conditional attributes, one real-valued decision feature and ten objects. Set $\tau = 0.2$. $C = \{M, N, P\}$. $D = \{K\}$.

The following tolerance classes are generated:

$U/F_{D,\tau} = \{\{o_1, o_2, o_4, o_5, o_6\}, \{o_3, o_7, o_8, o_9, o_{10}\}\}$,

$U/F_{C,\tau} = \{\{o_1\}, \{o_2, o_4\}, \{o_3, o_5\}, \{o_6, o_8, o_9\}, \{o_7, o_{10}\}\}$,

$U/F_{C-\{M\},\tau} = U/F_{\{N,P\},\tau} = \{\{o_1\}, \{o_2, o_4\}, \{o_3, o_5\}, \{o_6, o_8, o_9\}, \{o_7, o_{10}\}\}$,

$U/F_{C-\{N\},\tau} = U/F_{\{M,P\},\tau} = \{\{o_1\}, \{o_2, o_4\}, \{o_3, o_5\}, \{o_6, o_7, o_8, o_9, o_{10}\}\}$,

$U/F_{C-\{P\},\tau} = U/F_{\{M,N\},\tau} = \{\{o_1, o_2, o_4\}, \{o_3, o_5\}, \{o_6, o_8, o_9\}, \{o_7, o_{10}\}\}$,

$U/F_{C-\{M,N\},\tau} = U/F_{\{P\},\tau} = \{\{o_1\}, \{o_2, o_4\}, \{o_3, o_5\}, \{o_6, o_7, o_8, o_9, o_{10}\}\}$,

$U/F_{C-\{M,P\},\tau} = U/F_{\{N\},\tau} = \{\{o_1, o_2, o_4, o_6, o_8, o_9\}, \{o_3, o_5\}, \{o_7, o_{10}\}\}$,

$U/F_{C-\{N,P\},\tau} = U/F_{\{M\},\tau} = \{\{o_1, o_2, o_3, o_4, o_5\}, \{o_6, o_7, o_8, o_9, o_{10}\}\}$.

The reduct for DDTRS:

**Table 1.** Example dataset

| Objects | $M$ | $N$ | $P$ | $K$ |
|---------|-----|------|-----|------|
| $o_1$ | 9 | 0.12 | 1 | 0.4 |
| $o_2$ | 9 | 0.09 | 4 | 0.41 |
| $o_3$ | 10 | 0.3 | 9 | 1.3 |
| $o_4$ | 11 | 0.1 | 5 | 0.38 |
| $o_5$ | 12 | 0.28 | 9 | 0.39 |
| $o_6$ | 27 | 0.13 | 13 | 0.42 |
| $o_7$ | 28 | 0.2 | 14 | 1.4 |
| $o_8$ | 28 | 0.12 | 13 | 1.36 |
| $o_9$ | 29 | 0.11 | 14 | 1.32 |
| $o_{10}$ | 29 | 0.21 | 13 | 1.37 |

Set $\beta = 0.8$, $\beta$ - lower approximation of the decision classes are calculated as follows:

$$\underline{C_\tau^\beta}\{o_1,o_2,o_4,o_5,o_6\} = \underline{\{M,N,P\}_\tau^\beta}\{o_1,o_2,o_4,o_5,o_6\} = \{o_1,o_2,o_4\},$$

$$\underline{C_\tau^\beta}\{o_3,o_7,o_8,o_9,o_{10}\} = \underline{\{M,N,P\}_\tau^\beta}\{o_3,o_7,o_8,o_9,o_{10}\} = \{o_7,o_{10}\}.$$

Hence, the positive region can be constructed:

$$POS_{C,\tau}^\beta(D) = \cup_{Y_j \in U/F_{D,\tau}}^\beta \underline{C_\tau^\beta} Y_j = \underline{C_\tau^\beta}\{o_1,o_2,o_4,o_5,o_6\} \cup \underline{C_\tau^\beta}\{o_3,o_7,o_8,o_9,o_{10}\} =$$
$$\{o_1,o_2,o_4,o_7,o_{10}\}$$

The resulting degree of dependency is:

$$\gamma_{C,\tau}^\beta(D) = \frac{\left|POS_{C,\tau}^\beta(D)\right|}{|U|} = \frac{|\{o_1,o_2,o_4,o_7,o_{10}\}|}{|\{o_1,o_2,o_3,o_4,o_5,o_6,o_7,o_8,o_9,o_{10}\}|} = \frac{5}{10}.$$

For feature set $C - \{M\}$, the corresponding dependency degree is:

$$\gamma_{C-\{M\},\tau}^\beta(D) = \frac{\left|POS_{C-\{M\},\tau}^\beta(D)\right|}{|U|} = \frac{|\{o_1,o_2,o_4,o_7,o_{10}\}|}{|\{o_1,o_2,o_3,o_4,o_5,o_6,o_7,o_8,o_9,o_{10}\}|} = \frac{5}{10},$$

$$\gamma_{C-\{M\},\tau}^\beta(D) = \gamma_{\{N,P\},\tau}^\beta(D) = \gamma_{C,\tau}^\beta(D) = \frac{5}{10}.$$

Feature $M$ is deleted from feature set $C$. Similarly, the dependency degree of feature set $\{N,P\} - \{N\}$ is:

$$\gamma_{\{N,P\}-\{N\},\tau}^\beta(D) = \frac{\left|POS_{\{N,P\}-\{N\},\tau}^\beta(D)\right|}{|U|} = \frac{|\{o_1,o_2,o_4,o_6,o_7,o_8,o_9,o_{10}\}|}{|\{o_1,o_2,o_3,o_4,o_5,o_6,o_7,o_8,o_9,o_{10}\}|} = \frac{8}{10},$$

$$\gamma_{\{N,P\}-\{N\},\tau}^\beta(D) = \frac{8}{10} \neq \gamma_{C,\tau}^\beta(D) = \frac{5}{10}.$$

The reduct for the DDTRS algorithm is $\{N,P\}$.

Parameter analysis of DDTRS:

$X_1 = \{o_1\}$, $X_2 = \{o_2,o_4\}$, $X_3 = \{o_3,o_5\}$, $X_4 = \{o_6,o_8,o_9\}$, $X_5 = \{o_7,o_{10}\}$.
$Y_1 = \{o_1,o_2,o_4,o_5,o_6\}$, $Y_2 = \{o_3,o_7,o_8,o_9,o_{10}\}$.

$U/F_{D,\tau} = \{Y_1,Y_2\} = \{\{o_1,o_2,o_4,o_5,o_6\}, \{o_3,o_7,o_8,o_9,o_{10}\}\}$. $ID(X_1,Y_1) = \frac{|X_1 \cap Y_1|}{|X_1|} = 1$, $ID(X_1,Y_2) = \frac{|X_1 \cap Y_2|}{|X_1|} = 0$, so, the parameter cut-off point of $X_1$ with respect to $U/F_{D,\tau}$ is $\alpha_1 = Max(ID(X_1,Y_j))$ $(j = 1,2) = 1$, for $\beta \in (0.5,1]$, $\beta$ - lower approximation of $Y_j$ $(j = 1,2)$ with respect to $X_1$ is $X_1 = \{o_1\}$. Similarly, the parameter cut-off point of $X_2$ with respect to $U/F_{D,\tau}$ is $\alpha_2 = 1$, for $\beta \in (0.5,1]$, $\beta$ - lower approximation of $Y_j$ with respect to $X_2$ is $\{o_2,o_4\}$. The parameter cut-off point of $X_3$ with respect to $U/F_{D,\tau}$ is $\alpha_3 = 0.5$, for $\beta$,

$\beta$ - lower approximation of $Y_j$ with respect to $X_3$ is $\{o_3, o_5\}$. The parameter cut-off point of $X_4$ with respect to $U/F_{D,\tau}$ is $\alpha_4 = 0.667$, for $\beta \in (0.5, 0.667]$, $\beta$ - lower approximation of $Y_j$ with respect to $X_4$ is $\{o_6, o_8, o_9\}$ . The parameter cut-off point of $X_5$ with respect to $U/F_{D,\tau}$ is $\alpha_5 = 1$, for $\beta \in (0.5, 1]$, $\beta$ - lower approximation of $Y_j$ with respect to $X_5$ is $\{o_7, o_{10}\}$.

For $\beta \in (0.5, 0.667]$, $POS^{\beta}_{C,\tau}(D) = \cup_{Y_j \in U/F_{D,\tau}} \underline{C^{\beta}_{\tau}} Y_j = \{o_1\} \cup \{o_2, o_4\} \cup \{o_6, o_8, o_9\} \cup \{o_7, o_{10}\} = \{o_1, o_2, o_4, o_6, o_7, o_8, o_9, o_{10}\}$, $\gamma^{\beta}_{C,\tau}(D) = \frac{\left| POS^{\beta}_{C,\tau}(D) \right|}{|U|} = \frac{8}{10}$; For $\beta \in (0.667, 1]$, $POS^{\beta}_{C,\tau}(D) = \{o_1, o_2, o_4, o_7, o_{10}\}$, $\gamma^{\beta}_{C,\tau}(D) = \frac{5}{10}$.

According to feature selection method based on variable precision tolerance rough sets, for different $\beta$, the results are shown in Table 2. $A_1$ denotes conditional feature set. $A_2$ is the parameter cut-off point of $X_i$ with respect to $U/F_{D,\tau}$. $\beta$ - lower approximation of $Y_j$ $(j = 1, 2)$ with respect to $X_i$ is denoted by $A_3$. $A_4$ is parameter interval. $A_5$ and $A_6$ are dependency degree and positive region respectively.

**Table 2.** Results based on different $\beta$

| Parameter interval | Dependency degree | Reduct |
|---|---|---|
| $\beta \in (0.5, 0.667]$ | $\frac{8}{10}$ | $\{P\}$ or $\{N\}$ |
| $\beta \in (0.667, 0.8]$ | $\frac{5}{10}$ | $\{N, P\}$ |
| $\beta \in (0.8, 1]$ | $\frac{5}{10}$ | $\{P\}$ |

Different selection results can be got according to different interval of parameter $\beta$, as shown in Table 3. As listed in Table 3, selection results and dependency degree are different when $\beta$ is changed. For $\beta \in (0.5, 0.667]$, the reduct is $\{P\}$ or $\{N\}$. For $\beta \in (0.8, 1]$, the reduct is $\{P\}$. The dependency degree of $\beta \in (0.5, 0.667]$ are differ from that of $\beta \in (0.8, 1]$. The dependency degrees of $\beta \in (0.667, 0.8]$ and $\beta \in (0.8, 1]$ are $\frac{5}{10}$, while selection results are different.

## 6   Experiments

To evaluate the performance of the proposed algorithm, we applied it to one dataset from UCI database. There are eight conditional attributes and one decision feature, which are L-Surf, L-O2, L-Bp, Surf-Stbl, Bp-Stbl, Core-Stbl, Comfort and Adm-Decs. Surf-Stbl, Bp-Stbl, Core-Stbl and Adm-Decs are crisp-valued attributes. Others are real-valued features.

When there are missing values in dataset, these values are filled with mean values for continuous features and majority values for nominal features [8]. According to DDTRS, set $\tau = 0.2$, there are 70 tolerance classes of condition attributes and 3 tolerance classes of decision feature. Different selection results can be got according to different interval of parameter $\beta$. The results are shown in Table 4. $m_1$, $m_2$, $m_3$, $m_4$, $m_5$, $m_6$, $m_7$ and $m_8$ denote L-Core, L-Surf, L-O2, L-Bp, Surf-Stbl, Bp-Stbl, Core-Stbl and Comfort.

**Table 3.** Selection results based on different $\beta$

| $A_1$ | $A_2$ | $A_3$ | $A_4$ | $A_5$ | $A_6$ |
|---|---|---|---|---|---|
| $\{M, N, P\}$ | $a_1 = 1$<br>$a_2 = 1$<br>$a_3 = 0.5$<br>$a_4 = 0.667$<br>$a_5 = 1$ | $\{o_1\}$<br>$\{o_2, o_4\}$<br>$\{o_3, o_5\}$<br>$\{o_6, o_8, o_9\}$<br>$\{o_7, o_{10}\}$ | $\beta \in (0.5, 0.667]$<br>$\beta \in (0.667, 1]$ | $\frac{8}{10}$<br>$\frac{5}{10}$ | $\{o_1, o_2, o_4, o_6,$<br>$o_7, o_8, o_9, o_{10}\}$<br>$\{o_1, o_2, o_4, o_7, o_{10}\}$ |
| $\{N, P\}$ | $a_1 = 1$<br>$a_2 = 1$<br>$a_3 = 0.5$<br>$a_4 = 0.667$<br>$a_5 = 1$ | $\{o_1\}$<br>$\{o_2, o_4\}$<br>$\{o_3, o_5\}$<br>$\{o_6, o_8, o_9\}$<br>$\{o_7, o_{10}\}$ | $\beta \in (0.5, 0.667]$<br>$\beta \in (0.667, 1]$ | $\frac{8}{10}$<br>$\frac{5}{10}$ | $\{o_1, o_2, o_4, o_6,$<br>$o_7, o_8, o_9, o_{10}\}$<br>$\{o_1, o_2, o_4, o_7, o_{10}\}$ |
| $\{M, P\}$ | $a_1 = 1$<br>$a_2 = 1$<br>$a_3 = 0.5$<br>$a_4 = 0.8$ | $\{o_1\}$<br>$\{o_2, o_4\}$<br>$\{o_3, o_5\}$<br>$\{o_6, o_7, o_8,$<br>$o_9, o_{10}\}$ | $\beta \in (0.5, 0.8]$<br>$\beta \in (0.8, 1]$ | $\frac{8}{10}$<br>$\frac{3}{10}$ | $\{o_1, o_2, o_4, o_6,$<br>$o_7, o_8, o_9, o_{10}\}$<br>$\{o_1, o_2, o_4\}$ |
| $\{M, N\}$ | $a_1 = 0.667$<br>$a_2 = 0.5$<br>$a_3 = 0.667$<br>$a_4 = 1$ | $\{o_1, o_2, o_4\}$<br>$\{o_3, o_5\}$<br>$\{o_6, o_8, o_9\}$<br>$\{o_7, o_{10}\}$ | $\beta \in (0.5, 0.667]$<br>$\beta \in (0.667, 1]$ | $\frac{8}{10}$<br>$\frac{2}{10}$ | $\{o_1, o_2, o_4, o_6,$<br>$o_7, o_8, o_9, o_{10}\}$<br>$\{o_7, o_{10}\}$ |
| $\{P\}$ | $a_1 = 1$<br>$a_2 = 1$<br>$a_3 = 0.5$<br>$a_4 = 0.8$<br>$a_5 = 1$ | $\{o_1\}$<br>$\{o_2, o_4\}$<br>$\{o_3, o_5\}$<br>$\{o_6, o_7, o_8$<br>$, o_9, o_{10}\}$<br>$\{o_7, o_{10}\}$ | $\beta \in (0.5, 0.8]$<br>$\beta \in (0.8, 1]$ | $\frac{8}{10}$<br>$\frac{5}{10}$ | $\{o_1, o_2, o_4, o_6,$<br>$o_7, o_8, o_9, o_{10}\}$<br>$\{o_1, o_2, o_4, o_7, o_{10}\}$ |
| $\{M\}$ | $a_1 = 0.8$<br>$a_2 = 0.8$ | $\{o_1, o_2, o_3,$<br>$o_4, o_5\}$<br>$\{o_6, o_7, o_8,$<br>$o_9, o_{10}\}$ | $\beta \in (0.5, 0.8]$<br>$\beta \in (0.8, 1]$ | $\frac{10}{10}$<br>$\frac{0}{10}$ | $\{o_1, o_2, o_3, o_4, o_5,$<br>$o_6, o_7, o_8, o_9, o_{10}\}$<br>$\phi$ |
| $\{N\}$ | $a_1 = 0.667$<br>$a_2 = 0.5$<br>$a_3 = 1$ | $\{o_1, o_2, o_4,$<br>$o_6, o_8, o_9\}$<br>$\{o_3, o_5\}$<br>$\{o_7, o_10\}$ | $\beta \in (0.5, 0.667]$<br>$\beta \in (0.667, 1]$ | $\frac{8}{10}$<br>$\frac{2}{10}$ | $\{o_1, o_2, o_4, o_6,$<br>$o_7, o_8, o_9, o_{10}\}$<br>$\{o_7, o_{10}\}$ |

The parameter cut-off points are 0.571, 0.583, 0.6, 0.625, 0.667 and 0.714. With the change of parameter $\beta$, the dependency degree and reduct are changed.

**Table 4.** Selection results based on different $\beta$

| Parameter interval | Dependency degree | Reduct |
|---|---|---|
| $\beta \in (0.5, 0.571]$ | $\frac{78}{90}$ | $\{m_2, m_4, m_7\}$ or $\{m_1, m_5, m_6, m_8\}$ or $\{m_4, m_5, m_6, m_7\}$ |
| $\beta \in (0.571, 0.583]$ | $\frac{78}{90}$ | $\{m_2, m_3, m_5, m_6\}$ |
| $\beta \in (0.583, 0.6]$ | $\frac{78}{90}$ | $\{m_1, m_3, m_4, m_7\}$ or $\{m_3, m_5, m_6, m_7, m_8\}$ or $\{m_1, m_2, m_3, m_5, m_7, m_8\}$ |
| $\beta \in (0.6, 0.625]$ | $\frac{78}{90}$ | $\{m_4, m_8\}$ or $\{m_3, m_6, m_8\}$ or $\{m_1, m_3, m_4, m_5\}$ |
| $\beta \in (0.625, 0.667]$ | $\frac{78}{90}$ | $\{m_1, m_3, m_8\}$ or $\{m_4, m_6, m_7\}$ |
| $\beta \in (0.667, 0.714]$ | $\frac{72}{90}$ | $\{m_1, m_3, m_8\}$ |
| $\beta \in (0.714, 1]$ | $\frac{72}{90}$ | $\{m_1, m_2, m_3, m_4, m_5, m_7, m_8\}$ |

# 7 Conclusions

In this paper, we address feature selection method based on variable precision tolerance rough sets. This paper extends the research of traditional rough sets.

**Acknowledgments.** This paper is supported by The Youth project of National Social Science Foundation of china (no. 13CFX049) and Shanghai University Young Teacher Training Program (no. hdzf10008).

# References

1. Pawlak, Z.: Rough Sets. International Journal of Information Computer Science 11(5), 341–356 (1982)
2. Wang, G.Y.: Rough Set Theory and Knowledge Acquisition. Jiaotong University Press, Xi'an (2001) (in Chinese)
3. Miao, D.Q., Wang, J.: Information-Based Algorithm for Reduction of Knowledge. In: IEEE International Conference on Intelligent Processing Systems, pp. 1155–1158 (1997)
4. Grzymala-Busse, J.W.: Discretization of Numerical Attributes. In: Klösgen, W., Zytkow, J. (eds.) Handbook of Data Mining and Knowledge Discovery, pp. 218–225. Oxford University Press (2002)
5. Parthalin, N.M., Shen, Q.: Exploring The Boundary Region of Tolerance Rough Sets for Feature Selection. Pattern Recognition 42, 655–667 (2009)
6. Ziarko, W.: Variable precision rough set model. Journal of Computer and System Sciences 46, 39–59 (1993)
7. Zhang, H.Y., Leung, Y., Zhou, L.: Variable-precision-dominance-based rough set approach to interval-valued information systems. Information Sciences 244, 75–272 (2013)
8. Grzymala-Busse, J.W., Grzymala-Busse, W.J.: Handling Missing Attribute Values. In: Maimon, O., Rokach, L. (eds.) Handbook of Data Mining and Knowledge Discovery, pp. 37–57 (2005)

# Incremental Approaches to Computing Approximations of Sets in Dynamic Covering Approximation Spaces

Guangming Lang[1],*, Qingguo Li[2], Mingjie Cai[2], and Qimei Xiao[1]

[1] School of Mathematics and Computer Science, Changsha University of Science and
Technology, Changsha, Hunan 410114, P.R. China
langguangming1984@126.com, qimeixiao@sohu.com
[2] College of Mathematics and Econometrics,
Hunan University, Changsha, Hunan 410082, P.R. China
liqingguoli@aliyun.com, cmjlong@163.com

**Abstract.** In practical situations, it is of interest to investigate computing approximations of sets as an important step of attribute reduction in dynamic covering information systems. In this paper, we present incremental approaches to computing the type-1 and type-2 characteristic matrices of coverings with the variation of elements. Then we construct the second and sixth lower and upper approximations of sets by using incremental approaches from the view of matrices. We also employ examples to show how to compute approximations of sets by using the incremental and non-incremental approaches in dynamic covering approximation spaces.

**Keywords:** Rough sets, Covering information system, Boolean matrice, Characteristic matrice.

## 1 Introduction

Covering-based rough set theory [7], as a powerful mathematical tool for studying covering approximation spaces, has attracted a lot of attention of researchers in various fields of sciences. Especially, various kinds of approximation operators have been proposed for covering approximation spaces. Recently, Wang et al. [6] transformed the computation of approximations of a set into products of the characteristic matrices and the characteristic function of the set. However, it paid little attention to approaches to calculating the characteristic matrices. In practice, the covering approximation space varies with time due to the characteristics of data collection, and the non-incremental approach to constructing the characteristic matrices is often very costly or even intractable in dynamic covering approximation spaces. It is necessary to present effective approaches to computing characteristic matrices of dynamic coverings.

To the best of our knowledge, researchers [1–5, 8, 9] have focused on computing approximations of sets. For instance, Chen et al. [1, 2] constructed approximations of sets when coarsening or refining attribute values. Li et al. [3] computed approximations in dominance-based rough sets approach under the variation of attribute set.

---

* Corresponding author.

D. Miao et al. (Eds.): RSKT 2014, LNAI 8818, pp. 510–521, 2014.
DOI: 10.1007/978-3-319-11740-9_47 © Springer International Publishing Switzerland 2014

Luo et al. [4, 5] studied dynamic maintenance of approximations in set-valued ordered decision systems under the attribute generalization and the variation of object set. Zhang et al. [8,9] updated rough set approximations based on relation matrices and investigated neighborhood rough sets for dynamic data mining. These works demonstrate that incremental approaches are effective and efficient for computing approximations of sets. It motivates us to apply an incremental updating scheme to conduct approximations of sets by using characteristic matrices in dynamic covering approximation spaces, which will provide an effective approach to computing approximations of sets from the view of matrices.

The purpose of this paper is to compute approximations of sets by using incremental approaches in dynamic covering approximation spaces. First, we present incremental approaches to computing the type-1 and type-2 characteristic matrices in dynamic covering approximation spaces. We mainly focus on the situation: the variation of elements in coverings when adding and deleting objects. Furthermore, we provide incremental algorithms for constructing the second and sixth lower and upper approximations of sets based on the type-1 and type-2 characteristic matrices, respectively. We compare computation complexities of the incremental algorithms with those of non-incremental algorithms. Several examples are employed to illustrate that calculating approximations of sets is simplified greatly by utilizing the proposed approach.

The rest of this paper is organized as follows: Section 2 briefly reviews the basic concepts of covering-based rough set theory. In Section 3, we introduce incremental approaches to computing the type-1 and type-2 characteristic matrices with respect to immigration of elements of coverings when adding and deleting objects. We also present incremental algorithms of calculating the second and sixth lower and upper approximations of sets by using the type-1 and type-2 characteristic matrices, respectively. We also employ examples to show that how to compute approximations of sets by using incremental and non-incremental approaches in dynamic covering approximation spaces. We conclude the paper in Section 4.

## 2 Preliminaries

In this section, we review some concepts of covering-based rough sets.

**Definition 1.** *[7] Let $U$ be a finite universe of discourse, and $\mathscr{C}$ a family of subsets of $U$. $\mathscr{C}$ is called a covering of $U$ if none of elements of $\mathscr{C}$ is empty and $\bigcup\{C|C \in \mathscr{C}\} = U$.*

**Definition 2.** *[6] Let $U = \{x_1, x_2, ..., x_n\}$ be a finite universe, and $\mathscr{C} = \{C_1, C_2, ..., C_m\}$ a covering of $U$. For any $X \subseteq U$, the second and sixth upper and lower approximations of $X$ with respect to $\mathscr{C}$ are defined as follows:*
*(1) $SH_{\mathscr{C}}(X) = \bigcup\{C \in \mathscr{C}|C \cap X \neq \emptyset\}$, $SL_{\mathscr{C}}(X) = [SH_{\mathscr{C}}(X^c)]^c$;*
*(2) $XH_{\mathscr{C}}(X) = \{x \in U|N(x) \cap X \neq \emptyset\}$, $XL_{\mathscr{C}}(X) = \{x \in U|N(x) \subseteq X\}$,*
*where $N(x) = \bigcap\{C_i|x \in C_i \in \mathscr{C}\}$.*

**Definition 3.** *[6] Let $U = \{x_1, x_2, ..., x_n\}$ be a finite universe, $\mathscr{C} = \{C_1, C_2, ..., C_m\}$ a family of subsets of $U$, and $M_{\mathscr{C}} = (a_{ij})_{n \times m}$, where $a_{ij} = \begin{cases} 1, & x_i \in C_j; \\ 0, & x_i \notin C_j. \end{cases}$ Then*

*(1) $\Gamma(\mathscr{C}) = M_{\mathscr{C}} \cdot M_{\mathscr{C}}^T = (b_{ij})_{n \times n}$ is called the type-1 characteristic matrice of $\mathscr{C}$, where $b_{ij} = \bigvee_{k=1}^m (a_{ik} \cdot a_{jk})$;*

*(2) $\prod(\mathscr{C}) = M_{\mathscr{C}} \odot M_{\mathscr{C}}^T = (c_{ij})_{n \times n}$ is called the type-2 characteristic matrice of $\mathscr{C}$, where $c_{ij} = \bigwedge_{k=1}^m (a_{kj} - a_{ik} + 1)$.*

Especially, we have the characteristic function $X_X = \begin{bmatrix} a_1 & a_2 & ... & a_n \end{bmatrix}^T$ of $X \subseteq U$, where $a_i = \begin{cases} 1, & x_i \in X; \\ 0, & x_i \notin X. \end{cases}$ In what follows, we have another descriptions of the second and sixth lower and upper approximation operators.

**Definition 4.** *[6] Let $U = \{x_1, x_2, ..., x_n\}$ be a finite universe, $\mathscr{C} = \{C_1, C_2, ..., C_m\}$ a covering of $U$, and $X_X$ the characteristic function of $X$ in $U$. Then*

*(1) $X_{SH(X)} = \Gamma(\mathscr{C}) \cdot X_X$, $X_{SL(X)} = \Gamma(\mathscr{C}) \odot X_X$; (2) $X_{XH(X)} = \prod(\mathscr{C}) \cdot X_X$, $X_{XL(X)} = \prod(\mathscr{C}) \odot X_X$.*

## 3   Computing Approximations of Sets with Variations of Elements in Coverings

In this section, we introduce incremental approaches to computing the second and sixth lower and upper approximation of sets with the variation of coverings when adding and deleting objects.

**Definition 5.** *Let $(U, \mathscr{C})$ and $(U^+, \mathscr{C}^+)$ be covering approximation spaces, where $U = \{x_1, x_2, ..., x_n\}$, $U^+ = U \cup \{x_{n+1}\}$, $\mathscr{C} = \{C_1, C_2, ..., C_m\}$, $\mathscr{C}^+ = \{C_1^+, C_2^+, ..., C_m^+, C_{m+1}^+\}$, where $C_i^+ = C_i \cup \{x_{n+1}\}$ or $C_i$ $(1 \leq i \leq m)$, and $x_{n+1} \in C_{m+1}^+$. Then $(U^+, \mathscr{C}^+)$ is called a AE-covering approximation space.*

By Definition 5, $\mathscr{C}^+$ is referred to as a AE-covering. In practice, there are several types of coverings when adding objects. For simplicity, we only discuss the AE-coverings.

In what follows, we discuss how to construct $\Gamma(\mathscr{C}^+)$ based on $\Gamma(\mathscr{C})$. For convenience, we denote $M_{\mathscr{C}} = (a_{ij})_{n \times m}$, $M_{\mathscr{C}^+} = (a_{ij})_{(n+1) \times (m+1)}$, $\Gamma(\mathscr{C}) = (b_{ij})_{n \times n}$ and $\Gamma(\mathscr{C}^+) = (c_{ij})_{(n+1) \times (n+1)}$.

**Theorem 1.** *Let $(U^+, \mathscr{C}^+)$ be a AE-covering approximation space of $(U, \mathscr{C})$, $\Gamma(\mathscr{C})$ and $\Gamma(\mathscr{C}^+)$ the type-1 characteristic matrices of $\mathscr{C}$ and $\mathscr{C}^+$, respectively. Then*

$$\Gamma(\mathscr{C}^+) = \begin{bmatrix} \Gamma(\mathscr{C}) & 0 \\ 0 & 0 \end{bmatrix} \bigvee \begin{bmatrix} \triangle_1 \Gamma(\mathscr{C}) & (\triangle_2 \Gamma(\mathscr{C}))^T \\ \triangle_2 \Gamma(\mathscr{C}) & c_{(n+1)(n+1)} \end{bmatrix},$$

*where*

$$\triangle_1\Gamma(\mathscr{C}) = \begin{bmatrix} a_{1(m+1)} \ a_{2(m+1)} \ \cdots \ a_{n(m+1)} \end{bmatrix}^T \cdot \begin{bmatrix} a_{1(m+1)} \ a_{2(m+1)} \ \cdots \ a_{n(m+1)} \end{bmatrix};$$

$$\triangle_2\Gamma(\mathscr{C}) = \begin{bmatrix} a_{(n+1)1} \ a_{(n+1)2} \ \cdots \ a_{(n+1)(m+1)} \end{bmatrix} \cdot \begin{bmatrix} a_{11} & a_{12} & \cdots & a_{1(m+1)} \\ a_{21} & a_{22} & \cdots & a_{2(m+1)} \\ & & \cdots & \\ & & \cdots & \\ & & \cdots & \\ a_{n1} & a_{n2} & \cdots & a_{n(m+1)} \end{bmatrix}^T;$$

$$c_{(n+1)(n+1)} = \begin{bmatrix} a_{(n+1)1} \ a_{(n+1)2} \ \cdots \ a_{(n+1)(m+1)} \end{bmatrix} \cdot \begin{bmatrix} a_{(n+1)1} \ a_{(n+1)2} \ \cdots \ a_{(n+1)(m+1)} \end{bmatrix}^T.$$

**Proof.** By Definition 3, we get $\Gamma(\mathscr{C})$ and $\Gamma(\mathscr{C}^+)$ as follows:

$$\Gamma(\mathscr{C}) = M_{\mathscr{C}} \cdot M_{\mathscr{C}}^T = \begin{bmatrix} a_{11} & a_{12} & \cdots & a_{1m} \\ a_{21} & a_{22} & \cdots & a_{2m} \\ & & \cdots & \\ & & \cdots & \\ & & \cdots & \\ a_{n1} & a_{n2} & \cdots & a_{nm} \end{bmatrix} \begin{bmatrix} a_{11} & a_{12} & \cdots & a_{1m} \\ a_{21} & a_{22} & \cdots & a_{2m} \\ & & \cdots & \\ & & \cdots & \\ & & \cdots & \\ a_{n1} & a_{n2} & \cdots & a_{nm} \end{bmatrix}^T = \begin{bmatrix} b_{11} & b_{12} & \cdots & b_{1n} \\ b_{21} & b_{22} & \cdots & b_{2n} \\ & & \cdots & \\ & & \cdots & \\ & & \cdots & \\ b_{n1} & b_{n2} & \cdots & b_{nn} \end{bmatrix};$$

$$\Gamma(\mathscr{C}^+) = M_{\mathscr{C}^+} \cdot M_{\mathscr{C}^+}^T = \begin{bmatrix} a_{11} & a_{12} & \cdots & a_{1m} & a_{1(m+1)} \\ a_{21} & a_{22} & \cdots & a_{2m} & a_{2(m+1)} \\ & & \cdots & & \\ & & \cdots & & \\ a_{n1} & a_{n2} & \cdots & a_{nm} & a_{n(m+1)} \\ a_{(n+1)1} & a_{(n+1)2} & \cdots & a_{(n+1)m} & a_{(n+1)(m+1)} \end{bmatrix} \begin{bmatrix} a_{11} & a_{12} & \cdots & a_{1m} & a_{1(m+1)} \\ a_{21} & a_{22} & \cdots & a_{2m} & a_{2(m+1)} \\ & & \cdots & & \\ & & \cdots & & \\ a_{n1} & a_{n2} & \cdots & a_{nm} & a_{n(m+1)} \\ a_{(n+1)1} & a_{(n+1)2} & \cdots & a_{(n+1)m} & a_{(n+1)(m+1)} \end{bmatrix}^T$$

$$= \begin{bmatrix} c_{11} & c_{12} & \cdots & c_{1n} & c_{1(n+1)} \\ c_{21} & c_{22} & \cdots & c_{2n} & c_{2(n+1)} \\ & & \cdots & & \\ & & \cdots & & \\ & & \cdots & & \\ c_{n1} & c_{n2} & \cdots & c_{nn} & c_{n(n+1)} \\ c_{(n+1)1} & c_{(n+1)2} & \cdots & c_{(n+1)n} & c_{(n+1)(n+1)} \end{bmatrix}.$$

In the sense of the type-1 characteristic matrice, we see that $b_{ij} \vee (a_{i(m+1)} \cdot a_{j(m+1)}) = c_{ij}$ for $1 \le i, j \le n$. To compute $\Gamma(\mathscr{C}^+)$ on the basis of $\Gamma(\mathscr{C})$, we only need to compute $\triangle_1\Gamma(\mathscr{C})$, $\triangle_2\Gamma(\mathscr{C})$ and $c_{(n+1)(n+1)}$. Concretely, $\triangle_1\Gamma(\mathscr{C})$, $\triangle_2\Gamma(\mathscr{C})$ and $c_{(n+1)(n+1)}$ are computed as follows:

$$\triangle_1\Gamma(\mathscr{C}) = (a_{i(m+1)} \cdot a_{j(m+1)})_{(1 \le i, j \le n)} = \begin{bmatrix} a_{1(m+1)} \ a_{2(m+1)} \ \cdots \ a_{n(m+1)} \end{bmatrix}^T \cdot \begin{bmatrix} a_{1(m+1)} \ a_{2(m+1)} \ \cdots \ a_{n(m+1)} \end{bmatrix};$$

$$\triangle_2\Gamma(\mathscr{C}) = \begin{bmatrix} c_{(n+1)1} \ c_{(n+1)2} \ \cdots \ c_{(n+1)n} \end{bmatrix} = \begin{bmatrix} a_{1(m+1)} \ a_{2(m+1)} \ \cdots \ a_{n(m+1)} \end{bmatrix} \cdot \begin{bmatrix} a_{11} & a_{12} & \cdots & a_{1(m+1)} \\ a_{21} & a_{22} & \cdots & a_{2(m+1)} \\ & & \cdots & \\ & & \cdots & \\ a_{n1} & a_{n2} & \cdots & a_{n(m+1)} \end{bmatrix}^T;$$

$$c_{(n+1)(n+1)} = \begin{bmatrix} a_{(n+1)1} \ a_{(n+1)2} \ \cdots \ a_{(n+1)(m+1)} \end{bmatrix} \cdot \begin{bmatrix} a_{(n+1)1} \ a_{(n+1)2} \ \cdots \ a_{(n+1)(m+1)} \end{bmatrix}^T.$$

Therefore, we have $\Gamma(\mathscr{C}^+) = \begin{bmatrix} \Gamma(\mathscr{C}) & 0 \\ 0 & 0 \end{bmatrix} \vee \begin{bmatrix} \Delta_1\Gamma(\mathscr{C}) & (\Delta_2\Gamma(\mathscr{C}))^T \\ \Delta_2\Gamma(\mathscr{C}) & c_{(n+1)(n+1)} \end{bmatrix}$.

We show an incremental algorithm of computing the second lower and upper approximations of sets as follows.

Input: $(U, \mathscr{C})$, $(U^+, \mathscr{C}^+)$ and $X \subseteq U^+$. Output: $\mathcal{X}_{SH(X)}$ and $\mathcal{X}_{SL(X)}$.

Step 1: Calculating $\Gamma(\mathscr{C}) = M_{\mathscr{C}} \cdot M_{\mathscr{C}}^T$, where $M_{\mathscr{C}} = (a_{ij})_{n \times m}$;

Step 2: Computing $\Delta_1\Gamma(\mathscr{C})$ and $\Delta_2\Gamma(\mathscr{C})$, where

$$M_{\mathscr{C}^+} = (a_{ij})_{(n+1) \times (m+1)}; \alpha_1 = \begin{bmatrix} a_{1(m+1)} & a_{2(m+1)} & \cdots & a_{n(m+1)} \end{bmatrix}; \alpha_2 = \begin{bmatrix} a_{(n+1)1} & a_{(n+1)2} & \cdots & a_{(n+1)m} \end{bmatrix};$$

$$\Delta_1\Gamma(\mathscr{C}) = \alpha_1^T \cdot \alpha_1; \Delta_2\Gamma(\mathscr{C}) = \begin{bmatrix} \alpha_2 & a_{(n+1)(m+1)} \end{bmatrix} \cdot \begin{bmatrix} M_{\mathscr{C}} & \alpha_1^T \end{bmatrix}^T; c_{(n+1)(n+1)} = \begin{bmatrix} \alpha_2 & a_{(n+1)(m+1)} \end{bmatrix} \cdot \begin{bmatrix} \alpha_2 & a_{(n+1)(m+1)} \end{bmatrix}^T.$$

Step 3: Constructing $\Gamma(\mathscr{C}^+)$, where

$$\Gamma(\mathscr{C}^+) = (c_{ij})_{(n+1)(n+1)} = \begin{bmatrix} \Gamma(\mathscr{C}) & 0 \\ 0 & 0 \end{bmatrix} \vee \begin{bmatrix} \Delta_1\Gamma(\mathscr{C}) & (\Delta_2\Gamma(\mathscr{C}))^T \\ \Delta_2\Gamma(\mathscr{C}) & c_{(n+1)(n+1)} \end{bmatrix}.$$

Step 4: Obtaining $\mathcal{X}_{SH(X)}$ and $\mathcal{X}_{SL(X)}$, where $\mathcal{X}_{SH(X)} = \Gamma(\mathscr{C}^+) \cdot \mathcal{X}_X$; $\mathcal{X}_{SL(X)} = \Gamma(\mathscr{C}^+) \odot \mathcal{X}_X$.

The time complexity of computing the second lower and upper approximations of sets is less than $O(3n^2 + 2nm + 8n + m + 4)$ by using the incremental algorithm. Furthermore, $O((n + 1)^2 \cdot (m + 1))$ is the time complexity of the non-incremental algorithm. Therefore, the time complexity of the incremental algorithm is lower than that of the non-incremental algorithm.

The following example is employed to show the process of constructing approximations of sets by using the incremental algorithm.

*Example 1.* Let $U = \{x_1, x_2, x_3, x_4\}$, $U^+ = U \cup \{x_5\}$, $\mathscr{C} = \{C_1, C_2, C_3\}$, $\mathscr{C}^+ = \{C_1^+, C_2^+, C_3^+, C_4^+\}$, where $C_1 = \{x_1, x_4\}$, $C_2 = \{x_1, x_2, x_4\}$, $C_3 = \{x_3, x_4\}$, $C_1^+ = \{x_1, x_4, x_5\}$, $C_2^+ = \{x_1, x_2, x_4, x_5\}$, $C_3^+ = \{x_3, x_4\}$, $C_4^+ = \{x_3, x_5\}$, and $X = \{x_3, x_4, x_5\}$. By Definition 3, we first have that

$$\Gamma(\mathscr{C}) = M_{\mathscr{C}} \cdot M_{\mathscr{C}}^T = \begin{bmatrix} 1 & 1 & 0 \\ 0 & 1 & 0 \\ 0 & 0 & 1 \\ 1 & 1 & 1 \end{bmatrix} \cdot \begin{bmatrix} 1 & 1 & 0 \\ 0 & 1 & 0 \\ 0 & 0 & 1 \\ 1 & 1 & 1 \end{bmatrix}^T = \begin{bmatrix} 1 & 1 & 0 & 1 \\ 1 & 1 & 0 & 1 \\ 0 & 0 & 1 & 1 \\ 1 & 1 & 1 & 1 \end{bmatrix}.$$

Second, by Theorem 1, we get that

$$\Delta_1\Gamma(\mathscr{C}) = \begin{bmatrix} 0 & 0 & 1 & 0 \end{bmatrix}^T \cdot \begin{bmatrix} 0 & 0 & 1 & 0 \end{bmatrix} = \begin{bmatrix} 0 & 0 & 0 & 0 \\ 0 & 0 & 0 & 0 \\ 0 & 0 & 1 & 0 \\ 0 & 0 & 0 & 0 \end{bmatrix}; \Delta_2\Gamma(\mathscr{C}) = \begin{bmatrix} 1 & 1 & 0 & 1 \end{bmatrix} \cdot \begin{bmatrix} 1 & 1 & 0 & 0 \\ 0 & 1 & 0 & 0 \\ 0 & 0 & 1 & 1 \\ 1 & 1 & 1 & 0 \end{bmatrix}^T = \begin{bmatrix} 1 & 1 & 1 & 1 \end{bmatrix};$$

$$c_{55} = \begin{bmatrix} 1 & 1 & 0 & 1 \end{bmatrix} \cdot \begin{bmatrix} 1 & 1 & 0 & 1 \end{bmatrix}^T = 1.$$

Thus, we obtain that

$$\Gamma(\mathscr{C}^+) = (c_{ij})_{55} = \begin{bmatrix} \Gamma(\mathscr{C}) & 0 \\ 0 & 0 \end{bmatrix} \bigvee \begin{bmatrix} \Delta_1\Gamma(\mathscr{C}) & (\Delta_2\Gamma(\mathscr{C}))^T \\ \Delta_2\Gamma(\mathscr{C}) & c_{55} \end{bmatrix} = \begin{bmatrix} 1\,1\,0\,1\,0 \\ 1\,1\,0\,1\,0 \\ 0\,0\,1\,1\,0 \\ 1\,1\,1\,1\,0 \\ 0\,0\,0\,0\,0 \end{bmatrix} \bigvee \begin{bmatrix} 0\,0\,0\,0\,1 \\ 0\,0\,0\,0\,1 \\ 0\,0\,1\,0\,1 \\ 0\,0\,0\,0\,1 \\ 1\,1\,1\,1\,1 \end{bmatrix} = \begin{bmatrix} 1\,1\,0\,1\,1 \\ 1\,1\,0\,1\,1 \\ 0\,0\,1\,1\,1 \\ 1\,1\,1\,1\,1 \\ 1\,1\,1\,1\,1 \end{bmatrix}.$$

On the other hand, by Definition 3, we have $\Gamma(\mathscr{C}^+) = M_{\mathscr{C}^+} \cdot M_{\mathscr{C}^+}^T = \begin{bmatrix} 1\,1\,0\,1\,1 \\ 1\,1\,0\,1\,1 \\ 0\,0\,1\,1\,1 \\ 1\,1\,1\,1\,1 \\ 1\,1\,1\,1\,1 \end{bmatrix}.$

It is obvious that the time complexity is more than the incremental approach. By Definition 4, we have that

$$X_{SH(X)} = \Gamma(\mathscr{C}^+) \cdot X_X = \begin{bmatrix} 1\,1\,0\,1\,1 \\ 1\,1\,0\,1\,1 \\ 0\,0\,1\,1\,1 \\ 1\,1\,1\,1\,1 \\ 1\,1\,1\,1\,1 \end{bmatrix} \cdot \begin{bmatrix} 0 \\ 0 \\ 1 \\ 1 \\ 1 \end{bmatrix} = \begin{bmatrix} 1\,1\,1\,1\,1 \end{bmatrix}^T;$$

$$X_{SL(X)} = \Gamma(\mathscr{C}^+) \odot X_X = \begin{bmatrix} 1\,1\,0\,1\,1 \\ 1\,1\,0\,1\,1 \\ 0\,0\,1\,1\,1 \\ 1\,1\,1\,1\,1 \\ 1\,1\,1\,1\,1 \end{bmatrix} \odot \begin{bmatrix} 0 \\ 0 \\ 1 \\ 1 \\ 1 \end{bmatrix} = \begin{bmatrix} 0\,0\,1\,0\,0 \end{bmatrix}^T.$$

Therefore, $SH(X) = \{x_1, x_2, x_3, x_4, x_5\}$ and $SL(X) = \{x_3\}$.

In Example 1, there is a need to compute all elements in $\Gamma(\mathscr{C}^+)$ on the basis of the type-1 characteristic matrice. But we only need to calculate elements in $\Delta_1\Gamma(\mathscr{C})$, $\Delta_2\Gamma(\mathscr{C})$ and $c_{55}$ by Theorem 1. Thereby, the incremental algorithm is effective to compute the second lower and upper approximations of sets.

In practice, there is also a need to construct the type-2 characteristic matrices of AE-coverings for computing the sixth lower and upper approximations of sets. Subsequently, we construct $\prod(\mathscr{C}^+)$ based on $\prod(\mathscr{C})$. For convenience, we denote $\prod(\mathscr{C}) = (d_{ij})_{n \times n}$ and $\prod(\mathscr{C}^+) = (e_{ij})_{(n+1) \times (n+1)}$.

**Theorem 2.** *Let $(U^+, \mathscr{C}^+)$ be a AE-covering approximation space of $(U, \mathscr{C})$, $\prod(\mathscr{C})$ and $\prod(\mathscr{C}^+)$ the type-2 characteristic matrices of $\mathscr{C}$ and $\mathscr{C}^+$, respectively. Then*

$$\prod(\mathscr{C}^+) = \begin{bmatrix} \prod(\mathscr{C}) & 0 \\ 0 & 0 \end{bmatrix} \bigwedge \begin{bmatrix} \Delta_1 \prod(\mathscr{C}) & \Delta_3 \prod(\mathscr{C}) \\ \Delta_2 \prod(\mathscr{C}) & e_{(n+1)(n+1)} \end{bmatrix},$$

*where*

$$\triangle_1 \prod(\mathscr{C}) = \begin{bmatrix} a_{1(m+1)} & a_{2(m+1)} & \cdots & a_{n(m+1)} \end{bmatrix}^T \odot \begin{bmatrix} a_{1(m+1)} & a_{2(m+1)} & \cdots & a_{n(m+1)} \end{bmatrix};$$

$$\triangle_2 \prod(\mathscr{C}) = \begin{bmatrix} a_{(n+1)1} & a_{(n+1)2} & \cdots & a_{(n+1)(m+1)} \end{bmatrix} \odot \begin{bmatrix} a_{11} & a_{12} & \cdots & a_{1m} & a_{1(m+1)} \\ a_{21} & a_{22} & \cdots & a_{2m} & a_{2(m+1)} \\ \cdot & \cdot & \cdots & \cdot & \cdot \\ \cdot & \cdot & \cdots & \cdot & \cdot \\ \cdot & \cdot & \cdots & \cdot & \cdot \\ a_{n1} & a_{n2} & \cdots & a_{nm} & a_{n(m+1)} \end{bmatrix}^T;$$

$$\triangle_3 \prod(\mathscr{C}) = \begin{bmatrix} a_{11} & a_{12} & \cdots & a_{1m} & a_{1(m+1)} \\ a_{21} & a_{22} & \cdots & a_{2m} & a_{2(m+1)} \\ \cdot & \cdot & \cdots & \cdot & \cdot \\ \cdot & \cdot & \cdots & \cdot & \cdot \\ \cdot & \cdot & \cdots & \cdot & \cdot \\ a_{n1} & a_{n2} & \cdots & a_{nm} & a_{n(m+1)} \end{bmatrix} \odot \begin{bmatrix} a_{(n+1)1} & a_{(n+1)2} & \cdots & a_{(n+1)(m+1)} \end{bmatrix}^T;$$

$$e_{(n+1)(n+1)} = \begin{bmatrix} a_{(n+1)1} & a_{(n+1)2} & \cdots & a_{(n+1)(m+1)} \end{bmatrix} \odot \begin{bmatrix} a_{(n+1)1} & a_{(n+1)2} & \cdots & a_{(n+1)(m+1)} \end{bmatrix}^T.$$

**Proof.** By Definition 3, we have $\prod(\mathscr{C})$ and $\prod(\mathscr{C}^+)$ as follows:

$$\prod(\mathscr{C}) = M_{\mathscr{C}} \odot M_{\mathscr{C}}^T = \begin{bmatrix} a_{11} & a_{12} & \cdots & a_{1m} \\ a_{21} & a_{22} & \cdots & a_{2m} \\ \cdot & \cdot & \cdots & \cdot \\ \cdot & \cdot & \cdots & \cdot \\ a_{n1} & a_{n2} & \cdots & a_{nm} \end{bmatrix} \odot \begin{bmatrix} a_{11} & a_{12} & \cdots & a_{1m} \\ a_{21} & a_{22} & \cdots & a_{2m} \\ \cdot & \cdot & \cdots & \cdot \\ \cdot & \cdot & \cdots & \cdot \\ a_{n1} & a_{n2} & \cdots & a_{nm} \end{bmatrix}^T = \begin{bmatrix} d_{11} & d_{12} & \cdots & d_{1n} \\ d_{21} & d_{22} & \cdots & d_{2n} \\ \cdot & \cdot & \cdots & \cdot \\ \cdot & \cdot & \cdots & \cdot \\ d_{n1} & d_{n2} & \cdots & d_{nn} \end{bmatrix};$$

$$\prod(\mathscr{C}^+) = M_{\mathscr{C}^+} \odot M_{\mathscr{C}^+}^T = \begin{bmatrix} a_{11} & a_{12} & \cdots & a_{1m} & a_{1(m+1)} \\ a_{21} & a_{22} & \cdots & a_{2m} & a_{2(m+1)} \\ \cdot & \cdot & \cdots & \cdot & \cdot \\ \cdot & \cdot & \cdots & \cdot & \cdot \\ a_{n1} & a_{n2} & \cdots & a_{nm} & a_{n(m+1)} \\ a_{(n+1)1} & a_{(n+1)2} & \cdots & a_{(n+1)m} & a_{(n+1)(m+1)} \end{bmatrix} \odot \begin{bmatrix} a_{11} & a_{12} & \cdots & a_{1m} & a_{1(m+1)} \\ a_{21} & a_{22} & \cdots & a_{2m} & a_{2(m+1)} \\ \cdot & \cdot & \cdots & \cdot & \cdot \\ \cdot & \cdot & \cdots & \cdot & \cdot \\ a_{n1} & a_{n2} & \cdots & a_{nm} & a_{n(m+1)} \\ a_{(n+1)1} & a_{(n+1)2} & \cdots & a_{(n+1)m} & a_{(n+1)(m+1)} \end{bmatrix}^T$$

$$= \begin{bmatrix} e_{11} & e_{12} & \cdots & e_{1n} & e_{1(n+1)} \\ e_{21} & e_{22} & \cdots & e_{2n} & e_{2(n+1)} \\ \cdot & \cdot & \cdots & \cdot & \cdot \\ \cdot & \cdot & \cdots & \cdot & \cdot \\ e_{n1} & e_{n2} & \cdots & e_{nn} & e_{n(n+1)} \\ e_{(n+1)1} & e_{(n+1)2} & \cdots & e_{(n+1)n} & e_{(n+1)(n+1)} \end{bmatrix}.$$

By Definition 3, we see that $d_{ij} \vee (a_{j(m+1)} - a_{i(m+1)} + 1) = e_{ij}$ for $1 \le i, j \le n$. To compute $\prod(\mathscr{C}^+)$ on the basis of $\prod(\mathscr{C})$, we only need to compute $\triangle_1 \prod(\mathscr{C})$, $\triangle_2 \prod(\mathscr{C})$, $\triangle_3 \prod(\mathscr{C})$ and $e_{(n+1)(n+1)}$. Concretely, $\triangle_1 \prod(\mathscr{C})$, $\triangle_2 \prod(\mathscr{C})$, $\triangle_3 \prod(\mathscr{C})$ and $e_{(n+1)(n+1)}$ are constructed as follows.

$$\triangle_1 \prod(\mathscr{C}) = (a_{j(m+1)} - a_{i(m+1)} + 1)_{n \times n} = \begin{bmatrix} a_{1(m+1)} & a_{2(m+1)} & \cdots & a_{n(m+1)} \end{bmatrix}^T \odot \begin{bmatrix} a_{1(m+1)} & a_{2(m+1)} & \cdots & a_{n(m+1)} \end{bmatrix};$$

$$\Delta_2 \prod(\mathscr{C}) = \begin{bmatrix} e_{(n+1)1} & e_{(n+1)2} & \cdots & e_{(n+1)n} \end{bmatrix} = \begin{bmatrix} a_{(n+1)1} & a_{(n+1)2} & \cdots & a_{(n+1)m} & a_{(n+1)(m+1)} \end{bmatrix} \odot \begin{bmatrix} a_{11} & a_{12} & \cdots & a_{1m} & a_{1(m+1)} \\ a_{21} & a_{22} & \cdots & a_{2m} & a_{2(m+1)} \\ \cdot & \cdot & \cdots & \cdot & \cdot \\ \cdot & \cdot & \cdots & \cdot & \cdot \\ a_{n1} & a_{n2} & \cdots & a_{nm} & a_{n(m+1)} \end{bmatrix}^{T};$$

$$\Delta_3 \prod(\mathscr{C}) = \begin{bmatrix} e_{1(n+1)} & e_{2(n+1)} & \cdots & e_{n(n+1)} \end{bmatrix} = \begin{bmatrix} a_{11} & a_{12} & \cdots & a_{1m} & a_{1(m+1)} \\ a_{21} & a_{22} & \cdots & a_{2m} & a_{2(m+1)} \\ \cdot & \cdot & \cdots & \cdot & \cdot \\ \cdot & \cdot & \cdots & \cdot & \cdot \\ a_{n1} & a_{n2} & \cdots & a_{nm} & a_{n(m+1)} \end{bmatrix} \odot \begin{bmatrix} a_{(n+1)1} & a_{(n+1)2} & \cdots & a_{(n+1)m} & a_{(n+1)(m+1)} \end{bmatrix}^{T};$$

$$e_{(n+1)(n+1)} = \begin{bmatrix} a_{(n+1)1} & a_{(n+1)2} & \cdots & a_{(n+1)(m+1)} \end{bmatrix} \odot \begin{bmatrix} a_{(n+1)1} & a_{(n+1)2} & \cdots & a_{(n+1)(m+1)} \end{bmatrix}^{T}.$$

Therefore, we have $\prod(\mathscr{C}^{+}) = \begin{bmatrix} \prod(\mathscr{C}) & 0 \\ 0 & 0 \end{bmatrix} \wedge \begin{bmatrix} \Delta_1 \prod(\mathscr{C}) & \Delta_3 \prod(\mathscr{C}) \\ \Delta_2 \prod(\mathscr{C}) & e_{(n+1)(n+1)} \end{bmatrix}$.

By Theorem 2, we present an incremental algorithm of computing the sixth lower and upper approximations of sets as follows.

Input: $(U, \mathscr{C})$, $(U^{+}, \mathscr{C}^{+})$ and $X \subseteq U^{+}$. Output: $X_{XH(X)}$ and $X_{XL(X)}$.

Step 1: Constructing $\prod(\mathscr{C})$, where $\prod(\mathscr{C}) = M_{\mathscr{C}} \odot M_{\mathscr{C}}^{T}$.

Step 2: Computing $\Delta_1 \prod(\mathscr{C})$ and $\Delta_2 \prod(\mathscr{C})$, where

$$M_{\mathscr{C}^{+}} = \begin{bmatrix} M_{\mathscr{C}} & \alpha_1^{T} \\ \alpha_2 & a_{(n+1)(m+1)} \end{bmatrix}; \alpha_1 = \begin{bmatrix} a_{1(m+1)} & a_{2(m+1)} & \cdots & a_{n(m+1)} \end{bmatrix}; \alpha_2 = \begin{bmatrix} a_{(n+1)1} & a_{(n+1)2} & \cdots & a_{(n+1)m} \end{bmatrix};$$

$$\Delta_1 \prod(\mathscr{C}) = \alpha_1^{T} \odot \alpha_1; \Delta_2 \prod(\mathscr{C}) = \begin{bmatrix} \alpha_2 & a_{(n+1)(m+1)} \end{bmatrix} \odot M_{\mathscr{C}^{+}}^{T}; \Delta_3 \prod(\mathscr{C}) = M_{\mathscr{C}^{+}} \odot \begin{bmatrix} \alpha_1 & a_{(n+1)(m+1)} \end{bmatrix}^{T};$$

$$e_{(n+1)(n+1)} = \begin{bmatrix} \alpha_2 & a_{(n+1)(m+1)} \end{bmatrix} \odot \begin{bmatrix} \alpha_2 & a_{(n+1)(m+1)} \end{bmatrix}^{T}.$$

Step 3: Calculating $\prod(\mathscr{C}^{+})$, where $\prod(\mathscr{C}^{+}) = \begin{bmatrix} \prod(\mathscr{C}) & 0 \\ 0 & 0 \end{bmatrix} \wedge \begin{bmatrix} \Delta_1 \prod(\mathscr{C}) & \Delta_3 \prod(\mathscr{C}) \\ \Delta_2 \prod(\mathscr{C}) & e_{(n+1)(n+1)} \end{bmatrix}$.

Step 4: Getting $X_{XH(X)}$ and $X_{XL(X)}$, where $X_{XH(X)} = \prod(\mathscr{C}^{+}) \cdot X_X$; $X_{XL(X)} = \prod(\mathscr{C}^{+}) \odot X_X$.

By using the incremental algorithm, the time complexity of calculating the sixth lower and upper approximations of sets is $O(3n^2 + 2nm + 8n + m + 4)$. Additionally, $O((n+1)^2 \cdot (m+1))$ is the time complexity of the non-incremental algorithm. Therefore, the incremental algorithm is more effective than the non-incremental algorithm.

The following example illustrates that how to compute the sixth lower and upper approximations of set by using the incremental algorithm.

*Example 2.* (Continuation of Example 1) We obtain that

$$\prod(\mathscr{C}) = M_{\mathscr{C}} \odot M_{\mathscr{C}}^{T} = \begin{bmatrix} 1 & 1 & 0 \\ 0 & 1 & 0 \\ 0 & 0 & 1 \\ 1 & 1 & 1 \end{bmatrix} \odot \begin{bmatrix} 1 & 1 & 0 \\ 0 & 1 & 0 \\ 0 & 0 & 1 \\ 1 & 1 & 1 \end{bmatrix}^{T} = \begin{bmatrix} 1 & 0 & 0 & 1 \\ 1 & 1 & 0 & 1 \\ 0 & 0 & 1 & 1 \\ 0 & 0 & 0 & 1 \end{bmatrix}.$$

By Theorem 2, we have that

$$\Delta_1 \prod(\mathscr{C}) = \begin{bmatrix}0&0&1&0\end{bmatrix}^T \odot \begin{bmatrix}0&0&1&0\end{bmatrix}^T = \begin{bmatrix}1&1&2&1\\1&1&2&1\\0&0&1&0\\1&1&2&1\end{bmatrix}; \Delta_2 \prod(\mathscr{C}) = \begin{bmatrix}1&1&0&1\end{bmatrix} \odot \begin{bmatrix}1&1&0&0\\0&1&0&0\\0&0&1&1\\1&1&1&0\end{bmatrix} = \begin{bmatrix}0&0&0&0\end{bmatrix};$$

$$\Delta_3 \prod(\mathscr{C}) = \begin{bmatrix}1&0&0&0\\0&1&0&0\\1&1&1&0\\1&1&0&1\end{bmatrix} \odot \begin{bmatrix}1&1&0&1\end{bmatrix}^T = \begin{bmatrix}1\\1\\0\\0\end{bmatrix}; e_{55} = \begin{bmatrix}1&1&0&1&1\end{bmatrix} \odot \begin{bmatrix}1&1&0&1&1\end{bmatrix}^T = 1.$$

Thus, we have

$$\prod(\mathscr{C}^+) = \begin{bmatrix}\prod(\mathscr{C})&0\\0&0\end{bmatrix} \bigwedge \begin{bmatrix}\Delta_1\prod(\mathscr{C})&\Delta_3\prod(\mathscr{C})\\\Delta_2\prod(\mathscr{C})&e_{55}\end{bmatrix} = \begin{bmatrix}1&0&0&1&1\\1&1&0&1&1\\0&0&1&0&0\\0&0&0&1&0\\0&0&0&0&1\end{bmatrix}.$$

On the other hand, by Definition 3, we have $\prod(\mathscr{C}^+) = M_{\mathscr{C}^+} \odot M_{\mathscr{C}^+}^T = \begin{bmatrix}1&0&0&1&1\\1&1&0&1&1\\0&0&1&0&0\\0&0&0&1&0\\0&0&0&0&1\end{bmatrix}.$

It is obvious that the time complexity is more than the incremental approach. By Definition 4, we obtain

$$\mathcal{X}_{XH(X)} = \prod(\mathscr{C}^+) \cdot \mathcal{X}_X = \begin{bmatrix}1&0&0&1&1\\1&1&0&1&1\\0&0&1&0&0\\0&0&0&1&0\\0&0&0&0&1\end{bmatrix} \cdot \begin{bmatrix}0\\0\\1\\1\\1\end{bmatrix} = \begin{bmatrix}1&1&1&1&1\end{bmatrix}^T;$$

$$\mathcal{X}_{XL(X)} = \prod(\mathscr{C}^+) \odot \mathcal{X}_X = \begin{bmatrix}1&0&0&1&1\\1&1&0&1&1\\0&0&1&0&0\\0&0&0&1&0\\0&0&0&0&1\end{bmatrix} \odot \begin{bmatrix}0\\0\\1\\1\\1\end{bmatrix} = \begin{bmatrix}0&0&1&1&1\end{bmatrix}^T.$$

Therefore, $XH(X) = \{x_1, x_2, x_3, x_4, x_5\}$ and $XL(X) = \{x_3, x_4, x_5\}$.

In Example 2, we need to compute all elements in $\prod(\mathscr{C}^+)$ for constructing approximations of sets by Definition 3. By Theorem 2, we only need to calculate elements in $\Delta_1 \prod(\mathscr{C})$, $\Delta_2 \prod(\mathscr{C})$, $\Delta_3 \prod(\mathscr{C})$ and $e_{55}$. Thereby, the incremental algorithm is more effective to compute approximations of sets.

Below, we propose incremental approaches for computing the second and sixth lower and upper approximations of sets with emigration of elements of coverings when deleting objects.

**Definition 6.** *Let* $(U, \mathscr{C})$ *and* $(U^-, \mathscr{C}^-)$ *be covering approximation spaces, where* $U = \{x_1, x_2, ..., x_n\}$, $U^- = U - \{x_n\}$, $\mathscr{C} = \{C_1, C_2, ..., C_m\}$, $C_m = \{x_n\}$, $\mathscr{C}^- = \{C_1^-, C_2^-, ..., C_{m-1}^-\}$, where $C_i^- = C_i - \{x_n\}$ or $C_i$ $(1 \le i \le m - 1)$. Then $(U^-, \mathscr{C}^-)$ is called a DE-covering approximation space.

By Definition 6, $\mathscr{C}^-$ is referred to as a DE-covering of $\mathscr{C}$. On the basis of $\Gamma(\mathscr{C})$, we show how to construct $\Gamma(\mathscr{C}^-)$ for computing the second lower and upper approximations of sets. First, we discuss the relationship between $\Gamma(\mathscr{C})$ and $\Gamma(\mathscr{C}^-)$. For convenience, we denote $M_{\mathscr{C}} = (a_{ij})_{n \times m}$, $M_{\mathscr{C}^-} = (a_{ij})_{(n-1) \times (m-1)}$, $\Gamma(\mathscr{C}) = (b_{ij})_{n \times n}$ and $\Gamma(\mathscr{C}^-) = (c_{ij})_{(n-1) \times (n-1)}$.

**Theorem 3.** *Let $(U^-, \mathscr{C}^-)$ be a AE-covering approximation space of $(U, \mathscr{C})$, $\Gamma(\mathscr{C})$ and $\Gamma(\mathscr{C}^-)$ the type-1 characteristic matrices of $\mathscr{C}$ and $\mathscr{C}^-$, respectively. Then*

$$\Gamma(\mathscr{C}^-) = (b_{ij})_{(n-1)(n-1)}.$$

**Proof.** The proof is similar to that of Theorem 1. □

By Theorem 3, the time complexity of calculating the second lower and upper approximations of sets is $O(n^2 + 2mn - n + m)$. Additionally, the time complexity is $O((n-1)^2 \cdot (m-1))$ by Definition 3. Therefore, the proposed approach is more effective for computing the second lower and upper approximations of sets.

The following example is employed to show the process of constructing the second lower and upper approximations of sets when deleting an object.

*Example 3.* Let $U = \{x_1, x_2, x_3, x_4\}$, $U^- = \{x_1, x_2, x_3\}$, $\mathscr{C} = \{C_1, C_2, C_3, C_4\}$, $\mathscr{C}^- = \{C_1^-, C_2^-, C_3^-\}$, where $C_1 = \{x_1, x_4\}$, $C_2 = \{x_1, x_2, x_4\}$, $C_3 = \{x_3, x_4\}$, $C_4 = \{x_4\}$, $C_1^- = \{x_1\}$, $C_2^- = \{x_1, x_2\}$, $C_3^- = \{x_3\}$, and $X = \{x_2, x_3\}$. According to Example 1 and Theorem 3, we have that

$$\Gamma(\mathscr{C}^-) = \begin{bmatrix} 1 & 1 & 0 \\ 1 & 1 & 0 \\ 0 & 0 & 1 \end{bmatrix}.$$

Thus, we have that

$$X_{SH(X)} = \Gamma(\mathscr{C}^-) \cdot X_X = \begin{bmatrix} 1 & 1 & 0 \\ 1 & 1 & 0 \\ 0 & 0 & 1 \end{bmatrix} \cdot \begin{bmatrix} 0 \\ 1 \\ 1 \end{bmatrix} = \begin{bmatrix} 1 & 1 & 1 \end{bmatrix}^T;$$

$$X_{SL(X)} = \Gamma(\mathscr{C}^-) \odot X_X = \begin{bmatrix} 1 & 1 & 0 \\ 1 & 1 & 0 \\ 0 & 0 & 1 \end{bmatrix} \odot \begin{bmatrix} 0 \\ 1 \\ 1 \end{bmatrix} = \begin{bmatrix} 0 & 0 & 1 \end{bmatrix}^T.$$

Therefore, $SH(X) = \{x_1, x_2, x_3\}$ and $SL(X) = \{x_3\}$.

In Example 3, there is a need to compute all elements in $\Gamma(\mathscr{C}^-)$ by Definition 3. But we do not need to calculate elements in $\Gamma(\mathscr{C}^-)$ by Theorem 3.

Subsequently, we construct $\prod(\mathscr{C}^-)$ based on $\prod(\mathscr{C})$ for computing the sixth lower and upper approximations of sets. For convenience, we denote $\prod(\mathscr{C}) = (d_{ij})_{n \times n}$ and $\prod(\mathscr{C}^-) = (e_{ij})_{(n-1) \times (n-1)}$.

**Theorem 4.** *Let* $(U^-, \mathscr{C}^-)$ *be a DE-covering approximation space of* $(U, \mathscr{C})$, $\prod(\mathscr{C})$ *and* $\prod(\mathscr{C}^-)$ *the type-2 characteristic matrices of* $\mathscr{C}$ *and* $\mathscr{C}^-$, *respectively. Then*

$$\prod(\mathscr{C}^-) = (d_{ij})_{(n-1)(n-1)}.$$

**Proof.** The proof is similar to that of Theorem 2.                                    □

By Theorem 4, the time complexity of calculating the sixth lower and upper approximations of sets is $O(n^2 + 2mn - n + m)$. Additionally, the time complexity is $O((n-1)^2 \cdot (m-1))$ by Definition 3. Therefore, the proposed approach is more effective for computing the second lower and upper approximations of sets.

The following example illustrates that how to compute the sixth lower and upper approximations of set by Theorem 4.

*Example 4.* (Continuation of Example 3) According to Example 2 and Theorem 4, we obtain

$$\prod(\mathscr{C}^-) = \begin{bmatrix} 1 & 1 & 0 \\ 0 & 1 & 0 \\ 0 & 0 & 1 \end{bmatrix} \odot \begin{bmatrix} 1 & 1 & 0 \\ 0 & 1 & 0 \\ 0 & 0 & 1 \end{bmatrix}^T = \begin{bmatrix} 1 & 0 & 0 \\ 1 & 1 & 0 \\ 0 & 0 & 1 \end{bmatrix}.$$

Thus, we have

$$\mathcal{X}_{XH(X)} = \prod(\mathscr{C}^-) \cdot \mathcal{X}_X = \begin{bmatrix} 1 & 0 & 0 \\ 1 & 1 & 0 \\ 0 & 0 & 1 \end{bmatrix} \cdot \begin{bmatrix} 0 \\ 1 \\ 1 \end{bmatrix} = \begin{bmatrix} 0 & 1 & 1 \end{bmatrix}^T ;$$

$$\mathcal{X}_{XL(X)} = \prod(\mathscr{C}^-) \odot \mathcal{X}_X = \begin{bmatrix} 1 & 0 & 0 \\ 1 & 1 & 0 \\ 0 & 0 & 1 \end{bmatrix} \odot \begin{bmatrix} 0 \\ 1 \\ 1 \end{bmatrix} = \begin{bmatrix} 0 & 0 & 1 \end{bmatrix}^T .$$

Therefore, $XH(X) = \{x_2, x_3\}$ and $XL(X) = \{x_3\}$.

In Example 4, all elements in $\prod(\mathscr{C}^-)$ need to be computed based on $\prod(\mathscr{C}^-) = M_{\mathscr{C}^-} \odot M_{\mathscr{C}^-}^T$. By Theorem 4, there is no need to construct elements in $\prod(\mathscr{C}^-)$. Thereby, the proposed approach is more effective to compute the sixth lower and upper approximations of sets.

## 4    Conclusions

In this paper, we have provided effective approaches to constructing approximations of concepts in dynamic covering approximation spaces. Concretely, we have constructed type-1 and type-2 characteristic matrices of coverings with the incremental approaches. Incremental algorithms have been presented for computing the second and sixth lower and upper approximations of sets. Several examples have been employed to illustrate that computing approximations of sets could be reduced greatly by using the incremental approaches.

In the future, we will propose more effective approaches to constructing the type-1 and type-2 characteristic matrices of coverings. Additionally, we will focus on the development of effective approaches for knowledge discovery in dynamic covering approximation spaces.

**Acknowledgments.** We would like to thank the anonymous reviewers very much for their professional comments and valuable suggestions. This work is supported by the National Natural Science Foundation of China (NO. 11071061,11201137), the National Basic Research Program of China (NO.2011CB311808), the Science and Technology Program of Hunan,China(No.2013FJ4037) and the Scientific Research Fund of Hunan Provincial Education Department(No.14C0049).

# References

1. Chen, H.M., Li, T.R., Qiao, S.J., Ruan, D.: A rough set based dynamic maintenance approach for approximations in coarsening and refining attribute values. International Journal of Intelligent Systems 25(10), 1005–1026 (2010)
2. Chen, H.M., Li, T.R., Ruan, D.: Maintenance of approximations in incomplete ordered decision systems while attribute values coarsening or refining. Knowledge-Based Systems 31, 140–161 (2012)
3. Li, S.Y., Li, T.R., Liu, D.: Incremental updating approximations in dominance-based rough sets approach under the variation of the attribute set. Knowledge-Based Systems 40, 17–26 (2013)
4. Luo, C., Li, T.R., Chen, H.M.: Dynamic maintenance of approximations in set-valued ordered decision systems under the attribute generalization. Information Sciences (2013), http://dx.doi.org/10.1016/j.ins.2013.09.035
5. Luo, C., Li, T.R., Chen, H.M., Liu, D.: Incremental approaches for updating approximations in set-valued ordered information systems. Knowledge-Based Systems 50, 218–233 (2013)
6. Wang, S.P., Zhu, W., Zhu, Q.H., Min, F.: Characteristic matrix of covering and its application to boolean matrice decomposition and axiomatization, ArXiv: 207.0262v3
7. Zakowski, W.: Approximations in the space $(u, \pi)$. Demonstratio Mathematics 16, 761–769 (1983)
8. Zhang, J.B., Li, T.R., Ruan, D., Liu, D.: Rough sets based matric approaches with dynamic attribute variation in set-valued information systems. International Journal of Approximate Reasoning 53(4), 620–635 (2012)
9. Zhang, J.B., Li, T.R., Ruan, D., Liu, D.: Neighborhood rough sets for dynamic data mining. International Journal of Intelligent Systems 27(4), 317–342 (2012)

# Advances in Granular Computing

# Knowledge Approximations in Multi-scale Ordered Information Systems

Shen-Ming Gu, Yi Wu, Wei-Zhi Wu, and Tong-Jun Li

School of Mathematics, Physics and Information Science,
Zhejiang Ocean University, Zhoushan, Zhejiang 316022, P.R. China
{gsm,wuwz,litj}@zjou.edu.cn, 394294312@qq.com

**Abstract.** The key to granular computing is to make use of granules in problem solving. However, there are different granules at different levels of scale in data sets having hierarchical scale structures. And in real-world applications, there may exist multiple types of data in ordered information systems. Therefore, the concept of multi-scale ordered information systems is first introduced in this paper. The lower and upper approximations in multi-scale ordered information systems are then defined, and their properties are examined.

**Keywords:** Granular computing, Granules, Multi-scale information systems, Ordered information systems, Rough sets.

## 1 Introduction

The purpose of Granular computing (GrC) is to seek for an approximation scheme which can effectively solve a complex problem at a certain level of granulation [23]. The root of GrC comes from the concept of information granulation which was first introduced by Zadeh in 1979 [31, 32]. Ever since the introduction of the concept of "GrC", we have witnessed a rapid development and a fast growing interest in the topic [1, 2, 13–15, 18, 19, 22, 27–30].

With the view of GrC, a granule is a primitive notion which is a clump of objects drawn together by the criteria of indistinguishability, similarity or functionality [32]. The set of granules provides a representation of the unit with respect to a particular level of granularity [24]. An important and common used model for GrC is partition model proposed by Yao [29]. An equivalence relation allows us to model the passage from one level of detail to another, but does not, on its own, model more than two levels of details needed in practice. By employing the notion of labelled partition, Bittner and Smith [3] developed an ontologically motivated formal theory of granular partitions which is relatively comprehensive and useful for granular levels, but it does not address the types of aggregation commonly used with data mining. In order to represent hierarchical structure of data measured at different levels of granularities, Keet [12] explored a formal theory of granularity to build structure of the contents for different types of granularities. More recently, Wu and Leung [23] developed a new knowledge representation system, called multi-scale granular labelled partition structure,

D. Miao et al. (Eds.): RSKT 2014, LNAI 8818, pp. 525–534, 2014.
DOI: 10.1007/978-3-319-11740-9_48 © Springer International Publishing Switzerland 2014

in which data are represented by different scales at different levels of granulations having a granular information transformation from a finer to a coarser labelled partition. Hence, the multi-scale information system is a new and interesting topic in the theory of GrC. Recently, Wu and Leung [24] explored optimal scale selection in multi-scale decision tables from the perspective of granular computation. Gu et al. [6–8] proposed some methods for knowledge acquisition in consistent or inconsistent multi-scale decision systems. In [9], multi-granulation rough sets were discussed in multi-scale information systems.

Various methods of GrC concentrating on concrete models in specific contexts have been proposed over the years. Rough set theory is perhaps one of the most advanced areas that popularize GrC [10, 11, 14, 18, 19, 27–30]. It was originally proposed by Pawlak [17] as a formal tool for modelling and processing incomplete information. However, classical rough set theory is not effective for taking into account scaling criteria, that is, attributes with preference-ordered domains. To solve this problem, Greco et al. [5] developed an extension of Pawlak's rough set approach, which is called the Dominance-based Rough Set Approach (DRSA). Presently, work on dominance-based rough set model were developed rapidly. For example, Greco et al. [4] explored the dominance-based rough fuzzy model by introducing the concept of DRSA into the fuzzy environment. Shao and Zhang [21] further proposed an extension of the dominance relation in incomplete ordered information systems. Sai et al. [20] discussed data analysis and mining in ordered information tables. Xu et al. [25] introduced an approach to attribute reductions in inconsistent ordered information systems. Yang et al. [26] investigated dominance-based rough set in incomplete interval-valued information systems, which contain both incomplete and imprecise evaluations of objects.

This paper mainly focuses on the study of knowledge approximations in multi-scale ordered information systems. The organization of this paper is as follows. In the next section, we first introduce some basic notions related to ordered information systems. The concept of multi-scale ordered information systems and the corresponding knowledge approximations are explored in Section 3. In Section 4, we investigate some properties of granules in multi-scale ordered information systems. We then conclude the paper with a summary and outlook for further research in Section 5.

# 2   Ordered Information Systems

## 2.1   Information Systems

The notion of information systems provides a convenient tool for the representation of objects in terms of their attributes [16].

An information system is a pair $(U, A)$, where $U = \{x_1, x_2, \ldots, x_n\}$ is a non-empty, finite set of objects called the universe of discourse and $A = \{a_1, a_2, \ldots, a_m\}$ is a non-empty, finite set of attributes, such that $a : U \rightarrow V_a$ for any $a \in A$, i.e., $a(x) \in V_a$, $x \in U$, where $V_a = \{a(x) : x \in U\}$ is called the domain of $a$ [9].

Any non-empty attribute set $B \subseteq A$, it determines an indiscernibility relation on $U$ as follows:

$$R_B = \{(x, y) \in U \times U : a(x) = a(y), \forall a \in B\}. \tag{1}$$

Since $R_B$ is an equivalence relation on $U$, it partitions $U$ into a family of disjoint subsets $U/R_B$ of $U$:

$$U/R_B = \{[x]_B : x \in U\}, \tag{2}$$

where $[x]_B$ denotes the equivalence class determined by $x$ w.r.t. $B$, i.e., $[x]_B = \{y \in U : (x, y) \in R_B\}$.

Any $X \subseteq U$ and $B \subseteq A$, the lower and upper approximations of $X$ w.r.t. the equivalence relation $R_B$ are defined as follows:

$$\underline{R_B}(X) = \{x \in U : [x]_B \subseteq X\}, \tag{3}$$

$$\overline{R_B}(X) = \{x \in U : [x]_B \cap X \neq \emptyset\}. \tag{4}$$

## 2.2   Ordered Information Systems

For an information system $(U, A)$, $a \in A$, if the domain of attribute $a$ is ordered according to a decreasing or increasing preference, then $a$ is a criterion. It is assumed that the domain of a criterion $a \in A$ is completely pre-ordered by an outranking relation $\succeq_a$; $x \succeq_a y$ means that $x$ is at least as good as $y$ w.r.t. criterion $a$ [4].

For a subset of attributes $B \subseteq A$, if $\forall a \in B, x \succeq_a y$, then $x \succeq_B y$, that is, $x$ is at least as good as $y$ w.r.t. all attributes in $B$.

If all the attributes are criterions, then we say $(U, A)$ is an ordered information system [21].

For an ordered information system $(U, A)$ and $B \subseteq A$, a dominance relation can be defined as:

$$R_B^{\geq} = \{(y, x) \in U \times U : y \succeq_B x\}. \tag{5}$$

If $(y, x) \in R_B^{\geq}$, then $y$ dominates $x$ w.r.t. $B$. $R_B^{\geq}$ is reflexive, transitive, and antisymmetric, so it is not an equivalence relation.

The inverse relation of $R_B^{\geq}$ will be denoted by $R_B^{\leq}$.

$$R_B^{\leq} = \{(y, x) \in U \times U : x \succeq_B y\}. \tag{6}$$

Given $B \subseteq A$ and $B = B_1 \cup B_2$, where attributes set $B_1$ is according to increasing preference, and $B_2$ is according to decreasing preference, the granules of knowledge induced by the dominance relation $R_B^{\geq}$ are the set of objects dominating $x$,

$$
\begin{aligned}
[x]_B^{\geq} &= \{y \in U : (y, x) \in R_B^{\geq}\} \\
&= \{y \in U : f(y, a_1) \geq f(x, a_1)(\forall a_1 \in B_1) \\
&\qquad and\ f(y, a_2) \leq f(x, a_2)(\forall a_2 \in B_2)\} \\
&= \{y \in U : (x, y) \in R_B^{\leq}\}.
\end{aligned}
$$

And the granules of knowledge induced by the dominance relation $R_B^\leq$ are the set of objects dominated by $x$,

$$
\begin{aligned}
[x]_B^\leq &= \{y \in U : (y,x) \in R_B^\leq\} \\
&= \{y \in U : f(y,a_1) \leq f(x,a_1)(\forall a_1 \in B_1) \\
&\quad\; and\; f(y,a_2) \geq f(x,a_2)(\forall a_2 \in B_2)\} \\
&= \{y \in U : (x,y) \in R_B^\geq\}.
\end{aligned}
$$

$[x]_B^\geq$ and $[x]_B^\leq$ are called the $B$-dominating set and the $B$-dominated set w.r.t. $x \in U$, respectively.

For any $X \subseteq U$ and $B \subseteq A$, the lower and upper approximations of $X$ w.r.t. the dominance relation $R_B^\geq$ are defined as follows:

$$
\underline{R_B^\geq}(X) = \{x \in U : [x]_B^\geq \subseteq X\}, \tag{7}
$$

$$
\overline{R_B^\geq}(X) = \{x \in U : [x]_B^\geq \cap X \neq \emptyset\}. \tag{8}
$$

The lower approximation $\underline{R_B^\geq}(X)$ is the set of objects which dominating set is a subset of $X$. And the upper approximation $\overline{R_B^\geq}(X)$ is the set of objects which dominating set has a nonempty intersection with $X$.

Similarly, for $X \subseteq U$ and $B \subseteq A$, one can define the lower and upper approximations of $X$ w.r.t. the dominance relation $R_B^\leq$ as follows:

$$
\underline{R_B^\leq}(X) = \{x \in U : [x]_B^\leq \subseteq X\}, \tag{9}
$$

$$
\overline{R_B^\leq}(X) = \{x \in U : [x]_B^\leq \cap X \neq \emptyset\}. \tag{10}
$$

The lower approximation $\underline{R_B^\leq}(X)$ is the set of objects which dominated set is subset of $X$. And the upper approximation $\overline{R_B^\leq}(X)$ is the set of objects which dominated set has a nonempty intersection with $X$.

## 3    Multi-scale Ordered Information Systems

### 3.1    Multi-scale Information Systems

A multi-scale information system [23] is a tuple $S = (U, A)$, where $U = \{x_1, x_2, \ldots, x_n\}$ is a non-empty, finite set of objects called the universe of discourse, $A = \{a_1, a_2, \ldots, a_m\}$ is a non-empty, finite set of attributes, and each $a_j \in A$ is a multi-scale attribute, i.e., for the same object in $U$, attribute $a_j$ can take on different values at different scales. We assume that all the attributes have the same number $I$ of levels of granulations. Hence a multi-scale information system can be represented as a system $(U, \{a_j^k : k = 1, 2, \ldots, I, j = 1, 2, \ldots, m\})$, where $a_j^k : U \to V_j^k$ is a surjective function and $V_j^k$ is the domain of the $k$-th scale attribute $a_j^k$. For $1 \le k \le I - 1$, there exists a surjective function $g_j^{k,k+1} : V_j^k \to V_j^{k+1}$, such that $a_j^{k+1} = g_j^{k,k+1} \circ a_j^k$, i.e.

$$a_j^{k+1}(x) = g_j^{k,k+1}(a_j^k(x)), \tag{11}$$

where $g_j^{k,k+1}$ is called a granular information transformation function [23].

For $k \in \{1, 2, \ldots, I\}$, we denote $A^k = \{a_j^k : j = 1, 2, \ldots, m\}$. Then a multi-scale information system $S = (U, A)$ can be decomposed into $I$ information systems $S^k = (U, A^k), k = 1, 2, \ldots, I$.

## 3.2 Multi-scale Ordered Information Systems

For a multi-scale information system $S = (U, A)$, if $S^1 = (U, A^1)$ is an ordered information system, and the granular information transformation functions $g_j^{1,2}, j = 1, 2, \ldots, m$, are order preserving, then $S^2 = (U, A^2)$ is an ordered information system. Furthermore, if $g_j^{k,k+1}, j = 1, 2, \ldots, m, k = 1, 2, \ldots, I - 1$, are order preserving, then $S^k = (U, A^k), k = 2, 3, \ldots, I$, are ordered information systems. If $S^k = (U, A^k), k = 1, 2, \ldots, I$, are ordered information systems, $S = (U, A)$ is called a multi-scale ordered information system.

**Example 1.** Table 1 depicts a multi-scale ordered information system, where $U = \{x_1, x_2, \ldots, x_8\}$, $A = \{a_1, a_2\}$. From the table we have $S^k = (U, A^k), k = 1, 2, 3$, this information system has three levels of granulations.

**Table 1.** A multi-scale ordered information system

| $U$ | $a_1^1$ | $a_2^1$ | $a_1^2$ | $a_2^2$ | $a_1^3$ | $a_2^3$ |
|---|---|---|---|---|---|---|
| $x_1$ | 95 | 93 | 5 | 5 | 1 | 1 |
| $x_2$ | 88 | 91 | 4 | 5 | 1 | 1 |
| $x_3$ | 92 | 86 | 5 | 4 | 1 | 1 |
| $x_4$ | 83 | 82 | 4 | 4 | 1 | 1 |
| $x_5$ | 76 | 84 | 3 | 4 | 1 | 1 |
| $x_6$ | 54 | 65 | 1 | 2 | 0 | 1 |
| $x_7$ | 67 | 76 | 2 | 3 | 1 | 1 |
| $x_8$ | 52 | 56 | 1 | 1 | 0 | 0 |

Given a multi-scale ordered information system $(U, \{A^k : k = 1, 2, \ldots, I\})$, $1 \le k \le I$, $S^k = (U, A^k)$ is an ordered information system. For $B^k \subseteq A^k$ and $X \subseteq U$, the lower and upper approximations of $X$ w.r.t. the dominance relation $R_{B^k}^{\ge}$ are defined as follows:

$$\underline{R_{B^k}^{\ge}}(X) = \{x \in U : [x]_{B^k}^{\ge} \subseteq X\}, \tag{12}$$

$$\overline{R_{B^k}^{\ge}}(X) = \{x \in U : [x]_{B^k}^{\ge} \cap X \ne \emptyset\}. \tag{13}$$

And the lower and upper approximations of $X$ w.r.t. the dominance relation $R_{\overline{B}^k}^{\leq}$ are defined as follows:

$$R_{\overline{B}^k}^{\leq}(X) = \{x \in U : [x]_{\overline{B}^k}^{\leq} \subseteq X\}, \tag{14}$$

$$\overline{R_{\overline{B}^k}^{\leq}}(X) = \{x \in U : [x]_{\overline{B}^k}^{\leq} \cap X \neq \emptyset\}. \tag{15}$$

## 4  Properties of Approximations

**Proposition 1.** *Given a multi-scale ordered information system* $(U, A)=(U, \{A^k : k = 1, 2, \ldots, I\})$, *for* $B^k \subseteq A^k$ *and* $X, Y \subseteq U$, *the lower and upper approximations of* $X$ *w.r.t. the dominance relation* $R_{\overline{B}^k}^{\geq}$ *satisfy the following additional properties:*

*(GLD)* $R_{\overline{B}^k}^{\geq}(X) =\sim \overline{R_{\overline{B}^k}^{\geq}}(\sim X)$,

*(GUD)* $\overline{R_{\overline{B}^k}^{\geq}}(X) =\sim R_{\overline{B}^k}^{\geq}(\sim X)$,

*(GL1)* $R_{\overline{B}^k}^{\geq}(\emptyset) = \emptyset$,

*(GU1)* $\overline{R_{\overline{B}^k}^{\geq}}(\emptyset) = \emptyset$,

*(GL2)* $R_{\overline{B}^k}^{\geq}(U) = U$,

*(GU2)* $\overline{R_{\overline{B}^k}^{\geq}}(U) = U$,

*(GL3)* $R_{\overline{B}^k}^{\geq}(X \cap Y) = R_{\overline{B}^k}^{\geq}(X) \cap R_{\overline{B}^k}^{\geq}(Y)$,

*(GU3)* $\overline{R_{\overline{B}^k}^{\geq}}(X \cup Y) = \overline{R_{\overline{B}^k}^{\geq}}(X) \cup \overline{R_{\overline{B}^k}^{\geq}}(Y)$,

*(GL4)* $X \subseteq Y \Rightarrow R_{\overline{B}^k}^{\geq}(X) \subseteq R_{\overline{B}^k}^{\geq}(Y)$,

*(GU4)* $X \subseteq Y \Rightarrow \overline{R_{\overline{B}^k}^{\geq}}(X) \subseteq \overline{R_{\overline{B}^k}^{\geq}}(Y)$,

*(GL5)* $R_{\overline{B}^k}^{\geq}(X \cup Y) \supseteq R_{\overline{B}^k}^{\geq}(X) \cup R_{\overline{B}^k}^{\geq}(Y)$,

*(GU5)* $\overline{R_{\overline{B}^k}^{\geq}}(X \cap Y) \subseteq \overline{R_{\overline{B}^k}^{\geq}}(X) \cap \overline{R_{\overline{B}^k}^{\geq}}(Y)$,

*(GL6)* $R_{\overline{B}^k}^{\geq}(X) \subseteq X$,

*(GU6)* $X \subseteq \overline{R_{\overline{B}^k}^{\geq}}(X)$.

**Proposition 2.** *Given a multi-scale ordered information system* $(U, A)=(U, \{A^k : k = 1, 2, \ldots, I\})$, *for* $B^k \subseteq A^k$ *and* $X, Y \subseteq U$, *then the lower and upper approximations of* $X$ *w.r.t. the dominance relation* $R_{\overline{B}^k}^{\leq}$ *satisfy the following properties:*

*(LLD)* $R_{\overline{B}^k}^{\leq}(X) =\sim \overline{R_{\overline{B}^k}^{\leq}}(\sim X)$,

*(LUD)* $\overline{R_{\overline{B}^k}^{\leq}}(X) =\sim R_{\overline{B}^k}^{\leq}(\sim X)$,

*(LL1)* $R_{\overline{B}^k}^{\leq}(\emptyset) = \emptyset$,

*(LU1)* $\overline{R_{\overline{B}^k}^{\leq}}(\emptyset) = \emptyset$,

*(LL2)* $R_{\overline{B}^k}^{\leq}(U) = U$,

*(LU2)* $\overline{R_{\overline{B}^k}^{\leq}}(U) = U$,

*(LL3)* $R_{\overline{B}^k}^{\leq}(X \cap Y) = R_{\overline{B}^k}^{\leq}(X) \cap R_{\overline{B}^k}^{\leq}(Y)$,

(LU3) $\overline{R^{\leq}_{B^k}}(X \cup Y) = \overline{R^{\leq}_{B^k}}(X) \cup \overline{R^{\leq}_{B^k}}(Y)$,

(LL4) $X \subseteq Y \Rightarrow \underline{R^{\leq}_{B^k}}(X) \subseteq \underline{R^{\leq}_{B^k}}(Y)$,

(LU4) $X \subseteq Y \Rightarrow \overline{R^{\leq}_{B^k}}(X) \subseteq \overline{R^{\leq}_{B^k}}(Y)$,

(LL5) $\underline{R^{\leq}_{B^k}}(X \cup Y) \supseteq \underline{R^{\leq}_{B^k}}(X) \cup \underline{R^{\leq}_{B^k}}(Y)$,

(LU5) $\overline{R^{\leq}_{B^k}}(X \cap Y) \subseteq \overline{R^{\leq}_{B^k}}(X) \cap \overline{R^{\leq}_{B^k}}(Y)$,

(LL6) $\underline{R^{\leq}_{B^k}}(X) \subseteq X$,

(LU6) $X \subseteq \overline{R^{\leq}_{B^k}}(X)$.

The lower and upper approximations of $X$ w.r.t. the dominance relation $R^{\geq}_{B^k}$ satisfy all the properties of usual lower and upper approximations. Obviously, the same holds for the approximations induced by the reverse relation. Moreover, the multi-scale ordered approximations satisfy the following additional properties.

**Proposition 3.** *For a multi-scale ordered information system* $(U, \{A^k : k = 1, 2, \ldots, I\}), 1 \leq k \leq I - 1$, $x \in U$ *and* $B^k \subseteq A^k$,

(1) $[x]^{\geq}_{B^k} \subseteq [x]^{\geq}_{B^{k+1}}$,

(2) $[x]^{\leq}_{B^k} \subseteq [x]^{\leq}_{B^{k+1}}$.

Proof. (1) $\forall x \in U$, let $B^k = B^k_1 \cup B^k_2$, where attributes set $B^k_1$ is according to an increasing preference, and $B^k_2$ is according to a decreasing preference.

$$[x]^{\geq}_{B^k} = \{y \in U : \forall a^k_j \in B^k_1, a^k_j(y) \geq a^k_j(x) \text{ and } \forall a^k_j \in B^k_2, a^k_j(y) \leq a^k_j(x)\}$$
$$\subseteq \{y \in U : \forall a^k_j \in B^k_1, g^{k,k+1}_j(a^k_j(y)) \geq g^{k,k+1}_j(a^k_j(x))$$
$$\text{and } \forall a^k_j \in B^k_2, g^{k,k+1}_j(a^k_j(y)) \leq g^{k,k+1}_j(a^k_j(x))\}$$
$$= \{y \in U : \forall a^{k+1}_j \in B^{k+1}_1, a^{k+1}_j(y) \geq a^{k+1}_j(x)$$
$$\text{and } \forall a^{k+1}_j \in B^{k+1}_2, a^{k+1}_j(y) \leq a^{k+1}_j(x)\}$$
$$= [x]^{\geq}_{B^{k+1}}.$$

Therefore $[x]^{\geq}_{B^k} \subseteq [x]^{\geq}_{B^{k+1}}$.

(2) It is similar to the Proof of (1).

**Proposition 4.** *Given a multi-scale ordered information system* $(U, \{A^k : k = 1, 2, \ldots, I\})$, $1 \leq k \leq I - 1$, *for* $B^k \subseteq A^k$ *and* $X \subseteq U$, *the lower and upper approximations of* $X$ *w.r.t. the dominance relation* $R^{\geq}_{B^k}$ *satisfy the following properties:*

(1) $\underline{R^{\geq}_{B^{k+1}}}(X) \subseteq \underline{R^{\geq}_{B^k}}(X)$,

(2) $\overline{R^{\geq}_{B^{k+1}}}(X) \supseteq \overline{R^{\geq}_{B^k}}(X)$.

Proof. (1) $\forall x \in U$,

$$x \in \underline{R^{\geq}_{B^{k+1}}}(X) \Longleftrightarrow [x]^{\geq}_{B^{k+1}} \subseteq X$$
$$\Longrightarrow [x]^{\geq}_{B^k} \subseteq X$$
$$\Longleftrightarrow x \in \underline{R^{\geq}_{B^k}}(X).$$

Therefore $R_{\overline{B}^{k+1}}^{\geq}(X) \subseteq R_{\overline{B}^k}^{\geq}(X)$.

(2) $\forall x \in U$,

$$x \in \overline{R_{\overline{B}^k}^{\geq}}(X) \Longleftrightarrow [x]_{\overline{B}^k}^{\geq} \cap X \neq \emptyset$$
$$\Longrightarrow [x]_{\overline{B}^{k+1}}^{\geq} \cap X \neq \emptyset$$
$$\Longleftrightarrow x \in \overline{R_{\overline{B}^{k+1}}^{\geq}}(X).$$

Therefore $\overline{R_{\overline{B}^{k+1}}^{\geq}}(X) \supseteq \overline{R_{\overline{B}^k}^{\geq}}(X)$.

**Proposition 5.** *Given a multi-scale ordered information system* $(U, \{A^k : k = 1, 2, \ldots, I\})$, $1 \leq k \leq I - 1$, *for* $B^k \subseteq A^k$ *and* $X \subseteq U$, *the lower and upper approximations of* $X$ *w.r.t. the dominance relation* $R_{\overline{B}^k}^{\leq}$ *satisfy the following properties:*

*(1)* $\underline{R_{\overline{B}^{k+1}}^{\leq}}(X) \subseteq \underline{R_{\overline{B}^k}^{\leq}}(X)$,

*(2)* $\overline{R_{\overline{B}^{k+1}}^{\leq}}(X) \supseteq \overline{R_{\overline{B}^k}^{\leq}}(X)$.

## 5    Conclusion

In this paper, we have developed a new knowledge representation system called a multi-scale ordered information system. A multi-scale ordered information system can be used to represent data sets having hierarchical scale structures measured at different levels of granulations in which every attributes is criteria. By using dominance relations to construct the dominating classes, we also introduced knowledge approximations in multi-scale ordered information systems and examined some basic properties. Our future work will focus on new approaches to knowledge acquisition in multi-scale ordered information systems.

**Acknowledgments.** This work is supported by grants from the National Natural Science Foundation of China (Nos. 61272021, 61075120, 61202206 and 61173181), and the Zhejiang Provincial Natural Science Foundation of China (No. LZ12F03002), and the Open Foundation from Marine Sciences in the Most Important Subjects of Zhejiang (No. 20130109).

## References

1. Bargiela, A., Pedrycz, W.: Granular Computing: An Introduction. Kluwer Academic Publishers, Boston (2002)
2. Bargiela, A., Pedrycz, W.: Toward a theory of granular computing for human-centered information processing. IEEE Transactions on Fuzzy Systems 16, 320–330 (2008)
3. Bittner, T., Smith, B.: A theory of granular partitions. In: Duckham, M., Goodchild, M.F., Worboys, M.F. (eds.) Foundations of Geographic Information Science, pp. 117–151. Taylor & Francis, London (2003)

4. Greco, S., Inuiguchi, M., Slowinski, R.: Fuzzy rough sets and multiple-premise gradual decision rules. International Journal of Approximate Reasoning 41, 179–211 (2006)
5. Greco, S., Matarazzo, B., Slowinski, R.: Rough approximation by dominance relation. International Journal of Intelligent Systems 17, 153–171 (2002)
6. Gu, S.-M., Wu, W.-Z.: Knowledge acquisition in inconsistent multi-scale decision systems. In: Yao, J., Ramanna, S., Wang, G., Suraj, Z. (eds.) RSKT 2011. LNCS, vol. 6954, pp. 669–678. Springer, Heidelberg (2011)
7. Gu, S.M., Wu, W.Z.: On knowledge acquisition in multi-scale decision systems. International Journal of Machine Learning and Cybernetics. 4 (2013) 477–486
8. Gu, S.M., Wu, W.Z., Zheng, Y.: Rule acquisition in consistent multi-scale decision systems. In: 8th International Conference on Fuzzy System and Knowledge Discovery, pp. 390–393. IEEE Computer Society, Los Alamitos (2011)
9. Gu, S.M., Li, X., Wu, W.Z., Nian, H.: Multi-granulation rough sets in multi-scale information systems. In: Proceedings of the 2013 International Conference on Machine Learning and Cybernetics, Tianjin, pp. 108–113 (2013)
10. Hu, Q.H., Liu, J.F., Yu, D.R.: Mixed feature selection based on granulation and approximation. Knowledge-Based Systems 21, 294–304 (2008)
11. Inuiguchi, M., Hirano, S., Tsumoto, S.: Rough Set Theory and Granular Computing. Springer, Heidelberg (2002)
12. Keet, C.M.: A Formal Theory of Granularity, PhD Thesis, KRDB Research Centre, Faculty of Computer Science, Free University of Bozen-Bolzano, Italy (2008)
13. Leung, Y., Zhang, J.S., Xu, Z.B.: Clustering by scale-space filtering. IEEE Transactions on Pattern Analysis and Machine Intelligence 22, 1396–1410 (2000)
14. Lin, T.Y., Yao, Y.Y., Zadeh, L.A.: Data Mining, Rough Sets and Granular Computing. Physica, Heidelberg (2002)
15. Ma, J.-M., Zhang, W.-x., Wu, W.-Z., Li, T.-J.: Granular computing based on a generalized approximation space. In: Yao, J., Lingras, P., Wu, W.-Z., Szczuka, M.S., Cercone, N.J., Ślęzak, D. (eds.) RSKT 2007. LNCS (LNAI), vol. 4481, pp. 93–100. Springer, Heidelberg (2007)
16. Mi, J.S., Wu, W.Z., Zhang, W.X.: Approaches to knowledge reduction based on variable precision rough setsmodel. Information Sciences 159, 255–272 (2004)
17. Pawlak, Z.: Rough Sets: Theoretical Aspects of Reasoning about Data. Kluwer Academic Publishers, Boston (1991)
18. Qian, Y.H., Liang, J.Y., Dang, C.Y.: Knowledge structure, knowledge granulation and knowledge distance in a knowledge base. International Journal of Approximate Reasoning 50, 174–188 (2009)
19. Qian, Y.H., Liang, J.Y., Yao, Y.Y., Dang, C.Y.: MGRS: A multi-granulation rough set. Information Sciences 180, 949–970 (2010)
20. Sai, Y., Yao, Y.Y., Zhong, N.: Data analysis and mining in ordered information tables. In: Proceedings of the 2001 IEEE International Conference on Data Mining, pp. 497–504. IEEE, Los Alamitos (2001)
21. Shao, M.W., Zhang, W.X.: Dominance relation and rules in an incomplete ordered information system. International Journal of Intelligent Systems 20, 13–27 (2005)
22. Wu, W.-Z.: Rough set approximations based on granular labels. In: Sakai, H., Chakraborty, M.K., Hassanien, A.E., Ślęzak, D., Zhu, W. (eds.) RSFDGrC 2009. LNCS, vol. 5908, pp. 93–100. Springer, Heidelberg (2009)
23. Wu, W.Z., Leung, Y.: Theory and applications of granular labelled partitions in multi-scale decision tables. Information Sciences 181, 3878–3897 (2011)
24. Wu, W.Z., Leung, Y.: Optimal scale selection for multi-scale decision tables. International Journal of Approximate Reasoning 54, 1107–1129 (2013)

25. Xu, W.H., Li, Y., Liao, X.W.: Approaches to attribute reductions based on rough set and matrix computation in inconsistent ordered information systems. Knowledge-Based Systems 27, 78–91 (2012)
26. Yang, X.B., Yu, D.J., Yang, J.Y., et al.: Dominance-based rough set approach to incomplete interval-valued information system. Data & Knowledge Engineering 68, 1331–1347 (2009)
27. Yao, Y.Y.: Stratified rough sets and granular computing. In: Dave, R.N., Sudkamp, T. (eds.) 18th International Conference of the North American Fuzzy Information Processing Society, pp. 800–804. IEEE Press, New York (1999)
28. Yao, Y.Y.: Information granulation and rough set approximation. International Journal of Intelligent Systems 16, 87–104 (2001)
29. Yao, Y.: A Partition Model of Granular Computing. In: Peters, J.F., Skowron, A., Grzymała-Busse, J.W., Kostek, B.z., Swiniarski, R.W., Szczuka, M.S. (eds.) Transactions on Rough Sets I. LNCS, vol. 3100, pp. 232–253. Springer, Heidelberg (2004)
30. Yao, Y(Y.Y.), Liau, C.-J., Zhong, N.: Granular computing based on rough sets, quotient space theory, and belief functions. In: Zhong, N., Raś, Z.W., Tsumoto, S., Suzuki, E. (eds.) ISMIS 2003. LNCS (LNAI), vol. 2871, pp. 152–159. Springer, Heidelberg (2003)
31. Zadeh, L.A.: Fuzzy sets and information granularity. In: Gupta, N., Ragade, R., Yager, R.R. (eds.) Advances in Fuzzy Set Theory and Applications, pp. 3–18. North-Holland, Amsterdam (1979)
32. Zadeh, L.A.: Towards a theory of fuzzy information granulation and its centrality in human reasoning and fuzzy logic. Fuzzy Sets and Systems 90, 111–127 (1997)

# An Addition Strategy for Reduct Construction

Cong Gao and Yiyu Yao

Department of Computer Science, University of Regina
Regina, Saskatchewan, Canada S4S 0A2
{gao266,yyao}@cs.uregina.ca

**Abstract.** This paper examines an addition strategy for constructing an attribute reduct based on three-way classification of attributes. Properties of three-way classification of attributes are used to design an algorithm for constructing a reduct by using useful attributes. The algorithm makes sure that every attribute to be added, together with already added attributes, will form a partial reduct (i.e., a subset of a reduct). Based on the results of this paper, it is possible to study a wide class of addition based reduct construction algorithms. Finally, variations of the proposed algorithm are discussed.

**Keywords:** Three-way classification, reduct construction, addition strategy.

## 1 Introduction

Reduct construction is one of the most important tasks in rough set theory. It is typically assumed that a finite set of objects is described by a finite set of attributes. The values of objects on these attributes can be expressed by an information table, with rows representing objects and columns representing attributes. An attribute reduct is a minimal subset of attributes that provides the same descriptive ability as the entire set of attributes.

There may be more than one reduct in an information table. Many methods for finding the set of all reducts or a single reduct have been proposed [3, 4, 6–8, 15, 17]. Wong and Ziarko [11] pointed out that finding the set of all reducts or finding a reduct with a minimum number of attributes is an NP problem. Yao et al. [16] formulated reduct construction as a search problem and classified all reduct construction algorithms into three types according to the search strategies used: (1) deletion strategy, (2) addition-deletion strategy and, (3) addition strategy. Deletion strategy based algorithms try to construct a reduct from all attributes by deleting redundant attributes, addition-deletion based algorithms construct a reduct from the empty set or core set of attribute by first adding attributes and then deleting redundant attributes, and the third type, known as addition based algorithms, only has the addition phase without the deletion step.

The main objective of this paper is to revisit the addition strategy and to present a reduct construction algorithm from the view point of three-way classification of attributes. Three-way classification of attributes may be considered

D. Miao et al. (Eds.): RSKT 2014, LNAI 8818, pp. 535–546, 2014.
DOI: 10.1007/978-3-319-11740-9_49 © Springer International Publishing Switzerland 2014

as an example of three-way decisions [12, 13]. An attribute is called a core attribute if it is in all reducts, called a useful attribute if it is in at least one reduct, and called a useless attribute if it is in none of the reducts [14]. The sets of core attributes, useless attributes, and difference of useful attributes and core attributes form a three-way classification of attributes. Similar notions have been used by many authors. For example, Wei et al. [10] referred to the three classes of attributes as the sets of absolutely necessary attributes, relatively necessary attributes, and absolutely unnecessary attributes, respectively. Nguyen and Nguyen [2] named the sets of useful attributes and useless attributes as the set of reductive attributes and redundant attributes, respectively. Intuitively speaking, addition strategy for reduct construction explores the set of useful attributes. Based on the basic results of this paper, it is possible to study a wide class of addition based reduct construction algorithms.

## 2  Reducts in an Information Table

In many data analysis applications, information and knowledge is stored and represented in an *information table* [4].

**Definition 1.** *(Information table) An information table (also called an information system) is the following tuple:*

$$S = (U, AT, \{V_a \mid a \in AT\}, \{I_a \mid a \in AT\}),$$

*where $U$ is a finite nonempty set of objects, $AT$ is a finite nonempty set of attributes, $V_a$ is a nonempty set of values for every attribute $a$ in $AT$, and $I_a : U \to V_a$ is an information mapping. For every $x \in U$, an attribute $a \in AT$, and a value $v \in V_a$, $I_a(x) = v$ means that the object $x$ has the value $v$ on attribute $a$.*

Two objects are discernible if their values are different on at least one attribute. Skowron and Rauszer [7] suggested a matrix representation of discernibility, called a *discernibility matrix*, in which each cell is the set of attributes that discern a pair of objects.

**Definition 2.** *(Discernibility matrix) Given an information table $S$, its discernibility matrix $M = (m(x,y))$ is an $|U| \times |U|$ matrix, in which the element $m(x, y)$ for an object pair $(x, y) \in U \times U$ is defined by:*

$$m(x, y) = \{a \in AT \mid I_a(x) \neq I_a(y)\}.$$

Objects $x$ and $y$ can be distinguished by any attribute in $m(x, y)$. The discernibility matrix $M$ is a symmetric and square matrix, i.e., $m(x, y) = m(y, x)$, and all elements on principal diagonal are empty set, i.e., $m(x, x) = \emptyset$, $x \in U$. Therefore, it is sufficient to consider only the lower triangle or the upper triangle of $M$. As another formulation, $M$ also can be expressed as a set consisting of all distinct and nonempty elements, i.e., $M = \{m(x, y) \neq \emptyset \mid x, y \in U\}$.

Given a subset of attributes $A \subseteq AT$, an indiscernibility relation $\text{IND}(A) \subseteq U \times U$ is defined by:

$$\text{IND}(A) = \{(x, y) \in U \times U \mid \forall a \in A, I_a(x) = I_a(y)\}.$$

For any two objects $x, y \in U$, if $(x, y) \in \text{IND}(A)$, then $x$ and $y$ are indiscernible based on the attribute set $A$.

**Definition 3.** *(Reduct) An attribute set $R \subseteq AT$ of an information table $S$ is called an attribute reduct if $R$ satisfies the following two conditions:*

(i).     $\text{IND}(R) = \text{IND}(AT)$;

(ii).    *For any $a \in R$, $\text{IND}(R - \{a\}) \neq \text{IND}(AT)$.*

Condition (i) is called the jointly sufficient condition and condition (ii) the individually necessary condition. On the one hand, condition (i) indicates the joint sufficiency of the attribute set $R$. The object pairs that cannot be distinguished by $R$ also cannot be distinguished by $AT$, and vice versa. On the other hand, the condition (ii) indicates that each attribute in $R$ is individually necessary. In other words, there exists at least one object pair that cannot be distinguished when any attribute $a \in R$ is deleted from $R$, although the pair can be distinguished by $AT$. The set of all reducts of an information table $S$ is denoted by $\text{RED}(S)$.

Skowron and Rauszer [7] suggested an alternative characterization of a reduct in terms of the discernibility matrix as shown by the following theorem.

**Theorem 1.** *Given the discernibility matrix $M$ of an information table $S$. An attribute set $R$ is an attribute reduct if and only if*

(i).     $\forall (x, y) \in U \times U [m(x, y) \neq \emptyset \Rightarrow R \cap m(x, y) \neq \emptyset]$;

(ii).    $\forall a \in R, \exists (x, y) \in U \times U [m(x, y) \neq \emptyset \wedge ((R - \{a\}) \cap m(x, y) = \emptyset)]$.

Condition (i) shows that $R$ is jointly sufficient for distinguishing all discernible object pairs. In fact, the set of attributes formed by the union of all elements of the discernibility matrix satisfies condition (i). Condition (ii) shows that each attribute in $R$ is individually necessary. Although the result of Theorem 1 provides a criterion to test if a subset of attributes is a reduct, it does not directly offer a method to compute a reduct. Many authors proposed and studied various algorithms to construct a reduct based on the discernibility matrix [4, 7, 9, 17].

In the rest of the paper, we use $m$ for $m(x, y)$ and use the set form $M = \{m(x, y) \mid \forall x, y \in U \wedge m(x, y) \neq \emptyset\}$ for representing the discernibility matrix, if there is no confusion. The set of all reducts of an information table $S$ is also denoted by $\text{RED}(M)$.

*Example 1.* An information table, taken from [16], is given in Table 1. The table has five attributes and seven objects, that is, $U = \{o_1, o_2, o_3, o_4, o_5, o_6, o_7\}$, $AT = \{a, b, c, d, e\}$, $V_a = \{0, 1\}$, $V_b = \{0, 1, 2, 3\}$, $V_c = \{0, 1, 2\}$, $V_d = \{0, 1\}$, $V_e = \{0, 1, 2\}$, $I_b(o_6) = 3$ and so on. The discernibility matrix of Table 1 is given in Table 2. The set representation of the discernibility matrix is:

$$M = \{\{b, c\}, \{b, d\}, \{b, e\}, \{c, d\}, \{d, e\}, \{a, b, d\},$$
$$\{b, c, e\}, \{b, d, e\}, \{a, b, c, d\}, \{a, b, c, e\},$$
$$\{a, b, d, e\}, \{b, c, d, e\}, AT\}.$$

The advantage of the set representation is that we only consider distinct elements of the matrix.

**Table 1.** An information table

|       | a | b | c | d | e |
|-------|---|---|---|---|---|
| $o_1$ | 0 | 0 | 0 | 1 | 1 |
| $o_2$ | 0 | 1 | 2 | 0 | 0 |
| $o_3$ | 0 | 1 | 1 | 1 | 0 |
| $o_4$ | 1 | 2 | 0 | 0 | 1 |
| $o_5$ | 0 | 2 | 2 | 1 | 0 |
| $o_6$ | 0 | 3 | 1 | 0 | 2 |
| $o_7$ | 0 | 3 | 1 | 1 | 1 |

**Table 2.** Discernibility Matrix of Table 1

|       | $o_1$ | $o_2$ | $o_3$ | $o_4$ | $o_5$ | $o_6$ | $o_7$ |
|-------|-------|-------|-------|-------|-------|-------|-------|
| $o_1$ | -     |       |       |       |       |       |       |
| $o_2$ | $\{b, c, d, e\}$ | -     |       |       |       |       |       |
| $o_3$ | $\{b, c, e\}$ | $\{c, d\}$ | -     |       |       |       |       |
| $o_4$ | $\{a, b, d\}$ | $\{a, b, c, e\}$ | $AT$  | -     |       |       |       |
| $o_5$ | $\{b, c, e\}$ | $\{b, d\}$ | $\{b, c\}$ | $\{a, c, d, e\}$ | -     |       |       |
| $o_6$ | $\{b, c, d, e\}$ | $\{b, c, e\}$ | $\{b, d, e\}$ | $\{a, b, c, e\}$ | $\{b, c, d, e\}$ | -     |       |
| $o_7$ | $\{b, c\}$ | $\{b, c, d, e\}$ | $\{b, e\}$ | $\{a, b, c, d\}$ | $\{b, c, e\}$ | $\{d, e\}$ | -     |

## 3    Three-Way Classification of Attributes

Given an information table $S$ with the discernibility matrix $M$, according to the set of all reducts $\mathrm{RED}(M)$, attributes in $AT$ can be divided into three classes.

**Definition 4.** *(Three-way Classification of Attributes) Given an information table $S$, attributes in $AT$ can be divided into three pair-wise disjoint classes:*

$$\mathrm{CORE} = \bigcap \mathrm{RED}(M),$$
$$\mathrm{USEFUL\text{-}NC} = \bigcup \mathrm{RED}(M) - \bigcap \mathrm{RED}(M),$$
$$\mathrm{USELESS} = AT - \bigcup \mathrm{RED}(M).$$

*Attributes in* CORE, USEFUL-NC, USELESS *are called core, useful non-core and useless attributes, respectively. The set of attributes* USEFUL $= \bigcup \mathrm{RED}(M) =$ CORE $\cup$ USEFUL-NC *is the set of all useful attributes.*

The following properties hold:

(A1)  CORE $\cup$ USEFUL-NC $\cup$ USELESS $= AT$ and they are pair-wise disjoint;

(A2)  Every attribute in USEFUL-NC appears in at least one reduct, but not in all reducts;

(A3)  All attributes in USELESS does not appear in any reduct;

(A4)  Each attribute in CORE appears in every reduct;

(A5)  RED($M'$) = RED($M$), where $M' = \{m - \text{USELESS} \mid m \in M\}$.

**Proof.** Properties (A1) to (A4) are easy to prove, we only prove (A5). By definition of $M'$, every element $m' \in M'$ is calculated by $m' = m - \text{USELESS}$. Thus, $m' \subseteq m$.

RED($M'$) $\subseteq$ RED($M$): Assume $R \in$ RED($M'$). By condition (i) of Theorem 1, $R \cap m' \neq \emptyset$. Since $m' \subseteq m$, we get $R \cap m \neq \emptyset$. Thus, $R$ satisfies the condition (i) of Theorem 1 for $M$. Consider an attribute $a \in R$. By the condition (ii) of Theorem 1 for $M'$, we conclude that exists at least one $m' \in M'$ such that $(R - \{a\}) \cap m' = \emptyset$. According to (A3), we know $a \notin$ USELESS. Thus, $(R - \{a\}) \cap m = \emptyset$. That is, $R$ satisfies condition (ii) of Theorem 1 for $M$. Therefore, $R \in$ RED($M$).

RED($M$) $\subseteq$ RED($M'$): Assume $R \in$ RED($M$). By condition (i) of Theorem 1, we have $R \cap m \neq \emptyset$ for all $m \in M$. By (A3), $R \cap$ USELESS $= \emptyset$. We can conclude that $R \cap m' = R \cap (m - \text{USELESS}) \neq \emptyset$ for all $m' \in M'$. Thus, condition (i) of Theorem 1 holds for $R$ with respect to $M'$. Since $R$ satisfies condition (ii) of Theorem 1 for $M$, by $m' \subseteq m$, thus, $R$ satisfies condition (ii) for Theorem 1 for $M'$. This means that $R \in$ RED($M'$) $\qquad \square$

Many authors discussed the classification of attributes [4, 9, 10, 16, 17] in an explicit or implicit way. A widely used classification is the division of attributes into core attributes and non-core attributes. Wei et al. [10] classified attributes into three classes and proposed a method for attribute reduction in information tables.

## 4  Operations for Discernibility Matrix

Skowron and Rauszer [7] introduced absorption operations for a discernibility matrix. Yao and Zhao [15] applied the operations for simplifying the discernibility matrix.

**Definition 5.** *(Element absorption, Discernibility matrix absorption, Absorbed discernibility matrix) Given a discernibility matrix $M$, an element $m_1 \in M$ can absorb another element $m_2 \in M$, if $m_1 \subseteq m_2$. The discernibility matrix absorption is a sequence of all possible element absorptions. The result of discernibility matrix absorption on $M$ is a new matrix $M'$, which is obtained by replacing all the absorbed elements by the absorbing elements in $M$ and is called absorbed discernibility matrix of $M$.*

By definition, $\emptyset \neq M' \subseteq M$. In the set representation of discernibility matrix, elements can be removed from $M$ if they can be absorbed by any other elements in $M$. This removel will not affect the set of all reducts RED($M$) [7]. In other words, if $M'$ is the absorbed discernibility matrix of the original matrix $M$, then RED($M'$) = RED($M$). Yao and Zhao [15] suggested another operation called attributes deletion, which will affect the set of all reduct, but make sure the existance of at least one reduct as shown by the following lemma.

**Lemma 1.** *Given a discernibility matrix $M$, $W = \bigcup_{m \in M} m$ and $A \subseteq W$, if the set of attributes $W - A$ is jointly sufficient (i.e., $\forall m \in M, m \neq \emptyset \Rightarrow (W-A) \cap m \neq \emptyset$), then* RED($M'$) $\neq \emptyset$ *and* RED($M'$) $\subseteq$ RED($M$)*, where $M' = \{m - A \mid m \in M\}$.*

According to the lemma, there exists at least one reduct of $M$ in the new matrix $M'$ after the deleting the set of attributes.

**Definition 6.** *(An attribute-induced group of sets of attributes) Given a discernibility matrix $M$, for an attribute $a$ in $M$ (i.e., $a \in \bigcup_{m \in M} m$), the family of sets of attributes defined by:*

$$Group(a) = \{m \in M \mid a \in m\}.$$

*is called the $a$-induced group of sets of attributes*

By inducing absorption and attribute-induced group, we give the next lemma.

**Lemma 2.** *Given a discernibility matrix $M$, if $M'$ is the absorbed discernibility matrix of $M$, then $\bigcup_{m \in M'} m = $ USEFUL, i.e., every attribute in absorbed discernibility matrix is useful.*

**Proof.** $\bigcup_{m \in M'} m \supseteq$ USEFUL: By Definition 4, we know $\forall a \in$ USEFUL, and then $a \in \bigcup$RED($M$). Since RED($M'$) = RED($M$), we can get $a \in \bigcup$RED($M'$) $\subseteq \bigcup_{m \in M'} m$.

$\bigcup_{m \in M'} m \subseteq$ USEFUL: Assume $a \in \bigcup_{m \in M'} m$. Since $M'$ is the absorbed discernibility matrix, $\forall m \in (M' - Group(a))$, $\forall g \in Group(a)$, $m \not\subseteq g$, i.e., $m - g \neq \emptyset$. By expressing $g = \{a\} \cup A$, we get $m - A \neq \emptyset$. By Lemma 1, we get RED($M_1$) $\subseteq$ RED($M'$), where $M_1 = \{m - A \mid m \in M'\}$. After removing $A$ from $M$, we obtain $M_1$ and, the original element $g = \{a\} \cup A$ becomes a singleton subset $\{a\}$. By using $\{a\}$ to absorb elements of $Group(a)$ in $M_1$, all elements of $Group(a)$ can be absorbed by $\{a\}$. After this absorption, we denote the result discernibility matrix as $M_2$. Since this absorption will not change any reduct, we have RED($M_2$) = RED($M_1$).

Now, we prove $M_2 = M' \cup \{\{a\}\}$. Consider $M_2$, it can be expressed as the union of two part: $M_2 = P_1 \cup P_2$. Each element in $P_1$ includes attribute $a$ and each element in $P_2$ does not include $a$. In $P_1$, there is only one element $\{a\}$. In $P_2$, all elements have been deleted by $A$, i.e., $P_2 = \{m' - A \mid m' \in (M' - Group(a))\}$. Since $\forall m' \in (M' - Group(a))$ does not include $a$ and $g = \{a\} \cup A$, it follows $m' - A = m' - g$. Therefore, $P_2 = \{m' - A \mid m' \in (M' - Group(a))\} = \{m' - g \mid m' \in (M' - Group(a))\}$. Thus, $M_2 = P_1 \cup P_2 = \{\{a\}\} \cup P_2$. It can be concluded that $\text{RED}(\{\{a\}\} \cup P_2) = \text{RED}(M_2) = \text{RED}(M_1) \subseteq \text{RED}(M') = \text{RED}(M)$. Every attribute in singleton subset in discernibility matrix is a core attribute [7]. Therefore, $a$ is a core attribute in $(\{\{a\}\} \cup P_2)$, this means there exists $R \in \text{RED}(\{\{a\}\} \cup P_2)$, such that $a \in R$ and $R \in \text{RED}(M')$. Thus, $a$ is useful, i.e., $a \in \text{USEFUL}$. □

Lemma 2 states an important property of the set of useful attribute USEFUL. Since CORE is easy to compute by collecting all attributes in singleton subsets in $M$. Thus, Lemma 2 essentially tells us the property of USEFUL-NC. Based on its proof, the next lemma can be stated, which gives the condition for testing whether an attribute is useful.

**Lemma 3.** *Given a discernibility matrix $M$, an attribute $a \in \text{USEFUL}$ if and only if there exists a $g \in Group(a)$, such that $g$ cannot be absorbed by any other elements in $M$.*

**Proof.** Suppose $M'$ is the absorbed discernibility matrix of $M$.

$\Leftarrow$: If there exists a $g \in Group(a)$, such that $g$ cannot be absorbed by any other elements in $M$, then $g$ will be in $M'$. By Lemma 2, $a \in \text{USEFUL}$.

$\Rightarrow$: Assume $a \in \text{USEFUL}$. By Lemma 2, $a \in \bigcup_{g \in M'} g$. This means there exists a $g \in M'$ that $a \in g$. Since $M'$ is already absorbed, $g$ cannot be absorbed by any other elements in $M$. □

*Example 2.* Consider the discernibility matrix $M$ given in Table 2. We have:

$\{b, c\}$ absorbs $\{b, c, e\}, \{a, b, c, d\}, \{a, b, c, e\}, \{b, c, d, e\}$ and $AT$.

$\{b, d\}$ absorbs $\{a, b, d\}, \{b, d, e\}, \{a, b, c, d\}, \{a, b, d, e\}, \{b, c, d, e\}$ and $AT$.

$\{b, e\}$ absorbs $\{b, c, e\}, \{b, d, e\}, \{a, b, c, e\}, \{a, b, d, e\}, \{b, c, d, e\}$ and $AT$.

$\{c, d\}$ absorbs $\{a, b, c, d\}, \{b, c, d, e\}$ and $AT$.

$\{d, e\}$ absorbs $\{b, d, e\}, \{a, b, d, e\}, \{b, c, d, e\}$ and $AT$.

Therefore, the absorbed discernibility matrix is

$$M' = \{\{b, c\}, \{b, d\}, \{b, e\}, \{c, d\}, \{d, e\}\}.$$

The classifications of attributes are given by:

$$\text{CORE} = \emptyset, \text{USEFUL} = \{b, c, d, e\}, \text{USELESS} = \{a\}.$$

The results illustrate the basic ideas introduced in this section.

## 5 Addition Strategy for Reduct Construction

Given an information table or a discernibility matrix, the addition strategy for constructing a reduct starts from the empty set or the CORE and adds one attribute at a time sequentially.

## 5.1  Construction of a Partial Reduct by Addition of Attributes

To design an algorithm based on the addition strategy, simply knowing whether an attribute is useful or not is insufficient. Attributes may have some dependencies among each other. For instance, given a discernibility matrix $M$, if $\text{RED}(M) = \{\{b, d\}, \{a, c, e\}\}$, both attribute $b$ and $c$ are useful, but there does not exist a reduct containing $b$ and $c$. This means that we cannot simply add all useful attributes to construct a reduct, since some attributes added earlier may become unnecessary. Thus, an extra step of deletion must be performed [16].

**Definition 7.** *(Partial reduct, Super reduct) A set of attribute $R'$ is called a partial reduct, if there exists a reduct $R$, such that $R' \subseteq R$. A subset of attribute $R'$ is called a super reduct, if there exists a reduct $R$, such that $R' \supseteq R$.*

An important property of a partial reduct is that each attribute in it is necessary. To avoid attribute deletion, one must make sure that each addition of an attribute will form a partial reduct. Consequently, the deletion step is not needed.

Once an attribute $a \in \bigcup_{m \in M'} m$ is chosen to be added into a partial reduct, those elements including $a$ can be removed from $M'$, because those object pairs associated with elements including $a$ can be discerned by $a$. A pivotal step given by Theorem 2 is needed to ensure each added attribute will be in a partial reduct.

**Theorem 2.** *Given a discernibility matrix $M$, if an attribute $a \in AT$ is useful, then there exists $g \in Group(a)$ satisfying the following properties:*

(i).  $\forall m \in (M - Group(a))$  $[m \neq \emptyset \Rightarrow (m - g) \neq \emptyset]$;

(ii).  $\text{RED}(M_s \cup \{\{a\}\}) \subseteq \text{RED}(M)$, $M_s = \{m - g \mid m \in (M - Group(a))\}$.

**Proof.** (i): Assume property (i) does not hold, i.e., $\forall g \in Group(a)$, $\exists m \in (M - Group(a))$ $[m \neq \emptyset \wedge (m - g) = \emptyset]$. We now prove this leads to a contradiction. By assumption that $(m - g) = \emptyset$, we can obtain $m \subseteq g$. Thus $g$ can be absorbed by $m$. By the assumption, for every $g \in Group(a)$, there exists an $m \in (M - Group(a))$ absorbing it. Therefore, there is no element containing attribute $a$ in the absorbed discernibility matrix $M'$. Since $\text{RED}(M') = \text{RED}(M)$, there does not exist a reduct that contains attribute $a$. By the definition of USELESS, it can be concluded that $a$ is useless, which conflicts with the given condition that $a$ is useful. Hence, property (i) must hold.

(ii): According to the proof of Lemma 2, it is easy to prove that property (ii) holds.    □

Theorem 2 tells us that once a useful attribute $a \in AT$ has been chosen to construct a reduct, this attribute will be a singleton subset in the modified discernibility matrix $M_s \cup \{\{a\}\}$, in which $a$ is a core attribute that can be added into a partial reduct.

Since the two actions, removing $Group(a)$ from $M$ and deleting $A$ from $M$, will be taken after adding an useful attribute $a$ into a partial reduct, such actions imply that for all $R \in \text{RED}(M_s)$, $(R \cup \{a\}) \in \text{RED}(M)$. The procedure of converting $M$ into $M_s$ makes every added attribute be in a partial reduct. Thus, by continually performing this routine, one reduct in $\text{RED}(M)$ will be computed.

Furthermore, Theorem 2 is not limited to the absorbed discernibility matrix. An unabsorbed discernibility matrix also can be used to find a reduct by property (ii). Therefore, Theorem 2 provides a crucial criterion to design reduct construction algorithms based on addition strategy. Thereom 2 can also deal with a relative discernibility matrix. Algorithms designed by Thereom 2, when applied to a relative discernibility matrix, can produce a relative reduct. Miao et al. [1] discussed three kinds of criteria to produce respective relative discernibility matrices, which can be used to compute relative reducts by the proposed algorithms.

## 5.2   An Addition Strategy Based Algorithm

By the analysis so far, Theorem 2 suggests an algorithm for reduct construction based on an addition strategy. The detailed steps are given in Algorithm 1.

---

**Algorithm 1.** An Addition Strategy based Algorithm for Constructing a Reduct

---

**input**  : A discernibility matrix $M$.
**output**: A reduct $R$.

(1) $R = \emptyset$ or CORE, $CA = AT$ or $AT - \text{USELESS}$;
(2) **while** $CA \neq \emptyset$ and $M \neq \emptyset$ **do**
    (2.1) Compute fitness values of all attributes in $CA$ by using a fitness function $\sigma$;
    (2.2) Select an attribute $a \in CA$ according to its fitness;
    (2.3) $M = M - Group(a)$;
    (2.4) $CA = CA - \{a\}$;
    (2.5) **if** $a$ *is useful* **then**
        (2.5.1) $R = R \cup \{a\}$;
        (2.5.2) Compute fitness values of all elements in $Group(a)$ by using a fitness function $\xi$;
        (2.5.3) Sort $Group(a)$ according to fitness;
        (2.5.4) **for** $i = 1 : |Group(a)|$, $g_i \in Group(a)$ **do**
            (2.5.4.1) **if** $g_i$ can not be absorbed in $M$ **then**
                (2.5.4.1.1) $CA = CA - g_i$;
                (2.5.4.1.2) $M = \{m - g_i \mid m \in M\}$;
                (2.5.4.1.3) break;

(3) Output $R$.

---

There are two fitness functions used in Algorithm 1, that is, $\sigma : AT \to \Re$ and $\xi(a) : Group(a) \to \Re$, respectively. For fitness function, frequency-besed or entropy-based ones are widely used. For $\sigma$, a simple frequency-besed and an entropy-based fitness function can be defined as $\sigma(a) = |\{m \in M \mid a \in m\}|$ and $\sigma(a) = H(a) = -\sum_{v \in V_a} p(v) \log p(v)$, respectively. For $\xi$, it is different from $\sigma$. Fitness function $\xi$ is used to calculate fitness value of elements in attribute-induced group, rather than attribute itself. However, the frequency-based strategy and entropy-based strategy still can be used to design $\xi$. For example, a frequency-based has been given by Yao et al. [16]: $\xi(g_i) = |\{m \in M \mid m \cap g_i \neq \emptyset\}|$.

### 5.3   Variations of the Proposed Algorithm

Lemma 3 provides an approach to test whether an attribute is useless or not. It works on both the original discernibility matrix and absorbed discernibility matrix.

On the one hand, the procedure of Algorithm 1 can be simplified when we use absorbed discernibility matrix. If we absorb the discernibility matrix every time before adding an attribute into a partial reduct, according to Lemma 2, every attribute in absorbed discernibility matrix is a useful attribute, then step (2.5), (2.5.4), (2.5.4.1), and (2.5.4.1.3) can be removed from Algorithm 1. Actually, the simplified algorithm is essentially equal to an algorithm proposed by Zhao and Wang [17].

On the other hand, the given algorithm can be generalized by replacing the set representation of discernibility matrix to a table representation. The table representation of discernibility matrix is the one which is expressed by a table like Table 2. Compared with the table representation one, the set representation discernibility matrix loses the position information in table which stands for the object pair. Thus, by using table representation discernibility matrix, the reduct construction procedure is the simplification to discernibility matrix [15].

### 5.4   Time Complexity Analysis

The complexity of the proposed method depends on two parts, i.e., the construction of the discernibility matrix and construction of a reduct. The time complexity to calculate the discernibility matrix is $O(|AT||U|^2)$. For the proposed reduct construction algorithm, within the while loop of step (2), according to Lemma 3, to test whether an attribute $a$ is useless or not needs $|Group(a)| * |M - Group(a)| * |AT|$ comparisons. Thus the time complexity of step (2.5) is $O(|AT||U|^4)$. Similarly, the time complexity from step (2.5.4) to step (2.5.4.1.3) is $O(|AT||U|^4)$. Third, the while loop of step (2) will be executed for $Min(|AT|, |M|)$ number of times. Finally, the overall time complexity of Algorithm 1 is $O(Min(|AT|, |U|^2) * |AT||U|^4)$.

By using certain fitness functions of $\sigma$ and $\xi$, this algorithm may be improved. For example, when the fitness function $\xi$ always chooses the first $g_i \in Group(a)$ which satisfies step (2.5), and because during the calculation of step (2.5) we get such $g_i$, then the loop of step (2.5.4) can be removed. This modification reduces $|Group(a)| * |M - Group(a)| * |AT|$ comparisons. However, it does not change the time complexity of algorithm.

In comparison with the algorithm proposed in [17], Algorithm 1 has a lower time complexity and needs not to do absorption operation before adding an attribute. During the construction procedure, the discernibility matrix will be shrinked. Consequently, less number of elements needs to be compared. In comparison with the algorithm proposed in [15], Algorithm 1 needs less storage space, because it is unnecessary to store the entire discernibility matrix.

# 6 Conclusion

Reduct construction algorithms play an important role in data mining and data analysis. When the number of attributes is large and the number of attribute in a reduct is small, an addition strategy for reduction construction will be an ideal choice.

An important criterion in designing an algorithm based on an addition strategy is to make every added attribute necessary, that is, a partial reduct is constructed. The discernibility matrix absorption and element absorption ensure that every added attribute is necessary.

Based on three-way classification of attributes, we explore some properties of the set of useful attributes USEFUL. The results enable us to design an addition strategy based reduct construction algorithm. The proposed algorithm can be conveniently generalized. A wide range of algorithms based on addition strategy can be further studied.

# References

1. Miao, D.Q., Zhao, Y., Yao, Y.Y., Li, H.X., Xu, F.F.: Relative reducts in consistent and inconsistent decision tables of the Pawlak rough set model. Information Sciences 179(24), 4140–4150 (2009)
2. Nguyen, L.G., Nguyen, H.S.: On elimination of redundant attributes in decision tables. In: Federated Conference on Computer Science and Information Systems (FedCSIS), pp. 317–322 (2012)
3. Nguyen, H.S.: Approximate Boolean reasoning: foundations and applications in data mining. In: Peters, J.F., Skowron, A. (eds.) Transactions on Rough Sets V. LNCS, vol. 4100, pp. 334–506. Springer, Heidelberg (2006)
4. Pawlak, Z.: Rough Sets: Theoretical Aspects of Reasoning About Data. Kluwer Academic Publishers, Dordrecht (1991)
5. Pawlak, Z.: Rough sets. International Journal of Computer Information and Science 11(5), 341–356 (1982)
6. Quafafou, M.: $\alpha$-RST: a generalization of rough set theory. Information Sciences 124(1-4), 301–316 (2000)
7. Skowron, A., Rauszer, C.: The discernibility matrices and functions in information systems. In: Slowinski, R. (ed.) Intelligent Decision Support. Theory and Decision Library, pp. 331–362. Springer, Netherlands (1992)
8. Swiniarski, R.W.: Rough sets methods in feature reduction and classification. International Journal of Applied Mathematics and Computer Science 11(3), 565–582 (2001)

9. Wang, J., Wang, J.: Reduction algorithms based on discernibility matrix: the ordered attributes method. Journal of Computer Science and Technology 16(6), 489–504 (2001)
10. Wei, L., Li, H.R., Zhang, W.X.: Knowledge reduction based on the equivalence relations defined on attribute set and its power set. Information Sciences 177(15), 3178–3185 (2007)
11. Wong, S.K.M., Ziarko, W.: On optimal decision rules in decision tables. Bulletin of Polish Academy of Sciences 33, 693–696 (1985)
12. Yao, Y.: An outline of a theory of three-way decisions. In: Yao, J., Yang, Y., Słowiński, R., Greco, S., Li, H., Mitra, S., Polkowski, L. (eds.) RSCTC 2012. LNCS, vol. 7413, pp. 1–17. Springer, Heidelberg (2012)
13. Yao, Y.Y.: The superiority of three-way decisions in probabilistic rough set models. Information Sciences 181(6), 1080–1096 (2011)
14. Yao, Y.Y.: Duality in rough set theory based on the square of opposition. Fundamenta Informaticae 127(1), 49–64 (2013)
15. Yao, Y.Y., Zhao, Y.: Discernibility matrix simplification for constructing attribute reducts. Information Sciences 179(7), 867–882 (2009)
16. Yao, Y.Y., Zhao, Y., Wang, J.: On reduct construction algorithms. In: Gavrilova, M.L., Tan, C.J.K., Wang, Y., Yao, Y., Wang, G. (eds.) Transactions on Computational Science II. LNCS, vol. 5150, pp. 100–117. Springer, Heidelberg (2008)
17. Zhao, K., Wang, J.: A reduction algorithm meeting users' requirements. Journal of Computer Science and Technology 17(5), 578–593 (2002)

# Analysis of User-Weighted π Rough k-Means

Georg Peters[1] and Pawan Lingras[2]

[1] Munich University of Applied Sciences, Munich,
Germany & Australian Catholic University, Sydney, Australia
georg.peters@cs.hm.edu
[2] Saint Mary's University, Halifax, Canada
pawan.lingras@smu.ca

**Abstract.** Since its introduction by Lingras and West a decade ago, rough k-means has gained increasing attention in academia as well as in practice. A recently introduced extension, π rough k-means, eliminates need for the weight parameter in rough k-means applying probabilities derived from Laplace's Principle of Indifference. However, the proposal in its more general form makes it possible to optionally integrate user-defined weights for parameter tuning using techniques such as evolutionary computing. In this paper, we study the properties of this general user-weighted π k-means through extensive experiments.

**Keywords:** Rough k-Means, User-Defined Weights, Soft Clustering.

## 1 Introduction

Clustering is one of the most popular methods in data mining. The k-means clustering algorithm [3,10] is among the most frequently used approaches. In the meantime, derivatives of the k-means based on soft computing concepts have been proposed to address uncertainty. They include famous algorithms like fuzzy c-means [2] and Krishnapuram and Keller's possibilistic c-means [4].

In 1982, Pawlak [17] introduced rough set theory. Lingras and West [9] proposed rough k-means that is derived from the interval interpretation of rough sets [23]. Since its introduction it has gained increasing attention and has become an integral and important part of soft clustering [20].

Recently, Peters [19] proposed π rough k-means that utilizes Laplace's Principle of Indifference [5] to determine the weights in the mean function. So, in contrast to original rough k-means, π rough k-means in its standard case does not have any user-defined weights. However, in the general form of π rough k-means additional weights for the lower approximation and boundary region are integrated in the algorithm (we refer to it as *weighted* π rough k-means). These weights make it possible to optimize the clustering results by parameter tuning using techniques such as evolutionary computing (see, e.g., Mitra [14], Peters et al. [22] and Lingras [6] for evolutionary rough clustering approaches). Peters [19] already outlined the weighted π rough k-means algorithm. However, it was not studied in depth.

D. Miao et al. (Eds.): RSKT 2014, LNAI 8818, pp. 547–556, 2014.
DOI: 10.1007/978-3-319-11740-9_50 © Springer International Publishing Switzerland 2014

Therefore, the objective of the paper is to compare weighted and non-weighted $\pi$ rough k-means in more detail. Since the initial parameters of weighted $\pi$ rough k-means (weights and the threshold) are user-defined, it is crucial to have a good understanding of their impacts. Thus, we analyze their effects in experiments on real-life data and present the obtained results as performance maps.[1]

The remainder of the paper is organized as follows. In Section 2, we describe the fundamentals of rough clustering and briefly summarize some of its extensions. Then, in Section 3, we review $\pi$ rough k-means and discuss properties of weighted $\pi$ rough k-means. In the next section, we analyse results of experiments based on weighted $\pi$ rough k-means. The paper concludes with a summary in Section 5.

## 2    Fundamentals of Rough Clustering

In the rough clustering domain, Lingras and West [9] rough k-means and its derivatives are probably among the most popular algorithms. Their algorithmic structure is similar to Lloyd's k-means [10]. The main difference is that rough k-means has to deal with two approximations, the lower approximation and the boundary of a cluster. Lingras and West's rough k-means proceeds as follows:[2]

*Initialization*

- Setting of the number of clusters $K$, the weight for the lower approximations $\underline{w}$ (and the boundary weight $\widehat{w} = 1 - \underline{w}$) and the threshold $\zeta \geq 1$.
- Selection of the initial cluster centers (means) and assignment of the objects to the cluster as shown in Equations 2 to 4.

*Iteration*

- *Means*
  Compute the new means:

$$
m_k = \begin{cases}
\underline{w} \sum\limits_{x_n \in \underline{C_k}} \frac{x_n}{|\underline{C_k}|} + \widehat{w} \sum\limits_{x_n \in \widehat{C_k}} \frac{x_n}{|\widehat{C_k}|} & \text{for } \underline{C_k} \neq \emptyset \wedge \widehat{C_k} \neq \emptyset \\
\sum\limits_{x_n \in \underline{C_k}} \frac{x_n}{|\underline{C_k}|} & \text{for } \underline{C_k} \neq \emptyset \wedge \widehat{C_k} = \emptyset \\
\sum\limits_{x_n \in \widehat{C_k}} \frac{x_n}{|\widehat{C_k}|} & \text{for } \underline{C_k} = \emptyset \wedge \widehat{C_k} \neq \emptyset
\end{cases}
\tag{1}
$$

- *Approximations*
  Assign the objects to the approximations:
  (i) Determine the nearest mean of object $x_n$:

$$
d_h^{min} = d(x_n, m_h) = \min_{k=1,...,K} d(x_n, m_k)
\tag{2}
$$

  and assign $x_n$ to the upper approximation of cluster $h$:

$$
x_n \in \overline{C_h}.
\tag{3}
$$

---

[1] For the sake of simplicity, we assume the number of clusters as given.
[2] We present it with the well established relative threshold parameter $\zeta$.

(ii) Determine:

$$T = \left\{ t : \frac{d(x_n, m_t)}{d_h^{min}} \leq \zeta \wedge h \neq t \right\}. \tag{4}$$

IF $[T \neq \emptyset]$ THEN $[x_n \in \overline{C_t}, \forall t \in T]$ ELSE $[x_n \in \underline{C_h}]$.

– *Stopping Criterion*

IF [current upper approximations unchanged to previous .or. maximum iterations reached] THEN [stop] ELSE [repeat iteration].

Since its introduction, several extensions and derivatives of Lingras and West's original rough k-means have been proposed. They include, Peters' refined rough k-means [18], rough medoids [21] and evolutionary rough clustering [6,14,22] where the initial parameters are optimized with respect to cluster validity indexes. Furthermore, hybrid clustering, merging rough with fuzzy or possibilistic approaches have been proposed by, e.g., Mitra et al. [15], Maji and Pal [11,12] and Maji and Paul [13].

For recent surveys on rough clustering please see Lingras and Peters [7,8] and for the relationship of rough clustering to further soft clustering approaches Peters et al. [20].

## 3   Foundations of $\pi$ Rough k-Means

### 3.1   Fundamental Idea of $\pi$ Rough k-Means

Recently introduced $\pi$ rough k-means [19] addresses some challenges observed in the established rough k-means algorithms, in particular:

1. Some established algorithms are rather sensitive with respect to outliers. $\pi$ rough k-means is quite robust.
2. In established approaches a boundary object (unclear membership) has a higher impact on the means than an object in a lower approximation (sure membership). In $\pi$ rough k-means sure memberships have greater influence on the means than objects with unclear memberships.
3. In established algorithms the overall impact of a boundary object increases when its lack of clarity (the number of boundaries it belongs to) increases. In $\pi$ rough k-means the overall impact of boundary objects is independent of the number of boundaries it belongs to.
4. Established rough k-means algorithms require user-defined weights. $\pi$ rough k-means is free of user-defined weights; the weights are derived from probabilities.
5. The setting of user-defined weights is often subjective or even arbitrary. The weights in $\pi$ rough k-means are based on a well established Laplace's Principle of Indifference. Thus, they are unbiased and free from subjective elements.

To achieve these goals the mean function in $\pi$ rough k-means is modified as follows:

- In the mean function, an object $x_n$ is weighted by the reciprocal of the number $|B_{x_n}|$, where $B_{x_n}$ a set of clusters $x_n$ belongs to. This applies to all objects independent of their membership in either lower approximations or boundaries. For example, if object $x_{17}$ belongs to the boundaries of the clusters $C_2$, $C_3$ and $C_5$ we obtain: $B_{x_{17}} = \{C_2, C_3, C_5\}$ and $|B_{x_{17}}| = 3$; if $x_{26}$ belongs to the lower approximation of $C_4$ we get $B_{x_{26}} = \{C_4\}$ and $|B_{x_{26}}| = 1$. This weighting is derived from Laplace's Principle of Indifference [5]: Since we have no reasons for using any other probabilities, the probability of membership to a cluster is uniformly distributed and equals the reciprocal of the number of clusters an object belongs to. For the above examples we obtain: $p(x_{17}) = \dfrac{1}{|B_{x_{17}}|} = \dfrac{1}{3}$ and $p(x_{17}) = \dfrac{1}{|B_{x_{16}}|} = \dfrac{1}{1}$.

- In original rough k-means the mean function consists of the weighted sum of the sub-means, i.e., the mean of the lower approximation and the mean of the boundary of a cluster. Normally, the number of objects in a boundary is much smaller than the number of objects in a lower approximation. Hence, we observe an anomalous situation that the influence of a boundary object on means is much higher than that of an object in a lower approximation. In contrast to this indirect impact via sub-means, in $\pi$ rough k-means each object has a direct impact on the mean function – independent of its assignment to a lower approximation or a boundary.

### 3.2   Algorithmic Structure of $\pi$ Rough k-Means

In detail $\pi$ rough k-means in its standard case, i.e., without user-defined weights, proceeds as follows:

*Initialization*

- Setting of the number of clusters $1 < K < N$ ($k = 1, ..., K$) and the threshold parameter $\zeta \geq 1$.
- Selection of the initial means and assignment of the objects to the upper approximations of their nearest clusters.

*Iteration*

- *Means*
  Compute the new means:

$$m_k = \frac{\sum\limits_{x_n \in \overline{C_k}} \frac{x_n}{|B_{x_n}|}}{\sum\limits_{x_n \in \overline{C_k}} \frac{1}{|B_{x_n}|}} \tag{5}$$

– *Approximations*

Assign the objects to the approximations:

(i) For each cluster $k$, determine the object that is nearest to its mean

$$d_g^{min} = d(m_k, x_g) = \min_{n=1,...,N} d(m_k, x_n) \qquad (6)$$

and assign it to the corresponding upper approximation of cluster $k$:

$$x_g \in \overline{C_k} \qquad (7)$$
$$B_{x_g} = \{k\} \text{ and } |B_{x_g}| = 1 \qquad (8)$$

(ii) For the remaining objects $x_m$ with $m = 1, ..., N - K$:

  • Determine the nearest mean of object $x_m$:

$$d_h^{min} = d(x_m, m_h) = \min_{k=1,...,K} d(x_m, m_k) \qquad (9)$$

  • Determine similarly near means. Including the nearest mean we obtain:

$$B_{x_m} = T = \left\{ t : \frac{d(x_m, m_t)}{d_h^{min}} \leq \zeta \right\} \qquad (10)$$

  • Assign object $x_n$ to the upper approximations of the respective clusters:

$$x_m \in \overline{C_t}, \forall t \in T \qquad (11)$$

– *Stopping Criterion*

IF [current upper approximations unchanged to previous .or. maximum iterations reached] THEN [stop] ELSE [repeat iteration].

Note, step *Approximations (i)* is optional. It prevents the rather unlikely case of empty clusters and consequently a division by 0 in Equation 5.

## 3.3  User-Weighted $\pi$ Rough k-Means

Let us replace Equation 5 by a more general mean function, i.e., with user-defined weights [19]:

$$m_k = \frac{\underline{w} \sum\limits_{x_n \in \underline{C_k}} x_n + \widehat{w} \sum\limits_{x_n \in \widehat{C_k}} \frac{x_n}{|B_{x_n}|}}{\underline{w}|\underline{C_k}| + \widehat{w} \sum\limits_{x_n \in \widehat{C_k}} \frac{1}{|B_{x_n}|}} \qquad (12)$$

$$\text{with} \sum\limits_{x_n \in \widehat{C_k}} \dots = 0 \text{ for } \widehat{C_k} = \emptyset$$

For $\underline{w} = \widehat{w} = 0.5$ Equation 12 melts down to Equation 5. Since the weights $\underline{w}$ and $\widehat{w} = 1 - \underline{w}$ are user-defined they can be used for parameter tuning. In the next section, we perform experiments to analyse their impacts as well as the impact of the threshold parameter $\zeta$ on the clustering results.

# 4  Experiments and Discussion

## 4.1  Preliminaries

We use the Iris and the Wine data from the UCI database [1] and the Vowel data [16]. The data are normalized to the unity interval $[0, 1]$ before clustering. In the experiments, we vary the threshold parameter in a range from 1.1 to 2.0 in 0.1 steps ($\zeta \in \{1.1, 1.2, \ldots, 2.0\}$) and the weight for the lower approximation from 0.1 to 0.9 ($\underline{w} \in \{0.1, 0.2, \ldots, 0.9\}$).

Note, that our objective is not to obtain the best possible clustering results. Some of the previous approaches perform better than the results presented here. Our focus is on the relative performance of the algorithm with respect to the threshold parameter $\zeta$ and the weights $\underline{w}$, $\widehat{w} = 1 - \underline{w}$.

## 4.2  Experiments

*Wine Data Experiments.* The Wine data set consists of 178 samples with 13 features representing three different types of wine. The results are shown in Table 1. The representation of the results in Table 1 is to a certain degree

**Table 1.** Wine Data: Results

| $\zeta$ | \multicolumn{9}{c}{$w$} | | | | | | | | | | | | | | | | | |
|------|-----|-----|-----|-----|-----|-----|-----|-----|-----|-----|-----|-----|-----|-----|-----|-----|-----|-----|
|      | 0.1 | 0.2 | 0.3 | 0.4 | 0.5 | 0.6 | 0.7 | 0.8 | 0.9 | 0.1 | 0.2 | 0.3 | 0.4 | 0.5 | 0.6 | 0.7 | 0.8 | 0.9 |
|      | \multicolumn{9}{c}{Correctly Clustered} | | | | | | | | | \multicolumn{9}{c}{Incorrectly Clustered} | | | | | | | | |
| 1.1 | 151 | 159 | 161 | 161 | 161 | 161 | 162 | 162 | 162 | 14 | 6 | 4 | 4 | 4 | 4 | 4 | 4 | 4 |
| 1.2 | 53 | 143 | 152 | 152 | 154 | 154 | 154 | 154 | 154 | 48 | 6 | 4 | 4 | 2 | 2 | 2 | 2 | 2 |
| 1.3 | 60 | 60 | 139 | 141 | 142 | 145 | 146 | 146 | 146 | 50 | 32 | 2 | 2 | 1 | 1 | 1 | 1 | 1 |
| 1.4 | 2 | 2 | 59 | 59 | 135 | 134 | 134 | 137 | 136 | 1 | 1 | 13 | 8 | 1 | 1 | 1 | 1 | 1 |
| 1.5 | 2 | 2 | 2 | 2 | 2 | 55 | 55 | 129 | 130 | 1 | 1 | 1 | 1 | 1 | 1 | 1 | 1 | 1 |
| 1.6 | 2 | 2 | 2 | 2 | 2 | 2 | 2 | 2 | 124 | 1 | 1 | 1 | 1 | 1 | 1 | 1 | 1 | 1 |
| 1.7 | 2 | 2 | 2 | 2 | 2 | 2 | 2 | 2 | 2 | 1 | 1 | 1 | 1 | 1 | 1 | 1 | 1 | 1 |
| 1.8 | 2 | 2 | 2 | 2 | 2 | 2 | 2 | 2 | 2 | 1 | 1 | 1 | 1 | 1 | 1 | 1 | 1 | 1 |
| 1.9 | 2 | 2 | 2 | 2 | 2 | 2 | 2 | 2 | 2 | 1 | 1 | 1 | 1 | 1 | 1 | 1 | 1 | 1 |
| 2.0 | 2 | 2 | 2 | 2 | 2 | 2 | 2 | 2 | 2 | 1 | 1 | 1 | 1 | 1 | 1 | 1 | 1 | 1 |
|      | \multicolumn{9}{c}{Number in Lower Approximations} | | | | | | | | | \multicolumn{9}{c}{Number in Boundaries} | | | | | | | | |
| 1.1 | 165 | 165 | 165 | 165 | 165 | 165 | 166 | 166 | 166 | 13 | 13 | 13 | 13 | 13 | 13 | 12 | 12 | 12 |
| 1.2 | 101 | 149 | 156 | 156 | 156 | 156 | 156 | 156 | 156 | 77 | 29 | 22 | 22 | 22 | 22 | 22 | 22 | 22 |
| 1.3 | 110 | 92 | 141 | 143 | 143 | 146 | 147 | 147 | 147 | 68 | 86 | 37 | 35 | 35 | 32 | 31 | 31 | 31 |
| 1.4 | 3 | 3 | 72 | 67 | 136 | 135 | 135 | 138 | 137 | 175 | 175 | 106 | 111 | 42 | 43 | 43 | 40 | 41 |
| 1.5 | 3 | 3 | 3 | 3 | 3 | 56 | 56 | 130 | 131 | 175 | 175 | 175 | 175 | 175 | 122 | 122 | 48 | 47 |
| 1.6 | 3 | 3 | 3 | 3 | 3 | 3 | 3 | 3 | 125 | 175 | 175 | 175 | 175 | 175 | 175 | 175 | 175 | 53 |
| 1.7 | 3 | 3 | 3 | 3 | 3 | 3 | 3 | 3 | 3 | 175 | 175 | 175 | 175 | 175 | 175 | 175 | 175 | 175 |
| 1.8 | 3 | 3 | 3 | 3 | 3 | 3 | 3 | 3 | 3 | 175 | 175 | 175 | 175 | 175 | 175 | 175 | 175 | 175 |
| 1.9 | 3 | 3 | 3 | 3 | 3 | 3 | 3 | 3 | 3 | 175 | 175 | 175 | 175 | 175 | 175 | 175 | 175 | 175 |
| 2.0 | 3 | 3 | 3 | 3 | 3 | 3 | 3 | 3 | 3 | 175 | 175 | 175 | 175 | 175 | 175 | 175 | 175 | 175 |

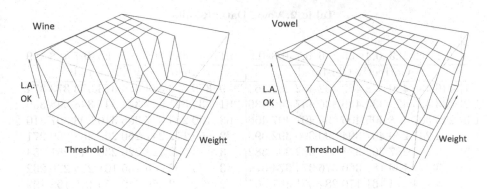

**Fig. 1.** Wine Vowel and Iris Data: Performance Maps

redundant: the sum of correctly and incorrectly clustered objects equals the number of objects in lower approximations and the number of boundary objects is the difference between the overall number of objects (178) and the objects in lower approximations. However, for clarity, Table 1 displays them separately. In Figure 1 (left), the number of correctly clustered objects (indicated by *L.A. OK*) is also depicted as a mesh.

As we can see good results are obtained for small thresholds. The results are balanced with respect to correctly and incorrectly clustered objects with a reasonably high number of objects in lower approximations. In particular, we observe that π rough k-means in its standard case ($\underline{w} = 0.5$) performs well. Its results with respect to correctly clustered objects for all thresholds are either the best or at least the second best. Figure 1 (left) also discloses a rather high and stable "platform" with a fairly small slope for small thresholds and/or high weights. The platform collapses for high thresholds and small weights. As we can see in Table 1, the number of objects in lower approximations goes down to the one object per cluster that is forced into a lower approximation by step *Approximations (i)* in π rough k-means.

*Vowel Data Experiments.* The Vowel data set consists of 871 objects with three features each. There are six different vowel types (classes). The classes have significant overlap. The clustering results for the Vowel data are shown in Table 2 and Figure 1 (right). Values with a leading apostrophe (') in Table 2 indicate that the algorithm terminated after the maximum number of iterations (set to 500) was reached. That is, in these cases the algorithm did not converge. Therefore, these results should be interpreted with even more caution than the remaining results.

In principle, the results are similar to the results for the Wine data. A main difference results from the different data qualities. The clusters of the Wine data are rather well separated while the clusters of the Vowel data have significant overlap. So, the percentage of correctly clustered objects is much lower and the percentage of the incorrectly clustered objects higher than for the Wine data.

**Table 2.** Vowel Data: Results

| ζ | w 0.1 | 0.2 | 0.3 | 0.4 | 0.5 | 0.6 | 0.7 | 0.8 | 0.9 | 0.1 | 0.2 | 0.3 | 0.4 | 0.5 | 0.6 | 0.7 | 0.8 | 0.9 |
|---|---|---|---|---|---|---|---|---|---|---|---|---|---|---|---|---|---|---|
| | Correctly Clustered | | | | | | | | | Incorrectly Clustered | | | | | | | | |
| 1.1 | '446 | 435 | 460 | 462 | 462 | 462 | 462 | 462 | 464 | '379 | 379 | 373 | 374 | 374 | 373 | 373 | 373 | 374 |
| 1.2 | 433 | 425 | 429 | 433 | 436 | 436 | 437 | 440 | 439 | 291 | 334 | 345 | 342 | 346 | 347 | 348 | 348 | 350 |
| 1.3 | '249 | 307 | 408 | 405 | 406 | 405 | 407 | 407 | 406 | '63 | 100 | 284 | 307 | 308 | 310 | 308 | 307 | 310 |
| 1.4 | '4 | 247 | 280 | 407 | 394 | 390 | 390 | 392 | 393 | '2 | 50 | 112 | 241 | 260 | 262 | 268 | 269 | 271 |
| 1.5 | 3 | 3 | 287 | 362 | 381 | 381 | 382 | 380 | 382 | 3 | 3 | 92 | 128 | 226 | 236 | 247 | 251 | 254 |
| 1.6 | '3 | '3 | 169 | 181 | 360 | 376 | 374 | 374 | 374 | '3 | '3 | 57 | 55 | 135 | 161 | 225 | 226 | 232 |
| 1.7 | 4 | 4 | 4 | 161 | 170 | 284 | 351 | 367 | 367 | 2 | 2 | 2 | 50 | 49 | 59 | 145 | 198 | 199 |
| 1.8 | 2 | 2 | 2 | 2 | 154 | 282 | 327 | 329 | 363 | 4 | 4 | 4 | 4 | 44 | 55 | 90 | 97 | 136 |
| 1.9 | 3 | 3 | 3 | 3 | 3 | 128 | 263 | 301 | 332 | 3 | 3 | 3 | 3 | 3 | 42 | 47 | 89 | 84 |
| 2.0 | 1 | 1 | 1 | 1 | 1 | 1 | 229 | 282 | 292 | 5 | 5 | 5 | 5 | 5 | 5 | 45 | 69 | 79 |
| | Number in Lower Approximations | | | | | | | | | Number in Boundaries | | | | | | | | |
| 1.1 | '825 | 814 | 833 | 836 | 836 | 835 | 835 | 835 | 838 | '46 | 57 | 38 | 35 | 35 | 36 | 36 | 36 | 33 |
| 1.2 | 724 | 759 | 774 | 775 | 782 | 783 | 785 | 788 | 789 | 147 | 112 | 97 | 96 | 89 | 88 | 86 | 83 | 82 |
| 1.3 | '312 | 407 | 692 | 712 | 714 | 715 | 715 | 714 | 716 | '559 | 464 | 179 | 159 | 157 | 156 | 156 | 157 | 155 |
| 1.4 | '6 | 297 | 392 | 648 | 654 | 652 | 658 | 661 | 664 | '865 | 574 | 479 | 223 | 217 | 219 | 213 | 210 | 207 |
| 1.5 | 6 | 6 | 379 | 490 | 607 | 617 | 629 | 631 | 636 | 865 | 865 | 492 | 381 | 264 | 254 | 242 | 240 | 235 |
| 1.6 | '6 | '6 | 226 | 236 | 495 | 537 | 599 | 600 | 606 | '865 | '865 | 645 | 635 | 376 | 334 | 272 | 271 | 265 |
| 1.7 | 6 | 6 | 6 | 211 | 219 | 343 | 496 | 565 | 566 | 865 | 865 | 865 | 660 | 652 | 528 | 375 | 306 | 305 |
| 1.8 | 6 | 6 | 6 | 6 | 198 | 337 | 417 | 426 | 499 | 865 | 865 | 865 | 865 | 673 | 534 | 454 | 445 | 372 |
| 1.9 | 6 | 6 | 6 | 6 | 6 | 170 | 310 | 390 | 416 | 865 | 865 | 865 | 865 | 865 | 701 | 561 | 481 | 455 |
| 2.0 | 6 | 6 | 6 | 6 | 6 | 6 | 274 | 351 | 371 | 865 | 865 | 865 | 865 | 865 | 865 | 597 | 520 | 500 |

For thresholds approximately between 1.5 and 1.6 and weights around 0.5 the ratio of correctly over incorrectly clustered objects is maximal at about 3.

*Iris Data Experiments.* The Iris data consist of three classes with 50 objects each. The characteristics of the Iris data are somehow in-between the Wine and Vowel data: two of the classes are overlapping while the third class is separated.

Since the results are in line with the results of the Wine and Vowel data we only show Figure 2 representing the correctly clustered objects. The performance map looks similar to the performance maps of the Wine and Vowel Data and, therefore, confirms these previous results.

### 4.3　Discussion

The experiments show that $\pi$ rough k-means performs well for small thresholds and high weights, independent of the characteristics of the data: rather separated (Wine), partly overlapping (Iris) or significantly overlapping (Vowel). Within a certain range of the initial parameters, the number of correctly clustered objects form a rather stable platform with a small angle of decline for increasing thresholds and decreasing weights. However, when a certain threshold, a certain weight

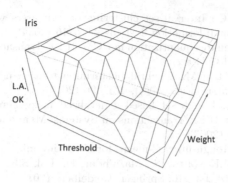

**Fig. 2.** Iris Data: Performance Map

is reached the platform ends in a cliff where the number of correctly clustered objects plunges and the number of boundary objects surges.

## 5 Conclusion

In the paper, we presented the $\pi$ rough clustering algorithm with user-defined weights in detail and discussed its properties. The weights give more freedom when the algorithm needs to be tuned with such techniques as evolutionary optimization. However, our experiments show that $\pi$ rough k-means without user-defined weights already performs well. So, in most conventional applications we would recommend to use $\pi$ rough k-means without user-defined weights. When evolutionary optimization needs to be implemented, other rough clustering algorithms that are more sensitive to the setting of the weights should be considered.

## References

1. Bache, K., Lichman, M.: UCI machine learning repository (2013)
2. Bezdek, J.C.: Pattern Recognition with Fuzzy Objective Algorithms. Plenum Press, New York (1981)
3. Jain, A.K.: Data clustering: 50 years beyond k-means. Pattern Recognition Letters 31, 651–666 (2010)
4. Krishnapuram, R., Keller, J.M.: A possibilistic approach to clustering. IEEE Transactions on Fuzzy Systems 1(2), 98–110 (1993)
5. Laplace, P.S.: Philosophical Essay on Probabilities. Dover Pub, New York (1951)
6. Lingras, P.: Evolutionary rough K-means clustering. In: Wen, P., Li, Y., Polkowski, L., Yao, Y., Tsumoto, S., Wang, G. (eds.) RSKT 2009. LNCS, vol. 5589, pp. 68–75. Springer, Heidelberg (2009)
7. Lingras, P., Peters, G.: Rough clustering. WIREs Data Mining and Knowledge Discovery 1, 64–72 (2011)
8. Lingras, P., Peters, G.: Applying rough set concepts to clustering. In: Peters, G., Lingras, P., Ślęzak, D., Yao, Y.Y. (eds.) Rough Sets: Selected Methods and Applications in Management and Engineering. Advanced Information and Knowledge Processing, pp. 23–37. Springer, London (2012)

9. Lingras, P., West, C.: Interval set clustering of web users with rough k-means. Journal of Intelligent Information Systems 23, 5–16 (2004)

10. Lloyd, S.P.: Least squares quantization in PCM. IEEE Transactions on Information Theory 28(2), 129–137 (1982)

11. Maji, P., Pal, S.K.: RFCM: A hybrid clustering algorithm using rough and fuzzy sets. Fundamenta Informaticae 80(4), 475–496 (2007)

12. Maji, P., Pal, S.K.: Rough set based generalized fuzzy c-means algorithm and quantitative indices. IEEE Transactions on Systems, Man, and Cybernetics - Part B 37(6), 1529–1540 (2007)

13. Maji, P., Paul, S.: Rough-fuzzy C-means for clustering microarray gene expression data. In: Kundu, M.K., Mitra, S., Mazumdar, D., Pal, S.K. (eds.) PerMIn 2012. LNCS, vol. 7143, pp. 203–210. Springer, Heidelberg (2012)

14. Mitra, S.: An evolutionary rough partitive clustering. Pattern Recognition Letters 25, 1439–1449 (2004)

15. Mitra, S., Banka, H., Pedrycz, W.: Rough-fuzzy collaborative clustering. IEEE Transactions on Systems, Man, and Cybernetics - Part B 36(4), 795–805 (2006)

16. Pal, S.K., Majumder, D.D.: Fuzzy sets and decision making approaches in vowel and speaker recognition. IEEE Transactions on Systems, Man, and Cybernetics 7, 625–629 (1977)

17. Pawlak, Z.: Rough sets. International Journal of Computer and Information Science 11, 341–356 (1982)

18. Peters, G.: Some refinements of rough k-means. Pattern Recognition 39, 1481–1491 (2006)

19. Peters, G.: Rough clustering utilizing the principle of indifference. Information Sciences 277, 358–374 (2014)

20. Peters, G., Crespo, F., Lingras, P., Weber, R.: Soft clustering - fuzzy and rough approaches and their extensions and derivatives. International Journal of Approximate Reasoning 54(2), 307–322 (2013)

21. Peters, G., Lampart, M.: A partitive rough clustering algorithm. In: Greco, S., Hata, Y., Hirano, S., Inuiguchi, M., Miyamoto, S., Nguyen, H.S., Słowiński, R. (eds.) RSCTC 2006. LNCS (LNAI), vol. 4259, pp. 657–666. Springer, Heidelberg (2006)

22. Peters, G., Lampart, M., Weber, R.: Evolutionary rough k-medoid clustering. In: Peters, J.F., Skowron, A. (eds.) Transactions on Rough Sets VIII. LNCS, vol. 5084, pp. 289–306. Springer, Heidelberg (2008)

23. Yao, Y.Y.: Two views of the theory of rough sets in finite universes. International Journal of Approximate Reasoning 15, 291–317 (1996)

# An Automatic Virtual Calibration
# of RF-Based Indoor Positioning with Granular Analysis

Ye Yin[1], Zhitao Zhang[1], Deying Ke[1], and Chun Zhu[2]

[1] College of Information, Mechanical and Electrical Engineering,
Shanghai Normal University, Shanghai 200234, China
yinye@188.com
[2] Shanghai Zenkore Co.,Ltd, Shanghai 201100, China
zhuchun@zenkore.com

**Abstract.** The positioning methods based received signal strength indicator (RSSI) is using the RSSI values to estimate the positions of the mobile. The RSSI positioning method based on propagation models, the system's accuracy depends on the adjustment of the propagation models parameters. In actual indoor environment, the propagation conditions are hardly predictable due to the dynamic nature of the RSSI, and consequently the parameters of the propagation model may change. In this paper, we propose and demonstrate an automatic virtual calibration technology of the propagation model that does not require human intervention; therefore, can be periodically performed, following the wireless channel conditions. We also propose the low-complexity Gaussian Filter (GF), Virtual Calibration Technology (VCT), Probabilistic Positioning Algorithm (PPA) , and Granular Analysis(GA) make the proposed algorithm robust and suitable for indoor positioning from uncertainty, self-adjective to varying indoor environment. Using MATLAB simulation, we study the calibration performance and system performance, especially the dependence on a number of system parameters, and their statistical properties. The simulation results prove that our proposed system is an accurate and cost-effective candidate for indoor positioning.

**Keywords:** Indoor positioning, Gaussian filter, Virtual calibration, Probabilistic localization algorithm, Granular analysis.

# 1   Introduction

The growing of wireless communications, mobile internet, smart sensors and smartphone has generated much commercial and research interests in statistical methods to track people and things, locations of devices and people have been considered to be quite valuable data. Inside stores, hospitals, high-rise buildings, warehouses, factories and underground parking lot, where Global Positioning System (GPS) [1] devices generally do not work. And Indoor Positioning System (IPS) [2] aims to provide location estimation for wireless devices, such as handheld devices and electronic badges. In this paper, we study the propagation model, characteristics of channel and people

D. Miao et al. (Eds.): RSKT 2014, LNAI 8818, pp. 557–568, 2014.
DOI: 10.1007/978-3-319-11740-9_51 © Springer International Publishing Switzerland 2014

pace model, propose a new method for indoor localization, which is a cost-effective for tracking objects.

The position estimation of a mobile can be achieved by using two different approaches [2], either rang-based or range–free. The former is defined by protocols that use absolute point to point distance estimates for calculating the location (see Tab.1). The later makes no assumption about the availability or validity of such information. As we know, the range-free solutions have a rough accuracy [3], so these techniques are unsuitable in applications where the location precision is one of the main requirements. Radio signal measurements are typical the RSSI, the angle of arrival (AOA), the time of arrival (TOA), the time difference arrival (TDOA) and the differential time differences of arrival (DTDoA). Although AOA, TDOA or DTDoA can achieve a high accuracy, they require a complex hardware, and the promotion of this approach is limited.

Due to advantages such as small size, low power consumption, low cost and easy deployment, the RFID [4] sensors are widely used to implement ubiquitous computing and smart city. We have successfully applied this technique to the 2010 Shanghai World Expo [5]. With the capability of providing RSSI, current advanced RFID systems are one of the potential candidates for indoor localization. In this work, we consider localization based on RSSI, since it does not require any special hardware and it is available in most of standard wireless devices. Several RFID based systems have been proposed for tracking objects in indoor environments. SpotON [2], [6] using an aggregation algorithm for three-dimensional localization. The tags use RSSI to obtain inter-tag distances based on empirical mapping between the two. SpotON assumes deterministic mapping between RSSI and distance and does not account for the range measurement uncertainty caused by the varying environment. LANDMARC [7] utilizes RSS measurement information to locate objects using k nearest reference tags. To diminish the uncertainty of the detected range caused by the varying environments, a large number of reference tags must be distributed in the environment. As we know, such positioning systems perform a preliminary calibration of the propagation

**Table 1.** Wireless technologies for indoor localization versus ranging techniques

| Ranging technique→<br>Wireless technology↓ | RSSI | TOA | TDOA | DTDOA | AOA |
|---|---|---|---|---|---|
| Ultrasonic | × | √ | √ | × | √ |
| Infrared | × | × | × | × | × |
| Bluetooth (IEEE 802.15.1) | √ | × | × | √ | × |
| RFID | √ | √ | √ | × | √ |
| WiFi (IEEE 802.11) | √ | √ | √ | √ | √ |
| IEEE 802.15.4a DSSS | √ | √ | × | × | √ |
| IEEE 802.15.4a UWB | √ | √ | √ | × | √ |
| IEEE 802.15.4a CSS | × | √ | × | × | × |
| 60 GHz | × | × | × | √ | × |

model parameters. The above two systems require the use of additional equipment [8] (such as reference tags, tags of this kind are mainly used to provide calibration signals), time-consuming human-based operations in order to carry out the calibration procedure. This limits its applications for most indoor scenarios, such as people tracking, fire disaster site, and so on.

In this paper, we propose an automatic calibration procedure (called virtual calibration) of the signal propagation model that is only based the RSSIs measured among readers and that can be executed periodically and automatically (i.e., without human intervention), this method is based on off-the-shelf active RFID technology. Based on the virtual calibration procedure, we use low-complexity GF to process the RSSIs before calibration procedure, and propose a PPA that uses RSSIs to estimate the locations of objects. The RSSIs have been processed by calibration procedure. This RF-based indoor positioning system is easy to deploy and cost-effective. Considering the uncertainties caused by the varying environment, we incorporate a probabilistic scheme based on Gaussian filter pretreatment, automatic virtual calibration and Bayesian inference to improve the localization accuracy. Bayesian inference was also used for traditional cellular and WLAN-based positioning system. But with a VCT, our proposed system can periodically and automatically calibrate the propagation model parameters without human intervention and additional equipment.

## 2   System Architecture

In our positioning system the off-the-shelf long distance active RFID system is used. The system works at the range of 2.4GHz frequency [9], with a minimal range of 0.5 meter and maximum range of 80 meters. The reader can not only receive RSSI from every tag within its range, but also broadcast RSSI. We exploit the later function in order to estimate the propagation model parameters using the RSSI exchanged among readers. Each reader can detect up to 200 tags simultaneously, each tag is pre-programmed with a unique 9-character ID (Identity) for identification by readers. In the next section, we will discuss the network layout of the system and details of each layer, such as composition and function of each layer.

### 2.1   Network Layout

The system consists of three network layers: sensing network layer, data collection layer and processing layer. The sensing network layer is used to measure the RSSI information from the readers to objects (each object wear an active tags) and to transmit the information to next layer. The data collection layer is used to collect RSSI information and to transmit the data to next layer. The processing layer receives RSSI information and processes the location information. In practice, the whole detection area may be covered by several servers. For simplicity, in Fig.1, we show the hierarchical architecture within the coverage of one server. Fig.4 shows the reader and tags used in our experiments.

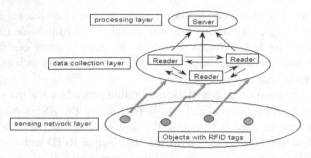

**Fig. 1.** System structure

**Fig. 2.** Reader    **Fig. 3.** Tags    **Fig. 4.** Reader and tags

## 2.2    Hardware Components

In our indoor positioning system, we mainly use three kinds of devices: RFID tags, RFID readers and servers. The followings are detailed descriptions of hardware features.

RFID tags: We use the active tags in our system. They are deployed in the sensing network layer. Each tracked object will be attached with a unique active RFID tag, called "object tag", used for identifying and tracking objects. Each tag has a unique ID, hence, we can distinguish objects by the corresponding ID number. Tags will periodically emit signals with their IDs, and the working hours of each tag is designed for 5 years.

RFID readers: The data collection layer consists of readers. Each reader also has a unique ID number. As mentioned above, our readers can not only receive RSSI, but also broadcast RSSI. The later function is an upgrade for the reader, this make it possible for virtual calibration. And every small detection area contains three readers. The whole detection area is covered by the data collection layer. Every object tag should be within the readable range of readers. The principle of readers' deployment should be satisfied that the distribution of the readers is not in one line in the space, and all readers' locations are known. A reader is responsible for: 1) Collecting and decoding the signals emitted by the active tag in its coverage; 2) Measuring the RSSI for each tag within its range; 3) Reporting tag ID, corresponding RSSI, and its own ID number to the server; 4) Broadcasting RSSI around, and receiving RSSI from other readers. To realize these functions, each reader is equipped with two interfaces: a RF interface that detects tags within its range, and a communication interface that transmits data to servers.

**Table 2.** Manufacturer and parameters of reader and tag

|  | Reader | Tag |
|---|---|---|
| Working frequency | 2.4GHz | |
| modulation mode | MSK | |
| communication distance | 0.5meter~80meter | |
| communication rate | 250kbps | |
| Working voltage | 12V | 1.5V |
| Communication interface | RJ45、RS232、RS485 | - |
| working temperature | -40℃~+80℃ | |
| Manufacturer: Shanghai Zhen Zhuo Electrical Technology Ltd. Co | | |

Servers: Each reader should be within the reach of at least one server. We can see this from Fig. 1. Readers communicate the measured RSSIs of the tags and readers with the server. A server is responsible for: 1) Collection RSSIs and IDs coming from readers; 2) Selecting the readers with high signal strength, which play the role of a coarse positioning; 3) Calculating the location of the object tags according to the positioning algorithm.

## 3    Virtual Calibration Procedure

The purpose of the calibration procedure is to adapt the propagation model to the actual environment. Due to the dynamic of the channel and the variation of wireless fading events, an automatic calibration procedure that can be periodically performed without human intervention may increase the accuracy of the localization system. The proposed calibration procedure achieves this goal, by exploiting the information exchanged from the readers. The parameters of the propagation model are: $\alpha_0$ (the RSSI at distance of 1 m), $\eta$ (the path loss exponent), $l_i$ (the attenuation of the wall of type i). $\alpha_0$ should be estimated in the free space, and it is not affected by the power of battery (see Fig.5 and Fig.6).

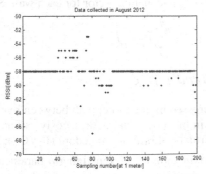
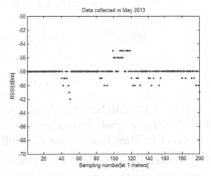

**Fig. 5.** Data collected in August 2013          **Fig. 6.** Data collected in February 2014

Therefore, in our method it is estimated a *priori*, it is not the object of our virtual calibration procedure. In this paper, we assume that $\alpha_0$ is estimated a *priori*, only calibrate the other two parameters $(\eta, l_i)$.

The rest of this section is organized as follows: First of all, we proposed two kind of virtual calibration procedures. Secondly, we evaluate the calibration results through comparing these two approaches with [8].

### 3.1    Global Virtual Calibration Procedure

During the virtual calibration procedure, we estimated all the required parameters $(\alpha_0, \eta, l_i)$. The Global Virtual Calibration Procedure (G-VCP) considers that $l_i$ of all types of walls are the same $(l_w)$. Therefore, replacing $d(i, j)$ with the actual distance between reader$_i$ and reader$_j$, $k(i, j)$ with the number of walls crossed by the direct path between reader$_i$ and reader$_j$, we get a estimation $RSSI'(i, j)$:

$$\text{RSSI}'_{(i,j)} = a_0 - 10\eta \log(d_{(i,j)}) - k_{(i,j)} l_w \tag{1}$$

$$\forall i, j : (reader_i, reader_j) \in C$$

where, we define $C = \{reader_i, reader_j\}$, reader$_i$ and reader$_j$ represent any two communicated readers. The estimated $RSSI'(i, j)$ differs from the actual measured $RSSI(i, j)$ by an error item $\varepsilon(i, j)$, we assume that all $\varepsilon(i, j)$ are independent and identically distributed. Keep in mind that $\alpha_0$ is a *priori*, according to [13], the approximation of the remaining two parameters $(\eta, l_w)$ can be achieved by a direct method that minimized the minimum mean square error (MMSE):

$$\varepsilon(i, j) = \left\| \text{RSSI}'_{(i,j)} - \text{RSSI}_{(i,j)} \right\|_2 \tag{2}$$

the computational overhead for direct method processing a linear MMS estimator problem is polynomial [13].

### 3.2    Per-wall Virtual Calibration Procedure

We estimate an individual attenuation factor for each wall between any pair of readers belonging to $C$. Let us assume that there are q types of walls in the positioning environment, define $L = \{l_1, l_2, l_3, \ldots, l_q\}$, L is the set of attenuation factor for each wall. So:

$$\text{RSSI}'_{(i,j)} = a_0 - 10\eta \log(d_{(i,j)}) - \sum_{d=1}^{q} k_{d(i,j)} l_d \tag{3}$$

$$\forall i, j : (reader_i, reader_j) \in C$$

where $k_{d(i,j)}$ is the number of wall of type $l_d$ crossed by the direct path between reader$_i$ and reader$_j$. The $\eta$ (path loss exponent) used in this procedure is previously estimated by the G-VCP. Therefore, in our Per-wall Virtual Calibration Procedure (P-VCP), we only estimated the parameters $l_i \in L$, the approximation of the remaining $q$ parameters $(l_1, l_2, l_3, \ldots, l_q)$.

### 3.3    Calibration Performance Analysis

In order to evaluate the performance of the calibration methods, in this section, we discuss the estimated distance error after VCT (G-VCP and P-VCP). To facilitate the analysis, we introduce the estimated distance of ideal propagation model for comparison, which is called realistic physical calibration procedure (R-PCP), details refer to [8]. Followings are the estimated distances of these three procedures. Each procedure has two equations, the first equation is the RSSI formula, and the second equation is the estimated distance.

I . R-PCP

$$RSSI = a_0 - 10\eta \log(d) - N(0, \sigma^2) \tag{4}$$

$$d_{R-procedure} = 10^{\frac{a_0 - RSSI - N(0,\sigma^2)}{10\eta}} \tag{5}$$

where $d$ is the distance between reader and tag, $N(0, \sigma^2)$ is the interference term of path loss model in [8], $a_0$ is the RSSI at distance of 1 meter, $\eta$ is the path loss exponent. $d_{R-procedure}$ is the estimated distance between reader$_i$ and tag$_j$.

II . G-VCP

$$RSSI = a_0 - 10\eta \log(d) - k_{(i,j)} l_w \tag{6}$$

$$d_{G-procedure} = 10^{\frac{a_0 - RSSI - k_{(i,j)} l_w}{10\eta}} \tag{7}$$

where $d$ is the distance between reader and tag, $k_{(i,j)} l_w$ is the interference term of path loss model in Section 3.1, $a_0$ is the RSSI at distance of 1 meter, $\eta$ is the path loss exponent. Keep in mind that $k_{(i,j)}$ is the number of walls crossed by the direct path between reader$_i$ and tag$_j$. $d_{G-procedure}$ is the estimated distance between reader$_i$ and tag$_j$.

III. P-VCP

$$RSSI = a_0 - 10\eta \log(d) - \sum_{d=1}^{q} k_{d(i,j)} l_d \tag{8}$$

$$d_{P-procedure} = 10^{\frac{a_0 - RSSI - \sum_{d=1}^{q} k_{d(i,j)} l_d}{10\eta}} \tag{9}$$

where $d$ is the distance between reader and tag, and $\sum_{d=1}^{q} k_{d(i,j)} l_d$ is the interference term of path loss model, $a_0$ is the RSSI at distance of 1 meter, $\eta$ is the path loss exponent. Keep in mind that $k_{(i,j)}$ is the number of walls crossed by the direct path between reader$_i$ and tag$_j$. $d_{P-procedure}$ is the estimated distance between reader$_i$ and tag$_j$.

After above analysis, next, we analyse the estimated distance error of the three methods. The main purpose of our error analysis is to obtain a distribution of difference value, and select a method with a smallest error. We define following formulas to describe the distance errors:

$$d_{error-R} = d_{practical} - d_{R-procedure} \qquad (10)$$

$$d_{error-G} = d_{practical} - d_{G-procedure} \qquad (11)$$

$$d_{error-P} = d_{practical} - d_{P-procedure} \qquad (12)$$

where $d_{error-R}$, $d_{error-G}$, $d_{error-P}$ are estimated distance errors, $d_{practical}$ is the actual distance between reader and tag.

Since each RSSI value corresponds to an estimate of the distance, also corresponds to an estimation error. We assume that each set of RSSI values is a uniformly distribution, and each set of estimation errors also is a uniformly distribution. In this part, we just analyse from the point of view of a single value, so as to distinguish it. Fig.7 is the flow chart of error analysis.

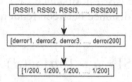

**Fig. 7.** Flow chart of error analysis

Based on the set of all the RSSIs measured between reader and tag, this set of RSSIs is $R_{2m}$. We evaluated the CPD of the estimated distance errors.

Fig.8 shows that there are three cumulative probability function curves, each curve representing a calibration method. This figure highlights that the P-VCP performs better than the G-VCP and R-PCP. Not surprisingly, the P-VCP outperforms the G-VCP, due to the better accuracy in the wall modeling. We also can know that virtual calibration procedure results in small estimated distance errors as the more expensive and complicated R-PCP. Tab.3 shows intuitive results of calibration performance.

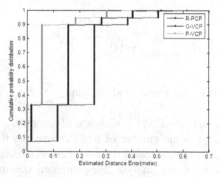

**Fig. 8.** CPD of estimated distance errors

**Table 3.** CPD of estimated distance errors

| CPD | P-VCP | G-VCP | R-PCP |
|-----|-------|-------|-------|
| 0.2 | 0     | 2cm   | 12cm  |
| 0.4 | 5cm   | 15cm  | 27cm  |
| 0.9 | 20cm  | 30cm  | 39cm  |

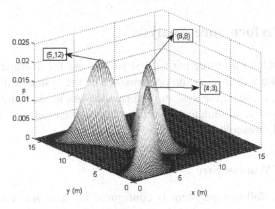

**Fig. 9.** Fit Model of three tracked tags

We tracked three objects in this room (see Fig.9), and the plane coordinates of these three objects have been marked out in the figure. From Fig.11, the result of positioning procedure is still very clear, even there are multiple targets in the same room.

At the end of algorithm, we evaluate the credibility of output by defining a confidence function $Bel'(x_k)$.

$$Bel'(x_k) = \frac{Bel(x_k)}{\max\left(Bel(x_k=(x_k,y_k))\right)} \tag{13}$$

We set a threshold $T$ ($0 \leq T \leq 1$) and only choose those grids with $Bel'(x_k) > T$. After one recursion, an estimated area is obtained.

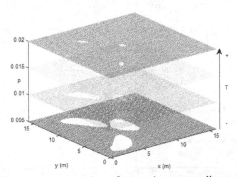

**Fig. 10.** Sectional drawing of space (corresponding to Fig.11)

Fig.10 shows sectional drawing of space corresponding to Fig.9. There are three objects in environment. The white areas are estimated areas of tracked objects. There are four pieces of sectional drawings with different thresholds, and from above figure, we can observe that estimated areas of three tracked objects are smaller as T increasing.

Therefore, we can achieve one more precise estimated area, if increasing the recursion time $R$ and threshold $T$.

# 4    System Performance Analysis

This section is divided into three parts. First we discuss the average positioning distance error. Then we will discuss the performance of the system with changing parameters. At last, we evaluate the system performance. As we known, the performance will be affected by several undetermined parameters, for example, recursion time $R$, window size $w$ and threshold $T$.

## 4.1    Impact of Window Size

Experiment 1: The following system is configured as: grid size $L$=40cm, threshold $T$=0.5, path loss exponent $\eta$=3, standard deviation $\sigma\sigma$=1.45, $\alpha_0\alpha$= -59.7581. Fig.33 plots APDE as a function of the window size $w$. In this part, we plan to evaluate the impact of window size on the localization accuracy. We perform the simulation for object 1, object 2 and object 3 as depicted in Fig.9.

From Fig.11, we observe that APDE decrease as the window size $w$ increase. In another words, the greater the window size $w$, the higher the localization accuracy. We also find that the APDE decrease rate depends on the window size. The APDE decrease rate will very high when the window size is relatively large.

From Fig.11, we can also observe a unique phenomenon, when the window size $w$ is very small, for example $w = 2$ or 4, the APDE decrease in early stage, but it slightly increase in later stage. The main reasons for this phenomenon are sample quantity and fault-tolerant rate. The fault-tolerant rate will increase with a less sample quantity. So, APDE will decrease with a bigger sample quantity (or called window size $w$).

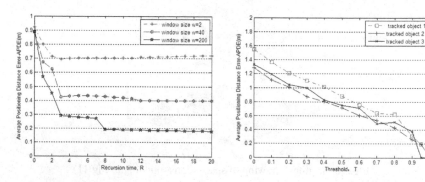

**Fig. 11.** Impact of window size $w$                **Fig. 12.** Impact of threshold $T$

### 4.2    Impact of Threshold, Path Loss Exponent

Experiment 2: The system configured as: grid size $L$=40cm, path loss exponent $\eta$=3, standard deviation $\sigma\sigma$=1.45, $\alpha_0\alpha$= -59.7581. In this experiment, we aim to evaluate the impact of threshold on the localization accuracy. We perform the simulation for object 1, object 2 and object 3. Fig.12 plots APDE as a function of the threshold $T$.

From Fig.12, we can get the conclusion that in most situation, the large the threshold $T$, the better the localization accuracy. We also can know that the relationship between threshold $T$ and localization accuracy is linearly decreasing. In other words, the localization accuracy can be fully decided by the threshold $T$, and the system will not appear the phenomenon of a steady state of APDE.

## 5    Conclusion

In this thesis, we presented a new method, an easy-setup and cost-effective indoor positioning method based on off-the-shelf active RFID technology combined with GF, VCT, PPA and PPM. Our indoor positioning system relies on a hierarchical architecture to cover an indoor environment. The proposed approach can automatic calibrate the propagation parameters according to the environment without additional overhead, reduce the uncertainty of localization, and at the same time obtain high positioning accuracy under conditions of indoor positioning. After we collected a large number of experimental data, using MATLAB simulations we have evaluated the performance of our proposal. In the destined system settings, the simulation results show that in 90% percent of the localization estimation, the system provides objects location with the APDE less than 57cm, in 80% percent of the localization estimation, the system provides objects location with the APDE less than 45cm, and in 70% percent of the localization estimation, the system provides objects location with the APDE less than 20cm. The simulation also shows that the system performance improves with the higher values of recursion time, window size, and path loss exponent. The simulation results can prove that the proposed system is an accuracy and cost-effective candidate for future indoor localization.

**Acknowledgements.** Authors are grateful to the anonymous reviewers for their helpful comments. They also wish to note that the research work is supported by the National Natural Science Foundation of China (NO. 61101209) and the Program of Shanghai Normal University (NO. A-7001-12-002006) and Special Fund of Software & IC Industry Development ( NO. 120493).

## References

1. Garmin Corporation, About GPS, http://www.garmin.com/aboutGPS/
2. Hui, L., Houshang, D., Pat, B., Jing, L.: Survey of wireless indoor positioning techniques and systems. IEEE Transactions on Systems, Man, and Cybernetics-Part C 37(6), 1067–1080 (2007)

3.  Giorgetti, G., Gupta, S., Manes, G.: Localization using signal strength: to range or not to range? In: Proceedings of the First ACM International Workshop on Mobile Entity Localization and Tracking in GPS-less Environments, New York (USA), pp. 91–96 (2008)
4.  Radio Frequency Identification (RFID) home page,
    http://www.aimglobal.org/technologies/rfid/
5.  Yin, Y., Zhou, J., Yin, J.: Design of World Expo tour sites guide system based on RFID technology. In: Proceedings of 2010 International Conference on Audio Language and Image Processing (ICALIP 2010), Shanghai (China), pp. 1026–1030 (2010)
6.  Hightower, J., Want, R., Borriello, G.: SpotON: An indoor 3d location sensing technology based on RF signal strength. UW-CSE 00-02-02, University of Washington, Department of Computer Science and Engineering, Seattle (USA). Thesis (2000)
7.  Ni, L., Liu, Y., Lau, Y.C., Patil, A.: Landmarc: Indoor location sensing using active RFID. In: Proceedings of the 1st IEEE International Conference on Pervasive Computing and Communications (PERCOM 2003), Dallas (USA), pp. 407–415 (2003)
8.  Xiao, L., Yin, Y., Wu, X.N., Wand, J.W.: A large-scale RF-based indoor localization system using low-complexity gaussian filter and improved bayesian inference. Radioengineering 22(1), 371–380 (2013)
9.  Zepernick, H.J., Wyscoki, T.A.: Multipath channel parameters for the indoor radio at 2.4 GHz ISMband. In: 1999 IEEE 49th Proceedings of Vehicular Technology Conference, Houston (USA), vol. 1, pp. 190–193 (1999)
10. Rappaport, T.S.: Wireless Communications Principles and Practices. Prentice-Hall Inc. (2002)
11. Green, E., Hata, M.: Microcellular propagation measurements in an urban environment. In: Proceedings of 1991 IEEE International Symposium on Personal, Indoor and Mobile Radio Communications, King's College London (U.K.), pp. 324–328 (1991)
12. Ito, K.: Gaussian filter for nonlinear filtering problems. In: Proceedings of the 39th IEEE Conference on Decision and Control, vol. 2, pp. 1218–1223 (2000)
13. Bjorck, A.: Solution of Equations in RN. In: Least Square methods: Handbook of Numerical Analysis, vol. 1, Elservier, NorthHolland (1990)
14. Madigan, D., Elnahrawy, E., Martin, R.: Bayesian indoor positioning systems. In: Proceedings of 24th Annual Joint Conference of the IEEE Computer and Communications Societies (INFOCOM 2005), Miami (USA), vol. 2, pp. 1217–1227 (2005)
15. Christopher, M.B.: Pattern Recognition and Machine Learning. Springer-Verlag Inc., New York (2006)

# Algebraic Structure of Fuzzy Soft Sets

Zhiyong Hong[1,2] and Keyun Qin[1]

[1] College of Mathematics, Southwest Jiaotong University,
Chengdu, Sichuan 610031, China
[2] School of Computer Science, Wuyi University, Jiangmen, Guangdong 529020, China
hongmr@163.com, keyunqin@263.net

**Abstract.** This paper is devoted to the discussion of algebraic structures of fuzzy soft sets. The fuzzy notion of soft equality relation on fuzzy soft sets is proposed and several related properties are investigated. Furthermore, MTL structures of fuzzy soft algebra and complex sample for affiliations and the mapping to the fuzzy soft quotient algebra are established.

**Keywords:** Fuzzy set, soft set, fuzzy soft equality, fuzzy soft algebra.

## 1 Introduction

To solve complicated problems in economics, engineering, environmental science and social science, methods in classical mathematics are not always successful because of various types of uncertainties presented in these problems. While probability theory, fuzzy set theory [1], rough set theory [2], and other mathematical tools are well-known and often useful approaches to describing uncertainty, each of these theories has its inherent difficulties as pointed out in [3,4]. Consequently, Molodtsov [3] proposed a completely new approach for modeling vagueness and uncertainty in 1999. This approach called soft set theory is free from the difficulties affecting existing methods.

Accordingly, works on soft set theory are progressing rapidly. Maji et al. [5] defined several algebraic operations on soft sets and conducted a theoretical study on the theory of soft sets. Based on [5], Ali et al. [6] introduced some new operations on soft sets and improved the notion of complement of soft set. They proved that certain De Morgan's laws with respect to these new operations hold in soft set theory. Qin et al. [7] introduced the notion of soft equality and established lattice structures and soft quotient algebras of soft sets. Maji et al. [8] initiated the study on hybrid structures involving soft sets and fuzzy sets. They proposed the notion of fuzzy soft set as a fuzzy generalization of classical soft sets and some basic properties were discussed. Afterwards, many researchers have worked on this concept. Various kinds of extended fuzzy soft sets have been proposed.

Algebraic structures play a fundamental role in many fields of mathematics. The lattice structure of some fuzzy algebraic systems such as fuzzy groups, G-fuzzy groups, some fuzzy ordered algebras and fuzzy hyperstructures were discussed in [9]. It is proved that under suitable conditions, these structures form a

D. Miao et al. (Eds.): RSKT 2014, LNAI 8818, pp. 569–576, 2014.
DOI: 10.1007/978-3-319-11740-9_52 © Springer International Publishing Switzerland 2014

distributive or modular lattice. The lattice structures of soft sets and fuzzy soft sets have been established [7,10]. This paper is devoted to a further study of the algebraic structure of fuzzy soft sets. The fuzzy soft equality relation between fuzzy soft sets is presented and the connections between fuzzy soft quotient algebras and nonclassical logic algebras are established.The paper is organized as follows:In Section 2, we recall some notions and properties of fuzzy sets, soft sets and fuzzy soft sets. In Section 3,we introduce the concept of fuzzy soft equality between fuzzy soft sets and established fuzzy soft quotient algebra of fuzzy soft sets. In Section 3,based on t-norm and its residuated implication, we establish the connections between soft algebras and nonclassical logic algebras. Finally, some conclusions are pointed out.

## 2   Preliminaries

This section presents a review of some fundamental notions of fuzzy sets, soft sets and fuzzy soft sets. We refer to [1,3,8] for details.

Fuzzy set theory initiated by Zadeh [1] provides an appropriate framework for representing and processing vague concepts by allowing partial memberships. Let $U$ be a nonempty set, called universe. A fuzzy set $\mu$ on $U$ is defined by a membership function $\mu : U \to [0,1]$. For $x \in U$, the membership value $\mu(x)$ essentially specifies the degree to which $x$ belongs to the fuzzy set $\mu$. The operations on fuzzy sets can be defined componentwise [1]. In what follows, the family of all subsets of $U$ and the family of all fuzzy sets on $U$ are denoted by $P(U)$ and $F(U)$ respectively.

In 1999, Molodtsov [3] introduced the concept of soft sets. Let $U$ be the universe set and $E$ the set of all possible parameters under consideration with respect to $U$. Usually, parameters are attributes, characteristics, or properties of objects in $U$. $(U, E)$ is called a soft space. Molodtsov defined the notion of a soft set in the following way:

**Definition 1.** *[3] A pair $(F, A)$ is called a soft set over $U$, where $A \subseteq E$ and $F$ is a mapping given by $F : A \to P(U)$.*

In other words, a soft set over $U$ is a parameterized family of subsets of $U$. For $e \in A$, $F(e)$ may be considered as the set of $e-$approximate elements of the soft set $(F, A)$.

Maji et al. [8] initiated the study on hybrid structures involving both fuzzy sets and soft sets. The notion of fuzzy soft sets was introduced as a fuzzy generalization of soft sets.

**Definition 2.** *[8] Let $(U, E)$ be a soft space. A pair $(F, A)$ is called a fuzzy soft set over $U$, where $A \subseteq E$ and $F$ is a mapping given by $F : A \to F(U)$.*

In the definition of fuzzy soft set, fuzzy sets in the universe $U$ are used as substitutes for the crisp subsets of $U$. Hence, every soft set may be considered as a fuzzy soft set. Based on operations on soft sets presented by Maji et al. [5] and Ali et al. [6], Qin et al. [10] defined several operations on fuzzy soft sets and established the lattice structure of fuzzy soft sets.

**Definition 3.** *Let $(F, A)$ and $(G, B)$ be two fuzzy soft sets over a common universe $U$.*

*(1) The extended union of $(F, A)$ and $(G, B)$, denoted by $(F, A) \cup_e (G, B)$, is the fuzzy soft set $(H, C)$, where $C = A \cup B$, and $H$ is given by:*

$$H(a) = \begin{cases} F(a), & \text{if} \quad a \in A - B, \\ G(a), & \text{if} \quad a \in B - A, \\ F(a) \cup G(a), & \text{if} \quad a \in A \cap B. \end{cases} \tag{1}$$

*(2) The extended intersection of $(F, A)$ and $(G, B)$, denoted by $(F, A) \cap_e (G, B)$, is the fuzzy soft set $(H, C)$, where $C = A \cup B$, and $H$ is given by:*

$$H(a) = \begin{cases} F(a), & \text{if} \quad a \in A - B, \\ G(a), & \text{if} \quad a \in B - A, \\ F(a) \cap G(a), & \text{if} \quad a \in A \cap B. \end{cases} \tag{2}$$

*(3) The restricted union of $(F, A)$ and $(G, B)$, denoted by $(F, A) \cup_r (G, B)$, is the fuzzy soft set $(H, C)$, where $C = A \cap B$, and $H(a) = F(a) \cup G(a)$ for every $a \in C$.*

*(4) The restricted intersection of $(F, A)$ and $(G, B)$, denoted by $(F, A) \cap_r (G, B)$, is the fuzzy soft set $(H, C)$, where $C = A \cap B$, and $H(a) = F(a) \cap G(a)$ for every $a \in C$.*

**Definition 4.** *(1) $(F, A)$ is called a relative null fuzzy soft set(with respect to the parameter set $A$), denoted by $\emptyset_A$, if $F(e) = \emptyset$ for all $e \in A$.*

*(2) $(G, A)$ is called a relative whole fuzzy soft set(with respect to the parameter set $A$), denoted by $U_A$, if $F(e) = U$ for all $e \in A$.*

**Definition 5.** *The relative complement of a fuzzy soft set $(F, A)$ is denoted by $(F, A)^r$ and is defined by $(F, A)^r = (F^r, A)$, where $F^r : A \to F(U)$ is a mapping given by $F^r(e) = \neg F(e)$ for all $e \in A$.*

Clearly, $((F, A)^r)^r = (F, A)$ holds.

We denote by $FS(U, E)$ the set of all fuzzy soft sets over the universe $U$ and the parameter set $E$, that is
$FS(U, E) = \{(F, A); A \subseteq E, F : A \to F(U)\}$.

**Theorem 1.** *[10]*
*(1) $(FS(U, E), \cup_e, \cap_r)$ is a bounded distributive lattice, $U_E$ and $\emptyset_\emptyset$ are the upper bound and lower bound respectively.*

*(2) Let $\leq$ be the ordering relation in lattice $(FS(U, E), \cup_e, \cap_r)$ and $(F, A)$, $(G, B) \in S(U, E)$. $(F, A) \leq (G, B)$ if and only if $A \subseteq B$ and $F(e) \subseteq G(e)$ for each $e \in A$.*

Let $A \subseteq E$, and $FS_A = \{(F, A); F : A \to F(U)\}$ be the set of all fuzzy soft sets over the universe $U$ and the parameter set $A$. It is trivial to verify that $(F, A) \cup_e (G, A), (F, A) \cap_r (G, A) \in FS_A$ for any $(F, A), (G, A) \in FS_A$. That is to say, $(FS_A, \cup_e, \cap_r)$ is a sublattice of $(FS(U, E), \cup_e, \cap_r)$, $U_A$ and $\emptyset_A$ are the upper bound and lower bound of $FS_A$ respectively.

# 3   Fuzzy Soft Equality Relation

Qin et al. [7] introduced the concept of soft equality between soft sets and established soft quotient algebra of soft sets. In this section, we extend the related concepts to fuzzy soft sets and discuss their basic properties.

**Definition 6.** *Let $(F, A)$ and $(G, B)$ be two fuzzy soft sets over a common universe $U$. $(F, A)$ is said to be fuzzy soft equal to $(G, B)$, denoted by $(F, A) \approx_{FS} (G, B)$, if $F(e) = G(e)$ for each $e \in A \cap B$, $F(e) = \emptyset$ for each $e \in A - B$ and $G(e) = \emptyset$ for each $e \in B - A$.*

**Theorem 2.** *Let $(F, A), (G, B) \in FS(U, E)$ be two fuzzy soft sets. $(F, A) \approx_{FS} (G, B)$ if and only if $(F, A) \cup_e (G, B) \approx_{FS} (F, A) \cap_r (G, B)$.*

*Proof.* Let $(F, A) \cup_e (G, B) = (H, A \cup B)$, $(F, A) \cap_r (G, B) = (L, A \cap B)$.

Suppose that $(F, A) \approx_{FS} (G, B)$. For any $e \in A \cap B$, we have $F(e) = G(e)$ and hence $H(e) = F(e) \cup G(e) = F(e) \cap G(e) = L(e)$. For any $e \in A \cup B - A \cap B = (A - B) \cup (B - A)$, if $e \in A - B$, then $F(e) = \emptyset$ and hence $H(e) = F(e) = \emptyset$; if $e \in B - A$, then $G(e) = \emptyset$ and hence $H(e) = G(e) = \emptyset$. Consequently, $(F, A) \cup_e (G, B) \approx_{FS} (F, A) \cap_r (G, B)$.

Conversely, assume that $(F, A) \cup_e (G, B) \approx_{FS} (F, A) \cap_r (G, B)$. For any $e \in A \cap B$, we have $F(e) \cup G(e) = H(e) = L(e) = F(e) \cap G(e)$ and consequently $F(e) = G(e)$. For any $e \in A - B$, we have $e \in A \cup B$ and $e \notin A \cap B$. It follows that $F(e) = H(e) = \emptyset$. For any $e \in B - A$, we have $e \in A \cup B$ and $e \notin A \cap B$. It follows that $G(e) = H(e) = \emptyset$. Thus $(F, A) \approx_{FS} (G, B)$.

Similarly, we have

**Theorem 3.** *Let $(F, A), (G, B) \in FS(U, E)$ be two fuzzy soft sets and $(F, A) \approx_{FS} (G, B)$. Then*
   (1) $(F, A) \cup_e (G, B) \approx_{FS} (F, A) \cap_e (G, B)$.
   (2) $(F, A) \cup_r (G, B) \approx_{FS} (F, A) \cap_r (G, B)$.

**Theorem 4.** *(1) $\approx_{FS}$ is an equivalence relation on $FS(U, E)$.*
   *(2) $(F, A) \approx_{FS} (G, B)$ if and only if $(F, A) \cup_r (G, B) \approx_{FS} (F, A) \cap_e (G, B)$.*

The following theorem shows that $\approx_{FS}$ is a congruence relation on $FS(U, E)$ with respect to $\cup_e$ and $\cap_r$.

**Theorem 5.** *Let $(F, A), (G, B), (H, C), (L, D) \in FS(U, E)$ and $(F, A) \approx_{FS} (G, B)$, $(H, C) \approx_{FS} (L, D)$. Then*
   (1) $(F, A) \cap_r (H, C) \approx_{FS} (G, B) \cap_r (L, D)$.
   (2) $(F, A) \cup_e (H, C) \approx_{FS} (G, B) \cup_e (L, D)$.

*Proof.* (1) Let $(F, A) \cap_r (H, C) = (M_1, A \cap C)$, $(G, B) \cap_r (L, D) = (M_2, B \cap D)$.

If $e \in (A \cap C) \cap (B \cap D)$, then $e \in A \cap B$, $e \in C \cap D$. It follows that $F(e) = G(e)$ and $H(e) = L(e)$. Consequently, $M_1(e) = F(e) \cap H(e) = G(e) \cap L(e) = M_2(e)$.

Let $e \in (A \cap C) - (B \cap D)$. It follows that $e \in A$, $e \in C$, $e \notin B \cap D$. If $e \notin B$, then $e \in A - B$ and hence $F(e) = \emptyset$. Hence we have $M_1(e) = F(e) \cap H(e) = \emptyset$. If $e \notin D$, then $e \in C - D$ and hence $H(e) = \emptyset$. Thus $M_1(e) = F(e) \cap H(e) = \emptyset$

Let $e \in (B \cap D) - (A \cap C)$. $M_2(e) = \emptyset$ can be proved similarly.

(2) Let $(F, A) \cup_e (H, C) = (T_1, A \cup C)$, $(G, B) \cup_E (L, D) = (T_2, B \cup D)$.

For any $e \in (A \cup C) \cap (B \cup D)$, we have $e \in A \cup C$, $e \in B \cup D$. Without losing generality, we assume that $e \in A$, $e \in D$. If $e \in B$, $e \in C$, then $e \in A \cap B$, $e \in C \cap D$. It follows that $F(e) = G(e)$ and $H(e) = L(e)$. Consequently, $T_1(e) = F(e) \cup H(e) = G(e) \cup L(e) = T_2(e)$. If $e \notin B$, $e \in C$, then $F(e) = \emptyset$, $H(e) = L(e)$ and hence $T_1(e) = F(e) \cup H(e) = H(e) = L(e) = T_2(e)$. If $e \in B$, $e \notin C$, then $F(e) = G(e)$, $L(e) = \emptyset$ and hence $T_1(e) = F(e) = G(e) = G(e) \cup L(e) = T_2(e)$. If $e \notin B$, $e \notin C$, then $F(e) = \emptyset$, $L(e) = \emptyset$ and hence $T_1(e) = \emptyset = T_2(e)$.

For any $e \in (A \cup C) - (B \cup D)$, we have $e \in A \cup C$, $e \notin B \cup D$. If $e \in A$, $e \in C$, then $e \in A - B$, $e \in C - D$. It follows that $F(e) = \emptyset$, $H(e) = \emptyset$ and hence $T_1(e) = F(e) \cup H(e) = \emptyset$. If $e \in A$, $e \notin C$, then $e \in A - B$, $F(e) = \emptyset$ and hence $T_1(e) = F(e) = \emptyset$. If $e \notin A$, $e \in C$, then $e \in C - D$, $H(e) = \emptyset$ and hence $T_1(e) = H(e) = \emptyset$.

$T_2(e) = \emptyset$ for any $e \in (B \cup D) - (A \cup C)$ can be proved similarly.

Let $(F, A)_{\approx_{FS}} = \{(G, B); (F, A) \approx_{FS} (G, B)\}$ be the congruence class containing $(F, A)$ and $FS(U, E)/ \approx_{FS} = \{(F, A)_{\approx_{FS}}; (F, A) \in FS(U, E)\}$. We define operations $\cup_{es}$ and $\cap_{rs}$ on $FS(U, E)/ \approx_{FS}$ as follows:

$(F, A)_{\approx_{FS}} \cup_{es} (G, B)_{\approx_{FS}} = ((F, A) \cup_e (G, B))_{\approx_{FS}}$,

$(F, A)_{\approx_{FS}} \cap_{rs} (G, B)_{\approx_{FS}} = ((F, A) \cap_r (G, B))_{\approx_{FS}}$.

These two operations are reasonable. $(FS(U, E)/ \approx_{FS}, \cup_{es}, \cap_{rs})$ is called fuzzy soft quotient algebra on soft space $(U, E)$.

**Theorem 6.** $(FS(U, E)/ \approx_{FS}, \cup_{es}, \cap_{rs})$ *is a distributive lattice.*

# 4  MTL Structure of Fuzzy Soft Algebra

In this section, based on t-norm and its residuated implication, we establish the connections between soft algebras and nonclassical logic algebras.

The fuzzy implication operator plays an important role in fuzzy logic and Zadeh's theory of approximate reasoning. A fuzzy implication operator is a binary operation $\rightarrow: [0, 1]^2 \rightarrow [0, 1]$ satisfying the following conditions [11]:

(I1) $0 \rightarrow a = 1$.

(2) $a \rightarrow 1 = 1$.

(3) $1 \rightarrow 0 = 0$.

(4) $a \rightarrow b$ is increasing with respect to $b$ and decreasing with respect to $a$.

In addition to these a number of other desirable properties associated with fuzzy implication operator have also been suggested [12].

Triangular norms (t-norms) are closely related to fuzzy implication operators. In what follows, the least upper bound (greatest lower bound) of a subset $G$ of $[0, 1]$ will be denoted by $\vee G$ ($\wedge G$), alternatively. A function $\otimes : [0, 1]^2 \rightarrow [0, 1]$ is said to be a t-norm if $\otimes$ is associative, commutative and satisfy the conditions $a \otimes 1 = a$ and that $a \leq b$ implies $a \otimes c \leq b \otimes c$ for all $a, b, c \in [0, 1]$. A t-norm $\otimes$ is left-continuous if $a \otimes \vee\{b_i; i \in I\} = \vee\{a \otimes b_i; i \in I\}$ holds where $a, b_i \in [0, 1] (i \in I)$, and $I$ is a nonempty index set.

**Theorem 7.** *[13] Suppose that $\otimes$ is a left-continuous t-norm, define $\to_\otimes$: $[0,1]^2 \to [0,1]$ as $a \to_\otimes b = \vee\{x \in [0,1]; a \otimes x \leq b\}$ for all $a, b \in [0,1]$. Then $\to_\otimes$ is a fuzzy implication operator and*

(1) $a \to_\otimes b = 1$ iff $a \leq b$.
(2) $a \leq b \to_\otimes c$ iff $b \leq a \to_\otimes c$.
(3) $a \to_\otimes (b \to_\otimes c) = b \to_\otimes (a \to_\otimes c)$.
(4) $1 \to_\otimes a = a$.
(5) $a \to_\otimes \wedge\{b_i; i \in I\} = \wedge\{a \to_\otimes b_i; i \in I\}$.
(6) $\vee\{b_i; i \in I\} \to_\otimes a = \wedge\{b_i \to_\otimes a; i \in I\}$.
(7) $a \otimes b \leq c$ iff $a \leq b \to_\otimes c$.

In this theorem, $\to_\otimes$ is called the residuated implication of $\otimes$.

*Example 1.* [13] The following are four left-continuous t-norms:

$$a \otimes_L b = (a + b - 1) \vee 0 \tag{3}$$

$$a \otimes_G b = a \wedge b \tag{4}$$

$$a \otimes_\pi b = ab \tag{5}$$

$$a \otimes_0 b = \begin{cases} a \wedge b, & \text{if } a + b > 1, \\ 0, & \text{if } a + b \leq 1. \end{cases} \tag{6}$$

The implication operators corresponding to the t-norms $\otimes_L$, $\otimes_G$, $\otimes_\pi$ and $\otimes_0$ are as follows (they are called Lukasiewicz operator, Godel operator, product operator, and $R_0$ operator, respectively):

$$a \to_{\otimes_L} b = (1 - a + b) \wedge 1 \tag{7}$$

$$a \to_{\otimes_G} b = \begin{cases} 1, & \text{if } a \leq b, \\ b, & \text{if } a > b. \end{cases} \tag{8}$$

$$a \to_{\otimes_\pi} b = \begin{cases} 1, & \text{if } a = 0, \\ \frac{b}{a} \wedge 1, & \text{if } a > 0. \end{cases} \tag{9}$$

$$a \to_{\otimes_0} b = \begin{cases} 1, & \text{if } a \leq b, \\ (1 - a) \vee b, & \text{if } a > b. \end{cases} \tag{10}$$

Based on t-norm and its residuated implication, we introduce $\otimes-$product and $\otimes-$implication operations on fuzzy soft sets.

**Definition 7.** *Let $\otimes$ be a left-continuous t-norm, $(F, E), (G, E) \in FS_E$. The $\otimes-$product of $(F, E), (G, E)$ is a fuzzy soft set $(H, E)$, denoted by $(F, E) \otimes (G, E) = (H, E)$, and is defined by $H(e)(x) = F(e)(x) \otimes G(e)(x)$ for any $e \in E$ and $x \in U$.*

**Definition 8.** *Let $\otimes$ be a left-continuous t-norm, $(F, E), (G, E) \in FS_E$. The $\otimes$-implication of $(F, E), (G, E)$ is a fuzzy soft set $(H, E)$, denoted by $(F, E) \rightarrow_\otimes (G, E) = (H, E)$, and is defined by $H(e)(x) = F(e)(x) \rightarrow_\otimes G(e)(x)$ for any $e \in E$ and $x \in U$.*

**Theorem 8.** *Let $(U, E)$ be a soft space and $\otimes$ a left-continuous t-norm, $(F, E), (G, E), (H, E) \in FS_E$. Then*
  (1) $(F, E) \otimes (G, E) = (G, E) \otimes (F, E)$.
  (2) $((F, E) \otimes (G, E)) \otimes (H, E) = (F, E) \otimes ((G, E) \otimes (H, E))$.
  (3) $U_E \otimes (F, E) = (F, E)$.
  (4) *If* $(F, E) \subseteq (G, E)$, *then* $(F, E) \otimes (H, E) \subseteq (G, E) \otimes (H, E)$.

**Theorem 9.** *Let $(U, E)$ be a soft space and $\otimes$ a left-continuous t-norm, $(F, E), (G, E), (H, E) \in FS_E$. Then $(F, E) \otimes (G, E) \subseteq (H, E)$ if and only if $(F, E) \subseteq (G, E) \rightarrow_\otimes (H, E)$.*

*Proof.* Suppose that $(F, E) \otimes (G, E) \subseteq (H, E)$. For any $e \in E$ and $x \in U$, we have $F(e)(x) \otimes G(e)(x) \leq H(e)(x)$ and hence $F(e)(x) \leq G(e)(x) \rightarrow_\otimes H(e)(x)$. It follows that $F(e) \subseteq G(e) \rightarrow_\otimes H(e)$ and consequently $(F, E) \subseteq (G, E) \rightarrow_\otimes (H, E)$.

Conversely, suppose that $(F, E) \subseteq (G, E) \rightarrow_\otimes (H, E)$. For any $e \in E$ and $x \in U$, we have $F(e)(x) \leq G(e)(x) \rightarrow_\otimes H(e)(x)$. It follows that $F(e)(x) \otimes G(e)(x) \leq H(e)(x)$ and hence $F(e) \otimes G(e) \subseteq H(e)$. Thus we have $(F, E) \otimes (G, E) \subseteq (H, E)$.

**Theorem 10.** *Let $(U, E)$ be a soft space and $\otimes$ a left-continuous t-norm, $(F, E), (G, E) \in FS_E$. Then $((F, E) \rightarrow_\otimes (G, E)) \cup_e ((G, E) \rightarrow_\otimes (F, E)) = U_E$.*

*Proof.* Let $((F, E) \rightarrow_\otimes (G, E)) \cup_e ((G, E) \rightarrow_\otimes (F, E)) = (H, E)$. For any $e \in E$ and $x \in U$, if $F(e)(x) \leq G(e)(x)$, then $F(e)(x) \rightarrow_\otimes G(e)(x) = 1$; if $F(e)(x) > G(e)(x)$, then $G(e)(x) \rightarrow_\otimes F(e)(x) = 1$. It follows that
$$H(e)(x) = (F(e)(x) \rightarrow_\otimes G(e)(x)) \vee (G(e)(x) \rightarrow_\otimes F(e)(x)) = 1.$$
Hence $H(e) = U$ and consequently $(H, E) = U_E$.

A MTL algebra is a bounded residuated lattice which satisfies the pre-linearity equation [12]. By Theorem 8, Theorem 9 and Theorem 10, we have the following theorem.

**Theorem 11.** $(FS_E, \cup_e, \cap_s, \otimes, \rightarrow_\otimes)$ *is a MTL algebra.*

We consider fuzzy soft quotient algebra $(FS(U, E)/ \approx_{FS}, \cup_{es}, \cap_{rs})$. For any fuzzy soft set $(F, A)$, there exists a unique fuzzy soft set $(F', E) \in FS_E$ such that $(F', E) \approx_{FS} (F, A)$. Namely, $F'(e) = F(e)$ for any $e \in A$ and $F'(e) = \emptyset$ for any $e \in E - A$. Thus there exists a one-to-one correspondence between $(FS(U, E)/ \approx_{FS}$ and $FS_E$. Consequently, we have

**Theorem 12.** $(FS(U, E)/ \approx_{FS}, \cup_{es}, \cap_{rs}, \otimes, \rightarrow_\otimes)$ *is a MTL algebra.*

# 5    Concluding Remarks

Soft sets and fuzzy soft sets are mathematical tools for dealing with uncertainties. This paper is devoted to the discussion of algebraic structures of fuzzy soft sets. The fuzzy notion of soft equality relation on fuzzy soft sets is proposed and several related properties are investigated. Furthermore, based on t-norm and its residuated implication operator, the product and implication operations on fuzzy soft sets are introduced. MTL structures of fuzzy soft algebra and fuzzy soft quotient algebra are established.

In further research, the connections between fuzzy soft algebras and nonclassical logic algebras, such as IMTL algebra, MV algebra and $R_0$ algebra is an important and interesting issue to be addressed.

**Acknowledgements.** This work has been supported by the National Natural Science Foundation of China (Grant No. 61175055, 61175044).

# References

1. Zadeh, L.A.: Fuzzy sets. Information and Control 8, 338–353 (1965)
2. Pawlak, Z.: Rough sets. International Journal of Computer and Information Science 11, 341–356 (1982)
3. Molodtsov, D.: Soft set theory-First results. Computers and Mathematics with Applications 37, 19–31 (1999)
4. Molodtsov, D.: The theory of soft sets, URSS Publishers, Moscow (2004) (in Russian)
5. Maji, P.K., Biswas, R., Roy, A.R.: Soft set theory. Computers and Mathematics with Applications 45, 555–562 (2003)
6. Ali, M.I., Feng, F., Liu, X., Min, W.K., Shabir, M.: On some new operations in soft set theory. Computers and Mathematics with Applications 57, 1547–1553 (2009)
7. Qin, K.Y., Hong, Z.Y.: On soft equality. Journal of Computational and Applied Mathematics 234, 1347–1355 (2010)
8. Maji, P.K., Biswas, R., Roy, A.R.: Fuzzy soft sets. The Journal of Fuzzy Mathematics 9, 589–602 (2001)
9. Borzooei, R.A., Bakhshi, M., Mashinchi, M.: Lattice structure on some fuzzy algebraic systems. Soft Computing 12, 739–749 (2008)
10. Qin, K., Zhao, H.: Lattice Structures of Fuzzy Soft Sets. In: Huang, D.-S., Zhao, Z., Bevilacqua, V., Figueroa, J.C. (eds.) ICIC 2010. LNCS, vol. 6215, pp. 126–133. Springer, Heidelberg (2010)
11. Yager, R.R.: On some new classes of implication operators and their role in approximate reasoning. Information Sciences 167, 193–216 (2004)
12. Esteva, F., Godo, L.: Monoidal t-norm based logic: towards a logic for left- continuous t-norms. Fuzzy Sets and Systems 124, 271–288 (2001)
13. Wang, G.J., Fu, L.: Unified forms of triple I method. Computers and Mathematics with Applications 49, 923–932 (2005)

# A Rapid Granular Method for Minimization of Boolean Functions

Zehua Chen[*], He Ma, and Yu Zhang

College of Information Engineering, Taiyuan University of Technology,
Taiyuan, Shanxi, P.R. China, 030024
chenzehua@tyut.edu.cn

**Abstract.** A rapid granular method for minimization of Boolean functions is proposed in this paper. Firstly, the Boolean function is changed into the sum of products. Secondly, truth table was got and statistic information under different knowledge space was computed as heuristic information for function minimization. Thirdly, information granules with different granularity were found according to the heuristic information. Finally, if all the terms in information granules have covered the universe, they will be the desired result. The algorithm was realized by *MATLAB* and experiments have shown its high efficiency.

**Keywords:** Granular computing, Boolean function, Boolean minimization.

## 1    Introduction

Boolean function is applied to describe the causal relationship between input and output logic variables. Minimization of Boolean function is greatly dependent on Boolean algebra operations, however, the complexity of minimization is dramatically increased as the number of input variables increased. Each Boolean function corresponds to a specific circuit structure, so the minimization of Boolean function is of great importance for simplification of digital logic circuits, so as to reduce the power cost and improve the security of the circuits [1].

The traditional methods for multivariable logic function reduction including algebraic reduction method, Karnaugh map(K-map) reduction method and $Q$-$M$ algorithm. However, the algebraic method need flexible application of the Boolean algebra laws, and there are no rules to follow; K-map method is not suitable for more than five input variables; the $Q$-$M$ algorithm and its improvements seems too complex with large loop iteration [1,2,3,4,5] and  some basic rough set method appeared [6,7,8]. Literature [9] proposed a granular matrix-based method for truth table reduction.

Granular computing(GrC) is an computable method for complex problems [10,11]. In this paper, a GrC-based Boolean function minimization algorithm was proposed from view of the statistic perspective and implemented by Matlab. The thought of

---

[*] Corresponding author.

D. Miao et al. (Eds.): RSKT 2014, LNAI 8818, pp. 577–585, 2014.
DOI: 10.1007/978-3-319-11740-9_53 © Springer International Publishing Switzerland 2014

granulation, operations under different granularity space and the way of solving the problem are consistent with the basic thought of GrC.

The paper was organized as follows: part 1, introduction, the background of the research; part 2, preliminaries, the basic concepts included in this paper; part 3, the description of the rapid granular algorithms for minimization of Boolean functions; and part 4, Discussion and Conclusions.

## 2    Preliminaries

In this part some basic definitions [1] will be given first for understanding the essence of the problem.

**Definition 1:** Boolean Function

A Boolean function can be described by a Boolean equation consisting of a binary variable identifying the function followed by an equal sign and a Boolean expression.

A single-output Boolean function is a mapping from each of possible combinations of value 0 and 1 on the function variables to value 0 or 1.

We only discuss the single-output Boolean function minimization in this paper.

Any arbitrary Boolean function can be expressed in the Sum-of-Product (SOP) forms.

**Definition 2:** Minterms

Product terms that have this property of all variables appearing exactly once (and, consequently, having the value of 1 for only one combination of values of the function variables) are called minterms. Let the complemented variables be represented with '0'and the uncomplemented ones be represented with '1'. We call such minterms the digital minterms.

**Example1:** Suppose there are 4 input variables, let minterms be $AB\overline{C}D$, $A\overline{B}C\overline{D}$, $A\overline{B}\,\overline{C}\,\overline{D}$ and $AB\overline{C}D$. Then the corresponding digital minterms are respectively 1011,1001,1000, 1010.

Minterms have some important properties as follows [1]:

1. There are $2^n$ minterms for $n$ Boolean variables. These minterms can be generated from the binary numbers from 0 to $2^n - 1$.
2. Any Boolean function can be expressed as SOP forms.
3. A function that includes all the $2^n$ minterms is equal to logic 1.

The property 2 shows that we can reduct SOP instead of minimizing Boolean function.

**Definition 3:** Reductive Minterm Group

If sum of minterms can be simplified to a term, we call group these minterms as reductive minterm group, represented with $RMG$.

The group of minterms $\overline{A}BCD$, $A\overline{B}CD$, $AB\overline{C}\overline{D}$, $ABC\overline{D}$ in example 1 can be grouped into a reductive minterm group that equals to the term $AB$.

Let $U$ be all the minterms in SOP, There are $p$ $RMG$ s, i.e. $U = \bigcup_{i=1}^{p} MG_i$ and $MG_i \cap MG_j \neq \varnothing (1 \leq i, j \leq p)$. It shows that all $RMG$ s form a coving.

**Definition 4:** Granule and Granularity

The common terms in $RMG$ are called granules. The variable number of granule is called granularity, represented by symbol $\omega$. The bigger the $\omega$, the finer the granlularity.

In example 1, the granule is $AB$, $\omega = 2$.

**Definition 5:** Granule Set

A collection of granules is called granule set, represented by $IGS$ and $IGS = \{Gr_1, Gr_2, ..., Gr_i, ...\}$, $Gr_i \in IGS$. $i$, the $i$th granule.

Different $IGS$ will be got in the process of Boolean function minimization. According to the different granularity, the minimal $IGS$ should be found to cover all the minterms, and that will be the result of Boolean function minimization.

***Theorem 1.*** For $m$ minterms with $n$ input variables, the sum of $m$ minterms will be 1 iff $m = 2^n$.

This is obviously correct according to the property 3.

***Theorem 2*** The sum of $RMG$ with $m$ minterms can be simplified as a term with $\omega$ variables iff $m = 2^{n-\omega}$.

**Proof:** Extract the common factors from minterms and the rest are the variables from other $n - \omega$ inputs. According to ***theorem 1***, if the rest can finally be reduced to 1 iff $m = 2^{n-\omega}$, i.e. if the $m$ minterms with $n$ inputs can be reducted as a common term, the necessary and sufficient condition is $m = 2^{n-\omega}$.

In example 1, $n = 4$, $\omega = 2$, $m = 2^{4-2} = 4$. So the $RMG$ can eventually be reducted as $A\overline{B}$.

Thus the Boolean function minimization based on Boolean logic relations is transformed into the reduction based on statistical information by thermo 2.

***Theorem 3.*** The information granules that got in coarser granularity are regarded as heuristic information granules. We do not calculate the information granules that contain the heuristic information granules.

***Theorem 3.*** corresponds to the absorption law of Boolean algebraic: $A + AB = A$.

For example, if $IGS$ contains heuristic information granule $A$ when $\omega = 1$, we don't consider the granules that contain $A$ in the after calculation, like $AB, AC, AB\overline{C}, A\overline{B}C\overline{D}$ and so on.

**Definition 7:** Don't-Care Conditions [11]

The 'don't care terms' contains the following two cases:

1. The input combinations never occur.
2. The input combinations are expected to occur, but the correspondent output are neglected.

The main work of this paper is Boolean function minimization (contains 'don't-care conditions').

# 3    The Rapid Granular Algorithm for Minimization of Boolean Function

To illustrate the algorithm, some symbol definition should be given first.

## 3.1    The Symbol Definition

$X$ : The input of Boolean function.

$Y$ : The output of Boolean function.

$G$ : All the combinations of inputs, $G = P(X)$  $G \neq \varnothing$.

$U_m$ : All the minterms generated by Boolean function.

$U_d$ : All the minterms generated by 'don't-care conditions'.

$RMG_\omega$ : All $RMG$ when the granularity is $\omega$, $RMG_\omega \subset RMG$.

$G_\omega$ : All combinations of inputs when the granularity is $\omega$, $G_\omega \subset G, G_{\omega i} \subset G_\omega$.

$V_\omega$ : All combinations of values of granules when the granularity is $\omega$, $V_{\omega i} \subset V_\omega$.

$N_{G_{\omega i}/V_{\omega i}}$ : The number of the minterms when the combination of inputs is $G_{\omega i}$ and the combination of values of inputs is $V_{\omega i}$.

$N$ : When the granularity is $\omega$, $N = 2^{n-\omega}$.

The ***theorem 2*** can be expressed as: $m$ minterms with $n$ inputs can be finally reduced as an information granule of iff $N_{G_{\omega i}/C_{\omega i}} = N$. This can be illustrated by example 2.

Example2, Let $n = 3, X = \{A, B, C\}$, then we have:

$G(X) = \{\{A\},\{B\},\{C\},\{A,B\},\{A,C\},\{B,C\},\{A,B,C\}\}$.

When $\omega = 1$, the granularity is coarsest.
$G_1 = \{\{A\},\{B\},\{C\}\}, V_1 = \{\{1\},\{0\}\}, N = 2^{(3-1)} = 4$.

When $\omega = 2$, the granularity becomes finer. $G_2 = \{\{A,B\},\{A,C\},\{B,C\}\}$,
$V_2 = \{\{0,1\},\{0,0\},\{1,0\},\{1,1\}\}, N = 2^{3-2} = 2$.

## 3.2    The Description of the Algorithm

The specific steps of the rapid granular algorithm for minimization of Boolean functions are described as follow:

*Step1*: Transform Boolean function into digital minterms in form of truth table.

*Step2*: Generate *IGS*. The following is how to generate *IGS*

---

**Procedure**

$\omega = 1$; % initialize granularity

for $\omega = 1$:n-1

   find $RMG_\omega$ and the corresponding $G_r$

   if $\bigcup RMG = U_m$

   break ;

end

check the minterms that are not included in *RMG*s and regard them as independent granules

delete the granules that are composed by $U_d$

output the final result, i.e. *IGS*

---

*Step3*: Output the *IGS* as the final result.

## 3.3    Example

To describe the algorithm more clearly, a concrete example is followed to explain the details of the proposed algorithm. The Boolean function can be easily transformed into SOP, so example 3 start from SOP.

**Example 3**: $X(A,B,C,D) = \Sigma m(0,5,6,7,9,13,15) + d(10,14)$

According to the algorithm in our paper, the specific procedure is followed.

**Step 1:** Transform the multivariable logic function (contains the 'don't care terms') into the digital minterms as shown in table1. In the second column, '*' distinguish the 'don't care terms' and '1'indicate the minterms.

As shown in table 1, the minterms are $\{m_0, m_5, m_6, m_7, m_9, m_{13}, m_{15}\}$ and the 'don't care terms' are $\{m_{10}, m_{14}\}$ .Obviously, the number of input variables is $n = 4$ .

**Table 1.** The Digital Minterms

| Minterms | Judge | $A$ | $B$ | $C$ | $D$ |
|:---:|:---:|:---:|:---:|:---:|:---:|
| $m_0$ | 1 | 0 | 0 | 0 | 0 |
| $m_5$ | 1 | 0 | 1 | 0 | 1 |
| $m_6$ | 1 | 0 | 1 | 1 | 0 |
| $m_7$ | 1 | 0 | 1 | 1 | 1 |
| $m_9$ | 1 | 1 | 0 | 0 | 1 |
| $m_{10}$ | * | 1 | 0 | 1 | 0 |
| $m_{13}$ | 1 | 1 | 1 | 0 | 1 |
| $m_{14}$ | * | 1 | 1 | 1 | 0 |
| $m_{15}$ | 1 | 1 | 1 | 1 | 1 |

**Step 2:** Generate $IGS$ .

When $\omega = 1$, $G_1 = \{\{A\},\{B\},\{C\},\{D\}\}. V_1 = \{\{0\},\{1\}\}$ $N = 2^{(4-1)} = 8$

$N_{A/0} = 4, N_{A/1} = 5; N_{B/0} = 3, N_{B/1} = 6; N_{C/0} = 4, N_{C/1} = 5; N_{D/0} = 4, N_{D/1} = 5;$

Because there is no number equals to $N = 2^{n-w} = 2^{4-1} = 8$, there is no information granule when $\omega = 1$.

When $\omega = 2$, Knowledge granularity space of the system becomes finer.

$G_2 = \{\{A,B\},\{A,C\},\{A,D\},\{B,C\},\{B,D\},\{C,D\}\}$ .

$V_1 = \{\{00\},\{01\},\{10\},\{11\}\}, N = 2^{(4-2)} = 4$ .

$N_{AB/00} = 1$ , $N_{AB/01} = 3$ , $N_{AB/11} = 3$ , $N_{AB/10} = 2$ ; $N_{AC/00} = 2$ , $N_{AC/01} = 2$ ,
$N_{AC/11} = 3$ , $N_{AC/10} = 2$ ; $N_{AD/00} = 2$ , $N_{AD/01} = 2$ , $N_{AD/11} = 3$ , $N_{AD/10} = 2$ ;
$N_{BC/00} = 2$ , $N_{BC/01} = 1$ , $N_{BC/11} = 4$ , $N_{BC/10} = 2$ ; $N_{BD/00} = 2$ , $N_{BD/01} = 1$ ,
$N_{BD/11} = 4, N_{BD/10} = 2; N_{CD/00} = 1; N_{CD/01} = 3, N_{CD/11} = 2, N_{CD/10} = 3$ .

$N_{BC/11} = N_{BD/11} = N . Gr_1 = BC , Gr_2 = BD$ . The corresponding $RMG$s are :

$R_1 = \{m_6, m_7, m_{14}, m_{15}\}$, $R_1 = \{m_5, m_7, m_{13}, m_{15}\}$.

The minterms included in *RMGs* do not cover the $U_m$. Continue to compute it.

When $\omega = 3, G_3 = \{\{A,B,C\},\{A,B,D\},\{A,C,D\},\{B,C,D\}\}$

$V_1 = \{\{000\},\{001\},\{010\},\{011\},\{100\},\{101\},\{110\},\{111\}\}$, $N = 2^{(4-3)} = 2$.

Because we get information granule $BC$ and $BD$ when $\omega = 2$, in the finer granularity, we do not need to compute the combinations that contain these granules. Like $N_{ABC/011}$, $N_{ABC/111}$, $N_{ABD/011}$, $N_{ABD/111}$, $N_{BCD/101}$, $N_{BCD/110}$, $N_{BCD/111}$. The search space is dramatically reduced.

$N_{ABC/000} = 1$ , $N_{ABC/001} = 0$ , $N_{ABC/010} = 1$ , $N_{ABC/100} = 1$ , $N_{ABC/101} = 1$ ,

$N_{ABC/110} = 1$ ; $N_{ABD/000} = 1$ , $N_{ABD/001} = 0$ , $N_{ABD/010} = 1$ , $N_{ABD/100} = 1$ ,

$N_{ABD/101} = 1$ ; $N_{ACD/110} = 1$ , $N_{ACD/000} = 1$ , $N_{ACD/001} = 1$ , $N_{ACD/010} = 1$ ,

$N_{ACD/011} = 1$ , $N_{ACD/100} = 0$ ; $N_{ACD/101} = 2$ , $N_{ACD/110} = 2$ , $N_{ACD/111} = 1$ ,

$N_{BCD/000} = 1, N_{BCD/001} = 1, N_{BCD/010} = 1, N_{BCD/011} = 0, N_{BCD/100} = 0$.

$N_{ACD/101} = N_{ACD/110} = N$ , $Gr_3 = A\overline{C}D$ , $Gr_4 = AC\overline{D}$ . The corresponding *RMGs* are $R_3 = \{m_9, m_{13}\}$, $R_4 = \{m_{10}, m_{14}\}$ . Now it has cover the universe, so stop computing.

So far, *RMGs* contains all minterms in $U_m$ except $m_0$. We consider $m_0$ as an independent granule, i.e. $Gr_5 = m_0 = \overline{A}\overline{B}\overline{C}\overline{D}$ . As $Gr_4$ is contained of the 'don't care terms', $Gr_4$ is invalid.

**Step 3:** $IGS = \{Gr_1, Gr_2, Gr_3, Gr_5\} = \{BC, BD, A\overline{C}D, \overline{A}\overline{B}\overline{C}\overline{D}\}$ . The final result is $Y = BC + BD + A\overline{C}D + \overline{A}\overline{B}\overline{C}\overline{D}$ .

### 3.4    Algorithm Complexity Analysis

The rapid granular algorithm for minimization of Boolean functions was programed by *MATLAB* and many examples are testified to prove its accuracy and efficiency.

The time complexity of this algorithm is d $(O(2^n))$. Because of the heuristic information, the time complexity will be greatly reduced and it is better than the $(O(3^n/3))$ of the *Q-M* algorithm.

## 4    Conclusions

Boolean function minimization is important not only in digital logic circuit design. SOP is the standard form of Boolean function. A rapid granular algorithm for minimization of Boolean functions is proposed in this paper. At first, the Boolean function is expressed in form of SOP, then, the basic theorem for minterms reduction is illustrated and proved. Secondly, we introduce the concepts of information granule and granularity. The static information was computed under different knowledge granularity space. And the stop condition was set to improve the algorithm efficiency. During this process, 'don't care terms' can also be considered and at last, output the final result. Experimental results show the accuracy and efficiency of the proposed algorithm.

The originality of this paper mainly lies that: 1.Introduce the concept of granularity and information granular to simplify the algorithm, the solution was searched under different knowledge granularity space.2.Find the statistical properties of minterms and use it to find the information granules.

The research finds a new way that totally different with the traditional ones to realize the Boolean function minimization with high efficiency, and on the other hand, enriches the application of granular computing. The follow-up work is still continuing.

**Acknowledgement.** This research is supported by Fund Program for the Scientific Activities of Selected Returned Overseas Professionals in Shanxi Province(2011-172) and Research Project Supported by Shanxi Scholarship Council of China (2013-031).

## References

[1]  Shi, Y.: Fundamentals of Digital Electronics, 5th edn. Higher Education Press, Beijing (2006) (in Chinese)

[2]  Zhijin, G., Yiqing, Z., Jianlin, Q.: Research of Best Cover for Logic Function with Big Number of Input and Output Variables. Computer Applications and Software 20(12), 11–13 (2004)

[3]  Bo, W.: An Improved Method to Extract Absolute Minimal Covers for Logic Function. Journal of Computer Aided Design and Graphics 11(2), 143–146 (1999)

[4]  Youlian, Z.: Research on Simplification Algorithm of Logical Functions with Computer. Journal of Nanjing University of Science and Technology: Natural Science Edition 27(4), 405–408 (2003)

[5]  Zongwei, L., Zhenghui, L.: Improved Algorithm to Multilevel Logic Optimization. Journal of Shanghai Jiaotong University 35(2), 209–211 (2001)

[6]  Yiqing, Z., Zhijin, G., Yanming, L.: An Algorithm of Combinatory Logic Optimization Based on Rough Set. Journal of Lanzhou University of Technology 33(1), 88–91 (2007)

[7]  Yiqing, Z., Yanming, L., Xun, L.: An Optimal Algorithm of Multiple Output Logic Function Based on Rough Set. Journal of Nantong University: Natural Science Edition 4(4), 59–61 (2006)

[8]  Yiqing, Z., Zhijin, G., Lixun: Expression in Rough Set for Logic Function and A Method of Minim I -ization. Heilongjiang University Journal of Natural Science 23(2), 265–268 (2006)

[9]  Zehua, C., Changqing, C., Gang, X.: Granular Matrix Based Rapid Reduction Algorithm for Multivariable Truth Table. Pattern Recognition and Artificial Intelligence 26(8), 745–750 (2013)

[10] Zadeh, L.A.: Some reflections on soft computing, granular computing and their roles in the conception, design and utilization of information/intelligent systems. Soft Computing (2), 23–25 (1998)

[11] Lin, T.Y.: Granular Computing: Practices, Theories, and Future Directions. In: Encyclopedia of Complexity and Systems Science, pp. 4339–4355 (2009)

# Big Data to Wise Decisions

# QoS-Aware Cloud Service Selection
# Based on Uncertain User Preference

Bin Mu, Su Li, and Shijin Yuan

School of Software Engineering, Tongji University, Shanghai, China
{binmu99,yuanshijin2003}@163.com lisu0901.tj@gmail.com

**Abstract.** With the growing number of alternative services being deployed by cloud service providers, and users usually can only provide uncertain QoS (Quality of Service) preferences to providers, it becomes difficult to select the most suitable service to satisfy users need. In this paper, we propose a novel model of cloud service selection which considers the uncertainty of user subjective and objective weight preferences. Based on this model, we first analyses the incompleteness and fuzziness of user preference, obtains the user subjective weight preference by intuitionistic fuzzy set and objective weight by attribute significance of rough set. Then we transform the uncertain user QoS preference-aware cloud service selection to a multiple attribute decision-making problem, further we use the technique of order the preference by similarity to an ideal solution to select best service for user. Lastly, we conduct a case study about cloud storage service selection to show the effectiveness and advantages of our approach.

**Keywords:** Cloud service selection, uncertain user preference, QoS, weight.

## 1 Introduction

Cloud computing [1] is a kind of scalable cheap distributed computing ability which can provide through internet. With the rapid development of cloud computing, more and more web services were deployed on the cloud platform to build SOA application and distributed cloud application [2]. Cloud services are so abundant and there are massive services with similar function but different non-functional attributions in the cloud, so it is deserve to think that how to select a suitable cloud service to fulfill user specific preference .User is the end beneficiary of service executing, so it is important to take a full consideration on the customized QoS preferences for service non-functional attribution so that we can improve user experience. It has become a hot point for research.

Nowadays, most of the service selection approaches based on user preference only considered complete and fuzzy QoS preference information; the following two things are not included in.

First is about the uncertainty of user weight preference. Usually when user are interacting with cloud service provider, it will often run into such situations that some

D. Miao et al. (Eds.): RSKT 2014, LNAI 8818, pp. 589–600, 2014.
DOI: 10.1007/978-3-319-11740-9_54 © Springer International Publishing Switzerland 2014

information was missed for itself or ignored by user's wrong recognition for its affection on QoS information. Moreover some preferences cannot be expressed clearly because of non-expert cloud user's limited specific technology. So all these situations will lead to uncertain QoS weight preference. In paper [3], the authors apply cloud model to compute the uncertainty of QoS and use mixed integer programming to identify the most suitable services, they only consider the uncertainty of QoS values but take no look into users' real preferences. Hongbing *et el.* [4] propose an approach which only consider the certainty QoS value preference to compute the similarity between descriptions candidates web service and user requirement. In paper [5] the MLOMSS local service selection model based on multi-dimensions service quality was presented, it applies the attribution importance of Fuzzy analytic hierarchy method to define the impact of QoS weight. But these models only focus on certain and complete QoS-aware service selection.

The second one is the ignorance of the importance of object QoS weight preference. Usually most of the existed papers do not consider the objective weight preference and other common evaluation with the same service requirement. In [6], authors apply the linear weighting method to assess the alternative services, they think that user subjective weights can reflect the relative importance of the objective function, however the results are somewhat not inconsistent with the actual situation and the weight standardization is too simple to reflect the complex scenarios of cloud service in the real world. Another discussion of integrated preferences for cloud service selection problem is using Analytic Hierarchy Process (AHP) [7] to solve this multi-attribution decision making problem. In [8], Godse *et el.* focus on the selection of software-as-a-service cloud based on AHP. It should be noted that all these considered factors except cost can hardly be quantified by an objective measure, thus their approach is still mainly based on subjective assessment.

Other approaches like paper [9] is too complicated to compute. As they first obtain the subjective weight preference by using the triangular fuzzy numbers method to convert language values and OWA operator to quantify; then adopt the information entropy method to determine the objective weights; lastly a synthetic parameters as the equilibrium factor is used to get the weight impact of the combining weight.

In this paper, different from all the existing models of cloud service selection, we present a novel model based on the aggregation of user uncertain weight preferences for QoS criteria. The main contributions of our paper can be summarized as follows.

- We address the problem of cloud service selection and demonstrate the influence of uncertainty of user preference on selection process.
- We propose a novel combination calculation method of subjective and objective user weight preference. We first utilize the intuitionistic fuzzy sets to handle the vague user subjective preference. Then according to the user history preference information for the same service request, we get the objective weight preference by the attribute significance of rough set. And last a compromise factor used to optimize the integrated weight preference.
- Based on the proposed model, we evaluate our approach by a real world case study and a comparison experiment with other method on this situation.

The rest of this paper is organized as follow. Section 2 is the model of our proposed cloud service selection. Section 3 introduces the QoS-aware service selection approach under the uncertain user weight preference. After that, a case study and experiment are presented in Section 4 to illustrate the effectiveness and advantages of our approach. We conclude this paper and discuss future work in Section 5.

## 2    Our Cloud Service Selection Decision Model

In this section, we describe the process of cloud service selection with uncertain user preferences as follows: given a set of candidate services $C$ which can meet the users' functional requirements firstly, $C= \{C_1, C_2,..., C_m\}$; according to the uncertain QoS weight preference information from cloud users, we can evaluate all the performance of QoS attributes for each service to choose the most suitable service for users. Therefore, this process can be defined as a multi-attributes decision model [10].

**Definition 1.** Cloud Service Selection Decision Model (**CSSDM**): Let CSSDM = (C, Q, V, $f$), where C is the set of candidate services and $C= \{C_1, C_2,..., C_m\}$; Q is the non-empty finite set of QoS attributes and $Q = \{q_1, q_2,..., q_n\}$; V is the QoS attributes value set of candidate services; The information function or approximation function is defined as $f$, where $f : S_x Q —> V$; W is the user preferred QoS attribute weight set and $W=\{w_1, w_2,...,w_n\}, \sum_{j=1}^{n}(w_j = 1) \ w_j > 0;$ DM express the decision matrix and DM = $(y_{ij})_{m \times n}$ , $y_{ij}$ represents the comprehensive evaluation value of the $j$ QoS attribute for the $i$ candidate service.

## 3    Cloud Service Selection Based on Uncertain User Preference

As a growing number of alternative value-added services have been deployed in the cloud to satisfy diversify requirements of organizations and individuals, the cloud service environment becomes more complicated for the various service fields and massive service amounts as well as the uncertainty of user preference. In this section, we propose a QoS-aware cloud service selection approach based on the uncertain user subjective and objective weight preferences.

The user subjective weight preference refers to the priority of one or a few QoS attributes must have special requirements and restraints according to user actual situation. However, usually user would often put forward uncertain and vague preferences due to the limitations of professional knowledge and thinking mindset. Also in real world situations, subjective assessment for cloud services and the importance weight for each QoS attribute are usually represented in the form of linguistic variables (e.g., "good" and "bad"). In order to deal with the inherent uncertainty of human languages, we apply a fuzzy simple additive weighting method in our model. Through this method, linguistic variables can be represented by fuzzy numbers for their fuzziness. And quantitative terms can also be represented in fuzzy number form. Then our model can effectively normalize and aggregate all different types of subjective preferences in real world situations. Before presenting the details of our model, some basic knowledge of intuitionistic fuzzy set will be introduced.

## 3.1    Subjective Weight Preference Based on Intuitionistic Fuzzy Set

Intuitionistic fuzzy set is a good description and illustration method for subjective uncertain information, it is originally proposed as the promotion of fuzzy set theory by Atanassov [11]. The fuzzy information is described via three parameters: Membership degree $\mu$, Non-membership degree $v$ and Uncertainty degree $\pi$.The ordered pair-intuition fuzzy number $[\mu, v]$ which composed by Membership degree and Non-membership degree usually used to represent a fuzzy concept.

Our paper will apply the mapping table between user linguistic described weight preference and intuitionistic fuzzy numbers which proposed in literature [12] to express and quantify subjective weight of a candidate for service. As shown in Table 1, the user linguistic weight preference for a certain QoS attribute is represented as an uncertainty degree parameter $L(\pi)$.

**Table 1.** Mapping table

| Linguistic Value | Intuitionistic Fuzzy Numbers |
|---|---|
| very important | $[0.9,0.1\text{-}\pi]$ |
| important | $[0.7,0.3\text{-}\pi]$ |
| medium | $[0.5,0.5\text{-}\pi]$ |
| unimportant | $[0.3,0.7\text{-}\pi]$ |
| very unimportant | $[0.1,0.9\text{-}\pi]$ |
| unknown | $[0,0]$ |

In order to apply to the multi-attribute decision making method, a conversion function is needed to transform the importance of a certain QoS attribute represented by $L(\pi)$ to weight coefficients. It is namely that the quantitative value is the remains of certain Membership degree subtract the uncertain portion of Non-membership degree, so the conversion function is:

$$\overline{w}(q) = \mu(q) - v(q) * \pi(q) = \mu(q) - (1 - \mu(q) - \pi(q)) * \pi(q)) \tag{1}$$

In addition, it is important to make sure that the total weight value equals 1 when specifying the importance degree of these $n$ QoS attributes. Therefore, a normalization function (2) is used to normalize the quantitative subjective weight preference to $[0, 1]$.

$$w_j^s = \overline{w}_j / \sum_{j=1}^{n} \overline{w}_j \tag{2}$$

For example, a potential cloud user marked a QoS attribute as important and the uncertainty degree is 0.1.So it can be expressed as $\pi = 0.1$, quantitative weight is: $w = 0.7 - (0.3 - 0.1) * 0.1 = 0.68$.

## 3.2    Objective Weight Preference Based on Rough Set

The user objective weight preference indicates a default preference when a user first time to use a service or he has no idea about the preference. Here we collect the user incomplete history preference information of same service request to get the objective weight preference by the QoS attribute significance of rough set.

Rough set theory [13] is convinced as an effective method that can be employed to analyze uncertain (including vague and incomplete) systems. Although it is less common than more traditional methods of probability, statistics and entropy, the key difference and unique strength of using classical rough set theory is that it provides an objective form of analysis. The rough set analysis requires no additional information, functions, grades, external parameters or subjective interpretations to determine set membership; it only uses the information presented within the given data. In this section, we will employ the attribute importance in rough set theory to process the incomplete user QoS preference.

**Definition 2.** Attribute Importance: Let $X \in U$ be a subset of attributes, $x_j \in U$ is an attribute, considering the importance of $x_j$ to $X$, namely that the definition enhancement after adding an attribute $x_j$ in set $X$, so the greater the enhancement, the more important that $x_j$ to $X$.

$$Sig_{x_i}(x_i) = 1 - \frac{|pox_{X-i}(U)|}{|pox_X(U)|} \tag{3}$$

Consider $P$ and $Q$ are the equivalence relation for set $U$, let $POS_p(Q)$ be the $P$ positive domain of $Q$, $POS_p(Q) = U_{X \in U/Q} PX$, the $P$ positive domain is a collection of all objects in set $U$ that can be exactly classified into $Q$ equivalence class relation according to the $\frac{U}{P}$ information. So the user objective QoS weight preference as follows:

$$w_j^o = \frac{Sig_{x_i}(x_i)}{\sum_{i=1}^n Sig_{x_i}(x_i)} \tag{4}$$

Obviously, it meets the basic requirements of weight as $0 \leq w_j^o \leq 1$ and $\sum_{j=1}^n w_j = 1$, therefore the objective and subjective weight preference of each user is:

$$w_j = \alpha w_j^s + (1 - \alpha)w_j^o \quad (0 \leq \alpha \leq 1, j = 1, 2, \dots n) \tag{5}$$

The $\alpha$ parameter indicates that importance of user subjective weight preference. The larger $\alpha$ is and the more obvious impact of the subjective weight on the integrated weight, usually the value of $\alpha$ is 0.8. According to the user preference and standardized QoS data, next we will introduce our approach to solve the service selection problem via the multi-attributes decision making method.

## 3.3    Our Proposed Approach

In this section, we present the details of our cloud service selection approach. Assume that a potential cloud user submits its request to the cloud service repository for finding the most suitable cloud service. After the preliminary selection according to the functional requirements and specific QoS value constraints, suppose that there are $m$ clouds left as the alternative clouds denoted by $\{C_i\}$, where $i = 1, \dots, m$. The final score of each alternative cloud is ranked by the close degree to the ideal positive solution according to the integrated user weight preferences. That namely multi-attributes decision making method-TOPSIS, it selects alternatives having the shortest distance from the positive ideal solution and the farthest distance from the negative ideal solution [14]. Here the positive ideal solution is a combination which composed by the best solution of properties in the candidate clouds; the negative ideal solution is inverse.

The detailed procedure of our approach is shown below:

**Step1: Preprocessing QoS attributes of candidate services.**

Firstly, choose the candidate cloud services which can meet all the functional and quantitative QoS value requirements to determine the initial QoS matrix . Assume that there are $m$ clouds left as the alternative clouds and each service has $n$ attributes. Then do preliminarily processing for the native QoS data based on its uncertainty and incompleteness. At last we get the normalization QoS matrix $P$, for details to see paper [15].

$$P = \begin{pmatrix} q_{11}' & \cdots & q_{1n}' \\ \vdots & \ddots & \vdots \\ q_{m1}' & \cdots & q_{mn}' \end{pmatrix}$$

**Step2: Constructing user preference weighted decision matrix.**

After we have got the normalization decision matrix $P$, then we should aggregate both user uncertain subjective and objective weight preferences to construct the weighted decision matrix.

- *Subjective QoS weight preference processing.*

As we have introduced before in our model, all the subjective attributes are represented in the form of linguistic variables. Here, we use the mapping illustrated in Table 1, which is frequently employed in research of service discovery and selection [16]. Each fuzzy number in Table 1 represents the fuzzy rating corresponding to the linguistic variable, suppose that the weight preference vector of cloud user for $n$ QoS attributes is $L(\pi)$.

$$L(\pi) = (L_1(\pi_1), L_2(\pi_2), \cdots, L_n(\pi_n)) \tag{6}$$

And $L_j(\pi_j)$ is the importance and uncertain degree of attribute $j$ for user, then we convert it to weight coefficient by the conversion function $\overline{w_j} = \mu_j - v_j * \pi_j$ ($j = 1, \dots, n$). Next we continue normalize the weighting coefficient to [0, 1] to get the final subjective weight preference:

$$w_j^s = \left( \frac{\overline{w_1}}{\sum_{j=1}^n \overline{w_j}}, \frac{\overline{w_2}}{\sum_{j=1}^n \overline{w_j}}, \cdots, \frac{\overline{w_n}}{\sum_{j=1}^n \overline{w_j}} \right) \tag{7}$$

- *Objective QoS weight preference processing.*

As we have introduced in 3.2, the history values often in the form of missing or incomplete but can be well processed by rough set. Equation (4) is for this situation when user can only provide limited subjective preferences, so the objective weight preference is $w_j^o$. Therefore, the integrated weight preference based on Equation (5) is:

$$W = \{w_1, w_2, \dots, w_n\} \tag{8}$$

The user preference weighted decision matrix $Y$ is:

$$Y = W^T * P = \begin{pmatrix} y_{11} & y_{12} & \cdots & y_{1n} \\ y_{21} & y_{22} & \cdots & y_{2n} \\ \cdots & \cdots & \cdots & \cdots \\ y_{m1} & y_{m2} & \cdots & y_{mn} \end{pmatrix} \tag{9}$$

**Step3: Choosing the positive and negative ideal cloud service.**
The purpose of service selection is to best meet the requirements of users. Thus the positive ideal solution is an ideal service that composed by the maximum value of properties $y_{ij}$ in the weighted decision matrix $Y$, on the contrary, the negative ideal solution is a service with the minimum value for properties $y_{ij}$ in weighted decision matrix $Y$. Assume they are respectively denoted by $A^+$ and $A^-$ as follows:

$$a_j^+ = \max(y_{ij}), a_j^- = \min(y_{ij}) \; (i = 1, \cdots, m \; and \; j = 1, ..., n)$$
$$A^+ = (a_1^+, a_2^+, \cdots, a_n^+) \quad A^- = (a_1^-, a_2^-, \cdots, a_n^-)$$

**Step4: Computing the distance and nearness degree between candidate services and ideal services.**
For each decision-maker $y_{ij}$ of a candidate cloud service $\{C_i\}$, the N-Dimensional Euclidean distance is employed to be the distance scale. Thus the distances of service between positive ideal solution are denoted by $d_i^+$ and the distances of service between negative ideal solution are denoted by $d_i^-$.

$$d_i^+ = \sqrt{\sum_{j=1}^n (y_{ij} - a_j^+)^2} \qquad d_i^- = \sqrt{\sum_{j=1}^n (y_{ij} - a_j^-)^2} \tag{10}$$

Then to calculate the relative nearness degree of candidate services to ideal solutions which is denoted by $c_i$.

$$c_i = \frac{d_i^-}{d_i^+ + d_i^-}, i = 1, ..., m \tag{11}$$

In the above equation (11), the closer of the value of nearness degree $c_i$ to 1, the more indicating that the candidate service is closer to the positive ideal solution, namely that it is more matching the preferences of user. Meanwhile, if the closer of the value of nearness degree $c_i$ to 0, the more indicating that the candidate service is closer to the negative ideal solution, namely that it is more off the preferences of user. If the value of nearness degree $c_i$ is 1, the candidate service actually is the ideal positive service. Finally, according to the final nearness degree values, all the alternative cloud services are ranked for the weight preference-aware selection for the potential cloud user.

# 4    Experimental Analysis

To evaluate the effectiveness of our modeling approach for handling uncertain user preferences and subsequent selection process, we apply it to a use case that aims select among cloud storage services, then we conduct a comparison with the MIP method used in [17] on the success ratio. All the experiments are conducted on the same computer with Intel(R) 2.66GHz processor, 2.0GB of RAM, Windows8.1.Both approaches are run independently for 100 times and all results are reported on average.

## 4.1   Case Study

Cloud storage services allow to easily upload data to Cloud datacenters, for instance for backup purposes or to share and access data on diverse devices. This use case is relevant for technical decision-makers in search for decision support to select Cloud storage services. Assume that after preliminary selection, a small business company which has a subscription plan among these seven cloud storage services: Dropbox, OneDrive, Box.com, iCloud, GoogleDrive, UbuntuOne and Wuala, which is denoted by $C_i$. The plan should consider the following six QoS attributes, including cost, reliability, reputation, safety, has free trial and response time. All these parameters of the QoS attributes are collected from the data which published online by service provider. The Table 2 shows the native QoS values of the cloud service candidates.

**Table 2.**   QoS Values of Service Candidates

| $C_i$ | Cost | Reliability (%) | Reputation | Safety | Has free trial | Response Time(ms) |
|-------|------|-----------------|------------|--------|----------------|-------------------|
| $C_1$ | 87 | 89 | 5 | Good | True | [220,260] |
| $C_2$ | 84 | 98 | 5 | Good | True | [180,230] |
| $C_3$ | 46 | 58 | 3 | Medium | True | [420,460] |
| $C_4$ | 85 | 85 | 4 | Good | True | [200,280] |
| $C_5$ | 65 | 75 | 4 | Good | True | [230,280] |
| $C_6$ | 44 | 55 | - | Medium | True | [320,400] |
| $C_7$ | 18 | 20 | 2 | Medium | False | [530,600] |

As the step 1 says, we firstly preprocess the QoS attributes data to obtain quantified and normalized QoS value matrix $P$. Next, in the subscription plan for the cloud storage service, the related QoS subjective weight preferences are illustrated in Table 3.

**Table 3.**   QoS Subjective Weight Preferences

| QoS Criteria | Weight Preferences | $\overline{w_s}$ | $W_s$ |
|--------------|--------------------|------------------|-------|
| Cost | M(0.2) | 0.44 | 0.1410 |
| Reliability | VI(0.0) | 0.9 | 0.2885 |
| Reputation | U(0.2) | 0.2 | 0.0641 |
| Safety | VI(0.0) | 0.9 | 0.2885 |
| Has free trial | - | - | - |
| Response Time | I(0.1) | 0.68 | 0.2179 |

In the Table 3, the company has marked the importance of "Cost" QoS attribute as "Medium" and the uncertain degree is 0.2,so this weight preference is expressed as

M(0.2); the same with the "Reliability" QoS attribute. The expression VI (0.0) means "very important" and the uncertainty is 0.According to the Equation (1) & (2), finally each QoS of subjective weight preference is listed in $\textbf{Ws}$ column.

As for this case study, we have launched a survey about what's preferred when difference background users consider using cloud storage service. We have collected abundant user historical preferences to generate the objective weight value $W_o$. By Equation (3) & (4), the objective weight preference of user is {0.0834, 0.3333, 0.0833, 0.3333, 0.0417, 0.125}.

Next we design an experiments to validate the effectiveness of our approach. If the potential cloud user totally has no ideal about the service requirements for limited technical knowledge, so the objective weight preference is recommended. On the other hand, if the company has a full understanding of his preferences, so he may adopt his subjective weight only. In other words, both scenarios are about uncertain user preferences which can be solved by assign corresponding values to the coordination parameter $\alpha$.Thus we consider three situations where $\alpha$ is 0, 1 and 0.8 to do evaluation.

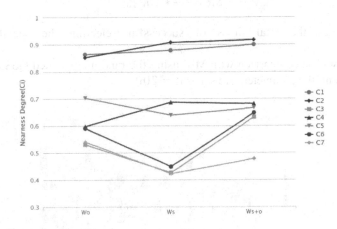

**Fig. 1.** Ws indicate the subjective weight preference; $W_{s+o}$ indicate an integrated weight preference and $W_o$ express the objective weight preference only.$c_i$ is the nearness value of each alternative cloud.

From Fig 1, after sorting by closeness degree value we can see that the cloud service C2 which ranked first in $Ws$ and $Ws+o$ scenarios, so it is the most suitable choice for this company's plan. This figure also shows that our approach can make a difference when dealing with the uncertainty of human languages user preferences. Also by considering objective preference assessments from historical cloud consumers, our approach takes into account some vital qualitative performance preferences which are usually ignored by users or providers or other approach in the selection process of a cloud service. Based on the analysis and experiment results in the above case study, our proposed approach is more effective to hit the right cloud service.

## 4.2    Comparison On Success Ratio

Given that there are no standard or authoritative cloud storage service selection test data sets available for researchers, as web service also is a kind of cloud component which can be integrated into cloud storage application for accessing information .So we adopt the **QWS Dataset**[1] by reclassified to derive our test data. We investigate the success ratio of our approach and MIP using the processed QWS Dataset which basically contains 120 services in real world.

The Success Ratio is how often the ratio of selecting the identical optimal service with users' QoS preference requirements during a service composition period. As In service composition system, an important goal of service selection is to select reliable services for service users. However, due to the uncertainty of QoS preferences, the selected service often deviates from the user expectations, which may lead to service composition failure in practical application. So in this comparison, we evaluate the success ratio of the two approaches. The Success Ratio (SR) can be calculated by

$$SR = \frac{srn}{n} * 100\% \tag{12}$$

$srn$ denotes the total times of successful selecting the specific service ing $n$ times.

Fig.2 shows the comparison with MIP using the random extended QoS datasets. In the experiment, the parameters are set as $n = 200$.

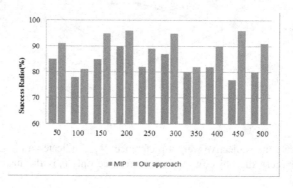

**Fig. 2.** Comparison on Success Ratio

According to the definition of SR, for a service selection approach, the higher its success ratio is, the better its performance. From Fig.2, with different number of service candidates, the success ratio of our approach is all higher than that of MIP. The overall success ratio of our approach is 90.95% on average, while that of MIP is only 82.08%.These experimental results indicate that our approach effectively reduces the influence of user provided uncertain QoS preference on the quality of composition service and greatly improves the reliability of service selection.

---

[1]  http://www.uoguelph.ca/~qmahmoud/qws/index.html

By using rough set and fuzzy set to process QoS preferences transactions, our approach effectively defense from services with large variances on its QoS, and reduce the difference between selected services and actual execution results. Hence, the reliability of service selection can be greatly improved.

## 5   Conclusion

In order to tackle cloud service selection challenges, we present a novel approach to describe and compute the uncertainty of cloud user QoS preference. Our approach applies the intuitionistic fuzzy sets to quantify the vague user subjective preference and the attribute significance of rough set to get the objective weight preference. Then it uses the multi-attribute decision theory to identify the most suitable cloud services. Finally, a case study demonstrates the advantages of our proposed model in cloud service selection. Furthermore, experimental results show that our model can help to improve reliability of service selection. Definitely this lets cloud users express their needs in linguistics terms which also brings a great comfort to them compared to systems that force users to assign exact weight for all preferences.

In our future work, we will continue to improve our model by considering the user hybrid QoS value and weight preferences in service selection. In addition, we plan to study the effect of different system factors to our model (e.g., de-fuzzification methods, sensitivity of member functions and mappings from linguistic variables to fuzzy numbers). Our goal is to help real-world service users find appropriate services according to their QoS requirements in the near future.

**Acknowledgments.** The work is financially supported by the 973 National Project "Research on the information service model and mechanism" (NO. 2010CB328106). We would like to show thanks to the sponsor.

## References

1. Armbrust, M., Fox, A., Griffith, R., et al.: A view of cloud computing. Communications of the ACM 53(4), 50–58 (2010)
2. Tsai, W.T., Sun, X., Balasooriya, J.: Service-oriented cloud computing architecture. In: 7th IEEE International Conference on Information Technology: New Generations (ITNG), pp. 684–689. IEEE Press (2010)
3. Wang, S., Zheng, Z., Sun, Q., Zou, H., Yang, F.: Cloud model for service selection. In: IEEE Conference on Computer Communications Workshops (INFOCOM WKSHPS), pp. 666–671. IEEE Press (2011)
4. Wang, H.B., Xu, J.X., Li, P.C., Patrick, H.: Incomplete preference-driven web Service selection. In: The Proceedings of IEEE International Conference on Service Computing (SCC 2008), vol. 1, pp. 75–82. IEEE Press (2008)
5. Hu, J.Q., Li, J.Z., Liao, G.P.: A Multi-Qos based Local Optimal Model of Service Selection. Chinese Journal of Computers 33(3), 526–534 (2010) (in Chinese)
6. Lo, C.C., Chen, D.Y., Tsai, C.F., Chao, K.M.: Service selection based on fuzzy TOPSIS method. In: 24th IEEE International Conference on Advanced Information Networking and Applications Workshops (WAINA), pp. 367–372. IEEE Press (2010)

7. Wanchun, D., Chao, L., Xuyun, Z., Chen, J.: A QoS-aware service evaluation method for co-selecting a shared service. In: IEEE International Conference on Web Services (ICWS), pp. 145–152. IEEE Press (2011)
8. Godse, M., Mulik, S.: An approach for selecting software-as-a-service (SaaS) product. In: IEEE International Conference on Cloud Computing (CLOUD 2009), pp. 155–158. IEEE Press (2009)
9. Fan, Z., Zhang, L., Shen, J., Wang, S.: A User's Preference Based Method for Web Service Selection. In: 2th IEEE International Conference onComputer Research and Development, pp. 39–45. IEEE Press (2010)
10. Cao, H., Feng, X., Sun, Y., et al.: A service selection model with multiple QoS constraints on the MMKP. In: IFIP International Conference onNetwork and Parallel Computing Workshops (NPC), pp. 584–589. IEEE Press (2007)
11. Atanassov, K.T.: Intuitionistic fuzzy sets. Fuzzy sets and Systems 20(1), 87–96 (1986)
12. Wang, P.: QoS-aware web services selection with intuitionistic fuzzy set under consumer's vague perception. Expert Systems with Applications 36(3), 4460–4466 (2009)
13. Pawlak, Z., Grzymala-Busse, J., Slowinski, R., Ziarko, W.: Rough sets. Communications of the ACM 38(11), 88–95 (1995)
14. Mahdavi, I., Mahdavi-Amiri, N., Heidarzade, A., Nourifar, R.: Designing a model of fuzzy TOPSIS in multiple criteria decision making. Applied Mathematics and Computation 206(2), 607–617 (2008)
15. Yau, S.S., Yin, Y.: Qos-based service ranking and selection for service-based systems. In: IEEE International Conference on Services Computing (SCC), pp. 56–63 (2011)
16. Torres, R., Astudillo, H., Salas, R.: Self-adaptive fuzzy QoS-driven web service discovery. In: IEEE International Conference on Services Computing (SCC), pp. 64–71 (2011)
17. Alrifai, M., Skoutas, D., Risse, T.: Selecting skyline services for QoS-based web service composition. In: Proceedings of the 19th International Conference on World Wide Web, pp. 11–20. ACM (2010)
18. Al-Masri, E., Mahmoud, Q.H.: Discovering the best web service. In: 16th International Conference on World Wide Web (WWW), pp. 1257–1258 (2007)

# Attribute Reduction in Decision-Theoretic Rough Set Model Using MapReduce

Jin Qian[1,2], Ping Lv[1,2], Qingjun Guo[1,2], and Xiaodong Yue[3]

[1] Key Laboratory of Cloud Computing & Intelligent Information Processing of
Changzhou City, Jiangsu University of Technology, 213001, P.R. China
[2] School of Computer Engineering, Jiangsu University of Technology,
Changzhou 213001, P.R. China
[3] School of Computer Engineering and Science, Shanghai University, Shanghai,
200444, P.R. China

**Abstract.** Attribute reduction is one of the most important research issues in decision-theoretic rough set model. This paper studies a new attribute measure preserving boundary region partition for a reduct. The relationships among the positive region, the probabilistic positive region and the indiscernibility object pairs for an equivalence class are analyzed. A heuristic attribute reduction algorithm framework using MapReduce in decision-theoretic rough set model is proposed. This study gives some insights into how to conduct attribute reduction in decision-theoretic rough set for big data.

## 1 Introduction

Rough set theory [8], proposed by Pawlak, is a useful mathematical framework to deal with uncertainty problems. It has been applied in many fields such as data analysis, data mining and machine learning. Using the lower approximations in Pawlak rough set model(PRS), one can derive the decision rules with certainty from the positive region. In practice, if considering little tolerance of errors, we may mine some latent useful knowledges from the boundary region. In recent years, some researchers begin to consider this issue and have extended PRS model into probabilistic approaches [13–15, 18]. All these proposals expand the positive and negative regions by introducing certain acceptable level of errors into the standard model, which promote the development and applications of rough set theory.

Attribute reduction is an important problem of rough set theory [8, 10, 12]. Prof. Yao proposed decision-theoretic rough set model(DTRS) and gave the corresponding attribute reduction algorithm in [15–17]. As the probabilistic negative region may not be the empty set, we cannot obtain the monotonicity of the probabilistic regions with respect to set inclusion of attributes [1, 5–7, 16, 17]. These monotonic attribute measures for positive region preservation in PRS model are inappropriate for DTRS model. Moreover, most existing algorithms can not deal with the massive data. Thus, a heuristic attribute reduction algorithm for DTRS model has not been considerably investigated so far for big data. In such

D. Miao et al. (Eds.): RSKT 2014, LNAI 8818, pp. 601–612, 2014.
DOI: 10.1007/978-3-319-11740-9_55 © Springer International Publishing Switzerland 2014

case, developing an efficient and effective approach to attribute reduction is very desirable.

In this paper, we investigate a monotonic attribute measure for the boundary region partition preservation in PRS model. We discuss the relationships among the positive region, the probabilistic positive region and the indiscernibility object pairs. Based on those results, we further propose an attribute reduction algorithm framework in DTRS model using MapReduce for big data.

## 2   Preliminaries

In this section, we will review attribute reduction in Pawlak rough set model and decision-theoretic rough set model. For a detailed description, please refer to papers [8, 10, 17].

### 2.1   Attribute Reduction in Pawlak Rough Set Model

In Pawlak rough set model, for a subset of attributes $A$, Pawlak defines that two objects in U are A-indiscernible if and only if they have the same values on all attributes in A. That is, $IND(A) = \{(x, y) \in U \times U | \forall a \in A, a(x) = a(y)\}$. The indiscernibility relation $IND(A)$ determines a partition of U, denoted by U/A or $\pi_A$. The equivalence class of U/A containing x is given by $[x]_A = \{y \in U | (x, y) \in IND(A)\}$. Let $\pi_D = \{D_1, D_2, ..., D_k\}$ be a partition of the universe U induced from the decision attribute D.

A reduct $A \subseteq C$ is called a positive region preserving reduct or P-reduct for short if and only if it requires that the positive region with respect to the decision attribute D is unchanged.

**Definition 1.** *For a decision table S, an attribute set $A \subseteq C$ is a Pawlak reduct of C with respect to D if it satisfies the following conditions:*
*(1) $POS(\pi_D | \pi_A) = POS(\pi_D | \pi_C)$;*
*(2) for any attribute $a \in A$, $POS(\pi_D | \pi_{A-\{a\}}) \neq POS(\pi_D | \pi_A)$.*

Based on this definition, one can derive the fact that the Pawlak positive region is monotonic, and the size of the positive region induced from the attribute set C is largest. A reduct A produces the same size of a positive region as what C does, and any subset of A cannot produce a larger positive region than A does [17]. Thus, we only check all the proper subsets A - {a} for $a \in A$.

A P-reduct also can be seen as a reduct preserving the quality of classification. The quality $\gamma(\pi_D | \pi_A)$ is defined as follows:

$$\gamma(\pi_D | \pi_A) = \frac{|POS(\pi_D | \pi_A)|}{|U|}. \tag{1}$$

Since the size of the positive region is monotonic with respect to set inclusion of attributes, one can derive the monotonicity of the $\gamma$ measure. Many researchers [9, 10, 12] use the $\gamma$ measure to construct some attribute reduction algorithms.

In general, any monotonic measure $f$ can be used to define a P-reduct if it satisfies the following condition

$$(f(\pi_D|\pi_A) = f(\pi_D|\pi_C)) \Longleftrightarrow (POS(\pi_D|\pi_A) = POS(\pi_D|\pi_C). \tag{2}$$

We consider a reduct preserving the boundary region. A subset of attributes $A \subseteq C$ is called a boundary region preserving reduct or a B-reduct for short if and only if it satisfies (1) $BND(\pi_D|\pi_A) = BND(\pi_D|\pi_C)$ and (2) for any attribute $a \in A$, $BND(\pi_D|\pi_A) \neq BND(\pi_D|\pi_{A-\{a\}})$. For a P-reduct $A \subseteq C$, since $NEG(D|A)$ is empty, we can have $POS(\pi_D|\pi_A) \cap BND(\pi_D|\pi_A) = \phi$ and $POS(\pi_D|\pi_A) \cup BND(\pi_D|\pi_A) = U$. Thus, the condition $POS(\pi_D|\pi_A) = POS(\pi_D|\pi_C)$ means that $BND(\pi_D|\pi_A) = BND(\pi_D|\pi_C)$ holds. The requirement of the same boundary region is implied in Definition 1. We have the following relationship between P-reduct and B-reduct: there exists a B-reduct B for a P-reduct A such that $A \subseteq B$ and there exists a P-reduct A for a B-reduct B such that $A \subseteq B$. Therefore, only considering the positive region for Pawlak rough set model is sufficient in the attribute reduction process.

## 2.2  Attribute Reduction in Decision-Theoretic Rough Set Model

Decision-theoretic rough set model introduces two tolerance threshold parameters $\alpha$ and $\beta(0 \leq \beta < \alpha \leq 1)$ to overcome the weakness in Pawlak rough set model, which can generate the probabilistic positive, boundary and negative regions based on the notion of expected lost from Bayesian decision theory. Each equivalence class $D_i \in \pi_D$ is called a decision class. The precision of an equivalence class $[x]_A \in \pi_A$ for predicting a decision class $D_i \in \pi_D$ is defined as:

$$p(D_i|[x]_A) = \frac{|[x]_A \cap D_i|}{|[x]_A|}. \tag{3}$$

where $|.|$ denotes the cardinality of a set.

**Definition 2.** *[17] For a decision table S, the three probabilistic regions can be defined as follows:*

$$POS_{(\alpha,\beta)}(\pi_D|\pi_A) = \{x \in U|p(D_{max}([x]_A)|[x]_A) \geq \alpha\}$$
$$BND_{(\alpha,\beta)}(\pi_D|\pi_A) = \{x \in U|\beta < p(D_{max}([x]_A)|[x]_A) < \alpha\} \tag{4}$$
$$NEG_{(\alpha,\beta)}(\pi_D|\pi_A) = \{x \in U|p(D_{max}([x]_A)|[x]_A) \leq \beta\}$$

*where $D_{max}([x]_A) \in \pi_D$ is a major decision class of the objects in $[x]_A$, i.e., $D_{max}([x]_A) = argmax_{D_i \in \pi_D}(\frac{|[x]_A \cap D_i|}{|[x]_A|})$ .*

Parallel to Pawlak's definition, a natural extension of attribute reduction for probabilistic positive region preserving reduct, or $P_{\{\alpha,\beta\}}$-reduct for short, can be defined as follows.

**Definition 3.** *For a decision table S, an attribute set $A \subseteq C$ is a probabilistic positive region-preserved reduct of C with respect to D if it satisfies the following conditions:*

*(1) $POS_{(\alpha,\beta)}(\pi_D|\pi_A) = POS_{(\alpha,\beta)}(\pi_D|\pi_C)$;*

*(2) for any $A' \subset A$, $POS_{(\alpha,\beta)}(\pi_D|\pi_{A'}) \neq POS_{(\alpha,\beta)}(\pi_D|\pi_A)$.*

In decision-theoretic rough set model, for a reduct $A \subseteq C$, as $NEG_{(\alpha,\beta)}(\pi_D|\pi_A)$ may be not empty set, we cannot have $POS_{(\alpha,\beta)}(\pi_D|\pi_A) \cup BND_{(\alpha,\beta)}(\pi_D|\pi_A) = U$. A reduct only for any probabilistic region preservation may change the other two probabilistic regions. In addition, we cannot obtain the monotonicity of the probabilistic positive region with respect to set inclusion of attributes. Therefore, we must check all the subset of a candidate reduct in condition 2 of Definition 3.

In decision-theoretic rough set model, the quality of a probabilistic classification $\gamma_{(\alpha,\beta)}$ is denoted as follows.

$$\gamma_{(\alpha,\beta)}(\pi_D|\pi_A) = \frac{|POS_{(\alpha,\beta)}(\pi_D|\pi_A)|}{|U|}. \tag{5}$$

As indicated in [16,17], since the probabilistic positive region is non-monotonic, $\gamma_{(\alpha,\beta)}$ is also non-monotonic. In other words, for $B \subset A$, we cannot obtain $\gamma_{(\alpha,\beta)}(\pi_D|\pi_B) < \gamma_{(\alpha,\beta)}(\pi_D|\pi_A)$. Note that the $\gamma_{(\alpha,\beta)}$ measure only reflects the size and cannot reveal the objects of the probabilistic positive region. If $\gamma_{(\alpha,\beta)}(\pi_D|\pi_A) = \gamma_{(\alpha,\beta)}(\pi_D|\pi_C)$, we can have $|POS_{(\alpha,\beta)}(\pi_D|\pi_A)| = |POS_{(\alpha,\beta)}(\pi_D|\pi_C)|$, but not guarantee that $POS_{(\alpha,\beta)}(\pi_D|\pi_A) = POS_{(\alpha,\beta)}(\pi_D|\pi_C)$ holds. On the other hand, $\gamma_{(\alpha,\beta)}$ also ignores the boundary region and negative region.

**Example 1:** Consider a simple decision table S shown in Table 1. Suppose $\alpha$=0.75, $\beta$=0.60, we can obtain the following probabilistic positive regions from some attribute sets.

$POS_{(\alpha,\beta)}(\pi_D|\pi_{\{c_1\}}) = \phi$

$POS_{(\alpha,\beta)}(\pi_D|\pi_{\{c_2\}}) = \phi$

$POS_{(\alpha,\beta)}(\pi_D|\pi_{\{c_5\}}) = \{o_3, o_4, o_5, o_9\}$

$POS_{(\alpha,\beta)}(\pi_D|\pi_{\{c_1,c_2\}}) = \{o_3\}$

$POS_{(\alpha,\beta)}(\pi_D|\pi_{\{c_1,c_5\}}) = \{o_4, o_7\}$

$POS_{(\alpha,\beta)}(\pi_D|\pi_{\{c_2,c_5\}}) = \{o_1, o_2, o_3, o_4, o_6, o_7, o_8\}$

$POS_{(\alpha,\beta)}(\pi_D|\pi_{\{c_1,c_2,c_5\}}) = \{o_1, o_3, o_4, o_7\}$

$POS_{(\alpha,\beta)}(\pi_D|\pi_C) = \{o_1, o_3, o_4, o_7\}$

Thus, we can obtain the probabilistic positive regions are non-monotonic with respect to set inclusion of attributes. Fig. 1 illustrates the relationship among the probabilistic positive regions and the set inclusion of attributes.

## 3   Attribute Reduction for the Boundary Region Partition Preservation

In decision-theoretic rough set model, the probabilistic positive regions contain all the consistent objects from the positive region and some objects from the boundary region in Pawlak rough set model. From Example 1, we can check that the

**Table 1.** An information table

| U | $c_1$ | $c_2$ | $c_3$ | $c_4$ | $c_5$ | $c_6$ | D |
|---|---|---|---|---|---|---|---|
| $o_1$ | 1 | 1 | 1 | 1 | 1 | 1 | 1 |
| $o_2$ | 1 | 0 | 1 | 0 | 1 | 1 | 1 |
| $o_3$ | 0 | 1 | 1 | 1 | 0 | 0 | 2 |
| $o_4$ | 1 | 1 | 1 | 0 | 0 | 1 | 2 |
| $o_5$ | 0 | 0 | 1 | 1 | 0 | 1 | 2 |
| $o_6$ | 1 | 0 | 1 | 0 | 1 | 1 | 3 |
| $o_7$ | 0 | 0 | 0 | 1 | 1 | 0 | 3 |
| $o_8$ | 1 | 0 | 1 | 0 | 1 | 1 | 3 |
| $o_9$ | 0 | 0 | 1 | 1 | 0 | 1 | 3 |

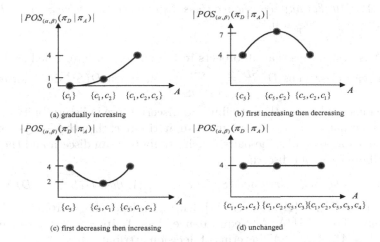

(a) gradually increasing   (b) first increasing then decreasing

(c) first decreasing then increasing   (d) unchanged

**Fig. 1.** The relationship between the size of probabilistic positive regions and the set inclusion of attributes

probabilistic positive region of $\{c_2, c_5\}, \{o_2, o_6, o_7, o_8\}$, is comprised of the positive region $\{o_7\}$ and the boundary region $\{o_2, o_6, o_8\}$ in PRS model. Moreover, the boundary region partitions of $\{c_1, c_2, c_5\}$ is the same as that of the whole attribute set $C$. Thus the boundary region partition plays a key role in the attribute reduction process for decision-theoretic rough set model. In what follows, we mainly consider how to construct the boundary region partition preservation.

### 3.1 Attribute Reduction for Boundary Region Partition Preservation in PRS

In PRS model, we can evaluate the size of indiscernibility object pairs from the boundary regions to construct a monotonic attribute measure.

**Definition 4.** *[10] Given a decision table S, let $A \subseteq C$ and $\pi_A = \{A_1, A_2, \ldots, A_r\}$, we get the number of indiscernibility pairs of objects that conditional attributes A can not discern.*

$$\widetilde{DIS_A^D} = \sum_{1 \le p \le r} \sum_{1 \le i < j \le k} n_p^i n_p^j \tag{6}$$

where $n_p^i (n_p^j)$ denotes the number of objects whose combinational value of conditional attributes is p and decision attribute value on d is i(j).

By Definition 4, one can check that the indiscernibility object pairs may be generated from the boundary regions with respect to the conditional attributes A. Thus, we can have the following property for the equivalence class in the positive region.

**Theorem 1.** *[10] For any equivalence class $A_p$ in a decision table S, $A_p \subseteq POS$ ($\pi_D | \pi_A$) iif $\sum\limits_{1 \le i < j \le k} n_p^i n_p^j = 0$.*

When A is $\phi$, it means that all objects in the universe are regarded as a special equivalence class, thus $\widetilde{DIS_\emptyset^D} = \sum\limits_{1 \le i < j \le k} n^i n^j$. When $\widetilde{DIS_A^D}$ does not equal $\widetilde{DIS_C^D}$, we must add some other attribute to discern those indiscernibility object pairs. Assume that the chosen attribute is an attribute c, then those object pairs remained are composed of some pairs which attribute c can discern and the rest which attribute c can not discern.

**Theorem 2.** *Given a decision table S, for $c \in C - A$, then $\widetilde{DIS_{A \cup c}^D} \le \widetilde{DIS_A^D}$.*

**Proof:** Suppose U/A = $\{A_1, A_2, \ldots, A_r\}$, the equivalence classes from $U/\{A \cup c\}$ are finer than those of U/A. Any equivalence class $A_p$ (p = 1, 2, ..., r) can be sub-divided as $A_p^1, A_p^2, \ldots, A_p^k$ in terms of decision attribute D.

For an equivalence class $A_p$, the number of the indiscernibility object pair $\widetilde{DIS_A^D}$ equals $\sum\limits_{1 \le i < j \le k} n_p^i n_p^j$. After adding attribute c into A, the number of the indiscernibility object pair $\widetilde{DIS_{A \cup c}^D}$ is the sum of $\sum\limits_{1 \le i < j \le k} n_{p,1}^i n_{p,1}^j$, $\sum\limits_{1 \le i < j \le k} n_{p,2}^i n_{p,2}^j$, ..., $\sum\limits_{1 \le i < j \le k} n_{p,m}^i n_{p,m}^j$, namely $\sum\limits_{1 \le i < j \le k} \sum\limits_{1 \le l \le m} n_{p,l}^i n_{p,l}^j$.

For any equivalence class $A_p$, suppose any two decision values i and j, we can have

$$n_{p,1}^i n_{p,1}^j + n_{p,2}^i n_{p,2}^j + \ldots + n_{p,m}^i n_{p,m}^j$$
$$\le (n_{p,1}^i + n_{p,2}^i + \ldots + n_{p,m}^i)(n_{p,1}^j + n_{p,2}^j + \ldots + n_{p,m}^j)$$
$$= n_p^i n_p^j.$$

Therefore, $\widetilde{DIS_{A \cup c}^D} \le \widetilde{DIS_A^D}$ holds. $\square$

**Theorem 3.** *Given a decision table S, $A \subseteq C$, if $\widetilde{DIS_A^D} = \widetilde{DIS_C^D}$, then BND ($\pi_D | \pi_A$) = $BND(\pi_D | \pi_C)$.*

**Proof:** According to Theorem 2, if $\widetilde{DIS_A^D} = \widetilde{DIS_C^D}$, we can have the equation $\pi_A = \pi_C$, thereby $BND(\pi_D | \pi_A) = BND(\pi_D | \pi_C)$ holds. $\square$

Theorem 2 and 3 indicates that this measure for evaluating the size of indiscernibility object pairs is monotonic and keeps the boundary region partition preservation unchanged.

## 3.2 Attribute Reduction for Boundary Region Partition Preservation in DTRS

Since the indiscernibility relation can keep the boundary region partition preservation in PRS, we employ the boundary region to discuss the probabilistic positive regions in DTRS.

**Theorem 4.** *Given a decision table $S$, $A \subseteq C$, if $BND(\pi_D|\pi_A) = BND(\pi_D|\pi_C)$, then $POS_{(\alpha,\beta)}(\pi_D|\pi_A) = POS_{(\alpha,\beta)}(\pi_D|\pi_C)$.*

**Proof:** Since $BND(\pi_D|\pi_A) = BND(\pi_D|\pi_C)$, $POS(\pi_D|\pi_A) = POS(\pi_D|\pi_C)$, thus we can conclude $POS_{(\alpha,\beta)}(\pi_D|\pi_A) = POS_{(\alpha,\beta)}(\pi_D|\pi_C)$.$\square$

As discussed above, the probabilistic positive regions may contain some objects from the boundary region, thus the number of the indiscernibility object pairs in a probabilistic positive region does not equal 0.

**Theorem 5.** *For any equivalence class $A_p$ in a decision table $S$, $A_p \subseteq POS_{(\alpha,\beta)}$ $(\pi_D|\pi_A)$ iif $\sum\limits_{1 \leq i < j \leq k} n_p^i n_p^j \leq \frac{1}{2}(1-\alpha)(1+\alpha-\frac{1}{|A_p|})|A_p|^2$.*

**Proof:** Suppose any equivalence class $A_p$ ($p = 1, 2, \ldots, r$) can be sub-divided as $A_p^1, A_p^2, \ldots, A_p^k$ in terms of decision attribute D. The numbers of these equivalence classes are denoted as $n_p^1, n_p^2, \ldots, n_p^k$, respectively. If $A_p \subseteq POS_{(\alpha,\beta)}(\pi_D|\pi_A)$ , there exists $A_p^h$ that $|A_p^h| \geq \alpha|A_p|$ holds. Therefore,

$$
\begin{aligned}
&\sum_{1 \leq i < j \leq k} n_p^i n_p^j \\
&= \frac{|A_p|^2 - \sum\limits_{1 \leq i \leq k} |A_p^i|^2}{2} \\
&= \frac{|A_p|^2 - |A_p^h|^2 - \sum\limits_{1 \leq i \leq k, i \neq h} |A_p^i|^2}{2} \\
&\leq \frac{|A_p|^2 - (\alpha|A_p|)^2 - (1-\alpha)|A_p|}{2} \\
&= \frac{1}{2}(1-\alpha)(1+\alpha-\frac{1}{|A_p|})|A_p|^2.\square
\end{aligned}
$$

Fig. 2 shows the main differences of three partitions. In Fig. 2(a) and (b), the boundary region partitions are the same. In Fig. 2(c), some objects from the boundary region and the objects from the positive region may be combined and form a new probabilistic positive region. Thus, during the attribute reduction process, we firstly acquire a reduct for boundary region partition preservation by indiscernibility relation, then further generate a reduct by deleting the redundant attribute in DTRS model.

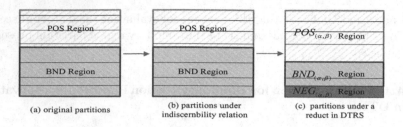

(a) original partitions

(b) partitions under indiscernibility relation

(c) partitions under a reduct in DTRS

**Fig. 2.** The main differences of three partitions

# 4 Attribute Reduction in Decision-Theoretic Rough Set Model Using MapReduce

## 4.1 Map Function for Computing the Equivalence Classes

Among classical attribute reduction algorithms, the most intensive calculation is to compute the equivalence classes. As we all know, it is obvious that the computation of one equivalence class is irrelevant to that of another equivalence class. Thus, different computations of the equivalence classes from a subset of attributes can be executed in parallel [11]. The pseudocode of map function is shown in Algorithm 1.

**Algorithm 1.** Map(key, value)
//Map phase of computing equivalence classes
//Parameters:Selected attributes A, a candidate attribute c
Input: Candidate Attributes $A \cup \{c\}$, a data split
Output:(EquivalenceClass, $[< d(x), 1 >]$)
Begin

**Step 1.** For an object x in a data split do {
**Step 2.** EquivalenceClass="";
**Step 3.** For any attribute $a \in A$ do
**Step 4.** {EquivalenceClass=EquivalenceClass + a(x) + " ";}
**Step 5.** EquivalenceClass=c(x)+ " " + EquivalenceClass;
**Step 6.** Output $< EquivalenceClass, < d(x), 1 >>.$}

End.

By Algorithm 1, we can compute the equivalence classes from candidate conditional attributes. In the next subsections, we mainly focus on how to compute the number of indiscernibility object pairs using MapReduce.

## 4.2 Reduce Function for Computing the Indiscernibility Object Pairs

In this subsection, we design a reduce function for computing the number of indiscernibility object pairs in cloud computing. The pseudocode for the reduce function is illustrated as algorithm 2.

**Algorithm 2.** Reduce(string EquivalenceClass, pairs[$< d1, n1 >$, $< d2, n2 >$, ...])

Input: (EquivalenceClass, pairs[$< d1, n1 >$, $< d2, n2 >$, ...])
Output:$< EquivalenceClass, NDis\_Num >$
Begin

**Step 1.** For any pair $< d, n > \in$ pairs [$< d1, n1 >$, $< d2, n2 >$, ...] do {
**Step 2.** Compute the frequencies $(n^1, n^2, ..., n^k)$ of decision value;}
**Step 3.** n= $\sum_{1}^{k} n^i$;
**Step 4.** $NDis\_Num = \sum_{1 \le i < j \le k} n^i n^j$;
**Step 5.** if $NDis\_Num > \frac{1}{2}(1-\alpha)(1+\alpha - \frac{1}{|n|})|n|^2$, then
**Step 6.** Output $< EquivalenceClass, NDis\_Num >$

End.

By algorithm 2, we can acquire the number of the indiscernibility object pairs generated from a candidate subset of attributes $A \cup \{c\}$ with respect to D, and store these different $< EquivalenceClass, NDis\_Num >$ pairs into the output files.

## 4.3 Attribute Reduction Algorithm in Decision-Theoretic Rough Set Model Using MapReduce

Parallel attribute reduction algorithm needs a MapReduce job to perform map and reduce functions. It reads the total number of indiscernibility object pairs from the output files in reduce phase, sums up them, and determines an attribute that will be added into a reduct. Thus, we can acquire the newly candidate subset of attributes for the next iteration. This iterative procedure must be executed serially as well. Here, we give a new definition of attribute significance as follows.

**Definition 5.** *For a decision table S, let $A \subseteq C$ and $a \in C - A$, then the significance of attribute a is defined by:*

$$sig(a, A, D) = \frac{\widetilde{DIS^D_{A \cup \{a\}}}}{\sum_{1 \le i < j \le k} n^i n^j} \tag{7}$$

As $\widetilde{DIS^D_\emptyset} \ge \widetilde{DIS^D_{A \cup \{a\}}} \ge 0$, we have $0 \le sig(a, A, D) \le 1$. Therefore, we can employ this attribute measure to construct attribute reduction algorithm using MapReduce. In DTRS model, we denote the number of the indiscernibility object pairs as $\widetilde{DIS^D_{A \cup \{a\}}}(\alpha)$. If $\alpha = 1$, we use $\widetilde{DIS^D_{A \cup \{a\}}}$ instead of $\widetilde{DIS^D_{A \cup \{a\}}}(1)$.

**Algorithm 3.** Attribute reduction algorithm in DTRS using MapReduce
Input: a decision table, S; a threshold parameter, $User\_\alpha$
Output: a reduct Red

Begin

**Step 1.** Let Red $= \emptyset$, $\alpha = 1$;

**Step 2.** Compute $\widetilde{DIS_C^D}$ by Algorithm 1 and 2;

**Step 3.** If $\widetilde{DIS_{Red}^D} = \widetilde{DIS_C^D}$ , then turn to Step 8;

**Step 4.** For each attribute $c \in C - Red$ do

**Step 5.** {Compute $\widetilde{DIS_{Red \cup c}^D}$ by Algorithm 1 and 2;}

**Step 6.** Let $\widetilde{DIS_{Red \cup c'}^D} = \min(\widetilde{DIS_{Red \cup c}^D})$;

**Step 7.** Red $=$ Red $\cup c'$, turn to step 3;

**Step 8.** $\alpha = User\_\alpha$;

**Step 9.** Compute $\widetilde{DIS_{Red}^D}(\alpha)$ by Algorithm 1 and 2;

**Step 10.** For each subset $A \subset Red$ do

**Step 11.** {Compute $\widetilde{DIS_A^D}(\alpha)$ by Algorithm 1 and 2;}

**Step 12.** if $\widetilde{DIS_A^D}(\alpha) < \widetilde{DIS_{Red}^D}(\alpha)$) then

**Step 13.** Red $=$ A, turn to Step 10;

**Step 14.** Output Red.

End.

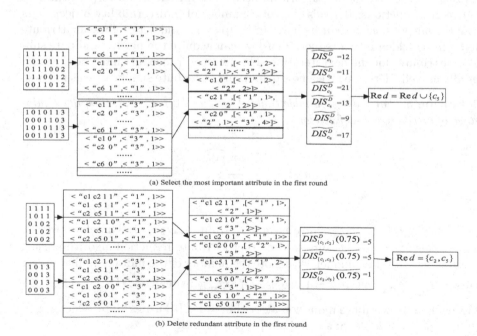

(a) Select the most important attribute in the first round

(b) Delete redundant attribute in the first round

**Fig. 3.** Attribute reduction process in DTRS model

By Algorithm 3, we first get a reduct that keeps the boundary region partition preservation by Step 1–7. Then we can acquire a reduct in DTRS through deleting the redundant attribute.

**Example 2:** (Continued from Example 1) Suppose $\alpha = 1$, the reduct is computed by Algorithm 3 as shown in Fig. 3. Suppose the whole dataset is divided into two data splits using MapReduce, we illustrate the attribute reduction process in DTRS model.

(1) Calculate $DIS_C^D = 3$. In the first round as shown in Fig. 3(a), we can select attribute $c_5$. Repeating the process, we can acquire a reduct $\{c_5, c_2, c_1\}$.

(2) Suppose $\alpha = 0.75$ and $\beta = 0.60$, $DIS_{\{c_5,c_2,c_1\}}^D(0.75) = 3$. We must check and compute the number of the indiscernibility object pairs of the subset of a reduct $\{c_5, c_2, c_1\}$ in DTRS model. In the first round as shown in Fig. 3(b), we can get a candidate reduct $\{c_5, c_2\}$. Since $DIS_{c_2}^D(0.75) = 11$ and $DIS_{c_5}^D(0.75) = 6$, we finally acquire the reduct $\{c_5, c_2\}$.

# 5   Conclusion

Many researchers focus on attribute reduction in the decision-theoretic rough set models in terms of decision preservation and region preservation. However, it is difficult to evaluate and interpret such a kind of subjective reductions. A new attribute measure for boundary region partition preservation is introduced. Moreover, an attribute reduction algorithm for decision-theoretic rough set models using MapReduce is designed in this paper. Heuristics and algorithms need to be further studied in our future work.

**Acknowledgment.** This work was partially sponsored by the National Natural Science Foundation of China under Grant No: 61103067, the Natural Science Foundation of Jiangsu Province of China under Grant No: BK20141152, the Key Laboratory of Cloud Computing and Intelligent Information Processing of Changzhou City under Grant No: CM20123004, Qing Lan Project of Jiangsu Province of China, the Key Laboratory of Embedded System and Service Computing, Ministry of Education under Grant No: ESSCKF201303, the Natural Science Foundation and Doctoral Research Foundation of Jiangsu University of Technology under Grant Nos: kyy12018, kyy13003.

# References

1. Beynon, M.: Reducts within the variable precision rough sets model: a further investigation. European Journal of Operational Research 134, 592–605 (2001)
2. Dean, J., Ghemawat, S.: MapReduce: Simplified data processing on large clusters. Communications of the ACM 51(1), 107–114 (2008)
3. Herbert, J.P., Yao, J.T.: Criteria for choosing a rough set model. Computers and Mathematics with Applications 57, 908–918 (2009)

4. Inuiguchi, M.: Attribute reduction in variable precision rough set model, International Journal of Uncertainty. Fuzziness and Knowledge-Based Systems 14, 461–479 (2006)
5. Jia, X.Y., Liao, W.H., Tang, Z.M., Shang, L.: Minimum cost attribute reduction in decision-theoretic rough set models. Information Sciences 219, 151–167 (2013)
6. Li, H.X., Zhou, X.Z., Zhao, J.B., Liu, D.: Attribute reduction in decision-theoretic rough set model: A further investigation. In: Klement, E.-P., Slany, W. (eds.) FLAI 1993. LNCS, vol. 695, pp. 466–475. Springer, Heidelberg (1993)
7. Mi, J.S., Wu, W.Z., Zhang, W.X.: Approaches to knowledge reduction based on variable precision rough set model. Information Sciences 159, 255–272 (2004)
8. Pawlak, Z.: Rough sets. International Journal of Computer and Information Sciences 11, 341–356 (1982)
9. Qian, Y.H., Liang, J.Y., Pedrycz, W., Dang, C.Y.: Positive approximation: An accelerator for attribute reduction in rough set theory. Artificial Intelligence 174, 597–618 (2010)
10. Qian, J., Miao, D.Q., Zhang, Z.H., Li, W.: Hybrid approaches to attribute reduction based on indiscernibility and discernibility relation. International Journal of Approximate Reasoning 52, 212–230 (2011)
11. Qian, J., Miao, D.Q., Zhang, Z.H., Yue, X.D.: Parallel attribute reduction algorithms using MapReduce. Information Sciences 279, 671–690 (2014)
12. Skowron, A., Rauszer, C.: The discernibility matrices and functions in information systems. In: Slowiński, R. (ed.) Intelligent Decision Support, Handbook of Applications and Advances of the Rough Sets Theory, pp. 331–362. Kluwer, Dordrecht (1992)
13. Slezak, D., Ziarko, W.: Attribute reduction in the Bayesian version of variable precision rough set model. Electronic Notes in Theoretical Computer Science 82, 263–273 (2003)
14. Yao, Y.Y.: Probabilistic approaches to rough sets. Expert Systems 20, 287–297 (2003)
15. Yao, Y.: Decision-theoretic rough set models. In: Yao, J., Lingras, P., Wu, W.-Z., Szczuka, M.S., Cercone, N.J., Ślęzak, D. (eds.) RSKT 2007. LNCS (LNAI), vol. 4481, pp. 1–12. Springer, Heidelberg (2007)
16. Yao, Y.Y., Zhao, Y.: Attribute reduction in decision-theoretic rough set models. Information Sciences 178, 3356–3373 (2008)
17. Zhao, Y., Wong, S.K.M., Yao, Y.Y.: A note on attribute reduction in the decision-theoretic rough set model. In: Peters, J.F., Skowron, A., Chan, C.-C., Grzymala-Busse, J.W., Ziarko, W.P. (eds.) Transactions on Rough Sets XIII. LNCS, vol. 6499, pp. 260–275. Springer, Heidelberg (2011)
18. Ziarko, W.: Probabilistic rough sets. In: Ślęzak, D., Wang, G., Szczuka, M.S., Düntsch, I., Yao, Y. (eds.) RSFDGrC 2005. LNCS (LNAI), vol. 3641, pp. 283–293. Springer, Heidelberg (2005)

# A Context-Aware Recommender System with a Cognition Inspired Model

Liangliang Zhao[1], Jiajin Huang[1], and Ning Zhong[1,2]

[1] International WIC Institute, Beijing University of Technology, P.R. China
{zlljianfeng,hjj}@emails.bjut.edu.cn
[2] Department of Life Science and Informatics,
Maebashi Institute of Technology, Japan
zhong@maebashi-it.ac.jp

**Abstract.** The development of information technologies raises the problem of information overload. Recommender systems aim to choose the best application or content from numerous applications or contents. And contextual information has been taken into account to improve the recommendation accuracy. Inspired by a cognitive architecture named ACT-R, this paper combines frequency and recency into contextual information to provide context-based recommendations for mobile applications. The experimental results show the ACT-R inspired method is effective in context-based recommendations.

## 1 Introduction

The purpose of recommender systems is to recommend items (information, products, services, etc.) which interest users to help them find useful items. It's one of the main methods to solve the problem of information overload. The classic methods of recommender systems can be divided into three kinds: collaborative filtering recommendation, content-based recommendation and hybrid recommendation which combines the two above methods. These methods mainly use relations between users and items to achieve recommendation. Adomavicius et al. pointed out that the use of contexts which describe the states of users or items, such as time, location, people around, device, etc. would increase the effect of recommendation [1], thus the context-aware recommender system [2] was presented. One of the important researches of context-aware recommender systems is to measure relations between contextual information and users' preferences. Jembere et al. presented a method to extract users' preferences for mobile applications using a strict partial order preference model [3]. Baltrunas et al. adopted collaborative filtering recommendation which pre-filters items' ratings to generate recommendations [4]. Su et al. recommended appropriate music for users based on their locations, moods, and health states [5]. Wang et al. proposed a recommender system which takes use of cognitive processing levels based on the cognitive psychology [6].

According to the ACT-R model proposed by Anderson, information used frequently and recently is easily extracted [7,8]. So what is used recently and

D. Miao et al. (Eds.): RSKT 2014, LNAI 8818, pp. 613–622, 2014.
DOI: 10.1007/978-3-319-11740-9_56 © Springer International Publishing Switzerland 2014

frequently becomes the major information. How to use these information to generate effective recommendation becomes a problem. The probabilistic model based on ACT-R's declarative memory retrieval mechanisms has been used in recommender systems. Fu et al. presented the SNIF-ACT model which achieves search purposes faster by evaluating ratings each link to the goal in the process of searching the Internet [9]. Stanley et al. used ACT-R's declarative memory retrieval mechanisms to generate predictions based on tags [10]. However, the above work did not consider the contextual information. In this paper, we use the activation equation combining measures of frequency and recency in ACT-R to discuss a context-based recommender system.

## 2   A Model

### 2.1   Model Review

ACT-R's spreading activation model pointed out that the likelihood and speed that memory information is extracted are determined by the strength of activation. The strength of activation mainly consists of two parts: the baseline level and the strength of association. The baseline level is mainly influenced by recency. Therefore, the more recently information is used, the more easily memory information is extracted. The activation of the baseline level determines the initial activation. The strength of association is the sum of association strength between each cue associated with the current context and the target. It reflects the current situation and adjusts the final activation level. The strength of association is mainly influenced by frequency. According to the spreading activation model, memory is characterized as a interconnected knowledge network. When items associated with the target arise in the context, the attention of the relevant items spread to the target items through cognitive structures to promote the extraction of target [7,8]. ACT-R calls these cognitive structures chunks.

In a context-aware recommender system, chunks can represent items and contextual information (such as time, location, people around, device, etc.). The strength of association between contextual information and items can be calculated based on users' historical data in the case of the previous contextual information. The more frequently that items are used in the contexts, the stronger is the strength of association. And users have more possibilities to use these items in the contexs.

We assume context $j$ ($j \in L$) in which item $i$ is used and $A_i$ which is the activation of item $i$. On the basis of Bayesian probability model in ACT-R [7,8], the possibility that item $i$ is used in the current contexts can be expressed as the following equation.

$$A_i = B_i + \sum_{j \in L} W_j S_{ji} \tag{1}$$

where $B_i$ is the baseline level activation of item $i$, $S_{ji}$ is the strength of association between context $j$ and item $i$, and $W_j$ reflects the attentional weight the model puts on context $j$. Equation (1) shows that $A_i$ is based on users' previous history and the current contexts.

Figure 1 shows how to use equation (1) to generate a recommended list. It consists of two parts: the offline computing and online recommendation. The offline computing consists of two sections: $B_i$ and $S_{ji}$. The online recommendation makes use of $A_i$ to rank items in a descending order and then recommends top-k items to users.

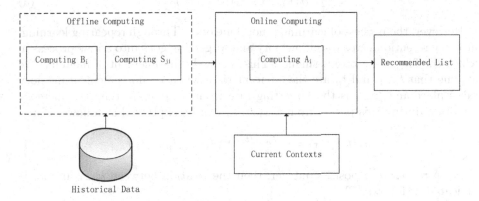

**Fig. 1.** The framework of model

Then how can we apply above activation equation in our mobile applications recommender system? From above, we can see activation $A_i$ is mainly influenced by recency and frequency, so we describe our model from the two aspects.

## 2.2 Recency

The baseline level activation $B_i$ is determined by recency that item is used. If an item is used recently, $B_i$ is higher. Human can clearly remember an event which is recent from numerous events. The recency can be modeled by a forgetting curve equation. There are two ways to express $B_i$ based on recency.

**The Basic Way.** We can use the expression of $B_i$ in the ACT-R model. $B_i$ can be obtained from the history that item $i$ was used [7,8].

$$B_i^{(1)} = ln \sum_{k=1}^{n} t_k^{-d} \tag{2}$$

where $t_k$ represents the subtraction between the $k$-th time that item $i$ was used and the current time and $d$ is a forgotten parameter whose value is a constant. $B_i$ is determined by recency that item $i$ is used. Taking the recency that item $i$ is used into account, we introduce $t_k$ and forgotten parameter $d$. The more recently item is used, the higher is the preference for the item and $B_i$.

The calculation for $B_i$ can be indicated through the following example. Supposing the history that item $i$ was used is listed as follows: (12.30, 2011), (1.14, 2013), (1.20, 2013), (1.20, 2013), (1.21, 2013), then $B_i$ in January 22, 2013 can be calculated as follows: $B_i = ln \left( 23^{-0.5} + 8^{-0.5} + 2^{-0.5} + 2^{-0.5} + 1^{-0.5} \right) = 1.157$.

**The Second Way.** The baseline level activation $B_i$ can also uses the following equation based on the forgetting curve [11].

We assume time $t$, forgetting rate $k$, then memory retention $p_a(t, k)$ can be indicated as follows [12]:

$$p_a(t, k) = e^{-kt}, t \in (0, \infty) \tag{3}$$

However, the process of learning is not done once. Through repeating learning, memory retention leaps to full memory and forgetting goes into a new process in which the forgetting rate is smaller. This is called reinforcement of memory. We assume that $t_{n-1}$ and $t_n$ are the adjacent time when reinforcement of memory takes place and $k_{n-1}$ is the forgetting rate from $t_{n-1}$ to $t_n$, then the memory retention during this period from $t_{n-1}$ to $t_n$ can be expressed as follows [12]:

$$p_a(t, k_{n-1}) = e^{-k_{n-1}(t-t_{n-1})}, t \in (t_{n-1}, t_n) \tag{4}$$

Let $\delta$ represent a positive integer, then the relation between $k_{n-1}$ and $k_n$ is indicated as follows [12]:

$$k_n = \frac{ln\left[1 + (\delta - 1)e^{-k_{n-1}(t_n - t_{n-1})}\right] - ln\delta}{t_{n-1} - t_n} \tag{5}$$

From the above equation, we can get forgetting rate $k_{n-1}$ and equation of forgetting curve after learning for $n$ times. The final function of forgetting curve after learning for $n$ times-$p_a(t, k_{n-1})$ can replace $B_i$ in ACT-R model:

$$B_i^{(2)} = p_a(t, k_{n-1}) = e^{-k_{n-1}(t-t_{n-1})}, t \in (t_{n-1}, \infty) \tag{6}$$

## 2.3   Frequency

$B_i$ is the baseline level activation for item $i$. $S_{ji}$ is used to express the strength of the association between item $i$ and context $j$ and it's mainly determined by frequency that item is used in the context. In the condition of context $j$, the more frequently that item $i$ is used, the greater is the strength of the association between item $i$ and context $j$. $S_{ji}$ can be indicated as follows:

$$S_{ji} = ln\left[\frac{prob(i|j)}{prob(i)}\right] \tag{7}$$

where $prob(i|j)$ is the conditional probability that item $i$ is used under the condition of context $j$ and $prob(i)$ is the probability that item $i$ is used.

The calculation for $S_{ji}$ can be indicated through the following example. Supposing the history that item 1, 2, 3 was used in the condition of context 1, 2, 3 is shown in figure 2, then $S_{ji}$ can be calculated as follows: $S_{11} = ln\left[\frac{1}{1} \div \frac{1}{3}\right] = ln3 = 1.099$, $S_{21} = ln\left[\frac{1}{2} \div \frac{1}{3}\right] = ln1.5 = 0.405$, $S_{31} = 0$.

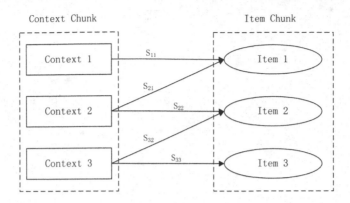

**Fig. 2.** The history that three items was used under three kind of condition

# 3 Experiments and Analysis

In recent years, with the upgrade of hardware, mobile phone, tablet and other portable mobile devices are playing an increasingly important role in people's life with convenience, widespread mobile network and easy operation. Mobile users have different preference in different contexts. How to effectively solve the problem to meet users' requirement for mobile network services is particularly urgent. Therefore, we'll verify the efficiency of our model in the mobile environment.

## 3.1 Data Set

Currently, under the mobile network environment, there still lacks public, authoritative and available mobile users' historical behavior data set that contains contextual information which tells us where and when one user used one mobile application for context-aware recommender systems. Borrowing the work in paper [3,4,6], we develop context generation rules and users' historical behavior generation rules. Then we construct a data set of users' historical behavior containing contextual information in mobile network environment to extract users' preference for mobile applications.

Firstly, we select a hundred kinds of Android applications which are downloaded frequently from Snap Pea applications center as mobile network application set and label them from 1 to 100. Then we select five hundred users as user set and label them from 1 to 500. Secondly, we generate context data set according to context generation rules which are shown in figure 3. Then, on the basis of mobile network application set and user set, we generate users' context behavior data set under the constraint of context data set. Users' context behavior data set tells us on which condition (such as time, location, people around, device, etc.) one user in user set used a mobile application in mobile network application set. Finally, we build users' historical behavior data set of five hundred users during a certain period (thirty days in this paper) according to generation process of users' context behavior data set.

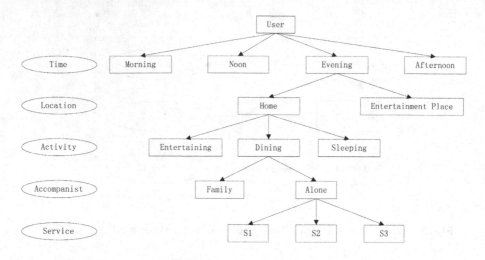

**Fig. 3.** An Example for spanning tree of context rule

The data set is generated gradually in accordance with constraint and dependency of tree structure. The spanning tree of context generation rules contains time, location, activity and accompanist from top to down. Each superior property should correspond to lower properties in line with the actual situation. For example, when "time" is morning, "location" can be office, transportation place and "activity" can be working, meeting, waiting, on-vehicle. When "time" is evening, "location" can be home, supermarket or entertainment place and "activity" can be dining, entertaining, sleeping, shopping. Part of spanning tree is shown in figure 3. From the figure, we can learn that users' behavior of using mobile applications is restricted by the state of context.

The final structure of the data set is shown as follows.

(1) The user set which contains five hundred users.

(2) The mobile network application set which contains one hundred applications.

(3) The context data set which contains time, location, activity and accompanist. Time is divided into morning, noon, afternoon and evening; location is divided into office, transportation place, home, restaurant, entertainment place and supermarket; activity includes working, meeting, waiting, dinning, entertaining, sleeping and shopping; accompanist includes colleagues, strangers, family, alone and friends.

(4) The users' historical behavior data set which contains context behavior of five hundred users during thirty days and records which mobile application was used on the condition of specific time, location, activity and accompanist.

The data set is divided into two parts: a training set and a test set. Eighty percent of the data is used as the training set to extract users' preference for mobile applications and twenty percent of the data is used as the test set for evaluating accuracy of the model.

## 3.2  Evaluation Criteria of Experiment Results

We adopt accuracy standard in this paper to evaluate the efficiency of our model. Firstly, we rank mobile applications in a descending order based on activation $A_i$. Then we choose top $N$ kinds of mobile applications as recommended list and compare the recommended list with actual data in the test set. The number of mobile applications in recommended list which equal to actual data in the test set divided by the number of actual data in the test set can get the accuracy of model. The higher accuracy of model is, the more effective is the model.

## 3.3  Procedure of Experiment and Analysis of Results

The hardware environment of experiment is CPU with Intel Pentium D 3.40 GHz and memory with DDR2 1.25 GB. The software environment of experiment is the operating system with Windows XP Professional SP3 and the development environment with JDK 1.7.0, Eclipse 3.6.

We have done two experiments. Experiment 1 is to verify the efficiency of our recommendation model and experiment 2 is to verify the efficiency of our model on data set which is affected strongly by one context-time. And the procedures of two experiments are shown as follows.

The first step: in the experiment 1, without considering the context, we generate data set randomly which means when users' historical behavior data is generated, we don't consider the fact that the use of mobile applications or services is constrained by context. In the experiment 2, we generate data set considering one kind of context-time. For example, when "time" is morning, we only generate the prior thirty kinds of mobile applications or services; when "time" is noon, we only generate the twenty-first mobile service to the fiftieth mobile service; when "time" is afternoon, we only generate the forty-first mobile service to the seventieth mobile service and when "time" is evening, we only generate the sixty-first mobile service to the hundredth mobile service.

The second step: the training set is input and we use our recommendation model to extract $N$ ($N$=3, 4, 5, 6) kinds of mobile applications whose preference values are higher as recommendation results. Meanwhile, we select $N$ kinds of mobile applications or services randomly as recommendation results.

The third step: we compare the recommendation results with the actual results of the test set and calculate recommendation accuracy of three methods (model combining $B_i^{(1)}$ and $S_{ji}$, model combining $B_i^{(2)}$ and $S_{ji}$, random method) separately.

The results of experiment 1 are shown in figure 4. The horizontal axis represents the number of recommended mobile applications and the vertical axis represents the accuracy of the each method. It's obvious that accuracy of recommendation model combining $B_i^{(1)}$ and $S_{ji}$ and recommendation model combining $B_i^{(2)}$ and $S_{ji}$ are much higher than the random method. The former two methods have the similar performance and model combining $B_i^{(2)}$ and $S_{ji}$ is a bit better than recommendation model combining $B_i^{(1)}$ and $S_{ji}$. As $N$ increases, the gap between the accuracy of our model and random method increases, so the advantages of our recommendation model are reflected.

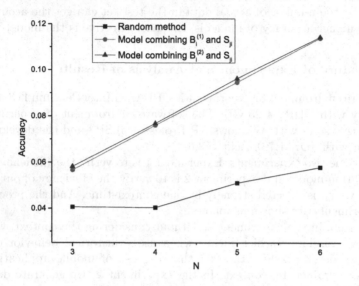

**Fig. 4.** Results of experiment 1

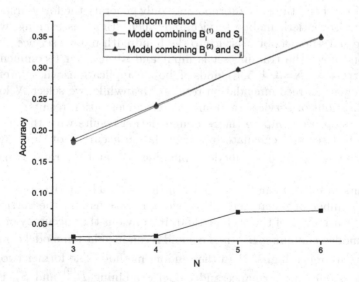

**Fig. 5.** Results of experiment 2

The results of experiment 2 are shown in figure 5. The horizontal axis represents the number of recommended mobile applications and the vertical axis represents the accuracy of three methods. We can learn from figure 5 that the accuracy of random method in experiment 2 has the similar performance with that in experiment 1, while the accuracy of recommendation model combining $B_i^{(1)}$ and $S_{ji}$ and recommendation model combining $B_i^{(2)}$ and $S_{ji}$ in experiment 2 improve greatly in comparison with that in experiment 1. This indicates that the performance of our recommendation model based on ACT-R is especially good when the use of mobile applications is influenced by the context.

## 4    Conclusions

In this paper, we combine the ACT-R cognitive architecture into users' preference for mobile applications or services from two elements namely frequency and recency. Then we develop an ACT-R inspired model and indicate that our model is efficient for mobile data set in the experiments, especially for the situation that users' behaviors of using applications or services are influenced strongly by contexts. In the future work, we'll add more contexts in and explore users' preference in the real social network instead of artificially generated data sets.

## References

1. Adomavicius, G., Sankaranarayanan, R., Sen, S., Tuzhilin, A.: Incorporating Contextual Information in Recommender Systems Using a Multidimensional Approach. ACM Transactions on Information System 23(1), 103–145 (2005)
2. Ricci, F., Rokach, L., Shapira, B., Kantor, P.B.: Recommender Systems Handbook. Springer, New York (2011)
3. Jembere, E., Adigun, M.O., Xulu, S.S.: Mining Context-Based User Preferences for M-Services Applications. In: Proceedings of the 2007 IEEE/WIC/ACM International Conference on Intelligent Agent Technology, pp. 753–763. IEEE Computer Society, New York (2007)
4. Baltrunas, L., Ricci, F.: Context-Based Splitting of Item Ratings in Collaborative Filtering. In: Proceedings of the 2009 ACM Conference on Recommender Systems, pp. 245–248. ACM Press, New York (2009)
5. Su, J., Yeh, H., Yu, P.S., Tseng, V.S.: Music Recommendation Using Content and Context Information Mining. IEEE Intelligent Systems 25(1), 16–26 (2010)
6. Wang, L.C., Meng, X.Y., Zhang, Y.J.: A Cognitive Psychology-Based Approach to User Preferences Elicitation for Mobile Network Services (in Chinese). Acta Electronica Sinica 39(11), 2547–2553 (2011)
7. Anderson, J.R., Bothell, D., Byrne, M.D., Douglass, S., Lebiere, C., Qin, Y.L.: An Integrated Theory of the Mind. Psychological Review 111, 1036–1060 (2004)
8. Anderson, J.R.: How Can the Human Mind Occur in the Physical Universe? Oxford University Press, New York (2007)
9. Fu, W., Pirolli, P.: SNIF-ACT: A Cognitive Model of User Navigation on World Wide Web. Human-Computer Interaction 22(4), 355–412 (2007)

10. Stanley, C., Byrne, M.D.: Predicting Tags for StackOverflow Posts. In: Proceedings of the 2013 International Conference on Cognitive Modeling, pp. 414–419. ICCM, Ottawa (2013)
11. Ebbinghaus, H.: A Contribution to Experimental Psychology, http://psy.ed.asu.edu/~classics/Ebbinghaus/index.htm
12. Yin, G.S., Cui, X.H., Ma, Z.Q.: A Collaborative Filtering Recommendation Model Based on Forgetting Curve (in Chinese). Journal of Harbin Engineering University 33(1), 85–90 (2012)

# Study on Fuzzy Comprehensive Evaluation Model of Education E-government Performance in Colleges and Universities

Fang Yu[1], Lijuan Ye[2], and Jiaming Zhong[2,*]

[1] Xiangnan University, associate professor, Chenzhou,
Hunan, China
yfjammy@qq.com
[2] Xiangnan University, professor, Chenzhou, Hunan, China
jmzhongcn@163.com

**Abstract.** First, it obtains the university education e-government performance evaluation index framework by making use of Delphi method. Then, it constructs the comprehensive quality evaluation hierarchy model by applying the analytic hierarchy process, to obtain the weight for each index, based on which to establish fuzzy comprehensive evaluation model, thus acquiring new method for university education e-government performance evaluation. Examples have proven the feasibility and effectiveness of this method.

**Keywords:** Education E-government, performance, fuzzy comprehensive evaluation, model.

## 1    Introduction

Education e-government construction is an important part of the national electronic government affairs and education information, is an important means of promoting the modernization of education. For the transformation of the mode of work and attitude, further improve the work quality and efficiency of the important task of the current colleges and universities to promote the building of e-government, and achieved some results, but overall, there is still a big investment, the problem such as high investment and low efficiency, small output of e-government construction of colleges and universities is far from meeting the requirement of the education reform and development of our country. International practices and studies have shown that the e-government project there is a high degree of risk, the probability of failure is extremely high, therefore, the western countries a large number of research institutions, consulting companies and scholars have carried out the study of e-government performance evaluation. To guide the healthy and orderly e-government's construction in colleges and universities, sustainable development, in-depth discussion and research

---

* Corresponding author.

D. Miao et al. (Eds.): RSKT 2014, LNAI 8818, pp. 623–630, 2014.
DOI: 10.1007/978-3-319-11740-9_57 © Springer International Publishing Switzerland 2014

to perfect, scientific education e-government performance evaluation index system of colleges and universities are imperative. [1-3]

## 2    Hierarchical Model of Education E-government Performance in Colleges and Universities

According to the principles of guidance, science, completeness, feasibility and development, it conducts several rounds of questionnaire and expert consultation by making use of Delphi method. Meanwhile, through several stages such as decomposition, convergence, test, revise, verification and perfection, it further analyzes and selects factors to be investigated, and then sequences these factors through expert judgment, to determine key factors to be investigated, thus obtaining the index framework. By making use of analytic hierarchy process, it obtains the weight for each index, thus obtaining the hierarchical model of comprehensive quality evaluation[4-11]. which is showed as Fig 1.

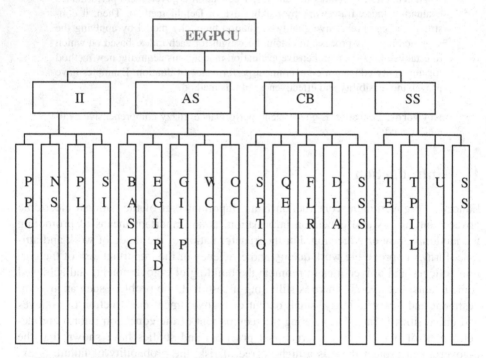

**Fig. 1.**    Hierarchical Model of Higher Learning Education E-government Performance

According to the analytic hierarchy process (AHP), we can get the **primary and secondary index weights**, it are as follows:

Primary index weight: $\omega_1 = 0.25, \omega_2 = 0.25, \omega_3 = 0.25, \omega_4 = 0.25$.
Secondary index weight: ,

$$\omega_{11} = 0.250, \omega_{12} = 0.250, \omega_{13} = 0.250, \omega_{14} = 0.250$$
$$\omega_{21} = 0.220, \omega_{22} = 0.220, \omega_{23} = 0.220, \omega_{24} = 0.220, \omega_{25} = 0.120 \quad ,$$
$$\omega_{31} = 0.197, \omega_{32} = 0.202, \omega_{33} = 0.197, \omega_{34} = 0.202, \omega_{35} = 0.202 \quad ,$$
$$\omega_{41} = 0.250, \omega_{42} = 0.250, \omega_{43} = 0.250, \omega_{44} = 0.250$$

# 3    Fuzzy Comprehensive Evaluation Model

## 3.1    Single Layer Fuzzy Comprehensive Evaluation Model

Education e-government performance evaluation system of evaluation index sets $A = \{a_1, a_2, \cdots, a_k\}$ ,according to the needs of evaluation decision, could be divided into such as evaluation criterion of $v_1, v_2, \cdots v_q$ q a level , the evaluation sets $V = \{v_1, v_2, \cdots v_q\}$.

If the ith indicators for single factor evaluation results $M_i = (m_{i1}, m_{i2}, \cdots, m_{iq})$, The evaluation index evaluation decision matrix k is:

$$M = (M_1, M_2, \cdots M_k)^T = \begin{pmatrix} m_{11} m_{12} \cdots m_{1q} \\ m_{21} m_{22} \cdots m_{2q} \\ \cdots\cdots\cdots\cdots \\ m_{k1} m_{k2} \cdots m_{kq} \end{pmatrix}.$$

Set the weight of each index distribution as follows: $\Omega = (\omega_1, \omega_2, \cdots \omega_k)$, The the comprehensive evaluation result is: $B = \Omega \circ M = (b_1, b_2, \cdots, b_q)$.

## 3.2    Multi-layer Fuzzy Comprehensive Evaluation Model

When the evaluation index in the system is no less than two layers, a multi-layer evaluation model is required, which should be built on the basis of single-layer evaluation model. The basic thought is as follows: first, it conducts single-layer comprehensive evaluation to the index at the bottom layer (or the most fundamental layer); then, by taking the evaluation results of this layer as the primary index of the upper layer, it evaluates the upper layer again, and so forth to the highest layer.

### 3.2.1    Index Classification
If the index set $A = \{a_1, a_2, \cdots, a_k\}$ is divided into p classes in accordance with the index attributes, i.e.:

$$A_i = \{a_{i1}, a_{i2}, \cdots, a_{ik_i}\}, i = 1, 2, \cdots, p$$

They will meet the following conditions:

(1) $k_1 + k_2 + \cdots + k_p = k$;

(2) $A_i \cap A_j = \varnothing (i \neq j)$.

(3) $A_1 \cup A_2 \cup \cdots \cup A_p = A$;

### 3.2.2   Establish Weight Vector

(1) Index Weight Vector

Let the weight for the index $A_i$ in the i class be $\omega_i (i = 1,2,\cdots, p)$, the index weight vector is:

$$\Omega = \{\omega_1, \omega_2, \cdots, \omega_p\}.$$

(2) Index Weight Vector

Let the weight of the index $a_{ij}$ in the i class be $\omega_{i\,j}$, the index weight vector is:

$$\Omega_i = \{\omega_{i1}, \omega_{i2}, \cdots, \omega_{ik_i}\}, i = 1,2, \cdots, p$$

### 3.2.3   Establish Multi-level Fuzzy Comprehensive Evaluation Model

Providing the fussy single-factor evaluation matrix of the subordinate index for certain index is:

$$M_i = \begin{pmatrix} m_{11}^{(i)} \cdots m_{1q}^{(i)} \\ m_{21}^{(i)} \cdots m_{2q}^{(i)} \\ \vdots \cdots \cdots \vdots \\ m_{k_i 1}^{(i)} \cdots m_{k_i q}^{(i)} \end{pmatrix}.$$

Then, the corresponding fussy comprehensive evaluation of each object is:

$$B_i = \Omega_i \circ M_i$$

$$= (\omega_{i1}, \omega_{i2}, \cdots, \omega_{ik_i}) \circ \begin{pmatrix} m_{11}^{(i)} \cdots m_{1q}^{(i)} \\ m_{21}^{(i)} \cdots m_{2q}^{(i)} \\ \vdots \cdots \cdots \vdots \\ m_{k_i 1}^{(i)} \cdots m_{k_i q}^{(i)} \end{pmatrix}.$$

$$= (b_{i1}, b_{i2}, \cdots, b_{iq})$$

Providing the secondary index comprehensive evaluation results are the elements, the fussy comprehensive evaluation matrix of the primary index is:

$$R = \begin{pmatrix} B_1 \\ B_2 \\ \cdots \\ B_p \end{pmatrix} = \begin{pmatrix} \Omega_1 \circ M_1 \\ \Omega_2 \circ M_2 \\ \cdots\cdots\cdots \\ \Omega_p \circ M_p \end{pmatrix},$$

Therefore, the fussy comprehensive evaluation of each object is

$$B = \Omega \circ M = (\omega_1, \omega_2, \cdots, \omega_p) \circ \begin{pmatrix} \Omega_1 \circ M_1 \\ \Omega_2 \circ M_2 \\ \cdots\cdots\cdots \\ \Omega_p \circ M_p \end{pmatrix} = (b_1, b_2, \cdots, b_q).$$

Therefore, we can evaluate the objects in line with the value of $b_1, b_2, \cdots, b_n$.

We can also divide the secondary indexes, to obtain the tertiary fussy comprehensive evaluation model or even the model with more layers. The multi-layer fussy comprehensive evaluation model can not only reflect the different layers of the evaluation index, but also avoid the difficulty in distributing weights because of too many indexes.

## 4    Evaluation Examples

According to the actual needs of evaluation decisions, the evaluation ranking standard can be divided into five classes, namely "excellent", "good", "medium", "qualified", and "disqualified".

v= {$v_1, v_2, v_3, v_4, v_5$}={excellent, good, medium, qualified, disqualified}

100 people, including personnel of teaching, administrative personnel, teaching supervisors, and students are invited to evaluate and mark each index of university A's education e-government performance in line with the defined evaluation ranking standards. The statistics results are shown as the Table 1.

**Table 1.** Score Table for university A's education e-government performance

| Primary Index | | Secondary Index | | Average score |
|---|---|---|---|---|
| Index Item | Weight | Index Item | Weight | |
| | | CPPC | 0.250 | 91.53 |
| II | 0.25 | CNS | 0.250 | 97.10 |
| | | NPL | 0.250 | 89.71 |
| | | ISI | 0.250 | 83.36 |

**Table 2.** (*Continued*)

|     |      | OBASC  | 0.220 | 85.19 |
|-----|------|--------|-------|-------|
|     |      | CEGIRD | 0.220 | 81.33 |
| AS  | 0.25 | CGIIP  | 0.220 | 79.15 |
|     |      | PWC    | 0.220 | 77.93 |
|     |      | SOC    | 0.120 | 71.56 |
|     |      | HSPTO  | 0.197 | 76.91 |
|     |      | OQE    | 0.202 | 77.53 |
| CB  | 0.25 | AFLR   | 0.197 | 73.11 |
|     |      | SDLA   | 0.202 | 62.31 |
|     |      | TSSS   | 0.202 | 81.79 |
|     |      | ITE    | 0.250 | 85.11 |
| SS  | 0.25 | ITPIL  | 0.250 | 81.33 |
|     |      | SU     | 0.250 | 75.10 |
|     |      | TSS    | 0.250 | 81.55 |

Construct the fuzzy membership functions

$$\mu_1(x) = \begin{cases} 0, 0 \le x \le 85 \\ \dfrac{x-85}{10}, 85 < x < 95 \\ 1, 95 \le x \le 100 \end{cases} \tag{1}$$

$$\mu_2(x) = \begin{cases} 0, 0 \le x \le 70 \\ \dfrac{x-70}{10}, 70 < x < 80 \\ 1, 80 \le x \le 90 \\ \dfrac{100-x}{10}, 90 < x \le 100 \end{cases} \tag{2}$$

$$\mu_3(x) = \begin{cases} 0, 0 \le x \le 55 \\ \dfrac{x-55}{10}, 55 < x < 65 \\ 1, 65 \le x \le 75 \\ \dfrac{85-x}{10}, 75 < x < 85 \\ 0, 85 \le x \le 100 \end{cases} \tag{3}$$

$$\mu_4(x) = \begin{cases} 0, 0 \leq x \leq 40 \\ \dfrac{x-40}{10}, 40 < x < 50 \\ 1, 50 \leq x \leq 60 \\ \dfrac{70-x}{10}, 60 < x < 70 \\ 0, 70 \leq x \leq 100 \end{cases} \tag{4}$$

$$\mu_5(x) = \begin{cases} 1, 0 \leq x \leq 40 \\ \dfrac{50-x}{10}, 40 < x < 50 \\ 0, 50 \leq x \leq 100 \end{cases} \tag{5}$$

According to Table 1 and formula (1)- (5), we can get the single-factor evaluation matrix of the secondary index as follows:

$$M_1 = \begin{pmatrix} 1 & 0.47 & 0 & 0 & 0 \\ 1 & 0.29 & 0 & 0 & 0 \\ 0 & 1 & 0.129 & 0 & 0 \\ 0 & 1 & 0.465 & 0 & 0 \end{pmatrix}, \quad M_2 = \begin{pmatrix} 0.019 & 1 & 0 & 0 & 0 \\ 0 & 1 & 0.367 & 0 & 0 \\ 0 & 0.915 & 0.585 & 0 & 0 \\ 0 & 0.793 & 0.707 & 0 & 0 \\ 0 & 0.155 & 1 & 0 & 0 \end{pmatrix}$$

$$M_3 = \begin{pmatrix} 0 & 0.691 & 0.809 & 0 & 0 \\ 0 & 0.753 & 0.747 & 0 & 0 \\ 0 & 0.311 & 1 & 0 & 0 \\ 0 & 0 & 0.731 & 0.769 & 0 \\ 0 & 1 & 0.321 & 0 & 0 \end{pmatrix}, \quad M_4 = \begin{pmatrix} 0.011 & 1 & 0 & 0 & 0 \\ 0 & 1 & 0.367 & 0 & 0 \\ 0 & 0.51 & 0.01 & 0 & 0 \\ 0 & 1 & 0.345 & 0 & 0 \end{pmatrix}$$

Let $\circ = (\cdot, +)$, using common matrix multiplication, Therefore,

$$B_1 = (0.5, 0.69, 0.149, 0, 0),$$
$$B_2 = (0.004, 0.835, 0.485, 0, 0)$$
$$B3 = (0, 0.5515, 0.7197912, 0.155338, 0), \quad B_4 = (0.003, 0.878, 0.179, 0, 0)$$
$$B = (0.127, 0.739, 0.383, 0.039, 0)$$

According to the maximum subordination principle, this university's education e-government performance is good.

Through the above mentioned analysis, we know that we can effectively evaluate nursing clinical teachers' comprehensive quality by utilizing analytic hierarchy method and fussy comprehensive evaluation method. Meanwhile, examples have proven that the evaluation is feasible, effective, and easily to be accepted and promoted. This model and algorithm have rigorous logical reasoning and theoretical basis, thus providing brand new methods and means to teachers' comprehensive quality evaluation.

**Acknowledgements.** This study is supported by the construct program of the key discipline in human province, Hunan Philosophy and social science program(NO: 13YBB205), Hunan science and technology program(NO:2014SK3229) , Hunan province ordinary university teaching reform project(xiangjiangtong[2013]NO:223), China.

# References

1. ZHanBoLei, XiaoFangJian: Reviews on the Performance Evaluation of E-government in China. Journal of Information 12, 13–17 (2006)
2. An, H., Gao, X., Zeng, D., Yun, P.: Application Case Study of Factor Analysis Method in the Education E-Government Comprehensive Evaluation. Transactions of Beijing Institute of Technology 27, 750–752 (2007)
3. Zhang, H., Kong, F.: Educational Information Evaluation. Publishing House of Electronics Industry, Beijing (2005)
4. Hu, B.-Q.: Basis of Fuzzy Theory. Press of Wuhan University, Wuhan (2004)
5. Du, D., Pang, Q.: Modern Comprehensive Evaluation Method and Case Selection. Tsinghua University Press, Beijing (2006)
6. Zhang, X.: Standardization Studies of the Indices Framework for Performance Evaluation of E-Goveernment. Terminology Standardization & Information Technology, 38–48 (2005)
7. Zhong, J., Li, D.: Comprehensive evaluation model integrated based on rough set and analytic hierarchy process. Engineering Journal of Wuhan University, 126–130 (2008)
8. Liu, Z., Zhong, J.: English Hearing Study Websties Evaluation Index System Analysis and Design. China Education Info., 27–30 (2009)
9. He, W., Liu, Z., Zhong, J.: On Comprehensive Evaluation of Educational Websites Based on Rough Set. Computer Applications and Software, 57–61 (2009)
10. Zhong, J., Li, D.: Research on Comprehensive Measurement Method of CAI Course Ware Based on Rough Set. Computer Engineering and Applications 44, 213–215 (2008)
11. Ye, L., Yu, F., Zhong, J.: Education E-government Performance Comprehensive Evaluation Index System of Analysis and Design in Colleges and Universities (2014)

# Parallel Attribute Reduction Based on MapReduce

Dachao Xi[1,2], Guoyin Wang[2,*], Xuerui Zhang[2], and Fan Zhang[2]

[1] Chongqing Key Laboratory of Computational Intelligence,
Chongqing University of Posts and Telecommunications, Chongqing 400065, China
[2] Institute of Electronic Information & Technology, Chongqing Institute of Green and
Intelligent Technology, Chinese Academy of Sciences,
Chongqing 401122, China
wangguoyin@cigit.ac.cn

**Abstract.** With the explosive increment of data, varieties of the parallel attribute reduction algorithm have been studied. To promote its efficiency, this paper proposes a new parallel attribute reduction algorithm based on MapReduce. It contains three parts, parallel computation of a simplified decision table, parallel computation of attribute significance and parallel computation of decision table. Data with different sizes are experimented. The experimental result shows that our algorithm has the ability of processing massive data with efficiency.

**Keywords:** Rough set, attribute reduction, MapReduce, parallel computing.

## 1 Introduction

With the fast-growing data in scientific and industrial areas, machine learning and data mining algorithms are facing the challenges from both the perspectives of data and computational complexity. Google come up with distributed file system (GFS) [1], parallel programming pattern (MapReduce) [2,3] and distributed and structured storage system (Bigtable) [4]. They become the basis of massive data processing. After that, many people begin to study data mining algorithm of massive data.

Rough set [5] is a classical tool it can process fuzzy and indeterminate problem. It has been widely used for the field of machine learning and data mining. Attribute reduction is one of very important study field of Rough Set. It is also the committed step of knowledge acquisition. Attribute reduction, on the condition that classification ability of knowledge base is unchanged, can delete data redundancy so that it can greatly improve knowledge definition of information system.

With the cloud computing coming, rough set theory models of massive data have become the popular in the field of machine learning and data mining. Zhang et al. [6,7] propose a parallel method for computing lower and upper approximations and a comparison of parallel large-scale knowledge acquisition using rough set theory. Zhang has proved the viability that classical rough set models can be paralleled. However, they just parallel the basis of rough set models. Qian et al. [8,12] come up with a parallel attribute reduction model, but it needs a special consistent decision table. There is another paper of Qian et al. [15], this paper propose a general parallel

D. Miao et al. (Eds.): RSKT 2014, LNAI 8818, pp. 631–641, 2014.
DOI: 10.1007/978-3-319-11740-9_58 © Springer International Publishing Switzerland 2014

algorithm for attribute reduction using MapReduce. It is a low efficiency of algorithm. Yang et al. [13] has a more efficient than Qian, but these reductions are not guaranteed to be the same as those discovered from the whole dataset since these sub-decision tables do not exchange the information for each other [15]. Yang el al. [14] shows us another algorithm. It can make sure correct of result, but it has less efficiency than Qian[15]. In this paper, a new parallel attribute reduction will propose. It has more efficiency and can make sure correctly.

The remaining of this paper is organized as fallows. Section 2 includes the elementary background introduction to rough sets and MapReduce. The algorithms of computing attribute reduction based on MapReduce are presented in Section 3. Experimental analysis is given in Section 4. The paper ends with conclusions and future work in Section 5.


# 2    Preliminaries

In this section, we will review the basic notions of rough set model in [9], and MapReduce programming model in cloud computing.


## 2.1    Rough sets

**Definition 1.** A decision table is an information system $S = (U, R, V, f)$, where $U = \{x_1, x_2, \cdots x_n\}$ is a non-empty finite set of objects, $R = C \cup D$ is a non-empty finite set of attributes. $V_a = \cup_{r \in R} V_r$ is a domain of the attribute a. $f : U \times R \to V$ is an information function such that $f(x, a) \in V_a$ for every $x \in U, a \in R$. An equivalence relation with respect to $B \subseteq R$ called the indiscernibility relation, denoted by $IND(B)$, is defined as $IND(B) = \{(x, y)|(x, y) \in U^2, \forall b \in B(b(x) = b(y))\}$

The equivalence relation $IND(B)$ partitions U into some equivalence classes given by $U/IND(B) = \{[x]_B | x \in U\}$

where $[x]_B$ denotes the equivalence class determined by $x$ with respect to $B$, $[x]_B = \{y \in U|(x, y) \in IND(B)\}$.

**Definition 2.** Let $X \subseteq U$ be a subset of $U$, and undefined relation $B$. The lower and upper approximations of $X$ are defined as

$$B_-(X) = \cup \{Y_i | (Y_i \in U/IND(B) \wedge Y_i \subseteq X\}$$
$$B^-(X) = \cup \{Y_i | (Y_i \in U/IND(B) \wedge Y_i \cap X \neq \emptyset\}$$

**Definition 3.** For a decision table $S$, $U/C = \{[u'_1]_c, [u'_2]_c, \cdots [u'_m]_c\}$ is a partition of $U$, $U' = \{u'_1, u'_2, \cdots, u'_m\}$ is a subsets of object, $POS_C(D) = [u'_{i_1}]_c \cup [u'_{i_2}]_c \cup \cdots \cup [u'_{i_t}]_c$, where $\forall u'_{i_s} \in U'$ and $|[u'_{i_s}]_c \backslash D| = 1 (s = 1, 2, \cdots, t)$. Let $U'_{pos} = \{[u'_{i_1}, u'_{i_2}, \cdots, u'_{i_t}\}$, $U'_{neg} = U' - U'_{pos}$, $U' = U'_{pos} \cup U'_{neg}$, $S' = (U', C \cup D, V, f)$ is a simplified decision table.

**Definition 4[11].** For a decision table $S$, $S' = (U', C \cup D, V, f)$ is a simplified decision table. $U'$ is the subset of table , P is a subset of attributes, $P \subseteq C$ and $\forall a \in (C - P)$, attribute significance of $a$ is: $sig_p(a) = |U'_{P \cup \{a\}} - U'_p|$

where $U'_p = \left\{\cup_{X \subseteq U' \backslash P \wedge X \in U'_{pos} \wedge |X \backslash D| = 1} X\right\} \cup \{\cup_{X \in U'/P \wedge X \subseteq U'_{neg}} X\}$.

## 2.2 The MapReduce model

MapReduce is a programming model and an associated implementation proposed by Google for processing and generating large data sets in a distributed computing environment that is amenable to a broad variety of real-world tasks. There are two steps of a MapReduce task: map and reduce. The computation takes a set of input key/value pairs, and produces a set of output key/value pair[10].

# 3    Parallel Attribute Reduction Based on MapReduce

### 3.1    A Traditional Attribute Reduction Algorithm

According to **Definitions 1- 4**, we present a Traditional Algorithm for Computation of Attribute Reduction (TACAR) which based on algorithm of Xu et al. [11]. The maximum of attributer significance is a reduction of the result. When the simplified decision table $U'$ is empty, we get the reduction of decision table. The algorithm is shown in **Algorithm 1** respectively.

---

**Algorithm 1.** TACAR

**Input:** A decision table $S = (U, R, V, f)$
**Output:** A reduction, Red

1. **begin**
2.   compute $U/C$, $U'$, $U'_{pos}$, $U'_{neg}$ // $U'$ is the simplified decision table
3.   $Red \leftarrow \emptyset$
4.   **while** $U' \mathrel{!=} \emptyset$ **do**
5.     **for** each $a \in C - R$ **do**
6.       compute $sig_R(a)$ // attribute significance of $a$
7.       // $B_R(a)$ is a collection that all objects of equivalence class are in $U'_{pos}$
       and
       their values of decision attribute are equal. $NB_R(a)$ is a collection that all
       object of equivalence class are in $U'_{neg}$.
8.       compute $B_R(a)$, $NB_R(a)$ and $U'/(R \cup \{a'\})$;
9.     **end**
10.    let $sig_R(a') \leftarrow \max sig_R(a)$;
11.    let $R \leftarrow R \cup \{a'\}$;
12.    let $U' \leftarrow U' - B_R a' - NB_R(a')$;
13.    let $U'_{pos} \leftarrow U'_{pos} - B_R(a')$;
14.    let $U'_{neg} \leftarrow U'_{neg} - NB_R(a')$;
15.  **end**
16.  $Red$;
17. **end**

---

## 3.2    Parallel Attribute Reduction Based on MapReduce

**3.2.1** Parallel Algorithm for Computation of a Simplified Decision Table.
**Definition 6.** Given a decision table $S = (U, R, V, f)$. Let $S = \cup_{i-1}^{m} S_i$, where $S_i = (U_i, R, V, f)$. It satisfies: (1) $U = \cup_{i-1}^{m} U_i$; (2) $U_j \cap U_k = \emptyset, \forall i, k \in \{1, \cdots, m\}$ and $j \neq k$. It means the decision tables S is divided into m sub-decision tables. Then we call $S_i$ is a sub-decision table of $S$.

According to **Definition 6,** each sub-decision table can compute equivalence classes independently. At the same time, the equivalence classes of different sub-decision tables can combine together if their information set is the same. The simplified decision table also can get through equivalence classes. Therefore, we can compute the simplified decision table in parallel. Here, we design a Parallel Algorithm for Computation of a Simplified Decision Table (PACSDT) based on MapReduce. The **Algorithm PACSDT** is divided into **Algorithm PACSDT-Map** and **Algorithm PACSDT-Reduce**, as outlined in **Algorithm 2 and 3.**

---

**Algorithm 2.** PACSDT-Map

**Input:** a decision table $S_i = (U_i, C \cup D, V, f)$, $S = \cup_{i-1}^{m} S_i$
**Output:** $< x\_C, x\_D >$ where $x\_C$ is condition of the object $x$, $x\_D$ is decision of the object $x$.

1. **begin**
2.    **for** each $x \in U_i$ **do**
3.        let $key \leftarrow x\_C$;
4.        let $value \leftarrow x\_D$;
5.        Output.collect *(key, value)*;
6.    **end**
7. **end**

---

After the **Algorithm 2**, the results of Map will be sorted order by the key of Map. The pair of Map will transmit to **Algorithm 3**.

---

**Algorithm 3.** PACSDT-Reduce

**Input:** $< x\_C, x\_D >$
**Output:** $< x\_C, x\_D + POS_C(D)\_flag + x\_No >$ where $POS_C(D)\_flag$ is a flag that $x \in POS_C(D)$ or $x \notin POS_C(D)$.

1.    **begin**
2.    let $num \leftarrow 1$ the number of key also the No of object
3.    let $value \leftarrow null$;
4.    **for** each $x\_C$ **do**
5.        let $key \leftarrow x\_C$;
6.        $ispos \leftarrow$ the flag of $POS_C(D)$;
7.        // ispos = 1 if all $x\_D$ has the same value, else ispos = 0
8.        let $value \leftarrow x\_D + ispos + num$;

9.    // for each $x\_C$ has many $x\_D$, but we just need one
10.    **let** $num \leftarrow num++$;
*11.*    Output.collect *(key, value)*;
**12. end**
**13. end**

We can get a new simplified decision table through **Algorithm 2** and **Algorithm 3**. It has a flag of $POS_C(D)$. This flag will play an important role in the computing of attribute significance.

### 3.2.2 Parallel Algorithm for Computation of Attribute Significance

The algorithm based on attribute significance has been widely used for classical attribute reduction. It is also used by Qian[8,12,15] and Yang[13,14] because of each attribute significance can be parallel computed. But in their algorithms each MapReduce can only get a significance of attribute[14,15]. In general, their algorithms employ the strategy in task parallel for the computations of attribute significance. Here, we design a Parallel Algorithm for Computation of Attribute Significance (PACAS). The algorithm can get all significances of attributes just only one MapReduce. The **Algorithm PACAS** is divided into **Algorithm PACAS-Map** and **Algorithm PACAS-Reduce** and **Algorithm PACAS**, as shown in **Algorithm 4**, **Algorithm 5** and **Algorithm 6**, respectively.

---

**Algorithm 4.** PACAS-Map

**Input:** A simplified decision table $S_i' = (U_i', C \cup D, V, f)$
**Output:** $< c + x\_c, x\_D + POS_C(D)\_flag + x\_No >$ where $c \in C$, $x\_c$ is the value of attribute c for each object $x$.

**1. begin**
2.    **for** each $x \in U_i'$ **do**
3.      **for** each $c \in C - R$ **do**
4.        **let** $key \leftarrow c + x\_c$;
5.        **let** $value \leftarrow x\_D + POS_c(D)\_flag + x\_No$;
6.        Output.collect *(key, value)*;
7.      **end**
8.    **end**
**9. end**

---

**Algorithm 5.** PACAS-Reduce

**Input:** $< c + x\_c, x\_D + POS_C(D)\_flag + x\_No >$
**Output:** $< c, sig_R(c) + B_R(c) + NB_R(c) >$ where $c \in C$, $sig_R(c)$ is a significance of attribute. $B_R(a)$ is a collection that all objects of equivalence class are in $U_{pos}'$ and their values of decision attribute are equal. $NB_R(a)$ is a collection that all objects of equivalence class are in $U_{neg}'$.

**1. begin**

2.     let $num \leftarrow 0$; //value of $|B_R(c)|$ and $|NB_R(c)|$

3.     let $posflag \leftarrow false$; // $POS_{cUR}(D)\_flag$ is equal or not in the same key

4.     let $B\_NB \leftarrow null$; //value of $B_R(c)$ and $NB_R(c)$

5.     let $valueofD \leftarrow false$; //$x\_D$ is equal or not in the same key

6.     **for each $c + x\_c$ do**

7.         **for** $Map\_value$ is not empty **do**

8.             let $num \leftarrow num++$;

9.             let $B\_NB \leftarrow x\_No$;

10.     **end**

11.     **if all** $POS_{cUR}(D)\_flags$ **are equal then**

12.         let $posflag \leftarrow true$;

13.     **end**

14.     **if all** $x\_Ds$ **are equal then**

15.         let $valueofD \leftarrow true$;

16.     **end**

17.     //is $B_R(c)$?

18.     **if** $posflag = true$ **and** $POS_{cUR}(D)\_flag = 1$ **and** $valueofD = true$ **then**

19.         let $key \leftarrow c$;

20.         let $value \leftarrow num+ B\_NB$;

21.     Output.collect $(key, value)$;

22.         let $num \leftarrow 0$;

23.         let $posflag \leftarrow false$;

24.     **end**

25.     //is $NB_R(c)$?

26.     **if** $posflag = true$ **and** $POS_{cUR}(D)\_flag = 0$ **then**

27.         let $key \leftarrow c$;

28.         let $value \leftarrow num+ B\_NB$;

29.     Output.collect$(key, value)$;

30.         let $num \leftarrow 0$;

31.         let $posflag \leftarrow false$;

32.     **end**

33. **end**

34. **end**

---

By **Algorithm 4** and **Algorithm 5**, we can get $B_R(c)$ and $B_R(c)$ of each attribute. Also can get $|B_R(c)|$ and $|NB_R(c)|$. The result has been stored on the HDFS. Read the file from HDFS and compute the max of significance. Then we can acquire one reduction. The algorithm is shown in **Algorithm 6** respectively.

---

**Algorithm 6. PACAS**

**Input:** A simplified decision table $S'_i = (U'_i, C \cup D, V, f)$

**Output:** one reduction

1. **begin**
2.   **let** *reduction* ← 0;
3.   Initiate a MapReduce job, compute $sig_R(c)$, $B_R(c)$ and $B_R(c)$ by executing
       Algorithm 4 and Algorithm 5;
4.   Read the result from the HDFS
5.   **let** *max* ← 0;
6.   **for** each $c \in C$ **do**
7.     **if** *max* < *sig(c)* **then**
8.       **let** *max* ← *sig(c)*;
9.       **let** *reduction* ← *c*;
10.    **end**
11.  **end**
12. **end**

### 3.2.3 Parallel Algorithm for Computation of Attribute Reduction Based on Attribute Significance

By **Algorithm 6**, we only get one reduction. So we need to execute **Algorithm 6** iterate to get the finally result. Before we execute **Algorithm 6** next, we need to delete redundancy of decision table. We design a Parallel Algorithm for Computation of Decision Table (PACDT). The algorithm is shown in **Algorithm 7** respectively.

**Algorithm 7.** PACDT-Map

**Input:** A simplified decision table $S'_i = (U'_i, C \cup D, V, f)$
**Output:** A new simplified decision table $S''_i = (U''_i, C \cup D, V, f)$

1.   **begin**
2.   **let** *Nos* ← $B_R(c)$ + $NB_R(c)$;
3.   **for** each $x \in U'_i$ **do**
4.     **if** *Nos* is not contains *x_No* **then**
5.       **let** *key* ← *x_C*;
6.       **let** *value* ← *x_D* + *ispos* + *num*;
7.       Output.collect*(key, value)*;
8.     **end**
9.   **end**
10.  **end**

By **Algorithm 7**, we get the new simplified decision table. Next reduction can be computed by it. And then, the whole algorithm of attribute reduction is shown in **Algorithm 8**. Parallel Algorithm for Computation of Attribute Reduction Based on Attribute Significance (PACARBAS) is summarized in **Algorithm 8**.

---

**Algorithm 8.** PACARBAS

**Input:** A decision table $S = (U, R, V, f)$
**Output:** *Reductions*

1. **begin**
2.    **let** *Reductions* ← ∅;
3.    **compute** a simplified decision table by Algorithm 2 and Algorithm 3;
4.    **while** $S'$ is not empty **do**
5.        **compute** a reduction by Algorithm 6;
6.        **let** *Reductions* ← *reduction;*
7.        **compute** a new simplified decision table by Algorithm 7;
8.    **end**
9.    *Reductions;*
10. **end**

---

## 4    Experimental Analysis

### 4.1    Experimental Setup

This section presents the experimental results of parallel algorithm for computing attribute reduction. We run the experiments on the Hadoop MapReduce platform [16], where Hadoop MapReduce is a programming model and software framework for developing applications that rapidly process massive data in parallel on large clusters of compute nodes, each of which has 8GB memory and uses Intel Core i7 CPU. Hadoop version 2.2.0 and java 1.7.45 are used as the MapReduce system for all experiments.

### 4.2    Data Sets

We utilize the large data set KDDCup-99 from the machine learning data repository, University of California at Irvine [17]. The data set KDDCup-99 consists of approximately five million records. Each record consists of 1decsion attribute and 41 condition attributes, where 6 are categorical and 35 are numeric. Since our parallel algorithms deal with categorical attributes, the 35 numeric attributes are discretized firstly. Besides, two synthetic data sets have been generated based on KDDCup-99. The data sets are outlined in **Table 1**.

**Table 1.** Description of the datasets

|   | Data Sets | Records | Attributes | Classes | Size(GB) |
|---|-----------|---------|------------|---------|----------|
| 1 | KDD99 | 4,898,421 | 41 | 23 | 0.48 |
| 2 | KDD-1.9G | 20,000,000 | 35 | 23 | 1.90 |
| 3 | KDD-3.9G | 40,000,000 | 41 | 23 | 3.80 |

### 4.3    Experimental Results

Our experimental results are as follows. We run the Algorithm of attribute reduction in 4 datanodes and 8 datanodes respectively. The running time of each dataset is shown in **Fig 2**. We can see that as the nodes increase, the running time of dataset is decline. But the time is not increase linear because of the network delay and data loading time.

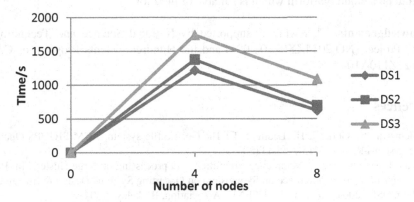

**Fig. 1.** Result with increasing nodes size

In **Fig 3**, we perform the speedup evaluation on datasets with different sizes and structures. The number of nodes varied from 1 to 8. As the result show, the proposed parallel algorithms have a good speedup performance. So this parallel algorithm can treat massive data efficiently.

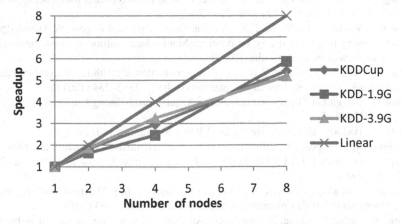

**Fig. 2.** Speadup

## 5    Conclusion

Traditional algorithm is not suitable for processing massive data. In this paper, we have presented a parallel algorithm of attribute reduction based on MapReduce.

It have been implemented the map and reduce functions for the new Simplified Decision Table and attribute significance. Experimental results present that the algorithm can deal with massive data better.

But when we run this algorithm of attribute reduction on Hadoop, there are many iterative computations. That means we have many IO operation with HDFS. When the dataset is small, most of time will waste for IO operation. So if we want to run the algorithm more efficiency, we can think about how to decrease iteration or run the algorithm on a cloud platform which is suitable of iteration.

**Acknowledgements.** This work is supported by National Science and Technology Major    Project (NO.2014ZX07104-006) and the Hundred Talents Program of CAS (NO.Y21Z110A10).

# References

1. Ghemawat, S., Gobioff, H., Leung, S.T.: The Google file system. ACM SIGOPS Operating Systems Review 37(5), 29–43 (2003)
2. Dean, J., Ghemawat, S.: Mapreduce: simplified data processing on large clusters. In: Proceedings of the 6th conference on Symposium on Opearting Systems Design & Implementation, OSDI 2004, vol. 6, p. 10. USENIX Association, Berkeley (2004)
3. Dean, J., Ghemawat, S.: MapReduce: simplified data processing on large clusters. Communications of the ACM 51(1), 107–113 (2008)
4. Chang, F., Dean, J., Ghemawat, S., et al.: Bigtable: A distributed storage system for structured data. ACM Transactions on Computer Systems (TOCS) 26(2), 4 (2008)
5. Pawlak, Z.: Rough set. International Journal of Computer and Information Sciences 11, 341–356 (1982)
6. Zhang, J., Li, T., Ruan, D., et al.: A parallel method for computing rough set approximations. Information Sciences 194, 209–223 (2012)
7. Zhang, J., Wong, J., Li, T., Li, P.Y.: A comparison of parallel large-scale knowledge acquisition using rough set theory on different MapReduce runtime systems. International Journal of Approximate Reasoning (2013)
8. Qian, J., Miao, D.Q., Zhang, Z.H.: Knowledge reduction algorithms in cloud computing. Jisuanji Xuebao (Chinese Journal of Computers) 34(12), 2332–2343 (2011)
9. Wang, G.: Rough Set Theory and knowledge Acquisition. Jiaotong University Press, Xi'an (2001) (in Chinese)
10. White, T.: Hadoop: The definitive guide. O'Reilly Media, Inc. (2012)
11. Zhangyan, X., Zuopeng, L., Bingru, Y., et al.: A quick attribute reduction algorithm with complexity of max {O (| C|| U|), O (| C| 2| U/C|)}. Chinese Journal of Computers 29(3), 391–399 (2006)
12. Qian, J., Miao, D.Q., Zhang, Z.H.: Research on Discernibility Matrix Knowledge Reduction Algorithm in Cloud Computing. Computer Science 38(8), 193 (2011)
13. Yang, Y., Chen, Z., Liang, Z., Wang, G.: Attribute reduction for massive data based on rough set theory and mapReduce. In: Yu, J., Greco, S., Lingras, P., Wang, G., Skowron, A. (eds.) RSKT 2010. LNCS, vol. 6401, pp. 672–678. Springer, Heidelberg (2010)
14. Yang, Y., Chen, Z.: Parallelized computing of attribute core based on rough set theory and mapReduce. In: Li, T., Nguyen, H.S., Wang, G., Grzymala-Busse, J., Janicki, R., Hassanien, A.E., Yu, H. (eds.) RSKT 2012. LNCS, vol. 7414, pp. 155–160. Springer, Heidelberg (2012)

15. Qian, J., Miao, D., Zhang, Z., et al.: Parallel attribute reduction algorithms using MapReduce. Information Sciences (2014)
16. Hadoop project develops open-source software for reliable, scalable, distribute computing, http://hadoop.apache.org
17. Newman, D., Hettich, S., Blake, C., Merz, C.: UCI Repository of Machine Learning Databases, University of California, Department of Information andComputer Science, Irvine, CA (1998), http://www.ics.uci.edu/~mlearn/MLRepository.html

# Dynamic Ensemble of Rough Set Reducts for Data Classification

Jun-Hai Zhai, Xi-Zhao Wang, and Hua-Chao Wang

Key Laboratory of Machine Learning and Computational Intelligence,
College of Mathematics and Computer Science, Hebei University, Baoding,
071002, Hebei, China
mczjh@126.com

**Abstract.** Ensemble learning also named ensemble of multiple classifiers is one of the hot topics in machine learning. Ensemble learning can improve not only the accuracy but also the efficiency of the classification system. Constructing the component classifiers in ensemble learning is crucial, because it has direct influence on the performance of the classification system. In the construction of component classifiers, it should be guaranteed that the constructed component classifiers possess certain accuracy and diversity. Based on the confidence degree of classifier, this paper presents an approach consisting of three steps to dynamically integrate rough set reducts. Firstly, multiple reducts are computed. Secondly, multiple component classifiers with certain diversity are trained on the different reducts. Finally, these component classifiers are integrated by adopting dynamic integration strategy. The experimental results show that the proposed algorithm is efficient and feasible.

**Keywords:** Ensemble learning, Dynamic ensemble, Component classifier, Rough set, Attribute reduct, Confidence degree.

## 1 Introduction

Ensemble learning, also known as ensemble of multiple classifiers, multi-classifier system, classifier integration or committee of experts, is to firstly train several component classifiers, and then integrate the outputs of these component classifiers using a strategy to classify unknown samples. It has been proved that ensemble learning outperforms single-classifier learning in accuracy [1]. One of the most representative works of ensemble learning is the Boosting method proposed by Schapire [2] in 1989. Freund and Schapire [3] improved this method and proposed the AdaBoost method in 1997. Another commonly used ensemble method named bagging is proposed by Breiman [4] in 1996. Many ensemble methods reported in the literature are developed from the two methods.

It is crucial to construct and select the component classifiers which should have certain diversity. It has been proved that ensemble of multiple component classifiers with certain diversity can improve the performance of integration system [5]. There are two commonly used methods to construct component classifiers with

D. Miao et al. (Eds.): RSKT 2014, LNAI 8818, pp. 642–649, 2014.
DOI: 10.1007/978-3-319-11740-9_59 © Springer International Publishing Switzerland 2014

certain diversity: the method based on sample subsets and the method based on attribute subsets. The former selects sample subsets from training set to train component classifiers. The representative works include Bagging [4], Boosting [2, 3]. The latter selects attribute subsets which have the same expressing ability as original attribute set to train component classifiers. The representative works include stochastic subspace method [6] and integrated rough subspace method [7].

Classifier integration and classifier selection are two strategies commonly used to classify the test samples in ensemble learning. Classifier integration classifies test samples by fusing all outputs of the trained component classifiers with some fusion strategies, for example, the averaging and the majority voting. There are two categories of classifier selection methods: static classifier selection and dynamic classifier selection. The static classifier selection chooses the best component classifier to classify the test samples. The algorithms of static classifier selection are simple and can be implemented easily, but the performances are not always ideal. While dynamic classifier selection chooses the appropriate component classifiers to classify different test samples. Based on the local accuracy, a dynamic classifier selection method was proposed in [8], which uses the estimation of local accuracy of each individual classifier in small regions of feature space surrounding the test sample. By combining classifier selection and classifier integration, the dynamic ensemble selection method was presented in [9], which dynamically integrates several classifiers by using some ensemble strategies. The method in [9] selects $k$ classifiers which are most similar to the oracle in the validation set, and then employ them as component classifiers to integrate and classify test samples.

If at least one component classifier can classify a test sample correctly, how to make the integrated classifier also classify this sample correctly? In order to deal with this problem, based on confidence degree of classifiers [10], this paper proposes an ensemble method which dynamically integrates rough set reducts. The basic idea of the proposed method is to employ confidence degree to measure the significance of a component classifier, and dynamically select the component classifiers used for integration for different test samples. If the confidence degree of one component classifier is much higher than the ones of other component classifiers, then we call this component classifier authoritative expert and it will be designated to classify the test samples. If there is no authoritative expert, then the classification results of the component classifiers will be integrated to decide the class label of the test samples.

The paper is organized as follows. In Section 2, the related basic concepts are briefly reviewed, the proposed method is presented in Section 3, the experimental results and analysis are given in Section 4. Finally Section 5 concludes this paper.

## 2  Preliminaries

In this section, we briefly review the basic concepts used in this paper including information system, decision information system, lower approximation, positive region, attribute reduct etc [11].

**Definition 1.** *An information system is a 4-tuple $IS = (U, A, V, f)$. Where $U$ is a non-empty finite set of objects, and $A$ is a non-empty finite set of attributes. $V = \bigcup_{a \in A} V_a$. $V_a$ is the domain of value of attribute $a \in A$. $f : U \times A \longrightarrow V$ is called information function which satisfy that $\forall a \in A$, $\forall x \in U$, $f(x, a) \in V_a$. If $A = C \cup D$, $C \cap D = \phi$, $C$ is the condition attribute set, $D$ is single decision attribute, then $IS$ is called decision information system or simply decision table, denoted by $DT$.*

**Definition 2.** *Given an information system $IS = (U, A, V, f)$, let $P \subseteq A$, $x, y \in U$, we call $x$ and $y$ are indiscernible to $P$, if and only if $\forall a \in P$, $f(x, a) = f(y, a)$. $\forall P \subseteq A$, $P$ is an equivalence relation on $U$, denoted by $IND(P)$ or $P$ in short. $\forall x \in U$, the equivalence class of $x$ generated by $P$ is $[X]_P$. Then equivalence relation $P$ partitions the set $U$ into disjoint subsets, denoted by $U/P = \{U_1, U_2, \ldots, U_k\}$.*

**Definition 3.** *Given a decision table $DT = (U, C \cup D, V, f)$, and $P \subseteq C$, $X \subseteq U$, the lower approximation of $X$ with respect to $P$ is defined as:*

$$\underline{P}(X) = \{x | (x \in U) \wedge ([x]_p \subseteq X)\}) \tag{1}$$

**Definition 4.** *Given a decision table $DT = (U, C \cup D, V, f)$, and $P \subseteq C$, $U/P = \{U_1, U_2, \ldots, U_k\}$ is partition of $U$ derive from $D$, the positive region of decision attribute $D$ with respect to $P$ is defined as:*

$$POS_P(D) = \bigcup_{i=1}^{k} \underline{P}(U_i) \tag{2}$$

**Definition 5.** *Given a decision table $DT = (U, C \cup D, V, f)$, and $P \subseteq C$, $\forall a \in P$, if $POS_P(D) = POS_{P-\{a\}}(D)$, then attribute $a$ is dispensable in $P$ with respect to $D$, otherwise attribute $a$ is indispensable. If $\forall a \in P$, attribute $a$ is indispensable in $P$ with respect to $D$, then $P$ is independent with respect to $D$.*

**Definition 6.** *Given a decision table $DT = (U, C \cup D, V, f)$, and $P \subseteq C$. $P$ is called a reduct of $C$ with respect to $D$, if the following conditions are satisfied: (1) $P$ is independent with respect to $D$; (2) $POS_P(D) = POS_C(D)$ The set of all reducts of $C$ with respect to $D$ is denoted by $RED_C(D)$.*

For a given decision table, classification rules can be extracted by calculating attribute reducts and value reducts with the following five steps:

**Step1**: Delete repeated objects in decision table;

**Step2**: Delete redundant condition attributes and obtain attribute reducts;

**Step3**: Delete redundant attribute values of each object and obtain value reducts;

**Step4**: Get the decision table reducts;

**Step5**: Obtain the classification rule according to the decision table reducts.

## 3  Dynamic Ensemble of Rough Set Reducts for Data Classification

In this section, we firstly present the definition of confidence degree, and then present the proposed algorithm.

For a given decision table, the samples in training set are categorized into $l$ classes, and the value domain of decision attribute $D$ is $\{d_1, d_2, \ldots, d_l\}$. Attribute reduct algorithm based on discernibility matrix [12] can calculate all reducts, denoted by $R_1, R_2, \ldots, R_s$ in this paper, $s$ classifiers can be trained with $s$ reducts. For convenience of description, the $s$ classifiers are also denoted by $R_1, R_2, \ldots, R_s$. Each classifier corresponds to a set of rules, and all the $s$ rule sets are also denoted by $R_1, R_2, \ldots, R_s$.

**Definition 7.** *Given rule set (or classifier) $R_j (1 \leq j \leq s)$, confidence degree of its $i$th rule is defined as:*

$$CD_{ij} = \frac{NC_{ij}}{NS_{ij}}, 1 \leq i \leq |R_j|, 1 \leq j \leq s \tag{3}$$

where $|R_j|$ denotes the number of rules in rule set $R_j$, $NC_{ij}$ denotes the number of samples matched correctly with $i$th rule of $j$th rule set $R_j$ (or classifier) in the training set, and $NS_{ij}$ denotes the number of samples matched with $i$th rule of $j$th rule set $R_j$ (or classifier) in training set.

**Definition 8.** *Given rule set (or classifier) $R_j (1 \leq j \leq s)$, its confidence degree is defined as:*

$$CD_j = \sum_{i=1}^{|R_j|} CD_{ij} \tag{4}$$

For a given $x$ and $R_j$, group the rules successfully matched with sample $x$ on $R_j$ by the second component of the rules (or class labels). Let $R_{jk}(1 \leq j \leq s, 1 \leq k \leq l)$ be the rule set corresponding to $k$th class, we define confidence degree of rule set $R_j$ (or classifier) by using the confidence degree of these rule corresponding to sample $x$.

**Definition 9.** *Given a test sample $x$ and classifier $R_j$, confidence degree of classifier $R_j$ corresponding to the sample $x$ is defined as:*

$$CD_j(x) = \sum_{k=1}^{l} \sum_{i=1}^{|R_{jk}|} CD_{ij} \tag{5}$$

If there is an authoritative expert, then the class label of the test sample $x$ is decided according to the following formulation.

$$d_{k_0}(x) = \arg\max_k \{ \sum_{i=1}^{|R_{jk}|} CD_{ij} \} \tag{6}$$

If there is no authoritative expert, we group all the classification rules of all the classifiers according to the second component of the rules (or class label) and the class label of the test sample $x$ is decided according to the following formulation.

$$d_{k_0}(x) = \arg\max_k \{CD_k(x)\} \tag{7}$$

where

$$CD_k(x) = \sum_{j=1}^{s} \sum_{i=1}^{|R_{jk}|} CD_{ij} \tag{8}$$

The proposed algorithm is described as follows.

**Algorithm DERSR**: Dynamic Ensemble Rough Set Reducts
   **Input**: Decision table, test sample $x$ and parameter $\delta (0 < \delta < 1)$
   **Output**: Classification result of $x$
   **STEP1**: Compute the attribute reducts $RED_C(D) = \{R_1, R_2, \ldots, R_s\}$ by the algorithm based on discernibility matrix.
   **STEP2**: For each reduct $R_j (1 \leq j \leq s)$, compute its value reduct and decision table reduct and obtain a set of classification rule (rule set or classifier), also denoted by $R_j (1 \leq j \leq s)$ .
   **STEP3**: For each rule set (classifier) $R_j (1 \leq j \leq s)$, firstly, compute confidence degree of its $i$th rule with (3), i.e. $CD_{ij}$, and then compute the confidence degree of $R_j$ with (4), i.e. $CD_j$.
   **STEP4**: For a given test sample $x$ and rule set (classifier) $R_j (1 \leq j \leq s)$, compute the confidence degree of $R_j$ on test sample $x$ with (5), i.e. $CD_j(x)(1 \leq j \leq s)$.
   **STEP5**: Decide the classification result of test sample $x$.
   If $CD_i(x) - \max\{CD_j(x)\} > CD_i(x) \times \delta$ (i.e. there exist authoritative expert), the classification result of test sample $x$ is decided by formulation (6), else, integrate the classification results of every classifier and decide the classification result of test sample $x$ by formulation (7).
   **STEP6**: Output the classification result of test sample $x$.

## 4    Experimental Results and Analysis

In order to verify the effectiveness of the proposed algorithm, we experimentally compared our algorithm with the one in [7] in test accuracy, and with the ID3 algorithm in three aspects: the number of rules, the number of attributes used for classification, and the test accuracy. We select 6 UCI data sets [13] for experiments, which are Monks-1, Monks-2, Monks-3, Car, Balance-scale and SPECT. In all experiments, we set $\delta = 0.3$, the experimental results are listed in table 1 and 2 respectively.

From the experimental results listed in table 1, it can be seen that the test accuracies of the proposed method DERSR are consistently higher than the ones of the method in [7]. The experimental results listed in table 2 show that the test

**Table 1.** The experimental results compared with the method in [7]

| Data set | Test Accuracy | |
|---|---|---|
| | DERSR | The method in [7] |
| Monks-1 | 0.8867 | 0.8516 |
| Monks-2 | 0.5871 | 0.5549 |
| Monks-3 | 0.8450 | 0.8361 |
| Car | 0.8334 | 0.8119 |
| Balance-scale | 0.6571 | 0.6506 |
| SPECT | 0.7138 | 0.6950 |

**Table 2.** The experimental results compared with the ID3 algorithm

| Data Set | #Rule | | #Attribute | | Test Accuracy | |
|---|---|---|---|---|---|---|
| | DERSR | ID3 | DERSR | ID3 | DERSR | ID3 |
| Monks-1 | 21 | 55 | 3 | 6 | 0.8867 | 0.7867 |
| Monks-2 | 236 | 98 | 6 | 6 | 0.5871 | 0.5347 |
| Monks-3 | 44 | 28 | 4 | 5 | 0.8450 | 0.8383 |
| Car | 235 | 274 | 6 | 6 | 0.8334 | 0.7585 |
| Balance-scale | 303 | 401 | 4 | 4 | 0.6571 | 0.3240 |
| SPECT | 707 | 40 | 18 | 23 | 0.7138 | 0.6663 |

accuracies of our method DERSR are also consistently higher than the ones of ID3. In addition, our method DERSR uses smaller attribute number than ID3, and sometimes smaller rule number. In the experiments, we find that almost all the rules extracted with ID3 are included in the rules generated with the proposed method DERSR. In order to further to verify this phenomenon, we also compared our approach with conventional rough set (RS in short) method, in this experiment, only one reduct is used for classification, similar results are obtained(see table 3). We analyzed the rules that are not included in the rule set generated with the method DERSR, and conclude that one reason is that some attributes are not included in reduct. Meanwhile, we also find an interesting phenomena that ID3 can generate a few rules with which there is no sample in training set can match. We explain this phenomenon by experiments on the Car data set.

For Car data set, decision rules are generated by DERSR and ID3 respectively. DERSR generates 235 rules, while ID3 method generates 274 rules and 4 of them are not included in rules generated by DERSR. There is no sample can match the 4 rules in the training set, see figure 1.

In figure 1, the number behind leaf node is class label. And the number behind class label is support degree of leaf node. It denotes the proportion of samples with leaf class label in partition data set. The 4 specific rules correspond to the 4 leaf node with 0 support degree in figure 1. Support degree of a leaf is 0 means that there is no sample match this leaf, namely the corresponding

**Table 3.** The experimental results compared with the RS method

| Data Set | #Rule | | #Attribute | | Test Accuracy | |
|---|---|---|---|---|---|---|
| | DERSR | RS | DERSR | RS | DERSR | RS |
| Monks-1 | 21 | 45 | 3 | 5 | 0.8867 | 0.8430 |
| Monks-2 | 236 | 94 | 6 | 5 | 0.5871 | 0.5519 |
| Monks-3 | 44 | 41 | 4 | 5 | 0.8450 | 0.8275 |
| Car | 235 | 269 | 6 | 6 | 0.8334 | 0.8029 |
| Balance-scale | 303 | 349 | 4 | 4 | 0.6571 | 0.6110 |
| SPECT | 707 | 38 | 18 | 17 | 0.7138 | 0.6941 |

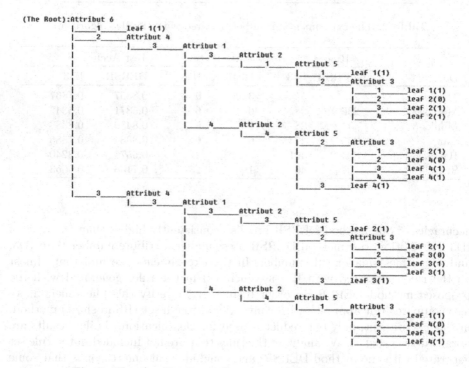

**Fig. 1.** The classification rules generated from Car with ID3

partition data set of this leaf is empty. The reason is that ID3 method partitions the sample set according to all attribute values of condition attribute, which results in some empty sample sets after partition and generates specific leaf node without support of training set.

## 5    Conclusion

Given a decision table, we can find multiple attribute reducts and each of them can have different contributions to the decision-making. If we only use one of them, even the most important one, to make decision, some important information will still be lost. In order to make full use of the information contributed by different reducts, this paper presents a novel algorithm to integrate these reducts based on the confidence degree of classifier with dynamic integration strategy, and some experiments are conducted to verify the effectiveness of the proposed method. Experimental results show that the proposed algorithm is efficient and feasible. As a by-product, an interesting phenomena on ID3 algorithm is found in our experiments. In the future, we will further study this interesting phenomena, and we will further investigate the impact of parameter $\delta$ on the performance of the integrated classification system.

**Acknowledgement.** This research is supported by the National Natural Science Foundation of China (71371063, 61170040), by the Natural Science Foundation of Hebei Province (F2013201110 and F2013201220), by the Key Scientific Research Foundation of Education Department of Hebei Province (ZD20131028), and the Scientific Research Foundation of Education Department of Hebei Province (Z2012101).

# References

1. Kittler, J., Hatef, M., Duin, R.P.W., et al.: On Combining Classifiers. IEEE Transactions on Pattern Analysis and Machine Intelligence 20(3), 226–239 (1998)
2. Schapire, R.E.: The strength of weak learnability. Machine Learning 5(2), 197–227 (1990)
3. Freund, Y., Schapire, R.E.: A decision-theoretic generalization of on-line learning and an application to boosting. Journal of Computer and System Sciences 55(1), 119–139 (1997)
4. Breiman, L.: Bagging Predictors. Machine Learning 6(2), 123–140 (1996)
5. Kuncheva, L.I., Whitaker, C.J.: Measures of diversity in classifier ensembles and their relationship with the ensemble accuracy. Machine Learning 51(2), 181–207 (2003)
6. Ho, T.K.: The random space method for constructing decision forests. IEEE Transactions on Pattern Analysis and Machine Intelligence 20(8), 832–844 (1998)
7. Hu, Q.H., Yu, D.R., Xie, Z.X., et al.: EROS: Ensemble rough subspaces. Pattern Recognition 40(12), 3728–3739 (2007)
8. Woods, K., Kegelmeyer, W.P., Bowyer, K.: Combination of multiple classifiers using local accuracy estimates. IEEE Transactions on Pattern Analysis and Machine Intelligence 19(4), 405–410 (1997)
9. Ko, A.H.R., Sabourin, R., Britto, A.S.: From dynamic classifier selection to dynamic ensemble selection. Pattern Recognition 41(5), 1718–1731 (2008)
10. Diintsch, I., Gediga, I.: Uncertainty measures of rough set prediction. Artificial Intelligence 106(1), 109–137 (1998)
11. Pawlak, Z.: Rough sets. International Journal of Information and Computer Sciences 11, 341–356 (1982)
12. Skowron, A., Rauszer, C.: The discernibility matrices and functions in information systems. In: Slowinski, R. (ed.) Intelligent Decision Support, Handbook of Applications and Advances of the Rough Sets Theory, Kluwer, Dordrecht (1992)
13. Blake, C.L., Merz, C.J.: UCI Repository of machine learning databases (1996), http://www.ics.uci.edu/~mlearn/MLRepository.html

# Rough Set Theory

# Intuitionistic Fuzzy Rough Approximation Operators Determined by Intuitionistic Fuzzy Triangular Norms

Wei-Zhi Wu[1,2], Shen-Ming Gu[1,2], Tong-Jun Li[1], and You-Hong Xu[1,2]

[1] School of Mathematics, Physics and Information Science,
Zhejiang Ocean University, Zhoushan, Zhejiang, 316022, P.R. China
[2] Chongqing Key Laboratory of Computational Intelligence,
Chongqing University of Posts and Telecommunications, Chongqing, 400065, China
{wuwz,gsm,litj,xyh}@zjou.edu.cn

**Abstract.** In this paper, relation-based intuitionistic fuzzy rough approximation operators determined by an intuitionistic fuzzy triangular norm $T$ are investigated. By employing an intuitionistic fuzzy triangular norm $T$ and its dual intuitionistic fuzzy triangular conorm, lower and upper approximations of intuitionistic fuzzy sets with respect to an intuitionistic fuzzy approximation space are first introduced. Properties of $T$-intuitionistic fuzzy rough approximation operators are then examined. Relationships between special types of intuitionistic fuzzy relations and properties of $T$-intuitionistic fuzzy rough approximation operators are further explored.

**Keywords:** Approximation operators, Intuitionistic fuzzy rough sets, Intuitionistic fuzzy sets, Intuitionistic fuzzy triangular norms, Rough sets.

## 1 Introduction

Rough set theory [8] is a new mathematical approach to deal with insufficient and incomplete information. The basic structure of rough set theory is an approximation space consisting of a universe of discourse and a binary relation imposed on it. Based on the approximation space, the notions of lower and upper approximation operators can be constructed. Using the concepts of lower and upper approximations in rough set theory, knowledge hidden in information tables may be unravelled and expressed in the form of decision rules.

One of the main directions in the research of rough set theory is naturally the generalization of concepts of Pawlak rough set approximation operators. Many authors have generalized the notion of rough set approximations by using non-equivalence binary relations. Other authors have also generalized the notion of rough set approximations into the fuzzy environment, and the results are called rough fuzzy sets (fuzzy sets approximated by a crisp approximation space) and fuzzy rough sets (fuzzy or crisp sets approximated by a fuzzy approximation space). As a more general case of fuzzy sets, the concept of intuitionistic fuzzy

D. Miao et al. (Eds.): RSKT 2014, LNAI 8818, pp. 653–662, 2014.
DOI: 10.1007/978-3-319-11740-9_60 © Springer International Publishing Switzerland 2014

(IF for short) sets, which was originated by Atanassov [1], has played a useful role in the research of uncertainty theories. Unlike a fuzzy set, which gives a degree of which element belongs to a set, an IF set gives both a membership degree and a nonmembership degree. Obviously, an IF set is more objective than a fuzzy set to describe the vagueness of data or information. The combination of IF set theory and rough set theory is a new hybrid model to describe the uncertain information and has become an interesting research issue over the years (see e.g. [2, 3, 5, 6, 9–12, 14–17]).

It is well-known that the dual properties of lower and upper approximation operators are of particular importance in the analysis of mathematical structures in rough set theory. The dual pairs of lower and upper approximation operators in the rough set theory are strongly related to the interior and closure operators in topological space, the necessity (box) and possibility (diamond) operators in modal logic, and the belief and plausibility functions in the Dempster-Shafer theory of evidence. On the other hand, we know that there are a lot of triangular norms which have been widely used in fuzzy set research. It should be noted that fuzzy inference results often depend upon the choice of the triangular norm. For analyzing uncertainty in complicated fuzzy systems, dual pairs of lower and upper fuzzy rough approximations defined by arbitrary triangular norms in rough set theory have been developed [7, 13]. According to this research line, the main objective of this paper is to present the study of IF rough sets determined by IF triangular norms. We will define a dual pair of lower and upper $T$-IF rough approximation operators and examine their essential properties.

## 2   Basic Notions Related to Intuitionistic Fuzzy Sets

In this section we recall some basic notions and previous results about intuitionistic fuzzy sets which will be used in the later parts of this paper.

Throughout this paper, $U$ will be a nonempty set called the universe of discourse. The class of all subsets (respectively, fuzzy subsets) of $U$ will be denoted by $\mathcal{P}(U)$ (respectively, by $\mathcal{F}(U)$). In what follows, $1_y$ will denote the fuzzy singleton with value 1 at $y$ and 0 elsewhere; $1_M$ will denote the characteristic function of a crisp set $M \in \mathcal{P}(U)$. For any $A \in \mathcal{F}(U)$, the complement of $A$ will be denoted by $\sim A$, i.e. $(\sim A)(x) = 1 - A(x)$ for all $x \in U$.

We first review a lattice on $[0, 1] \times [0, 1]$ originated by Cornelis et al. [4].

**Definition 1.** *Denote*

$$L^* = \{(x_1, x_2) \in [0, 1] \times [0, 1] \mid x_1 + x_2 \leq 1\}. \tag{1}$$

*A relation $\leq_{L^*}$ on $L^*$ is defined as follows:* $\forall (x_1, x_2), (y_1, y_2) \in L^*$,

$$(x_1, x_2) \leq_{L^*} (y_1, y_2) \iff x_1 \leq y_1 \text{ and } x_2 \geq y_2. \tag{2}$$

The relation $\leq_{L^*}$ is a partial ordering on $L^*$ and the pair $(L^*, \leq_{L^*})$ is a complete lattice with the smallest element $0_{L^*} = (0, 1)$ and the greatest element $1_{L^*} = (1, 0)$. The meet operator $\wedge$ and the join operator $\vee$ on $(L^*, \leq_{L^*})$ linked to the ordering $\leq_{L^*}$ are, respectively, defined as follows: $\forall (x_1, x_2), (y_1, y_2) \in L^*$,

$$
\begin{aligned}
(x_1, x_2) \wedge (y_1, y_2) &= (\min(x_1, y_1), \max(x_2, y_2)), \\
(x_1, x_2) \vee (y_1, y_2) &= (\max(x_1, y_1), \min(x_2, y_2)).
\end{aligned}
\tag{3}
$$

And for any index set $J$ and $a_j = (x_j, y_j) \in L^*$, $j \in J$, we define

$$
\begin{aligned}
\bigwedge_{j \in J} a_j &= \bigwedge_{j \in J} (x_j, y_j) = (\bigwedge_{j \in J} x_j, \bigvee_{j \in J} y_j), \\
\bigvee_{j \in J} a_j &= \bigvee_{j \in J} (x_j, y_j) = (\bigvee_{j \in J} x_j, \bigwedge_{j \in J} y_j).
\end{aligned}
\tag{4}
$$

Meanwhile, an order relation $\geq_{L^*}$ on $L^*$ is defined as follows: $\forall x = (x_1, x_2), y = (y_1, y_2) \in L^*$,

$$
(y_1, y_2) \geq_{L^*} (x_1, x_2) \iff (x_1, x_2) \leq_{L^*} (y_1, y_2),
\tag{5}
$$

and

$$
x = y \iff x \leq_{L^*} y \text{ and } y \leq_{L^*} x.
\tag{6}
$$

For $(x_1, x_2) \in L^*$, we define the complement element of $(x_1, x_2)$ in $L^*$ as follows:

$$
1_{L^*} - (x_1, x_2) = (x_2, x_1).
\tag{7}
$$

Since $\leq_{L^*}$ is a partial ordering, the order-theoretic definitions of conjunction and disjunction on $L^*$ called *IF triangular norm* (IF $t$-norm for short) and *IF triangular conorm* (IF $t$-conorm for short) are introduced as follows:

**Definition 2.** *An IF triangular norm (IF $t$-norm for short) on $L^*$ is an increasing, commutative, associative mapping $T : L^* \times L^* \to L^*$ satisfying $T(1_{L^*}, x) = x$ for all $x \in L^*$.*

**Definition 3.** *An IF triangular conorm (IF $t$-conorm for short) on $L^*$ is an increasing, commutative, associative mapping $S : L^* \times L^* \to L^*$ satisfying $S(0_{L^*}, x) = x$ for all $x \in L^*$.*

Obviously, the greatest IF $t$-norm (respectively, the smallest IF $t$-conorm) with respect to (w.r.t.) the ordering $\leq_{L^*}$ is min (respectively, max), defined by $\min(x, y) = x \wedge y$ (respectively, $\max(x, y) = x \vee y$) for all $x, y \in L^*$.

An IF $t$-norm $T$ and an IF $t$-conorm $S$ on $L^*$ are said to be *dual* if

$$
\begin{aligned}
T(x, y) &= 1_{L^*} - S(1_{L^*} - x, 1_{L^*} - y), \forall x, y \in L^*, \\
S(x, y) &= 1_{L^*} - T(1_{L^*} - x, 1_{L^*} - y), \forall x, y \in L^*.
\end{aligned}
\tag{8}
$$

Each IF $t$-norm $T$ can be associated two functions $T_1, T_2 : L^* \times L^* \to [0, 1]$ which are defined as follows:

$$
T(a, b) = (T_1(a, b), T_2(a, b)), \forall a, b \in L^*.
\tag{9}
$$

Likewise, from an IF $t$-conorm $S$ on $L^*$, we can derive two functions $S_1$, $S_2$ : $L^* \times L^* \to [0,1]$ which satisfy the following equation.

$$S(a,b) = (S_1(a,b), S_2(a,b)), \forall a, b \in L^*. \tag{10}$$

Since $T$ and $S$ are increasing, by Eq. (2) we can conclude

**Proposition 1.** *If $T$ is an IF $t$-norm on $L^*$ and $S$ the IF $t$-conorm on $L^*$ dual to $T$, then $T_1$ and $S_1$ are increasing and $T_2$ and $S_2$ are decreasing for both arguments.*

**Proposition 2.** *If $T$ is an IF $t$-norm on $L^*$, and $S$ is the IF $t$-conorm on $L^*$ dual to $T$. Then*

(1) $S_1(a,b) = T_2(1_{L^*} - a, 1_{L^*} - b)$, *for all $a,b \in L^*$.*
(2) $S_2(a,b) = T_1(1_{L^*} - a, 1_{L^*} - b)$, *for all $a,b \in L^*$.*
(3) $T_1(a,b) = S_2(1_{L^*} - a, 1_{L^*} - b)$, *for all $a,b \in L^*$.*
(4) $T_2(a,b) = S_1(1_{L^*} - a, 1_{L^*} - b)$, *for all $a,b \in L^*$.*

*Proof.* For $a = (a_1, a_2), b = (b_1, b_2) \in L^*$, by Eqs. (7) and (8), we have

$$\begin{aligned}
S(a,b) &= (S_1(a,b), S_2(a,b)) = 1_{L^*} - T(1_{L^*} - a, 1_{L^*} - b) \\
&= 1_{L^*} - (T_1(1_{L^*} - a, 1_{L^*} - b), T_2(1_{L^*} - a, 1_{L^*} - b)) \\
&= 1_{L^*} - (T_1((a_2, a_1), (b_2, b_1)), T_2((a_2, a_1), (b_2, b_1))) \\
&= (T_2((a_2, a_1), (b_2, b_1)), T_1((a_2, a_1), (b_2, b_1))) \\
&= (T_2(1_{L^*} - a, 1_{L^*} - b), T_1(1_{L^*} - a, 1_{L^*} - b)).
\end{aligned}$$

Thus (1) and (2) hold. Similarly, we can conclude (3) and (4).

**Definition 4.** [1] *Let a set $U$ be fixed. An IF set $A$ in $U$ is an object having the form*

$$A = \{\langle x, \mu_A(x), \gamma_A(x) \rangle \mid x \in U\},$$

*where $\mu_A : U \to [0,1]$ and $\gamma_A : U \to [0,1]$ satisfy $0 \le \mu_A(x) + \gamma_A(x) \le 1$ for all $x \in U$, and $\mu_A(x)$ and $\gamma_A(x)$ are, respectively, called the degree of membership and the degree of non-membership of the element $x \in U$ to $A$. The family of all IF subsets in $U$ is denoted by $\mathcal{IF}(U)$. The complement of an IF set $A$ is defined by $\sim A = \{\langle x, \gamma_A(x), \mu_A(x) \rangle \mid x \in U\}$.*

It can be observed that an IF set $A$ is associated with two fuzzy sets $\mu_A$ and $\gamma_A$. Here, we denote $A(x) = (\mu_A(x), \gamma_A(x))$, then it is clear that $A \in \mathcal{IF}(U)$ iff $A(x) \in L^*$ for all $x \in U$. Obviously, a fuzzy set $A = \{\langle x, \mu_A(x) \rangle \mid x \in U\}$ can be identified with the IF set of the form $\{\langle x, \mu_A(x), 1 - \mu_A(x) \rangle \mid x \in U\}$. Thus an IF set is indeed an extension of a fuzzy set.

Some basic operations on $\mathcal{IF}(U)$ are introduced as follows [1]: for $A, B, A_i \in \mathcal{IF}(U)$, $i \in J$, $J$ is an index set,

- $A \subseteq B$ iff $\mu_A(x) \le \mu_B(x)$ and $\gamma_A(x) \ge \gamma_B(x)$ for all $x \in U$,
- $A \supseteq B$ iff $B \subseteq A$,
- $A = B$ iff $A \subseteq B$ and $B \subseteq A$,
- $A \cap B = \{\langle x, \min(\mu_A(x), \mu_B(x)), \max(\gamma_A(x), \gamma_B(x)) \rangle \mid x \in U\}$,

- $A \cup B = \{\langle x, \max(\mu_A(x), \mu_B(x)), \min(\gamma_A(x), \gamma_B(x))\rangle \mid x \in U\},$
- $\bigcap\limits_{i \in J} A_i = \{\langle x, \bigwedge\limits_{i \in J} \mu_{A_i}(x), \bigvee\limits_{i \in J} \gamma_{A_i}(x)\rangle \mid x \in U\},$
- $\bigcup\limits_{i \in J} A_i = \{\langle x, \bigvee\limits_{i \in J} \mu_{A_i}(x), \bigwedge\limits_{i \in J} \gamma_{A_i}(x)\rangle \mid x \in U\}.$

For $(\alpha, \beta) \in L^*$, $\widehat{(\alpha, \beta)}$ will be denoted by the constant IF set: $\widehat{(\alpha, \beta)}(x) = (\alpha, \beta)$, for all $x \in U$. For any $y \in U$ and $M \in \mathcal{P}(U)$, IF sets $1_y$, $1_{U-\{y\}}$, and $1_M$ are, respectively, defined as follows: for $x \in U$,

$$\mu_{1_y}(x) = \begin{cases} 1, & \text{if} \quad x = y, \\ 0, & \text{if} \quad x \neq y. \end{cases} \qquad \gamma_{1_y}(x) = \begin{cases} 0, & \text{if} \quad x = y, \\ 1, & \text{if} \quad x \neq y. \end{cases}$$

$$\mu_{1_{U-\{y\}}}(x) = \begin{cases} 0, & \text{if} \quad x = y, \\ 1, & \text{if} \quad x \neq y. \end{cases} \qquad \gamma_{1_{U-\{y\}}}(x) = \begin{cases} 1, & \text{if} \quad x = y, \\ 0, & \text{if} \quad x \neq y. \end{cases}$$

$$\mu_{1_M}(x) = \begin{cases} 1, & \text{if} \quad x \in M, \\ 0, & \text{if} \quad x \notin M. \end{cases} \qquad \gamma_{1_M}(x) = \begin{cases} 0, & \text{if} \quad x \in M, \\ 1, & \text{if} \quad x \notin M. \end{cases}$$

The IF universe set is $U = 1_U = \widehat{(1,0)} = \widehat{1_{L^*}} = \{\langle x, 1, 0\rangle \mid x \in U\}$ and the IF empty set is $\emptyset = \widehat{(0,1)} = \widehat{0_{L^*}} = \{\langle x, 0, 1\rangle \mid x \in U\}.$

By using $L^*$, IF sets on $U$ can be represented as follows: for $A, B, A_j \in \mathcal{IF}(U)(j \in J, J$ is an index set), $x, y \in U$, and $M \in \mathcal{P}(U)$

- $A(x) = (\mu_A(x), \gamma_A(x)) \in L^*,$
- $U(x) = (1, 0) = 1_{L^*},$
- $\emptyset(x) = (0, 1) = 0_{L^*},$
- $x = y \Longrightarrow 1_y(x) = 1_{L^*}$ and $1_{U-\{y\}}(x) = 0_{L^*},$
- $x \neq y \Longrightarrow 1_y(x) = 0_{L^*}$ and $1_{U-\{y\}}(x) = 1_{L^*},$
- $x \in M \Longrightarrow 1_M(x) = 1_{L^*},$
- $x \notin M \Longrightarrow 1_M(x) = 0_{L^*},$
- $A \subseteq B \Longleftrightarrow A(x) \leq_{L^*} B(x), \forall x \in U \Longleftrightarrow B(x) \geq_{L^*} A(x), \forall x \in U,$
- $\left(\bigcap\limits_{j \in J} A_j\right)(x) = \bigwedge\limits_{j \in J} A_j(x) = \left(\bigwedge\limits_{j \in J} \mu_{A_j}(x), \bigvee\limits_{j \in J} \gamma_{A_j}(x)\right) \in L^*,$
- $\left(\bigcup\limits_{j \in J} A_j\right)(x) = \bigvee\limits_{j \in J} A_j(x) = \left(\bigvee\limits_{j \in J} \mu_{A_j}(x), \bigwedge\limits_{j \in J} \gamma_{A_j}(x)\right) \in L^*.$

For two IF sets $A, B \in \mathcal{IF}(U)$, we define two IF sets $A \cap_T B$ and $A \cup_S B$ as follows:

$$\begin{aligned} (A \cap_T B)(x) &= T(A(x), B(x)), \quad x \in U, \\ (A \cup_S B)(x) &= S(A(x), B(x)), \quad x \in U. \end{aligned} \tag{11}$$

It can easily be verified that

$$A \cup_S B = \sim ((\sim A) \cap_T (\sim B)). \tag{12}$$

# 3    $T$-Intuitionistic Fuzzy Rough Approximation Operators

In this section, by employing an IF $t$-norm $T$ and its dual IF $t$-conorm $S$ on $L^*$, we will define the lower and upper approximations of IF sets w.r.t. an arbitrary IF approximation space and discuss properties of $T$-IF rough approximation operators.

**Definition 5.** *Let $U$ and $W$ be two nonempty universes of discourse. A subset $R \in \mathcal{IF}(U \times W)$ is referred to as an IF binary relation from $U$ to $W$, namely, $R$ is given by*

$$R = \{\langle (x,y), \mu_R(x,y), \gamma_R(x,y) \rangle \mid (x,y) \in U \times W\}, \qquad (13)$$

*where $\mu_R : U \times W \to [0,1]$ and $\gamma_R : U \times W \to [0,1]$ satisfy $0 \le \mu_R(x,y) + \gamma_R(x,y) \le 1$ for all $(x,y) \in U \times W$. We denote the family of all IF relations from $U$ to $W$ by $\mathcal{IFR}(U \times W)$. An IF relation $R \in \mathcal{IFR}(U \times W)$ is said to be serial if $\bigvee_{y \in W} R(x,y) = 1_{L^*}$ for all $x \in U$. If $U = W$, $R \in \mathcal{IFR}(U \times U)$ is called an IF binary relation on $U$. $R \in \mathcal{IFR}(U \times U)$ is said to be reflexive if $R(x,x) = 1_{L^*}$ for all $x \in U$. $R$ is said to be symmetric if $R(x,y) = R(y,x)$ for all $x,y \in U$. $R$ is said to be $T$-transitive if $\bigvee_{y \in U} T(R(x,y), R(y,z)) \le_{L^*} R(x,z)$ for all $x, z \in U$, where $T$ is an IF $t$-norm.*

Throughout this section, we always assume that $T$ is an IF continuous $t$-norm on $L^*$ and $S$ the IF $t$-conorm dual to $T$.

**Definition 6.** *Let $U$ and $W$ be two non-empty universes of discourse and $R$ an IF relation from $U$ to $W$, then the triple $(U, W, R)$ is called a generalized IF approximation space. For $A \in \mathcal{IF}(W)$, the $T$-lower and $T$-upper approximations of $A$, denoted as $\underline{R}(A)$ and $\overline{R}(A)$, respectively, w.r.t. the approximation space $(U, W, R)$ are IF sets of $U$ and are, respectively, defined as follows:*

$$\underline{R}(A)(x) = \bigwedge_{y \in W} S(1_{L^*} - R(x,y), A(y)), \quad x \in U. \qquad (14)$$

$$\overline{R}(A)(x) = \bigvee_{y \in W} T(R(x,y), A(y)), \quad x \in U. \qquad (15)$$

*The operators $\underline{R}, \overline{R} : \mathcal{IF}(W) \to \mathcal{IF}(U)$ are, respectively, referred to as $T$-lower and $T$-upper IF rough approximation operators of $(U, W, R)$, and the pair $(\underline{R}(A), \overline{R}(A))$ is called the $T$-IF rough set of $A$ w.r.t. $(U, W, R)$.*

**Theorem 1.** *Let $(U, W, R_1)$ and $(U, W, R_2)$ be two IF approximation spaces, if $R_1 \subseteq R_2$, then*
  (1) *$\overline{R_1}(A) \subseteq \overline{R_2}(A)$ for all $A \in \mathcal{IF}(W)$.*
  (2) *$\underline{R_2}(A) \subseteq \underline{R_1}(A)$ for all $A \in \mathcal{IF}(W)$.*

*Proof.* It can be deduced directly from Definition 6.

**Definition 7.** *If $U, V, W$ are three nonempty sets, $R_1$ is an IF relation from $U$ to $V$, and $R_2$ is an IF relation from $V$ to $W$, we define an IF relation from $U$ to $W$, denoted $R_1 \circ R_2$, called the $T$-composition of $R_1$ and $R_2$ as follows:*

$$R_1 \circ R_2(x,z) = \bigvee_{y \in V} T(R_1(x,y), R_2(y,z)), \forall (x,z) \in U \times W. \qquad (16)$$

**Theorem 2.** *Let $(U, V, R_1)$ and $(V, W, R_2)$ be two IF approximation spaces, then*
(1) $\overline{R_1 \circ R_2}(A) = \overline{R_1}(\overline{R_2}(A))$ *for all $A \in \mathcal{IF}(W)$.*
(2) $\underline{R_1 \circ R_2}(A) = \underline{R_1}(\underline{R_2}(A))$ *for all $A \in \mathcal{IF}(W)$.*

*Proof.* (1) For any $u \in U$, we have

$$
\begin{aligned}
\overline{R_1}(\overline{R_2}(A))(u) &= \bigvee_{v \in V} T(R_1(u, v), \overline{R_2}(A)(v)) \\
&= \bigvee_{v \in V} T(R_1(u, v), \bigvee_{w \in W} T(R_2(v, w), A(w))) \\
&= \bigvee_{v \in V} \bigvee_{w \in W} T(R_1(u, v), T(R_2(v, w), A(w))) \\
&= \bigvee_{w \in W} \bigvee_{v \in V} T(T(R_1(u, v), R_2(v, w)), A(w)) \\
&= \bigvee_{w \in W} T(\bigvee_{v \in V} T(R_1(u, v), R_2(v, w)), A(w)) \\
&= \bigvee_{w \in W} T(R_1 \circ R_2(u, w), A(w)) \\
&= \overline{R_1 \circ R_2}(A)(u).
\end{aligned}
$$

Thus, $\overline{R_1 \circ R_2}(A) = \overline{R_1}(\overline{R_2}(A))$.
(2) It is similar to the proof of (1).

**Theorem 3.** *Let $(U, W, R)$ be an IF approximation space, $T$ an IF t-norm on $L^*$, and $S$ the IF t-conorm dual to $T$, then*
(IFL1)    $\underline{R}(A) = \sim \overline{R}(\sim A)$ *for all $A \in \mathcal{IF}(W)$.*
(IFU1)    $\overline{R}(A) = \sim \underline{R}(\sim A)$ *for all $A \in \mathcal{IF}(W)$.*

*Proof.* For any $A \in \mathcal{IF}(W)$ and $x \in U$, by Eq. (4) and Proposition 2 we have

$$
\begin{aligned}
\overline{R}(\sim A)(x) &= \bigvee_{y \in W} T(R(x, y), (\sim A)(y)) \\
&= \bigvee_{y \in W} (T_1(R(x, y), 1_{L^*} - A(y)), T_2(R(x, y), 1_{L^*} - A(y))) \\
&= (\bigvee_{y \in W} T_1(R(x, y), 1_{L^*} - A(y)), \bigwedge_{y \in W} T_2(R(x, y), 1_{L^*} - A(y))) \\
&= (\bigvee_{y \in W} S_2(1_{L^*} - R(x, y), A(y)), \bigwedge_{y \in W} S_1(1_{L^*} - R(x, y), A(y))) \\
&= 1_{L^*} - (\bigwedge_{y \in W} S_1(1_{L^*} - R(x, y), A(y)), \bigvee_{y \in W} S_2(1_{L^*} - R(x, y), A(y))).
\end{aligned}
$$

Thus

$$
\begin{aligned}
(\sim \overline{R}(\sim A))(x) &= 1_{L^*} - \overline{R}(\sim A)(x) \\
&= (\bigwedge_{y \in W} S_1(1_{L^*} - R(x, y), A(y)), \bigvee_{y \in W} S_2(1_{L^*} - R(x, y), A(y))) \\
&= \bigwedge_{y \in W} (S_1(1_{L^*} - R(x, y), A(y)), S_2(1_{L^*} - R(x, y), A(y))) \\
&= \bigwedge_{y \in W} S(1_{L^*} - R(x, y), A(y)) \\
&= \underline{R}(A)(x).
\end{aligned}
$$

Therefore, we conclude (IFL1). Similarly, we can prove that (IFU1) holds.

Properties (IFL1) and (IFU1) in Theorem 3 show that the $T$-IF rough approximation operators $\underline{R}$ and $\overline{R}$ are dual with each other. The following theorem presents some basic properties of $T$-IF rough approximation operators.

**Theorem 4.** *Let* $(U, W, R)$ *be an IF approximation space. Then the upper and lower $T$-fuzzy rough approximation operators defined in Definition 6 satisfy the following properties: For all* $A, B \in \mathcal{IF}(W)$, $A_j \in \mathcal{IF}(W)(\forall j \in J, J$ *is an index set),* $M \subseteq W$, $(x, y) \in U \times W$ *and all* $(\alpha, \beta) \in L^*$,

(IFL2)  $\underline{R}(\widehat{(\alpha, \beta)} \cup_S (A)) = \widehat{(\alpha, \beta)} \cup_S \underline{R}(A).$

(IFU2)  $\overline{R}(\widehat{(\alpha, \beta)} \cap_T A) = \widehat{(\alpha, \beta)} \cap_T \overline{R}(A).$

(IFL3)  $\underline{R}(\bigcap_{j \in J} A_j) = \bigcap_{j \in J} \underline{R}(A_j).$

(IFU3)  $\overline{R}(\bigcup_{j \in J} A_j) = \bigcup_{j \in J} \overline{R}(A_j).$

(IFL4)  $A \subseteq B \Longrightarrow \underline{R}(A) \subseteq \underline{R}(B).$

(IFU4)  $A \subseteq B \Longrightarrow \overline{R}(A) \subseteq \overline{R}(B).$

(IFL5)  $\underline{R}(\bigcup_{j \in J} A_j) \supseteq \bigcup_{j \in J} \underline{R}(A_j).$

(IFU5)  $\overline{R}(\bigcap_{j \in J} A_j) \subseteq \bigcap_{j \in J} \overline{R}(A_j).$

(IFL6)  $\underline{R}(W) = U.$

(IFU6)  $\overline{R}(\emptyset_W) = \emptyset_U.$

(IFL7)  $\underline{R}(1_{W - \{y\}})(x) = 1_{L^*} - R(x, y).$

(FU7)  $\overline{R}(1_y)(x) = R(x, y).$

(IFL8)  $\widehat{(\alpha, \beta)} \subseteq \underline{R}(\widehat{(\alpha, \beta)}).$

(IFU8)  $\overline{R}(\widehat{(\alpha, \beta)}) \subseteq \widehat{(\alpha, \beta)}.$

(IFL9)  $\underline{R}(1_M)(x) = \bigwedge_{y \notin M} (1_{L^*} - R(x, y)).$

(IFU9)  $\overline{R}(1_M)(x) = \bigvee_{y \in M} R(x, y).$

(IFL10)  $\underline{R}(1_{W - \{y\}} \cup_S \widehat{(\alpha, \beta)})(x) = S(1_{L^*} - R(x, y), (\alpha, \beta)).$

(IFU10)  $\overline{R}(1_y \cap_T \widehat{(\alpha, \beta)})(x) = T(R(x, y), (\alpha, \beta)).$

*Proof.* The proof for properties of the upper $T$-IF rough approximation operator can be found in [17], and properties of lower $T$-IF rough approximation operator can be deduced directly by employing the dual properties (IFL1) and (IFU1) in Theorem 3.

By using Theorem 5 in [17] and the dualities in Theorem 3, we can obtain following Theorems 5-8, which show that properties of some special IF relations, say serial IF relations, reflexive IF relations, symmetric IF relations, and $T$-transitive IF relations, can be equivalently characterized by properties of the $T$-IF rough approximation operators.

**Theorem 5.** *Let $(U, W, R)$ be an IF approximation space, then*

$$R \text{ is serial} \iff (\text{IFL0}) \ \underline{R}((\widehat{\alpha, \beta})) = (\widehat{\alpha, \beta}), \ \forall(\alpha, \beta) \in L^*.$$
$$\iff (\text{IFU0}) \ \overline{R}((\widehat{\alpha, \beta})) = (\widehat{\alpha, \beta}), \ \forall(\alpha, \beta) \in L^*.$$
$$\iff (\text{IFL0})' \ \underline{R}(\emptyset_W) = \emptyset_U.$$
$$\iff (\text{IFU0})' \ \overline{R}(W) = U.$$

**Theorem 6.** *Let $(U, R)$ be an IF approximation space (i.e. $R$ is an IF relation on $U$), then*

$$R \text{ is reflexive} \iff (\text{IFLR}) \ \underline{R}(A) \subseteq A, \quad \forall A \in \mathcal{IF}(U).$$
$$\iff (\text{IFUR}) \ A \subseteq \overline{R}(A), \quad \forall A \in \mathcal{IF}(U).$$

**Theorem 7.** *Let $(U, R)$ be an IF approximation space, then*

$$R \text{ is symmetric} \iff (\text{IFLS}) \ \underline{R}(1_{U-\{x\}})(y) = \underline{R}(1_{U-\{y\}})(x), \forall(x, y) \in U \times U.$$
$$\iff (\text{IFUS}) \ \overline{R}(1_x)(y) = \overline{R}(1_y)(x), \forall(x, y) \in U \times U.$$

**Theorem 8.** *Let $(U, R)$ be an IF approximation space, then*

$$R \text{ is } T - transitive \iff (\text{IFLT}) \ \underline{R}(A) \subseteq \underline{R}(\underline{R}(A)), \forall A \in \mathcal{IF}(U).$$
$$\iff (\text{IFUT}) \overline{R}(\overline{R}(A)) \subseteq \overline{R}(A), \forall A \in \mathcal{IF}(U).$$

## 4 Conclusion

We have studied a general type of relation-based intuitionistic fuzzy rough sets determined by IF triangular norms with their dual IF triangular conorms. We have introduced a dual pair of $T$-lower and $T$-upper IF rough approximation operators induced from a generalized IF approximation space. We have presented some properties of $T$-lower and $T$-upper IF rough approximation operators and have also examined essential properties of $T$-IF rough approximation operators corresponding to some special types of IF binary relations. For further study, we will investigate more mathematical structures of the $T$-IF rough approximation operators.

**Acknowledgement.** This work was supported by grants from the National Natural Science Foundation of China (Nos. 61272021, 61075120, 11071284, and 61173181), the Zhejiang Provincial Natural Science Foundation of China (Nos. LZ12F03002 and LY14F030001), and Chongqing Key Laboratory of Computational Intelligence (No. CQ-LCI-2013-01).

## References

1. Atanassov, K.: Intuitionistic Fuzzy Sets: Theory and Applications. Physica-Verlag, Heidelberg (1999)
2. Chakrabarty, K., Gedeon, T., Koczy, L.: Intuitionistic fuzzy rough set. In: Proceedings of 4th Joint Conference on Information Sciences (JCIS), Durham, NC, pp. 211–214 (1998)

3. Cornelis, C., Cock, M.D., Kerre, E.E.: Intuitionistic fuzzy rough sets: at the cross-roads of imperfect knowledge. Expert Systems 20, 260–270 (2003)
4. Cornelis, C., Deschrijver, G., Kerre, E.E.: Implication in intuitionistic fuzzy and interval-valued fuzzy set theory: construction, classification, application. International Journal of Approximate Reasoning 35, 55–95 (2004)
5. Huang, B., Li, H.-X., Wei, D.-K.: Dominance-based rough set model in intuitionistic fuzzy information systems. Knowledge-Based Systems 28, 115–123 (2012)
6. Jena, S.P., Ghosh, S.K.: Intuitionistic fuzzy rough sets. Notes on Intuitionistic Fuzzy Sets 8, 1–18 (2002)
7. Mi, J.-S., Leung, Y., Zhao, H.-Y., Feng, T.: Generalized fuzzy rough sets determined by a triangular norm. Information Sciences 178, 3203–3213 (2008)
8. Pawlak, Z.: Rough Sets: Theoretical Aspects of Reasoning about Data. Kluwer Academic Publishers, Boston (1991)
9. Radzikowska, A.M.: Rough approximation operations based on IF sets. In: Rutkowski, L., Tadeusiewicz, R., Zadeh, L.A., Żurada, J.M. (eds.) ICAISC 2006. LNCS (LNAI), vol. 4029, pp. 528–537. Springer, Heidelberg (2006)
10. Rizvi, S., Naqvi, H.J., Nadeem, D.: Rough intuitionistic fuzzy set. In: Proceedings of the 6th Joint Conference on Information Sciences (JCIS), Durham, NC, pp. 101–104 (2002)
11. Samanta, S.K., Mondal, T.K.: Intuitionistic fuzzy rough sets and rough intuitionistic fuzzy sets. Journal of Fuzzy Mathematics 9, 561–582 (2001)
12. Sun, B., Ma, W., Liu, Q.: An approach to decision making based on intuitionistic fuzzy rough sets over two universes. Journal of the Operational Research Society 64, 1079–1089 (2013)
13. Wu, W.-Z.: On some mathematical structures of $T$-fuzzy rough set algebras in infinite universes of discourse. Fundamenta Informaticae 108, 337–369 (2011)
14. Zhang, Z.M.: Generalized intuitionistic fuzzy rough sets based on intuitionistic fuzzy coverings. Information Sciences 198, 186–206 (2012)
15. Zhou, L., Wu, W.-Z.: On generalized intuitionistic fuzzy approximation operators. Information Sciences 178, 2448–2465 (2008)
16. Zhou, L., Wu, W.-Z.: Characterization of rough set approximations in Atanassov intuitionistic fuzzy set theory. Computers and Mathematics with Applications 62, 282–296 (2011)
17. Zhou, L., Wu, W.-Z., Zhang, W.-X.: On characterization of intuitionistic fuzzy rough sets based on intuitionistic fuzzy implicators. Information Sciences 179, 883–898 (2009)

# Covering Approximations
# in Set-Valued Information Systems

Yanqing Zhu and William Zhu

Lab of Granular Computing,
Minnan Normal University, Zhangzhou, China
williamfengzhu@gmail.com

**Abstract.** As one of three basic theories of granular computing, rough set theory provides a useful tool for dealing with the granularity in information systems. Covering-based rough set theory is a generalization of this theory for handling covering data, which frequently appear in set-valued information systems. In this paper, we propose a covering in terms of attribute sets in a set-valued information system and study its responding three types of covering approximations. Moreover, we show that the covering approximation operators induced by indiscernible neighborhoods and neighborhoods are equal to the approximation operators induced by the tolerance and similarity relations, respectively. Meanwhile, the covering approximation operators induced by complementary neighborhoods are equal to the approximation operators induced by the inverse of the similarity relation. Finally, by introducing the concept of relational matrices, the relationships of these approximation operators are equivalently represented.

**Keywords:** Covering, Rough set, Granular computing, Approximation operator, Set-valued information system.

## 1 Introduction

Rough set theory based on equivalence relations, as proposed by Pawlak [1–3] in the early 1980s, is a mathematical tool for dealing with the uncertainty, vagueness and granularity in information systems. Owing to the restrictions of the applications of equivalence relations, many researchers have presented various extensions [4–10]. For an equivalence relation, it partitions a universe into disjoint subsets. By extending partitions to coverings, covering-based rough sets [11–17] are investigated as the extensions of rough set theory.

An information system consists of a non-empty finite set of objects, a non-empty finite set of attributes, a domain of attributes and an information function. In many practical issues, some of the attribute values for an object are set-valued, which are always used to characterize uncertain information and missing information in the information system [18]. For an information system, if each attribute has a unique attribute value, then it is called a single-valued information system; otherwise, it is called a set-valued (multi-valued) information system [19, 20]. In a set-valued information system, covering is a common form of data representation. Covering-based rough sets are useful mathematical tools for dealing with covering data.

D. Miao et al. (Eds.): RSKT 2014, LNAI 8818, pp. 663–672, 2014.
DOI: 10.1007/978-3-319-11740-9_61 © Springer International Publishing Switzerland 2014

In this paper, we present a covering by attribute sets in a set-valued information system. Based on the covering, we connect three existing covering-based rough sets and two relation-based rough sets in this system. Moreover, we prove that the covering approximation operators induced by indiscernible neighborhoods and neighborhoods are equal to the approximation operators induced by the tolerance and similarity relations, respectively. Furthermore, the covering approximation operators induced by complementary neighborhoods are equal to the approximation operators induced by the inverse of a similarity relation. Finally, we redescribe the relationships of these approximation operators from the viewpoint of Boolean matrix. Especially, we propose the characteristic matrices of the three types of covering approximation operators.

The rest of this paper is arranged as follows. Section 2 reviews some fundamental concepts of covering-based rough sets and set-valued information systems. In Section 3, we present a covering in a set-valued information system. Based on the covering, we study relationships between three types of covering approximation operators and the approximation operators induced by two binary relations defined through attribute sets. Section 4 redescribes the relationships among the approximation operators from the viewpoint of Boolean matrix. Section 5 concludes this paper.

## 2    Preliminaries

In this section, we review some fundamental concepts and existing results of three types of covering-based rough sets and two types of relation-based rough sets in set-valued information systems. Throughout this paper, $U$ is a non-empty finite universe.

### 2.1    Three Types of Covering-Based Rough Sets

**Definition 1.**  *[16] Let $U$ be a universe of discourse and $C$ a family of subsets of $U$. If none subsets in $C$ is empty and $\bigcup C = U$, then $C$ is called a covering of $U$.*

Neighborhoods are important concepts in covering-based rough sets. The following definition will present three types of neighborhoods.

**Definition 2.**  *[21, 22] Let $C$ be a covering of $U$ and $x \in U$. $I_C(x) = \bigcup\{K \in C | x \in K\}$, $N_C(x) = \bigcap\{K \in C | x \in K\}$ and $M_C(x) = \{y \in U | x \in N_C(y)\}$ are called the indiscernible neighborhood, neighborhood and complementary neighborhood of $x$ with respect to $C$, respectively. When there is no confusion, we omit the subscript $C$.*

We often characterize data with a neighborhood granule derived from coverings. In covering-based rough sets, neighborhood-based approximations are proposed for describing any subset of a universe. There are three pairs of of neighborhood-based lower and upper approximation operators in the following definition.

**Definition 3.**  *[21–23] Let $C$ be a covering of $U$. For all $X \subseteq U$,*
*$SH_C(X) = \{x \in U | I_C(x) \cap X \neq \emptyset\}$, $SL_C(X) = \{x \in U | I_C(x) \subseteq X\}$,*
*$IH_C(X) = \{x \in U | N_C(x) \cap X \neq \emptyset\}$, $IL_C(X) = \{x \in U | N_C(x) \subseteq X\}$,*
*$CH_C(X) = \{x \in U | M_C(x) \cap X \neq \emptyset\}$, $CL_C(X) = \{x \in U | M_C(x) \subseteq X\}$,*
*are called the second, fifth and complementary upper and lower approximations of $X$ with respect to $C$, respectively. When there is no confusion, we omit the subscript $C$.*

In practical applications, much knowledge is redundant. Therefore, it is necessary to remove the redundancy and retain the essence.

**Definition 4.** *[16] Let $C$ be a covering of $U$ and $K \in C$. If $K$ is a union of some sets in $C - \{K\}$, then $K$ is said to be reducible; otherwise, $K$ is irreducible. The family of all irreducible elements of $C$ is called the reduct of $C$, denoted as $Reduct(C)$.*

## 2.2  Two Types of Relation-Based Rough Sets in Set-Valued Information Systems

An information system is a quadruple $IS = (U, A, V, f)$, where $U$ is a non-empty finite set of objects; $A$ is a non-empty finite set of attributes; $V = \bigcup_{b \in A} V_b$ and $V_b$ is a domain of attribute $b$; $f : U \times A \to 2^V$ is an information function such that $f(x, b) \in 2^{V_b}$ for every $x \in U, b \in A$. For an information system $IS = (U, A, V, f)$, if each attribute has a unique attribute value, then $IS$ with $V = \bigcup_{b \in A} V_b$ is called a single-valued information system; if an information system is not a single-valued information system, it is called a set-valued (multi-valued) information system (see Table 1).

**Table 1.** A set-valued information system

| $U$ | $b_1$ | $b_2$ | $b_3$ |
|---|---|---|---|
| $x_1$ | $\{0\}$ | $\{0\}$ | $\{1, 2\}$ |
| $x_2$ | $\{0, 1, 2\}$ | $\{0, 1, 2\}$ | $\{0\}$ |
| $x_3$ | $\{1, 2\}$ | $\{1\}$ | $\{1, 2\}$ |
| $x_4$ | $\{0, 1\}$ | $\{0, 2\}$ | $\{1\}$ |
| $x_5$ | $\{1, 2\}$ | $\{1, 2\}$ | $\{1\}$ |

**Definition 5.** *[19, 24] In a set-valued information system $IS = (U, A, V, f)$, for $b \in A$, the tolerance relation $T_b$ is defined as :*
$T_b = \{(x, y) | f(x, b) \cap f(y, b) \neq \emptyset\}$.
*For $B \subseteq A$, the tolerance relation $T_B$ is defined as follows:*
$T_B = \{(x, y) | \forall b \in B, f(x, b) \cap f(y, b) \neq \emptyset\} = \bigcap_{b \in B} T_b$.
*When $(x, y) \in T_B$, we call $x$ and $y$ are indiscernible or $x$ is tolerant with $y$ with respect to $B$. Let $T_B(x) = \{y | y \in U, y T_B x\}$. We call $T_B(x)$ the tolerance class of $x$ with respect to $T_B$.*

**Example 1.** A set-valued information system is presented in Table 1. Let $B = \{b_1, b_2\}$. The classes of objects in $U$ can be computed by Definition 5. Then $T_B(x_1) = \{x_1, x_2, x_4\}$, $T_B(x_2) = \{x_1, x_2, x_3, x_4, x_5\}$, $T_B(x_3) = T_B(x_5) = \{x_2, x_3, x_5\}$, $T_B(x_4) = \{x_1, x_2, x_4, x_5\}$.

**Definition 6.** *[24] Given a set-valued information system $IS = (U, A, V, f)$, $\forall X \subseteq U$, $B \subseteq A$, the lower and upper approximations of $X$ with respect to $T_B$ are defined as:*
$\underline{T_B}(X) = \{x \in U | T_B(x) \subseteq X\}$,
$\overline{T_B}(X) = \{x \in U | T_B(x) \cap X \neq \emptyset\}$.

**Definition 7.** *[25] In a set-valued information system $IS = (U, A, V, f)$, for $b \in A$, the relation $T_b'$ called similarity relation is defined as :*
$T_b' = \{(x, y) | f(x, b) \subseteq f(y, b)\}$.
*For $B \subseteq A$, the relation $T_B'$ is defined as follows:*
$T_B' = \{(x, y) | \forall b \in B, f(x, b) \subseteq f(y, b)\} = \bigcap_{b \in B} T_b'$.

When $(x, y) \in T_B'$, we call $x$ and $y$ are indiscernible with respect to $B$. Let $T_B'(x) = \{y | y \in U, x T_B' y\}$. We call $T_B'(x)$ the similarity class for $x$ with respect to $T_B'$.

*Example 2.* A set-valued information system is presented in Table 1. Let $B = \{b_1, b_3\}$. The classes of objects in $U$ can be computed by Definition 7. Then $T_B'(x_1) = \{x_1\}$, $T_B'(x_2) = \{x_2\}$, $T_B'(x_3) = \{x_3\}$, $T_B'(x_4) = \{x_4\}$, $T_B'(x_5) = \{x_3, x_5\}$.

**Definition 8.** *Given a set-valued information system $IS = (U, A, V, f)$, $\forall X \subseteq U$, $B \subseteq A$, the lower and upper approximations of $X$ with respect to $T_B'$ are defined as:*
$\underline{T_B}'(X) = \{x \in U | T_B'(x) \subseteq X\}$,
$\overline{T_B}'(X) = \{x \in U | T_B'(x) \cap X \neq \emptyset\}$.

## 3 Three Types of Covering Approximation Operators in Set-Valued Information Systems

In the above section, we introduce two binary relations $T_B$ and $T_B'$. In this section, as another form of data representation in this system, a covering is proposed through attributes.

**Definition 9.** *Let $(U, A, V, f)$ be a set-valued information system. For any attribute $b \in A$, suppose that $V_b = \{\alpha_1, \alpha_2, \cdots, \alpha_m\}$. We define $C_b$ as:*
$C_b = \{K_{\alpha_1}, K_{\alpha_2}, \cdots, K_{\alpha_m}\}$,
*where $K_{\alpha_i} = \{x | f(x, b) \cap \{\alpha_i\} \neq \emptyset\}$, where $i \in \{1, 2, \cdots, m\}$.*

**Proposition 1.** *Let $(U, A, V, f)$ be a set-valued information system and $b \in A$. Then $C_b$ forms a covering of $U$.*

*Proof.* For any $x \in U$, we have $f(x, b) \in 2^{V_b}$, i.e., $f(x, b) \cap V_b \neq \emptyset$. Hence, $K_\alpha \neq \emptyset$ for any $\alpha \in V_b$, and there exists $\alpha' \in V_b$ such that $x \in K_{\alpha'} \in C_b$. According to the arbitrariness of $x$, it is easy to see that $\cup_{\alpha \in V_b} K_\alpha = U$, i.e., $\cup C_b = U$. Thus, $C_b$ is a covering of $U$.

In a set-valued information system, for $B \subseteq A$, we have $T_B = \bigcap_{b \in B} T_b$ and $T_B' = \bigcap_{b \in B} T_b'$. For any attribute $b \in A$, it can induce a covering of $U$. However, different attributes generate different coverings. In order to address the issues about attribute sets, we firstly present a concept of intersection of two coverings induced by two different attributes in the following definition.

**Definition 10.** *Let $(U, A, V, f)$ be a set-valued information system. Suppose that $C_{b_1}$ and $C_{b_2}$ are two coverings of $U$ induced by $b_1$ and $b_2$, respectively. The intersection of $C_{b_1}$ and $C_{b_2}$ is defined as follows:*
$C_{b_1} \sqcap C_{b_2} = \{K_i \cap K_j | K_i \cap K_j \neq \emptyset, K_i \in C_{b_1}, K_j \in C_{b_2}\}$.

**Proposition 2.** *Let* $(U, A, V, f)$ *be a set-valued information system. Suppose that* $C_{b_1}$ *and* $C_{b_2}$ *are two coverings of* $U$ *induced by* $b_1$ *and* $b_2$, *respectively. Then* $C_{b_1} \sqcap C_{b_2}$ *is a covering of* $U$.

*Proof.* For any $x \in U$, since $C_{b_1}$ and $C_{b_2}$ are two coverings of $U$, there exist $K_i \in C_{b_1}$ and $K_j \in C_{b_2}$ such that $x \in K_i$ and $x \in K_j$. Thus, $x \in K_i \cap K_j \in C_{b_1} \sqcap C_{b_2}$. According to the arbitrariness of $x$, $C_{b_1} \sqcap C_{b_2}$ is a covering of $U$.

*Example 3.* From Table 1, we have $C_{b_1} = \{\{x_1, x_2, x_4\}, \{x_2, x_3, x_4, x_5\}, \{x_2, x_3, x_5\}\}$ and $C_{b_2} = \{\{x_1, x_2, x_4\}, \{x_2, x_3, x_5\}, \{x_2, x_4, x_5\}\}$. Then $C_{b_1} \sqcap C_{b_2} = \{\{x_1, x_2, x_4\}, \{x_2\}, \{x_2, x_4\}, \{x_2, x_5\}, \{x_2, x_3, x_5\}, \{x_2, x_4, x_5\}\}$. It is clear that $C_{b_1} \sqcap C_{b_2}$ is still a covering of $U$.

Based on the coverings induced by attributes, we can obtain the second, fifth and complementary upper and lower approximations of any subset of a universe. Moreover, we have the following three theorems.

**Theorem 1.** *Let* $(U, A, V, f)$ *be a set-valued information system and* $B \subseteq A$. *Denote* $C_B = \sqcap_{b \in B} C_b$, *where* $C_b$ *is a covering of* $U$ *induced by* $b$. *For all* $X \subseteq U$, *then*
*(1)* $SH_{C_B}(X) = \overline{T_B}(X)$,
*(2)* $SL_{C_B}(X) = \underline{T_B}(X)$.

*Proof.* For all $x, y \in U$, suppose that $(x, y) \in R_I^{C_b}$ if and only if $y \in I_{C_b}(x)$. We will proof this theorem by the following two steps.
(1) $T_b = R_I^{C_b}$.
For all $x, y \in U$, $(x, y) \in T_b \Leftrightarrow f(x, b) \cap f(y, b) \neq \emptyset \Leftrightarrow \exists \alpha_i \in V_b, s.t.\ x, y \in K_{\alpha_i} \Leftrightarrow y \in I_{C_b}(x) \Leftrightarrow (x, y) \in R_I^{C_b}$.
(2) $R_I^{C_B} = \sqcap_{b \in B} R_I^{C_b}$.
Firstly, we prove that $R_I^{C_{b_1} \sqcap C_{b_2}} = R_I^{C_{b_1}} \cap R_I^{C_{b_2}}$, where $b_1, b_2 \in A$. For all $x, y \in U$, $(x, y) \in R_I^{C_1} \cap R_I^{C_2} \Leftrightarrow ((x, y) \in R_I^{C_{b_1}}) \wedge ((x, y) \in R_I^{C_2}) \Leftrightarrow (y \in I_{C_{b_1}}(x)) \wedge (y \in I_{C_{b_2}}(x)) \Leftrightarrow (\exists K_1 \in C_{b_1}, s.t.x, y \in K_1) \wedge (\exists K_2 \in C_{b_2}, s.t.x, y \in K_2) \Leftrightarrow \exists K = K_1 \cap K_2 \in C_{b_1} \sqcap C_{b_2}, s.t.x, y \in K \Leftrightarrow y \in I_{C_{b_1} \sqcap C_{b_2}}(x) \Leftrightarrow (x, y) \in R_I^{C_{b_1} \sqcap C_{b_2}}$. Then $R_I^{C_{b_1} \sqcap C_{b_2}} = R_I^{C_{b_1}} \cap R_I^{C_{b_2}}$. Hence, it is clear that $R_I^{C_B} = \sqcap_{b \in B} R_I^{C_b}$.
To sum up, for all $X \subseteq U$, $SH_{C_B}(X) = \overline{R_I^{C_B}}(X) = \overline{T_B}(X)$ and $SL_{C_B}(X) = \underline{R_I^{C_B}}(X) = \underline{T_B}(X)$.

**Theorem 2.** *Let* $(U, A, V, f)$ *be a set-valued information system and* $B \subseteq A$. *Denote* $C_B = \sqcap_{b \in B} C_b$, *where* $C_b$ *is a covering of* $U$ *induced by* $b$. *For all* $X \subseteq U$, *then*
*(1)* $IH_{C_B}(X) = \overline{T'_B}(X)$,
*(2)* $IL_{C_B}(X) = \underline{T'_B}(X)$.

*Proof.* For all $x, y \in U$, suppose that $(x, y) \in R_N^C$ if and only if $y \in N_C(x)$. We will proof this theorem by the following two steps.
(1) $T'_b = R_N^{C_b}$.
For all $x, y \in U$, $(x, y) \in T'_b \Leftrightarrow f(x, b) \subseteq f(y, b) \Leftrightarrow \forall K \in C_b, x \in K \rightarrow y \in K \Leftrightarrow y \in N_{C_b}(x) \Leftrightarrow (x, y) \in R_N^{C_b}$.

(2) $R_N^{C_B} = \sqcap_{b \in B} R_N^{C_b}$.

Firstly, we prove that $R_N^{C_{b_1} \sqcap C_{b_2}} = R_N^{C_{b_1}} \cap R_N^{C_{b_2}}$, where $b_1, b_2 \in A$. For all $x, y \in U$, $(x, y) \in R_N^{C_{b_1}} \cap R_N^{C_{b_2}} \Leftrightarrow ((x, y) \in R_N^{C_{b_1}}) \wedge ((x, y) \in R_N^{C_{b_2}}) \Leftrightarrow (y \in N_{C_{b_1}}(x)) \wedge (y \in N_{C_{b_2}}(x)) \Leftrightarrow (\forall K_1 \in C_{b_1}, x \in K_1 \rightarrow y \in K_1) \wedge (\forall K_2 \in C_{b_2}, x \in K_2 \rightarrow y \in K_2) \Leftrightarrow \forall K = K_1 \cap K_2 \in C_{b_1} \sqcap C_{b_2}, x \in K \rightarrow y \in K \Leftrightarrow y \in N_{C_{b_1} \sqcap C_{b_2}}(x) \Leftrightarrow (x, y) \in R_N^{C_{b_1} \sqcap C_{b_2}}$. Then $R_N^{C_{b_1} \sqcap C_{b_2}} = R_N^{C_{b_1}} \cap R_N^{C_{b_2}}$. Hence, it is clear that $R_N^{C_B} = \sqcap_{b \in B} R_N^{C_b}$.

To sum up, for all $X \subseteq U$, $IH_{C_B}(X) = \overline{R_N^{C_B}}(X) = \overline{T_B'}(X)$ and $IL_{C_B}(X) = \underline{R_N^{C_B}}(X) = \underline{T_B'}(X)$.

**Theorem 3.** *Let $(U, A, V, f)$ be a set-valued information system and $B \subseteq A$. Denote $C_B = \sqcap_{b \in B} C_b$, where $C_b$ is a covering of $U$ induced by $b$. For all $X \subseteq U$, then*
*(1) $CH_{C_B}(X) = \overline{(T_B')^-}(X)$,*
*(2) $CL_{C_B}(X) = \underline{(T_B')^-}(X)$,*
*where $(T_B')^-$ is the inverse of $T_B'$.*

*Proof.* For all $x, y \in U$, suppose that $(x, y) \in R_M^C$ if and only if $y \in M_C(x)$. Similar to the proof of Theorem 2, we can obtain that $(T_b')^- = R_M^{C_b}$ and $R_M^{C_B} = \sqcap_{b \in B} R_M^{C_b}$ by the definition of $M_C(x)$. Hence, for all $X \subseteq U$, $CH_{C_B}(X) = \overline{R_M^{C_B}}(X) = \overline{(T_B')^-}(X)$ and $CL_{C_B}(X) = \underline{R_M^{C_B}}(X) = \underline{(T_B')^-}(X)$.

## 4    Matrix Representation of Covering Approximation Operators

In this section, we propose the characteristic functions of three pairs of covering approximation operators and discuss the reductions of covering. Firstly, we present the concepts of the matrix representation of a covering, the relational matrix of a binary relation and the characteristic function of a set.

**Definition 11.** *[26] Let $C = \{K_1, K_2, \cdots, K_m\}$ be a covering of $U = \{x_1, x_2, \cdots, x_n\}$. We define $M_C = ((M_C)_{ij})_{n \times m}$ as follows:*

$$(M_C)_{ij} = \begin{cases} 1, & x_i \in K_j, \\ 0, & x_i \notin K_j. \end{cases}$$

*$M_C$ is called a matrix representation of $C$, or a matrix representing $C$.*

Suppose $R$ is a relation on $U = \{x_1, x_2, \cdots, x_n\}$. Then its relational matrix $M_R = ((M_R)_{ij})_{n \times n}$ is defined as follows:

$$(M_R)_{ij} = \begin{cases} 1, & (x_i, x_j) \in R, \\ 0, & (x_i, x_j) \notin R. \end{cases}$$

**Definition 12.** *[27] Let $U = \{x_1, x_2, \cdots, x_n\}$, and $X$ a subset of $U$. The characteristic function $G(X) = (g_X(x_1), g_X(x_2), \cdots, g_X(x_n))^T$ (T denotes the transpose operation) is defined as:*

$$g_X(x_i) = \begin{cases} 1, & x_i \in X, \\ 0, & x_i \notin X. \end{cases}$$

*In other words $G(X)$ assigns 1 to an element which belongs to $X$ and 0 to an element which does not belong to $X$.*

In the following definition, we present a new operation between two Boolean matrices to represent covering-based rough sets in terms of the Boolean product.

**Definition 13.** *[26] Let $A = (a_{ij})_{n \times m}$ and $B = (b_{ij})_{m \times p}$ be two Boolean matrices. We define $C = A \odot B$ as follows: $C = (c_{ij})_{n \times p}$, $c_{ij} = \wedge_{k=1}^{m}(b_{kj} - a_{ik} + 1)$.*

$A$ and $B$ are Boolean matrices, but $A \odot B$ is not necessarily a Boolean matrix. We will illustrate this argument by the following example.

*Example 4.* Let $A = (0,0,0,0)$ and $B = (1,1,1,1)^T$ be two Boolean matrices, then $A \odot B = (2)_{1 \times 1}$.

In literature [26], the type-1 characteristic matrix $\Gamma(C)$ and type-2 characteristic matrix $\Pi(C)$ of $C$ were defined, where $\Gamma(C) = M_C \cdot M_C^T$ and $\Pi(C) = M_C \odot M_C^T$. Meanwhile, it was proved that the two types of characteristic matrices were the relational matrices of the relations induced by indiscernible neighborhoods and neighborhoods, respectively. Similarly, we present the type-3 characteristic matrix of $C$ through the following definition.

**Definition 14.** *Let $C$ be a covering of $U$ and $M_C$ a matrix representing $C$. Then $(M_C \odot M_C^T)^T$ is called the type-3 characteristic matrix of $C$, denoted as $\Theta(C)$.*

Combining the definition of complementary neighborhoods with the relational matrices of the relations induced by neighborhoods, we obtain that the type-3 characteristic matrix of a covering is the relational matrix of the relation induced by complementary neighborhoods.

**Proposition 3.** *Let $C$ be a covering of $U$. Then, for all $X \subseteq U$,*
*(1) $G(CH(X)) = \Theta(C) \cdot G(X)$,*
*(2) $G(CL(X)) = \Theta(C) \odot G(X)$.*

*Proof.* Denote $(M_C \odot M_C^T)^T = (t_{ij})_{n \times n}$. If $X = \emptyset$, then $G(CH(X)) = (M_C \odot M_C^T)^T \cdot (0,0,\cdots,0)^T = (0,0,\cdots,0)^T$ and $G(CL(X)) = (M_C \odot M_C^T)^T \odot (0,0,\cdots,0)^T = (0,0,\cdots,0)^T$. Hence $G(CH(X)) = G(CL(X)) = \emptyset$.
Otherwise, (1) $x_i \in CH(X) \Leftrightarrow g_{CH(X)}(x_i) = 1 \Leftrightarrow \vee_{k=1}^{m}(t_{ik} \wedge g_X(x_k)) = 1 \Leftrightarrow \exists k_0 \in \{1,2,\cdots,m\}$ such that $t_{ik_0} = g_X(x_{k_0}) = 1 \Leftrightarrow x_{K_0} \in M_C(x_i), x_{k_0} \in X \Leftrightarrow M_C(x_i) \cap X \neq \emptyset$. So $G(CH(X)) = \Theta(C) \cdot G(X)$.
(2) $x_i \in CL(X) \Leftrightarrow g_{CL(X)}(x_i) = 1 \Leftrightarrow \vee_{k=1}^{m}(g_X(x_k) - t_{ik} + 1) = 1 \Leftrightarrow if\ t_{ik} = 1, then\ g_X(x_k) = 1 \Leftrightarrow if\ x_k \in M_C(x_i), then\ x_k \in X \Leftrightarrow M_C(x_i) \subseteq X$. So $G(CL(X)) = \Theta(C) \odot G(X)$.

In the rest of this section, we propose the matrix representation of the covering approximation operators as shown in the above section and discuss their reductions.

**Theorem 4.** *Let $(U, A, V, f)$ be a set-valued information system. Denote $C_B = \sqcap_{b \in B} C_b$, where $C_b$ is a covering of $U$ induced by $b$. Then, for all $X \subseteq U$,*
*(1) $G(SH_{C_B}(X)) = \Gamma(C_B) \cdot G(X)$,*

$$G(SL_{C_B}(X)) = \Gamma(C_B) \odot G(X);$$
$$(2)\ G(IH_{C_B}(X)) = \Pi(C_B) \cdot G(X),$$
$$G(IL_{C_B}(X)) = \Pi(C_B) \odot G(X);$$
$$(3)\ G(CH_{C_B}(X)) = \Theta(C_B) \cdot G(X),$$
$$G(CL_{C_B}(X)) = \Theta(C_B) \odot G(X).$$

If $K$ is reducible in $C$, then $N_C(x) = N_{C-\{K\}}(x)$ for all $x \in U$. It is clear that $M_C(x) = M_{C-\{K\}}(x)$. Therefore, we have the following two propositions.

**Proposition 4.** *Let $C$ be a covering of $U$ and $K \in C$. If $K$ is reducible, then $\Theta(C) = \Theta(C - \{K\})$.*

**Proposition 5.** *Let $C$ be a covering of $U$. Then $\Theta(C) = \Theta(Reduct(C))$.*

Thus, we can easily obtain the following two theorems by removing smaller covering blocks and reducible elements, respectively.

**Theorem 5.** *Let $(U, A, V, f)$ be a set-valued information system, $C_b$ a covering of $U$ induced by an attribute $b \in A$ and $K \in C_b$. If there exists $K' \in C_b - \{K\}$ such that $K \subseteq K'$, for all $X \subseteq U$, then*
$$\Gamma(C_b) \cdot G(X) = \Gamma(C_b - \{K\}) \cdot G(X),$$
$$\Gamma(C_b) \odot G(X) = \Gamma(C_b - \{K\}) \odot G(X).$$

*Proof.* We need to prove that $\Gamma(C_b) = \Gamma(C_b - \{K\})$. Since $K \subseteq K'$, it is clear that $K \cup K' = K'$. According to the definition of indiscernible neighborhoods, $I_{C_b}(x) = I_{C_b-\{K\}}(x)$ for all $x \in U$. It is straightforward that $\Gamma(C_b) = \Gamma(C_b - \{K\})$.

**Theorem 6.** *Let $(U, A, V, f)$ be a set-valued information system and $C_b$ a covering of $U$ induced by an attribute $b \in A$. Then, for all $X \subseteq U$,*
$$(1)\ \Pi(C_b) \cdot G(X) = \Pi(Reduct(C_b)) \cdot G(X),$$
$$\Pi(C_b) \odot G(X) = \Pi(Reduct(C_b)) \odot G(X);$$
$$(2)\ \Theta(C_b) \cdot G(X) = \Theta(Reduct(C_b)) \cdot G(X),$$
$$\Theta(C_b) \odot G(X) = \Theta(Reduct(C_b)) \odot G(X).$$

*Proof.* (1) If $K$ is a reducible element in $C_b$, it is easy to see that $N_{C_b}(x) = N_{C_b-\{K\}}(x)$ for all $x \in U$. It is straightforward that $\Pi(C_b) \cdot G(X) = \Pi(C_b - \{K\}) \cdot G(X)$. Hence $\Pi(C_b) \cdot G(X) = \Pi(Reduct(C_b)) \cdot G(X)$.
(2) It is straightforward by Proposition 5.

# 5   Conclusions

In this paper, we proposed a covering in a set-valued information system. Meanwhile, based on the covering, three types of covering approximation operators were presented in this system. Moreover, we studied the relationships between these covering approximation operators and the approximation operators induced by two binary relations. Finally, we described them from the viewpoint of Boolean matrix.

**Acknowledgments.** This work is in part supported by National Nature Science Foundation of China under Grant Nos. 61170128 and 61379049, the Key Project of Education Department of Fujian Province under Grant No. JA13192, and the Zhangzhou Municipal Natural Science Foundation under Grant No. ZZ2013J03.

# References

1. Pawlak, Z.: Rough sets. International Journal of Computer and Information Sciences 11, 341–356 (1982)
2. Pawlak, Z.: Rough classification. International Journal of Man-Machine Studies 20, 469–483 (1984)
3. Orlowska, E., Pawlak, Z.: Representation of nondeterministic information. Theoretical Computer Science 29, 27–39 (1984)
4. Skowron, A., Stepaniuk, J.: Tolerance approximation spaces. Fundamenta Informaticae 27, 245–253 (1996)
5. Slowinski, R., Vanderpooten, D.: A generalized definition of rough approximations based on similarity. IEEE Transactions on Knowledge and Data Engineering 12, 331–336 (2000)
6. Liu, G., Zhu, W.: The algebraic structures of generalized rough set theory. Information Sciences 178, 4105–4113 (2008)
7. Yao, Y.Y.: On generalizing pawlak approximation operators. In: Polkowski, L., Skowron, A. (eds.) RSCTC 1998. LNCS (LNAI), vol. 1424, pp. 298–307. Springer, Heidelberg (1998)
8. Yao, Y.: Constructive and algebraic methods of theory of rough sets. Information Sciences 109, 21–47 (1998)
9. Li, T., Zhang, W.: Rough fuzzy approximations on two universes of discourse. Information Sciences 178, 892–906 (2008)
10. Deng, T., Chen, Y., Xu, W., Dai, Q.: A novel approach to fuzzy rough sets based on a fuzzy covering. Information Sciences 177, 2308–2326 (2007)
11. Zakowski, W.: Approximations in the space $(u, \pi)$. Demonstratio Mathematica 16, 761–769 (1983)
12. Bryniarski, E.: A calculus of rough sets of the first order. Bulletin of the Polish Academy of Sciences 36, 71–77 (1989)
13. Bonikowski, Z.: Algebraic structures of rough sets. In: Ziarko, W. (ed.) Rough Sets, Fuzzy Sets and Knowledge Discovery, Springer, pp. 243–247. Springer, Heidelberg (1994)
14. Bonikowski, Z., Bryniarski, E., Wybraniec-Skardowska, U.: Extensions and intentions in the rough set theory. Information Sciences 107, 149–167 (1998)
15. Zhu, W.: Relationship among basic concepts in covering-based rough sets. Information Sciences 179, 2478–2486 (2009)
16. Zhu, W., Wang, F.: Reduction and axiomization of covering generalized rough sets. Information Sciences 152, 217–230 (2003)
17. Qin, K., Gao, Y., Pei, Z.: On covering rough sets. In: Yao, J., Lingras, P., Wu, W.-Z., Szczuka, M.S., Cercone, N.J., Ślęzak, D. (eds.) RSKT 2007. LNCS (LNAI), vol. 4481, pp. 34–41. Springer, Heidelberg (2007)
18. Zhang, W., Ma, J., Fan, S.: Variable threshold concept lattices. Information Sciences 177, 4883–4892 (2007)
19. Guan, Y., Wang, H.: Set-valued information systems. Information Sciences 176, 2507–2525 (2006)
20. Qian, Y., Dang, C., Liang, J., Tang, D.: Set-valued ordered information systems. Information Sciences 179, 2809–2832 (2009)

21. Zhu, W.: Relationship between generalized rough sets based on binary relation and covering. Information Sciences 179, 210–225 (2009)
22. Ma, L.: On some types of neighborhood-related covering rough sets. International Journal of Approximate Reasoning 53, 901–911 (2012)
23. Pomykala, J.A.: Approximation operations in approximation space. Bulletin of the Polish Academy of Sciences 35, 653–662 (1987)
24. Zhang, W.: Incomplete information system and its optimal selections. Computers & Mathematics with Applications 48, 691–698 (2004)
25. Zhang, W., Yao, Y., Liang, Y.: Rough set and concept lattice. Xi'an Jiaotong University Press (2006)
26. Wang, S., Zhu, W., Zhu, Q., Min, F.: Characteristic matrix of covering and its application to boolean matrix decomposition. Information Sciences 263, 186–197 (2014)
27. Liu, G.: The axiomatization of the rough set upper approximation operations. Fundamenta Informaticae 69, 331–342 (2006)

# Rough Fuzzy Set Model for Set-Valued Ordered Fuzzy Decision System[*]

Zhongkui Bao[1,2], Shanlin Yang[1], and Ju Zhao[1]

[1] School of Management, Hefei University of Technology, Hefei, Anhui, China
[2] School of Mathematical Sciences, Anhui University, Heifei, Anhui, China
zkbao@ahu.edu.cn

**Abstract.** The classical rough set theory can not be directly used to reduce knowledge in set-valued ordered fuzzy decision system. Firstly, we propose a dominance relation-based rough fuzzy set model in set-valued ordered fuzzy decision system, and some important properties are investigated. Then, based on rough fuzzy set, the definitions of approximation consistent set and assignment consistent set are given. Judgment theorems of approximation consistent set and assignment consistent set are also obtained, meanwhile, attribute reduction approach based on discernibility matrices is proposed to eliminate redundant attributes that are not essential from the view of fuzzy decisions. Finally, an example is given to illustrate the effectiveness of the proposed method.

**Keywords:** Set-valued ordered fuzzy decision system, rough fuzzy set, knowledge reduction.

## 1 Introduction

Rough set theory, proposed by Pawlak [1, 2] in 1982, has been an effective tool to conceptualize and analyze various types of data. At present, the rough set theory has been successfully applied to artificial intelligence, data mining, decision analysis and intelligent information processing[3-5].

In many practical situations, some of the attribute values for an object may be set-valued, which are used to characterize uncertain and missing information in information systems [6]. Set-valued information systems are an important type of data tables, can be viewed as generalized models of single-valued information systems. Guan and Wang [7] defined tolerance relation and relative reduction in set-valued decision information systems, derived optimal decision rules. Dai et al. [8] proposed two new relations for set-valued information systems, and kinds of uncertainty measures of knowledge are defined in set-valued information systems. Qian et al. [9] proposed the method to queuing problems for objects and dominance-based rough set approach for set-valued ordered information systems in which attributes are criteria.

---

[*] The work is supported by the National Natural Science Foundation of China (No. 71201044, No. 71131002) and the Youth Science Research Foundation of Anhui University (No. 33050054).

D. Miao et al. (Eds.): RSKT 2014, LNAI 8818, pp. 673–682, 2014.
DOI: 10.1007/978-3-319-11740-9_62 © Springer International Publishing Switzerland 2014

The classical rough set theory can be used to deal with the clear concept, but in reality many concepts are fuzzy. So, it is necessary to generalize rough set theory in fuzzy environment. Fuzzy rough set and rough fuzzy set, first proposed by Dubois and Prade [10], have been made up the deficiencies of the traditional rough set theory in several aspects [11-15]. Dai and Tian [16] define a fuzzy relation and construct a fuzzy rough set model for set-valued information systems. Huang et al. [17, 18] applied rough fuzzy set approach to interval-valued fuzzy objective information system and interval-valued intuitionistic fuzzy information systems to simplify knowledge representation and extract nontrivial simpler decision rules. However, so far, how to reduce knowledge and make decision using rough fuzzy set model has scarcely been reported in set-valued ordered fuzzy decision system. In this paper, we first propose the dominance relation-based rough fuzzy set model in set-valued ordered fuzzy decision system, and then, the concepts of approximation consistent set and assignment consistent set are given. Finally, knowledge reduction based on discernibility matrices is investigated.

The remainder of the paper is organized as follows. Section 2 gives the concepts of set-valued ordered fuzzy decision system and dominance relation. In Section 3, we define the rough fuzzy set model while several properties of the model are also examined, and then present the definitions of approximation consistent set and assignment consistent set. In Section 4, the knowledge reduction of set-valued ordered fuzzy decision system is investigated. Finally, some concluding remarks are presented in Section 5.

## 2    Preliminaries

### 2.1    Set-Valued Ordered Fuzzy Decision System

**Definition 1.** A set-valued fuzzy decision system is a 4-tuple $S = <U, C \cup D, V, f>$. Where $U$ is a non-empty finite set of objects called universe, $C$ is a non-empty finite set of condition attributes, $D$ is a non-empty finite set of fuzzy decision attributes, $C \cap D = \varnothing$, $V = V_C \cup V_D$, $V_C$ and $V_D$ are the set of condition attribute value domains and decision attribute value domains, respectively. $f$ is a mapping from $U \times C$ to $V$ such that $f : U \times C \rightarrow 2^{V_c}$ is a set-valued mapping, and $f: U \times D \rightarrow [0,1]$ is a single-valued mapping, called information function.

As discussed in [7], set-valued can be divided into conjunctive set-valued and disjunctive set-valued. In conjunctive set-valued information system, the value of each attribute is all of values in the value domain, and in disjunctive set-valued information system, the value of each attribute is only one of values in the value domain. This paper we only discusses conjunctive set-valued.

**Definition 2.** A conjunctive set-valued fuzzy decision system is called a conjunctive set-valued ordered fuzzy decision system if all condition attribute are inclusion criterion[9].

**Example 1.** Table 1 is a conjunctive set-valued ordered fuzzy decision system, which is extended by a set-valued ordered decision system in [9]. Where $U=\{x_1, x_2, x_3, x_4, x_5, x_6\}$, $A=\{c_1, c_2, c_3, c_4\}=\{$Audition, Spoken Language, Reading, Writing$\}$ and $D=\{d_1, d_2\}=\{$Most likely to excellent student, Most likely to ordinary student$\}$. In the Table 1, $E$, $F$ and $G$ will stand for English, French and German, respectively.

**Table 1.** A set-valued ordered fuzzy decision system about language ability

| $U$ | $c_1$ | $c_2$ | $c_3$ | $c_4$ | $d_1$ | $d_2$ |
|-----|-------|-------|-------|-------|-------|-------|
| $x_1$ | $\{E\}$ | $\{E\}$ | $\{F,G\}$ | $\{F,G\}$ | 0.3 | 0.8 |
| $x_2$ | $\{E,G\}$ | $\{E,F\}$ | $\{F,G\}$ | $\{F,G\}$ | 0.8 | 0.2 |
| $x_3$ | $\{E,F\}$ | $\{E, G\}$ | $\{F,G\}$ | $\{F\}$ | 0.5 | 0.6 |
| $x_4$ | $\{F\}$ | $\{G\}$ | $\{F\}$ | $\{E,F\}$ | 0.2 | 0.8 |
| $x_5$ | $\{F,G\}$ | $\{G\}$ | $\{F,G\}$ | $\{F,G\}$ | 0.4 | 0.6 |
| $x_6$ | $\{E,F\}$ | $\{E,G\}$ | $\{F,G\}$ | $\{F,G\}$ | 0.8 | 0.4 |

## 2.2    Dominance Relation

Qian et al. [9] defined the following binary dominance relation between objects in set-valued ordered information system with all set-valued for conjunctive type.

**Definition 3.** Let $S=<U, C \cup D, V, f>$, $B \subseteq C$, dominance relation with respect to $B$ can be defined as:

$$R_B^{\geq} = \{(x,y) \in U \times U \mid \forall a \in B, f(x,a) \supseteq f(y,a)\} \tag{1}$$

$R_B^{\geq}$ is reflexive and transitive , symmetric is not satisfied. Denoted by $[x]_B^{\geq} = \{y \in U \mid (y,x) \in R_B^{\geq}\}$, $[x]_B^{\leq} = \{y \in U \mid (x,y) \in R_B^{\geq}\}$. $[x]_B^{\geq}$ describes objects that may dominate $x$ and $[x]_B^{\leq}$ describes objects that may be dominated by $x$ in terms of $B$.

# 3    Rough Fuzzy Set Model for Set-Valued Ordered Fuzzy Decision System

In practice, the type of data set is complicated, and classical rough set model can not deal with continuous or fuzzy data. So, based on dominance relation, we define the rough fuzzy set model for set-valued ordered fuzzy decision system.

**Definition 4.** Let $S=<U, C \cup D, V, f>$, $B \subseteq C$, $\forall d \in D$, $x \in U$, the lower and upper approximation of $x$ with respect to condition attribute subset $B$ are defined as follows:

$$\underline{B}(d)(x) = \inf\{f(y,d) : y \in [x]_B^{\geq}\} \tag{2}$$

$$\overline{B}(d)(x) = \sup\{f(y,d) : y \in [x]_B^{\geq}\} \tag{3}$$

$(\underline{B}(d), \overline{B}(d))$ is called rough fuzzy set of fuzzy set $d$ with respect to $B$. Compared with the classical rough set model, the presented lower and upper approximations of each object in set-valued ordered fuzzy decision system are fuzzy numbers. When $d$ is crisp set, rough fuzzy set $(\underline{B}(d), \overline{B}(d))$ is degenerated to Pawlak rough set based on dominance relation.

**Property 1.** Let $S=<U, C \cup D, V, f>$, $A \subseteq B \subseteq C$, $\forall x \in U$, then

(1)    $\underline{B}(d)(x) \leq \overline{B}(d)(x)$ ;        (2)    $\underline{A}(d)(x) \leq \underline{B}(d)(x)$ ;        (3)

$\overline{B}(d)(x) \leq \overline{A}(d)(x)$ ;

(4)   For $\forall y \in [x]_B^{\geq}$, then $\underline{B}(d)(x) \leq \underline{B}(d)(y)$, $\overline{B}(d)(y) \leq \overline{B}(d)(x)$.

**Example 2.** (Continued from Example 1) Compute the lower and upper approximations of all objects in Table 1, and the results are shown in Table 2.

Table 2. All the lower and upper approximations of objects with respect to $D$

| $U$ | $[x]_C^{\geq}$ | $\underline{C}(d_1)(x)$ | $\overline{C}(d_1)(x)$ | $\underline{C}(d_2)(x)$ | $\overline{C}(d_2)(x)$ |
|---|---|---|---|---|---|
| $x_1$ | $\{x_1, x_2, x_6\}$ | 0.3 | 0.8 | 0.2 | 0.8 |
| $x_2$ | $\{x_2\}$ | 0.8 | 0.8 | 0.2 | 0.2 |
| $x_3$ | $\{x_3, x_6\}$ | 0.5 | 0.8 | 0.4 | 0.6 |
| $x_4$ | $\{x_4, x_6\}$ | 0.2 | 0.8 | 0.4 | 0.8 |
| $x_5$ | $\{x_5\}$ | 0.4 | 0.4 | 0.6 | 0.6 |
| $x_6$ | $\{x_6\}$ | 0.8 | 0.8 | 0.4 | 0.4 |

**Definition 5.** Let $S=<U, C \cup D, V, f>$, $B \subseteq C$, $D = \{d_1, d_2, ..., d_r\}$, $\alpha \in [0,1]$ is a fuzzy number. Denoted as

$$L_B(x) = (\underline{B}(d_1)(x), \underline{B}(d_2)(x), ..., \underline{B}(d_r)(x)) \tag{4}$$

$$H_B(x) = (\overline{B}(d_1)(x), \overline{B}(d_2)(x), ..., \overline{B}(d_r)(x)) \tag{5}$$

$$L_B^{\alpha}(x) = \{d_j \mid \underline{B}(d_j)(x) \geq \alpha, 1 \leq j \leq r\} \tag{6}$$

$$H_B^{\alpha}(x) = \{d_j \mid \overline{B}(d_j)(x) \geq \alpha, 1 \leq j \leq r\} \tag{7}$$

(1) If $L_B(x) = L_C(x)$ [1], for any $x \in U$, then $B$ is called the lower approximation consistent set of $C$, and $L_A(x) \neq L_B(x)$, for any $A \subseteq B$, then $B$ is called the lower approximation reduction of $C$.

(2) If $H_B(x) = H_C(x)$ [2], for any $x \in U$, then $B$ is called the upper approximation consistent set of $C$, and $H_A(x) \neq H_B(x)$, for any $A \subseteq B$, then $B$ is called the upper approximation reduction of $C$.

---

[1]    $L_B(x) = L_C(x) \, iff \, \underline{B}(d_j)(x) = \underline{C}(d_j)(x), for \, \forall d_j \in D$.

[2]    $H_B(x) = H_C(x) \, iff \, \overline{B}(d_j)(x) = \overline{C}(d_j)(x), for \forall d_j \in D$.

(3) If $L_B^\alpha(x) = L_C^\alpha(x)$, for any $x \in U$, then $B$ is called $\alpha$ lower assignment consistent set of $C$, and $L_A^\alpha(x) \neq L_B^\alpha(x)$, for any $A \subseteq B$, then $B$ is called $\alpha$ lower assignment reduction of $C$.

(4) If $H_B^\alpha(x) = H_C^\alpha(x)$, for any $x \in U$, then $B$ is called $\alpha$ upper assignment consistent set of $C$, and $H_A^\alpha(x) \neq H_B^\alpha(x)$, for any $A \subseteq B$, then $B$ is called $\alpha$ upper assignment reduction of $C$.

The lower (upper) approximation consistent set of set-valued ordered fuzzy decision system is a subset of condition attribute set that preserves certainty (possible) membership degree of each object in each decision attribute, and $\alpha$ lower (upper) assignment consistent set is a subset of condition attribute set that preserves that certainty (possible) membership degree of each object in each decision attribute is at least $\alpha$.

# 4    Knowledge Reduction to Set-Valued Ordered Fuzzy Decision System

To acquire the approximation reduction set and assignment reduction set, the judgment theorem of approximation consistent set and assignment consistent set is given as follows.

**Theorem 1.** Let $S = <U, C \cup D, V, f>$, $B \subseteq C$, then

(1) $B$ is the lower approximation consistent set of $C \Leftrightarrow$ For $\forall x, y \in U$, if $L_C(y) \geq L_C(x)$ [3] is not satisfied, then $y \notin [x]_B^\geq$.

(2) $B$ is the upper approximation consistent set of $C \Leftrightarrow$ For $\forall x, y \in U$, if $H_C(y) \leq H_C(x)$ [4] is not satisfied, then $y \notin [x]_B^\geq$.

(3) $B$ is $\alpha$ lower assignment consistent set of $C \Leftrightarrow$ For $\forall x, y \in U$, if $L_C^\alpha(y) \supseteq L_C^\alpha(x)$ is not satisfied, then $y \notin [x]_B^\geq$.

(4) $B$ is $\alpha$ upper assignment consistent set of $C \Leftrightarrow$ For $\forall x, y \in U$, if $H_C^\alpha(y) \subseteq H_C^\alpha(x)$ is not satisfied, then $y \notin [x]_B^\geq$.

**Proof.**  (1) "$\Rightarrow$": If $y \in [x]_B^\geq$, then $[y]_B^\geq \subseteq [x]_B^\geq$. For $\forall j \in \{1, 2, \dots r\}$, $\underline{B}(d_j)(y) \geq \underline{B}(d_j)(x)$, by Definition 5, then $L_B(y) \geq L_C(x)$. $B$ is the lower approximation consistent set of $C$, then $L_B(x) = L_C(x)$. Therefore, $L_C(y) \geq L_C(x)$.

"$\Leftarrow$": For $y \in [x]_B^\geq$, then $L_C(y) \geq L_C(x)$. That is, for $\forall j \in \{1, 2, \dots r\}$, then $\underline{C}(d_j)(y) \geq \underline{C}(d_j)(x)$. By Definiton 4, $f(y, d_j) \geq \underline{C}(d_j)(y)$, therefore,

---

[3] $L_C(y) \geq L_C(x) \, iff \, \underline{C}(d_j)(y) \geq \underline{C}(d_j)(x), for \forall d_j \in D$.

[4] $H_C(y) \leq H_C(x) \, iff \, \bar{C}(d_j)(y) \leq \bar{C}(d_j)(x), for \forall d_j \in D$.

$\underline{B}(d_j)(x) = \inf\{f(y,d_j)\big|y \in [x]_B^\geq\} \geq \underline{C}(d_j)(y) \geq \underline{C}(d_j)(x)$. According to Property 1, for $B \subseteq C$, then $\underline{B}(d_j)(x) \leq \underline{C}(d_j)(x)$. So, for $\forall d_j$, $\underline{B}(d_j)(x) = \underline{C}(d_j)(x)$. That is, $L_B(x) = L_C(x)$, $B$ is the lower approximation consistent set of $C$.

(3) "$\Rightarrow$": If $y \in [x]_B^\geq$, then $[y]_B^\geq \subseteq [x]_B^\geq$. For $\forall j \in \{1,2,\ldots r\}$, $\underline{B}(d_j)(y) \geq \underline{B}(d_j)(x)$, then, for $\forall \alpha \in [0,1]$, $L_B^\alpha(y) \supseteq L_B^\alpha(x)$. $B$ is $\alpha$ lower assignment consistent set of $C$, then $L_B^\alpha(x) = L_C^\alpha(x)$. Therefore, $L_C^\alpha(y) \supseteq L_C^\alpha(x)$.

"$\Leftarrow$": For $B \subseteq C$, then, $\underline{B}(d_j)(x) \leq \underline{C}(d_j)(x)$, so, for $\forall \alpha \in [0,1]$, $L_B^\alpha(x) \subseteq L_C^\alpha(x)$. The next, we need to prove that $L_C^\alpha(x) \subseteq L_B^\alpha(x)$ is satisfied. Let $\forall d_j \in L_C^\alpha(x)$. For $y \in [x]_B^\geq$, $L_C^\alpha(y) \supseteq L_C^\alpha(x)$, then $\underline{C}(d_j)(y) \geq \underline{C}(d_j)(x) \geq \alpha$. In addition, $f(y,d_j) \geq \underline{C}(d_j)(y)$, therefore, $\underline{B}(d_j)(x) = \inf\{f(y,d_j)\big|y \in [x]_B^\geq\} \geq \underline{C}(d_j)(x) \geq \alpha$. So, $d_j \in L_B^\alpha(x)$, therefore, $L_C^\alpha(x) \subseteq L_B^\alpha(x)$. That is, $L_B^\alpha(x) = L_C^\alpha(x)$, $B$ is $\alpha$ lower approximation consistent set of $C$.

The proof of (2) and (4) is similar to the proof of (1) and (3).

The above theorem provides approach to judge whether a subset of condition attributes is consistent or not. We can further obtain practical approach to approximation reduction in set-valued ordered fuzzy decision system. First, the definitions of discernibility matrices are given as follows.

**Definition 6.** Let $S=<$U, C∪D, V, f$>$. Denoted as

$$\underline{D}(x,y) = \begin{cases} \{a \in C \big| (x,y) \notin R_a^\geq\}, otherwise \\ \varnothing \qquad\qquad , L_C(x) \geq L_C(y) \end{cases} \tag{8}$$

$$\bar{D}(x,y) = \begin{cases} \{a \in C \big| (x,y) \notin R_a^\geq\}, otherwise \\ \varnothing \qquad\qquad , H_C(x) \leq H_C(y) \end{cases} \tag{9}$$

$$\underline{D}^\alpha(x,y) = \begin{cases} \{a \in C \big| (x,y) \notin R_a^\geq\}, otherwise \\ \varnothing \qquad\qquad , L_C^\alpha(x) \supseteq L_C^\alpha(y) \end{cases} \tag{10}$$

$$\bar{D}^\alpha(x,y) = \begin{cases} \{a \in C \big| (x,y) \notin R_a^\geq\}, otherwise \\ \varnothing \qquad\qquad , H_C^\alpha(x) \subseteq H_C^\alpha(y) \end{cases} \tag{11}$$

$\underline{D} = \underline{D}(x,y)$, $\bar{D} = \bar{D}(x,y)$, $\underline{D}^\alpha = \underline{D}^\alpha(x,y)$ and $\bar{D}^\alpha = \bar{D}^\alpha(x,y)$ are called the lower approximation, the upper approximation, $\alpha$ lower assignment and $\alpha$ upper assignment discernibility matrices of set-valued ordered fuzzy decision system.

**Theorem 2.** Let S=$<$U, C$\cup$D, $V$, f$>$,  $B \subseteq C$,  $\forall x, y \in U$, then

(1) $B$ is the lower approximation consistent set of $C \Leftrightarrow \forall \underline{D}(x,y) \neq \varnothing$, then, $B \cap \underline{D}(x,y) \neq \varnothing$.

(2) $B$ is the upper approximation consistent set of $C \Leftrightarrow \forall \overline{D}(x,y) \neq \varnothing$, then, $B \cap \overline{D}(x,y) \neq \varnothing$.

(3) $B$ is $\alpha$ lower assignment consistent set of $C \Leftrightarrow \forall \underline{D}^{\alpha}(x,y) \neq \varnothing$, then, $B \cap \underline{D}^{\alpha}(x,y) \neq \varnothing$.

(4) $B$ is $\alpha$ upper assignment consistent set of $C \Leftrightarrow \forall \overline{D}^{\alpha}(x,y) \neq \varnothing$, then, $B \cap \overline{D}^{\alpha}(x,y) \neq \varnothing$.

**Proof.** (1) "$\Rightarrow$": $B$ is the lower approximation consistent set of $C$, for $\forall x, y \in U$, $\exists a \in B$,  such that $(x,y) \notin R_a^{\geq}$ while $L_C(x) \geq L_C(y)$ is not satisfied. Therefore, $a \in \underline{D}(x,y)$, $B \cap \underline{D}(x,y) \neq \varnothing$.

"$\Leftarrow$": If $\forall \underline{D}(x,y) \neq \varnothing$, then $B \cap \underline{D}(x,y) \neq \varnothing$. So, for $\forall x, y \in U$, $\exists a \in B$ and $a \in \underline{D}(x,y)$. while $L_C(x) \geq L_C(y)$ is not satisfied, $(x,y) \notin R_a^{\geq}$, i.e., $y \notin [x]_B^{\delta \geq}$. Therefore, $B$ is the lower approximation consistent set of $C$.

The proof of (2)-(4) is similar to the proof of (1).

By Theorem 2, we know that the lower (upper) approximation reduction and $\alpha$ lower (upper) assignment reduction of $C$ are minimum set $B$ which satisfy $B \cap \underline{D} \neq \varnothing$ ($B \cap \overline{D} \neq \varnothing$) and $B \cap \underline{D}^{\alpha} \neq \varnothing$ ($B \cap \overline{D}^{\alpha} \neq \varnothing$), which usually can be obtained by Boolean reasoning.

**Definition 7.** Let S=$<$U, C$\cup$D, V, f$>$, denoted by

$$\underline{\Delta} = \bigwedge_{(x,y) \in U \times U} \vee \underline{D}(x,y), \quad \overline{\Delta} = \bigwedge_{(x,y) \in U \times U} \vee \overline{D}(x,y) \qquad (12)$$

$$\underline{\Delta}^{\alpha} = \bigwedge_{(x,y) \in U \times U} \vee \underline{D}^{\alpha}(x,y), \quad \overline{\Delta}^{\alpha} = \bigwedge_{(x,y) \in U \times U} \vee \overline{D}^{\alpha}(x,y) \qquad (13)$$

$$\underline{\Delta}(x) = \bigwedge_{y \in U} \vee \underline{D}(x,y), \quad \overline{\Delta}(x) = \bigwedge_{y \in U} \vee \overline{D}(x,y) \qquad (14)$$

$$\underline{\Delta}^{\alpha}(x) = \bigwedge_{(x,y) \in U \times U} \vee \underline{D}^{\alpha}(x,y), \quad \overline{\Delta}^{\alpha}(x) = \bigwedge_{(x,y) \in U \times U} \vee \overline{D}^{\alpha}(x,y) \qquad (15)$$

$\underline{\Delta}(\overline{\Delta})$ and $\underline{\Delta}^{\alpha}(\overline{\Delta}^{\alpha})$ are called the lower (upper) approximation discernibility function and $\alpha$ lower (upper) assignment discernibility function. $\underline{\Delta}(x)(\overline{\Delta}(x))$ and $\underline{\Delta}^{\alpha}(x)(\overline{\Delta}^{\alpha}(x))$ are called the lower (upper) approximation discernibility function and $\alpha$ lower (upper) assignment discernibility function of object $x$.

**Theorem 3.** Let $S=<U,\ C \cup D,\ V,\ f>$, the minimal disjunctive normal forms of $\underline{\Delta}$ and $\overline{\Delta}$ are $\vee_{k=1}^{t} \wedge_{s=1}^{q_k} c_{i_s}$ and $\vee_{k=1}^{r} \wedge_{m=1}^{p_k} c_{i_m}$, denoted by $\underline{B}_k = \{c_{i_s} : s = 1, 2, \cdots, q_k\}$ ($k$=1, 2, $\cdots$, $r$), $\overline{B}_k = \{c_{i_m} : m = 1, 2, \cdots, p_k\}$ ($k$=1, 2, ..., $t$), then $\{\underline{B}_k : k = 1, 2, \cdots, t\}$ and $\{\overline{B}_k : k = 1, 2, \cdots, r\}$ are the set of all the lower and upper approximation reduction.

**Proof.** It follows directly from Theorem 2 and the definition of minimal disjunctive normal forms of discernibility functions.

Similarly, all $\alpha$ lower (upper) assignment reduction of set-valued ordered fuzzy decision system can be obtained from the minimal disjunctive normal forms of $\underline{\Delta}^{\alpha}$ and $\overline{\Delta}^{\alpha}$.

**Example 3.** (Continued from Example 2) Let $\alpha = 0.5$, $L_B(x)$, $H_B(x)$, $L_B^{\alpha}(x)$ and $H_B^{\alpha}(x)$ can be obtained, the results are shown in Table 3.

**Table 3.** All the result of $L_B(x)$, $H_B(x)$, $L_B^{0.5}(x)$ and $H_B^{0.5}(x)$

| $U$ | $L_B(x)$ | $H_B(x)$ | $L_B^{0.5}(x)$ | $H_B^{0.5}(x)$ |
|---|---|---|---|---|
| $x_1$ | (0.3, 0.2) | (0.8, 0.8) | $\varnothing$ | $\{d_1, d_2\}$ |
| $x_2$ | (0.8, 0.2) | (0.8, 0.2) | $\{d_1\}$ | $\{d_1\}$ |
| $x_3$ | (0.5, 0.4) | (0.8, 0.6) | $\{d_1\}$ | $\{d_1, d_2\}$ |
| $x_4$ | (0.2, 0.4) | (0.8, 0.8) | $\varnothing$ | $\{d_1, d_2\}$ |
| $x_5$ | (0.4, 0.6) | (0.4, 0.6) | $\{d_2\}$ | $\{d_2\}$ |
| $x_6$ | (0.8, 0.4) | (0.8, 0.4) | $\{d_1\}$ | $\{d_1\}$ |

Compute the approximation discernbility matrices and assignment discernibility matrices in Table 1. The results are shown as follows:

$$\underline{D} = \begin{pmatrix} \varnothing & c_1,c_2 & c_1,c_2 & c_1,c_2,c_4 & c_1,c_2 & c_1,c_2 \\ \varnothing & \varnothing & c_1,c_2 & c_1,c_2,c_4 & c_1,c_2 & c_1,c_2 \\ \varnothing & c_1,c_2,c_4 & \varnothing & \varnothing & c_1,c_4 & c_4 \\ C & C & c_1,c_2,c_3 & \varnothing & c_1,c_2,c_4 & C \\ \varnothing & c_1,c_2 & c_1,c_2 & \varnothing & \varnothing & c_1,c_2 \\ \varnothing & \varnothing & \varnothing & \varnothing & c_1 & \varnothing \end{pmatrix}$$

$$\overline{D} = \begin{pmatrix} \varnothing & c_1,c_2 & c_1,c_2 & \varnothing & c_1,c_2 & c_1,c_2 \\ \varnothing & \varnothing & \varnothing & \varnothing & c_1,c_2 & \varnothing \\ \varnothing & c_1,c_2,c_4 & \varnothing & \varnothing & c_1,c_4 & c_4 \\ \varnothing & C & c_1,c_2,c_3 & \varnothing & c_1,c_2,c_4 & C \\ \varnothing & c_1,c_2 & \varnothing & \varnothing & \varnothing & c_1,c_2 \\ \varnothing & c_1,c_2 & \varnothing & \varnothing & c_1 & \varnothing \end{pmatrix}$$

$$
\underline{D}^{0.5} = \begin{pmatrix}
\varnothing & c_1,c_2 & c_1,c_2 & c_1,c_2,c_4 & c_1,c_2 & c_1,c_2 \\
\varnothing & \varnothing & \varnothing & \varnothing & c_1,c_2 & \varnothing \\
\varnothing & \varnothing & \varnothing & \varnothing & c_1,c_4 & \varnothing \\
\varnothing & C & c_1,c_2,c_3 & \varnothing & c_1,c_2,c_4 & C \\
\varnothing & c_1,c_2 & c_1,c_2 & \varnothing & \varnothing & c_1,c_2 \\
\varnothing & \varnothing & \varnothing & \varnothing & c_1 & \varnothing
\end{pmatrix}
$$

$$
\overline{D}^{0.5} = \begin{pmatrix}
\varnothing & c_1,c_2 & \varnothing & \varnothing & c_1,c_2 & c_1,c_2 \\
\varnothing & \varnothing & \varnothing & \varnothing & c_1,c_2 & \varnothing \\
\varnothing & c_1,c_2,c_4 & \varnothing & \varnothing & c_1,c_4 & c_4 \\
\varnothing & C & \varnothing & \varnothing & c_1,c_2,c_4 & C \\
\varnothing & c_1,c_2 & \varnothing & \varnothing & \varnothing & c_1,c_2 \\
\varnothing & \varnothing & \varnothing & \varnothing & c_1 & \varnothing
\end{pmatrix}
$$

We have $\underline{\Delta} = \overline{\Delta} = c_1 \wedge c_4$, $\underline{\Delta}^{0.5} = c_1$, $\overline{\Delta}^{\alpha} = c_1 \wedge c_4$. Meanwhile, discernibility function of each object can be obtained. For example, $\underline{\Delta}(x_1) = \underline{\Delta}(x_2) = c_1 \vee c_2$, $\underline{\Delta}(x_3) = c_4$, $\overline{\Delta}(x_5) = c_1 \vee c_2$, $\overline{\Delta}(x_6) = c_1$, $\underline{\Delta}^{0.5}(x_2) = c_1 \vee c_2$, $\underline{\Delta}^{0.5}(x_3) = c_1 \vee c_4$, $\overline{\Delta}^{0.5}(x_4) = c_1 \vee c_2 \vee c_4$, $\overline{\Delta}^{0.5}(x_5) = c_1 \vee c_2$. The results of other objects can be derived similarly. Therefore, the proposed knowledge reduction method can eliminate redundant condition attributes and simplify the set-valued ordered fuzzy decision system.

## 5    Conclusion

Set-valued information systems are generalized models of single-valued information systems. In this paper, we firstly propose dominance relation-based rough fuzzy set model in set-valued ordered fuzzy decision system. Then, the concepts of approximation reduction and assignment reduction are given, and judgment theorems of approximation consistent set and assignment consistent set are obtained. Finally, attribute reduction approaches based on rough fuzzy set are investigated. Although the Boolean reasoning approach based on discernbility matrices can yield all attribute reduction sets, the complexity is high and grows exponentially with the attribute size. So efficient attribute reduction approaches are needed in the following work. Besides, we will develop fuzzy rough set model to those more complicated information systems.

## References

1. Pawlak, Z.: Rough sets. International Journal of Computer and Information Sciences 11, 341–356 (1982)
2. Pawlak, Z.: Rough sets: Theoretical Aspects of Reasoning about Data. Kluwer Academic Publishers, London (1991)

3. Pawlak, Z.: Rough set theory and its applications in data analysis. Cybernetics and Systems 29, 661–688 (1998)
4. Greco, S., Matarazzo, B., Slowinski, R.: Rough sets theory for multi-criteria decision analysis. European Journal of Operational Research 129, 1–47 (2001)
5. Pawlak, Z., Skowron, A.: Rough sets: some extensions. Information Sciences 177, 28–40 (2007)
6. Zhang, W.X., Leung, Y., Wu, W.Z.: Information Systems and Knowledge Discovery. Science Press, Beijing (2003)
7. Guan, Y.Y., Wang, H.K.: Set-valued information systems. Information Sciences 176, 2507–2525 (2006)
8. Dai, J.H., Tian, H.W., Liu, L.: Entropy measures and granularity measures for set-valued information systems. Information Sciences 240, 72–82 (2013)
9. Qian, Y.H., Dang, C.Y., Liang, J.Y., Tang, D.W.: Set-valued ordered information systems. Information Sciences 179, 2809–2832 (2009)
10. Dubois, D., Prade, H.: Rough fuzzy sets and fuzzy rough sets. International Journal of General System 17, 191–209 (1990)
11. Wu, W.Z., Mi, J.S., Zhang, W.X.: Generalized fuzzy rough sets. Information Sciences 151, 263–282 (2003)
12. Mi, J.S., Zhang, W.X.: An axiomatic characterization of a fuzzy generalization of rough sets. Information Sciences 160, 235–249 (2004)
13. Gong, Z.T., Sun, B.Z., Chen, D.G.: Rough set theory for the interval-valued fuzzy information systems. Information Sciences 178, 1968–1985 (2008)
14. Sun, B.Z., Gong, Z.T., Chen, D.G.: Fuzzy rough set theory for the interval-valued fuzzy information systems. Information Sciences 178, 2794–2815 (2008)
15. Hu, Q.H., Yu, D.R., Guo, M.Z.: Fuzzy preference based rough sets. Information Sciences 180, 2003–2022 (2010)
16. Dai, J.H., Tian, H.W.: Fuzzy rough set model for set-valued data. Fuzzy Sets and Systems 229, 54–68 (2013)
17. Huang, B.: Graded dominance interval-based fuzzy objective information systems. Knowledge-Based Systems 24, 1004–1012 (2011)
18. Huang, B., Wei, D.K., Li, H.X., Zhuang, Y.L.: Using a rough set model to extract rules in dominance-based interval-valued intuitionistic fuzzy information systems. Information Sciences 221, 215–229 (2013)

# Optimal-Neighborhood Statistics Rough Set Approach with Multiple Attributes and Criteria

WenBin Pei, He Lin*, and LingYue Li

School of Information Science and Engineering,
LanZhou University, Lanzhou 730000, P.R. China
`linhe@lzu.edu.cn`

**Abstract.** This paper focuses on the sorting problems with multiple types of attributes. About the attributes, in which are divided into qualitative attributes, quantitative attributes, qualitative criteria and quantitative criteria. Granules of knowledge are defined by applying four types of relations simultaneously: indiscernibility relation defined on qualitative attributes, similarity relation defined on quantitative attributes, dominance relation defined on qualitative criteria and quasi-partial order relation defined on quantitative criteria. To guarantee the tolerance of the system, the threshold is adjusted, resulting in a N-neighborhood system comes into being. The consistency measure which possess properties of monotonicity is regarded as the Likelihood Function, so the optimal threshold is obtained by Maximum Likelihood Estimation, as a result, N-neighborhood system is converted into optimal 1-neighborhood system. Therefore, we proposed the Optimal-Neighborhood Statistics Rough Set Approach with Multiple Attributes and Criteria.

**Keywords:** Quasi-partial order relation, Consistency measure, Maximum Likelihood Estimation, N-neighborhood system, Rough Set.

## 1 Introduction

Rough set theory[1-2] is a mathematic approach to process incertitude information. The lower approximation is the union of equivalent classes which are subset of the approximated set, so every object in the lower approximation is consistent with each other. The lower approximation might be an empty set, which is, however, should not be empty in fact, as some consistent objects may be treated as inconsistent objects improperly owing to inevitable noise interference. Thus some extensions of rough set model have been proposed, such as Variable Precision Rough Set[7-8] and Bayesian Rough Set[9-10], etc.

About the attributes, among which have been divided into qualitative attributes and quantitative attributes because it enables people to analyze the characteristics of the objects better(Greco et al.,2002). In this paper, among the criteria we distinguish between qualitative criteria and quantitative criteria. By exploiting Pansystems Theory[5] to explore the implied laws between the

---

* Corresponding author.

D. Miao et al. (Eds.): RSKT 2014, LNAI 8818, pp. 683–692, 2014.
DOI: 10.1007/978-3-319-11740-9_63 © Springer International Publishing Switzerland 2014

relation defined on qualitative criteria and the relation defined on quantitative criteria, the conclusion is discovered that the quasi-partial order relation could analyze the objects evaluated by quantitative criteria. In view of the existing situation that the attributes set of the decision express table contains qualitative attributes, quantitative attributes, qualitative criteria and quantitative criteria, granules of knowledge are defined by applying four types of relations simultaneously: the indiscernibility relation defined on qualitative attributes, the similarity relation defined on quantitative attributes, the dominance relation defined on qualitative criteria and the quasi-partial order relation defined on quantitative criteria.

As for quantitative attributes or criteria whose values exist by numerical values, it is acknowledged that the acquired attribute values might be incertitude, uncertain, or unknown owing to noise interference, inaccurate measurements, statistical errors and unclear definition, etc. So, nextly, this paper mainly discusses how to guarantee the tolerance of the system by threshold adjusting. Neighborhoods of a object change along with the change of threshold, thus a N-neighborhood system comes into being. The optimal threshold is obtain by the Maximum Likelihood Estimation, and the Likelihood Function adopts the consistency measure which possess properties of monotonicity.

This paper is organized as follows. Section 2 deducts the relation defined on quantitative criteria. Section 3 is devoted to present the Optimal-Neighborhood Statistics Rough Set Approach with Multiple Attributes and Criteria. In Section 4, an example is analyzed. In Section 5, the conclusion of this paper is drawn.

## 2   Relation Defined on Quantitative Criteria

The cognitive complicacy of the things stems from their uncertainties and diversities. The uncertainties mainly express in the uncertainties of knowledge discernment, denotation description(vague definition of some attribute values), and denotation numbers. However, the diversity is frequently noted as using multiple attributes of various types to describe objects, and a specialized type of attribute are analyzed accurately by a specialized relation.

Known from literature[4], qualitative attributes are analyzed by indiscernibility relation, while quantitative attributes are analyzed by similarity relation. As was admitted in previous study, the domain of a criterion has to be ordered by a preference. The qualitative criteria is the type of criteria whose values is described by the concepts or symbols, so the dominance relation is defined on them, see[4]. However, the values of the quantitative criteria are ordinal numbers, and then the relation is deduced to dispose the objects evaluated by quantitative criteria from the perspectives of properties of Pansystem relation.

**Theorem 1.** If $A$ is a nonempty set, $f \subset A \times A$, then $f - f^{-1}$ is anti-symmetric, $f \cap f^{-1}$ is symmetric.

Proof is omitted because the theorem can be easily proved.                    □

**Theorem 2**[5]. If $A$ is a nonempty set, $f \subset A \times A$, then $f$ can be divided into two relations $f_1$ and $f_2$, which are independent of each other, that is $f = f_1 \cup f_2$, where $f_1 = f - f^{-1}$ is anti-symmetric, $f_2 = f \cap f^{-1}$ is symmetric.

**Definition 1**[5]. If $A$ is a nonempty set, $f \subset A \times A$, the transitivity is defined as

$$T[A] = \{f | f_1 \circ f_1, f_1 \circ f_2, f_2 \circ f_1, f_2 \circ f_2 \subset f, f_1 = f - f^{-1}, f_2 = f \cap f^{-1}\} \quad (1)$$

and the definition of quasi-transitivity is

$$T_q[A] = \{f | f_1 \circ f_1, f_1 \circ f_2, f_2 \circ f_1 \subset f, f_1 = f - f^{-1}, f_2 = f \cap f^{-1}\} \quad (2)$$

Analysis of the above definition shows that the quasi-transitivity is the weakening of the transitivity when the condition $f_2 \circ f_2 \subset f$ is not satisfied.

As we all know, if $A$ is a nonempty set, $f \in T[A]$, $\forall x, y, z \in A$, $(x, y) \in f$ and $(y, z) \in f$, then $(x, z) \in f$, which can be expressed as:

(1).$(x, y) \in f_1$ and $(y, z) \in f_1 \Rightarrow (x, z) \in f_1 \circ f_1 \subset f \Rightarrow (x, z) \in f$,

(2).$(x, y) \in f_1$ and $(y, z) \in f_2 \Rightarrow (x, z) \in f_1 \circ f_2 \subset f \Rightarrow (x, z) \in f$,

(3).$(x, y) \in f_2$ and $(y, z) \in f_1 \Rightarrow (x, z) \in f_2 \circ f_1 \subset f \Rightarrow (x, z) \in f$,

(4).$(x, y) \in f_2$ and $(y, z) \in f_2 \Rightarrow (x, z) \in f_2 \circ f_2 \subset f \Rightarrow (x, z) \in f$.

However, if $f \in T[A]$ is replaced by $f \in T_q[A]$, without changing with (1),(2) and (3) because $f_1 \circ f_1$, $f_1 \circ f_2$, $f_2 \circ f_1$ also belong to $f$. While (4) does not hold becaese $f_2 \circ f_2$ is not necessarily belong to $f$, so $(x, y) \in f_2$ and $(y, z) \in f_2$, $(x, z)$ is not necessarily belong to $f$.

Let us give a example to interpret. For a given universe $A$, which is a nonempty set, $\forall a, b, c \in A$, $g = \{(a, b), (b, c), (a, c)\}$, $f = \{(a, b), (b, a), (a, c), (c, a), (c, b)\}$. It is a obvious fact that $g$ has transitivity as well as quasi-transitivity. After analysis, $f_1 = \{(c, b)\}$, $f_2 = \{(a, b), (b, a), (a, c), (c, a)\}$, and $f_1 \circ f_1 = \varnothing \subset f$, $f_1 \circ f_2 = \{(c, a)\} \subset f$, $f_2 \circ f_1 = \{(a, b)\} \subset f$ and $f_2 \circ f_2 = \{(a, a), (b, b), (b, c), (c, b), (c, c)\} \nsubseteq f$, so a conclusion that $f$ has quasi-transitivity is reached. In fact, the conclusion can also be interpreted by the following method, $(c, a) \in f_2$ and $(a, b) \in f_2 \Rightarrow (c, b) \in f$, however, $(a, c) \in f_2$ and $(b, a) \in f_2 \Rightarrow (b, c) \notin f$, so $f$ is not transitive.

**Theorem 3.** If $A$ is a nonempty set, $f \subset A \times A$, $f_1 = f - f^{-1}$, $f_2 = f \cap f^{-1}$, then $f_1 \circ f_1 \subset f$, $f_1 \circ f_2 \subset f$, $f_2 \circ f_1 \subset f$, $f_2 \circ f_2 \subset f \Longleftrightarrow f \circ f \subset f$.

The proof is omitted because the property can be easily proved. □

For a given universe $A$, which is a nonempty set, $f \subset A \times A$, $\forall x, y \in A$, if $(x, y) \in f$ and $(x, y) \notin f$, then $x$ is preferred to $y$ on $f$, denoted as $xD_f^k y$, where $k > 0$. If $(x, y) \in (f \cap f^{-1})$, then no difference is indicated between $x$ and $y$ on $f$, denoted as $xD_f^0 y$. If $(x, y) \in f$, but there is no way to verify whether $(y, x)$ belongs to $f$, then $x$ is not worse than $y$, denoted as $xD_f^k y$, where $k \succsim 0$.

**Theorem 4.** If $A$ is a nonempty set, $f \in T[A]$, $\forall x, y, z \in A$, then

(1)If $xD_f^k y$ and $yD_f^t z$ on $f$, $k > 0$, $t > 0$, then $xD_f^h z$, $h > 0$ on $f$.

(2)If $xD_f^k y$ and $yD_f^0 z$ on $f$, $k > 0$, then $xD_f^h z$, $h > 0$ on $f$.

(3)If $xD_f^0 y$ and $yD_f^k z$ on $f$, $k > 0$, then $xD_f^h z$, $h > 0$ on $f$.

(4)If $xD_f^0 y$ and $yD_f^0 z$ on $f$, then $xD_f^0 z$ on $f$.

**Proof.** (1) $\because$ On $f$, $xD_f^k y$ and $k > 0$ $\therefore$ $(x, y) \in f$ and $(y, x) \notin f$. $\because$ $yD_f^t z$ and $t > 0$ $\therefore$ $(y, z) \in f$ and $(z, y) \notin f$. Theorem 2 shows that $f = f_1 \cup f_2$, where $f_1 = f - f^{-1}$, $f_2 = f \cap f^{-1}$. Analysis of the above results in $(x, y) \in f_1$, $(y, z) \in f_1$. $\because$ $f \in T[A] \Rightarrow f_1 \circ f_1 \subset f$ $\therefore$ $(x, z) \in f$. Assuming that $(z, x) \in f$, then $(x, z) \in f_2$. $\because f_2 \circ f_1 \subset f$ $\therefore (z, x) \in f_2$ and $(x, y) \in f_1 \Rightarrow (z, y) \in f$, conflicts with the precondition $(z, y) \notin f$. $\therefore$ assumption is incorrect, $(z, x) \notin f$. In conclusion, $xD_f^h z$, $h > 0$ on $f$.    □

Proofs of (2) and (3) are similar to (1), so they had been omitted.    □

**Proof.** (4)On $f$, $\because xD_f^0 y$ $\therefore (x, y) \in f$ and $(y, x) \in f$ $\because yD_f^0 z$ $\therefore (y, z) \in f$ and $(z, y) \in f$. $\because f = f_1 \cup f_2$, $\therefore (x, y) \in f_2$, $(y, z) \in f_2$, similarly, $(y, x) \in f_2$, $(z, y) \in f_2$. $\because f \in T[A] \Rightarrow f_2 \circ f_2 \subset f$ $\therefore (x, z) \in f$, $(z, x) \in f$. So $xD_f^0 z$ on $f$. □

**Theorem 5**[5]. If $A$ is a nonempty set, $f \in T_q[A]$, $\forall x, y, z \in A$, then

(1)If $xD_f^k y$ and $yD_f^t z$ on $f$, $k > 0$, $t > 0$, then $xD_f^h z$, $h > 0$ on $f$.

(2)If $xD_f^k y$ and $yD_f^0 z$ on $f$, $k > 0$, then $xD_f^h z$, $h \succsim 0$ on $f$.

(3)If $xD_f^0 y$ and $yD_f^k z$ on $f$, $k > 0$, then $xD_f^h z$, $h \succsim 0$ on $f$.

(4)If $xD_f^0 y$ and $yD_f^0 z$ on $f$, then $xD_f^k z$ and $k$ is not necessarily equal to 0 on $f$. The only main difference between Theorem 4 and 5 lies in a difference between transitivity and quasi-transitivity which were analyzed previously. So the fact is realized that the information used to describe the objects is completely or strictly delivered by transitivity, however, quasi-transitivity delivers the information related to the objects partly. For example, if $x$ is preferred to $y$ and $y$ is indiscernible from $z$ on $f \in T[A]$, then $x$ is preferred to $z$ on $f$, which is supported by (2) of Theorem 4. However, about (2) from Theorem 5, if $x$ is preferred to $y$ and $y$ is indiscernible from $z$ on $f \in T_q[A]$, then $x$ is not worse than $z$ on $f$.

It was acknowledged that dominance relation is reflexive and transitive, however, quasi-partial order relation has reflexivity and quasi-transitivity. According to the definitions of the quantitative and qualitative criteria, the perceived fact is that the values of qualitative criteria are often accurate or stable, while imprecision or uncertainty is always along with the values of quantitative criteria. So the conclusion is obtained that dominance relation is defined to dispose the objects described by qualitative criteria, while quasi-partial order relation is defined on quantitative criteria, because it is too strict for dominance relation which has transitivity to analyze the objects described by quantitative criteria. The conclusion can be illustrated by the following example.

**Table 1.** Assessment of student achievement

| U | q1 | q2 |
|---|---|---|
| s1 | excellent | 91 |
| s2 | good | 88 |
| s3 | good | 87 |
| s4 | medium | 88 |

In table 1, there are shown four students(reference objects) described by two condition attributes, where $q1$ is a survey course and $q2$ is a examination course. It is obvious and acquired easily from the table that $q1$ is a qualitative criterion and $q2$ is a quantitative criterion. In general, it is hard for a qualitative criterion to be misjudged since the concepts are relatively stable and accurate. But it is common that some degree of errors come with the ordinal numbers, such as errors in scoring and statistics during the marking process, as well as existence of subjective interference during the fraction of the judgment.

The conclusion is realized that $q1$ is analyzed by dominance relation. For example, if $s1$ is better than $s2$ and $s2$ is indiscernible from $s3$ on $q1$, then $s1$ is better than $s3$ on $q1$. If $q2$ is also analyzed by dominance relation, then $s1$ is better than $s2$ and $s2$ is indiscernible from $s4$ on $q2$, so $s1$ is better than $s4$. In fact, it is a very real possibility that $s1$ is maybe indiscernible from $s2$ and $s2$ is maybe not indiscernible from $s4$ on $q2$ owing to some degree of errors. But it is normal that $s1$ is not worse than $s4$ on $q2$. By analysis above indicates that quasi-partial order relation is more appropriate than dominance relation to analyze the objects described by quantitative criteria.

# 3 Optimal-Neighborhood Statistics Rough Set Approach with Multiple Attributes and Criteria

Formally, by an decision express table we applied the 4-tuple $S = < U, A, V, f >$, where $U$ is a finite set of objects, called universe; $A = C \cup D$ is a finite set of attributes, set $C$ of condition attributes, set $D$ of decision criteria, $C$ and $D$ need to satisfy the equation of $C \cap D = \varnothing$; $V = \bigcup_{a \in C} V_a$, $V_a$ is the domain of the attribute $a \in C$; the information function $f$ specifies the attribute value of each object $x$. In S, $D = \{d\}$ makes a partition of the set of reference objects $U$ into a finite number of decision classes $Cl_1, Cl_2, \cdots, Cl_n$. Let $Cl = \{Cl_t, t \in T\}$, $T = \{1, 2, \cdots, n\}$, $\forall u, v \in T$, if $u > v$ such that the objects from $Cl_u$ are preferred to the objects from $Cl_v$, see[6]. In this paper, $C$ can be divided into $C^>$, $C^{\succeq}$, $C^=$, $C^\frown$, where $C^>$ is the set of qualitative criteria, $C^{\succeq}$ is the set of quantitative criteria, $C^=$ is the set of qualitative attributes, $C^\frown$ is the set of quantitative attributes, $C^>$, $C^{\succeq}$, $C^=$, $C^\frown$ need to satisfy equivalence of $C = C^> \cup C^{\succeq} \cup C^= \cup C^\frown$, and the intersection of any two subsets is empty. The sets to be approximated are t-upward union $Cl_t^{\geq} = \bigcup_{s \geq t} Cl_s$ and t-downward union $Cl_t^{\leq} = \bigcup_{s \leq t} Cl_s$.

For $\forall x, y \in U$, $R \subset U \times U$, if $xRy$, then $y$ is $R$-related to $x$. So, for every $x \in U$, the set of objects associated with $x$ is called the neighborhood of $x$, which is defined as $n(x) = \{y \in U : xRy\}$. The neighborhood system of $x$ is a nonempty set of all neighborhoods of $x$, marked as $US(x)$, see [11].

There are two general inconsistencies need to be considered. Firstly, the objects $x$ and $y$ have the identical descriptions on every attribute and criterion, but are assigned to two different decision classes. Secondly, object $x$ dominates object $y$ on every criterion and $x$ and $y$ have identical or similar descriptions on all considered attributes, but $x$ is assigned to a decision class worse than $y$[4].

In S, for any $P \subseteq C$, $P$ can be divided into $P^>$, $P^{\succeq}$, $P^=$, $P^\frown$, where $P^> = P \cap C^>$ is the subset of qualitative criteria, $P^{\succeq} = P \cap C^{\succeq}$ is the subset of quantitative criteria, $P^= = P \cap C^=$ is the subset of qualitative attributes, $P^\frown = P \cap C^\frown$ is the subset of quantitative attributes. For $\forall x, y \in U$, $\forall q \in P^>$, $x$ qualitatively dominates $y$ on $P^>$, denoted as $xD_P y$; $\forall q \in P^{\succeq}$, $x$ quantitatively dominates $y$ on $P^{\succeq}$, denoted as $xB_P^{u_1} y$, where $u_1$ is the dominant threshold. However, for any $P \subseteq P^=$, $x$ is indiscernible from $y$ on $P^=$, denoted as $xI_Py$; for any $P \subseteq P^\sim$, $x$ is similar to $y$ on $P^\sim$, denoted as $xR_P^{u_2}y$, where $u_2$ is the similarity threshold. It was concluded that the similarity between objects is represented by the similarity relation defined on the set of attribute values, that is, $\forall x, y \in U$, $\forall q \in P^\sim$, $xR_P^{u_2}y$ if and only if $\frac{|f(x,q)-f(y,q)|}{f(y,q)} \leq u_2$, see[4]. Similarly, $xB_P^{u_1}y$ if and only if $\frac{f(x,q)-f(y,q)}{f(y,q)} \geq u_1$.

Futhermore, for $\forall x, y \in U$, $\forall q \in P^>$, $f(x,q) \geq f(y,q)$ indicates that $x$ qualitatively dominates $y$ on $P^>$; $f(x,q) = f(y,q)$ indicates that $x$ is indiscernible from $y$ on $P^>$; $f(x,q) \leq f(y,q)$ indicates that $x$ is qualitatively dominated $y$ by on $P^>$. However, $\forall q \in P^\sim$, $\frac{f(x,q)-f(y,q)}{f(y,q)} \geq u_1$ indicates that $x$ quantitatively dominates $y$ on $P^\sim$; $\frac{f(y,q)-f(x,q)}{f(y,q)} < u_1$ illustrates that $x$ is not quantitatively dominated by $y$ on $P^\sim$; $\frac{|f(x,q)-f(y,q)|}{f(y,q)} < u_1$ shows that $x$ is indiscernible from $y$ on $P^\sim$, $\frac{f(y,q)-f(x,q)}{f(y,q)} \geq u_1$ shows that $x$ is quantitatively dominated by $y$ on $P^\sim$, where $0 \leq u_1 \leq 1$ is a dominate threshold. If $f[y,q] \neq 0$, then

(1) $\frac{f(x,q)-f(y,q)}{f(y,q)} \geq u_1 \Rightarrow f(x,q) \geq (1+u_1)f(y,q)$

(2) $\frac{f(y,q)-f(x,q)}{f(y,q)} < u_1 \Rightarrow f(x,q) > (1-u_1)f(y,q)$

(3) $\frac{|f(x,q)-f(y,q)|}{f(y,q)} < u_1 \Rightarrow (1-u_1)f(y,q) < f(x,q) < (1+u_1)f(y,q)$

(4) $\frac{f(y,q)-f(x,q)}{f(y,q)} \geq u_1 \Rightarrow f(x,q) \leq (1-u_1)f(y,q)$

**Theorem 6.** If $U$ is a nonempty set, for each $q \in P^>$, $\forall x, y, z \in U$, then(1)If $f(x,q) \geq f(y,q)$ and $f(y,q) \geq f(z,q)$ on $P^>$, then $f(x,q) \geq f(z,q)$ on $P^>$. (2)If $f(x,q) \geq f(y,q)$ and $f(y,q) = f(z,q)$ on $P^>$, then $f(x,q) \geq f(z,q)$ on $P^>$. (3)If $f(x,q) = f(y,q)$ and $f(y,q) \geq f(z,q)$ on $P^>$, then $f(x,q) \geq f(z,q)$ on $P^>$. (4)If $f(x,q) = f(y,q)$ and $f(y,q) = f(z,q)$ on $P^>$, then $f(x,q) = f(z,q)$ on $P^>$. The proof is omitted because the theorem can be easily proved.     □

**Theorem 7.** If $U$ is a nonempty set, for each $q \in P^\sim$, $\forall x, y, z \in U$, $f(y,q) \neq 0$, $f(z,q) \neq 0$, $0 \leq t \leq 1$, then

(1)If $f(x,q) \geq (1+t)f(y,q)$ and $f(y,q) \geq (1+t)f(z,q)$ on $P^\sim$, then $f(x,q) \geq (1+t)f(z,q)$ on $P^\sim$.

(2)If $f(x,q) \geq (1+t)f(y,q)$ and $(1-t)f(z,q) < f(y,q) < (1+t)f(z,q)$ on $P^\sim$, then $f(x,q) > (1-t)f(z,q)$ on $P^\sim$.

(3)If $(1-t)f(y,q) < f(x,q) < (1+t)f(y,q)$ and $f(y,q) \geq (1+t)f(z,q)$ on $P^\sim$, then $f(x,q) > (1-t)f(z,q)$ on $P^\sim$.

(4)If $(1-t)f(y,q) < f(x,q) < (1+t)f(y,q)$ and $(1-t)f(z,q) < f(y,q) < (1+t)f(z,q)$ on $P^\sim$, then there is not necessarily $(1-t)f(z,q) < f(x,q) < (1+t)f(z,q)$ on $P^\sim$.

The proof is omitted because the theorem can be easily proved.     □

As for quantitative attributes and criteria, neighborhoods of objects change along with the change of threshold, thus a N-neighborhood system is engendered. How to guarantee the optimal tolerance of the system? How to convert the N-neighborhood into the optimal 1-neighborhood.

Two binary relations are defined on $U$, denoted by $M_{uP}$ and $M^*_{uP}$, $\forall x, y \in U \bullet$ $xM_{uP}y$ if and only if $xD_Py$ for each $q \in P^>$, $xB_P^{u1}y$ for each $q \in P^{\succsim}$, $xI_Py$ for any $q \in P^=$, $yR_P^{u2}x$ for any $q \in P^\sim$.

• $xM^*_{uP}y$ if and only if $xD_Py$ for each $q \in P^>$, $xB_P^{u1}y$ for each $q \in P^{\succsim}$, $xI_Py$ for any $q \in P^=$, $xR_P^{u2}y$ for any $q \in P^\sim$.

From $M_{uP}$ and $M^*_{uP}$, four types of granules of knowledge can be generated:

(1)$M_{uP}^{L+}(x) = \{y \in U | yM_{uP}x\}$ is a set of objects $y$ qualitatively dominating $x$ on $P^>$, quantitatively dominating $x$ on $P^{\succsim}$ and indiscernible from $x$ on $P^=$, and $x$ is similar to $y$ on $P^\sim$.

(2)$M_{uP}^{U+}(x) = \{y \in U | yM^*_{uP}x\}$ is a set of objects $y$ qualitatively dominating $x$ on $P^>$, quantitatively dominating $x$ on $P^{\succsim}$ and indiscernible from $x$ on $P^=$, and $y$ is similar to $x$ on $P^\sim$.

(3)$M_{uP}^{L-}(x) = \{y \in U | xM^*_{uP}y\}$ is a set of objects $y$ qualitatively dominated by $x$ on $P^>$, quantitatively dominated by $x$ on $P^{\succsim}$ and indiscernible from $x$ on $P^=$, and $x$ is similar to $y$ on $P^\sim$.

(4)$M_{uP}^{U-}(x) = \{y \in U | xM_{uP}y\}$ is a set of objects $y$ qualitatively dominated by $x$ on $P^>$, quantitatively dominated by $x$ on $P^{\succsim}$ and indiscernible from $x$ on $P^=$, and $y$ is similar to $x$ on $P^\sim$.

Literature[8] proposed two types of consistency measures, that is, the gain-type consistency measure $f_x^p(y)$ and the cost-type consistency measure $g_x^p(y)$, and gave four types of monotonicity properties. For simplicity, this paper regards a cost-type consistency measure $\varepsilon_{Cl_t^\geq}^P(x) = \frac{card(D^+(x) \cap Cl_{t-1}^\leq)}{card(D^+(x))}$ as the Likelihood Function, which possess properties of monotonicity. In fact, you can consider any interestingness measures related to consistency and monotonicity. Probability of intersection of all random events that are independent of each other is:

$$P(\bigcap_{i=1}^n X_i = x_i) = \prod_{i=1}^n P(X_i = x_i) = \prod_{i=1}^n \varepsilon_{Cl_t^\geq}^P(x_i; u) \tag{3}$$

Let $L_P^{L+}(U) = \prod_{i=1}^n \varepsilon_{Cl_t^\geq}^P(x_i; u^{L+})$, the way to calculate the Maximum likelihood Estimation of parameter $u^{L+}$ can be converted into the question solved by computing the maximum value of Likelihood Function $L_P^{L+}(U)$. By analysis, $max\{L_P^{L+}(U)\} = max\{\prod_{i=1}^n (\frac{card(B_{u1}^+(x_i) \cap Cl_{t-1}^\leq)}{card(B_{u1}^+(x_i))} + \frac{card(R_{u2}^L(x_i) \cap (U-Cl_t))}{card(R_{u2}^L(x_i))})\} =$
$max\{\prod_{i=1}^n \frac{card(B_{u1}^+(x_i) \cap Cl_{t-1}^\leq)}{card(B_{u1}^+(x_i))}\} + max\{\prod_{i=1}^n \frac{card(R_{u2}^L(x_i) \cap (U-Cl_t))}{card(R_{u2}^L(x_i))}\} = max\{L_P^+(U_1)\} + max\{L_P(U_2^L)\}$. Similarly,
$L_P^{U+}(U) = \prod_{i=1}^n \varepsilon_{Cl_t^\geq}^P(x_i; u^{U+}) = \prod_{i=1}^n (\frac{card(B_{u1}^+(x_i) \cap Cl_{t-1}^\leq)}{card(B_{u1}^+(x_i))} + \frac{card(R_{u2}^U(x_i) \cap (U-Cl_t))}{card(R_{u2}^U(x_i))})$
$= \prod_{i=1}^n \frac{card(B_{u1}^+(x_i) \cap Cl_{t-1}^\leq)}{card(B_{u1}^+(x_i))} + \prod_{i=1}^n \frac{card(R_{u2}^U(x_i) \cap (U-Cl_t))}{card(R_{u2}^U(x_i))} = L_P^+(U_1) + L_P(U_2^U);$
$L_P^{L-}(U) = \prod_{i=1}^n \varepsilon_{Cl_t^\leq}^P(x_i; u^{L-}) = \prod_{i=1}^n (\frac{card(B_{u1}^-(x_i) \cap Cl_{t+1}^\geq)}{card(B_{u1}^-(x_i))} + \frac{card(R_{u2}^L(x_i) \cap (U-Cl_t))}{card(R_{u2}^L(x_i))})$
$= \prod_{i=1}^n \frac{card(B_{u1}^-(x_i) \cap Cl_{t+1}^\geq)}{card(B_{u1}^-(x_i))} + \prod_{i=1}^n \frac{card(R_{u2}^U(x_i) \cap (U-Cl_t))}{card(R_{u2}^U(x_i))} = L_P^-(U_1) + L_P(U_2^L);$

$$L_P^{U-}(U) = \prod_{i=1}^{n} \varepsilon_{Cl_t^{\geq}}^{P}(x_i; u^{U-}) = \prod_{i=1}^{n} \left( \frac{card(B_{u1}^{-}(x_i) \cap Cl_{t+1}^{\geq})}{card(B_{u1}^{-}(x_i))} + \frac{card(R_{u2}^{U}(x_i) \cap (U - Cl_t))}{card(R_{u2}^{U}(x_i))} \right)$$

$$= \prod_{i=1}^{n} \frac{card(B_{u1}^{-}(x_i) \cap Cl_{t+1}^{\geq})}{card(B_{u1}^{-}(x_i))} + \prod_{i=1}^{n} \frac{card(R_{u2}^{U}(x_i) \cap (U - Cl_t))}{card(R_{u2}^{U}(x_i))} = L_P^{-}(U_1) + L_P(U_2^{U}).$$

**Definition 2.** Given $P \subseteq C$, $t \in T$, the P-lower approximation of $Cl_t^{\geq}$ is

$$P_-(Cl_t^{\geq}) = \{x \in U | M_{uP}^{L+}(x) \subseteq Cl_t^{\geq}, u = [u_1, u_2^L]\} \qquad (4)$$

where, $u_1 = max\{L_P^+(U_1)\}$, $u_2^L = max\{L_P(U_2^L)\}$,
the P-upper approximation of $Cl_t^{\geq}$ is

$$P^-(Cl_t^{\geq}) = \{x \in Cl_t^{\geq} | M_{uP}^{U+}(x) \cap Cl_t^{\geq} \neq \varnothing, u = [u_1, u_2^U]\} \qquad (5)$$

where, $u_1 = max\{L_P^+(U_1)\}$, $u_2^U = max\{L_P(U_2^U)\}$,
and the P-boundary of $Cl_t^{\geq}$ is defined as

$$Bn_P(Cl_t^{\geq}) = P^-(Cl_t^{\geq}) - P_-(Cl_t^{\geq}) \qquad (6)$$

By the same token, given $P \subseteq C$, $t \in T$, the P-lower approximation of $Cl_t^{\leq}$ is $P_-(Cl_t^{\leq}) = \{x \in U | M_{uP}^{L-}(x) \subseteq Cl_t^{\leq}, u = [u_1, u_2^L]\}$, where $u_1 = max\{L_P^-(U_1)\}$, $u_2^L = max\{L_P(U_2^L)\}$. The P-upper approximation of $Cl_t^{\leq}$ is $P^-(Cl_t^{\leq}) = \{x \in Cl_t^{\leq} | M_{uP}^{U-}(x) \cap Cl_t^{\leq} \neq \varnothing, u = [u_1, u_2^U]\}$, where $u_1 = max\{L_P^-(U_1)\}$, $u_2^U = max\{L_P(U_2^U)\}$. The P-boundary of $Cl_t^{\leq}$ is $Bn_P(Cl_t^{\leq}) = P^-(Cl_t^{\leq}) - P_-(Cl_t^{\leq})$.

**Property 1.** Given $P \subseteq C$, $t \in T$, $P_-(Cl_t^{\geq}) \subseteq Cl_t^{\geq} \subseteq P^-(Cl_t^{\geq})$, $P_-(Cl_t^{\leq}) \subseteq Cl_t^{\leq} \subseteq P^-(Cl_t^{\leq})$.
The proof is omitted because the property can be easily proved. □

**Property 2.** Given $R \subseteq P \subseteq C$, $t \in T$, then

$$P^-(Cl_t^{\geq}) \subseteq R^-(Cl_t^{\geq}) \qquad R_-(Cl_t^{\geq}) \subseteq P_-(Cl_t^{\geq}) \qquad Bn_P(Cl_t^{\geq}) \subseteq Bn_R(Cl_t^{\geq})$$

$$P^-(Cl_t^{\leq}) \subseteq P^-(Cl_t^{\geq}) \qquad R_-(Cl_t^{\leq}) \subseteq P_-(Cl_t^{\leq}) \qquad Bn_P(Cl_t^{\leq}) \subseteq Bn_R(Cl_t^{\leq})$$

The proof is omitted because the property can be easily proved. □

**Definition 4.** Given $P \subseteq C$, $t \in T$, the accuracy of approximation of $Cl_t^{\geq}$ and $Cl_t^{\leq}$ are defined as $\partial_P(Cl_t^{\geq}) = \frac{card(P_-(Cl_t^{\geq}))}{card(P^-(Cl_t^{\geq}))}$, $\partial_P(Cl_t^{\leq}) = \frac{card(P_-(Cl_t^{\leq}))}{card(P^-(Cl_t^{\leq}))}$.

**Definition 5.** Given $P \subseteq C$, $t \in T$, the quality of approximation of partition $Cl$ is defined as $\lambda_P(Cl) = \frac{card(U - (\bigcup_{t \in T} Bn_P(Cl_t^{\geq})) \cup (\bigcup_{t \in T} Bn_P(Cl_t^{\leq})))}{card(U)}$.

**Definition 6.** Given $P \subseteq C$, $t \in T$, each minimal subset $R \subseteq P$ such that $\lambda_R(Cl) = \lambda_P(Cl)$, then $R$ is called a reduct of $Cl$, denoted by $RED_{Cl}(C)$. It is acknowledged that a decision table can have more than one reduct, the intersection of all reducts is called the core and is denoted by $CORE_{Cl}(C)$.

## 4   An Example

The following example(extending the one given by slowinski, 2002) illustrates the conclusions introduced above. In table 2, there are shown exemplary decisions of a decision maker concerning eight warehouses(reference objects) decribed by four types of condition attributes:

- $A_1$, capacity of the sales staff,
- $A_2$, maintenance cost,
- $A_3$, geographical region,
- $A_4$, area.

In the case of $C = \{A_1, A_2, A_3, A_4\}$ and $D = \{d\}$. By analysis the table can indicate that $A_1$ is a qualitative criterion, $A_2$ is a quantitative criterion, $A_3$ is a qualitative attribute, while $A_4$ is a quantitative attribute. About this example,

**Table 2.** Decision table with exemplary decision

| U | A1 | A2 | A3 | A4 | d |
|---|-----|-----|----|-----|--------|
| w1 | Medium | 750 | A | 500 | Loss |
| w2 | Good | 680 | A | 400 | Profit |
| w3 | Medium | 720 | A | 450 | Profit |
| w4 | Good | 650 | B | 400 | Loss |
| w5 | Good | 700 | B | 475 | Profit |
| w6 | medium | 710 | B | 425 | Profit |
| w7 | medium | 730 | B | 350 | Profit |
| w8 | medium | 750 | B | 350 | Loss |

the value of $U_1$ is only 0.01, 0.02 and 0.03. Nextly, the optimal threshold among them can be obtain by Maximum Likelihood Estimation.

When $Cl_2^{\geq} = \{w2, w3, w5, w6, w7\}$ is considered, then $max\{L_P(U1)\} = max\{\prod_{i=1}^{n} \frac{card(B_{u1}^+(x_i) \cap Cl_{t-1}^{\leq})}{card(B_{u1}^+(x_i))}\} = max\{\prod_{i=1}^{8} \frac{card(B_{u1}^+(x_i) \cap Cl_{\bar{i}}^{\leq})}{card(B_{u1}^+(x_i))}\}$. We get $u_1 = 0.03$ is the optimal dominant threshold among $u_1 = 0.01$, 0.02, 0.03. If the values of $U_2$ are 0.08, 0.1 and 0.12, $max\{L_P(U_2^L)\} = max\{\prod_{i=1}^{8} \frac{card(R_{u2}^L(x_i) \cap (U - Cl_t))}{card(R_{u2}^L(x_i))}\} = max\{\prod_{i=1}^{8} \frac{card(R_{u2}^L(x_i) \cap Cl_1)}{card(R_{u2}^L(x_i))}\}$, we can obtain $u_2^L = 0.12$ is the optimal similarity threshold among $u_2 = 0.08$, $u_2 = 0.1$ , $u_2 = 0.12$. While $max\{L_P(U_2^U)\} = max\{\prod_{i=1}^{8} \frac{card(R_{u2}^U(x_i) \cap (U - Cl_t))}{card(R_{u2}^U(x_i))}\} = max\{\prod_{i=1}^{8} \frac{card(R_{u2}^U(x_i) \cap Cl_1)}{card(R_{u2}^U(x_i))}\}$, we can receive $u_2^U = 0.1$ is the optimal similarity threshold among $u_2 = 0.08$, 0.1, 0.12.

From analysis introduced above, we obtain $M^{L+}(w1) = \{w1\}$, $M^{L+}(w2) = \{w2\}$, $M^{L+}(w3) = \{w1, w3\}$, $M^{L+}(w4) = \{w4\}$, $M^{L+}(w5) = \{w5\}$, $M^{L+}(w6) = \{w6\}$, $M^{L+}(w7) = \{w7\}$, $M^{L+}(w8) = \{w8\}$. While $M^{U+}(w1) = \{w1\}$, $M^{U+}(w2) = \{w2\}$, $M^{U+}(w3) = \{w3\}$, $M^{U+}(w4) = \{w4\}$, $M^{U+}(w5) = \{w5\}$, $M^{U+}(w6) = \{w6\}$, $M^{U+}(w7) = \{w7\}$, $M^{U+}(w8) = \{w8\}$.

So the C-lower approximation, the C-upper approximation and C-boundary of set of $Cl_2^{\geq}$ are, respectively, $C_-(Cl_2^{\geq}) = \{w2, w5, w6, w7\}$, $C^-(Cl_2^{\geq}) = \{w2, w3, w5, w6, w7\}$, $Bn_c(Cl_2^{\geq}) = \{w3\}$. By the similar way, the C-lower approximation, the C-upper approximation and C-boundary of set $Cl_1^{\leq} = \{w1, w4, w8\}$ are $C_-(Cl_1^{\leq}) = \{w1, w4, w8\}$, $C^-(Cl_1^{\leq}) = \{w1, w4, w8\}$, $Bn_c(Cl_1^{\leq}) = \varnothing$. Therefore, the accuracy of the approximation is equal to 0.8 for $Cl_2^{\geq}$ and to 1 for $Cl_1^{\leq}$, and the quality of approximation of partition $Cl$ is 0.875. There is only one reduct $RED_{Cl}(C) = \{A2, A3, A4\}$, which is also the core.

# 5  Conclusion

The paper proposed the Optimal-Neighborhood Statistics Rough Set Approach with Multiple Attributes and Criteria, which is based on the ideas that multiple types of attributes are needed to describe objects, and the specialized type of attribute is analyzed by the specialized relation. And the quasi-partial order relation was deduced to analyze the objects described by quantitative criteria. When the attributes set contains qualitative attributes, quantitative attributes, qualitative criteria and quantitative criteria, granules of knowledge are generated by applying four types of relations simultaneously:indiscernibility relation, similarity relation, dominance relation and quasi-partial order relation. Because of existence of quantitative attributes or criteria, in order to guarantee the optimal tolerance of the system, the optimal threshold can be obtaind by the Maximum Likelihood Estimation, and the Likelihood Function adopts the consistency measures which possesses properties of monotonicity.

**Acknowledgement.** The paper is supported by the Fundamental Research Funds for the Central Universities(lzujbky-2012-43), and thanks valued amendments which Professor Guoyin Wang, Weibin Deng raised.

# References

1. Pawlak, Z.: Rough sets. Interational Journal of Computer and Information Sciences 11, 341–356 (1982)
2. Pawlak, Z.: Rough sets. Theoretical Aspects of Reasoning About Data. Kluwer Academic Publishers (1991)
3. Slowinski, R., Vanderpooten, D.: A generalized definition of rough approximations based on similarity. IEEE Transactionson Knowledge and Data Engineering 12, 331–336 (2000)
4. Greco, S., Matarazzo, B., Slowinski, R.: Rough sets methodology for sorting problems in presence of multiple attributes and criteria. European Journal of Operational Research 138, 247–259 (2002)
5. Wu, X.M.: Views to world from Pansystem. The Renmin University of China Press, Beijing (1990)
6. Greco, S., Matarazzo, B., Slowinski, R.: Rough approximation of a preference relation by dominance relations. European Journal of Operational Research 117, 63–83 (1999)
7. Ziarko, W.: Variable precision Rough sets model. Journal of Computerand System Science 46, 39–59 (1993)
8. Blaszczynski, J., Greco, S., Slowinski, R., Szelag, M.: Monotonic Variable Consistency Rough Set Approaches. International Journal of Approximate Reasoning 50, 979–999 (2009)
9. Greco, S., Pawlak, Z., Slowinski, R.: Can Bayesian confirmation measures be useful for rough set decision rules? Engineering Applications of Artificial Intelligence 17, 345–361 (2004)
10. Greco, S., Matarazzo, B., Slowinski, R.: Rough membership and Bayesian confirmation measures for parameterized rough sets. International Journal of Approximate Reasoning 49, 285–300 (2008)
11. Yao, Y.Y.: Relational interpretations of neighborhood operators and rough set approximation operators. Information Sciences 111, 239–259 (1998)

# Thresholds Determination for Probabilistic Rough Sets with Genetic Algorithms

Babar Majeed, Nouman Azam, and Jing Tao Yao

Department of Computer Science, University of Regina, Canada S4S 0A2
{majeed2b,azam200n,jtyao}@cs.uregina.ca

**Abstract.** Probabilistic rough sets define the lower and upper approximations and the corresponding three regions by using a pair of $(\alpha, \beta)$ thresholds. Many attempts have been made to determine or calculate effective $(\alpha, \beta)$ threshold values. A common principle in these approaches is to combine and utilize some intelligent technique with a repetitive process in order to optimize different properties of rough set based classification. In this article, we investigate an approach based on genetic algorithms that repeatedly modifies the thresholds while reducing the overall uncertainty of the rough set regions. A demonstrative example suggests that the proposed approach determines useful threshold values within a few iterations. It is also argued that the proposed approach provide similar results to that of some existing approaches such as the game-theoretic rough sets.

## 1  Introduction

A fundamental issue in the application of probabilistic rough set models is the determination or estimation of $(\alpha, \beta)$ threshold values [12]. This issue is generally approached as an optimization or minimization of some properties or examining a tradeoff solution between multiple criteria [3]. The decision-theoretic rough sets (DTRS) minimize the overall cost associated with different actions of classifying objects to obtain the thresholds. The game-theoretic rough sets (GTRS) examine a tradeoff solution between multiple properties to determine the thresholds [4], [10]. The information-theoretic rough sets (ITRS) minimize the overall uncertainties of the rough set regions to estimate the thresholds [1], [2], [11].

Researchers have recently paid some attention in examining and combining iterative or repetitive methods with existing models (such as DTRS, GTRS and ITRS) to guide in learning and obtaining more effective thresholds. An adaptive learning cost function algorithm called Alcofa was combined with a DTRS model to learn the thresholds in [6]. The idea is to use a heuristic of overall decision cost associated with different regions to guide in learning the thresholds. The same study also utilize a genetic algorithm with the fitness function defined as the overall decision cost using the DTRS model. A gradient descent approach was combined with the ITRS model to iteratively reduce the uncertainty of the rough set regions in [3]. A learning method based on improving different proprieties of rough sets based classification was combined with the

D. Miao et al. (Eds.): RSKT 2014, LNAI 8818, pp. 693–704, 2014.
DOI: 10.1007/978-3-319-11740-9_64 © Springer International Publishing Switzerland 2014

GTRS model in [1]. In majority, if not in all cases, the iterative mechanisms introduced in these studies are specifically designed and implemented to be used with a certain model. This leads to a difficulty in making a direct comparison between iterative determination of thresholds using different models. Moreover, it sometimes becomes difficult to understand whether a certain performance gain is due to the iterative process or the model itself.

In this article, we make use of genetic algorithms (GA) to construct a general approach for determination of thresholds. The approach is demonstrated for iteratively reducing the overall uncertainty of rough set regions calculated using the ITRS model. The proposed approach provides benefits in at least two aspects. Firstly, it provides a framework for investigating further rough set models with the GA by realizing them in the form of different fitness functions. This facilitates and enables the comparison between different models thereby leading to a better insight into their relative performances. Secondly, it extends and enrich the ITRS model with capabilities of GA to improve the quality of the determined thresholds. A demonstrative example suggests that the proposed approach provide similar results to that of some existing approaches like the GTRS model.

## 2 Background Knowledge

A main result of probabilistic rough sets is that we can obtain the three rough set regions based on a pair of thresholds as,

$$\text{POS}_{(\alpha,\beta)}(C) = \{x \in U | P(C|[x]) \geq \alpha\}, \tag{1}$$

$$\text{NEG}_{(\alpha,\beta)}(C) = \{x \in U | P(C|[x]) \leq \beta\}, \tag{2}$$

$$\text{BND}_{(\alpha,\beta)}(C) = \{x \in U | \beta < P(C|[x]) < \alpha\}, \tag{3}$$

where and $U$ is the universe of objects and $P(C|[x])$ is the conditional probability of an object $x$ to be in $C$, given that $x \in [x]$. The three probabilistic regions are pair-wise disjoint and lead to a partition of the universe given by, $\pi_{(\alpha,\beta)} = \{\text{POS}_{(\alpha,\beta)}(C), \text{NEG}_{(\alpha,\beta)}(C), \text{BND}_{(\alpha,\beta)}(C)\}$. Another partition with respect to a concept $C$ is created as $\pi_C = \{C, C^c\}$. The uncertainty in $\pi_C$ with respect to the three probabilistic regions are [3]:

$$\begin{aligned} H(\pi_C | \text{POS}_{(\alpha,\beta)}(C)) = {} &- P(C | \text{POS}_{(\alpha,\beta)}(C)) \log P(C | \text{POS}_{(\alpha,\beta)}(C)) \\ &- P(C^c | \text{POS}_{(\alpha,\beta)}(C)) \log P(C^c | \text{POS}_{(\alpha,\beta)}(C)), \end{aligned} \tag{4}$$

$$\begin{aligned} H(\pi_C | \text{NEG}_{(\alpha,\beta)}(C)) = {} &- P(C | \text{NEG}_{(\alpha,\beta)}(C)) \log P(C | \text{NEG}_{(\alpha,\beta)}(C)) \\ &- P(C^c | \text{NEG}_{(\alpha,\beta)}(C)) \log P(C^c | \text{NEG}_{(\alpha,\beta)}(C)), \end{aligned} \tag{5}$$

$$\begin{aligned} H(\pi_C | \text{BND}_{(\alpha,\beta)}(C)) = {} &- P(C | \text{BND}_{(\alpha,\beta)}(C)) \log P(C | \text{BND}_{(\alpha,\beta)}(C)) \\ &- P(C^c | \text{BND}_{(\alpha,\beta)}(C)) \log P(C^c | \text{BND}_{(\alpha,\beta)}(C)). \end{aligned} \tag{6}$$

The above three equations may be viewed as the measure of uncertainty in $\pi_C$ with respect to $\text{POS}_{(\alpha,\beta)}(C)$, $\text{NEG}_{(\alpha,\beta)}(C)$ and $\text{BND}_{(\alpha,\beta)}(C)$ regions, respectively. The conditional probabilities in these equations, e.g., $P(C|\text{POS}_{(\alpha,\beta)}(C))$ is computed as,

$$P(C|POS_{(\alpha,\beta)}(C)) = \frac{|C \cap POS_{(\alpha,\beta)}(C)|}{|POS_{(\alpha,\beta)}(C)|}. \tag{7}$$

Equation (7) is interpreted as the portion of $POS_{(\alpha,\beta)}(C)$ that belongs to $C$. Conditional probabilities for other regions are similarly obtained. The overall uncertainty is computed as an average uncertainty of the regions [2], [3], i.e.,

$$\begin{aligned}
H(\pi_C|\pi_{(\alpha,\beta)}) &= P(POS_{(\alpha,\beta)}(C))H(\pi_C|POS_{(\alpha,\beta)}(C)) \\
&\quad + P(NEG_{(\alpha,\beta)}(C))H(\pi_C|NEG_{(\alpha,\beta)}(C)) \\
&\quad + P(BND_{(\alpha,\beta)}(C))H(\pi_C|BND_{(\alpha,\beta)}(C)).
\end{aligned} \tag{8}$$

The probability of a certain region, say, the positive region, is determined as, $P(POS_{(\alpha,\beta)}(C)) = |POS_{(\alpha,\beta)}(C)|/|U|$. The probabilities of other regions are similarly defined.

Equation (8) was reformulated in a more readable form in [1]. Considering $\Delta_P(\alpha,\beta)$, $\Delta_N(\alpha,\beta)$ and $\Delta_B(\alpha,\beta)$ as the uncertainties of the positive, negative and boundary regions, respectively, i.e.,

$$\Delta_P(\alpha,\beta) = P(POS_{(\alpha,\beta)}(C))H(\pi_C|POS_{(\alpha,\beta)}(C)), \tag{9}$$

$$\Delta_N(\alpha,\beta) = P(NEG_{(\alpha,\beta)}(C))H(\pi_C|NEG_{(\alpha,\beta)}(C)), \tag{10}$$

$$\Delta_B(\alpha,\beta) = P(BND_{(\alpha,\beta)}(C))H(\pi_C|BND_{(\alpha,\beta)}(C)). \tag{11}$$

Using Equations (9) - (11), Equation (8) is rewritten as,

$$\Delta(\alpha,\beta) = \Delta_P(\alpha,\beta) + \Delta_N(\alpha,\beta) + \Delta_B(\alpha,\beta), \tag{12}$$

which represents the overall uncertainty with respect to a particular $(\alpha,\beta)$ threshold pair. From Equation (12), it is noted that different $(\alpha,\beta)$ thresholds will lead to different overall uncertainties. An effective model is obtained by considering a learning mechanism that searches for effective thresholds by minimizing the overall uncertainty. We intend to use an approach based on GA for such a purpose.

# 3    A Genetic Algorithm Based Approach

Genetic algorithms are extensively applied in optimization problems [7], [8]. They utilize techniques inspired from natural evolution, such as inheritance, mutation, selection, and crossover to evolve an optimal solution. We briefly review the working of genetic algorithms and then examine its possible application in determining probabilistic thresholds.

## 3.1    Genetic Algorithms

The GA appear in different versions and variants in the literature. We consider the Holland's genetic algorithm which is also sometimes called as a simple or simplified genetic algorithm (SGA) [5], [9] .

---

**Algorithm 1.** A simple genetic algorithm

---

**Input:** Initial population
**Output:** Optimum solution
  1. Initialize population
  2. Evaluate population
  3. **While** termination criteria is not reached
  4.     Selecting next population using a fitness function
  5.     Perform crossover and mutation
  6. **End**

---

The working of the SGA is based on a population of some initial binary strings called chromosomes. Each string represents the encoding of a possible solution of an optimization problem. The algorithm repeatedly generates subsequent population generations from the current population using the genetic operators such as crossover and mutation. This process continues until a predetermined termination criteria is reached. The step by step procedure of genetic algorithm is presented as **Algorithm 1**.

### 3.2  Threshold Determination Using Genetic Algorithms

We introduce a five step framework or approach for threshold determination using the SGA which is adopted from the description in Section 3.1. A detailed description of these steps and their interpretation and implication in terms of probabilistic thresholds is elaborated and outlined below.

**Step 1.** Generating initial population.
The implementation of SGA starts with an encoding mechanism for representing variables of an optimization problem. We consider a binary encoding mechanism for the sake of simplicity. This means that the possible values of the thresholds $(\alpha, \beta)$ are represented using binary strings. The encoded binary strings are also referred to as chromosomes and each of them may represent a possible solution. Considering $0 \leq \beta < 0.5 \leq \alpha \leq 1$, we may encode possible values of $\alpha$ using two bits such that $00 = 0.7$, $01 = 0.8$, $10 = 0.9$ and $11 = 1.0$. In the same way, we may encode the values of $\beta$ with two bits such that $00 = 0.0$, $01 = 0.1$, $10 = 0.2$ and $11 = 0.3$. Increasing the number of bits will result in different encoding with a possibility of leading to a more accurate and exact solution. One may consider the data itself to obtain the initial encoding rather than initializing them subjectively.

A chromosome is represented by combining the encoding corresponding to one $\alpha$ value and one $\beta$ value. Each chromosome formed in this way corresponds to a threshold pair. For instance, the encoding 00 for $\alpha$ and the encoding 00 for $\beta$ are joined to obtained a chromosome 0000 which corresponds to a threshold pair of $(\alpha, \beta) = (0.70, 0.0)$. A set of chromosomes is known as population. From thresholds perspective, the population is a set or collection of threshold pairs including a possible optimal threshold pair. The encoding performed in this way allows for further operations to be performed on the population of chromosomes.

Another interesting issue in SGA is how to set or determine the initial population of chromosomes. A simple approach is to select them randomly. A better approach however is to utilize some domain specific knowledge. In case of threshold determination, we may inspect the quality of data to handle this issue. For instance, if the data contains minimum level of noise and provides precise information required to identify or specify a concept, a minimum size boundary region is may be expected. This means that certain decisions with high accuracy are possible while keeping $\alpha$ close to 1.0 and $\beta$ close to 0.0. An optimal solution is expected to be in close vicinity of threshold values $(\alpha, \beta) = (1, 0)$, i.e., the Pawlak model. An initial population is therefore selected based on threshold values in the neighbourhood of the Pawlak model.

**Step 2.** Evaluating population.
Each chromosome in the initial population generated in **Step 1** is evaluated by using a fitness function. The evaluation of a chromosome represents its fitness to survive and reproduce. The fitness function is also used in selecting chromosomes for the next stage or iteration. In this study, we used Equation (12), i.e., the overall uncertainty of rough set regions, as a fitness function. It should be noted that different fitness functions may be defined based on different rough set models. For instance, the fitness function measuring the overall decision cost may be obtained with the DTRS model [6]. Since chromosomes represent threshold pairs, the evaluation of chromosomes is essentially the evaluation of threshold pairs based on their associated levels of uncertainties. A threshold pair $(\alpha, \beta)$ will be highly evaluated compared to another threshold pair $(\alpha', \beta')$, if the latter provides better overall uncertainty, i.e., $\Delta(\alpha', \beta') < \Delta(\alpha, \beta)$. By selecting threshold pairs that minimizes the overall uncertainty, the process will iteratively guide towards effective thresholds.

**Step 3.** Termination conditions or criteria.
The iterative process of threshold optimization needs to be stopped at a suitable point in order to ensure effective and efficient performance. This may be approached in different ways, such as, a bound on the number of iterations or the evaluations reaching or crossing some limits or subsequent iterations does not provide any improvements in performance. In previous studies, the stop conditions are being defined in different ways. The stop condition of achieving certain levels of classification accuracy and precision were used in [4]. The decision cost associated with attributes was used to define the termination condition in [6]. The termination conditions of the boundary region becoming empty, the positive region size exceeds the prior probability of the concept $C$ and the total uncertainty of positive and negative regions exceeds that of the boundary region were used in [1].

In the SGA based approach, the fitness function may be used to define the termination conditions. For instance, the fitness value reaches a certain level or

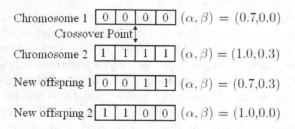

**Fig. 1.** Crossover

subsequent iterations do not significantly improve the fitness. These conditions are mathematically expressed as,

$$\Delta(\alpha, \beta) < \tau \ \text{ and } \ \Delta(\alpha, \beta)_i \approx \Delta(\alpha, \beta)_{i+1} \ \ ...., \tag{13}$$

where $\tau$ represents some predetermined threshold level and $\Delta(\alpha, \beta)_i$ denotes the fitness value at the $i$th iteration. In addition, one may consider the termination conditions utilized in the earlier studies as discussed above.

**Step 4.** Selecting new population.
The SGA generate new population from existing population by making use of a selection mechanism. The design principle in selection mechanisms is to provide more chances to chromosome with higher evaluations to be selected in the next generation. In the threshold determination perspective, this means that the threshold pairs (corresponding to chromosomes) associated with lesser uncertainty will be given higher priority for generating and obtaining further threshold pairs.

A common selection operator is based on the roulette-wheel selection mechanism. The chances of selecting a chromosome is seen as spinning a roulette wheel with the size of the slot for each chromosome as being proportional to its fitness or evaluation. The selection mechanism is formulated by determining the selection probability of each chromosome which is determined as a fitness value of a chromosome divided by the average fitness value of the chromosomes in the entire population. The values are then normalized to obtain selection probabilities. The selection probabilities defined in this way, reflect the sizes of the slots in the wheel. Now, similar to the roulette-wheel procedure, the chromosomes with higher fitness (slot sizes) will have more chance of being chosen. It is possible for one member to dominate all the others and get selected all the time.

**Step 5.** Performing crossover and mutation.
A crossover represents the exchange of genetic material between two parents to produce a unique offspring which has properties of both parents. A crossover can be either one-point or multi-point. The one-point crossover involves cutting the chromosomes of the parents at a randomly chosen common point and exchanging the right-hand-side sub-chromosomes. The multi-point involves cutting and exchanging of parent chromosomes from more than one point. For instance,

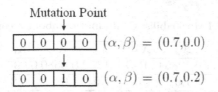

**Fig. 2.** Mutation

Chromosome $1 = 0000$ represents encoding of $(\alpha, \beta) = (0.7,0.0)$ and Chromosome $2 = 1111$ represents encoding of $(\alpha, \beta) = (1.0,0.3)$. The one-point crossover between Chromosome 1 and Chromosome 2 is shown in Fig. 1. The second bit is considered as the one-point crossover point in this case. After swapping the right-hand-side of chromosomes, the new offspring $1 = 0011$ and offspring $2 = 1100$, representing $(\alpha, \beta) = (0.7,0.3)$ and $(\alpha, \beta) = (1.0,0.0)$, respectively are being generated. Multi-point crossover is applied in the same way. For instance, considering two bit cross over with bit 2 and bit 3 as crossover points. New offspring in this case will be offspring $1 = 0010$ and offspring $2 = 1101$ which correspond to $(\alpha, \beta) = (0.7,0.2)$ and $(\alpha, \beta) = (1.0,0.1)$, respectively.

There are sometimes situations in which one may obtain similar initial and final chromosomes in a particular iteration. This may result in unnecessary iterations without improvements. To avoid such situations, mutation is applied to chromosomes by inverting a bit value. The position of the inverting bit in the bit string of a chromosome can be chosen randomly or by applying different probabilistic approaches. We consider the random approach in this article. For instance, in Fig. 2 we randomly choose the third bit of a chromosome and inverted its value from 0 to 1. This results in a new offspring.

The **Step 3** to **Step 5** are repeated until one of the termination conditions is satisfied. Finally, when the algorithm stops, the chromosome and its associated threshold pair having the best fitness (or least uncertainty) among the chromosomes in the final population is returned as the optimal solution.

## 4    A Demonstrative Example and Further Analysis

We further examine and analyze the applicability and usefulness of the proposed approach in this section. A demonstrative example and a comparison with an existing method is considered for this purpose.

### 4.1    An Example of Thresholds Determination

Considering probabilistic information about a concept $C$ which is presented in Table 1. Each $X_i$ in the table represents an equivalence class and each $P(C|X_i)$ is the conditional probability of $X_i$ given $C$. For ease in computations, the equivalence classes are written in decreasing order of $P(C|X_i)$. Moreover, for the sake of simplicity, we consider a majority-oriented model given by $0 \leq \beta < 0.5 \leq \alpha \leq 1$.

**Table 1.** Probabilistic information about a concept

|            | $X_1$  | $X_2$  | $X_3$  | $X_4$  | $X_5$  | $X_6$  | $X_7$  | $X_8$  |
|------------|--------|--------|--------|--------|--------|--------|--------|--------|
| $Pr(X_i)$  | 0.0177 | 0.1285 | 0.0137 | 0.1352 | 0.0580 | 0.0069 | 0.0498 | 0.1070 |
| $Pr(C/X_i)$| 1.0    | 1.0    | 1.0    | 1.0    | 0.9    | 0.8    | 0.8    | 0.6    |

|            | $X_9$  | $X_{10}$ | $X_{11}$ | $X_{12}$ | $X_{13}$ | $X_{14}$ | $X_{15}$ |
|------------|--------|----------|----------|----------|----------|----------|----------|
| $Pr(X_i)$  | 0.1155 | 0.0792   | 0.0998   | 0.1299   | 0.0080   | 0.0441   | 0.0067   |
| $Pr(C/X_i)$| 0.5    | 0.4      | 0.4      | 0.2      | 0.1      | 0.0      | 0.0      |

**Table 2.** Encoding scheme for values of $\alpha$ and $\beta$

| $\alpha$ | $\alpha$-Encoding | $\beta$ | $\beta$-Encoding |
|----------|-------------------|---------|------------------|
| 0.6      | 00                | 0.0     | 00               |
| 0.8      | 01                | 0.1     | 01               |
| 0.9      | 10                | 0.2     | 10               |
| 1.0      | 11                | 0.4     | 11               |

It should also be noted that the data in Table 1 is similar to the data considered in [1], [2] to facilitate comparisons with existing methods.

The **Step 1** to apply the SGA based approach is to encode the threshold pairs in binary bits as outlined in Section 3.2. For the sake of simplicity, we consider a two-bit encoding scheme presented in Table 2. This is a reasonable choice in this case, since based on Table 1 and the conditions associated with majority oriented model, the only values of interest for thresholds $\alpha$ and $\beta$ appear to be $\{1.0, 0.9, 0.8, 0.6\}$ and $\{0.0, 0.1, 0.2, 0.4\}$, respectively. In other words, there are equivalence classes associated only with these threshold values. Each threshold value is represented by using only two bits, e.g., $\alpha = 0.6$ is encoding as 00 and $\alpha = 1.0$ is encoded as 11. A population of size 4 is assumed in this example. To start the algorithm, we randomly choose four threshold pairs resulting in four chromosomes. This is presented in Table 3 under the column of Initial Population. The threshold pairs (0.6, 0.0), (0.8, 0.1), (0.6, 0.2) and (1.0, 0.4) with their corresponding chromosomes or binary encoded strings 0000, 0101, 0010 and 1111, respectively, are being chosen in this case.

The **Step 2** is to evaluate the chromosomes or the resulting population using the fitness function of Equation (12). To evaluate a certain chromosome or threshold pair, say $(\alpha, \beta) = (1.0, 0.0)$, we first determine the positive, negative and boundary regions (based on the data in Table 1 and) according to Equations (1)-(3). The $\text{POS}_{(1.0,0.0)}(C) = \bigcup\{X_1, X_2, X_3, X_4\}$, $\text{NEG}_{(1.0,0.0)}(C) = \bigcup\{X_{14}, X_{15}\}$ and $\text{BND}_{(1.0,0.0)}(C) = \bigcup\{X_3, X_4, ..., X_{13}\}$. The probability of the positive region is $P(\text{POS}_{(\alpha,\beta)}(C)) = P(X_1) + P(X_2) + P(X_3) + P(X_4) = 0.2951$. The probabilities for the negative and boundary regions are similarly obtained from Table 1 as 0.0508 and 0.6541, respectively.

**Table 3.** Initial configuration

| Initial Population | Thresholds | | Fitness Function |
|:---:|:---:|:---:|:---:|
| | $\alpha$ | $\beta$ | |
| 0000 | 0.6 | 0.0 | 0.6756 |
| 0101 | 0.8 | 0.1 | 0.6286 |
| 0010 | 0.6 | 0.2 | 0.6701 |
| 1111 | 1.0 | 0.4 | 0.6228 |

The conditional probability of $C$ with positive region is,

$$P(C|\text{POS}_{(1,0)}(C)) = \frac{\sum_{i=1}^{4} P(C|X_i) * P(X_i)}{\sum_{i=1}^{4} P(X_i)}$$

$$= \frac{1*0.0177+1*0.1285+1*0.0137+1*0.1352}{0.0177+0.1285+0.0137+0.1352} = 1. \quad (14)$$

The probability $P(C^c|\text{POS}_{(1,0)}(C))$ is computed as $1 - P(C|\text{POS}_{(1,0)}(C)) = 1 - 1 = 0$. The Shannon entropy of the positive region based on Equation (4) is therefore calculated as,

$$H(\pi_C|\text{POS}_{(1,0)}(C)) = -1 * log1 - (0 * log0) = 0. \quad (15)$$

The average uncertainty of the positive region according to Equation (9) is $\Delta_P(1,0) = P(\text{POS}_{(1,0)}(C))H(\pi_C|\text{POS}_{(1,0)}(C)) = 0$. In the same way, the uncertainties of the negative and boundary regions according to Equations (10) and (11) are determined as $\Delta_N(1,0) = 0$ and $\Delta_B(1,0) = 0.6537$, respectively. The total uncertainty according to Equation (12) is therefore $\Delta(1,0) = 0.6537$. The fitness values for all chromosomes are similarly obtained and shown in by column of Fitness Function in Table 3.

The algorithm enters into an iterative process once the initial configuration as presented in Table 3 is obtained. This means that **Steps 3** to **Step 5** of the of proposed approach will be executed until one of the termination conditions is reached. A single condition of 2 iterations is used in this example. It is just for the sake of ease and demonstration and should not be treated in any sense as limitation. One can continue the algorithm using the guidelines outlined in Section 3.2 to obtained more accurate results. Please be noted that the first iteration is not counted as it is just to setup the parameters of the algorithm.

Table 4 summarizes the results for the first iteration. The first column represents the initial population which is obtained from initial configuration in Table 3. Next, the selection operation is being applied to determine the role of each chromosome in the next generation. The selection probability is calculated for this purpose according the roulette-wheel principle described as **Step 4** in Section 3.2. This is determined in two sub-steps. First, divide each chromosome

**Table 4.** First iteration

| Initial Population | Thresholds | | Fitness Function | Fitness | Selection Probability | Crossover | Final Population |
|---|---|---|---|---|---|---|---|
| | $\alpha$ | $\beta$ | | | | | |
| 0000 | 0.6 | 0.0 | 0.6756 | 1.0405 | 0.26 | 2 | 0100 |
| 0101 | 0.8 | 0.1 | 0.6286 | 0.9682 | 0.24 | 1 | 0001 |
| 0010 | 0.6 | 0.2 | 0.6701 | 1.0321 | 0.26 | 4 | 0110 |
| 1111 | 1.0 | 0.4 | 0.6228 | 0.9592 | 0.24 | 3 | 1011 |

**Table 5.** Second iteration

| Initial Population | Thresholds | | Fitness Function | Fitness | Selection Probability | Crossover | Final Population |
|---|---|---|---|---|---|---|---|
| | $\alpha$ | $\beta$ | | | | | |
| 0100 | 0.8 | 0.0 | 0.6290 | 0.9898 | 0.25 | 4 | 0010 |
| 0001 | 0.6 | 0.1 | 0.6758 | 1.0635 | 0.27 | 3 | 0111 |
| 0110 | 0.8 | 0.2 | 0.6150 | 0.9678 | 0.24 | 2 | 0000 |
| 1011 | 0.9 | 0.4 | 0.6220 | 0.9788 | 0.24 | 1 | 1101 |

fitness value by the average fitness value of all chromosomes (which in this case is 0.6493). This is represented by the table column of Fitness in Table 4. Next, the selection probability of a chromosome is obtained as the normalized value of its fitness. This is shown in Table 4 under the column of Selection Probability. We set the condition for selecting and duplicating a particular chromosome if it is atleast 10% greater in value than the remaining selection probabilities. In this case, all the chromosomes are selected and none of them are being duplicated, since the difference in probabilities is not very significant.

A crossover between chromosomes is performed in **Step 5** of the SGA based approach. A random multi-point crossover on bit 1 and bit 3 is considered in this example. The column under the name crossover in Table 4 represents this. A crossover between chromosome 1 with chromosome 2 and chromosome 3 with chromosome 4 are being performed. The mutation operation is not performed in this case since, all chromosomes in initial population and final population are not the same.

The minimum fitness value at the end of the iteration for any chromosome is 0.6228 which corresponds to $(\alpha, \beta) = (1.0, 0.4)$. The **Step 3** to **Step 5** are repeated to obtain the results for the second iteration which are summarized in Table 5. The minimum value of uncertainty after second iteration is 0.6150 for the thresholds $(\alpha, \beta) = (0.8, 0.2)$ which represents the optimum values in this case.

## 4.2    Comparison with Game-Theoretic Rough Sets

We further analyze the results with the SGA based approach by considering its comparison with the GTRS based threshold determination. A GTRS based

**Table 6.** Payoff table

|  | | $D$ | |
|---|---|---|---|
|  | $s_1 = \alpha_\downarrow$ = 10% dec. $\alpha$ | $s_2 = \beta_\uparrow$ = 10% inc. $\beta$ | $s_3 = \alpha_\downarrow\beta_\uparrow$ = 10% (dec. $\alpha$ & inc. $\beta$) |
| $I$ | $s_1 = \alpha_\downarrow$ = 10% dec. $\alpha$ | (0.949,0.474) | (0.977,0.416) | **(0.946,0.480)** |
|  | $s_2 = \beta_\uparrow$ = 10% inc. $\beta$ | (0.977,0.416) | (0.944,0.490) | (0.923,0.542) |
|  | $s_3 = \alpha_\downarrow\beta_\uparrow$ = 10% (dec. $\alpha$ & inc. $\beta$) | (0.946,0.480) | (0.923,0.542) | (0.893,0.590) |

**Table 7.** Comparison with GTRS

| Approach | Determined thresholds | Solution type | Fitness or payoffs associated with thresholds |
|---|---|---|---|
| SGA based | (0.8,0.2) | Optimization | 0.6150 |
| GTRS based | (0.8,0.1) | Trade-off | (0.946,0.480) |

game discussed in [1] is considered for this purpose. The players in the game are
the immediate decision region (containing positive and negative regions) and the
deferred or boundary decision region. Each player can choose from three possible
strategies, i.e., $s_1 = \alpha_\downarrow = 10\%$ decrease in $\alpha$, $s_2 = \beta_\uparrow = 10\%$ increase in $\beta$ and
$s_3 = \alpha_\downarrow\beta_\uparrow = 10\%$ decrease in $\alpha$ and 10% increase in $\beta$. For the sake of briefness
we omit the details of the game. Interested reader is referred to reference [1].

Table 6 represents the payoff table corresponding to the game which is calculated based on the data in Table 1. The cell with bold values, i.e., **(0.946,0.480)**,
represents the game solution in this case. This solution corresponds to strategies
$(s_1, s_3)$ for the players which leads to thresholds $(\alpha, \beta) = (0.8, 0.1)$. This solution
is very similar to the one obtained with the SGA based approach, i.e., (0.8,0.2).
Table 7 summarizes the comparison between the two approaches.

The difference between the two solutions may be explained by considering the
nature of the solution provided by the two approaches. The solution provided by
the GTRS is based on the tradeoff between the uncertainties associated with different decision regions. The players compete and jointly determine the thresholds
which provide a balance between their personal interests. The solution provided
by the SGA based approach is based on the optimization of threshold values.
The approach iteratively utilize the genetic operators like selection, crossover
and mutation in guiding towards effective thresholds. Although the two solutions are very similar in this case, they may not be always the same due to
different objectives that are being employed in these approaches.

## 5   Conclusion

This article introduces an approach based on genetic algorithms for determining
thresholds in probabilistic rough sets. The proposed approach utilizes the overall

uncertainty of the rough set regions as a fitness function to evaluate the quality of obtained thresholds. The genetic operators such as selection, crossover and mutation are iteratively used to guide in the direction of optimal values. A demonstrative example suggests that the proposed approach provides useful threshold values within a few iterations. Moreover, it also provides similar results to that of the game-theoretic rough set model. These preliminary results advocates for the use of the suggested approach as an alternative method for threshold determination. It should also be noted that the proposed approach is also applicable if other fitness functions are being used.

# References

1. Azam, N., Yao, J.T.: Analyzing uncertainties of probabilistic rough set regions with game-theoretic rough sets. International Journal of Approximate Reasoning 55(1), 142–155 (2014)
2. Deng, X.F., Yao, Y.Y.: An information-theoretic interpretation of thresholds in probabilistic rough sets. In: Li, T., Nguyen, H.S., Wang, G., Grzymala-Busse, J., Janicki, R., Hassanien, A.E., Yu, H. (eds.) RSKT 2012. LNCS, vol. 7414, pp. 369–378. Springer, Heidelberg (2012)
3. Deng, X.F., Yao, Y.Y.: A multifaceted analysis of probabilistic three-way decisions (2013) (Manuscript)
4. Herbert, J.P., Yao, J.T.: Game-theoretic rough sets. Fundamenta Informaticae 108(3-4), 267–286 (2011)
5. Holland, J.H.: Adaptation in Natural and Artificial Systems. MIT Press, Cambridge (1992)
6. Jia, X.Y., Tang, Z.M., Liao, W.L., Shang, L.: On an optimization representation of decision-theoretic rough set model. International Journal of Approximate Reasoning 55(1), 156–166 (2014)
7. Kao, Y.T., Zahara, E.: A hybrid genetic algorithm and particle swarm optimization for multimodal functions. Applied Soft Computing 8(2), 849–857 (2008)
8. Leung, Y.W., Wang, Y.: An orthogonal genetic algorithm with quantization for global numerical optimization. IEEE Transactions on Evolutionary Computation 5(1), 41–53 (2001)
9. Srinivas, M., Patnaik, L.M.: Genetic algorithms: A survey. Computer 27(6), 17–26 (1994)
10. Yao, J.T., Herbert, J.P.: A game-theoretic perspective on rough set analysis. Journal of Chongqing University of Posts and Telecommunications (Natural Science Edition) 20(3), 291–298 (2008)
11. Yao, Y.Y.: Decision-theoretic rough set models. In: Yao, J.T., Lingras, P., Wu, W.-Z., Szczuka, M.S., Cercone, N.J., Ślęzak, D. (eds.) RSKT 2007. LNCS (LNAI), vol. 4481, pp. 1–12. Springer, Heidelberg (2007)
12. Yao, Y.Y.: Two semantic issues in a probabilistic rough set model. Fundamenta Informaticae 108(3-4), 249–265 (2011)

# Three-Way Decisions, Uncertainty, and Granular Computing

# Three-Way Weighted Entropies
# and Three-Way Attribute Reduction

Xianyong Zhang[1,2] and Duoqian Miao[1]

[1] Department of Computer Science and Technology, Tongji University,
Shanghai 201804, P.R. China
[2] College of Mathematics and Software Science, Sichuan Normal University,
Chengdu 610068, P.R. China
{zhang_xy,dqmiao}@tongji.edu.cn

**Abstract.** Rough set theory (RS-Theory) is a fundamental model of granular computing (GrC) for uncertainty information processing, and information entropy theory provides an effective approach for its uncertainty representation and attribute reduction. Thus, this paper hierarchically constructs three-way weighted entropies (i.e., the likelihood, prior, and posterior weighted entropies) by adopting a GrC strategy from the concept level to classification level, and it further explores three-way attribute reduction (i.e., the likelihood, prior, and posterior attribute reduction) by resorting to a novel approach of Bayesian inference. From two new perspectives of GrC and Bayesian inference, this study provides some new insights into the uncertainty measurement and attribute reduction of information theory-based RS-Theory.

**Keywords:** Rough set theory, uncertainty, granular computing, three-way decision, information theory, weighted entropy, Bayesian inference, attribute reduction.

## 1 Introduction

Rough set theory (RS-Theory) [1] is a fundamental model of granular computing (GrC) for uncertainty information processing. Information theory [2] is an important way to reflect information and measure uncertainty, and it was first introduced into RS-Theory for uncertainty representation and reduction measurement by Prof. Miao in 1997 [3]; a measurement called rough entropy was further put forward by Prof. Beaubouef in 1998 [4]. In the development of more than a decade, many systematic fruits [5-11] based on the information entropy, conditional entropy, and mutual information, have been widely used, especially for attribute reduction.

The Bayesian inference in machine learning [12] provides an effective approach for practical data processing, i.e., introducing the prior information into the likelihood function to produce the posterior probability. Following this approach, this paper mainly evolves the Bayesian probability formula in RS-Theory from a new perspective of weighted entropies, and it also explores relevant Bayesian expressions at different levels. When the weighted entropies are constructed from

D. Miao et al. (Eds.): RSKT 2014, LNAI 8818, pp. 707–719, 2014.
DOI: 10.1007/978-3-319-11740-9_65 © Springer International Publishing Switzerland 2014

the concept level to classification level, the GrC strategy is adopted, because GrC [13,14] is an effective structural methodology for dealing with hierarchical issues. Finally, the hierarchical weighted entropies are utilized to construct attribute reduction. In particular, three-way decision theory, proposed by Prof. Yao [15,16], plays a key role in decision making. Herein, from the three-way decision viewpoint, relevant Bayesian items and systemic reduction are also considered by using a longitudinal strategy, and we will concretely construct three-way weighted entropies and three-way reducts based on the likelihood, prior, and posterior items.

In summary, we mainly use two new perspectives of GrC and Bayesian inference to preliminary explore uncertainty measuring and attribute reduction. Thus, this study can provide some new insights into information theory-based RS-Theory. Moreover, the constructed three-way pattern regarding likelihood, prior, and posterior can partially enrich the three-way decision theory from a new perspective. Next, Section 2 provides preliminaries, Section 3 and 4 study the three-way weighted entropies at the concept and classification levels, respectively, Section 5 further discusses three-way attribute reduction, Section 6 finally provides conclusions.

## 2   Preliminaries

The decision table (D-Table) $(U, \mathcal{C} \cup \mathcal{D})$ serves as a main framework. Herein, $X \in U/IND(\mathcal{D}) = \{X_j : j = 1, .., m\}$, $\mathcal{A} \subseteq \mathcal{C}$, $[x]_{\mathcal{A}} \in U/IND(\mathcal{A}) = \{[x]_{\mathcal{A}}^i : i = 1, .., n\}$. $\mathcal{B} \subseteq \mathcal{A} \subseteq \mathcal{C}$ refers to the granulation relationship with a partial order - $\mathcal{A} \preceq \mathcal{B}$. If $\mathcal{A} \preceq \mathcal{B}$, then $\forall [x]_{\mathcal{B}} \in U/IND(\mathcal{B})$, $\exists k \in \mathbf{N}$, s.t., $\bigcup_{t=1}^{k} [x]_{\mathcal{A}}^t = [x]_{\mathcal{B}}$; thus, representative granular merging $\bigcup_{t=1}^{k} [x]_{\mathcal{A}}^t = [x]_{\mathcal{B}}$ can be directly utilized for verifying granulation monotonicity [17]. Moreover, $U/IND(\mathcal{B}) = \{U\}$ if $\mathcal{B} = \emptyset$, and let $\forall b \in \mathcal{B}$.

The conditional entropy and mutual information are

$$H(\mathcal{D}/\mathcal{A}) = - \sum_{i=1}^{n} p([x]_{\mathcal{A}}^i) \sum_{j=1}^{m} p(X_j/[x]_{\mathcal{A}}^i) \log p(X_j/[x]_{\mathcal{A}}^i)$$

and $I(\mathcal{A}; \mathcal{D}) = H(\mathcal{D}) - H(\mathcal{D}/\mathcal{A})$, respectively. Both uncertainty measures have granulation monotonicity, i.e., if $\mathcal{A} \preceq \mathcal{B}$ then $H(\mathcal{D}/\mathcal{A}) \leq H(\mathcal{D}/\mathcal{B})$ and $I(\mathcal{A}; \mathcal{D}) \geq I(\mathcal{B}; \mathcal{D})$, so they are used to construct two types of D-Table reduct, which are equivalent to the classical D-Table reduct based on regions [1,6,11].

(1) $\mathcal{B}$ is a region-based reduct of $\mathcal{C}$, if $POS_{\mathcal{B}}(\mathcal{D}) = POS_{\mathcal{C}}(\mathcal{D})$, $POS_{\mathcal{B}-\{b\}}(\mathcal{D}) \subset POS_{\mathcal{B}}(\mathcal{D})$.
(2) $\mathcal{B}$ is a conditional entropy-based reduct of $\mathcal{C}$, if $H(\mathcal{D}/\mathcal{B}) = H(\mathcal{D}/\mathcal{C})$, $H(\mathcal{D}/\mathcal{B}-\{b\}) > H(\mathcal{D}/\mathcal{B})$.
(3) $\mathcal{B}$ is a mutual information-based reduct of $\mathcal{C}$, if $I(\mathcal{B}; \mathcal{D}) = I(\mathcal{C}; \mathcal{D})$, $I(\mathcal{B} - \{b\}; \mathcal{D}) < I(\mathcal{B}; \mathcal{D})$.

Moreover, monotonous information entropy $H(\mathcal{A}) = -\sum_{i=1}^{n} p([x]_{\mathcal{A}}^{i})\log p([x]_{\mathcal{A}}^{i})$ is used to equivalently define the information-system reduct [1,3,7].

(1) $\mathcal{B}$ is a knowledge-based reduct of $\mathcal{C}$, if $U/IND(\mathcal{B}) = U/IND(\mathcal{C})$, $U/IND(\mathcal{B}-\{b\}) \neq U/IND(\mathcal{B})$.
(2) $\mathcal{B}$ is an entropy-based reduct of $\mathcal{C}$, if $H(\mathcal{B}) = H(\mathcal{C})$, $H(\mathcal{B}-\{b\}) < H(\mathcal{B})$.

## 3  Three-Way Weighted Entropies of a Concept

By evolving the Bayesian probability formula, this section mainly proposes three-way weighted entropies of a concept. For granule $[x]_{\mathcal{A}}$ and concept $X$, we first analyze the relevant causality mechanism of three-way probabilities, then discuss three-way entropies, and finally construct three-way weighted entropies.

Suppose $p(T) = \frac{|T|}{|U|}$ ($\forall T \in 2^{U}$), then $(U, 2^{U}, p)$ constitutes a probability space. Thus, there are four types of probability to construct the Bayesian formula

$$p([x]_{\mathcal{A}}/X) = \frac{p([x]_{\mathcal{A}}) \cdot p(X/[x]_{\mathcal{A}})}{p(X)}. \tag{1}$$

For given concept $X$, $p(X) = \frac{|X|}{|U|}$ becomes a constant, so the surplus three-way probabilities are worth analyzing.

From a causality viewpoint, concept $X$ represents a result while divided granule $[x]_{\mathcal{A}}$ means factors. Furthermore, from a Bayesian viewpoint, $\mathcal{A}$ can be viewed as a granulation parameter within a subset range of $\mathcal{C}$.

(1) $p(X/[x]_{\mathcal{A}}) = \frac{|X \cap [x]_{\mathcal{A}}|}{|[x]_{\mathcal{A}}|}$ is the likelihood probability for granulation parameter $\mathcal{A}$ to describe granular decision $X$.
(2) $p([x]_{\mathcal{A}}/X) = \frac{|X \cap [x]_{\mathcal{A}}|}{|X|}$ is the posterior probability to describe granulation parameters on a premise of result $X$.
(3) $p([x]_{\mathcal{A}}) = \frac{|[x]_{\mathcal{A}}|}{|U|}$ is the prior probability to describe cause parameter $\mathcal{A}$.

The three-way probabilities, which correspond to relative and absolute measures [18], respectively, exhibit different probability semantics and decision actions. In particular, likelihood $p(X/[x]_{\mathcal{A}})$ and posterior $p([x]_{\mathcal{A}}/X)$ directly reflect causality from the cause-to-effect and effect-to-cause viewpoints, respectively, so their relevant measures can thoroughly describe correlative relationships between the decision concept and its condition structures. Clearly, $p([x]_{\mathcal{A}}/X)$ is more perfect for reduction because reduction is a concrete effect-to-cause pattern, and its calculation is also more optimal. Moreover, prior $p([x]_{\mathcal{A}})$ mainly measures cause uncertainty by reflecting structural information of $\mathcal{A}$.

Our original intention is to describe the causality system regarding $\mathcal{A}$ and $X$ and to further study attribute reduction by constructing benign measures based on the three-way probabilities. In view of entropy's importance for measuring uncertainty, we next exhibit three-way entropies.

**Definition 1.** For concept $X$,

$$H^X(\mathcal{A}) = -\sum_{i=1}^{n} p([x]_\mathcal{A}^i)\log p([x]_\mathcal{A}^i), \tag{2}$$

$$H(\mathcal{A}/X) = -\sum_{i=1}^{n} p([x]_\mathcal{A}^i/X)\log p([x]_\mathcal{A}^i/X), \tag{3}$$

$$H(X/\mathcal{A}) = -\sum_{i=1}^{n} p(X/[x]_\mathcal{A}^i)\log p(X/[x]_\mathcal{A}^i) \tag{4}$$

are called prior, posterior, and likelihood entropies, respectively.

Definition 1 proposes three-way entropies of a concept. In fact, $H^X(\mathcal{A})$ and $H(\mathcal{A}/X)$ naturally measure uncertainty of granulation $\mathcal{A}$ without limitations and on promise $X$, respectively, because both $p([x]_\mathcal{A})$ and $p([x]_\mathcal{A}/X)$ form a probability distribution. Moreover, $H(X/\mathcal{A})$ is also formally proposed to measure the likelihood structure, though $\sum_{i=1}^{n} p(X/[x]_\mathcal{A}^i) \neq 1$. For $X$, $H^X(\mathcal{A})$ conducts absolute pre-evaluation, while $H(X/\mathcal{A})$ and $H(\mathcal{A}/X)$ make relative descriptions from two different causality directions. Thus, the three-way entropies, especially $H(X/\mathcal{A})$ and $H(\mathcal{A}/X)$, can measure causality between granulation parameter $\mathcal{A}$ and decision set $X$.

**Proposition 1.** If $\mathcal{A} \preceq \mathcal{B}$, then $H^X(\mathcal{A}) \geq H^X(\mathcal{B})$, $H(\mathcal{B}/X) \geq H(\mathcal{A}/X)$, but neither $H(X/\mathcal{A}) \geq H(X/\mathcal{B})$ nor $H(X/\mathcal{A}) \leq H(X/\mathcal{B})$ necessarily holds.

**Proof.** The results can be proved by entropy properties, because $p([x]_\mathcal{B}) = \sum_{t=1}^{k} p([x]_\mathcal{A}^t)$ and $p([x]_\mathcal{B}/X) = \sum_{t=1}^{k} p([x]_\mathcal{A}^t/X)$ but $p(X/[x]_\mathcal{B}) \neq \sum_{t=1}^{k} p(X/[x]_\mathcal{A}^t)$. □

Granulation monotonicity is an important feature for evaluating an entropy. Based on Proposition 1, the prior/posterior and likelihood entropies have granulation monotonicity and non-monotonicity, respectively. In particular, the following Example 1 illustrates the non-monotonicity of the likelihood entropy.

**Example 1.** Given $[x]_\mathcal{A}^1$, $[x]_\mathcal{A}^2$ and complementary $X_1$, $X_2$. Let $|[x]_\mathcal{A}^1| = 40 = |[x]_\mathcal{A}^2|, |X_1| = 29, |X_2| = 51$; moreover, $|[x]_\mathcal{A}^1 \cap X_1| = 1, |[x]_\mathcal{A}^2 \cap X_1| = 28$, so $|[x]_\mathcal{A}^1 \cap X_2| = 39, |[x]_\mathcal{A}^2 \cap X_2| = 12$. For $[x]_\mathcal{A}^1 \cup [x]_\mathcal{A}^2 = [x]_\mathcal{B}$ regarding $X_1$, $p(X_1/[x]_\mathcal{A}^1) = 0.025$, $p(X_1/[x]_\mathcal{A}^2) = 0.7$, $p(X_1/[x]_\mathcal{B}) = 0.3625$, so $-0.025\log0.025 - 0.7\log0.7 = 0.4932 < 0.5307 = -0.3625\log0.3625$; regarding $X_2$, $p(X_2/[x]_\mathcal{A}^1) = 0.975$, $p(X_2/[x]_\mathcal{A}^2) = 0.3, p(X_2/[x]_\mathcal{B}) = 0.6375$, so $-0.975\log0.975 - 0.3\log0.3 = 0.5567 > 0.4141 = -0.6375\log0.6375$. If $U/IND(\mathcal{A}) = \{[x]_\mathcal{A}^1, [x]_\mathcal{A}^2\}$ and $U/IND(\mathcal{B}) = \{[x]_\mathcal{B}\}$, then $\mathcal{A} \preceq \mathcal{B}$ but $H(X_1/\mathcal{A}) \leq H(X_1/\mathcal{B}), H(X_2/\mathcal{A}) \geq H(X_2/\mathcal{B})$. □

For the three-way probabilities, $p([x]_\mathcal{A}/X)$ and $p(X/[x]_\mathcal{A})$ reflect causality between $\mathcal{A}$ and $X$; for the three-way entropies, only $H^X(\mathcal{A})$ and $H(\mathcal{A}/X)$ exhibit necessary monotonicity. Thus, $p([x]_{\mathcal{A}}/X)$ and further $H(\mathcal{A}/X)$ hold important significance for describing $\mathcal{A}$ structure based on $X$. In fact, posterior entropy $H(\mathcal{A}/X)$ reflects the average information content of granulation $U/IND(\mathcal{A})$ for given concept $X$, and at the entropy level, only it has perfect value for measuring uncertainty of $\mathcal{A}$ for $X$. This posterior entropy's function underlies latter importance of the posterior weighed entropy and posterior attribute reduction.

Though the posterior entropy is valuable, however, there are no relationships for the three-way entropies. Thus, better three-way measures with granulation monotonicity are worth deeply mining to establish an essential connection. For this purpose, we first creatively evolve the Bayesian probability formula to naturally mine three-way weighted entropies, and we then explore their monotonicity and relationship.

**Theorem 1.** $-\sum_{i=1}^{n} p(X)p([x]_{\mathcal{A}}^{i}/X)\log p([x]_{\mathcal{A}}^{i}/X)$

$$= -\sum_{i=1}^{n} p(X/[x]_{\mathcal{A}}^{i})p([x]_{\mathcal{A}}^{i})\log p([x]_{\mathcal{A}}^{i}) - \sum_{i=1}^{n} p([x]_{\mathcal{A}}^{i})p(X/[x]_{\mathcal{A}}^{i})\log p(X/[x]_{\mathcal{A}}^{i})$$
$$+ p(X)\log p(X).$$

**Proof.** First, $p([x]_{\mathcal{A}}^{i}/X) = \frac{p([x]_{\mathcal{A}}^{i}) \cdot p(X/[x]_{\mathcal{A}}^{i})}{p(X)}$, $\forall i \in \{1, ..., n\}$. Thus,

$-p([x]_{\mathcal{A}}^{i}/X)\log p([x]_{\mathcal{A}}^{i}/X) = -\frac{p([x]_{\mathcal{A}}^{i}) \cdot p(X/[x]_{\mathcal{A}}^{i})}{p(X)}[\log p([x]_{\mathcal{A}}^{i}) + \log p(X/[x]_{\mathcal{A}}^{i}) - \log p(X)]$. Hence, $-p(X)p([x]_{\mathcal{A}}^{i}/X)\log p([x]_{\mathcal{A}}^{i}/X) = -p(X/[x]_{\mathcal{A}}^{i})p([x]_{\mathcal{A}}^{i}) \log p([x]_{\mathcal{A}}^{i}) - p([x]_{\mathcal{A}}^{i})p(X/[x]_{\mathcal{A}}^{i})\log p(X/[x]_{\mathcal{A}}^{i}) + p([x]_{\mathcal{A}}^{i})p(X/[x]_{\mathcal{A}}^{i})\log p(X)$. Furthermore, the result is obtained by summation, where $\sum_{i=1}^{n} p([x]_{\mathcal{A}}^{i})p(X/[x]_{\mathcal{A}}^{i})$

$\log p(X) = \sum_{i=1}^{n} p([x]_{\mathcal{A}}^{i} \cap X)\log p(X) = [\sum_{i=1}^{n} p([x]_{\mathcal{A}}^{i} \cap X)]\log p(X) = p(X)$ $\log p(X).$ □

Theorem 1 develops the Bayesian theorem in an entropy direction, and there is actually a core form containing both an entropy and weights, i.e., a weighted entropy. Thus, a weighted entropy plays an core role and can establish an equation. This entropy evolution inherits the Bayesian probability formula and inspires our following further works.

**Definition 2.** For probability distribution $(\xi, p_i)$ and weight $w_i \geq 0$, $H_W(\xi) = -\sum_{i=1}^{n} w_i p_i \log p_i$ is called the weighted entropy. In particular, the generalized weighted entropy has not constraint condition $\sum_{i=1}^{n} p_i = 1$.

The weighted entropy mainly introduces weights into the entropy, and weights usually reflect importance degrees for information receivers. In particular, it develops the entropy and degenerates into the latter by setting up $w_i = 1$. Herein, the generalized weighted entropy is mainly used in view of $\sum_{i=1}^{n} p(X/[x]_{\mathcal{A}}^{i}) \neq 1$.

**Definition 3.** For concept $X$,

$$H_W^X(\mathcal{A}) = -\sum_{i=1}^{n} p(X/[x]_\mathcal{A}^i)p([x]_\mathcal{A}^i)\log p([x]_\mathcal{A}^i), \tag{5}$$

$$H_W(\mathcal{A}/X) = -\sum_{i=1}^{n} p(X)p([x]_\mathcal{A}^i/X)\log p([x]_\mathcal{A}^i/X) = p(X)H(\mathcal{A}/X), \tag{6}$$

$$H_W(X/\mathcal{A}) = -\sum_{i=1}^{n} p([x]_\mathcal{A}^i)p(X/[x]_\mathcal{A}^i)\log p(X/[x]_\mathcal{A}^i) \tag{7}$$

are called prior, posterior, and likelihood weighted entropies, respectively.

The three-way weighted entropies originate from three-way entropies by adding probability-based weight coefficients. In fact, $H_W^X(\mathcal{A})$ improves upon absolute $H^X(\mathcal{A})$ by introducing relative $p(X/[x]_\mathcal{A}^i)$, while $H_W(\mathcal{A}/X)$ and $H_W(X/\mathcal{A})$ improve upon relative $H(\mathcal{A}/X)$ and $H(X/\mathcal{A})$ by introducing absolute $p(X)$ and $p([x]_\mathcal{A}^i)$, respectively. Thus, $H_W^X(\mathcal{A})$, $H_W(\mathcal{A}/X)$, $H_W(X/\mathcal{A})$ inherit uncertainty semantics by different probability weights, and they exhibit systematic completeness and superior stability from the double-quantitative perspective [18], so they can better describe the system regarding cause $\mathcal{A}$ and result $X$. Moreover, posterior weighted entropy $H_W(\mathcal{A}/X)$ has a simple and perfect structure, because it can be directly decomposed into a product of posterior entropy $H(\mathcal{A}/X)$ and constant $p(X)$. Note that weighted entropy symbol $H_W(.)$ is distinguished from information entropy symbol $H(.)$.

**Proposition 2.** If $\mathcal{A} \preceq \mathcal{B}$, then $H_W^X(\mathcal{A}) \geq H_W^X(\mathcal{B})$, $H_W(\mathcal{A}/X) \geq H_W(\mathcal{B}/X)$, $H_W(X/\mathcal{A}) \leq H_W(X/\mathcal{B})$.

**Proof.** Herein, we only provide the proof for the likelihood weighted entropy by utilizing granular merging $\bigcup_{t=1}^{k} [x]_\mathcal{A}^t = [x]_\mathcal{B}$. $f(u) = -u\log u$ $(u \in [0.1])$ is a concave function; thus, if $\sum_{t=1}^{k} \lambda_t = 1$, then $-\sum_{t=1}^{k} \lambda_t p_t \log p_t \leq -[\sum_{t=1}^{k} \lambda_t p_t]\log[\sum_{t=1}^{k} \lambda_t p_t]$.

$$-\sum_{t=1}^{k} p([x]_\mathcal{A}^t)p(X/[x]_\mathcal{A}^t)\log p(X/[x]_\mathcal{A}^t) = -\sum_{t=1}^{k} p([x]_\mathcal{B})\frac{|[x]_\mathcal{A}^t|}{|[x]_\mathcal{B}|}p(X/[x]_\mathcal{A}^t)\log p(X/[x]_\mathcal{A}^t)$$

$$= p([x]_\mathcal{B})[-\sum_{t=1}^{k} \frac{|[x]_\mathcal{A}^t|}{|[x]_\mathcal{B}|}p(X/[x]_\mathcal{A}^t)\log p(X/[x]_\mathcal{A}^t)]$$

$$\leq -p([x]_\mathcal{B})[\sum_{t=1}^{k} \frac{|[x]_\mathcal{A}^t|}{|[x]_\mathcal{B}|}p(X/[x]_\mathcal{A}^t)]\log \frac{\sum_{t=1}^{k} |[x]_\mathcal{A}^t \cap X|}{|[x]_\mathcal{B}|}$$

$$= -p([x]_\mathcal{B})\frac{|[x]_\mathcal{B} \cap X|}{|[x]_\mathcal{B}|}\log \frac{|[x]_\mathcal{B} \cap X|}{|[x]_\mathcal{B}|} = -p([x]_\mathcal{B})p(X/[x]_\mathcal{B})\log p(X/[x]_\mathcal{B}). \quad \square$$

Based on Proposition 2, three weighted entropies exhibit perfect granulation monotonicity and thus hold significance. In particular, $H_W(X/\mathcal{A})$ becomes monotonicity though $H(X/\mathcal{A})$ is non-monotonicity, and this monotonicity difficulty is proved by utilizing a concave feature of function $-ulogu$.

**Theorem 2.** $H_W(\mathcal{A}/X) = H_W^X(\mathcal{A}) + H_W(X/\mathcal{A}) + p(X)\log p(X)$
$= H_W^X(\mathcal{A}) - [-p(X)\log p(X) - H_W(X/\mathcal{A})]$, and $-p(X)\log p(X) - H_W(X/\mathcal{A}) \geq 0$.

Theorem 2 provides an important relationship for the three-way weighted entropies (where $-p(X)\log p(X)$ is a constant), i.e., the posterior weighted entropy becomes a linear translation of the sum of the prior and likelihood weighted entropies. Thus, the Bayesian probability formula can deduce essential relationships regarding not three-way entropies but three-way weighted entropies. Furthermore, $-p(X)\log p(X) - H_W(X/\mathcal{A})$ can be chosen as a new measure to simplify the fundamental equation by eliminating the translation distance.

**Definition 4.** $H_W^*(X/\mathcal{A}) = -p(X)\log p(X) - H_W(X/\mathcal{A})$.

**Corollary 1.** (1) If $\mathcal{A} \preceq \mathcal{B}$, then $H_W^*(X/\mathcal{A}) \geq H_W^*(X/\mathcal{B})$.
(2) $H_W(\mathcal{A}/X) = H_W^X(\mathcal{A}) - H_W^*(X/\mathcal{A})$.

Herein, $H_W^*(X/\mathcal{A})$ corresponds to $H_W(X/\mathcal{A})$ by a negative linear transformation, so it exhibits opposite granulation monotonicity. Furthermore, the posterior weighted entropy becomes the difference between prior weighted entropy $H_W^X(\mathcal{A})$ and $H_W^*(X/\mathcal{A})$, and the latter corresponds to the likelihood weighted entropy.

## 4    Three-Way Weighted Entropies of a Classification

Three-way weighted entropies are proposed for a concept in Section 3, and they will be further constructed for a classification in this section by a natural integration strategy of GrC. Moreover, they will be linked to the existing RS-Theory system with the information entropy, conditional entropy and mutual information. Next, classification $U/IND(\mathcal{D}) = \{X_1, .., X_m\}$ with $m$ concepts is given.

**Definition 5.** For classification $U/IND(\mathcal{D})$,

$$H_W^\mathcal{D}(\mathcal{A}) = \sum_{j=1}^m H_W^{X_j}(\mathcal{A}), \; H_W(\mathcal{A}/\mathcal{D}) = \sum_{j=1}^m H_W(\mathcal{A}/X_j), \; H_W(\mathcal{D}/\mathcal{A}) = \sum_{j=1}^m H_W(X_j/\mathcal{A})$$

are called prior, posterior, and likelihood weighted entropies, respectively. Moreover, let $H_W^*(\mathcal{D}/\mathcal{A}) = \sum_{j=1}^m H_W^*(X_j/\mathcal{A})$.

For decision classification $U/IND(\mathcal{D})$, the three-way weighted entropies are corresponding sum of concepts' weighted entropies regarding classification's internal concepts, because we naturally adopt a GrC strategy from an internal concept to its integrated classification. Thus, they inherit relevant causality mechanisms and hold corresponding functions for measuring uncertainty; moreover, they also inherit the essential monotonicity and mutual relationship.

**Proposition 3.** If $\mathcal{A} \preceq \mathcal{B}$, then $H_W^\mathcal{D}(\mathcal{A}) \geq H_W^\mathcal{D}(\mathcal{B})$, $H_W(\mathcal{A}/\mathcal{D}) \geq H_W(\mathcal{B}/\mathcal{D})$, $H_W(\mathcal{D}/\mathcal{A}) \leq H_W(\mathcal{D}/\mathcal{B})$, $H_W^*(\mathcal{D}/\mathcal{A}) \geq H_W^*(\mathcal{D}/\mathcal{B})$.

**Theorem 3 (Weighted Entropies' Bayesian Formula).**

$$H_W(\mathcal{A}/\mathcal{D}) = H_W^\mathcal{D}(\mathcal{A}) - [H(\mathcal{D}) - H_W(\mathcal{D}/\mathcal{A})] = H_W^\mathcal{D}(\mathcal{A}) - H_W^*(\mathcal{D}/\mathcal{A}).$$

Theorem 3 describes an important relationship of the three-way weighted entropies by introducing $H(\mathcal{D})$. Thus, the posterior weighted entropy is difference between the prior weighted entropy and $H_W^*(\mathcal{D}/\mathcal{A})$, and the latter is a linear transformation of the likelihood weighted entropy. In particular, Theorem 3 essentially evolves the Bayesian probability formula, so it is called by *Weighted Entropies' Bayesian Formula* to highlight its important values.

Next, we summarize the above GrC works via Table 1. There are three GrC levels which are located at the micro bottom, meso layer, and macro top, respectively.

(1) At Level (1), the three-way probabilities describe granule $[x]_{\mathcal{A}}$ and concept $X$, and the Bayesian probability formula holds by using premise $p(X)$.

(2) At Level (2), the three-way weighted entropies describe granulation $\mathcal{A}$ and concept $X$ and exhibit granulation monotonicity. In particular, $H_W(\mathcal{A}/X) = H_W^X(\mathcal{A}) - H_W^*(X/\mathcal{A})$ acts as an evolutive Bayesian formula, where $H_W^*(X/\mathcal{A})$ is a linear adjustment of $H_W(X/\mathcal{A})$ by using premise $-p(X)\log p(X)$.

(3) At Level (3), the three-way weighted entropies describe granulation $\mathcal{A}$ and classification $\mathcal{D}$ and inherit granulation monotonicity. In particular, $H_W(\mathcal{A}/\mathcal{D}) = H_W^{\mathcal{D}}(\mathcal{A}) - H_W^*(\mathcal{D}/\mathcal{A})$ acts as an evolutive Bayesian result, where $H_W^*(\mathcal{D}/\mathcal{A})$ is a linear adjustment of $H_W(\mathcal{D}/\mathcal{A})$ by using premise $H(\mathcal{D})$.

Thus, our GrC works establish an integrated description for $\mathcal{A}$ and $\mathcal{D}$ by using a bottom-top strategy, so they underlie the further discussion, especially for attribute reduction. In fact, attribute reduction is mainly located at Level (3), where $H(\mathcal{D})$ is a constant from the causality perspective.

**Table 1.** GrC-Based Weighted Entropies and Relevant Bayesian Formulas

| Level | Objects | Three-Way Measures | Relevant Bayesian Formulas |
|---|---|---|---|
| (1) | $[x]_{\mathcal{A}}$, $X$ | $p([x]_{\mathcal{A}})$, $p([x]_{\mathcal{A}}/X)$, $p(X/[x]_{\mathcal{A}})$ | $p([x]_{\mathcal{A}}/X) = \frac{p([x]_{\mathcal{A}}) \cdot p(X/[x]_{\mathcal{A}})}{p(X)}$ |
| (2) | $\mathcal{A}$, $X$ | $H_W^X(\mathcal{A})$, $H_W(\mathcal{A}/X)$, $H_W(X/\mathcal{A})$ (or $H_W^*(X/\mathcal{A})$) | $H_W(\mathcal{A}/X) = H_W^X(\mathcal{A}) - H_W^*(X/\mathcal{A})$ |
| (3) | $\mathcal{A}$, $\mathcal{D}$ | $H_W^{\mathcal{D}}(\mathcal{A})$, $H_W(\mathcal{A}/\mathcal{D})$, $H_W(\mathcal{D}/\mathcal{A})$ (or $H_W^*(\mathcal{D}/\mathcal{A})$) | $H_W(\mathcal{A}/\mathcal{D}) = H_W^{\mathcal{D}}(\mathcal{A}) - H_W^*(\mathcal{D}/\mathcal{A})$ |

Finally, we explain the novel system of the three-way weighted entropies by the previous system based on information theory, and we also analyze both systems' relationships.

**Theorem 4.** $H_W^{\mathcal{D}}(\mathcal{A}) = H(\mathcal{A})$, $H_W(\mathcal{A}/\mathcal{D}) = H(\mathcal{A}/\mathcal{D})$, $H_W(\mathcal{D}/\mathcal{A}) = H(\mathcal{D}/\mathcal{A})$, $H_W^*(\mathcal{D}/\mathcal{A}) = I(\mathcal{A};\mathcal{D})$.

**Proof.** (1) $H_W^{\mathcal{D}}(\mathcal{A}) = \sum_{j=1}^{m} H_W^{X_j}(\mathcal{A}) = -\sum_{j=1}^{m}\sum_{i=1}^{n} p(X_j/[x]_{\mathcal{A}}^i)p([x]_{\mathcal{A}}^i)\log p([x]_{\mathcal{A}}^i)]$

$= -\sum_{j=1}^{m}[p(X_j/[x]_{\mathcal{A}}^1)p([x]_{\mathcal{A}}^1)\log p([x]_{\mathcal{A}}^1) - ... - p(X_j/[x]_{\mathcal{A}}^n)p([x]_{\mathcal{A}}^n)\log p([x]_{\mathcal{A}}^n)]$

$= -[\sum_{j=1}^{m} p(X_j/[x]_{\mathcal{A}}^1)]p([x]_{\mathcal{A}}^1)\log p([x]_{\mathcal{A}}^1) - ... - [\sum_{j=1}^{m} p(X_j/[x]_{\mathcal{A}}^n)]p([x]_{\mathcal{A}}^n)\log p([x]_{\mathcal{A}}^n)$

$= -p([x]_{\mathcal{A}}^1)\log p([x]_{\mathcal{A}}^1) - ... - p([x]_{\mathcal{A}}^n)\log p([x]_{\mathcal{A}}^n) = H(\mathcal{A})$.

(2) $H_W(\mathcal{A}/\mathcal{D}) = \sum\limits_{j=1}^{m} H_W(A/X_j) = \sum\limits_{j=1}^{m} p(X_j)H(A/X_j)$

$= - \sum\limits_{j=1}^{m} p(X_j) \sum\limits_{i=1}^{n} p([x]_\mathcal{A}^i/X_j)\log p([x]_\mathcal{A}^i/X_j) = H(\mathcal{A}/\mathcal{D}).$

(3) $H_W(\mathcal{D}/\mathcal{A}) = H_W(X_1/A) + ... + H_W(X_m/A)$

$= [-p([x]_\mathcal{A}^1)p(X_1/[x]_\mathcal{A}^1)\log p(X_1/[x]_\mathcal{A}^1) - ... - p([x]_\mathcal{A}^n)p(X_1/[x]_\mathcal{A}^n)\log p(X_1/[x]_\mathcal{A}^n)] +$

$...$

$+ [-p([x]_\mathcal{A}^1)p(X_m/[x]_\mathcal{A}^1)\log p(X_m/[x]_\mathcal{A}^1) - ... - p([x]_\mathcal{A}^n)p(X_m/[x]_\mathcal{A}^n)\log p(X_m/[x]_\mathcal{A}^n)]$

$= -p([x]_\mathcal{A}^1)[p(X_1/[x]_\mathcal{A}^1)\log p(X_1/[x]_\mathcal{A}^1) + ... + p(X_m/[x]_\mathcal{A}^1)\log p(X_m/[x]_\mathcal{A}^1)] - ...$

$- p([x]_\mathcal{A}^n)[p(X_1/[x]_\mathcal{A}^n)\log p(X_1/[x]_\mathcal{A}^n) + ... + p(X_m/[x]_\mathcal{A}^n)\log p(X_m/[x]_\mathcal{A}^1)]$

$= - \sum\limits_{i=1}^{n} p([x]_\mathcal{A}^i) \sum\limits_{j=1}^{m} p(X_j/[x]_\mathcal{A}^i)\log p(X_j/[x]_\mathcal{A}^i) = H(\mathcal{D}/\mathcal{A}).$

Thus, $H_W^*(\mathcal{D}/\mathcal{A}) = H(\mathcal{D}) - H_W(\mathcal{D}/\mathcal{A}) = H(\mathcal{D}) - H(\mathcal{D}/\mathcal{A}) = I(\mathcal{A};\mathcal{D})$     $\square$.

**Theorem 5.** $H_W(\mathcal{A}/\mathcal{D}) = H_W^\mathcal{D}(\mathcal{A}) - H_W^*(\mathcal{D}/\mathcal{A})$ is equivalent to $H(\mathcal{A}/\mathcal{D}) = H(\mathcal{A}) - I(\mathcal{A};\mathcal{D})$.

Based on Theorem 4, three-way weighted entropies $H_W^\mathcal{D}(\mathcal{A})$, $H_W(\mathcal{A}/\mathcal{D})$, $H_W(\mathcal{D}/\mathcal{A})$ are equivalent to prior entropy $H(\mathcal{A})$, conditional entropy $H(\mathcal{A}/\mathcal{D})$, conditional entropy $H(\mathcal{D}/\mathcal{A})$, respectively; moreover, $H_W^*(\mathcal{D}/\mathcal{A})$ corresponds to mutual information $I(\mathcal{A};\mathcal{D})$. Furthermore, Theorem 5 reflects equivalence between $H_W(\mathcal{A}/\mathcal{D}) = H_W^\mathcal{D}(\mathcal{A}) - H_W^*(\mathcal{D}/\mathcal{A})$ and $H(\mathcal{A}/\mathcal{D}) = H(\mathcal{A}) - I(\mathcal{A};\mathcal{D})$, which are from two different systems. Thus, the weighted entropy system (including its Bayesian formula) has been explained/verified by the previous information theory system. In contrast, the former can thoroughly explain the latter as well. Therefore, both systems exhibit theoretical equivalence. However, the weighted entropy approach conducts a GrC construction, and it also emphasizes the causality semantics and application direction based on the Bayesian mechanism. Thus, the three-way weighted entropies hold at least two fundamental values. First, they construct, explain, and deepen the existing information system of RS-Theory by the GrC construction and Bayesian formula; moreover, they underlie systemic attribute reduction by the essential uncertainty measure and effective Bayesian inference.

## 5   Three-Way Attribute Reduction

The three-way weighted entropies (of a classification) and their monotonicity and relationship have been provided in Section 4. This section mainly uses them to systemically construct three-way attribute reduction, and the poster reduction will be emphasized via the Bayesian inference and causality theory.

**Definition 6.** $\mathcal{B}$ is called likelihood, prior, and posterior reducts of $\mathcal{C}$, if it satisfies the following three conditions, respectively.

(1) $H_W^*(\mathcal{D}/\mathcal{B}) = H_W^*(\mathcal{D}/\mathcal{C})$, $H_W^*(\mathcal{D}/\mathcal{B} - \{b\}) < H_W^*(\mathcal{D}/\mathcal{B})$
    (or $H_W(\mathcal{D}/\mathcal{B}) = H_W(\mathcal{D}/\mathcal{C})$, $H_W(\mathcal{D}/\mathcal{B} - \{b\}) > H_W(\mathcal{D}/\mathcal{B})$).
(2) $H_W^\mathcal{D}(\mathcal{B}) = H_W^\mathcal{D}(\mathcal{C})$, $H_W^\mathcal{D}(B - \{b\}) < H_W^\mathcal{D}(\mathcal{B})$.
(3) $H_W(\mathcal{B}/\mathcal{D}) = H_W(\mathcal{C}/\mathcal{D})$, $H_W(\mathcal{B} - \{b\}/\mathcal{D}) < H_W(\mathcal{B}/\mathcal{D})$.

**Theorem 6.**

(1) A likelihood reduct is equivalent to a D-Table reduct. Furthermore, a likelihood reduct based on $H_W^*(\mathcal{D}/\mathcal{A})$ or $H_W(\mathcal{D}/\mathcal{A})$ is equivalent to a D-Table reduct based on the mutual information or conditional entropy, respectively.
(2) A prior reduct is equivalent to an information-system reduct.
(3) A posterior reduct is different from both D-Table and information-system reducts.

**Proof.** (1) For the D-Table reduct, the region-based method is equivalent to the mutual information-based and conditional entropy-based ways [1,6,11]; furthermore, $I(\mathcal{A}; \mathcal{D})$ and $H(\mathcal{D}/\mathcal{A})$ correspond to $H_W^*(\mathcal{D}/\mathcal{A})$ and $H_W(\mathcal{D}/\mathcal{A})$, respectively, so the likelihood reduct is equivalent to the two information-based reducts and the classical region-based reduct. (2) For the information-system reduct, the prior reduct is equivalent to the entropy-based reduct and further knowledge-based reduct, because $H_W^{\mathcal{D}}(\mathcal{A}) = H(\mathcal{A})$. (3) The difference of the posterior reduct is verified by the following D-Table example.                    □.

**Example 2.** In D-Table $S = (U, \mathcal{C} \cup \mathcal{D})$ provided by Table 2,
$U = \{x_1, ..., x_{12}\}$, $C = \{a, b, c\}$, $D = \{d\}$, $U/\{d\} = \{X_1, X_2, X_3\}$,
$X_1 = \{x_1, ..., x_4\}$, $X_2 = \{x_5, ..., x_8\}$, $X_3 = \{x_9, ..., x_{12}\}$. Thus, $U/\{a, b, c\} =$
$U/\{a, b\} = \{\{x_2, x_3, x_6, x_{11}\}, \{x_4, x_8, x_{12}\}, \{x_1\}, \{x_5\}, \{x_7, x_{10}\}, \{x_9\}\}$,
$U/\{a\} = U/\{a, c\} = \{\{x_2, x_3, x_6, x_7, x_{10}, x_{11}\}, \{x_4, x_8, x_{12}\}, \{x_1\}, \{x_5\}, \{x_9\}\}$,
$U/\{b\} = \{\{x_2, x_3, x_6, x_{11}\}, \{x_4, x_8, x_{12}\}, \{x_1, x_7, x_{10}\}, \{x_5, x_9\}\}$,
$U/\{c\} = \{\{x_2, x_3, x_6, x_7, x_{10}, x_{11}\}, \{x_4, x_8, x_{12}\}, \{x_1, x_5, x_9\}\}$,
$U/\{b, c\} = \{\{x_2, x_3, x_6, x_{11}\}, \{x_4, x_8, x_{12}\}, \{x_1\}, \{x_5, x_9\}, \{x_7, x_{10}\}\}$.

**Table 2.** D-Table in Example 2

| $U$ | $a$ | $b$ | $c$ | $d$ | $U$ | $a$ | $b$ | $c$ | $d$ | $U$ | $a$ | $b$ | $c$ | $d$ |
|---|---|---|---|---|---|---|---|---|---|---|---|---|---|---|
| $x_1$ | 3 | 3 | 3 | 1 | $x_5$ | 4 | 4 | 3 | 2 | $x_9$ | 5 | 4 | 3 | 3 |
| $x_2$ | 1 | 1 | 1 | 1 | $x_6$ | 1 | 1 | 1 | 2 | $x_{10}$ | 1 | 3 | 1 | 3 |
| $x_3$ | 1 | 1 | 1 | 1 | $x_7$ | 1 | 3 | 1 | 2 | $x_{11}$ | 1 | 1 | 1 | 3 |
| $x_4$ | 2 | 2 | 2 | 1 | $x_8$ | 2 | 2 | 2 | 2 | $x_{12}$ | 2 | 2 | 2 | 3 |

First, there are only two D-Table reducts $\{a\}$, $\{c\}$ and one information-system reduct $\{a, b\}$. Herein, $H_W(\mathcal{C}/\mathcal{D}) = P(X_1)H(\mathcal{C}/X_1) + P(X_2)H(\mathcal{C}/X_2) + P(X_3) H(\mathcal{C}/X_3) = \frac{4}{12}[(-0.5\log 0.5 - 2 \times 0.25\log 0.25) - 4 \times 0.25\log 0.25 - 4 \times 0.25\log 0.25] = 1.8333 = H_W(\{b\}/\mathcal{D})$, $H_W(\{a\}/\mathcal{D}) = H_W(\{a, c\}/\mathcal{D}) = H_W(\{c\}/\mathcal{D}) = \frac{4}{12}(-0.5\log 0.5 - 2 \times 0.25\log 0.25) \times 3 = 1.500 < 1.8333$. Thus, $\{b\}$ becomes the sole posterior reduct and is neither the $(U, \mathcal{C} \cup \mathcal{D})$ reduct nor $(U, \mathcal{C})$ reduct; in contrast, neither $\{a\}$, $\{c\}$ nor $\{a, b\}$ is a posterior reduct. Note that the key granular merging regarding $\{x_1, x_7, x_{10}\}$ is allowed not for the other reducts but for the posterior reduct.                    □

The three weighted entropies can measure uncertainty, and they are used to naturally define the three-way reducts. In spite of measuring of all weighted entropies, the three-way reducts exhibit different reduction essence. The likelihood reduct and prior reduct, which are also related to the mutual information, conditional entropy and information entropy, mainly correspond to the qualitative reducts regarding D-Table and information-system, respectively. In contrast, the posterior reduct completely corresponds to a quantitative reduct. Thus, the posterior weighted entropy exhibit more essential metrizability, and the posterior reduct exhibits novelty and transcendence, so both are worth emphasizing.

Next, based on the posterior weighted entropy, we analyze important significance of the posterior reduct for D-Table reduct.

(1) The D-Table reduct usually uses likelihood information in the cause-to-effect (or condition-to-decision) direction. According to the Bayesian inference, the posterior weighted entropy adjusts the likelihood weighted entropy by strengthening the prior knowledge, so the posterior reduct improves upon the likelihood reduct by pursuing quantitative uncertainty rather than qualitative absoluteness. In fact, by considering the granulation distribution, the posterior reduct achieves the highest posterior uncertainty and lowest risk according to the granulation monotonicity and maximum entropy principle, respectively, so it can avoid the over-fitting problem due to its measurability, generality, and robustness.

(2) In D-Table $(U, \mathcal{C} \cup \mathcal{D})$, granulation $U/IND(\mathcal{A})$ and classification $\mathcal{D}$ correspond to the condition cause and decision effect, respectively. D-Table reduction aims to choose appropriate granulation parameters $\mathcal{A}$ to preserve specific decision information regarding $\mathcal{D}$, i.e., it mainly seeks condition parameters $\mathcal{A}$ on a stable premise of $\mathcal{D}$. Thus, from the causality viewpoint, posterior weighted entropy $H_W(\mathcal{A}/\mathcal{D})$ not only reflects the causality relationship between $C$ and $\mathcal{D}$ but also more adheres to the operational pattern of D-Table reduction, so the posterior reduct holds practical significance by adopting the cause-to-effect (or decision-to-condition) strategy.

In summary, within a new framework of Bayesian inference, the posterior weighted entropy bears important information of uncertainty distribution, and it also positively improves upon the likelihood weighted entropy by considering the prior information. Moreover, $H_W(\mathcal{A}/\mathcal{D})$ (i.e., condition entropy $H(\mathcal{A}/\mathcal{D})$) is simpler than $H_W^*(\mathcal{D}/\mathcal{A})$ (i.e., mutual information $I(\mathcal{A}; \mathcal{D})$) and $H_W(\mathcal{D}/\mathcal{A})$ (i.e., condition entropy $H(\mathcal{D}/\mathcal{A})$). Thus, the posterior reduct holds advantages regarding uncertainty semantics, causality directness and calculation optimization.

## 6   Conclusions

Based on the GrC technology and Bayesian inference approach, we construct three-way weighted entropies and three-way attribute reduction, and the relevant results deepen information theory-based RS-Theory, especially the GrC uncertainty measurement and attribute reduction. The three-way weighted entropies and three-way attribute reduction actually correspond to the likelihood, prior, posterior decisions, so they also enrich the three-way decision theory from a new viewpoint. In particular, hierarchies of three-way attribute reduction are worth deeply exploring, and the posterior weighted entropy and posterior attribute reduction need in-depth theoretical exploration and further practical verification.

**Acknowledgments.** This work was supported by the National Science Foundation of China (61273304 and 61203285), Specialized Research Fund for Doctoral Program of Higher Education of China (20130072130004), China Postdoctoral Science Foundation Funded Project (2013T60464 and 2012M520930), and Shanghai Postdoctoral Scientific Program (13R21416300).

# References

1. Pawlak, Z.: Rough sets. International Journal of Information and Computer Science 11(5), 341–356 (1982)
2. Shannon, C.E.: The mathematical theory of communication. The Bell System Technical Journal 27(3–4), 373–423 (1948)
3. Miao, D.Q.: Rough set theory and its application in machine learning (Ph. D. Thesis). Beijing, Institute of Automation, The Chinese Academy of Sciences (1997)
4. Beaubouef, T., Petry, F.E., Arora, G.: Information-theoretic measures of uncertainty for rough sets and rough relational databases. Information Sciences 109(1-4), 185–195 (1998)
5. Slezak, D.: Approximate entropy reducts. Fundamenta Informaticae 53, 365–390 (2002)
6. Miao, D.Q., Hu, G.R.: A heuristic algorithm for reduction of knowledge. Chinese Journal of Computer Research & Development 36, 681–684 (1999)
7. Miao, D.Q., Wang, J.: An information representation of the concepts and operations in rough set theory. Journal of Software 10(2), 113–116 (1999)
8. Liang, J.Y., Qian, Y.H., Chu, D.Y., Li, D.Y., Wang, J.H.: The algorithm on knowledge reduction in incomplete information systems. Fuzziness and Knowledge-Based Systems 10, 95–103 (2002)
9. Liang, J., Shi, Z., Li, D., Wierman, M.: Information entropy, rough entropy and knowledge granularity in incomplete information systems. International Journal of General Systems 35(6), 641–654 (2006)
10. Wang, G.Y., Yu, H., Yang, D.C.: Decision table reduction based on conditional information entropy. Chinese Journal of Computers 25, 759–766 (2002)
11. Wang, G.Y., Zhao, J., An, J.J., Wu, Y.: A comparative study of algebra viewpoint and information viewpoint in attribute reduction. Fundamenta Informaticae 68(3), 289–301 (2005)
12. Bishop, C.M.: Pattern Recognition and Machine Learning. Springer (2006)
13. Zadeh, L.A.: Towards a theory of fuzzy information granulation and its centrality in human reasoning and fuzzy logic. Fuzzy Sets and Systems 90, 111–127 (1997)
14. Lin, T.Y.: Granular computing: from rough sets and neighborhood systems to information granulation and computing with words. In: European Congress on Intelligent Techniques and Soft Computing, pp. 1602–1606 (1997)
15. Yao, Y.: An outline of a theory of three-way decisions. In: Yao, J., Yang, Y., Słowiński, R., Greco, S., Li, H., Mitra, S., Polkowski, L. (eds.) RSCTC 2012. LNCS, vol. 7413, pp. 1–17. Springer, Heidelberg (2012)

16. Yao, Y.Y.: Three-way decisions with probabilistic rough sets. Information Sciences 180, 341–353 (2010)
17. Miao, D.Q., Zhang, X.Y.: Change uncertainty of three-way regions in knowledge-granulation. In: Liu, D., Li, T.R., Miao, D.Q., Wang, G.Y., Liang, J.Y. (eds.) Three-Way Decisions and Granular Computing, pp. 116–144. Science Press, Beijing (2013)
18. Zhang, X.Y., Miao, D.Q.: Quantitative information architecture, granular computing and rough set models in the double-quantitative approximation space on precision and grade. Information Sciences 268, 147–168 (2014)

# Applying Three-way Decisions
# to Sentiment Classification with Sentiment Uncertainty

Zhifei Zhang[1,2] and Ruizhi Wang[1,3,*]

[1] Department of Computer Science and Technology,
Tongji University, Shanghai 201804, China
[2] Department of Computer Science and Operations Research,
University of Montreal, Quebec H3C 3J7, Canada
[3] Chongqing Key Laboratory of Computational Intelligence,
Chongqing University of Posts and Telecommunications, Chongqing 400065, China
ruizhiwang@tongji.edu.cn

**Abstract.** Sentiment uncertainty is a key problem of sentiment classification. In this paper, we mainly focus on two issues with sentiment uncertainty, i.e., context-dependent sentiment classification and topic-dependent sentiment classification. This is the first work that applies three-way decisions to sentiment classification from the perspective of the decision-theoretic rough set model. We discuss the relationship between sentiment classification rules and thresholds involved in three-way decisions and then prove it. The experiment results on real data sets validate that our methods are satisfactory and can achieve better performance.

**Keywords:** Sentiment classification, sentiment uncertainty, decision-theoretic rough sets, three-way decisions.

## 1 Introduction

Sentiment classification is the field of study that analyzes people's opinions, sentiments, evaluations, attitudes, and emotions towards entities and their attributes [1]. Sentiment classification has made considerable progress in the past more than ten yeas. However, the uncertainty due to the diversity of text content and form remains unsettled, and makes sentiment classification still difficult.

Sentiment uncertainty is reflected in domain-dependence, context-dependence, topic-dependence, and multi-label emotion [2]. The same two words may convey different sentiment in different context or topic even though they are in the same domain [3]. In this paper, we focus on context-dependent sentiment classification and topic-dependent sentiment classification with single-label.

Taking the context-dependent sentiment word "high" as an example, it is negative in "high cost" while positive in "high quality". However, as the word "low" in phrases like "low cost" and "low quality", the exact reverse is the case. Those context-dependent sentiment words can not be discarded in fact [4]. "high" and "low" form an antonym pair that is helpful in sentiment classification [5]. We can utilize antonym pairs to guide context-dependent sentiment classification.

D. Miao et al. (Eds.): RSKT 2014, LNAI 8818, pp. 720–731, 2014.
DOI: 10.1007/978-3-319-11740-9_66 © Springer International Publishing Switzerland 2014

Sentiment words are more important than topic words in sentiment classification [6], but topic words can assist with domain adaptation sentiment classification and aspect-based opinion mining [7]. The existing work is hardly aware that the topic can reflect the trend of sentiment. For example, the topic "one born in 1990s became a professor" indicates the positive trend, but "edible oil rose in price" indicates the negative trend. We can utilize topic information to guide topic-dependent sentiment classification.

There are two main kinds of methods for sentiment classification, i.e., lexicon-based unsupervised methods [8] and corpus-based supervised methods [9], while they are short of sentiment uncertainty analysis. Rough set theory [10], as an effective tool for uncertainty analysis, is seldom used in sentiment classification presently. Decision-theoretic rough set model [11] has been studied as a generalization of Pawlak rough set model. Three-way decisions interpret the positive, negative and boundary regions as acceptance, rejection and deferment respectively in a ternary classification [12], and provide a means for trading off different types of classification error in order to obtain a minimum cost classifier [13]. Many recent studies further investigated extensions and applications of three-way decisions [14–16]. We firstly apply three-way decisions to sentiment classification with sentiment uncertainty.

The rest of the paper is organized as follows. In Section 2, we review three-way decisions in decision-theoretic rough sets. In Section 3 and 4, we illustrate the context-dependent sentiment classification method and the topic-dependent sentiment classification method based on three-way decisions respectively. We provide the experiment results and analysis in Section 5. The concluding remarks are given in Section 6.

## 2    Three-way Decisions in Decision-theoretic Rough Sets

The essential idea of three-way decisions is described in terms of a ternary classification according to evaluations of a set of criteria [17]. Given a finite nonempty set $U$ and a finite set of criteria $C$, $U$ is divided into three pair-wise disjoint regions, POS, NEG, and BND respectively, based on $C$. We accept an object in POS as satisfying the set of criteria, reject the object in NEG as not satisfying the criteria. We neither accept nor reject the object in BND but opt for a noncommitment. The third option may also be referred to as a deferment decision that requires further information or investigation.

There are a set of two states $\Omega = \{X, \neg X\}$, indicating that an object is in $X$ and not in $X$ respectively, and a set of three actions $A = \{a_P, a_B, a_N\}$ for each state, where $a_P$, $a_B$, and $a_N$ mean classifying an object into three regions, namely, POS, BND and NEG respectively. The losses of three classification actions with respect to two states are given in Table 1, and they are under the following conditions [18].

$$
\begin{aligned}
&(c_1) \quad 0 \leq \lambda_{PP} < \lambda_{BP} < \lambda_{NP}, 0 \leq \lambda_{NN} < \lambda_{BN} < \lambda_{PN} \\
&(c_2) \quad (\lambda_{NP} - \lambda_{BP})(\lambda_{PN} - \lambda_{BN}) > (\lambda_{BP} - \lambda_{PP})(\lambda_{BN} - \lambda_{NN})
\end{aligned}
\tag{1}
$$

Given a pair of thresholds $(\alpha, \beta)$ with $0 \leq \beta < \alpha \leq 1$, the decision-theoretic rough set model makes the following three-way decisions for $X \subseteq U$:

$$
\begin{aligned}
\text{POS}_{(\alpha,\beta)}(X) &= \{x \in U \mid \Pr(X|[x]) \geq \alpha\} \\
\text{NEG}_{(\alpha,\beta)}(X) &= \{x \in U \mid \Pr(X|[x]) \leq \beta\} \\
\text{BND}_{(\alpha,\beta)}(X) &= \{x \in U \mid \beta < \Pr(X|[x]) < \alpha\}
\end{aligned}
\tag{2}
$$

**Table 1.** Losses for three-way decisions

| State \ Action | $a_P$ | $a_N$ | $a_B$ |
|---|---|---|---|
| $X$ | $\lambda_{PP}$ | $\lambda_{NP}$ | $\lambda_{BP}$ |
| $\neg X$ | $\lambda_{PN}$ | $\lambda_{NN}$ | $\lambda_{BN}$ |

where $[x]$ is the equivalence class containing x. The conditional probability $Pr(X|[x])$ and the thresholds $(\alpha, \beta)$ can be computed [18].

$$Pr(X|[x]) = \frac{|X \cap [x]|}{|[x]|} \tag{3}$$

$$\alpha = \frac{\lambda_{PN} - \lambda_{BN}}{(\lambda_{PN} - \lambda_{BN}) + (\lambda_{BP} - \lambda_{PP})}$$
$$\beta = \frac{\lambda_{BN} - \lambda_{NN}}{(\lambda_{BN} - \lambda_{NN}) + (\lambda_{NP} - \lambda_{BP})} \tag{4}$$

## 3 Context-dependent Sentiment Classification

### 3.1 Contextual Antonym Pairs

It is often the case that two sentiment words from an antonym pair share the same context yet opposite polarity.

**Definition 1 (Contextual Antonym Pairs).** *A contextual antonym pair is represented as $AP = \{sw, ws\}$, where $sw$ and $ws$ are ambiguous and antonymous adjectives. Within the same context that is simplified into a neighboring noun $nn$, we have*

$$Polarity(ws|nn) = -Polarity(sw|nn) \tag{5}$$

This paper focuses on 8 antonym pairs, i.e., {high, low}, {big, small}, {many, few}, {fast, slow}, {deep, shallow}, {long, short}, {light, heavy} and {thick, thin}, denoted as $APs$. The 16 context-dependent sentiment words from $APs$ are all one-character Chinese words and are frequently used in opinionated texts [4].

The collocations of context-dependent sentiment words and neighboring nouns in the training set are stored in a word polarity decision table.

**Definition 2 (Word Polarity Decision Table).** *A word polarity decision table is formalized as a quad $WDT = (U, C \cup D, V, f)$, where $U$ is a finite nonempty set of objects, i.e., the universal set; $C = \{n, w\}$ is a finite nonempty set of two condition attributes that respectively represent neighboring noun and context-dependent sentiment word; $D = \{l\}$ is a finite nonempty set of one decision attribute that represents the polarity of $w$; $V = V_n \cup V_w \cup V_l$, $V_n$ is a nonempty set of neighboring nouns, $V_w$ contains the 16 context-dependent sentiment words, and $V_l = \{1, 0, -1\}$; $f = \{f_a | f_a : U \rightarrow V_a\}$, $f_a$ is an information function that maps an object in $U$ to one value in $V_a$.*

## 3.2  Context-dependent Sentiment Three-way Decisions

Given a word polarity decision table $WDT = (U, C \cup D, V, f)$, $U$ is classified into three polarity groups based on $C$, i.e., $Positive = \{x \in U | f_l(x) = 1\}$, $Negative = \{x \in U | f_l(x) = -1\}$, and $Neutral = \{x \in U | f_l(x) = 0\}$. Each polarity group is divided into three disjoint regions, POS, NEG, and BND respectively.

Multi-category classification based on three-way decisions is usually transformed into multiple binary classification [19]. For example, $\Omega = \{Positive, \neg Positive\}$ represents that an object $x$ belongs to the positive polarity group and does not belong to this group respectively. According to Table 1, all three-way decision losses for three polarity groups are given by experiences [20].

**Table 2.** Three-way decision losses for three polarity groups ($u$ is the unit loss)

| Polarity \ Loss | $\lambda_{PP}$ | $\lambda_{BP}$ | $\lambda_{NP}$ | $\lambda_{PN}$ | $\lambda_{BN}$ | $\lambda_{NN}$ |
|---|---|---|---|---|---|---|
| Positive | 0 | $4u$ | $9u$ | $8u$ | $3u$ | 0 |
| Negative | 0 | $3u$ | $8u$ | $7u$ | $2u$ | 0 |
| Neutral | 0 | $u$ | $10u$ | $9u$ | $3.5u$ | 0 |

In view of Eq. (4), we have $\alpha_{Positive} = 0.556$, $\alpha_{Negative} = 0.625$, and $\alpha_{Neutral} = 0.846$. For an object $x$,

$$\begin{aligned}
&\text{If } \Pr(Positive|x) \geq \alpha_{Positive}, \text{Then } x \in \text{POS}(Positive) \\
&\text{If } \Pr(Negative|x) \geq \alpha_{Negative}, \text{Then } x \in \text{POS}(Negative)
\end{aligned} \tag{6}$$

where

$$\begin{aligned}
\Pr(Positive|x) &= \frac{count(n = f_n(x), w = f_w(x), l = 1)}{count(n = f_n(x), w = f_w(x))} \\
\Pr(Negative|x) &= \frac{count(n = f_n(x), w = f_w(x), l = -1)}{count(n = f_n(x), w = f_w(x))}
\end{aligned} \tag{7}$$

To determine the polarity of context-dependent sentiment words, we propose the following two rules: Bi-direction Rule and Uni-direction Rule.

**Proposition 1 (Bi-direction Rule).** *Suppose that we have a neighboring noun nn and a sentiment word sw, if there are two objects $x$ and $y$, $f_n(x) = f_n(y) = nn$, $f_w(x) = sw$, $f_w(y) = ws$, $\{sw, ws\} \in APs$, a bi-direction rule is made to obtain the polarity of sw given nn.*

$$Polarity(sw|nn) = \begin{cases} 1 & x \in \text{POS}(Positive) \wedge y \in \text{POS}(Negative) \\ -1 & x \in \text{POS}(Negative) \wedge y \in \text{POS}(Positive) \\ 0 & \text{otherwise} \end{cases} \tag{8}$$

**Theorem 1.** *Given two objects $x$ and $y$, if $f_n(x) = f_n(y) = nn \wedge \{f_w(x) = sw,$
$f_w(y) = ws\} \in APs$ holds, then $x \in \text{POS}(Positive) \wedge y \in \text{POS}(Negative)$ is
equivalent to $Polarity(sw|nn) = 1$.*

*Proof.* (1) $\because y \in \text{POS}(Negative) \therefore y \notin \text{POS}(Positive), \because x \in \text{POS}(Positive) \therefore$
$Polarity(sw|nn) = 1$.
(2) *Let* $\Pr(Positive|x) = a$ *and* $\Pr(Negative|y) = b$, *the expected losses incurred
by taking actions* $a_P$, $a_B$ *and* $a_N$ *respectively for $x$ and $y$, are listed in the following
equations:*

$$R(a_P|x, a_P|y) = 8u(1-a) + 7u(1-b), \ R(a_P|x, a_B|y) = 8u(1-a) + 3ub + 2u(1-b)$$

$$R(a_P|x, a_N|y) = 8u(1-a) + 8ub, \ R(a_B|x, a_P|y) = 4ua + 3u(1-a) + 7u(1-b)$$

$$R(a_B|x, a_B|y) = 4ua + 3u(1-a) + 3ub + 2u(1-b)$$

$$R(a_B|x, a_N|y) = 4ua + 3u(1-a) + 8ub, \ R(a_N|x, a_P|y) = 9ua + 7u(1-b)$$

$$R(a_N|x, a_B|y) = 9ua + 3ub + 2u(1-b), \ R(a_N|x, a_N|y) = 9ua + 8ub$$

*According to Eq.* (5), $\because Polarity(sw|nn) = 1 \therefore Polarity(ws|nn) = -1$. *Thus,*
$(a_P, a_P) = \underset{(D_1, D_2) \in \{a_P, a_B, a_N\}^2}{\arg\min} R(D_1|x, D_2|y)$ *need to hold. We have $a \geq 0.556, b \geq$*
$0.625$ *through pairwise comparison of expected losses. Obviously,* $\Pr(Positive|x) \geq$
$\alpha_{Positive}$, $\Pr(Negative|y) \geq \alpha_{Negative}$, *so* $x \in \text{POS}(Positive) \wedge y \in \text{POS}(Negative)$.

When two sentiment words from one contextual antonym pair do not share the same
context, e.g., only "high price" appears in the training set but "low price" does not,
the bi-direction rule is not suitable. We compute the $Z$-score statistic with one-tailed
test to perform the significant test. The statistical confidence level is set to 0.95, whose
corresponding $Z$-score is -1.645. If $Z$-score is greater than -1.645, the collocation of
one sentiment word and one neighboring noun is significant.

$$Z\text{-}score(x, X) = \frac{\Pr(X|x) - P_0}{\sqrt{\frac{P_0 \cdot (1 - P_0)}{|X \cap [x]_t|}}}, X \in \{Positive, Negative\} \tag{9}$$

where the hypothesized value $P_0$ is set to 0.7 [20], and $[x]_t$ is the equivalence class of
an object $x$ based on the condition attribute $t$.

**Proposition 2 (Uni-direction Rule).** *Suppose that we have a neighboring noun $nn$
and a sentiment word $sw$, if there is an object $x$ such that $f_n(x) = nn \wedge f_w(x) = sw$
but not an object $y$ with $f_n(y) = nn \wedge \{sw, f_w(y)\} \in APs$, then a uni-direction rule
is made to obtain the polarity of $sw$ given $nn$.*

$$Polarity(sw|nn) = \begin{cases} 1 & x \in \text{POS}(Positive) \wedge Z\text{-}score(x, Positive) > -1.645 \\ -1 & x \in \text{POS}(Negative) \wedge Z\text{-}score(x, Negative) > -1.645 \\ 0 & \text{otherwise} \end{cases} \tag{10}$$

**Theorem 2.** *In Uni-direction Rule, if $\Pr(Positive|x) \geq P_0$, then $Polarity(sw|nn) =$
1 holds.*

*Proof.* $\because \Pr(Positive|x) \geq P_0 > 0.556 \therefore x \in \text{POS}(Positive), \because \Pr(Positive|x) \geq$
$P_0 \therefore Z\text{-}score(x, Positive) \geq 0 > -1.645$, *So* $Polarity(sw|nn) = 1$.

# 4   Topic-dependent Sentiment Classification

## 4.1   Priori Sentiment

We believe that the first step in topic-dependent sentiment classification is to obtain the priori sentiment from topic information. When all sentences have been classified from the perspective of topic, the priori sentiment can be defined as follows.

**Definition 3** (**Priori Sentiment**). *Given $N_T$ sentences with topic $T$, among of which $N_{TP}$ are positive and $N_{TN}$ are negative, the priori sentiment of the topic $T$ is*

$$Priori(T) = \begin{cases} \frac{N_{TP}-N_{TN}}{N_T} & \left| \frac{N_{TP}-N_{TN}}{N_T} \right| \geq \frac{1}{2} \\ 0 & \text{otherwise} \end{cases} \tag{11}$$

If $Priori(T) > 0$, the topic $T$ tends to express the positive polarity. If $Priori(T) < 0$, the topic $T$ tends to express the negative polarity.

In order to describe the sentiment of a sentence, we introduce 6 features, $\{P, SU, SN, SP, ST, SD\}$, listed in Table 3 to represent it [2], and store in a sentence polarity decision table.

Table 3. Feature representation for sentences

| Feature | Range | Explanation |
| --- | --- | --- |
| $P$ | $\{-1, 0, 1\}$ | the polarity by counting positive(negative) words |
| $SU$ | $(0, 1]$ | the sentiment uncertainty measure |
| $SN$ | $\mathbb{N}$ | the number of single negation words |
| $SP$ | $\mathbb{N}$ | the number of single punctuation marks |
| $ST$ | $\{Q, D, O\}$ | the interrogative sentence or the exclamatory sentence |
| $SD$ | $\{T, F\}$ | whether the distinguishing adverb is contained |

**Definition 4** (**Sentence Polarity Decision Table**). *A sentence polarity decision table is formalized as a quad $SDT = (U, C \cup D, V, f)$, where $U$ is a finite nonempty set of sentences, i.e., the universal set; $C = \{P, SU, SN, SP, ST, SD\}$ is a finite nonempty set of six condition attributes; $D$ has only one decision attribute that represents the polarity of the sentence; $V = \cup V_a$, $V_a$ is a nonempty set of values of $a \in C \cup D$; $f = \{f_a | f_a : U \to V_a\}$, $f_a$ is an information function.*

**Definition 5** (**Positive Conditional Probability**). *Given a sentence polarity decision table $SDT = (U, C \cup D, V, f)$, the positive polarity group is $S_{Pos} = \{x \in U | D(x) = 1\}$, then the positive conditional probability of a sentence $x$ is*

$$Pos(x) = \Pr(S_{Pos} | [x]) = \frac{|S_{Pos} \cap [x]|}{|[x]|} \tag{12}$$

where $[x]$ is the equivalent class of a sentence $x$ based on all condition attributes $C$.

## 4.2 Topic-dependent Sentiment Three-way Decisions

We define three regions for the positive polarity group $S_{Pos}$ with a pair of thresholds $(\alpha, \beta)$, $0 \le \beta < \alpha \le 1$. The three regions respectively correspond to the positive, negative, and neutral polarity decisions.

$$\text{POS}(Positive) = \{x \in U | Pos(x) \ge \alpha\}$$
$$\text{NEG}(Positive) = \{x \in U | Pos(x) \le \beta\} \tag{13}$$
$$\text{BND}(Positive) = \{x \in U | \beta < Pos(x) < \alpha\}$$

By virtue of $SDT$, we can give a sentence $x$ the polarity label $Label(x)$ based on three-way decisions. Assume that $x$ belongs to the topic $T$, if $Label(x)$ and $Priori(T)$ are consistent, i.e., they are both greater or less than 0, the final decision is the same thing as $Label(x)$. But if they are contradictory, i.e., $Label(x) \times Priori(T) < 0$, the final polarity decision is made by the following weighted polarity $Pos_w$ or $Neg_w$.

**Definition 6 (Weighted Positive Polarity).** *Given a sentence $x$ with the topic $T$, if $Label(x) = 1$ but $Priori(T) < 0$, a weighted positive polarity measure is $Pos_w$:*

$$Pos_w = Priori(T) \times (1 - Pos(x)) + (1 + Priori(T)) \times Pos(x) \tag{14}$$

If $Pos_w > 0$ holds, $Pos(x)$ must be big enough, the final polarity is 1; If $Pos_w < 0$ holds, $Pos(x)$ must be small enough, the final polarity is $-1$.

**Theorem 3.** *Given a sentence $x$ with the topic $T$, if $Label(x)=1$ and $\alpha > -Priori(T) > 0$, then $Polarity(x) = 1$.*

*Proof.* $\because -Priori(T) > 0$ $\therefore Priori(T) < 0$. *Meanwhile,* $Label(x) = 1$, *and according to Eq. (14),* $Pos_w = Priori(T) + Pos(x)$ *holds.* $\because Label(x) = 1$ $\therefore x \in$ $\text{POS}(Positive)$, *i.e.,* $Pos(x) \ge \alpha$. $\because \alpha > -Priori(T)$ $\therefore Pos(x) > -Priori(T)$, *i.e.,* $Priori(T) + Pos(x) > 0$. *So* $Pos_w > 0$, *i.e.,* $Polarity(x) = 1$.

**Definition 7 (Weighted Negative Polarity).** *Given a sentence $x$ with the topic $T$, if $Label(x) = -1$ but $Priori(T) > 0$, a weighted negative polarity measure is $Neg_w$:*

$$Neg_w = Priori(T) \times Pos(x) + (Priori(T) - 1) \times (1 - Pos(x)) \tag{15}$$

If $Neg_w < 0$ holds, $Pos(x)$ must be small enough, the final polarity is $-1$; If $Neg_w > 0$ holds, $Pos(x)$ must be big enough, the final polarity is 1.

**Theorem 4.** *Given a sentence $x$ with the topic $T$, if $Label(x) = -1$ and $\beta < 1 - Priori(t) < 1$, then $Polarity(x) = -1$.*

*Proof.* $\because 1 - Priori(T) < 1$ $\therefore Priori(T) > 0$. *Meanwhile,* $Label(x) = -1$, *and according to Eq. (15),* $Neg_w = Priori(T) + Pos(x) - 1$ *holds.* $\because Label(x) = -1$ $\therefore$ $x \in \text{NEG}(Positive)$, *i.e.,* $Pos(x) \le \beta$. $\because \beta < 1 - Priori(T)$ $\therefore Pos(x) < 1 - Priori(T)$, *i.e.,* $Priori(T) + Pos(x) - 1 < 0$. *So* $Neg_w < 0$, *i.e.,* $Polarity(x) = -1$.

# 5  Experiments

## 5.1  Data Sets

To test our methods for context-dependent sentiment classification, two real-world data sets are used. COAE is from Task 1 of Chinese Opinion Analysis Evaluation 2012, and SEMEVAL is from Task 18 of Evaluation Exercises on Semantic Evaluation 2010.

To test our methods for topic-dependent sentiment classification, Weibo data set from NLP&CC 2012 Evaluation is used. We select three representative topics: "one born in 1990s became a professor", "ipad3", and "edible oil rose in price", simply marked as "Prof", "iPad", and "Oil" respectively.

**Table 4.** Descriptions of three data sets

|          | COAE | SEMEVAL | Weibo | | |
|----------|------|---------|-------|------|-----|
|          |      |         | Prof  | iPad | Oil |
| Positive | 598  | 1202    | 110   | 41   | 7   |
| Negative | 1295 | 1715    | 13    | 60   | 71  |
| Neutral  | 507  | 0       | 12    | 121  | 45  |
| **Total**| 2400 | 2917    | 480   |      |     |

## 5.2  Experiment Methods

For context-dependent sentiment classification, the evaluation criteria are $micro\text{-}F_1$ and $macro\text{-}F_1$, and the four methods are as follows.

**Baseline.** The method [8] discarding the context-dependent sentiment words.
**TWD.** Context-dependent sentiment three-way decisions for antonym pairs + Baseline.
**TWD-CE.** Context expansion (finding the synonyms for neighboring nouns) + TWD.
**TWD-DE.** Polarity expansion (finding the synonyms for context-dependent sentiment words) + TWD-CE.

For topic-dependent sentiment classification, the evaluation criterion is $micro\text{-}F_1$, and the six methods are as follows.

**S-RS.** Three-way decisions based polarity classification for sentences.
**T-S-RS.** Topic-dependent sentiment three-way decisions with standard topic classification + S-RS.
**T-S-RS(KNN/1).** Topic-dependent sentiment three-way decisions with KNN($k = 1$) topic classification + S-RS.
**T-S-RS(KNN/11).** Topic-dependent sentiment three-way decisions with KNN($k = 11$) topic classification + S-RS.
**T-S-RS(SVM).** Topic-dependent sentiment three-way decisions with SVM topic classification + S-RS.
**T-S-RS(Random).** Topic-dependent sentiment three-way decisions with random topic classification + S-RS.

## 5.3    Experiment Results for Context-dependent Sentiment Classification

The comparative results between Baseline and TWD are shown in Fig. 1. The performance on SEMEVAL is obviously improved in that those context-dependent sentiment words appear in about 97.6% of all sentences, but the performance on COAE is slightly improved due to the small percentage (29.5%). More sentences can be truly classified with polarity classification for contextual antonym pairs.

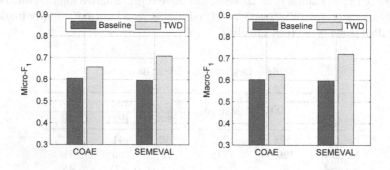

**Fig. 1.** Comparative results between Baseline and TWD

The behavior analysis of 16 context-dependent sentiment words is carried out. About 75% of the appearance is positive or negative, and the remainder is neutral for lack of context. We believe that classifying the context-dependent sentiment words truly is very important to sentiment classification.

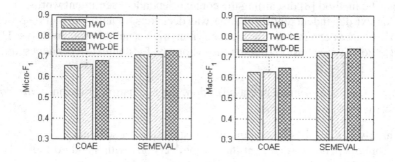

**Fig. 2.** Comparative results between TWD-CE, TWD-DE and TWD

When neighboring nouns or context-dependent sentiment words have been expanded with synonym identification, TWD-CE and TWD-DE are compared with TWD (see Fig. 2). Context expansion and polarity expansion can further improve the classification performance, because more contextual polarities are discovered. With the aid of double expansion, 26 additional neighboring nouns and 37 additional sentiment words on COAE are found, and 131 additional neighboring nouns and 60 additional sentiment words on SEMEVAL are found.

### 5.4   Experiment Results for Topic-dependent Sentiment Classification

The priori sentiment for each topic is: $Priori(\text{Prof}) = 0.72$, $Priori(\text{iPad}) = 0$, and $Priori(\text{Oil}) = -0.52$. The three values indicate that the topic "Prof" tends to express the positive polarity, but the topic "Oil" tends to express the negative polarity. The sentiment trend of the topic "iPad" is not significant.

T-S-RS combines S-RS with standard topic classification, which means each sentence on Weibo data set must be truly labelled. Figure 3 shows the comparative results between S-RS and T-S-RS.

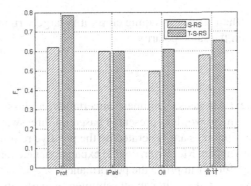

**Fig. 3.** Results of Weibo sentiment classification with standard topic classification

The standard topic can be helpful to sentiment classification. If the priori sentiment is equal to 0, the topic is useless, e.g., the performance of "iPad" is not improved. If the priori sentiment is not equal to 0, the bigger the absolute value, the greater the performance improvement, e.g., the improvement of "Prof" is more than that of "Oil".

If standard topic classification is not available, automatic topic classification, i.e., traditional text classification, should be done before sentiment classification. Assume that the topic for each sentence on Weibo is unknown, and four methods, i.e., SVM, KNN($k = 11$), KNN($k = 1$) and Random, will be used. The results of Weibo sentiment classification with automatic topic classification are illustrated in Fig. 4.

The performance of automatic topic classification is actually poor for Weibo texts with the characteristic of short length. For example, SVM assigns the topic "iPad" for all sentences. Because the priori sentiment of "iPad" is equal to 0, the performance of T-S-RS(SVM) is the same with S-RS. The five other methods are obviously poorer than T-S-RS. In this case, the performance of topic-dependent sentiment classification mainly depends on the performance of topic classification. KNN($k = 1$) achieves the best performance among KNN($k = 1$), KNN($k = 11$) and SVM, which are all better than Random. The short length of Weibo text results in sparseness. If both two texts have the

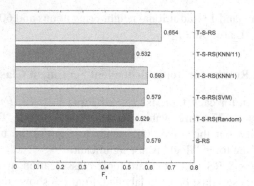

**Fig. 4.** Results of Weibo sentiment classification with automatic topic classification

same sentiment word, their distance is quite close, it is largely right that KNN($k = 1$) labels the two texts with the same polarity.

## 6  Conclusions

Seeing different context can cause sentiment uncertainty, a classification method for context-dependent sentiment words is presented based on three-way decisions. The bi-direction and uni-direction rules are generated by the positive regions of two sentiment words from each antonym pair. Moreover, double expansion, i.e., context expansion and polarity expansion, can further improve the performance.

Since different topics can also bring about sentiment uncertainty, a topic-dependent sentiment classification method is presented. The topic is transformed into the priori sentiment for classifying sentences together with three-way decisions. The relationship between the lower and upper-bound threshold and the priori sentiment is also proven.

In the future, we would like to dig into more accurate context information extraction which can help to filter noisy neighboring nouns. Better automatic topic classification will further improve the performance of topic-dependent sentiment classification.

**Acknowledgments.** This work is partially supported by the National Natural Science Foundation of China (Grant No.61273304, and No.61202170), the Specialized Research Fund for the Doctoral Program of Higher Education of China (Grant No.20130072130004), Chongqing Key Laboratory of Computational Intelligence, China (Grant No.CQ-LCI-2013-04), the Fundamental Research Funds for the Central Universities, and the State Scholarship Fund of China.

## References

1. Liu, B.: Sentiment analysis and opinion mining. Synthesis Lectures on Human Language Technologies 5(1), 1–167 (2012)
2. Zhang, Z.: Multi-granularity sentiment classification based on rough set theory. Ph.D. thesis, Tongji University, Shanghai, China (2014)

3. Ding, X., Liu, B., Yu, P.S.: A holistic lexicon-based approach to opinion mining. In: Proceedings of the 2008 International Conference on Web Search and Data Mining, pp. 231–240. ACM, New York (2008)
4. Wu, Y., Wen, M.: Disambiguating dynamic sentiment ambiguous adjectives. In: Proceedings of the 23rd International Conference on Computational Linguistics, pp. 1191–1199. ACL, Stroudsburg (2010)
5. Mohammad, S., Dorr, B., Hirst, G.: Computing word-pair antonymy. In: Proceedings of the 2008 Conference on Empirical Methods in Natural Language Processing, pp. 982–991. ACL, Stroudsburg (2008)
6. Hu, Y., Li, W.: Document sentiment classification by exploring description model of topical terms. Computer Speech and Language 25(2), 386–403 (2011)
7. Mei, Q., Ling, X., Wondra, M., Su, H., Zhai, C.: Topic sentiment mixture: modeling facets and opinions in weblogs. In: Proceedings of the 16th International Conference on World Wide Web, pp. 171–180. ACM, New York (2007)
8. Turney, P.D.: Thumbs up or thumbs down?: semantic orientation applied to unsupervised classification of reviews. In: Proceedings of the 40th Annual Meeting on Association for Computational Linguistics, pp. 417–424. ACL, Stroudsburg (2002)
9. Pang, B., Lee, L., Vaithyanathan, S.: Thumbs up?: sentiment classification using machine learning techniques. In: Proceedings of the 2002 Conference on Empirical Methods in Natural Language Processing, pp. 79–86. ACL, Stroudsburg (2002)
10. Pawlak, Z., Grzymala-Busse, J., Slowinski, R., Ziarko, W.: Rough sets. Communications of the ACM 38(11), 88–95 (1995)
11. Yao, Y., Wong, S.K.M.: A decision theoretic framework for approximating concepts. International Journal of Man-machine Studies 37(6), 793–809 (1992)
12. Yao, Y.: Two semantic issues in a probabilistic rough set model. Fundamenta Informaticae 108(3-4), 249–265 (2011)
13. Yao, Y.: Three-way decisions with probabilistic rough sets. Information Sciences 180(3), 341–353 (2010)
14. Yang, X., Yao, J.: Modelling multi-agent three-way decisions with decision-theoretic rough sets. Fundamenta Informaticae 115(2-3), 157–171 (2012)
15. Qian, Y., Zhang, H., Sang, Y., Liang, J.: Multigranulation decision-theoretic rough sets. International Journal of Approximate Reasoning 55(1), 225–237 (2014)
16. Liu, D., Li, T., Liang, D.: Incorporating logistic regression to decision-theoretic rough sets for classifications. International Journal of Approximate Reasoning 55(1), 197–210 (2014)
17. Yao, Y.: An outline of a theory of three-way decisions. In: Yao, J., Yang, Y., Słowiński, R., Greco, S., Li, H., Mitra, S., Polkowski, L. (eds.) RSCTC 2012. LNCS, vol. 7413, pp. 1–17. Springer, Heidelberg (2012)
18. Yao, Y.: The superiority of three-way decisions in probabilistic rough set models. Information Sciences 181(6), 1080–1096 (2011)
19. Zhou, B.: Multi-class decision-theoretic rough sets. International Journal of Approximate Reasoning 55(1), 211–224 (2014)
20. Zhang, L., Liu, B.: Identifying noun product features that imply opinions. In: Proceedings of the 49th Annual Meeting of the Association for Computational Linguistics: Human Language Technologies, pp. 575–580. ACL, Stroudsburg (2011)

# Three-Way Formal Concept Analysis

Jianjun Qi[1,3], Ling Wei[2,3], and Yiyu Yao[3]

[1] School of Computer Science & Technology, Xidian University,
Xi'an, 710071, P.R. China
qijj@mail.xidian.edu.cn
[2] School of Mathematics, Northwest University, Xi'an, 710069, P.R. China
wl@nwu.edu.cn
[3] Department of Computer Science, University of Regina,
Regina, Saskatchewan, Canada S4S 0A2
yyao@cs.uregina.ca

**Abstract.** In this paper, a novel concept formation and novel concept lattices are developed with respect to a binary information table to support three-way decisions. The three-way operators and their inverse are defined and their properties are given. Based on these operators, two types of three-way concepts are defined and the corresponding three-way concept lattices are constructed. Three-way concept lattices provide a new kind of model to make three-way decisions.

**Keywords:** Three-way decisions, Formal concept analysis, Three-way operators, Three-way concepts, Three-way concept lattices.

## 1 Introduction

Three-way decisions are widely used in real-word decision-making. They are used in different fields and disciplines by different names and notations. Observing this phenomenon, Yao proposed a unified framework description of three-way decisions [14]. The theory of three-way decisions is an extension of the common two-way decision model [4, 11–14]. Its applications and extensions are investigated by many recent studies [2, 4–7, 9, 15, 16].

The essential idea of three-way decisions is a ternary classification based on the notions of acceptance, rejection and noncommitment [14]. Its aim is to divide a universe into three pair-wise disjoint regions, called the positive, negative and boundary regions, written as POS, NEG and BND, respectively, according to evaluations of a set of criteria. The three regions are viewed as the regions of acceptance, rejection and noncommitment, respectively, in a ternary classification. Corresponding to the three regions, one may construct rules for three-way decisions. One can construct rules for acceptance from the positive region (inclusion method) and rules for rejection from the negative region (exclusion method). Whenever it is impossible to make an acceptance or a rejection decision, the third option of noncommitment is chosen [12].

Formal concept analysis (FCA) is proposed by Wille [8]. The basis of FCA is formal concepts and their ordered hierarchical structures, concept lattices.

D. Miao et al. (Eds.): RSKT 2014, LNAI 8818, pp. 732–741, 2014.
DOI: 10.1007/978-3-319-11740-9_67 © Springer International Publishing Switzerland 2014

Formal concepts can be formed through a pair of operators induced by a binary relation between an object universe and an attribute universe.

In FCA, formal concepts are from the classical view of concepts. A concept has two facets: extension (an object subset) and intension (an attribute subset). Each element in the extension possesses all elements in the intension and each element in the intension is shared by all elements in the extension. This implies two-way decisions. For a given concept, one can determine whether an object (an attribute) certainly possesses (is shared by) all elements in the intension (the extension) according to whether the object (the attribute) belongs to the extension (the intension). Such decision-making (inclusion method) can be supported in FCA.

In daily life, exclusion method is also used commonly when making decisions. One may want to determine whether an object (an attribute) does not possesses (is not shared by) any elements in the intension (the extension) of a concept. Combining inclusion method (acceptance) with exclusion method (rejection) induces a kind of three-way decisions. This is not supported by FCA. In order to support three-way decisions, there is a need for a new formal concept formation in which the extension or (and) intension of a concept should be an orthopair studied by Ciucci [1]. The paper presents our results on this topic.

The paper is organized as follows. Section 2 introduces some preliminaries about pairs of subsets. Section 3 discusses two-way and three-way operators in a binary information table and their properties. Section 4 presents two-way and three-way concept lattices defined by relevant operators. Section 5 gives an example. Finally, Section 6 concludes the paper.

## 2    Preliminaries

In this section, we introduce some operators about pairs of subsets which will be useful later.

Let $S$ be a non-empty finite set. We write $\mathcal{P}(S)$ to denote the power set, the set of all subsets, of $S$, and $\mathcal{DP}(S)$ to denote the set of all pairs of subsets of $S$, i.e. $\mathcal{DP}(S) = \mathcal{P}(S) \times \mathcal{P}(S)$. Set-theoretic operators on $\mathcal{DP}(S)$, intersection $\cap$, union $\cup$ and complement $c$, can be defined componentwise using standard set operators. For two pairs of subsets $(A, B), (C, D) \in \mathcal{DP}(S)$, we have

$$
\begin{aligned}
(A, B) \cap (C, D) &= (A \cap C, B \cap D), \\
(A, B) \cup (C, D) &= (A \cup C, B \cup D), \\
(A, B)^c &= (S - A, S - B) = (A^c, B^c).
\end{aligned}
\tag{1}
$$

These two pairs of subsets can be ordered by

$$
(A, B) \subseteq (C, D) \iff A \subseteq C \text{ and } B \subseteq D.
\tag{2}
$$

Obviously, $\mathcal{DP}(S)$ is a Boolean algebra.

# 3   Two-Way and Three-Way Operators with Binary Information Tables

## 3.1   Binary Information Tables

Let $U$ be a non-empty finite set of objects and $V$ a non-empty finite set of attributes. The relationship between objects and attributes can be formally defined by a binary relation $R$ from $U$ to $V$, $R \subseteq U \times V$. For a pair $u \in U$ and $v \in V$, if $uRv$, we say that object $u$ has attribute $v$, or alternatively, attribute $v$ is possessed by object $u$. The binary relation $R$ can also be conveniently represented as a binary information table, where the value of the attribute $v$ with respect to the object $u$ is 1 if $uRv$, otherwise, the value is 0. A binary information table $(U, V, R)$ is called a formal context in formal concept analysis [3].

For an object $u \in U$, the set of all attributes possessed by $u$ is called its attribute set, written as $uR$, that is,

$$uR = \{v \in V \mid uRv\}. \tag{3}$$

Similarly, for an attribute $v \in V$, the set of all objects having $v$ is called its object set, written as $Rv$, namely,

$$Rv = \{u \in U \mid uRv\}. \tag{4}$$

The complement of the binary relation $R$ is defined by [10]

$$R^c = \{(u, v) \mid \neg(uRv)\} = U \times V - R. \tag{5}$$

That is, $uR^cv$ if and only if $\neg(uRv)$. Similar to $R$, $uR^c = \{v \in V \mid uR^cv\}$ is the set of attributes not possessed by the object $u$, $R^cv = \{u \in U \mid uR^cv\}$ is the set of objects not having the attribute $v$.

## 3.2   Two-Way Operators

With respect to a binary information table $(U, V, R)$, the following pair of operators can be defined. For simplicity, both operators are denoted by the same symbol.

For $X \subseteq U$ and $A \subseteq V$, a pair of operators, $* : \mathcal{P}(U) \longrightarrow \mathcal{P}(V)$ and $* : \mathcal{P}(V) \longrightarrow \mathcal{P}(U)$, called positive operators, are defined by

$$
\begin{aligned}
X^* &= \{v \in V \mid \forall x \in X (xRv)\} \\
&= \{v \in V \mid X \subseteq Rv\}, \\
A^* &= \{u \in U \mid \forall a \in A (uRa)\} \\
&= \{u \in U \mid A \subseteq uR\}.
\end{aligned} \tag{6}
$$

This pair of operators are called derivation operators and denoted by symbol $'$ in formal concept analysis [3].

With $X$, the attribute set $V$ can be divided into two disjoint regions, $X^*$ and $V - X^*$. $X^*$ is the set of all attributes shared by all objects in $X$. And for any attribute in $V - X^*$, there must exist an object in $X$ which does not have that attribute. There maybe exist an empty region. We regard an empty region as a real region for convenience. Likewise, for $A \subseteq V$, the object set $U$ can also be divided into two disjoint regions, $A^*$ and $U - A^*$. $A^*$ is the set of all objects having all attributes in $A$. And for an object in $U - A^*$, there must exist an attribute in $A$ which is not possessed by that object. Due to such bipartition, this pair of operators are called two-way operators.

The two positive operators have the following properties [3]: if $X, Y \subseteq U$ are sets of objects and $A, B \subseteq V$ are sets of attributes, then

(C1) $X \subseteq X^{**}$ and $A \subseteq A^{**}$,
(C2) $X \subseteq Y \Longrightarrow Y^* \subseteq X^*$ and $A \subseteq B \Longrightarrow B^* \subseteq A^*$,
(C3) $X^* = X^{***}$ and $A^* = A^{***}$,
(C4) $X \subseteq A^* \Longleftrightarrow A \subseteq X^*$,
(C5) $(X \cup Y)^* = X^* \cap Y^*$ and $(A \cup B)^* = A^* \cap B^*$,
(C6) $(X \cap Y)^* \supseteq X^* \cup Y^*$ and $(A \cap B)^* \supseteq A^* \cup B^*$.

Through the positive operators, one can obtain a set of attributes common to the objects in a given object subset, or a set of objects having all attributes in a given attribute subset. On the other hand, there may exist some attributes which are not possessed by any object in a given object subset, and there may also exist some objects which do not have any attribute in a given attribute subset. With respect to this case, we can define another pair of operators, $\bar{*} : \mathcal{P}(U) \longrightarrow \mathcal{P}(V)$ and $\bar{*} : \mathcal{P}(V) \longrightarrow \mathcal{P}(U)$, called negative operators, as follows. For $X \subseteq U$ and $A \subseteq V$, we have

$$
\begin{aligned}
X^{\bar{*}} &= \{v \in V \mid \forall x \in X(\neg(xRv))\} \\
&= \{v \in V \mid \forall x \in X(xR^cv)\} \\
&= \{v \in V \mid X \subseteq R^cv\}, \\
A^{\bar{*}} &= \{u \in U \mid \forall a \in A(\neg(uRa))\} \\
&= \{u \in U \mid \forall a \in A(uR^ca)\} \\
&= \{u \in U \mid A \subseteq uR^c\}.
\end{aligned}
\tag{7}
$$

The same symbol is used for both operators. An attribute in $X^{\bar{*}}$ is not possessed by any object in $X$, and an object in $A^{\bar{*}}$ does not have any attribute in $A$. Clearly, the negative operators of $R$ are just the positive operators of the complement of $R$. Hence the properties that the former have are the same as the latter. The negative operators are also two-way operators.

### 3.3  Three-Way Operators

Combining the operators $*$ and $\bar{*}$, we can get the following pair of new operators, called three-way operators. For $X \subseteq U$ and $A \subseteq V$, the three-way operators,

$\prec : \mathcal{P}(U) \longrightarrow \mathcal{DP}(V)$ and $\prec : \mathcal{P}(V) \longrightarrow \mathcal{DP}(U)$, are defined by

$$X^{\preccurlyeq} = (X^*, X^{\bar{*}}),$$
$$A^{\preccurlyeq} = (A^*, A^{\bar{*}}). \tag{8}$$

The same symbol is again used for both operators.

According to the operator $\prec : \mathcal{P}(U) \longrightarrow \mathcal{DP}(V)$, for an object subset $X \subseteq U$, one can obtain a pair of subsets, $(X^*, X^{\bar{*}})$, of the attribute set $V$. By such pair, $V$ can be divided into the following three regions

$$\mathrm{POS}_X^V = X^*,$$
$$\mathrm{NEG}_X^V = X^{\bar{*}}, \tag{9}$$
$$\mathrm{BND}_X^V = V - (X^* \cup X^{\bar{*}}).$$

$\mathrm{POS}_X^V$ is the positive region, in which every attribute is definitely shared by all objects in $X$. $\mathrm{NEG}_X^V$ is the negative region, in which each attribute is not possessed definitely by any object in $X$. Those attributes possessed by some, but not all, objects in $X$ belong to the boundary region $\mathrm{BND}_X^V$.

If $X = \emptyset$, then $\mathrm{POS}_X^V = \mathrm{NEG}_X^V = V$ because that $X^* = X^{\bar{*}} = V$. Otherwise, $\mathrm{POS}_X^V \cap \mathrm{NEG}_X^V = \emptyset$. In this case, the three regions, $\mathrm{POS}_X^V$, $\mathrm{NEG}_X^V$ and $\mathrm{BND}_X^V$, are pair-wise disjoint. The positive region $\mathrm{POS}_X^V$ and the negative region $\mathrm{NEG}_X^V$ are given explicitly by the operator $\prec$, while the boundary region $\mathrm{BND}_X^V$ is given implicitly. One or two of the three regions may be empty, so the family of these regions may not be a partition of $V$. In this paper, the family of such three regions is still called a tripartition for convenience.

By the operator $\prec : \mathcal{P}(V) \longrightarrow \mathcal{DP}(U)$, for an attribute subset $A \subseteq V$, one can obtain a pair of subsets, $(A^*, A^{\bar{*}})$, of the object set $U$, and $U$ can be divided into the following three regions

$$\mathrm{POS}_A^U = A^*,$$
$$\mathrm{NEG}_A^U = A^{\bar{*}}, \tag{10}$$
$$\mathrm{BND}_A^U = U - (A^* \cup A^{\bar{*}}).$$

The positive region $\mathrm{POS}_A^U$ contains objects having all attributes in $A$. The negative region $\mathrm{NEG}_A^U$ is the set of objects not possessing any attribute in $A$. The boundary region $\mathrm{BND}_A^U$ includes objects possessing some, but not all, attributes in $A$.

Likewise, $\mathrm{POS}_A^U = \mathrm{NEG}_A^U = U$ if $A = \emptyset$. If $A \neq \emptyset$, then $\mathrm{POS}_A^U \cap \mathrm{NEG}_A^U = \emptyset$, and the family of the three regions, $\mathrm{POS}_A^U$, $\mathrm{NEG}_A^U$ and $\mathrm{BND}_A^U$, is a tripartition of $U$. In these regions, $\mathrm{POS}_A^U$ and $\mathrm{NEG}_A^U$ are given explicitly and $\mathrm{BND}_A^U$ is given implicitly.

In relation to the three-way operators, we can define their inverse, $\succ :$ $\mathcal{DP}(U) \longrightarrow \mathcal{P}(V)$ and $\succ : \mathcal{DP}(V) \longrightarrow \mathcal{P}(U)$, for $X, Y \subseteq U$ and $A, B \subseteq V$, as follows

$$(X, Y)^{\succ} = \{v \in V \mid v \in X^* \text{ and } v \in Y^{\bar{*}}\}$$
$$= X^* \cap Y^{\bar{*}},$$
$$(A, B)^{\succ} = \{u \in U \mid u \in A^* \text{ and } u \in B^{\bar{*}}\} \tag{11}$$
$$= A^* \cap B^{\bar{*}}.$$

The same symbol is again used for both operators.

Through the inverse operators, for a pair of subsets $(X, Y)$ of $U$, we can obtain a subset $X^* \cap Y^{\bar{*}}$ of $V$, in which each attribute is shared by all objects in $X$ and not possessed by any object in $Y$. Similarly, for a pair of subsets $(A, B)$ of $V$, one can get a subset $A^* \cap B^{\bar{*}}$ of $U$, in which each object has all attributes in $A$ and does not possess any attribute in $B$.

The following properties hold for the three-way operators and their inverse. For $X, Y, Z, W \subseteq U$ and $A, B, C, D \subseteq V$, we have

(E1) $X \subseteq X^{\lessdot\gtrdot}$ and $A \subseteq A^{\lessdot\gtrdot}$,

(E2) $X \subseteq Y \Longrightarrow Y^{\lessdot} \subseteq X^{\lessdot}$ and $A \subseteq B \Longrightarrow B^{\lessdot} \subseteq A^{\lessdot}$,

(E3) $X^{\lessdot} = X^{\lessdot\gtrdot\lessdot}$ and $A^{\lessdot} = A^{\lessdot\gtrdot\lessdot}$,

(E4) $X \subseteq (A, B)^{\gtrdot} \Longleftrightarrow (A, B) \subseteq X^{\lessdot}$,

(E5) $(X \cup Y)^{\lessdot} = X^{\lessdot} \cap Y^{\lessdot}$ and $(A \cup B)^{\lessdot} = A^{\lessdot} \cap B^{\lessdot}$,

(E6) $(X \cap Y)^{\lessdot} \supseteq X^{\lessdot} \cup Y^{\lessdot}$ and $(A \cap B)^{\lessdot} \supseteq A^{\lessdot} \cup B^{\lessdot}$.

(EI1) $(X, Y) \subseteq (X, Y)^{\gtrdot\lessdot}$ and $(A, B) \subseteq (A, B)^{\gtrdot\lessdot}$,

(EI2) $(X, Y) \subseteq (Z, W) \Longrightarrow (Z, W)^{\gtrdot} \subseteq (X, Y)^{\gtrdot}$
and $(A, B) \subseteq (C, D) \Longrightarrow (C, D)^{\gtrdot} \subseteq (A, B)^{\gtrdot}$,

(EI3) $(X, Y)^{\gtrdot} = (X, Y)^{\gtrdot\lessdot\gtrdot}$ and $(A, B)^{\gtrdot} = (A, B)^{\gtrdot\lessdot\gtrdot}$,

(EI4) $(X, Y) \subseteq A^{\lessdot} \Longleftrightarrow A \subseteq (X, Y)^{\gtrdot}$,

(EI5) $((X, Y) \cup (Z, W))^{\gtrdot} = (X, Y)^{\gtrdot} \cap (Z, W)^{\gtrdot}$
and $((A, B) \cup (C, D))^{\gtrdot} = (A, B)^{\gtrdot} \cap (C, D)^{\gtrdot}$,

(EI6) $((X, Y) \cap (Z, W))^{\gtrdot} \supseteq (X, Y)^{\gtrdot} \cup (Z, W)^{\gtrdot}$
and $((A, B) \cap (C, D))^{\gtrdot} \supseteq (A, B)^{\gtrdot} \cup (C, D)^{\gtrdot}$.

# 4 Two-Way and Three-Way Concept Lattices

## 4.1 Two-Way Concept Lattices

Property (C4) shows that the pair of positive operators set up a Galois connection between $\mathcal{P}(U)$ and $\mathcal{P}(V)$ and define a lattice of formal concepts [3].

**Definition 1.** [3] *Let $(U, V, R)$ be a binary information table. A pair $(X, A)$ of an object subset $X \subseteq U$ and an attribute subset $A \subseteq V$ is called a formal concept, for short, a concept, of $(U, V, R)$, if and only if $X^* = A$ and $A^* = X$. $X$ is called the extension and $A$ is called the intension of the concept $(X, A)$.*

The concepts of a binary information table $(U, V, R)$ are ordered by

$$(X, A) \leq (Y, B) \Longleftrightarrow X \subseteq Y \Longleftrightarrow B \subseteq A, \tag{12}$$

where $(X, A)$ and $(Y, B)$ are concepts. $(X, A)$ is called a sub-concept of $(Y, B)$, and $(Y, B)$ is called a super-concept of $(X, A)$. All the concepts form a complete lattice that is called the concept lattice of $(U, V, R)$ and written as $CL(U, V, R)$. The infimum and supremum are given by [3]

$$\begin{aligned} (X, A) \wedge (Y, B) &= (X \cap Y, (A \cup B)^{**}), \\ (X, A) \vee (Y, B) &= ((X \cup Y)^{**}, A \cap B). \end{aligned} \tag{13}$$

The pair of negative operators also define a lattice that is just the concept lattice defined by the positive operators of the complement of $R$. This lattice is called a complement concept lattice and denoted by $CL(U, V, R^c)$.

## 4.2    Object-Induced Three-Way Concept Lattices

From property (E4), it is easy to know that the operators $\lessdot : \mathcal{P}(U) \longrightarrow \mathcal{DP}(V)$ and $\gtrdot : \mathcal{DP}(V) \longrightarrow \mathcal{P}(U)$ form a Galois connection between $\mathcal{P}(U)$ and $\mathcal{DP}(V)$ and define a lattice of object-induced three-way concepts.

**Definition 2.** *Let $(U, V, R)$ be a binary information table. A pair $(X, (A, B))$ of an object subset $X \subseteq U$ and two attribute subsets $A, B \subseteq V$ is called an object-induced three-way concept, for short, an OE-concept, of $(U, V, R)$, if and only if $X^{\lessdot} = (A, B)$ and $(A, B)^{\gtrdot} = X$. $X$ is called the extension and $(A, B)$ is called the intension of the OE-concept $(X, (A, B))$.*

If $(X, (A, B))$ and $(Y, (C, D))$ are OE-concepts, then they can be ordered by

$$(X, (A, B)) \leq (Y, (C, D)) \Longleftrightarrow X \subseteq Y \Longleftrightarrow (C, D) \subseteq (A, B), \tag{14}$$

$(X, (A, B))$ is called a sub-concept of $(Y, (C, D))$, and $(Y, (C, D))$ is called a super-concept of $(X, (A, B))$. All the OE-concepts form a complete lattice, which is called the object-induced three-way concept lattice of $(U, V, R)$ and written as $OEL(U, V, R)$. The infimum and supremum are given by

$$\begin{aligned} (X, (A, B)) \wedge (Y, (C, D)) &= (X \cap Y, ((A, B) \cup (C, D))^{\gtrdot\lessdot}), \\ (X, (A, B)) \vee (Y, (C, D)) &= ((X \cup Y)^{\lessdot\gtrdot}, (A, B) \cap (C, D)). \end{aligned} \tag{15}$$

## 4.3    Attribute-Induced Three-Way Concept Lattices

Obviously, by property (EI4), the operators $\gtrdot : \mathcal{DP}(U) \longrightarrow \mathcal{P}(V)$ and $\lessdot : \mathcal{P}(V) \longrightarrow \mathcal{DP}(U)$ also set up a Galois connection between $\mathcal{DP}(U)$ and $\mathcal{P}(V)$ and define a lattice of attribute-induced three-way concepts.

**Definition 3.** *Let $(U, V, R)$ be a binary information table. A pair $((X, Y), A)$ of two object subsets $X, Y \subseteq U$ and an attribute subset $A \subseteq V$ is called an attribute-induced three-way concept, for short, an AE-concept, of $(U, V, R)$, if and only if $(X, Y)^{\gtrdot} = A$ and $A^{\lessdot} = (X, Y)$. $(X, Y)$ is called the extension and $A$ is called the intension of the AE-concept $((X, Y), A)$.*

If $((X,Y),A)$ and $((Z,W),B)$ are AE-concepts, then they can be ordered by

$$((X,Y),A) \leq ((Z,W),B) \Longleftrightarrow (X,Y) \subseteq (Z,W) \Longleftrightarrow B \subseteq A, \qquad (16)$$

$((X,Y),A)$ is called a sub-concept of $((Z,W),B)$, and $((Z,W),B)$ is called a super-concept of $((X,Y),A)$. All the AE-concepts also form a complete lattice, which is called the attribute-induced three-way concept lattice of $(U,V,R)$ and written as $AEL(U,V,R)$. The infimum and supremum are given by

$$\begin{aligned}((X,Y),A) \wedge ((Z,W),B) &= ((X,Y) \cap (Z,W),(A \cup B)^{<>}), \\ ((X,Y),A) \vee ((Z,W),B) &= (((X,Y) \cup (Z,W))^{><},A \cap B).\end{aligned} \qquad (17)$$

## 5 Example

Table 1 is a binary information table with $U = \{1,2,3,4\}$ and $V = \{a,b,c,d,e\}$. The corresponding concept lattice, complement concept lattice, object-induced three-way concept lattice and attribute-induced three-way concept lattice are shown as Figure 1 – 4, respectively. For the sake of simplicity, in the figures, we denote the sets by listing their elements. For example, the attribute subset $\{a,b,c\}$ is denoted by $abc$.

The figures show that the binary information table in this example has 6 formal concepts, 6 complement concepts, 8 OE-concepts and 11 AE-concepts. For every formal concept or every complement concept, there exists an OE-concept with the same extension and the different intension in the object-induced concept lattice. The intension of an OE-concept has two parts, so it can be used to divide the attribute universe $V$ into three regions in order to make three-way decisions. For example, the intension of the OE-concept $(13,(d,c))$ is $(d,c)$. Based on this intension, $V$ can be divided into the following regions: positive region $d$, negative region $c$ and boundary region $abe$. The same applies to the attribute-induced concept lattice.

## 6 Conclusions

As an extension of classic two-way decisions, three-way decisions are widely used in many fields and disciplines. Traditional formal concept analysis is good for two-way decisions, but not good for three-way decisions. In this paper, we proposed an extended model of formal concept analysis, three-way formal concept analysis, which is suitable for three-way decisions. We presented a new concept formation based on three-way operators, and constructed the object-induced concept lattices and the attribute-induced concept latticed. Three-way concept lattices can be used for not only normal three-way decisions but also granularity-based sequential three-way decisions [15].

**Acknowledgements.** This work was partially supported by the National Natural Science Foundation of China (Grant Nos. 11371014, 11071281 and 61202206), the Natural Science Basic Research Plan in Shaanxi Province of China (Program No. 2014JM8306) and the State Scholarship Fund of China. We thank Dr. Jianmin Ma, Dr. Hongying Zhang and Dr. Yanhong She for their discussions.

**Table 1.** A binary information table

| Object | $a$ | $b$ | $c$ | $d$ | $e$ |
|--------|-----|-----|-----|-----|-----|
| 1 | 1 | 1 | 0 | 1 | 1 |
| 2 | 1 | 1 | 1 | 0 | 0 |
| 3 | 0 | 0 | 0 | 1 | 0 |
| 4 | 1 | 1 | 1 | 0 | 0 |

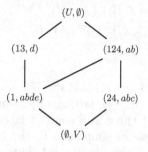

**Fig. 1.** The concept lattice $CL(U, V, R)$

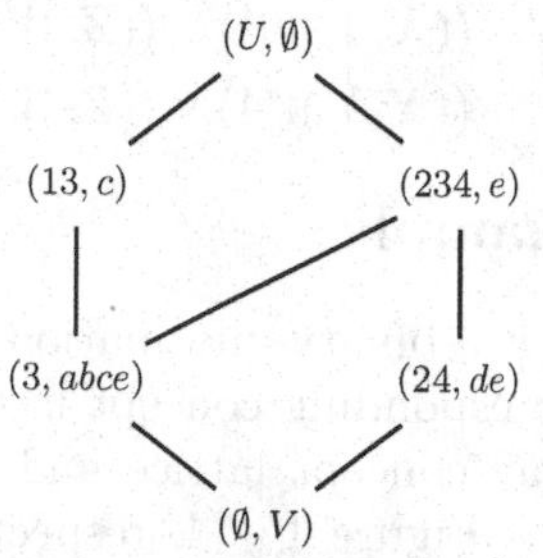

**Fig. 2.** The complement concept lattice $CL(U, V, R^c)$

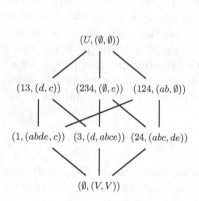

**Fig. 3.** The object-induced three-way concept lattice $OEL(U, V, R)$

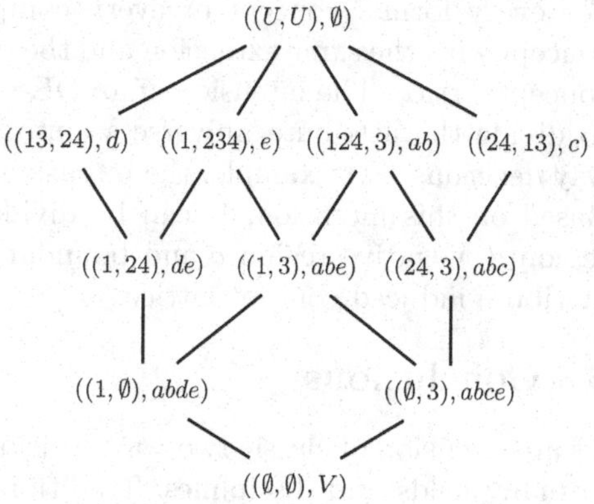

**Fig. 4.** The attribute-induced three-way concept lattice $AEL(U, V, R)$

# References

1. Ciucci, D.: Orthopairs: a simple and widely used way to model uncertainty. Fundamenta Informaticae 108, 287–304 (2011)
2. Deng, X.F., Yao, Y.Y.: Decision-theoretic three-way approximations of fuzzy sets. Information Sciences (2014), http://dx.doi.org/10.1016/j.ins.2014.04.022
3. Ganter, B., Wille, R.: Formal Concept Analysis. Mathematical Foundations. Springer, Heidelberg (1999)

4. Hu, B.Q.: Three-way decisions space and three-way decisions. Information Sciences (2014), http://dx.doi.org/10.1016/j.ins.2014.05.015

5. Li, H.X., Zhou, X.Z.: Risk decision making based on decision-theoretic rough set: A three-way view decision model. International Journal of Computational Intelligence Systems 4, 1–11 (2011)

6. Liu, D., Li, T.R., Ruan, D.: Probabilistic model criteria with decision-theoretic rough sets. Information Sciences 181, 3709–3722 (2011)

7. Liu, D., Li, T.R., Li, H.X.: A multiple-category classification approach with decision-theoretic rough sets. Fundamenta Informaticae 115, 173–188 (2012)

8. Wille, R.: Restructuring lattice theory: an approach based on hierarchies of concepts. In: Rival, I. (ed.) Ordered Sets, pp. 445–470. Reidel Publishing Company, Dordrecht-Boston (1982)

9. Yang, X.P., Yao, J.T.: Modelling multi-agent three-way decisions with decisiontheoretic rough sets. Fundamenta Informaticae 115, 157–171 (2012)

10. Yao, Y.: A Comparative Study of Formal Concept Analysis and Rough Set Theory in Data Analysis. In: Tsumoto, S., Słowiński, R., Komorowski, J., Grzymała-Busse, J.W. (eds.) RSCTC 2004. LNCS (LNAI), vol. 3066, pp. 59–68. Springer, Heidelberg (2004)

11. Yao, Y.: Three-Way Decision: An Interpretation of Rules in Rough Set Theory. In: Wen, P., Li, Y., Polkowski, L., Yao, Y., Tsumoto, S., Wang, G. (eds.) RSKT 2009. LNCS, vol. 5589, pp. 642–649. Springer, Heidelberg (2009)

12. Yao, Y.Y.: Three-way decisions with probabilistic rough sets. Information Sciences 180, 341–353 (2010)

13. Yao, Y.Y.: The superiority of three-way decisions in probabilistic rough set models. Information Sciences 181, 1080–1096 (2011)

14. Yao, Y.: An outline of a theory of three-way decisions. In: Yao, J., Yang, Y., Słowiński, R., Greco, S., Li, H., Mitra, S., Polkowski, L. (eds.) RSCTC 2012. LNCS, vol. 7413, pp. 1–17. Springer, Heidelberg (2012)

15. Yao, Y.: Granular Computing and Sequential Three-Way Decisions. In: Lingras, P., Wolski, M., Cornelis, C., Mitra, S., Wasilewski, P. (eds.) RSKT 2013. LNCS, vol. 8171, pp. 16–27. Springer, Heidelberg (2013)

16. Zhou, B.: Multi-class decision-theoretic rough sets. International Journal of Approximate Reasoning 55, 211–224 (2014)

# Three-Way Decision Based on Belief Function

Zhan'ao Xue[*], Jie Liu, Tianyu Xue, Tailong Zhu, and Penghan Wang

College of Computer and Information Engineering,
Henan Normal University, Xinxiang 453007, China
xuezhanao@163.com

**Abstract.** In this paper, the basic knowledge of three-way decision and the D-S evidence theory are reviewed, respectively. A new model of the three-way decision is proposed, which is based on a belief function. The probability function is replaced with the belief function in the classical three-way decision model. Besides the decision rules are proposed in this model, some properties are also discussed. Meanwhile, a universe is divided into three disjoint regions by the different values of the belief functions in this model and their decision rules. Finally, a comprehensive illustration is presented to verify the effectiveness and feasibility of this model.

**Keywords:** Three-way decision, belief function, D-S theory of evidence.

## 1 Introduction

Decision rough set model was proposed by Yao, Wong and Lingras in 1990 [1, 2]. Subsequently, Yao founded the three-way decision theory in 2010 [3, 4, 5]. The theory has greatly enriched the method and theory of granular computing, which can represent the probability of rough sets greatly and reflect the rough sets approximation principle accurately. The thought of three-way decision is widely used in real life, which is groundbreaking in its systematization, theorization and modeling. Three-way decision is a common problem solving strategy used in many disciplines. It has been used in email spam filtering [6], investment management [7], cluster analysis and data analysis [8]. Considering the values of loss function of the three-way decision model are either random numbers or interval numbers [9, 10].

The Dempster/Shafer (D-S) theory of evidence, which was first proposed by Dempster and was completed by Shafer. It had further profound impact in dealing with uncertainty [11]. The D-S theory of evidence is a mathematical tool to deal with uncertain information, which continues to receive attention in the research fields. Li [12] used the D-S theory of evidence to evaluate the maritime management. Liu and Hu [13] referred to such an issue as the application of decision support systems based on the D-S theory of evidence, and so on.

In recent years, the three-way decision theory is widely applied in many fields, the D-S theory of evidence is important for solving uncertain problems. At present, a

---

[*] Corresponding author.

D. Miao et al. (Eds.): RSKT 2014, LNAI 8818, pp. 742–752, 2014.
DOI: 10.1007/978-3-319-11740-9_68 © Springer International Publishing Switzerland 2014

research about the three-way decision based on the belief function is less. In this paper, a new three-way decision model based on the belief function is proposed. The probability function is replaced with the belief function in the classical three-way decision model, and the decision rules are proposed in this model. According to the new model and rules, a universe can be divided into three disjoint regions. Finally, by an example, the validity and feasibility of the presented method is demonstrated.

## 2    Preliminaries

For the sake of convenience for statement, we firstly summarize some related definitions and theorems which will be used in the following section.

**Definition 1 [8].** Let $U$ be a finite and nonempty set and $R$ be an equivalence relation on $U$. The pair $apr = (U, R)$ is called an approximation space. The equivalence relation $R$ induces a partition of $U$, denoted by $U / R$. For a subset $A \subseteq U$, its lower and upper approximations are defined by:

$$\underline{apr}(X) = \left\{ x \in U \mid [x]_R \subseteq X \right\}, \qquad \overline{apr}(X) = \left\{ x \in U \mid [x]_R \cap X \subseteq \phi \right\}. \tag{1}$$

The equivalence class containing object $x$ is denoted as $[x]_R$.

**Definition 2 [8].** Based on the rough set approximations of $X$, one can divide the universe $U$ into three disjoint regions, the positive region $POS(X)$, the boundary region $BND(X)$, and the negative region $NEG(X)$:

$$POS(X) = \underline{apr}(X) = \left\{ x \in U \mid [x]_R \subseteq X \right\},$$

$$BND(X) = \overline{apr}(X) - \underline{apr}(X) = \left\{ x \in U \mid [x]_R \cap X \neq \phi \wedge [x]_R \not\subseteq \phi \right\}, \tag{2}$$

$$NEG(X) = U - \overline{apr}(X) = \left\{ x \in U \mid [x]_R \cap X = \phi \right\}.$$

We can accept an object as it being an instance of a concept $X$ based on a positive rule, reject an object as it being an instance of $X$ based on a negative rule, and abstain based on a boundary rule.

Here are some necessary notions about D-S theory of evidence.

**Definition 3 [14].** Let $\Omega = \left( h_1, h_2, ..., h_n \right)$ be a finite set of $n$ hypotheses (frame of discernment). Constructing a mapping, $m : 2^\Omega \to [0,1]$, which is called a basic probability assignment and satisfies two axioms:

$$(1)\, m(\phi) = 0, \quad (\phi \text{ is empty set}),$$

$$(2)\, \sum_{A \subseteq \Omega} m(A) = 1.$$

A subset $A \subseteq \Omega$ is called a focal element of $m$, which satisfies the condition $m(A) > 0$. All the related focal elements are collectively called the body of evidence.

Definition 4 [14]. A belief function is a mapping from $2^{\Omega}$ to unit interval $[0,1]$, and satisfies the following axioms:

(1) $Bel(\phi) = 0$,

(2) $Bel(\Omega) = 1$,

(3) For every positive integer n and every collection $A_1, A_2, \ldots, A_n \subseteq \Omega$,

$$Bel\left(\bigcup_{i=1}^{n} A_i\right) \geq \sum_{\substack{I \subset \{1,2,\ldots,N\} \\ I \neq \phi}} (-1)^{|I|+1} Bel(\bigcap_{i \in I} A_i).$$

The dual of a belief function, called a plausibility function $Pl$, is defined by:

$$Pl(A) = 1 - Bel(\neg A).$$

Clearly, $Pl(A)$ represents the extent to which we fail to disbelieve $A$.

For any subset $A \subseteq \Omega$, $Bel(A) \leq Pl(A)$.

$[Bel(A), Pl(A)]$ constitutes the interval of support $A$ and can be seen as the lower and upper bounds of the probability which $A$ is supported, named as $m(\Omega) = Pl(A) - Bel(A)$.

Definition 5 [14]. Using the basic probability assignment, belief and plausibility of $A$ are expressed as:

$$Bel(A) = \sum_{D \subset A} m(D), \qquad Pl(A) = \sum_{A \cap D = \phi} m(D). \tag{3}$$

In a Pawlak rough set algebra, the qualities of lower and upper approximations of a subset $A \subseteq U$ are defined by [15, 16]:

$$q(X) = \left|\underline{apr}(X)\right| / |U|, \qquad \overline{q}(X) = \left|\overline{apr}(X)\right| / |U|. \tag{4}$$

Theorem 1 [15]. For a subset $X \subseteq U$ and $R$ is an equivalence relation on $U$, and its partition $U / R = \{X_1, X_2, \ldots, X_n\}$, the quality of lower approximation $q(X)$ is a belief function. The basic probability assignment is $m(x_i) = \dfrac{X_i}{U}$, $i = 1, 2, \ldots, n$ and

$m(A) = 0$, $A \notin U / R$, the quality of upper approximation $\overline{q}(X)$ is a plausibility function.

Theorem 2 [15]. Suppose $Bel$ is a belief function on $U$, which satisfies two conditions:

(1) the set of focal elements of $Bel$ is a partition of $U$,
(2) $m(A) = A / U$ for every focal element $A$ of $Bel$.

Where $m$ is the basic probability assignment of $Bel$. There exists a Pawlak rough set algebra such that $\underline{q}(X) = Bel(A)$, for every $A \subseteq U$.

Theorem 3[15]. For a decision table $A$, the finite set $U$ is divided into $Class_A(D) = \{X_{r(1)}, X_{r(2)}, ..., X_{r(n)}\}$ by the decision attribute $D$, $X_{r(i)} = \{x \in U, D(x) = r(i)\}$, $(i = 1, 2, ..., n)$. $r(i)$ is the value of decision, and $V_D = \{r(1), r(2), ..., r(n)\}$, for every focal element $A$ and $\forall \beta \in V_D$. The relationship between the rough set theory and evidence theory is as follows:

$$m(\phi) = 0, \qquad m(\beta) = \frac{|x \in U, D(x) = \beta|}{|U|},$$

$$Bel_A(\beta) = \frac{|\underline{A} \cup_{i \in \beta} X_i|}{|U|}, \qquad Pl_A(\beta) = \frac{|\overline{A} \cup_{i \in \beta} X_i|}{|U|}. \tag{5}$$

Theorem 4 [16]. The corresponding belief and plausibility functions of conditional set function $m*$ are as follows:

$$Bel^0(A/e) = \sum_{X \subseteq A} m^0(X/e), \qquad Pl^0(A/e) = \sum_{X \cap A = \phi} m^0(X/e). \tag{6}$$

When $A$ has a single element, $Bel^0(A/e) = m^0(X/e)$.

Property 1 [16]. Suppose $Bel(*/r_j)$ is a belief function on the frame of discernment $U$, $Bel(A) = \sum_{j=1}^{m} Bel(A/r_j)p(r_j)$ is a belief function.

# 3    Three-Way Decision Based on Belief Function Model

With respect to the model of three-way decision based on the belief function, we have a set of two states $\Omega = \{X, \neg X\}$, indicating that an element is in $X$ or not in $X$, respectively. The set of three actions is given by $A = \{a_P, a_B, a_N\}$, where $a_P$, $a_B$, and $a_N$ represent the three actions in classifying an object, deciding $x \in POS(X)$, deciding $x \in BND(X)$, and deciding $x \in NEG(X)$, respectively. Let $E(X|x)$ and $E(\neg X|x)$ be the belief function of an object in $X$ and not in $X$. $Bel(X)$ and $Pl(X)$ are the belief function and plausibility function of an object in $X$. $\lambda_{PP}$, $\lambda_{BP}$ and $\lambda_{NP}$ denote the losses incurred for taking actions $a_P$, $a_B$ and $a_N$ respectively, when an object belongs to $X$. And $\lambda_{PN}$, $\lambda_{BN}$, $\lambda_{NN}$ denote the losses incurred for taking the same actions when the object does not belong to $X$.

For every $x$, compute the expected loss associated with taking the individual actions, which the actions are $a_P$, $a_B$ and $a_N$ can be expressed as:

$$R(a_P | x) = \lambda_{PP} E(X | x) + \lambda_{PN} E(\neg X | x),$$

$$R(a_B \mid x) = \lambda_{BP} E(X \mid x) + \lambda_{BN} E(\neg X \mid x), \tag{7}$$

$$R(a_N \mid x) = \lambda_{NP} E(X \mid x) + \lambda_{NN} E(\neg X \mid x).$$

Therefore, we can also replace $(1 - Pl(X))$ with $E(\neg X \mid x)$. Namely $Bel(X)$ and $(1 - Pl(X))$ can instead of $E(X \mid x), E(\neg X \mid x)$.

$$R(a_P \mid x) = \lambda_{PP} Bel(X) + \lambda_{PN} (1 - Pl(X)),$$

$$R(a_B \mid x) = \lambda_{BP} Bel(X) + \lambda_{BN} (1 - Pl(X)), \tag{8}$$

$$R(a_N \mid x) = \lambda_{NP} Bel(X) + \lambda_{NN} (1 - Pl(X)).$$

The Bayesian decision procedure suggests the following minimum-risk decision rules:

(P):  $R(a_P \mid x) \le R(a_B \mid x)$ and $R(a_P \mid x) \le R(a_N \mid x)$, decide $x \in POS(X)$,

(B):  $R(a_B \mid x) \le R(a_P \mid x)$ and $R(a_B \mid x) \le R(a_N \mid x)$, decide $x \in BND(X)$,

(N):  $R(a_N \mid x) \le R(a_P \mid x)$ and $R(a_N \mid x) \le R(a_B \mid x)$, decide $x \in NEG(X)$.

Since $Pl(A) = m(\Omega) + Bel(A)$, we can simplify the rules only based on the belief function and the loss function $\lambda$. Consider a special kind of loss functions with:

$$0 \le \lambda_{PP} \le \lambda_{BP} < \lambda_{NP}, \qquad\qquad 0 \le \lambda_{NN} \le \lambda_{BN} < \lambda_{PN}. \tag{9}$$

That is, the loss of classifying an object $x$ belonging to $X$ into the positive region $POS(X)$, is less than or equal to the loss of classifying $x$ into the boundary region $BND(X)$, and both of these losses are strictly less than the loss of classifying $x$ into the negative region $NEG(X)$. The reverse order of the losses is used for classifying an object not in $X$. The action of conditional risk are $a_P$, $a_B$ and $a_N$, which can also be written as:

$$R(a_P \mid x) = \lambda_{PP} Bel(X) + \lambda_{PN} (1 - m(\Omega) - Bel(X)),$$

$$R(a_B \mid x) = \lambda_{BP} Bel(X) + \lambda_{BN} (1 - m(\Omega) - Bel(X)), \tag{10}$$

$$R(a_N \mid x) = \lambda_{NP} Bel(X) + \lambda_{NN} (1 - m(\Omega) - Bel(X)).$$

Under the rules (P), (B) and (N), the decision rules can be re-expressed as:
For the rule (P):

$$R(a_P \mid x) \le R(a_B \mid x) \Leftrightarrow Bel(X) \ge \frac{(1 - m(\Omega))(\lambda_{PN} - \lambda_{BN})}{(\lambda_{PN} - \lambda_{BN}) + (\lambda_{BP} - \lambda_{PP})},$$

$$R(a_P \mid x) \le R(a_N \mid x) \Leftrightarrow Bel(X) \ge \frac{(1 - m(\Omega))(\lambda_{PN} - \lambda_{NN})}{(\lambda_{PN} - \lambda_{NN}) + (\lambda_{NP} - \lambda_{PP})}.$$

For the rule (B):

$$R(a_B \mid x) \le R(a_P \mid x) \Leftrightarrow Bel(X) \le \frac{(1-m(\Omega))(\lambda_{PN} - \lambda_{BN})}{(\lambda_{PN} - \lambda_{BN}) + (\lambda_{BP} - \lambda_{PP})},$$

$$R(a_B \mid x) \le R(a_N \mid x) \Leftrightarrow Bel(X) \ge \frac{(1-m(\Omega))(\lambda_{BN} - \lambda_{NN})}{(\lambda_{BN} - \lambda_{NN}) + (\lambda_{NP} - \lambda_{BP})}.$$

For the rule $(N)$:

$$R(a_N \mid x) \le R(a_P \mid x) \Leftrightarrow Bel(X) \le \frac{(1-m(\Omega))(\lambda_{PN} - \lambda_{NN})}{(\lambda_{PN} - \lambda_{NN}) + (\lambda_{NP} - \lambda_{PP})},$$

$$R(a_N \mid x) \le R(a_B \mid x) \Leftrightarrow Bel(X) \le \frac{(1-m(\Omega))(\lambda_{BN} - \lambda_{NN})}{(\lambda_{BN} - \lambda_{NN}) + (\lambda_{NP} - \lambda_{BP})}.$$

Where the parameters $\alpha$, $\beta$ and $\gamma$ are defined as:

$$\alpha = \frac{(1-m(\Omega))(\lambda_{PN} - \lambda_{BN})}{(\lambda_{PN} - \lambda_{BN}) + (\lambda_{BP} - \lambda_{PP})},$$

$$\beta = \frac{(1-m(\Omega))(\lambda_{BN} - \lambda_{NN})}{(\lambda_{BN} - \lambda_{NN}) + (\lambda_{NP} - \lambda_{BP})}, \tag{11}$$

$$\gamma = \frac{(1-m(\Omega))(\lambda_{PN} - \lambda_{NN})}{(\lambda_{PN} - \lambda_{NN}) + (\lambda_{NP} - \lambda_{PP})}.$$

For the rule $(B)$: $\alpha > \beta$, then the loss function satisfies the condition $\frac{\lambda_{BP} - \lambda_{PP}}{\lambda_{PN} - \lambda_{BN}} < \frac{\lambda_{NP} - \lambda_{BP}}{\lambda_{BN} - \lambda_{NN}}$, as we all know, $\frac{b}{a} > \frac{d}{c} \Rightarrow \frac{b}{a} > \frac{b+d}{a+c} > \frac{d}{c} (a,b,c,d > 0)$, so $\frac{\lambda_{BP} - \lambda_{PP}}{\lambda_{PN} - \lambda_{BN}} < \frac{\lambda_{NP} - \lambda_{PP}}{\lambda_{PN} - \lambda_{NN}} < \frac{\lambda_{NP} - \lambda_{BP}}{\lambda_{BN} - \lambda_{NN}}$, namely $0 \le \beta < \gamma < \alpha \le 1$. The other decision rules can be re-expressed as:

$(P_1)$: If $Bel(X) \ge \alpha$, decide $x \in POS(X)$,

$(B_1)$: If $\alpha < Bel(X) < \beta$, decide $x \in BND(X)$,

$(N_1)$: If $Bel(X) \le \beta$, decide $x \in NEG(X)$.

The parameter $\gamma$ is no longer needed.

For the rule $(P_1)$: We accept $x$ as a member of $X$ if $Bel(X) \ge \alpha$.

For the rule $(B_1)$: We neither accept nor reject $x$ as a member of $X$ if $\alpha < Bel(X) < \beta$.

For the rule$(N_1)$: We reject $x$ as a member of $X$ if $Bel(X) \le \beta$.

When $\alpha = \beta$, we have $\alpha = \gamma = \beta$. In this case, we use the decision rules:

$(P_2)$: If $Bel(X) \ge \alpha$, decide $x \in POS(X)$,

$(B_2)$: If $Bel(X) = \alpha$, decide $x \in BND(X)$,

$(N_2)$: If $Bel(X) \leq \alpha$, decide $x \in NEG(X)$.

In this situation, the three-way decision can degenerate into traditional two-way decision.

## 4    Example

Suppose Table1 is given the following decision, which is a simplified diagnosis case library. When $A : \{a_1, ..., a_6\}$ is condition attributes, {temperature, permeability, the difference of pressure, porosity, pipe status} is replaced by $\{a_1, a_2, a_3, a_4, a_5, a_6\}$. $D$ is a decision attribute. $R_1$, $R_2$ and $R_3$ represent the different results of decision. According to the above five instances of condition attributes, decision result and the attributes of $c_6$, $c_7$. We can get a decision for $c_6$ and $c_7$ based on the new rules of three-way decision.

**Table 1.** Example of reservoir damage type diagnosis

| Attribute<br>U | $a_1$ | $a_2$ | $a_3$ | $a_4$ | $a_5$ | $a_6$ | D(decision) |
|---|---|---|---|---|---|---|---|
| $c_1$ | high | low | big | small | medium | ordinary | $R_1$ |
| $c_2$ | high | low | small | small | big | clean | $R_2$ |
| $c_3$ | low | middle | big | medium | small | dirty | $R_1$ |
| $c_4$ | low | high | medium | small | medium | ordinary | $R_3$ |
| $c_5$ | middle | low | small | big | medium | clean | $R_2$ |
| $c_6$ | middle | middle | small | big | small | ordinary | ? |
| $c_7$ | middle | low | medium | medium | small | ordinary | ? |

Here are the loss functions

$$\lambda_{PP} = 0.385, \quad \lambda_{BP} = 0.398, \quad \lambda_{NP} = 0.512;$$
$$\lambda_{NN} = 0.397, \quad \lambda_{BN} = 0.561, \quad \lambda_{PN} = 0.714.$$

We can regard the five instances information as prior experience. Combining with the rough sets, one can get some prior information which the theory of evidence needs, then the attributes of $c_6$ and $c_7$ are regarded as the known evidence.

(1)    In set $c_6$ $\{a_1, a_2, a_3, a_4, a_5, a_6\}$, the values of each attribute can be expressed as {middle, middle, small, big, small, ordinary}, respectively.

For $a_1$, according to the Theorem 3 and Theorem 4,

$$m_1(\{c_1, c_2\} = \text{high}) = 2/5, \quad m_1(\{c_5\} = \text{middle}) = 1/5, \quad m_1(\{c_3, c_4\} = \text{low}) = 2/5.$$

By Property 1, $m_1(R_1/\{\text{middle}\}) = 0$,

$m_1(R_2/\{\text{middle}\}) = 1/2$, $m_1(R_3/\{\text{middle}\}) = 0$, $m_1(\Omega) = 0$.

The same can be, for other five attributes $a_2 \sim a_6$, we can get the results as follows:

$a_2$ : $m_2(R_1/\{\text{middle}\}) = 1/2$ , $m_2(R_2/\{\text{middle}\}) = 0$ , $m_2(R_3/\{\text{middle}\}) = 0$ , $m_2(\Omega) = 0$.

$a_3$ : $m_3(R_1/\{\text{medium}\}) = 0$ , $m_3(R_2/\{\text{medium}\}) = 1$ , $m_3(R_3/\{\text{medium}\}) = 0$ , $m_3(\Omega) = 0$.

$a_4$ : $m_4(R_1/\{\text{big}\}) = 0$, $m_4(R_2/\{\text{big}\}) = 1/2$, $m_4(R_3/\{\text{big}\}) = 0$, $m_4(\Omega) = 3/5$.

$a_5$ : $m_5(R_1/\{\text{medium}\}) = 1/2$ , $m_5(R_2/\{\text{medium}\}) = 0$ , $m_5(R_3/\{\text{medium}\}) = 0$ , $m_5(\Omega) = 3/5$.

$a_6$ : $m_6(R_1/\{\text{ordinary}\}) = 1/4$ , $m_6(R_2/\{\text{ordinary}\}) = 0$ , $m_6(R_3/\{\text{ordinary}\}) = 0$ , $m_6(\Omega) = 0$.

According to the Dempster's rule of combination, $m(\Omega) = 3/5*3/5 = 9/25 = 0.360$, for the Theorem 4 and Property1, the results of evidence are shown as follows:

$Bel(R_1) = 1/2*1/5 + 1/2*1/5 + 1/4*2/5 = 3/10$,

$Bel(R_2) = 1/2*1/5 + 1*2/5 + 1/2*1/5 = 6/10$,

$Bel(R_3) = 0$.

(2) In set $c_7$ $\{a_1, a_2, a_3, a_4, a_5, a_6\}$, the values of each attribute can be expressed as $\{middle, low, medium, medium, small, ordinary\}$.

$a_1$: $m_1(R_1/\{\text{middle}\}) = 0$, $m_1(R_2/\{\text{middle}\}) = 1/2$, $m_1(R_3/\{\text{middle}\}) = 0$, $m_1(\Omega) = 0$.

$a_2$: $m_2(R_1/\{\text{low}\}) = 1/6$, $m_2(R_2/\{\text{low}\}) = 0$, $m_2(R_3/\{\text{low}\}) = 0$, $m_2(\Omega) = 0$.

$a_3$ : $m_3(R_1/\{\text{medium}\}) = 0$ , $m_3(R_2/\{\text{medium}\}) = 0$ , $m_3(R_3/\{\text{medium}\}) = 1$ , $m_3(\Omega) = 0$.

$a_4$ : $m_4(R_1/\{\text{medium}\}) = 1/2$ , $m_4(R_2/\{\text{medium}\}) = 0$ , $m_4(R_3/\{\text{medium}\}) = 0$ , $m_4(\Omega) = 3/5$.

$a_5$: $m_5(R_1/\{\text{small}\}) = 1/2$, $m_5(R_2/\{\text{small}\}) = 0$, $m_5(R_3/\{\text{small}\}) = 0$, $m_5(\Omega) = 3/5$.

$a_6$: $m_6(R_1/\{\text{ordinary}\}) = 1/4$, $m_6(R_2/\{\text{ordinary}\}) = 0$, $m_6(R_3/\{\text{ordinary}\}) = 0$, $m_6(\Omega) = 0$.

According to the Dempster's rule of combination $m(\Omega) = 3/5*3/5 = 9/25 = 0.360$

$Bel(R_1) = 1/6*3/5 + 1/2*1/5 + 1/2*1/5 + 1/4*2/5 = 4/10$,

$Bel(R_2) = 1/2*1/5 = 1/10$,

$Bel(R_3) = 1*1/5 = 1/5$.

According to Formula 11, so

$$\alpha = \frac{(1 - m(\Omega))(\lambda_{PN} - \lambda_{BN})}{(\lambda_{PN} - \lambda_{BN}) + (\lambda_{BP} - \lambda_{PP})} = \frac{(1 - 0.360) \times (0.714 - 0.561)}{(0.714 - 0.561) + (0.398 - 0.385)} = 0.590,$$

$$\beta = \frac{(1 - m(\Omega))(\lambda_{BN} - \lambda_{NN})}{(\lambda_{BN} - \lambda_{NN}) + (\lambda_{NP} - \lambda_{BP})} = \frac{(1 - 0.360) \times (0.561 - 0.397)}{(0.561 - 0.397) + (0.512 - 0.398)} = 0.378.$$

Through the above analysis and the rules (P1), (B1) and (N1), here we can get two different decisions for $c_6$, $c_7$.

For $c_6$, $R_2$ and $R_3$ are divided into the negative region, and $R_1$ is divided into the positive region according to the semantic interpretation of new rules, and in the end, it makes a final decision that $R_2$ is the most likely decision result of $c_6$.

For $c_7$, $R_2$ and $R_3$ are divided into the negative region, and $R_1$ is divided into the boundary region. According to the new rules, it draws a conclusion that $R_2$ and $R_3$ are the most unlikely decision results of $c_6$, so we can eliminate them immediately. According to the new model of three-way decision and rules, $R_1$ should take a decision of deferment. By observing more evidences, one can make a further judgment for $R_1$. The loss of decision can be reduced effectively.

## 5    Conclusion

In this paper, the new three-way decision model based on the belief function is mainly investigated. For the classical three-way decision, the universe is divided by the probability function. Instead of that, the probability function is replaced with the belief function in this new model. The universe is divided into three disjoint regions by different values of the belief functions in this model and those decision rules. Finally, the instance is presented to verify the availability and feasibility of the new model. The prospect for the further research is that the three-way decision theory can be combined with other measures, just like, Sugeno measure, necessity measure and so on.

**Acknowledgements.** This work is supported by the national natural science foundation of China under Grant No. 61273018, and Foundation and Advanced Technology Research Program of Henan Province of China under Grant No.132300410174, and the key scientific and technological project of Education Department of Henan Province of China under Grant No.14A520082, and the key scientific and technological project of XinXiang City of China under Grant No.ZG14020.

## References

1. Yao, Y., Wong, S.K.M.: A decision theoretic framework for approximating concepts. International Journal of Man-machine Studies 37, 793–809 (1992)
2. Ziako, W.: Variable precision tough set model. Journal of Computer and System Sciences 46, 39–59 (1993)

3. Yao, Y.: Three-way decision: An interpretation of rules in rough set theory. In: Wen, P., Li, Y., Polkowski, L., Yao, Y., Tsumoto, S., Wang, G. (eds.) RSKT 2009. LNCS, vol. 5589, pp. 642–649. Springer, Heidelberg (2009)
4. Yao, Y.: Three-way decisions with probabilistic rough sets. Information Sciences 180, 341–353 (2010)
5. Yao, Y.: The superiority of three-way decisions in probabilistic rough set models. Information Sciences 181, 1080–1096 (2011)
6. Zhou, B., Yao, Y., Luo, J.G.: A three-way decision approach to email spam filtering. In: Farzindar, A., Kešelj, V. (eds.) Canadian AI 2010. LNCS, vol. 6085, pp. 28–39. Springer, Heidelberg (2010)
7. Liu, D., Yao, Y., Li, T.R.: Three-way investment decisions with decision-theoretic rough sets. International Journal of Computational Intelligence Systems 4, 66–74 (2011)
8. Lingras, P., Chen, M., Miao, D.Q.: Rough cluster quality index based on decision theory. IEEE Transactions on Knowledge and Data Engineering 21, 1014–1026 (2009)
9. Liu, D., Li, T.R., Li, H.X.: Interval-valued decision-theoretic rough sets. Computer Science 39, 178–181 (2012) (in Chinese)
10. Liu, D., Li, T.R., Liang, D.C.: Fuzzy decision-theoretic rough sets. Computer Science 39, 25–29 (2012) (in Chinese)
11. Wang, W.S.: Principle and application of artificial intelligence. Publishing house of electronics industry, Beijing (2012) (in Chinese)
12. Li, M.: The comprehensive assessment to capabilities of maritime administrations based on DS theory of evidence. Wuhan University Press, Wuhan (2012) (in Chinese)
13. Liu, X.G., Hu, X.G.: Application of D-S evidence theory to DSS. Computer Systems and Applications 19, 112–115 (2010) (in Chinese)
14. Yao, Y., Lingras, P.J.: Interpretations of belief functions in the theory of rough sets. Information Sciences 104, 81–106 (1998)
15. Zhao, W.D., Li, Q.H.: Evidential reasoning of knowledge base system based on rough set theory. Computer Science 23, 447–449 (2002) (in Chinese)
16. Wu, J.Y.: The application and research of rough sets and evidence theory in medical intelligent diagnosis system. Hunan University, Changsha (2010) (in Chinese)
17. Zhou, Y.: Research on information fusion based on rough set and evidence theory. Central South University, Changsha (2008) (in Chinese)
18. Wang, Y.S., Yang, H.J.: Decision rule analysis of D-S theory of evidence. Computer Engineering and Application 40, 14–17 (2004) (in Chinese)
19. Dempster, A.P.: Upper and lower probabilities induced by a multi-valued mapping. The Annals of Mathematical Statistics 38, 325–339 (1967)
20. Dempster, A.P.: A generalization of Bayesian inference. Journal of the Royal Statistical Society Series B 30, 205–247 (1968)
21. Zhang, W.X.: Leung Yee, The uncertainty reasoning Principles, Xi'an: Xi'an Jiaotong University Press (1996)
22. Yao, J., Herbert, J.P.: Web-based support systems with rough set analysis. In: Kryszkiewicz, M., Peters, J.F., Rybiński, H., Skowron, A. (eds.) RSEISP 2007. LNCS (LNAI), vol. 4585, pp. 360–370. Springer, Heidelberg (2007)
23. Liu, D., Yao, Y., Li, T.R.: Three-way investment decision with decision-theoretic rough sets. International Journal of Computational Intelligence Systems 4, 66–74 (2011)
24. Liu, D., Li, T.R., Liang, D.C.: Three-way government decision analysis with decision - theoretic rough sets. International Journal of Uncertainty, Fuzziness and Knowledge-Based Systems 20, 119–132 (2012)

25. Zhao, W.Q., Zhu, Y.L.: An email classification scheme based on decision-theoretic rough set theory and analysis of email security. Proceedings of 2005 IEEE Region 10, 1–6 (2005)
26. Liu, D., Yao, Y., Li, T.R.: Three-way decision-theoretic rough sets. Computer Science 38, 246-250, in Chinese (2011)
27. Hu, B.Q.: Basic theory of fuzzy sets. Wuhan University Press, Wuhan (2010) (in Chinese)
28. Grabowski, A., Jastrzebska, M.: On the lattice of intervals and rough sets. Formalized Mathematics 17, 237–244 (2009)

# Semantically Enhanced Clustering in Retail Using Possibilistic K-Modes

Asma Ammar[1], Zied Elouedi[1], and Pawan Lingras[2]

[1] LARODEC, Institut Supérieur de Gestion de Tunis, Université de Tunis
41 Avenue de la Liberté, 2000 Le Bardo, Tunisie
asma.ammar@voila.fr, zied.elouedi@gmx.fr
[2] Department of Mathematics and Computing Science, Saint Mary's University
Halifax, Nova Scotia, B3H 3C3, Canada
pawan@cs.smu.ca

**Abstract.** Possibility theory can be used to translate numeric values into semantically more meaningful representation with the help of linguistic variables. The data mining applied to a dataset with linguistic variables can lead to results that are easily interpretable due to the inherent semantics in the representation. Moreover, the data mining algorithms based on these linguistic variables tend to orient themselves based on underlying semantics. This paper describes how to transform a real-world dataset consisting of numeric values using linguistic variables based on possibilistic variables. The transformed dataset is clustered using a recently proposed possibilistic k-modes algorithm. The resulting cluster profiles are semantically accessible with very little numerical analysis.

**Keywords:** K-modes method, possibility theory, retail databases, possibility distribution.

## 1 Introduction

The numeric values in real-world databases do not lend themselves well to semantic interpretations. Data mining based on numeric computations needs to be interpreted with the help of a domain expert. Fuzzy set theory provides semantic representation of numeric values using linguistic variables. For example, a numeric value of temperature is represented using a fuzzy linguistic variable with different fuzzy membership for values *cold, cool, warm,* and *hot.* Possibility theory proposed a more flexible use of these semantic variables. Data mining techniques based on these linguistic variables result in a semantically oriented knowledge discovery process and lead to easily interpretable results. This paper demonstrates such a process with a recently proposed possibilistic k-modes clustering algorithm for a real-world retail dataset.

Clustering methods have been widely applied in numerous fields including medicine, data mining, marketing, security, and banking. When clustering real-world objects (e.g. patients, customers, visitors) into similar groups, uncertainty is present at different levels. The main sources of uncertainty can be the lack of

D. Miao et al. (Eds.): RSKT 2014, LNAI 8818, pp. 753–764, 2014.
DOI: 10.1007/978-3-319-11740-9_69 © Springer International Publishing Switzerland 2014

information, incomplete or imprecise knowledge, defective systems, and unreliable sources of information. Uncertainty can be located in clusters i.e. an object is belonging to different groups (e.g. when a patient suffers from different sorts of diseases), or in the values of attributes if we do not have the exact value of the instance (e.g. when measuring blood pressure, temperature or humidity levels).

Many uncertainty theories have been applied to several clustering methods in order to handle imperfect knowledge. The fuzzy set and the belief function theories have been adapted to the c-means method respectively in the fuzzy c-means [6] and in the evidential c-means [16]. To handle uncertain and categorical data, the belief k-modes [5] was proposed as well. In addition, the rough set theory has been successfully used with different clustering methods such as the works presented in [3], [4], [8], and [14]. Possibility theory has also attracted attention of researchers and has provided interesting results when handling uncertain qualitative and quantitative data. We can mention the possibilistic c-means [12], the possibilistic k-modes [1], the k-modes using possibilistic membership [2].

In this work, we focus on categorical databases and use a modified version of the k-modes method called the possibilistic k-modes, denoted by PKM defined in our previous work [1]. We apply this approach to a retail store database which contains real-world transactions. Each attribute of the customer and the product datasets is represented using a possibility distribution. Through the use of these real-world datasets, we demonstrate the ability of the PKM to handle uncertainty in real-world situations. Furthermore, we show how this uncertain method can help us make better decisions by analyzing the behavior of customers in a retail store.

The rest of the paper is structured as follows: Section 2 gives the background relative to the possibility theory. Section 3 presents an overview of the k-modes method. Section 4 details the possibilistic k-modes. Section 5 presents and analyzes experimental results using the retail databases. Section 6 concludes the paper.

## 2   Possibility Theory

Possibility theory is a well-known uncertainty theory. It was proposed by Zadeh in [17] and further improved by Dubois and Prade [9]. It deals with imperfect pieces of information from complete knowledge to total ignorance. Numerous researchers [1], [2], [3] have applied possibility theory to several domains including data mining, medicine, and pattern recognition.

In [17], Zadeh defined the universe of discourse $\Omega$ as the set of states (or events) $\omega_i$. From this universe $\Omega = \{\omega_1, \omega_2, ..., \omega_n\}$, the possibility distribution function $\pi$ associates a value from the scale $L = [0, 1]$ to the event $\omega_i$.
Other concepts in possibility theory include the normalization and the extreme cases of knowledge.

1. The normalization is defined by:

$$max_i \{\pi(\omega_i)\} = 1. \tag{1}$$

2. The complete knowledge presented through $\exists$ a *unique* $\omega_0$, $\pi(\omega_0) = 1$, otherwise $\pi(\omega) = 0$.
3. The total ignorance illustrated by $\forall \omega \in \Omega, \pi(\omega) = 1$.

One of the well-known possibilistic similarity measures in possibility theory is the information affinity [7]. It measures the similarity between two normalized possibility distributions $\pi_1$ and $\pi_2$. Equation (2) presents the information affinity.

$$InfoAff(\pi_1, \pi_2) = 1 - 0.5[D(\pi_1, \pi_2) + Inc(\pi_1, \pi_2)]. \tag{2}$$

with $D(\pi_1, \pi_2) = \frac{1}{n}\sum_{i=1}^{n}|\pi_1(\omega_i) - \pi_2(\omega_i)|$, $n$ the number of objects and $Inc(\pi_1, \pi_2) = 1 - \max(\pi_1(\omega) Conj \pi_2(\omega))$ and $\forall \omega \in \Omega$, $\Pi_{Conj}(\omega) = \min(\Pi_1(\omega), \Pi_2(\omega))$ where $Conj$ denotes the conjunctive mode.

## 3    The K-Modes Method

The standard k-modes method, denoted in this work by SKM, is a clustering method defined in [10], [11]. It handles large categorical datasets. It is a modified version of the k-means algorithm [15]. The SKM was proposed to overcome the numeric limitation of the k-means method. It uses a simple matching dissimilarity measure and a frequency-based function in order to cluster objects into $k$ clusters. The resulting partition has to have a high intra-similarity and a low inter-similarity.

Given two objects with $m$ categorical attributes such as $X_1 = (x_{11}, x_{12}, ..., x_{1m})$ and $X_2 = (x_{21}, x_{22}, ..., x_{2m})$. The simple matching dissimilarity measure $(d)$ is defined by:

$$d(X_1, X_2) = \sum_{t=1}^{m} \delta(x_{1t}, x_{2t}). \tag{3}$$

$$\delta(x_{1t}, x_{2t}) = \begin{cases} 0 \text{ if } x_{1t} = x_{2t} \\ 1 \text{ if } x_{1t} \neq x_{2t} \end{cases}. \tag{4}$$

Thus, $d \in [0, m]$ has two extreme values. It is equal to 0 when all the attributes' values of $X_1$ and $X_2$ are similar. Otherwise, it is equal to $m$.

The frequency based method used to update the modes consists of computing the $fr(A_p) = \frac{n_{C_j}}{n}$ where $n_{C_j}$ is the number of objects having the same value of attribute $A_p$ relative to cluster $C_j$ and $1 \leq p \leq m$. The values that more frequently appear are kept as new mode values.

Generally, given $S = \{X_1, X_2, ..., X_n\}$ a set of $n$ objects with its k-modes $Q = \{Q_1, Q_2, ..., Q_k\}$ and $k$ clusters $C = \{C_1, C_2, ..., C_k\}$, we can cluster $S$ into $k \leq n$ clusters. The minimization problem of the clustering is illustrated as follows:

$$\min D(W, Q) = \sum_{j=1}^{k} \sum_{i=1}^{n} w_{i,j} d(X_i, Q_j). \tag{5}$$

where $W$ is an $n \times k$ partition matrix and $w_{i,j} \in \{0, 1\}$ presents the degree of belonging of $X_i$ in $C_j$. It is equal to 0 if the object $X_i$ is a member of the cluster

$C_j$ otherwise, it is equal to 1.

The k-modes algorithm is detailed as follows:

1. *Randomly select k initial modes, one mode for each cluster.*
2. *Calculate the dissimilarities between instances and modes using Equation (3).*
3. *Allocate objects to the most similar cluster.*
4. *Update the mode using the frequency-based function.*
5. *Retest the similarity and reallocate objects to appropriate clusters.*
6. *Repeat (2-5) until all clusters are stable.*

In spite of its successful application for clustering large categorical databases, the SKM has some issues. It faces the problem of the non-uniqueness of clusters' modes i.e. we can have several modes for a cluster and hence, we have to randomly select a mode for this cluster. Moreover, the SKM cannot handle uncertainty. It only deals with certain databases. These issues can make the SKM results inaccurate. To overcome these limitations, the possibilistic k-modes method (PKM) [1] has been developed.

## 4   The Possibilistic K-Modes: PKM

The possibilistic k-modes method [1] is a modified version of the SKM. It uses categorical values and deduces their degrees of uncertainty by presenting them using possibility degrees. As a result, the PKM benefits from the uncertainty located in the attribute values to provide better clustering results. In addition, the PKM keeps the advantages and overcomes the limitations of the SKM by overcoming the problem of the non-uniqueness of cluster mode and by dealing with the uncertainty in real-world databases.

### 4.1   Parameters

1. An uncertain training set: where each categorical attribute value in the customers and products databases has a degree of possibility representing the degree of uncertainty on this values.

2. The possibilistic similarity measure denoted by $IA$: is a modified version of the information affinity [7] defined in Equation (2). It is applied between two normalized possibility distributions. It provides the similarity between objects of the training set and the cluster modes given by:

$$IA(X_1, X_2) = \frac{\sum_{p=1}^{m} InfoAff(\pi_{1p}, \pi_{2p})}{m}. \tag{6}$$

3. The update of the clusters' modes: It computes the average of possibility distributions of objects belonging to the same cluster.

$$\forall \omega \in A_p, \pi_{pC}(\omega) = \frac{\sum_{i=1}^{n'} \pi_{ip}(\omega)}{n'}. \tag{7}$$

where $\pi_{ip}(\omega)$ is the possibility degree defined for the attribute $p$ related to the object $i$ in the cluster $C$, $n'$ is the number of objects, $A_p$ is the set of $t$ values of the attribute $p$, and $1 \leq p \leq m$.

## 4.2  Algorithm

The PKM algorithm takes as inputs: an uncertain database, $k$ the number of clusters and provides as a result groups of similar uncertain instances.

1. *Randomly select $k$ initial modes, one mode for each cluster.*
2. *Use Equation (6) to calculate the possibilistic similarity measure $IA$ applied between instances and modes.*
3. *Allocate an object to the most similar cluster to it based on $IA$.*
4. *Update the mode using Equation (7)*
5. *Retest the similarity and reallocate objects to appropriate clusters.*
6. *Repeat (2-5) until cluster assignments do not change.*

## 5  Experiments

### 5.1  The Framework

We used retail databases consisting of real-world transactions by customers when buying products from a small retail chain from 2005 to 2007. The database contains 15341 customers and 8987 products. Each customer is characterized by different attributes namely the number of visits, revenues, number of products bought, and the profits. The number of visits corresponds to the loyalty of the customers. Similarly, each product is represented by several attributes mainly the number of visits, the revenues and profits realized by this product, the quantity sold, and the number of customers that have bought the product. The number of visits that buy a particular product corresponds to its popularity. As there are many customers that do not often visit the retail store and many products are rarely bought, we only consider the top 1000 of the customers and top 500 of the products.

The absolute numerical values of spending, visits, number of products bought need to be put in context. For example, $15000 may be high spending in a computer store, but it will be be considered low expenditure in a car store. Therefore, we may want to represent the variable spending with linguistic values of *high, modest,* and *low.* These categories are too rigid. Therefore, we will later assign a possibilistic membership to these values. For example, a purchase of $2000 may have a possiblistic memberships given by $\{(low, 0.0), (modest, 0.3), (high, 0.9)\}$. Please note that unlike fuzzy memberships, the possibilistic values do not have to add up to 1 as explained in Equation (1). Using this principle, we propose the following representation for customers where the number of intervals depends on the initial values and frequency of each interval.

1. #Products (number of products bought): low, modest, and high.
2. Revenues: low, modest, high, and veryHigh.
3. Loyalty (number of visits): low, modest, high, and veryHigh.
4. Profits: low, modest, and high.

Similarly, the attribute values of the products dataset are categorized:

1. #Customers (number of customers who buy products): low, modest, and high.
2. Quantity: small, medium, and big.
3. Revenues: low, modest, high, and veryHigh.
4. Popularity (determined using number of visits): low, modest, high.
5. Profits: modest, high, and veryHigh.

We apply the equal-frequency interval discretization function on the retail datasets. For each attribute, we obtain categorical values described in Table 1-Table 4 for the customers datasets and in Table 5-Table 9 for the product dataset. We use the function summary to intervals $I$ to identify the possibility degrees.

**Table 1.** Equal-frequency discretization of the attribute number of products

| Number of products | | Summary function values | | |
|---|---|---|---|---|
| | | 1st Quartile (1st Qu.) | Median | 3rd Quartile (3rd Qu.) |
| $[5, 40)$ | low | 25 | 31.5 | 36 |
| $[40, 58)$ | modest | 43 | 46 | 51.25 |
| $[58, 546]$ | high | 67 | 78 | 102 |

**Table 2.** Equal-frequency discretization of the attribute revenues

| Revenues | | Summary function values | | |
|---|---|---|---|---|
| | | 1st Quartile (1st Qu.) | Median | 3rd Quartile (3rd Qu.) |
| $[57, 712)$ | low | 404.2 | 531.5 | 621 |
| $[712, 1041)$ | modest | 781 | 849 | 941 |
| $[1041, 1624)$ | high | 1188 | 1296 | 1439 |
| $[1624, 12614]$ | veryhigh | 1893 | 2325 | 3245 |

**Table 3.** Equal-frequency discretization of the attribute profits

| Profits | | Summary function values | | |
|---|---|---|---|---|
| | | 1st Quartile (1st Qu.) | Median | 3rd Quartile (3rd Qu.) |
| $[25, 287)$ | low | 160.2 | 214.5 | 251 |
| $[287, 428)$ | modest | 321 | 351 | 385 |
| $[428, 657)$ | high | 473 | 533 | 589 |
| $[657, 3484]$ | veryhigh | 767 | 935 | 1206 |

## 5.2    Possibilistic Representation of the Attribute Values

The use of possibility theory, in this step, allows the representation of the uncertainty that can occur in the values of attributes of the customer and product

**Table 4.** Equal-frequency discretization of the attribute loyalty

| Loyalty | | Summary function values | | |
|---|---|---|---|---|
| | | 1st Quartile (1st Qu.) | Median | 3rd Quartile (3rd Qu.) |
| [3, 16) | low | 11 | 12 | 14 |
| [16, 21) | modest | 17 | 18 | 19 |
| [21, 29) | high | 22 | 24 | 26 |
| [29, 289] | veryhigh | 33 | 38 | 52 |

**Table 5.** Equal-frequency discretization of the attribute number of customers

| Number of customers | | Summary function values | | |
|---|---|---|---|---|
| | | 1st Quartile (1st Qu.) | Median | 3rd Quartile (3rd Qu.) |
| [7, 32) | low | 20 | 24 | 29 |
| [32, 48) | modest | 35 | 38 | 42 |
| [48, 377] | high | 56 | 65 | 84 |

**Table 6.** Equal-frequency discretization of the attribute revenues

| Revenues | | Summary function values | | |
|---|---|---|---|---|
| | | 1st Quartile (1st Qu.) | Median | 3rd Quartile (3rd Qu.) |
| [98, 886) | low | 228 | 388 | 627.5 |
| [886, 1785) | modest | 1064 | 1268 | 1521 |
| [1785, 3305) | high | 2035 | 2403 | 2816 |
| [3305, 40800] | veryhigh | 4040 | 4993 | 7074 |

**Table 7.** Equal-frequency discretization of the attribute profits

| Profits | | Summary function values | | |
|---|---|---|---|---|
| | | 1st Quartile (1st Qu.) | Median | 3rd Quartile (3rd Qu.) |
| [6, 491) | low | 92 | 211 | 369 |
| [491, 1063) | modest | 606.5 | 741 | 894.2 |
| [1063, 20084] | high | 1354 | 1668 | 2425 |

**Table 8.** Equal-frequency discretization of the attribute popularity

| Popularity | | Summary function values | | |
|---|---|---|---|---|
| | | 1st Quartile (1st Qu.) | Median | 3rd Quartile (3rd Qu.) |
| [38, 58) | low | 45 | 49 | 52 |
| [58, 86) | modest | 65 | 70 | 76 |
| [86, 867] | high | 99 | 116 | 157 |

**Table 9.** Equal-frequency discretization of the attribute quantity

| Quantity | | Summary function values | | |
|---|---|---|---|---|
| | | 1st Quartile (1st Qu.) | Median | 3rd Quartile (3rd Qu.) |
| $[43, 73)$ | small | 52 | 57 | 65 |
| $[73, 120)$ | medium | 81 | 92 | 103 |
| $[120, 1395)$ | big | 136 | 169 | 224 |

datasets. For each attribute value, we defined a possibility distribution relative to each object expressing the level of uncertainty. Each value of attribute known with certainty is represented using a possibility degree equal to 1. It corresponds to the case of complete knowledge, whereas each uncertain value is represented using a high possibility degree (very possible value) or low possibility degree from (may be considered as nearly impossible). We define the possibility distributions for the attributes of the datasets with respect to their numeric values and using these possibilistic degrees.

Assume $X_i$ is a categorical attribute value, the *trapezoidal function* is used to represent $X_i$.

1. If we deal with the first interval:

$$d = \begin{cases} 0 & \text{if } X_i > 1stQu._{(I+1)} \\ \frac{1stQu._{(I+1)} - X_i}{1stQu._{(I+1)} - 3rdQu._I} & \text{if } 3rdQu._I \leq X_i \leq 1stQu._{(I+1)} \\ 1 & \text{if } X_i < 3rdQu._I \end{cases} \quad (8)$$

2. If we deal with the last interval:

$$d = \begin{cases} 0 & \text{if } X_i < 3rdQu._{(I-1)} \\ \frac{X_i - 3rdQu._{(I-1)}}{1stQu._I - 3rdQu._I} & \text{if } 3rdQu._{(I-1)} \leq X_i \leq 1stQu._I \\ 1 & \text{if } X_i > 1stQu._I \end{cases} \quad (9)$$

3. Otherwise: we are representing two intervals $(I - 1)$ and $(I + 1)$ with $I$ is the current interval:

$$d = \begin{cases} 0 & \text{if } X_i > 1stQu._{(I+1)} \text{ or } X_i < 3rdQu._{(I-1)} \\ \frac{X_i - 3rdQu._{(I-1)}}{1stQu._I - 3rdQu._{(I-1)}} & \text{if } 3rdQu._{(I-1)} < X_i < 1stQu._I \\ 1 & \text{if } 1stQu._I \leq X_i \leq 3rdQu._I \\ \frac{1stQu._{(I+1)} - X_i}{1stQu._{(I+1)} - 3rdQu._I} & \text{if } 3rdQu._I < X_i < 1stQu._{(I+1)} \end{cases} \quad (10)$$

The representations of the customers and products attributes' values using the *trapezoidal function* are illustrated in Figures 1 and 2.

## 5.3   Experimental Results

After experimentation with different values of $k$ for the customer dataset, we use $k=5$ and for the product dataset we use $k=3$ [13]. Table 10 shows the impact of the possibilistic structure on the execution time.

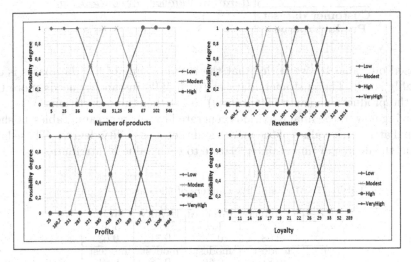

**Fig. 1.** Possibilistic degrees of uncertain values of attributes relative to customers data

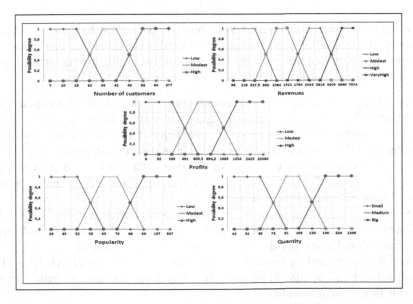

**Fig. 2.** Possibilistic degrees of uncertain values of attributes relative to products data

**Table 10.** The average of iteration number and execution time

| | *Iteration number* | *Execution time* |
|---|---|---|
| Customer dataset ‖ | 12 | 87 s |
| Product dataset ‖ | 8 | 54 s |

Based on Table 10, we notice that the running time of the databases is considerably small. The customers dataset (with 1000 instances) needs more time than the product one (with 500 objects) to get the stable partitions.

Moving now to the final partitions generated by our program, Tables 11 shows the final attributes' values of the modes and their possibility degrees. These latter explain the degree of uncertainty relative to each value of attribute.

**Table 11.** Modes of five clusters of the customers dataset

| Modes | #Products | Revenues | Loyalty | Profits |
|---|---|---|---|---|
| $cc_1$ | low 0.68 | low 0.75 | low 0.71 | low 0.63 |
| $cc_2$ | modest 0.81 | modest 0.69 | modest 0.78 | modest 0.81 |
| $cc_3$ | low 0.72 | low 0.71 | high 0.67 | modest 0.92 |
| $cc_4$ | low 0.85 | high 0.73 | high 0.64 | high 0.59 |
| $cc_5$ | high 0.66 | veryhigh 0.71 | veryhigh 0.85 | veryhigh 0.94 |

The PKM grouped customers into clusters based on their behaviors and characteristics. The possibility degrees assigned at the beginning to instances have ensured a high intra-similarity between objects belonging to the stable partitions. The values of modes suggest that we have five types of customers.

1. Type 1 (corresponding to mode 1): customers who rarely visits, and hence not loyal. They do not contribute significantly to revenues or profits .
2. Type 2 (corresponding to mode 2): customers that do not often visit the stores but, they spend enough money to make a modest profit.
3. Type 3 (corresponding to mode 3): customers that do not buy or spend a lot of money in the stores but, they frequently visit. Their visits could be considered as a means of entertainment.
4. Type 4 (corresponding to mode 4): customers who are considered loyal. They come often to the store, buy a lot of products and result in a reasonable profit for the store. Generally, they have a high socio-economic status.
5. Type 5 (corresponding to mode 5): customers who frequently visit the store indicating high loyalty. Moreover, they buy large number of products and result in significant profits.

Tables 12 shows the three final modes with their degrees of possibility.

As shown in Tables 12, we get three stable partitions. Each partition groups similar products having the same characteristics.

**Table 12.** Modes of products dataset

| Modes | #Customers | Revenues | Popularity | Profits | Quantity |
|-------|-----------|----------|------------|---------|----------|
| $cp_1$ | low | low | low | low | small |
|        | 0.92 | 0.645 | 0.814 | 0.894 | 0.908 |
| $cp_2$ | modest | modest | modest | modest | medium |
|        | 0.89 | 0.69 | 0.72 | 0.85 | 0.81 |
| $cp_3$ | high | high | high | veryhigh | big |
|        | 0.79 | 0.85 | 0.91 | 0.93 | 0.877 |

1. Partition 1: contains products that are least bought and least profitable. Customers seem to rarely need them.
2. Partition 2: presents products that are bought with modest quantity. They are not needed for everyday life but, they are bought by reasonable number of customers.
3. Partition 3: contains products which make a large profit for the store. These products are very popular as customers often buy them in large quantity.

The profiles of customers and products obtained from the PKM using linguistic variables based on possibility theory are semantically meaningful and can be created without further assistance from the domain expert using their tabular representations from Tables 11 and 12. The fact that the linguistic values were introduced prior to the application of the PKM algorithm, means that the semantics is taken into account during the clustering process.

# 6 Conclusion

This paper illustrates how introduction of semantics through the linguistics variables based on possibility theory can lead to a more meaningful data mining in a retail store. The numeric values from a transactional database are put in context using linguistic values for variables such as loyalty, popularity, and revenues. A recently proposed possibilistic k-modes (PKM) algorithm is used to create profiles of customers and products through a clustering process. The fact that the linguistic values are introduced prior to the clustering process ensures that the semantics is part of the clustering process. The modes of resulting clusters are easily translatable to meaningful customer and product profiles. We are currently conducting a detailed comparison of the profiles and cluster quality resulting from the PKM and traditional numerical K-means algorithm. The results will be presented in a forthcoming publication.

# References

1. Ammar, A., Elouedi, Z.: A New Possibilistic Clustering Method: The Possibilistic K-Modes. In: Pirrone, R., Sorbello, F. (eds.) AI*IA 2011. LNCS, vol. 6934, pp. 413–419. Springer, Heidelberg (2011)

2. Ammar, A., Elouedi, Z., Lingras, P.: K-modes clustering using possibilistic membership. In: Greco, S., Bouchon-Meunier, B., Coletti, G., Fedrizzi, M., Matarazzo, B., Yager, R.R. (eds.) IPMU 2012, Part III. Communications in Computer and Information Science, vol. 299, pp. 596–605. Springer, Heidelberg (2012)
3. Ammar, A., Elouedi, Z., Lingras, P.: RPKM: The Rough Possibilistic K-Modes. In: Chen, L., Felfernig, A., Liu, J., Raś, Z.W. (eds.) ISMIS 2012. LNCS, vol. 7661, pp. 81–86. Springer, Heidelberg (2012)
4. Ammar, A., Elouedi, Z., Lingras, P.: Incremental rough possibilistic K-modes. In: Ramanna, S., Lingras, P., Sombattheera, C., Krishna, A. (eds.) MIWAI 2013. LNCS, vol. 8271, pp. 13–24. Springer, Heidelberg (2013)
5. Ben Hariz, S., Elouedi, Z., Mellouli, K.: Selection Initial modes for Belief K-modes Method. International Journal of Applied Science, Engineering and Technology 4, 233–242 (2007)
6. Bezdek, J.C.: Pattern Recognition with Fuzzy Objective Function Algorithms. Kluwer Academic Publishers (1981)
7. Jenhani, I., Ben Amor, N., Elouedi, Z., Benferhat, S., Mellouli, K.: Information Affinity: a new similarity measure for possibilistic uncertain information. In: Proceedings of the 9th European Conference on Symbolic and Quantitative Approaches to Reasoning with Uncertainty, pp. 840–852 (2007)
8. Joshi, M., Lingras, P., Rao, C.R.: Correlating Fuzzy and Rough Clustering. Fundamenta Informaticae 115(2-3), 233–246 (2011)
9. Dubois, D., Prade, H.: Possibility theory: An approach to computerized processing of uncertainty. Plenum Press (1988)
10. Huang, Z.: Extensions to the k-means algorithm for clustering large datasets with categorical values. Data Mining and Knowledge Discovery 2, 283–304 (1998)
11. Huang, Z., Ng, M.K.: A note on k-modes clustering. Journal of Classification 20, 257–261 (2003)
12. Krishnapuram, R., Keller, J.M.: A possibilistic approach to clustering. IEEE Trans. Fuzzy System 1, 98–110 (1993)
13. Lingras, P., Elagamy, A., Ammar, A., Elouedi, Z.: Iterative meta-clustering through granular hierarchy of supermarket customers and products. Information Sciences 257, 14–31 (2014)
14. Lingras, P., West, C.: Interval Set Clustering of Web Users with Rough K-means. Journal of Intelligent Information Systems 23, 5–16 (2004)
15. MacQueen, J.B.: Some methods for classification and analysis of multivariate observations. In: Proceedings of the Fifth Berkeley Symposium on Mathematical Statistics and Probability. Statistics, vol. 1, pp. 281–297 (1967)
16. Masson, M.H., Denoeux, T.: ECM: An evidential version of the fuzzy c-means algorithm. Pattern Recognition 41, 1384–1397 (2008)
17. Zadeh, L.A.: Fuzzy sets as a basis for a theory of possibility. Fuzzy Sets and Systems 1, 3–28 (1978)

# A Three-Way Decisions Clustering Algorithm for Incomplete Data

Hong Yu, Ting Su, and Xianhua Zeng

Chongqing Key Laboratory of Computational Intelligence, Chongqing University of
Posts and Telecommunications, Chongqing, 400065, P.R. China
yuhong@cqupt.edu.cn

**Abstract.** Clustering is one of the most widely used efficient approaches
in data mining to find potential data structure. However, there are some
reasons to cause the missing values in real data sets such as difficulties
and limitations of data acquisition and random noises. Most of clustering
methods can't be used to deal with incomplete data sets for clustering
analysis directly. For this reason, this paper proposes a three-way decisions clustering algorithm for incomplete data based on attribute significance and miss rate. Three-way decisions with interval sets naturally
partition a cluster into positive region, boundary region and negative region, which has the advantage of dealing with soft clustering. First, the
data set is divided into four parts such as sufficient data, valuable data,
inadequate data and invalid data, according to the domain knowledge
about the attribute significance and miss rate. Second, different strategies are devised to handle the four types based on three-way decisions.
The experimental results on some data sets show preliminarily the effectiveness of the proposed algorithm.

**Keywords:** clustering, incomplete data, three-way decisions, attribute
significance.

## 1 Introduction

Cluster analysis is the task of grouping a set of objects in such a way that objects
in the same cluster are more similar to each other than to those in other clusters.
The study of cluster analysis not only has important theoretical significance, but
also has important engineering application value and humanistic value. Clustering has been widely applied in many fields, including pattern recognition, picture
processing, business intelligence and so on.

The missing values often appears in real data analysis. This may be due to
various reasons, such as difficulties and limitations of data acquisition, random
noises, data misunderstanding and data lost. Missing data is a big challenge that
data mining technology faced. For example, the benchmark database in machine
learning filed, UCI repository [9] even has more than 40% data sets with missing
data. Usually, a data set with missing data/values is also called as an incomplete
data set.

D. Miao et al. (Eds.): RSKT 2014, LNAI 8818, pp. 765–776, 2014.
DOI: 10.1007/978-3-319-11740-9_70 © Springer International Publishing Switzerland 2014

However, the traditional clustering approaches can not process incomplete data sets directly. Many scholars have achieved some meaningful results. Fuzzy clustering algorithms adapted to incomplete data perform well in recent studies. Li et al. [8] proposed an attribute weighted fuzzy c-means algorithm for incomplete data which introduce weighted Euclidean distance could get better clustering results; and then they [7] proposed a hybrid genetic algorithm with fuzzy c-means approach for the problem of incomplete data clustering. Himmelspach et al. [3] studied fuzzy clustering methods and compared the performance of them through experiences. Yamamoto et al. [11] considered the applicability of $\beta$-spread transformation for handling incomplete relational data in FCMdd-type linear clustering.

Despite there are some improved fuzzy clustering algorithms for incomplete data, there also exists some other clustering methods. For example, Wu et al. [10] proposed a new extended mean field annealing algorithm to solve the problem of clustering for incomplete data in the continuous-value state space; this algorithm uses a perturbation method to get the membership of each datum to a cluster and calculates the energy function after every perturbation. Lai et al. [5] used minimum description length concept to hierarchical clustering with incomplete data. Honda et al. [4] proposed a PCA-guided procedure for k-Means clustering of incomplete data sets, and the PCA-guided approach is more robust to initialization problems than the $k$-Means procedure using PDS.

In order to solve the problem of clustering for incomplete data, this paper presents a three-way decisions clustering algorithm for incomplete data. The concept of three-way decisions plays an important role in many real world decision-making problems. One usually makes a decision based on available information and evidence. In widely used two-way decision models, it is assumed that an object either satisfies the criteria or does not satisfy the criteria; we usually call that a positive decision or a negative decision. When the evidence is insufficient or weak, it might be impossible to make either a positive or a negative decision. One therefore chooses an alternative decision that is neither yes nor no. We called it the third-way decision, or a boundary decision, or a deferred decision; actually, it is a non-commitement. When the evidence becomes sufficient enough or strong enough, it might be possible to make either a positive or a negative decision.

The notion of three-way decisions represents a close relationship between rough set analysis, Bayesian decision analysis, and hypothesis testing in statistics [13]. Furthermore, Professor Yao had outlined a theory of three-way decisions in reference [12], and pointed out that we can describe and explain three-way decisions in many-valued logics and generalizations of set theory, including interval sets, rough sets, decision-theoretic rough sets, fuzzy sets, and shadowed sets. Therefore, we had proposed three-way decisions with interval sets in our previous work [14]. This three-way decisions method may have many real-world applications [1] [16] [6].

Specifically, the main idea of the proposed algorithm is to consider attribute significance as well as miss rate to deal with missing values. That is, the original data set is divided into four types such as sufficient data, valuable data, inadequate data and invalid data, according to the domain knowledge about the attribute significance and miss rate. Then, for sufficient data, the weighted distance between two incomplete objects and similar value estimation formula are defined, and a grid-based method is proposed to obtain an initial clustering result. For other types, the distance and membership between object and cluster are defined respectively, and three-way decisions rules are used to obtain the final clustering result. The experimental results on some data sets show preliminarily the effectiveness of the proposed algorithm.

## 2 Classify an Incomplete Data Set

### 2.1 Representation of Clustering

To define our framework, let a universe be $U = \{\mathbf{x}_1, \cdots, \mathbf{x}_n, \cdots, \mathbf{x}_N\}$, and the clustering result is $\mathbf{C} = \{C^1, \cdots, C^k, \cdots, C^K\}$, which is a family of clusters of the universe.

The universe can be represented as an information system $S = (U, A, V, F, W)$. $U = \{\mathbf{x}_1, \mathbf{x}_2, \cdots, \mathbf{x}_n, \cdots, \mathbf{x}_N\}$ and $A = \{a_1, a_2, \cdots, a_D\}$ are finite nonempty sets of objects and attributes respectively. $V = \{V_1, V_2, \cdots, V_D\}$ is the set of possible attribute values, $V_i$ is the possible attribute values of $a_i$, $f$ is an information function, $f : V_{ik} = f(\mathbf{x}_i, a_k) \in V_k$. $W = \{w_1, w_2, \cdots, w_D\}$ is a set of attribute weights, $w_i$ is the weight of $a_i$. The $\mathbf{x}_n$ is an object which has $D$ attributes, namely, $\mathbf{x}_n = (x_n^1, x_n^2, \cdots, x_n^d, \cdots, x_n^D)$. The $x_n^d$ denotes the value of $d$th attribute of object $\mathbf{x}_n$, where $n \in \{1, \cdots, N\}$, and $d \in \{1, \cdots, D\}$.

When there are some missing values, the information system $S$ will be an incomplete information system. Table 1 shows an example, which contains 10 objects, and each object has 9 attributes. The missing value is expressed by the symbol $*$.

**Table 1.** An Incomplete Information System

| U | $a_1$ | $a_2$ | $a_3$ | $a_4$ | $a_5$ | $a_6$ | $a_7$ | $a_8$ | $a_9$ |
|---|---|---|---|---|---|---|---|---|---|
| $\mathbf{x}_1$ | 3 | 2 | 1 | 25 | 5 | 1 | $*$ | 9 | $*$ |
| $\mathbf{x}_2$ | 2 | $*$ | 8 | 15 | 4 | 2 | 4 | 6 | 9 |
| $\mathbf{x}_3$ | $*$ | $*$ | $*$ | $*$ | $*$ | 6 | 5 | $*$ | 10 |
| $\mathbf{x}_4$ | 2 | $*$ | $*$ | 23 | 7 | 5 | $*$ | $*$ | $*$ |
| $\mathbf{x}_5$ | $*$ | 8 | 9 | 20 | 4 | 7 | 5 | 6 | $*$ |
| $\mathbf{x}_6$ | $*$ | $*$ | $*$ | $*$ | $*$ | 5 | 8 | 6 | 9 |
| $\mathbf{x}_7$ | $*$ | $*$ | $*$ | 19 | 20 | 2 | 4 | 9 | 4 |
| $\mathbf{x}_8$ | 2 | 3 | 9 | $*$ | $*$ | $*$ | 3 | 4 | 6 |
| $\mathbf{x}_9$ | 3 | 2 | 1 | 25 | 5 | $*$ | $*$ | $*$ | 2 |
| $\mathbf{x}_{10}$ | 3 | 5 | $*$ | $*$ | 4 | $*$ | $*$ | $*$ | $*$ |

We can look at the cluster analysis problem from a view of decisions making. For a hard clustering, it is a typical two-way decisions in some sense; and for a soft clustering, it is a kind of three-way decisions. The positive decisions decide objects into the positive region of a cluster definitely, the negative decisions decide objects into the negative region of a cluster definitely, and the boundary decisions decide objects into the boundary region of a cluster. Using interval sets to represent clusters can be more appropriate than crisp representations, which directly leads to an interpretation in three-way decisions for clustering. Let's review some basic concepts of clustering using interval sets [15].

With respect to the family of clusters, $\mathbf{C}$, we have the following family of clusters formulated by interval sets as follows.

$$\mathbf{C} = \{[\underline{C^1}, \overline{C^1}], \ldots, [\underline{C^k}, \overline{C^k}], \ldots, [\underline{C^K}, \overline{C^K}]\} \tag{1}$$

Therefore, the sets $\underline{C^k}$, $\overline{C^k} - \underline{C^k}$ and $U - \overline{C^k}$ formed by certain decision rules construct the three regions of the cluster $C^k$ as the positive region, boundary region and negative region, respectively. The three-way decisions rules are described as follows.

$$\begin{aligned} POS(C^k) &= \underline{C^k} \\ BND(C^k) &= \overline{C^k} - \underline{C^k} \\ NEG(C^k) &= U - \overline{C^k} \end{aligned} \tag{2}$$

According to Equation (2), the family of clusters $\mathbf{C}$ gives a three-way decisions clustering result. Namely, objects in $POS(C^k)$ belong to the cluster $C^k$ definitely, objects in $NEG(C^k)$ don't belong to the cluster $C^k$ definitely, and objects in the region $BND(C^k)$ might belong to the cluster or not. The $BND(C^k) \neq \emptyset$ means we need more information to help making decisions.

## 2.2    Classification of Incomplete Data

Few clustering methods consider the difference between different attributes in the process of clustering. For example, when Euclidean distance-based clustering algorithms calculate similarity between objects, there is an implicit assumption that each attribute contributes the same to clustering. However, different attribute produces different effect on clustering. The greater the attribute significance degree is, the more it can reveal the underlying characteristics of the data set.

In this paper, we suppose we get the attribute significance degree through some way in advance. The assumption is reasonable because we really have lots of expert knowledge in some cases. We can compute the attribute significance by some way before clustering if we have no expert knowledge. The proposed algorithm can still work well even if we don't know attribute significance, to say the least.

Therefore, we assume the descending order of attribute importance degree is $A = \{a_1, a_2, \cdots, a_m, \cdots, a_D\}$. Under a priori knowledge, $M = \{a_1, a_2, \cdots, a_m\}$ is the set of important attributes, $L = \{a_{m+1}, \cdots, a_D\}$ is the set of non-important attributes. Set $W = \{w_1, w_2, \cdots, w_k, \cdots, w_D\}$ be the set of attribute weights, and $w_1 \geq w_2 \geq \cdots \geq w_k \geq \cdots \geq w_D$.

The objective of this paper is incomplete data sets. However, the influence of missing important attributes and non-important attributes to data is different.

When missing values are with non-important attributes, but no with important attributes, the object retains important information so that it can be considered to be relatively complete. When both values of important attributes and non-important attributes are partially missing, the object retains some degree of important information. When all important attributes are missing, the quality of clustering will decline and the cost of deciding will increase.

Therefore, we use a two-tuples $I(\mathbf{x}_i) = (u, \varphi)$ to represent the complete degree of an object $\mathbf{x}_i$. $u = p/|M|$ represents the miss rate of important attributes, and $u \in [0,1]$, $|M|$ is the number of important attributes, $p$ is the number of missing important attributes of this object. $\varphi = q/|L|$ represents the miss rate of non-important attributes, and $\varphi \in [0,1]$, $|L|$ is the number of non-important attributes, $q$ is the number of missing non-important attributes.

According to the concept of complete degree, the data set can be divided into four types:

– sufficient data: when $0 \leq u \leq 0.2$, the object belongs to sufficient data,
– valuable data: when $0.2 < u \leq 0.5$ and $0 \leq \varphi \leq 0.5$, the object belongs to valuable data,
– inadequate data: when $0.2 < u \leq 0.5$ and $0.5 < \varphi \leq 1$, or $0.5 < u < 1$, the object belongs to inadequate data,
– invalid data: when $u = 1$, the object belongs to invalid data.

Generally speaking, attribute values are missed randomly in most cases, there exists a certain probability distribution of missing values especially for large data sets. So we take 0.5 as the division standard for miss rate of non-important attribute. On the other hand, missing more important attributes will seriously affect the reliability of clustering, so we choose 0.2 as the division standard for miss rate of important attribute. How to decide the two parameters automatically is a problem, it could be the further work.

Let's take Table 1 as the example again. According to the domain knowledge, the ordered important attributes by significance is $\{a_1, a_2, a_3, a_4, a_5\}$, and the set of non-important attributes is $\{a_6, a_7, a_8, a_9\}$.

The complete degree of objects are $I(\mathbf{x}_1) = (0, 0.5), I(\mathbf{x}_2) = (0.2, 0), I(\mathbf{x}_3) = (1, 0.25), I(\mathbf{x}_4) = (0.4, 0.75), I(\mathbf{x}_5) = (0.2, 0.25), I(\mathbf{x}_6) = (1, 0), I(\mathbf{x}_7) = (0.6, 0), I(\mathbf{x}_8) = (0.4, 0.25), I(\mathbf{x}_9) = (0, 0.75), I(\mathbf{x}_{10}) = (0.4, 1)$, respectively. So the set of sufficient data is $\{\mathbf{x}_1, \mathbf{x}_2, \mathbf{x}_5, \mathbf{x}_9\}$, the set of valuable data is $\{\mathbf{x}_8\}$, the set of inadequate data is $\{\mathbf{x}_4, \mathbf{x}_7, \mathbf{x}_{10}\}$, the set of invalid data is $\{\mathbf{x}_3, \mathbf{x}_6\}$.

# 3 Clustering Algorithm for Incomplete Data

## 3.1 Related Definitions

For sufficient data, a grid-based clustering method is proposed to obtain an initial underlying structure of the data set $U$. Thus, we first introduce some notions about grid-based clustering. For the information system $S$, we set $G = V_1 \times V_2 \times \cdots \times V_D$. We can partition the data space $G$ into non-overlapping rectangle units by dividing the every dimension into equal intervals, and the $Rth$ is the length of the unit. Usually, a unit is called as a grid.

Given an ordered sequence $\{r_1, r_2, \cdots, r_D\}$, every element of the sequence denotes the order number of the corresponding dimension, and we say the sequence denotes the position of a grid. A D-dimensional object $\mathbf{x}_n = (x_n^1, x_n^2, \cdots, x_n^d, \cdots, x_n^D)$ is contained in a grid having the position $\{r_1, r_2, \cdots, r_D\}$, if and only if $(r_j - 1) \times Rth \leq x_n^j < r_j \times Rth, 1 \leq j \leq D$. The density of a grid $g$ is defined as the number of objects contained in it, denoted as $\rho(g)$. A grid $g$ is called dense if $\rho(g) \geq minpoints$, or sparse if $\rho(g) < minpoints$, where $minpoints$ is a density threshold.

**Definition 1. Neighbour Grid:** Neighbor grids of a grid $g$ are those grids with the common boundary or the common vertexes with $g$.

**Definition 2. Adjacent Grid:** Adjacent grids of a grid $g$ include direct and indirect adjacent grids. Direct adjacent grids are these neighbor grids of $g$, and indirect adjacent grids are these neighbor grids of $g$'s neighbors. In this paper, the weight of attribute is taken into account to improve the quality of clustering. So we define a weighted distance function based on the partial distance [2] to calculate the distance under the environment of incomplete data.

**Definition 3. Weighted Distance between Two Objects:** $\forall \mathbf{x}_i, \mathbf{x}_j \in U$, the weighted distance between two objects is defined as follows.

$$Dist(\mathbf{x}_i, \mathbf{x}_j) = \frac{1}{\sum_{d=1}^{D} I_d w_d} (\sum_{d=1}^{D} (x_i^d - x_j^d)^2 I_d w_d^2)^{1/2}, \ where \ I_d = \begin{cases} 1 & x_i^d \neq *, x_j^d \neq * \\ 0 & else \end{cases}$$

(3)

When the objects are complete, Formula (3) is the Euclidean distance formula. Because an object is more similar to its nearest-neighbor than others, it is reasonable that to estimate the object's missing values by using its neighbor's features. Thus, we have the following.

**Definition 4. $h$ Nearest-Neighbors:** Nearest-neighbors of object $\mathbf{x}_i$ in a universe $U$ is expressed as follows.

$$N_h(\mathbf{x}_i, U) = \{\mathbf{x}_1, \mathbf{x}_2, \cdots, \mathbf{x}_h\}$$

(4)

where $h$ is the number of nearest-neighbors, and $\{\mathbf{x}_1, \mathbf{x}_2, \cdots, \mathbf{x}_h\} \in U$.

It's easy to achieve $h$ nearest-neighbors by using Formula (3). And, the missing attribute value of an object can be estimated by the average value of nearest-neighbors which don't miss this attribute value.

**Definition 5. Similar Value Estimation:** The similar value of the attribute $x_i^d$ for the object $\mathbf{x}_i$ is estimated by the follow formula.

$$Sve_h(x_i^d) = \frac{\sum_{x_z \in N_h(\mathbf{x}_i, R)} x_z^d I_z^d}{\sum_{z=1}^{h} I_z^d},$$

(5)

where

$$I_z^d = \begin{cases} 1 & x_z^d \neq * \\ 0 & x_z^d = * \end{cases}$$

and $1 \leq d \leq D, N_h(\mathbf{x}_i, U)$ is the $h$ nearest-neighbors of in $U$.

The distance between objects can be used to express the similarity of them, the greater the distance, the less the similarity. When deciding which cluster the object belongs to, the decision is based on the similarity between the object and clusters. The similarity between object and cluster is often calculated by the distance between the object and cluster center. However, this method is suitable for spherical clusters, not for arbitrary shape. Therefore, this paper calculates the similarity by the average distance value of the object with all data points in the cluster.

**Definition 6. Distance between Object and Cluster:** For $\mathbf{x}_i$ is an object, $C^k$ is a cluster, and $\mathbf{x}_i \notin C^k$, the distance between the object and cluster is defined as follows.

$$DisObjClu(\mathbf{x}_i, C^k) = \frac{\sum\limits_{\mathbf{x}_j \in C^k} Dist(\mathbf{x}_i, \mathbf{x}_j)}{|C^k|} \tag{6}$$

where $Dist(\mathbf{x}_i, \mathbf{x}_j)$ is the distance between $\mathbf{x}_i$ and $\mathbf{x}_j$, $|C^k|$ is the cardinality of the cluster.

According to the above definition, it's easy to define the membership between them. The higher the membership is, the more similar they are. So the decision is transferred to find the membership between the object and clusters.

**Definition 7. Membership between Object and Cluster:** For $\mathbf{x}_i$ is an object, $C^k$ is a cluster, and $\mathbf{x}_i \notin C^k$, the membership of an object to a cluster is defined as follows.

$$\mu(\mathbf{x}_i | C^k) = 1 - \frac{DisObjClu(\mathbf{x}_i, C^k)}{\sum\limits_{1 \leq k \leq K} DisObjClu(\mathbf{x}_i, C^k)} \tag{7}$$

## 3.2  Description of the Algorithm

The algorithm has four phases: (1) classifying the data set into four subsets according to the classification in Subsetion 2.2; (2) filling the missing values in sufficient data; (3) obtaining the initial cluster partitions from sufficient data; (4) obtaining the final clustering. The three-way decisions clustering algorithm for incomplete data is described in Algorithm 1, shorted by TWD-CI.

The first phase, it is easy to get the four subsets corresponding to the four types defined in Subsetion 2.2. We call them sufficient data, valuable data, inadequate data and invalid data. The second phase, for the sufficient data, we first estimate missing values according to Definition 5. Since sufficient data retain most of important information, it is reasonable to think they are relatively complete. Then, the grid-based clustering method is used to obtain a general underlying structure.

The fourth phase, the basic idea is to cluster them using corresponding three-way decisions strategy. The useful information in valuable data, inadequate data

---

**Algorithm 1.** TWD-CI: Three-way Decisions Clustering Algorithm for Incomplete Data

---

**Input**: $U$, $W$, the number of important attributes $R$, the number of nearest-neighbors $h$, the density threshold $points$, the length of interval $Rth$, the thresholds $\alpha$ and $\beta$.

**Output**: Final clusters.

**begin**

Step 1: classifying the data set into four subsets according to the classification rules in Subsetion 2.2.

Step 2: for every objects with missing values in sufficient data:

2.1 using Formula (3) to calculate the distance to other objects, then finding its $h$ nearest-neighbors.

2.2 using Formula (5) to fill the missing values.

Step 3: obtaining the initial cluster partitions from sufficient data:

3.1 dividing the data space into grids as introduced in Subsection 3.1.

3.2 mapping each object to the corresponding grid.

3.3 for unsolved dense grids: assign all objects in an unsolved dense grid into a new cluster, and use the bread first searching to find its adjacent grids; if the adjacent grid is dense, to assign objects in this grid into the cluster, and label the dense grid with solved.

3.4 for unsolved sparse grids: if there exists dense neighbor grids, then assign objects of this sparse grid to the cluster with the maximum dense neighbor grids, label the sparse grids with solved.

3.5 for unsolved grids: assign objects of this grid to the cluster with the maximum sparse neighbor grids.

Step 4: for every object $\mathbf{x}_i$ in valuable data, inadequate data and invalid data:

4.1 mapping each object $\mathbf{x}_i$ to a set of grids by considering the attribute with no missing values.

4.2 if objects in the set of grids belong to the same cluster $C_k$, then using Formula (7) to calculate the membership $\mu(\mathbf{x}_i|C^k)$; if $\mu \geq \alpha$, then the object belongs to $POS(C_k)$; if $\beta < \mu < \alpha$, then the object belongs to $BND(C_k)$.

4.3 if objects in the set of grids belong to more than one cluster, then using Formula (7) to calculate the membership $\mu$; find the cluster $C_k$ with the maximum of membership; if $\mu \geq \alpha$, then the object belongs to $POS(C_k)$; if $\beta < \mu < \alpha$, then the object belongs to $BND(C_k)$.

**end**

---

and invalid data is descending. Thus, clustering valuable data first, then we cluster other types orderly. Because an object in these three types data has missing values in a certain extent, it is not reasonable to assign it to a sole grid unit; we assign it to the units according to corresponding attributes with no missing values. Then according to the membership of an object to a cluster, we assign the object to the positive or the boundary region. That is, the three-way decisions rules used here are as follows.

$$
\begin{aligned}
&if \ \mu(\mathbf{x}_i|C^k) \geq \alpha, \ decide \ the \ object \ to \ POS(C^k); \\
&if \ \beta < \mu(\mathbf{x}_i|C^k) < \alpha, \ decide \ the \ object \ to \ BND(C^k); \\
&if \ \mu(\mathbf{x}_i|C^k) \leq \beta, \ decide \ the \ object \ to \ NEG(C^k).
\end{aligned}
\tag{8}
$$

## 4  Experiments

### 4.1  Experiment 1

The purpose of this experiment is to observe the statistical distribution of the incomplete data set, in order to verify the reasonability of the classification we introduced in Section 2.

First, we set some attribute values missed randomly on data sets. Then the experiment tests three times under miss rate 5%, 10%, 15% and 20%, respectively. In addition, the incomplete data set must meet the following conditions: (1) any object must keep at least one complete attribute; (2) any attribute in the data set must keep at least a complete value. That is to say, we don't keep the row or column with all missing values in the test.

Then, we count the proportion of every type of the data under a miss rate, where the data sets from UCI repository [9] except Face is a synthetic data set. Table 2 records the statistical result, $|U|$, $D$ and $M$ denote the number of objects in the data set, the number of attributes, the number of important attributes, respectively. The column "Miss" records the different missing rate. The "Sufficient", "Valuable", "Inadequate" and "Invalid" means the four types of data.

From Table 2 we can see that the number of sufficient data is reduced with the increasing of miss rate. Even if the miss rate reaches to 20%, the proportion of sufficient data is still more than 50%; which indicates the rationality of the proposed algorithm based on sufficient data. The proportion of valuable data is much less than sufficient data, but more than that of inadequate data and invalid data. Seeds and Face only has sufficient data and invalid data, because Seeds only has one important attribute and all attributes in Face are important. The experimental result shows the rationality of classification proposed in Section 2.

### 4.2  Experiment 2

In order to evaluate the validity of the thought that considers the attribute significance as well as the miss rate when classifying data set into four types, we

**Table 2.** The Proportion of Four Types Data

| Data Set | $|U|$ | D | M | Miss | Sufficient | Valuable | Inadequate | Invalid |
|---|---|---|---|---|---|---|---|---|
| Iris | 150 | 4 | 2 | 5% | 89.33% | 10.76% | 0.00% | 0.00% |
| | | | | 10% | 83.33% | 14.67% | 0.00% | 2.00% |
| | | | | 15% | 72.67% | 24.67% | 0.66% | 2.00% |
| | | | | 20% | 65.33% | 29.34% | 2.00% | 3.30% |
| Banknote Authentication | 1372 | 4 | 2 | 5% | 90.89% | 8.97% | 0.00% | 0.15% |
| | | | | 10% | 79.96% | 18.59% | 0.22% | 1.24% |
| | | | | 15% | 70.92% | 25.87% | 0.51% | 2.70% |
| | | | | 20% | 62.90% | 32.14% | 1.38% | 3.57% |
| Seeds | 210 | 7 | 1 | 5% | 97.62% | 0.00% | 0.00% | 2.38% |
| | | | | 10% | 87.14% | 0.00% | 0.00% | 2.86% |
| | | | | 15% | 95.24% | 0.00% | 0.00% | 4.76% |
| | | | | 20% | 94.29% | 0.00% | 0.00% | 5.71% |
| Wilt | 4839 | 5 | 3 | 5% | 85.45% | 13.78% | 0.74% | 0.02% |
| | | | | 10% | 73.13% | 23.77% | 2.87% | 0.23% |
| | | | | 15% | 61.67% | 30.89% | 7.07% | 0.37% |
| | | | | 20% | 51.00% | 37.53% | 10.66% | 0.81% |
| Face | 800 | 2 | 2 | 5% | 90.00% | 10.00% | 0.00% | 0.00% |
| | | | | 10% | 80.00% | 20.00% | 0.00% | 0.00% |
| | | | | 15% | 69.37% | 30.63% | 0.00% | 0.00% |
| | | | | 20% | 60.00% | 40.00% | 0.00% | 0.00% |

**Table 3.** The Results of Two Strategies

| Data Set | Miss Rate | Parameters | | | Strategy 1 | | Strategy 2 | |
|---|---|---|---|---|---|---|---|---|
| | | $Rth$ | $minpoints$ | $h$ | Accuracy | Time | Accuracy | Time |
| Iris | 5% | 0.8 | 3 | 6 | 93.33% | 0.188 | 88.67% | 0.156 |
| | 10% | 0.8 | 6 | 6 | 93.33% | 0.188 | 90.00% | 0.172 |
| | 15% | 0.8 | 6 | 6 | 91.33% | 0.250 | 86.67% | 0.312 |
| | 20% | 0.8 | 6 | 6 | 91.33% | 0.218 | 87.33% | 0.219 |
| Banknote Authentication | 5% | 3.5 | 51 | 10 | 77.41% | 2.562 | 63.63% | 1.688 |
| | 10% | 3.5 | 45 | 10 | 70.34% | 4.015 | 68.07% | 1.250 |
| | 15% | 2.7 | 14 | 10 | 69.39% | 4.781 | 68.07% | 2.141 |
| | 20% | 3.1 | 22 | 10 | 68.80% | 6.625 | 68.37% | 3.047 |
| Seeds | 5% | 1 | 6 | 10 | 86.67% | 1.984 | 85.71% | 1.562 |
| | 10% | 1 | 7 | 10 | 85.24% | 1.375 | 84.29% | 1.250 |
| | 15% | 1 | 7 | 10 | 84.76% | 1.328 | 77.62% | 1.188 |
| | 20% | 1 | 6 | 10 | 84.76% | 2.14 | 84.29% | 0.890 |
| Wilt | 5% | 28 | 21 | 20 | 84.21% | 284.404 | 83.43% | 296.562 |
| | 10% | 28 | 20 | 20 | 81.81% | 448.984 | 79.40% | 323.219 |
| | 15% | 28 | 19 | 20 | 70.74% | 277.594 | 70.74% | 244.248 |
| | 20% | 28 | 17 | 20 | 67.95% | 224.344 | 67.53% | 286.434 |
| Face | 5% | 26 | 21 | 10 | 79.25% | 0.750 | 79.25% | 0.750 |
| | 10% | 26 | 18 | 10 | 76.00% | 1.453 | 76.00% | 1.421 |
| | 15% | 24 | 16 | 10 | 78.88% | 1,703 | 78.88% | 1.734 |
| | 20% | 26 | 16 | 10 | 72.00% | 2.578 | 72.00% | 2.578 |

compare the proposed algorithm with the one that only considers the miss rate when classifying.

The proposed algorithm TWD-CI is called Strategy 1, another one is called Strategy 2. That is, the input data set of Strategy 1 is the data set ordered by attribute significance; and the input data set of Strategy 2 is the original data set or with the random order attributes, in other words, the weights of attributes are equal. The other processing of two strategies are same. We set the threshold $\alpha = 0.7, \beta = 0.5$ in experiment, and the result is shown in Table 3.

Observe Table 3, the accuracy of Strategy 1 is better than that of Strategy 2 in most cases. For Face data set, because two attribute weights are same, the advantage of Strategy 1 is not very evident. Strategy 1 is roughly the same as Strategy 2 in time. The results indicate that the proposed algorithm TWD-CI is adapted to the environment of attributes with different importance.

## 5  Conclusions

In this paper, we presented a three-way decisions clustering algorithm for incomplete data, which considers the attribute significance as well as miss rate. The algorithm divides the data set into four types based on attribute significance and the miss rate of the data. The four types are sufficient data, valuable data, inadequate data and invalid data. For sufficient data, grid-based clustering method is used to obtain a basic underlying structure of data. The processing for the rest of the three parts follows the order of valuable data, inadequate data and invalid data; and the three-way decisions rules are adopted to decide which cluster the datum belongs to.

The essential idea of the approach is to classify data set into four subsets, and the preliminary experimental results show that it is effective that clustering based on the sufficient data. However, there is still some problems need to solve, for example, how to automatically divide original incomplete data set into the four parts no depending on the parameters. Making better of Step 3 and Step 4 of the proposed algorithm will produce an important effect to improve the clustering performance and to reduce the time, which is also of the further work.

**Acknowledgments.** This work was supported in part by the China NSFC grant (No.61379114 & No.61272060).

## References

1. Azam, N.: Formulating Three-way decision making with game-theoretic rough sets. In: Proceedings of CanadainConference on Electrical and Computer Engineering (CCECE 2013), pp. 695–698. IEEE Press (2013)
2. Dixon, J.K.: Pattern recognition with partly missing data. IEEE Transactions on Systems, Man, and Cybernetics 9, 617–621 (1979)
3. Himmelspach, L., Hommers, D., Conrad, S.: Cluster tendency assessment for fuzzy clustering of incomplete data. In: Proceedings of the 7th conference of the European Society for Fuzzy Logic and Technology, pp. 290–297. Atlantis Press (2011)

4. Honda, K., Nonoguchi, R., Notsu, A., Ichihashi, H.: PCA-guided k-Means clustering with incomplete data. In: 2011 IEEE International Conference on Fuzzy Systems (FUZZ), pp. 1710–1714. IEEE Press (2011)
5. Lai, P.H., O'Sullivan, J.A.: MDL hierarchical clustering with incomplete data. In: Information Theory and Applications Workshop (ITA), pp. 1–5. IEEE Press (2010)
6. Liang, D.C., Liu, D.: A novel risk decision-making based on decision-theoretic rough sets under hesitant fuzzy information. J. IEEE Transactions on Fuzzy Systems (2014)
7. Li, D., Gu, H., Zhang, L.Y.: A hybrid genetic algorithm-fuzzy c-means approach for incomplete data clustering based on nearest-neighbor intervals. J. Soft Computing. 17, 1787–1796 (2013)
8. Li, D., Zhong, C.Q., Li, J.H.: An attribute weighted fuzzy c-means algorithm for incomplete data sets. In: 2012 International Conference on System Science and Engineering (ICSSE), pp. 449–453. IEEE Press (2012)
9. UCIrvine Machine Learning Repository: http://archive.ics.uci.edu/ml/
10. Wu, J., Song, C.H., Kong, J.M., Lee, W.D.: Extended mean field annealing for clustering incomplete data. In: International Symposium on Information Technology Convergence, pp. 8–12. IEEE Press (2007)
11. Yamamoto, T., Honda, K., Notsu, A., Ichihashi, H.: FCMdd-type linear fuzzy clustering for incomplete non-Euclidean relational data. In: 2011 IEEE International Conference on Fuzzy Systems (FUZZ), pp. 792–798. IEEE Press (2011)
12. Yao, Y.: An outline of a theory of three-way decisions. In: Yao, J., Yang, Y., Słowiński, R., Greco, S., Li, H., Mitra, S., Polkowski, L. (eds.) RSCTC 2012. LNCS, vol. 7413, pp. 1–17. Springer, Heidelberg (2012)
13. Yao, Y.Y.: Three-way decisions with probabilistic rough sets. J. Information Sciences 180, 341–353 (2010)
14. Yu, H., Liu, Z.G., Wang, G.Y.: An automatic method to determine the number of clusters using decision-theoretic rough set. International Journal of Approximate Reasoning 55, 101–115 (2014)
15. Yu, H., Wang, Y.: Three-way decisions method for overlapping clustering. In: Yao, J., Yang, Y., Słowiński, R., Greco, S., Li, H., Mitra, S., Polkowski, L. (eds.) RSCTC 2012. LNCS, vol. 7413, pp. 277–286. Springer, Heidelberg (2012)
16. Zhou, B., Yao, Y.Y., Luo, J.G.: Cost-sensitive three-way email spam filtering. Journal of Intelligent Information Systems 42, 19–45 (2013)

# Sentiment Analysis with Automatically Constructed Lexicon and Three-Way Decision

Zhe Zhou, Weibin Zhao, and Lin Shang

State Key Laboratory for Novel Software Technology, Department of Computer
Science and Technology, Nanjing University, Nanjing 210023, China
zhouzhenjucs@gmail.com zhaowb@njupt.edu.cn shanglin@nju.edu.cn

**Abstract.** An unsupervised sentiment analysis method is presented to
classify user comments on laptops into positive ones and negative ones.
The method automatically extracts informative features in testing dataset
and labels the sentiment polarity of each feature to make a domain-
specific lexicon. The classification accuracy of this lexicon will be compared to that with an existing general sentiment lexicon. Besides, the
concept of three-way decision will be applied in the classifier as well,
which combines lexicon-based methods and supervised learning methods together. Results indicate that the overall performance can reach
considerable improvements with three-way decision.

**Keywords:** sentiment analysis, opinion mining, sentiment lexicon, three-
way decision.

## 1 Introduction

Sentiment analysis, also known as opinion mining, refers to detecting the sentiment polarity or sentiment strength of a given piece of text[1]. Nowadays people
can freely post their opinions and comments on the Internet and receive others'
views at the same time[2]. Therefore, sentiment analysis becomes popular and
urgent for some particular groups of Internet users. For example, commodity
producers may collect reviews written by consumers and try to obtain the overall sentiment tendency in order to know whether their products are popular or
not and what advantages and disadvantages they have[3, 4]. On the other hand,
consumers can as well search their peers' opinions and reviews in order to know
whether the product they want is worth buying[5]. In such cases, techniques
similar to traditional topic-based classification algorithms can be used to automatically assign sentiment labels to product reviews. However, such methods
may run into difficulty due to the speciality of product review sentiment analysis. Firstly, a major difference between traditional topic-based classification and
sentiment analysis is that sentiment is often expressed in a subtle way, which will
pose challenges in the classification work[6]. Besides, reviews on products or services often pay attention to detailed features or aspects[3, 4, 7], so feature-level
analysis must be taken into consideration in the analysis process.

D. Miao et al. (Eds.): RSKT 2014, LNAI 8818, pp. 777–788, 2014.
DOI: 10.1007/978-3-319-11740-9_71 © Springer International Publishing Switzerland 2014

In this article, we will design an unsupervised sentiment analysis algorithm to deal with these problems. The algorithm can help to understand the sentiment of product reviews better by utilizing the feature-level information. Also, we will apply a three-way-decision-like concept into the scheme in order to boost its performance and get a finer-grained system. The proposed scheme will be applied on a dataset made up by laptop reviews to test its efficiency. The remainder of this article is organized as follows. Sect. 2 will list some related work about sentiment analysis. Sect. 3 will explain the proposed method in detail. Sect. 4 will list and analyse the experiment results. Sect. 5 will conclude the whole work above and look into the future work.

## 2    Related Work

There have been many contributions studying text sentiment analysis during the past decade. Pang et al.[6] collected over a thousand movie reviews for binary sentiment classification and compared performances of three different machine learning algorithms including Naive Bayesian, Max Entropy and the Support Vector Machine. These movie reviews have been one of the most well-known benchmark datasets for sentiment analysis since this contribution. Besides, Pang and Lee[8] also focused on extracting only the subjective sentences for feature selection aiming to improve the performance of sentiment analysis. The process of subjectivity summarization was based on the minimum-cut algorithm and proved to be beneficial to the classifier's performance. Recent works include Hu et al.[9], who provided a supervised sentiment analysis approach in microblogging by utilizing users' social relation information to tackle noises. Besides those works based on supervised learning, there have been enormous unsupervised-learning-based contributions as well. Turney[10] calculates the semantic orientations of a large number of 2-grams with the help of search engines and use them to classify the given text. Li and Liu[11] introduced an clustering-based approach in sentiment analysis and obtained satisfying results by applying TF-IDF weighting, voting mechanism and important term scores. Taboada et al.[12] used different general sentiment lexicons in their lexicon-based sentiment analysis approaches and made a comparison between those lexicons. Hogenboom et al.[13] manually created a lexicon consisting emoticons to aid the traditional classification work. With the problem that current expressions on social media are usually unstructured and informal, Hu et al.[14] incorporated emotional signals into the unsupervised sentiment analysis framework and experimental results proved the effectiveness of emotional signals.

Current sentiment analysis works mostly focus on adjective features as adjectives are believed to be more informative in indicating sentiment orientations. However, Zhang and Liu[15] pointed out that in sentiment analysis on product reviews, it is often necessary to apply domain-specific noun features into the feature space. There have been many contributions concerning this aspect. For example, Yin and Peng[3] build semantic orientations between product features and sentiment words in reviews written in Chinese. Hu and Liu[7] addressed sentiment analysis on customer reviews by extracting frequent product features and using them to summarize the sentiment of the whole review. Similarly, Zhang et al.[4] addressed feature-level sentiment analysis by combining association rules and point-wise mutual information together to extract product features and identify their sentiment orientations. Riloff et al.[16] proposed a method to extract subjective nouns, which again proved that noun features can help sentiment analysis quite well, especially in product reviews.

# 3    Proposed Method: Feature Lexicon Construction and Three-Way Decision

In this section we will explain the proposed scheme to tackle the potential difficulty in traditional sentiment analysis methods. We extract informative patterns from the dataset and calculate sentiment scores of those patterns with the help of a general lexicon. The newly formed product feature lexicon and the general lexicon will be separately used in a lexicon-based sentiment analysis algorithm and return two different results. Finally, we will introduce the concept of three-way decision and use a similar method to reach a better classification accuracy.

## 3.1    Data Preprocessing

Our work aims to run a sentiment analysis on comments written in Chinese, so the most crucial parts in preprocessing step will be word segmentation and part-of-speech tagging. Word segmentation refers to cutting every sentence into its component words and part-of-speech tagging means using natural language processing techniques to obtain the part of speech of each word.

## 3.2    Lexicon-Based Sentiment Analysis

A sentiment lexicon, which can be considered as a special kind of dictionary, is a data structure containing different words and their sentiment orientations. Typically, the sentiment orientation is represented by a numerical value. A value greater than zero refers to a positive orientation and a value smaller than zero indicates a negative one. At the moment there are plenty of public sentiment lexicons on the Internet. Those general lexicons are integrated by other people's manual work and can be applied into sentiment analysis works of any domain.

In our proposed scheme, a general lexicon of HowNet[17], which contains about 9000 Chinese words and their sentiment polarities as positive or negative, will be utilized as the lexicon to classify a piece of text into two sentiment categories. The pseudo-code of classification algorithm is shown below.

As [1, 6] have mentioned, sometimes lexicon-based sentiment analysis may encounter a large amount of ties where the sentiment score will be 0. It's usually because the times of occurrences of positive and negative words are equal (usually both are 0 when the text is not long enough). According to [12], people tend to favour positive expressions when they make comments and negative languages are often expressed in a obscure and indirect way which is actually hard to detect.

```
Input: text, lexicon
Output: sentiment label for text

score = 0
negation = 1
for i = 0 : text.length
    if text[i] is "不"        //"not"
        negation *= -1
    if text[i] is punctuation
        negation = 1
    if text[i] is in lexicon
        polarity = lexicon.get(text[i]) == positive ? 1 : -1
        score += negation * polarity
if score > 0
    classify text as positive
else
    classify text as negative

return score
```

**Fig. 1.** Lexicon-based algorithm for sentiment analysis

This phenomenon makes negative oriented comments much easier to be ignored and incorrectly classified. As a solution to the "positive bias", [12] gives negative words additional weights to reach a balance. For the same reason, out tie-breaker will always label tie comments as negative ones.

### 3.3   Automatically Constructed Feature Lexicon

General sentiment lexicons can make contributions to many sentiment analysis works, but our work focuses on the categorization of comments on laptops, which will be more challenging due to their unique traits. For example, the comment "the cost-performance is low" expresses a negative sentiment, although neither "cost-performance" nor "low" can be found in general sentiment lexicons. From the example, it's easy to see that when customers comment on electronic devices, they tend to express their opinions on product features rather than directly use sentiment-carrying words, especially when they want to show their negative views. So it will be inefficient to run analysis only with general lexicons. In order to solve this problem, we design an algorithm to automatically extract the product-feature-related phrases out of the whole corpus. These phrases together form a laptop feature lexicon.

The process of constructing a laptop feature lexicon is based upon an assumption that is consistent with people's general intuitions: the sentiment expressed by a word is to some extent correlated with the sentiment of words co-occurring with it. In [10], similar assumptions were used to label target phrases by calculating mutual information between the phrase and seed words. The detailed

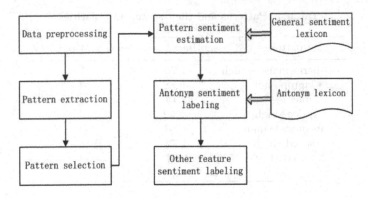

**Fig. 2.** Steps of building a laptop feature lexicon

scheme is shown below. The whole process consists of two main steps: pattern extraction and sentiment polarity assignment.

In the patterns extraction step, the algorithm detects all the "n-adj" patterns (a "n-adj" pattern is a noun and an adjective with no punctuations between them) from the corpus and stores them in pre-defined data structure. Afterwards, if a pattern's noun part occurs at least 100 times in the corpus and the occurrence of the pattern itself is greater than 10, then it will be put into the laptop feature lexicon. Otherwise the pattern will be removed.

**Table 1.** Patterns after selection

| pattern | noun freq | n.+adj. freq |
|---|---|---|
| cost-performance+high | 685 | 233 |
| laptop+hot | 181 | 67 |
| camera+clear | 200 | 14 |
| price+high | 632 | 33 |
| memory+small | 491 | 67 |
| speed+high | 873 | 168 |
| speed+low | 873 | 321 |
| ... | ... | ... |

Table 1 shows part of the terms in the final laptop feature lexicon. Of course in our experiment all those patterns are written in Chinese. Then our algorithm will automatically assign a sentiment label to every pattern in the laptop feature lexicon. When a pattern is found in the pattern extraction process, our algorithm will extract the n words before and following its noun part and store them in another data structure (in this work we let n=5). We call those words "neighbourhood sentences". For every pattern, we extract all the neighbourhood sentences near its occurrences and use the same method as Fig. 2 to compute their sentiment scores. After that we will be able to get the average sentiment

**Table 2.** Patterns and their sentiment polarities

| pattern | sentiment score | sentiment polarity |
| --- | --- | --- |
| cost-performance+high | 1.70 | 1 |
| laptop+hot | -0.88 | -1 |
| camera+clear | 1.14 | 1 |
| price+high | -0.52 | -1 |
| memory+small | -0.91 | -1 |
| speed+high | 1.72 | 1 |
| speed+low | 0.07 | -1 |
| ... | ... | ... |

score of all the neighbourhood sentences, we see that as an estimation of the sentiment score of the pattern itself. Table 2 shows part of the selected patterns, their corresponding average sentiment scores are shown in the middle column.

Lastly we transform every sentiment score to +1/-1. This is done with the help of an antonym lexicon which contains a large number of antonym pairs (see Table 3). We traverse through all the frequent patterns to find the pattern pairs whose noun parts are the same but adjective parts are antonyms. When such pairs are encountered, the pattern with larger sentiment score is given the sentiment polarity of +1 and the other -1. The polarity of remaining patterns will be decided by the sign of their sentiment scores.

**Table 3.** An antonym lexicon

| word1 | word2 |
| --- | --- |
| front | back |
| forward | backward |
| high | low |
| public | private |
| cold | hot |
| die | alive |
| ... | ... |

After all the patterns are given a sentiment polarity label (see the last column in Table 2), the laptop feature lexicon is finally constructed. Now we can run the algorithm in Fig. 2 again with the laptop feature lexicon in place of the general HowNet lexicon and expect the new algorithm to reach satisfying results.

## 3.4  Three-Way Decision

Most sentiment analysis problems are treated as binary classification tasks[3, 4, 6–8, 10–12, 14, 15], in which a piece of text is either labelled as "positive" or "negative". Such idea is simple and direct, but sometimes can not reflect the nature of the real world. Take a real world problem into consideration, if a classifier is trained to predict whether an incoming Email is a spam one or not, it will encounter some Emails which are hard to classify into either of the two categories. For example, assume the classifier is trained by logistic regression, and there is an Email whose probability of being negative (spam mail) is estimated as 0.55. Of course the Email will be classified as a spam, but the classifier will take great risks doing so because the probability of the mail being legitimate is up to 0.45 as well. Therefore, it is necessary to introduce a third way of decision into the classification task, where the classifier can refuse to classify emails if it's not confident enough of the emails' categorization. A rejected email will be labelled as "suspicious" and presented to the user, who will make his own judgements whether it's a spam or not. Such concept is called three-way decision.

Three-way decision has been widely studied in previous contributions and is usually associated with the rough set theory, as Yao have introduced in [18, 19]. According to the three regions in the rough set, Yao concluded three rules shown in (1)-(3) for decision. The values of alpha and beta are computed from six predefined losses when different actions are taken on different objects. Zhou et al.[20] put the three-way decision into application by designing a Email spam filtering algorithm. In their work, an email may be rejected if the risk of labelling it as either "legitimate" or "spam" is high enough. From experimental results, the three-way classifier reached a better weighted accuracy than a normal one.

$$\text{If } \Pr\left(X|[x]\right) \geq \alpha, \text{decide } x \in POS\left(X\right) \tag{1}$$

$$\text{If } \beta < \Pr\left(X|[x]\right) < \alpha, \text{decide } x \in BND\left(X\right) \tag{2}$$

$$\text{If } \Pr\left(X|[x]\right) \leq \beta, \text{decide } x \in NEG\left(X\right) \tag{3}$$

In this work we will use a method which is similar to the process of three-way decision to provide another sentiment classification algorithm on the given dataset, which is combined by the two lexicon-based methods introduced above. First, we apply the algorithm based on general lexicon and the algorithm based on laptop feature lexicon separately on the dataset and get two different results about the sentiment polarity of every piece of comment. Then we combine the two results together and let them vote for a final one. The rule is simple: if two results are the same, then the sentiment polarity of the comment will be the same with the two results; if two algorithms return different sentiment labels, then the comment will be put into the rejection set for further decision, which means we are not assigning a sentiment label to the comment at the moment. This allows us to put aside comments with which the classifiers are not confident enough and thus can reach a better accuracy on the comments who are given an exact sentiment label.

Our last aim in this work is to deal with the comments in the rejection set in order to complete the whole three-way decision concept. The idea is shown below in Fig. 3. We use the supervised learning method to classify the unlabelled comments in the rejection set, and the training data in supervised learning is made up by the comments that are previously labelled by the two lexicon-based algorithms. As [12, 21] have mentioned, the two main weaknesses of supervised learning are that it's difficult to find abundant labelled data for training and the classifier's performance may drop harshly when applied into a new domain of topic. However, our work extracts training data from the unlabelled dataset itself (see Fig. 3), which can solve the two problems at the same time.

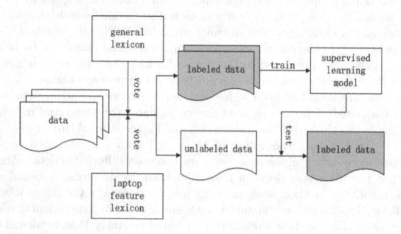

**Fig. 3.** A hybrid algorithm for sentiment analysis

Obviously, the idea has its drawbacks as part of the training data may be wrongly classified by lexicon-based algorithms and thus have an incorrect sentiment label, which may reduce the accuracy of supervised learning. But it has been indicated in [21] that the supervised method can still reach considerable accuracy provided that a large amount of training data have their labels assigned correctly. For this reason, we can expect our supervised learning algorithm to provide a good performance, as [21] have shown in their work.

# 4   Experimental Results

## 4.1   Dataset

Our test dataset is called ChnSentiCorp-nb-4000, which is collected by S.Tan[1]. The dataset consists of 3993 different comments on laptops, 1996 of which is positive and others negative. The average length of comments is around 60 Chinese characters. We will run the sentiment classification scheme introduced in Sect. 3 on the dataset, predicting the sentiment polarity of each comment.

---

[1] http://www.searchforum.org.cn/tansongbo/senti_corpus.jsp

## 4.2   Performance Measure

We will generally evaluate the performance of our method by its classification accuracy. When the general sentiment lexicon and the laptop feature lexicon is independently applied to the dataset, the calculation of Accuracy is presented below, where TP means true positives, TN means true negatives and ALL means the number of comments in the dataset.

$$\text{Accuracy} = \frac{TP + TN}{ALL} \tag{4}$$

After two lexicon-based methods are combined together, the calculation of Accuracy is shown in (5), where TIE means number of comments that are rejected by the classifier because the vote results in a tie. Besides, we will introduce a new measure called Reject (or tie rate) to represent the percentage of comments that are rejected.

$$\text{Accuracy} = \frac{TP + TN}{ALL - TIE} \tag{5}$$

$$\text{Reject} = \frac{TIE}{ALL} \tag{6}$$

When the three-way decision is applied, the calculation of Accuracy is shown in (7). TP2 and TN2 mean the number of true positives and true negatives that are obtained by the tie-breaker (in our work we use Naive Bayesian Classifier as the tie-breaker). 3rdWayAccuracy represents the accuracy our tie-breaker reached on the unassigned comments. For comparison, we set a a baseline where we randomly "guess" a sentiment label for each rejected comment. It is obvious that theoretically the 3rdWayAccuracy for random-choice strategy will be 50%, so (9) represents the baseline accuracy, which our method must be superior to.

$$\text{Accuracy} = \frac{TP + TN + TP2 + TN2}{ALL} \tag{7}$$

$$\text{3rdWayAccuracy} = \frac{TP2 + TN2}{TIE} \tag{8}$$

$$\text{Baseline} = \frac{TP + TN + 0.5 * TIE}{ALL} \tag{9}$$

## 4.3   Results and Analysis

Before presenting our experimental results, we will firstly introduce a previous work which is applied on the same dataset. The previous work is done by Yu et al.[22], in which parallelized sentiment classification algorithms are ran with different weighting methods, feature selection methods and supervised classifiers. The results in [22] is shown below and the average accuracy is around 80.2%.

**Table 4.** Experimental results by Yu et al.

| feature selection method | weighting method | classifier | accuracy |
|---|---|---|---|
| Bigram | Boolean | Knn | 84% |
| Bigram | Boolean | Rocchio | 82.9% |
| Bigram | Tf-idf | Knn | 78.4% |
| Bigram | Tf-idf | Rocchio | 87.7% |
| Sentimen Lexicon | Boolean | Knn | 78.9% |
| Sentimen Lexicon | Boolean | Rocchio | 78.4% |
| Sentimen | Tf-idf Lexicon | Knn | 74.1% |
| Sentimen | Tf-idf Lexicon | Rocchio | 81.6% |
| Substring | Boolean | Knn | 79.5% |
| Substring | Boolean | Rocchio | 68.8% |
| Substring | Tf-idf | Knn | 82.9% |
| Substring | Tf-idf | Rocchio | 85.3% |
| average | | | 80.2% |

Then our experimental results is shown below in Table 5. The first two rows shows the classification accuracy of general lexicon HowNet and the laptop feature lexicon constructed in our work. Results indicate that the laptop feature lexicon can do almost as well as the general sentiment lexicon which is publicly available on the Internet. This suggests that our method of extracting patterns describing product features and estimate their sentiment scores can actually make contributions to sentiment analysis.

Furthermore, when the two lexicons are combined together to make a vote system, the result is shown in the third row. 27.35% of all the reviews are rejected with rule (2) but those not rejected can reach the accuracy of 85.66%. Taking rejection as a third way of decision will provide a more subtle view in classification problems, which can reflect the true state of nature better[20].

According to [20], additional information is needed to deal with the undecided samples. So we apply the algorithm in Fig. 3, use supervised learning as a tie-breaker to classify the unlabelled data. The comparison result is shown in the forth and fifth row in Table 5. Our proposed scheme reached accuracy of 84.90%, which outperforms the baseline of random guess strategy and either of the two lexicons. Also, when comparing Table 4 with Table 5 it is easy to see our proposed scheme is better than [22] in most cases and is superior to its average accuracy as well. Unlike [22], our proposed scheme is unsupervised (as shown in Fig. 3, the "training data" in the supervised process is part of the testing data itself), which again proves its effectiveness. Those results suggest that our method can return satisfying classification results while maintaining its unsupervised feature.

## 5   Conclusion and Future Work

In this work, an unsupervised sentiment analysis scheme is ran on a dataset made up by customers' comments on laptops. A new laptop feature lexicon

**Table 5.** Experimental results

| algorithm | accuracy | reject rate |
|---|---|---|
| General Lexicon | 77.54% | - |
| Feature Lexicon | 74.28% | - |
| GL + FL | 85.66% | 27.35% |
| 3-way: baseline | 75.91% | - |
| 3-way: proposed | 84.90% | - |

which is generated from the dataset itself is introduced to provide an extra view in sentiment categorization. Also, three-way decision methods are used as well in order to get better classification accuracy. Experiment results show that the laptop feature lexicon can do almost as well as a general lexicon in classification accuracy. Besides, when the two lexicons are combined together with the three-way decision method, the classifier can reach a great improvement in its performance.

In the future, we aim to apply our feature lexicon construction methods to other domains to test its validity. Besides, the three-way decision model used in the proposed scheme is simple and intuitive. In the future, we hope to build a three-way classifier which is more theoretically precise in order to make the work more convincing.

# References

1. Pang, B., Lee, L.: Opinion mining and sentiment analysis. Foundations and Trends in Information Retrieval 2(1-2), 1–135 (2008)
2. Cambria, E., Schuller, B., Xia, Y., et al.: New avenues in opinion mining and sentiment analysis (2013)
3. Yin, C., Peng, Q.: Sentiment Analysis for Product Features in Chinese Reviews Based on Semantic Association. In: International Conference on Artificial Intelligence and Computational Intelligence, AICI 2009, vol. 3, pp. 81–85. IEEE (2009)
4. Zhang, H., Yu, Z., Xu, M., et al.: Feature-level sentiment analysis for Chinese product reviews. In: 2011 3rd International Conference on Computer Research and Development (ICCRD), vol. 2, pp. 135–140. IEEE (2011)
5. Feldman, R.: Techniques and applications for sentiment analysis. Communications of the ACM 56(4), 82–89 (2013)
6. Pang, B., Lee, L., Vaithyanathan, S.: Thumbs up?: sentiment classification using machine learning techniques. In: Proceedings of the ACL 2002 Conference on Empirical Methods in Natural Language Processing, vol. 10, pp. 79–86. Association for Computational Linguistics (2002)
7. Hu, M., Liu, B.: Mining and summarizing customer reviews. In: Proceedings of the Tenth ACM SIGKDD International Conference on Knowledge Discovery and Data Mining, pp. 168–177. ACM (2004)
8. Pang, B., Lee, L.: A sentimental education: Sentiment analysis using subjectivity summarization based on minimum cuts. In: Proceedings of the 42nd Annual Meeting on Association for Computational Linguistics, p. 271. Association for Computational Linguistics (2004)

9. Hu, X., Tang, L., Tang, J., et al.: Exploiting social relations for sentiment analysis in microblogging. In: Proceedings of the Sixth ACM International Conference on Web Search and Data Mining, pp. 537–546. ACM (2013)
10. Turney, P.D.: Thumbs up or thumbs down?: semantic orientation applied to unsupervised classification of reviews. In: Proceedings of the 40th Annual Meeting on Association for Computational Linguistics, pp. 417–424. Association for Computational Linguistics (2002)
11. Li, G., Liu, F.: A clustering-based approach on sentiment analysis. In: 2010 International Conference on Intelligent Systems and Knowledge Engineering (ISKE), pp. 331–337. IEEE (2010)
12. Taboada, M., Brooke, J., Tofiloski, M., et al.: Lexicon-based methods for sentiment analysis. Computational Linguistics 37(2), 267–307 (2011)
13. Hogenboom, A., Bal, D., Frasincar, F., et al.: Exploiting emoticons in sentiment analysis. In: Proceedings of the 28th Annual ACM Symposium on Applied Computing, pp. 703–710. ACM (2013)
14. Hu, X., Tang, J., Gao, H., et al.: Unsupervised sentiment analysis with emotional signals. In: Proceedings of the 22nd International Conference on World Wide Web, International World Wide Web Conferences Steering Committee, pp. 607–618 (2013)
15. Zhang, L., Liu, B.: Identifying noun product features that imply opinions. In: Proceedings of the 49th Annual Meeting of the Association for Computational Linguistics: Human Language Technologies: Short Papers, vol. 2, pp. 575–580. Association for Computational Linguistics (2011)
16. Riloff, E., Wiebe, J., Wilson, T.: Learning subjective nouns using extraction pattern bootstrapping. In: Proceedings of the Seventh Conference on Natural Language Learning at HLT-NAACL 2003, vol. 4, pp. 25–32. Association for Computational Linguistics (2003)
17. Dong, Z., Dong, Q.: HowNet (2000)
18. Yao, Y.: Three-way decisions with probabilistic rough sets. Information Sciences 180(3), 341–353 (2010)
19. Yao, Y.: Three-way decision: An interpretation of rules in rough set theory. In: Wen, P., Li, Y., Polkowski, L., Yao, Y., Tsumoto, S., Wang, G. (eds.) RSKT 2009. LNCS, vol. 5589, pp. 642–649. Springer, Heidelberg (2009)
20. Zhou, B., Yao, Y., Luo, J.: A three-way decision approach to email spam filtering. In: Farzindar, A., Kešelj, V. (eds.) Canadian AI 2010. LNCS, vol. 6085, pp. 28–39. Springer, Heidelberg (2010)
21. Tan, S., Wang, Y., Cheng, X.: Combining learn-based and lexicon-based techniques for sentiment detection without using labeled examples. In: Proceedings of the 31st Annual International ACM SIGIR Conference on Research and Development in Information Retrieval, pp. 743–744. ACM (2008)
22. Yu, Y., Xiang, X., Shang, L.: Research on Parallelized Sentiment Classification Algorithms. Computer Science 40(6), 206–210 (2013)

# A Novel Intelligent Multi-attribute Three-Way Group Sorting Method Based on Dempster-Shafer Theory

Baoli Wang[1,2] and Jiye Liang[1,*]

[1] Key Laboratory of Computational Intelligence and Chinese Information Processing of Ministry of Education, School of Computer and Information Technology of Shanxi University, Taiyuan 030006, China
[2] Department of Applied Mathematics, Yuncheng University, Yuncheng 044000, China
ljy@sxu.edu.cn,
pollycomputer@163.com

**Abstract.** Multi-attribute group sorting (MAGS) has become a popular subject in multi-attribute decision making fields. The optimization preference disaggregation method and the outranking relation method are frequently used to solve this kind of problems. However, when faced with a MAGS with more attributes and alternatives, these methods show their limitations such as the intensive computations and the difficulty to determine the necessary parameters. To overcome these limitations, we here propose an intelligent three-way group sorting method based on Dempster-Shafer theory for obtaining a more credible sorting result. In the proposed method, decision evidences are constructed by computing the fuzzy memberships of an alternative belonging to the decision classes; the famous Dempster combination approach is further used to aggregate these evidences for making the final group sorting. In the end, a simulation example is employed to show the effectiveness of the new method.

**Keywords:** Three-way decision, Dempster-Shafer theory, Fuzzy preference relation, Evidence combination, Group sorting.

## 1 Introduction

Multi-attribute decision making (MADM) is an important and familiar decision activity that usually occurs in the fields such as economics, management, medicine, pattern recognition [9,25].There are mainly four tasks [28] in MADM problems:

- to construct a rank of alternatives from the worst to the best [31];
- to select the best alternative(s) from the given alternative set [29];

---

* Corresponding author.

D. Miao et al. (Eds.): RSKT 2014, LNAI 8818, pp. 789–800, 2014.
DOI: 10.1007/978-3-319-11740-9_72 © Springer International Publishing Switzerland 2014

– to identify the major distinguish attributes and perform their description based on these attributes [22,23];
– to sort the alternatives into predefined homogenous groups [10].

These decision making problems have been studied in variety of fields from different aspects[9,13,26,32]. In recent years, multi-attribute sorting has attracted several researches' attention and become a hot topic in decision making.

Several representative multi-attribute sorting techniques have been proposed for solving practical problems. Multi-attribute utility theory models the decision maker's preferential system as a value function and determines the cut points for making sorting decision [16,18]. Outranking relation theory is a prominent approach to introduce the incomparable relation to the multi-attribute decision making. ELECTRE (ELimination Et Choix Traduisant la REalité)[27] and PROMETHEE (Preference Ranking Organization METHod for Enrichment Evaluations) [2] are two families of outranking relation approaches. Several improved outranking relation based methods [11,15,19] were proposed for specially solving the multi-attribute sorting problems. Case-based disaggregation methods are designed to analyze the actual decisions taken by the decision maker so that an appropriate model can be constructed representing the decision maker system of preferences as consistently as possible [5,6,9,12,13].

The above mentioned methods are all designed for the multi-attribute sorting problem with a single decision maker. However, a decision making model with a single decision maker can not adapt to the progressively complex socio-economic environment, and more and more practical decision making problems are in need of a group of decision makers to participate in the decision processes in order to consider different aspects of a subject. To the best of our knowledge, only a few researchers have addressed the multi-attribute group sorting problems. Bregar et al. [3] presented a ELECTRE-like group sorting procedure that captures preferential information in the form of the threshold model. Cai et al. [4] proposed a interactive method which considers the inconsistency to assist a group of decision makers with different priorities. Damart et al. [7] developed a ELECTRE-TRI based method to solve the group sorting problem by using a decision support system IRIS, which is considered to preserve the consistency of the sorting results both at the individual level and at the group level. Jabeur et al. [17] proposed an ordinal group sorting method to integrate the individual preferences expressed in partial pre-orders by taking into the importance of experts. Kadziński et al. [20] introduced the concept of a representative value function and proposed an optimization model which extends Doumpos' individual decision maker's preference disaggregation method to aggregate the preferences of several decision makers.

These approaches have made beneficial explorations for solving multi-attribute group sorting. However, Refs. [4] and [20] are done by constructing a optimization models; and Refs. [3] and [4] required some necessary parameters to be provided before the methods being executed, and it is usually quite hard to articulate these parameters by the decision makers. Then, it is urgent to design a new and easy understandable approach. A nature thought is to apply the method, which involves a single decision maker, plus voting for multi-attribute group sorting, while there might exist too many ties in the voting process as analyzed in [24]. Therefore, it is also interesting to introduce advanced aggregation method to solve the multi-attribute group sorting problems. In this paper, the famous Dempster-Shafer theory will be used to help to aggregate the decision makers' opinions in the following sections.

Three-way (positive, negative, deferment) decision is one of the most common used sorting in real decision-making problems [21,34,35]. In an invited lecture in International Joint Rough Set Symposium, Professor Yao has proposed a formal framework for three-way decision theory and pointed out that it is necessary to study the theory because of the three-way decision widely existing in human's daily life [36]. In this paper, we aim to propose an Dempster-Shafer theory based method for solving the multi-attribute group sorting in which each decision maker provide the primary three-way evaluation information by considering the samples' measurements under the condition attributes. Two main stages are involved in the proposed approach. In the first Stage, given a new alternative and $L$ sets of three-way decision case, we construct $L$ groups of pair of basic belief assignments (decision evidences), i.e., the memberships of alternative belonging to different decision classes from inner and outer perspectives. Hence, $2 \times L$ basic belief assignments are obtained with plenty of fuzzy sorting information. Then, the famous Dempster's combination method is used to combine $2 \times L$ basic belief assignments to make the final decision in the second stage. Since the multi-perspective decision evidences are considered in the new method, the final decision making result is more credible. In order to show the effectiveness of the method, we simulate a multi-attribute group sorting process by using the open WEKA data set PASTURE. The illustrative example shows that if each decision maker can provide the exact sorting results for the selected sample set, a more credible aggregation decision result can be got by using the proposed intelligent method.

The remainder of this paper is organized as follows. Section 2 describes the multi-attribute three-way group sorting problems, defines some basic concepts and reviewed some basic concepts of the D-S theory. The intelligent multi-attribute group sorting method is proposed in Section 3. In Section 4, an illustrative example is given to show the effectiveness of the new method. The paper is concluded in Section 5.

## 2   Preliminaries

### 2.1   Description of the Problem

The fundamental components involved in a multi-attribute three-way group sorting model are listed as follows.

- A finite set of alternatives: $A = \{a_1, a_2, \ldots, a_N\}(N > 1)$, under evaluation.
- A finite set of attributes: $C = \{c_1, c_2, \ldots, c_M\}(M > 1)$, under which each object in $A$ is measured; a decision attribute: $d$.

- A finite set of decision makers: $E = \{e_1, e_2 \ldots, e_L\}(L > 1)$, who deliver their primary decisions under $d$.
- A non-empty subset $A^k$ is chosen from $A$ to be evaluated by $e_k (k = 1, 2, ..., L)$.
- The primary decisions are expressed as the three-way decisions: Positive $(P)$, Boundary $(B)$, Negative $(N)$.

Let $S = (A, C, f)$ be a multi-attribute information system, where $f : A \times C \to R$ is a measure function and $f(a_i, a_j) = v_{ij} \in R$. Suppose $S^k = (A^k, C, d, f^k)$ is a multi-attribute evaluation decision system provided by $e_k$ $(k = 1, 2, ..., L)$, where $f^k : A^k \times \{d\} \to \{P, B, N\}$ is decision information function expressed by $e^k$, and $P$ denotes the positive opinion, $B$ the boundary or hesitant opinion, and $N$ the negative opinion. The objective of this problem is to decide the group decision results, i.e., the sorting classes for all alternatives in $A$.

In this study, we assume that each attribute in $C$ are profit, and take it for granted that the primary decision $P$ is preferred to $B$ $(P \succ B)$ and $B$ is preferred to $N$ $(P \succ N)$ under $d$. At the same time, the indifference relation is denoted by $\sim$, i.e., $P \sim P$, $B \sim B$ and $N \sim N$. "$\succeq$" means "$\succ$ or $=$". Given a multi-attribute evaluation system $S = (A, C, d, f)$. The measurement of $a_i$ under an attribute $c \in C$ is denoted as $f(a_i, c)$. The greater and less fuzzy preference relations over $A$ under an condition attribute $c$ are computed by Logsig sigmoid transfer function respectively as

$$r^{>c}(a_i, a_j) = \frac{1}{1 + e^{-\mu(f(a_i, c) - f(a_j, c))}} \tag{1}$$

and

$$r^{<c}(a_i, a_j) = \frac{1}{1 + e^{\mu(f(a_i, c) - f(a_j, c))}} \tag{2}$$

where $\mu > 0$ is a fuzzy factor. Actually, it is easy to prove that $r^{>c}(a_i, a_j) + r^{<c}(a_j, a_i) = 1$ and $r^{>c}(a_i, a_j) = r^{<c}(a_j, a_i)$. In many practical information system, the assessment values are usually in different domains, the values should be normalized into [0,1] before computing the fuzzy preference relations by the logsig function. After obtaining $r^{>c_j}(a_i, a_j)(r^{<c_j}(a_i, a_j))$ for individual attribute, we get the overall fuzzy preference relation under $C$ as

$$r^{>C}(a_i, a_j) = \min\{r^{>c_l}(a_i, a_j \mid l = 1, 2, ..., M)\} \tag{3}$$

and

$$r^{<C}(a_i, a_j) = \min\{(r^{<c_l}(a_i, a_j) \mid l = 1, 2, ..., M\} \tag{4}$$

for the attribute set $C$. There are some other approaches for calculating the overall fuzzy preference relations such as by weighting aggregation techniques [14].

## 2.2  Dempster-Shafer Theory

The Dempster-Shafer theory of evidence (D-S theory), also called the "evidence theory", has been proposed as a generalization of the subjective probability theory and as a model that allows representing the total ignorance case [1,8,30].

**Definition 1.** [8] *Let $\Theta$ be a nonempty finite set, called the frame of discernment, and $2^{\Theta}$ be the set of all subsets of $\Theta$. A set function $m : 2^{\Theta} \to [0,1]$ is referred to as a basic belief assessment (BBA) if it satisfies axioms (M1) and (M2):*

$$\text{(M1)} \ m(\emptyset) = 0, \quad \text{(M2)} \ \sum_{X \subseteq \Theta} m(X) = 1. \tag{5}$$

The quantity $m(X)$, called the belief mass of subset $X$, represents the partial belief that $X$ is true, but does not express any belief for the proper subset of $X$. $X$ is called a focal element when $m(X) \neq 0$. The family of all focal elements of $m$ is denoted by $\mathcal{M}$. Associated with each belief structure, a pair of belief and plausibility functions[30]can be defined as

$$Bel(X) = \sum_{B \subseteq X} m(B), \ \forall X \in 2^{\Theta}, \tag{6}$$

and

$$Pl(X) = \sum_{B \cap X \neq \emptyset} m(B), \ \forall X \in 2^{\Theta}. \tag{7}$$

Dempster's combination rule is one of the most widely used combination rules. It demands that the BBAs are obtained from the independent sources.

Let $m_1$ and $m_2$ be two independent BBAs. Dempster's rule is defined for a set $X \in 2^{\Theta}$ as follows:

$$m(X) = m_1 \oplus m_2(X) = \begin{cases} \frac{1}{1-K} \sum_{B \cap C = X} m_1(B) \cdot m_2(C), & X \neq \emptyset, \\ 0, & X = \emptyset, \end{cases} \tag{8}$$

where $K = \sum_{B \cap C = \emptyset} m_1(B) \cdot m_2(C)$ is the belief mass that the combination assigns to the empty set.

Dempster's combination operator are commutative and associative, therefore, the combination result of several BBAs is irrespective of the combination order.

Let $m$ be an aggregated BBA. The BBA based decision making rule is given as follows. Let $X_1, X_2 \in 2^{\Theta}$. If

$$m(X_1) = \max\{m(X), X \in 2^{\Theta}\}. \tag{9a}$$

$$m(X_2) = \max\{m(X), X \in 2^{\Theta}, X \neq X_1\}, \tag{9b}$$

$$\begin{cases} m(X_1) - m(X_2) > \varepsilon_1, \\ m(\Theta) < \varepsilon_2, \\ m(X_1) > m(\Theta), \end{cases} \tag{9c}$$

then $X_1$ is the result of judgement. $\varepsilon_1$ and $\varepsilon_2$ are the predefined thresholds.

## 3   Intelligent Group Three-Way Sorting Decision

Due to different experiences, knowledge and personalities, different decision makers might assign the same alternative into different class, and even if assign the same alternative into the same class, the memberships of the alternative to the class are usually diverse for the individual attitudes. In Subsection 3.1, the fuzzy decision evidences of alternatives are obtained from the inner and outer respective, and use Dempster's combination rule to aggregate the evidences for make the final decision. Subsection 3.2 proposes the algorithm for solving the multi-attribute three-way group sorting problem.

## 3.1    Evidences of Three-Way Decision for Alternatives

In practical decision making, one can judge the membership of an alternative to an decision class from inner and outer aspects. Take a paper reviewing decision for example, to evaluate the quality of a paper and decide whether to accept it, some reviewers compare the new manuscript with the worst accepted paper (inner perspective) to make their final decision, and others might contrast it with the best not-accepted paper (outer perspective) to express the final decision. In what follows, we obtain the decision evidences for an alternative belonging to the three-way decisions from the inner and outer perspectives.

Given a three-way evaluation decision system $S^k = (A^k, C, d, f^k)$, we denote $Z_{\overline{k}}^{\succeq} = \{a | f^k(a, d) \succeq Z\}$ and $Z_{\overline{k}}^{\preceq} = \{a | f^k(a, d) \preceq Z\}$ as the upward and downward classes of $Z$, respectively, where $Z \in \{P, B, N\}$.

**Definition 2.** *Let $S = (A, C, f)$ be a description information system and $S^k = (A^k, C, d, f^k)(A^k \subseteq A)$ be an evaluation decision system provided by $e^k$. For any $x \in A$ and $Z \in \{P, B\}$, the inner and outer memberships of $x$ belonging to the upward (downward) class $Z_{\overline{k}}^{\succeq}$ are defined as*

$$\hat{Z}_{\overline{k}}^{\succeq}(x) = \max_{y \in Z_{\overline{k}}^{\succeq}} r^{\succ c}(x, y) \quad (\hat{Z}_{\overline{k}}^{\preceq}(x) = \max_{y \in Z_{\overline{k}}^{\preceq}} r^{\prec c}(x, y)), \tag{10}$$

*and*

$$\check{Z}_{\overline{k}}^{\succeq}(x) = \min_{y \notin Z_{\overline{k}}^{\succeq}} r^{\succ c}(x, y) \quad (\check{Z}_{\overline{k}}^{\preceq}(x) = \min_{y \notin Z_{\overline{k}}^{\preceq}} r^{\prec c}(x, y)), \tag{11}$$

*respectively.*

From Definitions 2, $\hat{Z}_{\overline{k}}^{\succeq}(x)$ (respectively, $\hat{Z}_{\overline{k}}^{\preceq}(x)$) indicates that the membership of $x$ to the upward (respectively, downward) class of $Z_k$ depends on the sample that is in $Z_{\overline{k}}^{\succeq}$ (respectively, $Z_{\overline{k}}^{\preceq}$) and produce the maximum greater (less) preference over $x$. While, $\check{Z}_{\overline{k}}^{\succeq}(x)$ (respectively, $\check{Z}_{\overline{k}}^{\preceq}(x)$) indicates that the membership of $x$ to the upward (respectively, downward) class of $Z_k$ depends on the sample that is not in $Z_{\overline{k}}^{\succeq}$ (respectively, $Z_{\overline{k}}^{\preceq}$) and produce the minimum greater (less) preference over $x$.

**Proposition 1.** *Let $S = \{A, C, f\}$ be a description information system and $S^k = \{A^k, C, d, f^k\}(A^k \subseteq A)$ be an evaluation decision system provided by $e^k$. Given an object $x$ in $A$, we have*
   (1) $\hat{P}_{\overline{k}}^{\succeq}(x) \leq \hat{B}_{\overline{k}}^{\succeq}(x) \leq \hat{N}_{\overline{k}}^{\succeq}(x)$;
   (2) $\check{P}_{\overline{k}}^{\succeq}(x) \leq \check{B}_{\overline{k}}^{\succeq}(x) \leq \check{N}_{\overline{k}}^{\succeq}(x)$;

(3) $\hat{P}_k^{\preceq}(x) \geq \hat{B}_k^{\preceq}(x) \geq \hat{N}_k^{\preceq}(x)$;

(4) $\check{P}_k^{\preceq}(x) \geq \check{B}_k^{\preceq}(x) \geq \check{N}_k^{\preceq}(x)$.

*Proof.* (1) Since $P \succ B \succ C$, if $f(a,d) \succ P$, then $f(a,d) \succ B$, we have $P_k^{\succeq} \subseteq B_k^{\succeq}$. $\hat{P}_k^{\succeq}(x) = \max\limits_{y \in P_k^{\succeq}} r^{\succ c}(x,y) \leq \max\limits_{y \in B_k^{\succeq}} r^{\succ c}(x,y) = \hat{B}_k^{\succeq}(x)$. Similarly, we have $\hat{B}_k^{\succeq}(x) \leq \hat{N}_k^{\succeq}(x)$.

(2) Since $P \succ B \succ C$, if $f(a,d) \succ P$, then $f(a,d) \succ B$, we have $P_k^{\succeq} \subseteq B_k^{\succeq}$ and $\overline{B_k^{\succeq}} \subseteq \overline{P_k^{\succeq}}$. $\check{P}_k^{\succeq}(x) = \min\limits_{y \notin P_k^{\succeq}} r^{\succ c}(x,y) = \min\limits_{y \in \overline{P_k^{\succeq}}} r^{\succ c}(x,y) \leq \min\limits_{y \in \overline{B_k^{\succeq}}} r^{\succ c}(x,y) = \min\limits_{y \notin B_k^{\succeq}} r^{\succ c}(x,y) = \check{B}_k^{\succeq}(x)$. Similarly, we have $\check{B}_k^{\succeq}(x) \leq \check{N}_k^{\succeq}(x)$.

The proofs of (3) and (4) are similar to that of (1) and (2).

In what follows, we continue to obtain the inner and outer evidences from the memberships calculated from the above two definitions. Begin with identifying decision class belief assignments for each object, i.e. we set the frame of discernment as $\Theta = \{P, B, N\}$. Note that a proposition is then the assignment an object $x$ to a decision class $P$, $B$, or $N$.

For any $x \in A$, $m_x(Z)$ $(Z \in \Theta)$ be a basic probability mass representing the degree of which the expert supports the hypothesis that the decision class of $x$ is $Z$. The remaining probability mass is assigned to $m_x(\Theta)$, which also represents the uncertainty degree of the BBA.

Let $S = (A^k, C, d, f^k)$ be an evaluation system provided by $e^k$ and $Z \in \Theta = \{P, B, N\}$. We determine the inner and outer BBAs on $\Theta$ for an object $x \in A$ as follows.

(1) The inner BBA.

$$\hat{m}_k^x(Z) = \frac{1}{I_k^x} \hat{Z}_k^{\preceq}(x) \cdot \hat{Z}_k^{\succeq}(x), \tag{12a}$$

$$\hat{m}_k^x(\Theta) = 1 - \sum_{Z \in \Theta} \hat{m}_x(Z), \tag{12b}$$

where

$$I_k^x = \max\{1, \sum_{Z \in \Theta} \hat{Z}_k^{\preceq}(x) \cdot \hat{Z}_k^{\succeq}(x)\}. \tag{13}$$

The coefficient $I_k^x$ is used to normalize the believe degrees. If $\sum_{Z \in \Theta}(\hat{Z}_k^{\preceq}(x) \cdot \hat{Z}_k^{\succeq}(x)) \geq 1$, then $I_k^x = \sum_{Z \in \Theta}(\hat{Z}_k^{\preceq}(x) \cdot \hat{Z}_k^{\succeq}(x))$ and $\hat{m}_k^x(\Theta) = 0$, which means the evidence is certain. If $\sum_{Z \in \Theta} \hat{Z}_k^{\preceq}(x) \cdot \hat{Z}_k^{\succeq}(x) < 1$, then $I_k^x = 1$ and $\hat{m}_k^x(\Theta) = 1 - \sum_{Z \in \Theta}(\hat{Z}_k^{\preceq}(x) \cdot \hat{Z}_k^{\succeq}(x)) > 0$, which means there exists uncertainty in the evidence.

(2) The outer BBA.

$$\check{m}_k^x(Z) = \frac{1}{O_k^x} \check{Z}_k^{\preceq}(x) \cdot \check{Z}_k^{\succeq}(x), \tag{14a}$$

$$\check{m}_k^x(\Theta) = 1 - \sum_{Z \in \Theta} \check{m}_x(Z), \tag{14b}$$

where

$$O_k^x = \max\{1, \sum_{Z \in \Theta} \check{Z}_k^{\preceq}(x) \cdot \check{Z}_k^{\succeq}(x)\}. \tag{15}$$

Aggregation of decision makers' decision judgments is a vital process in group decision making. Then, we continue to combine the derived evidences into a unique one and make the final sorting decision. After obtaining the decision evidences, the next taks is to combine the evidences to the comprehensive BBA and make the final three-way sorting decision by the overall BBA.

### 3.2   A Novel Multi-criteria Three-Way Group Sorting Method Based on D-S Theory

This subsection provides an algorithm to show the detail computation processes of the proposed multi-criteria group sorting method.

---

**Algorithm 1.** A novel multi-criteria group sorting method based on D-S theory

---

**Input**: A evaluation information system $S = (A, C, d, f)$; $L$ decision systems $S^k = (A^k, C, d, f^k)(k = 1, 2, ..., L)$ provided by $L$ different decision makers.
For each alternative $x$ in $A$, execute the following procedures.
**Output**: The final three-way decision $f(x, d)$ for each alternative $x$.

Step 1: Compute the fuzzy preference relation of $x$ with respect to each alternative in $A^k$ by Eq. (3) and Eq. (4).
Step 2: For every $Z \in \Theta$, calculate $\hat{Z}_k^{\succeq}(x)$, $\hat{Z}_k^{\preceq}(x)$, $\check{Z}_k^{\succeq}(x)$, and $\check{Z}_k^{\preceq}(x)$ by (10)-(11).
Step 3: Compute the normalizing coefficients $I_x^k$ and $O_x^k$ by (13) and (15).
Step 4: Derive $\hat{m}_k^x(Z)$ and $\hat{m}_k^x(\Theta)$ by (12a) and (12b).
Step 5: Derive $\check{m}_k^x(Z)$ and $\check{m}_k^x(\Theta)$ by (14a) and (14b).
Step 6: Compute $\hat{m}^x = \hat{m}_1^x \oplus \hat{m}_2^x \oplus ... \oplus \hat{m}_L^x$ and $\check{m}^x = \check{m}_1^x \oplus \check{m}_2^x \oplus ... \oplus \check{m}_L^x$ ($\forall x \in A$) by (8).
Step 7: Calculate $m_x = \hat{m}_x \oplus \check{m}_x$ ($\forall x \in A$) by (8).
Step 8: Use $m$ to make the final overall decision of $x$ ($\forall x \in A$) by (9).

---

## 4   An Illustrative Example

In this section, we simulate an three-way group sorting process to show the effective of the proposed method. A three-way sorting data set, Pasture Production, was download from the WEKA web site (http://www.cs.waikato.ac.nz/ml/weka/). The data set was collected and organized by Dave Barker for predicting pasture production from a variety of biophysical features. Vegetation and soil variables from areas of the North Island hill country with different management histories (1973-1994) were measured and subdivided into 36 paddocks (the evaluation alternatives denoted as $a_1, a_2, ...$ and $a_{36}$), which are characterized by 22 attributes. The 22 attributes are: Fertilizer , Slope ($c_1$), Aspect-dev-NW ($c_2$), OlsenP ($c_3$), MinN ($c_4$),

TS ($c_5$), Ca-Mg ($c_6$), LOM ($c_7$), NFIX-mean ($c_8$), EWorms-main-3 ($c_9$), Eworms-No-species ($c_{10}$), KUnSat ($c_{11}$), OM, Air-Perm ($c_{12}$), Porosity ($c_{13}$), HFRG-pct-mean ($c_{14}$), Legume-yield ($c_{15}$), OSPP-pct-mean ($c_{16}$), Jan-Mar-mean-TDR ($c_{17}$), Annual-Mean-Runoff ($c_{18}$), Root-surface-area ($c_{19}$), Leaf-P ($c_{20}$). The pasture productions are classified into three groups: Low ($N$), Median($B$), High($P$). The first criterion is enumerated, and the thirteenth is with no information for all 36 paddocks being with 0. Therefore, we remove the first and the thirteenth criteria.

As discussed in Ref. [10], the model validation of decision making is usually implemented by interacting with the decision makers. Doumpos et al. also indicated that the validation mechanism of multi-criteria group decision making could be referred to that of statistical learning. In this study, we assign three decision makers with different samples and suppose they can make the exact sorting decisions for the alternatives in their individual sample sets. Then an existing method is used to make the final decisions. Finally, we compare the decision result with the actual decision to show the validation of the model.

We start with a multi-criteria information system $S = (A, C, f)$, where $A = \{a_1, a_2, ..., a_{36}\}$, $C = \{c_1, c_2, ..., c_{20}\}$ and $f$ is shown as in the pasture data set. Before the decision process, we first normalized the measurements under criteria by the approach given in Ref. [33]. Three decision makers $e_1, e_2, e_3$ take part in the decision process. Suppose $S^k = \{A^k, C, d, f^k\}$ is provided by $e_k$ ($k = 1, 2, 3$). Assume $e_k$ chooses $A^k = \{a_l | l \mod 6 = k+1\}$ and makes the accurate decisions with $f^k(a_i, d)$ being equal to the real label in the pasture data set. In what follows, we use $a_1$ under the perspective of $e_1$ to detail the calculation process.

Step 1: Compute the fuzzy preference relation
$(r^{\succ c}(a_1, a_1), r^{\succ c}(a_1, a_7), ..., r^{\succ c}(a_1, a_{34})) = (0.500, 0.003, ..., 0.002)$ and
$(r^{\prec c}(a_1, a_1), r^{\prec c}(a_1, a_7), ..., r^{\prec c}(a_1, a_{34})) = (0.500, 0.017, ..., 0.005)$ by Eq. (3) and Eq. (4).

Step 2: By using Eqs. (10), (11), we get

$$\hat{P}_1^{\succ}(a_1) = 0.008, \hat{B}_1^{\succ}(a_1) = 0.500, \hat{N}_1^{\succ}(a_1) = 0.500,$$

$$\hat{P}_1^{\prec}(a_1) = 0.500, \hat{B}_1^{\prec}(a_1) = 0.500, \hat{N}_1^{\prec}(a_1) = 0.1641,$$

$$\check{P}_1^{\succ}(a_1) = 0.003, \check{B}_1^{\succ}(a_1) = 0.0004, \check{N}_1^{\succ}(a_1) = 1,$$

$$\check{P}_1^{\prec}(a_1) = 1, \check{B}_1^{\prec}(a_1) = 0.0046, \check{N}_1^{\prec}(a_1) = 0.0046.$$

Step 3: By using Eqs. (13) and (15), we compute the normalizing coefficients as $I_x^k = 1$ and $O_x^k = 1$.

Step 4: From (12a) and (12b), we get $\hat{m}_1(P) = \hat{P}_1^{\succ}(a_1) \times \hat{P}_1^{\prec}(a_1) = 0.009 \times 0.500 = 0.0004$, $\hat{m}_1(B) = \hat{B}_1^{\succ}(a_1) \times \hat{B}_1^{\prec}(a_1) = 0.25$, $\hat{m}_1(N) = 0.0821$ and $\hat{m}_1(\Theta) = 1 - (\hat{m}_1(p) + \hat{m}_1(B) + \hat{m}_1(N)) = 0.6675$.

Step 5: Similarly, $\check{m}_1(P) = 0.0003$, $\check{m}_1(B) = 0.0000$, $\check{m}_1(N) = 0.0046$ and $\check{m}_1(\Theta) = 0.9951$.

The other two groups of decision evidences are calculated as

$$\hat{m}_2(P) = 0.0001, \hat{m}_2(B) = 0.0004, \hat{m}_2(N) = 0.0054, \hat{m}_2(\Theta) = 0.9941,$$

$\check{m}_2(P) = 0.0001$, $\check{m}_2(B) = 0.0000$, $\check{m}_2(N) = 0.0010$, $\check{m}_2(\Theta) = 0.9989$,

$\hat{m}_3(P) = 0.0002$, $\hat{m}_3(B) = 0.0004$, $\hat{m}_3(N) = 0.0012$, $\hat{m}_3(\Theta) = 0.9982$,

$\check{m}_3(P) = 0.0001$, $\check{m}_3(B) = 0.0000$, $\check{m}_3(N) = 0.0014$, $\check{m}_3(\Theta) = 0.9985$.

Step 6: $\hat{m}^{a_1}(P) = 0.0007$, $\hat{m}^{a_1}(B) = 0.2493$, $\hat{m}^{a_1}(N) = 0.0865$ and $\hat{m}^{a_1}(\Theta) = 0.6636$ are calculated from combination of three decision makers' inner evidences.

Step 7: $\check{m}^{a_1}(P) = 0.0005$, $\check{m}^{a_1}(B) = 0.0000$, $\check{m}^{a_1}(N) = 0.0070$ and $\check{m}^{a_1}(\Theta) = 0.9925$ are calculated from combination of three decision makers' outer evidences.

Step 8: Combine the inner and outer evidences: $m^{a_1}(P) = 0.0095$, $m^{a_1}(B) = 0.2479$, $m^{a_1}(N) = 0.0912$ and $m^{a_1}(\Theta) = 0.6599$. According to (9), the final integrated sorting decision is $B$ by setting the parameters $\varepsilon_1$ and $\varepsilon_2$ as 0.001 and 0.8, respectively.

Other alternatives can be similarly obtained. There are 24 out of 36 paddocks being evaluated exactly. In many group decision making problems, majority voting is a good procedure to aggregate the views of different decision makers' sorting. Here we compare our method with the two sorting methods: "SVM+vote" and "UTIDIS+vote". The decision conditions are the same as the above. The first "SVM+vote" which with the Gaussian kernel function present the 21 correct sorting results out of 36 paddocks, and the second one carry out 12 correct sorting results out of 36 paddocks. Therefore, the decision result got from our method is better than the other two, so the result is more credible.

## 5     Conclusion

In this paper, we propose a novel Dempster-Shafer theory based multi-attribute three-way group sorting method for obtaining more credible sorting results. The new method has two characteristics different from the methods in literature: (1) only one parameter, fuzzy factor, is needed in the method; (2) from each decision maker's sample set, the soft decision, i.e., basic belief assignment of an alternative belonging to three-way decision classes is constructed along with the consideration of the uncertainty degree. The superiority of the proposed method is also shown in an illustrative example.

**Acknowledgments.** This work was supported by the National Natural Science Foundation of China (Nos. 71031006, 61273294), the Shanxi Provincial Foundation for Returned Scholars (No. 2013-101) and the Construction Project of the Science and Technology Basic Condition Platform of Shanxi Province (No. 2012091002-0101).

## References

1. Boujelben, M.A., De Smet, Y., Frikha, A., Chabchoub, H.: A ranking model in uncertain, imprecise and multi-experts contexts: The application of evidence theory. International Journal of Approximate Reasoning 52(8), 1171–1194 (2011)

2. Brans, J.P., Mareschal, B.: Promethee-v-mcdm problems with segmentation constraints. INFOR 30(2), 85–96 (1992)
3. Bregar, A., Györkös, J., Jurič, M.B.: Interactive aggregation/disaggregation dichotomic sorting procedure for group decision analysis based on the threshold model. Informatica 19(2), 161–190 (2008)
4. Cai, F., Liao, X., Wang, K.: An interactive sorting approach based on the assignment examples of multiple decision makers with different priorities. Annal of Operation Research 197(1), 87–108 (2012)
5. Chen, Y., Hipel, K.W., Kilgour, D.M.: Multiple-criteria sorting using case-based distance models with an application in water resources management. IEEE Transactions on Systems, Man and Cybernetics-Part A 37(5), 680–691 (2007)
6. Chen, Y., Kilgour, D.M., Hipel, K.W.: Using a benchmark in case-based multi-criteria ranking. IEEE Transactions on Systems, Man and Cybernetics-Part A 39(2), 358–368 (2009)
7. Damart, S., Dias, L.C., Mousseau, V.: Supporting groups in sorting decisions: methodology and use of a multi-criteria aggregation/disaggregation dss. Decision Support Systems 43(4), 1464–1475 (2007)
8. Dempster, A.P.: Upper and lower probabilities induced by a multivalued mapping. The Annals of Mathematical Statistics, 325–339 (1967)
9. Doumpos, M., Zopounidis, C.: Multicriteria decision aid classification methods. Kluwer Academic Publishers (2002)
10. Doumpos, M., Zopounidis, C.: Preference disaggregation and statistical learning for multicriteria decision support: A review. European Journal of Operational Research 209(3), 203–214 (2011)
11. Figueira, J., Mousseau, V., Roy, B., et al.: Multiple Criteria Decision Analysis: State of the Art Surveys. Springer (2005)
12. Greco, S., Inuiguchi, M., Słowiński, R.: Dominance-based rough set approach using possibility and necessity measures. In: Alpigini, J.J., Peters, J.F., Skowron, A., Zhong, N. (eds.) RSCTC 2002. LNCS (LNAI), vol. 2475, pp. 85–92. Springer, Heidelberg (2002)
13. Greco, S., Matarazzo, B., Slowinski, R.: Rough approximation of a preference relation by dominance relations. European Journal of Operational Research 117(1), 63–83 (1999)
14. Hu, Q., Yu, D., Guo, M.: Fuzzy preference based rough sets. Information Sciences 180(10), 2003–2022 (2010)
15. Hu, Y.C., Chen, C.J.: A promethee-based classification method using concordance and discordance relations and its application to bankruptcy prediction. Informaiton Sciences 181(22), 4959–4968 (2011)
16. Andrews, C.J.: Sorting out a consensus-analysis in support of multiparty decisions. Environment and Planning B-Planing & Design 19(2), 189–204 (1992)
17. Jabeur, K., Martel, J.M.: An ordinal sorting method for group decision-making. European Journal of Operational Research 180(3), 1272–1289 (2007)
18. Jacquet-Lagrèze, E., Siskos, Y.: Assessing a set of additive utility functions for multicriteria decision-making, the uta method. European Journal of Operational Research 10(2), 151–164 (1982)
19. Janssen, P., Nemery, P.: An extension of the flowsort sorting method to deal with imprecision. 4OR- A Quanter Journal Operations Research 11(2), 171–193 (2013)
20. Kadziński, M., Greco, S., Słowiński, R.: Selection of a representative value function for robust ordinal regression in group decision making. Group Decision Negotiation 22(3), 429–462 (2013)

21. Liu, D., Yao, Y., Li, T.: Three-way investment decisions with decision-theoretic rough sets. International Journal of Computational Intelligence Systems 4(1), 66–74 (2011)

22. Ma, J., Fan, Z., Huang, L.: A subjective and objective integrated approach to determine attribute weights. European Journal of Operational Research 112(2), 397–404 (1999)

23. Ma, J., Fan, Z., Wei, Q.: Existence and construction of weight-set for satisfying preference orders of alternatives based on additive multi-attribute value model. IEEE Transactions on Systems, Man and Cybernetics, Part A: Systems and Humans 31(1), 66–72 (2001)

24. Madani, K., Read, L., Shalikarian, L.: Voting under uncertainty: A stochastic framework for analyzing group decision maing problems. Water Resources Management 28(7), 1839–1856 (2014)

25. Pang, J., Liang, J.: Evaluation of the results of multi-attribute group decision-making with linguistic information. Omega 40(3), 294–301 (2012)

26. Qian, Y., Liang, J., Pedrycz, W., Dang, C.: Positive approximation: An accelerator for attribute reduction in rough set theory. Artificial Intelligence 174(9), 597–618 (2010)

27. Roy, B.: Classement et choix en présence de point de vue multiples: Le méthode electre. Revue Francaise d'Informatique et de Recherche Opérationnelle 8(1), 57–75 (1968)

28. Roy, B.: Méthodologie Multicritère d' Aide à la Décision. Economica, Paris (1985)

29. Satty, T.L.: The analytic hierarchy process. McGrawHill, New York (1980)

30. Shafer, G.: A mathematical theory of evidence, vol. 1. Princeton University Press, Princeton (1976)

31. Song, P., Liang, J., Qian, Y.: A two-grade approach to ranking interval data. Knowledge-Based Systems 27, 234–244 (2012)

32. Wu, W., Leung, Y.: Optimal scale selection for multi-scale decision tables. International Journal of Approximate Reasoning 54(8), 1107–1129 (2013)

33. Xu, Z.: Uncertain multiple attribute decision making: methods and applications (2004)

34. Yao, Y.: Three-way decisions with probabilistic rough sets. Information Sciences 180(3), 341–353 (2010)

35. Yao, Y.: The superiority of three-way decisions in probabilistic rough set models. Information Sciences 181(6), 1080–1096 (2011)

36. Yao, Y.: An outline of a theory of three-way decisions. In: Yao, J., Yang, Y., Słowiński, R., Greco, S., Li, H., Mitra, S., Polkowski, L. (eds.) RSCTC 2012. LNCS (LNAI), vol. 7413, pp. 1–17. Springer, Heidelberg (2012)

# Dynamic Maintenance of Three-Way Decision Rules

Chuan Luo[1,2], Tianrui Li[1], and Hongmei Chen[1]

[1] School of Information Science and Technology, Southwest Jiaotong University,
Chengdu, 610031, China
luochuan@my.swjtu.edu.cn, {trli,hmchen}@swjtu.edu.cn
[2] Department of Computer Science, University of Regina, Regina, Saskatchewan,
S4S 0A2, Canada
luo256@cs.uregina.ca

**Abstract.** Decision-theoretic rough sets provide a three-way decision framework for approximating a target concept, with an error-tolerance capability to handle uncertainty problems by using a pair of thresholds on probability. The three-way decision rules of acceptance, rejection and deferment decisions can be derived directly from the three regions implied by rough set approximations. The decision environment is prone to dynamic instead of static in reality. With the data changed continuously, the three regions of a target decision will be changed inevitably, while the induced three-way decision rules will be changed avoidably. In this paper, we discuss the dynamic maintenance principles of three-way decision rules based on the variation of three regions with an incremental object. Decision rules can be updated incrementally without re-computing rule sets from the very beginning when a new object is added up to an information system.

**Keywords:** three-way decisions, decision-theoretic rough sets, incremental object.

## 1 Introduction

Rough sets theory (RST) is a relatively new mathematical tool to represent and reason imprecision and uncertain information emphasized in decision making [12]. For a target concept, RST provides a method for approximating concept according to three pair-wise disjoint regions, namely, the positive, boundary and negative regions. The three-way decision rules of acceptance, rejection and deferment decisions are derived from the above three regions [16]. In the Pawlak RST, the decisions of acceptance and rejection are made without any error, *i.e.*, the classification must be fully correct or certain which is too restrictive in practical applications. In order to resolve this situation, probabilistic RST with probabilistic information, which is a generalization of Pawlak RST, was proposed by allowing acceptable tolerance of errors [17]. By considering the costs of different types of classification decisions, Yao et al. proposed a more

D. Miao et al. (Eds.): RSKT 2014, LNAI 8818, pp. 801–811, 2014.
DOI: 10.1007/978-3-319-11740-9_73 © Springer International Publishing Switzerland 2014

general probabilistic RST, called decision theoretic rough sets (DTRS), which can derive various existing probabilistic RST models through setting different pairs of thresholds $\alpha$ and $\beta$, such as 0.5 probabilistic RST, variable precision RST and Bayesian RST [18]. Nowadays, DTRS has been applied in a wide variety of applications, such as email filtering [21], investment decision [9] and cluster analysis [20].

In real-world applications, data in an information system does not usually remain a stable condition [13]. However, most of RST-based data analysis methods in a batch mode cannot be effectively applied when data are collected sequentially. Recent years have witnessed great success of incremental learning techniques in improving the data analysis algorithms based on RST. Current studies on incremental RST-based data analysis can be classified into three categories: rough approximation, attribute reduction, and rule induction.

(1) Rough approximation

Li et al. proposed an incremental method for updating rough approximations in incomplete information system under the characteristic relation [7]. Chen et al. presented dynamic maintenance approach for computing approximations with the refining and coarsening of attribute values [1]. Luo et al. proposed incremental approaches for updating approximations in the set-valued information systems [5,6].

(2) Attribute reduction

Wang et al. developed a dynamic attribute reduction approach based on information entropy with dynamically-increasing attributes [15]. Liang et al. presented a group incremental attribute reduction algorithm with the addition of multiple objects [11]. Shu et al. proposed an incremental positive region-based attribute reduction algorithm with the variation of attributes [14].

(3) Rule induction

Fan et al. developed an incremental rule-extraction algorithm to efficiently handle added-in data [2]. Guo et al. proposed an incremental method for extracting rules based on the discernibility matrix and search tree [3]. Huang et al. proposed dynamic alternative rule induction algorithms with incremental objects [4].

Recently, three-way decisions have been paid close attentions. Many attempts have been made to apply the three-way decisions in a dynamical environment. Yao presented sequential three-way decisions from granular computing perspective, and demonstrated its superior in the context of multiple levels of information granularity [19]. Li et al. proposed a cost-sensitive sequential three-way decision model by considering the costs of decision result and decision process in sequential decision making [10]. Liu et al. investigated a dynamic DTRS model to deal with the dynamic change of loss functions [8]. To deal with the dynamic data set, in this paper, we propose a dynamic maintenance framework for three-way decisions, capable of dealing with an incremental object each time.

The remainder of the paper is organized as follows. In Section 2, some basic concepts of the three-way decisions based on DTRS are briefly reviewed. Principles of dynamic maintenance of three-way decision rules with an incremental object are analyzed in Section 3. The paper ends with conclusions and further research topics in Section 4.

## 2  Three-Way Decisions with DTRS

Basic concepts, notions and results of three-way decisions based on DTRS are outlined in this section [16,18].

**Definition 1.** *An information system is defined by a tuple:*

$$S = (U, AT, \{V_a | a \in AT\}, \{I_a : U \to V_a | a \in AT\}), \tag{1}$$

*where $U$ is a finite nonempty set of objects called the universe, $AT$ is a finite nonempty set of attributes, $V_a$ is the domain of an attribute $a$ and $I_a$ is a description function that assign a value from $V_a$ to each object.*

**Definition 2.** *Given a subset of attributes $A \subseteq AT$, let $ind(A)$ denote an equivalence relation on a nonempty and finite set of objects $U$, which can be defined as follows:*

$$x \ ind(A) \ y \iff \forall a \in A \ [I_a(x) = I_a(y)], \tag{2}$$

*where $I_a(x)$ denotes the value of an object $x \in U$ on an attribute $a \in A$.*

The equivalence class containing $x$ is given by $[x] = \{y \in U | x \ ind(A) \ y\}$. The partition $U/ind(A)$ induced by the equivalence relation $ind(A)$ contains all equivalence classes which are the building blocks to construct rough set approximations.

**Definition 3.** *Given a subset of universe $C \subseteq U$, $\forall x \in U$, the conditional probability of $x$ belonging to $C$ can be simply estimated as follows:*

$$Pr(C|[x]) = |C \cap [x]|/|[x]|, \tag{3}$$

*where $| \bullet |$ denotes the cardinality of a set.*

In terms of conditional probability, a pair of thresholds $(\alpha, \beta)$ should be chosen to introduce three probabilistic regions with acceptable tolerance of errors. The required thresholds can be derived from DTRS based on Bayesian decision procedure [18].

**Definition 4.** *Given a pair of thresholds $\alpha$ and $\beta$ with $\alpha > \beta$, the $(\alpha, \beta)$-probabilistic positive, boundary and negative regions are defined as follows:*

$$\begin{aligned}
POS_{(\alpha, \bullet)}(C) &= \{x \in U | Pr(C|[x]) \geq \alpha\}, \\
BND_{(\alpha, \beta)}(C) &= \{x \in U | \beta < Pr(C|[x]) < \alpha\}, \\
NEG_{(\bullet, \beta)}(C) &= \{x \in U | Pr(C|[x]) \leq \beta\}.
\end{aligned} \tag{4}$$

*The $(\alpha, \beta)$-probabilistic lower and upper approximations are defined by:*

$$\underline{apr}_{(\alpha,\bullet)}(C) = \{x \in U | Pr(C|[x]) \geq \alpha\}, \tag{5}$$

$$\overline{apr}_{(\bullet,\beta)}(C) = \{x \in U | Pr(C|[x]) > \beta\}, \tag{6}$$

*where $\alpha$ and $\beta$ can be systematically computed from a risk or loss function based on the Bayesian decision procedure.*

According to the three probabilistic regions, one can make three-way decisions based on the following positive, boundary and negative rules:

$$\Re_P(C): \quad Des([x]) \to Des(C), \, for \, [x] \subseteq POS_{(\alpha,\bullet)}(C),$$
$$\Re_B(C): \quad Des([x]) \to Des(C), \, for \, [x] \subseteq BND_{(\alpha,\beta)}(C),$$
$$\Re_N(C): \quad Des([x]) \to Des(C), \, for \, [x] \subseteq NEG_{(\bullet,\beta)}(C),$$

where $Des([x])$ denotes the logic formula defining the equivalence class $[x]$, which is typically a conjunction of attribute-value pairs in an information system, and $Des(C)$ is the name of the concept. For each decision rules, the conditional probability $Pr(C|[x])$ can be associated as a probabilistic measure called the accuracy or confidence of the rule.

## 3 Dynamic Three-Way Decisions Based on Incremental Object

Given a dynamic information system with incremental object, principles for maintaining and updating the three-way decision rules based on DTRS are proposed in this section. To describe a dynamic maintenance process from time $t$ to time $t + 1$, we denote the notions at time $t$ with the same superscript $(t)$, and those at time $t + 1$ with the same superscript $(t + 1)$. Specially, the conditional probabilities are denote as $Pr(t)$ and $Pr(t+1)$ at time $t$ and $t+1$ for convenience.

With the additional of an object $\mathbf{x}$ to a given information system, the equivalence class and target concept will be updated as follows:

$$[x]^{(t+1)} = \begin{cases} [x]^{(t)} \cup \{\mathbf{x}\}, & \mathbf{x} \, blongs \, to \, the \, equivalence \, class \, [x]; \\ [x]^{(t)}, & otherwise. \end{cases}$$

$$C^{(t+1)} = \begin{cases} C^{(t)} \cup \{\mathbf{x}\}, & \mathbf{x} \, blongs \, to \, the \, target \, concept \, C; \\ C^{(t)}, & otherwise. \end{cases}$$

From the definition of conditional probability in Definition 3, $Pr(C|[x])$ may non-monotonically change due to the variation of intersection of $[x]$ and $C$ which is shown in Table 1.

Based on the above four updating patterns of the conditional probability, in the following, we focus on the implementation of the dynamic maintenance of

**Table 1.** Updating patterns of the conditional probability

| | Patterns | $[x]$ | $C$ | $Pr(C\|[x])$ |
|---|---|---|---|---|
| 1. | $\mathbf{x} \in [x] \wedge \mathbf{x} \in C$ | $[x] \cup \{\mathbf{x}\}$ | $C \cup \{\mathbf{x}\}$ | $\frac{\|C \cap [x]\| + 1}{\|[x]\| + 1}$ |
| 2. | $\mathbf{x} \notin [x] \wedge \mathbf{x} \in C$ | $[x]$ | $C \cup \{\mathbf{x}\}$ | $\frac{\|C \cap [x]\|}{\|[x]\|}$ |
| 3. | $\mathbf{x} \in [x] \wedge \mathbf{x} \notin C$ | $[x] \cup \{\mathbf{x}\}$ | $C$ | $\frac{\|C \cap [x]\|}{\|[x]\| + 1}$ |
| 4. | $\mathbf{x} \notin [x] \wedge \mathbf{x} \notin C$ | $[x]$ | $C$ | $\frac{\|C \cap [x]\|}{\|[x]\|}$ |

three-way decision rules with an incremental object.

**Case 1: $\mathbf{x} \in [x]$ and $\mathbf{x} \in C$;**
In this case, for the conditional probability, we have:

$$Pr(t+1) \geq Pr(t)$$

(1) If $[x] \subseteq POS_{(\alpha,\bullet)}^{(t)}(C)$, i.e., $Des([x]) \to Des(C) \in \Re_P(C)$, then

$$POS_{(\alpha,\bullet)}^{(t+1)}(C) = POS_{(\alpha,\bullet)}^{(t)}(C) \cup \{\mathbf{x}\}.$$

Furthermore, since $Des(\{\mathbf{x}\}) = Des([x])$, the three-way decision rules for the target concept $C$: $\Re_P(C)$, $\Re_B(C)$ and $\Re_N(C)$ will remain constant.

(2) If $[x] \subseteq BND_{(\alpha,\bullet)}^{(t)}(C)$, i.e., $Des([x]) \to Des(C) \in \Re_B(C)$, we have:
(a) if $Pr(t+1) \geq \alpha$, then

$$POS_{(\alpha,\bullet)}^{(t+1)}(C) = POS_{(\alpha,\bullet)}^{(t)}(C) \cup ([x] \cup \{\mathbf{x}\}),$$
$$BND_{(\alpha,\bullet)}^{(t+1)}(C) = BND_{(\alpha,\bullet)}^{(t)}(C) - [x].$$

Furthermore, $Des([x])$ will be included into the conditions of positive rule $\Re_P(C)$, and will be excluded from the conditions of boundary rule $\Re_B(C)$. The negative rule $\Re_N(C)$ will remain constant.
(b) if $Pr(t+1) < \alpha$, then

$$BND_{(\alpha,\bullet)}^{(t+1)}(C) = BND_{(\alpha,\bullet)}^{(t)}(C) \cup \{\mathbf{x}\}.$$

The three-way decision rules for the target concept $C$: $\Re_P(C)$, $\Re_B(C)$ and $\Re_N(C)$ will remain constant.

(3) If $[x] \subseteq NEG_{(\alpha,\bullet)}^{(t)}(C)$, i.e., $Des([x]) \to Des(C) \in \Re_N(C)$, we have:
(a) if $Pr(t+1) \geq \alpha$, then

$$POS_{(\alpha,\bullet)}^{(t+1)}(C) = POS_{(\alpha,\bullet)}^{(t)}(C) \cup ([x] \cup \{\mathbf{x}\}),$$
$$NEG_{(\alpha,\bullet)}^{(t+1)}(C) = NEG_{(\alpha,\bullet)}^{(t)}(C) - [x].$$

Furthermore, $Des([x])$ will be included into the conditions of positive rule $\Re_P(C)$, and will be excluded from the conditions of negative rule $\Re_N(C)$. The boundary rule $\Re_B(C)$ will remain constant.

(b) if $\beta < Pr(t+1) < \alpha$, then

$$BND_{(\alpha,\bullet)}^{(t+1)}(C) = BND_{(\alpha,\bullet)}^{(t)}(C) \cup ([x] \cup \{\mathbf{x}\}),$$

$$NEG_{(\alpha,\bullet)}^{(t+1)}(C) = NEG_{(\alpha,\bullet)}^{(t)}(C) - [x].$$

Furthermore, $Des([x])$ will be included into the conditions of boundary rule $\Re_B(C)$, and will be excluded from the conditions of negative rule $\Re_N(C)$. The positive rule $\Re_P(C)$ will remain constant.

(c) if $Pr(t+1) \leq \beta$, then

$$NEG_{(\alpha,\bullet)}^{(t+1)}(C) = NEG_{(\alpha,\bullet)}^{(t)}(C) \cup \{\mathbf{x}\}.$$

Furthermore, the three-way decision rules for the target concept $C$: $\Re_P(C)$, $\Re_B(C)$ and $\Re_N(C)$ will remain constant.

**Case 2:** $\mathbf{x} \notin [x]$ and $\mathbf{x} \in C$;

In this case, for the conditional probability, we have:

$$Pr(t+1) = Pr(t)$$

Therefore, the three-way decision rules for the target concept $C$: $\Re_P(C)$, $\Re_B(C)$ and $\Re_N(C)$ will always remain constant.

**Case 3:** $\mathbf{x} \in [x]$ and $\mathbf{x} \notin C$;

In this case, for the conditional probability, we have:

$$Pr(t+1) < Pr(t)$$

(1) If $[x] \subseteq POS_{(\alpha,\bullet)}^{(t)}(C)$, i.e., $Des([x]) \to Des(C) \in \Re_P(C)$, we have:

(a) if $Pr(t+1) \geq \alpha$, then

$$POS_{(\alpha,\bullet)}^{(t+1)}(C) = POS_{(\alpha,\bullet)}^{(t)}(C) \cup \{\mathbf{x}\}.$$

Since $Des(\{\mathbf{x}\}) = Des([x])$, the three-way decision rules: $\Re_P(C)$, $\Re_B(C)$ and $\Re_N(C)$ will always remain constant.

(b) if $\beta < Pr(t+1) < \alpha$, then

$$BND_{(\alpha,\bullet)}^{(t+1)}(C) = BND_{(\alpha,\bullet)}^{(t)}(C) \cup ([x] \cup \{\mathbf{x}\}),$$

$$POS_{(\alpha,\bullet)}^{(t+1)}(C) = POS_{(\alpha,\bullet)}^{(t)}(C) - [x].$$

Furthermore, $Des([x])$ will be included into the conditions of boundary rule $\Re_B(C)$, and will be excluded from the conditions of positive rule $\Re_P(C)$. The negative rule $\Re_N(C)$ will remain constant.

(c) if $Pr(t+1) \leq \beta$, then

$$NEG_{(\alpha,\bullet)}^{(t+1)}(C) = NEG_{(\alpha,\bullet)}^{(t)}(C) \cup ([x] \cup \{\mathbf{x}\}),$$

$$POS_{(\alpha,\bullet)}^{(t+1)}(C) = POS_{(\alpha,\bullet)}^{(t)}(C) - [x].$$

Furthermore, $Des([x])$ will be included into the conditions of negative rule $\Re_N(C)$, and will be excluded from the conditions of positive rule $\Re_P(C)$. The boundary rule $\Re_B(C)$ will remain constant.

(2) If $[x] \subseteq BND^{(t)}_{(\alpha,\bullet)}(C)$, i.e., $Des([x]) \to Des(C) \in \Re_B(C)$, we have:
   (a) if $Pr(t+1) > \beta$, then

$$BND^{(t+1)}_{(\alpha,\bullet)}(C) = BND^{(t)}_{(\alpha,\bullet)}(C) \cup \{x\}.$$

The three-way decision rules for the target concept $C$: $\Re_P(C)$, $\Re_B(C)$ and $\Re_N(C)$ will always remain constant.
   (b) if $Pr(t+1) \leq \beta$, then

$$NEG^{(t+1)}_{(\alpha,\bullet)}(C) = NEG^{(t)}_{(\alpha,\bullet)}(C) \cup ([x] \cup \{\mathbf{x}\}),$$
$$BND^{(t+1)}_{(\alpha,\bullet)}(C) = BND^{(t)}_{(\alpha,\bullet)}(C) - [x].$$

Furthermore, $Des([x])$ will be included into the conditions of negative rule $\Re_N(C)$, and will be excluded from the conditions of boundary rule $\Re_B(C)$. The positive rule $\Re_P(C)$ will remain constant.
(3) If $[x] \subseteq NEG^{(t)}_{(\alpha,\bullet)}(C)$, i.e., $Des([x]) \to Des(C) \in \Re_N(C)$, we have:

$$NEG^{(t+1)}_{(\alpha,\bullet)}(C) = NEG^{(t)}_{(\alpha,\bullet)}(C) \cup \{x\}.$$

Furthermore, the three-way decision rules for the target concept $C$: $\Re_P(C)$, $\Re_B(C)$ and $\Re_N(C)$ will remain constant.

**Case 4: $\mathbf{x} \notin [x]$ and $\mathbf{x} \notin C$;**
   In this case, for the conditional probability, we have:

$$Pr(t+1) = Pr(t)$$

Therefore, the three-way decision rules for the target concept $C$: $\Re_P(C)$, $\Re_B(C)$ and $\Re_N(C)$ will always remain constant.

## 4   An Illustrative Case Study

A given information system $S = (U, AT, \{V_a | a \in AT\}, \{I_a : U \to V_a | a \in AT\})$ shown in Table 2 is used for exemplifying the dynamic maintenance of three-way decision rules with an incremental object.

The equivalence relation $ind(AT)$ partitions the universe into six equivalence classes: $[x_1] = \{x_1\}$, $[x_2] = \{x_2, x_6, x_8\}$, $[x_3] = \{x_3\}$, $[x_4] = \{x_4\}$, $[x_5] = \{x_5, x_9\}$, $[x_7] = \{x_7\}$. Furthermore, the decision attribute $d$ partitions the universe into three decision classes: $M = \{x_1, x_2\}$, $Q = \{x_3, x_4, x_5\}$, $F = \{x_6, x_7, x_8, x_9\}$. Suppose the two thresholds $\alpha = 0.75$ and $\beta = 0.60$ are calculated from the loss functions for the three states regarding $M$, $Q$ and $F$. The three regions of the decision $F$ can be obtained as follows:

$$POS^{(t)}_{(\alpha,\bullet)}(F) = \{x_7\};$$
$$BND^{(t)}_{(\alpha,\beta)}(F) = \{x_2, x_6, x_8\};$$

**Table 2.** The original information system of case study

| $U$ | $a_1$ | $a_2$ | $a_3$ | $a_4$ | $a_5$ | $a_6$ | $d$ |
|-----|-------|-------|-------|-------|-------|-------|-----|
| $x_1$ | 1 | 1 | 1 | 1 | 1 | 1 | $M$ |
| $x_2$ | 1 | 0 | 1 | 0 | 1 | 1 | $M$ |
| $x_3$ | 0 | 1 | 1 | 1 | 0 | 0 | $Q$ |
| $x_4$ | 1 | 1 | 1 | 0 | 0 | 1 | $Q$ |
| $x_5$ | 0 | 0 | 1 | 1 | 0 | 1 | $Q$ |
| $x_6$ | 1 | 0 | 1 | 0 | 1 | 1 | $F$ |
| $x_7$ | 0 | 0 | 0 | 1 | 1 | 0 | $F$ |
| $x_8$ | 1 | 0 | 1 | 0 | 1 | 1 | $F$ |
| $x_9$ | 0 | 0 | 1 | 1 | 0 | 1 | $F$ |

$$NEG^{(t)}_{(\bullet,\beta)}(F) = \{x_1, x_3, x_4, x_5, x_9\}.$$

The three-way decision rules for the decision $F$ can be obtained as follows:

$$\Re_P(F): Des([x_7]) \rightarrow Des(F);$$
$$\Re_B(F): Des([x_2]) \rightarrow Des(F);$$
$$\Re_N(F): Des([x_1] \cup [x_3] \cup [x_4] \cup [x_5]) \rightarrow Des(F).$$

In the following, to illustrate the dynamic maintenance process of three-way decision rules, we assume that object $x_{10}$, $x_{11}$, and $x_{12}$ in Table 3 are inserted into Table 2, respectively.

**Table 3.** The incremental objects of case study

| $U$ | $a_1$ | $a_2$ | $a_3$ | $a_4$ | $a_5$ | $a_6$ | $d$ |
|-----|-------|-------|-------|-------|-------|-------|-----|
| $x_{10}$ | 1 | 0 | 1 | 0 | 1 | 1 | $F$ |
| $x_{11}$ | 1 | 0 | 1 | 0 | 1 | 1 | $M$ |
| $x_{12}$ | 0 | 0 | 0 | 1 | 1 | 0 | $F$ |

(1) The insertion of object $x_{10}$.

For the equivalence classes induced by $U/ind(AT)$, we have $x_{10} \in [x_2] \wedge x_{10} \in F$. Then, according to **Case 1**, since $[x_2] \subseteq BND_{(\alpha,\beta)}(F)$ and $Pr(t+1) = 0.75 \geq \alpha$, we have

$$POS^{(t+1)}_{(\alpha,\bullet)}(F) = POS^{(t)}_{(\alpha,\bullet)}(F) \cup [x_2] \cup \{x_{10}\} = \{x_2, x_6, x_7, x_8, x_{10}\};$$
$$BND^{(t+1)}_{(\alpha,\bullet)}(F) = BND^{(t)}_{(\alpha,\bullet)}(F) - [x_2] = \emptyset.$$

For the equivalence classes $[x_i]$, $i \neq 2$, we have $x_{10} \notin [x_i] \wedge x_{10} \in F$, according to **Case 2**, $Des([x_i])$ in the three-way decision rules remains constant.

Finally, the three-way decision rules for the decision $F$ are updated as follows:

$\Re_P(F)$: $Des([x_2] \cup [x_7]) \rightarrow Des(F)$;
$\Re_B(F)$: $\emptyset$;
$\Re_N(F)$: $Des([x_1] \cup [x_3] \cup [x_4] \cup [x_5]) \rightarrow Des(F)$.

(2) The insertion of object $x_{11}$.

For the equivalence classes induced by $U/ind(AT)$, we have $x_{11} \in [x_2] \wedge x_{11} \notin F$. Then, according to **Case 3**, since $[x_2] \subseteq BND_{(\alpha,\beta)}(F)$ and $Pr(t+1) = 0.50 < \beta$, we have

$$NEG_{(\alpha,\bullet)}^{(t+1)}(C) = NEG_{(\alpha,\bullet)}^{(t)}(C) \cup [x_2] \cup \{x_{10}\} = \{x_1, x_2, x_3, x_4, x_5, x_6, x_8, x_9\};$$
$$BND_{(\alpha,\bullet)}^{(t+1)}(C) = BND_{(\alpha,\bullet)}^{(t)}(C) - [x_2] = \emptyset;$$

For the equivalence classes $[x_i]$, $i \neq 2$, we have $x_{10} \notin [x_i] \wedge x_{10} \notin F$, according to **Case 4**, $Des([x_i])$ in the three-way decision rules remains constant.

Finally, the three-way decision rules for the decision $F$ are updated as follows:

$\Re_P(F)$: $Des([x_7]) \rightarrow Des(F)$;
$\Re_B(F)$: $\emptyset$;
$\Re_N(F)$: $Des([x_1] \cup [x_2] \cup [x_3] \cup [x_4] \cup [x_5]) \rightarrow Des(F)$.

(3) The insertion of object $x_{12}$.

For the equivalence classes induced by $U/ind(AT)$, we have $x_{12} \in [x_7] \wedge x_{12} \in F$. Then, according to **Case 1**, since $[x_7] \subseteq POS_{(\alpha,\beta)}(F)$, we have

$$POS_{(\alpha,\bullet)}^{(t+1)}(F) = POS_{(\alpha,\bullet)}^{(t)}(F) \cup \{x_{12}\} = \{x_7, x_{12}\};$$

For the equivalence classes $[x_i]$, $i \neq 2$, we have $x_{12} \notin [x_i] \wedge x_{12} \in F$, according to **Case 2**, $Des([x_i])$ in the three-way decision rules remains constant.

Finally, since $Des(\{x_{12}\}) = Des([x_7])$, the three-way decision rules for the decision $F$ remain constant.

# 5    Conclusion

The three-way decisions implied by rough set approximations has attracted increasing attentions. In this paper, we have focused on the dynamic maintenance of three-way decision rules with an incremental object. When a new object is added up, the conditional probability may non-monotonically change, which represents four different updating patterns. Based on the change of conditional probability, four different maintenance strategies of the three-way decision rules based on the three regions were implemented, respectively. The proposed approach can update decision rules by modifying partial original rule sets without forgetting prior knowledge. An illustrative case study was presented to demonstrate the feasibility of the proposed approach. In our future study, the experimental analysis will be carried out to verify the effectiveness of the proposed approach in real applications.

**Acknowledgements.** This work is supported by the National Science Foundation of China (Nos. 61175047, 71201133, 61100117 and 61262058) and NSAF (No. U1230117), the Youth Social Science Foundation of the Chinese Education Commission (Nos. 10YJCZH117, 11YJC630127), the Fundamental Research Funds for the Central Universities (Nos. SWJTU11ZT08, SWJTU12CX091, SWJTU12CX117), the Doctoral Innovation Foundation of Southwest Jiaotong University (No. 2014LC), and the National Scholarship for Building High Level Universities, China Scholarship Council (No. 201307000040). The authors would like to thank Professor Yiyu Yao for useful comments on this study.

# References

1. Chen, H.M., Li, T.R., Ruan, D.: Maintenance of approximations in incomplete ordered decision systems while attribute values coarsening or refining. Knowledge-Based Systems 31, 140–161 (2012)
2. Fan, Y.N., Tseng, T.L., Chern, C.C., Huang, C.C.: Rule induction based on an incremental rough set. Expert Systems with Applications 36(9), 11439–11450 (2009)
3. Guo, S., Wang, Z.Y., Wu, Z.C., Yan, H.P.: A novel dynamic incremental rules extraction algorithm based on rough set theory. In: Proceedings of 4th International Conference of Machine Learning and Cybernatics, pp. 1902–1907 (2005)
4. Huang, C.C., Tseng, T.L., Fan, Y.N., Hsu, C.H.: Alternative rule induction methods based on incremental objet using rough set theory. Applied Soft Computing 13(1), 372–389 (2013)
5. Luo, C., Li, T.R., Chen, H.M.: Dynamic maintenance of approximations in set-valued ordered decision systems under the attribute generalization. Information Sciences 257, 210–228 (2014)
6. Luo, C., Li, T.R., Chen, H.M., Liu, D.: Incremental approaches for updating approximations in set-valued ordered information systems. Knowledge-Based Systems 50, 218–233 (2013)
7. Li, T.R., Ruan, D., Geert, W., Song, J., Xu, Y.: A rough sets based characteristic relation approach for dynamic attribute generalization in data ming. Knowledge-Based Systems 20, 485–494 (2007)
8. Liu, D., Li, T., Liang, D.: Three-way decisions in dynamic decision-theoretic rough sets. In: Lingras, P., Wolski, M., Cornelis, C., Mitra, S., Wasilewski, P. (eds.) RSKT 2013. LNCS, vol. 8171, pp. 291–301. Springer, Heidelberg (2013)
9. Liu, D., Yao, Y.Y., Li, T.R.: Three-way investment decisions with Decision-theoretic rough sets. International Journal of Computational Intelligence Systems 4, 66–74 (2011)
10. Li, H., Zhou, X., Huang, B., Liu, D.: Cost-sensitive three-way decision: A sequential strategy. In: Lingras, P., Wolski, M., Cornelis, C., Mitra, S., Wasilewski, P. (eds.) RSKT 2013. LNCS, vol. 8171, pp. 325–337. Springer, Heidelberg (2013)
11. Liang, J.Y., Wang, F., Dang, C.Y., Qian, Y.H.: A group incremental approach to feature selection applying rough set technique. IEEE Transactions on Knowledge and Data Engineering 26(2), 194–308 (2014)

12. Pawlak, Z.: Rough Sets. International Journal of Computer and Information Sciences 11, 341–356 (1982)
13. Raghavan, V., Hafez, A.: Dynamic data mining. In: Logananthara, R., Palm, G., Ali, M. (eds.) IEA/AIE 2000. LNCS (LNAI), vol. 1821, pp. 220–229. Springer, Heidelberg (2000)
14. Shu, W.H., Shen, H.: Updating attribute reduction in incomplete decision systems with the variation of attribute set. International Journal of Approximate Reasoning 55, 867–884 (2014)
15. Wang, F., Liang, J.Y., Qian, Y.H.: Attribute reduction: A dimension incremental strategy. Knowledge-Based Systems 39, 95–108 (2013)
16. Yao, Y.: Three-way decision: An interpretation of rules in rough set theory. In: Wen, P., Li, Y., Polkowski, L., Yao, Y., Tsumoto, S., Wang, G. (eds.) RSKT 2009. LNCS (LNAI), vol. 5589, pp. 642–649. Springer, Heidelberg (2009)
17. Yao, Y.Y.: Probabilistic rough set approximations. International Journal of Approximate Reasoning 49(2), 255–271 (2008)
18. Yao, Y.: Decision-theoretic rough set models. In: Yao, J., Lingras, P., Wu, W.-Z., Szczuka, M.S., Cercone, N.J., Ślęzak, D. (eds.) RSKT 2007. LNCS (LNAI), vol. 4481, pp. 1–12. Springer, Heidelberg (2007)
19. Yao, Y.: Granular computing and sequential three-way decisions. In: Lingras, P., Wolski, M., Cornelis, C., Mitra, S., Wasilewski, P. (eds.) RSKT 2013. LNCS (LNAI), vol. 8171, pp. 16–27. Springer, Heidelberg (2013)
20. Yu, H., Liu, Z.G., Wang, G.Y.: An automatic method to determine the number of clusters using decision-theoretic rough set. International Journal of Approximate Reasoning 55(1), 101–115 (2014)
21. Zhou, B., Yao, Y., Luo, J.: A three-way decision approach to email spam filtering. In: Farzindar, A., Kešelj, V. (eds.) Canadian AI 2010. LNCS (LNAI), vol. 6085, pp. 28–39. Springer, Heidelberg (2010)

# An Overview of Function Based Three-Way Decisions

Dun Liu[1] and Decui Liang[2]

[1] School of Economics and Management, Southwest Jiaotong University
Chengdu 610031, P.R. China
newton83@163.com
[2] School of Management and Economics, University of Electronic Science and
Technology of China
Chengdu 610054, P.R. China
decuiliang@126.com

**Abstract.** By considering the various of studies on loss functions with three-way decisions, a function based three-way decisions is proposed to generalize the existing models. A "four-level" approach with granular perspective is built, and the existing models can be categorized to a "four-level" framework through different decision criteria. Our work provides a novel "granularity" viewpoint on the current three-way decision researches.

**Keywords:** Three-way decisions, loss functions, decision-theoretic rough sets, decision-making.

## 1 Introduction

Three-way decisions (TWD), a new perspective of probabilistic rough sets, which were proposed by Y.Y. Yao, have drawn more and more attentions in nearly five years [38, 39, 42, 43]. A theory of three-way decisions is constructed based on the notions of the acceptance, rejection or noncommitment, which can be directly generated by the three regions of probabilistic rough sets. The rules generated by the positive region are used to make a decision of acceptance, the rules generated by the negative region are used to make a decision of rejection, the rules generated by the boundary region are used for making a decision of noncommitment [38, 40]. In general, three-way decisions describe the human cognitive process during decision making, and establish an intimate connection between rough sets and decision theory.

With carefully investigate current studies, there are three main research directions on three-way decisions. (1). The extended models of three-way decisions. By considering the key ingredient in three-way decisions is the loss functions, some researches adopted stochastic numbers [21], intervals [14, 23], fuzzy intervals [24, 27], triangular fuzzy numbers [12], intuitionistic fuzzy numbers [15], hesitant fuzzy numbers [13], shadow sets [5] to estimate the losses in three-way decisions,

D. Miao et al. (Eds.): RSKT 2014, LNAI 8818, pp. 812–823, 2014.
DOI: 10.1007/978-3-319-11740-9_74 © Springer International Publishing Switzerland 2014

which are the extension of losses under uncertain decision environments. Another extending of three-way decisions emphasized on the methodologies, e.g., multiple-classification three-way decisions [20, 46, 47], multi-agent three-way decisions [33], cost-sensitive three-way decisions [30], information-theoretic based three-way decisions [4], sequential three-way decisions [41], dynamic three-way decisions [28], game-theoretic based three-way decisions [1, 7], clusters based three-way decisions [16, 44], three-way decisions with two universes [29], etc. (2). The attributes reduction methods and rules acquisitions approaches of three-way decisions. The main attribute reduction methods in three-way decisions include probabilistic attribute reduct [37, 45], non-monotonicity of probabilistic positive region reduct [10], minimum cost attribute reduction [8], cost-sensitive attribute reduction [11] and some machine learning based attribute reduction [3]. (3). The applications of three-way decisions. The essential ideas of three-way decisions are commonly used in many domains, e.g., information sciences, engineering, management sciences, medical decision-making, etc [17, 26]. The aforementioned studies indicate that the three-way decisions has gradually been a hot research topic in granular computing and rough sets.

In this paper, we focus on investigating the loss functions in three-way decisions. The remainder of this paper is organized as follows: Section 2 provides the basic concepts of three-way decisions and it's extensions. A generalized three-way decision model with functional perspective is proposed in Section 3. Then, the similarity and difference of existing three-way decision models are carefully analyzed in Section 4, and a "four level" structure model for three-way decisions is built. Section 5 concludes the paper and outlines the future work.

## 2 Preliminaries

Basic concepts, notations and results of three-way decisions are briefly reviewed in this section [2, 18, 22, 31, 32, 34–36, 38, 48].

As Yao stated in [42, 43], many generalizations of sets have been proposed and studied with three-way decisions, including interval sets and three-valued logic, Pawlak rough sets, Decision-theoretic rough sets (DTRS), three-valued approximations in many-valued logic, fuzzy sets and shadowed sets. In our following discussions, we mainly discuss three-way decisions in DTRS.

The DTRS model is inspired by the Bayesian decision theory, a well known theorem in decision analysis [6]. It considers 2 states $\Omega = \{X, \neg X\}$ and 3 actions $\mathcal{A} = \{a_P, a_B, a_N\}$ during decision process. The set of states is given by $\Omega$ indicating that an object is in $X$ and not in $X$, respectively. Meanwhile, $a_P$, $a_B$, and $a_N$ in $\mathcal{A}$ represent the three actions in classifying an object $x$, namely, deciding $x \in \mathrm{POS}(X)$, deciding $x$ should be further investigated $x \in \mathrm{BND}(X)$, and deciding $x \in \mathrm{NEG}(X)$, respectively. The loss function $\lambda$ regarding the risk or cost of actions in different states is given by the $3 \times 2$ matrix:

|  | $X\ (P)$ | $\neg X\ (N)$ |
|---|---|---|
| $a_P$ | $\lambda_{PP}$ | $\lambda_{PN}$ |
| $a_B$ | $\lambda_{BP}$ | $\lambda_{BN}$ |
| $a_N$ | $\lambda_{NP}$ | $\lambda_{NN}$ |

In the matrix, $\lambda_{PP}$, $\lambda_{BP}$ and $\lambda_{NP}$ denote the losses incurred for taking actions of $a_P$, $a_B$ and $a_N$, respectively, when an object belongs to $X$. Similarly, $\lambda_{PN}$, $\lambda_{BN}$ and $\lambda_{NN}$ denote the losses incurred for taking the same actions when the object belongs to $\neg X$. $Pr(X|[x])$ is the conditional probability of an object $x$ belonging to $X$ given that the object is described by its equivalence class $[x]$. For an object $x$, the expected loss $R(a_i|[x])$ associated with taking the individual actions can be expressed as:

$$R(a_P|[x]) = \lambda_{PP} Pr(X|[x]) + \lambda_{PN} Pr(\neg X|[x]),$$
$$R(a_B|[x]) = \lambda_{BP} Pr(X|[x]) + \lambda_{BN} Pr(\neg X|[x]),$$
$$R(a_N|[x]) = \lambda_{NP} Pr(X|[x]) + \lambda_{NN} Pr(\neg X|[x]).$$

The Bayesian decision procedure suggests the following minimum-cost decision rules:

(P) If $R(a_P|[x]) \leq R(a_B|[x])$ and $R(a_P|[x]) \leq R(a_N|[x])$, decide $x \in POS(X)$;

(B) If $R(a_B|[x]) \leq R(a_P|[x])$ and $R(a_B|[x]) \leq R(a_N|[x])$, decide $x \in BND(X)$;

(N) If $R(a_N|[x]) \leq R(a_P|[x])$ and $R(a_N|[x]) \leq R(a_B|[x])$, decide $x \in NEG(X)$.

Since $Pr(X|[x]) + Pr(\neg X|[x]) = 1$, we simplify the rules based only on the probability $Pr(X|[x])$ and the loss function. By considering a reasonable kind of loss functions with $\lambda_{PP} \leq \lambda_{BP} < \lambda_{NP}$ and $\lambda_{NN} \leq \lambda_{BN} < \lambda_{PN}$, the decision rules (P)-(N) can be expressed concisely as:

(P) If $Pr(X|[x]) \geq \alpha$ and $Pr(X|[x]) \geq \gamma$, decide $x \in POS(X)$;

(B) If $Pr(X|[x]) \leq \alpha$ and $Pr(X|[x]) \geq \beta$, decide $x \in BND(X)$;

(N) If $Pr(X|[x]) \leq \beta$ and $Pr(X|[x]) \leq \gamma$, decide $x \in NEG(X)$.

The thresholds values $\alpha$, $\beta$, $\gamma$ generated by DTRS are given by:

$$\alpha^{DTRS} = \frac{(\lambda_{PN} - \lambda_{BN})}{(\lambda_{PN} - \lambda_{BN}) + (\lambda_{BP} - \lambda_{PP})};$$
$$\beta^{DTRS} = \frac{(\lambda_{BN} - \lambda_{NN})}{(\lambda_{BN} - \lambda_{NN}) + (\lambda_{NP} - \lambda_{BP})};$$
$$\gamma^{DTRS} = \frac{(\lambda_{PN} - \lambda_{NN})}{(\lambda_{PN} - \lambda_{NN}) + (\lambda_{NP} - \lambda_{PP})}. \tag{1}$$

In addition, as a well-defined boundary region, the conditions of rule (B) suggest that $\alpha > \beta$, which implies $0 \leq \beta < \gamma < \alpha \leq 1$. $\forall X \subseteq U$, the $(\alpha, \beta)$-lower approximation, $(\alpha, \beta)$-upper approximation of DTRS are defined as follows:

$$\underline{apr}_{(\alpha,\beta)}^{DTRS}(X) = \{x \in U | Pr(X|[x]) \geq \frac{(\lambda_{PN} - \lambda_{BN})}{(\lambda_{PN} - \lambda_{BN}) + (\lambda_{BP} - \lambda_{PP})}\};$$
$$\overline{apr}_{(\alpha,\beta)}^{DTRS}(X) = \{x \in U | Pr(X|[x]) > \frac{(\lambda_{BN} - \lambda_{NN})}{(\lambda_{BN} - \lambda_{NN}) + (\lambda_{NP} - \lambda_{BP})}\}. \tag{2}$$

The three regions of $(\alpha, \beta)$- positive, boundary and negative regions in DTRS can be written as:

$$\text{POS}^{DTRS}_{(\alpha,\beta)}(X) = \{x \in U \mid Pr(X|[x]) \geq \frac{(\lambda_{PN} - \lambda_{BN})}{(\lambda_{PN} - \lambda_{BN}) + (\lambda_{BP} - \lambda_{PP})}\},$$

$$\text{BND}^{DTRS}_{(\alpha,\beta)}(X) = \{x \in U \mid \frac{(\lambda_{BN} - \lambda_{NN})}{(\lambda_{BN} - \lambda_{NN}) + (\lambda_{NP} - \lambda_{BP})} < Pr(X|[x])$$
$$< \frac{(\lambda_{PN} - \lambda_{BN})}{(\lambda_{PN} - \lambda_{BN}) + (\lambda_{BP} - \lambda_{PP})}\},$$

$$\text{NEG}^{DTRS}_{(\alpha,\beta)}(X) = \{x \in U \mid Pr(X|[x]) \leq \frac{(\lambda_{BN} - \lambda_{NN})}{(\lambda_{BN} - \lambda_{NN}) + (\lambda_{NP} - \lambda_{BP})}\}.$$

Specially, if the assumption "$\alpha > \beta$" does not hold, the three-way decisions convert to two-way decisions, and it can be rewritten as:

$$\text{POS}^{DTRS}_{(\gamma,\gamma)}(X) = \{x \in U \mid Pr(X|[x]) \geq \frac{(\lambda_{PN} - \lambda_{NN})}{(\lambda_{PN} - \lambda_{NN}) + (\lambda_{NP} - \lambda_{PP})}\},$$

$$\text{NEG}^{DTRS}_{(\gamma,\gamma)}(X) = \{x \in U \mid Pr(X|[x]) < \frac{(\lambda_{PN} - \lambda_{NN})}{(\lambda_{PN} - \lambda_{NN}) + (\lambda_{NP} - \lambda_{PP})}\}.$$

According to above discussions, DTRS gives a brief semantics explanation with minimum decision risks when comparing with other rough set models (e.g., probabilistic rough set model). Note that, the two thresholds $\alpha$ and $\beta$ in DTRS are not setting in advance by human's experiments, and they are associated to decision risk with different loss functions [19, 38, 39, 42, 43].

## 3    A Function Based Perspective of Three-Way Decisions

In this section, we propose a more generalized model of three-way decisions. As we stated in Section 2, the decision rules generated by the three regions are closely related with loss functions. In DTRS, the values of losses are precise real numbers. In some cases, one can directly utilize the precise value (i.e., money, energy and time, etc) to estimate the costs [19]. However, in most of cases, it is rather difficult to utilize one value to illustrate the costs [12–14, 21, 23, 24, 27]. Inspired by the deficiencies, we propose a function based three-way decisions. In this model, we use two groups of loss functions $f(\lambda_{PP})$, $f(\lambda_{BP})$, $f(\lambda_{NP})$ and $f(\lambda_{PN})$, $f(\lambda_{BN})$, $f(\lambda_{NN})$ to instead of six single values, and the matrix refers to the cost of actions in different states can be rewritten as:

|       | $X$ $(P)$ | $\neg X$ $(N)$ |
|-------|-----------|----------------|
| $a_P$ | $f(\lambda_{PP})$ | $f(\lambda_{PN})$ |
| $a_B$ | $f(\lambda_{BP})$ | $f(\lambda_{BN})$ |
| $a_N$ | $f(\lambda_{NP})$ | $f(\lambda_{NN})$ |

In the matrix, $\lambda_{\bullet\bullet}$ ($\bullet = P, B, N$) is not a fixed number, but an independent variable. Given a fixed $\lambda_{\bullet\bullet}$, by considering the losses of classifying an object $x$

belonging to $X$ into the positive region $POS(X)$ is less than or equal to the loss of classifying $x$ into the boundary region $BND(X)$, and both of these losses are strictly less than the loss of classifying $x$ into the negative region $NEG(X)$. The reverse order of losses is used for classifying an object not in $X$. Therefore, the following two conditions should be considered:

$$f(\lambda_{PP}) \leq f(\lambda_{BP}) < f(\lambda_{NP}),$$
$$f(\lambda_{NN}) \leq f(\lambda_{BN}) < f(\lambda_{PN}). \tag{3}$$

Under (3) and $Pr(X|[x]) + Pr(\neg X|[x]) = 1$, the thresholds values $\alpha$, $\beta$, $\gamma$ with a function based three-way decisions can be calculated as:

$$\alpha^{DTRS} = \frac{f(\lambda_{PN}) - f(\lambda_{BN})}{(f(\lambda_{PN}) - f(\lambda_{BN})) + (f(\lambda_{BP}) - f(\lambda_{PP}))};$$

$$\beta^{DTRS} = \frac{f(\lambda_{BN}) - f(\lambda_{NN})}{(f(\lambda_{BN}) - f(\lambda_{NN})) + (f(\lambda_{NP}) - f(\lambda_{BP}))};$$

$$\gamma^{DTRS} = \frac{f(\lambda_{PN}) - f(\lambda_{NN})}{(f(\lambda_{PN}) - f(\lambda_{NN})) + (f(\lambda_{NP}) - f(\lambda_{PP}))}. \tag{4}$$

Similarly, suppose $0 \leq \beta < \gamma < \alpha \leq 1$, the three regions of $(\alpha, \beta)$- positive, boundary and negative regions with a function based three-way decisions can be written as:

$$POS_{(\alpha,\beta)}^{DTRS}(X) = \{x \in U \mid Pr(X|[x]) \geq \frac{f(\lambda_{PN}) - f(\lambda_{BN}))}{(f(\lambda_{PN}) - f(\lambda_{BN})) + (f(\lambda_{BP}) - f(\lambda_{PP}))}\},$$

$$BND_{(\alpha,\beta)}^{DTRS}(X) = \{x \in U \mid \frac{f(\lambda_{BN}) - f(\lambda_{NN})}{(f(\lambda_{BN}) - f(\lambda_{NN}) + f(\lambda_{NP}) - f(\lambda_{BP}))} < Pr(X|[x])$$
$$< \frac{f(\lambda_{PN}) - f(\lambda_{BN})}{(f(\lambda_{PN}) - f(\lambda_{BN})) + (f(\lambda_{BP}) - f(\lambda_{PP}))}\},$$

$$NEG_{(\alpha,\beta)}^{DTRS}(X) = \{x \in U \mid Pr(X|[x]) \leq \frac{f(\lambda_{BN}) - f(\lambda_{NN})}{(f(\lambda_{BN}) - f(\lambda_{NN})) + (f(\lambda_{NP}) - f(\lambda_{BP}))}\}.$$

Specially, if the assumption "$\alpha > \beta$" does not hold, the three-way decisions converts to two-way decisions, and it can be rewritten as:

$$POS_{(\gamma,\gamma)}^{DTRS}(X) = \{x \in U \mid Pr(X|[x]) \geq \frac{f(\lambda_{PN}) - f(\lambda_{NN})}{(f(\lambda_{PN}) - f(\lambda_{NN})) + (f(\lambda_{NP}) - f(\lambda_{PP}))}\},$$

$$NEG_{(\gamma,\gamma)}^{DTRS}(X) = \{x \in U \mid Pr(X|[x]) < \frac{f(\lambda_{PN}) - f(\lambda_{NN})}{(f(\lambda_{PN}) - f(\lambda_{NN})) + (f(\lambda_{NP}) - f(\lambda_{PP}))}\}.$$

## 4   The Framework and Comparison of Existing Three-Way Decision Models

In the following, we construct a "four-level" framework of existing three-way decision models, which can be regarded as the "granular structure" of three-way decisions with loss functions.

Level 1: The basic expression of three-way decisions

In this basic level, the loss function $f(\lambda_{\bullet\bullet})$ ($\bullet = P, B, N$) is set as a precise value number: $f(\lambda_{\bullet\bullet}) = \lambda_{\bullet\bullet}$ ($\bullet = P, B, N$), and the function based three-way decisions degenerate into precise-value based three-way decisions. Obviously, the basic model of three-way decisions is DTRS [34, 35]. Furthermore, due to the single valued losses in DTRS, the precise-value based three-way decisions can be regarded as static and certain decision model.

Level 2: Three-way decisions with uncertainty

As we known, "uncertainty" in decision making means the lack of certainty, it used to describe the state of having limited knowledge where it is impossible to exactly describe the existing state, a future outcome, or more than one possible outcome. The main mathematical theories to describe uncertainty are probability theory, fuzzy set theory, rough set theory, interval set theory, inclusion degree theory, evidence theory, etc. The former three theorems are most famous uncertainty methodologies and have been widely used in human's decision process. (1). Probability theory is the branch of mathematics concerned with probability, the analysis of random phenomena, it utilizes "stochastic number" to express uncertainty and indicates that a particular subject is seen from point of view of randomness. The uncertainty in probability theory means one thing cannot definite happened, but may happen with a probability. The central objects of probability theory are random variables, stochastic processes, and events: mathematical abstractions of non-deterministic events or measured quantities that may either be single occurrences or evolve over time in an apparently random fashion. (2). The uncertainty in fuzzy set theory comes from the fuzzification, which is used to describe the unclear classification for a concept. (e.g., the concept of "young"). The fuzzification comprises the process of transforming crisp values into grades of membership for linguistic terms of fuzzy sets. (3). The uncertainty in rough set theory comes from the inaccuracy for boundary sets. A rough set is a formal approximation of a crisp set in terms of a pair of sets which give the lower and the upper approximation of the original set, and can deal with inaccuracy, inconsistent, incomplete information in a decision system.

With above discussions, we investigate three types of uncertain three-way decision models as follows.

• Interval three-way decisions

In this scenario, the loss function $f(\lambda_{\bullet\bullet})$ ($\bullet = P, B, N$) can be rewritten as: $f(\lambda_{\bullet\bullet}) = [\lambda_{\bullet\bullet}^{-}, \lambda_{\bullet\bullet}^{+}]$ ($\bullet = P, B, N$), $\lambda_{\bullet\bullet}^{-}$ and $\lambda_{\bullet\bullet}^{+}$ are the lower bound and the upper bound of $\lambda_{\bullet\bullet}$. Liu et al. firstly introduced interval-valued loss functions to DTRS and carefully discussed the corresponding propositions and criteria of interval-valued three-way decisions [23]. Furthermore, Liu et al. used a fuzzy interval number $[\tilde{\lambda}_{\bullet\bullet}^{-}, \tilde{\lambda}_{\bullet\bullet}^{+}]$ and proposed fuzzy interval-valued three-way decisions [27]. Liang and Liu did systematic studies on three-way decisions with interval-valued decision-theoretic rough sets, they derived three-way decisions with the aid of two conventional methods and proposed a new optimization method in the viewpoint of the flexibility of information granularity [14].

• Fuzzy three-way decisions

In this scenario, the loss function $f(\lambda_{\bullet\bullet})$ ($\bullet = P, B, N$) can be rewritten as: $f(\lambda_{\bullet\bullet}) = \widetilde{\lambda}_{\bullet\bullet}$ ($\bullet = P, B, N$). Liu et al. firstly introduced fuzzy loss functions to DTRS and investigated the corresponding propositions and criteria of fuzzy three-way decisions [24]. Liang et al. considered the triangular fuzzy loss functions $\widetilde{\lambda}_{\bullet\bullet} = (l_{\bullet\bullet}, m_{\bullet\bullet}, u_{\bullet\bullet})$ ($\bullet = P, B, N$), and discussed 5 ranking functions for measure triangular fuzzy number [12]. In addition, Liang and Liu further took into account the losses of DTRS with hesitant fuzzy elements ($\widetilde{\lambda}_{\bullet\bullet} = h_E(\lambda_{\bullet\bullet})$) and proposed a new model of hesitant fuzzy three-way decisions, a novel risk decision-making method with the aid of hesitant fuzzy DTRS was developed in their work [13]. As to intuitionistic fuzzy sets, Liang and Liu used the degree of membership $\mu$ and degree of non-membership $\nu$ to depict the fuzzification [15]. The loss function for the intuitionistic fuzzy three-way decisions can be set as: $\widetilde{\lambda}_{\bullet\bullet} = (\lambda_{\bullet\bullet}, \mu(\lambda_{\bullet\bullet}), \nu(\lambda_{\bullet\bullet}))$. The decision rules generated by intuitionistic fuzzy three-way decisions both considered the membership and non-membership functions. In addition, Deng and Yao investigated mean-value-based three-way shadowed sets, they introduced a generalized decision-theoretic shadowed set model using the mean value [5]. In a word, all the aforementioned work have solid contributions to develop fuzzy three-way decision theorems.

• Stochastic three-way decisions

In this scenario, the loss function $f(\lambda_{\bullet\bullet})$ ($\bullet = P, B, N$) can be rewritten as: $f(\lambda_{\bullet\bullet}) = \lambda_{\bullet\bullet}^\varepsilon$ ($\varepsilon$ denotes the loss function $\lambda_{\bullet\bullet}^\varepsilon$ is a stochastic number). Liu et al. introduced stochastic loss functions to DTRS and proposed stochastic decision-theoretic rough set theory [21]. In their studies, a model of stochastic three-way decisions was built with respect to the minimum bayesian expected risk, and they further investigated two special stochastic three-way models under uniform distribution and normal distribution, respectively.

To sum up, Level 2 focuses on the decision of uncertainty, but these models are still within a static and close decision environment.

Level 3: Three-way decisions with multi-stage decision making

Level 3 begins to discuss the dynamic decision environment, and the rules acquisition by three-way decisions are not a single step decision, but a multi-stage process. Liu et al. considered the dynamic change of loss functions $f(\lambda_{\bullet\bullet}) = g(\lambda_{\bullet\bullet}^t)$ ($\bullet = P, B, N$) in DTRS with the time $t$, and proposed dynamic three-way decisions [28]. Yang and Yao proposed a multi-agent three-way decision model, the new model is utilized to seek for synthesized or consensus decisions when there are multiple decision preferences and criteria adopted by different agents [33]. Yao and Deng discussed sequential three-way decisions when adding attributes [41]. It suggested that if a decision of acceptance or rejection with certain tolerable levels of errors can be made at a higher level, it is not necessary to move to a lower level. This work enabled the decision maker to consider both the cost of various mis-classifications and the cost of obtaining the necessary evidence for making a classification decision. Liu et al. investigated multiple-category classification problems with three-way decisions [20], and further proposed a two-stage method to choose the best candidate classification. Zhou also provided a new formulation of multi-class three-way decision model, which can be well suited for cost-sensitive classification tasks where different types of classification errors have different costs [46, 47]. In summary, all the above mentioned studies make effective interpretations on dynamic three-way decisions.

Level 4: Function based three-way decisions

As we stated in Levels 1-3, Level 4 is the generation level for three-way decisions because the loss function in former three levels can be treated as a special case of $f(\lambda_{\bullet\bullet})$. In Level 4, $f(\lambda_{\bullet\bullet})$ can be treated as power function, exponeneial function, nonlinear function, etc. It is the generalization level for three-way decisions.

For simplicity, we choose three dimensions (characters): certain or uncertain decisions, single-stage or multi-stage decisions, static or dynamic decisions, to illustrate the similarity and difference of different three-way decision models. Figure 1 outlines the framework of the four-level model in three-way decisions. Table 1 summarizes the characters for the models in Figure 1.

**Table 1.** The characters for different three-way decision models

| Level | Model | | Character 1 | Character 2 | Character 3 |
|---|---|---|---|---|---|
| Level 1 | Precise-value (DTRS) | | Certain | Single-stage | Static |
| Level 2 | Interval | Interval-valued | Uncertain | Single-stage | Static |
| | | Fuzzy interval | Uncertain | Single-stage | Static |
| | | Interval sets | Uncertain | Single-stage | Static |
| | Fuzzy | Triangular | Uncertain | Single-stage | Static |
| | | Intuitionistic | Uncertain | Single-stage | Static |
| | | Hesitant | Uncertain | Single-stage | Static |
| | | Shadow | Uncertain | Single-stage | Static |
| | Stochastic | Uniform | Uncertain | Single-stage | Static |
| | | Normal | Uncertain | Single-stage | Static |
| | | Other | Uncertain | Single-stage | Static |
| Level 3 | Dynamic | | Uncertain | Multi-stage | Dynamic |
| | Multi-agent | | Uncertain | Multi-stage | Dynamic |
| | Sequential | | Uncertain | Multi-stage | Dynamic |
| | Multiple-category | | Uncertain | Multi-stage | Dynamic |
| Level 4 | Function | | Both | Both | Dynamic |

In Figure 1 and Table 1, Level 1 and Level 2 are based on the static three-way decisions, Level 3 and Level 4 are based on the dynamic three-way decisions. If we consider the variation of loss functions, the static models can change to dynamic models. In addition, Level 1 is the specialization level, and Level 4 is the generalization level. Between Level 1 and Level 4, the middle two levels respectively consider the decision risk and decision stages. With the insightful gain from the top-down or bottom-up perspective in granular computing, DTRS model in three-way decisions corresponds finest granularity, and the function based three-way decision model corresponds the coarsest granularity. In general, one can choose a special model by using different decision criterion or viewpoint in a real decision problem.

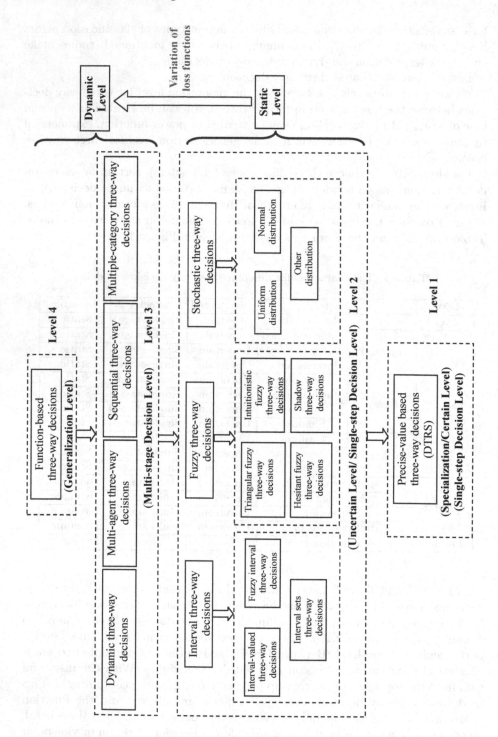

**Fig. 1.** The framework of the four-level model in three-way decisions

# 5 Conclusions

In this paper, we systematically study the loss functions with three-way decisions in nearly two decades and propose a function based three-way decision model. The current models of three-way decisions are carefully investigated, then classified into a generalization research framework via three different dimensions. A "four-level" model in three-way decisions is built, which draw an intuitive and clear impression for different three-way decision models. Our future researches will focus on properties for different types of function based three-way decisions, the group decision method in three-way decisions will be our another future research topic.

**Acknowledgements.** This work is partially supported by the National Science Foundation of China (No. 71201133), the Youth Social Science Foundation of the Chinese Education Commission (No. 11YJC630127), the Research Fund for the Doctoral Program of Higher Education of China (No. 20120184120028) and the Fundamental Research Funds for the Central Universities of China (No. SWJTU12CX117).

# References

1. Azam, N., Yao, J.T.: Analyzing uncertainties of probabilistic rough set regions with game-theoretic rough sets. International Journal of Approximate Reasoning 55, 142–155 (2014)
2. Abd El-Monsef, M.M.E., Kilany, N.M.: Decision analysis via granulation based on general binary relation. International Journal of Mathematics and Mathematical Sciences, article ID 12714 (2007), doi:10.1155/2007/12714
3. Chebrolu, S., Sanjeevi, S.G.: Attribute reduction in decision-theoretic rough set models using genetic algorithm. In: Panigrahi, B.K., Suganthan, P.N., Das, S., Satapathy, S.C. (eds.) SEMCCO 2011, Part I. LNCS, vol. 7076, pp. 307–314. Springer, Heidelberg (2011)
4. Deng, X., Yao, Y.: An information-theoretic interpretation of thresholds in probabilistic rough sets. In: Li, T., Nguyen, H.S., Wang, G., Grzymala-Busse, J., Janicki, R., Hassanien, A.E., Yu, H. (eds.) RSKT 2012. LNCS, vol. 7414, pp. 369–378. Springer, Heidelberg (2012)
5. Deng, X.F., Yao, Y.Y.: Mean-value-based decision-theoretic shadowed sets. In: 2013 Joint IFSA World Congress and NAFIPS Annual Meeting, Edmonton, Canada, pp. 1381–1387 (2013)
6. Duda, R.O., Hart, P.E.: Pattern classification and scene analysis. Wiley Press, New York (1973)
7. Herbert, J.P., Yao, J.T.: Game-theoretic rough sets. Fundamenta Informaticae 108, 267–286 (2011)
8. Jia, X.Y., Liao, W.H., Tang, Z.M., Shang, L.: Minimum cost attribute reduction in decision-theoretic rough set models. Information Sciences 219, 151–167 (2012)
9. Li, H.X., Zhou, X.Z.: Risk decision making based on decision-theoretic rough set: A three-way view decision model. International Journal of Computational Intelligence Systems 4, 1–11 (2011)

10. Li, H.X., Zhou, X.Z., Zhao, J.B., Liu, D.: Non-monotonic attribute reduction in decision-theoretic rough sets. Fundamenta Informaticae 126, 415–432 (2013)
11. Li, H., Zhou, X., Huang, B., Liu, D.: Cost-sensitive three-way decision: A sequential strategy. In: Lingras, P., Wolski, M., Cornelis, C., Mitra, S., Wasilewski, P. (eds.) RSKT 2013. LNCS, vol. 8171, pp. 325–337. Springer, Heidelberg (2013)
12. Liang, D.C., Liu, D., Pedrycz, W., Hu, P.: Triangular fuzzy decision-theoretic rough sets. International Journal of Approximate Reasoning 54, 1087–1106 (2013)
13. Liang, D.C., Liu, D.: A novel risk decision-making based on decision-theoretic rough sets under hesitant fuzzy information. IEEE Transactions on Fuzzy Systems (2014), doi:10.1109/TFUZZ.2014.2310495
14. Liang, D.C., Liu, D.: Systematic studies on three-way decisions with interval-valued decision-theoretic rough sets. Information Sciences 276, 186–203 (2014)
15. Liang, D.C., Liu, D.: Deriving three-way decisions from intuitionistic fuzzy decision-theoretic rough sets. Information Sciences (manuscript)
16. Lingras, P., Chen, M., Miao, D.Q.: Rough cluster quality index based on decision theory. IEEE Transaction on Knowledge and Data Engineering 21, 1014–1026 (2009)
17. Liu, D., Li, H.X., Zhou, X.Z.: Two decades' research on decision-theoretic rough sets. In: Proceeding of 9th IEEE International Conference on Cognitive Informatics, pp. 968–973 (2010)
18. Liu, D., Yao, Y.Y., Li, T.R.: Three-way investment decisions with decision-theoretic rough sets. International Journal of Computational Intelligence Systems 4, 66–74 (2011)
19. Liu, D., Li, T.R., Ruan, D.: Probabilistic model criteria with decision-theoretic rough sets. Information Sciences 181, 3709–3722 (2011)
20. Liu, D., Li, T.R., Li, H.X.: A multiple-category classification approach with decision-theoretic rough sets. Fundamenta Informaticae 115, 173–188 (2012)
21. Liu, D., Li, T., Liang, D.: Decision-theoretic rough sets with probabilistic distribution. In: Li, T., Nguyen, H.S., Wang, G., Grzymala-Busse, J., Janicki, R., Hassanien, A.E., Yu, H. (eds.) RSKT 2012. LNCS, vol. 7414, pp. 389–398. Springer, Heidelberg (2012)
22. Liu, D., Li, T.R., Liang, D.C.: Three-way government decision analysis with decision-theoretic rough sets. International Journal of Uncertainty, Fuzziness and Knowledge-Based Systems 20, 119–132 (2012)
23. Liu, D., Li, T.R., Li, H.X.: Interval-valued decision-theoretic rough sets. Computer Science 39(7), 178–181 (2012) (in chinese)
24. Liu, D., Li, T.R., Liang, D.C.: Fuzzy decision-theoretic rough sets. Computer Science 39(12), 25–29 (2012) (in chinese)
25. Liu, D., Li, T.R., Liang, D.C.: Incorporating logistic regression to decision-theoretic rough sets for classifications. International Journal of Approximate Reasoning 55, 197–210 (2014)
26. Liu, D., Li, T.R., Liang, D.C.: Rough Set Theory: A three-way decisions perspective. Journal of Nanjing University: Nat. Sci. Ed. 49(5), 574–581 (2013) (in chinese)
27. Liu, D., Li, T.R., Liang, D.C.: Fuzzy interval decision-theoretic rough sets. In: Proc. of 2013 IFSA World Congress NAFIPS Annual Meeting, pp. 1315–1320 (2013)
28. Liu, D., Li, T., Liang, D.: Three-way decisions in dynamic decision-theoretic rough sets. In: Lingras, P., Wolski, M., Cornelis, C., Mitra, S., Wasilewski, P. (eds.) RSKT 2013. LNCS (LNAI), vol. 8171, pp. 291–301. Springer, Heidelberg (2013)
29. Ma, W.M., Sun, B.Z.: On relationship between probabilistic rough set and Bayesian risk decision over two universes. International Journal of General Systems 41, 225–245 (2012)

30. Min, F., He, H.P., Qian, Y.H., Zhu, W.: Test-cost-sensitive attribute reduction. Information Sciences 181, 4928–4942 (2011)
31. Pawlak, Z.: Rough sets. International Journal of Computer and Information Science 11, 341–356 (1982)
32. Ślęzak, D., Ziarko, W.: The investigation of the bayesian. International Journal of Approximate Reasoning 40, 81–91 (2005)
33. Yang, X.P., Yao, J.T.: Modelling multi-agent three-way decisions with decision-theoretic rough sets. Fundamenta Informaticae 115, 157–171 (2012)
34. Yao, Y.Y., Wong, S.K.M.: A decision theoretic framework for approximating concepts. International Journal of Man-Machine Studies 37, 793–809 (1992)
35. Yao, Y.: Decision-theoretic rough set models. In: Yao, J., Lingras, P., Wu, W.-Z., Szczuka, M.S., Cercone, N.J., Ślęzak, D. (eds.) RSKT 2007. LNCS (LNAI), vol. 4481, pp. 1–12. Springer, Heidelberg (2007)
36. Yao, Y.Y.: Probabilistic rough set approximations. International Journal of Approximate Reasoning 49, 255–271 (2008)
37. Yao, Y.Y., Zhao, Y.: Attribute reduction in decision-theoretic rough set models. Information Sciences 178, 3356–3373 (2008)
38. Yao, Y.Y.: Three-way decisions with probabilistic rough sets. Information Sciences 180, 341–353 (2010)
39. Yao, Y.Y.: The superiority of three-way decision in probabilistic rough set models. Information Sciences 181, 1080–1096 (2011)
40. Yao, Y.Y.: Two semantic issues in a probabilistic rough set model. Fundamenta Informaticae 108, 249–265 (2011)
41. Yao, Y.Y., Deng, X.F.: Sequential three-way decisions with probabilistic rough sets. In: Proc. 10th IEEE International Conference on Cognitive Informatics and Cognitive Computing, pp. 120–125 (2011)
42. Yao, Y.Y.: Three-way decisions using rough sets. In: Peters, G., et al. (eds.) Rough Sets: Selected Methods and Applications in Management and Engineering, Advanced Information and Knowledge Processing, pp. 79–93 (2012)
43. Yao, Y.: An outline of a theory of three-way decisions. In: Yao, J., Yang, Y., Słowiński, R., Greco, S., Li, H., Mitra, S., Polkowski, L. (eds.) RSCTC 2012. LNCS (LNAI), vol. 7413, pp. 1–17. Springer, Heidelberg (2012)
44. Yu, H., Liu, Z.G., Wang, G.Y.: An automatic method to determine the number of clusters using decision-theoretic rough set. International Journal of Approximate Reasoning 55, 101–115 (2014)
45. Zhao, Y., Wong, S.K.M., Yao, Y.: A note on attribute reduction in the decision-theoretic rough set model. In: Peters, J.F., Skowron, A., Chan, C.-C., Grzymala-Busse, J.W., Ziarko, W.P. (eds.) Transactions on Rough Sets XIII. LNCS, vol. 6499, pp. 260–275. Springer, Heidelberg (2011)
46. Zhou, B.: A new formulation of multi-category decision-theoretic rough sets. In: Yao, J., Ramanna, S., Wang, G., Suraj, Z. (eds.) RSKT 2011. LNCS (LNAI), vol. 6954, pp. 514–522. Springer, Heidelberg (2011)
47. Zhou, B.: Multi-class decision-theoretic rough sets. International Journal of Approximate Reasoning, 211–224 (2014)
48. Ziarko, W.: Probabilistic approach to rough set. International Journal of Approximate Reasoning 49, 272–284 (2008)

# Multicost Decision-Theoretic Rough Sets Based on Maximal Consistent Blocks

Xingbin Ma[1], Xibei Yang[1,2,*], Yong Qi[3], Xiaoning Song[1],
and Jingyu Yang[2,4]

[1] School of Computer Science and Engineering, Jiangsu University of Science and
Technology, Zhenjiang, Jiangsu 212003, P.R. China
[2] Key Laboratory of Intelligent Perception and Systems for High-Dimensional
Information, Nanjing University of Science and Technology, Ministry of Education,
Nanjing, Jiangsu 210094, P.R. China
yangxibei@hotmail.com
[3] School of Economics and Management, Nanjing University of Science and
Technology, Nanjing, Jiangsu 210094, P.R. China
[4] School of Computer Science and Technology, Nanjing University of Science and
Technology, Nanjing, Jiangsu 210094, P.R. China

**Abstract.** Decision-theoretic rough set comes from Bayesian decision
procedure, in which a pair of the thresholds is derived by the cost matrix for the construction of probabilistic rough set. However, classical
decision-theoretic rough set can only be used to deal with complete information systems. Moreover, it does not take the property of variation
of cost into consideration. To solve above two problems, the maximal
consistent block is introduced into the construction of decision-theoretic
rough set by using multiple cost matrixes. Our approach includes optimistic and pessimistic multicost decision-theoretic rough set models.
Furthermore, the whole decision costs of optimistic and pessimistic multicost decision-theoretic rough sets are calculated in decision systems.
This study suggests potential application areas and new research trends
concerning decision-theoretic rough set.

**Keywords:** Decision-theoretic rough set, incomplete information system, multicost, maximal consistent block.

## 1 Introduction

Rough set [1] is a mathematical tool which can be used to characterize the
uncertainty by the difference between the lower and upper approximations. In
recent years, this theory has been widespread concerned by research scholars
[2–5] in many areas. The traditional rough set method can be used to deal with
complete information systems with discrete attribute values. However, due to
the error of the measurement data, the data acquisition constraints and other
factors, we often face the incomplete data in information systems. To deal with
such problems, many scholars have made great efforts [6–8].

---

* Corresponding author.

D. Miao et al. (Eds.): RSKT 2014, LNAI 8818, pp. 824–833, 2014.
DOI: 10.1007/978-3-319-11740-9_75 © Springer International Publishing Switzerland 2014

Obviously, the above rough set models do not take the cost into consideration. Moreover, in the field of machine learning and data mining, the cost sensitive learning occupies a significant position. Currently, many rough set scholars have explored many rough set problems related to the test cost and the misclassification cost. For instance, Yang et al. [9] proposed test cost-sensitive multigranulation rough set; Min et al. [10–13] proposed the concept of the test cost system and discussed the related problem of attribute reduction. For the misclassification cost, Yao [14] proposed decision-theoretic rough set in 1990, which used cost matrix to obtain a pair thresholds to construct the probabilistic approximation, from this point of view, Liu et al. [15] studied the multi-classification decisions which based on rough sets.

However, Yao's decision-theoretic rough set is based on complete information systems, therefore, introducing the concept of decision-theoretic into incomplete information systems can bring a new solution to deal with incomplete data. Furthermore, Yao's decision-theoretic rough set model only uses a cost matrix to describe the costs, which have some limitations in dealing with practical problems. For example, for the same medical diagnosis, different countries or regions may execute different compensation standards, using the same standards to evaluate all countries is unreasonable. This phenomenon indicates that Yao's decision-theoretic rough set did not consider costs inherent diversity and variability.

Therefore, we will discuss the concept of maximal consistent blocks and multiple cost matrixes in decision-theoretic rough set model. Furthermore, we will propose new rough set models based on multiple cost matrixes. In the models, the optimistic and pessimistic models will be discussed.

## 2  Preliminary Knowledge on Rough Sets

### 2.1  Maximal Consistent Block in Rough Set

Formally, an information system can be considered as a pair $I = < U, AT >$, in which $U$ is a non-empty finite set of the objects called the universe; $AT$ is a non-empty finite set of the attributes. $\forall a \in AT$, $V_a$ is the domain of attribute $a$. $\forall x \in U$, $a(x)$ denotes the value that $x$ holds on $a(\forall a \in AT)$.

It may happen that some of attribute values for an object are missing. We call it null value on this paper. For at least one attribute $a \in AT$, $V_a$ contains null value, then $I$ is called an incomplete information system, otherwise it is complete. In this paper, the null value is denoted by $*$.

**Definition 1.** *[6] Let $I$ be an incomplete information system in which $A \subseteq AT$, the tolerance relation in terms of $A$ is denoted by $TOL(A)$ where*

$$TOL(A) = \{(x, y) \in U^2 : \forall a \in AT, a(x) = a(y) \vee a(x) = * \vee a(y) = *\}.$$

Though the tolerance relation has been widely used to deal with incomplete information system, it has the following limitations[17]:

1. Firstly, different two tolerance classes may have inclusion relation.

2. Secondly, for all objects in $TOL_A(x) = \{y \in U : (x,y) \in TOL(A)\}$, they may have no common values.

From discussion above, a more reasonable classification analysis in incomplete information system has become a necessity. To solve such problem, Leung et al.[8] proposed the maximal consistent block technique, which is adopted from discrete mathematics.

**Definition 2.** *[8] Let $I$ be an incomplete information system in which $A \subseteq AT$, the set of all maximal consistent block in terms of $A$ is denoted by $\mu(A)$ where*

$$\mu(A) = \{Y \subseteq U : X^2 \subseteq TOL(A) \wedge (\forall x \notin Y \to (Y \cup \{x\})^2) \nsubseteq TOL(A))\}. \quad (1)$$

**Definition 3.** *[8] Let $I$ be an incomplete information system in which $A \subseteq AT$, the lower and upper approximations of $X$ in terms of $\mu(A)$ are denoted by $\underline{\mu_A}(X)$ and $\overline{\mu_A}(X)$, respectively, where*

$$\underline{\mu_A}(X) = \cup\{Y \in \mu(A) : Y \subseteq X\}; \quad (2)$$
$$\overline{\mu_A}(X) = \cup\{Y \in \mu(A) : Y \cap X \neq \emptyset\}. \quad (3)$$

The pair $[\underline{\mu_A}(X), \overline{\mu_A}(X)]$ is referred to as the rough set of $X$ in terms of the maximal consistent blocks in $\mu(A)$.

## 2.2   Maximal Consistent Block in Rough Set

For a Bayesian decision procedure, a finite set of the states can be denoted by $Q = \{w_1, w_2, \ldots, w_s\}$, a finite set of $t$ possible actions can be denoted by $B = \{b_1, b_2, \ldots, b_t\}$. $\forall x \in U$, let $Pr(w_j|x)$ be the conditional probability of object $x$ being in state $w_j$, and $\lambda(b_i|w_j)$ be the loss, or cost for taking action $b_i$ when the state is $w_j$. Suppose that we take the action $b_i$ for object $x$, then the expected loss is $R(b_j|x) = \sum_{i=1}^{s} \lambda(b_i|\omega_j) \cdot Pr(\omega_j|x)$.

For Yao's decision-theoretic rough set model, the set of states is composed by two classes such that $P = \{X, \sim X\}$, it can be used to indicate that an object is in class $X$ or out of class $X$; the set of actions is given by $E = \{e_P, e_B, e_N\}$, in which $e_P$; $e_B$ and $e_N$ express three actions: $e_P$ means that $x$ is classified into positive region of $X$, i.e., $POS(X)$; $e_B$ means that $x$ is classified into boundary region of $x$, i.e., $BND(X)$; $e_N$ means that $x$ is classified into negative region of $X$, i.e., $NEG(X)$. The loss function regarding the costs of three actions in two different states is given in Tab. 1. Obviously, Tab. 1 is a $3 \times 2$ matrix and it is denoted by $\mathbf{M}$ in this paper.

In Tab. 1, $\lambda_{PP}$; $\lambda_{BP}$ and $\lambda_{NP}$ are the losses for taking actions of $e_P$; $e_B$ and $e_N$, respectively, when stating $x$ is included into class $X$. $\lambda_{PN}$; $\lambda_{BN}$ and $\lambda_{NN}$ are the losses for taking actions of $e_P$; $e_B$ and $e_N$, respectively, when stating $x$ is out of class $X$. $\forall x \in U$, by using the conditional probability $Pr(X|Y)$, in which $Y \subset \mu_A(x)$, $\mu_A(x) = \{y \in \mu(A) : x \in Y\}$, the expected losses associated with three different actions.

**Table 1.** Cost matrix of decision-theoretic rough set

|         | $X$           | $\sim X$      |
| ------- | ------------- | ------------- |
| $e_P$   | $\lambda_{PP}$ | $\lambda_{PN}$ |
| $e_B$   | $\lambda_{BP}$ | $\lambda_{BN}$ |
| $e_N$   | $\lambda_{NP}$ | $\lambda_{NN}$ |

We can assume a reasonable loss function with the conditions in decision-theoretic rough set model, such that $0 \le \lambda_{PP} \le \lambda_{BP} \le \lambda_{NP}$ and $0 \le \lambda_{NN} \le \lambda_{BN} \le \lambda_{PN}$, then in an incomplete information system, $\forall x \in U$, by the maximal consistent block technique decision rules (P), (B) and (N) can be expressed as follows:

(P) if $\exists Y \subset \mu_A(x)$,s.t. $Pr(X|Y) \ge \alpha$ and $Pr(X|Y) \ge \gamma$, then $x \in POS(X)$;

(B) if $\exists Y \subset \mu_A(x)$, s.t. $Pr(X|Y) < \alpha$ and $Pr(X|Y) < \beta$, then $x \in BND(X)$;

(N) if $\exists Y \subset \mu_A(x)$, s.t. $Pr(X|Y) < \gamma$ and $Pr(X|Y) \le \beta$, then $x \in NEG(X)$;

where $\alpha = \frac{(\lambda_{PN}-\lambda_{BN})}{(\lambda_{PN}-\lambda_{BN})+(\lambda_{BP}-\lambda_{PP})}$; $\beta = \frac{(\lambda_{BN}-\lambda_{NN})}{(\lambda_{BN}-\lambda_{NN})+(\lambda_{NP}-\lambda_{BP})}$;

$\gamma = \frac{(\lambda_{PN}-\lambda_{NN})}{(\lambda_{PN}-\lambda_{NN})+(\lambda_{NP}-\lambda_{PP})}$.

$\forall x \in U$, since $0 \le \beta < \gamma < \alpha \le 1$, then we have

(P) if $\exists Y \subset \mu_A(x)$, s. t. $Pr(X|Y) \ge \alpha$, then $x \in POS(X)$;

(B) if $\exists Y \subset \mu_A(x)$, s. t. $\beta < Pr(X|Y) < \alpha$, then $x \in BND(X)$;

(N) if $\exists Y \subset \mu_A(x)$, s. t. $Pr(X|Y) \le \beta$, then $x \in NEG(X)$.

From discussions above, the lower approximation, the upper approximation of $X$ are

$$\underline{A}_{DT}(X) = \{x \in U : \exists Y \subset \mu_A(x), \text{s.t.} Pr(X|Y) \geqslant \alpha\};$$
$$\overline{A}_{DT}(X) = \{x \in U : \exists Y \subset \mu_A(x), \text{s.t.} Pr(X|Y) > \beta\}.$$

The pair $[\underline{A}_{DT}(X), \overline{A}_{DT}(X)]$ is referred to as a decision-theoretic rough set of $X$ in incomplete information system, $POS^A_{DT} = \underline{A}_{DT}(X)$ is referred to as the decision-theoretic positive region of $X$, $BND^A_{DT} = \overline{A}_{DT}(X) - \underline{A}_{DT}(X)$ is referred to as the decision-theoretic boundary region of $X$, $NEG^A_{DT} = U - \overline{A}_{DT}(X)$ is referred to as the decision-theoretic negative region of $X$.

## 3    Multicost Based Decision-Theoretic Rough Sets

For Yao's classical decision-theoretic rough set, one and only one $3 \times 2$ cost matrix is used. Yao assumed that such cost matrix comes from the experts evaluation. However, as what have been argued in Introduction part, it is clear that a single cost matrix is too subjective to help us make a better or a more suitable decision. Therefore, multiple cost matrixes and related decision-theoretic rough set approach will be explored in this section.

### 3.1    Decision Rules Based on Multicost

Suppose that $\mathbf{M}_1, \mathbf{M}_2, \ldots, \mathbf{M}_m$ are $m$ different cost matrixes, $\forall i = 1, \ldots, m$, the $i$-th loss function regarding the costs of three actions in two different states is given in Tab. 2.

**Table 2.** $i$-th cost matrix

|       | $X$            | $\sim X$       |
|-------|----------------|----------------|
| $e_P$ | $\lambda^i_{PP}$ | $\lambda^i_{PN}$ |
| $e_B$ | $\lambda^i_{BP}$ | $\lambda^i_{BN}$ |
| $e_N$ | $\lambda^i_{NP}$ | $\lambda^i_{NN}$ |

In Tab. 2, $\lambda^i_{PP}$; $\lambda^i_{BP}$ and $\lambda^i_{NP}$ are the $i$-th losses for taking actions of $e_P$; $e_B$ and $e_N$, respectively, when $x$ is included into class $X$. $\lambda^i_{PN}$; $\lambda^i_{BN}$ and $\lambda^i_{NN}$ are the $i$-th losses for taking actions of $e_P$; $e_B$ and $e_N$, respectively, when $x$ is out of class $X$. Similar to classical decision theoretic rough set, for each cost matrix, we may obtain three minimum-risk decision rules. For example, $\forall x \in U$, by considering the $i$-th cost matrix, the corresponding minimum-risk decision rules are:

(P) if $\exists Y \subset \mu_A(x)$, s.t. $Pr(X|Y) \geq \alpha_i$, then $x \in POS(X)$;

(B) if $\exists Y \subset \mu_A(x)$, s.t. $\beta_i < Pr(X|Y) < \alpha_i$, then $x \in BND(X)$;

(N) if $\exists Y \subset \mu_A(x)$, s.t. $Pr(X|Y) \leq \beta_i$, then $x \in NEG(X)$;

where $\alpha_i = \frac{(\lambda^i_{PN}-\lambda^i_{BN})}{(\lambda^i_{PN}-\lambda^i_{BN})+(\lambda^i_{BP}-\lambda^i_{PP})}$; $\beta_i = \frac{(\lambda^i_{BN}-\lambda^i_{NN})}{(\lambda^i_{BN}-\lambda^i_{NN})+(\lambda^i_{NP}-\lambda^i_{BP})}$.

Therefore, the immediate problem is to fuse these $m$ different (P), (B) or (N) rules when facing $m$ cost matrixes. To achieve such goal, we will present the following two fusion strategies: one is the optimistic strategy and the other is the pessimistic strategy.

1. Firstly, we may consider the minimal values of $m$ different values of $\alpha$ or $\beta$ and then the corresponding fusions of risk decision rules are:

(IP) if $\exists Y \subset \mu_A(x)$, s.t. $Pr(X|Y) \geq min_{i=1}^m \alpha_i$, then $x \in POS(X)$;

(IB) if $\exists Y \subset \mu_A(x)$, s.t. $min_{i=1}^m \beta_i < Pr(X|Y) < min_{i=1}^m \alpha_i$, then $x \in BND(X)$;

(IN) if $\exists Y \subset \mu_A(x)$, s.t. $Pr(X|Y) \leq min_{i=1}^m \beta_i$, then $x \in NEG(X)$.

Take for instance (IP), we can see that if the probability $Pr(X|Y)$ is greater than or equal to the minimum of $m$ values of $\alpha$, then $x$ should be included into the positive region of $X$. The explanations of (IB) and (IN) are similar to that of (IP). These semantic explanations show us an optimistic view of the fusion of $m$ decision rules.

2. Secondly, we may consider the maximal values of $m$ different values of $\alpha$ or $\beta$ and then the corresponding fusions of risk decision rules are:

(IIP) if $\exists Y \subset \mu_A(x)$, s.t. $Pr(X|Y) \geq max_{i=1}^m \alpha_i$, then $x \in POS(X)$;

(IIB) if $\exists Y \subset \mu_A(x)$, s.t. $max_{i=1}^m \beta_i < Pr(X|Y) < max_{i=1}^m \alpha_i$, then $x \in BND(X)$;

(IIN) if $\exists Y \subset \mu_A(x)$, s.t. $Pr(X|Y) \leq max_{i=1}^m \beta_i$, then $x \in NEG(X)$.

Take for instance (IIP), we can see that if the probability $Pr(X|Y)$ is greater than or equal to the maximum of $m$ values of $\alpha$, then $x$ can be included into the positive region of $X$. The explanations of (IIB) and (IIN) are also similar to that of (IIP). These semantic explanations show us a pessimistic view of the fusion of $m$ decision rules.

## 3.2 Optimistic and Pessimistic Decision-Theoretic Rough Sets

From the above two fusion strategies of risk decision rules, it is not difficult to present the following two multicost based decision-theoretic rough sets, we call them optimistic and pessimistic decision-theoretic rough sets, respectively.

**Definition 4.** *Let* $I \ =<\ U, AT >$ *be an information system in which* $A \subseteq AT$, $\mathbf{M}_1, \mathbf{M}_2, \ldots, \mathbf{M}_m$ *are m different cost matrixes,* $\forall X \subseteq U$, *the optimistic decision-theoretic lower approximation, upper approximation and boundary region of X are defined as*

$$\underline{A}_{ODT}(X) = \{x \in U : \exists Y \subset \mu_A(x), s.t. Pr(X|Y) \geqslant \min_{i=1}^{m} \alpha_i\}; \quad (4)$$

$$\overline{A}_{ODT}(X) = \{x \in U : \exists Y \subset \mu_A(x), s.t. Pr(X|Y) > \min_{i=1}^{m} \beta_i\}; \quad (5)$$

$$BND_{ODT}(X) = \{x \in U : \exists Y \subset \mu_A(x), s.t. \min_{i=1}^{m} \beta_i < Pr(X|Y) < \min_{i=1}^{m} \alpha_i\}. \quad (6)$$

*the pessimistic decision-theoretic lower approximation, upper approximation and boundary region of X are defined as*

$$\underline{A}_{PDT}(X) = \{x \in U : \exists Y \subset \mu_A(x), s.t. Pr(X|Y) \geqslant \max_{i=1}^{m} \alpha_i\}; \quad (7)$$

$$\overline{A}_{PDT}(X) = \{x \in U : \exists Y \subset \mu_A(x), s.t. Pr(X|Y) > \max_{i=1}^{m} \beta_i\}; \quad (8)$$

$$BND_{PDT}(X) = \{x \in U : \exists Y \subset \mu_A(x), s.t. \max_{i=1}^{m} \beta_i < Pr(X|Y) < \max_{i=1}^{m} \alpha_i\}. \quad (9)$$

**Proposition 1.** *Let* $I \ =<\ U, AT >$ *be an information system in which* $A \subseteq AT$, $\mathbf{M}_1, \mathbf{M}_2, \ldots, \mathbf{M}_m$ *are m different cost matrixes,* $\forall X \subseteq U$, *the optimistic decision-theoretic lower approximation, upper approximation and boundary region of X are defined as*

$$\underline{A}_{ODT}(X) \supseteq \underline{A}_{PDT}(X); \tag{10}$$

$$\overline{A}_{ODT}(X) \supseteq \overline{A}_{PDT}(X). \tag{11}$$

*Proof.* $\forall X \in \underline{A}_{PDT}(X)$, by Eq.(7), we have $Pr(X|\mu_A(x)) \geq max_{i=1}^{m}\alpha_i$ and then $Pr(X|\mu_A(x)) \geq min_{i=1}^{m}\alpha_i$ holds obviously, it follows that $X \in \underline{A}_{ODT}(X)$, i.e., $\underline{A}_{ODT}(X) \supseteq \underline{A}_{PDT}(X)$, Similarity, it is not difficult to prove that $\overline{A}_{ODT}(X) \supseteq \overline{A}_{PDT}(X)$.

Proposition 1 shows the relationships between optimistic decision-theoretic lower/upper approximations and pessimistic decision-theoretic lower/upper approximations, respectively. It tells us that optimistic decision-theoretic lower and upper approximations include pessimistic decision-theoretic lower and upper approximations, respectively.

**Proposition 2.** *Let $I =< U, AT >$ be an information system in which $A \subseteq AT$, $M_1, M_2, \ldots, M_m$ are $m$ different cost matrixes, $\forall X \subseteq U$, we have*

$$\underline{A}_{ODT}(X) = \cup_{i=1}^{m} \underline{A}_{DT}^i(X); \tag{12}$$

$$\overline{A}_{ODT}(X) = \cup_{i=1}^{m} \overline{A}_{DT}^i(X); \tag{13}$$

$$\underline{A}_{PDT}(X) = \cap_{i=1}^{m} \underline{A}_{DT}^i(X); \tag{14}$$

$$\overline{A}_{PDT}(X) = \cap_{i=1}^{m} \overline{A}_{DT}^i(X). \tag{15}$$

*where $\underline{A}_{DT}^i(X)$ and $\overline{A}_{DT}^i(X)$ are classical decision-theoretic lower and upper approximations constructed by the $i$-th cost matrix.*

*Proof.* $\forall x \in U$, by Eq. (4), we have $x \in \underline{A}_{ODT}(X) \Leftrightarrow Pr(X|Y) \geq min_{i=1}^{m} \alpha_i$
$\Leftrightarrow Pr(X|Y) \geq \alpha_1 \vee Pr(X|Y) \geq \alpha_2 \vee \ldots \vee Pr(X|Y) \geq \alpha_m$
$\Leftrightarrow x \in \underline{A}_{DT}^1(X) \vee x \in \underline{A}_{DT}^2(X) \vee \ldots \vee x \in \underline{A}_{DT}^m(X)$
$\Leftrightarrow x \in \bigcup_{i=1}^{m} \underline{A}_{DT}^i(X).$
That completes the proof of $\underline{A}_{ODT}(X) = \bigcup_{i=1}^{m} \underline{A}_{DT}^i(X)$. Similarity, it is not difficult to prove Eqs. (13), (14) and (15).

Proposition 2 shows the relationships among multicost based two decision-theoretic rough sets and classical decision-theoretic rough set. The details are: (1) optimistic decision-theoretic lower approximation is the union of $m$ classical decision-theoretic lower approximations; (2) optimistic decision-theoretic upper approximation is the union of $m$ classical decision-theoretic upper approximations; (3) pessimistic decision-theoretic lower approximation is the intersection of $m$ classical decision-theoretic lower approximations; (4) pessimistic decision-theoretic upper approximation is the intersection of $m$ classical decision-theoretic upper approximations.

## 3.3   Costs in Decision System

The end result of rough set theory is to derive decision rules from decision system. A decision system is a special information system $I =< U, AT \cup D >$, in which $AT$ is the set of the condition attributes, while $D$ is the set of the decision attributes. To simplify our discussions, we only consider one decision attribute $d$ and then the decision system is denoted by $I =< U, AT \cup d >$ in the context of this paper.

Generally speaking, decision attribute $d$ can partition the universe into a set of the equivalence classes such that $U/IND(d) = \{D_1, D_2, \ldots, D_n\}$. Then, a decision rule can be regarded as the relationship between descriptions on condition and decision attributes. In decision-theoretic rough set theory, three important decision rules should be considered, they are positive, boundary and negative rules. In Section 3.1 we have presented the rules. Since we have presented optimistic and pessimistic decision-theoretic rough sets in Section 3.2, and then it is interesting to discuss the costs of different decision rules, which are supported by objects in our optimistic and pessimistic decision-theoretic rough sets.

For optimistic decision-theoretic rough set, we may consider following three optimistic costs of decision rules which are supported by objects in lower approximation, boundary region and negative region, respectively.

- Optimistic cost of positive rule which is supported by $x \in \underline{A}_{ODT}(D_j)$, where $Pr(D_j|Y) = max\{Y \subset \mu_A(x) : Pr(D_j|Y) \geq min_{i=1}^m \alpha_i\}$,

$$Pr(D_j|Y) \cdot min_{i=1}^m \lambda_{PP}^i + (1 - Pr(D_j|Y)) \cdot min_{i=1}^m \lambda_{PN}^i;$$

- Optimistic cost of boundary rule which is supported by $x \in BND_{ODT}(D_j)$, where $Pr(D_j|Y) = max\{Y \subset \mu_A(x) : min_{i=1}^m \beta_i < Pr(D_j|Y) \geq min_{i=1}^m \alpha_i\}$,

$$Pr(D_j|Y) \cdot min_{i=1}^m \lambda_{BP}^i + (1 - Pr(D_j|Y)) \cdot min_{i=1}^m \lambda_{BN}^i;$$

- Optimistic cost of negative rule which is supported by $x \notin \overline{A}_{ODT}(D_j)$, where $Pr(D_j|Y) = max\{Y \subset \mu_A(x) : Pr(D_j|Y) > min_{i=1}^m \beta_i\}$,

$$Pr(D_j|Y) \cdot min_{i=1}^m \lambda_{NP}^i + (1 - Pr(D_j|Y)) \cdot min_{i=1}^m \lambda_{NN}^i .$$

Optimistic costs mean that we are expecting to get the lowest possible costs in terms of $m$ different cost matrixes. By considering three optimistic costs we mentioned above, it is not difficult to present the following optimistic cost in a decision system.

$$COST_A^{OPT} =$$
$$\Sigma_{j=1}^n \Sigma_{x \in \underline{A}_{ODT}(D_j)} Pr(D_j|Y) \cdot min_{i=1}^m \lambda_{PP}^i + (1 - Pr(D_j|Y)) \cdot min_{i=1}^m \lambda_{PN}^i$$
$$+ \Sigma_{j=1}^n \Sigma_{x \in BND_{ODT}(D_j)} Pr(D_j|Y) \cdot min_{i=1}^m \lambda_{BP}^i + (1 - Pr(D_j|Y)) \cdot min_{i=1}^m \lambda_{BN}^i$$
$$+ \Sigma_{j=1}^n \Sigma_{x \notin \overline{A}_{ODT}(D_j)} Pr(D_j|Y) \cdot min_{i=1}^m \lambda_{NP}^i + (1 - Pr(D_j|Y)) \cdot min_{i=1}^m \lambda_{NN}^i .$$

Similar to optimistic case, for pessimistic decision-theoretic rough set, we may also consider following three pessimistic costs of decision rules which are supported by objects in lower approximation, boundary region and negative region, respectively.

- Pessimistic cost of positive rule which is supported by $x \in \underline{A}_{PDT}(D_j)$, where $Pr(D_j|Y) = min\{Y \subset \mu_A(x) : Pr(D_j|Y) \geq max_{i=1}^m \alpha_i\}$,

$$Pr(D_j|Y) \cdot max_{i=1}^m \lambda_{PP}^i + (1 - Pr(D_j|Y)) \cdot max_{i=1}^m \lambda_{PN}^i;$$

- Pessimistic cost of boundary rule which is supported by $x \in BND_{PDT}(D_j)$, where $Pr(D_j|Y) = min\{Y \subset \mu_A(x) : max_{i=1}^m \beta_i < Pr(D_j|Y) \geq max_{i=1}^m \alpha_i\}$,

$$Pr(D_j|Y) \cdot max_{i=1}^m \lambda_{BP}^i + (1 - Pr(D_j|Y)) \cdot max_{i=1}^m \lambda_{BN}^i;$$

- Pessimistic cost of negative rule which is supported by $x \notin \overline{A}_{PDT}(D_j)$, where $Pr(D_j|Y) = min\{Y \subset \mu_A(x) : Pr(D_j|Y) > max_{i=1}^m \beta_i\}$,

$$Pr(D_j|Y) \cdot max_{i=1}^m \lambda_{NP}^i + (1 - Pr(D_j|Y)) \cdot max_{i=1}^m \lambda_{NN}^i .$$

Pessimistic costs mean that we are expecting to get the highest possible costs in terms of $m$ different cost matrixes. Similar to the optimistic case, we may also present the following pessimistic cost in a decision system:

$$COST_A^{PES} =$$
$$\Sigma_{j=1}^n \Sigma_{x \in \underline{A}_{PDT}(D_j)} Pr(D_j|Y) \cdot max_{i=1}^m \lambda_{PP}^i + (1 - Pr(D_j|Y)) \cdot max_{i=1}^m \lambda_{PN}^i$$
$$+ \Sigma_{j=1}^n \Sigma_{x \in BND_{PDT}(D_j)} Pr(D_j|Y) \cdot max_{i=1}^m \lambda_{BP}^i + (1 - Pr(D_j|Y)) \cdot max_{i=1}^m \lambda_{BN}^i$$
$$+ \Sigma_{j=1}^n \Sigma_{x \notin \overline{A}_{PDT}(D_j)} Pr(D_j|Y) \cdot max_{i=1}^m \lambda_{NP}^i + (1 - Pr(D_j|Y)) \cdot max_{i=1}^m \lambda_{NN}^i.$$

We have presented the optimistic and pessimistic costs in a decision system, which can be computed by objects in different regions of the optimistic and pessimistic decision theoretic rough sets, respectively. Since optimistic and pessimistic decision-theoretic rough sets were proposed based on multiple cost matrixes, then it is interesting to analyze the differences between costs come from decision-theoretic rough sets with respect to multiple cost matrixes and single cost matrix. From such point of view, we will compute four different types of costs, they are optimistic cost, pessimistic cost, maximal cost and minimal costs. The mathematical expressions of maximal and minimal costs are presented as follows.

$$COST_{MAX} = max_{i=1}^{m} COST_A^i;$$
$$COST_{MIN} = min_{i=1}^{m} COST_A^i.$$

where $COST_A^i$ is the decision cost for the $i$-th cost matrix, i.e., the cost derived from Yao's decision-theoretic rough set, which is constructed based on the $i$-th cost matrix. Maximal/Minimal costs are the greatest/least values of $m$ decision costs, which are derived from $m$ Yao's decision theoretic rough sets since each cost matrix supports one Yao's decision-theoretic rough set.

$$COST_A^i = \Sigma_{j=1}^{n} \Sigma_{x \in \underline{A}_{DT}(D_j)} Pr(D_j|Y) \cdot \lambda_{PP}^i + (1 - Pr(D_j|Y)) \cdot \lambda_{PN}^i$$
$$+ \Sigma_{j=1}^{n} \Sigma_{x \in \overline{A}_{DT}(D_j) - \underline{A}_{DT}(D_j)} Pr(D_j|Y) \cdot \lambda_{BP}^i + (1 - Pr(D_j|Y)) \cdot \lambda_{BN}^i$$
$$+ \Sigma_{j=1}^{n} \Sigma_{x \notin \overline{A}_{DT}(D_j)} Pr(D_j|Y) \cdot \lambda_{NP}^i + (1 - Pr(D_j|Y)) \cdot \lambda_{NN}^i$$

## 4    Conclusions

According to the real world applications, generalising classical rough set model is important to rough set theory. This paper introduced the decision-theoretic rough set into incomplete information systems, and applied maximal consistent blocks to construct decision-theoretic rough set model. Furthermore, by considering the cost's diversity and variability, the notions of optimistic and pessimistic multicost decision-theoretic rough set models were proposed. Finally, the whole decision costs of optimistic and pessimistic multicost decision-theoretic rough sets have been further analyzed. The further research will focus on the attribute reduction of multicost decision-theoretic rough set and propose a more general model as far as possible.

**Acknowledgments.** This work is supported by the Natural Science Foundation of China (Nos. 61100116, 61272419, 61305058), Natural Science Foundation of Jiangsu Province of China (Nos. BK2011492, BK2012700, BK20130471), Qing Lan Project of Jiangsu Province of China, Postdoctoral Science Foundation of China (No. 2014M550293), Key Laboratory of Intelligent Perception and Systems for High-Dimensional Information (Nanjing University of Science and Technology), Ministry of Education (No. 30920130122005), Natural Science Foundation of Jiangsu Higher Education Institutions of China (Nos. 13KJB520003, 13KJD520008).

# References

1. Pawlak, Z.: Rough Sets-theoretical aspects of reasoning about data. Kluwer Academic, Dordrecht (1991)
2. Hu, Q.H., Che, X.J., Zhang, L., et al.: Rank entropy based decision trees for monotonic classification. IEEE Transactions on Knowledge and Data Engineering 24(11), 2052–2064 (2012)
3. Hu, Q.H., Pan, W.W., Zhang, L., et al.: Feature selection for monotonic classification. IEEE Transactions on Fuzzy Systems 20(1), 69–81 (2012)
4. Luo, G.Z., Yang, X.B.: Limited dominance-based rough set model and knowledge reductions in incomplete decision system. Journal of Information Science and Engineering 26(6), 2199–2211 (2010)
5. Yang, X.B., Yang, J.Y.: Incomplete information system and rough set theory: Models and attribute reductions. Science Press, Beijing (2012)
6. Kryszkiewicz, M.: Rough set approach to incomplete information systems. Information Sciences 112(1-4), 39–49 (1998)
7. Kryszkiewicz, M.: Rule in Incomplete Information Systems. Information Sciences 113(3-4), 271–292 (1999)
8. Leung, Y., Li, D.Y.: Maximal consistent block technique for rule acquisition in incomplete information systems. Information Sciences 153, 85–106 (2003)
9. Yang, X.B., Qi, Y.S., Song, X.N., Yang, J.Y.: Test cost sensitive multigranulation rough set: Model and minimal cost selection. Information Sciences 250, 184–199 (2013)
10. Min, F., He, H.P., Qian, Y.H., Zhu, W.: Test-cost-sensitive attribute reduction. Information Sciences 181(22), 4928–4942 (2011)
11. Min, F., Liu, Q.H.: A hierarchical model for test-cost-sensitive decision systems. Information Sciences 179(14), 2442–2452 (2009)
12. Min, F., Zhu, W.: Attribute reduction of data with error ranges and test costs. Information Sciences 211, 48–67 (2012)
13. Min, F., Hu, Q.H., Zhu, W.: Feature selection with test cost constraint. International Journal of Approximate Reasoning 55(1), 167–179 (2014)
14. Yao, Y.Y., Wong, S.K.M., Lingras, P.: A decision-theoretic rough set model. In: 5th International Symposium on Methodologies for Intelligent Systems, pp. 17–25 (1990)
15. Liu, D., Li, T.R., Li, H.X.: A multiple-category classification approach with decision-theoretic rough sets. Fundamenta Informaticae 115(2-3), 173–188 (2012)
16. Jia, X.Y., Liao, W.H., Tang, Z.M., et al.: Minimum cost attribute reduction in decision-theoretic rough set models. Information Sciences 219, 151–167 (2013)
17. Guan, Y.Y., Wang, H.K.: Set-valued information systems. Information Sciences 176, 2507–2525 (2006)

# A Method to Reduce Boundary Regions in Three-Way Decision Theory

Ping Li[1], Lin Shang[1], and Huaxiong Li[2]

[1] Key Laboratory for Novel Software Technology,
Department of Computer Science and Technology,
Nanjing University, Nanjing 210023, China
[2] School of Management and Engineering,
Nanjing University, Nanjing 210093, China
lipingnju@gmail.com, {shanglin,huaxiongli}@nju.edu.cn

**Abstract.** A method for dealing the boundary region in three-way decision theory is proposed. In the three-way decision theory, all the elements are divided into three regions: positive region, negative region and boundary region. Positive region makes a decision of acceptance, negative region makes a decision of rejection. They can generate certain rules. However, boundary region makes a decision of abstaining. They generate uncertain rule. In classification, we always do with the boundary region. In this paper, we propose a method based on tri-training algorithm to reduce the boundary region. In the tri-training algorithm, we build up three classifiers based on three-way decision. We divide all the data into three parts randomly, aiming to keep the three classifiers different. We adopt a voting mechanism to label test samples. Experiments have shown that in most cases, tri-training algorithm is not only benefit for reducing boundary regions but also for improving classification precision. We also find some rules about the parameters alpha and beta how to affect boundary regions and classification precision.

**Keywords:** Boundary region, three-way decision theory, tri-training algorithm.

## 1 Introduction

Since $Yao$ proposed Decision-theoretic rough set model($DTRS$)[1], many researchers have concentrated on the study of the theory. $DTRS$ is probabilistic rough set model and mainly makes two contributions to rough set theory [2]. On one hand, it provides the semantic interpretation of positive regions and negative regions which are used in the rough set models. On the other hand, compared with other probabilistic rough set model[3,4,5], it provides a theoretic framework to calculate the thresholds.

Based on Decision-theoretic rough set model, $Yao$ proposed three-way decision theory[6]. It is an extension of two-way decision theory. In traditional two-way decision theory, there are only two actions being taken: one is accepting, which means divide the sample into positive regions; another is rejecting,

D. Miao et al. (Eds.): RSKT 2014, LNAI 8818, pp. 834–843, 2014.
DOI: 10.1007/978-3-319-11740-9_76 © Springer International Publishing Switzerland 2014

which means divide the sample into negative regions. The theory requires the decision makers to make decision actions immediately regardless of lack of information or not. Thus, it may result in wrong decisions when the information is not enough. In this situation, we can take three-way decision strategy. This strategy is similar to human decision procedure in practical decision problems. When the information is enough, human will make decisions immediately; however, when the available information is limited, human will wait and see. Thus, the three-way decision consists of three regions: positive regions, negative regions and boundary regions.

All the time, the idea of three-way decision is widely used in many fields such as medical diagnosis [7,8,9,10], social judgment theory [11], management theory [12,13,14], paper review [15] and etc. After three-way decision theory was proposed by $Yao$, Many researchers also applied this theory in many applications, e.g. spam filtering [16, 17], text classification[18] and etc.

The main superiority of three-way decision compared to two-way decision is the utility of the boundary decision. However, it is also a disadvantage of three-way decision. Because we need to deal with the boundary regions. In those applications, they usually don't handle the boundary regions, just leave them to decision makers. If the boundary regions are large, decision makers may spend a lot of time dealing with them. In this case, how to reduce the boundary regions is a new problem. Aiming to solve this problem, we propose a method to reduce the boundary regions in three-way decisions. We adopt the idea of tri-training algorithm [19] and put forward a tri-training algorithm based on three-way decisions. Experimental results show the method we proposed is good at reducing the boundary regions.

The paper is organized as follows: In section 2, we introduce the basic theory of three-way decisions. In section 3, we introduce tri-training algorithm based on three-way decisions in detail. In section 4, we conduct several groups of comparative experiments and analyze the experimental results. In the last, we draw a conclusion.

## 2  Preliminaries

In this section, we will review some basic notions of three-way decision theory[6].

### 2.1  Basic Knowledge of Three-Way Decision

In the three-way decision, $\Omega = (X, \bar{X})$, which represents the actual state of an object. $A = \{a_P, a_N, a_B\}$ means action set,where $a_P, a_N, a_B$ represent the actions that decide an object to $POS(X)$, $NEG(X)$ and $BND(X)$ respectively, where $POS(X)$, $NEG(X)$, $BND(X)$ represent positive region, negative region and boundary region respectively. Table 1 presents all costs for three-way decisions. The cost $\lambda_{ij}$ forms a matrix denoted as $(\lambda_{ij})_{2*3}$, where $i \in \{P, B, N\}$ which represents three actions; and $j \in \{P, N\}$ which represents actual states of objects.

Normally, we have $\lambda_{PP} \leq \lambda_{BP} \leq \lambda_{NP}$ and $\lambda_{NN} \leq \lambda_{BN} \leq \lambda_{PN}$. Because the costs of right decisions are less than wrong decisions. Moreover, we make a reasonable assumption is that $\lambda_{PP}$ and $\lambda_{NN}$ are equal to zero.

**Table 1.** Decision cost Matrix

| Actual States | Decide POS(X) | Decide BND(X) | Decide NEG(X) |
|:---:|:---:|:---:|:---:|
| $X$ | $\lambda_{PP}$ | $\lambda_{BP}$ | $\lambda_{NP}$ |
| $\bar{X}$ | $\lambda_{PN}$ | $\lambda_{BN}$ | $\lambda_{NN}$ |

According to Bayesian decision theory, we compare all decision costs of $A = \{a_P, a_N, a_B\}$ and select out the action which has the minimum expected decision cost. So we first define the expected decision cost of action $a_P, a_N, a_B$ as follows:

$$R\left(a_P|\left[x\right]\right) = \lambda_{PP}P\left(X|\left[x\right]\right) + \lambda_{PN}P\left(\bar{X}|\left[x\right]\right). \tag{1}$$

$$R\left(a_N|\left[x\right]\right) = \lambda_{NP}P\left(X|\left[x\right]\right) + \lambda_{NN}P\left(\bar{X}|\left[x\right]\right). \tag{2}$$

$$R\left(a_B|\left[x\right]\right) = \lambda_{BP}P\left(X|\left[x\right]\right) + \lambda_{BN}P\left(\bar{X}|\left[x\right]\right). \tag{3}$$

According to the Bayesian decision procedure, the minimum risk decision rules are shown as follows:

$$if R\left(a_P|\left[x\right]\right) \leq R\left(a_N|\left[x\right]\right) \&\& R\left(a_P|\left[x\right]\right) \leq R\left(a_B|\left[x\right]\right), then\, decide\, POS\left(X\right). \tag{4}$$

$$if R\left(a_N|\left[x\right]\right) \leq R\left(a_P|\left[x\right]\right) \&\& R\left(a_N|\left[x\right]\right) \leq R\left(a_B|\left[x\right]\right), then\, decide\, NEG\left(X\right). \tag{5}$$

$$if R\left(a_B|\left[x\right]\right) \leq R\left(a_P|\left[x\right]\right) \&\& R\left(a_B|\left[x\right]\right) \leq R\left(a_N|\left[x\right]\right), then\, decide\, BND\left(X\right). \tag{6}$$

As defined above, $\lambda_{PP} \leq \lambda_{BP} \leq \lambda_{NP}$ and $\lambda_{NN} \leq \lambda_{BN} \leq \lambda_{PN}$, and $P\left(X|\left[x\right]\right) + P\left(\bar{X}|\left[x\right]\right) = 1$, we can get same equivalent formulas as follows:

$$if P\left(X|\left[x\right]\right) \geq \lambda \&\& P\left(X|\left[x\right]\right) \geq \alpha, then\, decide\, POS\left(X\right). \tag{7}$$

$$if P\left(X|\left[x\right]\right) \leq \lambda \&\& P\left(X|\left[x\right]\right) \leq \beta, then\, decide\, NEG\left(X\right). \tag{8}$$

$$if P\left(X|\left[x\right]\right) \geq \beta \&\& P\left(X|\left[x\right]\right) \leq \alpha, then\, decide\, BND\left(X\right). \tag{9}$$

where
$$\alpha = \frac{\lambda_{PN}-\lambda_{NN}}{(\lambda_{PN}-\lambda_{NN})+(\lambda_{NP}-\lambda_{PP})}, \beta = \frac{\lambda_{BN}-\lambda_{NN}}{(\lambda_{BN}-\lambda_{NN})+(\lambda_{NP}-\lambda_{BP})}, \gamma = \frac{\lambda_{PN}-\lambda_{NN}}{(\lambda_{PN}-\lambda_{NN})+(\lambda_{NP}-\lambda_{PP})}$$

Because $\alpha \geq \beta \geq \gamma$, the variations of decision rules are shown as follows:

$$if P\left(X|\left[x\right]\right) \geq \alpha, then\, decide\, POS\left(X\right). \tag{10}$$

$$if P\left(X|\left[x\right]\right) \leq \beta, then\, decide\, NEG\left(X\right). \tag{11}$$

$$if P\left(X|\left[x\right]\right) \geq \beta \&\& P\left(X|\left[x\right]\right) \leq \alpha, then\, decide\, BND\left(X\right). \tag{12}$$

## 2.2  The Calculated Way of Three Regions

As all we know, when we want to classify a sample into one of these three regions, we need to compute $P\left(X|\left[x\right]\right)$. However, the posterior probability $P\left(X|\left[x\right]\right)$ is not always easy to get from data directly. So in paper[17], it use a monotonic transformation of the posterior probability to construct an equivalent classifier. For the transformation, the decision rules are as follows:

$$POS\left(X\right) = \{x| \sum_{i=1}^{n} log \frac{P\left(x_i|X\right)}{P\left(x_i|\bar{X}\right)} \geq \alpha'\}. \tag{13}$$

$$BND\left(X\right) = \{\beta' \prec \sum_{i=1}^{n} log \frac{P\left(x_i|X\right)}{P\left(x_i|\bar{X}\right)} \prec \alpha'\}. \tag{14}$$

$$NEG\left(X\right) = \{x| \sum_{i=1}^{n} log \frac{P\left(x_i|X\right)}{P\left(x_i|\bar{X}\right)} \leq \beta'\}. \tag{15}$$

where $\alpha' = log \frac{P(\bar{X})}{P(X)} + log \frac{\lambda_{PN} - \lambda_{BN}}{\lambda_{BP} - \lambda_{PP}}$, $\beta' = log \frac{P(\bar{X})}{P(X)} + log \frac{\lambda_{BN} - \lambda_{NN}}{\lambda_{NP} - \lambda_{BP}}$
We know from the formulas, the parameters $\alpha'$ and $\beta'$ are not in $[0, 1]$ anymore.

According to formulas above, all the factors are easy to get from data, so in this paper, we use these formulas above to compute probabilities.

## 3  The Method to Reduce Boundary Regions in Three-Way Decisions

In this section, we will show the tri-training algorithm based three-way decisions for reducing the boundary regions.

Tri-training algorithm[19] was proposed by Zhou in 2005 for semi-supervised learning. It is a new co-training[20] style algorithm. Compared with traditional co-training algorithm, tri-training algorithm does not require sufficient and redundant views, nor does it require different classifiers. Zhou firstly trained three initial classifiers from data sets which are labeled. Then, for any classifier, an unlabeled example can be labeled for it as long as the other two classifiers agree on the labeling of this example.

According to the idea, we propose the tri-training algorithm based on three-way decisions for reducing boundary regions. Firstly, we select three splits of training data via random sampling. Then we train three classifiers based on three-way decisions on these splits. Because these splits are different, there are some differences between the classifiers. When a test example comes in, we use these three classifiers to make a decision. If more than two of these three classifiers give the same classification result, then the example will be given the result label. For example, two classifiers label it as POS(NEG or BND), then the example will be given POS(NEG or BND) respectively. If exists this situation like that one classifier label it as POS, one classifier label it as NEG and the last classifier label it as BND, then the example will be given boundary label finally.

The detail of our algorithm is presented as below.

---

**Algorithm 1.** Three-way decision classifier

**Input:**
   Decision table $S = (U, C, D, V, f)$.
**Output:**
   a set named $bnd$, means all the samples in the boundary regions.
1. $bnd = \phi$
2. initialize $\lambda_{PP}, \lambda_{PN}, \lambda_{PB}, \lambda_{NN}, \lambda_{NP}, \lambda_{NB}$, then compute parameters $\alpha, \beta$; or set up $\alpha, \beta$ artificially.
3. make statistic of training data, preparing for probability calculation.
4. for every test sample, compute $p = \sum_{i=1}^{n} log \frac{P(x_i|X)}{P(x_i|\bar{X})}$ according to the statistic results.
   if $p \geq \alpha$, then divide the test sample into positive class.
   if $p \leq \beta$, then divide the test sample into negative class.
   if $\beta \prec p \prec \alpha$, then divide the test sample into boundary region, add the sample into $bnd$
5. output $bnd$

---

Algorithm 1 describes a classifier based on three-way decision theory in detail. Firstly, we compute the probability of each test sample, according to which we divide the sample into respective regions.

---

**Algorithm 2.** Tri-training algorithm based on three-way decisions

**Input:**
   Decision table $S = (U, C, D, V, f)$.
**Output:**
   a set named $bnd$, means all the samples in the boundary regions.
1. $bnd = \phi$
2. generate $s_i, i = 1...3$ randomly, $s_i$ presents a split of training data for each three-way decision classifier.
3. for $(i = 1...3)$
   train the three-way decision classifier according to $s_i$.
4. for each test sample, three classifiers give the classification result respectively.
   if more than two classifiers give the same results, then the test sample will be classified according to the result.
   if one classifier label it as POS, one classifier label it as NEG, and another classifier label it as BND, then add the sample into $bnd$.
5. output $bnd$

---

Algorithm 2 describes the tri-training algorithm based on three-way decisions in detail. Each classifier is a three-way decision classifier.

# 4    Experiments

We carry out several experiments to show that the tri-training algorithm are effective to reduce boundary regions. At the same time, we find some rules about the parameters alpha and beta how to affect boundary regions and classification precision.

## 4.1    Experimental Settings

Our experimental environment is as follows: Hardware environment: Inter(R) Core(TM) 2 Duo, 2.4GHz, 4GB memory; Software environment: the operation system is *windows*7; development environment is Kepler Release 2013, *JDK*7.

We choose two data sets from *UCI* machine learning repository[21]. Table 2 shows the details of the data sets.

**Table 2.** Data sets from *UCI*

| Data | Data Description | Number of Training data | Number of Testing data | Attributes | Classes |
|------|------------------|-------------------------|------------------------|------------|---------|
| *wdbc* | Breast Cancer Wisconsin(Diagnostic) | 469 | 100(pos:77, neg:33) | 32 | 2 |
| *wpdc* | Breast Cancer Wisconsin(Prognostic) | 198 | 28(pos:10, neg:18) | 34 | 2 |

In first dataset *wdbc*, we set up different $\alpha'$ and $\beta'$ by changing decision cost. With these different parameters, we conduct four groups of comparative experiments between three-way decision classifier and tri-training algorithm based on three-way decision classifier. In second dataset *wpdc*, we change $\alpha'$ and $\beta'$ artificially. We also conduct two groups of comparative experiments.

## 4.2    Experimental Results

The experimental results are shown in Table 3, Fig 1, Table 4, and Fig 2. TW means three-way decision classifier, TR means tri-training algorithm based on three-way decision classifier.

In Table 3, we conduct four groups of experiments. Group 1 is that we set up PN/NP={10,50,100}, and parameter $\beta'$ is not changed, but $\alpha'$ grows gradually; Group 2 is that we set up NP/PN={10,50,100}, and parameter $\alpha'$ is not changed, but $\beta'$ declines gradually; Group 3 and Group 4 are that we set up BP/BN={10,20,30} and BN/BP={10,20,30} respectively, and the parameter $\alpha'$ declines but $\beta'$ grows gradually. PN, NP, BP and BN represent $\lambda_{PN}, \lambda_{NP}, \lambda_{PB}, \lambda_{BN}$ respectively.

**Table 3.** Experimental results on dataset *wdbc*

| Group | Decision Cost | $\alpha'$ and $\beta'$ | preOfTW | preOfTR | bndOfTW | bndOfTR |
|---|---|---|---|---|---|---|
| 1 | PN=100.0, BN=5.0, BP=5.0, NP=10.0 | $\alpha'$=1.10805, $\beta'$=-0.17069 | 0.92 | 0.93 | 1 | 1 |
| 1 | PN=500.0, BN=5.0, BP=5.0, NP=10.0 | $\alpha'$=1.82493, $\beta'$=-0.17069 | 0.92 | 0.92 | 1 | 3 |
| 1 | PN=1000.0, BN=5.0, BP=5.0, NP=10.0 | $\alpha'$=2.12815, $\beta'$=-0.17069 | 0.92 | 0.92 | 1 | 3 |
| 2 | PN=10.0, BN=5.0, BP=5.0, NP=100.0 | $\alpha'$=-0.17069, $\beta'$=-1.44944 | 0.92 | 0.94 | 3 | 1 |
| 2 | PN=10.0, BN=5.0, BP=5.0, NP=500.0 | $\alpha'$=-0.17069, $\beta'$=-2.16633 | 0.92 | 0.94 | 3 | 1 |
| 2 | PN=10.0, BN=5.0, BP=5.0, NP=1000.0 | $\alpha'$=-0.17069, $\beta'$=-2.46954 | 0.92 | 0.94 | 3 | 1 |
| 3 | PN=500.0, BN=5.0, BP=50.0, NP=500.0 | $\alpha'$=0.82493, $\beta'$=-2.12493 | 0.92 | 0.93 | 4 | 2 |
| 3 | PN=500.0, BN=5.0, BP=100.0, NP=500.0 | $\alpha'$=0.52390, $\beta'$=-2.07378 | 0.92 | 0.93 | 3 | 2 |
| 3 | PN=500.0, BN=5.0, BP=150.0, NP=500.0 | $\alpha'$=0.34781, $\beta'$=-2.01579 | 0.92 | 0.93 | 3 | 2 |
| 4 | PN=500.0, BN=50.0, BP=5.0, NP=500.0 | $\alpha'$=1.78354, $\beta'$=-1.16633 | 0.92 | 0.92 | 3 | 4 |
| 4 | PN=500.0, BN=100.0, BP=100.0, NP=5.0 | $\alpha'$=1.73239, $\beta'$=-0.86530 | 0.92 | 0.92 | 3 | 4 |
| 4 | PN=500.0, BN=150.0, BP=5.0, NP=500.0 | $\alpha'$=1.67440, $\beta'$=-0.689210 | 0.92 | 0.92 | 2 | 4 |

According to the results in Table 3, we generate 4 line charts in Fig 1, where alpha and beta means $\alpha'$ and $\beta'$; preOfTW and preOfTR means classification precision of three-way decision classifier and that of tri-training classifier; bndOfTW and bndOfTR means the number of boundary regions of three-way classifier and that of tri-training classifier.

From Table 3 and Fig 1, we can obtain:
(1)When the parameters $\alpha'$ and $\beta'$ are same, the classification precision of tri-training classifier is higher than three-way decision classifier.
(2)When $\beta'$ is not changed, $\alpha'$ grows gradually, there is no change in boundary regions of three-way decision classifier; however, the boundary regions of tri-training classifier grows a little.
(3)When $\alpha'$ is not changed, $\beta'$ declines gradually, there is no change in boundary regions of three-way decision classifier and tri-training classifier, and at the same time, the number of boundary regions of tri-training classifier is less than three-way decision classifier.
(4)When $\alpha'$ declines and $\beta'$ grows, the boundary regions in three-way decision classifier declines; however, there is no change of tri-training classifier.
(5)In most cases, the boundary regions of tri-training classifier is less than three-way decision classifier.

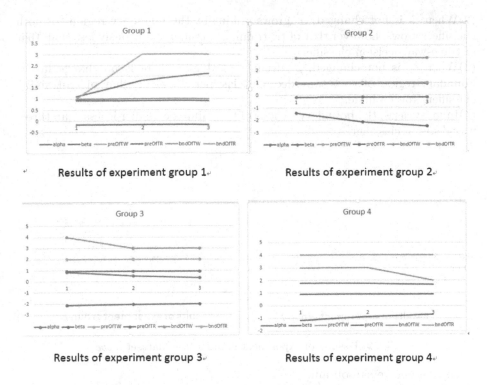

Results of experiment group 1.    Results of experiment group 2.

Results of experiment group 3.    Results of experiment group 4.

**Fig. 1.** Results of experiment group 1-4 on dataset *wdbc*

In Table 4, we set up parameters $\alpha'$ and $\beta'$ artificially. We conduct two groups of comparative experiments. One is that keeping $\beta'$ unchangeable, $\alpha'$ grows gradually; another is that keeping $\alpha'$ unchangeable, $\beta'$ declines gradually.

**Table 4.** Experimental results on dataset *wpbc*

| Group | $\alpha'$ and $\beta'$ | preOfTW | preOfTR | bndOfTW | bndOfTR |
|-------|------------------------|---------|---------|---------|---------|
| 1 | $\alpha'$ =0.3, $\beta'$ =0 | 0.6428 | 0.75 | 2 | 1 |
| 1 | $\alpha'$ =0.5, $\beta'$ =0 | 0.6428 | 0.75 | 3 | 1 |
| 1 | $\alpha'$ =1.0, $\beta'$ =0 | 0.5714 | 0.67 | 7 | 3 |
| 2 | $\alpha'$ =0.3, $\beta'$ =-0.2 | 0.6428 | 0.75 | 3 | 1 |
| 2 | $\alpha'$ =0.3, $\beta'$ =-0.5 | 0.6428 | 0.7142 | 3 | 2 |
| 2 | $\alpha'$ =0.3, $\beta'$ =-1.0 | 0.6428 | 0.6428 | 3 | 6 |

According to the results, we also generate 2 line charts in Fig 2.

From Table 4 and Fig 2, we can obtain:

(1)When the parameters $\alpha'$ and $\beta'$ are same, the classification precision of tri-training classifier is higher than three-way decision classifier.

(2)When $\beta'$ is not changed, $\alpha'$ grows gradually, the boundary regions of both classifiers grows, however that of tri-training classifier is obviously less than that of three-way decision classifier.

(3)When $\alpha'$ is not changed, $\beta'$ declines gradually, there is no change in the boundary region of three-way decision classifier, however, that of tri-training classifier grows a little.

(4)In most cases, the boundary regions of tri-training classifier is less than three-way decision classifier.

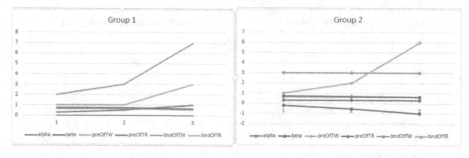

Results of experiment group 1          Results of experiment group 2

**Fig. 2.** Results of experiment group 1-2 on dataset *wpbc*

To sum up, we can obtain:

(1)When the parameters $\alpha'$ and $\beta'$ are same, the classification precision of tri-training classifier is higher than three-way decision classifier.

(2)The parameters $\alpha'$ and $\beta'$ will affect the boundary regions. With the change of these two parameters, the boundary regions of tri-training algorithm is not always superior to that of three-way decision classifier; however, in most cases, it is superior to that of three-way decision classifier.

## 5    Conclusion

In this paper, we proposed a method to reduce boundary regions in three-way decision. We adopted the idea of tri-training. In the tri-training algorithm, we built up three classifiers based on three-way decision. For a test sample, we used a voting mechanism to label it. We conducted several groups of comparative experiments between three-way decision and tri-training algorithm. The experimental results show that compared with three-way decision classifier, the classification precision of tri-training algorithm is higher; and in most cases, the boundary regions are less. Thus, tri-training algorithm based on three-way decision is benefit for reducing boundary regions and improving classification precision.

**Acknowledgements.** This work is supported by the National Natural Science Foundation of China (NSFC No.61170180, NSFC No.61035003, NSFC No.71201076) and Natural Science Foundation of Jiangsu, China(BK2011564).

# References

1. Yao, Y.Y., Wong, S.K.M., Lingras, P.: A decision-theorectic rough set model. Methodologies for Intelligent Systems 5, 17–24 (1990)
2. Pawlak, Z.: Rough sets. International Journal of Computer and Information Science 11(5), 341–356 (1982)
3. Pawlak, Z.: Theoretical Aspects of Reasoning about Data. Kluwer Academic Publishers, Dordrecht (1991)
4. Ziarko, W.: Variable precision rough set model. Journal of Computer and System Science 46, 39–59 (1993)
5. Slezak, D., Ziarko, W.: The investigation of the Bayesian rough set model. International Journal of Approximation Reasoning 40, 81–91 (2005)
6. Yao, Y.Y.: Three-way decision: An interpretation of rules in rough set theory. Rough Sets and Knowledge Technology, 642–649 (2009)
7. Chan, A., Gilon, D., Manor, O., et al.: Probabilistic reasoning and clinical decision-making: do doctors overestimate diagnostic probalities. An International Journal of Medicine 96, 763–769 (2003)
8. Lurie, J.D., Sox, H.C.: Principles of medical decision making. Spine 24, 493–498 (1999)
9. Pauker, S.G., Kassirer, J.P.: The threshold approach to clinical decision making. The New England Journal of Medicine 302, 1109–1117 (1980)
10. Van Der Gaag, L.C., Coupe, V.M.H.: Sensitive analysis for threshold decision making with Bayesian belief network. In: Advances in Artifical Intelligence. LNCS, vol. 1792, pp. 37–48 (2000)
11. Sherif, M., Hovland, C.I.: Social judgement: assimilation and constrast effects in communication and attitude change. Yale University Press, New Haven (1961)
12. Forster, M.R.: Key concepts in model selection performance and generalizability. Journal of Mathematical Psychology 44, 205–231 (2000)
13. Goudey, R.: Do statistical inferences allowing three alternative decisions give better feedback for environmentally precautionary decision-making. Journal of Environment Management 85, 338–334 (2007)
14. Woodward, P.W., Naylor, J.C.: An application of Bayesian methods in SPC. The Staatistician 42, 461–469 (1993)
15. Weller, A.C.: Editorial peer review: Its strengths and weakness. Information Today, Inc., Medford (2001)
16. Jia, X., Zheng, K., Li, W., Liu, T., Shang, L.: Three-way decisions solution to filter spam email: An empirical study. In: Yao, J., Yang, Y., Słowiński, R., Greco, S., Li, H., Mitra, S., Polkowski, L. (eds.) RSCTC 2012. LNCS, vol. 7413, pp. 287–296. Springer, Heidelberg (2012)
17. Zhou, B., Yao, Y.Y., Luo, J.G.: Cost-sensitive three-way email spam filtering. Journal of Intelligent Information Systems (2013)
18. Li, W., Miao, D.Q., Wang, W.L., Zhang, N.: Hierarchical rough decision theoretic framework for text classification. In: Proceedings of the 9th International Conference on Congnitive Informatics, pp. 484–489 (2010)
19. Zhou, Z.H., Li, M.: Tri-training: Exploiting unlabeled data using three classifiers. IEEE Transactions Knowledge and Data Engineering, 1529–1541 (2005)
20. Blum, A., Mitchell, T.: Combining Labeled and Unlabeled Data with Co-Training. In: Computational Learning Theory, pp. 92–100 (1998)
21. UCI Machine Learning Repository, http://archive.ics.uci.edu/ml/

# An Integrated Method for Micro-blog Subjective Sentence Identification Based on Three-Way Decisions and Naive Bayes

Yanhui Zhu, Hailong Tian, Jin Ma, Jing Liu, and Tao Liang

School of Computer and Communication, Hunan University of Technology,
Zhuzhou 412008, Hunan, China
{swayhzhu,tianhailongbmg}@163.com

**Abstract.** Microblog's subjective sentence recognition is the basis of it's public opinion analysis further research .Therefore, its recognition accuracy is crucial for future research work. Owing to the imprecision or incomplete of information, the precision of traditional SVM, NB and other machine learning algorithms that for microblog's subjective sentence recognition is not ideal. Presents a method based on the integrated of three-way decision and Bayesian algorithms to distinguish microblog's subjective sentence. Compared with traditional Bayesian algorithms, Experimental results show that the proposed integrated approach can significantly improve the accuracy of subjective sentence's recognition .

**Keywords:** Three-way decision, Bayes, micro-blog, subjective sentence.

## 1    Introduction

With the rapid development of Internet, for people to obtain information , micro-blog has become an important channel. Statistical Report ,which was on Internet Development 33rd China Internet Network [1] shows that by the end of December 2013,China micro-blog users reached 281 million, and in   netizen the Micro-blog unilization ratio was 45.5%. In recent years, many scholars launched the study of micro-blog identificational subjective sentences, Yang Wu et al [2] by using Bayes classifier, researched the classification of subjective and objective to micro-blog statement. Firstly, they analyzed the main differences between micro-blog text and other texts, and extract some features of subjective and objective clues for the characteristics of expression about micro-blog text. Then, they researched to 2-POS mode on the best select way, finally, as semantic features by feature words and the objective and subjective clues, as grammatical features by 2-POS mode, to study their impact on the classification results using Naïve Bayes classifier. Experimental results show that the method of taking into account semantic features and the structural characteristics of grammatical is better than the mehtod of only considering a feature better. Its classification precision was 81.9%, the recall 80.5%, F-Measure 81.2%.

D. Miao et al. (Eds.): RSKT 2014, LNAI 8818, pp. 844–855, 2014.
DOI: 10.1007/978-3-319-11740-9_77 © Springer International Publishing Switzerland 2014

Alexander Pak et al [3] to select N-Gram and speech tagging of micro-blog as feature, identification research on subjective sentences of micro-blog using Naïve Bayes classifier, compare with the two kinds of classifiers that SVM(support vector machine) and CRFs, expermental results show that, subject sentence identification of micro-blog based on Naïve Bayes was the best.

It's a binary classification problem about identification of subjective sentence. Since micro-blog has characteristics of short text, colloquial, randomness, etc, the subjective information is often imprecise,  incomple, and the classification accuracy rate is not high enough .This paper presents the method of subjective sentence recognition that micro-blog based on Three-Way Decision and machine learning algorithms integrated, the method using decision-rough sets and Three-Way Decision as theoretical basis, achieve identification that subjective sentence of micro-blog through integration with Bayes classifier that many previous studies and the best effect. The experiments show that the method can significantly impove the accuracy of subjective sentence recongnition.

## 2    Three-Way Decision Theory

Three-Way Decision theory [4-6] is proposed in rough sets [7-8] and decision-rough sets [9-13] by Yao, based on this model, Yao study on the semantic of positive region, negative region, boundary region in the rough sets theory and proposed to explain the rough sets rules extraction problem from the perspective of three-way decisions.

Decision-rough set is the expand of Pawlak algebra rough sets and 0.5-probability rough sets[10].the core of rough sets is the definition of the upper and lower approximation.

Definition of the upper and lower approximation of Pawlak algebra rough sets:

$$\overline{apr}(X) = \{x \in U \mid [x] \cap X \neq \Phi\},$$
$$\underline{apr}(X) = \{x \in U \mid [x] \subseteq X\}. \tag{1}$$

Definition of positive region $POS(X)$, negative region $NEG(X)$, boundary region $BND(X)$ based on the upper and lower approximation:

$$POS(X) = \underline{apr}(X) = \{x \in U \mid [x] \subseteq X\},$$
$$NEG(X) = U - \overline{apr}(X) = \{x \in U \mid [x] \cap X = \Phi\},$$
$$BND(X) = \overline{apr}(X) - \underline{apr}(X) = \{x \in U \mid [x] \cap X \neq \Phi \wedge \neg([x] \subseteq X)\}. \tag{2}$$

For the set of states $\Omega = \{X, \neg X\}$, there is the probability of conditions ,following:

$$p(X \mid [x]) = \frac{|X \cap [x]|}{|[x]|},$$

$$p(\neg X \mid [x]) = \frac{|\neg X \cap [x]|}{|[x]|} = 1 - p(X \mid [x]).$$

(3)

Which $\left| \ . \ \right|$ represents a potential of collection, namely the number of collection,$[X]$ said equivalence class. The three domains of Pawlak algebra rough sets is described by using the probability as follows:

$$POS(X) = \{x \in U \mid p(X \mid [x]) \geq 1\},$$
$$BND(X) = \{x \in U \mid 0 < p(X \mid [x]) < 1\},$$
$$NEG(X) = \{x \in U \mid p(X \mid [x]) \leq 0\}.$$

(4)

The rough sets mode only using two extreme values of probability, it lack of fault tolerance when applied to classification decisions. Based on this, Yao et al proposed a decision-rough set model .Suppose $0 \leq \beta < \alpha \leq 1$, as a pair of thresholds, and define three domain as follows:

$$POS_{(\alpha,\beta)}(X) = \{x \in U \mid p(X \mid [x]) \geq \alpha\},$$
$$BND_{(\alpha,\beta)}(X) = \{x \in U \mid \beta < p(X \mid [x]) < \alpha\},$$
$$NEG_{(\alpha,\beta)}(X) = \{x \in U \mid p(X \mid [x]) \leq \beta\}.$$

(5)

When object $X$ belongs $X$, orders $\lambda_{pp}$, $\lambda_{np}$, $\lambda_{bp}$ as the loss function of $x$ divided into $POS(X)$, $NEG(X)$, $BND(X)$, when $X$ belongs to $\neg X$, appropriate orders $\lambda_{pn}$, $\lambda_{nn}$, $\lambda_{bn}$ as the loss function of divided into the same three domains. Shown in the following table:

**Table 1.** Table of Loss Function

|        | $POS(X)$      | $BND(X)$      | $NEG(X)$      |
|--------|---------------|---------------|---------------|
| $X$    | $\lambda_{pp}$ | $\lambda_{bp}$ | $\lambda_{np}$ |
| $\neg X$ | $\lambda_{pn}$ | $\lambda_{bn}$ | $\lambda_{nn}$ |

The risk of equivalence class divided into three domains is defined as:

$$R(POS(X) \mid [x]) = \lambda_{pp} p(X \mid [x]) + \lambda_{pn} p(\neg X \mid [x]),$$
$$R(BND(X) \mid [x]) = \lambda_{bp} p(X \mid [x]) + \lambda_{bn} p(\neg X \mid [x]),$$
$$R(NEG(X) \mid [x]) = \lambda_{np} p(X \mid [x]) + \lambda_{nn} p(\neg X \mid [x]).$$

(6)

Given the minimum rish decision rule by Bayes decision theory:

If $R(POS(X) \mid [x]) \leq R(BND(X) \mid [x])$ and $R(POS(X) \mid [x]) \leq R(NEG(X) \mid [x])$ then $X \in POS(X)$;

If $R(BND(X) \mid [x]) \leq R(POS(X) \mid [x])$ and $R(BND(X) \mid [x]) \leq R(NEG(X) \mid [x])$, then $X \in BND(X)$;

If $R(NEG(X) \mid [x]) \leq R(POS(X) \mid [x])$ and $R(NEG(X) \mid [x]) \leq R(BND(X) \mid [x])$ then $X \in NEG(X)$.

For $R(X \mid [x]) + R(\neg X \mid [x]) = 1$, orders

$$\lambda_{pp} \leq \lambda_{bp} < \lambda_{np},$$
$$\lambda_{nn} \leq \lambda_{bn} < \lambda_{pn},$$
$$(\lambda_{pn} - \lambda_{bn})(\lambda_{np} - \lambda_{bp}) > (\lambda_{bp} - \lambda_{pp})(\lambda_{bn} - \lambda_{nn}),$$
$$\alpha = \frac{(\lambda_{pn} - \lambda_{bn})}{(\lambda_{pn} - \lambda_{bn}) + (\lambda_{bp} - \lambda_{pp})},$$
$$\gamma = \frac{(\lambda_{pn} - \lambda_{nn})}{(\lambda_{np} - \lambda_{pp}) + (\lambda_{pn} - \lambda_{nn})},$$
$$\beta = \frac{(\lambda_{bn} - \lambda_{nn})}{(\lambda_{bn} - \lambda_{nn}) + (\lambda_{np} - \lambda_{bp})}. \tag{7}$$

The above rules has the following equivalent description:

If $p(X \mid [x]) \geq \alpha$ then $X \in POS(X)$;

If $\beta < p(X \mid [x]) < \alpha$ then $X \in BND(X)$;

If $p(X \mid [x]) \leq \beta$ then $X \in NEG(X)$.

## 3    Three-Way Decision Approach to Microblog Subjective Sentence Recognition

To the positive , negative and boundary region of decision-theoretic rough set,Yao proposed positive rules, negative rules, boundary rules of three-way decisions .Sets said as follows:

positive    rules : $R(X \mid [x]) \geq \alpha, [x] \subseteq POS_{(\alpha,\beta)}(X)$

negative    rules : $R(X \mid [x]) \leq \beta, [x] \subseteq NEG_{(\alpha,\beta)}(X)$

boundary    rules : $\beta < R(X \mid [x]) < \alpha, [x] \subseteq BND_{(\alpha,\beta)}(X)$

That is: $\forall x \in U$, the rate of $R X \mid [x])$ is greater than or equal to the $\alpha$ .Then divided $[x]$ into positive region of $X$,At the same time make positive region decision, Other formula explained in the same way. And $U$ is a finite set of objects.

Use $p( L \mid x)$ represent the rate that $x$ is belong to subjective sentences category, It microblog whether belongs to the subjective sentences three-way decisions representation are as follows:

$p( L \mid x) \geq \alpha$, Decide $x$ is subjective sentences;

$p( L \mid x) \leq \beta$, Decide $x$ is not subjective sentences

$\beta < p( L \mid x) < \alpha$, It microblog whether belongs to the subjective sentences is uncertain, made by artificial processing decisions.

Three-way decisions model divided microblog into subjective sentences classes, not subjective sentences classes and boundary classes and compared with two classification model, not simply added a category, transformed the two classification problems into the multi-classification problems, But building on the basis of the decision-theoretic rough set for the target concept collection of positive ,negative and boundary region depiction.

## 4     Evaluation Standard

For positive class (category of subjective sentences), and negative class (category of nor subjective sentences) of the three-way decisions, this paper uses four evaluation indexes which are   namely ,precision rate ( $P$ ), recalling rate ( $R$ ), $F$ and macro-average. The table as follows:

**Table 2.**   Contingency Table of Categories

|  | Actually belong to positive classes documents | Actually belong to negative   classes documents |
|---|---|---|
| Judged belong to positive classes documents | $a_{pp}$ | $a_{pn}$ |
| Judged belong to negative classes documents | $a_{np}$ | $a_{nn}$ |
| judged belong to boundary classes documents | $a_{bp}$ | $a_{bn}$ |

Then, there are evaluation indexes of positive classes and negative classes, as follows:

$$P_p = \frac{a_{pp}}{a_{pp} + a_{pn}} \qquad R_p = \frac{a_{pp}}{a_{pp} + a_{np} + a_{bp}},$$

$$P_n = \frac{a_{nn}}{a_{nn} + a_{np}} \qquad R_n = \frac{a_{nn}}{a_{nn} + a_{pn} + a_{bn}},$$

$$F_p = \frac{2 \times P_p \times R_p}{P_p + R_p} \qquad F_n = \frac{2 \times P_n \times R_n}{P_n + R_n},$$

$$P_{avg} = \frac{P_p + P_n}{2} \qquad R_{avg} = \frac{R_p + R_n}{2} \qquad F_{avg} = \frac{F_p + F_n}{2}. \tag{8}$$

In the formula, $P_p$, $P_n$ and $P_{avg}$ denote the precision for the positive classes ,negative class and macro-average, Similarly, other the same subscript do the corresponding explanation.

## 5    Experimental Design

### 5.1    Microblog Feature Extraction

This paper uses the method on the base of combination of dictionary and statistical analysis to select microblog's candidate subjective features, and information gain (IG) method to extract features.   All steps are as follows:

Step1: Structuring basic domain of view dictionary. Using the positive and negative emotion words of HowNet [14]. For positive and negative evaluation words, there are 8746 words after removing duplicate words.

Step2: Using multiple word segmentation system [15] to build custom thesaurus for the recognition of the domain of view word. To get 5820 custom thesaurus words .

Step3: Constructing view words in candidate domain. Combined the word in step1 and step2, so as to remove duplicate words, and then there will be 13827 words of the candidate domain of view words.

Step4: Expanding the domain of view words by the conjunctions word dictionary. To get 14064 words of the candidate domain of view words.

Step5: Statistics corpus question mark and exclamation point are occupying a larger proportion. Besides that there is differences of its proportion from its subjective sentence and non-subjective sentence. So the question mark and exclamation point can be candidate subjective features.

Step6: Building the candidate subjective features. Emerged step4 and step5, and see this combination as custom thesaurus to separate Corpus's words and added into segmentation system of ICTCLAS2014[16]. Extracted the same words from segmentation corpus and custom thesaurus, and see it as candidate subjective features. Finally get 6232 candidate subjective features.

Step7: Using IG to extract candidate subjective features, the extracting number is determined by experiment .

## 5.2    Threshold—An Explanation

Make the following assumptions:

$$\lambda_{pp} = \lambda_{nn} = 0,$$
$$\lambda_{np} = \eta\lambda_{bp},$$
$$\lambda_{pn} = \eta\lambda_{bn},$$
$$\lambda_{bn} = 2\lambda_{bp}.$$

(9)

Because in the massive micro-blog text environment, subjective sentence recognition is a consideration of sensitive issues. The cost of subjective micro-blog into non-subjective micro-blog is smaller than non- subjective micro-blog divided into subjective micro-blog. There are:

$$\alpha = \frac{(\lambda_{pn} - \lambda_{bn})}{(\lambda_{pn} - \lambda_{bn}) + (\lambda_{bp} - \lambda_{pp})} = \frac{2\eta - 2}{2\eta - 1} = 1 - \frac{1}{2\eta - 1},$$

$$\beta = \frac{(\lambda_{bn} - \lambda_{nn})}{(\lambda_{bn} - \lambda_{nn}) + (\lambda_{np} - \lambda_{bp})} = \frac{2}{\eta + 1},$$

$$\gamma = \frac{(\lambda_{pn} - \lambda_{nn})}{(\lambda_{np} - \lambda_{pp}) + (\lambda_{pn} - \lambda_{nn})} = \frac{2}{3}.$$

(10)

$\eta > 2$, as $\alpha > \gamma > \beta$. The final value for $\eta$ is determined by experiment.

## 5.3    The Three-Way Decision Classifier Design Based on NB

Using the prior probability and class-conditional estimate probability $p(L \mid x)$ [17]:

$$p(L \mid x) = \frac{p(L) \times p(x \mid L)}{p(x)}.$$

(11)

(1)   NB probability estimation model uses the Bernoulli model :

$p(L)$ = The total number of documents subjective sentence / The total number of documents in the training text .

$p(x_i \mid L)$ =(The number of contain $L$ of subjective sentences and $x_i$ of features +1) / ( The total number of documents $L$ of subjective sentence +2) .

(2)   Assume that each feature is independent to each other .

(3)   Total Probability : $p(x) = \sum_{L_i} p(x \mid L_i)p(L_i)$.

## 5.4    Experimental Result and Analysis

Experimental corpus as "Chinese micro-blog sentiment analysis evaluation data set" [18] , Choose one of 3415 in the corpus , After feature selection, feature extraction and feature weighting, using the vector space model for text representation. With constructing micro-blog decision table, in the experiment shows that the number of features being the 3000, based on a subjective sentence micro-blog NB best recognition performance. With this micro-blog decision table, Threshold test, Experimental set the interval [2,22], In steps of 0.5 threshold value of the experiment. Experimental results shown in Figure 1, Figure 2, Figure 3.

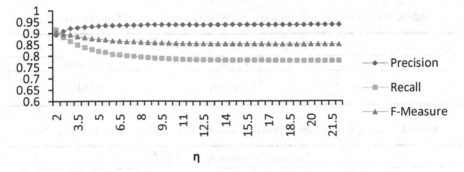

**Fig. 1.** Parameter $\eta$ experimental result for the positive class

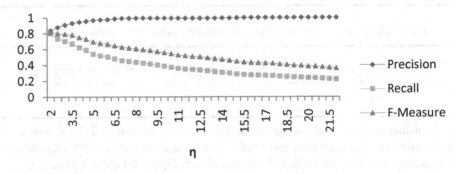

**Fig. 2.** Parameter $\eta$ experimental result for the negative class

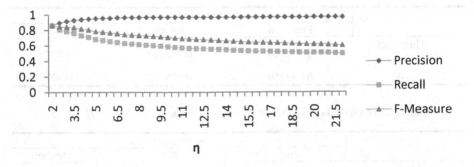

**Fig. 3.** Parameter $\eta$ experimental result for the macro-average

By comparison chart shows the experimental results, The overall trend with increasing values of $\eta$, precision rate on the rise, The recall and F-measure show a downward trend. Precision rate of the positive class is highest at $\eta = 14.5$, Precision rate of the negative class is 1 at $\eta = 15.5$, Precision rate of the macro-average is highest at $\eta = 15.5$. The results shown in Table 3, Table 4, Table 5:

**Table 3.** Three-way decisions improve precision rate effectively of the positive class

| The positive class $\eta = 14.5$ | | |
|---|---|---|
| Precision | Recall | F-Measure |
| 0.937398 | 0.780598 | 0.8518 |

**Table 4.** Three-way decisions improve precision rate effectively of the negative class

| The negative class $\eta = 15.5$ | | |
|---|---|---|
| Precision | Recall | F-Measure |
| 1 | 0.28039 | 0.437984 |

**Table 5.** Three-way decisions improve precision rate effectively of the macro-average

| The macro-average $\eta = 15.5$ | | |
|---|---|---|
| Precision | Recall | F-Measure |
| 0.968699 | 0.53049 | 0.645 |

Considering the precision, accuracy, recall and F-measure, $\eta = 3$ shows the best results of the experiment and results were compared with the NB classification. The results are shown in Table 6, Table 7, Table 8, Figure 4, Figure 5, Figure 6:

**Table 6.** Comparison of experimental results of the positive class at $\eta = 3$

| | The positive class | | |
|---|---|---|---|
| | Precision | Recall | F-Measure |
| Three-way | 0.921294 | 0.864914 | 0.8923 |
| NB | 0.823259 | 0.975521 | 0.8929 |

**Table 7.** Comparison of experimental results of the negative class at $\eta = 3$

| | The negative class | | |
|---|---|---|---|
| | Precision | Recall | F-Measure |
| Three-way | 0.906852 | 0.70058 | 0.79048 |

| NB | 0.932584 | 0.61786 | 0.74328 |

**Table 8.** Comparison of experimental results of the macro-average at $\eta = 3$

| | The macro-average | | |
| | Precision | Recall | F-Measure |
| Three-way | 0.914073 | 0.78275 | 0.842 |
| NB | 0.877922 | 0.79669 | 0.818 |

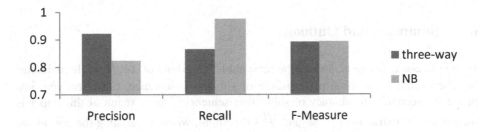

**Fig. 4.** Comparison of the positive class results

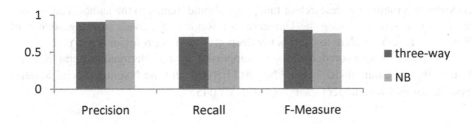

**Fig. 5.** Comparison of the negative class results

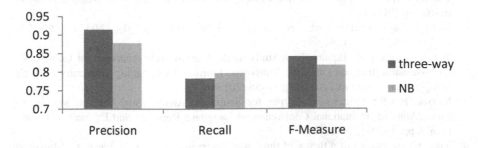

**Fig. 6.** Comparison of the macro-average results

By comparing the experimental results, compared with the NB method, Precision is improved by nearly 10%, F-Measure basically unchanged, Recall rate decreased slightly according to Three-Way Decision, which means the macro-average experimental result was significantly higher than NB. It described the Three-Way Decision method ,meanwhile maintaining the overall identification performance of subjective sentence, can significantly improve the accuracy of classification. There are a lot of subjective micro-blog statements in the mass micro-blog text environment. It plays a very important role for subsequent emotional research and improves the accuracy of the analysis of public opinion that improving the recognition accuracy of subjective sentence.

## 6    Summary and Outlook

Using three-way decision, and set the reasonable Threshold of $\alpha, \beta$ ,while maintaining the overall recognition performance of subjective sentence, can effectively improve the recognition accuracy of subjective sentence. The next job of this paper is mainly on the extraction of $\alpha$ and $\beta$'s threshold, while researching the consideration of microblog's three decisions. And probing a more effective evaluation criteria which based on three-way decision.

**Acknowledgment.** The research of this paper should thanks to the author Yanhui Zhu as a visiting scholar during the University of Regina in Canada in the guidance of Professor Yiyu Yao ,Here to express my deep gratitude to Professor Yiyu Yao.

This paper is sponsored by Project supported by the National Natural Science Foundation of China under Grant（No. 61170102）and the National Social Science Foundation of China under Grant（No. 12BYY045）.

## References

1. China Internet Network Information Center. The 33th China Internet Development StatisticsReport [EB/OL],
   `http://www.cnnic.net.cn/hlwfzyj/hlwxzbg/hlwtjbg/201301/P0201`
   `40221376266085836.pdf`
2. Wu, Y., Jingjing, S., Jiqiang, T.: A Study on the Classification Approach for Chinese MicroBlog Subjective and Objective Sentences. Journal of Chongqing University of Technology (Natural Science) 2013(01), 51–56 (2013)
3. Patrick, P.A.P.: Twitter as a Corpus for Sentiment Analysis and Opinion Mining. In: Proceedings of International Conference on Language Resource and Evaluation. Lisbon: [s. n. ], pp. 1320–1326 (2010)
4. Yao, Y.: An outline of a theory of three-way decisions. In: Yao, J., Yang, Y., Słowiński, R., Greco, S., Li, H., Mitra, S., Polkowski, L. (eds.) RSCTC 2012. LNCS, vol. 7413, pp. 1–17. Springer, Heidelberg (2012)
5. Yao, Y.Y.: Three-way decisions with probabilistic rough sets. Information Sciences 180, 341–353 (2010)

6. Yao, Y.Y.: The superiority of three-way decisions in probabilistic rough set models. Information Sciences 181, 1080–1096 (2011)
7. Pawlak, Z.: Rough sets. International Journal of Computer and Information Sciences 11(5), 341–356 (1982)
8. Pawlak, Z.: Rough set:theoretical aspects of reasonsing about data. Kluwer Academic Publishers, Dordrecht (1991)
9. Yao, Y.Y., Wong, S.K.M., Lingras, P.: A decision –theoretic rough set model. In: The 5th International Symposium on Methodologies for Intelligent Systems (1990)
10. Yao, Y.Y., Wong, S.K.M.: A decision theoretic framework for approximating concepts. International Journal of Man-Machine Studies 37, 793–809 (1992)
11. Yao, Y.Y.: Probabilistic approaches to rough sets. Expert System 20, 287–297 (2003)
12. Yao, Y.Y.: Probabilistic rough set approximations. International Journal of Approximate Reasoning 49, 255–271 (2008)
13. Wong, S.K.M., Ziarko, W.: A Probabilistic Model of Approximate Classification and Decision Rules with Uncertainty in Inductive Learning. Technical Report CS-85-23. Department of Computer Science. University of Regina (1985)
14. HowNet, http://www.keenage.com/html/c_index.html
15. Yanhui, Z., Yeqiang, X., Wenhua, W., et al.: Research on Opinion Extraction of Chinese Review. In: The Third Chinese Opinion Analysis Evaluation Proceedings, pp. 126–135. Conference Publishing, Shandong (2011)
16. ICTCLAS 2014 (2014), http://ictclas.nlpir.org/
17. Xiuyi, J., Lin, S., Xianzhong, Z., et al.: Threey-way decisions theory and its applications, pp. 61–79. Nanjing University Press, Nanjing (2012)
18. China Computer Federation. The Chinese Micro-Blog Emotional Analysis and Evaluation: Sample Data Sets [EB/OL] (July 01 2012),
    http://tcci.ccf.org.cn/conference/2012/pages/page04_eva.html

# Probabilistic Rough Set Model
# Based on Dominance Relation

Wentao Li[1] and Weihua Xu[1,2,*]

[1] School of Mathematics and Statistics,
Chongqing University of Technology, Chongqing, 400054, P.R. China
[2] Key Laboratory of Intelligent Perception and Systems for High-Dimensional
Information, Ministry of Education,
Nanjing University of Science and Technology, Nanjing, 210094, P.R. China
liwentao@2012.cqut.edu.cn,
chxuwh@gmail.com

**Abstract.** Unlike Pawlak rough set, probabilistic rough set models allow
a tolerance inaccuracy in lower and upper approximations. Dominance relation cannot establish probability measure space for the universe. In this
paper, the basic set assignment function, namely partition function is introduced into our work, which can transform the non-probability measure
generated by dominance relation into a probability measure space. The
probabilistic rough set model is established based on dominance relation,
and explained clearly through an example.

**Keywords:** Dominance relation, Probabilistic rough set, Partition function, Probability space.

## 1 Introduction

The notion of probabilistic rough set approximations was first introduced by Wong
and Ziarko [11], expressed through a pair of lower and upper approximations. The
acceptance of probabilistic rough sets is merely due to the fact that they are defined
by using probabilistic information and are more general and flexible. The introduction of probability enables the models to treat the universe of objects as samples
from a much larger universe [8]. Yao presented a decision making method based on
probabilistic rough set, which is called decision-theoretic rough set, where decision
rules obtained from positive region, negative region and boundary region [15], [17].
Essentially, the decision-theoretic rough set is a special case of probabilistic rough
set. The two thresholds in the probabilistic rough set model can be directly and
systematically calculated by minimizing the decision costs with Bayesian decision
procedure, which gives a brief semantics explanation in practical applications with
minimum decision risks. Bayesian decision theory deals with making decisions with

---

* Corresponding author is W.H. Xu. He is a Ph. D and a Prof. of Chongqing University of Technology. His main research fields are rough set, fuzzy set and artificial
intelligence.

D. Miao et al. (Eds.): RSKT 2014, LNAI 8818, pp. 856–863, 2014.
DOI: 10.1007/978-3-319-11740-9_78 © Springer International Publishing Switzerland 2014

minimum risk based on observed evidence. The probabilistic rough set has much more wider application after introducing the Bayesian decision principle.

Since the decision-theoretic rough set was proposed by Yao in 1990 [16], it has attracted much more attentions. Yao gave a decision theoretic framework for approximating concepts in 1992 [14]. Azam et al. proposed a threshold configuration mechanism for reducing the overall uncertainty of probabilistic regions in the probabilistic rough sets [1]. Jia et al. developed an optimization representation of decision-theoretic rough set model and raised an optimization problem [5]. Yu et al. applied decision-theoretic rough set model for automatically determining the number of clusters with much smaller time cost [18]. Liu et al. combined the logistic regression and the decision-theoretic rough set into a new classification approach [7].

The original rough set theory does not consider attributes with preference ordered domain, relations in the rough set theory are not equivalence relations. It is vital to propose an extension rough set theory called the dominance-based rough set approach [3] to take account into the ordering properties of criteria. The innovation is mainly based on substitution of the indiscernibility relation by a dominance relation. Recently, several studies have been made on properties and algorithmic implementations of dominance-based rough set approach [10]. Nevertheless, with the dominance-based rough set approach proposed by Greco et al. [3], only a limited number of methods use dominance-based rough set approach to acquire knowledge from inconsistent ordered information systems , but they did not clearly point out the semantic explanation of unknown values. Then Shao et al. further explored an extension of the dominance relation in an inconsistent ordered information system [9]. Many researchers have enriched the ordered theories and obtained many achievements. For instance, Xu et al. constructed a method of attribute reduction based on evidence theory in ordered information system [12], and others [2], [13].

Probabilistic rough set is based on an equivalence relation. However, in real life, one may often consider the rank of attributes. So we need to extend the probabilistic rough set theory by considering dominance relation. Relevantly, Greco et al. discussed a Bayesian decision theory for dominance-based rough set model in 2007 [4]. Kusunoki et al. studied an empirical risk associated with the classification function [6]. These approach want to take account into costs of misclassification in fixing parameters of the dominance-based rough set approach, while didn't transact the essence of issue about how to establish a probability measure space through a dominance relation. When we use the probabilistic rough set theory by considering a dominance relation, we may be face with problems that the dominance relation can't induce probability measure spaces. It is important that one solves this issue. Our objective is to explore how to establish probabilistic rough set model based on dominance relation. The rest of this paper is organized as follows. Some preliminary concepts about the rough set model based on dominance relation and probabilistic rough set are briefly reviewed in Section 2. In Section 3, we developed the probabilistic rough set based on dominance relation by using the partition function. Finally, Section 4 gets the conclusions.

## 2   Preliminaries

In this section, we review some basic concepts about rough sets based on dominance relation [3], probabilistic approaches to rough set theory.

A partial relation from $U$ to $U$ meets reflexivity, antisymmetry and transitivity, including decreasing preference $R^{\preceq}$ and increasing preference $R^{\succeq}$. As the decreasing preference can be converted to increasing preference, in this paper we only consider the increasing preference, namely the dominance relation $R^{\succeq}$ without any loss of generality.

Let $U$ be a universe of discourse, and $R^{\succeq}$ be a dominance relation on $U$. $U/R^{\succeq}$ is the set of dominance classes induced by a dominance relation $R^{\succeq}$, and $[x]_{R^{\succeq}}$ is called dominance class containing $x$. For an arbitrary set $X \subseteq U$, one can characterize $X$ by a pair of lower and upper approximations which are defined as follows.

$$\underline{R^{\succeq}}(X) = \{x \in U | [x]_{R^{\succeq}} \subseteq X\},$$

$$\overline{R^{\succeq}}(X) = \{x \in U | [x]_{R^{\succeq}} \cap X \neq \emptyset\}.$$

The pair $(\underline{R^{\succeq}}(X), \overline{R^{\succeq}}(X))$ is called the dominance-based rough set of $X$ with respect to $(U, R)$. If $\underline{R}(X) \neq \overline{R}(X)$, then $X$ is said to be a dominance-based rough set.

One can define $P$ as probability measure if the set-valued function $P$ maps from $2^U$ to $[0, 1]$, which can satisfy the two conditions: $P(U) = 1$; if $A \cap B = \emptyset$, then $P(A \cup B) = P(A) + P(B)$. And then $P$ is a probability measure of $\sigma - algebra$ which is combined by the family subset of $U$.

Mathematically, one may introduce a probability function on $\sigma - algebra$ of a universal set to construct a probabilistic approximation space, with which relationships between concepts can be defined in probabilistic terms. We can estimate the conditional probability of a set given an equivalence class. With probabilistic theory, an equivalence class is in the lower approximation if and only if an element in the equivalence class has a high probability (i.e., greater than or equal to a threshold) to be in the set.

Given $U$ as a non-empty and finite set of objects, where $R$ is an equivalence relation in $U$. Denote $[x]_R$ as the equivalence class with respect to $x$. And $P$ is a probability measure of $\sigma - algebra$ which is combined by the family subset of $U$. The triple $A_P = (U, R, P)$ is called probability approximation space.

**Definition 2.1.** [14] Let $0 \leq \beta < \alpha \leq 1$, for any $X \subseteq U$, the lower and upper approximations based on thresholds $\alpha, \beta$ with respect to $A_P = (U, R, P)$ are defined as follows

$$\underline{pr}_R^{(\alpha,\beta)}(X) = \{x \in U | P(X|[x]_R) \geq \alpha\},$$

$$\overline{pr}_R^{(\alpha,\beta)}(X) = \{x \in U | P(X|[x]_R) > \beta\}.$$

If $\underline{pr}_R^{(\alpha,\beta)}(X) = \overline{pr}_R^{(\alpha,\beta)}(X)$, then $X$ is a definable set, otherwise $X$ is a rough set.

Accordingly, the probabilistic positive, negative and boundary region are

$$pos(X) = \underline{pr}_R^{(\alpha,\beta)}(X) = \{x \in U | P(X|[x]_R) \geq \alpha\};$$

$$neg(X) = U - \overline{pr}_R^{(\alpha,\beta)}(X) = \{x \in U | P(X|[x]_R) \leq \beta\};$$

$$bn(X) = \overline{pr}_R^{(\alpha,\beta)}(X) - \underline{pr}_R^{(\alpha,\beta)}(X) = \{x \in U | \beta < P(X|[x]_R) < \alpha\}.$$

The parameters $\alpha, \beta$ in the probabilistic rough set theory above can be determined by special methods according to some additional conditions.

Based on the well-established Bayesian decision procedure, the decision-theoretic rough set model is derived from probability. That is to say, the decision-theoretic rough set model is a kind of probabilistic rough set model. The decision-theoretic rough set provides systematic methods for deriving the required thresholds on probabilistic rough set.

In real application of the probabilistic rough set models, we can obtain the thresholds $\alpha, \beta$ based on an intuitive understanding the levels of tolerance for errors. Just like we confirm the value of parameters $\alpha$ and $\beta$ included in the Section 3. And the calculation methods of the conditional probability can also meet for demands in application.

## 3  Probabilistic Rough Set Model Based on Dominance Relation

Probabilistic rough set models allow a tolerance inaccuracy in lower and upper approximations, or equivalently in the probabilistic positive, negative and boundary regions. When the relations are never equivalence relations but dominance relations, they will not produce the probability measure space. Here one can handle the dominance classes induced by the dominance relation with an operator to transform the non-probability measure into a probability measure space.

$R_A^{\succeq}$ is a dominance relation, $[x]_{R_A^{\succeq}}$ is the dominance class containing $x$. And $P(X|Y)$ is the conditional probability of whether concept $X$ happens or not depends on $Y$. We get the following definition.

**Definition 3.1.** Let $R_A^{\succeq}$ be a dominance relation. The basic set assignment function $j$ is from $2^U$ to $2^U$, is defined as

$$j(X) = \{x \in U | [x]_{R_A^{\succeq}} = X\}, X \in 2^U.$$

Obviously, $x \in j(X) \Leftrightarrow [x]_{R_A^{\succeq}} = X$.

The basic set assignment function $j([x]_{R_A^{\succeq}})$ contains these two properties:

- $\bigcup_{X \subseteq U} j(X) = U;$
- $For X \neq Y$, then $j(X) \bigcap j(Y) = \emptyset.$

It is easy to notice that the function $j([x]_{R_A^{\succcurlyeq}})$ is a partition function of the universe $U$, one can also call the partition function as set-valued mapping approximation operator. Accordingly, this operator transforms the triple $A_P = (U, R_A^{\succcurlyeq}, P)$, which is not a probability approximation space into probability measure approximation space.

**Definition 3.2** Let $R_A^{\succcurlyeq}$ be a dominance relation. Set $0 \leq \beta < \alpha \leq 1$, for any $X \subseteq U$, the lower and upper approximations based on parameters $\alpha, \beta$ with respect to $A_P = (U, R_A^{\succcurlyeq}, P)$ are defined as follows

$$\underline{jpr}_{R_A^{\succcurlyeq}}^{(\alpha,\beta)}(X) = \{x \in U | P(X|j([x]_{R_A^{\succcurlyeq}})) \geq \alpha\},$$

$$\overline{jpr}_{R_A^{\succcurlyeq}}^{(\alpha,\beta)}(X) = \{x \in U | P(X|j([x]_{R_A^{\succcurlyeq}})) > \beta\}.$$

If $\underline{jpr}_{R_A^{\succcurlyeq}}^{(\alpha,\beta)}(X) = \overline{jpr}_{R_A^{\succcurlyeq}}^{(\alpha,\beta)}(X)$, then $X$ is a definable set, otherwise $X$ is a rough set.

Accordingly, the probabilistic positive, negative and boundary region are

$$pos(X) = \underline{jpr}_{R_A^{\succcurlyeq}}^{(\alpha,\beta)}(X) = \{x \in U | P(X|j([x]_{R_A^{\succcurlyeq}})) \geq \alpha\};$$

$$neg(X) = U - \overline{jpr}_{R_A^{\succcurlyeq}}^{(\alpha,\beta)}(X) = \{x \in U | P(X|j([x]_{R_A^{\succcurlyeq}})) \leq \beta\};$$

$$bn(X) = \overline{jpr}_{R_A^{\succcurlyeq}}^{(\alpha,\beta)}(X) - \underline{jpr}_{R_A^{\succcurlyeq}}^{(\alpha,\beta)}(X) = \{x \in U | \beta < P(X|j([x]_{R_A^{\succcurlyeq}})) < \alpha\}.$$

An example is employed to present the probabilistic rough sets based on dominance relation.

**Example 3.1** In Table 1, $U = \{x_1, x_2, \cdots, x_7\}$ is a universe which consists of 7 objects, $a_1, a_2, a_3, a_4$ are the conditional attributes. One uses A, B, C, D to denote the values of these attributes. Moreover, A $\geq$ B $\geq$ C $\geq$ D.

**Table 1.** An information table

| $U$ | $a_1$ | $a_2$ | $a_3$ | $a_4$ |
|-----|-------|-------|-------|-------|
| $x_1$ | B | C | C | D |
| $x_2$ | C | B | B | A |
| $x_3$ | B | B | C | B |
| $x_4$ | A | D | A | C |
| $x_5$ | C | B | B | A |
| $x_6$ | B | A | D | B |
| $x_7$ | B | C | C | D |

Here we consider all of these four conditions: $a_1, a_2, a_3, a_4$, accordingly, $R^{\succcurlyeq}$ is the dominance relation induced by these four attributes. Then one can obtain that the dominance classes are as following

$[x_1]_{R^{\succcurlyeq}} = \{x_1, x_3, x_7\}$, $[x_2]_{R^{\succcurlyeq}} = \{x_2, x_5\}$, $[x_3]_{R^{\succcurlyeq}} = \{x_3\}$, $[x_4]_{R^{\succcurlyeq}} = \{x_4\}$, $[x_5]_{R^{\succcurlyeq}} = \{x_2, x_5\}$, $[x_6]_{R^{\succcurlyeq}} = \{x_6\}$, $[x_7]_{R^{\succcurlyeq}} = \{x_1, x_3, x_7\}$.

It is obvious that these seven classes form a covering of the universe, but not a partition. Accordingly, one may use the partition function $j$. Then we can get

$$j(X_1) = \{x_1, x_7\},$$
$$j(X_2) = \{x_2, x_5\},$$
$$j(X_3) = \{x_3\},$$
$$j(X_4) = \{x_4\},$$
$$j(X_5) = \{x_6\}.$$

These five sets, namely $j(X_1)$, $j(X_2)$, $j(X_3)$, $j(X_4)$ and $j(X_5)$ form a partition of the universe $U$.

Given $X = \{x_2, x_3, x_5\}$. We assume that $\alpha = 2/3, \beta = 1/4$. Conditional probability is $P(X|Y)$, where

$$P(X|Y) = \frac{|X \cap Y|}{|Y|}.$$

Then the conditional probabilities with respect to $R^\succ$ are shown as following:
$P(X|j([x_1]_{R^\succ})) = 1/3$, $P(X|j([x_7]_{R^\succ})) = 1/3$,
$P(X|j([x_2]_{R^\succ})) = 1$, $P(X|j([x_5]_{R^\succ})) = 1$,
$P(X|j([x_3]_{R^\succ})) = 1$,
$P(X|j([x_4]_{R^\succ})) = 0$,
$P(X|j([x_6]_{R^\succ})) = 0$.
The lower and upper approximations based on parameters $\alpha, \beta$ with respect to $A_P = (U, R^\succ, P)$ are computed as

$$\underline{jpr}_{R^\succ}^{(\frac{2}{3}, \frac{1}{4})}(X) = \{x \in U | P(X|j([x]_{R^\succ})) \geq 2/3\} = \{x_2, x_3, x_5\},$$
$$\overline{jpr}_{R^\succ}^{(\frac{2}{3}, \frac{1}{4})}(X) = \{x \in U | P(X|j([x]_{R^\succ})) > 1/4\} = \{x_1, x_2, x_3, x_5, x_7\}.$$

And then the probabilistic positive, negative and boundary region are

$$pos(X) = \underline{jpr}_{R^\succ}^{(\frac{2}{3}, \frac{1}{4})}(X) = \{x_2, x_3, x_5\};$$
$$neg(X) = U - \overline{jpr}_{R^\succ}^{(\frac{2}{3}, \frac{1}{4})}(X) = \{x_4, x_6\};$$
$$bn(X) = \overline{jpr}_{R^\succ}^{(\frac{2}{3}, \frac{1}{4})}(X) - \underline{jpr}_{R^\succ}^{(\frac{2}{3}, \frac{1}{4})}(X) = \{x_1, x_7\}.$$

Through the basic set assignment function, namely the partition function $j$, one can easily achieve the probability approximation space.

## 4    Conclusions

By considering the probabilistic rough sets based on dominance relation, the basic set assignment function, namely partition function is introduced into our work. The dominance relation results in a non-probability measure space. By the basic set assignment function, we can transact the covering of universe $U$ induced by a dominance relation into a partition of the universe $U$. This paper presents a partition function to construct a probability measure combining the probability and rough set theory, and proposes the probabilistic rough set based on dominance relation. In the future work, we can do further and relevant studies about the probabilistic rough set model based on dominance relation.

**Acknowledgements.** This work is supported by Natural Science Foundation of China (No. 61105041), National Natural Science Foundation of CQ CSTC (No. cstc 2013jcyjA40051), and Key Laboratory of Intelligent Perception and Systems for High-Dimensional Information (Nanjing University of Science and Technology), Ministry of Education (No. 30920140122006).

# References

1. Azam, N., Yao, J.T.: Analyzing uncertainty of probabilistic rough set region with game-theoretic rough sets. International Journal of Approximate Reasoning 55, 142–155 (2014)
2. Benferhat, S., Lagrue, S., Papini, O.: Reasoning with partially ordered information in a possibilistic logic framework. Fuzzy Set and Systems 144, 25–41 (2014)
3. Greco, S., Matarazzo, B., Slowinski, R.: Rough approximation by dominance relations. International Journal of Intelligent Systems 17, 153–171 (2002)
4. Greco, S., Słowiński, R., Yao, Y.: Bayesian decision theory for dominance-based rough set approach. In: Yao, J., Lingras, P., Wu, W.-Z., Szczuka, M.S., Cercone, N.J., Ślęzak, D. (eds.) RSKT 2007. LNCS (LNAI), vol. 4481, pp. 134–141. Springer, Heidelberg (2007)
5. Jia, X.Y., Tang, Z.M., Liao, W.H., Shang, L.: On an optimization representation of decision-theoretic rough set model. International Journal of Approximate Reasoning 55, 156–166 (2014)
6. Kusunoki, Y., Błaszczyński, J., Inuiguchi, M., Słowiński, R.: Empirical risk minimization for variable precision dominance-based rough set approach. In: Lingras, P., Wolski, M., Cornelis, C., Mitra, S., Wasilewski, P. (eds.) RSKT 2013. LNCS, vol. 8171, pp. 133–144. Springer, Heidelberg (2013)
7. Liu, D., Li, T.R., Li, H.X.: A multiple-category classification approach with decision-theoretic rough sets. Fundamenta Informaticae 115, 173–188 (2012)
8. Qian, Y.H., Zhang, H., Sang, Y.L., Liang, J.Y.: Multigranulation decision-theoretic rough sets. International Journal of Approximation Reasoning 55, 225–237 (2014)
9. Shao, M.W., Zhang, W.X.: Dominance relation and rules in an incomplete ordered information system. International Journal of Intelligent System 20, 13–27 (2005)
10. Susmaga, R., Slowinski, R., Greco, S., Matarazzo, B.: Generation of reducts and rules in multi-attributes and multi-criteria classification. Control and Cybernetics 4, 969–988 (2000)
11. Wong, S.K.M., Ziarko, W.: Comparison of the probabilistic approximate classification and the fuzzy set model. Fuzzy Sets and Systems 21, 357–362 (1987)
12. Xu, W.H., Zhang, X.Y., Zhong, J.M., Zhang, W.X.: Attribute reduction in ordered information system based on evidence theory. Knowledge Information Systems 25, 169–184 (2010)
13. Yang, X.B., Yang, J.Y., et al.: Dominance-based rough set approach and knowledge reductions in incomplete ordered information system. Information Sciences 178, 1219–1234 (2008)

14. Yao, Y.Y.: A decision theoretic framework for approximating concepts. International Journal of Man-Machine Studies 37, 793–809 (1992)
15. Yao, Y.Y.: Probabilistic approaches to rough sets. Expert Systems 20, 287–297 (2003)
16. Yao, Y.Y., Wong, S.K., Lingras, P.: A decision-theoretic rough set model. Methodlogies for Intelligent Systems 5, 17–24 (1990)
17. Yao, Y.Y.: Three-way decisions with probabilistic rough sets. Information Sciences 180, 341–353 (2010)
18. Yu, H., Liu, Z.G., Wang, G.Y.: An automatic method to determine the number of clusters using decision-theoretic rough set. International Journal of Approximation Reasoning 55, 101–115 (2014)

# Author Index

Ammar, Asma 753
Anitha, K. 69
Aszalós, László 399
Azam, Nouman 693

Bae, Hae-Young 149
Bai, Xiaoming 272
Bao, Zhongkui 673
Bi, Zhongqin 89

Cai, Mingjie 510
Chen, Degang 3
Chen, Hongmei 801
Chen, Lin 260
Chen, Peiqiu 364
Chen, Yufei 260
Chen, Zehua 577
Cheng, Xin 240, 332
Ciucci, Davide 15
Csajbók, Zoltán Ernő 15

Deng, Weihui 411
Du, Xiuquan 249
Duan, Huixian 229
Duan, Jie 121

Elouedi, Zied 753
El-Zekey, Moataz 27

Fan, Anjing 101
Feng, JiangFan 149
Feng, Tao 111
Fuh, Gene 376

Gao, Can 77
Gao, Cong 535
Gao, Zhen 376
Gautam, Vinay 39
Gu, Shen-Ming 525, 653
Guo, Qingjun 601
Guo, Yishuai 411

Han, Banghe 184
He, Xu 194
Hong, Zhiyong 569
Hsu, Ping-Yu 444

Hu, Qinghua 121
Huang, Jiajin 613
Huang, Jie 376
Huang, Li-Dong 295

Jiang, Tao 295
Jiao, Na 500
Jing, Liping 352
Jiyuan, Chen 285

Ke, Deying 557
Kothari, Ashwin G. 216

Lang, Guangming 510
Lei, Jingsheng 89
Li, Huaxiong 834
Li, LingYue 683
Li, Ping 834
Li, Qingguo 510
Li, Shaoyong 173
Li, Su 589
Li, Tianrui 173, 801
Li, Tong-Jun 525, 653
Li, Wei 260
Li, Wentao 856
Li, Xiaonan 184
Li, Xinrui 249
Lian, Hao 423
Liang, Decui 812
Liang, Jiye 789
Liang, Tao 844
Lin, He 683
Lin, Zhongda 272
Lingras, Pawan 547, 753
Liu, Caihui 59
Liu, Dun 812
Liu, Jie 742
Liu, Jing 844
Liu, Lu 272
Liu, Na 229
Liu, Yujiang 59
Lu, Qiang 149
Luo, Chuan 801
Lv, Ping 601
Lv, Renjie 423

Ma, He    577
Ma, Jin    844
Ma, Qiang    386
Ma, Xingbin    824
Majeed, Babar    693
Mei, Lin    229
Mi, Jusheng    111
Miao, Duoqian    77, 332, 707
Mihálydeák, Tamás    15, 399
Min, Fan    194
Mu, Bin    589

Niu, Junxia    129

Pan, Wei    432
Pei, WenBin    683
Peters, Georg    547

Qi, Jianjun    732
Qi, Yong    467, 824
Qian, Jin    601
Qin, Keyun    569
Qiu, Taorong    272

Raghavan, K.S. Sreenivasa    216

Shang, Lin    777, 834
Shao, Yabin    386
She, Kun    432
Shen, Yuan-Hong    444
Shuf, Yefim    376
Singh, Anupam K.    39
Song, Jingjing    467
Song, Lei    229
Song, Xiaoning    467, 824
Su, Ting    765

Tian, Hailong    844
Tian, Sheng    309
Tiwari, Kanchan S.    216
Tiwari, S.P.    39

Wang, Baoli    789
Wang, Guoyin    411, 631
Wang, Hanli    364
Wang, Hua-Chao    642
Wang, Jun    229
Wang, Lei    332
Wang, Meizhi    59
Wang, Min    59

Wang, Penghan    742
Wang, Ruizhi    240, 720
Wang, Xia    139
Wang, Xiaoxue    121
Wang, Xin    295
Wang, Xi-Zhao    642
Wang, Yinglong    454
Wang, Yongsheng    206
Wang, Zhicheng    260
Wei, Lai    320
Wei, Ling    732
Wei, Pengyuan    432
Wei, Zhihua    309, 364
Wu, Wei-Zhi    139, 525, 653
Wu, Yi    525

Xi, Dachao    631
Xia, Ying    149
Xian, Yikun    376
Xiao, Qimei    510
Xie, Xiang-An    444
Xu, Feifei    89
Xu, Weihua    856
Xu, You-Hong    653
Xue, Tianyu    742
Xue, Zhan'ao    742

Yadav, Vijay K.    39
Yan, Yuan-Ting    343
Yang, Bin    489
Yang, Jingyu    467, 824
Yang, Liu    352
Yang, Shanlin    673
Yang, Xibei    467, 824
Yao, JingTao    161, 693
Yao, Yiyu    535, 732
Ye, Lijuan    623
Yin, Ye    557
Yu, Dingjun    364
Yu, Fang    623
Yu, Hong    765
Yu, Hualong    467
Yu, Jian    352
Yu, Ying    454
Yuan, Shijin    589
Yue, Xiaodong    49, 77, 601

Zeng, Kai    432
Zeng, Xianhua    423, 765
Zha, Yongliang    249

Zhai, Jun-Hai    642
Zhang, Fan    631
Zhang, Lingjun    121
Zhang, Nan    49, 479
Zhang, Wei    77
Zhang, Xianyong    707
Zhang, Xiaoxia    3, 386
Zhang, Xuerui    411, 631
Zhang, Yan-Ping    249, 343
Zhang, Yi-Wen    343
Zhang, Yu    577
Zhang, Zehua    479
Zhang, Zhifei    720
Zhang, Zhitao    557
Zhao, Hong    101, 129
Zhao, Ju    673

Zhao, Liangliang    613
Zhao, Shu    249
Zhao, Weibin    777
Zhao, Weidong    260
Zheng, Xuefeng    206
Zhihua, Wei    285
Zhong, Jiaming    623
Zhong, Ning    613
Zhou, Ying    161
Zhou, Zhe    777
Zhu, Chun    557
Zhu, Tailong    742
Zhu, William    101, 129, 194, 489, 663
Zhu, Yanhui    844
Zhu, Yanqing    663

Printed in the United States
By Bookmasters